M. Goeldner, R. Givens (Eds.)
Dynamic Studies in Biology
Phototriggers, Photoswitches
and Caged Biomolecules

Further Titles of Interest

S. Daunert, S. K. Deo (Eds.)

Photoproteins in Bioanalysis

2005. ISBN 3-527-31016-9

P. G. Wang, T. B. Cai, N. Taniguchi (Eds.)

Nitric Oxide Donors for Pharmaceutical and Biological Applications

2005. ISBN 3-527-31015-0

W. R. Briggs, J. L. Spudich (Eds.)

Handbook of Photosensory Receptors

2005. ISBN 3-527-31019-3

E. Keinan (Ed.)

Catalytic Antibodies

2004. ISBN 3-527-30688-9

G. Kraus

Biochemistry of Signal Transduction and Regulation
3rd Edition

2003. ISBN 3-527-30591-2

Dynamic Studies in Biology

Phototriggers, Photoswitches and Caged Biomolecules

Edited by
Maurice Goeldner, Richard Givens

WILEY-VCH Verlag GmbH & Co. KGaA

Editors

Prof. Dr. Maurice Goeldner
Lab. De Chimie Bioorganique
UMR 7514 CNRS
Université L. Pasteur Strasbourg
74, route du Rhin
F-67401 Illkirch Cedex
France

Prof. Dr. Richard Givens
Department of Chemistry
University of Kansas
5023 Malott Hall
Lawrence, KS 66045
USA

Original Issue

■ All books published by Wiley-VCH are carefully produced. Nevertheless, authors, editors, and publisher do not warrant the information contained in these books, including this book, to be free of errors. Readers are advised to keep in mind that statements, data, illustrations, procedural details or other items may inadvertently be inaccurate.

Library of Congress Card No.: Applied for

British Library Cataloging-in-Publication Data:
A catalogue record for this book is available from the British Library

Bibliographic information published by Die Deutsche Bibliothek
Die Deutsche Bibliothek lists this publication in the Deutsche Nationalbibliografie; detailed bibliographic data is available in the Internet at http://dnb.ddb.de.

© 2005 WILEY-VCH Verlag GmbH & Co. KGaA, Weinheim

All rights reserved (including those of translation into other languages). No part of this book may be reproduced in any form – nor transmitted or translated into machine language without written permission from the publishers. Registered names, trademarks, etc. used in this book, even when not specifically marked as such, are not to be considered unprotected by law.

Printed in the Federal Republic of Germany
Printed on acid-free paper

Cover M. Gannon, J. Busse (Technical PhotoGraphic Solutions), Lawrence, Kansas
Typesetting K+V Fotosatz GmbH, Beerfelden
Printing betz-druck GmbH, Darmstadt
Bookbinding Litges & Dopf Buchbinderei GmbH, Heppenheim

ISBN-10 3-527-30783-4
ISBN-13 978-3-527-30783-8

Contents

Foreword *XVII*

Preface *XXI*

List of Authors *XXIII*

1	**Photoremovable Protecting Groups Used for the Caging of Biomolecules** *1*	
1.1	2-Nitrobenzyl and 7-Nitroindoline Derivatives *1*	
	John E. T. Corrie	
1.1.1	Introduction *1*	
1.1.1.1	Preamble and Scope of the Review *1*	
1.1.1.2	Historical Perspective *2*	
1.1.2	Synthetic Considerations *3*	
1.1.3	Survey of Individual Caged Compounds and Caging Groups *5*	
1.1.3.1	2-Nitrobenzyl Cages *5*	
1.1.3.1.1	Mechanistic Aspects of Photocleavage and By-Product Reactions of 2-Nitrobenzyl Cages *6*	
1.1.3.1.2	Representative Survey of Nitrobenzyl-Caged Compounds *11*	
1.1.3.2	7-Nitroindoline Cages *21*	
1.1.3.2.1	Mechanistic and Structural Aspects of Photochemical Cleavage of 1-Acyl-7-nitroindolines *21*	
1.1.3.2.2	Survey of 7-Nitroindoline Caged Compounds *22*	
1.1.4	Conclusion *24*	
	References *25*	
1.2	Coumarin-4-ylmethyl Phototriggers *29*	
	Toshiaki Furuta	
1.2.1	Introduction *29*	
1.2.2	Spectroscopic and Photochemical Properties *30*	
1.2.2.1	Overview *30*	
1.2.2.2	Phototriggers *33*	
1.2.2.2.1	MCM Groups: 7-Alkoxy-Substituted Coumarins *33*	
1.2.2.2.2	DMCM Groups: 6,7-Dialkoxy-Substituted Coumarins *34*

1.2.2.2.3	Bhc Groups: 6-Bromo-7-alkoxy-Substituted Coumarins *34*
1.2.2.2.4	DEACM Groups: 7-Dialkylamino-Substituted Coumarins *35*
1.2.2.3	Target Molecules *36*
1.2.2.3.1	Phosphates *36*
1.2.2.3.2	Carboxylates and Sulfates *41*
1.2.2.3.3	Amines *43*
1.2.2.3.4	Alcohols and Phenols *44*
1.2.2.3.5	Carbonyl Compounds *46*
1.2.3	Synthesis *46*
1.2.3.1	Synthesis of Precursor Molecules (Caging Agents) *46*
1.2.3.1.1	Phosphates *49*
1.2.3.1.2	Carboxylates *50*
1.2.3.1.3	Amines *50*
1.2.3.1.4	Alcohols and Phenols *50*
1.2.4	Applications *50*
1.2.5	Conclusion and Perspective *53*
	References *54*
1.3	*p*-Hydroxyphenacyl: a Photoremovable Protecting Group for Caging Bioactive Substrates *55*
	Richard S. Givens and Abraham L. Yousef
1.3.1	Introduction and History *55*
1.3.2	*p*-Hydroxyphenacyl *57*
1.3.2.1	General Physical and Spectroscopic Properties *59*
1.3.2.2	Synthesis of pHP-Caged Substrates *60*
1.3.3	Mechanistic Studies *64*
1.3.3.1	A Triplet "Photo-Favorskii" Rearrangement *64*
1.3.3.2	Role of the Triplet Phenol *65*
1.3.3.3	Correlation of the ^3pK_a with the Quantum Efficiency *65*
1.3.4	Applications *66*
1.3.4.1	Neurotransmitter Release *66*
1.3.4.2	Peptide Release *68*
1.3.4.3	Nucleotide Release *70*
1.3.4.4	Enzyme Photoswitches *72*
1.3.5	Advantages and Limitations *74*
	References *75*
1.4	Caging of ATP and Glutamate: a Comparative Analysis *76*
	Maurice Goeldner
1.4.1	Introduction *76*
1.4.2	General Properties for Caging Groups *76*
1.4.3	Caged-ATP *78*
1.4.3.1	Introduction *78*
1.4.3.2	Syntheses of the P^3-caged ATP Derivatives *78*
1.4.3.3	General Properties of the Caged-ATP Molecules *80*
1.4.3.4	Conclusion *84*
1.4.4	Caged Glutamate *84*

1.4.4.1	Introduction *84*
1.4.4.2	Syntheses of the Caged Glutamate Derivatives *86*
1.4.4.3	General Properties of the Caged-Glutamate Molecules *91*
1.4.4.4	Conclusion *92*
	Abbreviations *92*
	References *93*
2	**Mechanistic Overview of Phototriggers and Cage Release** *95*
	Richard S. Givens, Mani B. Kotala, and Jong-Ill Lee
2.1	Introduction *95*
2.2	Major Photoremovable Protecting Groups *96*
2.2.1	2-Nitrobenzyl (2-NB) *96*
2.2.1.1	The Reactive Excited State *98*
2.2.1.2	The *aci*-Nitro Intermediate *99*
2.2.1.3	Decay of the *aci*-Nitro Intermediate *100*
2.2.1.4	The Role of the Leaving Group: Alcohols vs ATP *101*
2.2.2	Benzoin (*Bnz*) *101*
2.2.2.1	Excitation: Singlet vs Triplet State Reactivity *103*
2.2.2.2	Heterolysis vs Homolysis *105*
2.2.3	Phenacyl (pHP) *109*
2.2.3.1	Singlet vs Triplet Excited States *111*
2.2.3.2	DFT Calculations on Excited Triplet State Reactivity *113*
2.2.3.3	Deprotonation Step *114*
2.2.3.4	Fate of the Spirodienedione Intermediate *116*
2.2.3.5	Effect of Meta Substituents *117*
2.2.3.6	Quantum Efficiency, Rate Constant, and Solvent Isotope Effect Determinations *117*
2.2.4	Benzyl (Bz), Arylmethyl, and Coumaryl (Cou) *118*
2.2.4.1	The Meta/Ortho-Meta Effect *118*
2.2.4.2	Heterolysis vs Homolysis *121*
2.2.4.3	Arylmethyl and Coumaryl (Cou) Derivatives *122*
2.3	Conclusions *125*
	Abbreviations *127*
	References *128*
3	**Caged Compounds and Solid-Phase Synthesis** *131*
	Yoshiro Tatsu, Yasushi Shigeri, and Noboru Yumoto
3.1	Introduction *131*
3.2	Solid-Phase Synthesis and Photolysis of Peptides *131*
3.2.1	General Overview of Solid-Phase Peptide Synthesis *131*
3.2.1.1	The Protecting Group in Solid-Phase Peptide Synthesis *132*
3.2.1.2	Formation of Peptide Fragments *133*
3.2.1.3	Purification and Identification *134*
3.2.2	Photolysis of Peptides *135*
3.2.2.1	Photocleavable Protecting Groups in Peptide Synthesis *135*

3.2.2.1.1	Amino Protection *135*
3.2.2.1.2	Protection of Other Functional Groups *137*
3.2.2.1.3	Photocleavable Linkers *139*
3.2.2.1.4	Photoremovable Scaffolds for Acyl Transfer *141*
3.2.2.2	Photolysis of 2-Nitrobenzyl Groups Bound to Peptides *143*
3.3	Synthesis of Caged Peptides *144*
3.3.1	Design of Caged Peptides *144*
3.3.2	Case Studies on Solid-Phase Synthesis of Caged Peptides *144*
3.3.2.1	Neuropeptide Y *145*
3.3.2.2	RS-20 *147*
3.3.2.3	AIP *147*
3.3.2.4	Nuclear Localization Signal *147*
3.3.2.5	S-Peptides *147*
3.3.2.6	Sperm-Activating Peptides *148*
3.3.2.7	Caged Phosphopeptides *148*
3.3.3	Caged Peptides Using Other Synthetic Methods *149*
3.4	Synthesis of Other Photoactive Biomolecular Compounds *150*
3.4.1	Thioxo Peptides *150*
3.4.2	Photochromic Peptides *151*
3.4.3	Photoaffinity Peptides *151*
3.5	Conclusions and Perspective *153*
	References *153*
4	**Control of Cellular Activity** *155*
4.1	Photochemical Release of Second Messengers – Caged Cyclic Nucleotides *155*
	Volker Hagen, Klaus Benndorf, and U. Benjamin Kaupp
4.1.1	Introduction *155*
4.1.2	Overview of Phototriggers for cNMPs *156*
4.1.2.1	*ortho*-Nitrobenzyl Derivatives *156*
4.1.2.2	(Coumarin-4-yl)methyl Derivatives *160*
4.1.2.3	Others *165*
4.1.2.4	Selection of the Most Useful Phototriggers *166*
4.1.3	Applications of Caged Cyclic Nucleotides *168*
4.1.3.1	Cyclic Nucleotide-Gated (CNG) Channels *168*
4.1.3.2	HCN Channels *172*
4.1.3.3	Dorsal Root Ganglion Cells *172*
4.1.3.4	cGMP-Signaling Pathway in Sperm *173*
4.1.3.5	Miscellaneous Systems *174*
	References *176*
4.2	Photochemical Release of Second Messengers – Caged Nitric Oxide *178*
	Christopher M. Pavlos, Hua Xu, and John P. Toscano
4.2.1	Introduction *178*
4.2.2	Photosensitive Precursors to Nitric Oxide *179*

4.2.2.1	Endogenous Photosensitive Precursors to Nitric Oxide	179
4.2.2.1.1	S-Nitrosothiols	180
4.2.2.1.2	N-Nitrosoamines (Nitrosamines)	181
4.2.2.2	Inorganic Photosensitive Precursors to Nitric Oxide	182
4.2.2.2.1	Nitric Oxide-Metalloprotein Adducts	182
4.2.2.2.2	Sodium Nitroprusside (SNP)	182
4.2.2.2.3	Roussin's Red and Black Salts (RRS, RBS)	184
4.2.2.2.4	Other Iron-based Photosensitive Precursors	186
4.2.2.2.5	Ruthenium-based Photosensitive Precursors	186
4.2.2.2.6	Chromium-based Photosensitive Precursors	188
4.2.2.2.7	Molybdenum-based Photosensitive Precursors	189
4.2.2.3	Organic Photosensitive Precursors to Nitric Oxide	189
4.2.2.3.1	Organic Nitrites	189
4.2.2.3.2	Aromatic Nitro Compounds	190
4.2.2.3.3	Furoxans	191
4.2.2.3.4	Linsidomine (SIN-1)	192
4.2.2.3.5	Phenyl N-tert-Butyl Nitrone	192
4.2.2.3.6	Dialkylamino-based Diazeniumdiolates	193
4.2.2.3.7	Carbon-based Diazeniumdiolates	198
4.2.2.3.8	Bis-N-nitroso-p-phenylenediamine Derivatives	198
4.2.3	Conclusions	200
	References	201
4.3	Photochemical Release of Neurotransmitters – Transient Kinetic Investigations of Membrane-bound Receptors on the Surface of Cells in the Microsecond-to-Millisecond Time Region	205
	George P. Hess	
4.3.1	Introduction	205
4.3.2	Cell Flow Technique	209
4.3.3	Photochemical Properties of Caged Neurotransmitters	212
4.3.4	Laser Pulse Photolysis Technique	213
4.3.5	Determination of the Rate and Equilibrium Constants of the Channel-Opening Mechanism	219
4.3.6	Comparison of the Values Obtained for the Rate and Equilibrium Constants of the Channel-Opening Process Using Independent Techniques	219
4.3.7	Summary	224
	References	226
4.4	Caged Neurotransmitters for Probing Neuronal Circuits, Neuronal Integration, and Synaptic Plasticity	232
	Deda C. Gillespie, Gunsoo Kim, and Karl Kandler	
4.4.1	Introduction	232
4.4.2	Functional Mapping of Neuronal Connections	232
4.4.2.1	Example: Uncaging Glutamate to Map Local Connections in the Mammalian Cortex	235

4.4.2.2	Example: Use of Caged Glutamate for Probing Refinement of Neuronal Connectivity *238*	
4.4.2.3	Conclusions and Future Challenges *239*	
4.4.3	Probing Neuronal Integration *239*	
4.4.3.1	Examples *240*	
4.4.3.2	Conclusions and Future Challenges *243*	
4.4.4	Investigating Synaptic Plasticity *243*	
4.4.5	Conclusions and Future Challenges *246*	
	References *246*	
5	**Photoregulation of Proteins** *253*	
5.1	Light-activated Proteins: An Overview *253*	
	Sandra Loudwig and Hagan Bayley	
5.1.1	Introduction *253*	
5.1.2	The Properties of Caged Proteins *254*	
5.1.2.1	Extents of Caging and Photoactivation *254*	
5.1.2.2	The Rate of Uncaging: Continuous Irradiation *254*	
5.1.2.3	The Rate of Uncaging: Breakdown of Intermediates *270*	
5.1.2.4	The Effects of pH on Photolysis *270*	
5.1.2.5	Problems with the Released Caging Group *271*	
5.1.2.6	Photochemistry Peculiar to Proteins *272*	
5.1.2.7	Polypeptide Chain Cleavage *273*	
5.1.2.8	Dominant Negative Effect *274*	
5.1.3	Sites of Modification in Caged Proteins *274*	
5.1.3.1	Random Modification *274*	
5.1.3.2	Active-site Directed Modification *275*	
5.1.3.3	The Use of Crosslinkers *276*	
5.1.3.4	Caging at Cys Residues *277*	
5.1.3.5	Sites other than Cys for Site-specific Modification *279*	
5.1.3.6	Reagents for Caging by Protein Modification *280*	
5.1.3.7	Determining the Extent of Protein Modification *281*	
5.1.3.8	True Residual Activity *282*	
5.1.3.9	Incomplete Uncaging *282*	
5.1.3.10	Means of Caging other than Chemical Modification *283*	
5.1.4	Applications of Caged Proteins *284*	
5.1.5	Future Prospects *284*	
5.1.5.1	Turning Proteins On *and* Off *284*	
5.1.5.2	Genetically Encoded Caged Proteins and Caging within Cells *294*	
5.1.5.3	Improved Spatial Control *295*	
5.1.5.4	Temporal Control *296*	
5.1.5.5	Medical Applications *297*	
	References *299*	

5.2	Photochemical Enzyme Regulation using Caged Enzyme Modulators *304*	
	Ling Peng and Maurice Goeldner	
5.2.1	Introduction *304*	
5.2.2	Photoregulation of ATPases *305*	
5.2.2.1	ATPases using Caged ATP Molecules *305*	
5.2.2.1.1	Na,K-ATPase *305*	
5.2.2.1.2	Transport ATPases and Carriers *305*	
5.2.2.1.3	Molecular Motors *306*	
5.2.2.2	Ca^{2+}-ATPase *307*	
5.2.2.2.1	Ca^{2+}-ATPase using Caged ATP *307*	
5.2.2.2.2	Ca^{2+}-ATPase using Caged Inhibitor *307*	
5.2.2.2.3	Ca^{2+}-ATPase using Caged Ca^{2+} *308*	
5.2.3	Photoregulation of GTPases: Ras Protein *308*	
5.2.3.1	Time-resolved Crystallographic Studies with Caged GTP *308*	
5.2.3.2	Time-resolved FTIR Studies with Caged GTP *309*	
5.2.4	Photoregulation of Kinases *310*	
5.2.4.1	cAMP-dependent Protein Kinase *310*	
5.2.4.1.1	PKA using Caged cAMP *311*	
5.2.4.1.2	PKA using Caged Inhibitor *312*	
5.2.4.2	PKC *312*	
5.2.4.2.1	PKC using Caged Activators *312*	
5.2.4.2.2	PKC using Caged Fluorescent Substrates *312*	
5.2.4.3	Calmodulin-dependent Protein Kinases *313*	
5.2.4.3.1	Myosin Light Chain Kinase (MLCK) using Caged Peptide Inhibitors *313*	
5.2.4.3.2	Calmodulin-dependent Protein Kinase II (CaMKII) using Caged Peptide Inhibitors *313*	
5.2.5	Photoregulation of Alcohol Dehydrogenases *314*	
5.2.5.1	Isocitrate Dehydrogenase (IDH): Caged NAD and NADP *314*	
5.2.5.2	Glucose-6-Phosphate Dehydrogenase (G6PDH): Caged Glucose 6-Phosphate and Caged NADP *315*	
5.2.5.3	7a-Hydroxy Steroid Dehydrogenase (7-HSDH): Caged Cholic Acid *315*	
5.2.5.4	Caged Benzyl Alcohol *317*	
5.2.6	Photoregulation of Cholinesterases *317*	
5.2.6.1	Caged Choline Derivatives *317*	
5.2.6.2	Caged Cholinesterase Substrate *318*	
5.2.7	Miscellaneous Examples of Caged Enzyme Modulators *320*	
5.2.7.1	Caged Nitric Oxide Synthase (NOS) Inhibitor *320*	
5.2.7.2	Caged Substrate of Quinol-oxidizing Enzymes *321*	
5.2.7.3	Caged Urease Substrate *322*	
5.2.7.4	Caged Ribonuclease Reductase Substrate *322*	
5.2.8	Conclusions and Perspectives *323*	
	References *323*	

5.3	The Use of Caged Proteins in Cell-based Systems *325*
	John S. Condeelis and David S. Lawrence
5.3.1	Introduction *325*
5.3.2	Cell Motility *326*
5.3.2.1	β-Thymosin *326*
5.3.2.2	Cofilin *328*
5.3.3	Electrical Conductance *330*
5.3.3.1	Antibodies *330*
5.3.3.2	Ion Channels *332*
5.3.4	Embryogenesis *335*
5.3.5	Protein Trafficking *336*
5.3.6	Signal Transduction *337*
5.3.7	Bacterial Exotoxins *338*
5.3.8	Summary and Conclusions *339*
	References *340*

6	**Photoremovable Protecting Groups in DNA Synthesis and Microarray Fabrication** *341*
	Michael C. Pirrung and Vipul S. Rana
6.1	Introduction *341*
6.2	Photoremovable Groups used in Conventional Nucleic Acid Synthesis *341*
6.3	The Photolithographic Method for Microarray Fabrication *348*
6.3.1	Background – Polypeptide Microarrays *350*
6.3.2	Early DNA Microarray Studies *351*
6.3.3	Methods to Assess Chemical Reactions on DNA Microarrays *353*
6.3.4	Protecting Groups *356*
6.3.4.1	Nitrobenzyl Groups *357*
6.3.4.2	Dimethoxybenzoin Carbonate Groups *358*
6.3.4.3	(Nitrophenylpropyloxy)carbonyl Groups *360*
6.3.4.4	Other Photoremovable Protecting Groups *362*
6.3.4.5	Photoacids with DMTr Groups *363*
6.4	The Future *364*
6.4.1	Multiphoton Sensitive Protecting Groups *365*
6.4.2	Direct Write Microarray Fabrication Systems *365*
	References *367*

7	**Analytical Time-resolved Studies using Photochemical Triggering Methods** *369*
7.1	Time-resolved IR Spectroscopy with Caged Compounds: An Introduction *369*
	Andreas Barth
7.1.1	IR Spectroscopy *369*
7.1.1.1	Introduction *369*
7.1.1.2	Information that can be Derived from the IR Spectrum *371*

7.1.1.3	Instrumentation *371*	
7.1.1.3.1	Fourier Transform IR (FTIR) Spectroscopy *371*	
7.1.1.3.2	Dispersive IR Spectrometers *372*	
7.1.1.3.3	IR Samples *373*	
7.1.1.4	Time-resolved IR spectroscopy *373*	
7.1.1.4.1	Overview *373*	
7.1.1.4.2	The Rapid Scan Technique *373*	
7.1.1.4.3	The Step Scan Technique *373*	
7.1.1.4.4	Single-wavelength Measurements *374*	
7.1.2	IR Spectroscopy of Caged Compounds *374*	
7.1.2.1	Introduction *374*	
7.1.2.2	The IR Absorbance Spectrum of NPE-caged ATP *375*	
7.1.2.3	Difference Spectroscopy *376*	
7.1.2.4	How to Interpret IR Spectra *378*	
7.1.2.5	The Difference Spectrum of NPE-caged ATP Photolysis *378*	
7.1.2.6	Side-reactions of the Photolysis Byproduct *381*	
7.1.2.7	Measuring Product Release Rates with IR Spectroscopy *383*	
7.1.2.8	IR Spectroscopy for Mechanistic Studies *384*	
7.1.3	Summary and Conclusions *387*	
	References *398*	
7.2	IR Spectroscopy with Caged Compounds: Selected Applications *400*	
	Vasanthi Jayaraman	
7.2.1	Introduction *400*	
7.2.2	Selected Examples *401*	
7.2.2.1	Allosteric Mechanism of Activation of the Glutamate Receptor *401*	
7.2.2.2	GTP Hydrolysis by Ras: Nature of the Intermediate *403*	
7.2.2.3	Active Transport of Ca^{2+} by Ca-ATPase *407*	
7.2.3	Conclusions *409*	
	References *409*	
7.3	New Perspectives in Kinetic Protein Crystallography using Caged Compounds *410*	
	Dominique Bourgeois and Martin Weik	
7.3.1	Introduction: Principles of Kinetic Crystallography *410*	
7.3.1.1	Protein Activity and Conformational Changes in Crystals *410*	
7.3.1.2	Timescales and Activation States *412*	
7.3.1.3	Accumulating Unstable Species in the Crystal *413*	
7.3.2	Real-time Crystallography and the Laue Technique *414*	
7.3.2.1	A Favorable Case: Isocitrate Dehydrogenase *415*	
7.3.2.2	The Case of Ha-*ras* p21: From Laue to Monochromatic Diffraction and Further Difficulties with Caged Compounds in Kinetic Crystallography *418*	
7.3.3	Trapping of Intermediate States *420*	
7.3.3.1	Cryo-photolysis of NPE-caged Compounds *422*	
7.3.3.2	Temperature-dependent Enzymatic Activity and Dynamical Transitions *427*	

7.3.3.3	Potential Applications of Cryo-photolysis Combined with Temperature-controlled Protein Crystallography *427*
7.3.3.4	Cryo-photolysis Wavelength and Cryo-radiolysis of NPE-caged Compounds *428*
7.3.3.5	X-ray Data Collection Strategy *429*
7.3.4	Conclusion *431*
	References *432*

8 Multiphoton Phototriggers for Exploring Cell Physiology *435*
Timothy M. Dore

8.1	Introduction and History *435*
8.2	Theory *436*
8.3	The Two-photon Action Cross-section, δ_u *438*
8.4	Chromophores for Two-photon Release of Small Organic Ligands or Metal Ions *440*
8.5	Applications *444*
8.5.1	Neurotransmitters *444*
8.5.2	Signal Transduction Molecules *446*
8.5.3	Fluorophores *450*
8.5.4	PDT and Gene Inactivation *453*
8.5.5	Drug Delivery *455*
8.6	Conclusion *456*
	References *456*

9 New Challenges *461*

9.1	Laser-Induced *T*-Jump Method: A Non-conventional Photoreleasing Approach to Study Protein Folding *461*
	Yongjin Zhu, Ting Wang, and Feng Gai
9.1.1	Introduction *461*
9.1.2	Laser-induced *T*-jump IR Technique *463*
9.1.3	Helix–Coil Transition Kinetics *466*
9.1.4	The Folding Mechanisms of β-Hairpins *470*
9.1.5	Ultrafast Folding of Helical Bundles *472*
9.1.5.1	Folding Kinetics of $1prb_{7-53}$ *472*
9.1.5.2	Folding Kinetics of α_3D *473*
9.1.6	Summary and Conclusions *474*
	References *475*
9.2	Early Kinetic Events in Protein Folding: The Development and Applications of Caged Peptides *479*
	Sunney I. Chan, Joseph J.-T. Huang, Randy W. Larsen, Ronald S. Rock, and Kirk C. Hansen
9.2.1	Introduction *479*
9.2.2	An Ultrafast Photochemical Triggering System *480*
9.2.3	Caging Strategies and the Development of a Photolabile Linker *481*
9.2.4	Synthesis of Cages and Caging of a Protein/Peptide *484*

9.2.4.1	DMB, Fmoc-Asp(DMB)-OH and Fmoc-Glu (DMB)-OH *484*	
9.2.4.2	3′,5′-Bis(caboxymethoxy)benzoin (BCMB) *485*	
9.2.4.3	BrAc-CMB *485*	
9.2.4.4	Protein Head-to-Side-chain Cyclization by BrAc-CMB *486*	
9.2.5	Monitoring the Refolding Kinetics of Protein Structural Motifs by Laser Flash Photolysis of "Caged Peptides" *487*	
9.2.6	Early Kinetic Events in Protein Folding *489*	
9.2.6.1	Side-chain Caging of the GCN4-p1 Leucine Zipper *489*	
9.2.6.2	Head-to-Side-chain Cyclization β-Sheets *490*	
9.2.6.3	Head-to-Side-chain Caging of the α-Helical Villin Headpiece Subdomain *492*	
9.2.7	Summary and Conclusions *493*	
	References *494*	
9.3	Photocontrol of RNA Processing *495*	
	Steven G. Chaulk, Oliver A. Kent, and Andrew M. MacMillan	
9.3.1	Introduction *495*	
9.3.2	Caging the Hammerhead Ribozyme *496*	
9.3.3	Splicing *498*	
9.3.3.1	Photocontrol of Splicing using Caged pre-mRNAs *498*	
9.3.3.2	ATP Requirements for Splicing *503*	
9.3.3.3	Phosphatase Activity and Spliceosome Assembly *504*	
9.3.4	Future Prospects *505*	
9.3.5	Experimental *506*	
9.3.5.1	Caged Hammerhead Ribozyme *506*	
9.3.5.1.1	Synthesis of 2′-*O*-(2-nitrobenzyl)adenosine Phosphoramidite *506*	
9.3.5.1.2	2′-*O*-(2-nitrobenzyl)adenosine *506*	
9.3.5.1.3	2′-*O*-(2-nitrobenzyl)-N^6-benzoyladenosine *506*	
9.3.5.1.4	5′-*O*-(4,4′-dimethoxytrityl)-2′-*O*-(2-nitrobenzyl)-N^6-benzoyladenosine-3′-*O*-(cyanoethyl-*N*,*N*-diisopropylamino) Phosphoramidite *507*	
9.3.5.1.5	Nucleoside Model Studies *507*	
9.3.5.1.6	T7 Transcription *508*	
9.3.5.1.7	Automated Synthesis *508*	
9.3.5.1.8	Nucleoside Composition Analysis *509*	
9.3.5.1.9	RNA Photolysis *509*	
9.3.5.1.10	Catalytic Ribozyme Cleavage Reaction *509*	
9.3.5.1.11	Saturating Ribozyme Kinetics *510*	
9.3.5.1.12	Measurement of Equilibrium Dissociation Constants for Ribozyme–RNA Complexes *510*	
9.3.5.2	Caged Pre-mRNA *511*	
9.3.5.2.1	Preparation of Caged Pre-mRNA *511*	
9.3.5.2.2	TLC Analysis of Pre-mRNA *511*	
9.3.5.2.3	Splicing and Photolysis of Caged Pre-mRNA *511*	
	References *512*	

9.4	Light Reversible Suppression of DNA Bioactivity with Cage Compounds	513
	W. Todd Monroe and Frederick R. Haselton	
9.4.1	Introduction	513
9.4.2	Evidence for DNA Caging	514
9.4.3	Reversible Blockade of Restriction Enzyme Activity	517
9.4.4	Reversible Blockade of *In Vitro* Transcription	518
9.4.5	Reversible Blockade of Hybridization	519
9.4.6	Reversible Inhibition of Antisense Activity	521
9.4.7	Reversible Blockade of GFP Expression in HeLa Cells	523
9.4.8	Reversible Blockade of Plasmid Expression in Cultured Corneas	524
9.4.9	Reversible Blockade of Plasmid Expression *In Vivo*	526
9.4.10	Reversible Blockade of mRNA for Expression Control *In Vivo*	526
9.4.11	Caged Hormones as Spatial Regulators of Gene Expression	527
9.4.12	Concerns for Caging to Control Gene Expression	527
9.4.13	Examples of Site-specific Caging of DNA and RNA	529
9.4.14	Future Directions and Conclusions	530
	References	531
9.5	Photoactivated Gene Expression through Small Molecule Inducers	532
	Sidney B. Cambridge	
9.5.1	Introduction	532
9.5.2	Tamoxifen and Estrogen Receptor (ER)	533
9.5.3	Ecdysone	534
9.5.4	Tetracycline	536
9.5.5	Summary and Conclusions	538
	References	538

Subject Index 539

Foreword

I came to the United States at the end of 1975 and joined the laboratory of Professor Joseph Hoffman at Yale. This was a particularly fortunate time for someone who had trained as an organic chemist and biophysicist and was interested in applying a chemical-based approach to interesting physiological problems. I began a collaboration with Biff Forbush, and for the next couple of years we embarked on trying to make a caged ATP that would efficiently release ATP rapidly on illumination under physiological conditions. The driving force for this approach came from the idea that in order to study an ion pump, or P-type ATPase, in a sided preparation, it would be ideal to be able to generate ATP within a closed compartment. The particular experiments that had been originally planned for this new tool (if it worked) have in fact still not been carried out, as their significance was made redundant by our increasing knowledge about the sodium pump.

As with most synthetic chemical approaches, the literature provided encouragement. Work on the photorelease of carboxylic acids from nitrobenzyl esters by Baltrop had shown that lengthy illumination in the ultraviolet range of these esters in acetone produced quantities of the free carboxylic acids. Not surprisingly, this was also an area (among so many) that had received attention from R. B. Woodward, who had demonstrated the utility of photodeprotection of nitrobenzyl phosphate esters. Indeed, photodeprotection had been appreciated as a potentially useful strategy in peptide and nucleic acid synthesis.

Biff Forbush and I then spent much of the next two years in the dark! Our initial synthetic efforts provided a clue that caged ATP was an achievable goal as our multi-step synthesis provided very small quantities of a long wavelength-absorbing material that yielded ATP on photolysis. Eventually we improved the synthetic route and could produce reasonable quantities of our new reagent. One of the first benefactors of this approach was David Trentham, who had recently come to the United States. He used some of our early samples in initial studies in skeletal muscle fibers. These led to studies over the next several years, from the laboratories of David Trentham and Yale Goldman among others, that have made significant steps forward in our understanding of the details of the contractile machinery, and also introduced David and his colleagues, including John Corrie, to the photorelease area in which they have made major mechanistic advances.

Dynamic Studies in Biology. Edited by M. Goeldner, R. Givens
Copyright © 2005 WILEY-VCH Verlag GmbH & Co. KGaA, Weinheim
ISBN: 3-527-30783-4

I then initiated studies of sodium pump partial reactions using the photorelease of caged ATP inside resealed red cell ghosts. These proved to be quite difficult and technically demanding. This led to me giving a series of seminars over the next couple of years on the "potential" of the caged ATP strategy, without being able to demonstrate any new biological information derived from its use. Eventually we were successful and were able to put to good use our new tool. During this time it became apparent to me that there would now be many variations on the caged ATP theme, as the chemistry was apparently relatively straightforward and photorelease could be applied to many different biological substrates and many different biological situations. Sure enough, the photorelease strategy was subsequently applied to other nucleotides, sugars, cyclic nucleoside monophosphates, neurotransmitters, etc. etc., using the same photochemical approach. Several of these approaches and applications are discussed in this monograph, and the reader is encouraged to discover the new developments in substrate release that are presented here.

Around this time it was becoming abundantly clear that the regulation and rapid alteration of cellular Ca levels was an essential and widely used factor in a large number of intracellular signaling systems. Furthermore, rapid changes in intracellular Ca are used as a trigger to activate a variety of physiological reactions. Many scientists were engaged in designing and evaluating reagents to reliably monitor intracellular Ca concentrations in a number of physiological systems. I realized that a reagent that could rapidly release Ca and generate rapid signaling transients or elevations in Ca might be a very useful tool in cellular and molecular biophysical studies. Obviously this is conceptually somewhat different than cleaving off a covalent protecting group from a substrate, but as usual the key elements were already in the literature. It was known that EDTA was a very effective chelator of polyvalent cations such as Ca, Mg etc., with dissociation constants in the nanomolar range. I also knew that iminodiacetic acid, half of the EDTA coordinating center, bound the same cations, but much more weakly (with a dissociation constant in the millimolar range). Thus if we could use light to bifurcate the Ca-EDTA complex to two iminodiacetic acid molecules, the bound Ca would be released. The synthetic chemistry proved to be very challenging, and, some ten years later, with the input of a post-doctoral Fellow, Graham Ellis-Davies, DM-nitrophen was successfully synthesized and had just the predicted properties for a caged Ca reagent base on EDTA. The kinetics of release turned out to be fairly rapid, and increases in Ca were achieved in around 60 microseconds. The selectivity of these chelators mirrored that of the parent compound EDTA, so they could as easily be used as photochemical sources of Ba, Mg etc. Subsequently, a similar approach led us to a molecule with higher selectivity for Ca, based on EGTA. Since then Graham Ellis-Davies has made some refinements to these initial caged Ca reagents. While Ca and other ion release using caged chelators are not a subject covered in this monograph, the reader is directed to a leading reference on this topic [G. C. R. Ellis-Davies "Development and application of calcium cages" in "Biophontonics" (Academic Press) eds. G. Marriott and I. Parker, Meth. En. 2003 360A, 226–238].

The particular advantages of photorelease from caged compounds as an experimental strategy are many. The rapidity with which substrates are released and hence reactions initiated now can be accomplished in the millisecond and sub-millisecond time range. The activating signal, high intensity light, is instantly available (using flash lamps or lasers) and can be readily switched off. Since prior to activation the caged substrates are relatively inert, they can be used to activate processes that are within closed compartments (vesicles or ghosts) or in ordered structures (muscle fibers or crystals). They can also be employed to synchronize arrays of proteins or enzymes. The technique provides high spatial and temporal resolution, and by virtue of the skills and imagination of a number of synthetic chemists, has a very broad versatility. There have been several recent and important developments that have extended the scope of this technology; these have included the introduction of multiphoton approaches, the application to "caged" fluorescence, the use of caged peptides or proteins, and the continuing development of newer photochromic moieties with more desirable (faster or longer wavelength) chemical or photochemical properties.

This volume brings together a group of expert practitioners in this field who have contributed greatly to its broader application and potential. This is the first time that a single volume has been dedicated to this important experimental strategy. I believe it will prove to be of considerable value to many biological scientists who already employ this approach as well as to those who are considering its application to their particular systems. I hope that it will also trigger the imagination of experimentalists and enable them to carry out studies that are made more accessible by access to caged compounds and photoactivation.

July 1, 2004

Jack H. Kaplan PhD, FRS
Benjamin Goldberg Professor & Head
Biochemistry & Molecular Genetics
University of Illinois at Chicago
900 S Ashland Ave
Chicago IL 60607

Preface

Cage compounds have been known in biochemistry and physiology for more than two decades [1]. The past few years have witnessed a considerable increase in interest in caged derivatives because of their high degree of temporal and spatial controllability. Light-initiated cage release is now extensively employed to investigate molecular processes in biochemistry and biophysics as well as to initiate other physiological phenomena. More recently, these same approaches have been extended to the release of substances in building or isolating libraries in combinatorial chemistry, in photolithography, and in the photorelease of caged reagents in chemical transformations. In parallel with recent developments for biological and medicinal technological innovations, the miniaturization of devices for delivery of reagents in analysis and chemical processing and in real-time spectral and physical measurements is a rapidly developing technology. These developments have placed increased demands on the need for even greater spatial and temporal control on processes including chemical transformations. Consequently, the use of light-activated processes becomes much more appealing to those working in the life sciences [2].

As will be discussed in this volume, a very extensive range of substrates and reagents has been delivered by the photolysis of photoactivated protecting groups. This monograph both reviews the recent accomplishments in the field of caged compounds and also looks forward to future inroads into other fields.

Much of what is presented here, in fact, refers to processes in many other fields where the control of reagent or substrate delivery is a key element in a biological study. Initially in this field, only small, low-molecular-weight compounds were "releasable". Now, whole proteins and oligonucleotides are the object of "caging" applications that can range from molecular beacons to enzyme switches.

In spite of the increased interest, the number and range of useful and practical cages remains very limited, with less than a handful that have found sustained interest within the life sciences. In fact, over 80% of the published discoveries employing photoremovable protecting groups are based on a single photochemical process, the photoredox chemistry of the 2-nitrobenzyl chromophore [3]. While this photochemical reagent has proven to be versatile, robust, and generally efficient, it has well-documented limitations, such as the slow rate of substrate release and generation of a reactive nitroso functionality, that significantly restrict its application in biological studies. These limitations have stimulated the design and development of cages, leading to an expansion in the availability of usable caging groups. Caging chromophores must have several key properties or attributes. Among these are a strong absorption in the near UV-vis region, an efficient photorelease process, a hypsochromic shift of the absorption spectrum due to the photoproduct, ease of attachment of the chromophore without introducing new added stereocenters, and a cage and photoproduct that are biologically inert. There are currently four principal chromophores, used in caged reactions, that are included in this monograph. These four are evaluated and their applications presented. The newer applications of multiphoton decaging have also been included. These have shown particular promise because of the improvement in spatial resolution for controlled release.

This monograph provides a timely assessment and overview of the state of the field and recent investigations of cage photorelease for a wide variety of applications. It remains to be discovered what new cages will replace or augment those chromophores now in use and what expanded repertoire of applications will emerge. Future research will undoubtedly extend both areas and require a reinvestigation of the accomplishments achieved with the aid of photoremovable protecting groups.

References

1 (a) KAPLAN, J. H., FORBUSH, B. I., HOFFMAN, J. F., Rapid photolytic release of adenosine 5′-triphosphate from a protected analogue: Utilization by the Na:K pump of human red blood cell ghosts. Biochemistry **1978**, *17*, 1929–1935. (b) ENGELS, J., SCHLAEGER, E.-J., Synthesis, structure, and reactivity of adenosine cyclic 3′,5′-phosphate benzyl triesters. *J. Med. Chem.* **1977**, *20*, 907–911.

2 G. MARRIOTT, Ed. *Methods in Enzymology*, Vol. 291, Academic Press, San Diego, **1998**.

3 WALKER, J. W., REID, G. P., MCCRAY, J. A., TRENTHAM, D. R., 1-(2-Nitrophenyl)ethyl phosphate esters of adenine nucleotide analogues: Synthesis and mechanism of photolysis. *J. Am. Chem. Soc.* **1988**, *110*, 7170–7178.

Strasbourg, France, July 2004　　　　　　　　　　　　　　　　　　Maurice Goeldner
Lawrence, Kansas, USA, July 2004　　　　　　　　　　　　　　　　Rich Givens

List of Authors

ANDREAS BARTH
Stockholm University
Department of Biochemistry
and Biophysics
Arrhenius Laboratories for Natural
Sciences
10691 Stockholm
Sweden

HAGAN BAYLEY
University of Oxford
Department of Chemistry
Mansfield Road
Oxford, OX1 3TA
UK

KLAUS BENNDORF
Friedrich-Schiller-Universität Jena
Institut für Physiologie II
07740 Jena
Germany

DOMINIQUE BOURGEOIS
Institut de Biologie Structurale
LCCP, UMR 5075
41 avenue Jules Horowitz
38027 Grenoble Cedex 1
France
and
ESRF
6 rue Jules Horowitz, BP 220
38043 Grenoble Cedex
France

SIDNEY. B. CAMBRIDGE
Max-Planck-Institute of Neurobiology
Department of Cellular and Systems
Neurobiology
Am Klopferspitz 18
82152 Munich-Martinsried
Germany

SUNNEY I. CHAN
Academia Sinica
Institute of Chemistry
Tapei 115
Taiwan
and
California Institute of Technology
Arthur Amos Noyes Laboratory
of Chemical Physics
Pasadena, CA 911125
USA

STEVEN G. CHAULK
University of Alberta
Department of Biochemistry
Edmonton, Alberta
T6G 2H7
Canada

JOHN S. CONDEELIS
The Albert Einstein College
of Medicine
Department of Anatomy
and Structural Biology
1300 Morris Park Ave.
Bronx, NY 10461
USA

JOHN E. T. CORRIE
National Institute
for Medical Research
Department of Physical Biochemistry
The Ridgeway, Mill Hill
London NW7 1AA
UK

TIMOTHY M. DORE
University of Georgia
Department of Chemistry
Athens, GA 30602-2556
USA

TOSHIAKI FURUTA
Toho University
Faculty of Science
Department of Biomolecular Science
2-2-1 Miyama,
Funabashi, Chiba 274-8510
Japan

FENG GAI
University of Pennsylvania
Department of Chemistry
231 South, 34th Street
Philadelphia, PA 19104
USA

DEDA C. GILLESPIE
University of Pittsburgh
School of Medicine
Department of Neurobiology
3500 Terrace Street
Pittsburgh, PA 15261
USA

RICHARD S. GIVENS
University of Kansas
Mallot Hall
Department of Chemistry
1251 Wescoe Hall Dr.
Lawrence, KS 66045-7582
USA

MAURICE GOELDNER
Université Louis Pasteur Strasbourg
Faculté de Pharmacie
Laboratoire de Chimie Bioorganique
UMR 7514 CNRS
74, route du Rhin
67401 Illkirch Cedex
France

VOLKER HAGEN
Forschungsinstitut für Molekulare
Pharmakologie
Robert-Rössle-Straße 10
13125 Berlin
Germany

KIRK C. HANSEN
Arthur Amos Noyes Laboratory
of Chemical Physics
California Institute of Technology
Pasadena, CA 91125
USA

FREDERICK R. HASELTON
Vanderbilt University
Department of Biomedical
Engineering
Box 1510 Station B
Nashville, TN 37232
USA

GEORGE P. HESS
Cornell University
Molecular Biology and Genetics
216 Biotechnology Building
Ithaca, NY 14853-2703
USA

JOSEPH J.-T. HUANG
Academia Sinica
Institute of Chemistry
Tapei 115
Taiwan

VASANTHI JAYARAMAN
University of Texas Houston
Health Science Center at Houston
Department of Integrative Biology
and Pharmacology
6431 Fannin, MSB 4.106
Houston, TX 77030
USA

KARL KANDLER
University of Pittsburgh
School of Medicine
Department of Neurobiology
3500 Terrace Street
Pittsburgh, PA 15261
USA

JACK H. KAPLAN
University of Illinois
Department of Biochemistry
and Molecular Genetics
Chicago, IL 60607
USA

U. BENJAMIN KAUPP
Forschungszentrum Jülich
Institut für Biologische
Informationsverarbeitung
52425 Jülich
Germany

OLIVER A. KENT
University of Alberta
Department of Biochemistry
Edmonton, Alberta
T6G 2H7
Canada

GUNSOO KIM
University of Pittsburgh School
of Medicine
Department of Neurobiology
3500 Terrace Street
Pittsburgh, PA 15261
USA

MANI B. KOTALA
University of Kansas
Malott Hall
The Center for Chemical
Methodology
and Library Development
Lawrence, KS 66045-0046
USA

RANDY W. LARSEN
University of South Florida
Department of Chemistry
Tampa, FL 33620
USA

DAVID S. LAWRENCE
The Albert Einstein College
of Medicine
of Yeshiva University
Department of Biochemistry
1300 Morris Park Ave.
Bronx, NY 10461-1602
USA

SANDRA LOUDWIG
University of Oxford
Department of Chemistry
Manfield Road
Oxford, OX1 3TA
UK

JONG-III LEE
Malott Hall
The Center for Chemical
Methodology
and Library Development
Lawrence, KS 66045-0046
USA

ANDREW M. MACMILLAN
University of Alberta
Department of Biochemistry
Edmonton, Alberta
T6G 2H7
Canada

W. Todd Monroe
Louisiana State University
Department of Biological
and Agricultural Engineering
Room 149 E.B Doran Building
Baton Rouge, LA 70803
USA

Christopher M. Pavlos
John Hopkins University
Department of Chemistry
3400 N. Charles Street
Baltimore, MD 21218
USA

Ling Peng
Université de Marseille
Département de chimie
AFMB CNRS UMR 6098
163, Av. de Luminy
13288 Marseille cedex 09
France

Michael C. Pirrung
University of California Riverside
Department of Chemistry
92521 Riverside, CA
USA

Vipul S. Rana
University of California Riverside
Department of Chemistry
92521 Riverside, CA
USA

Ronald S. Rock
California Institute of Technology
Arthur Amos Noyes Laboratory
of Chemical Physics
Pasadena, CA 911125
USA

Yasushi Shigeri
National Institute of Advanced
Industrial Science
and Technology (AIST)
1-8-31 Midorigaoka, Ikeda,
Osaka 563-8577
Japan

Yoshiro Tatsu
National Institute of Advanced
Industrial Science
and Technology (AIST)
1-8-31 Midorigaoka, Ikeda,
Osaka 563-8577
Japan

John P. Toscano
John Hopkins University
Department of Chemistry
3400 N. Charles Street
Baltimore, MD 21218
USA

Ting Wang
University of Pennsylvania
Department of Chemistry
231 South, 34th Street
Philadelphia, PA 19104
USA

Martin Weik
Institut de Biologie Structurale
LBM, UMR 5075
41 avenue Jules Horowitz
38027 Grenoble Cedex 1
France

Hua Xu
John Hopkins University
Department of Chemistry
3400 N. Charles Street
Baltimore, MD 21218
USA

ABRAHAM YOUSEF
University of Kansas
Malott Hall
Department of Chemistry
1251 Wescoe Hall Dr.
Lawrence, KS 66045-7582
USA

NOBORU YUMOTO
National Institute of Advanced
Industrial Science
and Technology (AIST)
3-11-46 Nakouji, Amagasaki
Hyogo 661-0974
Japan

YONGJIN ZHU
University of Pennsylvania
Department of Chemistry
231 South, 34th Street
Philadelphia, PA 19104
USA

1
Photoremovable Protecting Groups Used for the Caging of Biomolecules

1.1
2-Nitrobenzyl and 7-Nitroindoline Derivatives

John E. T. Corrie

1.1.1
Introduction

1.1.1.1 Preamble and Scope of the Review

This chapter covers developments with 2-nitrobenzyl (and substituted variants) and 7-nitroindoline caging groups over the decade from 1993, when the author last reviewed the topic [1]. Other reviews covered parts of the field at a similar date [2, 3], and more recent coverage is also available [4–6]. This chapter is not an exhaustive review of every instance of the subject cages, and its principal focus is on the chemistry of synthesis and photocleavage. Applications of individual compounds are only briefly discussed, usually when needed to put the work into context. The balance between the two cage types is heavily slanted toward the 2-nitrobenzyls, since work with 7-nitroindoline cages dates essentially from 1999 (see Section 1.1.3.2), while the 2-nitrobenzyl type has been in use for 25 years, from the introduction of caged ATP **1** (Scheme 1.1.1) by Kaplan and co-workers [7] in 1978.

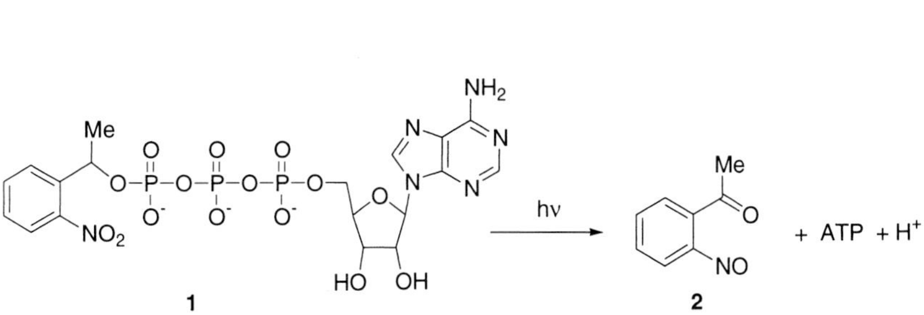

Scheme 1.1.1 Overall photolysis reaction of NPE-caged ATP **1**.

Dynamic Studies in Biology. Edited by M. Goeldner, R. Givens
Copyright © 2005 WILEY-VCH Verlag GmbH & Co. KGaA, Weinheim
ISBN: 3-527-30783-4

1.1.1.2 Historical Perspective

The pioneering work of Kaplan et al. [7], although preceded by other examples of 2-nitrobenzyl photolysis in synthetic organic chemistry, was the first to apply this to a biological problem, the erythrocytic Na:K ion pump. As well as laying the foundation for the field, the paper contains some early pointers to difficulties and pitfalls in the design of caged compounds. Specifically, it was shown that 2-nitrobenzyl phosphate **3** and its 1-(2-nitrophenyl)ethyl analog **4** both released inorganic phosphate in near-quantitative yield upon prolonged irradiation. However, when the same two caging groups were used on ATP, namely **1** and its non-methylated analog **5**, the maximum yield of released ATP from **5** was only ~25%, whereas that from **1** was at least 80%. It was suggested that the 2-nitrosobenzaldehyde by-product released from **5**, in contrast to nitrosoketone **2** released from **1**, might react with the liberated ATP to render it inactive. This hypothesis has not been further studied, but the observations provide an indication that different substituents on the caging group may have unexpected effects. We return to this in later sections that consider rates and mechanisms of caged-compound photolysis.

The use of nitrobenzyl-caged compounds to investigate millisecond time scale biological processes began when laser flash photolysis was used to release ATP from **1**. Initial experiments studied solution interactions between actin and myosin [8], but were soon extended to related work in skinned muscle fibers [9]. These early flash photolysis studies for release of active compounds were contemporaneous with work by the Lester group, who used *cis-trans* photoisomerization of azobenzene derivatives to manipulate pharmacological activity of receptor ligands [10, 11]. Although the latter work involves different photochemistry to that reviewed here, it has been a significant contributor to the adoption of the flash photolysis technique in biology.

In contrast to 2-nitrobenzyl photochemistry, which has its roots in 100-year old observations by Ciamician and Silber on the photochemical isomerization of 2-nitrobenzaldehyde [12], photocleavage of 1-acyl-7-nitroindolines has a much shorter history, dating from work by Patchornik and co-workers in 1976 [13]. They found that 1-acyl-5-bromo-7-nitroindolines (**6**) were photolyzed in organic solvents containing a low proportion of water, alcohols, or amines to yield the nitroindoline **7** and a carboxylic acid, ester or amide, depending on the nucleo-

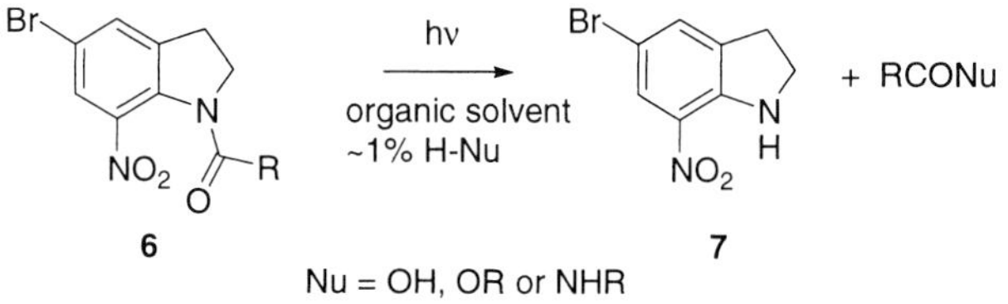

Scheme 1.1.2 Photolysis reaction of 1-acyl-nitroindolines in aprotic organic solvent containing a low proportion of water, an alcohol, or an amine.

phile present (Scheme 1.1.2). The work was briefly examined for its potential in peptide synthesis [14] but was unused for ~20 years, until our group began to use compounds of this type for the release of neuroactive amino acids (see Section 1.1.3.2). In view of the recent nature of most work on 7-nitroindolines, this review includes discussion of unpublished results to illustrate both the scope and limitations of this cage.

1.1.2
Synthetic Considerations

The development of an effective caged compound involves multiple factors, irrespective of the particular cage group employed. Not only must the chemical synthesis be achieved, but photochemical, physicochemical, and pharmacological properties must fit the intended application. Photochemical properties principally concern the efficiency and rate of photolysis upon pulse illumination by light of a suitable wavelength (normally >300 nm to minimize damage to proteins and nucleic acids). There is confusion in the literature about efficiency and rate, arising from different applications of this photochemistry. For use in time-resolved studies of bioprocesses, the relevant rate is of product release following a light flash of a few ns duration, and governs the time resolution with which the bioprocess can be observed. Other workers, using continuous illumination in synthetic photodeprotection applications, apply the term rate as a measure of conversion per unit time. While both uses are legitimate, only the former is generally relevant to studies of rapid bioprocesses. Efficiency of photoconversion, namely the percentage of caged compound converted to the effector species by the light pulse, is influenced by a combination of extinction coefficient at the irradiation wavelength (the proportion of incident light absorbed) and quantum yield (the proportion of molecules that undergo photolysis to form the desired photoproduct after absorbing a photon). Quantum yield is a property governed by the molecule itself and is not readily manipulated, so the investigator only has flexibility to vary the extinction coefficient. This can be achieved, at least for aromatic nitro compounds, by adding electron-donating groups to the aromatic ring. However, data from at least one such substituent study indicate that this can be an unpredictable exercise, sometimes with conflicting effects on extinction coefficient and quantum yield [15].

Important physicochemical properties are water solubility and resistance to hydrolysis. The significance of the latter is to avoid background hydrolytic release prior to the light flash, although stability during storage of aqueous solutions is also an issue. Finally, pharmacological properties need to be considered. In some cases attachment of a caging group can block an effector's actions but not necessarily eliminate its binding to a target protein. For example, caged ATP **1** retains affinity for actomyosin and, although not hydrolyzed by the enzyme, inhibits the shortening velocity of muscle fibers [16]. Similarly, nitroindoline-caged GABA and glycine (but not glutamate) bind to their respective receptors, in each case blunting the response to photoreleased amino acid [17].

The newly emerging field of 2-photon photolysis, first demonstrated by Webb and colleagues [18], offers the potential for highly localized photorelease (within a volume of a few μm^3) but places additional demands on optical properties of the cage. Existing 2-nitrobenzyl cages have very small 2-photon cross-sections and require light doses that cause significant photodamage to live tissue [19]. The Bhc (6-bromo-7-hydroxycoumarin-4-ylmethyl) cage is one option with a more useful 2-photon cross section [20], and some progress has also been made with the 4-methoxy-7-nitroindoline cage [21–23]. A recent paper described 1-(2-nitrophenyl)ethyl ethers of 7-hydroxycoumarins that had surprisingly high 2-photon cross-sections [25]. The underlying mechanism and generality of this approach remains to be determined.

Synthetic routes are specific to particular compounds, but some general points and matters of interest can be brought out. Often, synthesis of caged compounds is achieved by derivatizing the native effector species rather than by *de novo* synthesis. This has advantages in that the chirality present in most biological molecules is preserved, avoiding a need for asymmetric synthesis. The usual preparative method for caged ATP **1** involves treatment of ATP with 1-(2-nitrophenyl)diazoethane (**8**) in a 2-phase system, with the aqueous phase at pH ∼4 [24]. This method is applicable to phosphates in general and esterifies only the weakly acidic hydroxyl group. However, it has been found that either of the strongly acidic hydroxyls of the pyrophosphate in cyclic ADP-ribose (**9**) can be esterified when the aqueous phase is at a lower pH [26], and similar pyrophosphate esterification in nicotinamide adenine dinucleotide (NAD) has been achieved by treating its anhydrous tributylammonium salt with 1-(4,5-dimethoxy-2-nitrophenyl)diazoethane in DMF [27]. An interesting method developed by several groups in recent years has been to phosphorylate alcohols using phosphoramidite reagents that already incorporate the nitrobenzyl cage, so introducing a caged phosphate group in one synthetic step [28–31]. Examples are given in Section 1.1.3.

In most 2-nitrobenzyl-type caged compounds, the nitro group is introduced together with the rest of the cage moiety. In contrast, for many of the nitroindoline-caged amino acids, it has been necessary to couple an indoline with a protected amino acid and later introduce the nitro group. This is because the unreactive amino group of 7-nitroindolines can be acylated only under harsh conditions, incompatible with protecting groups on some amino acids, particularly glutamate, which has been a primary interest (see Section 1.1.3.2.2). Introducing the nitro group at a late stage is facilitated by the reactivity of the 1-acylindoline system but does require some care in the selection of reaction conditions and choice of protecting groups in the acyl side chain.

Finally, a desirable goal in this area would be to be able to cause sequential release of different effectors using different types of caging chemistry and light of two different wavelengths. Some progress has been made in recent papers, principally for selective differentiation of protecting groups in organic synthesis [32, 33], including an ingenious recent use of a kinetic isotope effect on abstraction of the benzylic proton to enhance differentiation between differently substituted 2-nitrobenzyl groups, of which one was dideuterated at the benzylic position [34]. Despite the gradual improvements being made in this area, application to caging chemistry remains as yet an unfulfilled aspiration.

1.1.3
Survey of Individual Caged Compounds and Caging Groups

1.1.3.1 2-Nitrobenzyl Cages
The main structural types considered here are the parent 2-nitrobenzyl (NB) (**10**), its 4,5-dimethoxy analog (DMNB) (**11**), 1-(2-nitrophenyl)ethyl (NPE) (**12**)

and its 4,5-dimethoxy analog (DMNPE) (**13**), and the α-carboxy-2-nitrobenzyl (CNB) (**14**) groups. Minor variants of these are discussed in context below. Substantial work has also been done with 2-(2-nitrophenyl)ethyl derivatives, which release products by a β-elimination mechanism, but information on the product release rate following flash photolysis is lacking [35]. These compounds have been used as photolabile precursors in chemical synthesis, largely for the generation of oligonucleotide microarrays [36–38].

1.1.3.1.1 Mechanistic Aspects of Photocleavage and By-Product Reactions of 2-Nitrobenzyl Cages

There has been much past work on aspects of 2-nitrobenzyl photolysis (reviewed in [1]). Yip and colleagues have shown evidence for the involvement of both singlet and triplet states in the photochemistry of various compounds [39, 40]. However, most mechanistic work has focused on the dark chemistry subsequent to the photo-induced process. An interesting recent study describes observation of mechanistic steps at the single-molecule level using a DMNB derivative tethered inside the pore of a modified hemolysin. Changes in single-channel currents monitored progress of different reaction stages [41]. The study of the photolysis of caged ATP **1** by Walker et al. [24] has largely been taken as a paradigm for caged compounds. An important aspect of the mechanism was that decay of the intermediate *aci*-nitro anion **16** (Scheme 1.1.3) was the rate-determining step for ATP release, and direct measurement by time-resolved infrared spectroscopy has confirmed this [42]. It was noted that the *aci*-nitro decay rate was proportional to proton concentration below pH 9: the lower pH limit of this proportionality was not determined [24]. The pH dependence was attributed, without specific evidence, to protonation of a non-bridging oxygen in the triphosphate chain. However, recent computational and experimental studies [43, 44] suggest an alternative explanation that the *aci*-nitro anion must reprotonate to allow closure to the benzisoxazoline **18** and subsequent reaction. Thus a complete reaction scheme is more reasonably formulated as in Scheme 1.1.3, shown for a general case where R is any substituent and X is the caged species.

The photolytic process itself consists in transfer of the benzylic proton to an oxygen of the nitro group to give the Z-nitronic acid **15**. Ionization of this species (from 2-nitrotoluene) was observable in pure water ($k \approx 2 \times 10^7 \text{ s}^{-1}$) but in the presence of buffer salts the ionization was within the 25-ns laser flash [44]. Subsequent reprotonation to give the isomeric E-nitronic acid **17** allows cyclization to the N-hydroxybenzisoxazoline **18**. Calculations indicate that direct cyclization of anion **16** to the conjugate base of **18** is prohibitively endothermic [43]. Concurrently, this reprotonation of **16** more rationally explains dependence of the decay rate on proton concentration. So far, no direct evidence for **18** has been found, and this intermediate is assumed to decay immediately to end products, these being the nitrosocarbonyl compound **20** and the released effector species (X^- in Scheme 1.1.3). The study by Walker et al. [24] specifically excluded an alternative collapse of **18** to hemiacetal **19**, at least to the extent that the latter did not accumulate as a stable intermediate. However, recently it has

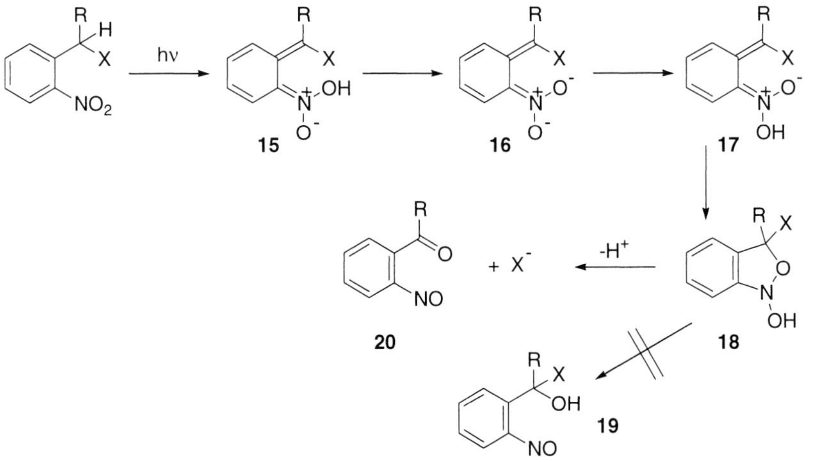

Scheme 1.1.3 Detailed reaction scheme for photolysis of generalized nitrobenzyl-caged compounds. The scheme incorporates results of recent studies by Wirz and colleagues [43, 44].

unequivocally been shown for NPE-caged alcohols such as **21** that **19** is a rate-limiting intermediate, where the hemiacetal decay rate (pH 7, 2 °C) is 0.11 s^{-1} for the compound where R=CH$_2$OPO$_2$OMe$^-$, while the *aci*-nitro decay rate is ~5000 s^{-1} [45]. Some evidence has been presented for a corresponding long-lived aminol in the photolysis of caged amides (see Section 1.1.3.1.2.1). Interestingly, the data for compounds like **21** strongly suggest that, in addition to the normal photolytic pathway via an *aci*-nitro intermediate as shown in Scheme 1.1.3, the major part of the reaction flux involves a direct, very rapid path from the initial nitronic acid (analogous to **15**) to hemiacetal **19** [45]. This anomalous, major pathway appears to operate for NPE-caged alcohols but not for analogous NB-caged alcohols. See Section 1.1.3.1.2.5 for additional comment on these compounds.

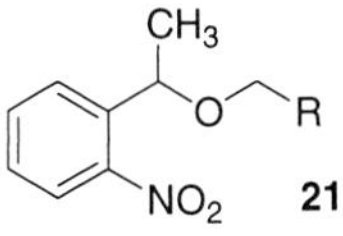

One difficulty of the caged compound field is that little available information exists to allow prediction of reaction rates or efficiencies. For example, measured *aci*-nitro decay rates of phosphates **3** and **4** (25 mM MOPS, 150 mM KCl, pH 7.0, 20 °C) were 660 and 34 300 s^{-1} respectively. Upon monomethylation of the phosphate in these compounds, the rate for the compound derived from **4** (NPE-caged methyl phosphate **23**) fell to 160 s^{-1}, while that for NB-caged methyl phosphate **22** became biphasic, with rate constants of ~1000 and 3 s^{-1} (relative amplitudes ~1:4) (J. E. T. Corrie and D. R. Trentham, unpublished data). The occurrence of biphasic rate constants for the decay of *aci*-nitro transients is not uncommon (see, for example, Refs. [46–48]) and has sometimes been attributed to possible *E/Z*-isomerism of the intermediates, as in structures **24** and **25**. Recent computational studies suggest that this is unlikely, at least in intermediates from the NB series, where the *E*-isomer **24** is calculated to be of very much lower energy [45, 49], so the presence of a significant proportion of *Z*-isomer is improbable.

22 R = H
23 R = Me
24
25

A further complication of 2-nitrobenzyl photochemistry arises from the formation of radical species, initially observed by chance during photolysis of NPE-caged ATP **1** solutions in an EPR spectrometer. The radical species was first assigned as a radical anion of caged ATP and was estimated to constitute approximately 10% of the reaction flux under the experimental conditions (10 mM caged ATP, 10 mM DTT, 200 mM buffer in the pH range 6–9, 1.5 °C) [50]. The species had non-exponential decay kinetics that were relatively insensitive to pH (times for decay to half the maximum intensity were 0.3, 0.4 and 1.2 s at pH 6, 7 and 9 respectively). The structure was later corrected to the cyclic nitroxyl **26**, and a mechanism for its formation was proposed [51]. NPE-caged methyl phosphate **23** gave a comparable EPR spectrum, indicating that only the cage group was involved in the formation of the radical [51]. No further reports of "long-lived" radical species from other caged compounds have appeared, but probably all such 2-nitrobenzyl systems generate a proportion of a radical species upon photolysis. Evidently the formation of low amounts of such radicals does not usually have deleterious effects.

26

The nitrosoarylcarbonyl by-product **20** is mentioned in many papers as being of high reactivity and potentially damaging to biological systems. In fact there is scant reported evidence of such problems, apart from the archetypal case of caged ATP where an added thiol was found necessary to block enzyme inhibition [7] and a recent example in which 2-nitrosoacetophenone was postulated to bind to alkaline phosphatase, although definitive proof of this was lacking [52]. Later follow-up work suggests that the original interpretation was incorrect (L. Zhang and R. Buchet, personal communication) so the evidence for by-product interference remains minimal. Nevertheless, a protective thiol is frequently incorporated during

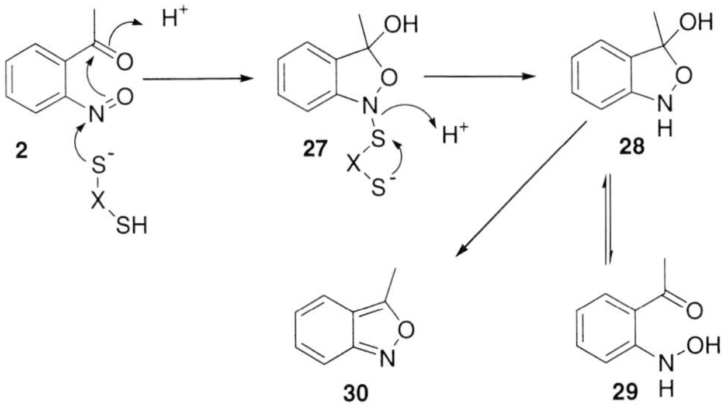

Scheme 1.1.4 Reactions of 2-nitrosoacetophenone **2** with a dithiol such as dithiothreitol. Note discussion in the text that the scheme does not apply to reactions with a monothiol such as glutathione, where the hydroxylamine product is not obtained.

photolysis, or the presence of reduced glutathione within cells ensures the same protection. The chemistry of the reaction of thiols with 2-nitrosoacetophenone (**2**) (the by-product from photolysis of NPE-caged compounds) has been clarified (Scheme 1.1.4) [42].

With DTT as the thiol, rapid-scan FTIR spectroscopy was just able to observe the intermediate **27** (no carbonyl absorption), which was rapidly reduced to a tautomeric mixture of **28** and **29**. These would be the predominant species present in a typical caged compound experiment. Dehydration to 3-methylanthranil (**30**) took place on a much slower time scale ($t_{1/2}$ 9 min at pH 7, 35 °C) [42]. However, this study of the thiol reaction of 2-nitrosoacetophenone cannot necessarily be applied to by-products from other types of cage or with other thiols. For example, Chen and Burka [53] reported that 2-nitrosobenzaldehyde, the by-product of the NB cage, reacts with a monothiol such as N-acetylcysteine to give 2-aminobenzaldehyde and not the hydroxylamine analogous to **28**/**29**. We re-examined this work and found that the product from the aldehyde and the ketone depends upon the thiol. With a monothiol (N-acetylcysteine in our work), both compounds give the corresponding amine, while, with a dithiol (DTT), both give the corresponding hydroxylamine (J. E. T. Corrie and V. R. N. Munasinghe, unpublished data). The different products presumably reflect operation of distinct pathways, controlled by kinetic competition after initial addition of thiol to the nitroso group. For a discussion of these reaction pathways, see Ref. [54]. An ingenious alternative method to remove the reactive nitroso by-product was described by Pirrung and colleagues [55], who prepared compounds incorporating

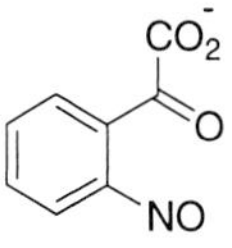

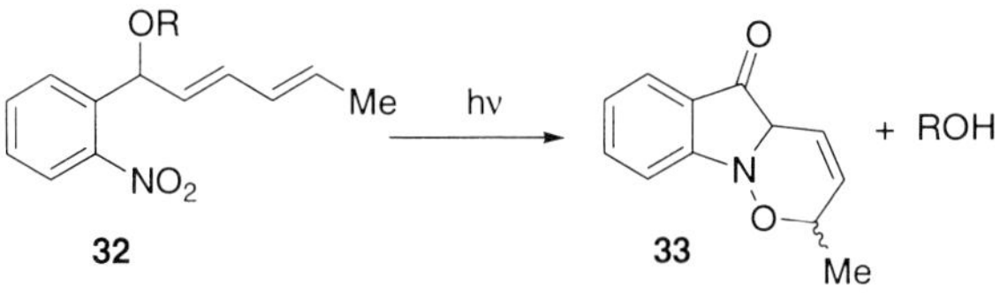

Scheme 1.1.5 Overall photolysis reaction of **32**, in which the diene side chain traps the nitrosoketone by-product of photolysis as a Diels-Alder adduct.

a pentadienyl system (such as **32**) that was able to trap the nitroso group as an intramolecular Diels-Alder adduct (**33**) (Scheme 1.1.5). Applications of this method have not yet been reported.

Many papers on the CNB cage **14** show the by-product as the nitrosopyruvate **31**, but there is little evidence for the existence of this compound, nor has its reaction with thiols been characterized. Indeed, FTIR data from a range of CNB caged compounds show that there is significant decarboxylation upon photolysis (J. E. T. Corrie and A. Barth, unpublished data), as might be expected from published data on simple nitrophenylacetates [56] and a related 2-nitrophenylglycine derivative [57]. The strong absorption band at 2343 cm^{-1}, characteristic for CO_2 in water, occurs in a region missing from the only published IR study of flash photolysis of a CNB-caged compound [58]. Current work (J. E. T. Corrie and A. Barth) aims to quantify the extent of CO_2 loss from CNB-caged compounds and to provide more definite evidence on the formation of **31**.

To summarize, investigators should be cautious in extrapolating from the relatively well-understood photocleavage of NPE-caged ATP **1** to other systems. A striking example is given by a caged diazenium diolate, that was assumed to undergo normal photocleavage of its NB group and ultimately to generate nitric oxide [59]. Subsequent studies showed that photolysis proceeded by an entirely different mechanism. The nitro group had essentially no effect on the reaction course, and formation of nitrous oxide was a minor pathway [60]. This case may be the most divergent from usual expectations but underscores the danger of unverified extrapolations. In cases where the product release rate is critical, simple measurements of *aci*-nitro decay rates may not be adequate to define the rate of product release (see the above discussion of NPE-caged alcohols **21** for a striking example). The identity of intermediates and by-products may be of less concern to most investigators, although, for example, in cases where fluorescence recordings are made after triggering a bioprocess by flash photolysis of a caged compound, awareness of light absorption of intermediates or by-products may be necessary to avoid optical artifacts (see Ref. [61] for an example of correcting a fluorescence signal for the transient inner filter effect of an *aci*-nitro intermediate).

1.1.3.1.2 Representative Survey of Nitrobenzyl-Caged Compounds

The following section is not a comprehensive coverage of the literature, but aims to pick out points of particular interest as a guide to general principles and opportunities for future work. The different types of 2-nitrobenzyl cage are covered in this section, which is mostly subdivided into different types of functional groups that have been caged. The first two sections deal with particular classes of caged species, as this seems a more rational classification.

Nucleotides As mentioned above, photolysis of NPE-caged ATP has been studied by FTIR difference spectroscopy, confirming directly that the *aci*-nitro decay rate is rate determining for release of ATP [42, 62]. The IR spectral assignments are discussed elsewhere in this volume. Work on caged cyclic nucleotides since 1993 has described NPE- and DMNB-caged versions of 8-bromo-cAMP and -cGMP and reinforces previous conclusions that the axial isomers of these caged compounds are significantly more resistant to hydrolysis than the equatorial isomers, as well as that electron-donating groups in the cage group increase the hydrolysis rate [63–64]. Caged cyclic nucleotides where the NB group bears a charged substituent that has electron-withdrawing properties (for example cGMP derivative **34**) have high stability and solubility in water [65]. Responses to photorelease of cGMP from either NPE-caged cGMP or **34** have been compared [66].

Extensive studies have led to nicotinamide coenzymes caged in various ways. NAD was caged on its pyrophosphate group (as for cyclic ADP-ribose, discussed in Section 1.1.2) and NADP on the 2′-phosphate of its adenosine moiety, in both cases with the DMNPE cage [27]. The amide nitrogen of the nicotinamide in both nucleotides was CNB-caged by a mix of enzymatic and chemical steps [27]. The caged NADP **35** had an *aci*-nitro decay rate of only 30 s^{-1}, very much slower than reported for most other CNB-caged compounds. These caged NADP compounds were used in time-resolved crystallographic studies of isocitrate dehydrogenase [67]. In related work, several NADP compounds caged on the nicotinamide were synthesized with NB, DMNB, CNB, NPE and a novel 4-carboxy-NPE group. These papers [68, 69] include useful synthetic detail on 2-nitrobenzyl chemistry and show evidence for long-lived intermediates in the photolysis of some of the caged NADP derivatives, postulated to be aminols analogous to the hemiacetal described in Section 1.1.3.1.1. The available data suggest that these putative aminols have lifetimes of hundreds of seconds (pH 7.3) [69], so the compounds are unlikely to be useful in applications requiring high time-resolution. It is noteworthy that there appears to be some discrepancy between the data for similar CNB-caged NAD/NADP compounds [27, 67, 69], and further investigation may be appropriate. These studies again underline the importance of reliably establishing product release rates for new caged compounds.

In addition to studies with simple mono- and dinucleotides, oligonucleotide sequences can be photomanipulated either by simple uncaging or by generating strand breaks. Thus, caged RNA sequences have been accessed from an NB-caged precursor such as **36**, that can be used in automated oligonucleotide synthesis [70, 71]. Photolysis releases intact RNA. These site-specific caged sequences contrast with non-specific modification of the backbone phosphates in DNA or RNA by a 4-coumarinylmethyl cage [72]. In complementary work, single- and double-strand breaks at specific sites in DNA sequences can be photogenerated by caged linkers introduced into the sequence during chemical synthesis [73, 74]. Structure **36** is an example that generates both 3'- and 5'-phosphate-terminated strand breaks directly by photolysis of each cage group. In an alternative approach, DNA oligomers with a 2-nitrobenzyl group at C-5' can directly generate breaks terminated in a 5'-phosphate but need base treatment to liberate the 3'-end [75]. Creation of strand breaks in specific locations is expected to aid studies of nucleic acid repair processes.

Peptides and proteins Current reviews [76, 77] detail much past work in this field, so only recent studies with features of particular interest are covered here. A further very recent review covers synthesis of caged proteins by biosynthetic incorporation of unnatural amino acids, which allows caging at specific sites with a flexibility impossible to achieve by chemical modification of whole proteins [78]. Thus, caging of peptides and proteins is moving from early (often non-specific) modifications to more highly designed blocking. In one such study, self-assembly of an amyloidogenic peptide was blocked by attachment of a pentacationic modified peptide joined to the amyloidogenic peptide by a photocleavable nitrobenzyl-type linker. Irradiation cleaved the inhibitory unit and initiated fibril formation [79]. Photocleavage kinetics were not determined, but, based on related data [80], are probably $\sim 2000\ s^{-1}$ (pH 7.5, 20 °C). A new promising strategy is peptide caging by a 2-nitrobenzyl group on a backbone amide, so the caging position is not limited by the presence of particular side chains [81, 82]. The *aci*-nitro decay rate is rapid ($\sim 27\,000\ s^{-1}$ at pH 7, ambient temperature), although the observation of long-lived aminols from other caged amides [69] raises caution about the actual rate of product release. A novel methionine-caged protein (horse heart cytochrome *c*) was prepared by alkylation at pH 1.5, where only the methionine side chains were reactive [83]. This strategy could only be applicable to proteins that refold after exposure to such low pH but may be useful for methionine-containing peptides. In a different strategy, the Michael acceptor **37** has been used for specific alkylation of thiol groups, for example, to prepare a caged papain [84]. Lastly, serine, threonine, and tyrosine phosphopeptides have been prepared in NPE-caged form, using an NPE-phosphoramidite reagent for direct introduction of the caged phosphate group during solid-phase peptide synthesis [30].

MeO   NO₂

MeO   NO₂   **37**

Caged carboxylates NB- and NPE-caged carboxylate groups generally have slow *aci*-nitro decay rates, limiting their value in studies of rapid processes (see Ref. [85] for the recent example of NPE-caged isocitrate, $k \approx 60\ s^{-1}$ at pH 7, 25 °C). One means to overcome this slow rate has been with the CNB group, extensively studied as CNB esters of neuroactive amino acids including L-glutamate [86], GABA [87], NMDA [88], and glycine [89]. Recently, a modified CNB group with an additional *para*-carboxylate (DCNB) has been used as a more hydrophilic cage on D-aspartate (compound **38**) [90]. *Aci*-nitro decay rates for these CNB esters are typically biphasic (although the relative amplitude of the phases varies between compounds), with the major component having rates (pH ~ 7, ambient temperature) in the range $20\,000$–$150\,000\ s^{-1}$. The fast component for DCNB ester **38** was slightly slower, at $16\,500\ s^{-1}$. Time-resolved infrared measurements indicate that the release rate of L-glutamate from its CNB γ-ester parallels the *aci*-nitro decay rate [58]. This compound had the least biphasic *aci*-nitro decay of all those described (90% of the amplitude at $33\,000\ s^{-1}$), so the IR measurements do not help interpret the release rates from compounds with a higher proportion of a slow phase in the decays. Reported spontaneous hydrolysis rates for CNB esters close to neutral pH and at ambient temperature are ~1% in 24 h, which in general requires some care to minimize contamination by free amino acid, but all these compounds have been used in experiments with appropriate isolated neuronal cells. It was reported that CNB-caged GABA had no significant pharmacological effects [87], but a recent study has found significant and novel pharmacology for this compound [91].

Apart from the nitroindoline cage discussed in Section 1.1.3.2, two other cages for carboxylates have been described. One is the 2,2′-dinitrobenzhydryl group, used as the β-ester of NMDA [92], where the *aci*-nitro decay rate was 165 000 s^{-1} and cleanly monophasic (pH 7, 22 °C), as well as being ~5-fold faster than for the CNB ester of the same carboxylate. The compound was reported to be stable to hydrolysis under these conditions, but the analogous caged glycine was rapidly hydrolyzed at pH 7 [93], as would be expected for the ester of an α-amino acid, so this cage group must be applied with caution. The second cage is a hydroxylated NPE structure, as in caged arachidonic acid **39**. This paper noted that CNB-caged fatty acids were chemically unstable and that the hydroxy-NPE compounds had *aci*-nitro decay kinetics significantly faster than for NPE-caged carboxylates [94].

In applications outside the neuroscience area, a range of caged strong acids, such as trichloroacetic acid, has been described for photorelease in anhydrous solvents in conjunction with photodirected oligonucleotide synthesis [95]. Modified lipids have been described that have additional carboxylate groups at the terminus of the normal fatty acid chains. When these carboxylates were caged (with NB, DMNB, NPE, or DMNPE groups) the lipids could be assembled into liposomes. Irradiation to release these carboxylates destabilized the liposomes and allowed leakage of their contents [96]. Similar results were obtained for liposomes with a DMNB-carbamate-caged amino group of a phosphatidylethanolamine [97].

Caged amines and amides Amines can be caged in two principal forms, either by direct attachment of a caging group, in which case the charge state of the amine is maintained, or indirect attachment via a carbonyl group to form a carbamate, in which case the amine, normally cationic at physiological pH, becomes a neutral species. Recent examples of each strategy applied to sphingolipids have been reviewed [98] and are of particular interest for examples of chemoenzymatic synthesis of caged compounds. Otherwise, little new work on direct amine cages has appeared in the last decade (apart from caged Ca^{2+} reagents, see below). Note that unpublished data (J. E. T. Corrie and A. Barth) indicate that compounds such as N-2-nitrobenzylglycine (**40**) undergo substantial decarboxylation upon photolysis. Work is in hand to quantify the extent of this side reaction, which is also observed with the isomeric 4-nitro compound, indicating that it operates by a different photochemical process than the normal cleavage reaction.

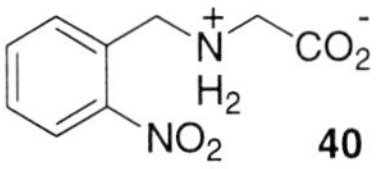

Carbamate derivatives of amines involve two stages for amine release: photocleavage of the caging group to leave a carbamate salt, which then undergoes non-photochemical decarboxylation (Scheme 1.1.6). Either step can be rate-limiting for overall release of the amine. This has been explored for two cages which illustrate rate limitation by different processes. Thus the NPE-caged glutamate **41** and related carbamate-caged amino acids are rate limited by decay of the *aci*-nitro intermediate, the process which generates the carbamate salt [99]. In contrast, the dimethoxybenzoin derivative **43** generates the carbamate salt on a sub-µs time scale, and release of the amine is controlled by the thermal decarboxylation (150 s^{-1} at pH 7, 21 °C) [100]. The decarboxylation rate will vary with the pK of the particular amino group but will always be an upper limit on the amine release rate from any carbamate derivative. Strategies that focus on rapid photocleavage of the cage group from carbamates, as from the CNB-derivative **42** [101], cannot avoid this limitation.

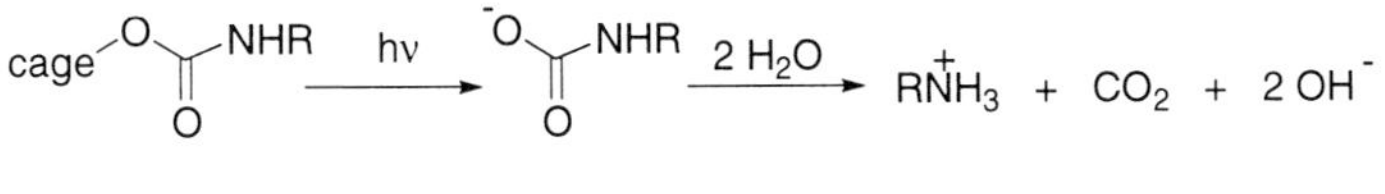

Scheme 1.1.6 Generalized scheme for photolysis of a carbamate-caged amine, showing the separate stages of photocleavage of the cage and thermal decarboxylation of the resulting carbamate salt.

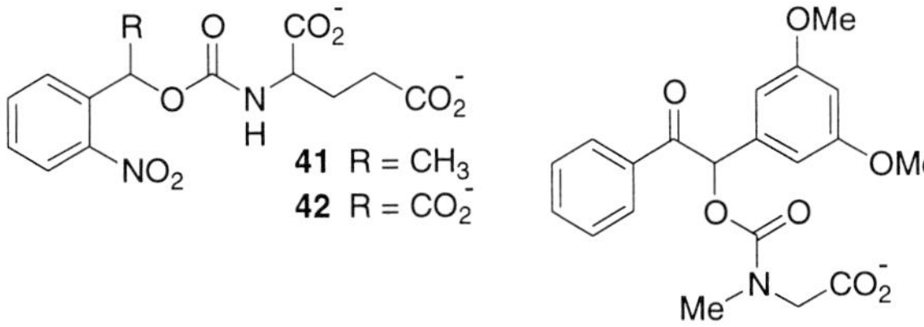

Most new work on caged amides has been described above in relation to caged NAD and NADP and has a particular caveat discussed in that Section about long-lived intermediates in the photocleavage process that limit the product release rate. In other applications, a series of NB, NPE and CNB-caged derivatives of asparagine, glutamine, GABA amide, and glycinamide were investigated [102]. Photolysis rates of the NB-, NPE- and CNB-caged glutamines were 385, 1925 and 900 s^{-1} respectively (pH 7.5, ambient temperature), where relevant being for the faster phase of biphasic traces. The suggested mechanism shows an aminol intermediate, but no evidence was presented for this or for its possible effect on the product release rate. Finally, the caged biotins **44** (R=H or CH$_3$) have been used as a means to control assembly of arrays of biological molecules on surfaces [103, 104]. By uniformly coating a surface with a caged biotin and irradiating defined areas through a mask, patterns of tethered biotin can be created which subsequently bind avidin or streptavidin and allow immobilization of species via conventional biotin/avidin layers.

Caged alcohols and phenols A major finding for NPE-caged alcohols has been discussed in Section 1.1.3.1.1 in relation to a long-lived hemiacetal intermediate that causes much slower product release than inferred from the *aci*-nitro decay rate [45]. This rate limitation appears to be general for NPE-caged alcohols, so applies to the NPE-caged choline reported by Peng and Goeldner [47], invalidating the μs time scale claimed for choline release. The NPE-caged 2-deoxyglucose **45b** has been described [105] but without data on release kinetics. Other caged sugars include the NB-glucoside and ether derivatives **45a** [105] and **46a,b** [46, 106]. As noted in Section 1.1.3.1.1, these and other NB-caged alcohols studied to date appear not to involve a long-lived hemiacetal intermediate in their photocleavage step, so product release is at the *aci*-nitro decay rate. However, a very recent detailed study of 2-nitrobenzyl methyl ether by Wirz and co-workers [107] did find evidence for a rate-limiting hemiacetal ($t_{1/2} \approx 2$ s at pH 7, 23 °C), and subsequent studies of 2-nitrobenzyloxyacetic acid gave similar results (J. Wirz, personal communication). In contrast, biological responses to release of glucose from the NB-caged glucose **46a** were faster [106], suggesting either a significantly faster decay rate or non-involvement of a hemiacetal. On balance, it seems probable that the process of product release from NB- and NPE-caged alcohols always proceeds via a hemiacetal. However, particularly in the NB case, the extent to which this intermediate accumulates and so limits the rate of product release is apparently very sensitive both to the precise compound and to the solution conditions. Further work to clarify aspects of this reaction is in progress (J. Wirz and J. E. T. Corrie, unpublished data).

A novel 4-nitro-CNB cage is present in the caged diacyl glycerol **47** [108], which has been used in a number of physiological studies. The additional nitro group was not commented upon but probably facilitates the synthetic route. As well as the free acid **47**, the corresponding methyl ester was prepared, from which the photolysis by-product **48** was isolated and characterized. Notably, in view of the discussion of decarboxylation during photolysis of other CNB compounds (Section 1.1.3.1.1), it was reported that the free acid corresponding to **48** could not be isolated, although this was attributed to instability [108]. A related caged diacyl glycerol using the Bhc cage [20] has been prepared to enable 2-photon release at different sites within a tissue preparation [109].

Caged phenols are largely represented by derivatives of phenylephrine **49**, an α-adrenergic agonist. Walker et al. described NB-, NPE-, DMNB- and CNB-derivatives caged on the phenolic oxygen, with *aci*-nitro rates (pH 7, 22 °C) between $\sim 2\,s^{-1}$ (NB) and $\sim 2000\,s^{-1}$ (CNB) [110]. Other workers prepared NB-, DMNB- and CNB-derivatives caged on the amino group [111, 112] and concluded that the *N*-linked NB-caged compound best blocked pharmacological activity [112]. The *N*-linked compounds tended to have faster *aci*-nitro decay rates than the corresponding *O*-linked compounds.

The CNB-caged phenolic carbonate **50** was prepared to study effects of 2,5-di-*t*-butylhydroquinone, a reversible inhibitor of the sarcoplasmic Ca^{2+} ATPase [113]. As for the carbamates discussed above (Scheme 1.1.6), release of the end product involves both photochemical and thermal cleavage steps. Here the *aci*-nitro decay rate was biphasic with its major component at $\sim 3800\,s^{-1}$, while the thermal loss of CO_2 was estimated at $\sim 130\,s^{-1}$ so was rate limiting for release of the phenol. It is noteworthy that the rate of CO_2 loss from an aliphatic monoalkyl carbonate, as formed on photolysis of a caged alcohol such as the Bhc-caged diacyl glycerol mentioned above [109], is predicted to be very much slower on the basis of previous studies of this decarboxylation [114], with an estimated half-time of $\sim 30\,s$ (pH 7) for release of the final product. This may not be important for diacyl glycerols, which probably diffuse slowly from the photo-

Caged phosphates Most work related to caged phosphates has been described in earlier sections relating to general synthetic methods or to nucleotides. A substantial advance mentioned in Section 1.1.2 is the use of phosphoramidite reagents to introduce a caged phosphate group as a complete entity, as opposed to modification of an existing phosphate [28–31]. Applications include synthesis of a caged sphingosine phosphate [29] and caged phosphopeptides [30]. An interesting use has been to prepare inositol phosphates caged on specific phosphate groups [28, 115], in contrast to the method originally used for IP$_3$, where caging was distributed among the three phosphate groups and relied on separation to obtain the biologically inert P^4- or P^5-caged isomers [116]. Note that regiospecific caging has also been achieved for a thiophosphate analog, 1-D-*myo*-inositol 1,4-bisphosphate 5-phosphorothioate, where the *S*-caged derivative **51** was formed by alkylating the free thiophosphate with 1-bromo-1-(2-nitrophenyl)-ethane [117]. In the context of inositol phosphates, an alternative approach to a caged IP$_3$ analog is **52**, caged on a free hydroxyl group and derivatized on the phosphates to make the compound cell-permeant. It can be loaded into cells where it is hydrolyzed by non-specific esterases to generate the caged IP$_3$ analog within the cell [118]. Previous uses of intracellular caged IP$_3$ have generally involved patch-clamped single cells, where the reagent was introduced via the patch pipette (for example, see Ref. [119]). Finally, in this Section, caging of phosphates has been extended to caged phosphoramides to prepare potential prodrugs of phosphoramide mustards [120].

Caged protons Several classes of caged compound, including nitrobenzyl-type cages, release a proton during the photolytic process (Scheme 1.1.1), but the focus of interest is usually directed to the other released species, as in all the compounds described above. Depending on the pK of the released effector compound, the proton may remain free or be taken up again as the effector is released. In most cases where the proton remains free, it is rapidly buffered by cellular or external medium buffers, but in experimental situations with minimal buffering, the released proton induces rapid acidification. To allow study of rapid proton-mediated events alone, it is desirable to use a reagent from which the other released component is not bioactive. One such is 2-nitrobenzaldehyde, which photoisomerizes to 2-nitrosobenzoic acid [12, 121, 122]. Other compounds more directly related to the general 2-nitrobenzyl type are the NPE-caged hydroxyphenyl phosphate **53** [123] and NPE-caged sulfate **54** [124]. The pH jumps that can be achieved by these compounds are restricted by the pK of the other released species, which will buffer the released proton. Thus **53** can achieve acidification to about pH 5, while **54** can reach pH ~2 (if sufficient photolysis can be achieved to generate ~10 mM proton concentration). For each of these compounds, liberation of the proton upon photolysis should occur within a ~25-ns laser flash [44].

Caged calcium Caged calcium reagents are unique among caged compounds, since release of Ca^{2+} depends on a change in the affinity of a photolabile chelating agent rather than direct rupture of a covalent bond between the calcium ion and the cage. The process is illustrated for DM-nitrophen (**55**) in Scheme 1.1.7, which shows rupture of the intact high-affinity chelator to fragments of much lower affinity. A very recent comprehensive review discusses interplay of the calcium affinity before and after photolysis, the desired free Ca^{2+} level before and after photolysis, and how photochemical parameters of different available cages influence the choice of reagent. It also covers progress with 2-photon photolysis of caged calcium reagents [125]. The main point of interest not in the review concerns details of the photochemical cleavage, which appears not solely to be as depicted in Scheme 1.1.7. In unpublished work (J. E. T. Corrie and A. Barth), we have observed substantial CO_2 formation on photolysis of DM-nitrophen, with or without Ca^{2+} present. Work now in hand aims to quantify the extent of this decarboxylation process and to determine to what extent it may affect modeling of the Ca^{2+} transients [126] induced by flash photolysis.

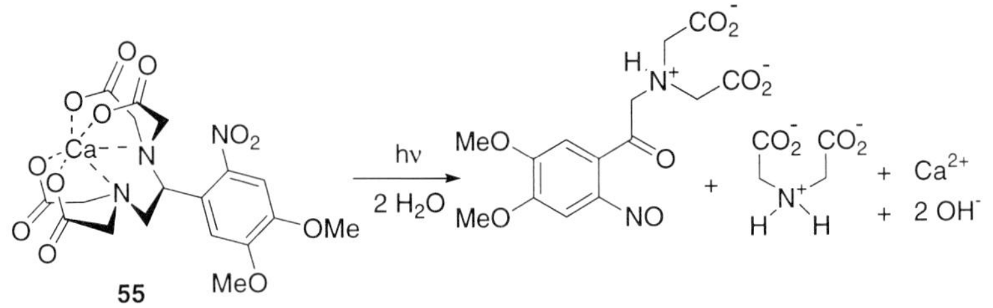

Scheme 1.1.7 Release of Ca^{2+} upon cleavage of a photolabile Ca^{2+} chelating agent (DM-nitrophen).

Caged fluorophores General strategy in caging fluorophores is to perturb the electronic structure by attachment of the caging group to make the molecule either non-fluorescent or very weakly so. Photolysis allows the delocalized electronic structure of the free fluorophore to re-establish, thereby regenerating fluorescence. An early example of the approach is caged fluorescein (**56**), in which alkylation of both phenolic groups locks the compound into the non-fluorescent lactone form [127]. The field is dominated by applications of the reagents rather than by particular chemical innovation. Most chemical and cell biological work with caged fluorophores has been by the Mitchison group in a series of elegant studies and has been recently reviewed [128]. A variant of a caged Q-rhodamine linked to dextran for cell lineage measurements has been described [129]. None of the work focuses on release rates of the fluorophores, as the cell biology applications have been more concerned with spatial definition of the released fluorescence than with high time resolution. Outside cell biology, the caged hydroxypyrenetrisulfonate **57** has been used to define the amplitude and/or temporal stability of concentration jumps of biologically active compounds [17, 130]. This is a useful technique for photolysis in small volumes, such as on a microscope stage, where the photolysis light is focused only on part of the field, and direct measurement of the released concentration is difficult. The surprising 1- and 2-photon photochemistry of NPE-caged 7-hydroxycoumarins is another recent approach but needs further work to establish its generality [25].

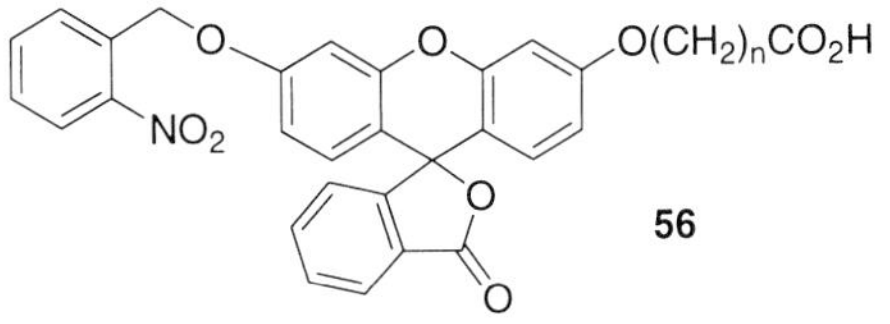

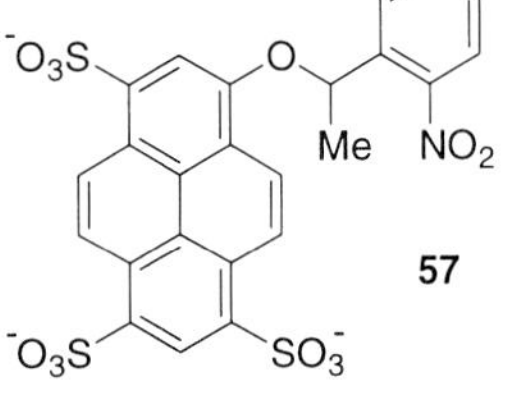

1.1.3.2 7-Nitroindoline Cages

As mentioned in Section 1.1.2, interest in this caging group was revived only in 1999, so this Section contains much less material than that on 2-nitrobenzyl cages. Applications to date are restricted to photorelease of neuroactive amino acids, but photochemical synthesis of amides in organic solutions has also been described, where an acyl group on the indoline nitrogen is transferred to a different primary or secondary amine [131–134].

1.1.3.2.1 Mechanistic and Structural Aspects of Photochemical Cleavage of 1-Acyl-7-nitroindolines

Unlike 2-nitrobenzyl compounds, photocleavage of nitroindoline cages gives different by-products in organic and aqueous media. The original work by Amit and colleagues [13] in aprotic solvent plus 1% water gave products shown in Scheme 1.1.2, where the water present in the solvent was incorporated into the released carboxylic acid. The same process enables the photochemical amide syntheses mentioned above. However, in 100% aqueous solution, the reaction gives not a nitroindoline but a nitrosoindole (Scheme 1.1.8), as well as the carboxylic acid: solvent water is not incorporated into the products [135]. There is a smooth transition from the nitroindoline to the nitrosoindole by-product as the proportion of water in the reaction solvent increases from 1 to 100% [136]. The carboxylate release rate in fully aqueous medium is $\sim 5 \times 10^6 \, s^{-1}$, and photocleavage involves a triplet state of the nitroindoline [136]. It was inferred that a common intermediate, the carboxylic nitronic anhydride **58**, was formed in either organic or aqueous solution, with partition into the different reaction products being determined by the ionizing power of the solvent and to some extent by the pK of the departing carboxylate group.

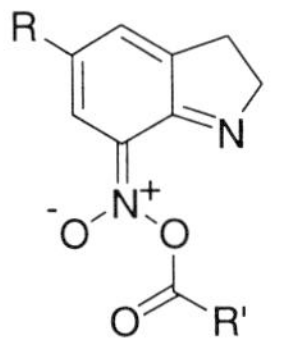

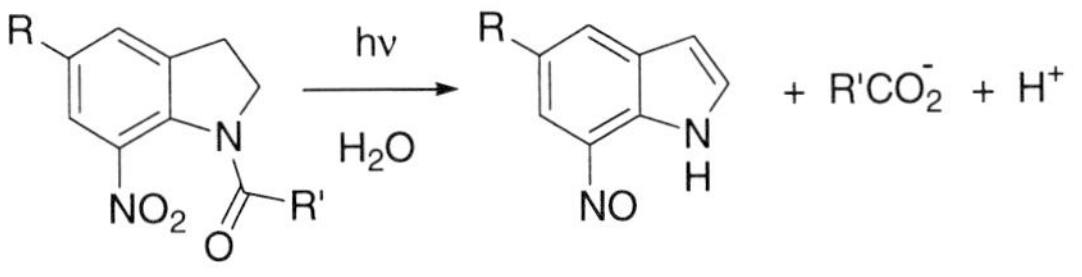

Scheme 1.1.8 Photolysis reaction of 1-acyl-nitroindolines in aqueous solution, showing formation of the nitrosoindole by-product.

In the recrudescence of the nitroindoline cage, we first replaced the 5-bromo substituent used by Amit [13] by a substituted alkyl group, as in **59** [135]. This led to improved solubility and an ~2-fold gain in photoefficiency. A substituent at C5 is useful to prevent nitration there, which otherwise takes place in competition with C7 nitration. In later work, different substituents were used, aiming to improve photosensitivity by enhanced near-UV absorption. Of three substituted species investigated, the 4-methoxy compound **60** had the best gain in sensitivity over **59**, although during synthesis a varying proportion of the 5-nitro isomer was always obtained [137, 138]. Blocking the 5-position in **61** to avoid 5-nitration led to decreased photosensitivity, probably because steric effects reduce overlap of the 4-substituent with the aromatic ring. Compound **62**, substituted with the strongly electron-releasing 4-dimethylamino group, was inert upon irradiation, probably because a competing process involving electron transfer from the tertiary amino substituent quenches an excited state [137].

60 R = H, R' = OMe
61 R = Me, R' = OMe
62 R = H, R' = NMe$_2$

1.1.3.2.2 Survey of 7-Nitroindoline Caged Compounds

The discussion focuses on successful applications of the nitroindoline cage, but also, to illustrate its limitations, describes unpublished cases where the methodology has failed. Our work has been directed principally to synthesis of caged neuroactive amino acids, largely L-glutamate, GABA, and glycine. Initial work produced nitroindoline-caged versions of these compounds, namely **63–65**. Note that for synthetic reasons the side chain ester at C5 of the indoline in the caged glycine was hydrolyzed to the free acid, but this did not affect photochemical release [17, 135]. Significant points about these compounds were their very high resistance to hydrolysis, so no leakage from the caged species occurs upon incubation near neutral pH [135], and the fact that their photochemical release rates [136] are more than adequate to mimic physiological release. The glutamate conjugate **63** had no detectable pharmacological activity before photolysis at concentrations up to at least 1 mM, but this was not so for the GABA **64** and glycine **65** conjugates, which showed evidence of binding to the relevant receptors [17]. At least in the case of GABA, this result, together with reported pharmacological activity of CNB-caged GABA [91], shows that an optimal caged GABA remains a challenge.

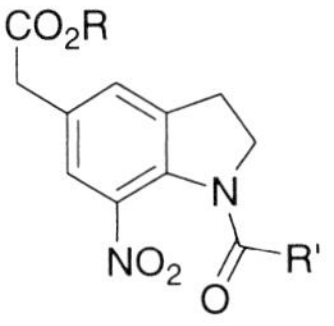

63 R = Me, R' = (CH$_2$)$_2$CH(NH$_3^+$)CO$_2^-$
64 R = Me, R' = (CH$_2$)$_3$NH$_3^+$
65 R = H, R' = CH$_2$NH$_3^+$

As mentioned above, later studies showed that a 4-methoxy substituent is beneficial to photolysis efficiency and the L-glutamate analog of **60** has been used in several physiological applications [17, 139–142], including work where localized release was achieved by 2-photon photolysis [21–23]. These results, together with the synthetic applications of nitroindoline photolysis described above [131–134] indicate that the strategy has broad applicability to caging carboxylic acids. However, in the course of attempts to extend applicability of the method, we have encountered examples where this cage was not useful (J. E. T. Corrie and G. Papageorgiou, unpublished data). Studies of the nitroindoline-caged Ca^{2+} chelating agent **66** were ultimately frustrated as the compound did not photolyze either in the presence or absence of Ca^{2+}. This is similar but not identical to data of Adams and colleagues, who reported that the dinitroindoline derivative **67** photolyzed only when excess Ca^{2+} was present [143]. We also prepared the EDTA derivative **68** and found that it photolyzed cleanly in the presence of a large excess of Ca^{2+}, but without the metal ion it decomposed to a complex, uncharacterized mixture, accompanied by CO$_2$ formation. We speculatively attribute the failure of **66** to photolyze to a quenching of the excited state by the electron-rich aryl rings of the BAPTA moiety, although this does not fully explain the disparate reactivity of **66** and **67**. Some additional evidence for possible quenching came from the nitroindolinyl carbamate derivative **69**, which was synthesized as a potential alternative to the caged fluorophore **57**. Compound **69** was also resistant to photolysis. The behavior of the EDTA derivative **68** is broadly consistent with electron transfer from the aliphatic tertiary amino group(s) to an excited state of the nitroindoline. Such single-electron transfer in this compound would be expected to cause decarboxylation, well precedented for cation radicals of α-amino acids [144].

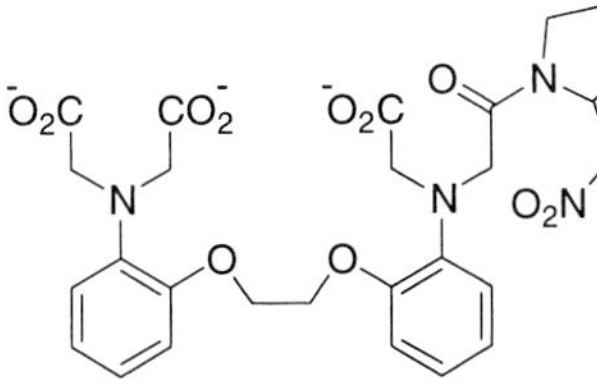

66 R = Me
67 R = NO$_2$

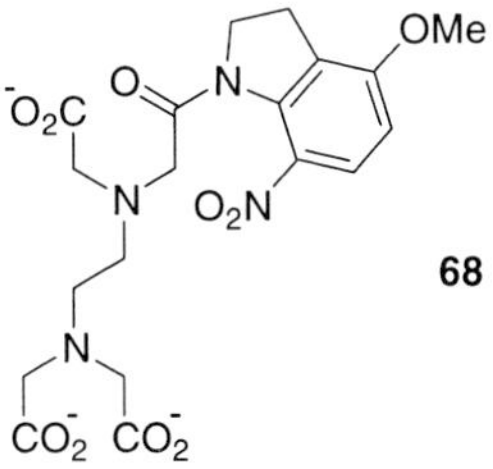

68

69

Despite these isolated failures, we continue to develop the strategy, notably in pursuit of higher photolytic efficiency and elimination of residual pharmacological activity in derivatives such as the caged GABA **64**. Very recently, we have shown that intramolecular triplet sensitization by an attached benzophenone can significantly enhance photosensitivity [145], and other results of these studies will be reported in the future.

1.1.4
Conclusion

Although caged compounds of the 2-nitrobenzyl type have been in use for 25 years, opportunities for modification of existing cages and extension to individual new caged effectors continue to arise. Hand in hand with this effort are mechanistic studies of particular compounds that clarify release rates of effectors and underline that the *aci*-nitro decay rate does not always correspond to the rate of product release. Development of the 7-nitroindoline cage has provided new opportunities that have begun to impact on neurophysiological research. Nevertheless, our understanding of many nuances of these photocleavage processes remains incomplete, for example, the failure adequately to explain the frequently observed biphasic rates of *aci*-nitro decay processes, the implications of the decarboxylation reactions observed for CNB cages, and the still somewhat unpredictable reactivity of the nitroindoline system. Several examples described in this chapter indicate that more attention should be paid to direct determination of product release rates, especially in cases where this is a critical parameter in physiological experiments. Progress with localized release by multi-photon uncaging makes this particularly important. If an end product is only released after a delayed series of dark reactions, the advantages of localized photolysis will be lost by diffusion of the photolytically generated intermediate. To conclude, cooperation between groups with separate expertise in chemistry, physiology, and optics is likely to deliver further improvements in reagents for the future.

References

1 J.E.T. Corrie, D.R. Trentham, In *Bioorganic Photochemistry*, H. Morrison, Ed., Wiley, New York, Vol. 2, pp 243–305, **1993**.
2 S.R. Adams, R.Y. Tsien, *Annu. Rev. Physiol.*, **1993**, *55*, 755–784.
3 J.H. Kaplan, *Annu. Rev. Physiol.*, **1990**, *52*, 897–914.
4 G. Marriott, Ed. *Methods in Enzymology*, Vol. 291, Academic Press, San Diego, **1998**.
5 A.P. Pelliccioli, J. Wirz, *Photochem. Photobiol. Sci.*, **2002**, *1*, 441–458.
6 R.S. Givens, P.G. Conrad, A.L. Yousef, J.I. Lee, In *CRC Handbook of Organic Photochemistry and Photobiology*, W. Horspool, F. Lenci, Eds., 2nd Edition, CRC Press, Boca Raton, pp 69.1–69.46, **2003**.
7 J.H. Kaplan, B. Forbush, J.F. Hoffman, *Biochemistry*, **1978**, *17*, 1929–1935.
8 J.A. McCray, L. Herbette, T. Kihara, D.R. Trentham, *Proc. Natl. Acad. Sci. USA*, **1980**, *77*, 7237–7241.
9 Y.E. Goldman, M.G. Hibberd, J.A. McCray, D.R. Trentham, *Nature*, **1982**, *300*, 701–705.
10 H.A. Lester, J.M. Nerbonne, *Ann. Rev. Biophys. Bioeng.*, **1982**, *11*, 151–175.
11 A.M. Gurney, H.A. Lester, *Physiol. Rev.*, **1987**, *67*, 583–617.
12 G. Ciamician, P. Silber, *Chem. Ber.*, **1901**, *34*, 2040–2046.
13 B. Amit, D.A. Ben-Efraim, A. Patchornik, *J. Am. Chem. Soc.*, **1976**, *98*, 843–844.
14 S. Pass, B. Amit, A. Patchornik, *J. Am. Chem. Soc.*, **1981**, *103*, 7674–7675.
15 C.P. Holmes, *J. Org. Chem.*, **1997**, *62*, 2370–2380.
16 H. Thirlwell, J. Sleep, M.A. Ferenczi, *J. Muscle Res. Cell Motil.*, **1995**, *16*, 131–137.
17 M. Canepari, L. Nelson, G. Papageorgiou, J.E.T. Corrie, D. Ogden, *J. Neurosci. Methods*, **2001**, *112*, 29–42.
18 W. Denk, J.H. Strickler, W.W. Webb, *Science*, **1990**, *248*, 73–76.
19 N.I. Kiskin, R. Chillingworth, J.A. McCray, D. Piston, D. Ogden, *Eur. Biophys. J.*, **2002**, *30*, 588–604.
20 T. Furuta, S.S.H. Wang, J.L. Dantzker, T.M. Dore, W.J. Bybee, E.M. Callaway, W. Denk, R.Y. Tsien, *Proc. Natl. Acad. Sci. USA*, **1999**, *96*, 1193–1200.
21 M. Matsuzaki, G.C.R. Ellis-Davies, T. Nemoto, Y. Miyashita, M. Iino, H. Kasai, *Nat. Neurosci.*, **2001**, *4*, 1086–1092.
22 M.A. Smith, G.C.R. Ellis-Davies, J.C. Magee, *J. Physiol.*, **2003**, *548*, 245–258.
23 M. Matsuzaki, N. Honkura, G.C.R. Ellis-Davies, H. Kasai, *Nature*, **2004**, *429*, 761–766.
24 J.W. Walker, G.P. Reid, J.A. McCray, D.R. Trentham, *J. Am. Chem. Soc.*, **1988**, *110*, 7170–7177.
25 Y. Zhao, Q. Zheng, K. Dakin, K. Xu, M.L. Martinez, W.H. Li, *J. Am. Chem. Soc.*, **2004**, *126*, 4653–4663.
26 R. Aarhus, K. Gee, H.C. Lee, *J. Biol. Chem.*, **1995**, *270*, 7745–7749.
27 B.E. Cohen, B.L. Stoddard, D.E. Koshland, *Biochemistry*, **1997**, *36*, 9035–9044.
28 J. Chen, G.D. Prestwich, *Tetrahedron Lett.*, **1997**, *38*, 969–972.
29 L. Qiao, A.P. Kozikowski, A. Olivera, S. Spiegel, *Bioorg. Med. Chem. Lett.*, **1998**, *8*, 711–714.
30 D.M. Rothman, M.E. Vázquez, E.M. Vogel, B. Imperiali, *Org. Lett.*, **2002**, *4*, 2865–2868.
31 C. Dinkel, O. Wichmann, C. Schultz, *Tetrahedron Lett.*, **2003**, *44*, 1153–1155.
32 C.G. Bochet, *Tetrahedron Lett.*, **2000**, *41*, 6341–6346.
33 A. Blanc, C.G. Bochet, *J. Org. Chem.*, **2002**, *67*, 5567–5577.
34 A. Blanc, C.G. Bochet, *J. Am. Chem. Soc.*, **2004**, *126*, 7174–7175.
35 S. Walbert, W. Pfleiderer, U.E. Steiner, *Helv. Chim. Acta*, **2001**, *84*, 1601–1611.
36 A. Hasan, K.P. Stengele, H. Giegrich, P. Cornwell, K.R. Isham, R.A. Sachleben, W. Pfleiderer, R.S. Foote, *Tetrahedron*, **1997**, *53*, 4247–4264.
37 M. Beier, A. Stephan, J.D. Hoheisel, *Helv. Chim. Acta*, **2001**, *84*, 2089–2095.

38 M.C. Pirrung, L. Wang, M.P. Montague-Smith, *Org. Lett.*, **2001**, *3*, 1105–1108.
39 R.W. Yip, D.K. Sharma, R. Giasson, D. Gravel, *J. Phys. Chem.*, **1985**, *89*, 5328–5330.
40 R.W. Yip, Y.X. Wen, D. Gravel, R. Giasson, D.K. Sharma, *J. Phys. Chem.*, **1991**, *95*, 6078–6081.
41 T. Luchian, S.H. Shun, H. Bayley, *Angew. Chem. Int. Ed.*, **2003**, *42*, 1926–1929.
42 A. Barth, J.E.T. Corrie, M.J. Gradwell, Y. Maeda, W. Mäntele, T. Meier, D.R. Trentham, *J. Am. Chem. Soc.*, **1997**, *119*, 4149–4159.
43 Y.V. Il'ichev, J. Wirz, *J. Phys. Chem. A*, **2000**, *104*, 7856–7870.
44 M. Schwörer, J. Wirz, *Helv. Chim. Acta*, **2001**, *84*, 1441–1457.
45 J.E.T. Corrie, A. Barth, V.R.N. Munasinghe, D.R. Trentham, M.C. Hutter, *J. Am. Chem. Soc.* **2003**, *125*, 8546–8554.
46 J.E.T. Corrie, *J. Chem. Soc., Perkin Trans. 1*, **1993**, 2161–2166.
47 L. Peng, M. Goeldner, *J. Org. Chem.*, **1996**, *61*, 185–191.
48 G.C.R. Ellis-Davies, J.H. Kaplan, R.J. Barsotti, *Biophys. J.*, **1996**, *70*, 1006–1016.
49 I.R. Dunkin, J. Gebicki, M. Kiszka, D. Sanin-Leira, *J. Chem. Soc., Perkin Trans. 2*, **2001**, 1414–1425.
50 J.E.T. Corrie, J. Baker, E.M. Ostap, D.D. Thomas, D.R. Trentham, *J. Photochem. Photobiol. A*, **1998**, *115*, 49–55.
51 J.E.T. Corrie, B.C. Gilbert, V.R.N. Munasinghe, A.C. Whitwood, *J. Chem. Soc., Perkin Trans. 2*, **2000**, 2483–2491.
52 L. Zhang, R. Buchet, G. Azzar, *Biophys. J.*, **2004**, *86*, 3873–3881.
53 L.J. Chen, L.T. Burka, *Tetrahedron Lett.*, **1998**, *39*, 5351–5354.
54 C. Diepold, P. Eyer, H. Kampffmeyer, K. Reinhardt, *Adv. Exp. Biol. Med.*, **1982**, *136B*, 1173–1181.
55 M.C. Pirrung, Y.R. Lee, K. Park, J.B. Springer, *J. Org. Chem.*, **1999**, *64*, 5042–5047.
56 J.D. Margerum, C.T. Petrusis, *J. Am. Chem. Soc.*, **1969**, *91*, 2467–2472.
57 C.D. Woodrell, P.D. Kehayova, A. Jain, *Org. Lett.*, **1999**, *1*, 619–621.
58 V. Jayaraman, S. Thiran, D.R. Madden, *FEBS Lett.*, **2000**, *475*, 278–282; Q. Cheng, M.G. Steinmetz, V. Jayaraman, *J. Am. Chem. Soc.*, **2002**, *124*, 7676–7677.
59 L.R. Makings, R.Y. Tsien, *J. Biol. Chem.*, **1994**, *269*, 6282–6285.
60 A. Srinivasan, N. Kebede, J.E. Saavedra, A.V. Nikolaitchik, D.A. Brady, E. Yourd, K.M. Davies, L.K. Keefer, J.P. Toscano, *J. Am. Chem. Soc.*, **2001**, *123*, 5465–5472.
61 Z.H. He, R.K. Chillingworth, M. Brune, J.E.T. Corrie, M.R. Webb, M.A. Ferenczi, *J. Physiol.*, **1999**, *517*, 839–854.
62 A. Barth, K. Hauser, W. Mäntele, J.E.T. Corrie, D.R. Trentham, *J. Am. Chem. Soc.*, **1995**, *117*, 10311–10316.
63 V. Hagen, C. Djeza, S. Frings, J. Bendig, E. Krause, U.B. Kaupp, *Biochemistry*, **1996**, *35*, 7762–7771.
64 V. Hagen, C. Djeza, J. Bendig, I. Baeger, U.B. Kaupp, *J. Photochem. Photobiol. B*, **1998**, *42*, 71–78.
65 L. Wang, J.E.T. Corrie, J.F. Wootton, *J. Org. Chem.*, **2002**, *67*, 3474–3478.
66 J. Pollock, J.H. Crawford, J.F. Wootton, J.E.T. Corrie, R.H. Scott, *Neurosci. Lett.*, **2003**, *338*, 143–146.
67 B.L. Stoddard, B.E. Cohen, M. Brubaker, A.D. Mesecar, D.L. Koshland, *Nat. Struct. Biol.*, **1998**, *5*, 891–897.
68 C.P. Salerno, M. Resat, D. Magde, J. Kraut, *J. Am. Chem. Soc.*, **1997**, *119*, 3403–3404.
69 C.P. Salerno, D. Magde, A.P. Patron, *J. Org. Chem.*, **2000**, *65*, 3971–3981.
70 S.G. Chaulk, A.M. MacMillan, *Nucl. Acids Res.*, **1998**, *26*, 3173–3178.
71 S.G. Chaulk, A.M. MacMillan, *Angew. Chem. Int. Ed.*, **2001**, *40*, 2149–2152.
72 H. Ando, T. Furuta, R.Y. Tsien, H. Okamoto, *Nat. Genet.*, **2001**, *28*, 317–325.
73 P. Ordoukhanian, J.S. Taylor, *Bioconjug. Chem.*, **2000**, *11*, 94–103.

74 K. Zhang, J. S. Taylor, *Biochemistry*, **2001**, *40*, 153–159.

75 A. Dussy, C. Meyer, E. Quennet, T. A. Bickle, B. Giese, A. Marx, *Chembiochem*, **2002**, *3*, 54–60.

76 Y. Shigeri, Y. Tatsu, N. Yumoto, *Pharmacol. Ther.*, **2001**, *91*, 85–92.

77 G. Marriott, P. Roy, K. Jacobson, *Methods Enzymol.*, **2003**, *360*, 274–288.

78 E. J. Petersson, G. S. Brandt, N. M. Zacharias, D. A. Dougherty, H. A. Lester, *Methods Enzymol.*, **2003**, *360*, 258–273.

79 C. J. Bosques, B. Imperiali, *J. Am. Chem. Soc.*, **2003**, *125*, 7530–7531.

80 D. Ramesh, R. Wieboldt, A. P. Billington, B. K. Carpenter, G. P. Hess, *J. Org. Chem.*, **1993**, *58*, 4599–4605.

81 Y. Tatsu, T. Nishigaki, A. Darszon, N. Yumoto, *FEBS Lett.*, **2002**, *525*, 20–24.

82 M. C. Pirrung, S. J. Drabik, J. Ahamed, H. Ali, *Bioconjug. Chem.*, **2000**, *11*, 679–681.

83 T. Okuno, S. Hirota, O. Yamauchi, *Biochemistry*, **2000**, *39*, 7538–7545.

84 R. Golan, U. Zehavi, M. Naim, A. Patchornik, P. Smirnoff, M. Herchmann, *J. Protein Chem.*, **2000**, *19*, 117–122.

85 M. J. Brubaker, D. H. Dyer, B. Stoddard, D. E. Koshland, *Biochemistry*, **1996**, *35*, 2854–2864.

86 R. Wieboldt, K. R. Gee, L. Niu, D. Ramesh, B. K. Carpenter, G. P. Hess, *Proc. Natl. Acad. Sci. USA*, **1994**, *91*, 8752–8756.

87 K. R. Gee, R. Wieboldt, G. P. Hess, *J. Am. Chem. Soc.*, **1994**, *116*, 8366–8367.

88 K. R. Gee, L. Niu, K. Schaper, G. P. Hess, *J. Org. Chem.*, **1995**, *60*, 4260–4263.

89 C. Grewer, J. Jäger, B. K. Carpenter, G. P. Hess, *Biochemistry*, **2000**, *39*, 2063–2070.

90 K. Schaper, S. A. M. Mobarekeh, C. Grewer, *Eur. J. Org. Chem.*, **2002**, 1037–1046.

91 P. Molnár, J. V. Nadler, *Eur. J. Pharmacol.*, **2000**, *391*, 255–262.

92 K. R. Gee, L. Niu, V. Jayaraman, G. P. Hess, *Biochemistry*, **1999**, *38*, 3140–3147.

93 S. Ueno, J. Nabekura, H. Ishibashi, N. Akaike, T. Mori, M. Shiga, *J. Neurosci. Methods*, **1995**, *58*, 163–166.

94 J. Xia, X. Huang, R. Sreekumar, J. W. Walker, *Bioorg. Med. Chem. Lett.*, **1997**, *7*, 1243–1248.

95 P. J. Seranowski, P. B. Garland, *J. Am. Chem. Soc.*, **2003**, *125*, 962–965.

96 K. Yamaguchi, Y. Tsuda, T. Shimakage, A. Kusumi, *Bull. Chem. Soc. Jpn.*, **1998**, *71*, 1923–1929.

97 Z. Y. Zhang, B. D. Smith, *Bioconjug. Chem.*, **1999**, *10*, 1150–1152.

98 R. H. Scott, J. Pollock, A. Ayer, N. M. Thatcher, U. Zehavi, *Methods Enzymol.*, **2000**, *312*, 387–400.

99 J. E. T. Corrie, A. De Santis, Y. Katayama, K. Khodakhah, J. B. Messenger, D. C. Ogden, D. R. Trentham, *J. Physiol.*, **1993**, *465*, 1–8.

100 G. Papageorgiou, J. E. T. Corrie, *Tetrahedron*, **1997**, *53*, 3917–3932.

101 F. M. Rossi, M. Margulis, C. M. Tang, J. P. Y. Kao, *J. Biol. Chem.*, **1997**, *272*, 32933–32939.

102 D. Ramesh, R. Wieboldt, A. P. Billington, B. K. Carpenter, G. P. Hess, *J. Org. Chem.*, **1993**, *58*, 4599–4605.

103 S. A. Sundberg, R. W. Barrett, M. Pirrung, A. L. Lu, B. Kiangsoontra, C. P. Holmes, *J. Am. Chem. Soc.*, **1995**, *117*, 12050–12057.

104 M. C. Pirrung, C. Y. Huang, *Bioconjug. Chem.*, **1996**, *7*, 317–321.

105 S. Watanabe, R. Hirokawa, M. Iwamura, *Bioorg. Med. Chem. Lett.*, **1998**, *8*, 3375–3378.

106 R. Lux, V. R. N. Munasinghe, F. Castellano, J. W. Lengeler, J. E. T. Corrie, S. Khan, *Mol. Biol. Cell*, **1999**, *10*, 1133–1146.

107 Y. V. Il'ichev, M. A. Schwörer, J. Wirz, *J. Am. Chem. Soc.*, **2004**, *126*, 4581–4595.

108 R. Sreekumar, Y. Q. Pi, X. P. Huang, J. W. Walker, *Bioorg. Med. Chem. Lett.*, **1997**, *7*, 341–346.

109 V. G. Robu, E. S. Pfeiffer, S. L. Robia, R. C. Balijepalli, Y. Q. Pi, T. J. Kamp, J. W. Walker, *J. Biol. Chem.*, **2003**, *278*, 48154–48161.

110 J.W. WALKER, H. MARTIN, F.R. SCHMITT, R.J. BARSOTTI, *Biochemistry*, **1993**, *32*, 1338–1345.

111 S. MURALIDHARAN, J.M. NERBONNE, *J. Photochem. Photobiol. B*, **1995**, *27*, 123–137.

112 W.A. BOYLE, S. MURALIDHARAN, G.M. MAHER, J.M. NERBONNE, *J. Photochem. Photobiol. B*, **1997**, *41*, 233–244.

113 F.M. ROSSI, J.P.Y. KAO, *J. Biol. Chem.*, **1997**, *272*, 3266–3271.

114 Y. POCKER, B.L. DAVIDSON, T.L. DEITS, *J. Am. Chem. Soc.*, **1978**, *100*, 3564–3567.

115 C. DINKEL, C. SCHULTZ, *Tetrahedron Lett.*, **2003**, *44*, 1157–1159.

116 J.W. WALKER, J. FEENEY, D.R. TRENTHAM, *Biochemistry*, **1989**, *28*, 3272–3280.

117 J.F. WOOTTON, J.E.T. CORRIE, T. CAPIOD, J. FEENEY, D.R. TRENTHAM, D.C. OGDEN, *Biophys. J.*, **1995**, *68*, 2601–2607.

118 W.H. LI, J. LLOPIS, M. WHITNEY, G. ZLOKARNIK, R.Y. TSIEN, *Nature*, **1998**, *392*, 936–941.

119 D. OGDEN, T. CAPIOD, *J. Gen. Physiol.*, **1997**, *109*, 741–756.

120 R. REINHARD, B.F. SCHMIDT, *J. Org. Chem.*, **1998**, *63*, 2434–2441.

121 G. BONETTI, A. VECLI, C. VIAPPIANI, *Chem. Phys. Lett.*, **1997**, *269*, 268–273.

122 S. ABBRUZZETTI, C. VIAPPIANI, J.R. SMALL, L.J. LIBERTINI, E.W. SMALL, *Biophys. J.*, **2000**, *79*, 2714–2721.

123 S. KHAN, F. CASTELLANO, J.L. SPUDICH, J.A. MCCRAY, R.S. GOODY, G.P. REID, D.R. TRENTHAM, *Biophys. J.*, **1993**, *65*, 2368–2382.

124 A. BARTH, J.E.T. CORRIE, *Biophys. J.*, **2002**, *83*, 2864–2871.

125 G.C.R. ELLIS-DAVIES, *Methods Enzymol.*, **2003**, *360*, 226–238.

126 G.C.R. ELLIS-DAVIES, J.H. KAPLAN, R.J. BARSOTTI, *Biophys. J.*, **1996**, *70*, 1006–1016.

127 G.A. KRAFFT, W.R. SUTTON, R.T. CUMMINGS, *J. Am. Chem. Soc.*, **1988**, *110*, 301–303.

128 T.J. MITCHISON, K.E. SAWIN, J.A. THERIOT, K. GEE, A. MALLAVARAPU, Chapter 4 of Ref. [4].

129 K.R. GEE, E.S. WEINBERG, D.J. KOZLOWSKI, *Bioorg. Med. Chem. Lett.*, **2001**, *11*, 2181–2183.

130 R. JASUJA, J. KEYOUNG, G.P. REID, D.R. TRENTHAM, S. KHAN, *Biophys. J.*, **1999**, *76*, 1706–1719.

131 K.C. NICOLAOU, B.S. SAFINA, N. WINSSINGER, *Synlett*, **2001**, *SI*, 900–903.

132 C. HELGEN, C.G. BOCHET, *Synlett*, **2001**, 1968–1970.

133 K. VIZVARDI, C. KREUTZ, A.S. DAVIS, V.P. LEE, B.J. PHILMUS, O. SIMO, K. MICHAEL, *Chem. Lett.*, **2003**, *32*, 348–349.

134 C. HELGEN, C.G. BOCHET, *J. Org. Chem.*, **2003**, *68*, 2483–2486.

135 G. PAPAGEORGIOU, D.C. OGDEN, A. BARTH, J.E.T. CORRIE, *J. Am. Chem. Soc.*, **1999**, *121*, 6503–6504.

136 J. MORRISON, P. WAN, J.E.T. CORRIE, G. PAPAGEORGIOU, *Photochem. Photobiol. Sci.*, **2002**, *1*, 960–969.

137 G. PAPAGEORGIOU, J.E.T. CORRIE, *Tetrahedron*, **2000**, *56*, 8197–8205.

138 G. PAPAGEORGIOU, J.E.T. CORRIE, *Synth. Commun.*, **2002**, *32*, 1571–1577.

139 M. CANEPARI, G. PAPAGEORGIOU, J.E.T. CORRIE, C. WATKINS, D. OGDEN, *J. Physiol.*, **2001**, *533*, 765–772.

140 M. CANEPARI, D. OGDEN, *J. Neurosci.*, **2003**, *23*, 4066–4071.

141 G. LOWE, *J. Neurophysiol.*, **2003**, *90*, 1737–1746.

142 M. CANEPARI, C. AUGER, D. OGDEN, *J. Neurosci.*, **2004**, *24*, 3563–3573.

143 S.R. ADAMS, J.P.Y. KAO, R.Y. TSIEN, *J. Am. Chem. Soc.*, **1989**, *111*, 7957–7968.

144 K.O. HILLER, B. MASLOCH, M. GÖBL, K.D. ASMUS, *J. Am. Chem. Soc.*, **1981**, *103*, 2734–2743.

145 G. PAPAGEORGIOU, M. LUKEMAN, P. WAN, J.E.T. CORRIE, *Photochem. Photobiol. Sci.*, **2004**, *3*, 366–373.

1.2
Coumarin-4-ylmethyl Phototriggers

Toshiaki Furuta

1.2.1
Introduction

Coumarin-4-ylmethyl groups are newly developed phototriggers that have been used to make caged compounds of phosphates, carboxylates, amines, alcohols, phenols, and carbonyl compounds. Coumarin, the parent compound of the chromophore, is the common name of benzo-α-pyrone, highlighted in bold with numbering in Scheme 1.2.1. The photochemistry typifying coumarin-4-ylmethyls can be realized as that of a member of arylalkyl-type photo-removable protecting groups [1]. Upon photolysis, the C-heteroatom bond (mostly oxygen) between C-4 methylene and X (leaving groups) is cleaved to produce an anion of the leaving group and a solvent-trapped coumarin as a photo by-product.

This chapter presents an overview of spectroscopic and photochemical properties, and synthetic methods of the reported coumarin-type phototrigger structural variants. The application of coumarin phototriggers remains limited in number, but the importance of coumarins as potential replacements for conventional 2-nitrobenzyls has been accepted widely. References that have appeared through January 2004 will be addressed.

Studies of coumarins have mainly addressed the utilization of their fluorescent properties [2]: fluorescent labeling agents, fluorogenic enzyme substrates, and laser dyes. About 20 years ago, Givens and Matuszewski first noticed that a phosphate ester of (coumarin-4-yl)methanol is photosensitive [3]. A benzene solution of (7-methoxycoumarin-4-yl)methyl ester of diethylphosphate (MCM-DEP) was photolyzed with a quantum yield of 0.038. They also demonstrated its use as a fluorescent labeling agent for nucleophilic molecules (Nu) including proteins, suggesting generation of an electrophilic coumarin-4-ylmethyl cation upon photolysis (Scheme 1.2.2).

About 10 years later, the reaction was reinvestigated as a replacement for the 2-nitrobenzyl-type cages [4]. Photochemical properties of MCM-cAMP were compared to those of the 1-(2-nitrophenyl)ethyl and desyl ester of cAMP under a simulated physiological environment. MCM-cAMP was photolyzed to produce

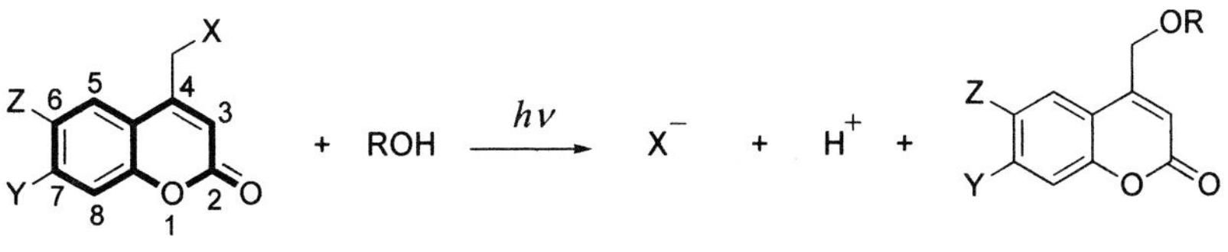

Scheme 1.2.1

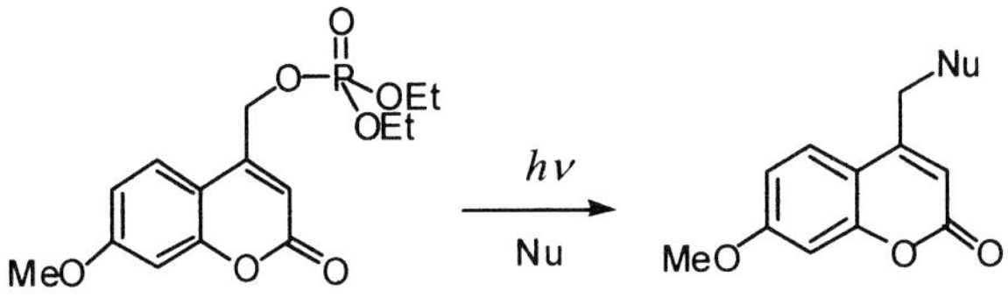

Scheme 1.2.2

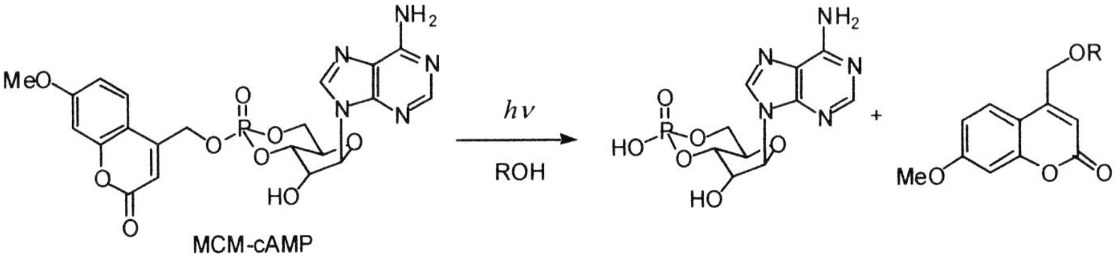

MCM-DEP

Scheme 1.2.3

MCM-cAMP

the parent cAMP with nearly quantitative yield upon 340-nm irradiation (Scheme 1.2.3). It also offers advantages over those reported previously, including improved stability in the dark and high photolytic efficiency [5]. From that time, MCM group and their structural variants, with their improved properties, have been synthesized and employed to produce several caged compounds, such as second messengers, neurotransmitters and DNA/RNA. Some of those caged compounds have been applied successfully to the investigation of cell chemistry.

1.2.2
Spectroscopic and Photochemical Properties

1.2.2.1 Overview

Fig. 1.2.1 shows structures of coumarin phototriggers reported thus far. They can be grouped into (1) 7-alkoxy group [3–18], (2) 6,7-dialkoxy group [14, 15, 18–21], (3) 6-bromo-7-alkoxy group [18, 22–29], and (4) 7-dialkylamino group

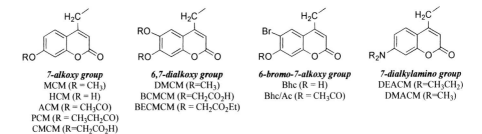

7-alkoxy group
MCM (R = CH$_3$)
HCM (R = H)
ACM (R = CH$_3$CO)
PCM (R = CH$_3$CH$_2$CO)
CMCM (R=CH$_2$CO$_2$H)

6,7-dialkoxy group
DMCM (R=CH$_3$)
BCMCM (R=CH$_2$CO$_2$H)
BECMCM (R = CH$_2$CO$_2$Et)

6-bromo-7-alkoxy group
Bhc (R = H)
Bhc/Ac (R = CH$_3$CO)

7-dialkylamino group
DEACM (R=CH$_3$CH$_2$)
DMACM (R=CH$_3$)

Fig. 1.2.1 Structures and acronyms of coumarin-4-ylmethyl phototriggers.

[14, 15, 18, 30, 31] in view of their structural similarity. Photochemical and photophysical properties as well as physical properties of coumarin-caged compounds depend on the nature of the structure of attached molecules. Especially important are the types of functional groups to be protected. For that reason, the properties of each compound will be collected and their differences assessed one by one for each functional group. Notwithstanding, it would be useful to overview spectroscopic and photochemical properties of the four groups to emphasize their distinctions and differences.

Tab. 1.2.1 summarizes the reported photochemical and photophysical properties of the four groups. The absorption properties of the groups differ remarkably. Absorption spectra of the four coumarin-4-ylmethanols – (7-methoxycoumarin-4-yl)methanol (MCM-OH), (6,7-dimethoxycoumarin-4-yl)methanol (DMCM-OH), (6-bromo-7-hydroxycoumarin-4-yl)methanol (Bhc-OH) and (7-diethylaminocoumarin-4-yl)methanol (DEACM-OH) – which represent the four groups as well as 4,5-dimethoxy-2-nitrobenzyl alcohol (NVOC-OH) were recorded in a simulated physiological environment (Fig. 1.2.2; M. Kawamoto, T. Watanabe, unpublished results). The parent coumarin-4-ylmethanol has its absorption maximum at 310 nm [15], whereas an electron-donating substitution on the C6 or C7 of the coumarin ring shifts the absorption maximum to longer wavelengths. The dialkylamino substitution on the C7 produces a strong red

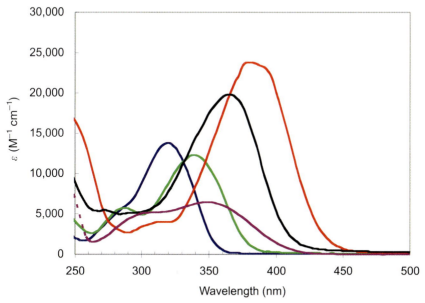

Fig. 1.2.2 UV/Vis spectra of coumarin-4-ylmethanols in KMOPS (pH 7.2). Red: DEACM-OH; Black: Bhc-OH; Green: DMCM-OH; Blue: MCM-OH; Purple: NVOC-OH, 337: N_2 laser (337 nm), 355: YAG laser (355 nm), 364: Ar laser (364 nm), 405: blue laser (405 nm). BP 330–385: transparent wavelength range of a wide band pass filter BP 330–385.

Tab. 1.2.1 Spectroscopic and photochemical properties of the coumarin phototriggers

Groups[a]	7-Alkoxy MCM, HCM, ACM, PCM and CMCM	6,7-Dialkoxy DMCM, BCMCM and BECMCM	6-Bromo-7-alkoxy Bhc and Bhc/Ac	7-Dialkylamino DEACM and DMACM
λ_{max}[b]	325	345	375[c]	395
ε_{max}[d]	4000–12 000	10 000–12 000	13 000–19 000	16 000–20 000
Φ_{chem}[e]	ca. 0.1	ca. 0.1	ca. 0.1	ca. 0.3
k[f]	ca. 10^8			ca. 10^9
λ_{em}[g] (Φ_f[h])	394 (0.65)	438 (0.59)	465 (0.61)	484 (0.082) 491 (0.21)
Light sources[i]	Xe, Hg, 337	Xe, Hg, 355, 364	Xe, Hg, 355, 364, 405	Xe, Hg, 364, 405
Functional groups[j]	P, C, S, Al	P, Al	P, C, Am, Al, Ph	P, Al
Comments	ACM and PCM are membrane permeable. CMCM has high water solubility	BECMCM is membrane permeable. BCMCM has high water solubility	Bhc/Ac is membrane permeable. Bhc has high 2-photon absorption cross-sections	Can be activated by longer wavelength UV/vis with improved efficiency
Refs.	3–18	14, 15, 18–21	18, 22–29	14, 15, 18, 30–32

a) The acronyms and full IUPAC nomenclature are: MCM, (7-methoxycoumarin-4-yl)methyl; HCM, (7-hydroxycoumarin-4-yl)methyl; ACM, (7-acetoxycoumarin-4-yl)methyl; PCM, (7-propionyloxycoumarin-4-yl)methyl; CMCM, (7-carboxymethoxycoumarin-4-yl)methyl; BCMCM, [6,7-bis(carboxymethoxy)coumarin-4-yl]methyl; BECMCM, [6,7-bis(ethoxycarbonylmethoxy)coumarin-4-yl]methyl; DMCM, (6,7-dimethoxycoumarin-4-yl)methyl; DMACM, (7-dimethylaminocoumarin-4-yl)methyl; DEACM, (7-diethylaminocoumarin-4-yl)methyl; and Bhc, (6-bromo-7-hydroxycoumarin-4-yl)methyl.
b) Absorption maximum (nm).
c) at pH 7. The value depends on a pH.
d) Molar absorptivity ($M^{-1}cm^{-1}$).
e) Quantum yields for disappearance of starting materials upon irradiation.
f) Rate constants of photolysis (s^{-1}).
g) Emission maxima (nm) of the MCM-OH, DMCM-OH, Bhc-OH, DEACM-OH, and DMACM-OH.
h) Fluorescence quantum yield.
i) Possible light sources. Xe: Xe lamps, Hg: Hg lamps, 337: N_2 laser (337 nm), 355: YAG laser (355 nm), 364: Ar laser (364 nm), 405: blue laser (405 nm)
j) Functional groups that have been caged. P: phosphates, C: carboxylates, S: sulfates, Al: alcohols, Am: amines, Ph: phenols.

shift of more than 80 nm. Bhc group was designed to lower the pKa of the C7 hydroxyl group so that the O-H bond is deprotonated at pH 7, causing a red shift of the absorption maximum by 60 nm.

Substantial overlaps between emission profiles of light sources and absorption profiles of phototriggers are necessary to achieve maximum photolysis efficiency. Therefore, ideal wavelength regions for excitations are 300–340 nm for the MCM, 320–360 nm for the DMCM, 330–420 nm for the Bhc, and 350–450 nm for the DEACM groups. Regarding cell biological applications, photoactivation is usually done by fluorescent microscopes equipped with an extra light source for photoactivation, including a Xe lamp with a band pass UV filter (for example, BP 330–385) and a UV laser. Light sources for epi-illumination can also be used for photoactivation. Tab. 1.2.1 shows possible light sources for each group and wavelengths.

Most coumarins have strong fluorescence, which may overlap with emission spectra of some fluorescent probes and cause difficulties in monitoring specific effects by fluorescence imaging after photoactivation. Tab. 1.2.1 summarizes emission maximum wavelengths and fluorescence quantum yields of the representative photo-byproducts 4-hydroxymethylcoumarins.

Overall, coumarin-type phototriggers offer the following advantages over other phototriggers: (1) large extinction coefficient at the wavelength greater than 350 nm; (2) high photolysis efficiency upon UV irradiation; (3) acceptable stability in the dark; (4) fast photolysis kinetics; and (5) practically useful 2-photon excitation cross-sections. Furthermore, the absorption properties, membrane permeability, and water solubility can be optimized easily by the nature of substitutions on the coumarin ring. In addition, diverse collections of phototriggers are already available.

1.2.2.2 Phototriggers

1.2.2.2.1 MCM Groups: 7-Alkoxy-Substituted Coumarins

Each group has a membrane-permeable and a water-soluble structural variant. They are designed to facilitate methods to incorporate the compounds into biological systems. The MCM family comprises the MCM, ACM, PCM, HCM, and CMCM groups. Most compounds reported to date are phosphate esters. Groups can also cage carboxylates, sulfates, and alcohols. The ACM and the PCM groups are designed to improve membrane permeability. After incorporation into live cells by simple diffusion, acetyl (or propionyl) moiety would be hydrolyzed by intrinsic esterases to produce a more polar HCM group that has almost no membrane permeability and might accumulate inside cells [6, 7]. Carboxylic acid in the CMCM group is almost fully ionized at a physiological pH. Consequently, CMCM-caged compounds should be highly water-soluble [14, 16]. All members have a single alkoxy substitution at the C7 position. They also have similar absorption and photochemical properties as summarized in Tab. 1.2.1. Typical values of the absorption maximum wavelength are around 325 nm with molar absorptivity of 4000 to 12 000 $M^{-1}cm^{-1}$; the value varies depending on molecular structures. Photolysis quantum yields were as high as 0.1, implying that the efficiency of photolysis, $\Phi\varepsilon$, the product of photolysis quantum yield and molar absorptivity, would be up to 1000. That value is more than one order of magnitude larger than those of 2-nitrobenzyls when photolysis is done at around 325 nm. Early examples using MCM-type phototriggers proved advantageous over others [5, 11–13]. Nevertheless, the absorption maximum at 325 nm is not ideal for cell biological applications. For that reason, new structural variants with longer absorption maxima have been developed: longer than 350 nm is desirable.

1.2.2.2.2 DMCM Groups: 6,7-Dialkoxy-Substituted Coumarins

The family consists of BCMCM, BECMCM, and DMCM groups, and is utilized to make caged compounds of phosphates and alcohols [14, 15, 18–21]. The BCMCM group has two carboxylic acids and renders the corresponding caged compounds highly water soluble [14]. The BECMCM is the membrane-permeable version of BCMCM. The two carboxylic acids in BCMCM are converted to ethyl esters in BECMCM, and the negative charges are thereby masked. Hydrolysis of the ethyl esters by intrinsic esterases is postulated after the compound is incorporated into live cells [20]. Tab. 1.2.1 summarizes the spectroscopic and photochemical properties of the groups. Longer-wavelength absorption maxima were red shifted by 20 nm compared to those of the MCM-type. Single alkoxy substitution on C6 must be responsible for the red shift because Eckardt et al. reported that introduction of an electron-donating substitution on the C6 position, not C7, causes a large red shift of the absorption maxima [15]. For unknown reasons, photolysis quantum yields of the DMCM groups are always lower than those of the corresponding MCM analogs. These results show that overall photochemical properties closely resemble, or are even worse than, those of the MCM groups. Therefore, no compelling reason exists for using the DMCM groups instead of the MCM groups.

1.2.2.2.3 Bhc Groups: 6-Bromo-7-alkoxy-Substituted Coumarins

The Bhc group has been applied to cage carboxylates, amines, phosphates, alcohols, and carbonyl compounds [18, 22–29]. It has already been proved to be a replacement for conventional 2-nitrobenzyl-type phototriggers. The Bhc group can add a substantial amount of water solubility to corresponding caged compounds because most C7 phenolic hydroxyl moiety in the Bhc is ionized at physiological pH. On the other hand, caged compounds having Bhc/Ac, in which the C7 hydroxyl is masked by acetylation, accumulate remarkably inside live cells, as in the case of their debromo analog, ACM. Results show that the Bhc group can render a target molecule either water soluble or membrane permeable [28, 29]. These are advantages of the Bhc group over other phototriggers, including other members of coumarins, because no single phototrigger, except for Bhc, satisfies all the following criteria: (1) has a strong absorption band at more than 350 nm; (2) has a substantially high photolysis quantum yield; (3) has a practically usable stability; (4) renders a target molecule water soluble; and (5) adds membrane permeability by simple modification. Another advantage that must be noted is its two-photon chemistry. Two-photon-induced uncaging action cross-sections of the Bhc caged glutamates were reported to be almost 1 GM upon 740-nm irradiation, which is more than two orders of magnitude larger than those of the 2-nitrobenzyls [22].

The Bhc group was designed to solve problems observed with the HCM group. The HCM group has two absorption maxima: one at around 325 nm corresponds to the protonated form of C7 hydroxyl moiety; the other, at 375 nm, corresponds to the ionized form. The protonated form is predominant at a physiological pH (pH 7) because the pKa of the C7 hydroxyl is 7.9. Consequently, its absorption maximum at 325 nm is larger than that at 375 nm. Introduction of an electron-withdrawing group into C6 should enhance acidity of the C7 phenolic oxygen through an inductive effect. In fact, the introduction of bromo substitution on C6 caused the lowering of the pKa by almost two units. Therefore, the Bhc group has a single absorption maximum at around 375 nm with a large molar absorptivity (ca. 19 000 $M^{-1}cm^{-1}$) under a simulated physiological environment (Fig. 1.2.2). Moreover, the bromine atom is known to accelerate the rate of intersystem crossing (heavy atom effect), leading to the increase in the fraction of excited triplet state. We compared photolysis quantum yields of the four types of coumarin-caged acetates with or without halogen substitutions: HCM-OAc, Bhc-OAc, Chc-OAc, and tBhc-OAc (see Fig. 1.2.5 for structures). Single bromine substitution increased the photolysis quantum yield by 150% (HCM vs Bhc), whereas a chloro substitution, which has almost no heavy atom effect, decreased it by 40% (HCM vs Chc). These results suggest that: (1) introduction of an electron-withdrawing substitution on HCM is unfavorable for the photolysis reaction, probably because it interferes with a through-bond electron transfer from the C7 oxygen to the C2 carbonyl at an electronically excited state (see a proposed mechanism of the reaction); (2) a heavy-atom effect of a bromine atom might compensate for this unfavorable electronic interaction, suggesting the existence of a triplet state as a reactive excited state. The existence of this state is further supported by the fact that 3,6,8-tribromo-7-hydroxycoumarin-4-ylmethyl (tBhc) acetate was photolyzed with a quantum yield of 0.065, which corresponds to 260% enhancement (HCM vs tBhc).

The absorption maximum of the Bhc group would be blue shifted by 40–50 nm if photolysis were performed under an environment where the protonated form is predominant. This blue shift would occur because the large absorption band at 375 nm comes from the ionized form of the C7 hydroxyl moiety, which has a pKa of 6.2. That consequent blue shift of the group might be disadvantageous if the compound is applied to an acidic compartment such as those in mitochondria.

1.2.2.2.4 DEACM Groups: 7-Dialkylamino-Substituted Coumarins

Dialkylamino substitution on C7 improved the spectroscopic and the photochemical properties remarkably. The absorption maxima are at 390–400 nm with a molar absorptivity of 20 000 $M^{-1}\ cm^{-1}$. Photolysis quantum yields are as high as 0.3, which is the highest among the reported coumarin cages. No remarkable difference is apparent between the dimethylamino (DMACM) and the diethylamino (DEACM) variants, except that the fluorescence intensity of the DEACM-OH $\Phi_f=0.082$) is considerably smaller than that of the DMACM-OH $\Phi_f=0.21$) [15]. Reported applications of the group to caging chemistry were limited to phosphates [14, 15, 30–32] and alcohols [18]. However, the observed spectroscopic and photochemical properties were highly desirable. For those reasons, the group must be considered as a potential replacement for conventional 2-nitrobenzyl phototriggers. Taking account of the structural similarity to the Bhc group, the DEACM group must be used to protect carboxylates, amines, diols, and carbonyl compounds. Although no structural variants other than DEACM and DMACM are reported at present, modification to enhance water solubility or improve membrane permeability is possible.

1.2.2.3 Target Molecules

1.2.2.3.1 Phosphates

Caged phosphates are the most successful applications of coumarin phototriggers. Hagen, Bendig, and Kaupp's group has contributed generously to the knowledge of coumarin-caged phosphate chemistry. Their studies have mainly focused on caged cyclic nucleotides; related topics will be discussed in other sections. In this chapter, representative examples of each structural variant will be examined to elucidate their differences and similarities. Reported caged phosphates are cAMP [5–7, 13–15, 19–21, 28, 29], cGMP [13, 14, 20, 21, 28, 29], 8-Br-cAMP [10, 12, 32], 8-Br-cGMP [10–12, 32], cytidine-5′-diphosphate (CDP) [30], adenosine-5′-monophosphate (AMP), adenosine-5′-diphosphate (ADP), adenosine-5′-triphosphate (ATP) [31], DNA, and RNA [23] (Fig. 1.2.3). Tab. 1.2.2 summarizes selected examples of spectroscopic and photochemical properties.

For cyclic nucleotides, all variations are available. Slight but acceptable differences were observed between axial and equatorial isomers: the axial isomers have larger photolytic quantum yields and better stabilities in an aqueous solution in most cases. The CMCM-cNMPs [14] showed highest water solubility (200–1000 µM); the second carboxylate in the BCMCM contributes almost nothing to solubility. DEACM-cNMPs [14] have the largest photolysis quantum yields (Φ of ca. 0.21) among the coumarin-caged cNMPs. Moreover, they provide fast kinetics for the cNMP release ($k > 10^9$ s^{-1}). Bhc-cNMPs [28] showed good photosensitivity under one- and two-photon excitation conditions and a certain water solubility. In addition, they can be converted to membrane-permeable derivatives by acetylation. One can choose compounds having appropriate properties for a variety of situations. Suggested guidelines for the selection of suitable compounds could include the following:

(1) photo-activated by Xe or Hg lamp with wide-band U-filter (330–385 nm): desirable – Bhc and DEACM
(2) photo-activated by YAG (355 nm) laser: desirable – Bhc and DEACM
(3) photo-activated by 405 nm laser: desirable – DEACM; usable – Bhc

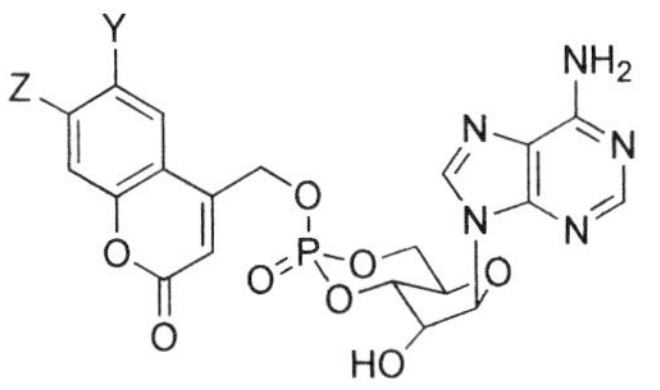

1.2 Coumarin-4-ylmethyl Phototriggers

MCM-cAMP (Y = H, Z = OCH$_3$)
ACM-cAMP (Y = H, Z = O$_2$CCH$_3$)
PCM-cAMP (Y = H, Z = O$_2$CCH$_2$CH$_3$)
HCM-cAMP (Y = H, Z = OH)
DMCM-caged cAMP (Y = Z = OCH$_3$)
CMCM-caged cAMP (Y = H, Z = OCH$_2$CO$_2$H)
BCMCM-caged cAMP (Y = Z = OCH$_2$CO$_2$H)
BECMCM-caged cAMP (Y = Z = OCH$_2$CO$_2$Et)
DEACM-caged cAMP (Y = H, Z = NEt$_2$)
DMACM-caged cAMP (Y = H, Z = NMe$_2$)
Bhc-cAMP (Y = Br, Z = OH)
Bhc-cAMP/Ac (Y = Br, Z = O$_2$CCH$_3$)

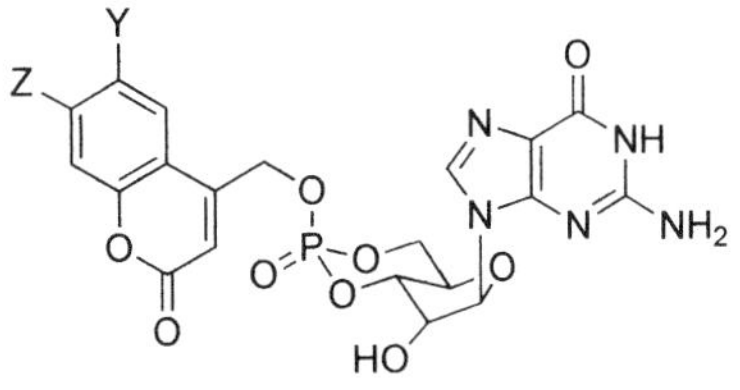

MCM-cged cGMP (Y = H, Z = OCH$_3$)
CMCM-caged cGMP (Y = H, Z = OCH$_2$CO$_2$H)
BCMCM-caged cGMP (Y = Z = OCH$_2$CO$_2$H)
BECMCM-caged cGMP (Y = Z = OCH$_2$CO$_2$Et)
DEACM-caged cGMP (Y = H, Z = NEt$_2$)
DMACM-caged cGMP (Y = H, Z = NMe$_2$)
Bhc-cGMP (Y = Br, Z = OH)
Bhc-cGMP/Ac (Y = Br, Z = O$_2$CCH$_3$)

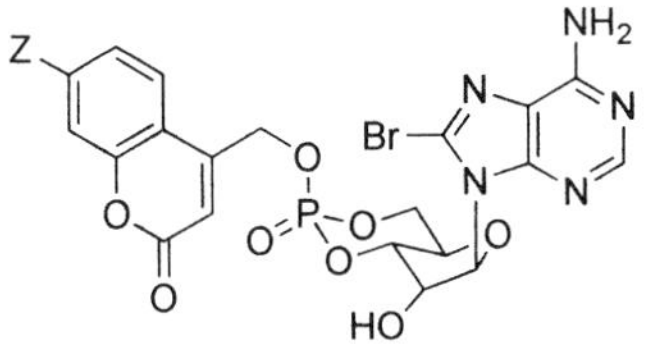

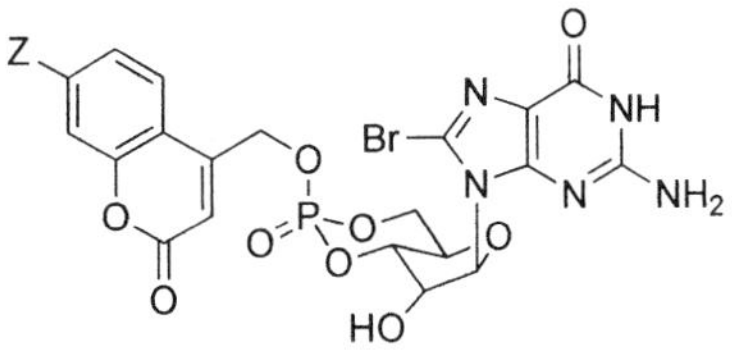

MCM-caged 8-Br-cGMP (Z = OCH$_3$)
DEACM-caged 8-Br-cGMP (Z = NEt$_2$)
DMACM-caged 8-Br-cGMP (Z = NMe$_2$)

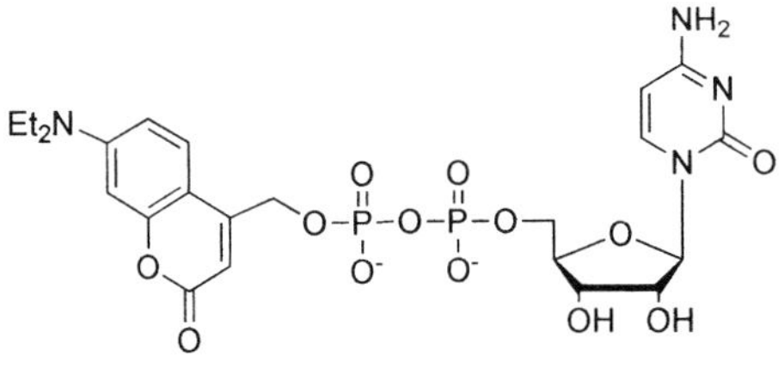

MCM-caged 8-Br-cAMP (Z = OCH$_3$)
DEACM-caged 8-Br-cAMP (Z = NEt$_2$)
DMACM-caged 8-Br-cAMP (Z = NMe$_2$)

DEACM-caged CDP

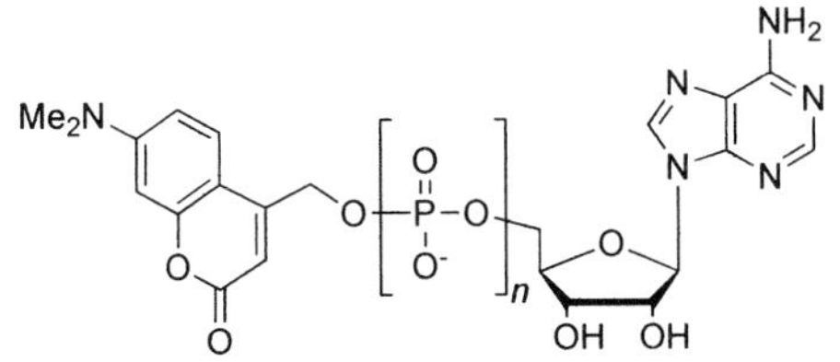

DMACM-caged ATP (n = 3)
DMACM-caged ADP (n = 2)
DMACM-caged AMP (n = 1)

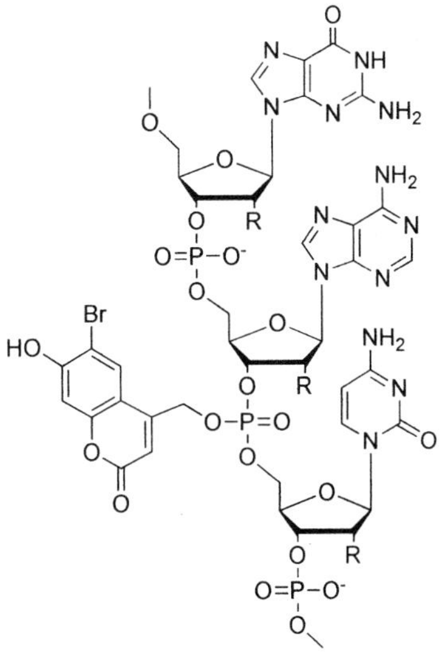

Bhc-mRNA (R = OH)
Bhc-DNA (R = H)

Fig. 1.2.3 Coumarin-caged phosphates.

Tab. 1.2.2 Spectroscopic and photochemical properties of coumarin-caged phosphates

Compounds	λ_{max} (ε)	Φ_{dis} [a]	Φ_{app} [b]	k [c]	$t_{1/2}$ [d]	s [e]	Comments	Refs.
MCM-cAMP (ax)	325[f] (13 300)	0.12[g]	0.10[g]	4.2	1000[g]			5, 13
MCM-cGMP (ax)[h]	327 (13 300)	0.21						13
MCM-cGMP (eq)[h]	325 (13 300)	0.092						13
MCM-8-Br-cAMP (ax)	326[h] (13 500)	0.14[h]		0.25	>400[h]	20[i]		12
ACM-cAMP	313[f] (7650)	0.056[g]					membrane permeable	7
PCM-cAMP	313[f] (6500)	0.054[g]					membrane permeable	7
HCM-cAMP	326[f] (16800)	0.062[g]			28.8 (a) 18.9 (e)			7
CMCM-caged cAMP (ax)[j]	326 (12 500)	0.12				900	equatorial isomer is less soluble (200 µM)	14
CMCM-caged cGMP (ax)[j]	326 (11 700)	0.16				350	equatorial isomer is more soluble (>1000 µM)	14
DMCM-caged cAMP (ax)[j]	349 (11 000)	0.04						15
BCMCM-caged cAMP (ax)[j]	346 (10 700)	0.10				500	equatorial isomer is more soluble (1000 µM)	14
DEACM-caged cAMP (ax)	402[i] (18 600)	0.21[h]		>1		135[i]	equatorial isomer is less soluble (15 µM)	14
DEACM-CDP[j]	392 (16 200)	0.029		0.2				30
DMACM-ATP[j]	385 (14 300)	0.072		1.6				31
DMACM-ADP[j]	385 (15 000)	0.063						31
DMACM-AMP[j]	385 (15 300)	0.072						31
Bhc-mRNA[k]	333, 383							23

Bhc-cAMP (eq)[1]	374 (16 300)	0.11	0.10	90	axial isomer is more stable ($t_{1/2}$ = 260 h)	28
Bhc-cGMP (eq)[1]	374 (15 300)	0.12	0.12	420	axial isomer is more stable ($t_{1/2}$ = 1240 h)	28

a) Quantum yields for disappearance of starting materials upon irradiation.
b) Quantum yields for appearance of the starting material.
c) Rate constants of photolysis (×10^9 s^{-1}).
d) Half-life (h) in the dark.
e) Solubility (µM).
f) In CH$_3$OH.
g) In DMSO/Ringer's solution (1/99) pH 7.4.
h) In CH$_3$OH/HEPES-KCl buffer (20/80) pH 7.2.
i) CH$_3$CN/HEPES-KCl buffer (5/95) pH 7.2.
j) In HEPES-KCl buffer pH 7.2.
k) In Tris-HCl pH 7.4.
l) In DMSO/KMOPS (0.1/99.9) pH 7.2.

(4) photo-activated by Ti-Sapphire laser (720–820 nm): Bhc
(5) highly water soluble: desirable – CMCM and BCMCM; usable – Bhc
(6) membrane permeable: Bhc/Ac, PCM, ACM and BECMCM

Hagen and Bendig's group (Fig. 1.2.4) proposed a mechanism for photolytic cleavage of MCM-, DMCM-, and DEACM-caged phosphates [13–15]. The mechanism involves heterolysis of the CH_2-OP bond from the lowest excited singlet state (S_1), an escape of the resulting ion pairs from the solvent cage, and trapping of the coumarin-4-ylmethyl cation by the solvent. Several pieces of evidence collected for photolysis of MCM-caged phosphates supported the mechanism. (1) A photoproduct aside from the parent phosphate was a (7-methoxycoumarin-4-yl)methanol (MCM-OH) when the photolysis was performed in an aqueous solution. (2) Photolysis of MCM-DEP in ^{18}O-labeled water resulted in incorporation of ^{18}O only in the MCM-OH, confirming that the reaction proceeded via the photo S_N1 mechanism and not by the photo solvolysis reaction. (3) No phosphorescence was detected from any of the MCM-caged compounds. (4) Only traces (<0.5%) of 4-methyl-7-methoxycoumarin, which is derived from homolysis pathway, were detected.

The fluorescence quantum yield of the MCM-cAMP (ax) is 0.030, whereas that of the MCM-OH is 0.65. Therefore, a strong fluorescence enhancement was observed as photolysis proceeded. Similar behavior was observed for other coumarin phototriggers. Measurement of the increased fluorescence would allow estimation of the amount of photochemically released cAMP because the fluorescence intensity is directly proportional to the molar fraction of the liberated MCM-OH (and therefore the liberated cAMP). However, fluorescence of MCM-OH was quenched completely in HEK 293 cells. Consequently, it could not be used for estimation.

The strong fluorescent property of the liberated coumarin-4-ylmethanol allows estimation of the rate constant of photolysis reaction. Thereby, the rate constants for the photolysis of the DEACM and the DMACM caged cyclic nucleotides were determined as $k=10^9$ s^{-1}, i.e., the concentration jump of the parent cyclic nucleotides occurred within a nanosecond.

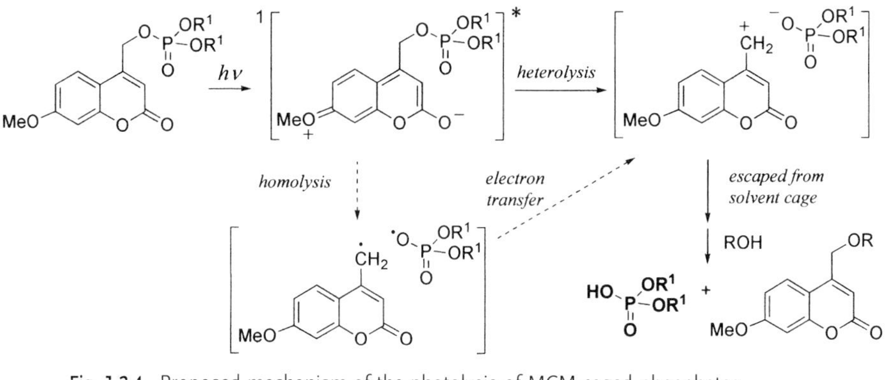

Fig. 1.2.4 Proposed mechanism of the photolysis of MCM-caged phosphates.

1.2.2.3.2 Carboxylates and Sulfates

Although coumarin-4-ylmethyl esters of carboxylates and sulfates should undergo analogous photochemistry to that of the phosphate esters, only a few examples have been reported so far (Fig. 1.2.5), partly because of the reported low photolysis efficiencies of the MCM esters of simple aliphatic acids. For example, the photolysis quantum yield for the MCM ester of heptanoic acid (MCM-OHep) is 0.0043 [13], which is more than one order of magnitude smaller than that of MCM-phosphates. The first successful application was the Bhc-caged glutamate [22]. The photolytic quantum yield of the Glu(γ-Bhc) was 0.019. The overall uncaging efficiency ($\Phi\varepsilon$) was more than one order of magnitude larger than that of the CNB- and the DMNPE-Glu. This difference indicates that the uncaging light intensity could be reduced 10-fold when Bhc-caged glutamates were used. The 2-photon uncaging action cross-sections of the Glu(γ-Bhc) were 0.89 GM at 740 nm and 0.42 GM at 800 nm, which are both more than two orders of magnitude larger than those of the conventional 2-nitrobenzyl caged compounds. Tab. 1.2.3 summarizes selected examples of spectroscopic and photochemical properties.

Low photosensitivity of the MCM ester can be overcome when a carboxylic acid with higher acidity is used. MCM and CMCM esters of α-amino acid derivatives were photolyzed with reasonably high quantum yields: 0.06 for MCM-TBOA and 0.04 for CMCM-TBOA [16], which is ten times as large as that of the MCM-OHep. A typical pKa value of α-carboxylate in α-amino acid is around two, which is lower than that of a simple aliphatic acid by two units: it approaches the pK_a of a phosphate. Differing acidities among simple aliphatic and α-amino acids must partially explain the variations in the quantum yields. Another matter for consideration is solvent polarity. Photolytic efficiencies of coumarin-caged phosphates [13] and carbonates [18] are less favorable when photo-

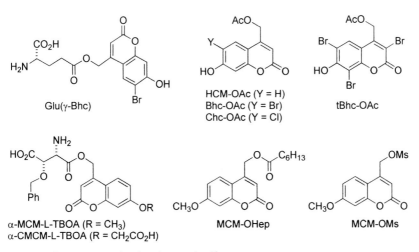

Fig. 1.2.5 Coumarin-caged carboxylates and sulfate.

Tab. 1.2.3 Spectroscopic and photochemical properties of coumarin-caged compounds

Compounds	λ_{max} (ε)	Φ_{dis} [a]	Φ_{app} [b]	k [c]	$t_{1/2}$ [d]	δ_u [e]	Refs.
Carboxylates							
MCM-OHep	324[f] (13 500)	0.0043[g]		0.0030			13
Glu(γ-Bhc)[h]	369 (19 550)	0.019			50	0.89 (740)	22
						0.42 (800)	
HCM-OAc[h]	325 (11 600)	0.025					22
Bhc-OAc[h]	370 (15 000)	0.037			183	1.99 (740)	22
						0.42 (800)	
Chc-OAc[h]	370 (16 000)	0.01					22
tBhc-OAc[h]	397 (15 900)	0.065			218	0.96 (740)	22
						3.1 (800)	
α-MCM-L-TBOA[i]	325 (4300)	0.06					16
α-CMCM-L-TBOA[i]	325 (4600)	0.04					16
Sulfates							
MCM-OMs	325[f] (13 000)	0.081[g]		0.41			13
Amines							
Bhcmoc-Glu[h]	368 (17 470)	0.019			stable after 35 h	0.95 (740) 0.37 (800)	22
MCM-NBu$_3$[j]	344 (11 700)	0.39					9
Alcohols and Phenols							
MCMoc-Gal[h]	322 (12 100)	0.020					18
DMCMoc-Gal[h]	344 (10 800)	0.0065					18
DEACM-Gal[h]	396 (17 300)	0.0058					18
Bhcmoc-Gal[h]	374 (15 000)	0.015					18
Bhcmoc-diC$_8$	343[h] (11 600)	0.014[h]			37[h]		18
	380[k] (15 000)	0.2[k]					27
Tyr(Bhcmoc)-OMe[h]	372 (13 900)	0.022	0.020		38		18
Bhcmoc-Adenosine[h]	373 (13 700)	0.012	0.010		467		18
Bhcmoc-Phenol[h]	372 (17 900)	0.067					18
Bhc-acetal[l]		0.028			stable after 2 weeks		24
Carbonyls							
Bhc-diol-acetophenone[h]	370 (18 000)	0.030				1.23 (740)	26
Bhc-diol-benzaldehyde[h]	370 (18 000)	0.057				0.90 (740)	26

a) Quantum yields for disappearance of starting materials upon irradiation.
b) Quantum yields for appearance of the starting material.
c) Rate constants of photolysis ($\times 10^9$ s^{-1}).
d) Half-life (h) in the dark.
e) Two-photon uncaging action cross sections ($\times 10^{-50}$ cm^4 s/photon, GM).
f) In CH$_3$CN/HEPES-KCl buffer (30/70) pH 7.2.
g) In CH$_3$OH/HEPES-KCl buffer (20/80) pH 7.2.
h) In DMSO/KMOPS (0.1/99.9) pH 7.2.
i) In PBS (+) solution.
j) In benzene.
k) In ethanol/Tris (50/50) pH 7.4.
l) In CH$_3$OH/HEPES-KCl (50/50) pH 7.4.

lysis is performed in an organic solvent. The solvent effect must be another reason for the lower quantum yield observed for MCM-OHep. The observed quantum yield of 0.0043 would improve if the reaction were performed in a 100% aqueous environment because the reaction of MCM-OHep was investigated in the presence of 20% methanol. In fact, HCM-OAc has a photolytic quantum yield of 0.025 in an aqueous buffer [22]. The reactions of the MCM- and the CMCM-TBOAs were performed in PBS. Therefore, photochemical properties of MCM esters of carboxylates must be reinvestigated under a simulated physiological environment.

1.2.2.3.3 Amines

Both carbamate and alkyl ammonium salts have been reported (Fig. 1.2.6). Although the examples remain limited to the Bhcmoc (6-bromo-7-hydroxycoumarin-4-ylmethoxycarbonyl) group [22, 25], coumarin-4-ylmethoxycarbonyls can serve as phototriggers of an amino group, such as aliphatic amines, amino acids, and nucleotides, and should have broader applicability than the alkyl ammonium salts. Photochemical properties of the Bhcmoc-Glu were almost identical to those of the carboxylate counterpart, Glu(γ-Bhc). A large $\Phi\varepsilon$ value of the Bhcmoc-Glu compared to that of the CNB-Glu was also seen for activation of GluR channels in rat brain slices [22]. Overall, coumarin-carbamates are highly photosensitive upon 1- and 2-photon excitation, applicable to protect primary and secondary amines, and stable under a simulated physiological environment. Tab. 1.2.3 summarizes selected examples of spectroscopic and photochemical properties.

A potential drawback is the relatively slow kinetics in comparison to those of phosphates and carboxylates. Photolysis is expected to proceed via a photo-induced cleavage followed by decarboxylation of the resultant carbamic acid. The rate constant for the decarboxylation reaction is estimated to be 2×10^2 s^{-1} at pH 7.2 [33]. Consequently, the rate constant for release of amines is no larger than 2×10^2 s^{-1}, which could pose a serious obstacle when a faster kinetic analysis, say that within a microsecond time window, is needed.

Neckers [8, 9] and Giese [17] reported the photocleavage of C-N bond in MCM amines. Homolytic cleavage of the C4-N bond in a radical anion intermediate was proposed. Such cleavage might be initiated by photo-induced electron transfer from appropriate electron donors (borates, amines, or thiols) to the C2 carbonyl in the coumarin ring (Fig. 1.2.8).

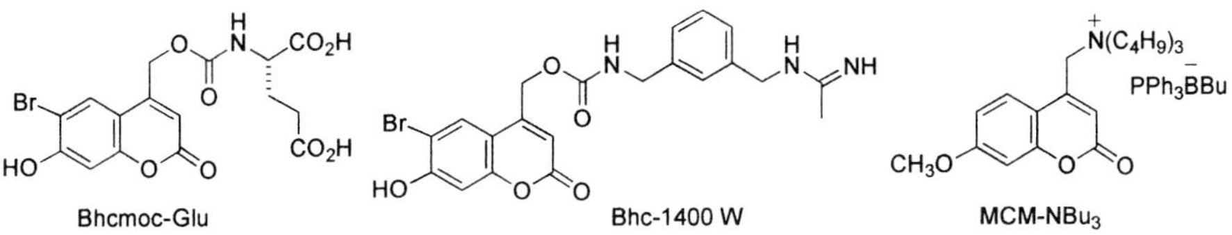

Fig. 1.2.6 Coumarin-caged amines.

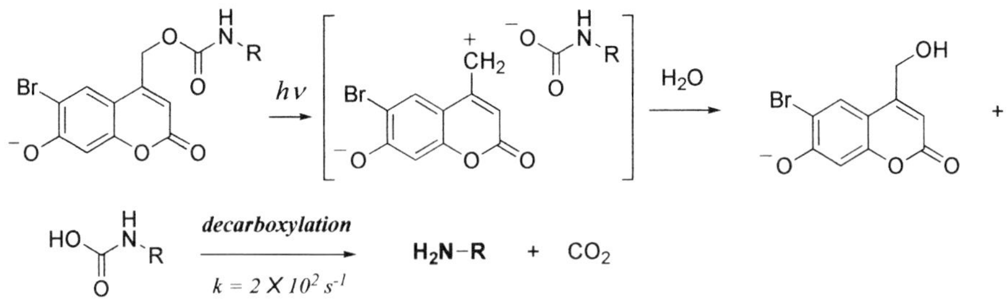

Fig. 1.2.7 Photolysis of Bhcmoc-caged amines.

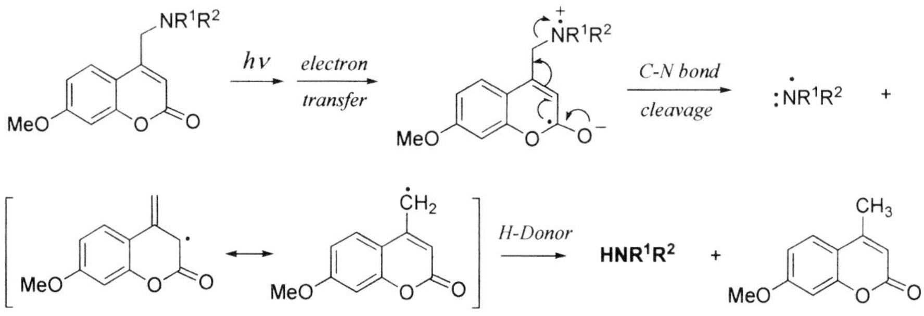

Fig. 1.2.8 Proposed mechanism of photolysis of coumarinyl amines.

1.2.2.3.4 Alcohols and Phenols

Coumarin-4-ylmethyl carbonates are photolabile. They release parent alcohols or phenols upon irradiation. All four types of coumarin-caged carbonates having the same leaving group were synthesized [18]. Their photochemical properties were investigated. Overall uncaging efficiencies, $\Phi\varepsilon$, of all coumarin derivatives were more than ten times better than that of the NVOC phototrigger at 350 nm. Surprisingly, a remarkable difference in photolytic quantum yield was observed between the MCMoc-Gal ($\Phi=0.020$) and the DMCMoc-Gal ($\Phi=0.0065$). Photolytic quantum yields decreased markedly when photolysis was performed in the presence of organic solvents: the quantum yield of MCMoc-Gal in 50% THF-H_2O was only 0.0029. The DEACMoc and the Bhcmoc groups showed similar reactivity. Both compounds must be favorable for making caged compounds of alcohols by converting them to the corresponding carbonates. The Bhcmoc-phenol has a higher photolytic quantum yield than those of the aliphatic alcohols, probably because the carbonic acid anion derived from phenol has a better leaving-group ability. Therefore, the Bhcmoc group was applied to produce caged compounds of dioctanoyl glycerol (Bhcmoc-diC_8), tyrosine methyl ester (Tyr(Bhcmoc)-OMe) and adenosine (5′-Bhcmoc-adenosine). Like coumarin carbamates, coumarin carbonates suffer from relatively slow kinetics of decarboxylation reactions from an intermediate carbonic acid.

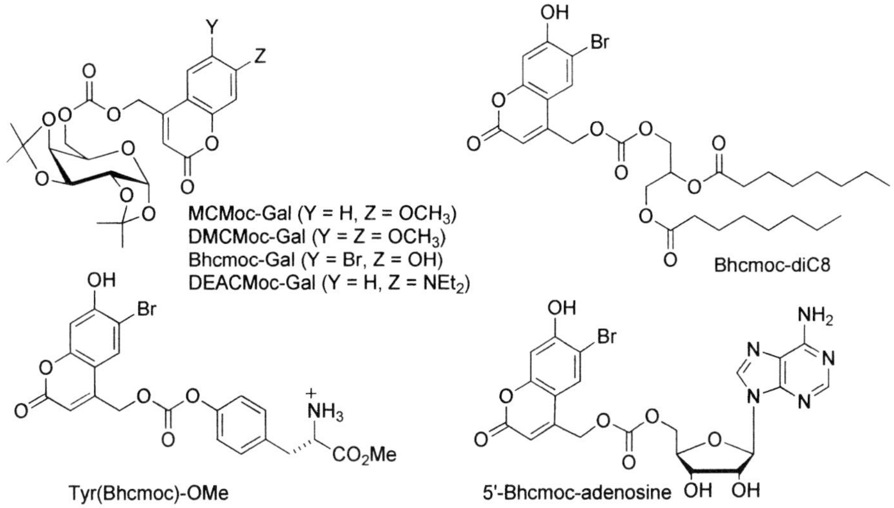

Fig. 1.2.9 Coumarin-caged alcohols and phenols.

Although no kinetic study is available, the rate constant of an alcohol release must not exceed 2×10^2 s^{-1} (Fig. 1.2.10).

Aliphatic ethers do not deprotect upon photolysis. Lin and Lawrence [24] proposed an interesting strategy for construction of caged 1,2-diols. They found that 1,3-dioxolanes of the 4-formylated Bhc were photolyzed to release parent 1,2-diols with quantum yields of 0.0278–0.0041. Fig. 1.2.11 shows a proposed mechanism of photocleavage. The adjacent oxygen in Bhc-acetal assists with the departure of the alkoxide anion, thereby stabilizing the resultant coumarin-4-yl-methyl cation. Interestingly, the six-membered counterparts, 1,3-dioxanes, were not photosensitive. The presence of a tertiary amine caused a seven-fold increase of the photolytic quantum yield, probably because the tertiary amine serves as a proton donor during acetal hydrolysis, which eventually accelerates the overall photorelease reaction.

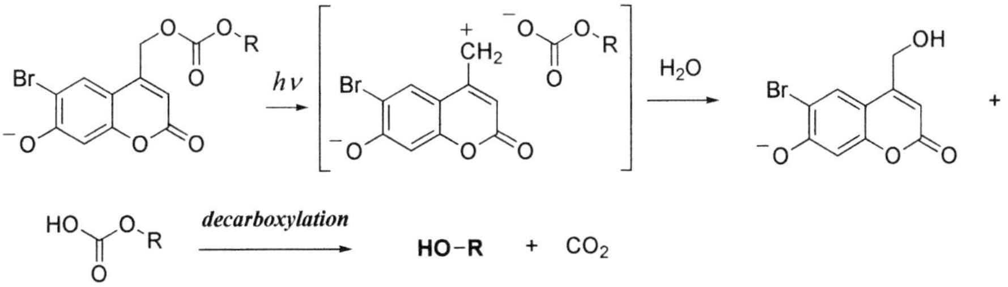

Fig. 1.2.10 Photolysis of Bhcmoc-caged alcohols and phenols.

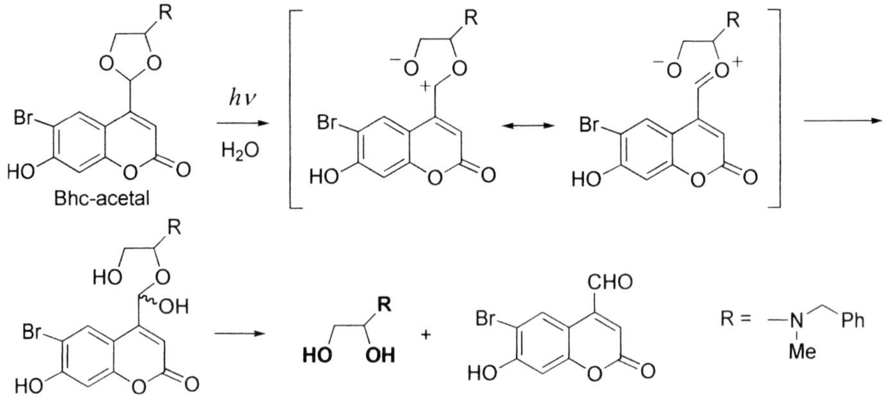

Fig. 1.2.11 Photolysis of Bhc-acetal.

1.2.2.3.5 Carbonyl Compounds

Dore and colleagues [26] found that aldehydes and ketones can be caged by 6-bromo-4-(1,2-dihydroxyethyl)-7-hydroxycoumarin (Bhc-diol) as corresponding acetals. The possible reaction mechanism shown in Fig. 1.2.12 involves the zwitterionic intermediate in which C4 methylene cation must be stabilized by the adjacent alkoxymethyl substitution. Photolytic quantum yields ($\Phi = 0.030–0.057$) were comparable to those of the other Bhc-caged compounds. Two photon uncaging cross-sections, δ_u, were measured to be 0.51–1.23 GM. These results are the only example of a phototrigger capable of releasing aldehydes and ketones by a two-photon excitation condition. They must have great potential for cell biological applications.

1.2.3
Synthesis

1.2.3.1 Synthesis of Precursor Molecules (Caging Agents)

Introduction of coumarin-type "phototriggers" into molecules of interest has been achieved using five types of precursor molecules having reactive functional

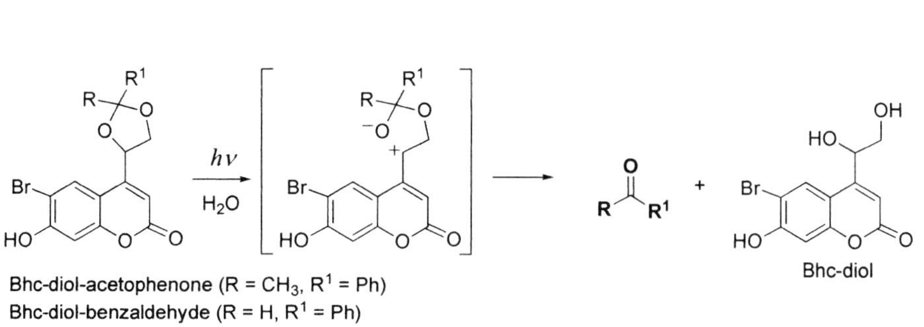

Bhc-diol-acetophenone (R = CH$_3$, R^1 = Ph)
Bhc-diol-benzaldehyde (R = H, R^1 = Ph)

Fig. 1.2.12 Photolysis of Bhc-diol-protected aldehydes and ketones.

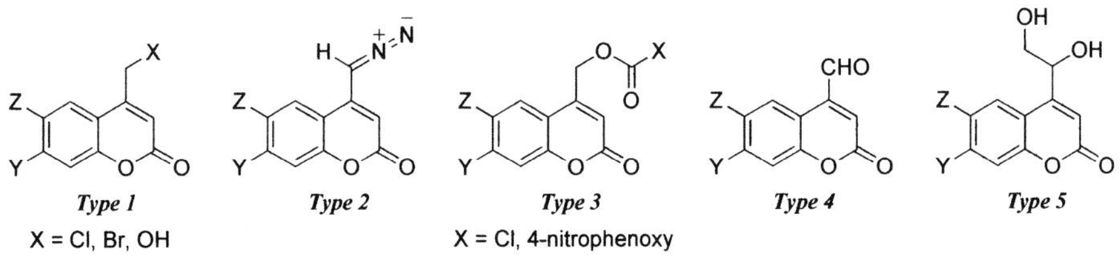

1.2 Coumarin-4-ylmethyl Phototriggers

Fig. 1.2.13 Precursor molecules for synthesis of coumarin-caged compounds.

groups on the C-4 position (Fig. 1.2.13). The halo- or hydroxymethyl (type 1) and the diazomethyl (type 2) precursors can be used to prepare corresponding coumarin-4-ylmethyl esters of phosphates, carboxylates, and sulfates. Amino- and hydroxyl-containing molecules can be converted into corresponding carbamates and carbonates using chloroformate or 4-nitrophenylcarbonate precursors (type 3). The type 4 precursor is used for caged 1,2-diols and type 5 for caged aldehydes and ketones.

Extensive studies of the synthesis of substituted coumarins have been made [34]; some substituted coumarins are commercially available. Fig. 1.2.14 shows the structures of commercially available coumarins that have been used to prepare caged compounds. Br-Mmc [35] and Br-Mac [36], which were originally developed as fluorescent labeling agents of carboxylates and alcohols, can be used for the synthesis of MCM- and ACM-caged phosphates and carboxylates. The remaining compounds also have the desired substitution pattern in which electron-donating substituents such as alkoxy and amino groups are located on the C7 position and a methyl group on the C4 position. Donor substitutions at C7 help the chromophore to be ionized via through-bond electron movement when the coumarins are excited electronically, facilitating cleavage of the CH_2-X bond [13]. Alkyl substitutions at the C4 position can undergo allylic oxidation. In addition, further modification to introduce appropriate functional groups is possible. These compounds have been utilized to produce caged compounds having MCM, DMCM, and DEACM phototriggers.

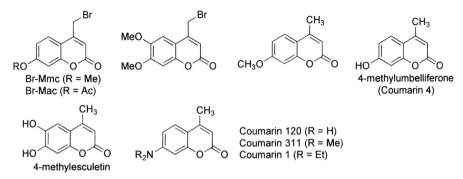

Fig. 1.2.14 Structures of commercially available coumarins with common names.

Type 1 precursors having the 4-bromomethyl group were prepared by radical bromination (NBS) of 4-methylcoumarins [15]; 4-hydroxymethyl group were prepared by hydrolysis of corresponding acetates [15] or chlorides [22].

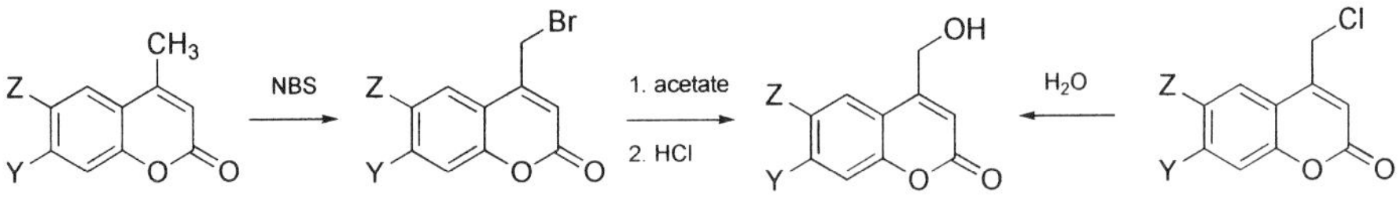

Scheme 1.2.4

Type 2 and 4 precursors were synthesized from 4-methylcoumarins according to the procedure of Ito [37]: the C4-methyl group was oxidized by selenium dioxide (or manganese dioxide) followed by tosylhydrazone formation and base-induced diazo formation.

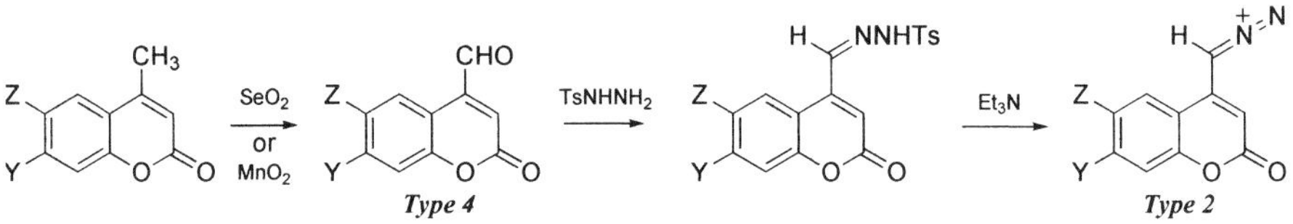

Scheme 1.2.5

For cases where an appropriate starting material is not commercially available, coumarin frameworks must be constructed starting from substituted phenols, either by the Pechmann reaction [38] or the Knoevenagel condensation [39] (Scheme 1.2.6).

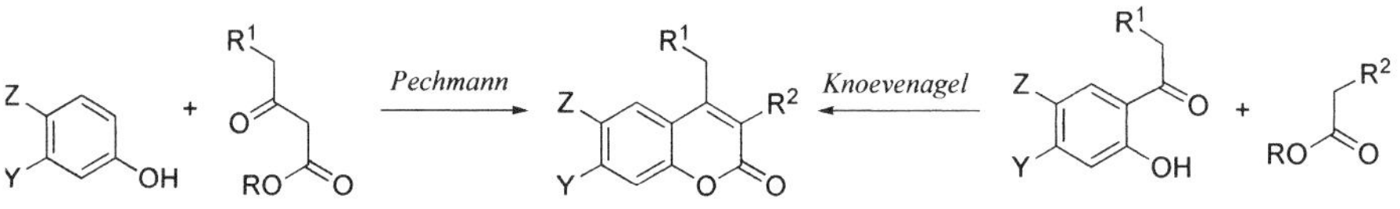

Scheme 1.2.6

Synthesis of the Bhc group has been performed using the Pechmann reaction. Scheme 1.2.7 illustrates synthetic routes toward the type 1 [22], 3 [18] and 5 [26] precursors of the Bhc group. The Bhc frameworks were constructed by the acid-catalyzed condensation of 4-bromoresorcinol and ethyl acetoacetate derivatives. Other type 3 precursors can be prepared in a similar manner [18].

In principle, conventional synthetic methods for esters, carbamates, carbonates, and acetals can be applied to the synthesis of coumarin-caged com-

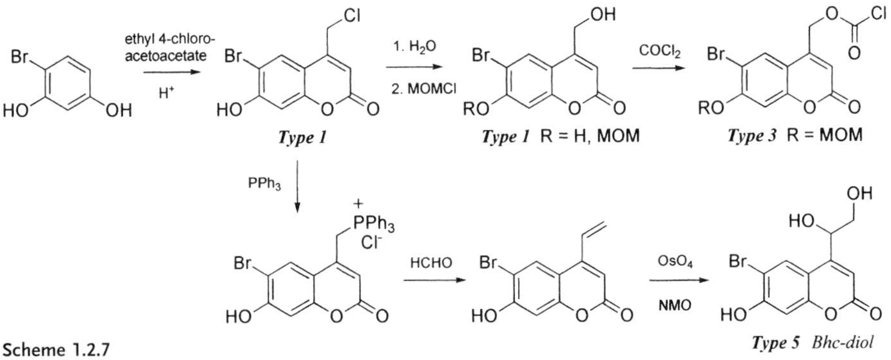

Scheme 1.2.7

pounds. Examples in subsequent sections will be limited to biologically relevant molecules. Their synthetic strategies are discussed briefly.

1.2.3.1.1 **Phosphates**
The coumarin phototriggers can be introduced into acidic molecules such as phosphates, carboxylates, and sulfates using type 1 and the type 2 precursors. Multistep synthesis using protection-deprotection chemistry is required when a target compound has more than a single functional group. An exception is synthesis of caged nucleotides. For example, a free phosphoric acid of cAMP was esterified with Br-Mmc or Br-Mac in the presence of silver (I) oxide [38] to yield MCM-cAMP in 44% yield (ax/eq=7/3) and ACM-cAMP in 14% yield (ax/eq=1/1), respectively [5–7]. The MCM-cAMP was also prepared by reaction of the tetra-*n*-butylammonium salt of cAMP with Br-Mmc (64% yield, ax/eq=8/2) [13]. Analogously, DMCM-caged cAMP was prepared either by the silver oxide method (18.7%, ax/eq=55/45) or the ammonium salt method (57.3%, ax/eq=85/15) [15]. Improved yields and better axial selectivity were observed in the ammonium salt method. In these reactions, protections of other functional groups are not necessary. Attempts to apply the methods to make HCM-cAMP and MCM-caged cGMP have failed [7, 13]. For that reason, the methods seem to be limited to the MCM, ACM and DCMCM groups, which have no free acidic hydroxy functional group on the coumarin ring. In contrast, HCM-, DEACM-, DMACM- and Bhc-cNMPs were synthesized using corresponding 4-diazomethylcoumarins, type 2 precursors, in 10–30% isolated yields [7, 14, 15, 28]. The 4-diazomethyl precursors are superior to other methods for preparation of caged cyclic nucleotides. Analogously, Bhc-caged RNAs (encode EGFP or a transcription factor en2) were prepared by the reaction of 4-diazomethyl-7-hydroxycoumarin (Bhc-diazo) and synthetic full-length mRNAs in DMSO [23].

1.2.3.1.2 Carboxylates

The synthesis of Glu(γ-Bhc) involves DBU-promoted esterification of the appropriately protected glutamate with the 4-chloromethyl-7-hydroxycoumarin [22]. The DBU method was developed for the synthesis of optically active amino acid esters [39]. No racemization was detected during the reaction. However, Shimamoto and colleagues [16] observed copious epimerization in the synthesis of caged L-TBOA. Therefore, they used the corresponding 4-diazomethylcoumarins to prepare MCM- and CMCM-L-TBOAs. Ito [37] and Goya [40] have studied the reaction of 4-diazomethylcoumarins with various carboxylic acids. Most carboxylic acids can be esterified by 4-diazomethylcoumarins without adding any catalysts or additives. However, dramatic acceleration was observed in the presence of either silica gel (known to catalyze the generation of carbene intermediates from diazoalkanes) or tetrafluoroborate (a strong Lewis acid).

1.2.3.1.3 Amines

Precursor molecules that can be used are 4-nitrophenyl carbonates and chloroformates (type 3). Introduction of the Bhcmoc group into the amino moiety of the protected glutamate was achieved via the 4-nitrophenyl carbonate intermediate with 31% isolated yield [22]. The yield is acceptable, but needs to be optimized. A potential precursor is the corresponding chloroformate. Use of (6-bromo-7-methoxymethoxycoumarin-4-yl)methyl chloroformate (MOM-Bhcmoc-Cl) [18] greatly improved yields of amino protections (T. Watanabe, unpublished results).

1.2.3.1.4 Alcohols and Phenols

Introduction of the Bhcmoc group into alcohols and phenols proceeded with almost quantitative yield when the MOM-Bhcmoc-Cl was used as a precursor [18]. Synthesis of Bhcmoc-diC8 (91% overall yield in 2 steps) and Tyr(Bhcmoc)-OMe (100% overall yield in 2 steps) were successful examples. Chloroformate precursors are so reactive that chemoselective protection of a hydroxyl moiety failed in the presence of an aromatic amino group. Therefore, the 4-nitrophenyl-carbonate precursor was used in synthesis of 5'-Bhcmoc-adenosine.

1.2.4
Applications

Applications of coumarin phototriggers to cell chemistry remain limited in number. However, the results are extremely promising. This section is intended to give an overview of reported applications of coumarin-caged compounds. Most of the reported live cell applications will be collected, including biological tests. To avoid repetition of information in related chapters, this section will specifically address the reaction conditions including cell types, methods of application to cells, types of the light sources, and monitoring methods (Tab. 1.2.4). The types of the compounds were second messengers (cAMP, cGMP, diC$_8$), a

Tab. 1.2.4 The reported biological applications of coumarin-caged compounds

Compounds	Cell types	Methods of application to cells	Light sources	Monitoring methods	Observed effects	Refs.
MCM-cAMP ACM-cAMP PCM-cAMP	Fish melanophores	Bathing	Xe lamp	Optical microscope	Pigment dispersion	5–7
MCM-8-Br-cAMP and cGMP	HEK 293 cells	Patch pipette and bathing	100 W Hg lamp and N_2 laser	Patch clamp and Ca^{2+} imaging	CNG channel opening	11, 12
MCM-8-Br-cGMP	Bovine sperm cells	Bathing	N_2 laser	Ca^{2+} imaging	CNG channel opening	10
DEACM-cAMP	HEK 293 cells	Patch pipette	405 nm laser	Patch clamp	CNG channel opening	14
BCMCM-cAMP	HEK 293 cells	Patch pipette	Xe lamp	Patch clamp	CNG channel opening	14
BECMCM-cAMP	Sea urchin and starfish sperms	Bathing	Xe flash lamp	Ca^{2+} imaging and motility	Ca^{2+} influx and movement	20, 21
BCMCM-cAMP	Rat olfactory receptor neuron	Patch pipette	Xe lamp	Patch clamp	CNG channel opening	19
DMACM-8-Br-cAMP and -cGMP DEACM-8-Br-cGMP	HEK 293 cells	Patch pipette and bathing	100W Hg lamp and Ar laser	Patch clamp and Ca^{2+} imaging	CNG channel opening	32
DMACM-caged ATP	Mouse astrocytes and brain slices	Bathing	Ar laser	Ca^{2+} imaging	transient $[Ca^{2+}]_i$ increase	31
Bhcmoc-diC8	Rat ventricular myocytes	Bathing	Ti/Sapphire laser	CLSM	twitch response	27
Bhcmoc-Glu	Rat brain slices	Bathing	N_2 and Ti/Sapphire lasers	Patch clamp	GluR opening	22

Tab. 1.2.4 (continued)

Compounds	Cell types	Methods of application to cells	Light sources	Monitoring methods	Observed effects	Refs.
Bhc-1400W	Purified proteins		366 nm and Ti/Sapphire laser		iNOS inhibition	25
α-CMCM-L-TBOA	MDCK cells	Bathing	365 nm	[^{14}C]Glu uptake	Blocks Glu transporters	16
Bhc-mRNA/DNA	Zebrafish embryos	Microinjection	355 nm	Phenotype	Ectopic expression of proteins	23
Bhc-cAMP	Newt olfactory receptor cells	Patch pipette	Xe lamp	Patch clamp	CNG channel opening	28
Bhc-cAMP/Ac	Fish melanophores	Bathing	Xe lamp	Optical microscope	Pigment dispersion	28
Bhc-cAMP/Ac	Sea urchin sperms	Bathing	Xe flash lamp	Ca^{2+} imaging	CNG channel opening	29
Bhc-cGMP/Ac						

neurotransmitter (glutamate), an isoform-specific enzyme inhibitor (1400W), a blocker for glutamate transporters (L-TBOA), a receptor ligand (ATP), DNA and mRNAs. Three compounds, Bhcmoc-Glu, Bhcmoc-diC$_8$ and Bhc-1400W, were photo-activated under two-photon excitation conditions.

1.2.5
Conclusion and Perspective

Progress in the field of caged compounds depends largely on the development of new compounds. Although *in vitro* measurement has shown that coumarin-4-yl-methyl groups have desirable properties as phototriggers for several types of biologically relevant molecules, few biological examples have been reported to date. One reason is that only limited types of compounds were synthesized compared to 2-nitrobenzyl type phototriggers; none of the coumarin-caged compounds are commercially available. The following target molecules may be useful to dissect cellular processes by taking advantage of DEACM and Bhc-type phototriggers.

Molecules that play a crucial role in signaling pathways are important targets. Intracellular signaling pathways depend largely on spatial and temporal distribution of signaling molecules. Photo-regulation of signaling molecules offers an ideal method to mimic a function of the molecule under a given physiological condition. Collections of small-molecular-weight caged compounds should be expanded by extension of the already existing chemistry of coumarin phototriggers. In addition to neurotransmitters and second messengers, the compounds that would be useful to be caged must include receptor ligands that can activate gene expressions, subtype specific inhibitors of various kinases, and small synthetic molecules that can alter specific cellular functions.

Photochemical regulation of a protein function is a goal for caging chemistry. Toward this end, caged compounds of mRNAs, DNAs, and proteins must be considered further. The Bhc-caged mRNA technology paves the way for controlling intracellular concentration of proteins in live animals in a spatially and temporally regulated manner. The method is applicable to produce caged plasmid DNAs which would provide a method to photo-control protein functions in mammalian cells. Several approaches can be considered for construction of caged proteins. Recent advances in methods to incorporate non-natural amino acids into proteins enable the construction of caged proteins in which an active site amino acid residue is modified by a coumarin phototrigger. Alternatively, multiple introductions of phototriggers can perturb the protein conformation, thereby inactivating the protein function. In this case, it is not necessary to introduce a phototrigger into an amino acid that is critical to an activity. For this reason, this method is applicable to all proteins even if the active site amino acid is unknown. However, the method is not practical when a 2-nitrobenzyl-type phototrigger is used because cleavage of multiply introduced phototriggers requires prolonged irradiation time to restore the original activity. Multiple modifications can be feasible when highly improved photosensitivity of coumarin phototriggers is utilized.

The multi-photon excitation technique is the most promising and elegant advance in the field because it offers true three-dimensional resolution, which is useful for photo-activation reaction on tissue slice samples and in whole organisms. The Bhc group has already proved itself useful for two-photon excitation. However, only a limited example is available for both *in vitro* and *in vivo* excitation conditions. Systematic investigation of two-photon absorption properties is needed, especially of various phototriggers and Bhc groups' uncaging action cross-sections and wavelength dependencies. Such obtained data would form the basis for designing new phototriggers that are favorably photo-activated under multi-photon excitation conditions.

References

1 S. A. FLEMING, J. A. PINCOCK, in *Molecular and Supramolecular Photochemistry 3 (Organic Molecular Photochemistry)*, V. RAMAMURTHY and K. S. SCHANZE, Eds. Marcel Dekker: New York, **1999**, pp 211–281.

2 For example, P. S. MUKHERJEE, H. T. KARNES, *Biomed. Chromatogr.* **1996**, *10*, 193–204.

3 R. S. GIVENS, B. MATUSZEWSKI, *J. Am. Chem. Soc.* **1984**, *106*, 6860–6861.

4 T. FURUTA, H. TORIGAI, T. OSAWA, M. IWAMURA, *Chem. Lett.* **1993**, 1179–1182.

5 T. FURUTA, H. TORIGAI, M. SUGIMOTO, M. IWAMURA, *J. Org. Chem.* **1995**, *60*, 3953–3956.

6 T. FURUTA, A. MOMOTAKE, M. SUGIMOTO, M. HATAYAMA, H. TORIGAI, M. IWAMURA, *Biochem. Biophys. Res. Commun.* **1996**, *228*, 193–198.

7 T. FURUTA, M. IWAMURA, *Methods Enzymol.* **1998**, *291*, 50–63.

8 A. M. SARKER, Y. KANEKO, A. V. NIKOLAITCHIK, D. C. NECKERS, *J. Phys. Chem. A* **1998**, *102*, 5375–5382.

9 A. M. SARKER, Y. KANEKO, D. C. NECKERS, *J. Photochem. Photobiol. A* **1998**, *117*, 67–74.

10 B. WIESNER, J. WEINER, R. MIDDENDORFF, V. HAGEN, U. B. KAUPP, I. WEYAND, *J. Cell. Biol.* **1998**, *142*, 473–484.

11 B. WIESNER, V. HAGEN, *J. Photochem. Photobiol. B* **1999**, *49*, 112–119.

12 V. HAGEN, J. BENDIG, S. FRINGS, B. WIESNER, B. SCHADE, S. HELM, D. LORENZ, U. B. KAUPP, *J. Photochem. Photobiol. B* **1999**, *53*, 91–102.

13 B. SCHADE, V. HAGEN, R. SCHMIDT, R. HERBRICK, E. KRAUSE, T. ECKARDT, J. BENDIG, *J. Org. Chem.* **1999**, *64*, 9109–9117.

14 V. HAGEN, J. BENDIG, S. FRINGS, T. ECKARDT, S. HELM, D. REUTER, U. B. KAUPP, *Angew. Chem. Int. Ed.* **2001**, *40*, 1046–1048.

15 T. ECKARDT, V. HAGEN, B. SCHADE, R. SCHMIDT, C. SCHWEITZER, J. BENDIG, *J. Org. Chem.* **2002**, *67*, 703–710.

16 K. TAKAOKA, Y. TATSU, N. YUMOTO, T. NAKAJIMA, K. SHIMAMOTO, *Bioorg. Med. Chem. Lett.* **2003**, *13*, 965–970.

17 R. O. SCHOENLEBER, B. GIESE, *Synlett.* **2003**, 501–504.

18 A. Z. SUZUKI, T. WATANABE, M. KAWAMOTO, K. NISHIYAMA, H. YAMASHITA, M. IWAMURA, T. FURUTA, *Org. Lett.* **2003**, *5*, 4867–4870.

19 J. BRADLEY, D. REUTER, S. FRINGS, *Science* **2001**, *294*, 2176–2178.

20 U. B. KAUPP, J. SOLZIN, E. HILDEBRAND, J. E. BROWN, A. HELBIG, V. HAGEN, M. BEYERMANN, F. PAMPALONI, I. WEYAND, *Nat. Cell. Biol.* **2003**, *5*, 109–117.

21 M. MATSUMOTO, J. SOLZIN, A. HELBIG, V. HAGEN, S. UENO, O. KAWASE, Y. MARUYAMA, M. OGISO, M. GODDE, H. MINAKATA, U. B. KAUPP, M. HOSHI, I. WEYAND, *Dev. Biol.* **2003**, *260*, 314–324.

22 T. FURUTA, S. S.-H. WANG, J. L. DANTZKER, T. M. DORE, W. J. BYBEE, E. M. CALLAWAY, W. DENK, R. Y. TSIEN, *Proc. Natl. Acad. Sci. USA* **1999**, *96*, 1193–1200.

23 H. ANDO, T. FURUTA, R. Y. TSIEN, H. OKAMOTO, *Nat. Genet.* **2001**, *28*, 317–325.

24 W. Lin, D. S. Lawrence, *J. Org. Chem.* **2002**, *67*, 2723–2726.
25 H. J. Montgomery, B. Perdicakis, D. Fishlock, G. A. Lajoie, E. Jervis, J. G. Guillemette, *Bioorg. Med. Chem.* **2002**, *10*, 1919–1927.
26 M. Lu, O. D. Fedoryak, B. R. Moister, T. M. Dore, *Org. Lett.* **2003**, *5*, 2119–2122.
27 V. G. Robu, E. S. Pfeiffer, S. L. Robia, R. C. Balijepalli, Y. Pi, T. J. Kamp, J. W. Walker, *J. Biol. Chem.* **2003**, *278*, 48154–48161.
28 T. Furuta, H. Takeuchi, M. Isozaki, Y. Takahashi, M. Sugimoto, M. Kanehara, T. Watanabe, K. Noguchi, T. M. Dore, T. Kurahashi, M. Iwamura, R. Y. Tsien, *ChemBioChem.* **2004**, *5*, 1119–1128.
29 T. Nishigaki, C. D. Wood, Y. Tatsu, N. Yumoto, T. Furuta, D. Ellias, K. Shiba, S. A. Baba, A. Darszon, *Dev. Biol.* **2004**, *272*, 376–388.
30 R. O. Schoenleber, J. Bendig, V. Hagen, B. Giese, *Bioorg. Med. Chem.* **2002**, *10*, 97–101.
31 D. Geisler, W. Kresse, B. Wiesner, J. Bendig, H. Kettenmann, V. Hagen, *ChemBioChem*, **2003** *4*, 162–170.
32 V. Hagen, S. Frings, B. Wiesner, S. Helm, U. B. Kaupp, J. Bendig, *ChemBioChem*, **2003** *4*, 434–442.
33 F. M. Rossi, M. Margulis, C.-M. Tang, J. P. Y. Kao, *J. Biol. Chem.* **1997**, *272*, 32933-32939.
34 This seminal review still gives us useful information on the synthesis of substituted coumarins. S. M. Sethna, N. M. Shah, *Chem. Rev.* **1945**, *36*, 1–62.
35 W. Duenges, *Anal. Chem.* **1977**, *49*, 442–445.
36 H. Tsuchiya, T. Hayashi, H. Naruse, N. Takagi, *J. Chromatog.* **1982**, *234*, 121–130.
37 K. Ito, J. Maruyama, *Chem. Pharm. Bull.* **1983**, *31*, 3014–3023.
38 T. Furuta, H. Torigai, T. Osawa, M. Iwamura, *J. Chem. Soc., Perkin Trans. 1* **1993**, 3139–3142.
39 N. Ono, T. Yamada, T. Saito, K. Tanaka, A. Kaji, *Bull. Chem. Soc. Jpn.* **1978**, *51*, 2401–2404.
40 A. Takadate, T. Tahara, H. Fujino, S. Goya, *Chem. Pharm. Bull.* **1982**, *30*, 4120–4125.

1.3
p-Hydroxyphenacyl: a Photoremovable Protecting Group for Caging Bioactive Substrates

Richard S. Givens and Abraham L. Yousef

1.3.1
Introduction and History

One of the newer and more promising photoremovable protecting groups is the *p*-hydroxyphenacyl (*p*-HOC$_6$H$_4$COCH$_2$, p**HP**) chromophore. The introduction of pHP as a cage for bioactive substrates began less than a decade ago with the demonstrated release of ATP from pHP-ATP [Eq. (1)] [1, 2]. Although the pHP group was discovered only recently, several reviews have already noted its applications in biological studies [3–5]. This review will be limited, however, to the general photochemistry, the mechanism of this reaction, the synthetic methodology required to protect a substrate, and a few recent applications of the pHP series. Other sections of this monograph will more fully address the earlier applications.

[Scheme for Eq. (1): pHP-ATP → (hν, Φ = 0.37, k_r = 5.5 × 10^8 s^−1, TRIS Buffer) → ATP + pHPA]

$$\text{pHP-ATP} \xrightarrow[\text{TRIS Buffer}]{\substack{h\nu \\ \Phi = 0.37 \\ k_r = 5.5 \times 10^8 \text{ s}^{-1}}} \text{ATP} + \text{pHPA} \qquad (1)$$

Historically, the parent phenacyl group ($C_6H_5COCH_2$) was first suggested as a photoremovable protecting group by Sheehan and Umezawa [6], who employed the p-methoxyphenacyl derivative for the release of derivatives of glycine. An earlier contribution from Anderson and Reese [7] reported the photoreactions of p-methoxy and p-hydroxy phenacyl chlorides, which rearranged, in part, to methyl phenylacetates when irradiated in methanol [Eq. (2)]. None of these earlier investigations, however, led to the development or exploited the potential of these chromophores as photoremovable protecting groups.

$$\text{X-PH-Cl} \xrightarrow[\text{-HCl}]{\substack{h\nu \\ \text{MeOH}}} \mathbf{A} = \text{acetophenones} + \text{X-PA methyl esters} \qquad (2)$$

(X = p-OH; p-OMe; o-OMe)

The p-hydroxyphenacyl group has several properties that make it appealing as a photoremovable protecting group. Ideally, such groups should have good aqueous solubility for biological studies. The photochemical release must be efficient, e.g., $\Phi \approx 0.10$–1.0, and the departure of the substrate from the protecting group should be a primary photochemical process occurring directly from the excited state of the cage chromophore. All photoproducts should be stable to the photolysis environment, and the excitation wavelengths should be greater than 300 nm with a reasonable absorptivity (a). The caged compounds as well as the photoproduct chromophore should be inert or at least benign with respect to the media and the other reagents and products. A general, high yielding synthetic procedure for attachment of the cage to the substrate including the separation of caged and uncaged derivatives must be available.

Of these properties, the p-hydroxyphenacyl photoremovable protecting group (ppg) satisfies most of them. It does have weak absorptivity above 300 nm. However, the chromophore can be enhanced by modifying the aromatic substituents through appending methoxy or carboxyl groups at the *meta* position or through extending the aryl ring (e.g., naphthyl, indole, etc.).

The "photo-Favorskii" rearrangement of the acetophenone chromophore to form a p-hydroxyphenylacetic acid derivative is shown in Eqs. (1) and (2) [8]. The realignment of the connectivity of the carbonyl and aryl groups results in a significant hypsochromic shift of the chromophore; this has beneficial effects on the overall photochemistry, especially on the conversion to products [8]. Also, according to Reese and Anderson [7], rearrangement to methyl phenylacetates occurred for only the o- or p-methoxy- and p-hydroxy-substituted phenacyl chlorides, and the ratio of rearrangement to direct reduction to the acetophenone was approximately 1:1. All other substituents examined as well as the parent phenacyl chloride gave only the acetophenone reduction product that preserved the intact chromophore. As such, the acetophenone products would effectively compete for the incident light whenever such a reaction is carried to a modest or high conversion.

1.3.2
p-Hydroxyphenacyl

Following the leads provided by Sheehan [6] and by Reese and Anderson [7], we surveyed additional p-substituted phenacyl phosphate derivatives for their efficacy toward releasing phosphate [1, 2]. Among the substituents examined, the p-acetamido, methyl p-carbamoyl, and n-butyl p-carbamoyl groups proved untenable because they gave a plethora of products from the chromophore, most of which resulted from coupling or reduction of an intermediate phenacyl radical [9, 10]. Tab. 1.3.1 gives the disappearance quantum efficiencies for several p-substituted phenacyl phosphates, from which it is evident that photoreactions of the acetamido and carbamoyl derivatives are also very efficient. However, the large array of products of the phototrigger discouraged further development of these two electron-donating groups.

The methoxy substituent showed a much cleaner behavior, yielding only two products, the acetophenone and a rearrangement product, methyl or t-butyl (R=CH_3 or ^tBu) p-methoxyphenylacetate [Eq. (3)]. Most unexpectedly, the p-hydroxyphenacyl diethyl phosphate (pHP-OPO(OEt)$_2$) gave exclusively the rearranged p-hydroxyphenylacetic acid when photolyzed in aqueous buffers or in H_2O [Eq. (4)]. In fact, of all of the groups examined, only p-hydroxy and p-methoxy produced any rearranged phenylacetates.

Tab. 1.3.1 Disappearance and product efficiencies for ammonium salts of p-substituted phenacyl phosphates (pX-P–OPO(OEt)$_2$) in pH 7.2 Tris buffer at 300 nm

p-Substitutent (pX-)	Φ_{dis}	$\Phi_{pX\text{-}PA}$	$\Phi_{pX\text{-}A}$	Φ_{other}
NH_2	<0.05	0.0	<0.05	na
CH_3CONH	0.38	0.0	0.11	dimers
CH_3OCONH	0.34	0.0	nd	2 unknowns
CH_3O [a]	0.42	0.20	0.07	na
HO [b]	0.38	0.12	0.0	0.0

a) Solvent was MeOH and diethyl phosphate was the leaving group.
b) The diammonium salt of the mono ester. 10% CH_3CN was added to the solvent.

[Reaction scheme 3: p-MeO P-OPO(OEt)₂ + hν, ROH → pMeO A + pMeO PA + HO-P(=O)(OEt)₂] (3)

[Reaction scheme 4: pHP-OPO(OEt)₂ + hν, H₂O → pHPA + HO-P(=O)(OEt)₂] (4)

Extending the substrates to include p-hydroxyphenacyl ATP (pHP-ATP) demonstrated that ATP and p-hydroxyphenylacetic acid (pHPA) were the *only* photoproducts formed, and this occurred with a good quantum efficiency of 0.37 ± 0.01 and a rate constant of $5.5 \pm 1.0 \times 10^8 \, \text{s}^{-1}$ for appearance of ATP [Eq. (1)]. The photorelease of γ-aminobutyrate (GABA) and L-glutamate (L-Glu) was also observed by photolysis of the corresponding p-hydroxyphenacyl esters [Eq. (5)] [11].

[Reaction scheme 5: pHP-GABA (R=H), pHP-L-Glu (R=CO₂⁻) + hν, H₂O or buffer → pHPA + GABA (or Glu)] (5)

α-Amino acids have proven to be less amenable to the pHP methodology in aqueous environments, because these esters slowly hydrolyzed at pH 7.0 over a 24 h period, thus compromising the usefulness of the pHP group under these conditions. At pH 7, the protected α-ammonium ion must enhance the nucleophilic attack of H₂O at the ester carbonyl. This rationale is supported by the observation that pHP-protected dipeptide Ala-Ala and oligopeptides such as bradykinin, which lack the exposed α-amino group, are quite stable to hydrolysis [Eq. (6)] [11].

pHP-Ala-Ala → (hv, H₂O or buffer) → pHPA + Ala-Ala (6)

Release of Ala-Ala from pHP-Ala-Ala has served as a model for further studies on the release of oligopeptides. The most dramatic example of release of a small oligopeptide was the release of synthetic bradykinin [Eq. (7)] [8]. Applications using caged second messengers have aided our understanding of the biological functions of astrocytes [12].

pHP-Bk (OC-Arg.Phe.Pro.Ser.Phe.Gly.Pro.Pro.Arg–NH₂) $\xrightarrow[-H_2O]{h\nu}$ pHPA + Bradykinin, $\Phi = 0.22$ (7)

1.3.2.1 General Physical and Spectroscopic Properties

Substrate-caged pHP and its derivatives are generally isolated as water-soluble crystalline solids. They are stable for long periods of time when stored in the dark under cold, dry conditions, and in neutral aqueous buffered media they exhibit hydrolytic stability for 24 h or longer (for exceptions, see above) [2]. The p-hydroxyacetophenone (pHA) chromophore is characterized by a moderately strong absorption band (λ_{max}) at 282 nm ($\varepsilon = 1.4 \times 10^4$ M^{-1}cm^{-1}) and significant absorptivity from 282 to 350 nm (shoulder at $\lambda_{max} \approx 330$ nm) at pH 7.0. In hydroxylic solvents, the conjugate base (pHA$^-$) is in equilibrium with pHA and at higher pH shifts the λ_{max} to 330 nm [see Eq. (8) and Fig. 1.3.1]. Most reported studies of photochemical activation have used 300–340 nm light for excitation of pHP.

pHA ⇌ pHA$^-$, $pK_a = 7.9$ (8)

Placement of electron-donating methoxy groups at the meta positions of the aromatic ring effectively extend the absorption range of the pHP chromophore toward the visible region ($\lambda > 375$ nm) [13]. In contrast, electron-withdrawing carboxylic ester and carboxamide groups have little effect on the absorption properties of the pHP chromophore. However, the choice of substituents appears to have a significant effect on the quantum efficiency for release of the substrate

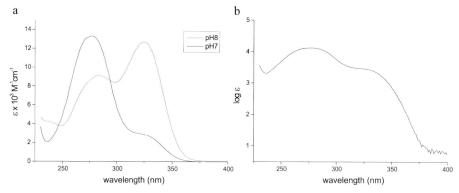

Fig. 1.3.1 UV spectral data for pHA. a Spectra (absorptivity as a function of λ) of a 0.13 mM solution of pHA in NaH$_2$PO$_4$ buffer (3% CH$_3$CN) at pH 7 and 8 showing the shift in absorption from the phenol to the conjugate base. b Absorbance (log ε vs wavelength (λ)) for pHA at pH 7.0.

Tab. 1.3.2 Spectral and photochemical properties of substituted pHP GABA derivatives

Substituted pHP	R	R'	λ_{max} (log ε)[a]	Φ_{rel}[b]	k_r (10^8 s^{-1})
(R'-, HO-, R- substituted phenyl-CO-CH$_2$-GABA)	H	H	282 (4.16) 325sh	0.21	1.9
	OCH$_3$	H	279 (3.97) 307 (3.90) 341sh	0.04	26.1
	OCH$_3$	OCH$_3$	303 (3.90) 355 (3.55)	0.03	0.22
	CONH$_2$	H	324 (4.15)	0.38	0.7
	CO$_2$Me	H	272 (4.09) 310 (3.22) 330 (3.94)	0.31	13
	CO$_2$H	H	281 (4.23)	0.04	0.68

a) In H$_2$O or aqueous buffered media.
b) For the appearance of GABA at pH 7.0. sh=shoulder.

(Φ_{rel}), as seen in Tab. 1.3.2. In fact, electron-withdrawing groups gave the best quantum efficiencies, whereas the opposite was true for electron-donating groups. Overall, the release of the substrate is an efficient photochemical process and occurs within tens of nsec (i.e., $k_{release} \approx 10^8$ s^{-1}).

1.3.2.2 Synthesis of pHP-Caged Substrates

The pHP chromophore can be attached to many functional groups, including carboxylic acids, phosphates, and thiols, with relative ease using a stoichiometric equivalent of 2-bromo-4'-hydroxyacetophenone (pHP-Br) with the nucleophilic

1.3 p-Hydroxyphenacyl: a Photoremovable Protecting Group for Caging Bioactive Substrates

function on the substrate. pHP-Br is most conveniently synthesized by bromination of commercially available pHA with copper(II) bromide in refluxing chloroform/ethyl acetate [14]. Substrates are normally used as their chemically protected derivatives with only the key nucleophilic group exposed. These are reacted with pHP-Br in an aprotic solvent in the presence of a mild base. The base-catalyzed S_N2 displacement of bromide from pHP-Br is a facile process, because of the enhanced reactivity of the α-haloketone. The increased reactivity is illustrated by a comparison of the rates for displacement of chloride from phenacyl chloride and n-C_4H_9Cl with potassium iodide. A rate enhancement of 10^5 s^{-1} is reported for this pair of halides [15]. Removal of the other protecting group(s) on the fully caged ester furnishes the corresponding photoprotected pHP substrate. Scheme 1.3.1 illustrates a specific example in which the Boc-protected dipeptide Ala-Ala was coupled with pHP-Br in 1,4-dioxane in the presence of 1,8-diazabicyclo-[5.4.0]undec-7-ene (DBU) to give Boc-protected pHP-Ala-Ala in 62% yield. Removal of the Boc group with TFA provided pHP-Ala-Ala in 78% yield [8].

A similar synthetic strategy was used to protect the C-terminal carboxylate group of the nonapeptide bradykinin (Bk). In this approach, the peptide sequence of bradykinin (Bk) was constructed first by a solid-phase Merrifield strategy, attaching of the C-terminus protected arginine of Bk to the 2-chlorotrityl resin. Fmoc, tBOC, and Pbf protection chemistry was employed [9, 10]. The resin-bound, fully protected bradykinin was then cleaved from the resin surface with dichloromethane/methanol/acetic acid (8:1:1), providing a protected Bk with only the C-terminal carboxylate exposed. Derivatization of the C-terminus with pHP-Br and DBU in DMF was followed by a two-step sequence to remove the N-Boc-, *tert*-butyl-, and N-Pbf-protecting groups employing a deprotection cocktail of 88% TFA, 7% thioanisole, and 5% water, as shown in Scheme 1.3.2. The pHP-Bk obtained in this sequence gave an overall yield of 60% after purification by RP-HPLC [8]. The strategy employed for pHP bradykinin should serve as a general one for C-terminus pHP protection of any oligopeptide.

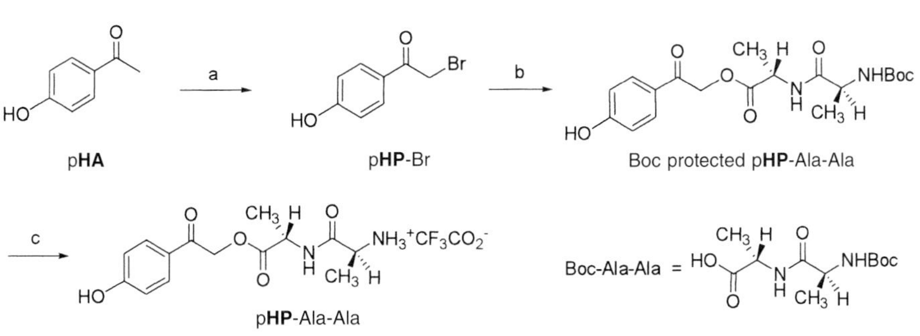

(a) $CuBr_2$, $CHCl_3$/EtOAc, 90 °C, 72%; (b) Boc-Ala-Ala, DBU, 1,4-dioxane, 15 °C, 62%; (c) TFA, 4 h, 0 °C, 78%

Scheme 1.3.1 Synthesis of pHP-Ala-Ala.

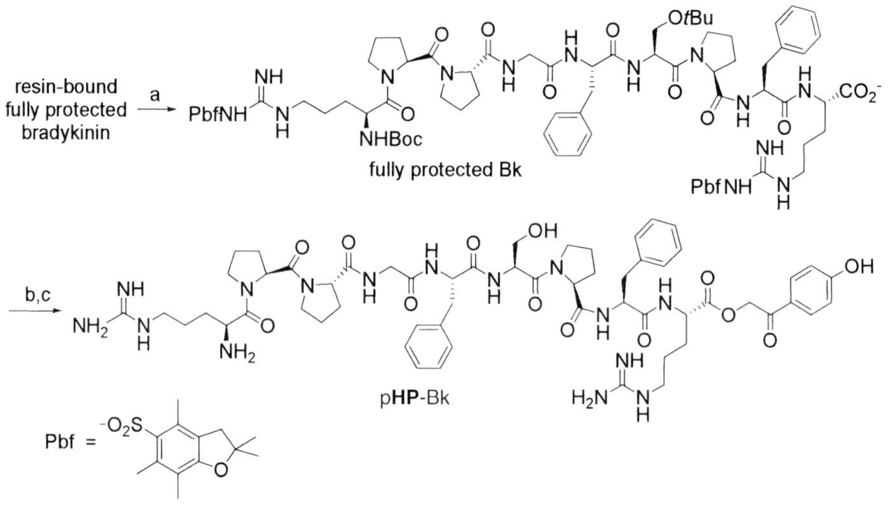

(a) CH$_2$Cl$_2$, CH$_3$OH, CH$_3$CO$_2$H (8:1:1), 2 h, quantitative; (b) pHP-Br, DBU, DMF, rt, 24 h, 72%; (c) 88% TFA, 7% thioanisole, 5% H$_2$O, 2.5 h, 84%

Scheme 1.3.2 Synthesis of p-hydroxyphenacyl bradykinin (pHP-Bk).

While the synthetic techniques outlined above are generally robust and applicable to a variety of oligopeptides, they are not appropriate for all substrates. For example, the synthesis of pHP-ATP required a modified approach in which pHP-OPO$_3^{-2}$ was coupled to ADP using carbonyl diimidazole (CDI) as shown in Scheme 1.3.3 [2, 16]. The photoprotected phosphate pHP-OPO$_3^{-2}$ was synthesized from alkylation of dibenzyl tetramethylammonium phosphate with pHP-Br, followed by removal of the benzyl groups by hydrogenolysis of the acetal-protected dibenzyl phosphate.

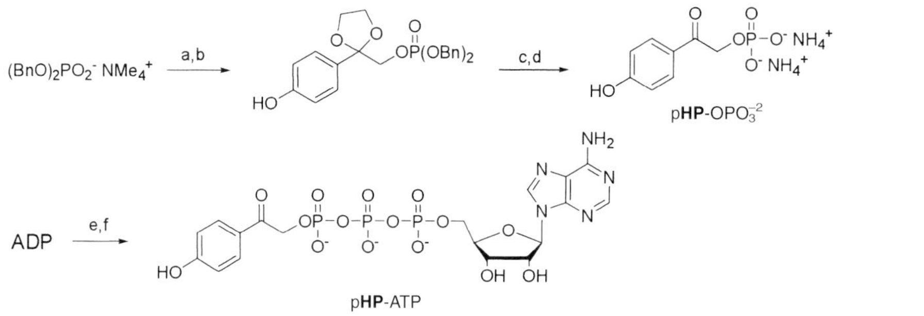

(a) pHP-Br, benzene, reflux, 85%; (b) ethylene glycol, p-TsOH, 82%; (c) Pd/C, H$_2$; (d) 1% HCl, then DEAE, Sephadex, NH$_4$OAc, 95%; (e) CDI; (f) pHP-OPO$_3^{-2}$, HMPA, then DEAE-Cellulose, NH$_4^+$HCO$_3^-$, 42%

Scheme 1.3.3 Synthesis of p-hydroxyphenacyl ATP (pHP-ATP).

1.3 p-Hydroxyphenacyl: a Photoremovable Protecting Group for Caging Bioactive Substrates

In certain cases it may be necessary or desirable to avoid the use of organic solvents in the protection scheme. For example, if the substrate is an enzyme or large oligopeptide (>35 amino acids), its conformational structure may be irreversibly altered in non-aqueous media. In such cases, the caging reaction can be carried out in aqueous buffered solution, provided that there exists one predominant nucleophilic site in the protein. An elegant study by Bayley et al. [17] targeted the catalytic subunit of protein kinase A through a thiophosphorylated threonine residue using pHP-Br as the caging reagent. The caging reaction was accomplished by reacting the enzyme, $P_ST^{197}C_a$, with an ethanolic solution of pHP-Br in an aqueous buffer containing 25 mM Tris·HCl, 200 mM NaCl, 2 mM EDTA, and 1% Prionex (pH 7.3) for 15 min at 25 °C in the dark, resulting in essentially quantitative derivatization [Eq. (9)].

$$P_ST^{197}C_\alpha \xrightarrow[25\ °C,\ 15\ min]{pHP\text{-}Br,\ buffer\ (pH\ 7.3)} pHP\text{-}P_ST^{197}C_\alpha \quad (9)$$

buffer = 25 mM Tris HCl, 200 mM NaCl, 2 mM EDTA, 1% Prionex

Pei et al. used a similar strategy to cage the active site of three protein tyrosine phosphatases (PTPs) with pHP-Br [18]. The caged phosphatase enzymes were PTP1B, SHP-1 (a phosphatase containing an Src homology 2 domain), and SHP-1(ΔSH2), the catalytic domain of SHP-1. Micromolar concentrations of each enzyme were incubated in HEPES buffer solution containing excess pHP-Br in DMF, yielding caged PTP [Eq. (10)]. Matrix-assisted laser desorption ionization mass spectroscopy of trypsin-digested fragments of inactivated pHP-SHP-1(ΔSH2) confirmed that derivatization occurred at the active site Cys-453 residue.

$$PTP \xrightarrow[rt,\ <10\ min]{pHP\text{-}Br,\ buffer\ (pH\ 7.4)} pHP\text{-}PTP \quad (10)$$

buffer: 50 mM HEPES (*N*-(2-hydroxyethyl)piperazine-*N*-(2-ethanesulfonic acid))

1.3.3
Mechanistic Studies

Recent experimental and theoretical studies provide mechanistic information on the photorelease of substrates from the p-hydroxyphenacyl group. However, there are questions remaining regarding the molecular changes involved in the photorelease processes and the identity of the intermediate(s). A brief general description of the mechanism based in part on laser flash photolysis (LFP) studies and density functional theory calculations is presented here. For a more detailed discussion, the reader is encouraged to consult Part II of this monograph.

1.3.3.1 A Triplet "Photo-Favorskii" Rearrangement

A proposed mechanism is outlined in Scheme 1.3.4. After initial excitation of the chromophore to its singlet excited state, rapid singlet-triplet (ST) intersystem crossing ($k_{ST} = 2.7 \times 10^{11}$ s^{-1}) quantitatively generates the triplet state [19]. The triplet excited phenolic group rapidly undergoes an adiabatic proton transfer to solvent, generating triplet phenoxide anion ^{3}pHP-X$^-$. It is speculated that ^{3}pHP-X$^-$ is the precursor to the rate-limiting release of the substrate and the putative spirodienedione intermediate **DD**. **DD** may also be in equilibrium with the oxyallyl valence isomer **OA**. Hydrolysis of **DD** leads to p-hydroxyphenylacetic acid (pHPA), the only major photoproduct of the p-hydroxyphenacyl chromophore.

Previous mechanistic hypotheses involving the direct heterolytic cleavage of the pHP-substrate bond from the triplet phenol ^{3}pHP-X, or homolytic cleavage followed by rapid electron transfer also suggested intermediate **DD** [20, 21] but did not incorporate the seminal role of the phenolic hydroxy group. In a few instances, a minor amount of photohydrolysis of the substrate competes and results in the appearance of <10% of 2,4′-dihydroxyacetophenone among the photolysis products of pHP [8].

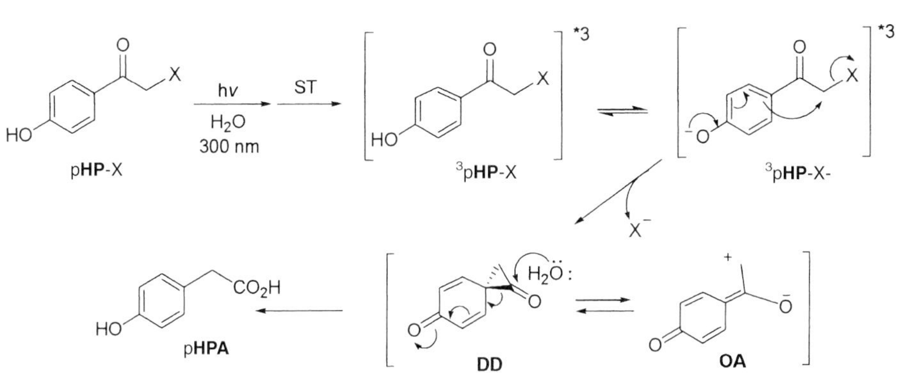

Scheme 1.3.4 Proposed mechanism for photorelease of substrates from pHP-X.

1.3.3.2 Role of the Triplet Phenol

The importance of the triplet phenoxide ion ^{3}pHP-X$^-$ in the photorelease mechanism is supported by LFP studies of pHA in aqueous acetonitrile [19]. The lower pK_a of ^{3}pHA (5.15) is a 60-fold increase in acidity compared with the ground state (7.93) [22]. This increased acidity was also determined for several meta-substituted pHA chromophores listed in Tab. 1.3.3. Thus the adiabatic ionization of the phenol on the triplet excited state energy surface drives the release of the conjugate base of the leaving group.

1.3.3.3 Correlation of the ^{3}pK_a with the Quantum Efficiency

Further evidence of the important role of the conjugate base is found in the correlation between the ^{3}pK_as and the efficiency of reaction shown in Fig. 1.3.2.

Tab. 1.3.3 Ground state and excited state pK_as for substituted pHP derivatives [22]

Substituted pHP	R	R'	pK_a [a]	^{3}pK_a [b]
(R' on phenyl ring, HO- and R positions shown)	H	H	7.93	5.15
	OCH$_3$	H	7.85	6.14
	OCH$_3$	OCH$_3$	7.78	6.16
	CONH$_2$	H	6.15	3.75
	CO$_2$Me	H	7.66	4.28
	CO$_2$H	H	6.28	4.33

a) Measured in water containing 10% methanol at 22–23 °C with an ionic strength of 0.10 M (NaCl).
b) Determined from phosphorescence measurements in ethanol (5% methanol and isopropanol) in liquid nitrogen. (We thank Professor Wirz, University of Basel, for the ^{3}pK_a measurements.)

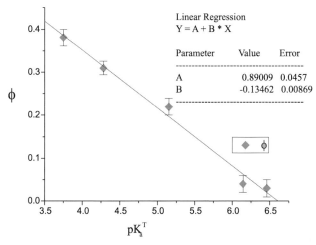

Fig. 1.3.2 Quantum efficiency for release of GABA as a function of the ^{3}pKa of the pHP derivative [22].

This trend suggests that the efficiency of release of the substrate hinges upon the chromophore's ability to form the corresponding phenoxide anion in the triplet excited state, suggesting a crucial role for the *p*-hydroxy function in the photorelease and photorearrangement mechanism.

1.3.4
Applications

1.3.4.1 Neurotransmitter Release

The rapid release rates of substrates from the pHP group ($k_r = 10^8$–10^9 s^{-1}) permit the study of the early, rapid events in neurotransmission and signal transduction processes. Caged derivatives of neurotransmitters, for example pHP-L-Glu [Eq. (11)], have been used successfully to probe the various aspects of chemical synaptic transmission in isolated brain cells. For example, Kandler et al. [23] employed pHP-L-Glu to study postsynaptic long-term depression (LTD) in CA1 hippocampal pyramidal cells. The changes in synaptic efficacy within such cells are thought to be an integral part of learning and memory. Caged glutamate was administered via bath application (50–200 µM) to hippocampal CA1 pyramidal cells in 300 µm rat brain slices. Tetrodotoxin (5 µM) blocked presynaptic transmission to isolate the responses to only the postsynaptic cell changes. Photolysis was conducted with time-resolved pulsed (20 ms) UV from a 100 W mercury arc lamp directed through two 10 µm diameter optical fibers to provide spatial control. Rapid glutamatergic currents, monitored by patch clamp electrodes, were elicited by the photorelease of glutamate (Fig. 1.3.3). The currents were suppressed by glutamate receptor antagonists, e.g., 6-cyano-7-nitroquinoxaline-2,3-dione (CNQX) and D-aminophosphonovalerate (D-APV), demonstrating that the observed responses were due to activation of ionotropic glutamate receptors. The rapid concentration jump in glutamate resulting from localized photolysis of pHP-L-Glu, combined with depolarization, led to LTD of the glutamate receptors.

$$\text{pHP-L-Glu} \xrightarrow{h\nu} \text{L-Glu} + \text{pHPA} \tag{11}$$

Caged bradykinin (pHP-Bk) was deployed by Haydon et al. to explore possible avenues for the connectivity among signaling astrocytes in hippocampal preparations [12]. Mice brain slices were bathed in calcium indicator solution (X-rhod-1 or fluo-$_4$AM) along with 1 nM Bk to identify the bradykinin-responsive cells in the stratum radiatum (Fig. 1.3.4 A, B). The astrocytes responsive to Bk trigger the release of calcium within the cell upon binding free Bk. These cells were identified as targets for initiating the signal transduction of neighboring

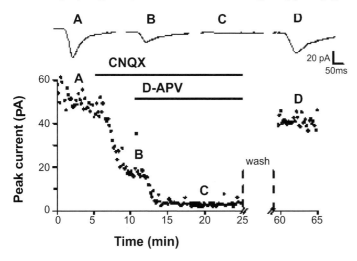

Fig. 1.3.3 Focal photolysis of pHP-L-Glu in CA1 hippocampal pyramidal cells. The elicited current (**A**) is attenuated upon bath application of CNQX, an AMPA-receptor antagonist (**B**); addition of the NMDA antagonist D-APV completely diminishes the current (**C**). Washing the tissue sample free of the antagonists restores the current (**D**). (With permission from Nature Neuroscience.)

cells. The preparation was then incubated in a 10 nM solution of caged pHP-Bk. Focal photolysis from a nitrogen-pulsed laser at 337 nm or a continuous-wave argon laser at 351 and 364 nm (1–20 ms pulses) photoreleased Bk from pHP-Bk, inducing the calcium signal in the Bk-responsive cell. Calcium signals were subsequently monitored by confocal imaging of the fluorescence of the calcium-bound indicator with an appropriate wavelength (e.g. 488 nm for X-rhod-1) from a second source. The initial fluorescent burst in the photostimulated cell was followed by delayed responses in neighboring cells six to twenty seconds after the initial response (Fig. 1.3.4 C, D). Since bath application of Bk only elicited a response from the cell that was directly photostimulated, it was concluded that the delayed calcium responses were not simply due to diffusion of photoreleased Bk. Furthermore, the pattern of the responses was asymmetric and did not include all astrocyte cells in the immediate vicinity of the focal cell. Simple diffusion of the signal transducer or any other diffusive activation mechanism is apparently ruled out by this observation. Thus, these results suggest that the astrocytes may be functionally connected. The cell did not respond in the presence of glutamate receptor antagonists CNQX, D-AP$_5$, and (s)-α-methyl-4-carboxyphenylglycine (MCPG), indicating that the glutamate that is naturally released in response to elevated calcium levels in astrocytes [24] is not responsible for the intercellular signaling mechanism.

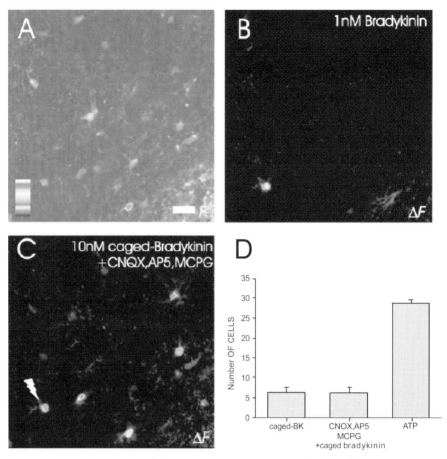

Fig. 1.3.4 Astrocyte intercellular calcium signaling upon focal photolysis of p**HP**-Bk. **A** A fluorescence image of the Bk-responsive astrocytes in the stratum radiatum, identified upon addition of 1 nM Bk. **B** A separate field of view showing only one bradykinin-responsive cell. **C** Delayed calcium responses observed upon photolysis of p**HP**-Bk near the Bk-responsive cell (lightning bolt) in the presence or absence of the glutamate receptor antagonists CNQX, D-AP$_5$, and MCPG. **D** Graph showing the number of cells for which delayed calcium responses were observed upon the photolytically induced calcium increase in the Bk-responsive cell. (With permission granted by the Journal of Neuron Glia Biology.)

1.3.4.2 Peptide Release

The p**HP** group can be used for the protection and photorelease of optically active peptides. Givens et al. [8] reported that the caged dipeptide p**HP**-Ala-Ala cleanly and efficiently releases Ala-Ala upon photolysis at 300 nm in D_2O [Eq. (12), Tab. 1.3.4]. The optical rotation of the released dipeptide was identical to that of free Ala-Ala. Similar conclusions were reported for the photolysis of p**HP**-Bk under analogous conditions, in which the released Bk CD spectrum

1.3 p-Hydroxyphenacyl: a Photoremovable Protecting Group for Caging Bioactive Substrates

matched that of an authentic sample [Eq. (7), Fig. 1.3.5]. These results demonstrate that the chiral integrity of optically active peptide substrates is preserved throughout the synthetic protection/photolytic deprotection cycle.

pHP-Ala-Ala, $[\alpha]_{20}^D = -39.73$ (c = 0.59, H_2O)

$\xrightarrow{h\nu,\ 300\ nm}_{D_2O,\ 18\ min}$ pHPA +

Ala-Ala, $[\alpha]_{20}^D = -14.77$ (c = 0.13, H_2O)

(100% retention of optical activity; $[\alpha]_{20}^D = -14.75$ for starting dipeptide)

(12)

Complete conversion of the starting pHP esters was observed. For example, pHP-Ala-Ala was photolyzed to 100% conversion as shown by ^1H NMR [8]. The exceptional conversions possible for pHP derivatives result from the transparent pHPA, the blue-shifted photoproduct formed by the rearrangement, a key feature of the pHP photochemistry.

Thiol-containing peptides have also been successfully used in the pHP protection strategy, as was demonstrated recently by Goeldner et al. [25], in which glutathione (GSH) was converted to pHP-SG in 92% yield [Eq. (13)]. As demonstrated from the study by Pei et al. discussed earlier, cysteines can be selectively modified in the presence of other functional groups in a protein upon direct reaction with pHP-Br.

GSH $\xrightarrow[\text{1:1 ethanol/buffer}]{\text{pHP-Br}}$ pHP-SG, 92%

buffer: 1N HEPES pH 7.0

(13)

Tab. 1.3.4 Quantum efficiencies for pHP-Ala-Ala and pHP-Bk in D_2O [8]

Entry	Φ_{dis} [c]	Φ_{app} [d]	Φ_{rear} [e]
pHP-Ala-Ala [a]	0.27 (0.04)	0.25 (0.05)	0.24 (0.04)
pHP-Bk [b]	0.20 (0.02)	0.22 (0.02)	0.19 (0.01)

a) Determined by ^1H NMR and HPLC (averaged).
b) Determined by HPLC.
c) For the disappearance of the starting ester.
d) For the appearance of the released substrate.
e) For the appearance of pHPA. Standard deviations are given in parentheses.

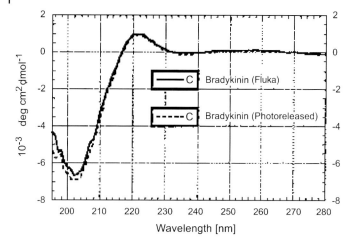

Fig. 1.3.5 CD Comparison of photoreleased Bk with authentic Bk. The broken line " --- " represents Bk isolated by HPLC from the photolysis of pHP-Bk; the continuous line " —— " represents Bk (Fluka) [Θ_{max} = 954 deg cm^2 dmol^{-1} (222 nm), Θ_{min} = –6705 deg cm^2 dmol^{-1} (202 nm). For Bk (photoreleased): Θ_{max} = 918 deg cm^2 dmol^{-1} (222 nm), Θ_{min} = –6965 deg cm^2 dmol^{-1} (202 nm), where Θ = mean residue ellipticity]. (With permission of the Journal of the American Chemical Society.)

Photolysis of pHP-SG in TRIS-buffer solution resulted in ~70% release of free GSH, accompanied by 23% of the corresponding thioester, in addition to pHPA and pHA, as shown in Eq. (14). The thioester was likely the result of nucleophilic attack of the thiol on the spirodienedione intermediate **DD** (see above). The undesired side reaction with **DD** may not be as likely at a protein-binding site where the intermediate(s) may readily escape and not remain in close proximity to the released thiol.

$$\text{pHP-SG} \xrightarrow[\text{Tris-HCl buffer, pH 7.2}]{h\nu,\ 312\ \text{nm}} \text{GSH} + \text{pHPA} + \text{pHA} \quad (14)$$

1.3.4.3 Nucleotide Release

Caged nucleotides have also been deployed as probes for physiological mechanism studies such as ion transport. Fendler et al. [26] used **pHP**-ATP to investigate the intracellular transport of sodium and potassium ions mediated by Na$^+$,K$^+$-ATPase. Lipid membranes to which were added membrane fragments containing Na$^+$,K$^+$-ATPase from pig kidney, along with pHP-ATP (10–300 µM) were exposed to light pulses (10 ns) from an XeCl excimer laser at 308 nm. The rapid jump in ATP concentration [see Eq. (1)] led to activation of the Na$^+$,K$^+$-

ATPase, which hydrolyzes ATP resulting in the generation of a transient current.

The release of ATP from pHP-ATP could be monitored using time-resolved Fourier transform infrared (TR-FTIR) spectroscopy. Among the most prominent changes observed were the disappearance of the 1270 cm^{-1} γ-PO$_3^-$ diester band of pHP-ATP and the appearance of the free 1129 cm^{-1} γ-PO$_3^{-2}$ band of the released ATP, shown in Fig. 1.3.6A. To kinetically resolve the release of ATP, the absorbance changes at selected difference bands were tracked (Fig. 1.3.6B). The positive signal observed at 1140 cm^{-1}, corresponding to the released ATP, and a sharp negative signal at 1270 cm^{-1}, representing pHP-ATP, were observed within the first few microseconds after the laser pulse. The signal at 1200 cm^{-1} was attributed to an artifact from the changes in transient absorption of water resulting from the energy absorbed by the sample. Fig. 1.3.6C shows the corrected ATP release signal (solid line) obtained after subtraction of the artifact signal at 1200 cm^{-1}. The integrated stray light signal of the laser pulse (dotted line) used to monitor the detector response time is also shown. Together these data show that the relaxation rate for ATP release occurs with a lower limit of 10^6 s^{-1}. Additional details on these studies are available in Chapter 7.1 of this book.

In a similar study, pHP-GTP [Eq. (15)] was used as a probe in a study of the mechanism of hydrolysis of GTP by Ras, a protein that is involved in cell signaling [27]. Upon binding GTP, the resulting Ras-GTP complex acts as an "on" switch in the signaling pathway. Hydrolysis of the bound GTP to GDP extinguishes the signal. In this particular study, photolytic release of ^{18}O-labeled caged GTP was used in conjunction with time-resolved FTIR spectroscopy to probe the nature of the transition structure for the rate-determining step in the hydrolysis of Ras-bound GTP. Buffer solutions containing the Ras protein were

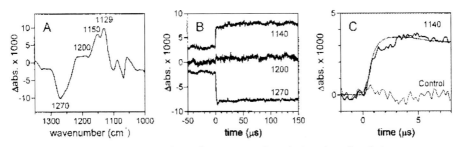

Fig. 1.3.6 TR-FTIR spectroscopic analysis of the release of ATP from pHP-ATP. A FTIR difference spectrum of the photolysis of pHP-ATP. The absorbance was measured 10 ms to 11 s after the photolysis flash and subtracted from the absorbance immediately prior to the flash. B The infrared absorbance changes at 1140, 1200, and 1270 cm^{-1} measured with a dispersive infrared spectrometer. C A closer look at the infrared absorbance change at 1140 cm^{-1} upon subtraction of the smoothed 1200 cm^{-1} signal. Bold line: ATP release signal (first flash), thin line: heat signal (second flash without measuring light), dotted line: integrated stray light signal of the laser pulse. (With permission of the Biophysical Journal.)

mixed with threefold equivalents of p**HP**-GTP, forming the caged GTP-Ras complex. A Tris·HCl:MgCl$_2$ buffer solution (pH 8.0) of the lyophilized protein was irradiated with a 308 nm pulsed laser for 30 s, and the subsequent changes in infrared absorption spectra of the phosphoryl groups were measured as a function of time. Monitoring of the hydrolysis of β-^{18}O$_3$ labeled GTP resulted in no detectable positional isotope exchange, suggesting a concerted hydrolysis mechanism. Further details on this application are available in Chapter 7.2 of this book.

$$\text{p}\textbf{HP}\text{-GTP} \xrightarrow[\text{Tris·HCl} \atop \text{MgCl}_2,\ \text{pH 8.0}]{h\nu,\ 308\ \text{nm}} \text{GTP} + \text{p}\textbf{HPA}$$

(15)

1.3.4.4 Enzyme Photoswitches

Normally, covalent attachment of inhibitors to key residues in an active site of an enzyme creates a non-reversible, inactive protein. However, photolabile groups attached at the active site become reversible inhibitors and thus act as switches that can alternately turn the enzyme off and reactivate it upon exposure to light. To date, two enzymes have been successfully caged and reactivated using the p**HP** group. The first enzyme, a protein tyrosine phosphatase (PTP), a class of enzymes that catalyze the hydrolysis of phosphotyrosine residues, was directly caged with p**HP**-Br. In this particular instance, the active site involves catalysis by a cysteine residue that participates in nucleophilic attack on a phosphate group of a bound phosphotyrosine substrate. This leads to a phosphorylated cysteinyl group at the active site, which is subsequently hydrolyzed by water [28]. Little is known about the specific role of these enzymes, other than as regulators of tyrosine phosphorylation. Pei et al. [18] explored a number of α-halophenacyl derivatives as potential PTP inhibitors by reacting a series of PTPs with p**HP**-Br and other phenacyl analogs. Derivatization of the key cysteine residue in the PTP active site with p**HP**-Br takes place in under 10 min in aqueous buffered media at room temperature, as shown previously in Eq. (10). Inactivation of the enzyme could be observed spectrophotometrically by using an assay with p-nitrophenyl phosphate and monitoring the change in absorbance at 405 nm as a function of time for varying concentrations of p**HP**-Br. Irradiation of the dormant PTPs at 350 nm resulted in reactivation of the enzymatic activity [Eq. (16)]. The extent of reactivation ranged from approximately 30% to as high as 80% of the original activity of the untreated enzyme, depending on the enzyme and the cage (Fig. 1.3.7).

1.3 p-Hydroxyphenacyl: a Photoremovable Protecting Group for Caging Bioactive Substrates

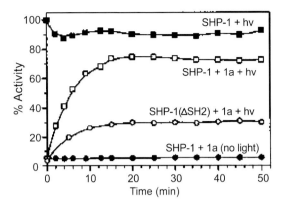

Fig. 1.3.7 Reactivation of pHP-caged SHP-1 and SHP-1(ΔSH2) enzymes [see Eq. (11) and accompanying text for details]. In the graph, **1a** represents pHP-Br [18]. (With permission from the Journal of the American Chemical Society.)

$$\text{pHP-PTP} \xrightarrow[\substack{\text{HEPES buffer} \\ 4\,°\text{C}}]{h\nu,\ 350\ \text{nm}} \text{PTP (free enzyme)} + \text{pHPA} \quad (16)$$

Protein kinase A (PKA) is a cell signaling protein that contains a pair of regulatory and catalytic (C) subunits. The regulatory subunits bind cAMP, which in turn induces a conformational change in the protein leading to release of an active monomeric C subunit. The C subunit of PKA contains two lobes as illustrated in Fig. 1.3.8. The small lobe is made up of the N-terminal sequences of the enzyme where ATP binding takes place, while the large lobe houses the C-terminal sequences with the residues necessary for substrate binding and catalysis. A key threonine residue (Thr-197) in the large lobe serves as a biological switch that, when phosphorylated, engages the enzyme in an active conformation [17].

An unphosphorylated C subunit of PKA was expressed from *Escherichia coli* and thiophosphorylated at Thr-197 with 3-phosphoinositide-dependent kinase 1. Bayley and coworkers [17] alkylated the resulting thiophosphate with pHP-Br to give the pHP(thio)-protected enzyme [Eq. (9)]. The modified enzyme pHP-$P_S T^{197} C_a$ exhibited a 17-fold reduction in activity toward the substrate kemptide, based on a ^{32}P assay. The covalently bound pHP group likely prevents the enzyme from achieving a fully active conformation. Photolysis of pHP-$P_S T^{197} C_a$ with near UV light in buffer solution released the free enzyme with a product quantum efficiency of 0.21 and an estimated deprotection yield of 85–90%, based on capture of photoreleased $P_S T^{197} C_a$ with 5-thio-2-nitrobenzoic acid (TNB)-thiol agarose [Eq. (17)].

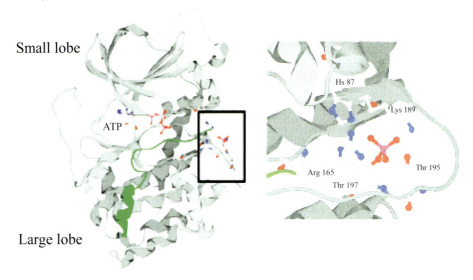

Fig. 1.3.8 Structure of the C subunit of PKA. Both the large and small lobes of the C subunit are shown. ATP binding takes place at the small lobe; the large lobe houses the substrate, in this example a peptide inhibitor (green). The key residues involved in catalysis, including Thr-197, are highlighted in the expanded view [17]. (With permission from the Journal of the American Chemical Society.)

$$\text{pHP-P}_S\text{T}^{197}\text{C}\alpha \xrightarrow[\text{Tris·HCl buffer (pH = 7.3)}]{h\nu,\ 280\text{-}370\ \text{nm}} \text{P}_S\text{T}^{197}\text{C}\alpha \qquad (17)$$

1.3.5
Advantages and Limitations

The **pHP** group is a highly versatile photoremovable protecting group that is amenable to a wide range of applications in both synthetic and mechanistic bioorganic chemistry. The advantages are the ease of synthesis and the rate and efficiency of release of a variety of substrates. The current limitations are the excitation wavelengths required (<400 nm) and the requirement that the leaving group be the conjugate base of an acid (phosphate, carboxylate, and thiolate, for example). These limitations can be remedied by proper design of the chromophore and manipulation of the leaving group attachments. Research is in progress that addresses these issues.

Acknowledgements

The authors thank the Center for Excellence in Chemical Methodologies and Library Design and the University of Kansas for financial support. The collaboration of the many research groups who have demonstrated the applications of pHP and its derivatives and the research efforts of the Photochemistry Group at the University of Kansas have been invaluable.

References

1 GIVENS, R.S., PARK, C.-H. *Tetrahedron Lett.*, **1996**, *35*, 6259–6266.
2 PARK, C.-H., GIVENS, R.S. *J. Am. Chem. Soc.*, **1997**, *119*, 2453–2463.
3 PELLICCIOLI, A.P., WIRZ, J. *Photochem. Photobiol. Sci.*, **2002**, *1*, 441–458.
4 BOCHET, C.G. *J. Chem. Soc., Perkin Trans. 1*, **2002**, 125–142.
5 ADAMS, S.R., TSEIN, R.Y. *Annu. Rev. Physiol.*, **2000**, *18*, 755–784.
6 SHEEHAN, J.C., UMEZAWA, K. *J. Org. Chem.*, **1973**, *38*, 3771–3774.
7 ANDERSON, J.C., REESE, C.B. *Tetrahedron Lett.*, **1962**, *1*, 1–4.
8 GIVENS, R.S., WEBER, J.F.W., CONRAD II, P.G., OROSZ, G., DONAHUE, S.L., THAYER, S.A. *J. Am. Chem. Soc.*, **2000**, *122*, 2687–2697.
9 BARLOS, K., GATOS, D., STAUROS, K., PAPAPHOTIV, G., SCHAFER, W., WENQUING, Y. *Tetrahedron Lett.*, **1989**, *30*, 3947–3950.
10 BODI, J., SULI-VARGHA, H., LUDANYI, K., VEKEY, K., OROSZ, G. *Tetrahedron Lett.*, **1997**, *38*, 3293–3296.
11 GIVENS, R.S., JUNG, A., PARK, C.-H., WEBER, J., BARTLETT, W. *J. Am. Chem. Soc.*, **1997**, *119*, 8369–8370.
12 SUL, J.-Y., OROSZ, G., GIVENS, R.S., HAYDON, P.G. *Neuron Glia Biology*, **2004**, *1*, 3–11.
13 CONRAD II, P.G., GIVENS, R.S., WEBER, J.F.W., KANDLER, K. *Org. Lett.*, **2000**, *2*, 1545–1547.
14 CONRAD II, P.G. Department of Chemistry, PhD Dissertation, University of Kansas: Lawrence, **2001**.
15 HINE, J. *Physical Organic Chemistry*, McGraw-Hill Book Company, Inc.: New York **1962**.
16 GIVENS, R.S., WEBER, J.F.W., JUNG, A.H., PARK, C.-H. In *Methods Enzymol*, GERARD MARRIOTT, ed.; Academic Press, **1998**; Vol. *291*, pp. 1–29.
17 ZOU, K., CHELEY, S., GIVENS, R.S., BAYLEY, H. *J. Am. Chem. Soc.*, **2002**, *124*, 8220–8229.
18 ARABACI, G., GUO, X-C., BEEBE, K.D., COGGESHALL, K.M., PEI, D. *J. Am. Chem. Soc.*, **1999**, *121*, 5085–5086.
19 CONRAD II, P.G., GIVENS, R.S., HELLRUNG, B., RAJESH, C.S., RAMSEIER, M., WIRZ, J. *J. Am. Chem. Soc.*, **2000**, *122*, 9346–9347.
20 GIVENS, R.S., CONRAD, P.G. II, YOUSEF, A.L., LEE, J.-I. In *CRC Handbook of Organic Photochemistry and Photobiology*, W. HORSPOOL, LENCI, F., ed.; CRC Press: New York, **2004**.
21 GIVENS, R.S., LEE, J.-I. *J. Photosci.*, **2003**, *10*, 37–48.
22 WIRZ, J., GIVENS, R.S. unpublished results, **2002**.
23 KANDLER, K., KATZ, L.C., KAUER, J.A. *Nat. Neurosci.*, **1998**, *1*, 119–123.
24 PARPURA, V., HAYDON, P.G. *PNAS, USA*, **2000**, *97*, 8629–8634.
25 SPECHT, A., LOUDWIG, S., PENG, L., GOELDNER, M. *Tetrahedron Lett.*, **2002**, *43*, 8947–8950.
26 GEIBEL, S., BARTH, A., AMSLINGER, S., JUNG, A.H., BURZIK, C., CLARKE, R.J., GIVENS, R.S., FENDLER, K. *Biophys. J.*, **2000**, *79*, 1346–1357.
27 DU, X., FREI, H., KIM, S.-H. *J. Biol. Chem.*, **2000**, *275*, 8492–8500.
28 DIXON, J.E., ZHANG, Z.-Y. *Adv. Enzymol. Relat. Areas Mol. Biol.*, **1994**, *68*, 1–36.

1.4
Caging of ATP and Glutamate: a Comparative Analysis

Maurice Goeldner

1.4.1
Introduction

Light-sensitive biomolecules (caged compounds) enable the study of rapid biological processes by fast and efficient photochemical triggering of the release of the bioactive compound. The caging of many substances including second messengers, neurotransmitters, peptides, and functional proteins and their modulators will be described in detail in this book. However, a precise comparative analysis of different caging groups for a prototypical bioactive molecule has not been described thus far. Consequently, such analyses could shed new light on the opportunity to develop new caging groups for these reference molecules. ATP and glutamate each represent typical molecules for which numerous photoremovable protecting groups have been described, justifying such analyses. A second reason to select these two substances is that the incorporation of the photochemical group involves the chemical modification on phosphates or carboxylates, chemical functions which are found in many biological active molecules. As a consequence, the synthetic methods as well as the chemical and photochemical properties described here will most likely be directly applicable to these related substances.

This chapter will list exhaustively the different caging groups which have been described for these two molecules. It will delineate briefly the key steps for their syntheses and emphasize their respective physico-chemical and photochemical properties defining the overall quality of the caging process. These properties will be summarized in tables to facilitate a direct comparison and to help select the cage which would be best adapted for a potential use in a given biological experiment. This chapter will not describe the biological applications for which caged ATP and caged Glu derivatives have been used, not only because the list would be fairly long, but mainly because biologists usually use commercially available substances. I believe it is more important, in this chapter, to inform biologists about the existing substances by describing the properties of the different caging groups in an unbiased manner and possibly stimulate collaborations among chemists and biologists.

1.4.2
General Properties for Caging Groups

What are the chemical and photochemical properties required for a caging group?
- The chemical synthesis represents the first element that confronts an experimentalist, knowing that only a few caged biomolecules are commercially available. Clearly a simple and efficient synthetic pathway is preferred. For

the syntheses of more complex molecules, i.e. caged peptides, proteins, or oligonucleotides, a direct transformation of the biomolecule into a caged derivative will be necessary and raises the question of the selectivity of the chemical modification in a multifunctional molecule.
- Properties in the dark: the caged molecule should display hydrolytic stability and aqueous solubility. Most importantly, the caged biomolecule should be biologically inactive. Different enzymes and receptors are involved in individual biological events: therefore, each system will use its own criteria to define the biological inertness.
- Photochemical properties: these involve several parameters:
  - Irradiation wavelengths should be greater than 300 nm to avoid the absorption of light by proteins and nucleic acids; therefore, a high absorbance of the caging group above 300 nm is desirable.
  - The photolytic reaction should occur with high quantum efficiency, a value which defines the intrinsic quality of the photolytic reaction (i.e. ratio of molecule released to photons absorbed). Two types of quantum efficiencies will be considered: if possible the quantum efficiency of product formation (Φ_{ATP} or Φ_{Glu}) and/or the quantum efficiency of disappearance of the caged biomolecule (Φ_{Dis}).
  - Efficacy of photorelease in a photolytic reaction is determined by a combination of the absorption coefficient (ε) at the wavelength of irradiation and the product quantum efficiency (Φ). The uncaging action cross-section, which is defined by the product $\varepsilon_\lambda \times \Phi$ at the irradiation wavelength λ, therefore represents a quantification of the photolytic efficacy. The values shown in Tabs. 1.4.2 and 1.4.4 are usually not described; they have been deduced, when possible, from data described in the literature. These values allow a useful comparative analysis among the different caging groups, at different irradiation wavelength. When high concentrations (i.e. mM concentrations) of the released biomolecules are necessary, the higher ε values can become detrimental for the photolytic reaction (i.e. non-uniform release of the product within the sample); however this drawback, in most instances, can be easily counteracted by selecting a different irradiation wavelength.
  - The side product of the photolytic reaction should not interfere with the ongoing reaction by absorbing the light at the excitation wavelength, and in addition this compound should not interfere either chemically or biologically with the biological system.
  - Last, but not least, the release of the biomolecule should be very fast in comparison to the time course of the biological process to be studied. The rate constant for release of the biomolecule will be given, and in its absence the rate constant of disappearance of the caged substrate.

1.4.3
Caged-ATP

1.4.3.1 Introduction

ATP (adenosine triphosphate) represents an ubiquitous carrier of chemical energy in most energy-requiring processes in living cells. ATP hydrolysis has been involved in active transports, in muscle contraction, in endo- and exocytosis, in cytoplasmic streaming, in ciliary movements, in conformational changes of proteins, and in many other dynamic processes. It is not surprising that ATP was one of the first bioactive molecules to be targeted for caging studies, as will be discussed in the introduction by Jack Kaplan, one of the conceivers of the concept [1].

For many years, most biologists used for their studies the caged-ATP molecules that were commercially available, that is, the P^3-2-nitrobenzyl and the P^3-1-(2-nitrophenyl)ethyl caged derivatives (NB- and NPE-caged ATP, respectively) [2–4]. The need for more powerful caging groups for ATP prompted the syntheses of ATP substituted by different photoremovable protecting groups. In the nitrobenzyl series, several derivatives were developed such as the a-carboxy-2-nitrobenzyl (CNB-caged ATP) [5] and the 1-(4,5-dimethoxy-2-nitrophenyl)ethyl (DMNPE-caged ATP) [6]. Alternatively, different chemical series were investigated, such as the 3,5-dinitrophenyl derivative (DNP-caged ATP) [7], the desyl (desyl-caged ATP) [8–10] and the related 3′,5′-dimethoxy derivative (DMB-caged ATP) [11, 12], the p-hydroxyphenacyl (pHP-caged ATP) [10, 13], and more recently the [7-(dimethylamino)coumarin-4-yl]methyl (DMACM-caged ATP) [14] derivatives. These molecules, listed in Tab. 1.4.1, are all substituted at the P^3 phosphate group of ATP to ensure that, as caged derivatives, they lack biological activity. The same phosphate ester photo-protecting groups have also been used for the blocking of phosphate groups of other relevant nucleotides including AMP, ADP, cAMP, and cGMP, as will be illustrated by V. Hagen and colleagues in Part IV of this monograph [15].

1.4.3.2 Syntheses of the P^3-caged ATP Derivatives

Different strategies were developed to synthesize these derivatives. In most cases, these involve the synthesis of the caged monophosphates, which were subsequently coupled to ADP in the presence of an activating reagent such as a carbodiimide or carbonyl diimidazole (Scheme 1.4.1). Different methods were developed to synthesize the monophosphate derivatives, starting usually from the corresponding alcohol. The first derivatives (NB- and NPE-caged ATP) [2] used anhydrous phosphoric acid in CCl_3CN as the phosphorylating reagent and generated the expected derivatives in low yields. The hydroxyl group of t-butyl 2-nitromandelate was phosphorylated with phosphorus oxychloride followed by hydrolysis to give the phosphate precursor [5]. A series of improvements targeted the synthesis of phosphate esters as direct precursors of the phosphates, allowing an easier purification in organic medium. The NPE- and the DMB-caged ATP used a biscyanoethyl phosphoramidite reagent to couple with opti-

Tab. 1.4.1 Caged ATP molecules: chemical formulas and abbreviations

Chemical structure	Formulas – Cage acronyms	References
Cage—O—P—O—P—O—P—O—[adenosine]	Cage—ATP	
2-nitrobenzyl-ATP	R=H: NB R=CH$_3$: NPE R=CO$_2$H: CNB	[2] [2–4] [5]
4,5-dimethoxy-2-nitrobenzyl-ATP	R=CH$_3$: DMNPE	[6]
3,5-dinitrophenyl-ATP	DNP	[7]
desyl-ATP	R=H: Desyl R=OCH$_3$: DMB	[10] [7–9, 12]
p-hydroxyphenacyl-ATP	pHP	[10, 13]
DMACM-ATP	DMACM	[15]

cally active (R)- or (S)-1-(2-nitrophenyl)ethanol [4] or a protected benzoin derivative [11] before oxidation and deprotection to 1-(2-nitrophenyl)ethyl phosphate or 3′,5′-dimethoxybenzoin phosphate, respectively. The DMACM derivative used a bis-t-butyl phosphoramidite reagent [14] to couple with the 4-hydroxymethyl coumarin derivative before oxidation and deprotection to the DMACM phosphate derivative.

For the synthesis of the desyl-caged ATP, a symmetrical dioxaphosphole intermediate was used to generate a desyl monophosphate [16], while, for the synthesis of the pHP derivative, the 4-hydroxyphenacyl bromide was converted into the dibenzylphosphate derivative, a precursor of the 4-hydroxyphenacyl phosphate [13]. It appears that the use of a phosphoramidite coupling reagent and subsequent conversion to the phosphate group does offer a better control of the reaction in organic media and should be applicable to the synthesis of most caged ATP molecules.

Alternatively, an efficient and direct coupling of ATP with nitrophenyl diazoethane derivatives, under controlled pH in a biphasic medium, was described for the synthesis of the NPE and DMNPE probes (Scheme 1.4.2) [3, 6]. A detailed analysis of this coupling methodology can be found in Ref. [7]. However, this method gave disappointing results for the synthesis of the DMACM-caged derivative [14]. Although very short and appealing, it does not represent a general synthetic method for the caging of ATP or other nucleotide molecules.

1.4.3.3 General Properties of the Caged-ATP Molecules

To our knowledge, all the known P^3-caged ATP molecules that have been published are listed in Tab. 1.4.1. Their structures are shown and their commonly used acronyms are included. Tab. 1.4.2 summarizes the chemical and photochemical properties including UV absorption data, photochemical quantum yields of ATP formation or disappearance of the caged derivative, the uncaging action cross-section at a given wavelength, the rate constants of photolysis, and finally, when described, hydrolytic stability and aqueous solubility. As might be anticipated, not all these data for each caged ATP are available from the literature. However, most published data do provide two essential factors, the quantum efficiencies and the rates of photolysis. The fragmentation mechanisms of the different protecting groups will be discussed in detail in several chapters of this monograph and are summarized in Part II [17].

For many years, the *ortho*-nitro benzyl series, i.e. NPE-caged ATP, was frequently employed for biological experiments. It was the best probe available at that time, its excellent quantum efficiency for ATP release ($\Phi_{ATP}=0.63$) compensated for its low absorptivity above 300 nm. Nevertheless this probe displayed several drawbacks such as a moderate rate of photolysis ($\sim 80\ s^{-1}$) and the formation of *ortho*-nitroso acetophenone, a highly absorbing product above 300 nm, which has been demonstrated to be chemically reactive toward proteins and cysteines in particular [18]. Unfortunately, the search for new derivatives in this series did not produce the expected improvements. For example, the probe substituted by

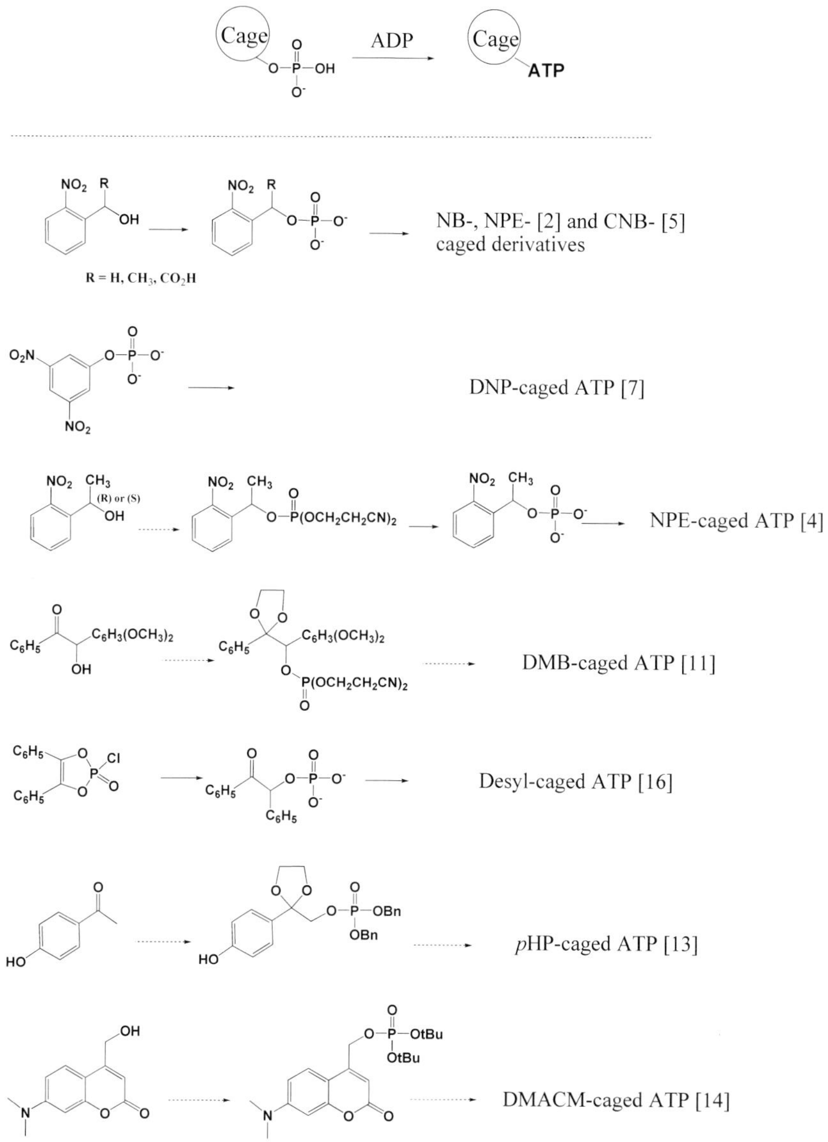

1.4 Caging of ATP and Glutamate: a Comparative Analysis

Scheme 1.4.1 Syntheses of caged-ATP derivatives: coupling of caged monophosphates to ADP

a carboxylic group at the benzylic position (CNB-caged ATP) showed increased water solubility. To our knowledge, the photochemical properties of this derivative are not available [5]. The 3,4-dimethoxy-2-nitrophenyl derivative (DMNPE-caged ATP), which displayed the expected UV-shift, showed a dramatic deficit in photochemical properties, i.e. a rather low quantum efficiency ($\Phi_{ATP}=0.07$) together with a very slow rate of photolysis (18 s^{-1}).

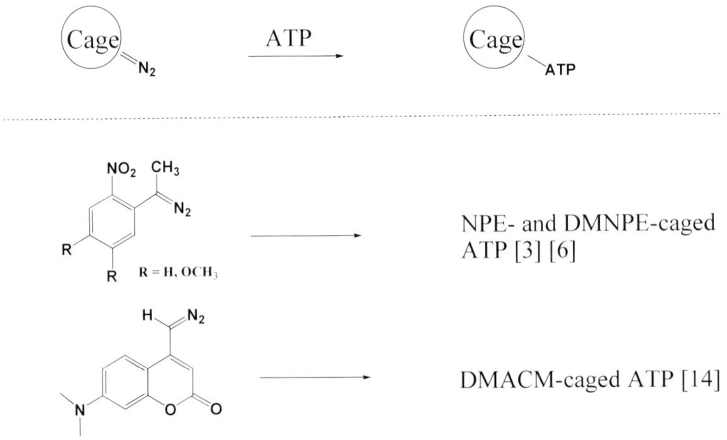

Scheme 1.4.2 Syntheses of caged-ATP derivatives: diazo-coupling approach

The quest for rapid photolytic reactions prompted the search for different photo-protecting groups for ATP, leading to the development of desyl-, DMB-, pHP- and DMACM-caged ATP derivatives (Tabs. 1.4.1 and 1.4.2). These molecules showed consistent improvement in chemical and photochemical properties. Clearly, the first two compounds to be developed, the desyl and the DMB-probe, despite fast rates of photolysis, displayed low photolytic efficiencies together with poor aqueous solubility, while the DMB-probe showed moderate hydrolytic stability. These probes have not been used further for biological experiments.

The pHP-probe displayed major improvements, high quantum yields ($\Phi_{ATP}=0.37$), fast fragmentation kinetics ($>10^7$ s^{-1}) together with hydrolytic stability. The photolytic conversion of the pHP-caged ATP generates exclusively p-hydroxyphenylacetic acid as side-product [13], which does not interfere with the ongoing photolytic reaction. The only limitation of this probe was a moderate photolytic efficiency, around 350 nm. Finally, the very recent description of the DMACM-probe [14] allowed extremely fast fragmentation kinetics ($>10^9$ s^{-1}) to be depicted together with a strong absorptivity above 350 nm of the coumarin chromophore to compensate for rather low quantum yields ($\Phi_{ATP}=0.086$) and ensure high photolytic efficiencies (Tab. 1.4.2). However, the formation of the strongly absorbing hydroxymethyl derivative during photolysis limits high conversions to wavelengths ≥ 360 nm. While showing hydrolytic stability, its water solubility, which represents a major drawback for the coumarin derivatives in general [20], has been improved by the presence of the dimethylamino substituent.

Tab. 1.4.2 Caged ATP derivatives: chemical and photochemical properties

Caged – ATP derivatives Abbreviations	λ_{max}; nm (ε: $M^{-1}\,cm^{-1}$)	Quantum efficiencies Φ_{ATP} or Φ_{Dis}[a]	Uncaging action cross-section $\Phi \cdot \varepsilon$ (λ)[b]	Rate constants (s^{-1}) of release of ATP[c]	Stability and/or solubility in buffered media	References
NB-caged ATP	260 (26 600)	$\Phi_{ATP} \leq 0.19$[c]	<130 (347 nm)			[2]
NPE-caged ATP	260 (26 600)	Φ_{ATP} 0.63	~410 (347 nm)	86 (pH 7.1)		[2, 3]
CNB-caged ATP						[5]
DMNPE-caged ATP	249 (17 200) 350 (5100)	Φ_{ATP} 0.07	~350 (347 nm)	18 (pH 7.0)		[6]
DNP-caged ATP		$\Phi_{ATP} \leq 0.007$	Low			[7]
Desyl-caged ATP		Φ_{ATP} 0.3		fast[e]	Sol.: Poor	[10]
DMB-caged ATP	256 (25 600)	Φ_{ATP} 0.3	~50 (347 nm)	>10^5	Stab.: Medium $24 \cdot 10^{-6}\,h^{-1}$ pH 7.0	[7–9, 12]
pHP-caged ATP	286 (14 600)	Φ_{ATP} 0.3; Φ_{Dis} 0.38		~10^7	Stab.: Stable >24 h pH 6.5 and 7.2	[10, 13]
DMACM-caged ATP	385 (15 300)	Φ_{ATP} 0.086	~645 (347 nm)	>$1.6 \cdot 10^9$	Stab.: Stable <0.5% hydrol · 24 h pH 7.2	[14]

a) Product quantum efficiency (Φ_{ATP}) or quantum efficiency of the disappearance of the caged derivative (Φ_{Dis}).
b) Photolytic efficiency see text.
c) For discussion see text.
d) This value has been deduced from Fig. 4 in Ref. [2].
e) The kinetics are presumably very fast (~$10^8\,s^{-1}$) according to data described on related compounds (see Ref. [17]).

1.4.3.4 Conclusion

Not all of the criteria for ideal caging of a biomolecule are ever completely satisfied for a given probe in a specific situation. Inevitably, a choice of the photoprotecting group will depend on the biological application. The DMACM-caged ATP represents at this time the most powerful caging probe for ATP at wavelengths ≥ 360 nm, especially if the release of the biological molecule has to be extremely fast ($\tau_{1/2} \ll 1$ µs). Using shorter wavelengths (300–340 nm) the pHP-caged ATP represents a useful alternative. The syntheses of these two probes are well described, reinforcing their potential for biological applications. It should be pointed out that the user-friendly (i.e., commercially available) NPE-caged ATP derivative, because of its well documented utilization, should be restricted to the study of slower ATP-dependent biological processes or situations where the rates of the biological processes are not germane to the study.

1.4.4
Caged Glutamate

1.4.4.1 Introduction

Glutamate is the main excitatory neurotransmitter in the mammalian central nervous system, mediating neurotransmission across most excitatory synapses. The postsynaptic signal is transduced by three classes of glutamate-gated ion channel receptors, the AMPA, the kainate and the NMDA receptors, respectively. Glutamate has therefore become a primary target for caging studies, opening up new opportunities for a better understanding of the organization of neuronal networks. Besides its usage for rapid chemical kinetic investigations, uncaging of glutamate has been used to mimic synaptic input as well as to map the glutamate sensitivity of dendritic arbors of single cells. An efficient uncaging can be used to generate action potentials with precise spatial resolution, allowing mapping of the location of neurons connected to a single cell. Several of these aspects will be described in detail in Part IV of this monograph by G. Hess [20] and K. Kandler [21], respectively. Clearly, several of these investigations require high spatial resolution, for which two-photon optical methods are recommended, as are also discussed in Part VIII of this monograph by T. Dore [22].

Many caged glutamate derivatives have been described in the literature, illustrating the attractiveness of the target as well as the continuing quest for more powerful probes. Tab. 1.4.3 gives the structures and the common acronyms for 24 caged glutamate derivatives. Most of the caging groups are similar to those described for ATP, the modification being the replacement of the phosphate ester group with a carboxylate ester group. The glutamate molecule offers three different caging possibilities, two carboxylic acid functions, α and γ respectively, in addition to the photochemical masking of the amino group of the neurotransmitter. There is one example in which both α and γ carboxylic acid groups are protected [29]. As with ATP, the most frequently used probe has been the *ortho*-nitrobenzyl protecting groups [23–29]. The *p*-hydroxy phenacyl series [30, 31], the desyl group [32], and the coumarinyl series [33] are also described. The *ortho*-nitro indoline series

Tab. 1.4.3 Caged glutamate molecules: chemical formulas and abbreviations

Chemical structure	Formula	Cage acronyms	Refs.
(structure 1)	$R=CH_3$, $R_1=H$	γ-NPE	[23]
	$R=CO_2H$, $R_1=H$	γ-CNB	[23]
	$R=CF_3$, $R_1=OCH_3$	γ-DMNPT	[24]
	$R=H$, $R_1=H, OCH_2CH=CH_2$	γ-ANB	[25]
(structure 2)	$R=CO_2H$, $R_1=H$	α-CNB	[23]
	$R=H$, $R_1=OCH_3$	α-DMNB	[26]
	$R=CH_3$, $R_1=OCH_3$	α-DMNPE	[26]
(structure 3)	$R=CH_3$	N-NPEOC	[27]
	$R=CO_2H$	N-Nmoc	[28]
(structure 4)	$R=H$	N-NB	[23]
	$R=CH_3$	N-NPE	[23]
	$R=CO_2H$	N-CNB	[23]
(structure 5)		Bis-α,γ-CNB	[29]
(structure 6)	R, R' = H	γ-pHP	[30]
	R = H, R' = OCH_3	γ-pHPM	[31]
	R, R' = OCH_3	γ-pHPDM	[31]
(structure 7)		γ-O-Desyl	[32]
(structure 8)		α-O-Desyl	[32]
(structure 9)		N-Desyl	[32]

Tab. 1.4.3 (cont.)

Chemical structure	Formula	Cage acronyms	Refs.
(bromo-hydroxycoumarin carbamate of glutamate)		N-Bhc	[33]
(bromo-hydroxycoumarin ester of glutamate)		γ-BhcMe	[33]
(nitroindoline glutamate)	R = CH$_2$CO$_2$Me, R' = H	γ-NI	[34, 35]
	R = H, R' = OCH$_3$	γ-MNI	[35]
(dimethoxy-nitrophenylpropyl glutamate)		γ-DMNPP	[37]

has been introduced more recently by Corrie and colleagues [34–36] and is described by the author in this monograph [38]. Finally we have recently synthesized a glutamate derivative [37] in the *ortho*-nitrophenethyl series developed by Steiner and colleagues [39].

1.4.4.2 Syntheses of the Caged Glutamate Derivatives

All caged glutamate derivatives required either *N-t*-Boc (amino) or *t*-butyl (acid) group protection during the synthetic attachment of the chromophore. These protected intermediates, reagents **A**, **B**, **C**, and **D** respectively, were subsequently coupled to the appropriate caging group and the protecting groups removed under acidic conditions (Scheme 1.4.3). This general strategy does allow the purification of the protected precursors in organic medium before the acidic deprotection treatment (i.e. a solution of TFA in methylene chloride) and a final HPLC purification of the caged glutamates.

The different coupling methods which have been used for the syntheses of the caged glutamate precursors are listed in Scheme 1.4.3. Most of the probes used an S$_N$2 reaction by displacement of an activated halogen derivative by the protected glutamates **A**, **B**, **C**, or **D**, in different reaction conditions. The α- or γ-esters of caged-Glu (CNB- [23]; *p*HP- [30]; *p*M-HP-; *p*HPDM- [31]; desyl- [32]) and the coumarin-caged Glu [33] were synthesized, in the presence of DBU, in apolar solvent, e.g., benzene or a benzene/THF mixture for the bis-α, γ-CNB [29]. More polar solvents, e.g., CH$_3$CN, were used for the synthesis of the *N*-derived molecules in the presence of potassium carbonate (*N*-NPE-; *N*-CNB- [23]) or diisopropyl ethylamine

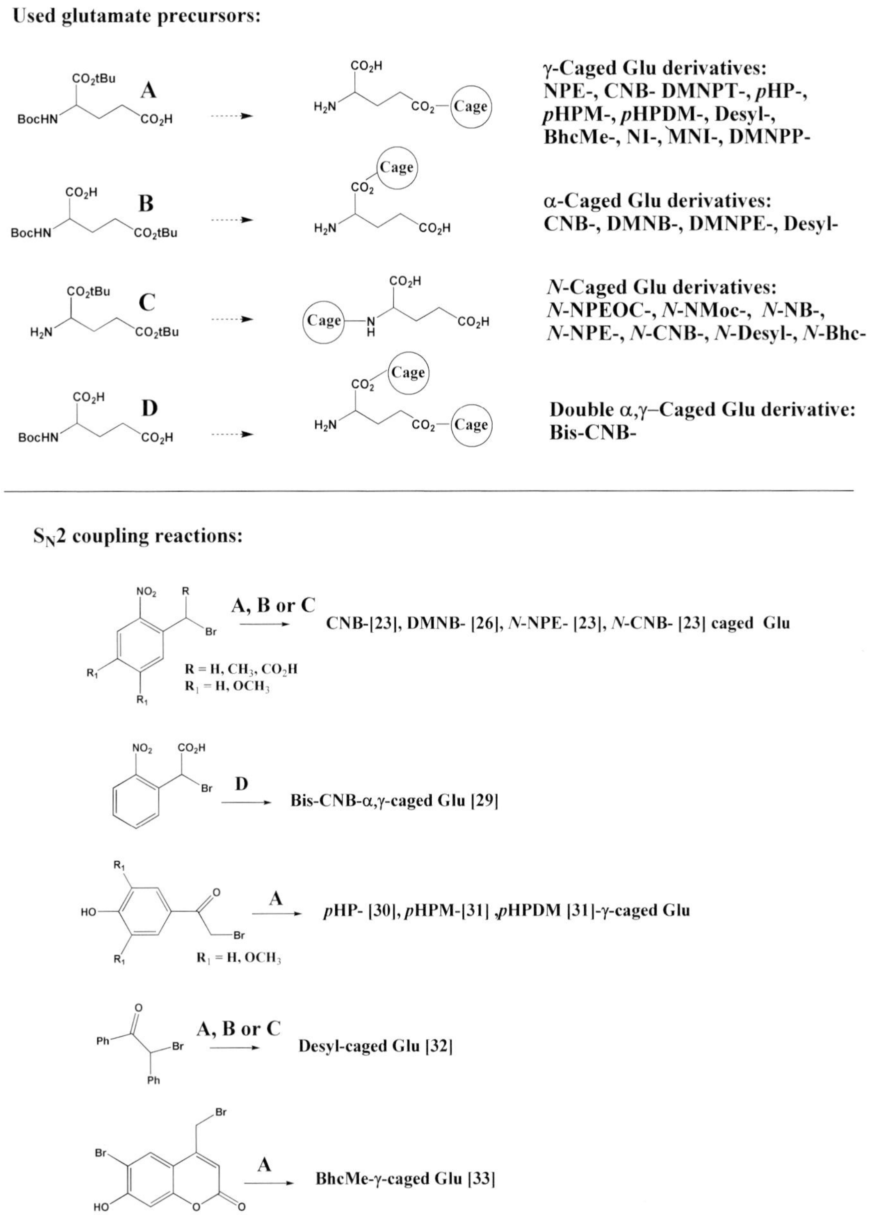

1.4 Caging of ATP and Glutamate: a Comparative Analysis

Used glutamate precursors:

[Structure A: BocHN-CH(CO$_2$tBu)-CH$_2$CH$_2$-CO$_2$H] → [H$_2$N-CH(CO$_2$H)-CH$_2$CH$_2$-CO$_2$-Cage]

γ-Caged Glu derivatives:
NPE-, CNB- DMNPT-, *p*HP-, *p*HPM-, *p*HPDM-, Desyl-, BhcMe-, NI-, MNI-, DMNPP-

[Structure B: BocHN-CH(CO$_2$H)-CH$_2$CH$_2$-CO$_2$tBu] → [H$_2$N-CH(CO$_2$-Cage)-CH$_2$CH$_2$-CO$_2$H]

α-Caged Glu derivatives:
CNB-, DMNB-, DMNPE-, Desyl-

[Structure C: H$_2$N-CH(CO$_2$tBu)-CH$_2$CH$_2$-CO$_2$tBu] → [Cage-NH-CH(CO$_2$H)-CH$_2$CH$_2$-CO$_2$H]

N-Caged Glu derivatives:
N-NPEOC-, N-NMoc-, N-NB-, N-NPE-, N-CNB-, N-Desyl-, N-Bhc-

[Structure D: BocHN-CH(CO$_2$H)-CH$_2$CH$_2$-CO$_2$H] → [H$_2$N-CH(CO$_2$-Cage)-CH$_2$CH$_2$-CO$_2$-Cage]

Double α,γ-Caged Glu derivative:
Bis-CNB-

S$_N$2 coupling reactions:

[o-NO$_2$-benzyl bromide with R, R$_1$ substituents] — A, B or C → CNB-[23], DMNB- [26], N-NPE- [23], N-CNB- [23] caged Glu
R = H, CH$_3$, CO$_2$H
R$_1$ = H, OCH$_3$

[o-NO$_2$-α-bromo phenylacetic acid] — D → Bis-CNB-α,γ-caged Glu [29]

[HO-aryl ketone with R$_1$ substituents, α-bromo] — A → *p*HP- [30], *p*HPM-[31], *p*HPDM [31]-γ-caged Glu
R$_1$ = H, OCH$_3$

[Ph-CO-CHBr-Ph (desyl bromide)] — A, B or C → Desyl-caged Glu [32]

[6-Bromo-4-(bromomethyl)-7-hydroxycoumarin] — A → BhcMe-γ-caged Glu [33]

Scheme 1.4.3 Syntheses of caged glutamate derivatives: protected glutamate derivatives (A, B, C, and D) and synthetic methods

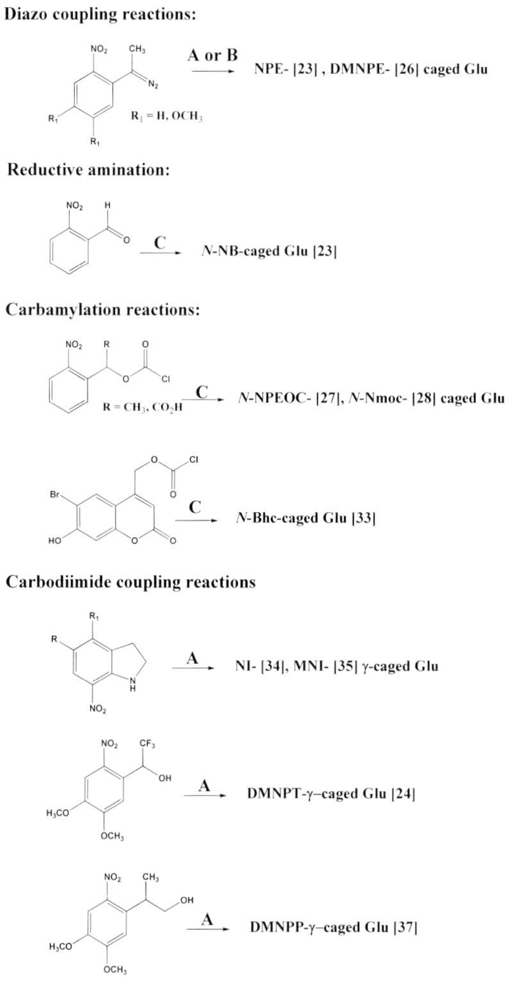

Diazo coupling reactions:

[structure] →(A or B) NPE- [23], DMNPE- [26] caged Glu

R_1 = H, OCH$_3$

Reductive amination:

[structure] →(C) N-NB-caged Glu [23]

Carbamylation reactions:

[structure, R = CH$_3$, CO$_2$H] →(C) N-NPEOC- [27], N-Nmoc- [28] caged Glu

[structure] →(C) N-Bhc-caged Glu [33]

Carbodiimide coupling reactions

[indoline structure] →(A) NI- [34], MNI- [35] γ-caged Glu

[structure] →(A) DMNPT-γ–caged Glu [24]

[structure] →(A) DMNPP-γ–caged Glu [37]

Scheme 1.4.3 (cont.)

(N-desyl-caged probes [32]). The synthesis of the precursor of the DMNB-caged α-Glu [26] used KF in refluxing acetone. The diazo precursors, developed for the caging of ATP, were used also for the syntheses of the γ- or α-NPE- and DMNPE-caged Glu respectively [23, 26]. One example of reductive amination was described on *ortho*-nitrobenzaldehyde for the synthesis of N-NB-caged Glu [23]. The syntheses of N-NPEOC- [27], N-Nmoc- [28] and Bhc- [33] caged Glu used an activated chloroformate, carbamate, or carbonate, respectively, to couple with the bis-*t*-butyl esters of glutamate **C** to generate the corresponding carbamates. The probes in the nitroindoline series (NI- and MNI-caged γ-Glu [34, 35]) provide an amide group in order to cage glutamate, using EDC-DMAP to couple the protected glutamic acid **A** onto the substituted indoline derivatives. Nitration and deprotection were achieved in a sub-

Tab. 1.4.4 Caged glutamate derivatives: chemical and photochemical properties

Caged Glu derivatives abbreviations	λ_{max}: nm (ε: M^{-1} cm^{-1})	Quantum efficiencies Φ_{Glu} or Φ_{Dis} [a]	Uncaging action cross section $\Phi \cdot \varepsilon$ (λ) [b]	Two-photon uncaging action cross section $\delta u(\lambda)$ (GM) [c]	Rate constant (s^{-1}) of release of Glu [d]	Stability and/or solubility in buffered medium	References
γ-NPE-caged Glu					~9 (pH 7.0)		[23]
γ-CNB-caged Glu	262 (5100)	Φ_{Glu} 0.14	~25 (365 nm) [33]		33×10^3 (pH 7.0)	Stab. Good <2% pH 7.0	[23]
γ-DMNPT-caged Glu						Unstable	[24]
γ-ANB-caged Glu							[25]
α-CNB-caged Glu		Φ_{Glu} 0.16			8.7×10^3 (pH 7.0)	Stab. Fair $\tau_{1/2}$ ~15 h pH 7.4	[23]
α-DMNB-caged Glu	345 (5900)	Φ_{Glu} 0.006 [33]	~30 (365 nm) [33]				[26]
α-DMNPE-caged Glu		Φ_{Glu} 0.65			~93 (pH 5.5)		[26]
N-NPEOC-caged Glu					~14 (pH 7.0)		[27]
N-Nmoc-caged Glu	265 (5700) 308 (2000)	Φ_{Dis} 0.11	~220 (308 nm)		~145 (pH 7.2)	Stable	[28]
N-NB-caged Glu		Φ_{Glu} 0.06			~210 (pH 7.0)		[23]
N-NPE-caged Glu		Φ_{Glu} 0.044			~2.2×10^3 (pH 7.0)		[23]
N-CNB-caged Glu [e]		Φ_{Glu} 0.036			~410 (pH 7.0) ~365 (pH 7.0)		[23]
Bis-α,γ-CNB-caged Glu [f]		Φ_{Glu} 0.16 Φ_{Glu} 0.14			8.7×10^3 (pH 7.0) 33×10^3 (pH 7.0)		[29]

Tab. 1.4.4 (continued)

Caged Glu derivatives abbreviations	λ_{max}: nm (ϵ: M^{-1} cm^{-1})	Quantum efficiencies Φ_{Glu} or Φ_{Dis} [a]	Uncaging action cross section $\Phi \cdot \epsilon$ (λ) [b]	Two-photon uncaging action cross section $\delta u(\lambda)$ (GM) [c]	Rate constant (s^{-1}) of release of Glu [d]	Stability and/or solubility in buffered medium	References
γ-pHP-caged Glu		Φ_{Glu} 0.08 Φ_{Dis} 0.12			~7×10^7 (pH 7.0)	Stable	[30]
γ-pHPM-caged Glu	279 (9310) 307 (7930)	Φ_{Glu} 0.035	~280 (307 nm)				[31]
γ-pHPDM-caged Glu	304 (11730)	Φ_{Glu} 0.035	~410 (304 nm)		~2×10^7 (pH 7.0)		[31]
		Φ_{Glu} 0.14 Φ_{Dis} 0.30			~10^7 (CH$_3$CN/H$_2$O)	Poor solub.	[32]
γ-O-Desyl-caged Glu		No Glu formed					[32]
α-O-Desyl-caged Glu N-Desyl-caged Glu		No Glu formed					[32]
N-Bhc-caged Glu	368 (17470)	Φ_{Glu} 0.019	~360 (365 nm)	~0.95 (740 nm) ~0.35 (800 nm)			[33]
γ-BhcMe-caged Glu	369 (19550)	Φ_{Glu} 0.019	~330 (365 nm)	~0.9 (740 nm) ~0.4 (800 nm)			[33]
γ-NI-caged Glu		Φ_{Dis} 0.043	~120 (347 nm)			Stable	[34, 35]
γ-MNI-caged Glu		Φ_{Dis} 0.085	~380 (347 nm)	~0.06 (730 nm)	~2.7×10^3 (pH 7.0)	Stable	[35, 36]

a) Product quantum efficiency (Φ_{Glu}) or quantum efficiency of the disappearance of the caged derivative (Φ_{Dis}).
b) Corresponds to the photolytic efficiency at a given wavelength: see text.
c) GM = Göppert-Mayer (10^{-50} cm$^4 \cdot$ s/photon).
d) For discussion of the "rate constant for release", see text.
e) Refer to the values of the two diastereoisomers.
f) The quantum yields and the rate constants values refer to the single α- and γ-CNB-caged derivatives respectively [23].

sequent step. The DMNPT ester [24] as well as the DMNPP ester [37] used the DCC, DMAP conditions which have been described for the synthesis of caged glycine [40] and which give excellent yields using 1-(o-nitrophenyl-[2,2,2]-trifluoro)ethanol [41] and 2-(4,5-dimethoxy-2-nitrophenyl)propanol [39] respectively.

1.4.4.3 General Properties of the Caged-Glutamate Molecules

In contrast to ATP, glutamate has been caged at three different positions generating distinct probes for the same photo-protecting group, leading to more complicated comparisons, a vs γ positions for the carboxylic acid functions and carboxylic acid vs amine functions. Another general observation relates to the impressive number of probes which have been synthesized at this point (twenty-four to our knowledge), emphasizing the interest in and need for powerful probes as well as the desire to find the ideal probe for glutamate. Tab. 1.4.4 summarizes the chemical and photochemical properties of the different probes. In the *ortho*-nitrobenzyl series (thirteen derivatives), overall slow kinetics are observed, the fastest derivative being the γ-CNB-caged Glu ($k = 33 \times 10^3$ s^{-1}). The slow rates may still be satisfactory for studies of phenomena involving slower glutamate receptors such as the NMDA receptor. For the N-caged carbamate derivatives (*N*-NPEOC, *N*-Nmoc, *N*-Bhc), the fragmentation kinetics are governed by the rates (ms time range) of decarboxylation of the intermediate carbamic acids. For the N-substituted nitrobenzyl derivatives (*N*-NB, *N*-NPE and *N*-CNB), caution should be taken that the published rate constant values, which refer to the decomposition of the *aci*-nitro intermediates are related to the actual glutamate release. Recent studies on the photochemical decomposition of related *o*-nitrobenzyl ether derivatives [42, 43] demonstrated that the rate-limiting step for the release of the alcohol was the decomposition of an *o*-nitroso hemiacetal intermediate which display much slower kinetics. Similar intermediates (i.e. *o*-nitroso hemiaminals) are likely to occur during the fragmentations of N-substituted nitrobenzyl derivatives.

Noticeably, all the caged glutamate derivatives which show fast fragmentation kinetics ($\tau_{1/2} \ll 1$ µs), i.e., the *p*HP, the desyl and the coumarin derivatives, have lower product quantum efficiencies, which in some cases (coumarin series) are compensated by a strong absorptivity around 350 nm. In the *p*-hydroxyphenacyl series, the photolytic efficiency above 350 nm could not be substantially improved by the addition of *meta*-methoxy substituents, mainly because of a decrease in the product quantum efficiencies [31]. The desyl series was disappointing; only the γ-substituted derivative showed the expected photo-fragmentation reaction [32] but with a low photolytic efficiency at higher wavelength and a recurrent poor water solubility. The derivatives in the nitro-indoline series, the NI- and MNI-caged γ-Glu probes [34–36], are very stable caged probes. These derivatives require photolytic cleavage of an amide bond. They exhibit an excellent photolytic efficiency due mainly to a strong absorptivity around 350 nm and are not hindered by the formation of the nitroso indole product during the photolysis. Nevertheless, the observed fragmentation kinetics restrict their use to biological events occurring in the sub-ms to ms time range.

1.4.4.4 Conclusion

This survey on ATP and glutamate, very likely the mostly targeted biomolecules for caging studies, underlines the fact that none of the described molecules can be quoted as the perfect caging group even though several of them had very satisfactory overall properties. Not all of the necessary data are available from the literature, and therefore a complete and precise comparison cannot be made. Tabs. 1.4.2 and 1.4.4 are intended to gather together the maximum available information for the reader to make an unbiased comparison among the probes. As already pointed out, the selection of a caged derivative will depend mainly on the biological experiment for which it is intended. Selection of a probe will have to take into account all the important properties: aqueous stability and solubility, photochemical kinetics, and efficiencies. In addition, for the caged glutamate derivatives or other caged neurotransmitters, neurobiological applications may be able to address the study of cellular events where two photon-uncaging techniques provide the improved spatial resolution during photoactivation [33, 44]. These techniques require high photochemical efficiencies around 350 nm [45] and do represent one of the key issues in the search for new caged neurotransmitters as will be discussed in detail in Part VIII of this monograph [22].

Abbreviations

ATP Derivatives

CNB-caged ATP: Adenosine triphosphate P^3-(a-carboxy-2-nitrobenzyl) ester.
DMACM-caged ATP: Adenosine triphosphate P^3-[(7-dimethylaminocoumarin-4-yl)methyl] ester.
DMB-caged ATP: Adenosine triphosphate P^3-[1-(3,5-dimethoxyphenyl)-2-phenyl-2-oxo]ethyl ester.
O-Desyl-caged ATP: Adenosine triphosphate P^3-(1,2-diphenyl-2-oxo)ethyl ester.
DMNPE-caged ATP: Adenosine triphosphate P^3-1-(4,5-dimethoxy-2-nitrophenyl)-ethyl ester.
DNP-caged ATP: Adenosine triphosphate P^3-(3,5-dinitrophenyl) ester.
pHP-caged ATP: Adenosine triphosphate P^3-(p-hydroxyphenacyl) ester.
NB-caged ATP: Adenosine triphosphate P^3-(2-nitrobenzyl) ester.
NPE-caged ATP: Adenosine triphosphate P^3-1-(2-nitrophenyl)ethyl ester.

Glutamate Derivatives

γ-ANB-caged Glu: L-Glutamic acid, γ-(5-allyloxy-2-nitrobenzyl) ester.
N-Bhc-caged Glu: N-(6-bromo-7-hydroxycoumarin-4-ylmethoxycarbonyl)-L-glutamic acid.
γ-BhcM-caged Glu: L-Glutamic γ-[(6-bromo-7-hydroxycoumarin-4-yl)methyl] ester.
Bis-a,γ-CNB-caged Glu: L-Glutamic acid, bis-a,γ-(a-carboxy-2-nitrobenzyl) diester.
a-CNB-caged Glu: L-Glutamic acid, a-(a-carboxy-2-nitrobenzyl) ester.

γ-CNB-caged Glu: L-Glutamic acid, γ-(α-carboxy-2-nitrobenzyl) ester.
N-CNB-caged Glu: N-(α-carboxy-2-nitrobenzyl)-L-glutamic acid.
α-O-Desyl-caged Glu: L-Glutamic α-[(1,2-diphenyl-2-oxo)ethyl] ester.
γ-O-Desyl-caged Glu: L-Glutamic γ-[(1,2-diphenyl-2-oxo)ethyl] ester.
N-Desyl-caged Glu: N-(1,2-Diphenyl-2-oxo)ethyl-L-glutamic acid.
γ-DMpHP-caged Glu: L-Glutamic γ-(m-dimethoxy-p-hydroxyphenacyl) ester.
α-DMNB-caged Glu: L-Glutamic acid, α-(4,5-dimethoxy-2-nitrobenzyl) ester.
α-DMNPE-caged Glu: L-Glutamic acid, α-[1-(4,5-dimethoxy-2-nitrophenyl)ethyl] ester.
γ-DMNPP-caged Glu: L-Glutamic acid, γ-[2-(4,5-dimethoxy-2-nitrophenyl)propyl] ester.
γ-DMNPT-caged Glu: L-Glutamic acid, γ-[1-(4,5-dimethoxy-2-nitrophenyl)-2,2,2-trifluoroethyl] ester.
γ-pHP-caged Glu: L-Glutamic γ-(p-hydroxyphenacyl) ester.
γ-pHMP-caged Glu: L-Glutamic γ-(p-hydroxy-m-methoxy-phenacyl) ester.
γ-MNI-caged Glu: 1-[S-(4-amino-4-carboxybutanoyl)]-4-methoxy-7-nitroindoline: (4-Methoxy-1-acyl-7-nitroindoline) derivative.
N-NB-caged Glu: N-(2-Nitrobenzyl)-L-glutamic acid.
γ-NI-caged Glu: Methyl 1-[S-(4-amino-4-carboxybutanoyl)]-7-nitroindoline-5-acetate: (1-acyl-7-nitroindoline) derivative.
N-Nmoc-caged Glu: N-(o-Nitromandelyl)oxycarbonyl-L-glutamic acid.
γ-NPE-caged Glu: L-Glutamic acid, γ-[1-(2-nitrophenyl)ethyl] ester.
N-NPE-caged Glu: N-(2-nitrophenyl ethyl)-L-glutamic acid.
N-NPEOC-caged Glu: N-1-(2-Nitrophenyl)ethoxycarbonyl-L-glutamic acid.

References

1 J. H. Kaplan, "Introduction" of this monograph.
2 J. H. Kaplan, B. Forbush III, J. F. Hoffman, *Biochemistry* **1978**, *17*, 1929–1935.
3 J. W. Walker, G. P. Reid, J. A. McCray, D. R. Trentham, *J. Am. Chem. Soc.* **1988**, *110*, 7170–7177.
4 J. E. T. Corrie, G. P. Reid, D. R. Trentham, M. B. Hursthouse, M. A. Mazid, *J. Chem. Soc. Perkin Trans. I* **1992**, 1015–1019.
5 R. P. Haugland, K. R. Gee, US Patents 5 635 608 and 5 888 829.
6 J. W. Wootton, D. R. Trentham, *NATO ASI Ser., Ser. C* **1989**, *272*, 277–296.
7 J. E. T. Corrie, D. R. Trentham, "Caged Nucleotides and Neurotransmitters" in *Biological Applications of Photochemical Switches*, H. Morrison Ed., John Wiley & Sons, New York, **1994**, pp 243–305.
8 H. Thirlwell, J. E. Corrie, G. P. Reid, D. R. Trentham, M. A. Ferenzi, *Biophys. J.* **1994**, *67*, 2436–2447.
9 V. S. Sokolov, H.-J. Apell, J. E. T. Corrie, D. R. Trentham, *Biophys. J.* **1998**, *74*, 2285–2298.
10 R. S. Givens, C.-H. Park, *Tetrahedron Lett.* **1996**, *37*, 6259–6262.
11 J. E. T. Corrie, D. R. Trentham, *J. Chem. Soc. Perkin Trans. I* **1992**, 2409–2417.
12 J. E. T. Corrie, Y. Katayama, G. P. Reid, M. Anson, D. R. Trentham, *Philos. Trans. R. Soc. Ser. A* **1992**, *340*, 233–244.
13 C.-H. Park, R. S. Givens, *J. Am. Chem. Soc.* **1997**, *119*, 2453–2463.
14 D. Geissler, W. Kresse, B. Wiesner, J. Bendig, H. Kettenmann, V. Hagen, *ChemBioChem* **2003**, *4*, 162–170.
15 V. Hagen, B. Kaupp, "Caged Cyclic Nucleotides", Chapter 5.1, this monograph.

16 R. S. Givens, P. S. Athey, B. Matuszewski, L. W. Kueper, J.-Y. Xue, T. Fister, *J. Am. Chem. Soc.* **1993**, *115*, 6001–6012.
17 "Mechanistic Overview of Phototriggers and Cage Release" Part II, this monograph.
18 A. Barth, J. E. T. Corrie, M. J. Gradwell, Y. Maeda, W. Mäntele, T. Meier, D. Trentham, *J. Am. Chem. Soc.* **1997**, *119*, 4149–4159.
19 V. Hagen, J. Bendig, S. Frings, T. Eckard, S. Helm, D. Reuter, U. Benjamin Kaupp, *Angew. Chem. Int. Ed.* **2001**, *40*, 1046–1048.
20 G. Hess, "Photochemical Release of Neurotransmitters" Chapter 6.1, this monograph.
21 G. Hess, "Caged Neurotransmitters for Probing Neuronal Circuits, Neuronal Integration and Synaptic Plasticity" Chapter 6.2, this monograph.
22 T. Dore, "Two-photon and multiphoton phototriggers" Part VIII, this monograph.
23 R. Wieboldt, K. R. Gee, L. Niu, D. Ramesh, B. K. Carpenter, G. P. Hess, *Proc. Natl. Acad. Sci. USA* **1994**, *91*, 8752–8756.
24 S. Guillou, M. Goeldner, Unpublished.
25 *Soc. Neurosci.* **1995**, Abstract 21, 579, abstract #238.11.
26 M. Wilcox, R. W. Viola, K. W. Johnson, A. P. Billington, B. K. Carpenter, J. A. McCray, A. P. Guzikowski, G. P. Hess, *J. Org. Chem.* **1990**, *55*, 1585–1589.
27 J. E. T. Corrie, A. Desantis, Y. Katayama, K. Khodakhah, J. B. Messenger, D. C. Ogden, D. R. Trentham, *J. Physiol.* **1993**, *465*, 1–8.
28 F. M. Rossi, M. Margulis, C.-M. Tang, J. P. Y. Kao, *J. Biol. Chem.* **1997**, *252*, 32933–32939.
29 D. L. Pettit, S. S.-H. Wang, K. R. Gee, G. J. Augustine, *Neuron* **1997**, *19*, 465–471.
30 R. S. Givens, A. Jung, C.-H. Park, J. Weber, W. Bartlett, *J. Am. Chem. Soc.* **1997**, *119*, 8369–8370.
31 P. G. Conrad II, R. S. Givens, J. F. W. Weber, K. Kandler, *Org. Lett.* **2000**, *2*, 1545–1547.
32 K. R. Gee, L. W. Kueper, J. Barnes, G. Dudley, R. S. Givens, *J. Org. Chem.* **1996**, *61*, 1228–1233.
33 T. Furuta, S. S.-H. Wang, J. L. Dantzker, T. M. Dore, W. J. Bybee, E. M. Callaway, W. Denk, R. Y. Tsien, *Proc. Natl. Acad. Sci. USA* **1999**, *96*, 1193–2000.
34 G. Papageorgiou, D. C. Ogden, A. Barth, J. E. T. Corrie, *J. Am. Chem. Soc.* **1999**, *121*, 6503–6504.
35 M. Canepari, L. Nelson, G. Papageorgiou, J. E. T. Corrie, D. Ogden, *J. Neurosci. Methods* **2001**, *112*, 29–42.
36 G. Papageorgiou, J. E. T. Corrie, *Tetrahedron* **2001**, *56*, 8197–8205.
37 W. Wittayanan, M. Goeldner, Unpublished
38 "Photoremovable Protecting Groups Used for the Caging of Biomolecules" Part I, this monograph.
39 S. Walbert, W. Pfleiderer, U. E. Steiner, *Helv. Chim. Acta* **2001**, *84*, 1601–1611.
40 C. Grewer, J. Jäger, B. K. Carpenter, G. P. Hess, *Biochemistry* **2000**, *39*, 2063–2070.
41 A. Specht, M. Goeldner, *Angew. Chem. Int. Ed. Engl.* **2004**, *43*, 2008–2012.
42 J. E. T. Corrie, A. Barth, V. R. N. Munasinghe, D. R. Trentham, M. C. Hutter, *J. Am. Chem. Soc.* **2003**, *125*, 8546–8554.
43 Y. Il'ichev, M. A. Schwörer, J. Wirz, *J. Am. Chem. Soc.* **2004**, *126*, 4581–4595.
44 O. D. Fedoryak, T. M. Dore, *Org. Lett.* **2002**, *4*, 3419–3422.
45 M. Matsuzaki, G. C. R. Ellis-Davies, T. Nemoto, Y. Miyashita, M. Iino, H. Kasai, *Nature Neurosci.* **2001**, *4*, 1086–1092.

2
Mechanistic Overview of Phototriggers and Cage Release

Richard S. Givens, Mani B. Kotala, and Jong-Ill Lee

Mechanisms help us understand how cage release reactions work, the advantages and limitations of a particular cage, and how cages can be applied. With the resurgence of interest in photoremovable protecting groups since their initial introduction for the release of ATP by Kaplan and cAMP by Engels in the late 1970s, there has also been a renewed interest in understanding the mechanisms for photoactivated release of substrates for biological studies. An overview of studies on these mechanisms has not been published since 1993. It is timely to provide this information to the biological and chemical communities. The scope of this review is limited to recent developments on those major functional groups that have been successfully employed as photoremovable cage chromophores. While not intended to be comprehensive, this review will focus on several selected examples of studies on successfully deployed protecting groups. The limiting rates for substrate release and the quantum efficiencies will be the primary focus of each review. In addition, details concerning the overall mechanism and the methods available to study the reaction will also be included. An attempt will be made to list the advantages and disadvantages for the photodeprotection reaction of each photoremovable protecting group.

2.1
Introduction

Light-activated removal of protecting groups ("cages") has provided biologists, biochemists, and physiologists with the ability to control triggering events that, through the release of substrates, initiate biological responses. The control is expressed in terms of the location, intensity, and duration of a light pulse that can be finely tuned to meet experimental protocols and demands. Researchers have made use of this ability by employing "cages" or photoremovable protecting groups (ppgs) that are chemically covalently bound to the substrates [1]. These photolabile precursors are biologically interesting molecules, such as caged second messengers, neurotransmitters, and nucleotides, and have even been employed to cage the active sites of enzymes, thereby serving as initiators for dynamic processes in biology and chemistry [2].

Dynamic Studies in Biology. Edited by M. Goeldner, R. Givens
Copyright © 2005 WILEY-VCH Verlag GmbH & Co. KGaA, Weinheim
ISBN: 3-527-30783-4

Several potential chromophores have been examined for cage applications, but very few have met the demands imposed by experimental protocols or successfully developed [1]. The constraints required for a successful chromophore include both mechanistic requirements and changes in physical properties that accompany the attachment of a chromophore to a substrate. An important limiting factor is the ease of synthesis of the substrate-cage complex; the users often find that organic syntheses are very challenging, thus limiting selection to those cages that are commercially available. The efficiency, the rate-limiting step for release of substrate, and the chemical and photochemical yields of the biological substrate are also principal features that must be considered and may have important and dramatic implications on the course of a study. Only a handful of prospective chromophores survive rigorous scrutiny of these constraints and only these are further considered. The four most robust candidates currently enjoying widest application, grouped according to the principal chromophore, are: (a) 2-nitrobenzyl (**2-NB**), (b) benzoin (**Bnz**), (c) *p*-hydroxyphenacyl (**pHP**), and (d) arylmethyl derivatives including the benzyl (**Bz**) and coumaryl (**Cou**) chromophores.

This chapter reviews the general mechanisms and relevant physical properties of the chromophores and includes a few examples. The discussion of each chromophore includes spectroscopic absorption and emission properties, the current understanding of the reaction mechanism including reactive excited state(s), the nature and timing of the bond-breaking process, and a discussion of the rate-limiting step(s) for the release of substrate.

2.2
Major Photoremovable Protecting Groups

2.2.1
2-Nitrobenzyl (2-NB)

Over 80% of the publications on caged compounds are applications of the 2-**NB** chromophore and its derivatives.[1] It is noteworthy that in spite of the intense interest in the 2-**NB** cage, many of the mechanistic details have yet to be determined. In fact, several very recent studies by Wirz [3] and Corrie [4] have improved our understanding of the events that occur between the absorption of light by the 2-**NB** chromophore and the emergence of the unfettered substrate HX [Eq. (1)].

$$\text{2-NB} \xrightarrow{h\nu\ (>300\ \text{nm})} [\text{aci-nitro intermediate}] \longrightarrow \text{product} + HX \tag{1}$$

---

1 According to a survey of SciFinder references to caged compounds.

2.2 Major Photoremovable Protecting Groups

The overall reaction appears at first to be an unremarkable photoinduced intramolecular redox reaction of a nitro ester to form a nitrosoketone [e.g., 2-nitrosoacetophenone, R=CH$_3$, Eq. (1)]. However, the events between light absorption and the eventual release of HX are far more complex than first imagined [4b]. Several photophysical, photochemical, and ground state events have been identified, which are outlined in the overall mechanistic sequence shown in Scheme 2.1. Supporting evidence for the individual steps is included in the mechanistic discussion.

To better understand the significance of the factors that affect the rates and mechanism for the 2-NB chromophore, the discussion of the mechanism has been subdivided into the five major steps: excitation, photoredox to the *aci*-nitro intermediate, cyclization to the oxazole, ring-opening to the hemiacetal, and collapse of the hemiacetal with release of the substrate. The complexity of the reaction and the competing kinetic pathways make it important that those investigating dynamic effects using caged substrates become particularly cognizant of the limiting kinetics of the key step: the release of substrate from the hemiacetal intermediate. Several of the intermediates along the reaction pathway are UV active and therefore have been used to determine rate constants for formation and decay of these transients. The rate constants vary with pH, media, and the nature of the substrate – variations that are discussed in detail below.

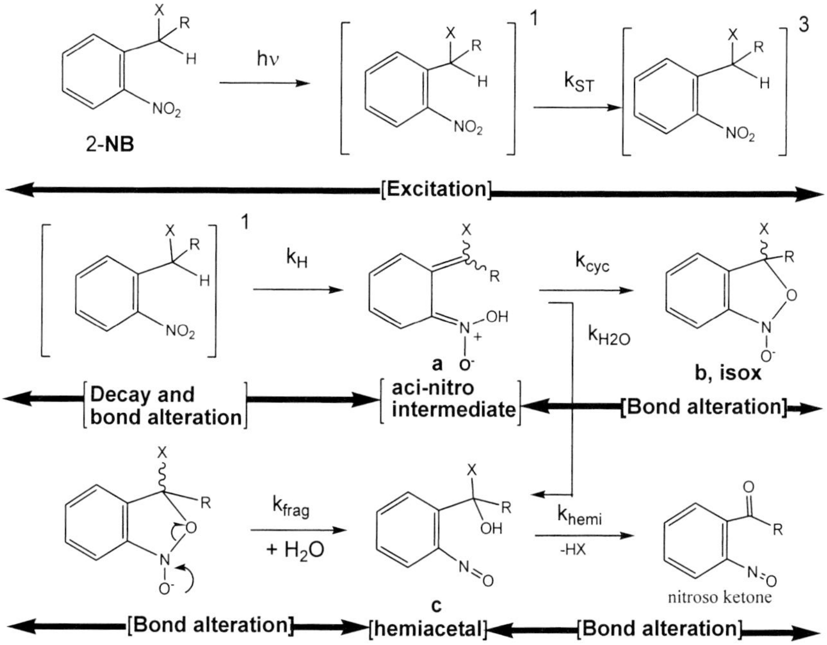

Scheme 2.1 General mechanism for 2-nitrobenzyl photorelease of a substrate.

X=substrate. Rate constants are: k_{ST}=single-triplet crossing; k_H=hydrogen abstraction; k_{cyc}=cyclization to the isoxazole; k_{H_2O}=direct hydrolysis to the isoxazole; k_{frag}=fragmentation of the isoxazole; k_{hemi}=hemiacetal or ketal hydrolysis. Intermediates **a**, **b**, and **c** are discussed in the text.

2.2.1.1 The Reactive Excited State

The 2-**NB** chromophore has a strong UV absorption (log $\varepsilon = 3.7$, λ_{max} 275 nm) that tails into the near UV as shown for the 2-**NB** methyl ether [3] (Fig. 1). The photoproduct also absorbs even more strongly in the same region as 2-**NB**. This can have a deleterious effect on the progress of the reaction especially at moderate to high conversions. The absorptivity of 2-**NB** chromophore can be enhanced by aryl substitutions (e.g., CH_3O). It has been found, however, that electron-donating groups on the chromophore significantly reduce the reaction efficiency. This may be the result of a lowering of the excited state energy [5] available for the carbon-hydrogen bond abstraction process.

The excited singlet of 2-**NB** may either cross over to the triplet state (k_{ST}) or undergo the hydrogen abstraction redox reaction (k_H). A priori, either the singlet or triplet excited state could undergo photoredox hydrogen abstraction. The first order rates for these two processes (k_{ST} and k_H, respectively) must be very fast. The rate constants for k_{ST} for nitrobenzenes generally cluster around $10^9\,s^{-1}$, whereas the hydrogen atom abstraction rate constants are less well known and are subject to greater variability. That the singlet redox reaction might be competitive with the singlet-triplet crossing can also be anticipated from the intramolecular nature of this redox process.

The rate constant for hydrogen atom transfer will vary with the C-H bond strength, the groups attached to the benzylic carbon, and the nature of the excited state (singlet or triplet). Though suggestive of a quantitative structure-reactivity relationship that might relate these parameters, such studies have not been reported. Nevertheless, it is generally safe to assume that abstraction rate constants are competitive with singlet-triplet crossing rate constants. The singlet lifetimes (τ_S) of 2-**NB** derivatives are generally very short, on the order of a few nanoseconds or less.

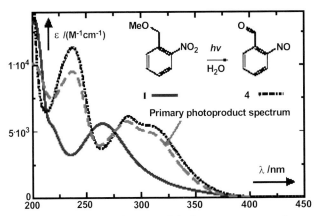

Fig. 2.1 UV-Vis spectra of 2-**NB** methyl ether (**1**, ⋯), 2-nitrosobenzaldehyde (**4**, − − − −) and the primary photoproduct of **1** in 1 mM aq. $HClO_4$. From Il'ichev, Schwoerer, and Wirz [3] and reprinted with the permission of the American Chemical Society.

The earlier studies by Yip and Gravel [5, 6] on 2-nitrotoluenes did demonstrate that both the singlet and triplet states are capable of abstracting an ortho methyl hydrogen atom. More recently, however, Wirz [3] has assigned the reactive state for the 2-**NB** methyl ether as the singlet.

2-**NB** triplet states are generally much longer lived, having lifetimes (τ_T) of 10^{-6} to 10^{-9} s, and are limited by the rate of decay of the triplet to the ground state 2-**NB** or by other competing chemical processes. The triplet intramolecular hydrogen atom transfer process does not appear to be an important one, at least not for the 2-**NB** ethers [3].

2.2.1.2 The *aci*-Nitro Intermediate

The product of the H-atom transfer is the non-aromatic *aci*-nitro intermediate that is stable enough to be observed as a long-lived, ground state transient through spectroscopic and kinetic studies. The transient decay is often monitored as the signature for the release of HX. In fact, as will be illustrated later, the release of free HX occurs at a much later stage in the mechanism, and the rate-determining step for release potentially can be any one of the intervening steps, including the decay of the penultimate hemi-acetal/ketal. Thus, it should not be assumed that the rate of decay of the *aci*-nitro intermediate is equivalent to the rate for HX release.

This mechanistic ambiguity has caused mistaken assignments of the release rate in the 2-**NB** cage literature. Unfortunately, missed assignments of the release rate constant are not restricted to just the 2-**NB** series. Similar situations have arisen for other cages (see below and accompanying references). Therefore, the importance of correctly defining the release step and determining the rate-limiting microscopic rate constants for the sequence leading to the release of the substrate cannot be overemphasized.

Lifetime and kinetic rate studies surrounding the formation and disappearance of the *aci*-nitro intermediate for 2-**NB** methyl ether (X=OCH$_3$, R=H), ethyl ether (X=OCH$_3$, R=CH$_3$), and ATP ester have been elaborately detailed in a recent series of studies by Wirz and his coworkers [3]. Employing time-resolved ps pump probe, ns laser flash, and time-resolved infrared spectroscopy, they have been able to define rate constants for many of the intermediate steps including the formation and decay of the *aci*-nitro intermediates for 2-**NB** methyl and ethyl ethers (Scheme 2.1). Interestingly, all the transient intermediates are sensitive to the pH of the media, including the penultimate hemi-acetal/ketal intermediate, which itself is subject to general acid catalytic decay to the aldehyde or ketone. Almost all of the processes have minimal rates at or near physiological pH as shown in their pH-rate profiles (Fig. 2.2). Thus, the reactions are slowest in the region of most interest to biochemists and physiologists! The dependence of the rate on pH also serves as a warning regarding mechanistic and kinetic investigations in that they must be performed at the physiological pH of interest to be meaningful.

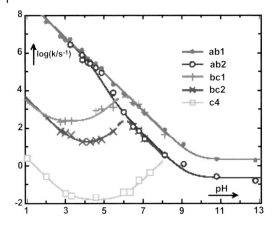

Fig. 2.2 pH rate profiles for decomposition of the isomeric *aci*-nitro intermediates (**ab1** and **ab2**), the isomeric isoxazoles (**isox**, **bc1**, and **bc2**) and the hemi-acetal (**c4**), the intermediate leading to the free aldehyde or ketone and methanol in the photolysis of 2-**NB** methyl ether. From Il'ichev, Schwoerer, and Wirz [3] and reprinted with the permission of the American Chemical Society.

2.2.1.3 Decay of the *aci*-Nitro Intermediate

There are several isomeric and tautomeric forms of the *aci*-nitro intermediate itself that further complicate our understanding of the reaction mechanism. Wirz [3] was able to sort out the predominant reactive *aci*-nitro isomer and to provide spectroscopic and kinetic information on its decay. The E,E isomer cyclizes directly to the benzisoxazole (**isox**, eq. 2). In highly aqueous media, Wan and Corrie [4] have suggested that hydrolysis to the hemiketal or hemiacetal may compete for *aci*-nitro intermediates depending on the conditions and the structure of the *aci*-nitro intermediate (Scheme 2.1).

aci-nitro
E,E
isomer

isox

(2)

The benzisoxazole intermediate (**isox**) is silent in the near UV-Vis, making kinetic investigations more difficult. Its formation has been inferred in order to account for the time delay between disappearance of the 400 nm band for the *aci*-nitro transient and the later emergence of a transient band at 310–330 nm assigned to the hydrated nitroso hemiacetal intermediate. This new transient's rate of formation did not conform to the decay of the 400 nm *aci*-nitro transient. However, time-resolved infrared (TRIR) spectra demonstrated that the decay of the *aci*-nitro bands at 1100–1400 and 1540–1650 cm^{-1} occurred with concomitant appearance of a new band at 1080 cm^{-1} in spectra that lacked both carbonyl and nitroso group absorptions. Upon decay of this transient, a prominent nitroso group absorption band at 1500 cm^{-1} appeared, but the new species still lacked a carbonyl absorption (Fig. 2.3), thus clearly demonstrating the stepwise process leading to the intermediacy of a hemiacetal or hemiketal.

The final step, hydrolysis of the hemi-acetal or -ketal, is signaled by the emergence of the free carbonyl monitored by fast-scan IR at 1700 cm^{-1}. The full IR spectrum becomes coincident with that of 2-nitrosobenzaldehyde as does the UV-Vis spectrum (see Fig. 2.1). Furthermore, the concomitant formation of methanol and loss of the hemi-acetal accounted for the changes in the 3200–3600 cm^{-1} OH region, which was transformed from a broad, diffuse band from the intramolecular hydrogen-bonded hemiacetal to the much narrower OH absorption of an uncomplexed CH_3OH.

The five-step mechanism, outlined in Scheme 2.1, is very sensitive to the solvent and pH as well as the nature of the leaving group. The pH-dependent rate constants for the formation of several intermediates encountered in the 2-**NB** methyl ether photolysis are depicted in Fig. 2.2. It is noteworthy that the slowest process of this sequence at physiological pH is the general acid-catalyzed breakdown of the hemi-acetal (green curve —), requiring a period of seconds to complete! Similar results were found for the 1-(2-nitrophenyl)ethyl methyl ether. In both cases, the limiting rate for methanol release was the hydrolysis of the hemiacetal/ketal.

2.2.1.4 The Role of the Leaving Group: Alcohols vs ATP

Wirz [3] examined the pH-dependent release of ATP from 1-(2-nitrophenyl)-ethyl ATP, the classic substrate for the caged 2-**NB** series [7]. In this case, the limiting rate for release of ATP, a superior leaving group to methanol, is the decay of the *aci*-nitro intermediate at pH >6. However, at pH <6, the isoxazole decomposition becomes rate limiting, as shown in Fig. 2.4. The hemiacetal is very unstable, hydrolyzing as quickly as it is formed.

From the Wirz study, the signatures of each of the significant transient intermediates could be detected and in most instances their formation and decay rate constants determined. The complex interplay of structure, media, pH, and other media factors can alter these rate constants and in some instances even the designation of the rate-determining step for substrate release. Such an exhaustive study in other 2-**NB** derivatives is unlikely and perhaps unwarranted. However, gauging the rate of substrate release should be an important objective for mechanistic studies in biology and will require measuring the rate of carbonyl or substrate formation. This will be the only true measure of the rate of release. Techniques for measuring these latter rates are very limited; it would appear that TR-FTIR is the most promising.

2.2.2
Benzoin (*Bnz*)

The benzoin or desyl group (**Bnz**) has regained prominence among photoremovable protecting groups during the past decade chiefly because of the chromophore's strong absorptivity in the near UV as shown for 3′,5′-dimethoxyben-

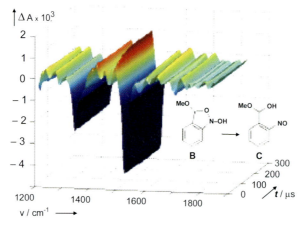

Fig. 2.3 The TRIR difference spectra of 2-**NB** methyl ether conversion of **isox** (**B**) to the **hemiacetal** (**C**) in CD3CN with 75 mM D$_2$SO$_4$. Time evolution is from front to back. From Il'ichev, Schwoerer, and Wirz [3] and reprinted with the permission of the American Chemical Society.

zoin acetate in Fig. 2.5, the good efficiencies for photoheterolysis, and the inert nature of the by-product. Benzoins are an attractive alternative to 2-**NB** as caging groups for a number of applications, primarily because syntheses of benzoin-caged derivatives are accomplished in good-to-excellent yields by straightforward procedures. Early studies by Sheehan [8, 9] demonstrated that benzoin acetates rearranged upon photolysis to 2-arylbenzofurans [Eq. (3)]. Recently, others [2, 10–12] have found that benzoin derivatives can serve as a good protecting group in a number of biologically active substrates despite the difficulties imposed by an added stereocenter and the diminished aqueous solubility of the caged derivative.

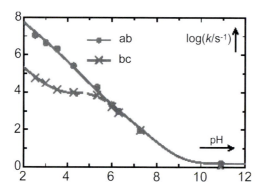

Fig. 2.4 pH rate profile for the *aci*-nitro intermediate (**ab**) and the isoxazole intermediate (**b**) after laser flash photolysis of 1-(2-nitrophenyl)-ethyl ATP (2-**NB** ATP) at 355 nm. From Il'ichev, Schwoerer, and Wirz [3] and reprinted with the permission of the American Chemical Society.

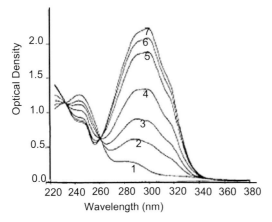

Fig. 2.5 Course of the photolysis of 3′,5′-dimethoxybenzoin acetate (**Bnz** OAc) to 2-phenyl-5,7-dimethoxybenzofuran in acetonitrile irradiated at 360 nm: 1, 0 s; 2, 20 s; 3, 40 s; 4, 80 s; 5, 180 s; 6, 260 s; 7, 420 s. From Sheehan, Wilson, and Oxford [9] and reprinted with the permission of the American Chemical Society.

$$\textbf{Bnz OAc} \xrightarrow[-\text{HOAc}]{h\nu} \text{2-phenylfuran} \qquad (3)$$

2.2.2.1 Excitation: Singlet vs Triplet State Reactivity

Both unsubstituted and methoxy-benzoin esters have been the subject of mechanistic investigations, and both have provided insights regarding electronic effects on the photochemistry [8–14]. Studies by Sheehan and Wilson [8, 9] showed that methoxy substitution on the benzyl ring gave excellent yields of the 2-phenylbenzofuran, and they assigned the reactive state as either a singlet or an unquenchable, short-lived triplet. On the other hand, unsubstituted benzoin derivatives have been shown to undergo very rapid ST crossing to the triplet, and both the lowest excited singlet and triplet states show significant n,π* character. Givens and Wirz [13] reported that benzoin phosphates photocyclize exclusively by a triplet pathway (see below).

Employing nanosecond laser flash photolysis (ns LFP) studies, Wan [14] proposed a singlet intramolecular exciplex mechanism for m,m'-**DM Bnz** OAc photolysis (Scheme 2.2). He also suggested that a short-lived cationic intermediate was generated. Two absorption bands at 330 and 420 nm were assigned to the m,m'-**DM Bnz** OAc triplet that decayed with rates of $\sim 10^6$ s^{-1}. A third band at 485 nm was assigned to the intermediate cyclized cation **Cat**$^+$ because dioxy-

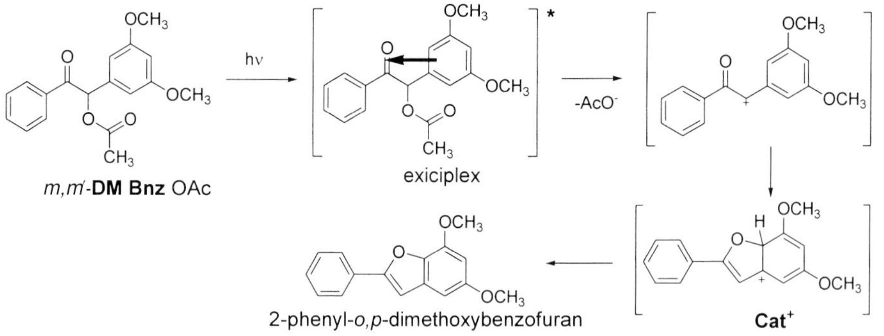

Scheme 2.2 Wan's mechanism for benzoin photoheterolysis [14].

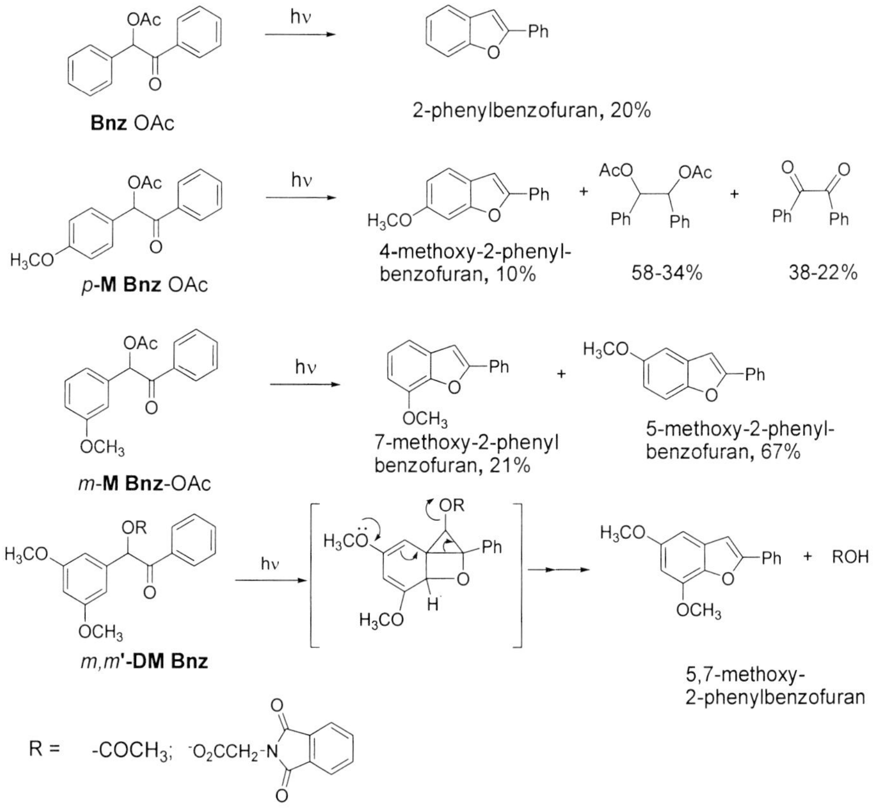

Scheme 2.3 The effects of methoxy substitution on benzoin acetate photochemistry [8, 9, 14].

gen had no effect on its rate of formation or decay and therefore its formation must have occurred through the singlet manifold.

Wan [14] proposed that an initial charge transfer intramolecular exciplex was formed between the electron-rich dimethoxybenzyl ligand and the electron-deficient $^1n,\pi^*$ excited carbonyl. Subsequent formation of the cyclized protonated furan followed by proton loss gave 2-phenylbenzofuran.

The placement of the methoxy substituents on the benzyl group was found to be crucial. For example, photolysis at 366 nm converted the *m*-methoxybenzyl and *m*,*m*′-dimethoxybenzyl analogs to the corresponding 2-phenylbenzofurans in excellent yields (88% and 94%, respectively) [8, 14], whereas the para isomer gave only 10% of 4-methoxy-2-phenylbenzofuran. Furthermore, the para isomer was beset with a mixture of other coupling byproducts and gave poor overall yields. The high quantum efficiency of 0.64 observed for *m*,*m*′-DM Bnz OAc and the short triplet lifetime ($\tau^3 < 10^{-10}$ s) indicated that the photorelease was both rapid and efficient.

2.2.2.2 Heterolysis vs Homolysis

Quenching studies on the unsubstituted or *p*-methoxy derivatives suggested a triplet state with a lifetime (τ^3) of ca. 10^{-10} s [12]. In clear contrast to these results, *m*,*m*′-dimethoxybenzoin acetate (*m*,*m*′-**DM Bnz OAc**) could not be quenched, suggesting that this derivative either reacts via its singlet excited state or possibly through a very short-lived, unquenchable triplet. A reactive exciplex was discounted because of the absence of any detectable new absorption band.

Chan [11] observed solvent incorporation for the photochemistry of 3′,5′-bis(carboxymethoxy)benzoin acetate (*m*,*m*′-**BCM Bnz OAc**) in aqueous media but not in methanol. The estimated lifetime of a putative intermediate cation **DMB**$^+$ formed under these conditions must be at least 5 ps in order for solvent to compete with cyclization. Alternatively, the solvent incorporation could have occurred from the cyclized intermediate as shown in Scheme 2.4.

DMB$^+$

A proposed biradical intermediate (**B OAc**) that preceded acetoxy migration and was followed by rearomatization to the benzofuran also has been suggested. This process was in direct competition with attack by water on the cyclized intermediate. The highly strained **oxetane**, originally proposed by Sheehan [9], was avoided through this rationale. The literature does not provide enough information for one to adequately sort through these disparate mechanisms for the dimethoxybenzoin series, and the question of a general mechanistic profile for this ppg remains unresolved at this time.

Scheme 2.4

Proposed mechanism for aqueous photorelease of acetate from *m,m'*-BCM Bnz OAc [11].

B-OAc

Oxetane

In contrast, Givens and Wirz [13] have provided sufficient information to construct a general reaction profile for the unsubstituted benzoin ppg. They have reported that the photocyclization of unsubstituted benzoin phosphates occurs exclusively through the triplet. As shown in Fig. 2.6, two short-lived transients are observed upon direct photolysis in aqueous acetonitrile solutions (Fig. 6B).

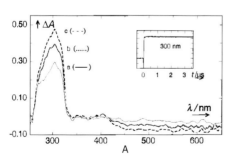
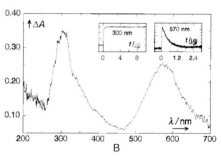

Fig. 2.6 **A** Difference spectrum recorded ca. 30 ns after laser excitation (248 nm) of a solution of benzoin diethyl phosphate in acetonitrile (**a**) after one flash, (**b**) after two flashes, and (**c**) after three flashes. The inset shows a kinetic trace of the absorbance at 300 nm due to the formation of 2-phenylbenzofuran. **B** Transient absorption spectrum recorded ca. 30 ns after laser excitation (248 nm) of a solution of benzoin diethyl phosphate in water containing 4% acetonitrile. The insets show kinetic traces of the absorbances at 300 and 570 nm. From Rajesh, Givens, and Wirz [13] and reprinted with the permission of the American Chemical Society.

The absorbance at 300 nm from the difference spectra recorded in acetonitrile about 30 ns after excitation (Fig. 2.6 A) is attributed to the formation of 2-phenylbenzofuran (also Fig. 2.5). The transient absorption spectrum in water (Fig. 2.6 B) revealed another absorption at 570 nm that decayed with a rate constant of 2.3×10^6 s^{-1}, and the decay rate constant increased in the presence of oxygen, suggesting that it was the reactive triplet. Naphthalene did not affect the decay rate but did reduce the amplitude by quenching the ^{3}**Bnz** O$_2$P(OEt)$_2$ with a rate constant of 3.7×10^9 M^{-1} s^{-1}. The product studies in various solvents, quenching studies with azide, bromide or chloride, and DFT calculations further suggested that the 570 nm transient was the triplet cation (Scheme 2.5).

Two competing triplet pathways are necessary, however, in order to account for the results of the nanosecond and picosecond laser flash studies. Partitioning of the reaction pathways begins at the initial excitation step, which freezes the two interconverting ground state conformations. Ring closure of the n,π* triplet to a diradical intermediate requires that the carbonyl group and the α-phenyl group be in close proximity as shown for the gauche structure **Bnz** g. Such a conformation may be disfavored in solvents that form strong hydrogen bonds to the carbonyl group. It is suggested that this conformer forms furan concomitant with or rapidly following the loss of phosphate. The anti conformation **Bnz** a, favored by hydrogen bonding, is confined to the s-trans conformation and heterolysis generates an extended benzyl cation **Bnz**$^+$. The barrier to conformational reorganization necessary for cyclization from this conformer is apparently much higher than that for attack by the solvent.

The rate-determining cyclization and phosphate loss to the furan were assumed to occur within 20 ns at room temperature based on the lifetime of ^{3}**Bnz** g in a picosecond pump-probe study. However, nucleophilic substitution can become a major pathway when trifluoroethanol is the solvent. The triplet benzoin cation ^{3}**BC**$^+$ was detected by LFP at 570 nm based on its reduced lifetime in the presence of oxygen. This relatively long-lived transient had a 430 ns lifetime in aqueous acetonitrile, which increased to 660 ns in trifluoroethanol. Thus, it was assigned as a triplet cation (^{3}**BC**$^+$). The decay of triplet to the ground state singlet occurred before the nucleophilic attack by trifluoroethanol. DFT calculations provided the energy-minimized structures for the singlet and triplet structures shown in Scheme 2.5 with an energy gap between singlet and triplet states of only ~1.7 kcal/mol.

In view of the advantageous characteristics of high quantum efficiencies and fast release rates, it would appear that benzoin has significant potential as a chromophore for biological applications. There are, however, major drawbacks. Its intrinsic chirality leads to diastereomers with chiral substrates, and incident-light competition by the cyclized byproducts lowers the yield of the decaged substrate. During synthesis, azlactone formation can also lower the caging yields when carbamates are the cage targets, as shown in Scheme 2.6 [15]. In a few instances, benzoin esters tautomerize at higher pH, switching the position of the carbonyl [Eq. (4)]. Biological applications are further limited by the insolubility of most benzoin-caged compounds, and the lack of product solubility further ex-

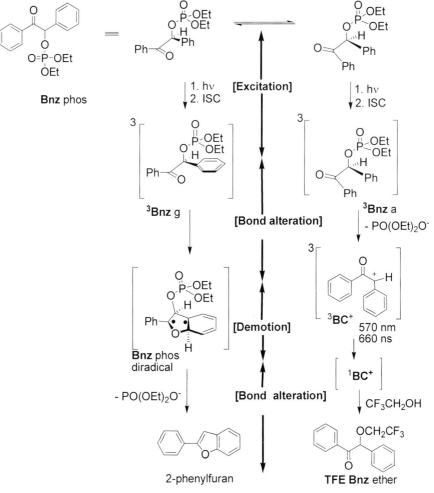

Scheme 2.5 Wirz and Givens mechanism for photosolvolysis of benzoin phosphates [13].

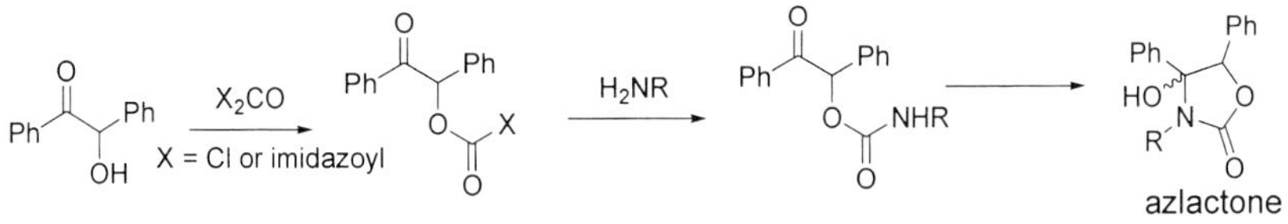

Scheme 2.6 Competitive by-product formation from benzoin carbamate synthesis [15].

acerbates the photochemical outcome since most 2-phenylbenzofurans are insoluble in aqueous media.

$$\text{(4)}$$

2.2.3
Phenacyl (pHP)

The newest addition to the repertoire of ppgs is the *p*-hydroxyphenacyl (pHP) chromophore [1b–e]. The discovery and development of the pHP group began in the mid 1990s and continues today. Because of its recent entry into this research area, the number of applications at this time is limited. Examples that are available have been informative and have demonstrated the potential of pHP protection and release in terms of the efficiency, the rapid rate of substrate release, and the high conversion of the released substrate that are possible. These attributes evolve from two important features of the photochemistry of the *p*-hydroxyphenacyl group: (1) the very rapid, primary photochemical heterolytic cleavage of the substrate-phenacyl bond and (2) the skeletal rearrangement of the conjugated chromophore leading to a transparent intermediate during or following heterolysis [Eq. (5)]. The resulting structure lacks a strong absorbance above 300 nm. Heterolytic release of the substrate occurs in a nanosecond time frame ($k_{release} > 10^8 \text{ s}^{-1}$) with efficiencies of 0.1–0.4, a result of the primary photolytic process. The rearrangement assures that complete conversion of the caged moiety to free substrate is possible.

$$\text{(5)}$$

The overall reaction is best described as an excited state "Favorskii-like" rearrangement [1d, 2b, 16–18], a process in which the substrate release step is part of the "Favorskii" cyclopropanone ring formation [19, 20]. The resulting spirodienedione (sDD), a putative intermediate in this rearrangement, must be rapidly hydrolyzed in aqueous media forming p-hydroxyphenylacetic acid (Scheme 2.7). The photochemical Favorskii process was first suggested many years ago by Reese and Anderson [16] for the photorearrangement of 2-chloro-4'-hydroxyacetophenones that included the putative cyclopropanone intermediate susceptible to rapid methanolysis. Key aspects of this process are the concomitant release of chloride ion (a heterolytic fragmentation process) and the formation of the spirodienedione intermediate. Other pathways were proposed.

Several features of this reaction make it unique among all photoremovable protecting groups. The most notable are that (1) the chromophore's absorption undergoes a bathochromic shift with increasing pH due to the phenol-phenolate equilibrium, (2) the photorelease sequence is adiabatic, occurring exclusively on the triplet manifold, (3) the reaction efficiencies are good to excellent, even in aerated solutions, and (4) the Favorskii rearrangement is the exclusive pathway for most of the pHP family of derivatives as long as the reactions are carried out in aqueous media. All of these features combine to make pHP a noteworthy candidate for applications for *in vivo* or *in vitro* aqueous studies, especially for critical time-resolution studies and for situations where conversion to products must be quantitative and complete. Other examples along with specific details on the studies cited here appear in other chapters in this monograph.

The generally understood mechanism for the photorelease and rearrangement, given in Scheme 2.7, is presented according to the following principal features: (1) identification of the reactive excited state, (2) DFT calculations on the excited state reactivity, (3) the deprotonation step, (4) fate of the spirodiene-

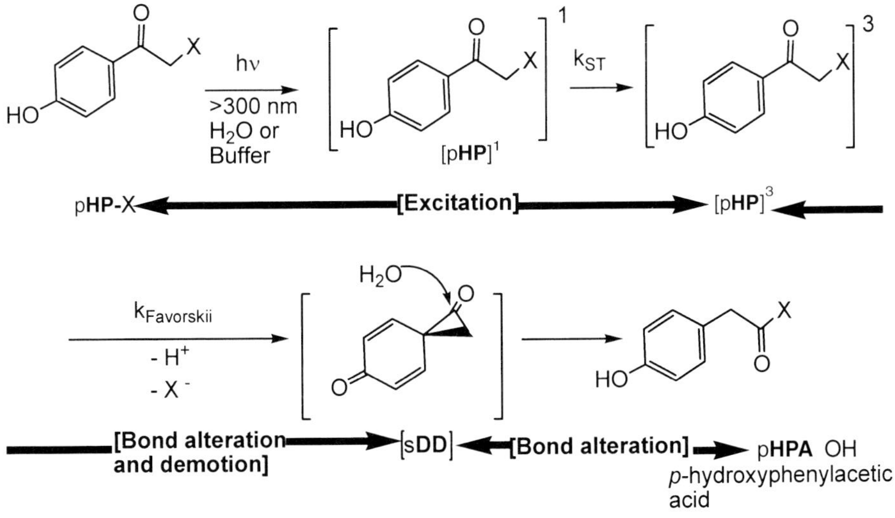

Scheme 2.7 Mechanism for photorelease of substrates from the p-hydroxyphenacyl chromophore.

dione (sDD) intermediate, (5) effect of substituents, and (6) quantum efficiency, rate constant, and solvent isotope effect determinations.

2.2.3.1 Singlet vs Triplet Excited States

As shown in Fig. 2.7, *p*-hydroxyacetophenone, the pHP chromophore, absorbs well beyond 300 nm in neutral aqueous solution. In basic media, the phenoxide contribution moves the absorption band even farther into the near UV, bordering on the visible spectral region. Initial excitation at neutral pH in aqueous media produces a singlet excited state that rapidly crosses to the lowest π,π^* triplet (k_{ST}=2.7 to 5×10^{11} s^{-1}) [13, 24]. The singlet, which is described as a mix of the n,π^* and π,π^* configurations, neither fluoresces under ambient conditions nor undergoes any reaction. Thus, singlet-triplet crossing appears to be the only exit channel available, resulting in unit efficiency for formation of the triplet excited state.

Low-temperature phosphorescence emission spectra (frozen ethanol/ether/isopropanol glasses, 77 K) have also been reported for several pHP derivatives ($E_T \approx 69$ to 71 kcal/mol) [10], although this does not appear to be a major exit channel at room temperature [13]. Instead, the primary exit channel from the triplet is deprotonation and release of substrate.

Alternative mechanisms, proposed by Corrie and Wan [21], suggest that the cleavage occurs from the pHP excited singlet. In one mechanism, the phenolic proton is effectively transported by a series of hydrogen-bond shifts in aqueous media from the phenolic OH to the pHP carbonyl oxygen forming a *p*-quinone

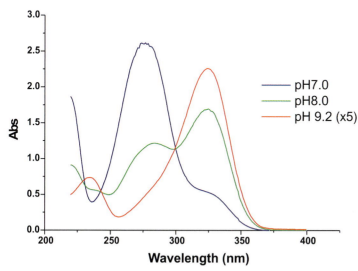

Fig. 2.7 Bathochromic shift in the UV-Vis spectra of pHP GABA as a function of pH in buffered solutions. (Unpublished results, A. Yousef, 2004.)

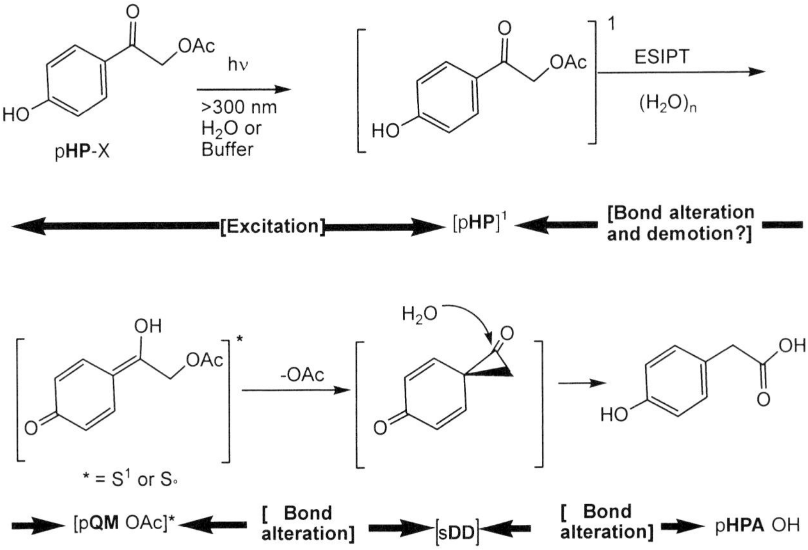

Scheme 2.8 Excited singlet ESIPT mechanism for the pHP photorearrangement [21, 22].

methide (pQM) (Scheme 2.8). This is described by the authors as a formal Excited-State Intramolecular Proton Transfer (ESIPT) process to the *p*-quinone methide (pQM) intermediate, for which there is ample evidence from other photosolvolysis reactions [22]. Further bond reorganization during photosolvolysis generates the spirodienedione (sDD) that is hydrolyzed to *p*-hydroxyphenylacetic acid (pHPA OH).

The intermediate pQM OAc might either react while still in the excited singlet state (an adiabatic ESIPT process) or after relaxation to the ground state. Included in these mechanisms is the reverse proton transfer, thereby explaining the nonproductive decay of some excited pHP derivatives and a possible source of the lower than unity quantum efficiencies (see below).

A variant mechanism suggested by the same authors is direct deprotonation in the singlet excited state of the phenolic hydrogen to solvent followed by departure of the leaving group and rearrangement to sDD. The latter step is similar to that proposed in Scheme 2.9 for the triplet pHP chemistry.

The significant differences between these mechanisms and that discussed below are the singlet reactivity and the intermediacy of the pQM. Experimental evidence from the low-temperature phosphorescence emission and the TR-LFP studies for a number of pHP esters established that the triplet state is the reactive state. Secondly, while there is no concrete evidence for the intermediacy of pQM (for that matter, the same is true for sDD), DFT calculations indicate that rearrangement from pQM to sDD would be endergonic in the ground state. Furthermore, the pQM has been detected for pHA's adiabatic proton tautomerizations in the triplet manifold [21] but has not been observed for any other pHP derivative studied thus far. However, the lack of evidence for a pQM may be due to the very rapid rearrangement channels available to pHP X.

2.2.3.2 DFT Calculations on Excited Triplet State Reactivity

The mechanistic interpretation relies on a bond-breaking process from the π,π^* triplet that occurs by a spin-allowed release of the conjugate base of the leaving substrate concomitant with the generation of a triplet oxyallyl 1,3-biradical intermediate [23] (see Fig. 2.8). This process could be either an adiabatic or a diabatic decay process that eventually yields the ground state oxyallyl intermediate. Triplet-singlet crossing is required at this point in the mechanism, however, since no further excited triplet intermediates have been detected. An initial adiabatic formation of the oxyallyl intermediate has the advantage of avoiding a violation of a spin-restricted triplet going directly to two singlet ground state moieties.

This mechanism is consonant, however, with the observed heterolysis of the substrate-carbon bond. The DFT calculations [23] do support formation of an open oxyallyl triplet, a species that is more stable than the spirodienedione triplet. It is interesting to note that the calculated energy difference between the planar triplet oxyallyl (**P0°**) and the 90° twisted triplet intermediate (**T90°**) is only ~6 kcal/mol. Therefore, both of these triplets are possible excited-state intermediates on the pathway to the ground state spirodienedione (**sDD**). The most likely pathway based on our calculations appears to be the rapid solvolysis of the triplet pHP conjugate base to a planar oxyallyl triplet (**P0°**) that rotates to the 90° twisted intermediate (**T90°**) during substrate-chromophore bond breaking. Closure to **sDD** may actually not take place until or while the molecule relaxes to the ground state surface. DFT calculations, however, show that the triplet energy of **sDD** is actually considerably higher than the triplet diradical energies of **P0°** and **T90°**. Thus, the formation of **sDD** is driven toward closure concomitant with T → S crossing to the ground state sDD ($\Delta E = E_0^{S\,(ground\,state)} - E^{T(twisted)} \approx -10$ kcal/mol). **sDD** is then attacked by H_2O to complete the "Favorskii" transformation.

Phillips [24] has examined the structure of the excited triplet of pHP OAc by ns and ps time-resolved Resonance Raman spectroscopy (TR^3). This approach has provided good evidence for a torsional rotation to a 15° out-of-plane twisted $(\pi,\pi^*)^3$ triplet in CH_3CN. The carbonyl C_8–O_{10} bond lengthens by 0.1 Å (see Fig. 2.8 C) whereas the C_6(ring)–C_8(carbonyl) bond contracts by 0.06 Å. These changes are in accord with an evolving formation of the cyclopropanone ring through formation of the C_6–C_9 bond. The leaving group (L=acetate) also rotates from a ground state anti conformation (176°) with respect to the plane of the phenol ring into a gauche orientation (103°), that is now positioned properly for back-side displacement from the electron-rich phenoxide's ipso carbon. The observation of a concomitant elongation of the C_9-leaving group bond by 0.04 Å also preludes the concerted cyclopropanone formation as acetate departs.

A.

meta sub (X)	a_{ipso}	q_{ipso}
H	0.017	0.100
OCH$_3$	0.016	0.098
F	0.025	0.101
CN	0.092	0.110

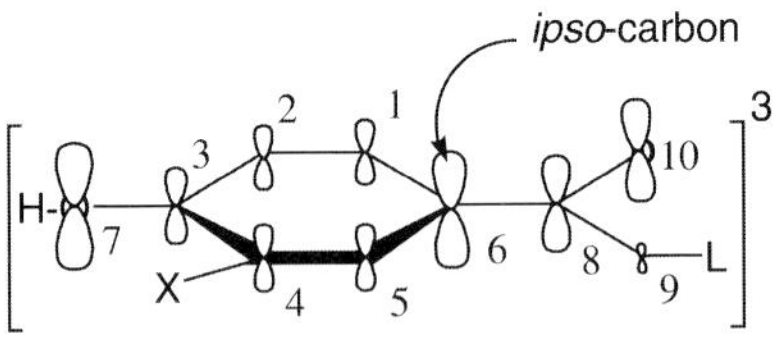

^{a} a_{ipso} = spin density on ipso-carbon
q_i = Charge density on ipso-carbon

B.

meta sub (X)	a_{ipso}	q_{ipso}
H	0.359	0.105
OCH3	0.318	-0.097
F	0.360	0.099
CN	0.379	0.096

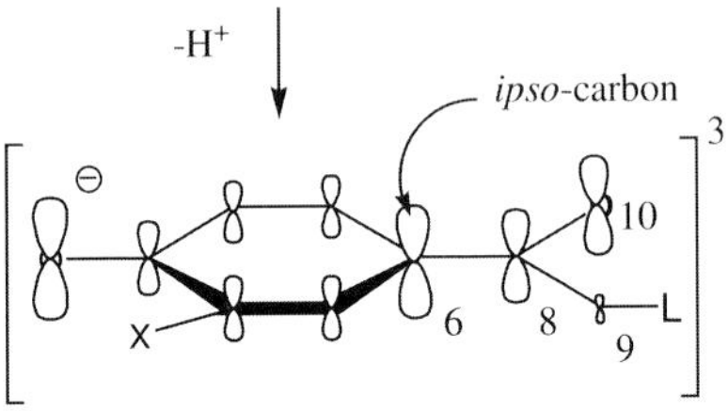

p-hydroxyphenacyl (pHP) triplet
E = -288665.3 kcal/mol

$-H^+$ ↓

pHP conjugate base triplet
E = -288329.6 kcal/mol

C.

−L (substrate)

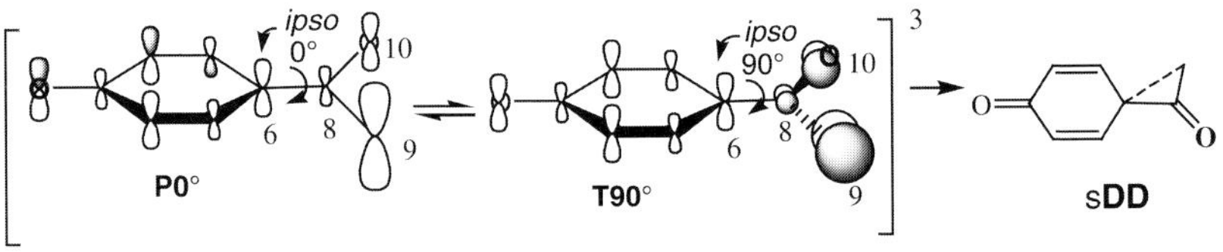

PO° ⇌ T90° → sDD

E = -287909.7 kcal/mol   E = -287915.7 kcal/mol

Triplet Oxyallyl Diradical	Spin density a_{ipso}	Charge density q_{ipso}
Planar conformer	0.408818	0.049146
Twisted conformer	0.397146	0.040913

Fig. 2.8 Charge (q_i) and spin (a_i) densities at the ipso positions on p-hydroxyphenacyl triplet states for pHP and three meta-substituted pHP cages. pHP (**A**), its conjugate base (**B**), and the diradical intermediate (**C**) [23].

2.2.3.3 Deprotonation Step

In addition to the energy considerations and the spin density build-up at the ipso carbon, two other factors are important for the successful expulsion of the leaving group in this mechanistic pathway. The decrease in the ^{3}pK_a of the phenolic hydroxyl and a good functional leaving group (L) are critical.

Deprotonation of the phenolic OH group from the triplet excited state of the p-hydroxyphenyl group in H$_2$O is a key initial step. Wirz and Givens [13] have reported a rate constant of 9×10^6 s^{-1} for deprotonation along with a diminution of the pK_a from 7.9 to 5.5 for the parent p-hydroxyacetophenone triplet. Thus, in aqueous media under neutral or near neutral conditions, the triplet rapidly deprotonates, exposing an electron-rich phenoxide ion. The concomitant twisting of the acetyl substituent as discussed above sets up the 1,3-alignment of the electron-rich aryl moiety juxtaposed a and antiparallel to the leaving group, the molecular equivalent of the ground state Favorskii rearrangement to the cyclopropanone intermediate as proposed by the Loftfield [25] and shown in Scheme 2.9.

For comparison, the mechanism of ground state Favorskii rearrangement [19, 20] is shown in Scheme 2.9. It begins with a rapid, base-catalyzed deprotonation of an a-hydrogen, an equilibrium process that is followed by rate-limiting loss of chloride. Two pathways have been proposed for this process: (1) a direct S_N2-like displacement of chloride to directly form the cyclopropanone intermediate (the Loftfield mechanism [25]) or (2) a π-assisted S_N1-like ionization to the oxyallyl zwitterions (zAO, the Dewar mechanism [26]). The zwitterion subsequently closes to cyclopropanone, which opens to the rearranged acid, ester, or amide on attack by the nucleophilic solvent. Both theoretical [27] and experimental evidence [28] favor the stepwise Dewar oxyallyl zwitterionic pathway. Bordwell [28] has demonstrated through MeO⁻/MeOD exchange experiments that the enolate anion is reversibly generated prior to the loss of chloride. Substituent effects (Hammett ρ value of −4.97), solvent effects (a Grunewald-Winstein m value of 0.647), a positive salt effect, and the absence of a common-ion effect support a rate-determining formation of the oxyallyl zwitterion followed by closure to the cyclopropanone. This nicely parallels the photochemical Favorskii rearrangement mechanism discussed earlier.

For the photochemical process, the removal of the vinylogous phenolic proton is equivalent to the a-proton in Bordwell's mechanism, generating a phenolate intermediate. In the ground state process, electron-donating substituents (i.e., p-MeO >> H >> p-NO$_2$) increase the rate of rearrangement by several orders of magnitude. For the photochemical version, the electron-rich phenolate ion greatly accelerates the π-assisted loss of the leaving group to form the zwitterionic oxyallyl intermediate (zOA).

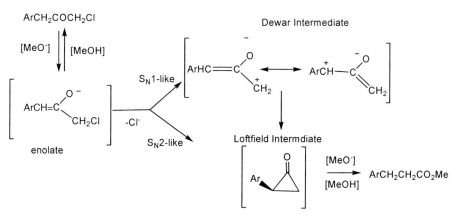

Scheme 2.9 The Favorskii rearrangement: the ground state mechanisms.

2.2.3.4 Fate of the Spirodienedione Intermediate

Evidence for the spirodienedione sDD is sparse. In fact, its existence is essentially inferred by the changes in bond connectivity, since it has eluded detection in all experimental protocols reported to date. It is, therefore, presumed to be a highly reactive intermediate with a lifetime of less than a nanosecond. The cyclopropanone ring opening is driven not only by relief of ring strain but also by the simultaneous rearomatization of the dienone (see Schemes 2.7 and 2.8).

Product analysis from several photochemical studies of substituted phenacyl chlorides, phosphates, carboxylates, and other leaving groups suggest that the rearrangement is general as long as the photoreactions are run under rather tightly defined conditions. For example, the absence of the *p*-hydroxy group leads to competing photoreductive cleavage of the leaving group, yielding the unrearranged acetophenone [17, 18, 29]. Anderson and Reese [16], in their initial report in 1962, demonstrated this with a series of phenacyl chlorides [Eq. (6) and Tab. 2.1]. In their study, the Favorskii rearrangement was observed for electron donating *p*-hydroxy, *p*-methoxy and *o*-methoxy groups. Attempts to extend the nature of the electron-donating groups even further to *p*-NH$_2$, *p*-NHCOCH$_3$, and *p*-NHCO$_2$CH$_3$ did not lead to Favorskii rearrangement behavior (see Tab. 1.3.1 in Chapter 1.3) [10]. The most reliable structural predictor has been the *p*-hydroxyphenacyl framework.

$$\text{X-PH Cl} \xrightarrow[\text{- HCl}]{\substack{h\nu \\ \text{ROH} \\ R = H, Me}} \text{X-acetophenone, X-AP} + \text{Methyl X-phenylacetate, X-PA OMe} \quad (6)$$

Tab. 2.1 Results from photomethanolysis of substituted phenacyl chlorides [16]

X-PH Cl	% yield X-PA OMe	% yield X-AP
p-OH	32	26
p-OMe	32	30
o-OH	–	3
o-OMe	32	16
p-Me	4	58
H	–	53
p-CO$_2$Me	–	48
p-Cl	–	55
o-Cl	–	45
m-OMe	–	15

2.2.3.5 Effect of Meta Substituents

A selected number of meta-substituted pHP derivatives have also been examined (see Tab. 1.3.2 in Chapter 1.3). Ortho substituents have not. The rich photochemistry known for the o-substituted aceto- and benzophenones including intramolecular hydrogen abstraction could dominate the photochemistry of these derivatives [30, 31]. Nevertheless, they deserve future scrutiny.

Electron-donating meta substituents do shift the absorption range toward the visible, but do so at the expense of lowered quantum efficiencies. Electron acceptors, on the other hand, do not lower the efficiency and are, in fact, more efficient. They do not markedly shift the absorption range, however (see Tab. 1.3.1 in Chapter 1.3), but they do provide a site for further synthetic elaboration for multi-functional applications of the pHP caging group. An empirical correlation was determined between the efficiency for the release of GABA ($\Phi_{release}$) and the triplet state pK_a^3 of the phenolic proton for the meta series, which is discussed in Chapter 1.3 (Fig. 1.3.2 and Tab. 1.3.3). No correlation has thus far been found between the pK_a^3 and the rate constants for release.

2.2.3.6 Quantum Efficiency, Rate Constant, and Solvent Isotope Effect Determinations

Rate constants for release determined by conventional Stern-Volmer quenching techniques and by ns and ps Time-Resolved Laser Flash Photolysis (ns- and ps-TR-LFP) have been reported for several pHP derivatives [32]. Solvent isotope effects [33] have also been determined along with solvent effects, and the effects have led to a working mechanistic hypothesis for this reaction (see Scheme 2.8).

Tab. 2.2 Physical and photochemical properties of pHP derivatives. Variation of the leaving group

Released substrate	λ_{max} (nm) (log ε)[a]	% yield release[b]	$\Phi_{Release}$	k_r (10^8 s^{-1})	Refs.
GABA	282 (4.16)	Quant.	0.21	1.9	35
Glutamate	273 (3.94)	77[c]	0.14	1.9[d]	34, 36, 37
OPO_3^{2-}	280 (4.48)	Quant.	0.38	n/a	17, 18
OPO_3Et_2	271 (4.18)	n/a	0.94	24	17, 18
GTP	n/a	n/a	n/a	n/a	38
ATP	260 (4.48); 286 (4.16)	Quant.	0.30	24	17, 18
Ala Ala	282 (4.12)	Quant.	0.25	2.3	39
Bradykinin	282 (4.07)	n/a	0.22	1.8	38, 40

a) In H_2O.
b) Product yield determined by ^1H NMR or HPLC. Quant. = quantitative yield.
c) Yield of pPA OH.
d) Assumed to be the same as GABA, the model ester.

The efficiencies and rate constants are consistently high (Tab. 2.2), a clear indication that the heterolysis is the primary channel out of the excited triplet state of the **pHP** chromophore. The results of a proton inventory assessment for the reaction conducted in mixtures of H_2O and D_2O suggest that the phenolic proton is donated to a neighboring solvent H_2O molecule at the rate-determining transition state [33, 34].

The photochemistry of the **pHP** chromophore is a good platform for the rapid and efficient photorelease of substrates if they are relatively strong conjugate bases. The excitation wavelength range is limited to $\lambda < 400$ nm, but other properties such as aqueous solubility, synthetic accessibility, and relative stability are excellent.

2.2.4
Benzyl (Bz), Arylmethyl, and Coumaryl (Cou)

Among the currently employed ppgs, benzyl (**Bz**) predates all others for release of amino acids. Barltrop [41] reported the photochemical deprotection of glycine from N-benzyloxycarbamoyl glycine (**Bz** CO_2NH gly) in aqueous media at 254 nm [Eq. (7)]. This was followed four years later by Chamberlain's study of m,m'-dimethoxybenzyl chromophores, which have now become the most popular substituted benzyl derivatives in the series [42].

Bz CO_2NH gly [R, R' = H] glycine
m,m'**DMBz** CO_2NH R' = H (85%) glycine, R' = CH_2SH (60%) cysteine, R' = Ph (66%) phenylalanine, R' = CH_2OH (72%) serine, R' = $(CH_2)_3$NH^{t}BOC (42%) BocGABA
    [R = OCH_3]

(7)

2.2.4.1 The Meta/Ortho-Meta Effect [43]

Chamberlain's work was seminal, for it demonstrated the importance of electron-donating groups in the meta position, maximizing heterolytic pathways for benzyl and subsequently other arylmethyl ppg platforms. Photosolvolysis reactions of benzyl esters, for which the reactive excited state has been characterized, react through their excited singlet states by heterolysis. For this reason the rate constants for heterolysis must be high, i.e., 10^8 to 10^9 s^{-1}, in order to compete with the other decay modes of a short-lived singlet. However, the actual release rate of the substrate from any carbamate ester will be much slower, $\sim 10^5$ s^{-1}, because of the slow loss of CO_2 from the resulting amine carbonate which yields the free substrate.

The earlier studies on the mechanism of benzyl photosolvolysis reactions were the classical studies by Havinga [44, 45]. However, it was not until the molecular-orbital-based mechanistic interpretations of Zimmerman [43] on benzyl and *m*-methoxybenzyl derivatives that an understanding of the enhanced reactivity with *meta*-methoxy substituents was forthcoming [46, 47]. In fact, Zimmerman showed that electron-donating groups in either the ortho or meta positions caused preferential photoheterolysis and gave higher efficiencies and rates relative to the photoreactions of the para derivatives. This structure-reactivity association runs contrary to the established relationships for ground state solvolysis reactions of substituted benzyl derivatives. Furthermore, the para-substituted analogs proceed by pathways that result principally in radical coupling products.

Zimmerman's mechanistic interpretations for benzyl acetates have become known as the "meta effect" in photosolvolysis chemistry. The observed preference for heterolysis is purportedly driven by the landscape of the interacting excited- and ground-state energy surfaces. The meta isomer's excited singlet surface crosses the ground state surface close to an energy minimum for the ground state surface of an ion pair (Fig. 2.9a and b). This surface crossing region can be likened to an excited state surface "funnel" through which the ester exits from the upper excited state surface to the ground state ion pair surface (Fig. 2.9; Scheme 2.10) [47].

Zimmerman [43] further reported that the *m,m'*-dimethoxybenzyl acetate (m,m'M Bz OAc) cleanly produces the photosolvolysis product in 77% yield with an efficiency of 0.10 [Eq. (8)], reinforcing the conclusion reached from Chamberlain's results [42].

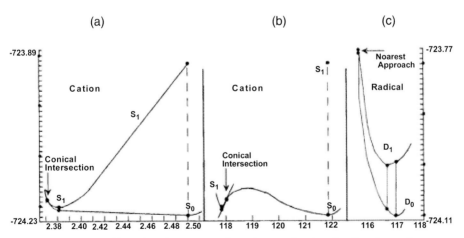

Fig. 2.9 Heterolytic and homolytic dissociation processes with a conical intersection in the heterolysis case. Cross sections: (a) uses bond lengths as the horizontal coordinate; (b) and (c) use bond angles. Energies are in hartrees. From Zimmerman [47] and reprinted with the permission of the American Chemical Society.

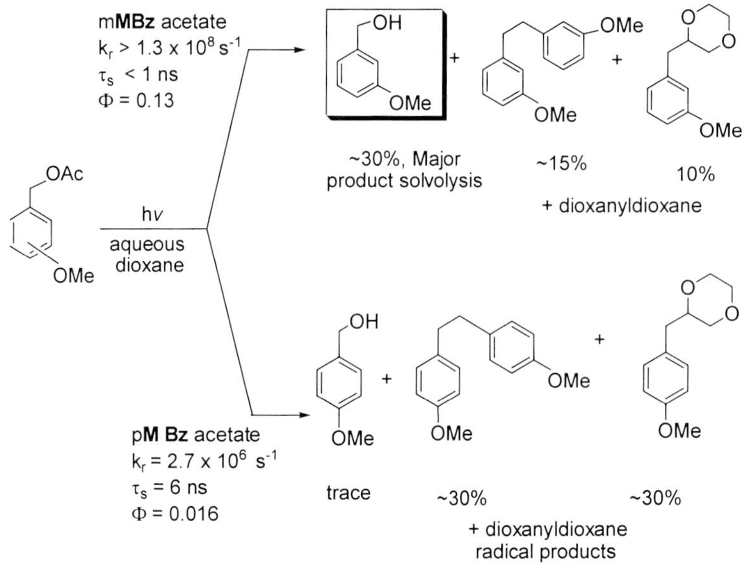
mMBz acetate
k_r > 1.3 x 10^8 s^-1
τ_S < 1 ns
Φ = 0.13
OH
OMe
OMe
OMe
O
O
OMe
~30%, Major product solvolysis
~15%
10%
+ dioxanyldioxane
OAc
OMe
hν
aqueous dioxane
OH
OMe
O
O
pM Bz acetate
k_r = 2.7 x 10^6 s^-1
τ_S = 6 ns
Φ = 0.016
OMe
OMe
OMe
trace
~30%
~30%
+ dioxanyldioxane
radical products

Scheme 2.10 The Zimmerman and Sandel origins of the "meta" effect [43].

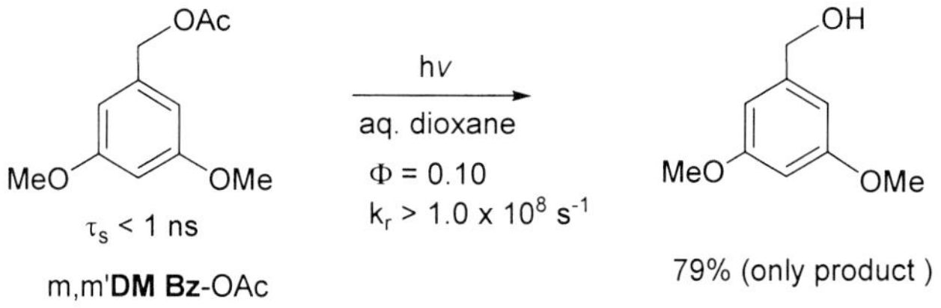
OAc
MeO
OMe
τ_S < 1 ns
m,m'DM Bz-OAc
hν
aq. dioxane
Φ = 0.10
k_r > 1.0 x 10^8 s^-1
OH
MeO
OMe
79% (only product)

(8)

Wan [48] has extended the benzyl acetate structure-reactivity relationship with three additional substituents, F, Cl, and CH_3 (Tab. 2.3). Quantum efficiencies for heterolysis for ortho and meta derivatives were generally higher than those for para, i.e., ortho~meta>>para. Multiple methoxy substitution further enhanced the reactivity in the order ortho,ortho' > ortho,meta>meta,meta' ($\Phi = 0.31$, 0.18, 0.0125, respectively). This latter discovery that substituent effects are nearly additive in the excited state solvolyses is significant for designing future chromophore candidates.

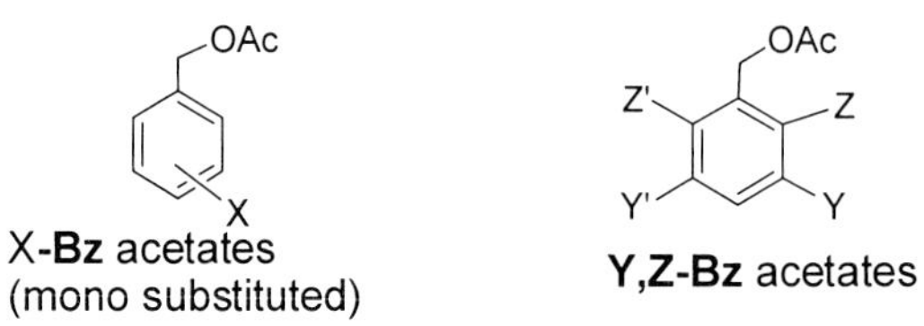
OAc
X
X-Bz acetates
(mono substituted)
OAc
Z'
Z
Y'
Y
Y,Z-Bz acetates

Tab. 2.3 Quantum efficiencies for mono-substituted benzyl acetates (X-**Bz**) [48]

Substituent (X)	Φ_{ortho}	Φ_{meta}	Φ_{para}
CH_3	0.15	0.094	0.042
F	0.078	0.075	0.001
Cl	0.018	0.014	0.005
OCH_3	0.16	0.13	0.016

2.2.4.2 Heterolysis vs Homolysis

A large body of evidence supports Zimmerman's meta (or ortho/meta) effect. However, this mechanism has also been a source of some controversy. Zimmerman [43, 46, 47] proposed the direct heterolysis of the C-O bond through a π,π^* singlet excited state, yielding an "ion-pair" that proceeded on to the ionic products. This process is in competition with homolysis, which leads to the "radical-derived" products. Thus, the product mixture from substituted benzyl acetates are "fingerprints" of the partitioning of the π,π^* singlet excited state between the heterolytic and homolytic pathways, i.e., they provide the ratio of "ion-derived" to "radical-derived" pathways, which are controlled by the type and position of the substituent.

Pincock [49, 50] proposed an alternative mechanism for substituent effects in which chromophore-substrate bonds undergo homolytic cleavage from the π,π^* singlet excited state, leading directly to radical pairs. The ratio of "ionic-derived" to "radical-derived" products in this mechanism is governed by a competition between electron transfer and radical reactions for these radical pairs, i.e., the ratio of the products is controlled by the redox potentials of the radical pair. Within the radical pair, a further competition between decarboxylation and other radical reactions also occurs. This competition serves as a "radical clock" – an internal probe that monitors the rates of other competing processes since the absolute rate constants are known for the decarboxylation reactions of many carboxy radicals. The product ratios thus provide relative rates for the various pathways. Factors that affect the electron-transfer versus decarboxylation rates include the solvent polarity, the ease of decarboxylation, and the availability of abstractable hydrogen atoms (Scheme 2.11).

An overarching factor is the ease of oxidation of the arylmethyl radical to the corresponding cation. Thus, in testing this hypothesis, Pincock [49, 50] examined esters known to produce carboxy radicals that had known decarboxylation rates, such as the phenylacetoxy and pivoyloxy radicals. From the decarboxylation rate constants and the radical/ion product ratios, he was able to assign the rates of electron transfer for the oxidation of a series of substituted arylmethyl esters. A Marcus correlation of the electron transfer rate constants to the known oxidation potentials was discovered which further supported the hypothesis that the ionic-derived products did in fact arise from a very rapid electron transfer process.

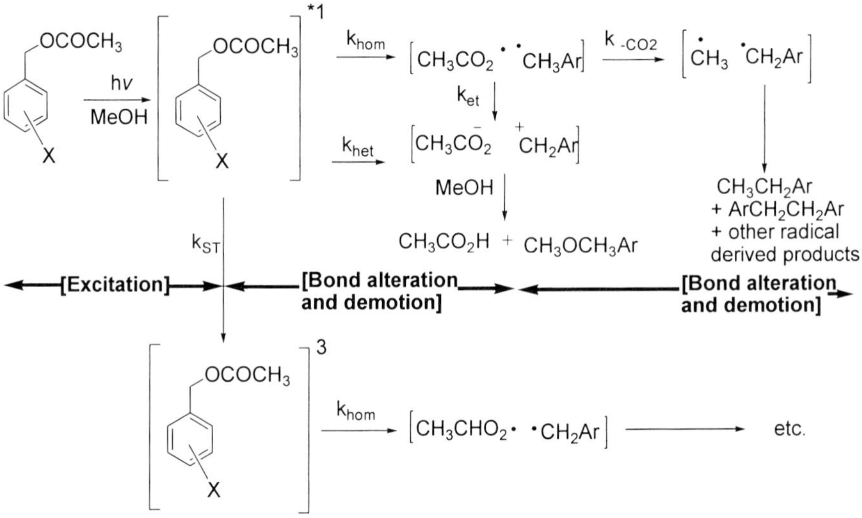

where rate constans k_{hom} = homolysis, k_{het} = heterolysis, k_{ST} = singlet triplet crossing, k_{et} = electron transfer and k_{-CO2} = decarboxylation

Scheme 2.11 Pincock mechanism for the photolysis of benzyl carboxylates [49].

Zimmerman [47] countered Pincock's conclusion with the surface crossing analysis (shown in Fig. 2.9 above) and further pointed out that certain very rapid decarboxylation reactions, i.e., the thermal decomposition of dipivoyl peroxide in particular, may undergo a concerted decarboxylation from its π,π^* singlet excited state rather than the stepwise homolysis sequence. At the present time, the photosolvolysis mechanism leading directly to the ion pair is accepted as the general mechanism.

2.2.4.3 Arylmethyl and Coumaryl (Cou) Derivatives

Less is known about other arylmethyl ppgs, and few applications are currently available. Preliminary mechanistic work on naphthylmethyl analogs [51] indicates that the photochemical reactions parallel the benzyl and methoxybenzyl chemistry. Most reactions have been shown to be singlet state primary photochemical processes where the heterolytic pathway dominates. These reactions generally require the use of polar, protic solvents, otherwise radical pathways become significant. The efficiencies tend to be lower than those for the benzyl analogs, chiefly because of the lower energy of naphthalene singlet.

Coumaryl analogs (**Cou**) are an exception and have generated much more interest as caging groups for several important reasons. The stability of the excited coumaryl moiety under intense photolytic conditions has been demonstrated through the many laser dye applications. Large numbers of new coumaryl derivatives have been synthesized for these and other purposes, and their photostability has been extensively tested under a variety of conditions. Many are highly fluorescent, a feature that can be both useful and deleterious. For these, a wide range of chromophores with known physical and photophysical properties are available. While the coumaryl (**Cou**) group and its hydroxy analog (**HCou**) are fluorescent [52], with good fluorescence efficiencies (ϕ_f) of 0.1–0.4, many of the caged substrates are only weakly fluorescent ($\phi_f \ll 0.1$). This change in fluorescence efficiency has been exploited to monitor the extent of reaction during photorelease studies. Recent work has shown that the 6-bromo-7-hydroxycoumarylmethyl group has a reasonably good two-photon cross-section [53] and can be employed in two-photon decaging studies. Finally, the synthesis of coumaryl-caged compounds is relatively straightforward. Details on these aspects can be found in Chapters 4.1 (Hagen) and 8 (Dore) in this volume.

A mechanistic scheme has been proposed by Bendig and Hagen [51, 54, 55] that outlines the major processes in the release of substrate from the π,π^* excited singlet state. As shown in Scheme 2.12, excitation of the coumaryl chromophore to its singlet excited state is followed by C-O fragmentation. The lack of evidence on the initial cleavage step precludes an unambiguous assignment of the homolytic vs heterolytic cleavage. However, by analogy with the benzyl series, the heterolysis pathway (k_{het}) is the more attractive at this time. The relative energetics for the two pathways are similar to those for the benzyl and naphthylmethyl analogs studied so thoroughly by Pincock and Zimmerman. Thus, the Zimmerman [43, 46, 47] mechanism most likely obtains here as well.

Bendig and Hagen include ion pair return, an often-overlooked process. Nevertheless, it is an important pathway and must be considered when measuring efficiencies and the rate constants for the "free" solvated substrate. The competition between ion pair return and escape from the solvent-separated ion pair to the fully solvated ions is nearly the rate of diffusion. Thus, the rate-controlling step is most likely k_{het}, which should be $\sim 10^9 \, s^{-1}$.

A series of substituted derivatives have been examined for the effect of substituents at the 7 position of the coumaryl chromophore. To assure reasonable hydrophilicity, the hydroxyl (**HCou**), 7-carboxymethoxy (**CM Cou**) and the 7-diethylamino coumaryl (**DEA Cou**) cages were synthesized. The three cages were tested and their release efficiencies determined using phosphate leaving groups, i.e., diethyl phosphate, cAMP, and cGMP. The **DEA Cou** derivative extended the excitation wavelength range above 405 nm without compromising the efficiency ($\phi_{release} \approx 0.20$–$0.25$). **BCM Cou** was the most hydrophilic and had the best thermal stability (>1000 h) among all the derivatives tested; its quantum efficiencies for photosolvolysis ranged from 0.08 to 0.24. The rate constant for phosphate cleavage was determined to be $\sim 10^9 \, s^{-1}$ in all cases. The **CM Cou** derivatives had much narrower effective wavelength ranges, <365 nm, and were less efficient ($\phi_{release} \approx 0.08$–$0.16$).

The photorelease efficiency of a series of leaving groups from 7-methoxycoumarylmethyl cages by the Hagen and Bendig group [54, 55] is given in Tab. 2.4. Greater than 95% yields were realized when the leaving groups were n-heptanoate, sulfonate, and diethyl phosphate. The rate constants for release and the

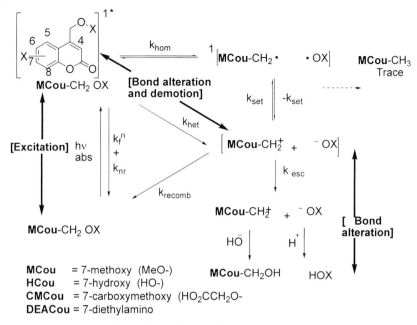

Scheme 2.12 Hagen and Bendig's mechanism for coumaryl ppgs [51, 54, 55].

relevant pK_as of the acids are also shown in Tab. 2.4. Interestingly, the efficiencies and rates increase as the leaving group pK_a decreases, which was taken as a measure of a leaving group effect on the reactivity. A correlation emerged between the quantum efficiency and the pK_a of the leaving group (Fig. 2.10), but the mechanistic insight regarding the nature of the bond-breaking step is not clear.

Furuta et al. [52, 56, 57] have also examined the photosolvolysis of 7-substituted coumarins. Electron-donating substituents in the 6- and 7-positions and electron-withdrawing groups at the 3-position resulted in strong bathochromic shifts in the absorption spectra. The authors' recent work in this area has led to

Tab. 2.4 Rate constants and efficiencies for a series of MCou substrates [54]

No.	Substrate (conj. Base)	pK_a (of the conj. Acid)	φ_{Cou} ×10^{-2}	k_r ×10^6 s^{-1}
3a	Hexanoate	4.89	4.3	3.0
3b	p-MeOC$_6$H$_4$CO$_2^-$	4.41	4.5	n/a
3c	C$_6$H$_5$CO$_2^-$	3.99	5.2	n/a
3d	p-NCC$_6$H$_4$CO$_2^-$	3.54	6.4	n/a
4	CH$_3$SO$_3^-$	−1.54	8.1	410
5	(EtO)$_2$PO$_2^-$	0.71	3.7	190

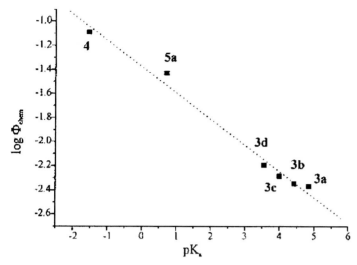

Fig. 2.10 Relationship between photochemical quantum efficiency ($\phi_{release}$) and the pK_a values of the corresponding acids. See Tab. 2.8 for the structures of **3a–d**, **4**, and **5**. From Schade, Hagen, Schmidt, Herbrich, Krause, Eckardt, and Bendig [54] and reprinted with the permission of the American Chemical Society.

the development of 6-bromo-7-hydroxy-coumarinyl-4-methyl (**BH Cou**) as a protecting group for two-photon activated photosolvolysis [Eq. (9)]. **BH Cou** esters and carbamates efficiently release carboxylates and amines upon photolysis, by both one- and two-photon processes, and are discussed in Section 1.2 of this monograph.

MCou cAMP R_1 = NH_2; R_2 = H
MCou cGMP R_1 = OH; R_2 = NH_2

(9)

2.3
Conclusions

The mechanisms afforded by these four classes of photoremovable protecting groups range from the relatively direct photosolvolysis of benzyl-, arylmethyl-, and coumaryl-protected substrates to the very complex rearrangements of 2-nitrobenzyl, benzoin, and p-hydroxyphenacyl derivatives. The 2-**NB** group and its derivatives undergo an initial excited redox process followed by a series of slower ground state steps before final release of the substrate. The rate constant for the final release step of a substrate can vary from as fast as 10^6 s^{-1} to less than 10^{-2} s^{-1} at the slow end of the reactivity spectrum, depending on a number of environmental and structural features. Each application has to be considered with respect to the particular experimental configuration and molecular parameters.

The efficiencies of the 2**NB** series are quite good and are usually reliably high because the photoredox process is irreversible. Thus, the *aci*-nitro intermediate must cascade to the eventual release of the substrate, but the pathway can be tortuous and slow, or it can be barrier free and lead to a quite rapid reaction rate. The structure of the *aci*-nitro derivative and the ability of the leaving group to depart do influence the rate of reaction. Benzoin, p-methoxyphenacyl, benzyl, and coumaryl all undergo direct S_N1 photosolvolysis from their excited singlet states. Rate constants approach 10^9 s^{-1} in order to complete with the other singlet decay routes. These rates are predictably faster than in the 2-**NB** series, but the efficiencies are variable depending on the leaving-group ability of the substrate and may be further reduced by the competition from ion pair return. This latter parameter has not been adequately examined in the past, and little is known about its importance in photosolvolysis reactions.

Finally, the p**HP** groups are unique in several ways. They react through their triplet excited states via a deep-seated rearrangement to phenylacetic acids, an advantage in terms of the hypsochromic shift of the product chromophore relative to the reactant p**HP** absorption. Conversion is not limited by competition from product absorption. Finally, it appears that substrate release occurs during the rearrangement and is in competition with the triplet state decay process. Again, this requires that release occurs at $\sim 10^9$ s^{-1} to be competitive, i.e., to be efficient under aerated conditions at room temperature. The mechanism of this rearrangement, a "Favorskii-like" process, is analogous to the photosolvolysis reactions discussed earlier and requires that substrates be reasonably good leaving groups. The excitation spectra for all the current p**HP** examples are limited to <400 nm for one-photon triggered reactions. As the wavelength of a chromophore is augmented by substitution, the energies of the excited singlet and triplet states are correspondingly reduced. At some point, lowering the available energy for bond breaking leads to an inefficient reaction.

A current technique that shows great promise for overcoming some of these limitations is multiphoton-induced photochemistry. Multiphoton excitations using lower energy infrared frequencies also provide much greater spatial resolution, since both lateral and depth convergence of the exciting photon flux are required. The need for robust, efficient, rapid cages that are biologically benign and can take advantage of this technology remains high.

Acknowledgements

The authors thank the University of Kansas Center for Chemical Methodology and Library Development (NIIH, KU-CMLD, P50 GM069663) and the University of Kansas for financial support. The collaboration of the many research groups who have demonstrated the applications of pHP and its derivatives and the research efforts of the Photochemistry Group at the University of Kansas have been invaluable.

Abbreviations

BH Cou	6-bromo-7-hydroxycoumaryl
Bnz a	anti conformation of benzoin
Bnz g	gauche conformation of benzoin
BMC Bnz	bis(carboxymethoxy)benzoin
Bnz	benzoin
Bz	benzyl
Cou	coumaryl
CM Cou	7-carboxymethoxycoumaryl
DEA Cou	7-diethylaminocoumaryl
DFT	density function theory
DM Bnz	dimethoxybenzoin
ESIPT	excited-state intramolecular proton transfer
HCou	7-hydroxycoumaryl
Isox	benzisoxazole
M Bnz	methoxybenzoin
2-NB	2-nitrobenzyl
ns- (ps-) LFP	nano- (pico-) second laser flash photolysis
OAc	acetate
P0°	planar intermediate
pHA	p-hydroxyacetophenone
pHP	p-hydroxyphenacyl
pHPA OH	p-hydroxyphenylacetic acid
ppg	photoremovable protecting group
pQM	p-quinone methide
sDD	spirodienedione
T90°	90° twisted intermediate
TRIR, TR FTIR	time-resolved (Fourier transform) infrared
TR LFP	time-resolved laser flash photolysis
TR'	time-resolved resonance Raman
zOA	zwitterionic oxyallyl intermediate

References

1 For recent reviews, see: (a) Adams, S. R., Tsien, R. Y. *Annu. Rev. Physiol.* **2000**, *18*, 755–784. (b) Pelliccioli, A. P. and Wirz, J., *Photochem. Photobiol. Sci.* **2002**, *1*, 441–458. (c) Bochet, C. G., *J. Chem. Soc. Perkin Trans.* **2002**, *1*, 125–142. (d) Givens, R. S., Lee, J.-I., *J. Photoscience* **2002**, *10*, 37-48. (e) Givens, R. S., Conrad, G. P. II, Yousef, A., Lee, J.-L. "Photoremovable Protecting Groups", Ch. 69 in *Handbook of Organic Photochemistry and Photobiology*, 2nd Edn., W. Horspool and F. Lenci, eds.

2 (a) Corrie, J. E. T., Trentham, D. R. *Biological Applications of Photochemical Switches*, J. Wiley and Sons: New York, **1993**. (b) Givens, R. S., Kueper, L. W. III, *Chem. Rev.* **1993**, *93*, 55–66.

3 Il'ichev, Y. V., Schwoerer, M. A., Wirz, J. *J. Am. Chem. Soc:* **2004**, *126*, 4581–4595.

4 (a) Papageorgiou, G. L., Matthew, L., Wan, P., Corrie, J. E. T., *Photochem. Photobiol. Sci.* **2004**, *3*, 366–373. (b) See also, Walker, J. W., Reid, G. P., McCray, J. A., Trentham, D. R. *J. Am. Chem. Soc.* **1988**, *110*, 7170–7177.

5 Yip, R. W., Sharma, D. K., Giasson, R., Gavel, D., *J. Phys. Chem.* **1985**, *89*, 5328–5330.

6 Yip, R. W., Sharma, D. K., Giasson, R., Gavel, D., *J. Phys. Chem:* **1984**, *88*, 5770–5772.

7 (a) Kaplan, J. H., Forbush, B. I., Hoffman, J. F., *Biochemistry* **1978**, *17*, 1929-1935. (b) Engels, J.; Schlaeger, E.-J., *J. Med. Chem.* **1977**, *20*, 907–911.

8 Sheehan, J. C., Wilson, R. M. *J. Am. Chem. Soc.* **1964**, *86*, 5277–5281.

9 Sheehan, J. C., Wilson, R. M., Oxford, A. W. *J. Am. Chem. Soc.* **1971**, *93*, 7222–7228.

10 Givens, R. S., Athey, P. S., Matuszewski, B., Kueper, L. W. III, Xue, J. Y., Fister, T. *J. Am. Chem. Soc.* **1993**, *115*, 6001–6012.

11 Rock, R. S., Chan, S. I. *J. Am. Chem. Soc.* **1998**, *120*, 10766–10767.

12 Pirrung, M. C., Bradley, J. C. *J. Org. Chem.* **1995**, *60*, 1116.

13 Rajesh, C. S., Givens, R. S., Wirz, J. *J. Am. Chem. Soc.* **2000**, *122*, 611–618.

14 Shi, Y., Corrie, J. E. T., Wan, P. *J. Org. Chem.* **1997**, *62*, 8278–8279.

15 Cameron, J. F., Willson, C. G., Frechet, J. M. J., *J. Chem. Soc., Perkin Trans. 1* **1997**, 2429.

16 Anderson, J. C., Reese, C. B., *Tetrahedron Lett.* **1962**, *1*, 1–4.

17 Givens, R. S., Park, C.-H., *Tetrahedron Lett.* **1996**, *37*, 6259–6262.

18 Park, C.-H., Givens, R. S., *J. Am. Chem. Soc.* **1997**, *119*, 2453–2463.

19 Favorskii, A. E., *J. Prakt. Chem.* **1894**, *51*, 559.

20 Kende, A. S., *Org. React.* **1960**, *11*, 261–316.

21 Zhang, K., Corrie, J. E. T., Munasinghe, V. R. N., Wan, P. *J. Am. Chem. Soc.* **1999**, *121*, 5625–5632.

22 Fischer, M., Wan, P., *J. Am. Chem. Soc.* **1998**, *120*, 2680.

23 Lee J.-I. and Givens R. S., Unpublished results.

24 Philips, D. L., Ma, C., Chan, W. S., Kwok, W. M., Zuo, P., *J. Phys. Chem. B* **2004**, *108*, 9264–9276.

25 Loftfield, R. G., *J. Am. Chem. Soc.* **1951**, *73*, 4707.

26 Burr, J. G., Jr., Dewar, M. J., *J. Chem. Soc.* **1954**, 2101.

27 Castillo, R. A. J., Moliner, V., *J. Phys. Chem. B* **2001**, *105*, 2453–2460.

28 Bordwell, F. G., Scameborn, R. G., Springer, W. R., *J. Am. Chem. Soc.* **1969**, *91*, 2087–2093.

29 Banerjee, A., Falvey, D. E., *J. Am. Chem. Soc.* **1998**, *120*, 2965–2966.

30 Wagner, P. J. "Yang Photocyclization: Coupling of Biradicals formed by Intramolecular Hydrogen Abstraction of Ketones", Ch 58 in *"Handbook of Organic Photochemistry and Photobiology"* 2nd Edn., W. Horspool and F. Lenci, (eds.) **2003**, 58-1 to 58–70.

31 Hasagawa, T. "Norrish Type II Processes of Ketones: Influence of Environment", Ch 55 in *"Handbook of Organic Photochemistry and Photobiology"* 2nd Edn., W. Horspool and F. Lenci, (eds.) **2003**, 55-1 to 55-14.

32 CONRAD II, P.G., GIVENS, R.S., HELL-RUNG, B., RAJESH, C.S., RAMSEIER, M., WIRZ, J., *J. Am. Chem. Soc.* **2000**, *122*, 9346-9347.

33 GIVENS, R., WIRZ, J. and MATA, J., Unpublished Results.

34 SWAIM, C.G., KUHN, D.A., SCHOWEN, R.L., *J. Am. Chem. Soc.* **1965**, *89*, 1553–1561.

35 GIVENS, R.S., JUNG, A., PARK, C.-H., WEBER, J., BARTLETT, W., *J. Am. Chem. Soc.* **1997**, *119*, 8369–8370.

36 KANDLER, K., GIVENS, R.S., KATZ, L.C., *Photostimulation with caged glutamate*, in Imaging Neurons, YUSTE, R., LANNI, F., KONNERTH, A. (eds.) (**2000**), Cold Spring Harbor Laboratory Press, Cold Spring Harbor, NY, 27/1–27/9.

37 CONRAD II, P.G., GIVENS, R.S., WEBER, J.F.W., KANDLER, K., *Org. Lett.* **2000**, *2*, 1545–1547.

38 DU, X., FREI, H., KIM, S.-H., *J. Biol. Chem.* **2000**, *275*, 8492–8500.

39 GIVENS, R.S., WEBER, J.F.W., CONRAD II, P.G., OROSZ, G., DONAHUE, S.L., THAYER, S.A., *J. Am. Chem. Soc.* **2000**, *122*, 2687–2697.

40 SUL, J.-Y, OROSZ, G., GIVENS, R.S., HAYDON, P.G., *Neuron Glia Biol.* **2004**, *1*, 3–10.

41 BARLTROP, J.A., SCHOFIELD, P., *Tetrahedron Lett.* **1962**, 697–699.

42 CHAMBERLAIN, J.W., *J. Org. Chem.* **1966**, *31*, 1658–1660.

43 ZIMMERMAN, H.E., SANDEL, V.R., *J. Am. Chem. Soc.* **1963**, *85*, 915–922.

44 HAVINGA, E., CORNELISSE, J., *Chem. Rev.* **1975**, *75*, 353.

45 HAVINGA, E., KRONENBERG, M.E., *Pure Appl. Chem.* **1968**, *16*, 137.

46 ZIMMERMAN, H.E., *J. Am. Chem. Soc.* **1995**, *117*, 8988–8991.

47 ZIMMERMAN, H.E., *J. Phys. Chem. A* **1998**, *102*, 5616–5621.

48 WAN, P., CHAK, B., LI, C., *Tetrahedron Lett.* **1986**, *27*, 2937–2940.

49 PINCOCK, J.A., *Acc. Chem. Res.* **1997**, *30*, 43–49.

50 HILBORN, J.W.M.E., PINCOCK, J.A., WEDGE, P.J., *J. Am. Chem. Soc.* **1994**, *116*, 3337–3346.

51 (a) GIVENS, R.S., MATUSZEWSKI, B., LEVI, N., LEUNG, D., *J. Am. Chem. Soc.* **1977**, *99*, 1896–1903. (b) PINCOCK, J.A. "The Photochemistry of Esters of Carboxylic Acids" Ch 66 in: Handbook of Organic Photochemistry and Photobiology 2nd Edn., W. HORSPOOL and F. LENCI (eds.), **2003**, 66-1 to 66-5.

52 ECKARDT, T., HAGEN, V., SCHADE, B., SCHMIDT, R., SCHWEITZER, C., BENDIG, J., *J. Org. Chem.* **2002**, *67*, 703–710.

53 FURUTA, T., WANG, S.S.H., DANTZKER, J.L., DORE, T.M., BYBEE, W.J., CALLAWAY, E.M., DENK, W., TSIEN, R.Y., *Proc. Natl. Acad. Sci. USA* **1999**, *96*, 1193–2000.

54 SCHADE, B., HAGEN, V., SCHMIDT, R., HERBRICH, R., KRAUSE, E., ECKARDT, T., BENDIG, J., *J. Org. Chem.* **1999**, *64*, 9109–9117.

55 HAGEN, V., BENDIG, J., FRINGS, S., ECKARDT, T., HELM, S., REUTER, D., KAUPP, U.B., *Angew. Chem. Int. Ed.* **2001**, *40*, 1045–1048.

56 FURUTA, T., IWAMURA, M. *New Caged Groups: 7-Substituted Coumarinylmethyl Phosphate Esters*, Academic Press, New York **1998**.

57 FURUTA, T., TORIGAI, H., SUGIMOTO, M., IWAMURA, M., *J. Org. Chem.* **1995**, *60*, 3953–3956.

3
Caged Compounds and Solid-Phase Synthesis

Yoshiro Tatsu, Yasushi Shigeri, and Noboru Yumoto

3.1
Introduction

Many proteins and peptides are known to function both temporally and spatially to regulate important biological processes. Caged proteins and peptides are promising tools for elucidating the mechanisms of such processes. However, difficulties in the preparation of caged peptides and proteins limit their use. Recently, several groups have made improvements to the traditional chemical modification method, as well as developing new methods to prepare caged peptides and proteins: a nonsense codon suppression technique and a solid-phase peptide synthesis.

In this chapter, the solid-phase synthesis of caged compounds, and in particular the synthesis of peptides, is described. Peptides are built up by combination of 20 amino acids, whose side chains have various functional groups. In the chemical synthesis of peptides, various protections are required, and photocleavable protection has been used as an orthogonal protection. The combination of photocleavable protection and solid-phase synthesis enables the facile synthesis of peptides carrying photocleavable protection that can act as caged peptides. The background and development of solid-phase synthesis of caged peptides is documented.

3.2
Solid-Phase Synthesis and Photolysis of Peptides

3.2.1
General Overview of Solid-Phase Peptide Synthesis

Solid-phase synthesis has been widely recognized as a facile method to synthesize compounds. The main advantage of solid-phase synthesis is ease of separation of target compounds from the reaction mixture. A building block for the target compound is anchored to polymer beads, and the second building block is then made to react. After completion of the reaction, the product is separated by filtration and washing on a funnel. Similarly, the cleavage of a protecting group or the conversion of a functional group can be accomplished by soaking the beads in the reaction cocktail, followed by separation on a funnel. A third building block can then be attached to the naked functional group. In the final step, all the protecting groups and the linkage with the polymer beads are cleaved, and the target compound is obtained. The purity of the product is not always high, mainly because of the difficulty of the purification process on the polymer beads, but an acceptable purity is usually achieved by high-pressure liquid chromatography (HPLC) purification.

3.2.1.1 The Protecting Group in Solid-Phase Peptide Synthesis

The most important aspect of solid-phase synthesis is orthogonality: each protecting group and linkage can be cleaved independently. Peptide synthesis proceeds in a one-dimensional direction, and at least two types of orthogonal protection are required. In standard solid-phase peptide synthesis, the protection of the N-terminal amino group is "temporary", and the C-terminal carboxyl or amide group and side chain functional groups are "permanent". The orthogonality is based on the relative stability of the protecting group against acids and bases.

There have been two major methods developed for solid-phase peptide synthesis: the Boc and the Fmoc methods. Boc (or Boc/benzyl) chemistry uses the t-butoxycarbonyl group as protection for the N-α-terminal amino group and benzyl group for side chain, respectively. N-α-Boc-protected amino acids are used as building blocks and made to react with a linker on polystyrene beads. The Boc group is cleaved by trifluoroacetic acid (TFA), whereas the benzyl group is chemically stable against TFA. In the final cleavage, hydrogen fluoride, a strong acid, is used to cleave the linker at the C-terminal and the protection on the side chains. Fmoc (or Fmoc/tBu) chemistry uses the fluorenylmethyloxycarbonyl (Fmoc) group for the N-α-terminal amino group and the t-butyl group as protection for the side chain, respectively. N-α-Fmoc-protected amino acids are used as building blocks and made to react with a linker on polystyrene beads. The Fmoc group is cleaved by bases, such as piperidine and 1,8-diazabicyclo[5.4.0]undec-7-ene (DBU). The t-butyl group is chemically stable against bases. In the final cleavage, TFA is used to cleave the linker at the C-terminal and the protection on the side chains. Representative protecting groups are summarized in Tab. 3.1. The 2-nitrobenzyl group is a representative caging group, and its susceptibility to base and acid attack is similar to that of the benzyl group. The benzyl group is orthogonal, and is stable to TFA. Therefore, it is assumed that a combination of Fmoc and 2-nitrobenzyl groups will provide 2-nitrobenzylated peptides that would serve as caged peptides. In fact, most caged peptides have been synthesized using Fmoc chemistry, and a case study will be discussed later.

Tab. 3.1 Chemical stability of representative protecting groups

	Protection	TFA	HF	Piperidine
$N\text{-}\alpha\text{-}NH_2$	Fmoc	–		+
	Boc	+	+	–
$N\text{-}\varepsilon\text{-}NH_2$	Boc	+	+	–
	Z	–	+	–
OH	Butyl	+	+	–
	benzyl	–	+	–
SH	Trt	+	+	–
	benzyl	–	+	–
NH_2	Boc	+	+	–
	Z	–	+	–
COOH	t-Bu	+	+	–
	benzyl	–	+	+

(+): cleaved; (–): stable
Trt: trityl, Z: benzyloxycarbonyl, *t*-Bu: t-butyl

3.2.1.2 Formation of Peptide Fragments

The elongation of peptide sequences in solid-phase synthesis almost universally proceeds stepwise from the C-terminal to the N-terminal (the reverse of natural synthesis). This is because N-urethane-protected amino acids, such as Boc and Fmoc, are relatively free of racemization and other side reactions, which may be caused by the formation of oxazolone during the activation of the carboxyl group. The completion of the coupling reaction is also ensured by use of excess N-urethane-protected amino acids.

There are several types of coupling techniques in solid-phase peptide synthesis, such as carbodiimide, preformed active esters, preformed anhydrides, amino acid halides, and coupling reagents. Among them, coupling reagents, such as benzotriazolyloxy-tris-pyrrolidino-phosphonium hexafluorophosphate (PyBOP), are the most popular choice in practice, owing to their simplicity and rapidity. The other coupling techniques are sometimes used for synthetic reasons, such as their higher reactivity. Cleavage of the N-urethane group is accomplished by a weak acid such as TFA for the Boc group or by a base such as piperidine for the Fmoc group. Repeated cycling of the coupling of N-urethane-protected amino acids and the cleavage of the urethane groups gives the desired amino acid sequence.

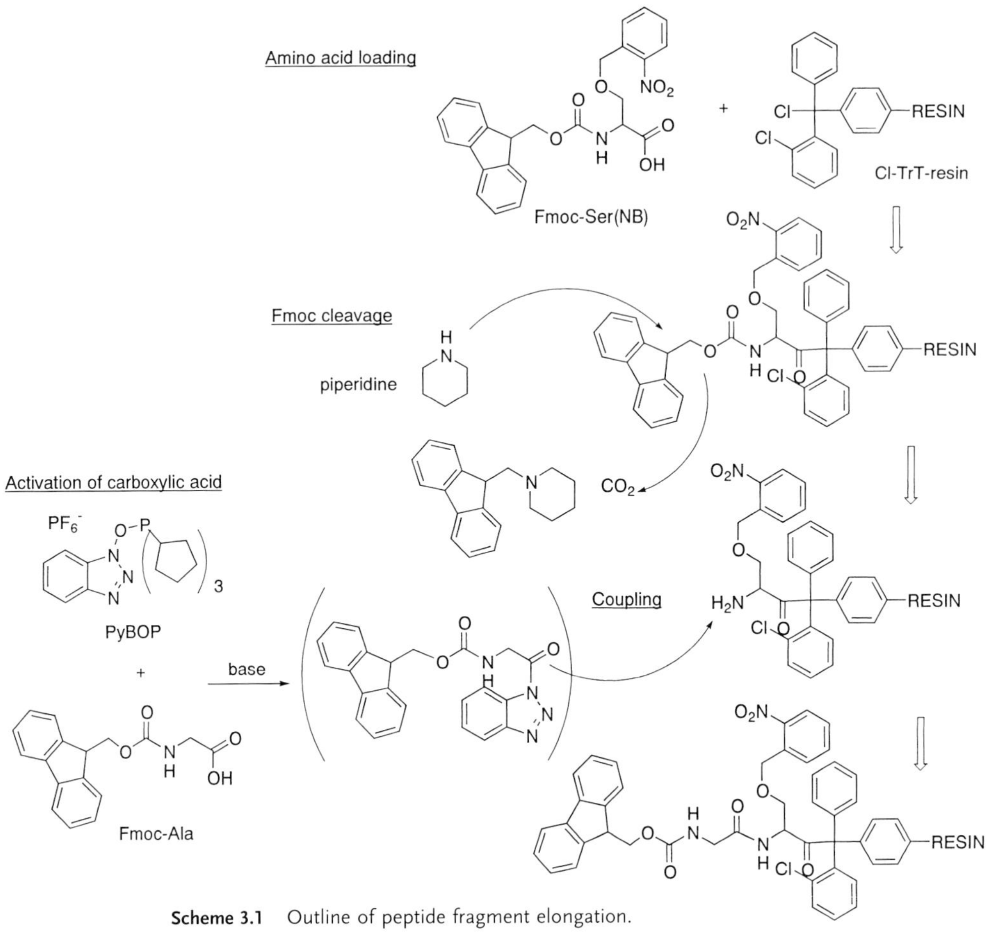

Scheme 3.1 Outline of peptide fragment elongation.

During the final stage, the C-terminal linkage is cleaved by a strong acid such as hydrogen fluoride for the tBoc strategy and by a weak acid such as TFA for the Fmoc strategy. The benzyl group is also cleaved by hydrogen fluoride. Caged peptides containing 2-nitrobenzyl groups are not compatible using a Boc strategy.

3.2.1.3 Purification and Identification

Synthetic peptides are usually purified by reversed phase HPLC, since the obtainable resolution is high. In many cases, deleted sequences, where one amino acid is deleted, have different retention times in HPLC chromatography. Usually, caged peptides have a higher hydrophobicity, and the elution from reversed-phase HPLC is slower than for the intact form. Therefore, contamination with the intact peptide is excluded in the HPLC purification. Although the caging groups might be decomposed by the UV detector used in HPLC, it is possible to fractionate the target peptide in the retention time without use of a detector after a preliminary experiment.

The fractionated peptide is generally identified using a mass spectrometer (MS). Electro spray ionization (ESI) and matrix-assisted laser ionization (MALDI) are applicable in the mass spectrometry of peptides. ESI is convenient, because the elution from HPLC can be directly connected to the MS, and the identification and fractionation can be performed immediately. However, multivalent ions and solvent adducts are often observed in ESI. In the MALDI technique, the parent ion is the dominant species observed, and identification is straightforward. In the case of caged peptides, however, care is needed in identifying the ions, because laser ionization also causes photolysis of the caging moiety.

3.2.2
Photolysis of Peptides

3.2.2.1 Photocleavable Protecting Groups in Peptide Synthesis

This section describes photocleavable protection in peptide synthesis including solution phase techniques. Peptides have a variety of functional moieties. Various orthogonal protections have been developed for each functional moiety, since orthogonal protection is useful for removing the protection at a targeted functional moiety among similar moieties. Photocleavable protection is one of the ultimate orthogonal protections, since it does not use chemical reagents. The photocleavable protecting group described in this section is not originally used as the cage, but the methods contain useful information related to the synthesis of caged peptides for biological usage. The 2-nitrobenzyl group is a commonly used photocleavable protecting group and has been used as an orthogonal protection in peptide synthesis.

3.2.2.1.1 Amino Protection

The most common protecting group for amines is carbamate. O-Alkyl cleavage of carbamate releases carbamic acids, which are unstable and decompose to give carbon dioxide and free amine. The N-2-nitrobenzyloxycarbonyl group, Z(2-NO$_2$), is used as a protection for the amino group, and the release of carbamic acids is initiated by UV irradiation [1]. N-Z(2-NO$_2$)-protected amino acids and peptides are normally prepared via the reaction of the free amine with 2-nitrobenzylchloroformate, which can be prepared by reacting 2-nitrobenzyl alcohol and phosgene or phosgene dimer. Z(2-NO$_2$), like the benzyloxycarbonyl group (Z), is stable to acids (100% TFA) and bases (30% piperidine), but unstable to catalytic hydrogenation. The installation and removal of the 2-nitroveratryloxycarbonyl group (NVOC) is similar to the installation and removal of Z(2-NO$_2$). The absorption maximum of NVOC (λ_{max}=ca. 350 nm) is longer than that of Z(2-NO$_2$) (λ_{max}=ca. 270 nm). A caged peptide based on the protection of the N-α-amino group with Z(2-NO$_2$) was synthesized by Iwamura via solution phase synthesis [2]. N-α-Protected amino acids can also be applied in a parallel synthesis.

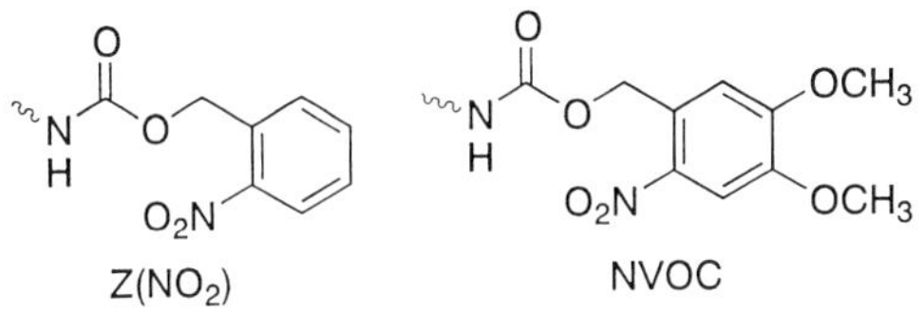

3 Caged Compounds and Solid-Phase Synthesis

Scheme 3.2 Photocleavable protection for amines.

N-α-Fmoc-N-ε-NVOC lysine has been used to prepare a disulfide-bridged peptide conjugate [3]. Since maleimide, used as a linkage for conjugation, is unstable under the conditions used for peptide synthesis, it should be introduced to the peptide after the cleavage from the resin. For the selective introduction of maleimide to the amino group at N-terminus, another amino group at the side chain of the lysine must be protected under both the SPPS and maleimide introduction. NVOC is stable against TFA, and the desired conjugate can be obtained by removal of NVOC group with UV irradiation ($\lambda > 280$ nm, in Pyrex) for 30 min.

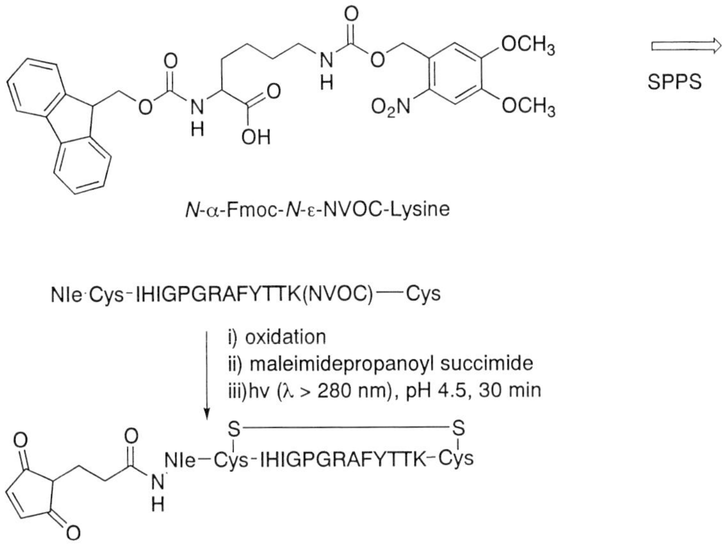

Scheme 3.3 Protection of the amine and photolysis.

Photoirradiation of an amine protected with benzyloxycarbonyl (Z), which has no nitro group, also gives free amine. Photolysis of Z-Gly-OH ($\lambda = 254$ nm) gives Gly, as well as various side products, under basic conditions. It was later found out that substitution by the methoxy group at the *meta* position enhanced the reactivity and yield. 3,5-Dimethoxycarbonyl on the α-amino group of lysine was selectively photocleaved in the presence of Z-protection at the ε-amino group [4]. 3,5-Dimethoxycarbonyl protection was applied in the dipeptide synthesis.

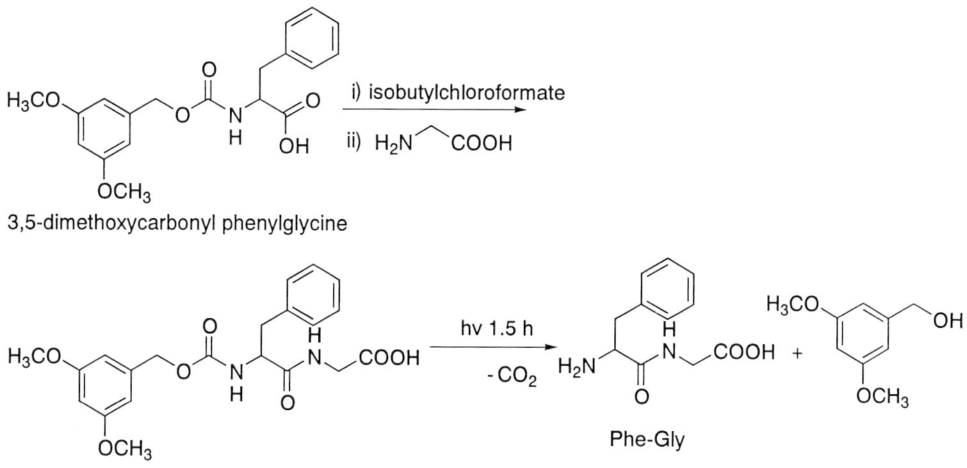

Scheme 3.4 Protection with 3,5-dimethoxycarbonyl and photolysis.

Another photocleavable protecting group is a phenacyl (2-oxo-2-phenylethyl) group. Church et al. showed that 4-methoxy phenacyloxycarbonyl (Phenoc) was stable to acids and bases, and that the phenacyl group can be cleaved in neutral conditions [5]. *In situ* deprotection and coupling were also shown. Boc-Val-Osu, which is an activated ester, was mixed with Phenoc-Ala-OMe and irradiated for 15 h. The released amine was coupled with activated ester to give the dipeptide.

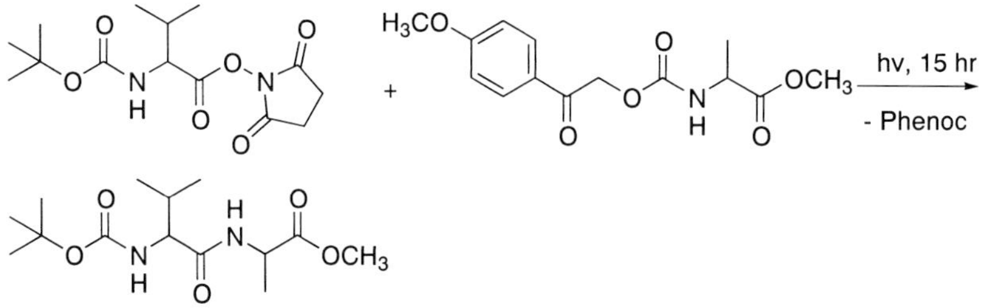

Scheme 3.5 Protection with phenacyl and photolysis.

3.2.2.1.2 Protection of Other Functional Groups

Carboxyl group The 2-nitrobenzyl group can also be used as a photocleavable protecting group for the carboxyl group. The mechanism of photocleavage is the same for amine protection, except that the carboxyl group is formed directly. The ester bond is not stable under nucleophilic conditions, similarly to the benzyl group.

Carboxyamide Henriksen et al. used a 2-nitrophenylglycine as a photocleavable protecting group for peptide amide synthesis [6]. 2-Nitrophenylglycine was introduced by transacylation catalyzed by carboxypeptidase Y with alanine at the C-terminal of the peptide. Photocleavage of the nitrophenyl glycine provided the C-terminal amide.

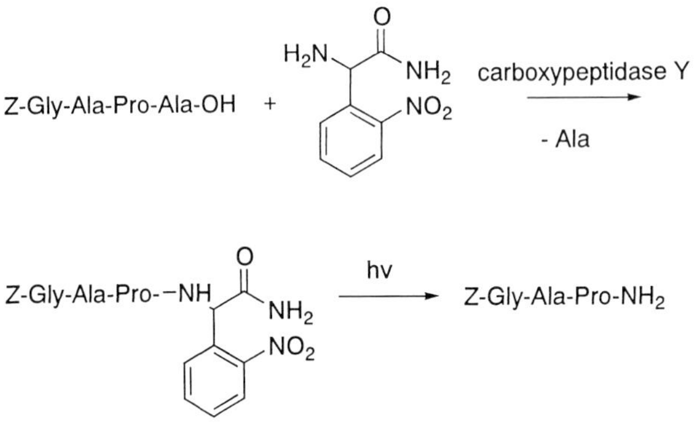

Scheme 3.6 Protection of the carboxyl group and photolysis.

Hydroxyl group Protection of hydroxyl groups with benzyl groups forms an ether bond, which is chemically stable. The 4,5-dimethoxy-2-nitrobenzyl group has been incorporated into the side chain of serine by a Lewis acid-catalyzed ring opening of Fmoc-aziridine-2-carboxylate benzyl ester with dimethoxy-2-nitrobenzyl alcohol [7].

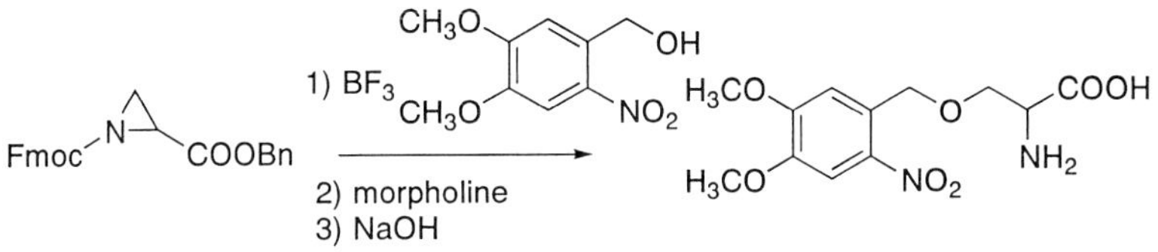

Scheme 3.7 o-4,5-Dimethoxybenzyl-serine.

The 2-nitrobenzyl group has been incorporated into the side chain of Z-tyrosine methyl ester using 2-nitrobenzyl chloride and sodium methoxide [8]. The ether bond was stable against bases and acids, but was photolyzed to give the free hydroxyl group. The dipeptide, N-Z-O-2-nitrobenzyl-tyrosyl-glycine ethyl ester, has been prepared, and the free hydroxyl group was also generated by irradiation.

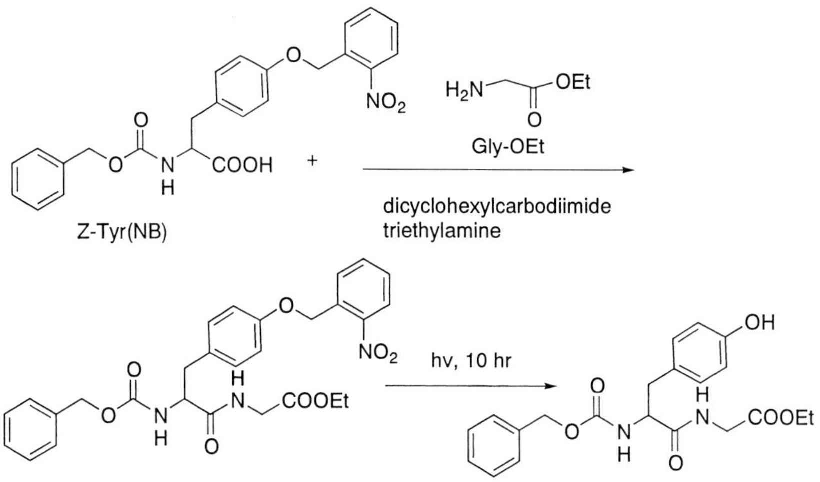

3.2 Solid-Phase Synthesis and Photolysis of Peptides

Scheme 3.8 Protection of the hydroxyl group and photolysis.

Imidazole The 2-nitrobenzyl group has also been incorporated by reaction of the silver salt of Boc-histidine methyl ester with 2-nitrobenzyl bromide in refluxed benzene solution [9]. The 2-nitrobenzyl group can be removed by photo-irradiation.

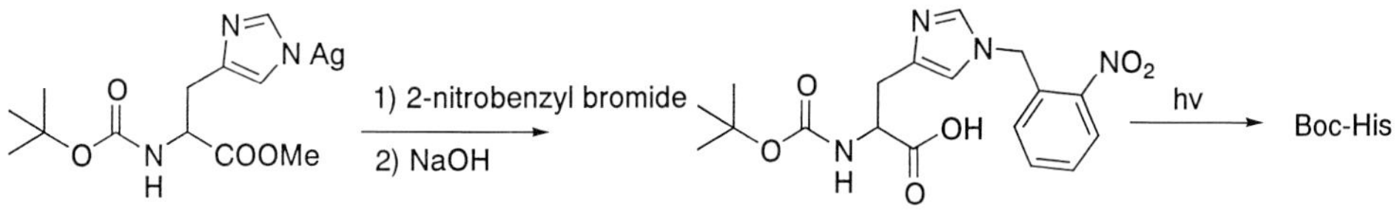

Scheme 3.9 N-α-Boc-N-im-nitrobenzyl-histidine.

Guanidine The 4,5-dimethoxy-2-nitrobenzyl group has been incorporated into the amine of ornithine to give a photolabile arginine residue, and this was applied in caged peptide synthesis [10].

3.2.2.1.3 Photocleavable Linkers

Photolysis is a mild, non-invasive technique, and much work has been done to develop photocleavable linkers. In principle, photolysis is compatible with the major protecting groups used in the Fmoc and Boc strategies, and it is also useful in obtaining peptide fragments where the side chains are protected. The most widely studied photocleavable linkers are based on the 2-nitrobenzyl moiety and the phenacyl moiety.

The 2-nitrobenzyl moiety has been introduced by nitration of chloromethylated polystyrene beads. The synthesized peptide on the resin was photoreleased, but the resin showed poor swelling properties due to the nitration. The introduction of a photolabile linker overcame this problem, ensuring that only the required number of nitrobenzyl moieties was loaded on the resin [11]. A drawback of photocleavable linkers based on 2-nitrobenzyl is the nitrosobenzaldehyde moiety photoproduct. This is thought to be a consequence of the formation of azo compounds, which act as a light filter. It was also pointed out that the oxidation of methionine into methionine sulfoxide during the photolysis occurs. These drawbacks were overcome by a-methylation [12].

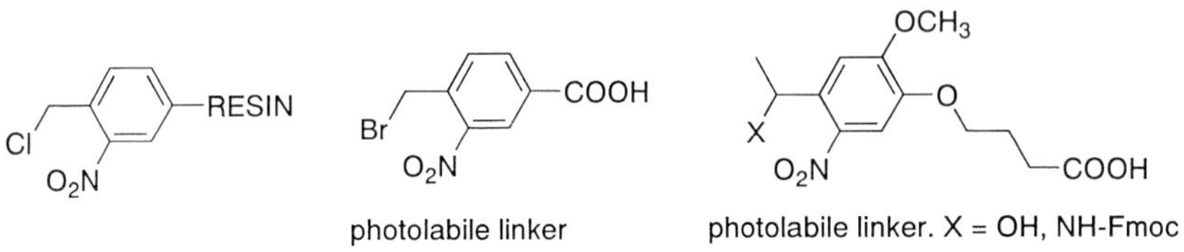

Scheme 3.10 Photocleavable linker.

A photolabile linker has also been introduced into polyethylene glycol (PEG) [13]. The soluble polymer acted not only as a protecting group for the C-terminal, but also increased solubility during the chemical reaction, and allowed for separation by filtering the peptide precipitated by the ether.

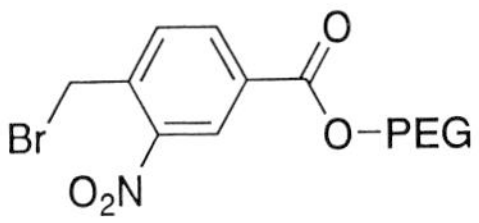

Scheme 3.11 Photocleavable support for peptide synthesis.

Peptide-bound beads were also used directly in the screening of a peptide library [14]. A nine-mer, 442 368 member, peptide library was constructed on polymer beads carrying a photolabile linker using the partitioning-mixing method. The beads were mixed with agarose and poured on plates containing melanophore cells. After the photolysis of the beads, the beads above the response area were retrieved and sequenced.

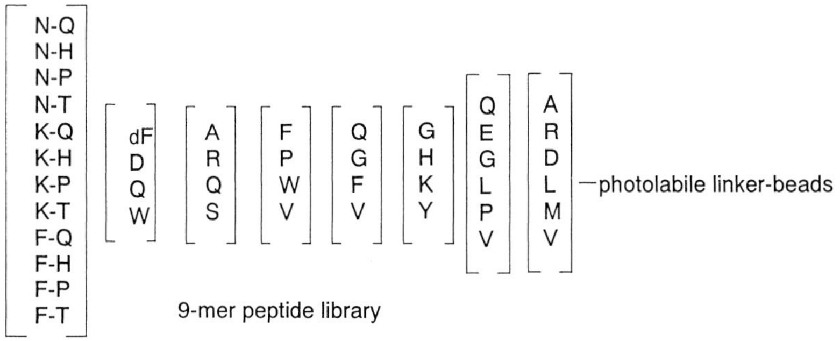

3.2 Solid-Phase Synthesis and Photolysis of Peptides | 141

[N-Q, N-H, N-P, N-T, K-Q, K-H, K-P, K-T, F-Q, F-H, F-P, F-T] [dF, D, Q, W] [A, R, Q, S] [F, P, W, V] [Q, G, F, V] [G, H, K, Y] [Q, E, G, L, P, V] [A, R, D, L, M, V] —photolabile linker-beads

9-mer peptide library

Scheme 3.12 Construction of peptide library.

A photolabile linker based on 3-methoxybenzoin has been incorporated into a peptide sequence [15]. The Fmoc building block for the linker was photochemically inert, since the benzoyl carbonyl was protected with dithiane, which can be removed under mild condition such as (trifluoroacetoxy)iodobenzene (TFAIB). The dithiane "safety-catch" would be useful to prevent the photolysis during the synthesis. The photocleavage of the backbone of a peptide based on the linker may be useful in applications involving caging techniques and folding studies.

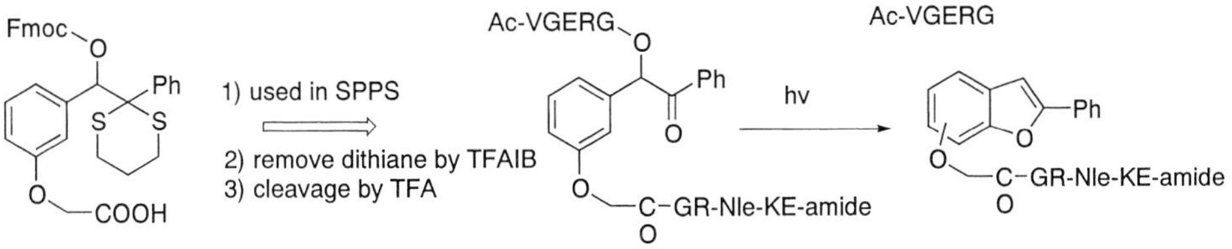

Scheme 3.13 Photocleavable linker within peptide fragments.

3.2.2.1.4 Photoremovable Scaffolds for Acyl Transfer

Long peptides Chemical ligation is a promising method to overcome the length limitation in the solid-phase synthesis of peptides, as well as semisynthesis using biologically prepared peptides. This peptide bond formation occurs via an S-N acyl shift and is quite convenient for polypeptide synthesis, because protecting groups are not required, and the reaction can be carried out in neutral aqueous solutions. Chemoselective reactions proceed between a thioester at the C-terminus of the N-terminal fragment and a cysteine residue at the N-terminus of the C-terminal fragment. Thioester is exchanged by thiol group in the cysteine residue and S-N acyl shift results in the formation of a native peptide bond. Thiol-containing linkers can be used to replace the cysteine residue. To overcome the limitation of a cysteine residue at the condensation site, a photoremovable auxiliary has been developed [16]. The N-terminus of serine was oxidized with periodate to give an N-glyoxyloyl group, and a 2-mercapto-1-(2-nitrophenyl)ethylamine group was attached. Then, the peptide was ligated to a peptide thioester. The auxiliary was removed by UV irradiation to give the ligated peptide, which had a native peptide bond, although the ligated sequence was Xaa-Gly.

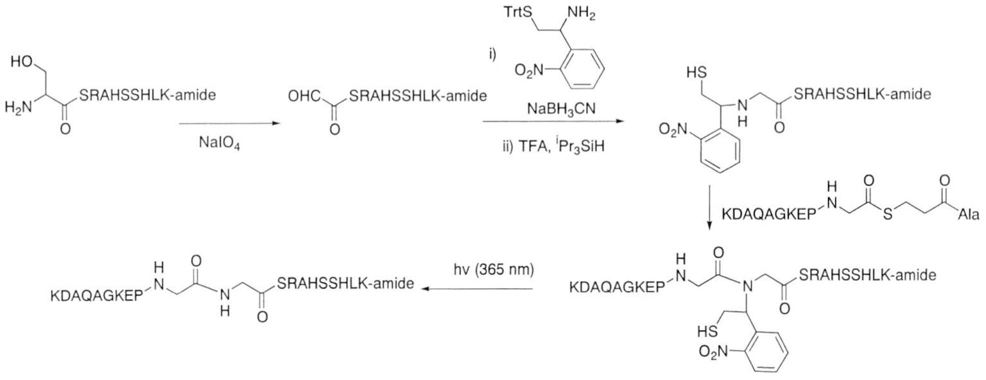

Scheme 3.14 Photoremovable scaffold for acyl transfer.

Cyclic peptides The synthesis of linear peptides generally proceeds well, but head-to-tail cyclization is sometimes ineffective. Peptide bonds prefer to adopt a trans conformation, and peptide fragments prefer to be extended. This may cause the deleterious separation of the head and tail, which should be linked. To facilitate such a reaction, a photoremovable auxiliary carrying a hydroxyl group, the 2-hydroxy-6-nitrobenzyl (HnB) group, has been introduced at the N-terminus and in the middle of the peptide amide [17, 18]. As the N-alkylation of the amide bonds lowers the trans-cis amide bond barrier, the auxiliary promotes cis amide bonds to facilitate the ring contraction. Activation of the C-terminus results in the formation of a cyclic ester. An O-N-acyl transfer results in a ring formation, and the auxiliary is removed by UV irradiation to give the cyclic peptide. Following this route, an all-L cyclic tetrapeptide was successfully prepared.

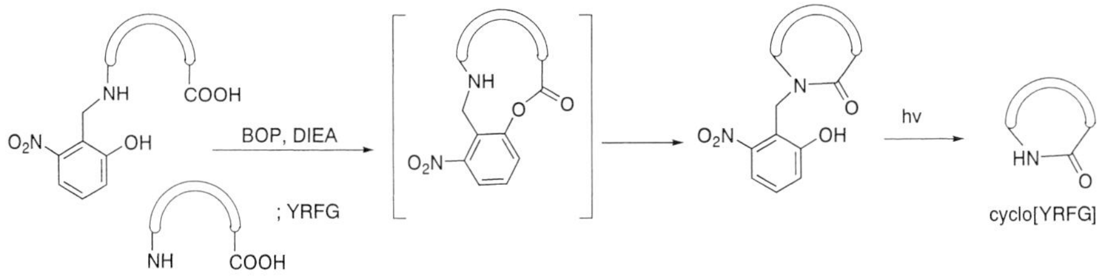

Scheme 3.15 Photolysis of 2-nitrobenzyl group

3.2.2.2 Photolysis of 2-Nitrobenzyl Groups Bound to Peptides

The photolytic process of the 2-nitrobenzyl group is thought to proceed through a mechanistic pathway that includes: (i) a photoinduced oxygen transfer from the nitro group to the benzylic carbon position and the formation of the *aci*-nitro compound, and (ii) a hemiacetal decomposition and production of a decaged compound.

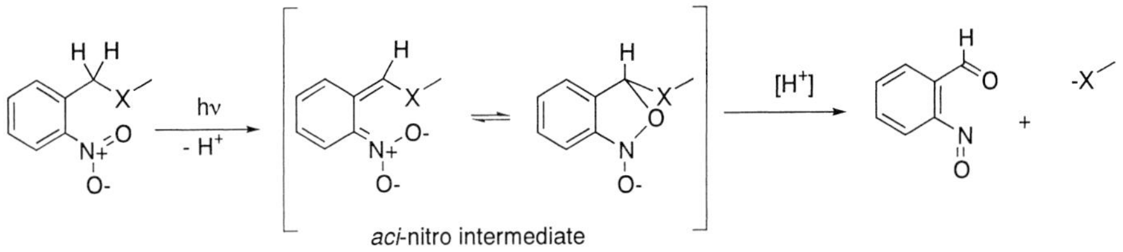

Scheme 3.16 Photolysis of the nitrobenzyl moiety.

While the completion of the photocleavage is required in the chemical syntheses mentioned above, rapidity is also required in caged compounds. Temporal analysis of the biological response upon photolysis of caged compounds requires the rapid release of the intact compound. The rate of decaging can be estimated using the flash photolysis technique. Irradiation of the 2-nitrobenzyl group with pulsed UV light causes a rapid increase in the absorption at $\lambda = 420$ nm, which can be ascribed to the formation of the *aci*-nitro intermediate, and the decay curve of the absorption corresponds to the decomposition of the *aci*-nitro intermediate. The dependence of the photolysis on the residue to which the 2-nitrobenzyl group is bound has been examined for model peptides (see Tab. 3.2) [19]. The decomposition rates of the exiting groups were in the order: $HOCH_2 > NHCO > HOC_6H_5 > HSCH_2 > HOOCNH$. The lower the pH in the reaction medium, the faster was the rate observed for all the model peptides studied, showing that acetal decomposition catalyzed by acid is involved in the

Tab. 3.2 Photolysis of 2-nitrobenzyl group bound to peptide

R-	$\tau_{1/2}$ (s)
N-a-Z(NO$_2$)-Ala-	0.61
N-a-Z(NO$_2$)-Gly-	0.52
N-a-Ac-N-ε-Z(NO$_2$)Lys-	0.27
N-a-Ac-S-nitrobenzyl-Cys-	0.17
N-a-Ac-O-nitrobenzyl-Tyr-	0.026
N-a-Ac, N-a-2-nitrobenzyl-Gly-	0.0073
N-a-Ac-O-nitrobenzyl-Ser-	0.0061

The photolabile residue, R-, is attached to 12mer peptide, GGPPPPPPPPPP-amide.
$\tau_{1/2}$, half-life of *aci*-nitro intermediate, was measured for the peptide dissolved in sodium phosphate buffer (0.05 M, pH 7.0).

photolysis. The rate and mechanism of decaging in the case of peptides is similar to those of other caged compounds.

3.3
Synthesis of Caged Peptides

3.3.1
Design of Caged Peptides

One of the indispensable properties for useful caged peptides is that the caged peptide precursor should be biologically inactive, or at least several orders of magnitude less active than the photolysis product, since the caged peptides are added prior to photolysis.

There have been a number of reports concerning site-directed mutagenesis in biologically active peptides, and the relationship between the amino acid sequence and the activity has been elucidated. The amino acid residues that are not involved in the activity can be used for the introduction of probes, such as biotin and fluorescent probes. The amino acid residues that are involved in the activity can be used for the introduction of the caging group, which is expected to interrupt the interaction between the peptide and the target biomolecules. When there is no such information about the structure-activity relationship of the peptide, it is necessary to survey the residue before synthesis of the caged derivatives. A sequential substitution, such as an alanine scan, in which each amino acid is substituted with alanine, may be useful to reveal the sequence-activity relationship. A caged amino acid scan is a more direct method. However, at present, there is a limitation in the introduction of a caging group into amino acid residues, such as Ser, Lys, Arg, Tyr, Cys, and N-α-amino residues. Furthermore, plural residues are usually involved in the biological activity of peptides. Therefore, it is not always the case that the introduction of a caging group into biologically active peptides gives the corresponding caged peptide whose biological activity is satisfactorily decreased.

3.3.2
Case Studies on Solid-Phase Synthesis of Caged Peptides

There have been several methods reported for the synthesis of caged peptides: solution-phase synthesis, chemical modification, and *in vitro* translation using amber *t*-RNA. The advantages of the solid-phase process are: rapidity of synthesis, selective incorporation of caged amino acids, and accessibility for biochemists. Scheme 3.17 summarizes the building blocks for caged amino acid residues. By following the standard protocol of the Fmoc solid-phase synthesis described above, caged peptides can be obtained. The drawbacks of the solid-phase synthesis are, however, limitation in the length of peptide sequences and insufficiency of the set of building blocks for caged amino acid residues. Therefore, synthesis of caged peptides should be undertaken using the appropriate method.

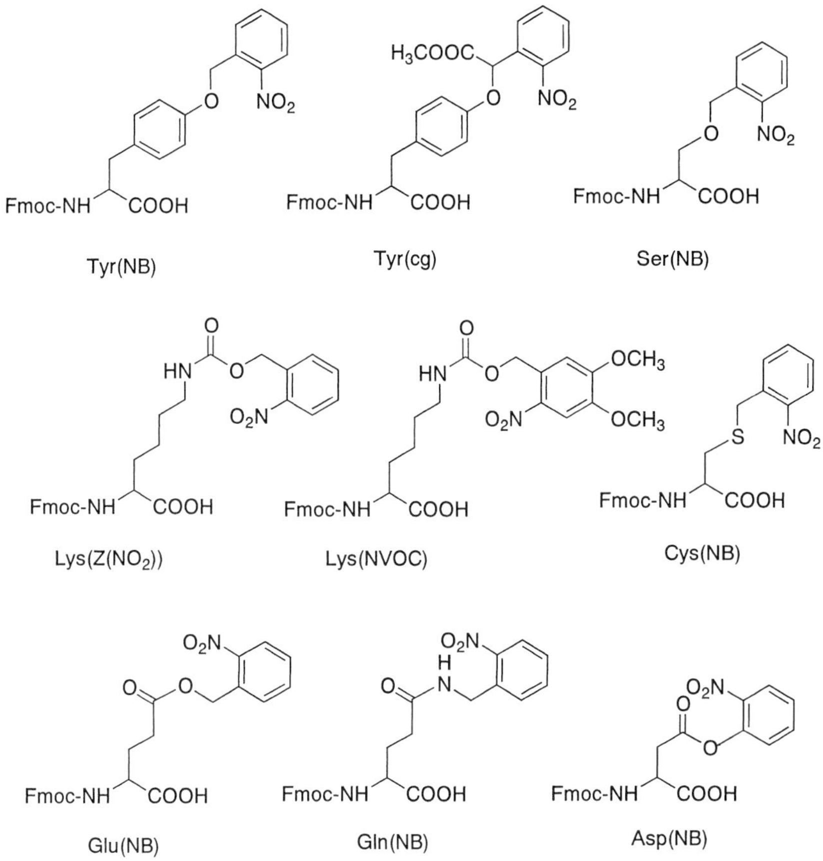

Scheme 3.17 Building block for caged peptides.

Although the synthesis of caged peptides is thought to be a promising route for the study of their function in living cells, there have been only a few reports in the literature on caged peptides. Caged peptides, which were prepared by Fmoc solid-phase peptide synthesis, are summarized in Tab. 3.3.

3.3.2.1 Neuropeptide Y [20]

Neuropeptide Y (NPY) is a 36 amino acid polypeptide found in both the central and peripheral nervous systems. NPY has been suggested in a wide variety of potential roles, including regulation of blood pressure, anxiety, circadian rhythms, and in feeding behavior. NPY contains tyrosine residues at both the N- and C-termini. These residues are essential for the activation of the NPYY1-type receptor. The introduction of a nitrobenzyl group at the N- or C-terminal tyrosine residue reduces the activity by more than one order of magnitude. The extent of the reduction was almost the same for each terminus, and an additive ef-

3 Caged Compounds and Solid-Phase Synthesis

Tab. 3.3 Caged peptides synthesized by solid phase synthesis

Peptide	Sequence	Caged residue	Biological activity	Refs.
Neuro-peptide Y	YPSKPDNPGEDAPA EDLARYYSALRHYI NLITRQRY-amide	Intact	$IC_{50} = 2$ nM	20
		Tyr(NB)-1	$IC_{50} = 100$ nM	
		Tyr(NB)-36	$IC_{50} = 55$ nM	
		Tyr(NB)-1,36	$IC_{50} = 1000$ nM	
			Y1 receptor binding	
RS20	ARRKYQKTGHAVRAIGRLSS	Intact	$IC_{50} = 2$ μM	21
		Tyr(cg)-5	$IC_{50} = 100$ μM	
			MLCK activity	
AIP	KKALRRQEAVDAL	Intact	$IC_{50} = 3.2 \times 10^{-8}$ M	22
		Lys(Z(NO$_2$))-1,2	$IC_{50} = 1.2 \times 10^{-6}$ M	
			Inhibitory activity	
NLS	GGGPKKKRKVGGGC	Intact	microscope observation	23
		Lys(NVOC)-5		
S-peptides	KETAAAKFERQH-NIe-DS	Intact	$V_i = 1.02 \times 10^{-5}$ M/min	24
		Lys(Z(NO$_2$))-1	$V_i = 1.01 \times 10^{-5}$ M/min	
		Glu(NB)-2	$V_i = 0.96 \times 10^{-5}$ M/min	
		Lys(Z(NO$_2$))-7	$V_i = 1.10 \times 10^{-5}$ M/min	
		Glu(NB)-9	$V_i = 0.99 \times 10^{-5}$ M/min	
		Gln(NB)-11	$V_i = 0$ M/min	
		Asp(NB)-14	$V_i = 0.29 \times 10^{-5}$ M/min	
			activity for reconstituted RNase	
Speract	GFDLSGGGVG	Intact	$IC_{50} = 0.67$ nM	25
		N-NB-Gly-6	$IC_{50} = 950$ nM	
	GYDLSGGGVG	Tyr-2 as intact	$IC_{50} = 2.78$ nM	
		Tyr(NB)-2	$IC_{50} = 179$ nM	
	GFDLSGGGVG	intact	$IC_{50} = 1.05$ nM	
		Ser(NB)-7	$IC_{50} = 71.8$ nM	
			receptor binding	
14-3-3 Binding peptide	Ac-RL-Dana-R-pSLPA-amide	pS	$K_D = 700 \pm 80$ nM	26
		cpS	>100 fold worse binding to 14-3-3 protein	

fect was observed upon the introduction of nitrobenzyl tyrosine at both termini. UV irradiation of the nitrobenzyl derivatives of NPY resulted in the recovery of their binding affinity for the NPYY1-type receptor.

3.3.2.2 RS-20 [21]

Another type of building block for Tyr is the α-carboxy-2-nitrobenzyl group (cg). RS-20, a 20-amino acid calcium-calmodulin binding domain of smooth muscle myosin light-chain kinase (MLCK), binds tightly to calcium-bound calmodulin (CaM) and inhibits the signaling function in cell motility. A structure-activity study showed that substitution of Trp-5 and Leu-18 with Glu reduced this activity. The substitution of Trp-5 residues with Tyr(cg) resulted in inactivation. Photolysis resulted in an increase in activity of about 50-fold. However, the substitution of both Trp-5 and Leu-18 with Tyr(cg) resulted in inactive peptides both before and after photolysis. With use of caged RS-20, it was indicated that both the action of calcium-calmodulin and MLCK, and by inference myosin II, are required for the amoeboid locomotion of eosinophil cells.

3.3.2.3 AIP [22]

A 13-amino acid peptide, AIP (autocamtide-2-related inhibitory peptide), was found to be a highly specific inhibitory peptide of CaMKII. A previous study indicated the importance of the first two lysine residues in the inhibitory activity, and IC_{50} values of des[Lys^{1,2}]AIP and AIP were 1.0×10^{-6} M and 3.2×10^{-8} M, respectively. Therefore, the Z(NO$_2$) group was introduced into a side chain of Lys^1 and Lys^2. The IC_{50} value for [Lys(Z(NO$_2$))^{1,2}]AIP (1.2×10^{-6} M) was found to be almost the same as that for des[Lys^{1,2}]AIP. By irradiation of the aqueous solution of the caged peptides with UV light, the inhibitory activity was restored to the same level as that of the intact AIP.

3.3.2.4 Nuclear Localization Signal [23]

Most nuclear protein import is mediated by a consensus sequence of basic amino acids called the nuclear localization signal (NLS). One lysine of NLS was substituted with Lys(NVOC), and the peptide was conjugated with fluorescent-labeled BSA, which enabled fluorescence observation. It was shown that following UV irradiation, the caged NLS conjugate translocated into and accumulated in the nucleus.

3.3.2.5 S-Peptides [24]

Ribonuclease S (RNase S) consists of two peptide fragments: an S-peptide (1–20) and an S-protein (21–124). The separated S-peptide readily rebounds to separated S-protein, and restores the native structure and activity. Although His-12 and His-19 are revealed to be active sites, caged His is unstable, and the other residues were substituted with caged amino acids. Six mutants, Lys(Z(NO$_2$))-1, Glu(NB)-2, Lys(Z(NO$_2$))-7, Glu(NB)-9, Gln(NB)-11, and Asp(NB)-14 were synthesized. The substitution at the 1, 2, 7, and 9 positions did not affect the activity of the reconstituted RNase, but substitution at the 11 and 14 positions resulted in a perfect suppression and a 30% suppression, respectively. The side chain of Gln-11 stabilizes the phosphate of an RNA substrate in the active site via a hydrogen bond, and it is presumed that the nitrobenzyl group in Gln-11 masks the active site.

3.3.2.6 Sperm-Activating Peptides [25]

Sperm-activating peptides (SAP or speract, GFDLQGGGVG) are diffusible components of the echinoderm egg jelly coat that show various biological effects on homologous spermatozoa. Speract transiently binds to its receptor on the sperm flagella activating a guanylate cyclase, and induces changes in the intracellular pH and Ca^{2+} ion concentration. Since the 3-D structure of the receptor bound with the ligand is not obtained, the structure-activity relationship of speract has been studied extensively. Gly-6 is considered to be essential for the activity of this peptide, because only this position is conserved among all the speract isoforms (>50) purified from various species of sea urchin. Substitution and deletion experiments showed that the activity of [Pro^6]speract is very low, suggesting that the 5–6 amide bond is important for receptor binding. The 2-nitrobenzyl group was therefore introduced at the amide bond site. The nitrobenzyl group was introduced as N-Fmoc-N-2′-nitrobenzyl-Gly. Acylation of the imine of N-2′-nitrobenzyl-Gly using PyBOP was unsuccessful, but acylation using acyl fluorides or carbodiimides proceeded to reaction. The backbone amide-caged speract showed a vastly reduced affinity for its receptor (IC$_{50}$=950 nM), and UV irradiation of the caged speract photocleaved the 2-nitrobenzyl group, restoring its affinity (IC$_{50}$=0.67 nM). Based on the conserved residues among the isoforms, the two caged peptides were obtained by substitution with Tyr(NB)-2 and Ser(NB)-7. However, the difference in the activity between the caged forms and decaged forms was less than two orders of magnitude. The advantage of using the backbone amide-caged peptide is that this strategy provides various sites for the introduction of a photocleavable group, whereas the side-chain-caged peptide has an obvious limitation in the variety of amino acid residues for the introduction of a photocleavable group.

3.3.2.7 Caged Phosphopeptides [27]

Phosphoproteins mediate key events in cell regulatory pathways. Imperialli et al. have developed a building block for caged phosphopeptides. Phosphotriesters are susceptible to β-elimination in basic conditions, and the cyanoethyl groups are removed during the Fmoc cleavage, but the resulting phosphate is stable.

While the benzyl group in standard phosphopeptide synthesis is removed by acid treatment, the 2-nitrophenyl-ethyl group is stable against a 90% TFA treatment, and thus a caged phosphopeptide-carrying photocleavable group at phosphate can be obtained.

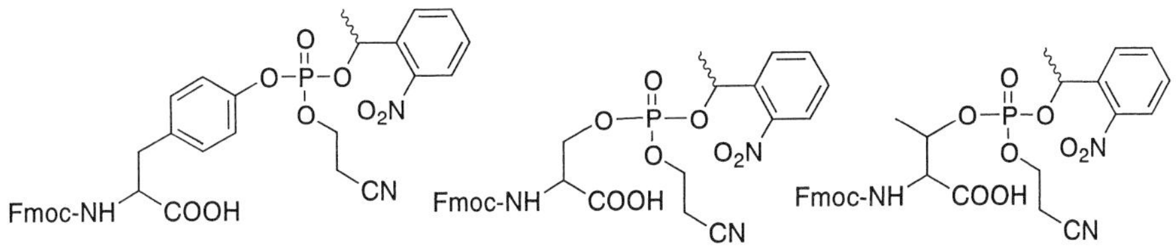

Scheme 3.18 Building block for caged phosphopeptides.

A fluorescent probe was incorporated into the caged phosphopeptide of a 14-3-3 binding peptide, which has an affinity for the 14-3-3 protein through a phosphorylation-dependent protein-protein interaction [26]. The intact 14-3-3 binding peptide exhibited a strong fluorescence, while the caged peptide exhibited a weaker fluorescence.

3.3.3
Caged Peptides Using Other Synthetic Methods

The easiest and commonest way to prepare caged peptides and proteins is by the chemical modification of their reactive amino acid residues. There are two methods, depending on the type of side chains available: the amino-reactive chloroformate [28] and the thiol-reactive bromide [29, 30] methods. One disadvantage of these chemical modification methods is that it is difficult to control the position and the number of modified sites, because most peptides and proteins contain multiple reactive groups.

A nonsense codon suppression technique is the most sophisticated way to prepare caged peptides and proteins. Schultz's group established the *in vitro* method [31, 32], and Lester's group improved on it to apply the method *in vivo* [33, 34]. By using the nonsense codon suppression technique, photolabile amino acids can be incorporated at a desired site in any peptide and protein. However, it is not easy for many laboratories to synthesize tRNA linked to a photolabile amino acid, and preparation of large quantities of the product in a cell-free translation system is usually very difficult.

A solution phase synthesis is suited for the preparation of large quantities of peptides, but not suited for a long peptide sequence. Caged peptides with a short amino acid sequence have been prepared using a solution phase technique: a caged apoptosis inducer bearing a Z(NO$_2$) on the N-terminal amino group [10], and a caged chemotactic peptide bearing a nitroveratryl group on the N-formyl amide group [35].

Tab. 3.4 Caged peptides and proteins other than solid-phase synthesis

Caged compounds	Caged residue	Refs.
Chemical modification		
G-actin	Lys	28
Heavy meromyocin	Cys	29
C-kemptide	Cys, thiophosphoryl	30
Kinase inhibitor	Arg	10
Nonsense codon suppression		
p21 ras protein	Asp	31
DNA polymerase	Ser	32
Acetylcholine receptor & K$^+$ channel	Gly	33
Acetylcholine receptor	Tyr	34
Solution phase synthesis		
Apoptosis inducer	N-terminal	2
fMLF peptide	N-fMet	35

3.4
Synthesis of Other Photoactive Biomolecular Compounds

3.4.1
Thioxo Peptides

Thioxo peptides, in which a thioamide linkage replaces a backbone amide linkage, have received attention recently, because they provide a way to synthesize analogs other than substitution with the 20 natural amino acids, which may be useful to increase enzymatic stability and activity, or selectivity of peptides. Although thioamides are a near isosteric replacement for amides, their absorption is higher than that of amides, and the photoswitch effect in cis-trans isomerization of an amide bond has been reported [36]. The thioxo bond of Tyr in Tyr-Pro-Xaa-Phe-amide peptide was introduced using a building block of thioxylated tyrosyl-6-nitrobenzotriazolide. The peptides were cleaved with SnCl$_4$ or

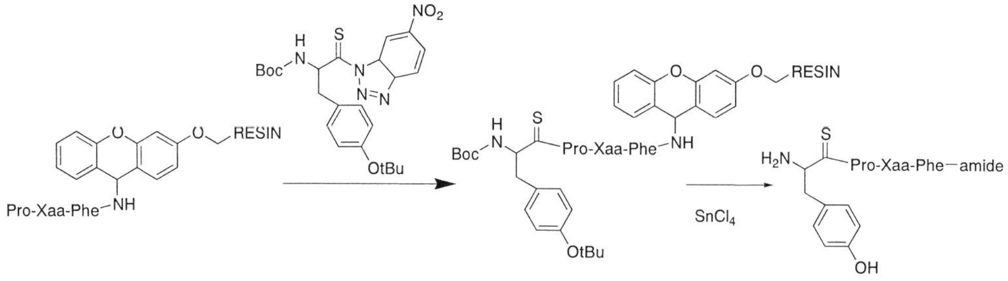

Scheme 3.19 Thioxo peptide.

ZnCl$_2$Et$_2$O to avoid acidolysis. Irradiation of the peptide at $\lambda = 337$ nm resulted in an increase in the cis content of the Tyr-φ[CS-N]-Pro bond. The photoisomerization was reversible without any photodecomposition.

3.4.2
Photochromic Peptides

A number of studies have been carried out to control the conformation of a polypeptide structure bearing photochromic moieties [37]. Most of the polypeptides studied were synthesized in the solution phase. Photocontrol of the lytic activity of melittin-carrying spiropyran has also been reported [38].

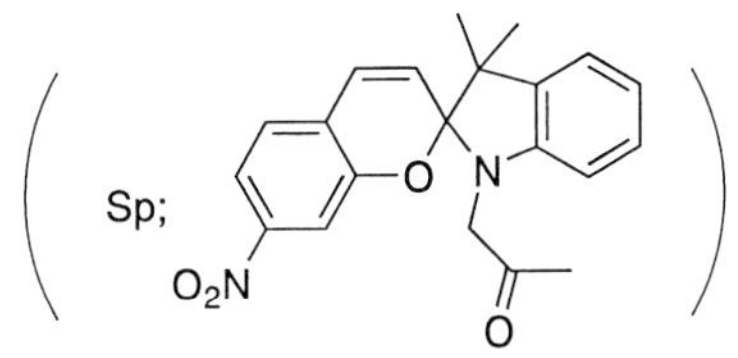

Scheme 3.20 Photoresponsive melittin.

3.4.3
Photoaffinity Peptides

Photoaffinity reagents are not categorized with caged compounds, but they are also photolabile compounds and useful tools to elucidate biomolecular interactions. Various photoaffinity peptides have been prepared more extensively than caged peptides. While the biological activity of a caged peptide is designed to be diminished in the photoaffinity peptide, the photolabile moiety is introduced on a site that does not affect biological activity. The photolabile moiety generates a reactive intermediate, such as carbene, upon photo-irradiation. The intermediate attacks an acceptor and forms a covalent bond. The acceptor may be a target biomolecule, such as a receptor that is in contact with the peptide, and thus the target biomolecule is conjugated with the peptide. The photoaffinity peptide generally has a probe, such as a radioisotope atom or biotin, to track the conjugate. Scanning the site of a photolabile moiety in the peptide and digestion of

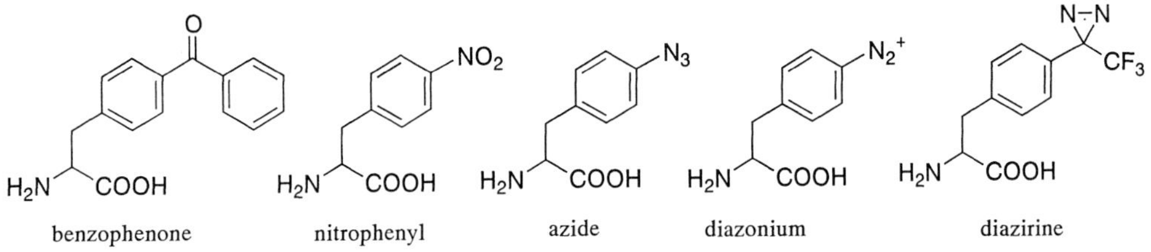

Scheme 3.21 Photoaffinity of amino acid residues.

Tab. 3.5 Photoaffinity peptides

Photoaffinity residue	Peptide	Synthesis	Refs.
Benzophenone			
Bpa	Nociceptin	Fmoc SPPS	39
Bpa	Parathyroid hormone	Boc SPPS	40
HO-Bpa	Substance P	Fmoc SPPS	41
Fla	Parathyroid hormone	Boc SPPS	42
Nitroaryl			
Phe(4-NO$_2$)	Substance P	Boc SPPS	43
Trp(6-NO$_2$)	CCK	Fmoc SPPS	44
Diazonium			
Phe(4-N$_2^+$)	Substance P	post-SPPS	43
Azide			
Phe(4-N$_3$)	Substance P	post-SPPS	43
Diazirine			
(Tmd)Phe	Neuropeptide Y	Fmoc SPPS	45

Bpa: Benzoylphenylalanine, HO-Bpa: (4-hydroxyl-benzoyl)phenylalanine
Fla: Fluorenyloxoalanine, (Tmd)Phe: 4-(3-trifluoromethyl)-3H-diazirin-3-ylphenylalanine

photo-conjugates may provide information on the contact sites between the peptide and receptor.

Benzophenone is extensively used in photoaffinity studies. UV irradiation results in an n–π* transition and generates a biradical. The biradical reacts with an accessible hydrocarbon and forms a new C–C bond. The reactivity is high against electron-rich centers such as the Cγ–H of Leu and the Cβ–H of Val. When there is no accessible hydrocarbon within the reactive volume, then the radical readily relaxes down to the ground state. Benzophenone is chemically stable under Boc and Fmoc chemistry, and several analogs have been developed and used in SPPS [39–42].

4-Nitrophenylalanine was originally used as a non-photoactivatable compound, but with the use of a strong light source it decomposes under a multiphoton process and decays into reactive intermediates. 4-Nitrophenylalanine is stable under Boc and Fmoc chemistry, and can be incorporated into peptides using SPPS [43]. 6-Nitrotryptophan has also been used as a photoaffinity probe [44].

The nitro group of nitrophenylalanine can also be used as a precursor for other photoactivatable groups. The nitro group is reduced to an amino group and is then converted to an azide group. Irradiation of phenyl azide generates nitrene. The nitrene prefers insertion into X–H bonds. The diazonium salt can be also prepared from a 4-nitrophenylalanine-containing peptide. Photo-irradiation generates an aryl cation, which inserts into the X–H bond.

Diazirines are photolyzed to generate carbenes. Carbene is more reactive than nitrene and results in a high reactivity with water. Diazirine is chemically stable toward strong acids and bases, and a residue possessing a diazirine group can be incorporated into peptides using SPPS [45].

3.5
Conclusions and Perspective

Caged peptides are promising tools for characterizing complex biological processes, but there are at least two important considerations in using them effectively. One is that a photolabile protecting group should be incorporated exactly into the active site to provide a caging effect. At present, it is very difficult to predict active sites in peptides solely from their amino acid sequences. Therefore, before the preparation of caged molecules, their structure-activity relationships should be clarified. The second consideration is to provide a complete set of photocleavable protections for the various amino acid residues in peptides and proteins. These procedures should be kept as simple as possible.

References

1 A. Patchornik, B. Amit, R. B. Woodward, *J. Am. Chem. Soc.* **1970**, *92*, 6333–6335.
2 M. Odaka, T. Furuta, Y. Kobayashi, M. Iwamura, *Biochem. Biophys. Res. Commun.* **1995**, *213*, 652–656.
3 V. K. Rusiecki, S. A. Warne, *Bioorg. Med. Chem. Lett.* **1993**, *3*, 707–710.
4 J. W. Chamberlin, *J. Am. Chem. Soc.* **1965**, *31*, 1658–1660.
5 G. Church, J. M. Ferland, J. Gauthier, *Tetrahedron Lett.* **1989**, *30*, 1901–1904.
6 D. B. Henriksen, K. Breddam, O. Buchardt, *Int. J. Pept. Protein Res.* **1993**, *41*, 169–180.
7 M. C. Pirrung, D. S. Nunn, *Bioorg. Med. Chem. Lett.* **1992**, *2*, 1489–1492.
8 B. Amit, E. Hazum, M. Fridkin, A. Patchornik, *Int. J. Pept. Protein. Res.* **1977**, *9*, 91–96.
9 S. M. Kalbag, R. W. Roeske, *J. Am. Chem. Soc.* **1975**, *97*, 440–441.
10 J. S. Wood, M. Koszelak, J. Liu, D. S. Lawrence, *J. Am. Chem. Soc.* **1998**, *120*, 7145–7146.
11 D. H. Rich, S. K. Gurwara, *J. Am. Chem. Soc.* **1975**, *97*, 1575–1579.
12 C. P. Holmes, *J. Org. Chem.* **1997**, *62*, 2370–2380.
13 F. S. Tjoeng, E. K. Tong, R. S. Hodges, *J. Org. Chem.* **1978**, *43*, 4190–4194.
14 C. K. Jayawickreme, H. Sauls, N. Bolio, J. Ruan, J. Moyer, W. Burkhart, B. Marron, T. Rimele, J. Shaffer, *J. Pharmacol. Toxicol.* **1999**, *42*, 189–197.
15 R. S. Rock, S. I. Chan, *J. Org. Chem.* **1996**, *61*, 1526–1529.
16 T. Kawakami, S. Aimoto, *Tetrahedron Lett.* **2003**, *44*, 6059–6061.
17 W. D. F. Meutermans, G. T. Bourne, S. W. Golding, D. A. Horton, M. R. Campitelli, D. Craik, M. Scanlon, M. L. Smythe, *Org. Lett.* **2003**, *5*, 2711–2714.
18 W. D. F. Meutermans, S. W. Golding, G. T. Bourne, L. P. Miranda, M. J. Dooley, P. F. Alewood, M. L. Smythe, *J. Am. Chem. Soc.* **1999**, *121*, 9790–9796.
19 Y. Tatsu, Y. Endo, N. Yumoto, *Peptide Science* **2002**, 405–406.

20 Y. Tatsu, Y. Shigeri, S. Sogabe, N. Yumoto, S. Yoshikawa, *Biochem. Biophys. Res. Commun.* **1996**, *227*, 688–693.
21 J.W. Walker, S.H. Gilbert, R.M. Drummond, M. Yamada, R. Sreekumar, R.E. Carraway, M. Ikebe, F.S. Fay, *Proc. Natl. Acad. Sci. USA* **1998**, *95*, 1568–1573.
22 Y. Tatsu, Y. Shigeri, A. Ishida, I. Kameshita, H. Fujisawa, N. Yumoto, *Bioorg. Med. Chem. Lett.* **1999**, *9*, 1093–1096.
23 Y. Watai, I. Sase, H. Shiono, Y. Nakano, *FEBS Lett.* **2001**, *488*, 39–44.
24 T. Hiraoka, I. Hamachi, *Bioorg. Med. Chem. Lett.* **2003**, *13*, 13–15.
25 Y. Tatsu, T. Nishigaki, A. Darszon, N. Yumoto, *FEBS Lett.* **2002**, *525*, 20–24.
26 M.E. Vazquez, M. Nitz, J. Stehn, M.B. Yaffe, B. Imperiali, *J. Am. Chem. Soc.* **2003**, *125*, 10150–10151.
27 D.M. Rothman, M.E. Vazquez, E.M. Vogel, B. Imperiali, *J. Org. Chem.* **2003**, *68*, 6795–6798.
28 G. Marriott, *Biochemistry* **1994**, *33*, 9092–9097.
29 G. Marriott, M. Heidecker, *Biochemistry* **1996**, *35*, 3170–3174.
30 P. Pan, H. Bayley, *FEBS Lett.* **1997**, *405*, 81–85.
31 S.K. Pollitt, P.G. Schultz, *Angew. Chem., Int. Ed.* **1998**, *37*, 2104–2107.
32 S.N. Cook, W.E. Jack, X.F. Xiong, L.E. Danley, J.A. Ellman, P.G. Schultz, C.J. Noren, *Angew. Chem., Int. Ed. Engl.* **1995**, *34*, 1629–1630.
33 P.M. England, H.A. Lester, N. Davidson, D.A. Dougherty, *Proc. Natl. Acad. Sci. USA* **1997**, *94*, 11025–11030.
34 J.C. Miller, S.K. Silverman, P.M. England, D.A. Dougherty, H.A. Lester, *Neuron* **1998**, *20*, 619–624.
35 M.C. Pirrung, S.J. Drabik, J. Ahamed, H. Ali, *Bioconjugate Chem.* **2000**, *11*, 679–681.
36 R. Frank, M. Jakob, F. Thunecke, G. Fischer, M. Schutkowski, *Angew. Chem., Int. Ed.* **2000**, *39*, 1120–1122.
37 O. Pieroni, A. Fissi, *J. Photochem. Photobiol. B-Biol.* **1992**, *12*, 125–140.
38 T. Ueda, K. Nagamine, S. Kimura, Y. Imanishi, *J. Chem. Soc.-Perkin Trans. 2* **1995**, 365–368.
39 C. Nakamoto, V. Behar, K.R. Chin, A.E. Adams, L.J. Suva, M. Rosenblatt, M. Chorev, *Biochemistry* **1995**, *34*, 10546–10552.
40 L. Mouledous, C.M. Topham, H. Mazarguil, J.C. Meunier, *J. Biol. Chem.* **2000**, *275*, 29268–29274.
41 C.J. Wilson, S.S. Husain, E.R. Stimson, L.J. Dangott, K.W. Miller, J.E. Maggio, *Biochemistry* **1997**, *36*, 4542–4551.
42 Y. Han, A. Bisello, C. Nakamoto, M. Rosenblatt, M. Chorev, *J. Pept. Res.* **2000**, *55*, 230–239.
43 E. Escher, R. Couture, G. Champagne, J. Mizrahi, D. Regoli, *J. Med. Chem.* **1982**, *25*, 470–475.
44 U.G. Klueppelberg, H.Y. Gaisano, S.P. Powers, L.J. Miller, *Biochemistry* **1989**, *28*, 3463–3468.
45 N. Ingenhoven, C.P. Eckard, D.R. Gehlert, A.G. Beck-Sickinger, *Biochemistry* **1999**, *38*, 6897–6902.

4
Control of Cellular Activity

4.1
Photochemical Release of Second Messengers – Caged Cyclic Nucleotides

Volker Hagen, Klaus Benndorf, and U. Benjamin Kaupp

4.1.1
Introduction

The cyclic nucleoside monophosphates (cNMPs) adenosine 3′,5′-cyclic monophosphate (cAMP) and guanosine 3′,5′-cyclic monophosphate (cGMP) control a variety of important cellular processes (for reviews see [1–3]). Their principal targets are cAMP- and cGMP-regulated phosphodiesterases (PDEs), cAMP- and cGMP-dependent kinases (cAKs and cGKs), cyclic nucleotide-gated (CNG) ion channels, hyperpolarization-activated and cyclic nucleotide-gated (HCN) channels, guanosine-nucleotide exchange factors (GEF), and bacterial transcription factors (CAP).

Caged compounds (biologically inert precursors) of cNMPs are very useful for studies of spatial- and time-dependent aspects of signaling pathways because these compounds enable one to produce rapid concentration jumps of cNMPs inside cells [4–10]. cAMP or cGMP are rendered biologically inactive (caged) by esterification of the free phosphate moiety with a photolabile protecting or caging group. The esterification yields axial and equatorial diastereomers.

The phototriggers must meet specific requirements. They should be sufficiently soluble in aqueous solutions (although sometimes membrane permeability is required), stable toward solvolysis (the ester bond may undergo hydrolysis in aqueous medium), and biologically inert. Furthermore, they should undergo fast and highly efficient photochemical reactions using long-wavelength UV/Vis activation. A high two-photon uncaging cross section would also be useful.

The most important caging groups for cAMP and cGMP are 2-nitrobenzyl and coumarinylmethyl derivatives. These two caging groups were also used for caging 8-bromo- and 8-(4-chlorophenylthio)-substituted cNMPs (8-Br-cAMP, 8-Br-cGMP, 8-pCPT-cAMP and 8-pCPT-cGMP). These substituted cNMPs are valuable derivatives because they are poorly hydrolyzed by phosphodiesterases [11,

Dynamic Studies in Biology. Edited by M. Goeldner, R. Givens
Copyright © 2005 WILEY-VCH Verlag GmbH & Co. KGaA, Weinheim
ISBN: 3-527-30783-4

12] and often display a higher biological efficacy [11–13]. Therefore, activation or inhibition of cAMP- or cGMP-dependent targets requires lower concentrations of cyclic nucleotides. This means that less intense light flashes are sufficient, which minimizes cell damages by UV light.

4.1.2
Overview of Phototriggers for cNMPs

4.1.2.1 *ortho*-Nitrobenzyl Derivatives

The 2-nitrobenzyl (NB) ester of cAMP, the first caged cNMP, was introduced by Engels and Schlaeger in 1977 [14]. Later the NB ester of cGMP [15], the 4,5-dimethoxy-2-nitrobenzyl (DMNB) [16] and 1-(2-nitrophenyl)ethyl (NPE) [17] esters of cAMP and cGMP, the 1-(4,5-dimethoxy-2-nitrophenyl)ethyl (DMNPE) ester of cAMP [17], and the 4,5-bis(carboxymethoxy)-2-nitrobenzyl ester (BCMNB) of cGMP (for this compound no physico-chemical data are available) [18] were described. Only recently, the 4-[N,N-bis(carboxymethyl)carbamoyl]-2-nitrobenzyl (BCMCNB) esters of cAMP and cGMP have been introduced [19]. DMNB and NPE esters were also prepared for 8-Br-cAMP and 8-Br-cGMP [20–22] as well as for 8-pCPT-cAMP and 8-pCPT-cGMP [23, 24] (Scheme 4.1.1).

The synthesis of caged compounds involves treatment of the free acids of cNMPs with the corresponding substituted nitrobenzyl diazoalkanes (yields about 50%, diastereomer ratios axial:equatorial ~ 55:45). However, the BCMCNB-caged compounds were synthesized by esterification of the tri-*n*-butylammonium salts of the cNMPs with 4-{N,N-bis-[2-(trimethylsilyl)ethoxycarboxymethyl]carbamoyl}-2-nitrobenzylbromide and subsequent deprotection with trifluoroacetic acid. The yield for the synthesis of BCMCNB-caged cAMP is 25–30% (diastereomer ratios axial:equatorial ~ 3:1). The diastereomers of BCMCNB-caged cGMP were formed in a similar ratio, but with much lower yield. Diastereomer separation was accomplished using preparative reverse-phase HPLC.

Irradiation of the nitrobenzyl esters of cNMPs by UV light in aqueous buffer results in the liberation of the free cyclic nucleotide, a 2-nitrosobenzaldehyde or 2-nitrosoacetophenone, and a proton (Scheme 4.1.2). The photolysis reaction pathway includes an *aci*-nitro intermediate and a cyclic benzisoxazoline intermediate (for details see [10, 25, 26]). The mechanism causes a relatively slow rate of photorelease of cNMPs. For example, the nitrobenzyl-caged cNMPs photolyze under physiological conditions with uncaging rate constants between 3000 s^{-1} (DMNB-caged derivatives) and 1.7 s^{-1} (BCMCNB-caged derivatives) [10, 19]; thus, the cNMPs are released within 0.33 and 590 ms, respectively. Therefore, nitrobenzyl caging groups are unsuitable for studying fast reactions on a millisecond or sub-millisecond time scale.

NPE and DMNB esters of cAMP and cGMP are superior to the corresponding NB derivatives with respect to their photochemical properties (higher quantum yields, higher rate of photorelease) [10]. The DMNPE ester of cAMP (the cGMP analog was not prepared) hydrolyzes so rapidly in aqueous buffer at pH 7 that it is of no practical use in biological research [17]. So the NPE-, DMNB- and

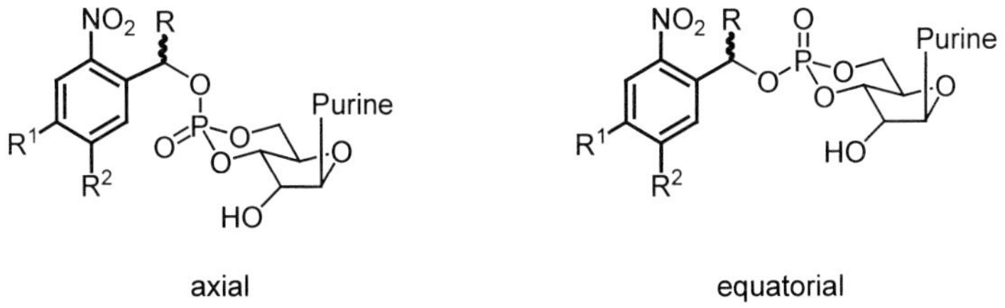

4.1 Photochemical Release of Second Messengers – Caged Cyclic Nucleotides

axial   equatorial

NB-caged cAMPs: Purine = adenin-9-yl, R = H, R^1 = R^2 = H
DMNB-caged cAMPs: Purine = adenin-9-yl, R = H, R^1 = R^2 = OCH$_3$
DMNB-caged cGMPs: Purine = guanin-9-yl, R = H, R^1 = R^2 = OCH$_3$
DMNB-caged 8-Br-cAMPs: Purine = 8-bromoadenin-9-yl, R = H, R^1 = R^2 = OCH$_3$
DMNB-caged 8-Br-cGMPs: Purine = 8-bromoguanin-9-yl, R = H, R^1 = R^2 = OCH$_3$
DMNB-caged 8-pCPT-cAMPs: Purine = 8-pCPT-adenin-9-yl, R = H, R^1 = R^2 = OCH$_3$
DMNB-caged 8-pCPT-cGMPs: Purine = 8-pCPT-guanin-9-yl, R = H, R^1 = R^2 = OCH$_3$
NPE-caged cAMPs: Purine = adenin-9-yl, R = CH$_3$, R^1 = R^2 = H
NPE-caged cGMPs: Purine = guanin-9-yl, R = CH$_3$, R^1 = R^2 = H
NPE-caged 8-Br-cAMPs: Purine = 8-bromoadenin-9-yl, R = CH$_3$, R^1 = R^2 = H
NPE-caged 8-Br-cGMPs: Purine = 8-bromoguanin-9-yl, R = CH$_3$, R^1 = R^2 = H
NPE-caged 8-pCPT-cGMPs: Purine = 8-pCPT-guanin-9-yl, R = CH$_3$, R^1 = R^2 = H
DMNPE-caged cAMPs: Purine = adenin-9-yl, R = CH$_3$, R^1 = R^2 = OCH$_3$
DMNPE-caged cGMPs: Purine = guanin-9-yl, R = CH$_3$, R^1 = R^2 = OCH$_3$
BCMNB-caged cGMPs: Purine = guanin-9-yl, R = H, R^1 = R^2 = OCH$_2$COOH
BCMCNB-caged cAMPs: Purine = adenin-9-yl, R = R^2 = H, R^1 = CON(CH$_2$COOH)$_2$
BCMCNB-caged cGMPs: Purine = guanin-9-yl, R = R^2 = H, R^1 = CON(CH$_2$COOH)$_2$

adenin-9-yl: X = H
8-bromoadenin-9-yl: X = Br
8-pCPT-adenin-9-yl: X = p-chlorophenylthio

guanin-9-yl: X = H
8-bromoguanin-9-yl: X = Br
8-pCPT-guanin-9-yl: X = p-chlorophenylthio

Scheme 4.1.1 Structures of the axial and equatorial isomers of nitrobenzyl-caged cNMPs.

BCMCNB-caged derivatives are the most important nitrobenzyl-caged cNMPs. Their chemical and photophysical properties in aqueous buffer are shown in Tab. 4.1.1. This table illustrates that the NPE- and BCMCNB-caged derivatives have high photochemical quantum yields, but their low absorptions at wavelengths >300 nm (see also Fig. 4.1.1) reduce the efficiency of photorelease at long-wavelength irradiation. The DMNB-caged compounds absorb at longer wavelengths (λ_{max} at 345 nm). However, their photochemical quantum yields are about 100- or 1000-fold smaller than those of the NPE-caged cNMPs, resulting in relatively low photoefficiencies. There are only small differences in photoefficiencies between axial and equatorial isomers. Taking the uncaging cross sections [these are the products of photolysis quantum yield (φ) and molar absorptivity (ε)] to compare the overall efficiency of the uncaging reaction, the NPE-, DMNB-, and BCMCNB-caged cNMPs are found to have relatively low values ($\varphi\varepsilon < 300$ M^{-1}cm^{-1}) at wavelengths >330 nm. As a consequence, relatively intense UV light is required to reach significant concentrations of the free

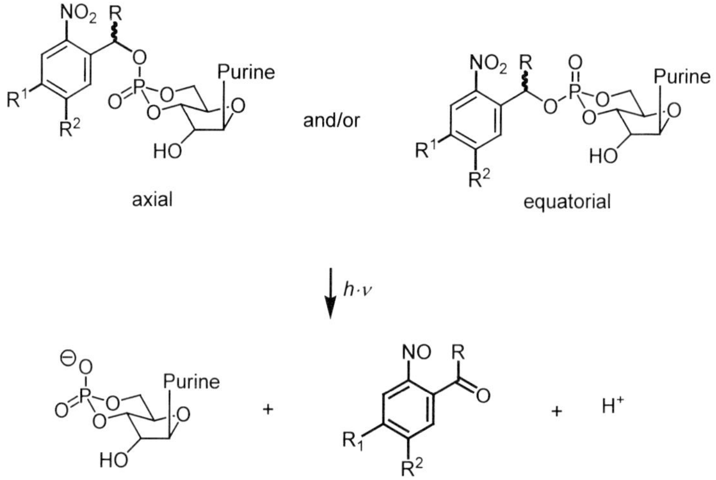

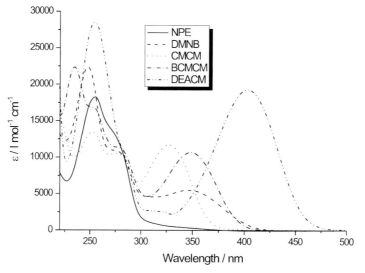

Scheme 4.1.2 Photolysis of the diastereomers of nitrobenzyl-caged cNMPs.

Fig. 4.1.1 UV/Vis spectra of the axial diastereomers of NPE-, DMNB-, CMCM-, BCMCM-, and DEACM-caged cGMPs.

cNMPs. Notwithstanding these unfavorable properties, NPE- and DMNB-caged cNMPs were successfully used in physiological studies (see Chapter 4.1.3).

Resistance of the DMNB-caged cNMPs to hydrolysis in aqueous buffer is relatively low. More resistant to hydrolysis are the NPE-caged compounds. Unexpectedly, the BCMCNB-caged compounds display an excellent hydrolytical stability. In all cases, the axial isomers are considerably more stable than the equatorial isomers.

Tab. 4.1.1 Long-wavelength absorption maximum λ_{max}, extinction coefficient ε at λ_{max}, photochemical quantum yield φ_{chem}, uncaging rate constant k, half-life $t_{1/2}$, and solubility s of selected 2-nitrobenzyl esters of cNMPs in aqueous buffer solutions, pH 7.0–7.2

Compound	λ_{max} (ε) [nm] ([M^{-1} cm^{-1}])	φ_{chem}	k [s^{-1}]	$t_{1/2}$ [h]	s [μM]	Refs.
DMNB-caged cAMP (axial)	345 (5800)[a]	0.05	>300[b]	24	>100[a]	10, 16, 17
DMNB-caged cAMP (equatorial)	346 (6000)[a]	0.05	>300[b]	4	100[a]	10, 16, 17
DMNB-caged cGMP (axial)	347 (5500)[a]	0.004	>3000[b]	43[a]	125[a]	10, 16, 17
DMNB-caged cGMP (equatorial)	347 (5500)[a]	0.003–0.009	>3000[b]	12[a]	n.d.	10, 16, 17
DMNB-caged 8-Br-cAMP (axial)	345 (6000)	0.05	n.d.[c]	60	120	20, 21
DMNB-caged 8-Br-cAMP (equatorial)	345 (6000)	0.05	n.d.[c]	8	80	20, 21
DMNB-caged 8-Br-cGMP (axial)	346 (5800)	0.005	n.d.[d]	50	100	20, 21
DMNB-caged 8-Br-cGMP (equatorial)	346 (5800)	0.005	n.d.[d]	5	10	20, 21
NPE-caged cAMP (axial)	259 (19300)[a]	0.39[b]	4.2[b]	231	160[b]	10, 17, 28
NPE-caged cAMP (equatorial)	260 (19300)[a]	0.39[b]	4.2[b]	39	160[b]	10, 17, 28
NPE-caged cGMP (axial)	255 (18300)[a]	n.d.[e]	n.d.[e]		225[a]	17
NPE-caged cGMP (equatorial)	255 (18900)[a]	n.d.[e]	n.d.[e]		n.d.	17
NPE-caged 8-Br-cAMP (axial)	265 (19500)	0.48[f]	n.d.[e]	600	225	22
NPE-caged 8-Br-cAMP (equatorial)	265 (19500)	0.49[f]	n.d.[e]	95	300	22
NPE-caged 8-Br-cGMP (axial)	264 (19400)	0.33[f]	n.d.[e]	300	35	22
NPE-caged 8-Br-cGMP (equatorial)	264 (19400)	0.27[f]	n.d.[e]	80	120	22
BCMCNB-caged cAMP (axial)	258 (21300)	0.16	1.7	3010	55000[g]	19
BCMCNB-caged cAMP (equatorial)	258 (21300)	n.d.[h]	n.d.[h]	1020	55000[g]	19
BCMCNB-caged cGMP (axial)	251 (20300)	0.24	1.7	3460	55000[g]	19
BCMCNB-caged cGMP (equatorial)	251 (20300)	n.d.[h]	n.d.[h]	990	55000[g]	19

a) V. Hagen, unpublished data. b) Measured for a mixture of the axial and equatorial isomer.
c) Likely to be similar to DMNB-caged cAMP. d) Likely to be similar to DMNB-caged cGMP.
e) Likely to be similar to NPE-caged cAMP. f) In MeOH-HEPES-KCl buffer (1:4), pH 7.2.
g) Solubility of the dipotassium salt. h) Likely to be similar to the axial isomer.

The DMNB- and NPE-caged cNMPs are hydrophobic substances and show limited aqueous solubility. In general the solubility of at least one of the diastereomers is, however, sufficient for their use in most biological systems. These compounds also permeate cell membranes. The BCMCNB-caged cNMPs and BCMNB-caged cGMP display very good aqueous solubility by virtue of two anionic centers. This allows the use of high concentrations of these caged cyclic nucleotides; therefore, it is expected that, upon photolysis, they give rise to spatially uniform increases in intracellular cNMPs [19].

A drawback of all nitrobenzyl-caged cNMPs is the production of highly reactive nitrosocarbonyl by-products during photolytic cleavage. These by-products may damage cells.

4.1.2.2 (Coumarin-4-yl)methyl Derivatives

Photoactivity of (7-methoxycoumarin-4-yl)methyl (MCM) esters was first reported in 1984 [27]. In the following years, MCM [28, 29], (7-hydroxycoumarin-4-yl)methyl- (HCM) as well as [7-(acyloxy)coumarin-4-yl]methyl (ACM) esters of cAMP were introduced as phototriggers for cAMP [29, 30], and MCM-caged cGMP [31], MCM-caged 8-Br-cAMP, and MCM-caged 8-Br-cGMP were described [32]. Recently, a number of coumarin-type cage analogs, including [7-(carboxymethoxy)coumarin-4-yl]methyl (CMCM) [33], (6,7-dimethoxycoumarin-4-yl)-methyl (DMCM) [34], [6,7-bis(carboxymethoxy)coumarin-4-yl]methyl (BCMCM) [33], [6,7-bis(ethoxycarbonylmethoxy)coumarin-4-yl]methyl (BECMCM) [35], [6,7-bis(methoxycarbonylmethoxy)coumarin-4-yl]methyl (BMCMCM) [24], [7-(diethylamino)coumarin-4-yl]methyl (DEACM) [33, 36], [7-(dimethylamino)coumarin-4-yl]methyl (DMACM) [34, 36], and (6-bromo-7-hydroxycoumarin-4-yl)methyl (BHCM, only for cAMP) [37, 38] have been employed to cage cNMPs. The structures of the phototriggers are shown in Scheme 4.1.3.

The coumarinylmethyl-caged cNMPs were synthesized by three different methods, involving the alkylation of the free acids of cNMPs with the substituted (4-diazomethyl)coumarins or the reaction of the respective 4-(bromomethyl)coumarins using either silver(I)oxide-activated cNMPs or the tetra-n-butylammonium salts of cNMPs. The reactions give a mixture of the axial and equatorial diastereomers. Using the diazo-based method (MCM, HCM, ACM, CMCM, BCMCM, BECMCM, DEACM, DMACM, BHCM esters), the overall yields are 7–35% (caged cAMPs, 8-Br-cAMPs, 8-Br-cGMPs as well as caged cGMPs; HCM, ACM and BHCM esters are only described for cAMP), and the diastereomers are formed in ratios of approximately 45:55 (axial/equatorial). Synthesis via the silver(I) oxide method gives yields of the caged cAMPs (MCM, HCM, ACM, DMCM esters) or caged 8-Br-cAMPs and 8-Br-cGMPs (MCM and DMCM esters) between 13 and 44%, and the diastereomer ratio is 65–55:35–45 (axial/equatorial). Only trace amounts of MCM- and DMCM-caged cGMPs were obtained using the silver(I) oxide method; HCM- and ACM-caged cGMPs were not described. The procedure involving tetrabutylammonium salts of the cNMPs (MCM, DMCM, DEACM esters) leads to the highest yields (30–90%) of the cor-

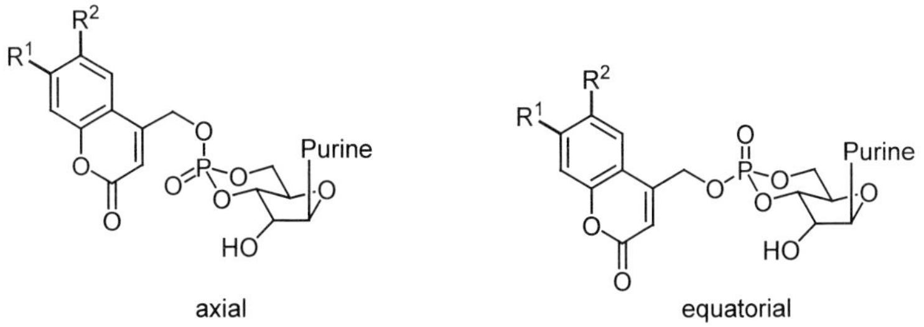

4.1 Photochemical Release of Second Messengers – Caged Cyclic Nucleotides

axial

equatorial

MCM-caged cAMPs: Purine = adenin-9-yl, R^1 = OCH_3, R^2 = H
MCM-caged cGMPs: Purine = guanin-9-yl, R^1 = OCH_3, R^2 = H
MCM-caged 8-Br-cAMPs: Purine = 8-bromoadenin-9-yl, R^1 = OCH_3, R^2 = H
MCM-caged 8-Br-cGMPs: Purine = 8-bromoguanin-9-yl, R^1 = OCH_3, R^2 = H
MCM-caged 8-pCPT-cAMPs: Purine = 8-pCPT-adenin-9-yl, R^1 = OCH_3, R^2 = H
MCM-caged 8-pCPT-cGMPs: Purine = 8-pCPT-guanin-9-yl, R^1 = OCH_3, R^2 = H
CMCM-caged cAMPs: Purine = adenin-9-yl, R^1 = OCH_2COOH, R^2 = H
CMCM-caged cGMPs: Purine = guanin-9-yl, R^1 = OCH_2COOH, R^2 = H
CMCM-caged 8-Br-cAMPs: Purine = 8-bromoadenin-9-yl, R^1 = OCH_2COOH, R^2 = H
CMCM-caged 8-Br-cGMPs: Purine = 8-bromoguanin-9-yl, R^1 = OCH_2COOH, R^2 = H
HCM-caged cAMPs: Purine = adenin-9-yl, R^1 = OH, R^2 = H
ACM-caged cAMPs: Purine = adenin-9-yl, R^1 = OAcyl, R^2 = H
BHCM-caged cAMPs: Purine = adenin-9-yl, R^1 = OH, R^2 = Br
DMCM-caged cAMPs: Purine = adenin-9-yl, R^1 = R^2 = OCH_3
DMCM-caged cGMPs: Purine = guanin-9-yl, R^1 = R^2 = OCH_3
DMCM-caged 8-Br-cAMPs: Purine = 8-bromoadenin-9-yl, R^1 = R^2 = OCH_3
DMCM-caged 8-Br-cGMPs: Purine = 8-bromoguanin-9-yl, R^1 = R^2 = OCH_3
BCMCM-caged cAMPs: Purine = adenin-9-yl, R^1 = R^2 = OCH_2COOH
BCMCM-caged cGMPs: Purine = guanin-9-yl, R^1 = R^2 = OCH_2COOH
BCMCM-caged 8-Br-cAMPs: Purine = 8-bromoadenin-9-yl, R^1 = R^2 = OCH_2COOH
BCMCM-caged 8-Br-cGMPs: Purine = 8-bromoguanin-9-yl, R^1 = R^2 = OCH_2COOH
BECMCM-caged cAMPs: Purine = adenin-9-yl, R^1 = R^2 = OCH_2COOEt
BECMCM-caged cGMPs: Purine = guanin-9-yl, R^1 = R^2 = OCH_2COOEt
BECMCM-caged 8-Br-cAMPs: Purine = 8-bromoadenin-9-yl, R^1 = R^2 = OCH_2COOEt
BECMCM-caged 8-Br-cGMPs: Purine = 8-bromoguanin-9-yl, R^1 = R^2 = OCH_2COOEt
BMCMCM-caged cAMPs: Purine = adenin-9-yl, R^1 = R^2 = OCH_2COOMe
BMCMCM-caged cGMPs: Purine = guanin-9-yl, R^1 = R^2 = OCH_2COOMe
DEACM-caged cAMPs: Purine = adenin-9-yl, R^1 = NEt_2, R^2 = H
DEACM-caged cGMPs: Purine = guanin-9-yl, R^1 = NEt_2, R^2 = H
DEACM-caged 8-Br-cAMPs: Purine = 8-bromoadenin-9-yl, R^1 = NEt_2, R^2 = H
DEACM-caged 8-Br-cGMPs: Purine = 8-bromoguanin-9-yl, R^1 = NEt_2, R^2 = H
DMACM-caged cAMPs: Purine = adenin-9-yl, R^1 = NMe_2, R^2 = H
DMACM-caged cGMPs: Purine = guanin-9-yl, R^1 = NMe_2, R^2 = H
DMACM-caged 8-Br-cAMPs: Purine = 8-bromoadenin-9-yl, R^1 = NMe_2, R^2 = H
DMACM-caged 8-Br-cGMPs: Purine = 8-bromoguanin-9-yl, R^1 = NMe_2, R^2 = H

Scheme 4.1.3 Structures of the axial and equatorial isomers of coumarinylmethyl-caged cNMPs.

responding caged cAMPs, 8-Br-cAMPs and 8-Br-cGMPs with diastereomer ratios axial:equatorial of 85–65:15–35%. Again the preparation of the coumarinylmethyl-caged cGMPs was not successful or gave significantly lower yields. All diastereomers were separated using preparative reverse phase HPLC.

The photocleavage of the coumarinylmethyl esters of the cNMPs is shown in Scheme 4.1.4. Irradiation of the axial and/or equatorial diastereomers in aqueous buffer with 330–440 nm light results in the liberation of the anion of the respec-

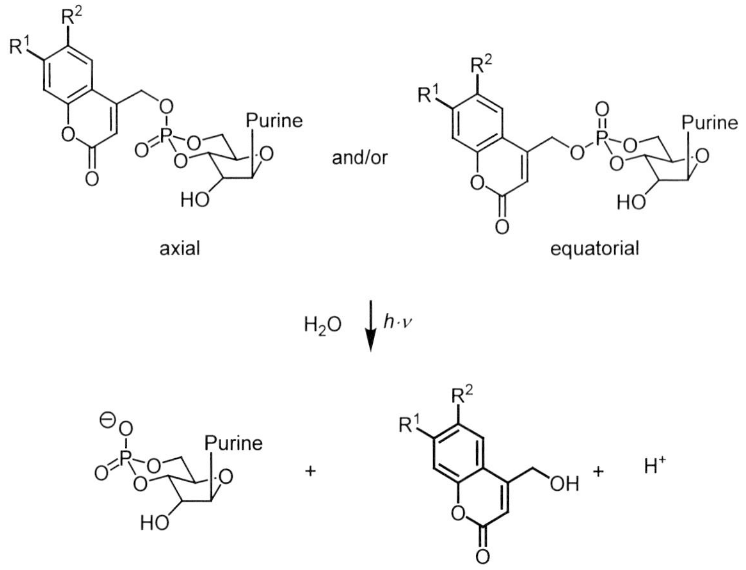

Scheme 4.1.4 Photolysis of the diastereomers of coumarinylmethyl-caged cNMPs.

tive cNMP, the corresponding 4-(hydroxymethyl)coumarin, and a proton. The mechanism for the photorelease of the cNMPs from their caged precursors was clarified by Schade et al. [31]. They detected a photo S_N1 reaction (solvent-assisted photoheterolysis), suggesting that the photocleavage proceeds via an excited singlet state, ion-pair formation, ion-pair separation by the polar solvent (formation of the cyclic nucleotide anion), and, finally, hydroxylation of the coumarinylmethyl carbocation [formation of the corresponding 4-(hydroxymethyl)coumarin]. This mechanism implies that the photorelease should occur very quickly. Indeed, using time-resolved fluorescence measurements [the photosolvolysis is accompanied by a large fluorescence enhancement caused by the formation of the strongly fluorescent 4-(hydroxymethyl)coumarin from the weakly fluorescent caged cNMP], it was shown that the 4-(hydroxymethyl)coumarins are formed with rate constants between 2.5×10^8 and 10^9 s^{-1}, that is, within a few nanoseconds [31–34, 36]. It is inferred that the cNMPs are formed with similar rates. Such high rate constants for uncaging are unique; the liberation of cyclic nucleotides is about 6 orders of magnitude faster than liberation from o-nitrobenzyl esters.

Tab. 4.1.2 summarizes the chemical and photophysical properties of a representative selection of the most important coumarinylmethyl-caged cNMPs. BHCM-caged cAMP was not included, because no data are available. The compound is expected to show a relatively high cross-section for two-photon photolysis [37]. Tab. 4.1.2 and Fig. 4.1.1 illustrate that the coumarinylmethyl-caged cNMPs absorb in the long-wavelength UV or visible region and the absorption coefficients are high. Compared to the MCM/CMCM caging group, the absorption of the

Tab. 4.1.2 Long-wavelength absorption maximum λ_{max} extinction coefficient ε at λ_{max}, photochemical quantum yield φ_{chem}, and solubility s of selected coumarinylmethyl esters of cNMPs in acetonitrile/HEPES-KCl buffer (5:95), pH 7.2

Compound	λ_{max} (ε) [nm] ([M^{-1} cm^{-1}])	φ_{chem}	s [μM]	Refs.
MCM-caged cAMP (axial)	328 (13200) [a, b]	0.12 [c]; 0.13 [a, b]	30 [b]	28, 29
MCM-caged cAMP (equatorial)	325 (13300) [a, b]	0.12 [c]; 0.07 [a, b]	10 [b]	28, 29
MCM-caged cGMP (axial)	327 (13300) [a]	0.21 [a]	10 [b]	31
MCM-caged cGMP (equatorial)	325 (13300) [a]	0.09 [a]	25 [b]	31
MCM-caged 8-Br-cAMP (axial)	325 (13300) [a]	0.14 [a]	20	32
MCM-caged 8-Br-cAMP (equatorial)	326 (13500) [a]	0.09 [a]	40	32
MCM-caged 8-Br-cGMP (axial)	328 (13400) [a]	0.20 [a]	40	32
MCM-caged 8-Br-cGMP (equatorial)	325 (13400) [a]	0.10 [a]	20	32
CMCM-caged cAMP (axial)	326 (12500)	0.12	900	33
CMCM-caged cAMP (equatorial)	324 (12400)	0.10	200	33
CMCM-caged cGMP (axial)	326 (11700)	0.16	350	33
CMCM-caged cGMP (equatorial)	325 (11200)	0.10	>1000	33
BCMCM-caged cAMP (axial)	346 (10700)	0.10	500	33
BCMCM-caged cAMP (equatorial)	347 (12400)	0.08	1000	33
BCMCM-caged cGMP (axial)	348 (11000)	0.14	550	33
BCMCM-caged cGMP (equatorial)	347 (11200)	0.09	>1000	33
BCMCM-caged 8-Br-cAMP (axial)	347 (10300)	0.10	650	24
BCMCM-caged 8-Br-cAMP (equatorial)	346 (9900)	0.08	1000	24
BCMCM-caged 8-Br-cGMP (axial)	349 (10200)	0.13	300	24
BCMCM-caged 8-Br-cGMP (equatorial)	346 (12700)	0.09	1000	24
BECMCM-caged cAMP (axial)	340 (10900) [a]	0.07 [a]	25	24, 35
BECMCM-caged cAMP (equatorial)	339 (10800) [a]	0.08 [a]	160	24, 35
BECMCM-caged cGMP (axial)	340 (11000) [a]	0.15 [a]	75	24, 35
BECMCM-caged cGMP (equatorial)	339 (11000) [a]	0.10 [a]	25	24, 35
DEACM-caged cAMP (axial)	402 (18600) [a]	0.21 [a]	135	33
DEACM-caged cAMP (equatorial)	396 (20200) [a]	0.23 [a]	15	33
DEACM-caged cGMP (axial)	403 (19300) [a]	0.25 [a]	120	33
DEACM-caged cGMP (equatorial)	396 (19300) [a]	0.26 [a]	15	33
DEACM-caged 8-Br-cAMP (axial)	401 (20100) [a]	0.23 [a]	3	36
DEACM-caged 8-Br-cAMP (equatorial)	395 (19600) [a]	0.24 [a]	10	36
DEACM-caged 8-Br-cGMP (axial)	406 (16800) [a]	0.27 [a]	120	36
DEACM-caged 8-Br-cGMP (equatorial)	396 (17300) [a]	0.28 [a]	4	36

a) In MeOH/HEPES-KCl buffer (1:4), pH 7.2.
b) V. Hagen, unpublished data.
c) Measured for a mixture of the axial and equatorial isomer in 1% DMSO in Ringer's solution.

BCMCM/BECMCM chromophore is red-shifted. The DEACM and the DMACM group exhibit the highest extinction coefficients and are further red-shifted, allowing effective uncaging between 350 and 440 nm. Long-wavelength activation of photorelease allows using non-damaging light conditions.

Photolysis quantum yields of most coumarinylmethyl-caged derivatives are acceptable; the quantum yields of DEACM-caged cNMPs are even relatively high. Uncaging cross-sections ($\varphi\varepsilon$ values) at the absorption maxima are 1300–1500 $M^{-1}cm^{-1}$ (MCM-, CMCM-caged cNMPs at about 325 nm), 1100–1400 $M^{-1}cm^{-1}$ (BCMCM-, BECMCM-caged cNMPs at about 350 nm) and 4200–4800 $M^{-1}cm^{-1}$ (DEACM-caged cNMPs at about 400 nm). The values of the DEACM-caged cNMPs as well as those of the corresponding DMACM-caged cNMPs (2500–4800 $M^{-1}cm^{-1}$) are extraordinarily high and illustrate the excellent photoefficiency of these caged compounds. The efficient photorelease of the cyclic nucleotides from their 7-(dialkylamino)coumarinylmethyl esters inside cells at long-wavelength excitation (irradiations at 405 nm or 436 nm) has also been demonstrated by experiments [33, 36].

The axial and equatorial diastereomers of the coumarinylmethyl-caged cNMPs are highly resistant to spontaneous hydrolysis in aqueous buffer solution at pH 7.2. Half-lives are >500–1000 h [31, 33, 36], and HPLC monitoring during a 24 h period revealed no measurable formation of the "free" cyclic nucleotide. Solubilities of the coumarinylmethyl-caged cNMPs in aqueous buffer vary between a few μM and a few mM (Tab. 4.1.2). The negatively charged CMCM- and BCMCM-caged compounds are highly soluble. The MCM-caged derivatives are poorly soluble, but because of the high biological activity of 8-Br-cAMP and 8-Br-cGMP, 8-bromosubstituted MCM-caged cNMPs were successfully used in physiological studies [31]. The solubilities of the axial diastereomers of the neutral BECMCM- and DEACM-caged cNMPs are sufficient for most applications.

The fluorescence quantum yields of the coumarinylmethyl-caged cNMPs in aqueous buffer are very small [31, 32, 34, 36], whereas the photoreleased 4-(hydroxymethyl)coumarins are strongly fluorescent. This difference is illustrated in Fig. 4.1.2, which compares the fluorescence spectra of BCMCM-caged cAMP and BCMCM-OH [6,7-bis(carboxymethoxy)-4-(hydroxymethyl)coumarin]. Therefore, upon photolysis of the caged compounds, the fluorescence is strongly enhanced and can be used as a tool for monitoring the progress of the photoreaction. In aqueous buffer, it is feasible to determine the concentration of the released 4-(hydroxymethyl)coumarins and thereby also the concentration of the cNMPs [32, 34, 36, 39]. However, in experiments with HEK293 cells, no fluorescence increase upon photolysis of MCM-, DMCM-, DMACM-, or DEACM-caged cNMPs was observed, because of the fluorescence quenching of the 4-(hydroxymethyl)coumarins [32, 36, 39]. In contrast, the hydrophilic BCMCM-OH is strongly fluorescent inside cells and the quantification of the release of cNMPs from their BCMCM esters inside cells is possible [39]. Fig. 4.1.6 shows the quantification of the concentration of cAMP following photolysis of BCMCM-caged cAMP in a HEK293 cell transfected with DNA encoding the CNGA2 channel. The measurement allowed the *in vivo* determination of the dose-response relation of a cellular reaction triggered by cyclic nucleotides.

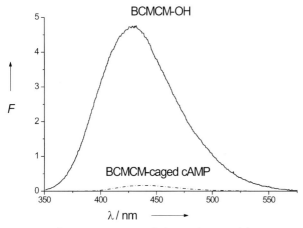

Fig. 4.1.2 Fluorescence spectra of 50 µM solutions of the equatorial isomer of BCMCM-caged cAMP and of BCMCM-OH in HEPES-KOH buffer at pH 7.2. F=relative fluorescence intensity. (Printed by permission from [39]).

4.1.2.3 Others

Desoxybenzoinyl (desyl) [40, 41], 2-naphthylmethyl [28], and 2-anthraquinonylmethyl groups [28] represent other photolabile caging groups that were used for caging cAMP (Scheme 4.1.5). Unfortunately, desyl-caged cAMP is very sensitive to solvolysis in aqueous buffer solution [28]. The other two caged cAMPs suffer

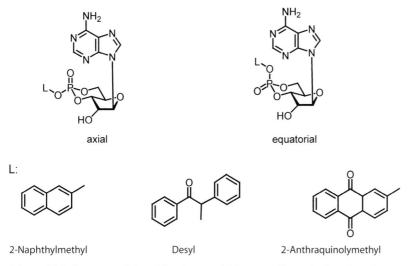

Scheme 4.1.5 Structures of the axial and equatorial isomers of non-nitrobenzyl- and non-coumarinylmethyl-caged cNMPs.

from low solubility in aqueous buffer [28]. No applications with these compounds have been described so far.

4.1.2.4 Selection of the Most Useful Phototriggers

The coumarinylmethyl-caged cNMPs bearing electron donor substituents in 7-position of the coumarinylmethyl moiety combine a set of favorable properties rendering them ideal tools for intra- and extracellular studies of cyclic nucleotide signaling. They show no background activity, are stable to solvolysis, and photolyze efficiently and extremely quickly using long-wavelength UV/Vis light. The hydrophilicity of these compounds can be regulated to a fair degree by substituents at the coumarinylmethyl moiety. The caged compounds are far superior to the 2-nitrobenzyl-caged cNMP analogs, and DMNB-, NPE- and BCMCNB-caged cNMPs should be used only when the slow photorelease and low photoefficiency do not matter or when the water solubility becomes a limiting factor (the BCMCNB-caged cNMPs are even more water-soluble than the highly soluble BCMCM-caged cNMPs, and the cell membrane-permeant NPE-caged cNMPs show a higher solubility in water than the BECMCM- or DEACM-caged cNMPs). Tab. 4.1.3 compares the efficiency of the photorelease of 8-Br-cGMP from different caged compounds at several wavelengths. The comparison illus-

Tab. 4.1.3 Comparison of the efficiency (in %) of the photorelease of 8-Br-cGMP from 25 µM solutions of caged compounds (axial isomers) in MeOH/HEPES-KCl buffer (1:4) with various irradiation times at different wavelengths

Compound	t (λ_{exc}=333 nm)			t (λ_{exc}=365 nm)			t (λ_{exc}=405 nm)			t (λ_{exc}=436 nm)		
	6 s	60 s	600 s	6 s	60 s	600 s	6 s	60 s	600 s	6 s	60 s	600 s
NPE-caged 8-Br-cGMP	<1	3	12	<1	<1	1	<1	<1	<1	<1	<1	<1
DMNB-caged 8-Br-cGMP	<1	4	30	<1	3	20	<1	1	2	<1	<1	<1
MCM-caged 8-Br-cGMP	17	77	>98	4	30	96	<1	2	10	<1	<1	<1
CMCM-caged 8-Br-cGMP [a]	16	75	>98	5	30	93	<1	1	8	<1	<1	<1
DMCM-caged 8-Br-cGMP	10	46	97	10	60	91	1	3	19	<1	1	3
BCMCM-caged 8-Br-cGMP [a]	9	45	96	13	66	96	1	2	18	<1	1	3
DMACM-caged 8-Br-cGMP	4	32	93	15	64	>98	20	90	>98	7	75	98
DEACM-caged 8-Br-cGMP	1	7	60	12	60	90	20	91	>98	12	80	>98

a) In HEPES-KCl buffer, pH 7.2.

trates the superiority of the coumarinylmethyl-caged compounds over the nitrobenzyl-caged derivatives and shows that the highest release of the cNMP is achieved at 333 nm with the MCM/CMCM-, at 365 nm with the DMCM/BCMCM/BECMCM- as well as the DMACM/DEACM-, and at 405 or 436 nm with the DMACM/DEACM-caged compounds.

The most useful coumarinylmethyl-caged derivatives are the DEACM-, BCMCM-, and BECMCM-caged compounds. Because of significant differences in solubility of the diastereomers, the pure isomers with the highest solubility (see Tab. 4.1.2) should be used for experiments. However, mixtures of the diastereomers or the less soluble isomer can also be successfully used. The equatorial BCMCM esters, because of their high aqueous solubility, allow the establishment of high concentrations of the phototriggers inside cells by means of patch pipettes (whole-cell configuration) or microinjection. They are expected to considerably stay inside cells, because they carry two negative charges and therefore cannot easily leave the cell across the membrane. The high concentration combined with the high photoefficiency allows large instantaneous steps of the cNMP concentration. The photorelease occurs efficiently at wavelengths between 330 and 380 nm (maximum at about 350 nm). Furthermore, an advantage of these compounds is the possibility to determine quantitatively the amount of the photoreleased cyclic nucleotide inside cells using fluorescence measurements. BECMCM-caged cNMPs are cell membrane-permeant analogs of the BCMCM-caged compounds. They can be introduced into cells by incubation and are the compounds of choice for studies of cell suspensions. The BECMCM derivatives are assumed to accumulate inside the cell by hydrolysis of the ester group at the coumarinyl moiety. Their photorelease occurs also very effectively at wavelengths between 330 and 380 nm. DEACM-caged cNMPs display the most favorable photochemical properties. Their ability to uncage upon long-wavelength irradiation (350 up to 440 nm) minimizes or even prevents cellular damage and chromophore bleaching by light. Furthermore, their high photoefficiency permits the generation of large jumps of the cNMP concentration. These caged compounds can be introduced into cells by incubation as well as by means of recording pipettes. However, the solubility of the compounds allows only the photorelease of cAMP, cGMP, and 8-Br-cGMP concentrations up to 100 µM or of 8-Br-cAMP concentrations up to 10 µM; for many applications this is sufficient. The development of DEACM-caged analogs of the cNMPs bearing anionic substituents is currently under way [24]. We assume that these compounds and the DEACM phototriggers will be the compounds of choice in studies of cyclic nucleotide-dependent processes using caged compounds. BHCM-caged cNMPs could be useful in studies using two-photon photolysis.

4.1.3
Applications of Caged Cyclic Nucleotides

Caged cyclic nucleotides have been used to unravel several cAMP- and cGMP-dependent signaling pathways. In the following, we will address each of these signaling pathways and their cAMP- or cGMP targets.

4.1.3.1 Cyclic Nucleotide-Gated (CNG) Channels

Cyclic nucleotide-gated channels play an important role in mediating the electrical response in rod and cone photoreceptors and olfactory neurons [3]. CNG channels are directly gated open by the binding of cAMP and cGMP to a binding site in the C-terminal domain of the channel polypeptide.

Karpen and his collaborators [42] were the first to study CNG channels using caged cGMP. Membrane patches excised from rod outer segments were exposed with their cytosolic side to a solution containing DMNB-caged cGMP. Upon a flash of UV light (≥ 14 ns), the CNG channels become activated with a time course that could be described with a single exponential.

The efficacy of the concentration jump method can be improved with the (coumarin-4-yl)methyl derivatives. The beneficial physico-chemical properties of these compounds permit larger concentration jumps than are possible with o-nitrobenzyl derivatives. With BCMCM-caged cGMP, the largest concentration jump we obtained was 57 µM cGMP. Fig. 4.1.3 shows experiments in which DEACM-caged cGMP was used to activate CNGA2-channel currents. The channels were heterologously expressed in *Xenopus* oocytes and the recordings were performed in excised patches. The concentration of free cGMP was determined from the ratio of the steady-state current to the steady-state current at saturating cGMP using the concentration-response relationship for these channels. In Fig. 4.1.3 A, DEACM-caged cGMP (30.0 µM) was used; the flash produced a concentration jump of 13.7 µM free cGMP. The highly resolved activation reveals a short initial delay followed by a time course that could be fitted with the sum of two exponentials, yielding the time constants τ_1 and τ_2. The time constants at +100 and –100 mV are slightly different. Before the flash, no current was present, indicative of extremely low levels of contamination of the substance with free cGMP. Fig. 4.1.3 B shows the initial activation phase of CNGA2 currents from another experiment at higher time resolution (10.0 µM DEACM-caged cGMP). The flash produced a concentration jump of 7.2 µM free cGMP. An initial delay phase is clearly visible. The complexity of the activation time course can be used to test the appropriateness of kinetic models for the activation of the channels.

Caged cGMP and cAMP were also employed to resolve an apparent discrepancy between the rapid and transient odorant-induced rise of cAMP in olfactory neurons (≤ 50 ms) [43] and the slow delayed electrical response of olfactory neurons (delay and duration several hundreds of milliseconds [44, 45]. These seemingly conflicting observations have led to the hypothesis that inherently slow channel

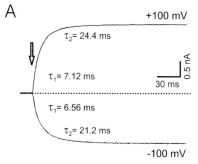 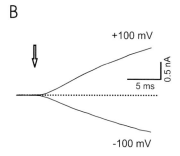

Fig. 4.1.3 Change of patch current following flash photolysis of DEACM-caged cGMP at +100 and –100 mV. The light flashes (arrows) were delivered from a Xenon flash lamp (light pulse duration ~150 µs) and guided directly to the experimental chamber (volume ~0.8 µl) via a light guide. Both bath and pipette contained symmetric K^+ (150 mM) solution. (A) Currents evoked with 30 µM DEACM caged cGMP. The time courses are fitted with the sum of two exponentials (curves superimposed), yielding the indicated time constants. (B) Initial delay phase of CNGA2 currents. The currents were evoked with 10 µM DEACM-caged cGMP. The traces were kindly provided by V. Nache, Jena.

kinetics determine the sluggish olfactory response [46]. However, release of cAMP in olfactory sensory neurons (OSNs) [47–49] or 8-Br-cGMP in a cell line expressing the A2 subunit of the olfactory CNG channel [20] produced currents with little or almost no delay (Fig. 4.1.4), which is in line with the short delay observed in excised patches as shown in Fig. 4.1.3. These results demonstrate that the activation of both CNGA1 and CNGA2 channels is dominated by an exponential time course and that only the rate of activation is slower in CNGA2 than in CNGA1 channels. Moreover, both the rise and recovery of electrical responses in OSNs evoked by odorants were significantly slower than those evoked by the release of cAMP from caged compounds [47–49]. These differences suggest that the waveform of the

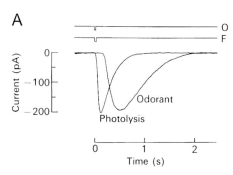 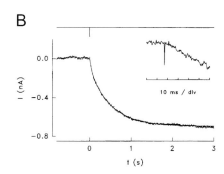

Fig. 4.1.4 Currents evoked by odorants and by photolysis of NPE-caged cAMP in olfactory neurons (A) and by DMNB-caged 8-Br-cGMP in HEK293 cells expressing the A2 channel (B). Printed by permission from Lowe and Gold [76] and Hagen et al. [20]. For experimental details see the original references.

odorant response is shaped by the synthesis and hydrolysis of cAMP and that the respective rates in preparations of isolated olfactory cilia may differ from those in intact olfactory neurons.

CNG channels produce mixed inward currents carried by Na^+ and Ca^{2+} ions. Caged cyclic nucleotides were particularly useful for the study of fractional Ca^{2+} permeability, i.e. the fraction of the total current carried by Ca^{2+} ions (designated P_f). The experiments involve the simultaneous measurement of the total current and the number of Ca^{2+} ions flowing through channels by a patch pipette and by a fluorescent Ca^{2+} indicator dye, respectively (Fig. 4.1.5 A). It is crucial that all Ca^{2+} ions entering the cell (1) are flowing through CNG channels

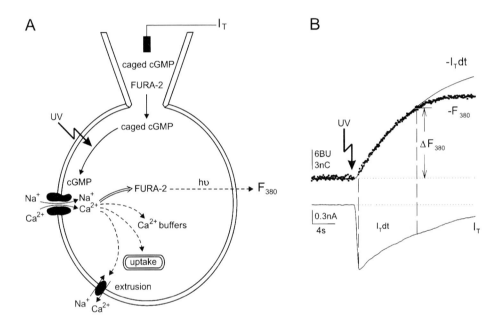

Fig. 4.1.5 Determination of the fractional Ca^{2+} current in CNG channels. (A) Experimental design: an HEK293 cell expressing CNG channel is loaded through a patch pipette in the whole-cell configuration with caged cGMP and FURA-2. Photorelease of cGMP by a UV flash leads to activation of channels that conduct both Na^+ and Ca^{2+}. An electrode inside the patch pipette records the total current, I_T, and Ca^{2+} entry is monitored by changes in the FURA-2 fluorescence, F_{380}, recorded by a photon counter. For quantitative recording of Ca^{2+} influx, FURA-2 is used at a concentration of 1–2 mM, which prevents loss of inflowing Ca^{2+} to cellular Ca^{2+} buffers and transport systems. (B) Evaluation of data recorded at –70 mV from a cell expressing bO. $[Ca^{2+}]_o$ was 0.3 mM; the pipette contained 70 µM caged cGMP and 1 mM FURA-2. The experiment was started with a 500 ms UV flash of 0.6 mV (arrow). Whole-cell current (I_T) and FURA-2 fluorescence (F_{380}) were recorded simultaneously. Fluorescence intensity is given in bead units (BU). Superposition of fluorescence and current integral ($-\int I_T dt$, given in nanoCoulomb) indicates proportionality within the time segment marked by the vertical lines. This segment is used to calculate the value of $f = \Delta F_{380} / \int I_T dt$ and to determine the fractional Ca^{2+} current, P_f. (Printed by permission from Dzeja et al. [52]).

and (2) are detected by the Ca^{2+} indicator. These requirements are readily met during the first few seconds after channel activation. At later times, the co-linear increase of current and Ca^{2+} concentration breaks down because of the uptake of Ca^{2+} into intracellular organelles and the active extrusion of Ca^{2+} from the cell (Fig. 4.1.5 B; for details see Refs. [50, 51]). The rapid and controlled activation of CNG channels inside cells is most easily accomplished using caged cyclic nucleotides. Fractional Ca^{2+} permeability using caged compounds has been determined for the A subunit of rod (A1), cone (A3), and OSNs (A2) channels [52] and for the native CNG channels in intact rod and cone photoreceptors [53]. It was found that CNG channels of cones and OSNs display larger P_f values than CNG channels of rods. In OSNs, under physiological conditions, the majority of the current – if not all – might be carried by Ca^{2+} ions. Although, the P_f values of rod and cone CNG channels are different ($P_f \cong 0.15$ versus 0.34), both are independent of membrane voltage [51]. This feature offers a functional advantage since it ensures that the kinetics of the photocurrent

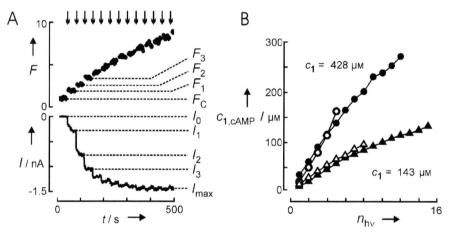

Fig. 4.1.6 Quantification of the concentrations of BCMCM-OH and cAMP following photolysis of BCMCM-caged cAMP.
(A) HEK293 cells were transfected with cDNA encoding the cAMP-gated cation channel (CNGA2) and loaded with 428 µM BCMCM-caged cAMP as well as with 35 µM BCMCM-OH for internal calibration. The fluorescence recording (upper trace) documents liberation of BCMCM-OH during a series of UV flashes ($t=20$ ms, $\lambda=365$ nm). The simultaneous current recording (lower trace) shows activation of the CNGA2 channels by photoreleased cAMP. Both fluorescence intensity and current amplitude can be used to calculate photolysis. (B) Calculation of photolytic liberation of BCMCM-OH from the increase in fluorescence intensity, and calculation of cAMP liberation from the normalized current $I_{norm}= I/I_{max}$ using the Hill equation. The results for quantification of photolysis obtained from fluorescence measurements (filled symbols) and from analysis of current recordings (open symbols) are in good agreement. While the current calibration is limited by the maximal channel activation at roughly 150 µM cAMP, the fluorescence-based calibration can be extended to higher concentrations. $I=$current $n_{hv}=$number of flashes; $c_{1,cAMP}=$concentration of photoreleased cAMP. (Printed by permission from Hagen et al. [39]).

(which depends on the intracellular Ca^{2+} concentration) is controlled by light and not by voltage.

For many applications it would be useful to determine quantitatively the degree of photolysis of the caged compound inside cells and, thereby, the amount of released cAMP or cGMP. This is feasible by fluorescence measurements using BCMCM-caged cAMP or cGMP (Fig. 4.1.6) [39]. The fluorescence quantum yields of the caged compounds are small, whereas the released alcohol (BCMCM-OH) is strongly fluorescent (see Fig. 4.1.2). The calibration procedure of the changes in fluorescence upon photolysis of BCMCM-caged cAMP has been validated using CNG channels. Calibrations using whole-cell current recordings and fluorescence measurements are in excellent agreement (Fig. 4.1.6 B) [39].

BCMCM-caged cAMP was used to study the inactivation of the olfactory CNG channel by Ca^{2+}/CaM [54]. Photolysis of BCMCM-caged cAMP rapidly activates CNG channels in isolated rat OSNs and heterologously expressed in HEK293 cells. Bradley and his collaborators [54] found that response times critically depended on the molecular composition of the channel. Inactivation by Ca^{2+}/CaM was fastest and matched inactivation of the native channel when all three subunits – A2, A4, and B1b – were present. This study thus demonstrates that all three subunits are involved in the binding of CaM and its action on the gating kinetics [54].

4.1.3.2 HCN Channels

Hyperpolarization-activated and cyclic nucleotide-gated (HCN) channels play an important role in endogenous rhythmic activity of cells, in sour taste, and in synaptic transmission (for reviews see [55, 56]). Unlike CNG channels, HCN channels do not open by binding of cAMP alone; instead, HCN channels are activated at hyperpolarized membrane potentials. Cyclic nucleotides enhance activation by shifting the P_o-V_m relation to more positive potentials. cAMP acts at significantly lower concentrations than cGMP.

Caged cAMP has been employed to show that the activation of HCN channels by cAMP (at negative V_m) occurs with almost no delay [57, 58], strengthening the notion that cAMP affects gating directly without involving phosphorylation steps [59]. However, the gating of HCN channels by cAMP is intrinsically slower than that of CNG channels. Using caged cyclic nucleotides also permitted recording of P_o-V_m relations in the absence and presence of cAMP in one and the same cell [58]. Caged cAMP was also employed in sino-atrial node cells of the rabbit [60]. Release of cAMP increased the firing frequency and induced an L-type Ca^{2+} current. The positive chronotropic response was attributed to the activation of Ca^{2+} channels rather than the activation of I_f (that is HCN channels).

4.1.3.3 Dorsal Root Ganglion Cells

Release of cGMP from NPE-caged cGMP in dorsal root ganglion (DRG) neurons evoked two different types of responses (inward current), a rapid response and a delayed response (delay of up to several minutes) [61, 62]. The instantaneous response might be produced by the activation of CNG channels, although there is no independent evidence for the expression of CNG channels in DRG neurons. The delayed response might be due to the activation of ryanodine receptor/Ca^{2+} release channels by a cyclic ADP-ribose-dependent mechanism. The formation of cyclic ADP-ribose is catalyzed by ADP-ribosyl cyclase [63]. cGMP-dependent phosphorylation upregulates the activity of ADP-ribosyl cyclase. Consistent with this interpretation, release of cAMP from a caged precursor does not evoke delayed responses [61]. Surprisingly, release of cGMP from the more water-soluble BCMCNB-caged cGMP congener did not evoke the instantaneous response, only the delayed response. The reason for this difference is not known [62].

4.1.3.4 cGMP-Signaling Pathway in Sperm

The egg and sperm meet through a process called chemotaxis. The egg releases a substance, called chemo-attractant, which forms a chemical gradient surrounding the egg. The sperm utilizes cues of the chemo-attractant gradient to reorient its swimming behavior and, thereby, locate the egg. During the past 30 years or so, a wealth of studies mainly in sea urchins has resulted in a detailed model of the cellular and physiological reactions underlying chemotaxis [64, 65]. The chemo-attractant, usually a short peptide, binds to a receptor protein (a guanylyl cyclase) on the surface of the sperm. Binding initiates a series of cellular events including the synthesis of cGMP and cAMP, which eventually cause the influx of Ca^{2+} from the outside. The increase of the Ca^{2+} concentration inside the cell changes the beating pattern of the sperm tail and thereby reorients the swimming trajectory. Recent evidence suggests that the model, except for the first step, may be not correct [35, 66]. Using BECMCM-caged cGMP and cAMP, Kaupp and his collaborators [35] demonstrated that cGMP either directly or indirectly promotes the influx of Ca^{2+} into the sperm (Fig. 4.1.7). The Ca^{2+} responses evoked by cGMP and the chemo-attractant are similar, whereas the Ca^{2+} response evoked by cAMP is distinctly different. Either cAMP and cGMP activate two different Ca^{2+}-entry pathways or one and the same Ca^{2+}-entry channel responds to both cAMP and cGMP in different ways. While it is tempting to speculate that the Ca^{2+}-entry channel in fact is a CNG channel, the significantly delayed onset (≥ 150 ms) of the rise in Ca^{2+} concentration does not support this idea. For example, activation by cAMP and cGMP of CNG channels from rods and OSNs, whether native or recombinant proteins, proceeds faster with almost no delay (Figs. 4.1.3 and 4.1.4) [20, 42]. A similar cGMP-signaling pathway and cGMP-dependent Ca^{2+} entry has been identified in starfish sperm using BECMCM-caged cAMP and cGMP [67]. Whatever the identity of the Ca^{2+}-entry channel and its mechanism of activation might be, these two studies using caged cyclic nucleotides clearly identify cGMP as the primary messenger of the Ca^{2+} response. A note of caution is necessary here. When comparing cGMP- and cAMP-dependent responses using caged compounds, it is often tacitly assumed that the concentrations of cyclic nucleotides inside the cell and the efficiency of photorelease are similar if not identical. While

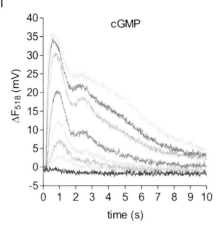

Fig. 4.1.7 Ca^{2+} response of sperm from the sea urchin *Arbacia punctulata* evoked by cyclic nucleotides. Ca^{2+} responses in sperm loaded with caged cGMP and Fluo-4 elicited by flashes of ultraviolet light of different energies. Relative light energies ranged from 2.5 to 100%. The energy of a full flash was 26.8 mJ cm^{-2}. For details see Ref. [35]. (Printed by permission from Kaupp et al. [35]).

the quantum yields and absorptivity of related congeners are rather similar, the membrane permeability of esterified compounds (for example BECMCM-caged), their subcellular distribution, and their hydrolysis by unspecific esterases may be different.

Caged cyclic nucleotides also allowed the identification of the motor response in free-swimming sperm [35]. In a shallow observation chamber, sperm swim in circles. Release of cGMP and cAMP from caged precursors changes the swimming trajectory in characteristic ways [35]. Periods of "turns" (enhanced curvature of the swimming path and enhanced asymmetry of flagellar waveform) alternate with periods of straight swimming (lower curvature of swimming path and reduced asymmetry of flagellar waveform).

4.1.3.5 Miscellaneous Systems

cAMP permeability of gap junctions Gap junctions provide direct intercellular pathways that allow exchange of ions and metabolites up to a molecular weight of about 1 kD. Bedner and his colleagues [68] used CNG channels to monitor the relative permeability of cAMP for gap junctions-coupled cell lines that were expressing a cAMP-sensitive mutant of the A2 channel [69] and were loaded with NPE-caged cAMP. cAMP was released in one cell by localized photolysis. Released cAMP reached the neighboring cell by diffusion across gap junctions and stimulated Ca^{2+}-entry through CNG channels.

Neuronal long-term depression Parallel fiber synapses onto Purkinje neurons undergo long-term depression (LTD) when presynaptic activity coincides with postsynaptic depolarization. LTD is also observed when cGMP and Ca^{2+} are photolytically released inside the Purkinje neurons [18]. The development of LTD requires precise tuning in the release of the two messengers: The increase must occur within 200–300 ms.

Growth cone guidance Gradients of pituitary adenylate cyclase-activating polypeptide (PACAP), a neuropeptide widely expressed in the developing nervous system, attract growth cones *in vitro* [70]. The attraction depends on localized cAMP signaling because it is blocked by global elevation of intracellular cAMP levels using forskolin, an activator of adenylyl cyclase. Local elevation of intracellular cAMP by flash photolysis of NPE-caged cAMP was sufficient to induce growth cone attraction. Attraction is mediated by cAMP-dependent protein phosphorylation because it is inhibited by protein kinase-specific inhibitors.

Munck and his collaborators [71] employed caged cAMP to generate pulsed cAMP gradients inside growth cones. Certain time patterns of cAMP release induce turning of growth cones, whereas a continuous increase in cAMP is ineffective.

Ciliary beating of ependymal cells Ependymal cells form a single-layered ciliated surface that lines the ventricular surface of vertebrate brain. Serotonin (5-HT) and ATP exert opposite effects on ciliary activity [72]. Stimulation by 5-HT enhances ciliary beat frequency, whereas stimulation by ATP decreases activity. Release of cAMP from DMNB-caged cAMP as well as superfusion of cells with forskolin both depress the beat frequency. These results suggest that 5-HT-mediated increase of $[Ca^{2+}]_i$ stimulates beat frequency, whereas a purinergic receptor-mediated increase of cAMP depresses beat frequency. Either ciliary activity is controlled by two independent different pathways or cAMP interferes with a component in the Ca^{2+}-mediated stimulation.

Regulation of Ca^{2+} oscillations by cAMP Megakaryocytes display agonist-induced oscillations of $[Ca^{2+}]_i$ that are inhibited by agents that elevate cAMP [73]. Using DMNB-caged cAMP, these authors demonstrate that cAMP directly inhibits IP_3-induced Ca^{2+} release in intact cells [74]. The inhibition requires phosphorylation by PKA.

Regulation of CFTR by cAMP A nice example of the use of caged cAMP is provided by Chen and collaborators [75]. The mechanism whereby cAMP stimulates Cl^- efflux through CFTR channels in secretory epithelia is controversial. One hypothesis is that cAMP triggers the translocation of the CFTR protein from an intracellular pool to the cell surface. These authors showed that cAMP released from NPE-caged cAMP activates CFTR currents of 100 pA yet does not change the membrane capacitance, which is a sensitive measure of the change in membrane surface that accompanies vesicle exocytosis or membrane endocytosis. This study supports the notion that the open probability of CFTR conductance is enhanced by a PKA-dependent phosphorylation of CFTR.

References

1. U. Walter, *Rev. Physiol. Biochem. Pharmacol.* **1979**, 113, 41–88.
2. T. M. Lincoln, T. L. Cornwell, *FASEB J.* **1993**, 7, 328–338.
3. U. B. Kaupp, R. Seifert, *Physiol. Rev.* **2002**, 82, 769–824.
4. H. A. Lester, J. M. Nerbonne, *Annu. Rev. Biophys. Bioeng.* **1982**, 11, 151–175.
5. J. M. Nerbonne, *Soc. Gen. Physiol. Ser.* **1986**, 40, 417–445.
6. A. M. Gurney, H. A. Lester, *Physiol. Rev.* **1987**, 67, 583–617.
7. J. H. Kaplan, A. P. Somlyo, *Trends Neurosci.* **1989**, 12, 54–59.
8. J. A. McCray, D. R. Trentham, *Annu. Rev. Biophys. Biophys. Chem.* **1989**, 18, 239–270.
9. J. P. Y. Kao, S. R. Adams, *Optical Microscopy: Emerging Methods and Applications* (B. Herman and J. L. Lemasters, Eds.), Academic Press Inc., San Diego **1993**, pp. 27–85.
10. J. E. T. Corrie, D. R. Trentham, *Biological Applications of Photochemical Switches* (H. Morrison, Ed.), John Wiley & Sons, Inc., **1993**, pp. 243–305.
11. A. L. Zimmerman, G. Yamanaka, F. Eckstein, D. A. Baylor, L. Stryer, *Proc. Natl. Acad. Sci. USA* **1985**, 82, 8813–8817.
12. K.-W. Koch, U. B. Kaupp, *J. Biol. Chem.* **1985**, 260, 6788–6800.
13. J. C. Tanaka, J. F. Eccleston, R. E. Furman, *Biochemistry* **1989**, 28, 2776–2784.
14. J. Engels, E. J. Schlaeger, *J. Med. Chem.* **1977**, 20, 907–911.
15. J. Engels, R. Reidys, *Experientia* **1978**, 34, 14–15.
16. J. M. Nerbonne, S. Richard, J. Nargeot, H. A. Lester, *Nature* **1984**, 310, 74–76.
17. J. F. Wootton, D. R. Trentham, *Photochemical Probes in Biochemistry*, NATO ASI Ser. C, (P. E. Nielsen, Ed.), Kluwer Academic, The Netherlands **1989**, 272, pp. 277–296.
18. V. Lev-Ram, T. Jiang, J. Wood, D. S. Lawrence, R. Y. Tsien, *Neuron* **1997**, 18, 1025–1038.
19. L. Wang, J. E. Corrie, J. F. Wootton, *J. Org. Chem.* **2002**, 67, 3474–3478.
20. V. Hagen, C. Dzeja, S. Frings, J. Bendig, E. Krause, U. B. Kaupp, *Biochemistry* **1996**, 35, 7762–7771.
21. U. B. Kaupp, C. Dzeja, S. Frings, J. Bendig, V. Hagen, *Methods Enzymol.* **1998**, 291, 415–430.
22. V. Hagen, C. Dzeja, J. Bendig, I. Baeger, U. B. Kaupp, *J. Photochem. Photobiol. B* **1998**, 42, 71–78.
23. B. Wiesner, J. Weiner, R. Middendorff, V. Hagen, U. B. Kaupp, I. Weyand, *J. Cell Biol.* **1998**, 142, 473–484.
24. V. Hagen, J. Bendig, Unpublished work.
25. J. W. Walker, G. P. Reid, J. A. McCray, D. R. Trentham, *J. Am. Chem. Soc.* **1998**, 110, 7170–7177.
26. M. Schwörer, J. Wirtz, *J. Helv. Chim. Acta* **2001**, 84, 1441–1458.
27. R. S. Givens, B. Matuszewski, *J. Am. Chem. Soc.* **1984**, 106, 6860–6861.
28. T. Furuta, H. Torigai, M. Sugimoto, M. Iwamura, *J. Org. Chem.* **1995**, 60, 3953–3956.
29. T. Furuta, M. Iwamura, *Methods Enzymol.* **1998**, 291, 50–63.
30. T. Furuta, A. Momotake, M. Sugimoto, M. Hatayama, H. Torigai, M. Iwamura, *Biochem. Biophys. Res. Commun.* **1996**, 228, 193–198.
31. B. Schade, V. Hagen, R. Schmidt, R. Herbrich, E. Krause, T. Eckardt, J. Bendig, *J. Org. Chem.* **1999**, 64, 9109–9117.
32. V. Hagen, J. Bendig, S. Frings, B. Wiesner, B. Schade, S. Helm, D. Lorenz, U. B. Kaupp, *J. Photochem. Photobiol. B* **1999**, 53, 91–102.
33. V. Hagen, J. Bendig, S. Frings, T. Eckardt, S. Helm, D. Reuter, U. B. Kaupp, *Angew. Chem. Int. Ed.* **2001**, 40, 1046–1048.
34. T. Eckhardt, V. Hagen, B. Schade, R. Schmidt, C. Schweitzer, J. Bendig, *J. Org. Chem.* **2002**, 67, 703–710.
35. U. B. Kaupp, J. Solzin, J. E. Brown, A. Helbig, V. Hagen, M. Beyermann, E. Hildebrand, I. Weyand, *Nature Cell Biology* **2003**, 5, 109–117.

36 V. Hagen, S. Frings, B. Wiesner, S. Helm, U. B. Kaupp, J. Bendig, ChemBioChem **2003**, 4, 434–442.

37 T. Furuta, S. S. Wang, J. L. Dantzker, J. L. Dore, T. M. Dore, W. J. Bybee, E. M. Callaway, W. Denk, R. Y. Tsien, Proc. Natl. Acad. Sci. USA **1999**, 96, 1193–1200.

38 R. Y. Tsien, T. Furuta, WO 00/31588. **2000**. PCT-Patent.

39 V. Hagen, S. Frings, J. Bendig, D. Lorenz, B. Wiesner, U. B. Kaupp, Angew. Chem. Int. Ed. **2002**, 41, 3625–3628.

40 R. S. Givens, P. S. Athey, L. W. I. Kueper, B. Matuszewski, J. Xue, J. Am. Chem. Soc. **1992**, 114, 8708–8710.

41 R. S. Givens, P. S. Athey, B. Matuszewski, L. W. I. Kueper, J. Xue, T. Fister, J. Am. Chem. Soc. **1993**, 115, 6001–6012.

42 J. W. Karpen, A. L. Zimmerman, L. Stryer, D. A. Baylor, Proc. Natl. Acad. Sci. USA **1988**, 85, 1287–1291.

43 H. Breer, I. Boekhoff, E. Tareilus, Nature **1990**, 345, 65–68.

44 S. Firestein, F. Werblin, Science **1989**, 244, 79–82.

45 S. Firestein, G. M. Shepherd, F. S. Werblin, J. Physiol. **1990**, 430, 135–158.

46 F. Zufall, H. Hatt, S. Firestein, Proc. Natl. Acad. Sci. USA **1993**, 90, 9335–9339.

47 T. Kurahashi, A. Menini, Nature **1997**, 385, 725–729.

48 G. Lowe, G. H. Gold, J. Physiol. **1993**, 462, 175–196.

49 G. Lowe, G. H. Gold, Proc. Natl. Acad. Sci. USA **1995**, 92, 7864–7868.

50 R. Schneggenburger, Z. Zhou, A. Konnerth, E. Neher, Neuron **1993**, 11, 133–143.

51 T. Ohyama, A. Picones, J. I. Korenbrot, J. Gen. Physiol. **2002**, 119, 341–354.

52 C. Dzeja, V. Hagen, U. B. Kaupp, S. Frings, EMBO J. **1999**, 18, 131–144.

53 T. Ohyama, D. H. Hackos, S. Frings, V. Hagen, U. B. Kaupp, J. I. Korenbrot, J. Gen. Physiol. **2000**, 116, 735–753.

54 J. Bradley, D. Reuter, S. Frings, Science **2001**, 294, 2176–2178.

55 U. B. Kaupp, R. Seifert, Annu. Rev. Physiol. **2001**, 63, 235–257.

56 R. B. Robinson, S. A. Siegelbaum, Annu. Rev. Physiol. **2003**, 65, 453–480.

57 R. Gauss, R. Seifert, U. B. Kaupp, Nature **1998**, 393, 583–587.

58 R. Seifert, A. Scholten, R. Gauss, A. Mincheva, P. Lichter, U. B. Kaupp, Proc. Natl. Acad. Sci. USA **1999**, 96, 9391–9396.

59 D. DiFrancesco, P. Tortora, Nature **1991**, 351, 145–147.

60 H. Tanaka, R. B. Clark, W. R. Giles, Proc. R. Soc. Lond. B **1996**, 263, 241–248.

61 J. H. Crawford, J. F. Wootton, G. R. Seabrook, R. H. Scott, J. Neurophysiol. **1997**, 77, 2573–2584.

62 J. Pollock, J. H. Crawford, J. F. Wootton, J. E. T. Corrie, R. H. Scott, Neurosci. Lett. **2003**, 338, 143–146.

63 N. Rusinko, H. C. Lee, J. Biol. Chem. **1989**, 264, 11725–11731.

64 A. Darszon, P. Labarca, T. Nishigaki, F. Espinosa, Physiol. Rev. **1999**, 79, 481–510.

65 A. Darszon, C. Beltrán, R. Felix, T. Nishigaki, C. L. Treviño, Dev. Biol. **2001**, 240, 1–14.

66 J. C. Kirkman-Brown, K. A. Sutton, H. M. Florman, Nat. Cell Biol. **2003**, 5, 93–96.

67 M. Matsumoto, J. Solzin, A. Helbig, V. Hagen, S.-I. Ueno, O. Kawase, Y. Maruyama, M. Ogiso, M. Godde, H. Minakata, U. B. Kaupp, M. Hoshi, I. Weyand, Dev. Biol. **2003**, 260, 314–324.

68 P. Bedner, H. Niessen, B. Odermatt, K. Willecke, H. Harz, Exp. Cell Res. **2003**, 291, 25–35.

69 W. Altenhofen, J. Ludwig, E. Eismann, W. Kraus, W. Bönigk, U. B. Kaupp, Proc. Natl. Acad. Sci. USA **1991**, 88, 9868–9872.

70 C. Guirland, K. B. Buck, J. A. Gibney, E. DiCicco-Bloom, J. Q. Zheng, J. Neurosci. **2003**, 23, 2274–2283.

71 S. Munck, P. Bedner, T. Bottaro, H. Harz, Eur. J. Neurosci. **2004**, 19, 791–797.

72 T. Nguyen, W.-C. Chin, J. A. O'Brien, P. Verdugo, A. J. Berger, J. Physiol. **2001**, 531, 131–140.

73 S. Teryshnikova, X. Yan, A. Fein, *J. Physiol.* **1998**, 512, 89–96.
74 S. Tertyshnikova, A. Fein, *Proc. Natl. Acad. Sci. USA* **1998**, 95, 1613–1617.
75 P. Chen, T.-C. Hwang, K. D. Gillis, *J. Gen. Physiol.* **2001**, 118, 135–144.
76 G. Lowe, G. H. Gold, *Nature* **1993**, 366, 283–286.

4.2
Photochemical Release of Second Messengers – Caged Nitric Oxide

Christopher M. Pavlos, Hua Xu, and John P. Toscano

4.2.1
Introduction

Recent research has demonstrated that nitric oxide (NO), a diatomic radical known previously as a noxious environmental pollutant, is involved in a wide range of critical bioregulatory processes including neurotransmission, blood clotting, and blood pressure control [1–3]. In addition, macrophages have been shown to kill cancerous tumor cells and intracellular parasites by releasing large amounts of NO [1–3].

Photosensitive precursors to NO (i.e. "caged NO") have been developed and will be discussed in this chapter. The focus of the discussion will be on the photochemistry of the various precursors. Brief mention will also be given to the use of these caged compounds in biology and medicine, which has centered mainly on the elucidation of neurophysiological roles of NO and the development of new agents for use in photodynamic therapy (PDT) applications. In addition, the photoproduction of NO in biological settings has often been demonstrated via vasorelaxation studies.

Neurophysiological studies using caged NO have aimed to clarify the role of NO in long-term potentiation and long-term depression, two neural processes that are thought to be involved in the mechanism of learning and memory. Since NO is reactive, with a half-life of a few seconds under physiological conditions [4], the use of caged compounds is a necessary but convenient way of producing rapid, precise NO concentration changes such that subsequent neurophysiological responses can be determined.

Proposed PDT applications using caged NO involve the potential of NO in the treatment of cancer. The role of NO in cancer is multifaceted; depending on the location, amount, and duration of release, NO can either kill tumor cells or cause them to proliferate [5–8]. In general, available concentrations of NO in the pico- to nanomolar concentration range are thought to enhance tumor cell growth, whereas concentrations approaching the micromolar range (and higher) are thought to be tumoricidal [8]. The mechanisms involved in the effects are still under active investigation, but appear (at least in part) to involve the regulation of angiogenesis and apoptosis [5–8]. Although NO itself is not generally effective against all tumor cell lines, recent studies have strongly suggested that NO donors can enhance clinically important modalities of cancer treatment including radiation therapy and chemotherapy [5–8].

A major deterrent to the effectiveness of cancer treatments like radiotherapy is the resistance of hypoxic (oxygen-deficient) tumor cells. Howard-Flanders first demonstrated in 1957 that NO radiosensitizes hypoxic bacteria [9]. More recently, NO and NO donors have been shown to be effective radiosensitizers of hypoxic mammalian cells, with an enhancement ratio of 2.4 [6, 10]. A major obstacle for the practical use of NO donors, however, is their influence on systemic blood pressure. Importantly, the use of caged precursors will make it possible to target the sensitizing effects of NO to tumor cells, avoiding unwanted systemic responses.

4.2.2
Photosensitive Precursors to Nitric Oxide

4.2.2.1 Endogenous Photosensitive Precursors to Nitric Oxide

Endothelial-derived nitric oxide synthase (eNOS) is a form of constitutive NOS that constantly produces small amounts of NO in order to regulate the constriction of vascular smooth muscle, and ultimately control blood flow [11, 12]. This constant production of NO results in formation of adducts (e.g., S-nitroso and N-nitroso compounds, nitrosylated metal centers) and NO end products (nitrite and nitrate) that may serve as endogenous NO stores. Furchgott et al. first reported the relaxation of vascular smooth muscle upon UV irradiation [13]. This "photorelaxation" effect is now thought to stem from electronic excitation of the above-mentioned NO stores, resulting in release of NO. Many studies have focused on the identification of the species responsible for this photorelaxation effect [14–24].

While previous studies are in general agreement that the potential identities of these NO stores are S- or N-nitroso compounds, nitrosylated metal centers, nitrite, or nitrate, Feelisch and co-workers have only recently clarified the degree of involvement of each of these compounds [25]. Concentrations of the above species present in rat thoracic aortae were quantified by HPLC methods. Nitrite (10 μM) and nitrate (42 μM) were found to be the major NO-derived species. Lower concentrations of S-nitroso (40 nM) and N-nitroso (33 nM) species were found, and no nitrosylated heme species were observed.

The action spectra for NO release were determined for S-nitrosoglutathione (GSNO) and S-nitrosoalbumin (as representative low- and high-molecular-weight S-nitroso compounds), N,N-dimethylnitrosamine (as a typical N-nitrosamine), nitrite (NO_2^-), and nitrate (NO_3^-). NO release was determined by chemiluminescence detection. A comparison of these data with the photorelaxation action spectrum of rat aortic tissue suggested that the photoreleased NO from the tissue originates mainly from low-molecular-weight S-nitrosothiols, with a small contribution from nitrite. (The photolysis of aqueous nitrite has been shown to result in the production of NO and hydroxyl radical [26, 27].) Mechanisms of bioactivation and potential physiological functions of these NO stores are under active investigation.

4.2.2.1.1 S-Nitrosothiols

The photochemistry of S-nitrosothiols has received significant attention. In 1965 Barrett et al. proposed that S-nitrosothiols undergo photochemical decomposition upon 365 nm irradiation to yield thiyl radical and NO [28]. Mutus and coworkers have reported that visible light photolysis of GSNO, while less efficient than UV photolysis, still significantly enhances NO release in comparison with the dark reaction [29]. The quantum yield of NO release for 545 nm photolysis was reported to be 0.056. Singh et al. have demonstrated that S-nitrosothiols (and also C-nitroso compounds) undergo efficient triplet sensitization with Rose Bengal and other photosensitizers and suggested such a strategy for the localized release of NO in phototherapy [30].

Wood et al. have further explored the photochemical decomposition of S-nitrosothiols in a recently reported laser flash photolysis study of S-nitrosoglutathione (GSNO) and the S-nitroso complex of bovine serum albumin (BSANO) [31]. Upon 355 nm photolysis of GSNO, unresolvable ground state bleaching is observed, attributed to photoinduced homolytic cleavage of the S-N bond producing NO and the glutathione thiyl radical. A slower, resolvable on the microsecond time scale, bleaching of the ground state is also observed. This slower depletion of GSNO is postulated to arise from the reaction of initially produced thiyl radical with GSNO to form a disulfide (GSSG) and a second equivalent of NO ($k=1.9\times10^9$ $M^{-1}s^{-1}$). Alternatively, the thiyl radical can react with oxygen to give glutathione peroxy radical. This transient species also reacts with ground state GSNO to give GSSG and NO. Thus, NO is produced not only via initial photoinduced homolysis, but also from the reactions of GSNO with the glutathione thiyl and peroxy radicals (Scheme 4.2.1). In contrast to these results, 355 nm excitation of BSANO led to photodecomposition, but no evidence for NO production was obtained. The production of BSA^- and NO^+ was suggested in this case.

Shishido and de Oliveira have reported reduced thermal and photochemical reactivity, including a lower quantum yield for NO production, of the S-nitroso adduct of N-acetylcysteine (SNAC) in a polyethylene glycol matrix as compared to that in the solution phase [32]. Additionally, GSNO and SNAC have been incorporated into F127 hydrogels for use as potential drug delivery systems that allow targeting of NO release for biomedical applications [33].

Tedesco and co-workers have also studied the photochemistry of GSNO and SNAC by laser flash photolysis and have utilized a system in which SNAC is photolyzed in the presence of zinc phthalocyanine to produce peroxynitrite [34]. Zinc phthalocyanine is known to produce superoxide anion upon photolysis [35, 36]. When both superoxide and NO are produced, their combination ($k=4\times10^9$ $M^{-1}s^{-1}$) leads ultimately to peroxynitrite formation.

Scheme 4.2.1 The photochemistry of S-nitrosoglutathione (GSNO).

Etchenique and co-workers have designed a NO photodelivery system utilizing S-nitrosothiols bound to a gold surface that has the potential to deliver NO in a controlled way to a very small area [37]. A monolayer of dithiothreitol was allowed to form on the surface prior to a transnitrosation reaction initiated by the addition of S-nitrosodithiothreitol. The newly formed surface was found to be thermally stable, but it released all bound NO upon 60 s of irradiation with a 30 W tungsten lamp.

Zhelyaskov and co-workers have employed GSNO and S-nitrosoacetylpenicillamine (SNAP), in the design of a system in which NO concentration can be kept relatively constant as a function of time [38]. Upon irradiation in aerated solution, NO is cleaved from the nitrosothiol. As mentioned above, the thiyl radical initially produced can react with oxygen to yield a peroxyradical. When the peroxyradical is photolyzed, superoxide is produced. As NO and superoxide react to form peroxynitrite, NO is lost from solution. Thus, the authors suggest that with use of the proper solution pH, nitrosothiol concentration, and light intensity, a constant concentration of NO can be maintained.

4.2.2.1.2 N-Nitrosoamines (Nitrosamines)

Nitrosamines, which are produced *in vivo* from the reaction of secondary amines with nitrite, are well-known carcinogens that are present in processed meats, cigarette smoke, and other foodstuffs. The photochemistry of nitrosamines has been reviewed previously by Chow [39]. Kikugawa and co-workers recently undertook a photochemical study of several nitrosamines to confirm NO release upon irradiation [40]. All of the nitrosamines studied, N-nitrosodimethylamine, N-nitrosodiethylamine, N-nitrosomorpholine, and N-nitrosopyrrolidine, release NO upon UV irradiation in phosphate buffer solution, analogous to S-nitrosothiol compounds discussed above. NO was quantified by spin trapping with EPR detection and by determination of nitrite and nitrate production. Streptozotocin (Fig. 4.2.1), which is known for inducing diabetes mellitus in experimental animals [41], has been also recently demonstrated to be a photochemical NO donor [42]. Because of the known toxicity of most nitrosamines, they are not suitable for most clinical applications.

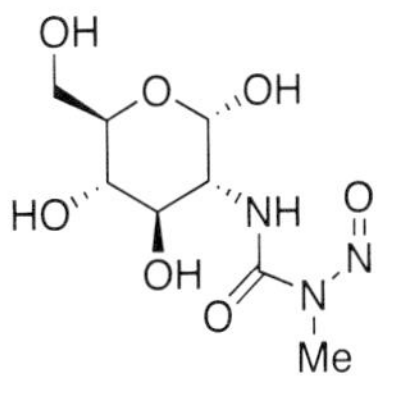

Fig. 4.2.1 The structure of streptozocin.

4.2.2.2 Inorganic Photosensitive Precursors to Nitric Oxide

4.2.2.2.1 Nitric Oxide-Metalloprotein Adducts

NO has been shown to bind to various metalloproteins *in vivo* to form metal-nitrosyl complexes. Consequently, different nitrosyl metalloporphyrins (Fe, Cr, Mn) have been investigated as potential photochemical NO donors. Photodecomposition of these complexes generally proceeds via initial loss of NO to form an intermediate solvento species which can either react with oxygen (yielding final products) or with NO (yielding starting material). The kinetics of recombination with NO can vary drastically depending on the coordinating ability of the solvent. A number of reviews have been written on the photochemistry and recombination kinetics of these systems [43–51].

4.2.2.2.2 Sodium Nitroprusside (SNP)

The photochemistry of the disodium salt of pentacyanonitrosylferrate, commonly known as sodium nitroprusside (SNP), has been investigated in numerous studies [52–56] dating back to the mid 1900s, but has only more recently received attention as a possible photochemical NO donor for clinical applications. Buxton et al. first proposed the ejection of nitric oxide and resultant formation of the aquated Fe(III) species as the primary photochemical process (Scheme 4.2.2) [57].

Wolfe and Swinehart [53] verified that NO and not the nitrosonium cation (NO^+), as previously suggested [52], was the species released upon photolysis. They additionally determined the quantum yield for production of the aquated species to be 0.35 and 0.18 for photolysis at 366 and 436 nm, respectively. Ford and co-workers [58] later determined by electrochemical methods that the production of the aquated species was stoichiometrically equivalent to the production of NO.

In addition to NO, both cyanide anion (CN^-) and cyanogen (C_2N_2) have been detected in solution after photolysis of SNP [59–61]. These products were initially thought to arise from photolysis of SNP. However, in 1995 de Oliveira et al. [62] proposed that CN^- and CN radical (which dimerizes to cyanogen) arise only from photolysis of the Fe(III) aquated species (Scheme 4.2.3). The production of these toxic species was later reported to be completely circumvented at wavelengths of photolysis greater than 480 nm [63].

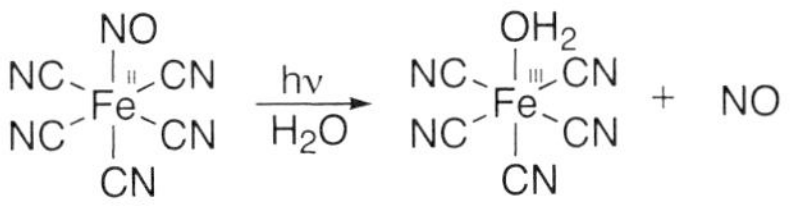

Scheme 4.2.2 The primary photochemical reaction of sodium nitroprusside (SNP).

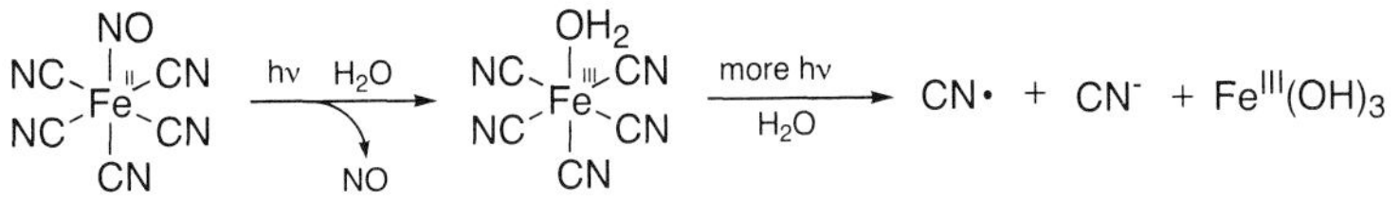

$$\text{NC}\underset{\text{CN}}{\overset{\text{NO}}{\underset{|}{\overset{|}{\text{Fe}}}}}\text{CN} \quad \xrightarrow[\text{H}_2\text{O}]{h\nu} \quad \text{NC}\underset{\text{CN}}{\overset{\text{OH}_2}{\underset{|}{\overset{|}{\text{Fe}}}}}\text{CN} \quad \xrightarrow[\text{H}_2\text{O}]{\text{more } h\nu} \quad \text{CN}\cdot \ + \ \text{CN}^- \ + \ \text{Fe}^{III}(\text{OH})_3$$

Scheme 4.2.3 The photochemistry of sodium nitroprusside (SNP).

The thermal production of NO from SNP is also well documented, and SNP has been used to this effect in many clinical applications [64]. However, irradiation of SNP greatly increases the rate of NO production. Unfortunately, research has also shown that SNP thermally releases CN^- *in vivo* [60].

Despite the complexity of the photochemistry of SNP and the potential CN^- toxicity concerns, it has been used in several applications requiring the introduction of NO in a controlled fashion. Specifically, SNP has been used to investigate the effects of both NO and cGMP on potassium channel activity in posterior pituitary nerve terminals [65]. Az-Ma and Yuge have irradiated SNP in the presence of various volatile anesthetics in order to study their interactions with NO [66].

Additionally, SNP has been utilized by Zhelyaskov and Godwin [67, 68] in the design of two systems that offer a controllable and constant concentration of nitric oxide for a given experiment. In the most recent example [68], an electronic system (utilizing negative feedback from an electrochemical NO sensor) which maintains a constant (within 2%) concentration of NO in a closed cell was described. A pre-determined concentration is set before NO is released photochemically from SNP. When the desired concentration is reached (as detected by the probe), a shutter blocks the light, and NO production stops. As the concentration of NO begins to decay (either by reaction with oxygen or by diffusion out of solution) the shutter is opened again. In this way a constant, controllable concentration of NO can be maintained in the cell.

A caged form of SNP itself has also been prepared by Kunkely and Vogler [69]. Photolysis of aqueous $[Fe(CN)_5(ONPh)]^{3-}$ was shown to produce SNP and benzene with a quantum yield of 0.03 at 546 nm. The authors propose that the MLCT state of the complex carries an additional negative charge on the nitrosobenzene ligand, expediting the protonation and subsequent decomposition to benzene and NO, which then recombines with the $Fe^{III}(CN)_5$ complex to form SNP (Scheme 4.2.4).

Stasicka and co-workers have explored the photochemistry of the SNP-thiolate system in search of a new form of caged NO that can be activated by light of longer wavelength [70, 71]. Solutions of SNP and various thiolates have been demonstrated to be in equilibrium, as shown in Scheme 4.2.5.

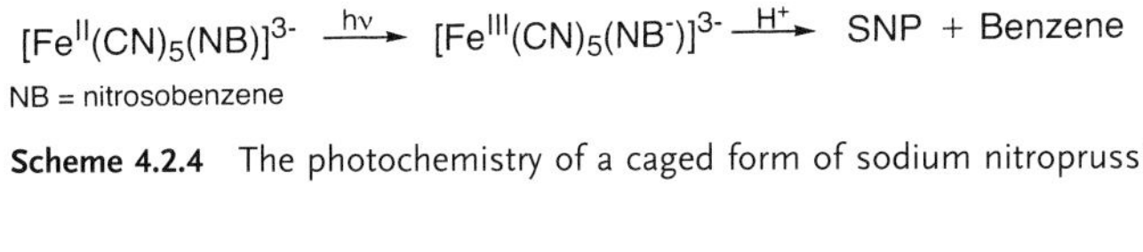

$$[Fe^{II}(CN)_5(NB)]^{3-} \xrightarrow{h\nu} [Fe^{III}(CN)_5(NB^-)]^{3-} \xrightarrow{H^+} SNP + Benzene$$

NB = nitrosobenzene

Scheme 4.2.4 The photochemistry of a caged form of sodium nitroprusside (SNP).

$[Fe(CN)_5(NO)]^{2-} + RS^- \rightleftharpoons [Fe(CN)_5N(O)SR]^{3-}$

Scheme 4.2.5 The equilibrium involving sodium nitroprusside and thiolates.

Although both SNP and the SNP-thiolate complexes are photoactive, photolysis of the latter results in the production of a nitrosothiol radical anion, ultimately yielding a nitrosothiol (via oxidation by oxygen). The use of this complex is advantageous because NO can be delivered ultimately (via thermal or photochemical decomposition of the nitrosothiol) by long wavelength photolysis ($\lambda > 500$ nm) [70].

As the system shown above is in equilibrium, the desired photoproduct can be chosen by the wavelength of irradiation (i.e., longer wavelength photolysis results in the production of RSNO, shorter wavelength photolysis yields NO directly). This system was further exploited recently and proposed for use as a molecular switch [71]. The equilibrium shown above can be manipulated by factors such as pH, temperature, ionic strength, and concentration of thiolate. Since two different primary photoproducts (NO or RSNO) can be produced depending on the species being irradiated, the ultimate observed product can be controlled by varying these factors accordingly.

4.2.2.2.3 Roussin's Red and Black Salts (RRS, RBS)

Although the first description of the anions of Roussin's black salts, $Fe_4S_3(NO)_7^-$ (RBS, Fig. 4.2.2), and of Roussin's red salt, $Fe_2S_2(NO)_4^{2-}$ (RRS, Fig. 4.2.2) dates back to the mid 19th century, their photochemistry has drawn attention only recently. Ford and co-workers have previously reviewed their photochemistry [45, 72].

RBS and RRS are highly colored compounds with strong absorption in the visible region. RBS has a broad absorption band at 580 nm which extends into the red. The strongest absorption band of RRS is centered at 380 nm and tails into the visible [45].

RBS is thermally stable in the dark and upon photolysis in deoxygenated solutions. However, upon photolysis in aerobic solutions, RBS slowly undergoes decomposition to give 5.9 equivalents of NO with a quantum yield of 0.0011 at

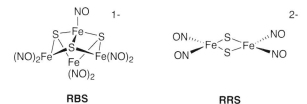

Fig. 4.2.2 The structures of Roussin's black salt (RBS) and Roussin's red salt (RRS).

365 nm [73, 74]. Photolysis in the presence of simple oxidants such as $Co(NH_3)_6^{3+}$ also results in the release of NO.

RRS is thermally unstable in aerated solution at neutral pH and reacts to give RBS over the course of a few hours in the dark. Upon photolysis, this process is significantly accelerated. The quantum yield for conversion of RRS to RBS is 0.14, independent of the irradiation wavelength [73]. Electrochemical NO analysis showed that 0.5 mol NO was released for each mol of RRS converted to RBS. Flash photolysis studies suggest that RRS photodissociates to NO and an intermediate species, $Fe_2S_2(NO)_3^{2-}$, that reacts competively with NO ($k_{NO} = 9.1\times10^8$ $M^{-1}s^{-1}$) to reform RRS and with O_2 ($k_{O_2} = 5.7\times10^7$ $M^{-1}s^{-1}$) to form a second intermediate, $Fe_2S_2(NO)_3^-$, which is likely the precursor of the eventual photoproduct RBS (Scheme 4.2.6). In deaerated aqueous solution, photolysis of the red salt also leads to the formation of RBS but with a much smaller quantum yield (0.0039) [72].

Despite being toxic to cell cultures [73], RBS has been used to enhance the effect of γ-irradiation in hypoxic V79 cells [73], to investigate the NO-dependent relaxation of smooth muscle in the guinea-pig [75], and to study the effect of NO on dopamine efflux in the rat brain [76].

Because neither RBS (because of its toxicity) nor RRS (because of its instability) is suitable for practical applications in complex systems, Roussin's red esters $Fe_2(SR)_2(NO)_4$ (RRE, Fig. 4.2.3) were developed very recently by Ford and coworkers [77].

RREs are thermally stable (in comparison to RRS) while remaining photoactive toward NO release. Continuous photolysis of these esters in aerated solutions leads to the release of 4 equivalents of NO. Quantum yields are moderate and vary over an order of magnitude (from 0.02 to 0.13) depending on the wavelength, light intensity, and solvent. LFP experiments indicate that the initial photoreaction is the reversible dissociation of NO. In the absence of O_2, the initial intermediate (I) reacts with NO to regenerate the parent compound ($k_{NO} = 1.1\times10^9$ $M^{-1}s^{-1}$). In the presence of O_2, the intermediate is oxidized ($k_{O_2} = 1.3\times10^7$ $M^{-1}s^{-1}$) to one or more new unidentified species (X) that ultimately decompose to final products (Scheme 4.2.7).

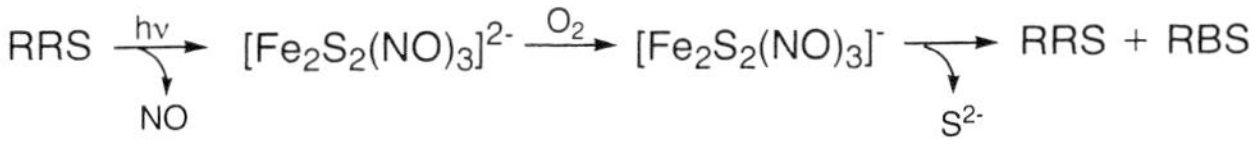

$$RRS \xrightarrow[NO]{hv} [Fe_2S_2(NO)_3]^{2-} \xrightarrow{O_2} [Fe_2S_2(NO)_3]^- \xrightarrow{S^{2-}} RRS + RBS$$

Scheme 4.2.6 The photochemistry of Roussin's red salt (RRS).

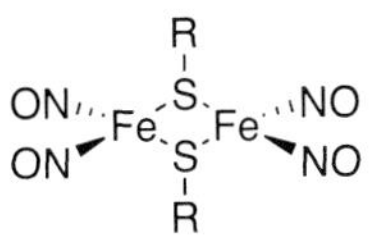

R = Me, Et, Bz, EtOH, EtSO₃

Fig. 4.2.3 The structure of Roussin's red esters (RRE).

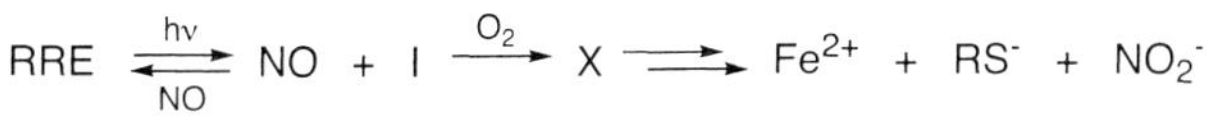

$$\text{RRE} \underset{\text{NO}}{\overset{h\nu}{\rightleftharpoons}} \text{NO} + \text{I} \xrightarrow{O_2} \text{X} \longrightarrow \text{Fe}^{2+} + \text{RS}^- + \text{NO}_2^-$$

Scheme 4.2.7 The photochemistry of Roussin's red ester (RRE).

4.2.2.2.4 Other Iron-based Photosensitive Precursors

Recently, Mascharak and co-workers reported the photochemical release of NO from a non-heme Fe(III) species [78]. (NO normally prefers Fe(II) centers.) When a solution of [Fe(PaPy$_3$)(CH$_3$CN)][ClO$_4$]$_2$ in acetonitrile is purged with NO gas, a red color is observed (attributed to the NO bound complex). (See Fig. 4.2.4 for the structure of PaPy$_3$.) When the synthesis is performed in methanol, red crystals suitable for X-ray analysis are obtained. This complex represents the first non-heme Fe(III) species to bind NO. The authors report that the complex is stable in acetonitrile solution, but loses NO upon photolysis (50 W tungsten lamp). Upon the addition of NO gas to a photolyzed solution, the NO bound complex is formed again.

Another Fe-based photochemical NO precursor has recently been developed by Kunkely and Vogler [79]. Iron(III) cupferronate is reported to undergo a photochemical redox reaction leading initially to the formation of Fe(II) and a cupferronate radical. The radical subsequently decays to nitrosobenzene and nitric oxide. Photolysis is efficient, as the quantum yield for disappearance of the compound is 0.03 at 366 nm.

4.2.2.2.5 Ruthenium-based Photosensitive Precursors

Nitrosyl chlorides Nitrosyl derivatives of ruthenium have been shown to be thermally stable, but photolabile, releasing NO on exposure to near-UV light

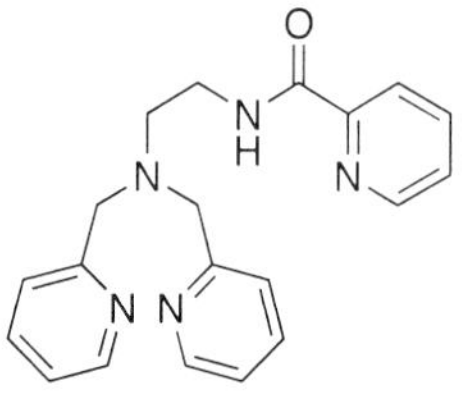

Fig. 4.2.4 The structure of PaPy$_3$.

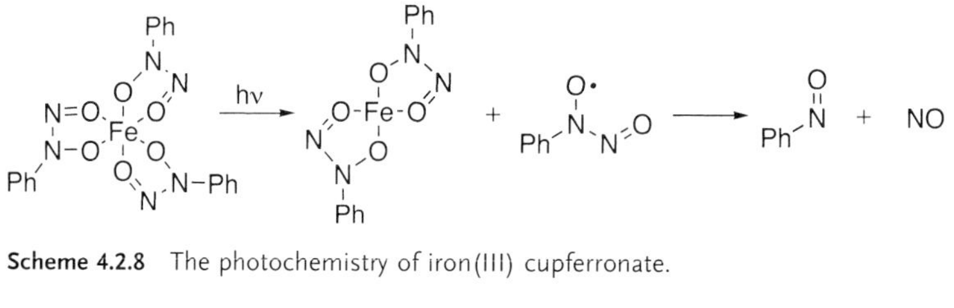

Scheme 4.2.8 The photochemistry of iron(III) cupferronate.

[80]. Two commercially available compounds, $K_2RuCl_5(NO)$ and $RuNOCl_3$ have been investigated by Trentham and co-workers [80]. $K_2RuCl_5(NO)$ has an absorption maximum at 253 nm ($\varepsilon = 11000$ M^{-1} cm^{-1}) and a shoulder band at 335 nm. Upon irradiation in the range 300–500 nm, NO is rapidly released (within 1 µs after the laser pulse) with a quantum yield of 0.06 [80]. Additionally, $RuNOCl_3$ was found to release NO with a quantum yield of 0.012. The authors suggested the use of $K_2RuCl_5(NO)$ in preference to $RuNOCl_3$ for clinical applications because of the higher efficiency of NO release. In addition to eliciting smooth muscle relaxation [80], $K_2RuCl_5(NO)$ has been used to study long-term potentiation of synaptic transmission in hippocampal slices [81, 82], interneuronal activity in mollusks [83], and the action of NO on rat neurohypophysial K^+ channels [84].

Ruthenium salen and salophen nitrosyls Ford and co-workers developed the salen-type ruthenium nitrosyl complexes Ru(R-salen)(X)(NO) and the related salophen-type complexes Ru(R-salophen)(X)(NO) (where R-salen is a derivative of the N,N'-ethylenebis(salicylideneiminato) dianion, R-salophen is a derivative of the N,N'1,2-phenylenebis(salicylideneiminato) dianion, and X=Cl or ONO, see Fig. 4.2.5) as a new class of ruthenium-based photosensitive NO precursors [85, 86]. Upon 365 nm photolysis in acetonitrile, these Ru(L)(X)(NO) compounds undergo NO dissociation to give the ruthenium(III) solvento products, Ru(L)(X)(Sol), with moderate quantum yields ranging from 0.055 to 0.13. The solvento species can undergo reaction with NO to reform the nitrosyl complexes. The rates of these back-reactions vary dramatically in different solvents. For example, the second-order rate constant k_{NO} is 5×10^{-4} M^{-1}s^{-1} in acetonitrile but 5×10^8 M^{-1}s^{-1} in cyclohexane [86].

A water-soluble derivative of these salen complexes, Ru(salen)(NO)(H$_2$O)$^+$, has also been prepared [86, 87]. The quantum yield for NO release from this derivative is 0.005 at 365 nm. Ru(salen)(H$_2$O)$_2^+$ was suggested to be the photoproduct based on UV-visible, EPR, and FTIR spectroscopic data.

Bis(2,2'-bipyridine)nitrosylruthenium(II) derivatives Callahan and Meyer first reported the photochemical release of NO from complexes of the type Ru^{II}X(bpy)$_2$(NO) (where bpy=bipyridine and X=N$_3^-$, Cl$^-$, NO$_2^-$, NH$_3$, pyridine, or CH$_3$CN) in 1977 [88]. The photochemistry of [Ru^{II}Cl(bpy)$_2$(NO)]$^{2+}$ has also been

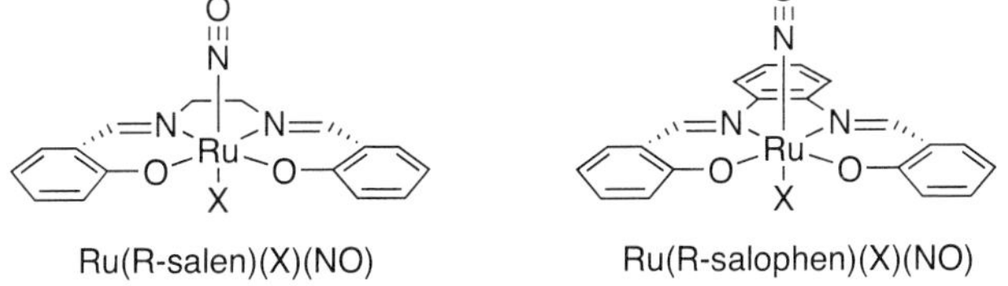

Ru(R-salen)(X)(NO)   Ru(R-salophen)(X)(NO)

Fig. 4.2.5 The structures of Ru(R-salen)(X)(NO) and Ru(R-salophen)(X)(NO) complexes.

studied more recently and NO release confirmed [89]. In aqueous pH 5.7 solution, the complex releases NO with a quantum yield of 0.98 following 355 nm photolysis. The photoproduct was identified as $[Ru^{III}Cl(bpy)_2(OH)]^+$ with a proposed mechanism of photodecomposition involving metal-to-ligand charge transfer, as previously reported [88], before ejection of NO.

Pyrazine-bridged nitrosyl ruthenium complexes An additional ruthenium-based form of caged NO has recently been developed by da Silva and co-workers [90]. In aqueous media, this dinuclear complex undergoes an electron transfer process upon visible light photolysis (532 nm). The resultant charge transfer state subsequently releases NO and associates with a water molecule (Scheme 4.2.9).

4.2.2.2.6 Chromium-based Photosensitive Precursors

The aqueous photochemistry of $trans$-$Cr(cyclam)(ONO)_2^+$ (cyclam = 1,4,8,11-tetraazacyclotetradecane) has been studied in detail [91, 92]. Although no photodecomposition takes place under an inert atmosphere, photolysis in aerated solution produces NO efficiently (quantum yield ~0.1). The proposed mechanism for photodecomposition proceeds via loss of NO and formation of an intermediate species. In the presence of oxygen, the intermediate is trapped via oxidation to give a stable Cr^V species, ultimately yielding permanent products. In the absence of oxygen, the back reaction with NO readily occurs ($k = 3.1 \times 10^6$ $M^{-1}s^{-1}$) and the starting material is reformed (Scheme 4.2.10). These complexes benefit from thermal stability that eludes other metal-nitrosyl complexes.

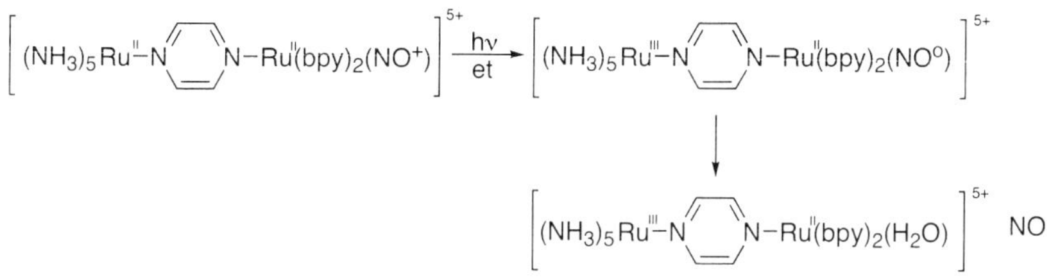

Scheme 4.2.9 The photochemistry of a pyrazine-based nitrosyl ruthenium complex.

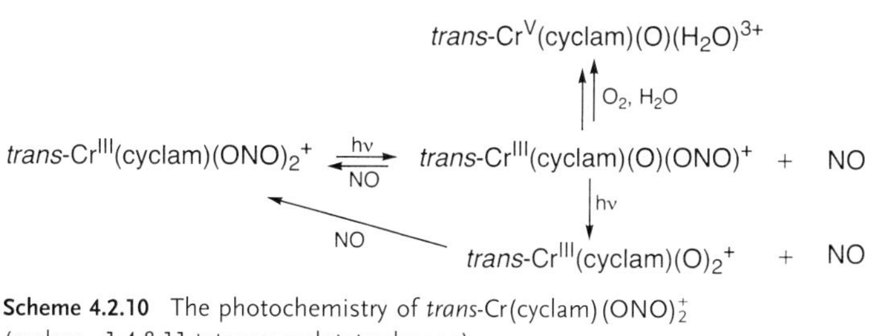

Scheme 4.2.10 The photochemistry of $trans$-$Cr(cyclam)(ONO)_2^+$ (cyclam = 1,4,8,11-tetraazacyclotetradecane).

4.2.2.2.7 Molybdenum-based Photosensitive Precursors

The photochemistry of a dinitrosyl Mo complex was recently studied by Yonemura and co-workers [93]. Solutions of the complex shown below in DMF are thermally stable but release one equivalent of NO upon exposure to ambient light as shown in Scheme 4.2.11. Crystal structures were obtained for both the mono and dinuclear complexes. The origin of the hydroxo bridges is unknown, but they are suggested to come from dissolved water or oxygen in the solvent.

4.2.2.3 Organic Photosensitive Precursors to Nitric Oxide

4.2.2.3.1 Organic Nitrites

In 1961, Barton and co-workers reported a new photochemical reaction for alkyl nitrites with sufficiently long alkyl chains (Scheme 4.2.12) [94]. The synthetic utility of this photochemical process, which leads ultimately to a γ-hydroxy oxime, has been reviewed recently by Majetich [95].

The primary photochemical process for organic nitrites is homolysis of the O-NO bond to give an alkoxy radical and NO. Subsequent processes for systems in which the Barton reaction is not possible include radical recombination and disproportionation. Ludwig and McMillan showed that the O-NO bond cleavage reaction has a temperature-dependent quantum yield ranging from 0.3 to 0.5 for isopropyl nitrite [96].

In the case of *tert*-butyl nitrite, the quantum yield for O-NO bond homolysis following photolysis at 366 nm is close to 1.0 [97]. However, radical recombina-

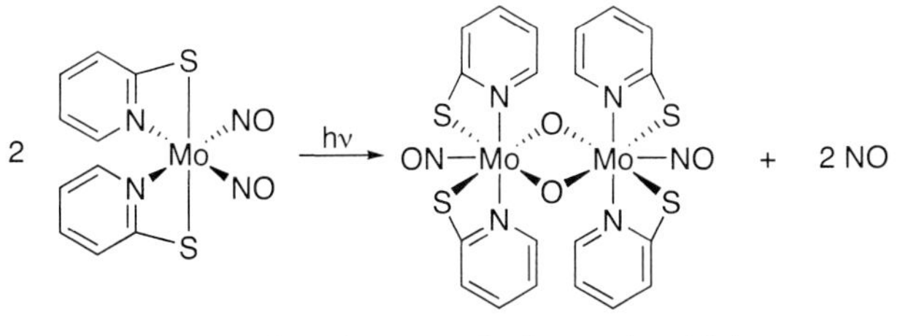

Scheme 4.2.11 The photochemistry of a dinitrosyl molybdenum complex.

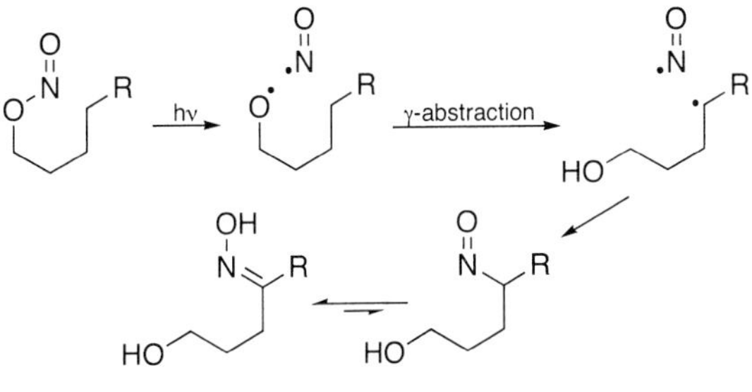

Scheme 4.2.12 The photochemistry of alkyl nitrites.

tion is completely efficient as disproportionation cannot take place. At lower wavelengths, *tert*-butoxy radical is proposed to undergo photochemical decomposition, leading to permanent production of NO.

Because (1) most alkyl nitrites do not lead ultimately to NO production, and (2) those that do simultaneously produce other radicals, as well as aldehydes, these compounds have not found utility in applications requiring the photochemical release of NO.

4.2.2.3.2 Aromatic Nitro Compounds

Although not all nitroarenes release NO upon photolysis, aromatic nitro compounds that do yield NO on irradiation share one common conformational feature: steric hindrance in each case causes the nitro group to be nearly perpendicular to the plane of the aromatic ring. In this out-of-plane geometry, substantial overlap exists between the p orbital of the oxygen atom and the p orbital of the aromatic ring in both the ground and excited states [98]. This overlap leads, in the excited state, to the formation of an oxaziridine ring, which reopens to form the nitrite. The aromatic nitrite, ultimately, undergoes O-N bond homolysis to release NO and a phenoxyl radical.

This mechanism was first proposed by Chapman and co-workers in a study of 9-nitroanthracene photochemistry (Scheme 4.2.13) [98]. A recent example is 6-nitrobenzo[a]pyrene (Fig. 4.2.6), which has been shown to release NO and

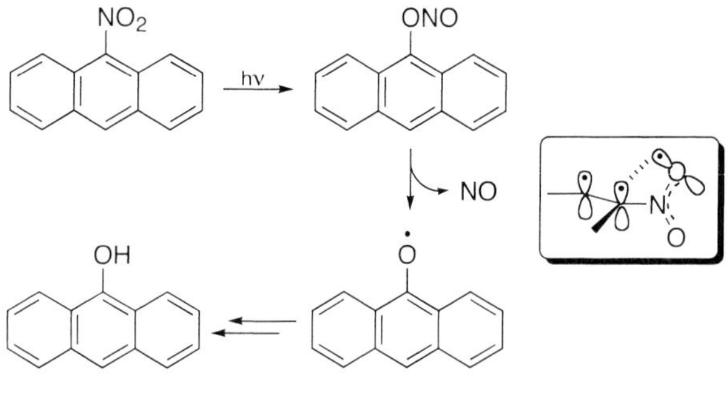

Scheme 4.2.13 The photochemistry of 9-nitroanthracene.

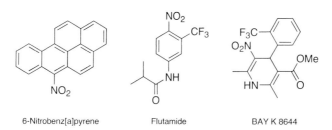

6-Nitrobenz[a]pyrene   Flutamide   BAY K 8644

Fig. 4.2.6 The structures of 6-nitrobenz[a]pyrene, flutamide, and BAY K 8644.

cause DNA damage [99]. The photoreactivity of the anticancer drug flutamide (Fig. 4.2.6) has recently been examined [100], and subsequently this compound has been immobilized on gold surfaces by incorporation into a novel self-assembled monolayer (SAM) [101]. These SAMs release NO upon irradiation ($\lambda > 380$ nm) and have been proposed to be useful for the delivery of a quantitative amount of NO to a target site. The aromatic nitro compound BAY K 8644 (1,4-dihydro-2,6-dimethyl-3-nitro-4-(2-trifluoromethylphenyl)-5-pyridinecarboxylic acid methyl ester, Fig. 4.2.6) has been used as a photochemical NO donor in a study of photorelaxation in the rat aorta [102].

4.2.2.3.3 Furoxans

Although 3,4-disubstituted furoxans (Fig. 4.2.7) were reported in 1995 to release NO through electron impact fragmentation [103, 104], only more recently have reports appeared indicating that irradiation of diarylfuroxans gives diaryacetylenes and NO. For example, photolysis of 3,4-bis-2'-chlorophenylfuroxan generates NO and bis-2-chlorophenylacetylene in 17% yield [105]. However, not all furoxans release NO photochemically, as irradiation of some unsymmetrical furoxans leads exclusively to isomerization [106, 107]. Auricchio and co-workers have recently clarified the issue, suggesting that the substituents R in the 3 and 4 positions control the outcome of the photochemical reaction [105]. Furoxans are known to be in equilibrium with the corresponding dinitroso species [106, 107], and interconversion can be effected thermally or photochemically. Additionally, the dinitroso species can exist in the form of a diazetine dioxide.

Auricchio et al. have calculated strain energies for several derivatives and found that the dinitroso species will be favored when R is alkyl or alkoxide. Thus, irradiation of these furoxan derivatives results first in the dinitroso species and ultimately leads to isomerization (Scheme 4.2.14).

However, dinitroso species where R=aryl were calculated to be more stable in the diazetine dioxide form. Therefore, upon irradation these furoxan derivatives form ultimately two equivalents of NO and an acetylene.

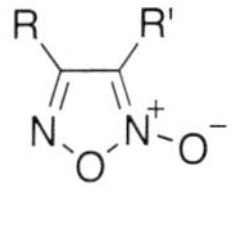

Fig. 4.2.7 The structure of 3,4-disubstituted furoxans.

R = Me
R' = OEt

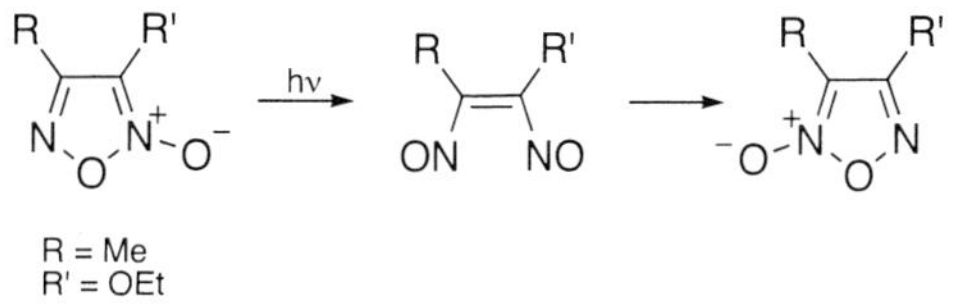

Scheme 4.2.14 The isomerization of 3,4-disubstituted furoxans.

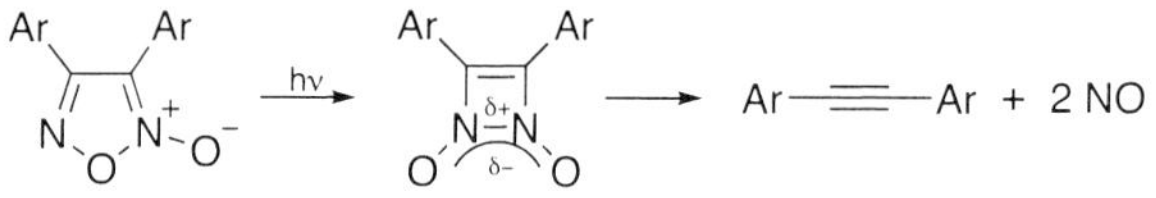

Scheme 4.2.15 The release of NO from 3,4-disubstituted furoxans.

Although furoxans have not enjoyed use as NO donors in clinical applications, Pashchenko and co-workers employed piridazinofuroxan to investigate the rate of NO trapping by the water-soluble iron complex Fe(II)-proline-dithiocarbamate [108]. The quantum yield for NO formation from this donor induced by 308 nm light was found to be 0.027.

4.2.2.3.4 Linsidomine (SIN-1)

SIN-1, which under aerobic conditions thermally decomposes to yield NO and also forms superoxide, is the active metabolite of the anti-anginal drug molsidomine. Ullrich et al. investigated the photochemical release of nitric oxide from SIN-1 in 1997 [109]. NO release was determined by quantification of nitrite and nitrate via the Griess assay, and was found to be enhanced by 61% as compared to the control (thermal release) upon irradiation (40 W tungsten lamp).

Tertyshnikova, Yan, and Fein utilized a caged form of SIN-1 for a study of inositol 1,4,5-triphosphate receptor-mediated calcium release in intact rat megakaryocytes [110]. Caged SIN-1 is thermally stable, but releases SIN-1 and ultimately NO within milliseconds after UV photolysis (Scheme 4.2.16).

4.2.2.3.5 Phenyl N-tert-Butyl Nitrone

NO has been reported to be produced upon photolysis of phenyl-N-tert-butyl nitrone (PBN), a commonly used spin trap in free radical research [111]. Aqueous photolysis of PBN results in the formation of tert-nitrosobutane (t-NB). Subsequent C-N bond homolysis provides NO and t-butyl radical. The free carbon-centered radical is then trapped by a second molecule of t-BN to yield the stable nitroxide radical, di-tert-butylnitroxide (Scheme 4.2.17).

NO production was verified by several methods, including iron-nitrosyl formation with EPR detection and nitrite determination via the Griess assay. Addition-

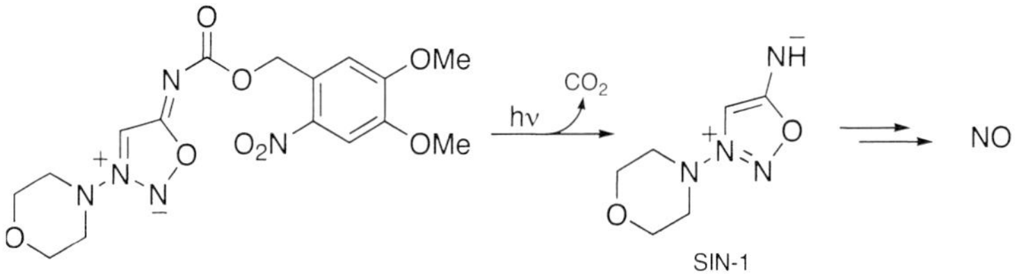

Scheme 4.2.16 The release of NO from a caged form of linsidomine (SIN-1).

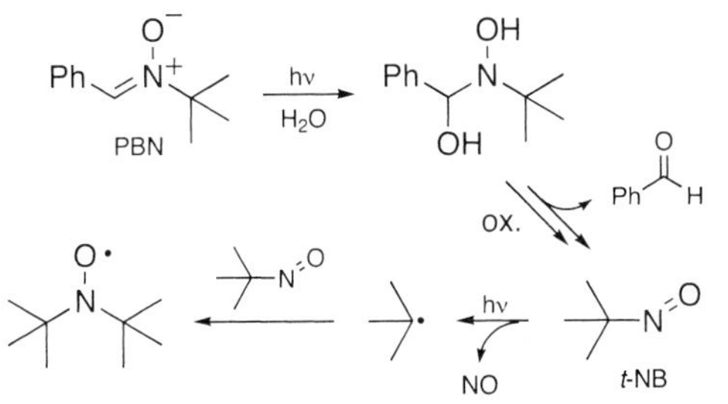

Scheme 4.2.17 The photochemistry of phenyl-N-tert-butyl nitrone (PBN).

ally, photolysis of PBN was shown to activate guanylate cyclase, causing an increase in the concentration of cGMP in solution.

4.2.2.3.6 Dialkylamino-based Diazeniumdiolates

Compounds containing the diazeniumdiolate [N(O)=NO]$^-$ functional group have proven useful as research tools in a variety of applications requiring spontaneous release of NO [112]. Anions such as 1-(N,N-dialkylamino)diazen-1-ium-1,2-diolates (1) are stable as solid salts, but release up to 2 mol of NO when dissolved in aqueous solution at physiologically relevant conditions (Scheme 4.2.18). (The formation of such compounds by the reaction of NO with nucleophiles such as amines has been known since the 1960s [113–116].) Keefer and co-workers have shown that the rate of NO release can be varied by modifying the substituents R, pH, or temperature, and have developed anions with half-lives in aqueous buffer at pH 7.4 and 37°C ranging from two seconds to 20 h [112]. In addition, diazeniumdiolate solution half-lives tend to correlate very well with their pharmacological durations of action, suggesting that they are minimally affected by metabolism [117]. These compounds have shown great potential in a variety of medical applications requiring either the rapid production or gradual release of NO [117, 118], and have allowed biological consequences of NO delivery rates to be probed [119].

Recent efforts to make diazeniumdiolates more effective pharmaceuticals have concentrated on using derivatives of such compounds to deliver NO specifically to a targeted site. A number of these efforts have focused on the development of photosensitive precursors to diazeniumdiolates, the first of which was reported in 1994 by Makings and Tsien, who synthesized O^2-2-nitrobenzyl diazeniumdiolate derivatives 2 (R'=H, OCH$_3$, OCH$_2$CO$_2$Et, OCH$_2$CO$_2^-$ K$^+$, OCH$_2$CO$_2$CH$_2$O$_2$CCH$_3$) and studied the kinetics of NO release [120]. Depending on the substituents R', extracellular or intracellular NO release could be targeted.

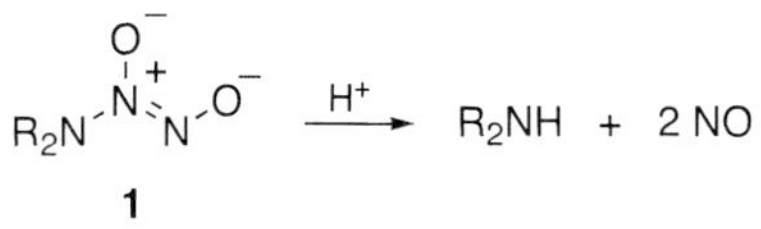

Scheme 4.2.18 The release of NO from 1-(N,N-dialkylamino)diazen-1-ium-1,2-diolates (1).

On the basis of known 2-nitrobenzyl chemistry, these potential prodrugs were *expected* to photodecompose rapidly to nitrosoaldehydes **3** and diazeniumdiolate **1** (R=Et) (Scheme 4.2.19), which has been shown [112] to release two moles of NO (along with diethylamine) with a half-life of 2.1 min at pH 7.4 and 37 °C. Instead, however, NO was produced much faster, within the 5 ms time resolution of Makings and Tsien's flash photolysis experiments. A potential rationalization of the observed rapid formation of NO is that the initially produced diazeniumdiolate is protonated (thus catalyzing its decomposition) as a result of the 2-nitrobenzyl uncaging mechanism. Quantum yields for NO formation were relatively low, ranging from 0.02 to 0.05; nonetheless, these phototriggered NO donors were used to inhibit thrombin-stimulated platelet aggregation [120], and were also used in a variety of biological applications, most notably to examine the induction of long-term depression in the cerebellum [121–123], and to study long-term potentiation in cultured hippocampal neurons [124–126].

Several years later, a photochemical study of O^2-alkylated diazeniumdiolates (including **2**, R'=H) was reported [127]. Product distributions and the mechanistically powerful tool of time-resolved infrared (TRIR) spectroscopy revealed two primary reaction pathways, the major (95%) of which involves the formation of potentially carcinogenic nitrosamine ($R_2NN=O$) and an oxygen-substituted nitrene (RON). The product distributions for O^2-2-nitrobenzyl derivative **2** (R'=H) are exactly analogous to those of O^2-benzyl derivative **4** (Scheme 4.2.20). For each compound examined, minor amounts of NO, potentially produced by secondary photolysis of the nitrosamine, were observed. Nanosecond TRIR experiments provided additional support for these reaction pathways and confirmed that nitrosamine is a primary photoproduct.

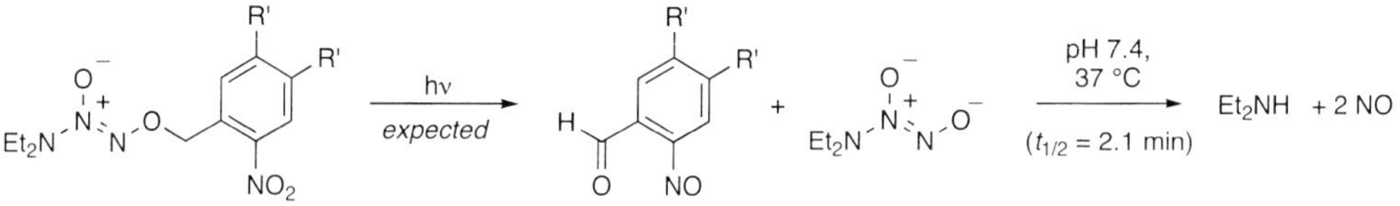

Scheme 4.2.19 Expected photochemistry of O^2-2-nitrobenzyl diazeniumdiolate derivatives **2**.

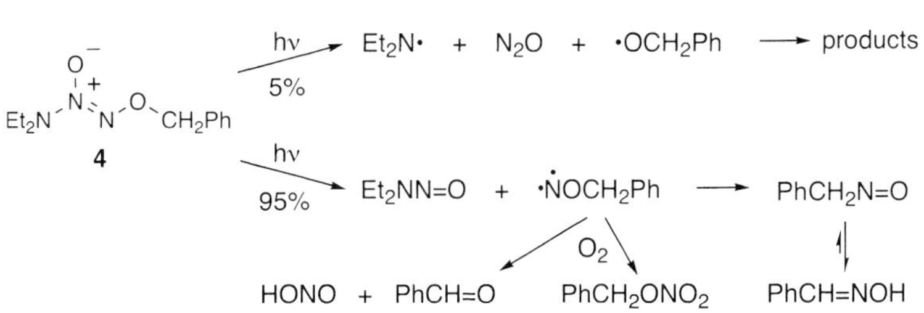

Scheme 4.2.20 Photochemistry of O^2-benzyl diazeniumdiolate derivative **4**.

Subsequent to the above studies, a series of *meta*-substituted O^2-benzyl-substituted diazeniumdiolates was reported as a potential class of photosensitive precursors to diazeniumdiolates [128]. These studies attempted to take advantage of the well-established meta effect of electron-donating and electron-withdrawing groups in benzylic systems [129–131]. Such substitution has been shown to favor the formation of ionic products in photochemical reactions. On the basis of this previous work, a series of O^2-benzyl-substituted diazeniumdiolates, **5–9**, was prepared and the effect of electron-donating meta substituents on the efficiency of NO release was examined [128]. Results concerning the photochemistry (Rayonet, 300 nm in 90% aqueous acetonitrile) of **5–9** were interpreted in terms of *Path A* (undesired) and *Path B* (desired), shown in Scheme 4.2.21. The relative contributions of *Path A* and *Path B* were found to be strongly dependent on the aromatic ring substitution pattern. The yield of nitrosamine decreases and that of NO increases with stronger π-donating meta substituents. The photodecomposition of O^2-substituted diazeniumdiolates **5–8** proceeds approximately 5, 10, 35, and 50% through *Path B*, respectively.

In order to differentiate NO production arising from **1** (R = Et) from non-diazeniumdiolate-derived NO, the pH dependence of diazeniumdiolate stability was utilized. At room temperature **1** (R = Et) has a lifetime of several hours at pH 11, but only several seconds at pH 3. Aqueous solutions at pH 11 of each reactant were irradiated and then analyzed for NO. Any NO detected under these conditions must arise from a non-diazeniumdiolate pathway. For all reactants examined, the amount of NO formed from such a pathway was negligible. In order to liberate the NO associated with photoreleased **1** (R = Et), solutions were acidified with 1.0 M H_2SO_4 to pH 2–3, and NO analysis was continued. Substantial amounts of NO were then observed and quantified.

Given the observation that the desired *Path B* is favored with strong π-donating meta substituents, dihydroxybenzyl derivative **9** was examined. Photoinduced deprotonation [132, 133] to yield an oxyanionic meta-substituted derivative was expected to result in significant enhancement of the relative contribution of *Path*

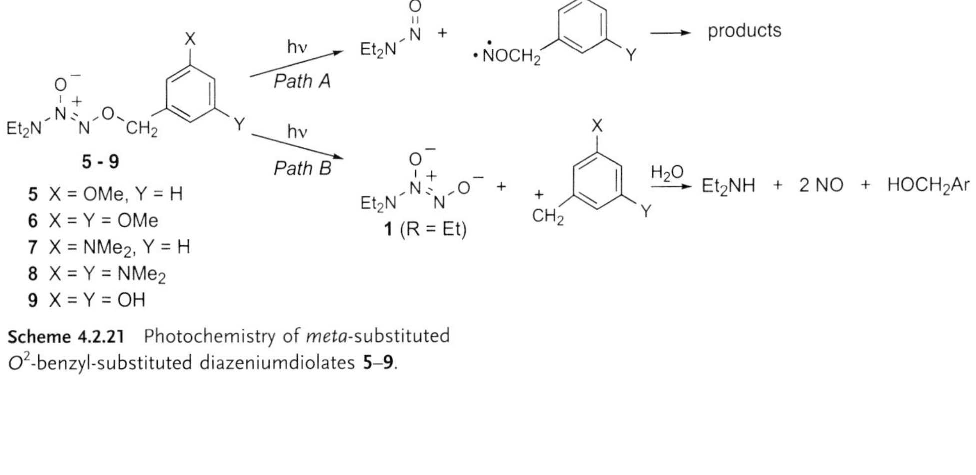

5 - 9

5 X = OMe, Y = H
6 X = Y = OMe
7 X = NMe$_2$, Y = H
8 X = Y = NMe$_2$
9 X = Y = OH

Scheme 4.2.21 Photochemistry of *meta*-substituted O^2-benzyl-substituted diazeniumdiolates **5–9**.

B to the observed photochemistry. Consistent with the expected large effect of oxyanionic meta substitution, when **9** was irradiated (Rayonet, 300 nm) in a pH 11.2 aqueous solution (containing 1% acetonitrile) to 5–35% conversion, the amount of NO detected corresponded to a yield of photoreleased **1** (R = Et) of approximately 92%. When **9** was irradiated (Rayonet, 300 nm) in pH 8.4 or 7.4 solutions, however, the yield of **1** (R = Et) was substantially reduced (to 18% and 16%, respectively), while that of nitrosamine was significantly enhanced. Excited state deprotonation must not be rapid enough to compete with *Path A* under these conditions. The pH 11.2 results are easily rationalized by simply considering that the pK_a of the phenolic protons of **9** is approximately 10. At pH 11.2, it is the oxyanion that is excited, and thus *Path B* becomes dominant. Consistent with this interpretation, the yield of nitrosamine is dependent on solution pH and showed a marked decrease at pH values above 10.

Since, as discussed above, 2-nitrobenzyl derivatives **2** have been reported to release NO much faster than expected (within 5 ms, rather than over the course of several minutes) upon photolysis [120], the NO release rate observed with **5**–**9** for photoreleased diazeniumdiolate **1** (R = Et) was confirmed to be identical to that observed upon normal thermal decomposition of **1** (R = Et).

Diazeniumdiolates **1** (R = alkyl) display absorption spectra with λ_{max} = ca. 250 nm that tail out to $A = 0$ at approximately 320 nm [134]. Thus, to avoid potential complications from secondary photolysis of photoreleased diazeniumdiolates, Bushan et al. examined a series of O^2-naphthylmethyl- and O^2-naphthylallyl-substituted diazeniumdiolates (**10–18**, Fig. 4.2.8) that can be photolyzed with light of wavelengths ≥ 350 nm [135].

Based on the work of Pincock and co-workers, who examined the photochemistry of a series of 1-naphthylmethyl esters and demonstrated efficient carboxylate photorelease for the 4-methyl derivative as well as for the unsubstituted parent compound [130, 136, 137], 1-naphthylmethyl derivatives **10** and **11** were examined. Unfortunately, these derivatives provided substantial amounts of nitro-

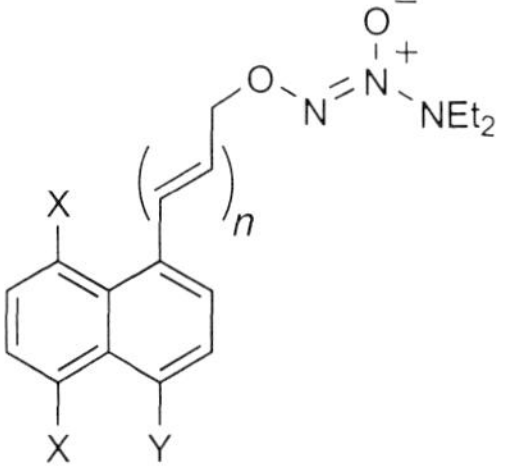

10 $n = 0$, X = H, Y = H
11 $n = 0$, X = H, Y = Me
12 $n = 1$, X = H, Y = H
13 $n = 0$, X = OMe, Y = H
14 $n = 1$, X = OMe, Y = H

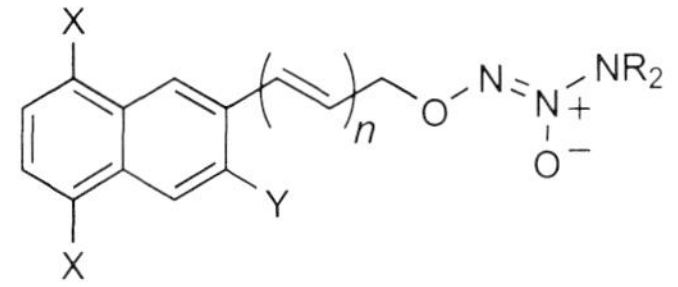

15 $n = 0$, X = OMe, Y = H, R = Et
16 $n = 0$, X = H, Y = OMe, R = Et
17 $n = 1$, X = H, Y = OMe, R = Et
18 $n = 1$, X = H, Y = OMe,
NR$_2$ = piperidinyl

Fig. 4.2.8 O^2-naphthylmethyl- and O^2-naphthylallyl-substituted diazeniumdiolates.

samine and very little (ca. 1%) photorelease of diazeniumdiolate **1**. Following the recent report of Rao and co-workers showing that arylallyl acetates undergo efficient ionic photodissociation in polar solvents [138], naphthylallyl derivative **12** was also examined; here, approximately 25% photorelease of diazeniumdiolate **1** was observed.

To determine what substitution patterns might enhance the photorelease of **1** in naphthylmethyl and naphthylallyl systems (as was accomplished by electron-donating meta substituents in the benzyl series), reported substituent effects on the excited state acidity of 1- and 2-naphthols were examined [133, 139–141]. Electron-*withdrawing* substituents in the 5 and 8 positions, and in the 3, 5, and 8 positions for 1- and 2-naphthol, respectively, have been shown to enhance the excited state acidity, indicating that these positions transmit electronic effects most effectively in the excited states of naphthalene derivatives. Thus, electron-*donating* substituents in these positions were proposed to enhance the formation of a naphthylmethyl or naphthylallyl cation, resulting in the desired photorelease of diazeniumdiolate anion **1** [135].

To test this hypothesis, methoxy-substituted compounds **13–17** were synthesized and, along with the parent compounds **10** and **12**, were analyzed for products following photolysis. Quantification of NO released upon photolysis was performed and was used to derive yields of photoreleased **1** (R=Et). Results indicated that the appropriate methoxy group substitution pattern has a significant effect on the efficiency of the photorelease of **1** (e.g., **10** (1% release) vs **13** (40% release), and **12** (25% release) vs **14** (95% release)). In addition, naphthylallyl derivatives perform better than their naphthylmethyl analogs (e.g., **10** (1% release) vs **12** (25% release), **13** (40% release) vs **14** (95% release), and **16** (20% release) vs **17** (50% release)). This latter trend may be the result of the production of a more stable naphthylallyl cation or may reflect greater transfer of electron density in the excited states of the naphthylallyl systems, as suggested by simple Hückel calculations which qualitatively reproduced the general trends observed for diazeniumdiolate photorelease from compounds **10–17**.

The most efficient NO-releasing compound of those examined in this study, in terms of both diazeniumdiolate photorelease (95%) and quantum yield of photodecomposition (0.66), is naphthylallyl derivative **14**. Importantly, this precursor overcomes the shortcomings of the previously studied benzyl derivatives [128]. The high efficiency of diazeniumdiolate photorelease is not pH dependent and can be initiated by long wavelength ($\geq$350 nm) light, making this diazeniumdiolate precursor potentially well suited for a range of biological applications. Moreover, since the rate of NO release from diazeniumdiolate **1** can be controlled by factors such as the substituents R, pH, and temperature, an additional advantageous feature of **14** (and the other derivatives discussed here) is that the flux of NO can likewise be controlled (and varied). This was demonstrated by simply changing the nature of the released diazeniumdiolate from a diethylamine derivative [**1** (R=Et)] to a piperidine derivative [**1** (NR$_2$=piperidinyl)] for **17** and **18**, respectively [135].

4.2.2.3.7 Carbon-based Diazeniumdiolates

Although all known naturally occurring diazeniumdiolates are carbon-based, these compounds have not received as much attention as their nitrogen-based counterparts since they are not known to decompose to NO thermally [142]. Hwu et al. first reported the generation of NO, along with an azoxybenzene derivative, upon photolysis of various cupferron (PhN(O)=NO$^-$) derivatives [143]. Following this report, Wang and co-workers developed O^2-substituted derivatives of cupferron [144] and neocupferron (NpN(O)=NO$^-$) [145] with substituents ranging from simple alkyl groups to the 2-nitrobenzyl photosensitive protecting group. All of these derivatives appear to give at least small amounts of NO upon 254 or 350 nm photolysis for cupferron and neocupferron, respectively (Scheme 4.2.22). NO release was verified by electrochemical methods as well as by oxyhemoglobin trapping. These compounds also benefit from increased thermal stability, which may be important since cupferron can be oxidized by certain enzymes to give NO and nitrosobenzene [146, 147].

4.2.2.3.8 Bis-N-nitroso-p-phenylenediamine Derivatives

Bis-N-nitroso-p-phenylenediamine (BNN) derivatives, developed by Fujimori and co-workers, have been shown recently to be a very promising class of NO donors [148]. They are thermally stable in aqueous solution (pH 7–9), as well as in non-acidic organic solvents, but upon photolysis yield two equivalents of NO plus a quinoimine byproduct (Scheme 4.2.23).

In general, BNNs have strong absorption maxima at ca. 300 nm (ε=13 500 M^{-1} cm^{-1}) tailing out to ca. 425 nm. These compounds benefit from very high NO releasing efficiency as the chemical yields of NO are approximately 2. Fujimori and co-workers performed initial laser flash photolysis experiments with BNN5 in pH 7.4 phosphate buffer solution. Upon 308 nm photolysis, a transient species having visible absorption (λ=405 nm) is observed within 20 ns. This transient species, which has been assigned to the radical resulting from loss of one NO molecule, can either recombine with NO to reform the starting material (k_r = 1.38×10^8 M^{-1}s^{-1}) or dissociate to a second equivalent of NO (k_d=2.96×10^4 s^{-1}) and the quinoimine byproduct (Scheme 4.2.24) [148]. Pacheco and co-workers recently reexamined these values and suggested k_r = 1.1×10^9 M^{-1}s^{-1} and k_d = 500 s^{-1} [149].

Because the quantum yield of NO is not significantly affected by air, direct photolytic N-N bond homolysis was initially proposed to occur from the singlet excited state [148]. However, observed quantum yields of nearly 2 for NO formation

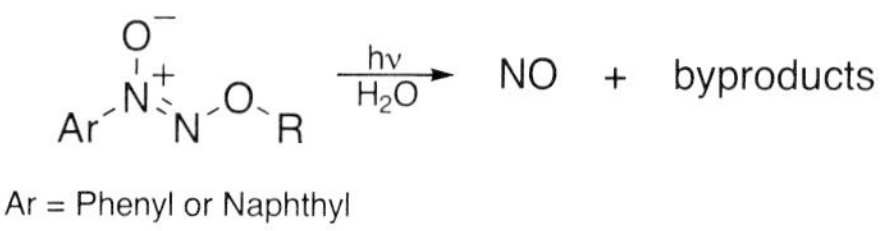

Ar = Phenyl or Naphthyl

Scheme 4.2.22 The production of NO from cupferron (Ar = phenyl) and neocupferron (Ar = naphthyl).

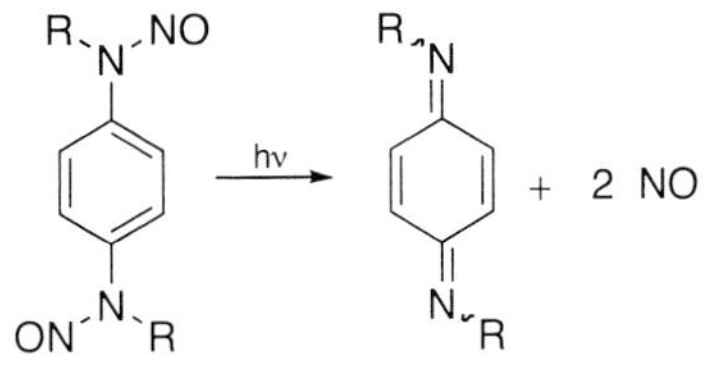

BNN3: R=CH$_3$
BNN5: R=CH$_2$CO$_2$H
BNN5M: R=CH$_2$CO$_2$CH$_3$

Scheme 4.2.23 The production of NO from bis-N-nitroso-p-phenylenediamine (BNN) derivatives.

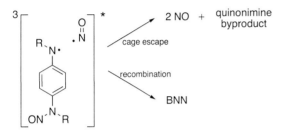

Scheme 4.2.24 Intermediates involved in the photochemistry of bis-N-nitroso-p-phenylenediamine (BNN) derivatives.

suggest that radical pair recombination to regenerate the starting material, a fate to be expected for a singlet radical pair in a solvent cage, is not an issue. Thus, the initial N-N bond homolysis was later proposed to occur from an excited triplet state, yielding a triplet radical pair whose fastest reaction path involves diffusion from the solvent cage rather than spin-forbidden recombination [150].

To clarify this issue, triplet-sensitized laser flash photolysis experiments were performed with biacetyl, Michler's ketone, and benzil [150]. Upon sensitization of BNN3, the transient absorption spectrum recorded at 2 µs after the laser shot was found to be identical to that observed upon direct photolysis with 308 nm light. The similar transient spectrum and decay profile suggests that both photolyses generate the same transient species. The rate of triplet energy transfer was slow with benzil ($k_{et}=4.91\times10^8$ M^{-1}s^{-1}); the triplet energy of BNN3 was estimated to be 230 kJ mol^{-1} based on the Sandros equation for endothermic energy transfer [150]. Although N-N bond dissociation from the excited singlet state of BNN was not completely eliminated, the excited singlet state of BNN3 formed upon direct photolysis was proposed to undergo intersystem crossing, yielding the triplet excited state of BNN3 from which the N-N bond homolysis occurs.

The potential advantage of the differently substituted BNNs, analogous to those derivatives reported by Makings and Tsien above for the 2-nitrobenzyl series [120], was demonstrated by carrying out photorelaxation experiments of rat aortic strips with a series of compounds [151]. Water soluble BNN5Na localizes in the water phase and is membrane impermeable, making it ideal for applications in which the NO donor must remain extracellular. BNN3 is readily incorporated into vascular smooth muscle cells, and, since it is not soluble in water, is localized at lipids in cells. BNN5M is muscle cell permeable and hydrolyzes to BNN5Na by cytosolic esterases. Once hydrolyzed, membrane impermeable BNN5Na remains in the cell. Rat aortic strips were loaded with a chosen BNN derivative prior to irradiation with UV light (300–360 nm). Photorelaxation was found to be ca. 23 times greater for strips loaded with a BNN as compared to the control. Thus, by choosing a suitable BNN derivative, NO can be delivered with precise time and space resolution at a specific compartment in a living system [151].

BNN5 was additionally used by Cabail and Pacheco to generate NO by laser flash photolysis in the presence of hydroxylamine oxidoreductase (HAO). Upon 308 nm laser photolysis an exponential growth was observed at 420, 523, and 552 nm which was consistent with reduction of HAO. No growth was observed in the absence of BNN5. NO was thus demonstrated to act as a one-electron reductant of HAO [152].

4.2.3
Conclusions

A variety of photosensitive precursors to NO have been developed. These precursors, as caged compounds, are able to deliver NO in a spatially and temporally controllable manner. In addition, some (i.e., diazeniumdiolate-based precursors) allow NO release rate profiles to be easily varied.

Although caged NO compounds have played a significant role in elucidating neurophysiological roles of NO, their use in other biological applications has been fairly limited. For example, caged NO compounds have been suggested to be promising new agents for photodynamic therapy applications, but only a few cases of their actual use have been reported. It is hoped that the future will see more biochemical and pharmacological applications of the caged NO compounds discussed in this chapter.

Abbreviations

NO	nitric oxide	SNP	sodium nitroprusside
NOS	nitric oxide synthase	RBS	Roussin's black salt
PDT	photodynamic therapy	RRS	Roussin's red salt
GSNO	S-nitrosoglutathione	RRE	Roussin's red ester
BSANO	S-nitroso bovine serum albumin	PBN	phenyl-N-tert-butyl nitrone
		t-NB	tert-nitrosobutane
SNAC	N-acetylcysteine	BNN	N,N'-dinitroso-p-phenylenediamine
SNAP	S-nitrosoacetylphenicillamine		
		HAO	hydroxylamine oxireductase

References

1 MURAD, F. *Angew. Chem., Int. Ed. Engl.* **1999**, *38*, 1856–1868.
2 FURCHGOTT, R. F. *Angew. Chem., Int. Ed. Engl.* **1999**, *38*, 1870–1880.
3 IGNARRO, L. J. *Angew. Chem., Int. Ed. Engl.* **1999**, *38*, 1882–1892.
4 THOMAS, D. D.; LIU, X.; KANTROW, S. P.; LANCASTER, J. R., Jr. *Proc. Natl. Acad. Sci. USA* **2001**, *98*, 355–360.
5 WINK, D. A.; VODOVOTZ, Y.; LAVAL, J.; LAVAL, F.; DEWHIRST, M. W.; MITCHELL, J. B. *Carcinogenesis* **1998**, *19*, 711–721.
6 WINK, D. A.; VODOVOTZ, Y.; COOK, J. A.; KRISHNA, M. C.; KIM, S.; COFFIN, D.; DEGRAFF, W.; DELUCA, A. M.; LIEBMANN, J.; MITCHELL, J. B. *Biochemistry (Moscow)* **1998**, *63*, 802–809.
7 MITCHELL, J. B.; COOK, J. A.; STEIN, W.; COFFIN, D.; ESPEY, M. G.; MIRANDA, K. M.; WINK, D. A. *Radiat. Res.* **2000**, *2*, 618–621.
8 HOFSETH, L. J.; HUSSAIN, S. P.; WOGAN, G. N.; HARRIS, C. C. *Free Radical Biol. Med.* **2003**, *34*, 955–968.
9 HOWARD-FLANDERS, P. *Nature* **1957**, *180*, 1191–1192.
10 MITCHELL, J. B.; WINK, D. A.; DEGRAFF, W.; GAMSON, J.; KEEFER, L. K.; KRISHNA, M. C. *Cancer Res.* **1993**, *53*, 5845–5848.
11 FLEMING, I.; BUSSE, R. *Am. J. Physiol.* **2003**, *284*, R1–R12.
12 BOO, Y. C.; JO, H. *Am. J. Physiol.* **2003**, *285*, C499–C508.
13 FURCHGOTT, R. F.; SLEATOR, W.; MCCAMAN, M. W.; ELCHLEPP, J. *J. Pharmacol. Exp. Ther.* **1955**, *113*, 22–23.
14 BUYUKAFSAR, K.; DEMIREL-YILMAZ, E.; GOCMEN, C.; DIKMEN, A. *J. Pharmacol. Exp. Ther.* **1999**, *290*, 768–773.
15 CHAUDHRY, H.; LYNCH, M.; SCHOMACKER, K.; BIRNGRUBER, R.; GREGORY, K.; KOCHEVAR, I. *Photochem. Photobiol.* **1993**, *58*, 661–669.
16 DAVE, K. C.; JINDAL, M. N.; KELKAR, V. V.; TRIVEDI, H. D. *Brit. J. Pharmacol.* **1979**, *66*, 197–201.
17 FURCHGOTT, R.; GREENBLATT, E.; EHRREICH, S. J. *J. Gen. Physiol.* **1961**, *44*, 499–519.
18 KARLSSON, J. O. G.; AXELSSON, K. L.; ELWING, H.; ANDERSSON, R. G. G. *J. Cyclic Nucleotide Protein Phosphorylation Res.* **1986**, *11*, 155–166.
19 KUBASZEWSKI, E.; PETERS, A.; MCCLAIN, S.; BOHR, D.; MALINSKI, T. *Biochem. Biophys. Res. Commun.* **1994**, *200*, 213–218.
20 LOVREN, F.; TRIGGLE, C. R. *Eur. J. Pharmacol.* **1998**, *347*, 215–221.
21 MATSUNAGA, K.; FURCHGOTT, R. F. *J. Pharmacol. Exp. Ther.* **1991**, *259*, 1140–1146.
22 MEGSON, I. L.; HOLMES, S. A.; MAGID, K. S.; PRITCHARD, R. J.; FLITNEY, F. W. *Brit. J. Pharmacol.* **2000**, *130*, 1575–1580.
23 VENTURINI, C. M.; PALMER, R. M. J.; MONCADA, S. *J. Pharmacol. Exp. Ther.* **1993**, *266*, 1497–1500.
24 WIGILIUS, I. M.; AXELSSON, K. L.; ANDERSSON, R. G. G.; KARLSSON, J. O. G.; ODMAN, S. *Biochem. Biophys. Res. Commun.* **1990**, *169*, 129–135.
25 RODRIGUEZ, J.; MALONEY, R. E.; RASSAF, T.; BRYAN, N. S.; FEELISCH, M. *Proc. Natl. Acad. Sci. USA* **2003**, *100*, 336–341.
26 TREININ, A.; HAYON, E. *J. Am. Chem. Soc.* **1970**, *92*, 5821–5828.
27 BILSKI, P.; CHIGNELL, C. F.; SZYCHLINSKI, J.; BORKOWSKI, A.; OLEKSY, E.; RESZKA, K. *J. Am. Chem. Soc.* **1992**, *114*, 549–556.
28 BARRETT, J.; DEBENHAM, D. F.; GLAUSER, J. *Chem. Commun.* **1965**, 248–249.
29 SEXTON, D. J.; MURUGANANDAM, A.; MCKENNEY, D. J.; MUTUS, B. *Photochem. Photobiol.* **1994**, *59*, 463–467.
30 SINGH, R. J.; HOGG, N.; JOSEPH, J.; KALYANARAMAN, B. *FEBS Lett.* **1995**, *360*, 47–51.
31 WOOD, P. D.; MUTUS, B.; REDMOND, R. W. *Photochem. Photobiol.* **1996**, *64*, 518–524.
32 SHISHIDO, S. M.; DE OLIVEIRA, M. G. *Photochem. Photobiol.* **2000**, *71*, 273–280.
33 SHISHIDO, S. M.; SEABRA, A. B.; LOH, W.; GANZAROLLI DE OLIVEIRA, M. *Biomaterials* **2003**, *24*, 3543–3553.
34 ROTTA, J. C. G.; LUNARDI, C. N.; TEDESCO, A. C. *Braz. J. Med. Biol. Res.* **2003**, *36*, 587–594.

35 Maree, M.D.; Nyokong, T. *J. Photochem. Photobiol. A* **2001**, *142*, 39–46.
36 Hadjur, C.; Wagnieres, G.; Ihringer, F.; Monnier, P.; vandenBergh, H. *J. Photochem. Photobiol. B* **1997**, *38*, 196–202.
37 Etchenique, R.; Furman, M.; Olabe, J.A. *J. Am. Chem. Soc.* **2000**, *122*, 3967–3968.
38 Zhelyaskov, V.R.; Gee, K.R.; Godwin, D.W. *Photochem. Photobiol.* **1998**, *67*, 282–288.
39 Chow, Y.L. *Acc. Chem. Res.* **1973**, *6*, 354–360.
40 Hiramoto, K.; Ohkawa, T.; Kikugawa, K. *Free Radical Res.* **2001**, *35*, 803–813.
41 Al-Achi, A.; Greenwood, R. *Drug Dev. Ind. Pharm.* **2001**, *27*, 465–468.
42 Kwon, N.S.; Lee, S.H.; Choi, C.S.; Kho, T.; Lee, H.S. *FASEB J.* **1994**, *8*, 529–533.
43 Coppens, P.; Novozhilova, I.; Kovalevsky, A. *Chem. Rev.* **2002**, *102*, 861–884.
44 Ford, P.C. *Int. J. Photoenergy* **2001**, *3*, 161–169.
45 Ford, P.C.; Bourassa, J.; Lee, B.; Lorkovic, I.; Miranda, K.; Laverman, L. *Coord. Chem. Rev.* **1998**, *171*, 185–202.
46 Hoshino, M.; Laverman, L.; Ford, P.C. *Coord. Chem. Rev.* **1999**, *187*, 75–102.
47 Wanat, A.; Wolak, M.; Orzel, L.; Brindell, M.; van Eldik, R.; Stochel, G. *Coord. Chem. Rev.* **2002**, *229*, 37–49.
48 Laverman, L.E.; Wanat, A.; Oszajca, J.; Stochel, G.; Ford, P.C.; van Eldik, R. *J. Am. Chem. Soc.* **2001**, *123*, 285–293.
49 Cao, W.; Christian, J.F.; Champion, P.M.; Rosca, F.; Sage, J.T. *Biochemistry* **2001**, *40*, 5728–5737.
50 Zavarine, I.S.; Kini, A.D.; Morimoto, B.H.; Kubiak, C.P. *J. Phys. Chem. B* **1998**, *102*, 7287–7292.
51 Adachi, H.; Sonoki, H.; Hoshino, M.; Wakasa, M.; Hayashi, H.; Miyazaki, Y. *J. Phys. Chem. A* **2001**, *105*, 392–398.
52 Mitra, R.P.; Jain, D.V.S.; Banerjee, A.K.; Chari, K.V.R. *J. Inorg. Nucl. Chem.* **1963**, *25*, 1263–1266.
53 Wolfe, S.K.; Swinehart, J.H. *Inorg. Chem.* **1975**, *14*, 1049–153.
54 Stochel, G. *Coord. Chem. Rev.* **1992**, *114*, 269–295.
55 Bloyce, P.E.; Hooker, R.H.; Lane, D.A.; Rest, A.J. *J. Photochem.* **1985**, *28*, 525–528.
56 Rusanov, V.; Stankov, S.; Trautwein, A.X. *Hyperfine Interact.* **2002**, *144*, 307–323.
57 Buxton, G.V.; Dainton, F.S.; Kalecinski, J. *Int. J. Radiat. Phys. Chem.* **1969**, *1*, 87–98.
58 Kudo, S.; Bourassa, J.L.; Boggs, S.E.; Sato, Y.; Ford, P.C. *Anal. Biochem.* **1997**, *247*, 193–202.
59 Spiegel, H.E.; Kucera, V. *Clin. Chem.* **1977**, *23*, 2329–2331.
60 Arnold, W.P.; Longnecker, D.E.; Epstein, R.M. *Anesthesiology* **1984**, *61*, 254–260.
61 Bisset, W.I.K.; Burdon, M.G.; Butler, A.R.; Glidewell, C.; Reglinski, J. *J. Chem. Res. S* **1981**, 299.
62 de Oliveira, M.G.; Langley, G.J.; Rest, A.J. *J. Chem. Soc., Dalton Trans.* **1995**, 2013–2019.
63 Shishido, S.M.; Ganzarolli de Oliveira, M. *Prog. React. Kinet. Mech.* **2001**, *26*, 239–261.
64 Butler, A.R.; Megson, I.L. *Chem. Rev.* **2002**, *102*, 1155–1165.
65 Klyachko, V.A.; Ahern, G.P.; Jackson, M.B. *Neuron* **2001**, *31*, 1015–1025.
66 Az-Ma, T.; Yuge, O. *Masui to Sosei* **1997**, *33*, 79–83.
67 Zhelyaskov, V.R.; Godwin, D.W. *Nitric Oxide* **1998**, *2*, 454–459.
68 Zhelyaskov, V.R.; Godwin, D.W. *Nitric Oxide* **1999**, *3*, 419–425.
69 Kunkely, H.; Vogler, A. *J. Photochem. Photobiol. A* **1998**, *114*, 197–199.
70 Szacilowski, K.; Oszajca, J.; Stochel, G.Y.; Stasicka, Z. *J. Chem. Soc., Dalton Trans.* **1999**, 2353–2358.
71 Szacilowski, K.; Stasicka, Z. *Coord. Chem. Rev.* **2002**, *229*, 17–26.
72 Bourassa, J.L.; Ford, P.C. *Coord. Chem. Rev.* **2000**, *200–202*, 887–900.
73 Bourassa, J.; DeGraff, W.; Kudo, S.; Wink, D.A.; Mitchell, J.B.; Ford, P.C. *J. Am. Chem. Soc.* **1997**, *119*, 2853–2860.

74 Bourassa, J.; Lee, B.; Bernard, S.; Schoonover, J.; Ford, P. C. *Inorg. Chem.* **1999**, *38*, 2947–2952.

75 Matthews, E. K.; Seaton, E. D.; Forsyth, M. J.; Humphrey, P. P. A. *Br. J. Pharmacol.* **1994**, *113*, 87–94.

76 Black, M. D.; Matthews, E. K.; Humphrey, P. P. A. *Neuropharmacology* **1994**, *33*, 1357–165.

77 Conrado, C. L.; Bourassa, J. L.; Egler, C.; Wecksler, S.; Ford, P. C. *Inorg. Chem.* **2003**, *42*, 2288–2293.

78 Patra, A. K.; Afshar, R.; Olmstead, M. M.; Mascharak, P. K. *Angew. Chem., Int. Ed.* **2002**, *41*, 2512–2515.

79 Kunkely, H.; Vogler, A. *Inorg. Chim. Acta* **2003**, *346*, 275–277.

80 Bettache, N.; Carter, T.; Corrie, J. E. T.; Ogden, D.; Trentham, D. R. *Methods Enzymol.* **1996**, *268*, 266–281.

81 Murphy, K. P. S. J.; Williams, J. H.; Bettache, N.; Bliss, T. V. P. *Neuropharmacology* **1994**, *33*, 1375–185.

82 Murphy, K. P. S. J.; Bliss, T. V. P. *J. Physiol.* **1999**, *515*, 453–462.

83 Gelperin, A. *Nature* **1994**, *369*, 61–63.

84 Ahern, G. P.; Hsu, S. F.; Jackson, M. B. *J. Physiol.* **1999**, *520 Pt 1*, 165–176.

85 Works, C. F.; Ford, P. C. *J. Am. Chem. Soc.* **2000**, *122*, 7592–7593.

86 Works, C. F.; Jocher, C. J.; Bart, G. D.; Bu, X.; Ford, P. C. *Inorg. Chem.* **2002**, *41*, 3728–3739.

87 Bordini, J.; Hughes David, L.; Da Motta Neto Joaquim, D.; da Cunha Carlos, J. *Inorg. Chem.* **2002**, *41*, 5410–5416.

88 Callahan, R. W.; Meyer, T. J. *Inorg. Chem.* **1977**, *16*, 574–581.

89 Togniolo, V.; Santana da Silva, R.; Tedesco, A. C. *Inorg. Chim. Acta* **2001**, *316*, 7–12.

90 Sauaia, M. G.; de Lima, R. G.; Tedesco, A. C.; da Silva, R. S. *J. Am. Chem. Soc.* **2003**, *125*, 14718–14719.

91 De Leo, M.; Ford, P. C. *J. Am. Chem. Soc.* **1999**, *121*, 1980–1981.

92 De Leo, M. A.; Ford, P. C. *Coord. Chem. Rev.* **2000**, *208*, 47–59.

93 Yonemura, T.; Nakata, J.; Kadota, M.; Hasegawa, M.; Okamoto, K.-i.; Ama, T.; Kawaguchi, H.; Yasui, T. *Inorg. Chem. Commun.* **2001**, *4*, 661–663.

94 Barton, D. H.; Beaton, J. M.; Geller, L. E.; Pechet, M. M. *J. Am. Chem. Soc.* **1961**, *83*, 4076–4083.

95 Majetich, G. *Tetrahedron* **1995**, *51*, 7095–7129.

96 Ludwig, B. E.; McMillan, G. R. *J. Am. Chem. Soc.* **1969**, *91*, 1085.

97 McMillan, G. R. *J. Phys. Chem.* **1963**, *67*, 931–932.

98 Chapman, O. L.; Heckert, D. C.; Reasoner, J. W.; Thackaberry, S. P. *J. Am. Chem. Soc.* **1966**, *88*, 5550–5554.

99 Fukuhara, K.; Kurihara, M.; Miyata, N. *J. Am. Chem. Soc.* **2001**, *123*, 8662–8666.

100 Sortino, S.; Giuffrida, S.; De Guidi, G.; Chillemi, R.; Petralia, S.; Marconi, G.; Condorelli, G.; Sciuto, S. *Photochem. Photobiol.* **2001**, *73*, 6–13.

101 Sortino, S.; Petralia, S.; Compagnini, G.; Conoci, S.; Condorelli, G. *Angew. Chem., Int. Ed.* **2002**, *41*, 1914–1917.

102 Lovren, F.; Triggle, C. R. *Eur. J. Pharmacol.* **1998**, *347*, 215–221.

103 Hwang, K. J.; Jo, I.; Shin, Y. A.; Yoo, S.; Lee, J. H. *Tetrahedron Lett.* **1995**, *36*, 3337–3340.

104 Lee, S. H.; Jo, I.; Lee, J. H.; Hwang, K. J. *B Kor. Chem. Soc.* **1997**, *18*, 1115–1117.

105 Auricchio, S.; Selva, A.; Truscello, A. M. *Tetrahedron* **1997**, *53*, 17407–17416.

106 Mallory, F. B.; Cammarata, A. *J. Am. Chem. Soc.* **1966**, *88*, 61–64.

107 Mallory, F. B.; Manatt, S. L.; Wood, C. S. *J. Am. Chem. Soc.* **1965**, *87*, 5433–5438.

108 Pashchenko, S. V.; Khramtsov, V. V.; Skatchkov, M. P.; Plyusnin, V. F.; Bassenge, E. *Biochem. Biophys. Res. Commun.* **1996**, *225*, 577–584.

109 Ullrich, T.; Oberle, S.; Abate, A.; Schroder, H. *FEBS Lett.* **1997**, *406*, 66–68.

110 Tertyshnikova, S.; Yan, X.; Fein, A. *J. Physiol.* **1998**, *512*, 89–96.

111 Chamulitrat, W.; Jordan, S. J.; Mason, R. P.; Saito, K.; Cutler, R. G. *J. Biol. Chem.* **1993**, *268*, 11520–11527.

112 Hrabie, J. A.; Keefer, L. K. *Chem. Rev.* **2002**, *102*, 1135–1154.

113 Drago, R.S.; Karstetter, B.R. *J. Am. Chem. Soc.* **1960**, *83*, 1819–1822.

114 Drago, R.S.; Paulik, F.E. *J. Am. Chem. Soc.* **1960**, *82*, 96–98.

115 Drago, R.S.; Ragsdale, R.O.; Eyman, D.P. *J. Am. Chem. Soc.* **1961**, *83*, 4337–4339.

116 Longhi, R.; Ragsdale, R.O.; Drago, R.S. *Inorg. Chem.* **1962**, *1*, 768–770.

117 Keefer, L.K. *Annu. Rev. Pharmacol. Toxicol.* **2003**, *43*, 585–607.

118 Saavedra, J.E.; Fitzhugh, A.L.; Keefer, L.K. *Nitric Oxide and the Cardiovascular System* **2000**, 431–446.

119 Mooradian, D.L.; Hutsell, T.C.; Keefer, L.K. *J. Cardiovasc. Pharmacol.* **1995**, *25*, 674–678.

120 Makings, L.R.; Tsien, R.Y. *J. Biol. Chem.* **1994**, *269*, 6282–6285.

121 Lev-Ram, V.; Makings, L.R.; Keitz, P.F.; Kao, J.P.; Tsien, R.Y. *Neuron* **1995**, *15*, 407–415.

122 Lev-Ram, V.; Nebyelul, Z.; Ellisman, M.H.; Huang, P.L.; Tsien, R.Y. *Learn Memory* **1997**, *4*, 169–177.

123 Lev-Ram, V.; Jiang, T.; Wood, J.; Lawrence, D.S.; Tsien, R.Y. *Neuron* **1997**, *18*, 1025–1038.

124 Arancio, O.; Kiebler, M.; Lee, C.J.; Lev-Ram, V.; Tsien, R.Y.; Kandel, E.R.; Hawkins, R.D. *Cell* **1996**, *87*, 1025–1035.

125 Arancio, O.; Lev-Ram, V.; Tsien, R.Y.; Kandel, E.R.; Hawkins, R.D. *J. Physiol.* **1997**, *90*, 321–322.

126 Lev-Ram, V.; Wong, S.T.; Storm, D.R.; Tsien, R.Y. *Proc. Natl. Acad. Sci. USA* **2002**, *99*, 8389–8393.

127 Srinivasan, A.; Kebede, N.; Saavedra, J.E.; Nikolaitchik, A.V.; Brady, D.A.; Yourd, E.; Davies, K.M.; Keefer, L.K.; Toscano, J.P. *J. Am. Chem. Soc.* **2001**, *123*, 5465–5472.

128 Ruane, P.H.; Bushan, K.M.; Pavlos, C.M.; D'Sa, R.A.; Toscano, J.P. *J. Am. Chem. Soc.* **2002**, *124*, 9806–9811.

129 Zimmerman, H.E. *J. Am. Chem. Soc.* **1995**, *117*, 8988–8991.

130 Pincock, J.A. *Acc. Chem. Res.* **1997**, *30*, 43–49.

131 Zimmerman, H.E. *J. Phys. Chem. A* **1998**, *102*, 5616–5621.

132 Bartok, W.; Lucchesi, P.J.; Snider, N.S. *J. Am. Chem. Soc.* **1962**, *84*, 1842–1844.

133 Tolbert, L.M.; Solntsev, K.M. *Acc. Chem. Res.* **2002**, *35*, 19–27.

134 Pavlos, C.M.; Cohen, A.D.; D'Sa, R.A.; Sunoj, R.B.; Wasylenko, W.A.; Kapur, P.; Relyea, H.A.; Kumar, N.A.; Hadad, C.M.; Toscano, J.P. *J. Am. Chem. Soc.* **2003**, *125*, 14934–14940.

135 Bushan, K.M.; Xu, H.; Ruane, P.H.; D'Sa, R.A.; Pavlos, C.M.; Smith, J.A.; Celius, T.C.; Toscano, J.P. *J. Am. Chem. Soc.* **2002**, *124*, 12640–12641.

136 Givens, R.S.; Matuszewski, B.; Neywick, C.V. *J. Am. Chem. Soc.* **1974**, *96*, 5547–5552.

137 DeCosta, D.P.; Pincock, J.A. *J. Am. Chem. Soc.* **1993**, *115*, 2180–2190.

138 Rao, G.V.; Reddy, M.J.R.; Srinivas, K.; Reddy, M.J.R.; Bushan, K.M.; Rao, V.J. *Photochem. Photobiol.* **2002**, *76*, 29–34.

139 Seiler, P.; Wirz, J. *Tetrahedron Lett.* **1971**, 1683–1686.

140 Shizuka, H. *Acc. Chem. Res.* **1985**, *18*, 141–147.

141 Agmon, N.; Rettig, W.; Groth, C. *J. Am. Chem. Soc.* **2002**, *124*, 1089–1096.

142 Hrabie, J.A.; Keefer, L.K. *Chem. Rev.* **2002**, *102*, 1135–1154.

143 Hwu, J.R.; Yau, C.S.; Tsay, S.C.; Ho, T.I. *Tetrahedron Lett.* **1997**, *38*, 9001–9004.

144 Hou, Y.; Xie, W.; Janczuk, A.J.; Wang, P.G. *J. Org. Chem.* **2000**, *65*, 4333–4337.

145 Hou, Y.; Xie, W.; Ramachandran, N.; Mutus, B.; Janczuk, A.J.; Wang, P.G. *Tetrahedron Lett.* **2000**, *41*, 451–456.

146 Alston, T.A.; Porter, D.J.T.; Bright, H.J. *J. Biol. Chem.* **1985**, *260*, 4069–4074.

147 McGill, A.D.; Zhang, W.; Wittbrodt, J.; Wang, J.Q.; Schlegel, H.B.; Wang, P.G. *Bioorg. Med. Chem.* **2000**, *8*, 405–412.

148 Namiki, S.; Arai, T.; Fujimori, K. *J. Am. Chem. Soc.* **1997**, *119*, 3840–3841.

149 Cabail, M.Z.; Lace, P.J.; Uselding, J.; Pacheco, A.A. *J. Photochem. Photobiol. A* **2002**, *152*, 109–121.

150 Yoshida, M.; Ikegami, M.; Namiki, S.; Arai, T.; Fujimori, K. *Chem. Lett.* **2000**, 730–731.

151 Namiki, S.; Kaneda, F.; Ikegami, M.; Arai, T.; Fujimori, K.; Asada, S.; Hama, H.; Kasuya, Y.; Goto, K. *Bioorg. Med. Chem.* **1999**, *7*, 1695–1702.

152 Cabail, M. Z.; Pacheco, A. A. *Inorg. Chem.* **2003**, *42*, 270–272.

4.3
Photochemical Release of Neurotransmitters – Transient Kinetic Investigations of Membrane-bound Receptors on the Surface of Cells in the Microsecond-to-Millisecond Time Region

George P. Hess

4.3.1
Introduction

The properties and use of photolabile precursors of neurotransmitters in transient kinetic investigations of chemical reactions mediated by membrane-bound neurotransmitter receptors in the millisecond-to-microsecond time region are described. The approach (reviewed previously in Ref. [1]; see also Ref. [2]) is expected to be applicable to investigations of many membrane-bound proteins, which are estimated to comprise a large fraction of all known proteins. The reactions mediated by neurotransmitter receptors regulate the transmission of signals between about 10^{12} cells of the mammalian central nervous system and between nerve and muscle cells, and are fundamental to the ability of the nervous system to receive, store, and process environmental information [3, 4]. They are mediated by a family of five structurally related proteins [5, 6] that initiate [the nicotinic acetylcholine (nAChR), glutamate, and serotonin type 3 receptors] or inhibit [the γ-aminobutyric acid (GABA$_A$) and glycine receptors] the signaling process. The name of the receptor protein derives from the chemical signal (neurotransmitter) that activates the receptor. Upon binding their specific neurotransmitter, the receptors transiently (in a few milliseconds) form channels through which small inorganic ions cross the membrane of neurons or muscle cells. Excitatory receptors form cation-conducting channels; inhibitory receptors form anion-conducting channels. *The combined action of excitatory and inhibitory receptors of a single cell leads to a transient voltage change across the cell membrane.* If the voltage change is of appropriate magnitude and sign ($\sim$+40 mV), an electrical signal is propagated along the axon of the neuron (Fig. 4.3.1 A). At the axonal terminal, the arriving electrical signal causes the release of a neurotransmitter, which diffuses across a gap (synapse) and binds to specific receptors on an adjacent cell.

The general reaction scheme that allows cells of the nervous system to communicate with one another is shown in Fig. 4.3.1 B; the nicotinic acetylcholine

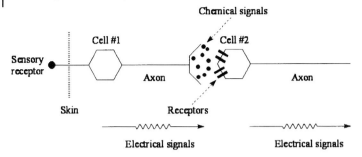

A

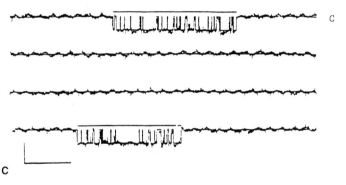

B

C

Fig. 4.3.1 A Transfer of signals between cells mediated by neurotransmitter receptors. The interaction between chemical signals (neurotransmitters) secreted by one cell and receptors on an adjacent cell leads to a transient change in transmembrane potential. When this change has the appropriate sign and magnitude ($\sim +40$ mV) an electrical signal is propagated along the axon of the cell and leads to the secretion of chemical signals. **B** A minimum mechanism for the neurotransmitter receptors. The reaction scheme is based on the original suggestion of Katz and Thesleff [6] for the nicotinic acetylcholine receptor and is consistent with the available kinetic information [reviewed 1].

The importance of the channel-opening equilibrium constant Φ^{-1} ($=k_{op}/k_{cl}$) was initially based on results of transient kinetic investigations with the nAChR [reviewed 1, 119] and subsequently on results obtained with all the other neurotransmitter receptors investigated [1, 101, 104]. L represents the neurotransmitter and A the active, non-desensitized receptor form. The chemical reaction contains several ligand-binding steps, characterized by the dissociation constant K_1, leading to receptor:ligand complexes, where the subscript represents the number of ligand molecules bound. Several transitions in protein conformation result: One change in the sub-ms time region leads to receptor serves as a model for the other receptors activated by neurotransmitters. The scheme is based on classical electrophysiological measurements, with only a second time resolution, made by Katz and Thesleff [6]. They determined that two ligand molecules must bind before the channel opens, and they observed a desensitization process in the seconds time region. Later, using rapid chemical kinetic (see Ref. [7] and Refs. [8, 9] for reviews) and modern electrophysiological (see Refs. [8–12] and Refs. [8, 13–15] for reviews) techniques, a receptor desensitization process that occurs in the millisecond time region was discovered [7]. *The properties of the receptor that desensitizes rapidly are quite different from those observed after this process has gone to completion* (see Refs. [1, 8] for reviews).

Only a few examples of the new information one can obtain about membrane-bound neurotransmitter receptors by use of transient kinetic techniques are available so far. Some of these examples are given at the end of this chapter. However, techniques to investigate rapid chemical reactions in solution have been available for many years [17, 18]. They have been invaluable in elucidating the specific elementary steps in such reactions and have provided essential insights into the mechanism by which enzymes accelerate the reactions, regulate the metabolism of essential cellular components, modify proteins involved in intercellular and intracellular signaling, and translate the nucleic acid codes. The techniques have been summarized in recent textbooks [19–21].

Electrophysiological techniques are essential in the use of rapid kinetic investigations of neurotransmitter receptors in cell membranes, as is discussed below. They were developed during the last twenty years [10, 22–25] and give essential information about the open-channel form of the receptors ($\overline{AL_2}$ in Fig. 4.3.1 B).

---

the formation of a transmembrane channel ($\overline{AL_2}$), characterized by the rate constants for channel opening (k_{op}) and closing (k_{cl}). In the presence of neurotransmitter, receptor proteins are also rapidly (15–200 ms) converted to inactive forms, I, with altered ligand-binding properties, characterized by the dissociation constant K_2. Katz and Thesleff [6] observed a slow (seconds) desensitization process in the muscle nAChR. It is now known that a slow desensitization process is associated with a minor muscle nAChR form in BC$_3$H1 cells and that the major form of the receptor desensitizes in the ms time [12]. It is assumed that a single K_1 value pertains to the dissociation of L from the AL and AL$_2$ forms. The equations used to evaluate the rate and equilibrium constants of the reaction scheme using rapid chemical reaction techniques are listed in Tab. 4.3.2. **C** Determination of P_0, the conditional probability that the channel is open when the receptor is in a nondesensitized state [24]. The value of P_0 is determined from single-channel current measurements. An example of a single-channel current recording of nAChRs on BC3H1 cells that have been activated by 200 µM carbamoylcholine is shown. The recording was made at 23 °C and at a membrane potential of –60 mV. Two bursts of channel activity are shown: a line has been drawn over each burst to denote the duration of the burst. Idealized data have been superimposed on the channel openings for identification purposes. A computer program was used to determine the duration of each burst and the sum of the duration of all channel openings in the burst. The procedure is repeated over many bursts to obtain the average fraction of time during a burst that the receptor channel is open. The horizontal and vertical calibration bars indicate measurements of 50 ms and 5 pA respectively (from Ref. [36]).

One can determine the current passing through a single channel (Fig. 4.3.1 C) and, therefore, the conductance or number of ions passing through the open channel per unit time and the lifetime of the open channel, from which one can determine the rate of channel closing, k_{cl}. With the excellent instruments now commercially available, designed originally by Neher and Sigworth [25], the current-recording technique is now simple to use. It has brought a wealth of information about all types of ion channels, including those modified by disease or by protein engineering (see Ref. [24] for review). The transient chemical kinetic techniques with a μs-ms time resolution described here (see Refs. [27–29] and Refs. [1, 8, 9, 30] for reviews) were developed to determine the rate and equilibrium constants of the minimum reaction scheme for the activation and desensitization of membrane-bound receptors (Fig. 4.3.1 B). The values of the constants are of interest for several reasons. For instance, to account for the integration of excitatory and inhibitory signals arriving at one cell and the resulting transmembrane voltage changes, one needs to know the receptor-controlled rate at which inorganic cations and anions cross the membrane through the receptor channels. The rates at which cations and anions move across the semi-permeable cell membrane are related to the transmembrane voltage [31, 32]. These rates depend not only on the conductance of the open receptor channels but also on the concentration of open receptor channels as a function of neurotransmitter concentration and time [33]. To determine the concentration of open channels as a function of neurotransmitter concentration and time, transient kinetic techniques are required. One needs to determine (i) the reaction path starting with free receptor and neurotransmitter and leading to the formation of neurotransmitter-receptor complexes and open receptor channels, (ii) the reaction path leading to desensitized (transiently inactive) receptor forms, and (iii) the rate and equilibrium constants associated with individual steps in the reaction. The reaction steps outlined in Fig. 4.3.1 B, like those of many enzyme-catalyzed reactions, occur in the μs-to-ms time domain and must, therefore, be investigated with techniques that have a μs-to-ms time resolution. Such techniques exist for investigating reactions mediated by soluble proteins [17–21, 34, 35]. The function of neurotransmitter receptors, however, must be studied in intact cells or vesicles with the receptor protein embedded in a membrane separating two solutions of different ionic composition. Kinetic techniques for studying cell surface receptors in the μs-to-ms time region, and over a wide range of reactant concentrations, only recently became available (see Ref. [28] and Refs. [1, 9, 30] for reviews), and the photochemical release of neurotransmitters plays a crucial role in these methods. Here I describe the kinetic techniques and refer to a few examples of the application of the new techniques using photolabile (caged) neurotransmitters in studies of the channel-opening process of neurotransmitter receptors [29] involved in drug-receptor interactions [36, 37] and receptor dysfunction in nervous system diseases (see, e.g., Ref. [104]).

Caged neurotransmitters have also been used in functional mapping [43, 44]. Some of these are described in another chapter in this monograph, as is the use of photolabile precursors of other molecules that play important roles in the nervous system, such as caged ATP, calcium, and nitric oxide. Caged neurotransmitters have also been used in voltammetry studies of neurotransmitter release and uptake [52, 53] and in studies of the modulation of AMPA and NMDA receptor function by Group III metabotropic receptors [54] and of synaptic inhibition by hypothalamic neuropeptides [55].

4.3.2
Cell Flow Technique

The discovery of rapid receptor desensitization (the reversible, transient inactivation of a receptor by its specific neurotransmitter) [7, 10] provided the impetus for developing techniques for rapidly equilibrating receptors with neurotransmitter in solutions flowing over cells. An ingenious device [11] for flowing solutions (in this case neurotransmitters) over a single cell was adapted [12]. The cell is attached to a recording electrode and the current due to opening of receptor channels can then be recorded by the whole-cell current-recording technique [57]. The theory of solution flowing over spherical objects as large as the cells that contain the neurotransmitter receptors predicts the existence of water layers (diffusion layers) adhering to the external cell membrane [58, 59] that can delay the access of ligands in the flowing solution to the cell surface proteins [28]. The equilibration of ligands in the flowing solutions with cell surface receptors can, therefore, be slow compared to receptor desensitization [8, 12]. In this case, the amplitude of the current recorded must be corrected to take account of desensitization that occurs during the equilibration of the neurotransmitter in the flowing solutions with the cell surface receptors [8, 12]. The thick solid line in Fig. 4.3.2 is the observed whole-cell current recorded when a BC_3H1 muscle cell containing acetylcholine receptors was equilibrated with 200 µM acetylcholine. The current first rises because of cations moving through receptor channels that open in response to the neurotransmitter. The falling phase of the current reflects receptor desensitization, in which the receptor channels close. The thin line in Fig. 4.3.2, which reaches an amplitude about twice that reached in the recorded current trace, is a calculated line [correction equation Eq. (A) – see below] and is the current corrected for desensitization that occurs during the equilibration of the receptors with neurotransmitters in the flowing solution [8, 12].

$$(I_A)_{t_n} = (e^{a\Delta t} - 1) \sum_{i=1}^{n}(I_{obs})\Delta t_i + (I_{obs})_{\Delta t_n}. \tag{A}$$

The correction equation, Eq. (A), was derived [8, 12] to calculate the current I_A that is associated with the active non-desensitized receptor form A (Fig. 4.3.1 B). To obtain the value of I_A from measurements of the observed current, I_{obs}, the current time course is divided into constant (1–5 ms) time intervals to take into account the equilibration time of small segments of the cell surface with ligand. The current is then corrected for the desensitization occurring during each time interval Δt. After n constant time intervals ($n\Delta t = t_n$), during each of which the current $(I_{obs})_{\Delta t}$ is measured, the corrected current is given by Eq. (A) [8, 12].

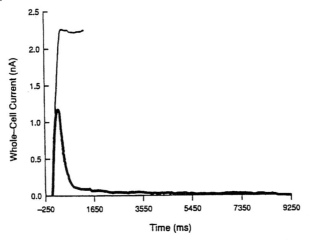

Fig. 4.3.2 A cell flow measurements with a BC$_3$H1 muscle cell containing nAChRs, pH 7.4, 23 °C, −60 mV transmembrane potential. A solution of 200 µM acetylcholine emerged from the cell flow device (Fig. 4.3.3 A) at a rate of 1 cm s^{-1}. The thick solid line represents the observed current; the thinner line represents the calculated current corrected for receptor desensitization using the correction equation [Eq. (A)] [8, 12].

$(I_{obs})_{\Delta t_i}$ is the observed current during the ith time interval, and t_n is equal to or greater than the current rise time, the time it takes the current to reach a maximum value. Under conditions of laminar flow of the neurotransmitter solution over the cell, the value of I_A can be determined with good precision (±10%) [12]. The correction equation, Eq. (A), is based on the theory of the flow of solutions over spherical objects [58, 59]. The BC$_3$H1 cells containing an embryonic form of the muscle nicotinic acetylcholine receptor [60] are nearly spherical when suspended by the current-recording electrode in solutions bathing the cell. When experiments are done with central nervous system neurons that are firmly attached to their substratum, it is possible to obtain spherical, tightly sealed receptor-containing membrane vesicles of about 15 µm diameter attached to the recording electrode [61, 62]. The vesicle can then be suspended in the fluid stream emerging from the flow device.

The time resolution of the laser pulse photolysis technique to be described next is two to three orders of magnitude higher than that of the cell flow technique. Experiments with the laser pulse photolysis technique employing photolabile precursors of the neurotransmitters (see below) verifies that the time it takes for the current to reach its maximum amplitude is much shorter than that observed in the cell flow experiment illustrated in Fig. 4.3.2. For the laser pulse photolysis method, biologically inert photolabile precursors of neurotransmitters (caged neurotransmitters) suitable for investigations in the µs time region of the excitatory acetylcholine, glutamate, and serotonin receptors and the inhibitory GABA and glycine receptors have been developed (Tab. 4.3.1) [29, 63–72].

Tab. 4.3.1 Photolytic properties of biologically inert, photolabile derivatives of neurotransmitters caged with an α-carboxy-2-nitrobenzyl group (pH 7.5, 22 °C)

α-Carboxy-2-nitrobenzyl group					
Caged neurotransmitter	Group caged	Photolysis $t_{1/2}$ (μs)	Product quantum yield	Target receptor	Refs.
Carbamoylcholine	Carbamate	45	0.8	Acetylcholine	29
Glutamate	γ-Carboxyl	21	0.14	Glutamate	65
Kainate	γ-Carboxyl	45	0.37	Kainate	64
Serotonin	Phenolic hydroxyl	16	0.08	Serotonin 5HT$_3$	72
γ-Aminobutyric acid (GABA)	γ-Carboxyl	19	0.16	GABA$_A$	69
Glycine	α-Carboxyl	60	0.38	Glycine	71

4.3.3
Photochemical Properties of Caged Neurotransmitters

Attempts to make chemical kinetic studies of acetylcholine receptors on cell surfaces were initiated in innovative studies by Bartels et al. [73], who synthesized photoisomerizable 3,3'-bis(trimethyl aminomethyl) azobenzene (Bis Q) and reported that the trans, but not the cis, form was an activating ligand for the acetylcholine receptor. In our development of a rapid-reaction method using light activation, we foresaw the need to measure any adverse effects of caged neurotransmitters and their photolysis products on the receptors; it became one of our strategic considerations [8, 12]. Using the rapid-mixing techniques we had developed for use with vesicles (see Ref. [7] for review) we found that "inactive" *cis*-Bis Q causes receptor desensitization and *trans*-Bis Q becomes a receptor inhibitor, even at low concentrations [74]. We therefore began to consider the synthesis of photolabile inert precursors of neurotransmitters.

Photocleavable protecting groups for biologically important compounds have many uses in cell biology. This is particularly true of cases where access of a compound to its reaction partner is slow but the induced reaction is fast (see Refs. [75, 76] and [9, 70, 76–79] for reviews). The most frequently used photocleavable protecting group is the 2-nitrobenzyl group, with various substituents. The photochemically induced transfer of oxygen in 2-nitrobenzyl derivatives by which the *ortho*-CH group becomes C-OH and the NO_2 group NO upon irradiation was first described by De Mayo [80]. Barltrop et al. [81] then protected the carboxy group of glycine and leucine using a 2-nitrobenzyl derivative. Subsequently, Patchornik et al. [82] used the 2-nitrobenzyloxycarbonyl group as a photosensitive blocking reagent in peptide synthesis. The use of this protecting group with biologically important phosphates was pioneered by Kaplan et al. [75] and McCray and Trentham [83], and led to widespread use of "caged compounds" in which the 2-nitrobenzyl group with and without substituents is the protecting group. When the α-amino group of glutamate was blocked by the 2-nitrobenzyl oxycarbonyl group [84], the photolysis rate was $17\ s^{-1}$ and the quantum yield 0.65 [78]. Several other caging groups that can be used to protect the amino group or carboxyl group of neurotransmitters have been reported, for example the 3',5-dimethoxybenzoin group [70, 84, 85] and the *p*-hydroxyphenacyl group [86]. Several caged phenylephrine derivatives [87–89] that are suitable for investigations of the α_1-adrenergic receptor have also been developed.

For transient kinetic investigations of neurotransmitter receptors on the surface of muscle cells and neurons, caged neurotransmitters must have several properties. In addition to being photolyzed in the µs time region so that photolysis is not rate limiting, they must also be (i) photolyzed to give the neurotransmitter with sufficient quantum yield to allow kinetic investigations to be made over a wide range of neurotransmitter concentration, (ii) water-soluble and sufficiently stable in aqueous solution before photolysis, (iii) photolyzed at a wavelength greater than 335 nm to avoid cell damage, and (iv) last but not least, neither the caged compound nor the photolysis products, with the exception of the liberated neurotransmitter, may modify the receptor-mediated reaction being studied.

All these criteria were kept in mind when we initiated the synthesis of photolabile inert precursors of neurotransmitters. We modified the well-known 2-nitrobenzyl group [29] by addition of a carboxy group and used the a-carboxy-2-nitrobenzyl group (a-CNB) as the photoremovable protecting group [29]. The aCBN group was then used to protect the amino group of the excitatory neurotransmitters carbamoylcholine [29, 90] (an analog of acetylcholine), glutamate [65], kainate [64], and N-methyl-D-aspartate [92], and of the inhibitory neurotransmitters glycine [93] and GABA [63, 69]. For the acetylcholine receptor, carbamoylcholine is used because it is a stable and well-characterized analog of acetylcholine, and, unlike acetylcholine, it contains a functional group, an amino group, to which the protecting group can be attached. The introduction of the a-carboxy-2-nitrobenzyl protecting group (a-CNB) [29] improved the photolysis rate of caged carbamoylcholine; the carbamate is photolyzed with a rate of $17\,000\ s^{-1}$ and a quantum yield of 0.8 [29]. It also improved the photolysis rate of the caged amino groups compared to the caging group used previously [63] for amino groups. However, although the N-protected photolabile precursors of glycine [93] and GABA [63, 69], and presumably of other amino-group-containing compounds, are photolyzed 35 times faster than the protecting groups previously used for amino groups [83], they are photolyzed 30 times slower than the caged carbamoylcholine we synthesized [29].

When we used the a-CNB group to protect the carboxyl group of the excitatory neurotransmitters glutamate [65], kainate [64] and serotonin [72] and the inhibitory neurotransmitter GABA [63] and glycine [71], we obtained compounds that meet all the criteria listed above for transient kinetic investigations. The photochemical properties of neurotransmitters caged with the a-CNB group [29] are listed in Tab. 4.3.1.

4.3.4
Laser Pulse Photolysis Technique

The apparatus used in laser pulse photolysis experiments is illustrated in Fig. 4.3.3 A. A cell attached to a current-recording electrode [57] is pre-equilibrated with a caged neurotransmitter. At zero time, a pulse of laser light photolyzes the caged compound in the μs time region. The liberated neurotransmitter binds to receptors on the cell surface and initiates the formation of transmembrane channels; the resulting current is measured (Fig. 4.3.3 B). The time resolution of the method is governed by the rate at which the neurotransmitter is liberated by a light pulse. Photolysis rates with $t_{1/2}$ values between 16 and 60 μs have been obtained with the compounds listed in Tab. 4.3.1. Caged neurotransmitters that proved unsuitable for transient kinetic measurements because they either photolyze too slowly, in the ms time region, or are not biologically inert [96], have also been synthesized [63, 67, 68, 93, 94, 96] but they are not listed in the Table. Many other caging groups suitable for caging neurotransmitter have been developed and

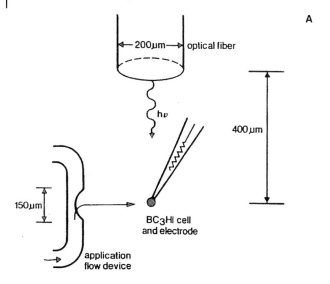
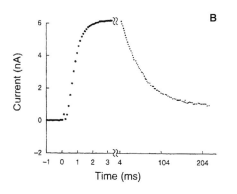
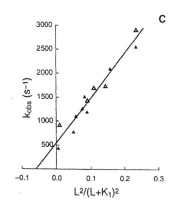
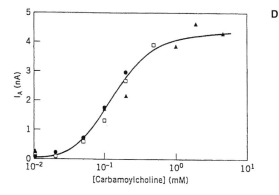

are described in other chapters in this book. Tab. 4.3.1 contains only those caged neurotransmitters that we have found to be biologically inert in our transient kinetic studies, that are photolyzed in the microsecond time domain with sufficient quantum yield for transient kinetic experiments, and that are water soluble and have sufficient thermal stability for the measurements to be conducted. The caged compounds must always be tested under the experimental conditions in which they will be used.

In laser pulse photolysis measurements, the concentration of neurotransmitter generated by photolysis of a caged precursor is measured by using the cell flow technique [12]. A known concentration of free neurotransmitter is applied to the cell and the amplitude of the resulting current, together with a dose-response curve is used to calibrate the concentration of neurotransmitter generated in the subsequent photolysis experiment. Cell flow experiments are also used at the end of the experiment to check if the laser pulse has damaged the cell or the receptors. The cell flow device (see Fig. 4.3.3 A and Ref. [66] for details) is used both to remove the neurotransmitter from the culture dish and to then add caged neurotransmitter.

Fig. 4.3.3 **A** A BC$_3$H1 cell carrying nicotinic acetylcholine receptors, of ~20 μm diameter, attached to an electrode for whole-cell current recording, was equilibrated with caged carbamoylcholine [27]. The side view of the cell flow device [11] used to equilibrate the cell surface with ligands is shown on the left. A Candela SLL500 dye laser was used. Rhodamine 640 or sulforhodamine 640 laser dye, together with a second harmonic generator, produced a wavelength of 328 nm. The laser beam was introduced from an optical fiber of diameter 200 μm. The fiber was adjusted to be ~400 μm away from the cell so that the area illuminated around the cell had a diameter of 300 to 400 μm. The energy of the laser pulse emerging from the fiber was ~500 μJ, and the pulse length was 600 ns. By projecting visible light through the optical fiber, the cell was illuminated and the fiber properly positioned. The wavelength of 328 nm was chosen to avoid cell damage at lower wavelengths and too low a product yield at higher wavelengths. Current was recorded in the whole-cell configuration [57]. Data were low-pass filtered (Krohn-Hite 3322) with a 1–10 kHz cutoff frequency (−3 dB point) and then digitized at a 2–20 Hz sampling frequency using a PDP 11/23 minicomputer; the data were then transferred to a Convex C210 computer where the constants for the rising and decaying phases of the whole-cell current were analyzed using the data analysis program PLOT. **B** Whole-cell current induced by 200 μM released carbamoylcholine at pH 7.4, 22–23 °C, and −60 mV [27]. The points represent the digitized current data. A first-order plot of the current rise time [Tab. 4.3.2, Eqs. (14) and (15)] yields an observed first-order rate constant of 2140 s^{-1} for the current rise. **C** Determination of k_{op}, k_{cl}, and K_1 for the opening of nAChR channels in BC$_3$H1 cells at pH 7.4, 23 °C, and −60 mV [27]. The values of k_{obs} determined from experiments shown in **B** are plotted according to $k_{obs} = k_{cl} + k_{op} L^2/(L+K_1)^2$ [Eq. (15), Tab. 4.3.2]. The values of k_{op}, k_{cl} and K_1 are given in Tab. 4.3.3. **D** Concentration dependence of the current amplitude corrected [see correction equation: Eq. (A)] for receptor desensitization, I_A. BC$_3$H1 muscle cells, pH 7.4, 22–23 °C, and −60 mV transmembrane potential. Data are from Udgaonkar and Hess [12] (●, single-channel current recordings; ▲, cell flow measurements) and Matsubara et al. [27] (□, laser pulse photolysis). The line indicating the concentration of open receptor channels was calculated from the constants pertaining to the channel-opening process determined in laser pulse photolysis experiments (Tab. 4.3.3)

Tab. 4.3.2 Equations for the reaction scheme shown in Fig. 4.3.1 B

$$\overline{(AL_2)}_0 = \frac{\overline{AL_2}}{A + AL + AL_2 + \overline{AL_2}} = \frac{L^2}{(L + K_1)^2 \Phi + L^2} = P_0 \tag{1}$$

$$I_A = I_M R_M \overline{(AL_2)}_0 \tag{2}$$

$$[I_M R_M (I_A)^{-1} - 1]^{1/2} = \Phi^{1/2} + \Phi^{1/2} K_1 [L_1]^{-1} \tag{3}$$

$$(I_{obs})_t - (I_{obs})_{t\infty} = [(I_{obs})_{t=0} - (I_{obs})_{t\infty}] e^{-at} \tag{4}$$

$$a = \Phi \left[\frac{L k_{43} + 2 K_2 k_{21}}{(L + 2K_2)\Phi} + \frac{(L k_{34} + 2 K_1 k_{12})L}{L^2(1 + \Phi) + 2 K_1 L \Phi + K_1^2 \Phi} \right] \tag{5}$$

$$a = \frac{k_{34} \Phi L^2}{(L + K_1)^2 \Phi + L^2} \tag{6}$$

$$I_A / I'_A = 1 + I_0 / K_1 \tag{7}$$

$$\frac{1}{K_I} = \frac{F_A}{(K_I)_1} + \frac{F_{AL}}{(K_I)_2} + \frac{F_{AL_2}}{(K_I)_3} + \frac{F_{\overline{AL_2}}}{(K_I)_4} \tag{8}$$

$$\frac{I_A}{I'_A} = 1 + \frac{I_0}{K_1} \frac{(2K_1 L + K_1^2) \Phi}{(L + K_1)^2 \Phi + L^2} \tag{9}$$

$$\frac{I_A}{I'_A} = 1 + \frac{I_0}{K_1} \overline{(AL_2)}_0 \tag{10}$$

$$\frac{I_A}{I'_A} = 1 + \frac{I_0}{K_I} + \frac{II_0}{K_{II}} \tag{11}$$

$$\frac{I_A}{I'_A} = 1 + \frac{I_0}{K_I} + \frac{II_0}{K_{II}} + \frac{I_0}{K_I} \frac{II_0}{K_{II}} = 1 + \frac{I_0}{K_I} + \frac{II_0}{K_{II}} \left(\frac{K_I + I_0}{K_I} \right) \tag{12}$$

$$\frac{K_I}{\overline{K_I}} = \frac{\Phi^{-1}}{\Phi_{I0}^{-1}} \quad \text{where} \quad \Phi^{-1} = \frac{k_{op}}{k_{cl}} \quad \text{and} \quad \Phi_{I0}^{-1} = \frac{k_{op*}}{k_{cl*}} \tag{13}$$

$$I_t = I_{max}[1 - \exp(-k_{obs} t)] \tag{14}$$

$$k_{obs} = k_{cl} + k_{op} \left(\frac{L}{L + K_I} \right)^2 \tag{15}$$

$$k_{obs} = k_{cl} \frac{K_I}{K_I + I_0} + k_{op} \left(\frac{L}{L + K_I} \right)^2 \tag{16}$$

$$k_{obs} = k_{cl} + k_{op} \left(\frac{L}{L + K_I} \right)^2 \frac{K_I}{K_I + I_0} \tag{17}$$

Tab. 4.3.2 (continued)

Eq. (1): $(\overline{AL_2})_0$ represents the fraction of receptor molecules in the open channel form. $\Phi^{-1} = k_{op}/k_{cl}$ is the channel-opening equilibrium constant, L represents the molar concentration of activating ligand, and P_0 is the conditional probability, determined in single-channel current recordings, that the receptor is in the open-channel form [8, 18]. All other symbols have been defined (Fig. 4.3.1 B).

Eq. (2): I_A is the current due to open receptor-channels in the cell membrane corrected for receptor desensitization. I_M is the current due to 1 mole of open receptor-channels, R_M represents the number of moles of receptors in the cell membrane.

Eq. (3): A linear version of Eq. (2) [101].

Eq. (4): a represents the rate coefficient for receptor desensitization obtained from the falling phase of the current in cell flow experiments [36]. I_{obs} represents the current during the rising phase of the current in laser pulse photolysis experiments. The subscripts t and $t=0$ refer to the time of measurement, and $t=0$ to the time when the current has reached its maximum value in laser pulse photolysis experiments.

Eq. (5): $a = 1/\tau_A$ where a is the rate coefficient for receptor.

Eq. (6): When k_{34} (Fig. 4.3.1 B) is the dominant rate constant, a simplified equation is obtained for the rate coefficient a for receptor desensitization [8, 12].

Eq. (7): The dissociation constant of the inhibitor from the non-desensitized receptor can be determined by both cell flow and photolysis measurements. In order to simplify the equations, we use a ratio method, I_A/I'_A, where I_A and I'_A represent the current maxima corrected for receptor desensitization in the absence and presence of inhibitor respectively. I_0 represents the inhibitor concentration and K_I the observed dissociation constant of the inhibitor from the receptor. It is assumed K_I is the same whether the receptor is monoliganded or biliganded.

Eq. (8): The relationship between the observed inhibitor dissociation constant K_I and the inhibitor dissociation constant for the A, AL, AL_2, and $\overline{AL_2}$ receptor forms and F_A, F_{AL}, F_{AL_2}, and $F_{\overline{AL_2}}$ which represent the fraction of receptors in forms A, AL, AL_2, and $\overline{AL_2}$.

Eq. (9): For a competitive inhibitor $1/K_I$ is multiplied by the fraction of receptor molecules in the A and the AL form.

Eq. (10): An inhibitor binding only to the open-channel form.

Eq. (11): Two inhibitors I_0 and II_0 binding to the same receptor site.

Eq. (12): Two inhibitors I_0 and II_0 binding to two different receptor sites.

Eq. (13): Equation pertaining to the cyclic mechanism in Fig 4.3.4 c.

$$\frac{K_I}{\overline{K_I}} = \frac{\Phi^{-1}}{\Phi_{IO}^{-1}} \quad \text{where} \quad \Phi^{-1} = \frac{k_{op}}{k_{cl}} \quad \text{and} \quad \frac{\Phi^{-1}}{\Phi_{IO}^1} = \frac{k_{op*}}{k_{cl*}}$$

Eq. (14): In the laser pulse photolysis experiments with BC$_3$H1 cells containing nicotinic acetylcholine receptors, the current rise time was observed to follow a single exponential rate law over 85% of the reaction [1]. In this equation I_t represents the observed current at time t and I_{max} represents the maximum current.

Eq. (15): The relationship between the observed rate constant for the current rise k_{obs}, and k_{op}, k_{cl}, K_1 of the reaction scheme (Fig. 4.3.1 B).

Eq. (16): k_{obs} in the presence of an inhibitor that binds only to the open channel form of the receptor.

Eq. (17): k_{obs} in the presence of an inhibitor that binds only to the closed channel form of the receptor. If the inhibitor binds both to the open and closed channel forms a combination of Eqs. (16) and (17) is obtained.

Tab. 4.3.3 Comparison of the values of the constants obtained in laser pulse photolysis (LaPP) experiments with BC$_3$H1 cells and those obtained by single-channel current recording (SCC) and cell flow techniques. Unless otherwise indicated, experiments were performed at 22–23 °C, pH 7.4, and a transmembrane voltage of –60 mV (from [27])

Constant	Method	Value of constant	Refs.
k_{op}	LaPP	9400 ± 3900 s^{-1}	27
	SCC	179 ± 89 s^{-1} [a]	56
k_{cl}	LaPP	$580 + 140$ s^{-1}	27
	SCC	350 ± 129 s^{-1}	27
$k_{op}(k_{op}+k_{cl})^{-1}$	LaPP	0.94	27
	Cell flow	0.84	27
K_1	LaPP	210 ± 90 μM	27
	Cell flow	240 μM	12, 28

[a] Two rate constants were evaluated from the measurements (made at 11 °C and a transmembrane voltage of –60 mV). A value of 11952 ± 2980 s^{-1} was considered not to reflect k_{op}.

In order to obtain laminar flow of the neurotransmitter solution over the cell, the cell must be suspended in the flowing solution and must be nearly spherical. In the case of BC$_3$H1 muscle cells, this is easily accomplished after making the whole-cell seal with the recording electrode and lifting the cells from the bottom of the culture dish [57, 95]. Alternatively, vesicles of about 10 μm diameter can be made from the membranes of central nervous system neurons. The method for obtaining vesicles has been described by Walstrom and Hess [62] and is similar to that described by Sather et al. [61]. A vesicle is obtained from a cell body by first making a whole-cell seal [57] and then gently lifting the recording electrode (glass pipette) until the membrane pinches off from the cell body, thus forming a vesicle. The vesicles typically have a diameter of about 10 μm and a capacitance of about 1–3 pF.

The methodology for measuring the photolysis rates and quantum yields of caged neurotransmitters and for characterizing the compounds is also described in a previous publication [96], as is the methodology for *in situ* photolysis with cells and subsequent current recording.

Contamination of the caged precursor by a small amount of the uncaged compound is one of the most frequently encountered problems in their use. Small amounts of uncaged compound can desensitize neurotransmitter receptors during equilibration with the caged compound. To avoid this, the caged compounds must be protected from light and stored in dry form. We store some caged compounds protected from light, under vacuum, and at –20 °C. Under these conditions, we have had no problems in storing our caged compounds for at least 6 months.

Before we use a caged compound, we test its inertness by using the cell flow technique [12] described above. Contamination of the caged compound by small amounts of free neurotransmitter can easily be detected by this technique. If necessary, we purify the caged compound by using HPLC. If HPLC is not available, column chromatography can be used.

4.3.5
Determination of the Rate and Equilibrium Constants of the Channel-Opening Mechanism

The nicotinic acetylcholine receptor in BC_3H1 muscle cells will be used as an example. Fig. 4.3.3 B shows some results obtained in a laser pulse photolysis experiment. The cell was equilibrated with caged carbamoylcholine. At zero time, the caged carbamoylcholine was photolyzed, liberating free carbamoylcholine in the µs time region. The time resolution of the technique allows one to observe three distinct phases of the reaction: a rising phase of the current reflecting the opening of acetylcholine receptor channels, a maximum current amplitude, a measure of the concentration of open receptor channels, and on a different and slower time scale the falling phase of the current, reflecting receptor desensitization. In experiments with the excitatory acetylcholine, glutamate and kainate receptors and the inhibitory $GABA_A$ and glycine receptors [27, 29, 62–68], the rise time of the current follows a single exponential rate equation over 85% of the reaction. In Fig. 4.3.3 C the dependence of the first-order rate constant for the current rise time k_{obs} on the carbamoylcholine concentration is demonstrated by using Eq. (15) (Tab. 4.3.2). The parameters in this equation, k_{op}, k_{cl}, L, and K_1 are defined in the legend to Fig. 4.3.1 B. The ordinate intercept of the line gives the value of the channel-closing rate constant, k_{cl}, and the slope the channel-opening rate constant, k_{op}. From the dependence of the maximum current amplitude on the concentration of carbamoylcholine one can obtain the value of the channel-opening equilibrium constant, $\Phi = k_{cl}/k_{op}$, and the value of K_1, the dissociation constant of the receptor site controlling channel opening (Fig. 4.3.1 B) by using Eq. (3) (Tab. 4.3.2). The values of K_1 and Φ ($\Phi = k_{cl}/k_{op}$) obtained from the effect of ligand concentration on the current amplitude can be compared to the values of K_1, k_{cl} and k_{op} obtained from the effect of ligand concentration on the observed rate constant, k_{obs}, for the rise time of the current (Fig. 4.3.3 B, C; Eqs. (4)–(6); Tab. 4.3.2).

The falling phase of the current gives information about the rate of receptor desensitization. Desensitization is slow compared to channel opening, occurs in a different time scale, and is investigated more conveniently by the cell flow method [8, 12].

4.3.6
Comparison of the Values Obtained for the Rate and Equilibrium Constants of the Channel-Opening Process Using Independent Techniques

The relationship between the concentration of open acetylcholine receptor channels and carbamoylcholine concentration, obtained by the laser pulse photolysis, cell flow [12], and single-channel current-recording [10, 23] techniques, is given in Fig. 4.3.3 D. As can be seen, when the single-channel current-recording and cell flow techniques are used at low ligand concentrations and in an appropriate time domain, there is excellent agreement between the results obtained with

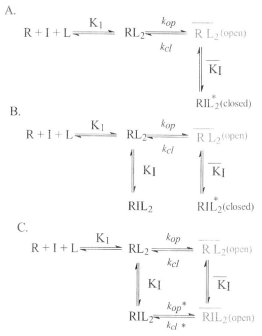

Fig. 4.3.4 Proposed mechanisms for inhibition of the nAChR by MK-801 and cocaine [101]. In each case, the upper line represents the minimum mechanism for the opening of the receptor channel. Receptor R binds the neurotransmitter L (or another activating ligand, for instance carbamoylcholine). RL_2 represents the closed-channel conformations. $\overline{RL}_2$ represents the open-channel conformation of the receptor that allows inorganic cations to cross the cell membrane. K_1 is the observed dissociation constant for the activating ligand. k_{op} and k_{cl} are the rate constants for channel opening and closing, respectively. Φ^{-1} ($=k_{op}/k_{cl}$) is the channel-opening equilibrium constant. The reactions shown occur in the microsecond-to-millisecond time region [1]. For clarity, the desensitization reaction, which in the case of the nAChR occurs in the 100–500 ms time region, and the binding of the inhibitor I to the unliganded receptor form are not shown. The relatively slow transitions of receptor/inhibitor complexes to non-conductive forms are also not shown. **A** Channel-blocking mechanism in which the inhibitor binds in the open channel and blocks it. **B** Extended channel-blocking mechanism. The inhibitor binds to the closed- and open-channel forms giving the non-conducting inhibited receptor forms IRL_2 and $I\overline{RL}_2$. K_I and $\overline{K}_I$ are the observed inhibitor dissociation constants pertaining to the closed- and open-channel form, respectively. **C** Proposed cyclic inhibition mechanisms involving a complex of the inhibitor with the open-channel conformation in which the open channel is not blocked by the inhibitor (i.e. it conducts ions). This minimum mechanism is based on chemical kinetic measurements. The principle of microscopic reversibility requires that the ratio $K_1/\overline{K}_1 = \Phi^{-1}/\Phi_{I0}^{-1}$ where $\Phi = k_{op}/k_{cl}$ and $\Phi_{I0}^{-1} = k_{op}^*/k_{cl}^*$. Therefore, compounds that bind to a regulatory site with higher affinity for the closed-channel conformation than the open-channel form will shift the equilibrium toward the closed-channel form and inhibit the receptor [101]. Compounds that bind to the open-channel conformation with equal or higher affinity than to the closed-channel form are not expected to change the channel-opening equilibrium constant unfavorably [101]. These compounds are, therefore, expected to displace inhibitors from the regulatory sites without inhibiting receptor activity.

these techniques and with the laser pulse photolysis technique. Such agreement with single-channel recordings is not observed at higher ligand concentrations, or when the measurements require longer periods, because of increased receptor desensitization.

A few examples of the use of photolabile neurotransmitters and the laser pulse photolysis technique in elucidating the mechanism of reactions mediated by membrane-bound receptors follow. Many therapeutic agents and abused drugs affect the function of neurotransmitter receptors [97]. Among important inhibitors of the nAChR are the anticonvulsant MK801 and the abused drug cocaine. Understanding the mechanism of inhibition of the nAChR has been a major goal in this field for over two decades [24, 98]. The laser pulse photolysis technique allows one to determine the rate constant of the current rise indicative of channel opening, and therefore to study the effects of inhibitors on k_{cl} and on k_{op} independently of one another. Before the laser pulse photolysis technique was developed, it was possible to determine the effects of inhibitors only on the channel-closing rate by using the single-channel current-recording technique. Mechanisms A and B in Fig. 4.3.4 were proposed prior to the application of transient kinetic techniques to this problem. In mechanism A, the inhibitor binds in the open channel and blocks it [24, 98]. In mechanism B, the inhibitor binds to both the open-channel and closed-channel forms; and the form to which the inhibitor binds cannot form an open channel [99]. When transient kinetic techniques were first applied to this problem, it was shown that the nAChR has two inhibitory sites [100]. This is illustrated with the anticonvulsant MK801 (Fig. 4.3.5 A) and the muscle type nAChR expressed in BC_3H1 cells. The first phase of the receptor inhibition occurs with a halftime of approximately 12 ms. This is followed by a second phase with a halftime of inhibition of about 300 ms, or about 25 times longer than the first phase of inhibition. The inhibition of the receptor by cocaine also has a fast and slow phase [37]. So far, only the rapid equilibrating inhibitory site has been characterized [100–102]. In laser pulse photolysis experiments the effects of inhibitors on both k_{op} and k_{cl} (Fig. 4.3.5 B and C) can be determined (see Ref. [1] for review). While the inhibition mechanisms shown in Fig. 4.3.4 A and B require that both k_{op} and k_{cl} decrease with increasing inhibitor concentration [100, 101], the laser pulse photolysis measurements indicate that in the case of the inhibitor MK801 k_{cl} actually increases with increasing inhibitor concentration while k_{op} remains essentially constant [100] (Fig. 4.3.5 B,C). On the basis of this and similar experiments using the laser pulse photolysis technique, we arrived at mechanism C (Fig. 4.3.4) [101]. In this mechanism the inhibitor binds to an allosteric site on both the closed- and open-channel forms of the receptor and shifts the channel-opening equilibrium constant towards the closed-channel form, thereby inhibiting the receptor. This inhibition process is followed by a relatively slow step to another inhibited form [100] not shown in mechanism C. Mechanism C and the equation pertaining to it [Eq. (13) in Tab. 4.3.2] predict that compounds that bind with equal affinity to the same inhibitory site on the closed- and open-channel form of the receptor, as do allosteric inhibitors such as MK801 and cocaine, will not change the channel-opening equilibrium but can replace inhibitors [101]. Both combinatorially synthesized RNA polymers [101] and small organic molecules [102] were found that fulfill this requirement. They alleviate cocaine inhibition of the embryonic muscle-type nAChR in BC$_3$H1 cells (Fig. 4.3.5 D). These results demonstrated for the first time that compounds ex-

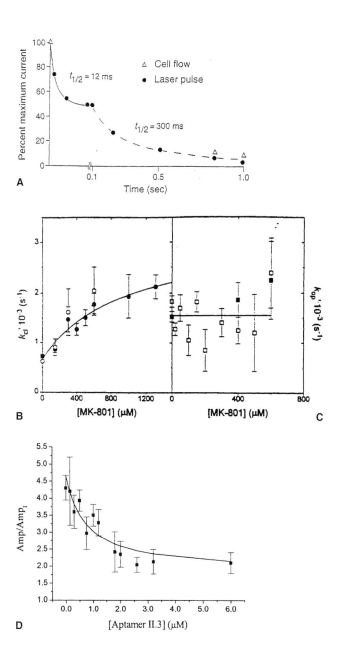

ist with which one can counteract the effect of allosteric inhibitors, like MK801 and cocaine, of the muscle-type nAChR. Furthermore, they demonstrate that knowledge of the mechanism is of crucial importance in finding such compounds.

The second example presented here concerns the dysfunction of a human $GABA_A$ receptor linked to generalized epilepsy with febrile seizures [38]. Epilepsy affects approximately 50 million people worldwide [38, 103]. In the mutation considered here, a lysine (K289) in the γ subunit has been mutated to a methionine residue, the K289M mutation [38]. This mutation when introduced into the rat γ_2 subunit of the $GABA_A$ receptor causes a roughly 50% decrease in the current response as compared to the wild-type receptor when exposed to the same concentration of GABA, comparable to the results reported for the human $GABA_A$ receptor [38]. Laser pulse photolysis experiments (Fig. 4.3.6) have traced this defect to a decrease in the channel-opening equilibrium measured by the rate constants k_{op}/k_{cl}. Fig. 4.3.6 shows a determination of k_{op} and k_{cl} for the wild-type (●) and mutated (▽) receptors. The laser pulse photolysis measurements are plotted according to Eqs. (4)–(6) (Tab. 4.3.2). The ordinate intercepts of the two lines in Fig. 4.3.6, measures of k_{cl}, are the same for the wild-type and mutated receptors. The slopes of the two lines, measures of k_{op}, are quite differ-

Fig. 4.3.5 **A** Laser pulse and cell flow experiments of the inhibition of the nAChR by 200 µM MK-801 in the presence of 100 µM carbamoylcholine, 22 °C, –60 mV, pH 7.4. Two phases of inhibition are shown. A fast process with a $t_{1/2}$ of 12 ms and a slow phase with a $t_{1/2}$ value of 300 ms. **B** Determination of k_{obs} for the current rise as a function of MK-801 concentration using the laser pulse photolysis technique. At the concentration of carbamoylcholine used (25 µM), k_{obs} reflects mainly k_{cl} (k'_{op} was calculated to be 90 s^{-1} at 25 µM carbamoylcholine). The cells were equilibrated with 400 or 800 µM caged carbamoylcholine with and without MK-801 (200 ms preincubation with MK-801). The solid symbols (●) represent the laser pulse photolysis experiments. Each data point represents the average of 3–10 experiments with at least two different cells. For comparison the inverse of the mean lifetimes of the open channel determined by single-channel recording are shown (○). The solid line represents the best fit according to Eq. (16) (Tab. 4.3.2). The value of k_{cl} was taken as 580 s^{-1}; the apparent value k'_{op} is 90 s^{-1} at this carbamoylcholine concentration. A nonlinear least-squares fitting program was used to obtain the values of k_{cl}, 3150 s^{-1}, and K_I, 950 µM, and to construct the dashed line. **C** Laser pulse photolysis experiments. Plot of $k'_{op} = (k_{obs} - k_{cl})$ at 160 µM released carbamoylcholine as a function of MK-801 concentration determined from the values obtained with 200 ms (□) and 4 s (■) preincubation. k_{cl} was obtained from k_{obs} at 25 µM released carbamoylcholine. Each data point represents the average of 3–10 experiments with at least two different cells. **D** Alleviation of cocaine inhibition of the nicotinic acetylcholine receptor [101]. The whole-cell current corrected for receptor desensitization was determined by using the cell flow technique [12] in the experiments shown. At a constant concentration (100 µM) of carbamoylcholine, the ratio of the maximum current amplitude obtained in the absence, A, and presence, A_1, of cocaine at a constant concentration (150 µM) was determined as a function of the concentration of RNA aptamer II-3 [101]. The BC_3H1 cell was pre-equilibrated with aptamer II-3 for 2 s at 22 °C, –60 mV, and pH 7.4. Each data point represents the average of two to three experiments using an average of two cells per point.

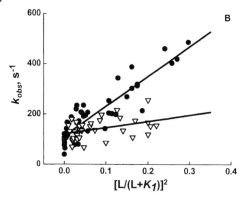

Fig. 4.3.6 Laser pulse photolysis experiments with the wild-type (●) and mutated (△) GABA$_A$ receptors at –60 mV, 22 °C, and pH 7.4. The cells were preincubated with 100 μM caged GABA for 400 ms prior to photolysis at time 0. The rising phase of the current (see Fig. 4.3.3 B for an example of a laser pulse experiment) can be fitted by a single exponential. The rate coefficients, k_{obs}, for the rising phase of the current obtained by the LaPP technique for the wild-type and mutant GABA$_A$ receptors are plotted as a function of $L/(L+K_1)^2$ [see Tab. 4.3.2, Eq. (15)], where L is the GABA concentration. For the wild-type receptor, the total number of measurements included in the fit is 49, and 20 cells were used for collecting these data: $k_{cl}=113\pm9$ s^{-1}, $k_{op}=1183\pm82$ s^{-1}, $K_1=34\pm13$ μM. For the mutant receptor, the total number of measurements included in the fit is 33, and 15 cells were used for collecting these data: $k_{cl}=121\pm11$ s^{-1}, $k_{op}=249\pm99$ s^{-1}, and $K_1=39\pm8$ μM.

ent. From the experiment in Fig. 4.3.6, we determined that the k_{op} value for the wild type is 1183 ± 82 s^{-1} and for the mutated receptor 249 ± 99 s^{-1}, or about fivefold lower. These results correspond to a channel-opening equilibrium constant of about 10 for the wild type and only about 2 for the mutated receptor, and this accounts for the observed dysfunction of the GABA$_A$ receptor [104]. The same experiments also showed that all other values pertaining to the channel-opening mechanisms of the GABA$_A$ receptor (Fig. 4.3.1 B) are the same for the two receptor types. In analogy with the experiments on alleviating the inhibition of the nAChR [101, 102], ligands that can shift the channel-opening equilibrium toward the open form are expected to alleviate the dysfunction of the mutated GABA$_A$ receptor.

4.3.7
Summary

I have described the properties of caged neurotransmitters that are used in transient kinetic measurements of reactions on cell surfaces [1, 70]. I have briefly described the laser pulse photolysis technique in which these caged neurotransmitters are used to obtain a time resolution of 60 μs or less in investigating membrane-bound protein-mediated reactions on the surface of cells. I have given just two examples in which transient kinetic techniques provide information about rapid reactions on cell surfaces that are not obtainable by previous techniques in which ligands are equilibrated with cell surface receptors relatively slowly. Other examples include measurement of the channel-opening kinetics of the kainate-activated glutamate receptor [105], the alpha (1) homooligomeric glycine receptor [106], the GluR1Q (flip) [107], and GluR2Q (flip) AMPA [108] receptors, and the GluR6 kainate receptor [109]. Caged neurotransmitters have also been used in transient kinetic studies of glutamate transport by the neuronal excitatory amino acid carrier (EAAC1) [110–114]. Fourier transform, infrared spectroscopy, and caged neurotransmitters are being used to characterize ligand-protein interactions [115–118]. While only a few examples that illustrate the use of transient kinetic techniques in investigating reactions of membrane-bound proteins are available so far, the impact of transient kinetic techniques on our understanding of biological reactions in solution is well documented [17–21].

Applications of transient kinetic measurements to investigations of membrane-bound proteins, such as the neurotransmitter receptors and transporters, are in their early stages. The mechanisms of a wealth of neurotransmitter receptors and transporters and their isoforms are not yet well understood. Many clinically important compounds (for example antidepressants, tranquilizers, anti-convulsants) and abused drugs (for example, cocaine) affect receptor function [97]. Dysfunction of neurotransmitter receptor-mediated reactions is implicated in diseases of the nervous system. The selection and use of combinatorially synthesized compounds, based on a knowledge of the mechanism, to alleviate receptor dysfunction has been illustrated. The results obtained so far suggest that the transient kinetic technique in which caged compounds play a central role is capable of providing essential information about these reactions on cell surfaces in the future.

Acknowledgments

I am grateful to the current and former members of the laboratory who obtained these results, and dedicate the chapter to the memory of Raymond Wieboldt, who made significant contributions to the development of the caged neurotransmitters used. The research reviewed here was supported by grants GM04842 and NS08527 awarded to G.P.H. by the National Institutes of Health Institute of General Medical Sciences, and Institute of Neurological Diseases and Stroke respectively.

References

1 Hess, G.P. and Grewer, C. (1998) Development and application of caged ligands for neurotransmitter receptors in transient kinetic and neuronal circuit mapping studies. *Methods Enzymol. Caged Compounds* **291**, 443–473.

2 Breitinger, H.G. (2001) Fast kinetic analysis of ligand-gated ion channels. *Neuroscientist* **7**, 95–103.

3 Crick, F. (1994) *The Astonishing Hypothesis. The Scientific Search for the Soul* (Charles Scribner & Sons, New York).

4 Betz, H. (1990) Ligand-gated ion channels in the brain: the amino acid receptor superfamily. *Neuron* **5**, 383–392.

5 Stroud, R.M., McCarthy, M.P. and Shuster, M. (1990) Nicotinic acetylcholine superfamily of ligand-gated ion channels. *Biochemistry* **29**, 11009–11023.

6 Katz, B. and Thesleff, S. (1957) A study of the "desensitization" produced by acetylcholine at the motor end-plate. *J. Physiol. (London)* **138**, 63–80.

7 Hess, G.P., Cash, D.J. and Aoshima, H. (1979) Acetylcholine receptor controlled ion fluxes in membrane vesicles investigated by fast reaction techniques. *Nature* **282**, 329–331.

8 Hess, G.P., Udgaonkar, J.B. and Olbricht, W.L. (1987) Chemical kinetic measurements of transmembrane processes using rapid reaction techniques. Acetylcholine receptor. *Annu. Rev. Biophys. Biophys. Chem.* **16**, 507–534.

9 Hess, G.P. (1993) Determination of the chemical mechanism of neurotransmitter receptor-mediated reactions by rapid chemical kinetic techniques. *Biochemistry* **32**, 989–1000.

10 Sakmann, B., Patlak, J. and Neher, E. (1980) Single acetylcholine-activated channels show burst-kinetics in presence of desensitizing concentrations of agonist. *Nature* **286**, 71–73.

11 Krishtal, O.A. and Pidoplichko, V.I. (1980) A receptor for protons in the nerve cell membrane. *Neuroscience* **5**, 2325–2327.

12 Udgaonkar, J.B. and Hess, G.P. (1987) Chemical kinetic measurements of a mammalian acetylcholine receptor using a fast reaction technique. *Proc. Natl. Acad. Sci. USA* **84**, 8758–8762.

13 Cash, D.J. and Subbarao, R. (1987) Channel opening of γ-aminobutyric acid receptor from rat brain: Molecular mechanisms of receptor response. *Biochemistry* **26**, 7562–7570.

14 Trussel, L.O. and Fischbach, G.D. (1989) Glutamate receptor desensitization and its role in synaptic transmission. *Neuron* **3**, 209–218.

15 Ochoa, W.L.M., Chattopadhyay, A. and McNamee, M.G. (1989) Desensitization of the nicotinic acetylcholine receptor. Molecular mechanism and effect of modulators. *Cell. Molec. Neurobiol.* **9**, 141–178.

16 Changeux, J.-P. (1990) Functional architecture and dynamics of the nicotinic receptor: An allosteric ligand-gated ion channel. *Fidia Research foundation Award Lectures* **4**, 21–168.

17 Hartridge, H. and Roughton, F.J.W. (1923) A method of measuring the velocity of very rapid chemical reactions. *Proc. R. Soc. (London)* **104**, 376–394.

18 Eigen, M. (1997) Immeasurably fast reactions. *Nobel Lectures in Chemistry 1969–1970* (Elsevier Publishing Co., Amsterdam), pp. 170–203.

19 Fersht, A. (1999) *Structure and Mechanism in Protein Science: A Guide to Enzyme Catalysis and Protein Folding* (W.H. Freeman and Co., New York).

20 Gutfreund, H. (1995) *Kinetics for the Life Sciences* (Cambridge University Press, Cambridge, UK).

21 Hammes, G.G. (2000) *Thermodynamics and Kinetics for the Biological Sciences* (John Wiley & Sons, New York).

22 Katz, B. and Miledi, R. (1972) The statistical nature of the acetylcholine potential and its molecular components. *J. Physiol. (London)* **224**, 665–669.

23 Neher, E. and Sakmann, B (1976) Single-channel currents recorded from membrane of denervated frog muscle fibres. *Nature* **260**, 779–802.

24 Sakmann, B. and Neher, E. (1995) (eds.) *Single-channel Recording* (2nd edn) (Plenum Press, New York).

25 SIGWORTH, F. (1983) Electronic design of the patch clamp. In *Single-Channel Recording* (eds. SAKMANN, B., NEHER, E.) (Plenum, NY), pp. 3–35.

26 MILLER, N. S. (1991) The *Pharmacology of Alcohol and Drugs of Abuse and Addiction*, Springer, New York.

27 MATSUBARA, N., BILLINGTON, A. P. and HESS, G. P. (1992) Laser pulse photolysis of caged carbamoylcholine in investigations of a mammalian nicotinic acetylcholine receptor in BC$_3$H1 cells: What can one learn from chemical kinetic measurements in the microsecond time region? *Biochemistry* **31**, 5507–5514.

28 UDGAONKAR, J. B. and HESS, G. P. (1987) Acetylcholine receptor: Channel opening kinetics evaluated by rapid chemical kinetic and single-channel current measurements. *Biophys. J.* **52**, 873–883.

29 MILBURN, T., MATSUBARA, N., BILLINGTON, A. P., UDGAONKAR, J. B., WALKER, J. W., CARPENTER, B. K., WEBB, W. W., MARQUE, J., DENK, W., MCCRAY, J. A. and HESS, G. P. (1989) Synthesis, photochemistry, and biological activity of a caged photolabile acetylcholine receptor ligand. *Biochemistry* **29**, 49–55.

30 HESS, G. P., NIU, L. and WIEBOLDT, R. (1995) Determination of the chemical mechanism of neurotransmitter receptor-mediated reactions by rapid chemical kinetic methods. *Ann. N.Y. Acad. Sci.* **757**, 23–39.

31 PLANCK, M. (1890) Über die Potentialdifferenz zwischen zweiverdünnten Lösungen binärer Electrolyte. *Ann. Phys. Chem.* **40**, 561–576.

32 GOLDMAN, D. E. (1943) Potential impedance and rectification in membranes. *J. Gen. Physiol.* **27**, 37–60.

33 HESS, G. P., KOLB, H.-A., LÄUGER, P., SCHOFFENIELS, E. and SCHWARZE, W. (1984) Acetylcholine receptor (from *Electrophorus electricus*): A comparison of single-channel current recordings and chemical kinetic measurements. *Proc. Natl. Acad. Sci. USA* **81**, 5281–5285.

34 EIGEN, M. (1967) Kinetics of reaction control and information transfer in enzymes, in *Fast Reactions and Primary Processes in Chemical Kinetics. Nobel Symp.* **5** (Ed. S. CLAESSON) (Interscience, New York) pp. 333–369.

35 JOHNSON, K. A. (1993) Transient-state kinetic analysis of enzyme reaction pathways, in *The Enzymes* **20**, 1–61 (Academic Press, New York).

36 NIU, L. and HESS, G. P. (1993) An acetylcholine receptor regulatory site in BC$_3$H1 cells: Characterized by laser pulse photolysis in the microsecond-to-millisecond time region. *Biochemistry* **32**, 3831–3855.

37 NIU, L., ABOOD, L. G. and HESS, G. P. (1995) Cocaine: Mechanism of inhibition of a muscle acetylcholine receptor studied by laser pulse photolysis. *Proc. Natl. Acad. Sci. USA* **92**, 12008–12012.

38 BAULAC., S., HUBERFELD, G., GOURFINKEL-AN, I., MITROPOULOU, G., BERANGER, A., PRUD'HOMME, J.-F., BAULAC, M., BRICE, A., BRUZZONE, R., and LEGUERN, E. (2001) First genetic evidence of GABA(A) receptor dysfunction in epilepsy: a mutant in the gamma 2-subunit gene. *Nat. Genet.* **28**, 46–48.

39 DENK, W., DELANEY, K. R., GELPERIN, A., KLEINFELD, D., STROWBRIDGE, B. W., TANK, D. W. and YUSTE, R. (1994) Anatomical and functional imaging of neurons using 2-photon laser scanning microscopy. *J. Neurosci. Methods* **54**, 151–162.

40 CALLAWAY, E. M. and KATZ, L. C. (1993) Photostimulation using caged glutamate reveals functional circuitry in living brain slices. *Proc. Natl. Acad. Sci. USA* **90**, 7661–7665.

41 KATZ, L. C. and DALVA, M. B. (1994) Scanning laser photostimulation. A new approach for analyzing brain circuits. *J. Neurosci. Methods* **54**, 205–218.

42 LI, H., AVERY, L., DENK, W. and HESS, G. P. (1997) Identification of chemical synapses in the pharynx of *Caenorhabditis elegans*. *Proc. Natl. Acad. Sci. USA* **94**, 5912–5916.

43 PETTIT, D. L. and AUGUSTINE, G. J. (2000) Distribution of functional glutamate and GABA receptors on hippocampal pyramidal cells and interneurons. *J. Neurophysiol.* **84**, 28–38.

44 PETTIT, D. L., HELMS, M. C., LEE, P., AUGUSTINE, G. J. and HALL, W. C. (1999) Local excitatory circuits in the inter-

mediate gray layer of the superior colliculus. *J. Neurophysiol.* **81**, 1424–1427.

45 PETTIT, D. L., WANG, S. S., GEE, K. R. and AUGUSTINE, G. J. (1997) Chemical two-photon uncaging: a novel approach to mapping glutamate receptors. *Neuron* **19**, 465–471.

46 WANG, S. S. and AUGUSTINE, G. J. (1995) Confocal imaging and local photolysis of caged compounds: dual probes of synaptic function. *Neuron* **15**, 755–760.

47 WANG, S. S., KHIROUG, L. and AUGUSTINE, G. J. (2000) Quantification of spread of cerebellar long-term depression with chemical two-photon uncaging of glutamate. *Proc. Natl. Acad. Sci. USA* **97**, 8635–8640.

48 AUGUSTINE, G. J. (1994) Combining patch-clamp and optical methods in brain slices. *J. Neurosci. Methods* **54**, 163–169.

49 KANDLER, K., KATZ, L. and KAUER, J. A. (1998) Focal photolysis of caged glutamate produces long-term depression of hippocampal glutamate receptors. *Nat. Neurosci.* **1**, 119–123.

50 SCHUBERT, D., KOTTER, R., ZILLES, K., LUHMANN, H. J. and STAIGER, J. F. (2003) Cell type-specific circuits of cortical layer IV spiny neurons. *J. Neurosci.* **23**, 2961–2967.

51 DODT, H. U., SCHIERLOH, A., EDER, M. and ZIEGLGANSBERGER, W. (2003) Circuitry of rat barrel cortex investigated by infrared-guided laser stimulation. *Neuroreport* **14**, 623–627.

52 LEE, T. H., GEE, K. R., DAVIDSON, C. and ELLINWOOD, E. H. (2002) Direct real-time assessment of dopamine release autoinhibition in the rat caudate-putamen. *Neuroscience* **112**, 647–654.

53 LEE, T. H., GEE, K. R., ELLINWOOD, E. H. and SEIDLER, F. J. (1998) Altered cocaine potency in the nucleus accumbens following 7-day withdrawal from intermittent but not continuous treatment: voltammetric assessment of dopamine uptake in the rat. *Psychopharmacology* **137**, 303–310.

54 TAVERNA, S. and PENNARTZ, C. M. (2003) Postsynaptic modulation of AMPA- and NMDA-receptor currents by Group III metabotropic glutamate receptors in rat nucleus accumbens. *Brain Res.* **976**, 60–68.

55 DAVIS, S. F., WILLIAMS, K. W., XU, W., GLATZER, N. R. and SMITH, B. N. (2003) Selective enhancement of synaptic inhibition by hypocretin (orexin) in rat vagal motor neurons: implications for autonomic regulation. *J. Neurosci.* **23**, 3844–3854.

56 SINE, S. M. and STEINBACH, J. H. (1986) Activation of acetylcholine receptors on clonal mammalian BC$_3$H1 cells by low concentrations of agonists. *J. Physiol. (London)* **373**, 129–162.

57 HAMILL, O., MARTY, A., NEHER, E., SAKMANN, B. and SIGWORTH, F. J. (1981) Improved patch clamp techniques for high-resolution current recordings from cells and cell-free membrane patches. *Pfluegers Arch.* **391**, 85–100.

58 LANDAU, V. G. and LIFSHITZ, E. M. (1959) *Fluid Mechanics* (Pergamon, Oxford), p. 219.

59 LEVICH, A. G. (1962) *Physicochemical Hydrodynamics* (Prentice-Hall, Englewood Cliffs, NJ).

60 SINE, S. and TAYLOR, P. (1979) Functional consequences of agonist-mediated staten transitions in the cholinergic receptor-studies in cultured muscle cells. *J. Biol. Chem.* **254**, 3315–3325.

61 SATHER, W., DIEUDONNE, S., MACDONALD, J. F. and ASCHER, P. (1992) *J. Physiol. (London)* **450**, 643–672.

62 WALSTROM, K. M. and HESS, G. P. (1994) On the mechanism for the channel-opening reaction of strychnine-sensitive glycine receptors on cultured embryonic mouse spinal cord cells. *Biochemistry* **33**, 7718–7730.

63 WIEBOLDT, R., RAMESH, D., CARPENTER, B. K. and HESS, G. P. (1994) Synthesis and photochemistry of photolabile derivatives of γ-aminobutyric acid for chemical kinetic investigations of the GABA receptor in the millisecond time region. *Biochemistry* **33**, 1526–1533.

64 NIU, L., GEE, K. R., SCHAPER, K. and HESS, G. P. (1996) Synthesis and photochemical properties of a kainate precursor, and activation of kainate and AMPA receptor channels on a microsecond time scale. *Biochemistry* **35**, 2030–2036.

65 WIEBOLDT, R., GEE, K. R., NIU, L., RAMESH, D., CARPENTER, B. K. and HESS, G. P. (1994) Photolabile precursors of glutamate: Synthesis, photochemical properties and activation of glutamate receptors on a microsecond time scale. *Proc. Natl. Acad. Sci. USA* **91**, 8752–8756.

66 NIU, L., GREWER, C. and HESS, G. P. (1996) Chemical kinetic investigations of neurotransmitter receptors on a cell surface in the µs time region. *Tech. Protein Chem.* **VII**, 139–149.

67 NIU, L., WIEBOLDT, R., RAMESH, D., CARPENTER, B. K. and HESS, G. P. (1996) Synthesis and characterization of a caged receptor ligand suitable for chemical kinetic investigations of the glycine receptor in the 3-µs time domain. *Biochemistry* **35**, 8136–8142.

68 RAMESH, D., WIEBOLDT, R., NIU, L., CARPENTER, B. K. and HESS, G. P. (1993) Photolysis of a new protecting group for the carboxyl function of neurotransmitters within 3 microseconds and with product quantum yield of 0.2. *Proc. Natl. Acad. Sci. USA* **90**, 11074–11078.

69 GEE, K. R., WIEBOLDT, R. and HESS, G. P. (1994) Synthesis and photochemistry of a new photolabile derivative of GABA. Neurotransmitter release and receptor activation in the µs time region. *J. Am. Chem. Soc.* **116**, 8366–8367.

70 GEE, K. R., CARPENTER, B. K. and HESS, G. P. (1998) Synthesis, photochemistry and biological characterization of photolabile protecting groups for carboxylic acids and neurotransmitters. *Methods Enzymol.* **291**, 30–50.

71 GREWER, C., JÄGER, J., CARPENTER, B. K. and HESS, G. P. (2000) A new photolabile precursor of glycine with improved properties: A tool for chemical kinetic investigations of the glycine receptor. *Biochemistry* **39**, 2063–2070.

72 BREITINGER, H.-G. A., WIEBOLDT, R., RAMESH, D., CARPENTER, B. K. and HESS, G. P. (2000) Synthesis and characterization of photolabile derivatives of serotonin for chemical kinetic investigations of the serotonin 5-HT3 receptor. *Biochemistry* **39**, 5500–5508.

73 BARTELS, E., WASSERMAN, N. H. and ERLANGER, B. F. (1971) Photochromic activators of the acetylcholine receptor. *Proc. Natl. Acad. Sci. USA* **68**, 1820–1823.

74 DELCOUR, A. H. and HESS, G. P. (1986) Chemical kinetic measurements of the effect of *trans*- and *cis*-3,3'bis[trimethylammonio)methyl]azobenzene bromide on acetylcholine receptor mediated ion translocation in *Electrophorus electricus* and *Torpedo californica*. *Biochemistry* **25**, 1793–1798.

75 KAPLAN, J. H., FORBUSH, B. and HOFFMAN, J. F. (1978) Rapid photolytic release of adenosine 5''-triphosphate from a protected analogue: Utilization by the Na:K pump of human red blood cell ghosts. *Biochemistry* **17**, 1929–1935.

76 MCCRAY, J. A., HERBETTE, L., KIHARA, T. and TRENTHAM, D. R. (1980) A new approach to time-resolved studies of ATP-requiring biological systems: Laser flash photolysis of caged ATP. *Proc. Natl. Acad. Sci. USA* **77**, 7237–7241.

77 ADAMS, R. S. and TSIEN, R. Y. (1993) Acetylcholine receptor kinetics. *Annu. Rev. Physiol.* **55**, 755–783.

78 CORRIE, J. E. T. and TRENTHAM, D. R. (1993) *Bioorg. Photochem.* **2** (Ed. MORRISON, H.) (Wiley, New York).

79 NERBONNE, J. M. (1996) Caged compounds: tools for illuminating neuronal responses and connections. *Curr. Opin. Neurobiol.* **6**, 379–386.

80 DEMAYO, P. (1960) Ultraviolet photochemistry of simple unsaturated systems. *Adv. Org. Chem.* **2**, 367–425.

81 BARLTROP, J. A., PLANT, P. J. and SCHOFIELD, P. (1966) Photosensitive protecting group. *J. Chem. Soc. Chem. Commun.* 822–823.

82 PATCHORNIK, A., AMIT, B. and WOODWARD, R. B. (1970) Photosensitive protecting groups. *J. Am. Chem. Soc.* **92**, 6333–6335.

83 MCCRAY, J. A. and TRENTHAM, D. R. (1989) Properties and uses of photoreactive caged compounds. *Annu. Rev. Biophys. Chem.* **29**, 239–270.

84 SHEEHAN, J. C., WILSON, R. M. and OXFORD, A. W. (1971) The photolysis of methoxy-substituted benzoin esters. A photosensitive protecting group for car-

boxylic acids. *J. Am. Chem. Soc.* **93**, 7222–7228.

85 Papageorgiou, G. and Corrie, J. E. T. (1997) Synthesis and properties of carbamoyl derivatives of photolabile benzoins. *Tetrahedron* **53**, 3917–3932.

86 Park, C. H. and Givens, R. S. (1997) New photoactivated protecting groups. 6. p-Hydroxyphenacyl: A phototrigger for chemical and biochemical probes. *J. Am. Chem. Soc.* **119**, 2453–2463.

87 Muralidharan, S. M., Mayer, G. M., Boyle, W. B. and Nerbonne, J. M. (1993) "Caged" phenylephrine: development and application to probe the mechanism of alpha-receptor mediated vasoconstriction. *Proc. Natl. Acad. Sci. USA* **90**, 5199–5203.

88 Walker, J. W., Martin, H., Schmitt, F. R. and Barsotti, R. J. (1993) Rapid release of alpha-adrenergic receptor ligand from photolabile analogues. *Biochemistry* **32**, 1338–1345.

89 Muralidharan, S. and Nerbonne, J. M. (1995) Photolabile "caged" adrenergic receptor agonists and related model compounds. *J. Photochem. Photobiol. (B)* **27**, 123–137.

90 Walker, J. W., McCray, J. A. and Hess, G. P. (1986) Photolabile protecting groups for an acetylcholine receptor ligand. Synthesis and photochemistry of a new class of o-nitrobenzyl derivatives and their effects on receptor function. *Biochemistry* **25**, 1799–1805.

91 Morrison, H. A. (1969) *The Chemistry of the Nitro and Nitroso Groups* (Feuer, H., Ed.) Part 1, pp. 165–213 (Wiley, New York).

92 Gee, K. R., Niu, L., Schaper, K. and Hess, G. P. (1995) Caged bioactive carboxylates, synthesis, photolysis studies, and biological characterization of a new caged N-methyl-D-aspartic acid (NMDA). *J. Org. Chem.* **60**, 4260–4263.

93 Billington, A. P., Walstrom, K. M., Ramesh, D., Guzikowski, A. P., Carpenter, B. K. and Hess, G. P. (1992) Synthesis and photochemistry of photolabile N-glycine derivatives, and effects of one on the glycine receptor. *Biochemistry* **31**, 5500–5507.

94 Banerjee, A., Grewer, C., Ramakrishnan, L., Jäger, J., Gameiro, A., Breitinger, H.-G. A., Gee, K. R., Carpenter, B. K. and Hess, G. P. (2003) Toward the development of new photolabile protecting groups that can rapidly release bioactive compounds upon photolysis with visible light. *J. Org. Chem.* **68**, 8361–8357.

95 Marty, A. and Neher, E. (1983) Tight-seal whole-cell recording. In *Single-Channel Recording* (eds. Sakmann, B. and Neher, E.) (Plenum, New York) pp. 107–122.

96 Billington, A. P., Matsubara, N., Webb, W. W. and Hess, G. P. (1992) Protein conformational changes in the µs time region investigated with a laser pulse photolysis technique. *Techniques in Protein Chemistry* **III**, 417–427.

97 Hardman, J. G., Limbird, L. E., Molinoff, P. B., Ruddon, R. W. and Gilman, A. G. (1996) *Goodman & Gilman's The Pharmacological Basis of Therapeutics* (9th Edn.) (McGraw-Hill, New York).

98 Arias, H. R. (1999) Role of local anesthetics on both cholinergic and serotonergic ionotropic receptors. *Neurosci. Biobehavioral Rev.* **23**, 817–843.

99 Adams, P. R. (1976) Drug blockage of open end-plate channels. *J. Physiol. (London)* **260**, 531–552.

100 Grewer, C. and Hess, G. P. (1999) On the mechanism of inhibition of a nicotinic acetylcholine receptor by the anticonvulsant MK-801 investigated by laser pulse photolysis in the microsecond-to-millisecond time region. *Biochemistry* **38**, 7837–7846.

101 Hess, G. P., Ulrich, H., Breitinger, H.-G., Niu, L., Gameiro, A. M., Grewer, C., Srivastava, S., Ippolito, J. E., Lee, S., Jayaraman, V. and Coombs, S. E. (2000) Mechanism-based discovery of ligands that prevent inhibition of the nicotinic acetylcholine receptor by cocaine and MK-801. *Proc. Natl. Acad. Sci. USA* **97**, 13895–13900.

102 Hess, G. P., Gameiro, A. M., Schoenfeld, R. C., Chen, Y., Ulrich, H., Nye, J. A., Sit, B., Carroll, F. I., and Ganem, B. (2003) Reversing the action of noncompetitive inhibitors (MK-801 and co-

caine) on a protein (nicotinic acetylcholine receptor)-mediated reaction. *Biochemistry* **42**, 6106–6114.
103 SCOTT, R. A., LHATOO, S. D. and SANDER, J. W. A. S. (2001) The treatment of epilepsy in developing countries: where do we go from here? *Bulletin of the World Health Organization* **79**, 344–351.
104 RAMAKRISHNAN, L., and HESS, G. P. (2004) On the mechanism of a mutated and abnormally functioning aminobutyric acid (A) receptor linked to epilepsy and alleviation of its dysfunction. *Biochemistry* **43**, 7534–7540.
105 JAYARAMAN, V. (1998) Channel-opening mechanism of a kainate-activated glutamate receptor: kinetic investigations using a laser pulse photolysis technique. *Biochemistry* **37**, 16735–16740.
106 GREWER, C. (1999) Investigation of the alpha(1)-glycine receptor channel-opening kinetics in the submillisecond time domain. *Biophys. J.* **77**, 727–738.
107 LI, G. and NIU, L. (2004) How fast does the GluR1Qflip AMPA receptor channel open? *J. Biol. Chem.* **279**, 3990–3997.
108 LI, G., PEI, W. and NIU, L. (2003) Channel-opening kinetics of GluR2(flip) AMPA receptor: a laser pulse photolysis study. *Biochemistry* **42**, 12358–12366.
109 LI, G., OSWALD, R. E. and NIU, L. (2003) Channel-opening kinetics of GluR6 kainate receptor. *Biochemistry* **42**, 12367–12375.
110 GREWER, C., WATZKE, N., WIESSNER, M. and RAUEN, T. (2000) Glutamate translocation of the neuronal glutamate transporter EAAC1 occurs within milliseconds. *Proc. Natl. Acad. Sci. USA* **97**, 9706–9711.
111 WATZKE, N., BAMBERG, E., and GREWER, C. (2001) Early intermediates in the transport cycle of the neuronal excitatory amino acid carrier EAAC1. *J. Gen. Physiol.* **117**, 547–562.
112 GREWER, C., MADANI MOBAREKEH, S. A., WATZKE, N., RAUEN, T., and SCHAPER, K. (2001) Substrate translocation kinetics of excitatory amino acid carrier 1 probed with laser pulse photolysis of a new photolabile precursor of D-aspartic acid. *Biochemistry* **40**, 232–240.
113 WATZKE, N. and GREWER, C. (2001) The anion conductance of the glutamate transporter EAAC1 depends on the direction of glutamate transport. *FEBS Lett.* **503**, 121–125.
114 GREWER, C., WATZKE, N., RAUEN, T. and BICHO, A. (2003) Is the glutamate residue Glu-373 the proton acceptor of the excitatory amino acid carrier 1? *J. Biol. Chem.* **278**, 2585–2592.
115 JAYARAMAN, V., THIRAN, S. and MADDEN, D. R. (2000) Fourier transform infrared spectroscopic characterization of a photolabile precursor of glutamate. *FEBS Lett.* **475**, 278–282.
116 JAYARAMAN, V., KEESEY, R. and MADDEN, D. R. (2000) Ligand-protein interactions in the glutamate receptor. *Biochemistry* **39**, 8693–8697.
117 MADDEN, D. R., THIRAN, S., ZIMMERMAN, H., ROMM, J. and JAYARAMAN, V. (2001) Stereochemistry of quinoxaline binding to a glutamate receptor investigated by Fourier transform infrared spectroscopy. *J. Biol. Chem.* **272**, 37821–37826.
118 CHENG, Q., STEINMETZ, M. G. and JAYARAMAN, V. (2002) Photolysis of gamma-(alpha-carboxy-2-nitrobenzyl)-L-glutamic acid investigated in the microsecond time scale by time-resolved FTIR. *J. Am. Chem. Soc.* **124**, 7676–7677.
119 HESS, G. P., CASH, D. J. and AOSHIMA, H. (1983). Acetylcholine receptor-controlled ion translocation: Chemical kinetic investigations of the mechanism. *Annu. Rev. Biochem. Biophys.* **225**, 500–504.

4.4
Caged Neurotransmitters for Probing Neuronal Circuits, Neuronal Integration and Synaptic Plasticity

Deda C. Gillespie, Gunsoo Kim, and Karl Kandler

4.4.1
Introduction

In recent years caged compounds have become important tools for investigating nervous system function; caged neurotransmitters in particular have been employed in almost every major area in the neurosciences where experiments can be performed *in vitro*. In this chapter we will highlight three areas in which caged neurotransmitters have been most successfully used to significantly enhance our understanding of brain function. We will focus first on an area where caged neurotransmitters have seen the most use, which is in the characterization of neuronal connectivity, where caged glutamate has been used to map inputs to individual neurons. A second area that has greatly profited from the use of caged neurotransmitters is neuronal integration; here, caged neurotransmitters have been employed to stimulate, with fine spatiotemporal control, small segments of a neuron's extensively branched dendritic tree. Finally, we will provide examples of how caged neurotransmitters have helped to advance our understanding of mechanisms involved in neuronal plasticity. We also want to alert the reader to a recent review that highlights somewhat different aspects of the use of caged biological substances in neuroscience [1].

4.4.2
Functional Mapping of Neuronal Connections

Understanding the organization and the wiring patterns of the mammalian brain has been a major challenge in neuroscience. Over the last century, researchers have come up with a vast number of different techniques for visualizing neuronal pathways. Currently, neuronal connectivity is most commonly revealed by injection of neuronal tracers into a small volume of brain tissue. These tracers are then taken up by neurons and their processes and are transported forward along axonal projections (anterograde transport) or backwards to neuronal cell bodies (retrograde transport). A great variety of substances can serve as neuronal tracers including fluorescent dyes or latex beads, enzymes or small organic molecules (reviewed in [2]), viruses [3–5], and genetically encoded proteins [6, 7]. Despite increasing specificity and sensitivity of these tracers and their visualization, anatomical tracing techniques have the limitation that physiological characteristics of the connections such as synaptic strength and reliability remain unknown. Photolytic cleavage (uncaging) of caged glutamate can overcome these limitations because, in addition to revealing the spatial organization of neuronal connections, it also provides information about the functional nature of these connections. Mapping neuronal connectivity by uncaging glutamate is based upon the fact that glutamate, the major excitatory neurotransmitter in the vertebrate central nervous system, excites nearly every neuron in the CNS. High intensity UV light flashes in the presence of caged glutamate can produce a rapid rise in the free glutamate concentration, which depolarizes neurons and, if the depolarization exceeds the firing threshold, generates action potentials. These action potentials propagate along axonal processes and into presynaptic terminals, where they cause the release of neurotransmitters, followed by synaptic currents in functionally connected neurons. Thus, by recording electrical responses from a single neuron in a living brain slice bathed in caged glutamate one can use locally applied UV "search" flashes to reveal the location of presynaptic neurons that functionally connect to the recorded cell (Fig. 4.4.1) [8]. In addition to providing data about neuronal connectivity on a single cell level, this method can also reveal functional properties of these connections such as neurotransmitter and receptor phenotype, physiological strength, and reliability. Finally, the recorded, postsynaptic neuron can be char-

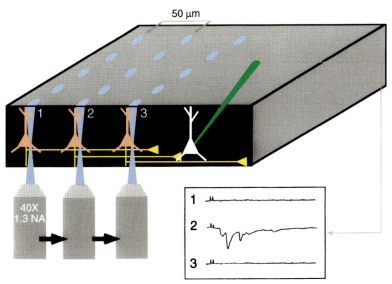

Fig. 4.4.1 Schematic illustration mapping functional neuronal connections with focal uncaging of glutamate. UV light is focused through the objective and a glass-bottomed chamber into a living brain slice. Membrane currents are recorded from a single neuron (white) via a whole-cell patch electrode (green). With caged glutamate in the bathing solution, short UV light flashes cause rapid increases in free glutamate, generating action potentials in cells near the light flash. Neurons that provide no synaptic input [1, 3] to the recorded cell will cause no change in current, while those that do provide input, such as neuron 2 in the figure, will create a postsynaptic current, as shown in the box. Laser scanning photostimulation can efficiently uncage glutamate at the many points in an array in order to construct the input map for a given neuron. (Adapted with permission from [8]).

acterized with regard to its physiological properties (active and passive membrane characteristics) and, by including dye in the recording pipette for intracellular cell-fills, with regard to its somatic, dendritic, and axonal morphology. More recently, simultaneous field potential recordings across multiple electrodes have also been combined with photostimulation to construct multiple connectivity maps [9]. However, because this method measures extracellular activity it cannot provide as full a morphological or physiological characterization of individual postsynaptic neurons as can intracellular recordings.

Figure 4.4.1 illustrates the laser scanning photostimulation technique. The tissue slice rests in a glass-bottomed perfusion chamber and is bathed in a caged glutamate-containing medium. A whole-cell patch electrode is used to record from an individual cell (white cell at right), while UV light flashes, focused through the objective, are delivered at numerous positions in a grid. Release of free glutamate causes action potentials in nearby neurons. If one of these active neurons is presynaptic to the neuron being recorded, a synaptic response will be measured through the patch electrode.

Three forms of caged glutamate have primarily been used for mapping neuronal connections: γ-CNB-caged glutamate (Molecular Probes [10–14]; see also Chapter 1.1 in this volume), Nmoc-caged glutamate [15, 16], and pHP-glutamate [17, 18] (see also Chapter 1.3 this volume). These caged compounds are relatively stable in physiological solutions, which is important for mapping studies that can take several hours. An increase of free glutamate by spontaneous hydrolysis can desensitize glutamate receptors and over time and at higher concentrations can have toxic effects. The speed of photolysis for these caged compounds is on the millisecond scale, which allows one to create steep enough glutamate concentration jumps to avoid the receptor desensitization and inactivation of fast voltage-gated sodium channels that can be created by gradual increases of glutamate and associated depolarizations. Finally, the quantum yield of these compounds is sufficiently high to circumvent phototoxicity and UV-induced changes in neurotransmitter receptor function due to excessive UV [19–23].

To achieve high spatial resolution, UV illumination and glutamate uncaging must be restricted to small tissue volumes, which has been accomplished by several different approaches. Most commonly, the light from a UV laser (He-Cd, nitrogen pulsed laser, or argon-ion laser) is focused into the brain slice through a 40× or 60× microscopic objective [8, 24]. In addition to lasers, other UV light sources such as xenon flash bulbs or mercury arc lamps have been proven suitable for rapid uncaging of glutamate in slices [13, 24–27] and turn-key systems are now commercially available [28, 29]. In place of a microscopic objective, several groups have used small-diameter (5–50 μm) or tapered optical quartz fibers to create focal spots of UV light [27, 30, 31], an approach that is both economical and technically simple but that is biased toward stimulation of superficial neurons in slice preparations. Recently, a modified confocal microscope [32] has also been used to focus the uncaging UV light. "Chemical two-photon uncaging," using glutamate coupled to two caging moieties [33], requires removal of both caging groups to activate the glutamate, and hence increases spatial resolution, particularly coaxial to the light beam. In addition, the double-caged glutamate is more stable than single-caged glutamate, and it does not require the more expensive lasers necessary for optical two-photon uncaging. Nevertheless, true two-photon uncaging is becoming more common with the recent development of new caging groups for two-photon glutamate uncaging [34, 35]. Multiphoton uncaging, which is discussed further in Chapter 8, allows for yet greater spatial resolution and deeper tissue stimulation with minimal UV toxicity.

In addition to the optics of the light delivery apparatus, light intensity, flash duration, and the concentration of caged glutamate, optical and biological properties of the preparation also determine the resolution of the uncaging system. UV light is easily scattered in brain tissue, and UV penetration into the slice depends on the optical properties of the tissue, which vary greatly among slices taken from different brain areas or developmental stages. Other significant factors are the effectiveness of neuronal or glial glutamate uptake systems, the dendritic geometry of the neurons to be stimulated, and their sensitivity to glutamate, all of which again vary with age and across brain regions. To control for these several sources of variability, the resolution of the uncaging technique must be determined experimentally for each preparation by measuring the area in which a given UV pulse elicits action potentials in neurons. Most mapping studies have reported a spike-eliciting uncaging radius of about 20–80 μm [18, 36–38].

4.4.2.1 Example: Uncaging Glutamate to Map Local Connections in the Mammalian Cortex

The mammalian neocortex exhibits a stereotyped three-dimensional architecture: the cortical plate consists of 6 horizontal layers and orthogonal columns spanning the 6-layered plate. The cortical layers, defined by specific neuronal types and by characteristic connectivity patterns, were initially described on the basis of anatomy. Interlaminar and intercolumnar connectivity patterns have been generally inferred from tracing studies, based on the assumption that axonal arbors make synaptic contacts with neurons with overlapping dendritic fields. However, cortical neurons exhibit a great deal of anatomical and functional diversity even within a layer, and it has become clear that the physical presence of axonal arbors alone provides insufficient information about what subset of neurons actually form synapses within the area.

Mapping neuronal connections with caged glutamate has provided significant insight into functional cortical connectivity by revealing that morphologically and functionally distinct types of cortical neurons receive distinctly different patterns of synaptic inputs from different types of presynaptic neurons. Moreover, even morphologically similar neurons can receive dramatically different laminar inputs, suggesting a greater diversity among cortical neurons than was implied by anatomical measures [39]. Laser scanning photostimulation is well suited to *in vivo-in vitro* studies in which cortical modularity is first assessed *in vivo* using intrinsic optical imaging, then living slices are cut from the same tissue in order to examine connectivity in the previously imaged brain [16, 40]. Callaway and colleagues have used laser scanning photostimulation in a series of detailed studies elucidating the circuitry of visual cortex. In acute brain slices, synaptic responses to computer-controlled laser photostimulation were obtained with whole-cell patch electrodes. The input maps constructed from these responses indicate the locations of cell bodies of neurons that provide synaptic input to the recorded neuron. Using this approach, Dantzker and Callaway [37] demonstrated that the excitatory input maps to pyramidal cells they recorded in layer 2/3 were similar, but that the excitatory inputs to various inhibitory interneurons were quite varied. For example, fast-spiking interneurons (Fig. 4.4.2 a) received strong inputs from layers 4 and 5 and weaker inputs from layer 2/3, whereas certain adapting interneurons (Fig. 4.4.2 b) received nearly all their excitatory inputs from layer 2/3. Other adapting interneurons appeared to belong to yet a different class exhibiting different input patterns. This approach has been used to examine patterns of synaptic input to morphologically distinct classes of layer 5 pyramidal cells [41] as well as to physically intermingled, but functionally distinct, sets of cortical projection neurons [42]. Results from this series of studies show that distinct patterns of synaptic input are able to distinguish morphologically distinct excitatory neurons, morphologically similar but functionally

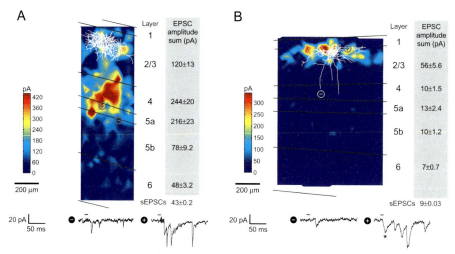

Fig. 4.4.2 Laser-scanning photostimulation reveals intrinsic synaptic connection in visual cortex. Input maps for a fast-spiking interneuron (**A**) and for an adapting interneuron (**B**) in cortical layer 2/3. The morphological reconstruction of the recorded and dye-filled neuron is superimposed in white. Maps are linear interpolations of the postsynaptic current amplitudes recorded after glutamate photolysis at different sites. Dark red areas indicate locations where glutamate application elicited maximal amplitude currents in the recorded neuron, and thus indicate the location of cell bodies of neurons strongly connected to the neuron. The fast-spiking neuron shown in **A** receives a strong input from layer 4 and a weaker but significant input from layer 2/3. The adapting interneuron in **B** receives the great majority of its input from layer 2/3. (Reproduced with permission from [37]).

distinct inhibitory interneurons, and excitatory projection neurons subserving different pathways in the visual stream.

Focal uncaging in other cortical areas [9, 14, 43, 44] and in the CA3-CA1 region of the hippocampus [45] has been used to elucidate connectivity and topography. Focal photolysis of caged glutamate has also allowed researchers to begin to identify intrinsic circuitry in the superior colliculus and suprachiasmatic nucleus, where the presence of fibers of passage had created difficulties for studies of local circuitry using traditional techniques such as electrical stimulation [46–48].

Finally, although the contributions of glia to neuronal processing have often been ignored, astrocytes can play a complex role in neuronal signaling, influencing neuronal activity through uptake and release of various molecules in the

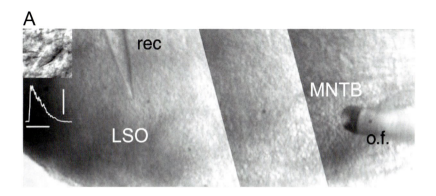

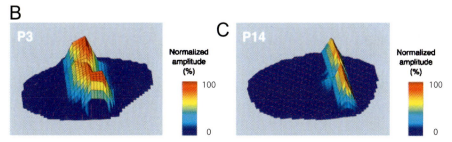

Fig. 4.4.3 Determining the developmental reorganization of inhibitory circuits with caged glutamate. **A** Whole-cell patch clamp recordings (recording electrode shown at "rec") were obtained from single LSO neurons in living brainstem slices while the input area of the MNTB was mapped with focal uncaging of PHP glutamate (100 µM). Small areas (diameter ~ 25 µM) were illuminated with short (5–50 ms) flashes of UV light using an optical quartz fiber (designated "o.f."; inner light-conducting diameter 20 µm). The spike-eliciting radius of this system was about 25 µM. Insets show an LSO neuron and its synaptic response to glutamate uncaging in the MNTB. **B** Example of an input map from MNTB to a single LSO neuron obtained in a P3 animal. **C** Input map obtained in a P14 animal. The input maps acquired in slices from P14 animals are narrower than those acquired in very young slices. Reproduced from [18].

extracellular space. A recent report [49] found that focal photolysis of caged glutamate activated metabotropic glutamate receptors on individual astrocytes and elicited a calcium response in the cell. This was followed by delayed calcium responses in other astrocytes, indicating the presence of a connected astrocytic circuit. Using caged bradykinin (see Chapter 1.3) to stimulate astrocytes in the presence of glutamate receptor antagonists, the authors further showed that astrocytes were not using glutamate to communicate among themselves. Rather, these astrocytic connections are probably mediated by ATP, and allow for waves of oscillatory calcium among astrocytes after synaptic stimulation (for a recent review see also Ref. [50]).

4.4.2.2 Example: Use of Caged Glutamate for Probing Refinement of Neuronal Connectivity

Mapping neuronal connections through photolysis of caged glutamate has been used to examine developmental reorganization of excitatory neuronal connections in the immature neocortex [36, 44]. Recently, our laboratory provided evidence that such functional reorganization also occurs among inhibitory connections by mapping functional connectivity in an inhibitory pathway in an auditory sound localization circuit [18]. In the mammalian auditory system, interaural intensity difference – a major clue for sound localization – is first processed in the lateral superior olive (LSO), a binaural brainstem nucleus that receives inhibitory inputs from the medial nucleus of the trapezoid body (MNTB). In mature animals, the MNTB-LSO pathway is precisely topographically organized. During development, this precise topographic connectivity pattern is achieved partly by pruning of axon collaterals [51]. To investigate whether and to what degree the connectivity in the MNTB-LSO pathway involves the reorganization of functional synaptic connections, functional MNTB-LSO connectivity was mapped in newborn rats during the period before hearing onset at two weeks, using an optical quartz fiber to deliver the uncaging light (Fig. 4.4.3).

Using this approach we found that, in newborn animals, individual LSO neurons receive functional inputs from an area that covers about 40% of the MNTB. During postnatal development, the size of these input maps decreases about fourfold such that in a two-week-old animal (the age at which physiological hearing begins), single LSO neurons receive functional inputs from only about 10% of the MNTB. Thus, during the first two postnatal weeks, LSO neurons lose most of their MNTB inputs by becoming disconnected from three out of four presynaptic partners. Functional elimination (silencing) of MNTB-LSO synapses occurs before hearing onset, while structural elimination of synapses by axonal pruning occurs after hearing onset, suggesting that the two processes are guided by different mechanisms.

4.4.2.3 Conclusions and Future Challenges

During the last 10 years, focal photolysis of caged glutamate has become an important tool for analyzing functional organization of both developing and mature neuronal circuits. The development of more stable and sensitive forms of caged glutamate, combined with a wider variety of approaches for delivering UV flashes to small tissue volumes, has played an important role in making this method more useful and user-friendly. A major limitation of current mapping methods is that their use is restricted to brain slices and thus to analysis of neuronal circuits that can be contained in a single 300–500 µm thick brain slice *in vitro*. This hampers the analysis of longer-range pathways, which are rarely restricted to a single plane. New methods for preparing nonplanar slices may extend the range of systems than can be analyzed *in vitro* [52]. The development of stable caging groups photolyzable by longer-wavelength light will also be a boon to these types of studies, as it will allow for deeper tissue penetration and thus, potentially, for use in more complete circuits.

4.4.3
Probing Neuronal Integration

Neurons receive thousands to tens of thousands of synaptic inputs from other neurons, and each neuron may in turn feed information to another 1000 neurons. The majority of synaptic inputs are located on geometrically complex, highly ramified processes, the dendrites. Temporal and spatial integration of these dendritic inputs eventually results in the generation or inhibition of action potentials, the final output signal of neurons. The effect of a single synapse on the postsynaptic neuron is a function of integration in the dendritic tree, and is determined by the dendrites' passive cable and active membrane properties (for reviews see Refs. [53–56]). Over the last 10 years, a number of technical advances such as dendritic patch clamp recordings, high-resolution optical imaging, and focal stimulation of small dendritic segments have resulted in a remarkable advance in our understanding of how information is processed in single neurons. Rapid and very focal uncaging of neurotransmitters at dendritic locations has aided us in understanding how a neuron integrates inputs over space. For example, since focal uncaging can induce local responses that in many respects mimic synaptic responses, uncaging has been used to create artificial presynaptic release sites. These artificial synapses can be positioned at different sites along the dendrite in order to study how synapse location influences the neuronal response [57–60].

The postsynaptic effect of a synapse is determined not only by its location, but also by the type and number of neurotransmitter receptors it activates. Most neurotransmitters can activate several distinct receptors, each with unique properties and unique responses. Traditionally, the cellular location and distribution of neurotransmitter receptors has been visualized using labeled agonists, toxins, or antibodies. Together these studies have provided evidence that neurotransmitter receptors are non-uniformly distributed along the dendritic length [61–65].

However, anatomically derived subcellular receptor maps cannot indicate whether receptors are functional, nor can they show the physiological consequences of specific receptor activation. In addition, where neuronal dendrites and somata overlap, as is usually the case in the brain, anatomical staining patterns may be difficult to interpret unless the cells are isolated *in vitro* or unless some specific structural specialization works in the researcher's favor. To overcome these limitations, iontophoresis [66] or pressure ejection of receptor agonists have been used to activate functional receptors. However, these methods are most practically limited to preparations such as acutely dissociated cells or cell cultures in which neurons are easily accessible.

Using two-photon microscopy and caged carbamoylcholine to activate nicotinergic acetylcholine receptors, Denk [67] provided the first demonstration that highly localized uncaging of neurotransmitters could be used to activate focally restricted populations of neurotransmitter receptors on cultured BC3H1 cells. Successive studies have applied uncaging of receptor agonists to neurons both in cultures and in the more complex environment of living brain slices. Receptor activation by the uncaged agonist produces a membrane current that can be recorded at the neuronal soma or along dendrites, and the magnitude of this current is a function of the number of receptors activated and of the electrical properties of the dendrites. Several groups have now used this approach to reveal the subcellular distribution of functional ionotropic glutamate receptors [33, 60, 68–71], GABA receptors [58, 69, 72], and nicotinergic acetylcholine receptors [73] in individual neurons from several brain regions of normal and mutant [74] mice. In general, these studies have emphasized a non-uniform distribution of neurotransmitter receptors along the dendritic tree, in some cases due to a smooth gradient of receptor density and in others due to the presence of high-density receptor clusters. Synaptic behavior is influenced not only by receptor distribution, but also by spatiotemporal patterns of neurotransmitter uptake by glutamate transporters located on neurons and glia. Recently, glutamate uncaging has been used to investigate glutamate uptake after stimulation of the climbing fiber-Purkinje cell synapse [75], and the synthesis of new caged glutamate transporter blockers may help to advance our understanding of synaptic glutamate transporter function yet further [76].

4.4.3.1 Examples

Pettit and colleagues [33] were the first to apply glutamate uncaging to map glutamate receptor distribution on individual dendrites of hippocampal pyramidal neurons, using a double-caged glutamate reagent in order to ensure a highly spatially restricted volume of neurotransmitter activation. Their initial finding was that AMPA receptors were more densely distributed on distal dendrites than on proximal dendrites. In a later study using both caged glutamate and caged GABA, the same laboratory [69] compared the distribution of functional glutamate and GABA receptors in two types of cells in the hippocampus. They uncaged neurotransmitters at many locations along the dendrites while recording the resulting membrane currents, and found not only that the current densities for GABA- and glutamate-mediated currents vary along the dendrites of a given neuron, but also that the current densities varied differently depending on the neurotransmitter and on the cell type as well. For example, in interneurons, glutamate receptor density, as inferred from current density, appeared to increase monotonically with distance from the soma (Fig. 4.4.4a). By contrast, GABA receptor density peaked at about 25 μm from the soma and decreased with increasing distance from the soma (Fig. 4.4.4b). Thus, in hippocampal interneurons, inhibitory inputs are more abundant fairly close to the soma, while glutamate receptors are found in abundance on the distal dendrites. In pyramidal cells apical dendrites, glutamate receptor density was lowest approximately 50 μm from the soma, while, in basal dendrites, density was highest near the soma and decreased with distance from the soma.

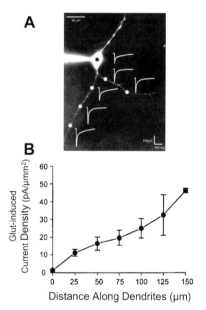

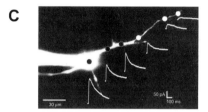

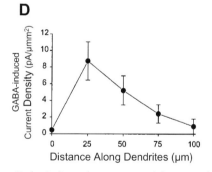

Fig. 4.4.4 Mapping cellular distributing of glutamate and GABA receptors in hippocampal interneurons. Density of receptor types along dendrites of hippocampal interneurons. **A** Volume-rendered image of an interneuron filled with a dye through the recording electrode. Circles indicate the position and diameter of the UV light spot used to uncage glutamate, and traces indicate averages of 5 evoked responses at each location. **B** Current density as a function of glutamate uncaging distance from the soma ($n=4$). **C** GABA responses in an interneuron. Circles indicate the position and diameter of the UV light spot used to uncage GABA, and nearby traces indicate averages of 3 responses evoked by uncaging at each location. **D** Current density as a function of GABA uncaging distance from the soma ($n=6$). Current density is the synaptic response recorded at the soma normalized by the dendritic area at the uncaging site. Peak amplitudes normalized to the amplitude of currents evoked at the cell body. (Reproduced with permission from [68]).

A number of studies from several laboratories have provided additional evidence of a non-uniform and specific subcellular distribution of neurotransmitter receptors. For example, glutamate uncaging onto layer V pyramidal neurons of somatosensory cortex, in conjunction with pharmacology to block specific glutamate receptor subtypes, supports a differential distribution of different glutamate receptors. In these neurons, NMDA receptor density is highest near the soma, while AMPA and kainite receptor densities appear to be highest on distal dendrites [68, 77]. Specific subcellular gradients have also been found for different types of inhibitory GABA receptors, with GABAB receptors more abundant proximally to the soma and GABAA receptors more abundant distally [72]. Lowe [58], in a careful study of GABA receptor sites, used focal GABA uncaging at numerous points along different dendrites to inhibit back-propagating action potentials in mitral cells, thereby determining with this approach the spatial spread of GABAergic inhibition (Fig. 4.4.5) and demonstrating a non-uniform distribution of GABA receptors throughout the dendritic tree. Differential distribution of other receptors, such as the nAChR, has also been seen on CA1 interneurons and pyramidal cells [73]. In this study, the dye-filled neuron was visualized on a confocal microscope, and caged carbachol was photolyzed in order to map $a7$-containing nAChRs in the 3-dimensional dendritic tree. An added benefit of uncaging was the temporal control that made measurement of the rapidly desensitizing $a7$-nAChR response possible. Focal photolysis with calcium imaging and intracellular recording has been used to study other aspects of dendritic computation such as compartmentalization in the dendritic tree and the contri-

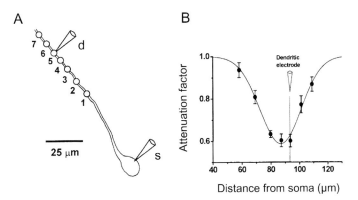

Fig. 4.4.5 Attenuation of back-propagating dendritic spikes in mitral cells by focal uncaging of GABA along the dendrite. Action potentials were evoked by current injection at the soma. A Cell with dendrite, showing location of somatic recording electrode (s), dendritic recording electrode (d), and sites of UV illumination along the dendrite. B Attenuation of dendritic action potentials induced by GABA uncaging as a function of distance from soma. Dendritic recording electrode position (near uncaging site 5) indicated by dashed line. Significant attenuation occurred over a range of about 20 μm, with attenuation asymmetric and greatest on the side proximal to the soma. (Reproduced with permission from [57]).

bution of different branches and of specific glutamate receptor subtypes to action potential generation [59]. Finally, two-photon MNI-glutamate uncaging was used in a study that showed a striking degree of variability in the glutamate responses at different dendritic spines on hippocampal neurons [35], demonstrating a relationship between synaptic response size and dendritic spine geometry.

4.4.3.2 Conclusions and Future Challenges

Focal uncaging of caged neurotransmitters has become an essential tool for deciphering dendritic function and integration. Mapping the subcellular distributions of various neurotransmitter receptor types has established the basis for a richer understanding of how individual synapses affect neuronal integration. The development of caged compounds to include specific receptor antagonists and modulators, as well as caging moieties sensitive to a range of wavelengths, will enable researchers to further dissect aspects of synaptic function and test hypotheses that have emerged from synaptic mapping studies and to probe for interactions between synapses. Focal excitation in subregions of the dendritic tree has given insight into propagation of action potentials. In the last ten years the importance of active membrane properties has become increasingly clear; caged antagonists to voltage-gated ion channels could provide a new approach for testing specific hypotheses about the function of these active properties.

4.4.4
Investigating Synaptic Plasticity

A hallmark of the nervous system is its ability to respond or adapt to novel situations, the brain also being able to create and store memories. Perhaps no other nervous system phenomenon has received as much attention (and controversy) as the twin phenomena of long-term potentiation (LTP) and long-term depression (LTD). LTP and LTD are changes in synaptic efficacy that occur subsequent to specific synaptic activity patterns and are generally considered to be the cellular phenomena that underlie the formation of memories. In addition, LTP and LTD have been strongly implicated as crucial mechanisms that direct the changes in circuit wiring found both in normal development and in such pathologies as addiction. Initially described in the CA3-CA1 region of the hippocampus [78], LTP, which can be induced by pairing postsynaptic activity with a high-intensity input, refers to the persistent (>1 h) facilitation of the synaptic response seen after pairing. Under certain stimulation conditions, such as low-frequency synaptic stimulation, LTD, a persistent weakening of synaptic connections, is induced. Decades of research, mostly in acute brain slice preparations, have established that induction of classical LTP or LTD requires postsynaptic activity, specifically a postsynaptic increase in the intracellular calcium concentration resulting from activation of NMDA receptors [79–82], and that the direction of plasticity depends on pairing of postsynaptic activity with specific activity patterns or specific pre- and post-synaptic timing [83–85]. Although LTP and LTD have been studied most extensively in the hippocampus, they are seen in other brain areas as well [86–90] and are generally believed to be a common feature of excitatory synapses.

The first use of uncaging of neurotransmitters to explore mechanisms of synaptic plasticity addressed the question of whether the primary locus of induction and expression of long-term synaptic plasticity in the hippocampal CA3-to-CA1 synapses is presynaptic or postsynaptic [27]. Instead of activating receptors on CA1 pyramidal neurons by stimulation of the presynaptic terminal, the authors blocked action potential-elicited neurotransmitter release and stimulated two separate small dendritic segments by focal photolysis of caged glutamate using small-diameter (10 μm) optical fibers. Pairing of glutamate uncaging at one of the dendritic sites with postsynaptic depolarization elicited a long-lasting depression of the response to glutamate (Fig. 4.4.6). Like synaptic LTD, uncaging-induced depression was input specific, as depression was observed only at the dendritic site at which uncaging was paired with depolarization, but not at the unpaired site. In addition, uncaging-induced depression required activation of NMDA receptors and protein phosphatases. Thus, glutamate uncaging was able to produce a purely "postsynaptic" LTD that shared many features of synaptically induced LTD. Interestingly, though the locus of LTP at this synapse is also believed to be postsynaptic, uncaging glutamate did not induce LTP, perhaps because a high enough glutamate concentration was not attained with focal photo-

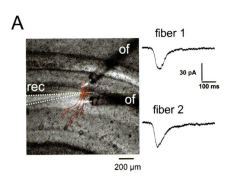
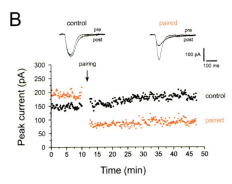

Fig. 4.4.6 Long-term depression of glutamate responses in hippocampal neurons induced by photolysis of caged glutamate. **A** Experimental approach. Membrane currents of pyramidal cell neuron were recorded (recording electrode shown at "rec") and two optical fibers (designated "o.f."; inner light-conducting diameter 10 μm) were positioned over two discrete dendritic sites. The biocytin-stained and reconstructed neuron from this experiment was superimposed on a video image of the original slice. Membrane responses elicited by 20 ms UV light flashes delivered through fiber 1 and fiber 2 are shown to the right. **B** Pairing of uncaging glutamate (1 Hz for 60 s) with strong depolarization of the recorded neuron (induced by current injection) produces a long-lasting depression of glutamate responses. This long-term depression (LTD) is specific for the site which experienced pairing (red). Sites which did not experience pairing (black) showed a stable response amplitude. Insets show responses before and after pairing for control sites (black) and paired sites (red). (Reproduced from [27]).

lysis. Subsequent studies using the uncaging technique found quite similar results for layer V pyramidal neurons in mammalian neocortex [91, 92]. That is, LTD, but not LTP, could also be induced in these neurons purely postsynaptically, by focal glutamate application.

A separate study by Wang et al. [93] took advantage of the spatial resolution with double-caged glutamate to investigate the spatial spread of LTD in cerebellar Purkinje cells. In these experiments, the authors combined calcium imaging and whole-cell patch recording with glutamate uncaging. Calcium imaging was used to determine the spatial spread of postsynaptic activation induced by parallel fiber activation. Pairing of parallel fiber activation with depolarization of the postsynaptic Purkinje cell induced LTD; synaptic currents declined by an average of 20%. After induction of synaptic LTD, the sensitivity of postsynaptic AMPA receptors was measured using photolysis of caged glutamate. Responses to glutamate uncaging decreased after LTD by at least as much as the synaptic response, indicating that LTD could be accounted for by the loss of AMPA receptor sensitivity. Simple rundown of AMPA receptors was ruled out because control cells (not depolarized) showed neither LTD nor decreased response to

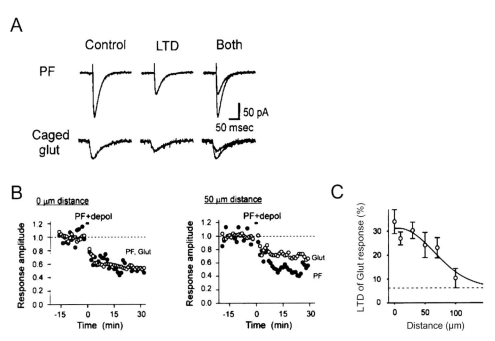

Fig. 4.4.7 In cerebellar Purkinje cells, depression of glutamate responses occurs within 100 μm of the site of LTD induction. **A** LTD was induced by pairing parallel fiber activation with postsynaptic depolarization. LTD was accompanied by a depression in the response to uncaged glutamate. **B** Depression of glutamate responses as a function of distance from LTD induction site (0 and 50 μm distance). **C** Percent depression at different distances from the site of LTD induction. Dashed line indicates average glutamate receptor rundown in the absence of LTD. (Reproduced from [92]).

glutamate uncaging. In order to quantify the spatial extent of decreased AMPA receptor sensitivity, the authors uncaged glutamate at different distances from the site of LTD induction. A reduction in AMPA receptor sensitivity was found within approximately 100 µm of the site of LTD reduction but not at greater distances. Thus, LTD is quite local, as only AMPA receptors near (<100 µm) the induction site are depressed, but the depression can nevertheless spread to nearby synapses that were inactive during the induction event. In addition to the studies mentioned above, glutamate uncaging has also been applied to investigate synaptic plasticity in the nucleus accumbens, a part of the brain in which LTP and LTD are associated with drug addiction [94], and has been used to investigate synaptic changes that are associated with epileptic brains [95, 96].

4.4.5
Conclusions and Future Challenges

Focal stimulation with caged neurotransmitters has helped to clarify some crucial issues in synaptic plasticity. Focal glutamate uncaging has unequivocally pointed to the postsynaptic neuron as the locus of LTD in the CA3-CA1 region, and focal uncaging has also been used to map the extent of plasticity. Further elucidation of mechanisms involved in synaptic plasticity may well involve the use of other caged compounds to test the involvement of neurotrophins [97], or to investigate downstream elements of NMDA receptor activation such as calcium, IP3, and other molecules known to participate in plasticity-related second messenger cascades.

References

1 CALLAWAY, E.M. & YUSTE, R. Stimulating neurons with light. Curr. Opin. Neurobiol. **12**, 587–592 (2002).
2 KOBBERT, C., APPS, R., BECHMANN, I., LANCIEGO, J.L., MEY, J. & THANOS, S. Current concepts in neuroanatomical tracing. Prog. Neurobiol. **62**, 327–351 (2000).
3 O'DONNELL, P., LAVIN, A., ENQUIST, L.W., GRACE, A.A. & CARD, J.P. Interconnected parallel circuits between rat nucleus accumbens and thalamus revealed by retrograde transynaptic transport of pseudorabies virus. J. Neurosci. **17**, 2143–2167 (1997).
4 UGOLINI, G., KUYPERS, H.G. & STRICK, P.L. Transneuronal transfer of herpes virus from peripheral nerves to cortex and brainstem. Science **243**, 89–91 (1989).
5 ENQUIST, L.W. & CARD, J.P. Recent advances in the use of neurotropic viruses for circuit analysis. Curr. Opin. Neurobiol. **13**, 603–606 (2003).
6 TSIEN, R.Y. The green fluorescent protein. Annu. Rev. Biochem. **67**, 509–544 (1998).
7 BRAZ, J.M., RICO, B. & BASBAUM, A.I. Transneuronal tracing of diverse CNS circuits by Cre-mediated induction of wheat germ agglutinin in transgenic mice. Proc. Natl. Acad. Sci. USA **99**, 15148–15153 (2002).
8 KATZ, L.C. & DALVA, M.B. Scanning laser photostimulation: a new approach for analyzing brain circuits. J. Neurosci. Methods **54**, 205–218 (1994).

9. STAIGER, J.F., KOTTER, R., ZILLES, K. & LUHMANN, H.J. Laminar characteristics of functional connectivity in rat barrel cortex revealed by stimulation with caged-glutamate. *Neurosci. Res.* **37**, 49–58 (2000).

10. WIEBOLDT, R., GEE, K.R., NIU, L., RAMESH, D., CARPENTER, B.K. & HESS, G.P. Photolabile precursors of glutamate: synthesis, photochemical properties, and activation of glutamate receptors on a microsecond time scale. *Proc. Natl. Acad. Sci. USA* **91**, 8752–8756 (1994).

11. SAWATARI, A. & CALLAWAY, E.M. Convergence of magno- and parvocellular pathways in layer 4B of macaque primary visual cortex. *Nature* **380**, 442–446 (1996).

12. KOTTER, R., STAIGER, J.F., ZILLES, K. & LUHMANN, H.J. Analysing functional connectivity in brain slices by a combination of infrared video microscopy, flash photolysis of caged compounds and scanning methods. *Neuroscience* **86**, 265–277 (1998).

13. SCHMID, S. & WEBER, M. Neurons of the superior olivary complex do not excite startle-mediating neurons in the caudal pontine reticular formation. *Neuroreport* **13**, 2223–2227 (2002).

14. SCHUBERT, D., KOTTER, R., ZILLES, K., LUHMANN, H.J. & STAIGER, J.F. Cell type-specific circuits of cortical layer IV spiny neurons. *J. Neurosci.* **23**, 2961–2970 (2003).

15. ROSSI, F.M., MARGULIS, M., TANG, C.M. & KAO, J.P. N-Nmoc-L-glutamate, a new caged glutamate with high chemical stability and low pre-photolysis activity. *J. Biol. Chem.* **272**, 32933–32939 (1997).

16. ROERIG, B. & KAO, J.P. Organization of intracortical circuits in relation to direction preference maps in ferret visual cortex. *J. Neurosci.* **19**, RC44 (1999).

17. GIVENS, R.S., WEBER, J.F., JUNG, A.H. & PARK, C.H. New photoprotecting groups: desyl and p-hydroxyphenacyl phosphate and carboxylate esters. *Methods Enzymol.* **291**, 1–29 (1998).

18. KIM, G. & KANDLER, K. Elimination and strengthening of glycinergic/GABAergic connections during tonotopic map formation. *Nat. Neurosci.* **6**, 282–290 (2003).

19. LESZKIEWICZ, D.N., KANDLER, K. & AIZENMAN, E. Enhancement of NMDA receptor-mediated currents by light in rat neurones in vitro. *J. Physiol.* **524 Pt 2**, 365–374 (2000).

20. LESZKIEWICZ, D.N. & AIZENMAN, E. A role for the redox site in the modulation of the NMDA receptor by light. *J. Physiol.* **545**, 435–440 (2002).

21. LESZKIEWICZ, D.N. & AIZENMAN, E. Reversible modulation of GABA(A) receptor-mediated currents by light is dependent on the redox state of the receptor. *Eur. J. Neurosci.* **17**, 2077–2083 (2003).

22. MIDDENDORF, T.R. & ALDRICH, R.W. Effects of ultraviolet modification on the gating energetics of cyclic nucleotide-gated channels. *J. Gen. Physiol.* **116**, 253–282 (2000).

23. CHANG, Y., XIE, Y. & WEISS, D.S. Positive allosteric modulation by ultraviolet irradiation on GABA(A), but not GABA(C), receptors expressed in Xenopus oocytes. *J. Physiol.* **536**, 471–478 (2001).

24. CALLAWAY, E.M. & KATZ, L.C. Photostimulation using caged glutamate reveals functional circuitry in living brain slices. *Proc. Natl. Acad. Sci. USA* **90**, 7661–7665 (1993).

25. ENGERT, F., PAULUS, G.G. & BONHOEFFER, T. A low-cost UV laser for flash photolysis of caged compounds. *J. Neurosci. Methods* **66**, 47–54 (1996).

26. DENK, W. Pulsing mercury arc lamps for uncaging and fast imaging. *J. Neurosci. Methods* **72**, 39–42 (1997).

27. KANDLER, K., KATZ, L.C. & KAUER, J.A. Focal photolysis of caged glutamate produces long-term depression of hippocampal glutamate receptors. *Nat. Neurosci.* **1**, 119–123 (1998).

28. RAPP, G. & GUTH, K. A low cost high intensity flash device for photolysis experiments. *Pflugers Arch.* **411**, 200–203 (1988).

29. RAPP, G. Flash lamp-based irradiation of caged compounds. *Methods Enzymol.* **291**, 202–222 (1998).

30. GODWIN, D.W., CHE, D., O'MALLEY, D.M. & ZHOU, Q. Photostimulation with caged neurotransmitters using fi-

ber optic lightguides. *J. Neurosci. Methods* **73**, 91–106 (1997).

31 PARPURA, V. & HAYDON, P. G. "Uncaging" using optical fibers to deliver UV light directly to the sample. *Croat. Med. J.* **40**, 340–345 (1999).

32 KORKOTIAN, E., ORON, D., SILBERBERG, Y. & SEGAL, M. Confocal microscopic imaging of fast UV-laser photolysis of caged compounds. *J. Neurosci. Methods* **133**, 153–159 (2004).

33 PETTIT, D. L., WANG, S. S., GEE, K. R. & AUGUSTINE, G. J. Chemical two-photon uncaging: a novel approach to mapping glutamate receptors. *Neuron* **19**, 465–471 (1997).

34 FURUTA, T., WANG, S. S., DANTZKER, J. L., DORE, T. M., BYBEE, W. J., CALLAWAY, E. M., DENK, W. & TSIEN, R. Y. Brominated 7-hydroxycoumarin-4-yl-methyls: photolabile protecting groups with biologically useful cross-sections for two photon photolysis. *Proc. Natl. Acad. Sci. USA* **96**, 1193–1200 (1999).

35 MATSUZAKI, M., ELLIS-DAVIES, G. C., NEMOTO, T., MIYASHITA, Y., IINO, M. & KASAI, H. Dendritic spine geometry is critical for AMPA receptor expression in hippocampal CA1 pyramidal neurons. *Nat. Neurosci.* **4**, 1086–1092 (2001).

36 DALVA, M. B. & KATZ, L. C. Rearrangements of synaptic connections in visual cortex revealed by laser photostimulation. *Science* **265**, 255–258 (1994).

37 DANTZKER, J. L. & CALLAWAY, E. M. Laminar sources of synaptic input to cortical inhibitory interneurons and pyramidal neurons. *Nat. Neurosci.* **3**, 701–707 (2000).

38 ROERIG, B., CHEN, B. & KAO, J. P. Different inhibitory synaptic input patterns in excitatory and inhibitory layer 4 neurons of ferret visual cortex. *Cereb. Cortex* **13**, 350–363 (2003).

39 SAWATARI, A. & CALLAWAY, E. M. Diversity and cell type specificity of local excitatory connections to neurons in layer 3B of monkey primary visual cortex. *Neuron* **25**, 459–471 (2000).

40 WELIKY, M. & KATZ, L. C. Functional mapping of horizontal connections in developing ferret visual cortex: experiments and modeling. *J. Neurosci.* **14**, 7291–7305 (1994).

41 BRIGGS, F. & CALLAWAY, E. M. Layer-specific input to distinct cell types in layer 6 of monkey primary visual cortex. *J. Neurosci.* **21**, 3600–3608 (2001).

42 YABUTA, N. H., SAWATARI, A. & CALLAWAY, E. M. Two functional channels from primary visual cortex to dorsal visual cortical areas. *Science* **292**, 297–300 (2001).

43 SCHUBERT, D., STAIGER, J. F., CHO, N., KOTTER, R., ZILLES, K. & LUHMANN, H. J. Layer-specific intracolumnar and transcolumnar functional connectivity of layer V pyramidal cells in rat barrel cortex. *J. Neurosci.* **21**, 3580–3592 (2001).

44 SHEPHERD, G. M., POLOGRUTO, T. A. & SVOBODA, K. Circuit analysis of experience-dependent plasticity in the developing rat barrel cortex. *Neuron* **38**, 277–289 (2003).

45 BRIVANLOU, I. H., DANTZKER, J. L., STEVENS, C. F. & CALLAWAY, E. M. Topographic specificity of functional connections from hippocampal CA3 to CA1. *Proc. Natl. Acad. Sci. USA* **101**, 2560–2565 (2004).

46 PETTIT, D. L., HELMS, M. C., LEE, P., AUGUSTINE, G. J. & HALL, W. C. Local excitatory circuits in the intermediate gray layer of the superior colliculus. *J. Neurophysiol.* **81**, 1424–1427 (1999).

47 HELMS, M. C., OZEN, G. & HALL, W. C. Organization of the intermediate gray layer of the superior colliculus. I. Intrinsic vertical connections. *J. Neurophysiol.* **91**, 1706–1715 (2004).

48 STRECKER, G. J., WUARIN, J. P. & DUDEK, F. E. GABAA-mediated local synaptic pathways connect neurons in the rat suprachiasmatic nucleus. *J. Neurophysiol.* **78**, 2217–2220 (1997).

49 SUL, J., OROSZ, G., GIVENS, R. S. & HAYDON, P. G. Astrocytic connectivity in the hippocampus. *Neuron Glia Biol.* **1**, 3–11 (2004).

50 FIELDS, R. D. The other half of the brain. *Sci. Am.* **290**, 54–61 (2004).

51 SANES, D. H. & SIVERLS, V. Development and specificity of inhibitory terminal ar-

borizations in the central nervous system. *J. Neurobiol.* **22**, 837–854 (1991).

52 WICKERSHAM, I. R., KLEINFELD, D., CALLAWAY, E. M. & GAZZANIGA, M. S. Nonplanar slices of living brain tissue. *Soc. Neurosci. Abstr.* **27**, 130.10 (2001).

53 SEGEV, I., RINZEL, J. & SHEPHERD, G. M. (eds.) The theoretical foundations of dendritic function: the collected papers of Wilfrid Rall with commentaries. MIT Press, Boston (1994).

54 JOHNSTON, D., MAGEE, J. C., COLBERT, C. M. & CRISTIE, B. R. Active properties of neuronal dendrites. *Annu. Rev. Neurosci.* **19**, 165–186 (1996).

55 STUART, G., SPRUSTON, N. & HAUSSER, M. Dendrites. Oxford University Press, Oxford, UK (1999).

56 MIGLIORE, M. & SHEPHERD, G. M. Emerging rules for the distributions of active dendritic conductances. *Nat. Rev. Neurosci.* **3**, 362–370 (2002).

57 SCHILLER, J., MAJOR, G., KOESTER, H. J. & SCHILLER, Y. NMDA spikes in basal dendrites of cortical pyramidal neurons. *Nature* **404**, 285–289 (2000).

58 LOWE, G. Inhibition of backpropagating action potentials in mitral cell secondary dendrites. *J. Neurophysiol.* **88**, 64–85 (2002).

59 WEI, D. S., MEI, Y. A., BAGAL, A., KAO, J. P., THOMPSON, S. M. & TANG, C. M. Compartmentalized and binary behavior of terminal dendrites in hippocampal pyramidal neurons. *Science* **293**, 2272–2275 (2001).

60 SMITH, M. A., ELLIS-DAVIES, G. C. & MAGEE, J. C. Mechanism of the distance-dependent scaling of Schaffer collateral synapses in rat CA1 pyramidal neurons. *J. Physiol.* **548**, 245–258 (2003).

61 CRAIG, A. M., BLACKSTONE, C. D., HUGANIR, R. L. & BANKER, G. Selective clustering of glutamate and gamma-aminobutyric acid receptors opposite terminals releasing the corresponding neurotransmitters. *Proc. Natl. Acad. Sci. USA* **91**, 12373–12377 (1994).

62 NUSSER, Z., ROBERTS, J. D., BAUDE, A., RICHARDS, J. G. & SOMOGYI, P. Relative densities of synaptic and extrasynaptic GABAA receptors on cerebellar granule cells as determined by a quantitative immunogold method. *J. Neurosci.* **15**, 2948–2960 (1995).

63 BAUDE, A., NUSSER, Z., MOLNAR, E., McILHINNEY, R. A. & SOMOGYI, P. High-resolution immunogold localization of AMPA type glutamate receptor subunits at synaptic and non-synaptic sites in rat hippocampus. *Neuroscience* **69**, 1031–1055 (1995).

64 LANDSEND, A. S., AMIRY-MOGHADDAM, M., MATSUBARA, A., BERGERSEN, L., USAMI, S., WENTHOLD, R. J. & OTTERSEN, O. P. Differential localization of delta glutamate receptors in the rat cerebellum: coexpression with AMPA receptors in parallel fiber-spine synapses and absence from climbing fiber-spine synapses. *J. Neurosci.* **17**, 834–842 (1997).

65 GULYAS, A. I., MEGIAS, M., EMRI, Z. & FREUND, T. F. Total number and ratio of excitatory and inhibitory synapses converging onto single interneurons of different types in the CA1 area of the rat hippocampus. *J. Neurosci.* **19**, 10082–10097 (1999).

66 TRUSSELL, L. O., THIO, L. L., ZORUMSKI, C. F. & FISCHBACH, G. D. Rapid desensitization of glutamate receptors in vertebrate central neurons. *Proc. Natl. Acad. Sci. USA* **85**, 4562–4566 (1988).

67 DENK, W. Two-photon scanning photochemical microscopy: mapping ligand-gated ion channel distributions. *Proc. Natl. Acad. Sci. USA* **91**, 6629–6633 (1994).

68 DODT, H. U., FRICK, A., KAMPE, K. & ZIEGLGANSBERGER, W. NMDA and AMPA receptors on neocortical neurons are differentially distributed. *Eur. J. Neurosci.* **10**, 3351–3357 (1998).

69 PETTIT, D. L. & AUGUSTINE, G. J. Distribution of functional glutamate and GABA receptors on hippocampal pyramidal cells and interneurons. *J. Neurophysiol.* **84**, 28–38 (2000).

70 FRICK, A., ZIEGLGANSBERGER, W. & DODT, H. U. Glutamate receptors form hot spots on apical dendrites of neocortical pyramidal neurons. *J. Neurophysiol.* **86**, 1412–1421 (2001).

71 LOWE, G. Flash photolysis reveals a diversity of ionotropic glutamate receptors

on the mitral cell somatodendritic membrane. *J. Neurophysiol.* **90**, 1737–1746 (2003).

72 EDER, M., RAMMES, G., ZIEGLGANSBERGER, W. & DODT, H. U. GABA(A) and GABA(B) receptors on neocortical neurons are differentially distributed. *Eur. J. Neurosci.* **13**, 1065–1069 (2001).

73 KHIROUG, L., GINIATULLIN, R., KLEIN, R. C., FAYUK, D. & YAKEL, J. L. Functional mapping and Ca^{2+} regulation of nicotinic acetylcholine receptor channels in rat hippocampal CA1 neurons. *J. Neurosci.* **23**, 9024–9031 (2003).

74 ANDRASFALVY, B. K., SMITH, M. A., BORCHARDT, T., SPRENGEL, R. & MAGEE, J. C. Impaired regulation of synaptic strength in hippocampal neurons from GluR1-deficient mice. *J. Physiol.* **552**, 35–45 (2003).

75 BRASNJO, G. & OTIS, T. S. Isolation of glutamate transport-coupled charge flux and estimation of glutamate uptake at the climbing fiber-Purkinje cell synapse. *Proc. Natl. Acad. Sci. USA* **101**, 6273–6278 (2004).

76 TAKAOKA, K., TATSU, Y., YUMOTO, N., NAKAJIMA, T. & SHIMAMOTO, K. Synthesis and photoreactivity of caged blockers for glutamate transporters. *Bioorg. Med. Chem. Lett.* **13**, 965–970 (2003).

77 EDER, M., BECKER, K., RAMMES, G., SCHIERLOH, A., AZAD, S. C., ZIEGLGANSBERGER, W. & DODT, H. U. Distribution and properties of functional postsynaptic kainate receptors on neocortical layer V pyramidal neurons. *J. Neurosci.* **23**, 6660–6670 (2003).

78 BLISS, T. V. & LOMO, T. Long-lasting potentiation of synaptic transmission in the dentate area of the anaesthetized rabbit following stimulation of the perforant path. *J. Physiol.* **232**, 331–356 (1973).

79 COLLINGRIDGE, G. L., KEHL, S. J. & MCLENNAN, H. The antagonism of amino acid-induced excitations of rat hippocampal CA1 neurones in vitro. *J. Physiol.* **334**, 19–31 (1983).

80 MALENKA, R. C., KAUER, J. A., ZUCKER, R. S. & NICOLL, R. A. Postsynaptic calcium is sufficient for potentiation of hippocampal synaptic transmission. *Science* **242**, 81–84 (1988).

81 KAUER, J. A., MALENKA, R. C. & NICOLL, R. A. A persistent postsynaptic modification mediates long-term potentiation in the hippocampus. *Neuron* **1**, 911–917 (1988).

82 SCHILLER, J., SCHILLER, Y. & CLAPHAM, D. E. NMDA receptors amplify calcium influx into dendritic spines during associative pre- and postsynaptic activation. *Nat. Neurosci.* **1**, 114–118 (1998).

83 MULKEY, R. M. & MALENKA, R. C. Mechanisms underlying induction of homosynaptic long-term depression in area CA1 of the hippocampus. *Neuron* **9**, 967–975 (1992).

84 MARKRAM, H., LUBKE, J., FROTSCHER, M. & SAKMANN, B. Regulation of synaptic efficacy by coincidence of postsynaptic APs and EPSPs. *Science* **275**, 213–215 (1997).

85 BI, G. Q. & POO, M. M. Synaptic modifications in cultured hippocampal neurons: dependence on spike timing, synaptic strength, and postsynaptic cell type. *J. Neurosci.* **18**, 10464–10472 (1998).

86 ITO, M., SAKURAI, M. & TONGROACH, P. Climbing fibre induced depression of both mossy fibre responsiveness and glutamate sensitivity of cerebellar Purkinje cells. *J. Physiol.* **324**, 113–134 (1982).

87 KOMATSU, Y., FUJII, K., MAEDA, J., SAKAGUCHI, H. & TOYAMA, K. Long-term potentiation of synaptic transmission in kitten visual cortex. *J. Neurophysiol.* **59**, 124–141 (1988).

88 LINDEN, D. J., DICKINSON, M. H., SMEYNE, M. & CONNOR, J. A. A long-term depression of AMPA currents in cultured cerebellar Purkinje neurons. *Neuron* **7**, 81–89 (1991).

89 CRAIR, M. C. & MALENKA, R. C. A critical period for long-term potentiation at thalamocortical synapses. *Nature* **375**, 325–328 (1995).

90 THOMAS, M. J., BEURRIER, C., BONCI, A. & MALENKA, R. C. Long-term depression in the nucleus accumbens: a neural correlate of behavioral sensitization to co-

caine. *Nat. Neurosci.* **4**, 1217–1223 (2001).

91 DODT, H., EDER, M., FRICK, A. & ZIEGLGANSBERGER, W. Precisely localized LTD in the neocortex revealed by infrared-guided laser stimulation. *Science* **286**, 110–113 (1999).

92 EDER, M., ZIEGLGANSBERGER, W. & DODT, H.U. Neocortical long-term potentiation and long-term depression: site of expression investigated by infrared-guided laser stimulation. *J. Neurosci.* **22**, 7558–7568 (2002).

93 WANG, S.S., KHIROUG, L. & AUGUSTINE, G.J. Quantification of spread of cerebellar long-term depression with chemical two-photon uncaging of glutamate. *Proc. Natl. Acad. Sci. USA* **97**, 8635–8640 (2000).

94 TAVERNA, S. & PENNARTZ, C.M. Postsynaptic modulation of AMPA- and NMDA-receptor currents by Group III metabotropic glutamate receptors in rat nucleus accumbens. *Brain Res.* **976**, 60–68 (2003).

95 MOLNAR, P. & NADLER, J.V. Mossy fiber-granule cell synapses in the normal and epileptic rat dentate gyrus studied with minimal laser photostimulation. *J. Neurophysiol.* **82**, 1883–1894 (1999).

96 WILLIAMS, P.A., WUARIN, J.P., DOU, P., FERRARO, D.J. & DUDEK, F.E. Reassessment of the effects of cycloheximide on mossy fiber sprouting and epileptogenesis in the pilocarpine model of temporal lobe epilepsy. *J. Neurophysiol.* **88**, 2075–2087 (2002).

97 KOSSEL, A.H., CAMBRIDGE, S.B., WAGNER, U. & BONHOEFFER, T. A caged Ab reveals an immediate/instructive effect of BDNF during hippocampal synaptic potentiation. *Proc. Natl. Acad. Sci. USA* **98**, 14702–14707 (2001)

5
Photoregulation of Proteins

5.1
Light-activated Proteins: An Overview

Sandra Loudwig and Hagan Bayley

5.1.1
Introduction

While there are a considerable number of publications describing small caged molecules for applications in biology (for reviews, see [1, 2] and this volume), there are still relatively few papers describing caged proteins (for earlier reviews, see [3–6]). This is perhaps because the preparation of such molecules requires an interdisciplinary approach and often presents considerable technical difficulties. Nevertheless, the potential applications of caged proteins are likely to be highly rewarding and an expansion of work in this area is certainly warranted. We define a caged protein as an inactive protein that can be activated by light (Fig. 5.1.1). The definition is best kept loose, and here we make several asides to discuss, for example, reversible activation and light-triggered inactivation.

For many applications, there is no alternative to the use of a caged protein. In other cases, the use of a caged small molecule might be considered instead. By comparison with small molecules, the advantages of caged proteins include the fact that the activity in question is precisely pinpointed and that in cellular experiments very low concentrations of a reagent can be used, avoiding, for example, problems with photolysis byproducts.

This chapter is focussed on how to make caged proteins, while that of Condeelis and Lawrence (Chapter 5.3) concentrates on how to use them *in vivo*. Proteins are usually caged by targeted chemical modification. We aim to point out the issues with which the experimentalist should be familiar and to illustrate them with examples; however, the examples do not in themselves offer a comprehensive review of the literature, which nowadays can be better obtained from databases.

Dynamic Studies in Biology. Edited by M. Goeldner, R. Givens
Copyright © 2005 WILEY-VCH Verlag GmbH & Co. KGaA, Weinheim
ISBN: 3-527-30783-4

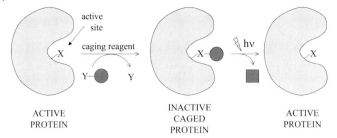

Fig. 5.1.1 Typical procedure for caging and uncaging a protein. A key residue (e.g. one in the active site of an enzyme) is derivatized with a caging reagent resulting in inactivation of the protein. The protein can be reactivated by photolysis of the caging group, which is released in an altered form.

5.1.2
The Properties of Caged Proteins

5.1.2.1 Extents of Caging and Photoactivation

Light-activated proteins are used to obtain control over the timing and location of activation of a biological process, and when necessary the extent of activation. If there are other biomolecules present, they should not be damaged. These requirements raise several important issues, the solutions to which will depend on the experimental system. While it is desirable to efficiently inactivate a protein with a caging reagent, and subsequently recover full activity upon irradiation, this cannot always be done. Therefore, the importance of the extents of caging and uncaging must be considered further, and will in turn affect the choice of caging reagent. For example, in the case of a caged signal transduction protein for injection into target cells (see Chapter 5.3), it is highly desirable to begin with a completely inactivated protein and even limited photoregeneration of activity within the cell (e.g. 5%) can be useful. In contrast, in many biophysical experiments, such as those involving time-resolved X-ray diffraction, circular dichroism or IR spectroscopy, a high recovery of activity is usually required (e.g. 95% or above), but complete inactivation is not usually as crucial (e.g. 5% of unmodified protein can often be tolerated).

5.1.2.2 The Rate of Uncaging: Continuous Irradiation

The rate of photorelease is often critical in biophysical experiments and should be fast in comparison to the timecourse of the event under observation. The half-time for activation of a caged reagent under continuous irradiation in dilute solution is given by [3]:

$$t_{1/2} \approx 0.3/\phi_P I_0 \varepsilon$$

where ϕ_P is the product quantum yield, I_0 is the light intensity and ε is the extinction coefficient. $t_{1/2}$, ε and I_0 are usually measured experimentally to find ϕ_P.

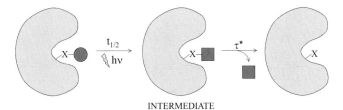

Fig. 5.1.2 Illustration of $t_{1/2}$ and τ^* for photolysis. During continuous photolysis at relatively low light intensity, the half-life of the caged protein is $t_{1/2}$, which is a function of the amount of light absorbed per unit time and the efficiency of conversion of the excited state to the uncaged product. Under these conditions, the rate at which the excited states (and intermediates when they exist) are converted to product, represented by τ^*, is not rate determining. After a short intense light flash, the half-time for formation of the uncaged product is given by τ^*.

This expression encompasses the rate at which a collection of molecules absorbs photons and the efficiency at which the excited states are converted to uncaged product. $t_{1/2}$ should be distinguished from the usually more rapid rate of breakdown of the excited states and intermediates (τ^*), which will be discussed later (Fig. 5.1.2). The relationship given above is valid for a single wavelength, as would be the case for laser irradiation, and it is a reasonable approximation for a narrow band of wavelengths. Where a wide band is used, e.g. as might be the case from an arc lamp, the variation of ε and I_0 with wavelength should be taken into account when determining ϕ_P for the prevailing conditions. This can be done conveniently by summing $I_0\varepsilon$ over the wavelength band by using, say, 10-nm intervals. To determine I_0 for each interval from the total lamp output (determined by actinometry), both the emission spectrum of the lamp and the filter characteristics must be considered [7].

Clearly, in most circumstances, I_0 can be increased far beyond the levels required to achieve photolysis in the desired time. The question is: what happens to other materials in the experimental system at high light intensities? The simple answer is that the consequences are usually deleterious. Most biological molecules are damaged by light, and the damage is exacerbated at short wavelengths and in the presence of oxygen. In a complex system, a wide variety of damage can occur, including crosslinking of nucleic acids and proteins. However, even in a simple biophysical experiment, involving a single protein, inter- and intramolecular crosslinks, polypeptide chain cleavage, and damage to susceptible residues such as tryptophan (Trp) can cause significant problems [8, 9].

At short photolysis times, a measure of the extent of uncaging relative to damage is given by:

$$E_{ud} = t_{1/2d}/t_{1/2u} = \phi_u\varepsilon_u/\phi_d\varepsilon_d$$

where the subscripts 'u' and 'd' indicate uncaging and damage, respectively. At longer photolysis times, as uncaging nears completion, damage will "catch up" with uncaging: a warning to irradiate for no longer than necessary.

Tab. 5.1.1 Caging groups for proteins

Caging group	Caging group characteristics	Caged functionality and/or molecule; caging reagent	Comments	Refs.
X–CH₂ group on 2-nitrobenzene (NO₂ ortho); labeled **NB**			2-Nitrobenzyl derivatives, initially described for the protection of carboxylates [141], are the most widely used caging groups. They have been used to cage thiols, thiophosphates, phosphates, carboxylates, amino groups and alcohols. Various modifications at the benzylic position (see NB, NPE, CNB and NPT) or on the aromatic ring (see DMNB, DMNPT, NVOC and MeNPoc) have been made in attempts to improve photochemical properties (see this table). The main drawback of the 2-nitrobenzyl group is the generation of a photolytic byproduct that reacts with thiol groups [25, 33, 34]. Further, when substituted in the benzylic position, the 2-nitrobenzyl group bears an asymmetric center.	25, 33, 34, 141
	$\phi_{(312;\ pH\ 6)} = 0.84$; $\phi_{(312;\ pH\ 8.5)} = 0.14$; $\phi\varepsilon_{(312;\ pH\ 6)} \approx 1090$; $\phi\varepsilon_{(312;\ pH\ 8.5)} \approx 180$	RCH₂S-NB Caged Cys in cAMP-dependent protein kinase (PKA). 2-nitrobenzyl bromide with Br at benzylic position, labeled **NBB**	A PKA mutant with a single Cys is modified with NBB. Caging with different 2-nitrobenzyl derivatives (BNPA, DMNBB and NBB) and photolysis at two pH values are compared (see this table). Modification with unsubstituted NBB gives the highest extent of inactivation and subsequent photoregeneration of activity.	65
	$\varepsilon_{312} = 1300$ [3]; $\phi_{(312;\ pH\ 5.8)} = 0.62$; $\phi_{(312;\ pH\ 7.2)} = 0.14$; $\phi\varepsilon_{(312;\ pH\ 5.8)} \approx 810$; $\phi\varepsilon_{(312;\ pH\ 7.2)} \approx 180$ [12]	RCH₂S-NB In the heptapeptide C-kemptide, the serine of kemptide is replaced with Cys. This residue is caged with NBB.	Caging with BNPA and DMNBB (see this table) and photolysis at two pH values are compared. Photolysis of C-kemptide modified with NBB gives the best quantum yield and a slightly higher yield of uncaged product: 70% at pH 5.8 against 62 and 67% for BNPA and DMNBB, respectively.	3, 12

Tab. 5.1.1 (continued)

Caging group	Caging group characteristics	Caged functionality and/or molecule; caging reagent	Comments	Refs.
	ε_{312} = 1300 [3] $\phi_{(312;\ pH\ 5.8)}$ = 0.23; $\phi_{(312;\ pH\ 7.2)}$ = 0.04 $\phi\varepsilon_{(312;\ pH\ 5.8)} \approx 300$; $\phi\varepsilon_{(312;\ pH\ 7.2)} \approx 50$ [12]	$$R-O-\overset{\overset{O}{\|}}{\underset{\underset{O^-}{\|}}{P}}-S\cdot NB$$ Caged thiophosphorylated serine in the hepta-peptide kemptide.	Caging with BNPA and DMNBB (see this table) and photolysis at two pH values are compared. Photolysis of thiophosphoserine modified with unsubstituted NBB gives the best quantum yield and the highest yield of uncaged product: 70% at pH 5.8 against 55% for DMNBB. At pH 4, thiophosphate can be selectively modified over Cys with NBB.	3, 12
	$\phi_{(312;\ pH\ 5.8)}$ = 0.37; $\phi_{(312;\ pH\ 7.3)}$ = 0.25 $\phi\varepsilon_{(312;\ pH\ 5.8)} \approx 480$; $\phi\varepsilon_{(312;\ pH\ 7.3)} \approx 325$	$$Ar-O-\overset{\overset{O}{\|}}{\underset{\underset{O^-}{\|}}{P}}-S\cdot NB$$ Caged thiophosphotyrosine in an 11 amino acid peptide.	A thiophosphotyrosine is modified with NBB. Caging with NBB and 4-hydroxyphenacyl bromide (HPB), and photolysis at two pH values are compared. Uncaging of HP gives the best quantum yields and a slightly better yield of uncaged product: 50–70% (at both pH 5.8 and 7.3) against 50–60% for NB (see below).	31
	λ_{max} = 272 nm ε_{272} = 6200 $\phi_{(300-350;\ pH\ 7)}$ = 0.05 $\phi\varepsilon_{312} \approx 65$ τ^* ND [11]	RPhO-NB Caged phenylephrine [11].	Photolysis of different 2-nitrobenzyl derivatives (NB, NPE, DMNB and CNB) are compared (see this table) [11]. Proteins have not been caged directly on hydroxyls, but catalytic serines, with enhanced nucleophilicity, might be selectively modified with NBB-like compounds. Also, caged tyrosines have been incorporated into proteins by unnatural amino acid mutagenesis [142, 143].	11, 142, 143
	λ_{max} = 262 nm ε_{262} = 5200 $\phi_{(300-350;\ pH\ 8)}$ = 0.25 τ^* (nitronate decay) = 1.7 ms (pH 7) $\phi\varepsilon_{312} \approx 325$ (pH 8)	ROC(O)NH-NB Caged carbamylcholine.	Proteins have not been caged as carbamates in this manner, but the work provides a useful study of the pH dependence of breakdown of the nitronate intermediate.	144

Tab. 5.1.1 (continued)

Caging group	Caging group characteristics	Caged functionality and/or molecule; caging reagent	Comments	Refs.
NPE (structure: benzene ring with CH(X)CH₃ and NO₂ substituents)	$\lambda_{max} = 272$ nm $\varepsilon_{272} = 6200$ $\phi_{(300-350;\ pH\ 7)} = 0.11$ τ^* (nitronate decay) = 300 ms (pH 7) $\phi\varepsilon_{312} \approx 140$	RPhO-NPE Caged phenyl-ephrine.	Proteins have not been caged directly in this way, but the work provides a useful comparison of various 2-nitrobenzyl derivatives. ϕ and hence $\phi\varepsilon$ values for NPE are about twice those for NB. The breakdown of the nitronate intermediate (Fig. 5.1.3) is slow in this case.	11
	$\lambda_{max} = 262$ nm $\varepsilon_{262} = 5200$ $\phi_{(300-350;\ pH\ 8)} = 0.25$ τ^* (nitronate decay) = 67 μs (pH 7) $\phi\varepsilon_{312} \approx 325$ (pH 8)	ROC(O)NH-NPE Caged carbamyl-choline.	Although the ϕ and $\phi\varepsilon$ values for carbamylcholine caged with NB and NPE are the same, the decay rate of the nitronate intermediate with NPE is much faster.	144
CNB (structure: benzene ring with CH(COOH)(X) and NO₂ substituents)	$\phi_{(312;\ pH\ 5.8)} = 0.21$ $\phi\varepsilon_{312} \approx 270$ For Cys in C-kemptide [12]	RCH₂S-CNB In the heptapeptide C-kemptide, the serine of kemptide is replaced with a Cys. This residue is caged with BNPA. Alternatively, the serine in kemptide is thiophosphorylated and then modified by BNPA [3, 12]. BNPA (structure: benzene ring with CH(COOH)(Br) and NO₂ substituents)	BNPA has been developed as a water-soluble reagent for the modification of nucleophiles in peptides and proteins. While 95% modification of the Cys in C-kemptide is achieved, only 10% modification of a thiophosphorylated serine in kemptide occurs under the same conditions. The ϕ value for Cys caged with CNB is much lower than for Cys caged with NB [3, 12]. Further, irradiation of CNB-Cys in PKA leads to 25–30% recovery of activity, while 80–100% recovery is found with PKA caged at Cys with NB [65]. BNPA has also been used for caging the pore-forming protein α-hemolysin [47].	3, 12, 47, 65
	$\lambda_{max} = 270$ nm $\varepsilon_{266} = 6000$ $\phi_{(300-350;\ pH\ 7)} = 0.28$ τ^* (nitronate decay) = 0.35 ms $\phi\varepsilon_{312} \approx 365$	RPhO-CNB Caged phenyl-ephrine.	The quantum yield for CNB is higher than for NB, NPE and DMNB. The lifetime of the nitronate is short compared to NPE and DMNB in the same context (τ^* not given for NB).	11

Tab. 5.1.1 (continued)

Caging group	Caging group characteristics	Caged functionality and/or molecule; caging reagent	Comments	Refs.
	$\lambda_{max} = 266$ nm $\varepsilon_{266} = 5200$ $\phi_{(300-350;\ pH\ 7)} = 0.8$ τ^* (nitronate decay) = 40 μs $\phi\varepsilon_{312} \approx 1040$ [144]	ROC(O)NH-CNB Caged carbamyl-choline [144].	ϕ is remarkably high and $\phi\varepsilon$ is thus favorable [144]. Also the rate of decay of the nitronate intermediate is faster than for NB and slightly faster than for NPE in the same context [10, 144].	10, 144
F$_3$C–X with NO$_2$, R substituents R = H, NPT R = OMe, DMNPT	R = H: $\phi_{(312;\ pH\ 7)} = 0.7$ $\tau^* = 2.2$ and 50 μs (biexponential rate for nitronate decay) $\phi\varepsilon_{312} \approx 910$ R = OMe: $\phi_{(312;\ pH\ 7)} = 0.43$ $\phi\varepsilon_{312} \approx 1935$	RO-NPT and RO-DMNPT Caged alcohols: ϕ values are for NPT-choline and DMNPT-arseno-choline.	ϕ for NPT is very high. ϕ for DMNPT can be as high as 0.62, e.g. for 4-DMNPT α-tolylgalactoside. The $\phi\varepsilon$ value for DMNPT is favorable. Proteins have not yet been caged in this manner.	145
X with NO$_2$, MeO, OMe substituents DMNB	$\varepsilon_{312} = 4500$ [3] $\phi_{(312;\ pH\ 5.8)} = 0.15$ $\phi\varepsilon_{312} \approx 675$ [12]	RCH$_2$S-DMNB The Cys in C-kemptide is caged with DMNBB [12]. Br, NO$_2$, MeO, OMe DMNBB	ϕ is only slightly lower than for BNPA, but much lower than for NBB. Nevertheless, since the ε value at 312 nm is higher, $\phi\varepsilon$ is about double that for the NB peptide. The yield of deprotection of 67% at pH 5.8 is similar to that of NB (70%) [12]. However, when DMNBB and NBB are used to cage a Cys in PKA, the photo-regeneration of activity with NB-PKA is 10–15 times greater than with DMNB-PKA [65].	3, 12, 65
	$\phi_{(312;\ pH\ 5.8)} = 0.06$ $\phi\varepsilon_{312} = 270$	$$R-O-\overset{\overset{O}{\|}}{\underset{\underset{O^-}{\|}}{P}}-S-DMNB$$ Caged thiophosphorylated serine in kemptide.	DMNBB is used to cage thiophosphorylserine in kemptide. ϕ for DMNB is lower than that for NB, but $\phi\varepsilon_{(312;\ pH\ 5.8)}$ values are about the same (300 for NB). ϕ values for NB and DMNB-caged thiophosphoserine are lower than ϕ values for caging a Cys residue.	3, 12

Tab. 5.1.1 (continued)

Caging group	Caging group characteristics	Caged functionality and/or molecule; caging reagent	Comments	Refs.
	$\lambda_{max} = 330$ nm $\varepsilon_{330} = 5000$ $\phi_{(300-350;\ pH\ 7)} = 0.13$ τ^* (nitronate decay) = 2.8 ms $\phi\varepsilon_{312} \approx 585$	RPhO-DMNB Caged phenyl-ephrine.	The ϕ value for DMNB-phenyl-ephrine is similar to that for NPE-phenylephrine, but $\phi\varepsilon$ is higher. The lifetime of the nitronate lies between the values for CNB and NPE.	11
	$\lambda_{max} = 350$ nm	$R-O-\overset{O}{\underset{O^-}{\overset{\|}{P}}}-O-DMNB$ Caged phosphates (cAMP and cGMP).	DMNB is first used with caged phosphate esters. Compared to NB, the absorption maximum is shifted to 350 nm and cyclic nucleotide photorelease is 200 times faster.	146
MeO–(Ar)–O–C(=O)–X, with MeO and NO$_2$ substituents **NVOC**	ε likely to be similar to DMNB $\phi_{(365;\ pH\ 7.2)} = 0.023$ [141]	NVOC–NH–C(=NH)–NH–R Caged arginine in a peptide inhibitor of protein kinase [141]. MeO–(Ar)–CH$_2$–O–C(=O)–Cl with MeO, NO$_2$ NVOC-Cl	In the case of caged Arg, the low value of ϕ is compensated by a high ε, as the NVOC λ_{max} is around 350 nm [141]. 6-Nitro-veratryloxycarbonyl chloride (NVOC-Cl) is used to randomly modify Lys residues in proteins (see Tab. 5.1.2).	147
MeO–(Ar)–CH(R)(X)–CH$_2$–NO$_2$ with MeO, NO$_2$	ε likely to be similar to DMNB	R–S–CH(–)–CH$_2$–NO$_2$ on dimethoxynitrophenyl Caged Cys in β-galactosidase. MeO–(Ar)–CH=CH–NO$_2$ with MeO, NO$_2$	1-(4,5-Dimethoxy-2-nitrophenyl)-2-nitroethene is used to modify one Cys residue per subunit in β-galactosidase.	66

Tab. 5.1.1 (continued)

Caging group	Caging group characteristics	Caged functionality and/or molecule; caging reagent	Comments	Refs.
MeNPoc	ε likely to be similar to DMNB	MeNPoc-OR Photoprotection of 5′ hydroxyl groups in deoxyribonucleotides. MeNPoc-Cl	1-(2-Nitro-4,5-methylenedioxy-phenyl)-ethyl-1-oxycarbonyl chloride is used for the protection of deoxyribonucleoside hydroxyl groups during the synthesis of oligonucleotide arrays. Proteins have not yet been caged in this manner, but the oxycarbonyl chloride might be used as an alternative to NVOC-Cl for caging by random modification of Lys residues.	148
(NO$_2$, OH, R, OH structure)	$\lambda_{max} = 262$ nm $\varepsilon_{259} = 5500$ (cyclohexanone ketal) $\lambda_{max} = 259$ nm $\varepsilon_{259} = 5300$ (glycol)	Caged aldehydes and ketones.	Aldehydes and ketones have been caged as dioxolane rings by using 2-nitrobenzyl derivatives. Ketones have been introduced into proteins by unnatural amino acid mutagenesis and it might be possible to cage them with diol reagents. Alternatively, caged ketones themselves could be introduced by unnatural amino acid mutagenesis.	149–153
NNMC	$\lambda_{max} = 259$ nm $\varepsilon_{259} = 7298$; $\varepsilon_{350} = 1289$ $\phi_{(380;\ pH\ 6.7)} = 0.63$ (nitroso formation) $\phi\varepsilon_{380} = 510$	Model compound. NNMCC	3-Nitro-2-naphthalenemethanol is irradiated as a model compound to obtain ϕ values. 3-Nitro-2-naphthyloxycarbonyl chloride (NNMCC) is then used for random caging of Lys residues in an immunoglobulin, despite the low water solubility of the compound.	50
NPPOC	$\lambda_{max} \approx 260$ nm $\varepsilon_{355} = 400$ $\phi_{(365;\ pH\ 7.2)} = 0.3$ [154] $\phi\varepsilon_{355} \approx 120$	NPPOC-OR Caged alcohol in thymidine [154].	The NPPOC group has been used in the synthesis of oligo-nucleotides microarrays [155]. Proteins have not yet been caged in this manner.	154, 155

Tab. 5.1.1 (continued)

Caging group	Caging group characteristics	Caged functionality and/or molecule; caging reagent	Comments	Refs.
MNI	R=OMe: $\lambda_{max} \approx 246$ nm $\varepsilon_{246} = 17500$ $\phi_{(347;\ pH\ 7.2)} = 0.085$ [156] Two photon uncaging (730 nm), $\delta_u = 0.06$ GM for a caged carboxylate [158]	RCO-MNI Caged carboxylates.	Photo-accelerated solvolysis of 1-acyl-7-nitroindoline derivatives is reported [156, 157]. ϕ is usually low, but ε is high, and photolysis occurs in high yield. Two-photon uncaging has been demonstrated [158]. Proteins have not yet been caged in this manner.	156–158
MNPCC	$\lambda_{max} = 262$ nm $\varepsilon_{(262,\ pH\ 7.4)} = 5000$ ϕ is low	MNPC-OR Caged alcohols.	N-Methyl-N-(2-nitrophenyl) carbamoyl chloride is used to cage various alcohols [159] and the catalytic serine of butyrylcholinesterase (see Tab. 5.1.2) [53].	53, 159
Benzoin	$R_1 = R_2 = $OMe: $\lambda_{max,\ MeOH} = 246$ nm $\varepsilon_{246} = 14250$; $\varepsilon_{282} = 3050$ $\phi_{(366,\ benzene)} = 0.64$ (benzofuran formation) $\tau^* \approx 0.1$ ns (quenching of triplet state) [161]	Photorelease of acetate [161]; benzoin-OC(O)OR, photorelease of alcohols [163]; benzoin-OC(O)NHR, photorelease of amines [164, 165].	The rate of product formation from the excited state is extremely fast and the quantum yield of benzofuran release is high [161]. Benzoins have been used to cage phosphates [162], alcohols [163] and amines [164, 165]. The chemically inert benzofuran byproduct has a huge absorbance above 300 nm (for $R_1 = R_2 = $OMe: $\lambda_{max,\ EtOH} = 301$ nm; $\varepsilon_{301} = 27480$) [161]. Proteins have not yet been caged as benzoins. Potential caging reagents might have solubility problems. The reagent presents a chiral center.	160–166

Tab. 5.1.1 (continued)

Caging group	Caging group characteristics	Caged functionality and/or molecule; caging reagent	Comments	Refs.
HP (4-hydroxyphenacyl: HO-C₆H₄-C(O)-CH₂-X)			4-Methoxyphenacyl derivatives were first reported by Sheehan in an attempt to simplify the benzoin protecting group [167]. Givens developed the 4-hydroxy derivative to improve aqueous solubility [17]. Like benzoin, the 4-hydroxy-phenacyl group offers a very short τ^* value. It does not contain or generate a chiral center and a predominant photolysis byproduct is usually released: 4-hydroxyphenylacetic acid. The byproduct, besides being chemically inert, does not absorb at the wavelengths of irradiation (> 300 nm) [168].	17, 167, 168
	$\phi_{(313;\ pH\ 7.2)} = 0.085$ $\phi\varepsilon_{310} \approx 670$ [169]	RS-HP Caged thiol in 3'-thio-2'-deoxythymidine phosphate [169]. HPB (HO-C₆H₄-C(O)-CH₂-Br)	The ϕ value is lower than for other 4-hydroxyphenacyl-caged functional groups and several photolysis byproducts are formed [169]. Nevertheless, 4-hydroxyphenacyl bromide (HPB) has proved successful for caging the catalytic Cys of a protein tyrosine phosphatase and about 70% activity is regained on uncaging [54].	54, 169
	$\phi_{(312;\ pH\ 5.8)} = 0.65$; $\phi_{(312;\ pH\ 7.3)} = 0.56$ $\phi\varepsilon_{310} \approx 5100$ (pH 5.8); $\phi\varepsilon_{310} \approx 4400$ (pH 7.3)	R-O-P(=O)(O⁻)-S-HP Caged thiophosphotyrosine in a peptide.	4-Hydroxyphenacyl bromide is used to cage a thiophosphorylated tyrosine in an unprotected peptide. Compared with the 2-nitrobenzyl (NB) group, HP exhibits better quantum yields. ϕ values are similar at pH 5.8 and 7.3.	31
	$\phi_{(312;\ pH\ 7.3)} = 0.21$ $\phi\varepsilon_{310} \approx 1650$	R-O-P(=O)(O⁻)-S-HP Caged thiophosphothreonine in PKA.	4-Hydroxyphenacyl bromide is employed to cage a thiophosphorylated threonine in PKA (see Tab. 5.1.2).	7

Tab. 5.1.1 (continued)

Caging group	Caging group characteristics	Caged functionality and/or molecule; caging reagent	Comments	Refs.
	$\lambda_{max} = 286$ nm $\varepsilon_{286} = 14600$ (H_2O/CH_3CN) [168] $\phi_{(300; pH\ 7.3)} = 0.37$ $\phi\varepsilon_{300} \approx 4400$ $\tau^* = 1.2$ ns (from quenching of triplet state) [17] values for caged phosphate	R–O–P(=O)(O⁻)–O–HP Caged phosphate (ATP) [17]. RCOO-HP Caged carboxylates [167].	The very short τ^* value is notable.	17, 16 168
MeO, HO, OMe (HDMP)	$\lambda_{max} = 370$ nm $\varepsilon_{304} = 11730$ $\phi_{(300-350;\ pH\ 7.2)} = 0.03–0.04$ $\tau^* = 69$ ns $\phi\varepsilon_{304} \approx 410$	RCOO-HDMP Caged carboxylate in GABA and glutamate.	The introduction of methoxy groups in positions 3 and 5 of the ring leads to a red-shift of the absorbance, but also to a drastic decrease of the quantum yield. The photolysis pathway appears to be altered and the release of the caging group as the 4-hydroxyphenylacetic acid is a minor pathway. The 3-monomethoxy reagent was also examined. Proteins have not yet been caged with methoxy substituted HP.	18
(coumarin structure with R and X substituents)			Coumarin derivatives usually do not display high quantum yields, but they do absorb strongly above 300 nm. They can also show an increase in fluorescence upon irradiation. A very interesting feature of several coumarins is susceptibility to two-photon irradiation with IR light, which enables spatial control of photolysis at the micometer level. A drawback is their low aqueous solubility.	15

Tab. 5.1.1 (continued)

Caging group	Caging group characteristics	Caged functionality and/or molecule; caging reagent	Comments	Refs.
HCM	$\lambda_{max} = 325$ nm; $\varepsilon_{(325, pH\ 7.2)} = 11600$; $\varepsilon_{(365, pH\ 7.2)} = 4100$; $\phi_{(365, pH\ 7.2)} = 0.025$; $\phi\varepsilon_{365} = 103$; Two-photon uncaging $\delta_u = 1.07$ GM (740 nm); $\delta_u = 0.13$ GM (800 nm) for caged acetate [15]	$R_1-O-\overset{O}{\underset{OR_2}{\overset{\|}{P}}}-O\ HCM$ Caged phosphate [162] cAMP [170] RCOO-HCM [15] RCNHCOO-HCM [15].	For caged acetate, the ϕ value is low, but the strong absorbance compensates for this [15]. Proteins have not yet been caged with the HCM group. Cys residues might be modified directly with 4-halomethyl-7-hydroxy-coumarins. Two-photon uncaging is demonstrated. The cross-sections are favorable.	15, 162, 170
TBHCM	$\lambda_{max} = 397$ nm; $\varepsilon_{(397, pH\ 7.2)} = 15\,900$; $\varepsilon_{(365, pH\ 7.2)} = 9\,700$; $\phi_{(365, pH\ 7.2)} = 0.065$; $\phi\varepsilon_{365} = 631$; $\delta_u = 0.96$ GM (740 nm); $\delta_u = 3.1$ GM (800 nm) values for caged acetate	AcCOO-TBHCM Caged carboxylate (acetate).	The introduction of bromine atoms on the coumarin ring of HCM increases both the quantum yield and the absorbance of caged carboxylates. The best results are achieved with three additional bromines, as shown. The two-photon cross section is high. Proteins have not yet been caged with the TBHCM group.	15
MCM	$\lambda_{max} = 327$ nm; $\varepsilon_{327} = 13\,300$; $\phi_{(333;\ pH\ 7.2)} = 0.21$; $\phi\varepsilon_{333} = 2793$ for caged cGMP [16]	$R_1-O-\overset{O}{\underset{OR_2}{\overset{\|}{P}}}-O\ MCM$ Caged phosphate (cAMP or cGMP).	Furuta et al. obtain $\phi\varepsilon_{340} = 670$ for caged cAMP in Ringer's solution [171]. Schade et al. obtain $\phi\varepsilon_{(333,\ pH\ 7.2)} = 1716$ for caged cAMP and $\phi\varepsilon_{333} = 2793$ for caged cGMP in 20% methanol/80% HEPES buffer [16]. Proteins have not yet been caged with the MCM group.	16, 171
DMACM	$\lambda_{max} = 394$ nm; $\varepsilon_{394} = 17\,200$; $\phi_{(333;\ pH\ 7.2)} = 0.28$ for caged AMP [173]	$R_1-O-\overset{O}{\underset{O^-}{\overset{\|}{P}}}-O\ DMACM$ Caged phosphates (cAMP, cGMP, ATP, ADP or AMP).	DMACM was first described for caging cAMP and cGMP [172]. λ_{max} is at a favorable wavelength.	172, 173

Tab. 5.1.1 (continued)

Caging group	Caging group characteristics	Caged functionality and/or molecule; caging reagent	Comments	Refs.
BHC-diol	$\lambda_{max} = 370$ nm $\varepsilon_{365} = 12600–19500$ $\phi_{(365;\ pH\ 7.2)} = 0.03–0.06$ $\phi\varepsilon_{365} = 403–1026$ $\delta_u = 0.51–1.23$ GM (740 nm)	Caged aldehydes and ketones.	BHC-diol is used to cage aldehydes and ketones. The quantum yields of uncaging are low, but the strong absorbance leads to high $\phi\varepsilon$ values. BHC-protected compounds can be uncaged by two-photon irradiation with a favorable cross section (δ_u).	41
BHQ	$\lambda_{max} = 369$ nm $\varepsilon_{369} = 2600$ $\phi_{(365;\ pH\ 7.2)} = 0.29$ $\phi\varepsilon_{365} \approx 750$ $\delta_u = 0.59$ GM (740 nm); $\delta_u = 0.087$ GM (780 nm) for BHQ-OAc	RCOO-BHQ Caged carboxylates.	BHQ-protected compounds can be photolysed by two-photon irradiation. The cross-section at 740 nm is quite good. Proteins have not been caged with members of this class of reagents.	40
AQMOC	$\lambda_{max} = 327$ nm $\varepsilon_{369} = 4100$; $\varepsilon_{350} = 1500$ $\phi_{350} = 0.1$ (disappearance of starting material) in 50% aqueous THF $\phi\varepsilon_{350} = 150$	AQMOC-OR Caged alcohol (galactose).	Anthraquinon-2-ylmethoxycarbonyl is used to cage galactose. The $\phi\varepsilon$ value is low. However, AQMOC proves to be superior to various arylmethyl carbonates (7-methoxycoumarinyl-4-methyloxycarbonyl, pyren-1-ylmethoxycarbonyl and phenanthren-9-ylmethoxycarbonyl) in terms of ϕ and $\phi\varepsilon$.	174
DANP	$R = Me_2N-$ $\lambda_{max} \approx 400$ nm $\varepsilon_{400} = 9077$ $\phi_{(308–360;\ pH\ 7.4)} = 0.03$; $\phi_{(450;\ pH\ 7.4)} = 0.002$ $\tau^* = 5$ μs (transient absorption change) [175]	DANP-OCOR' Caged carboxylate.	For $R = Me_2N-$, the quantum yield is very low, although the absorbance is high [175]. The ester is quite unstable in aqueous media (hydrolysis: $t_{1/2} = 99$ min at pH 7), but the stability is improved compared to the compound with $R = OMe$ (hydrolysis: $t_{1/2} = 6.1$ min at pH 7.1) [176]. The caging group is released as the corresponding phenol.	175, 176

Tab. 5.1.1 (continued)

Caging group	Caging group characteristics	Caged functionality and/or molecule; caging reagent	Comments	Refs.
CINN-OX (structure: Et₂N, OH, cinnamate with OX ester)		CINN-OR to modify various serine proteases (structure: Et₂N, OH, cinnamate with OR ester)	Cinnamate ester derivatives, CINN-OR have been used to modify the catalytic Ser of various proteases. The nature of the ester used depends on the substrate specificity of the protease. The photochemical data from the different proteases are compared below, together with a model compound. The absorbance and, especially, the quantum yield vary in the microenvironment of the protein.	177
	$\phi_{(trans\ to\ cis)} = 0.13$ $\lambda_{max} = 358$ nm $\varepsilon_{(366,\ pH\ 7.4)} = 19\,200$ $\phi\varepsilon_{366} \approx 2500$ [177]	CINN-OEt Model ester.		177
	$\phi_{(trans\ to\ cis)} = 0.17$ $\lambda_{max} = 372$ nm $\varepsilon_{(366,\ pH\ 7.4)} = 25\,700$ $\phi\varepsilon_{366} \approx 4370$ [177] $\tau^* = 199$ μs (cyclization after photoisomerization) [106]	CINN-O-chymotrypsin Caged catalytic serine.		106, 177
	$\phi_{(trans\ to\ cis)} = 0.23$ $\lambda_{max} = 358$ nm $\varepsilon_{(366,\ pH\ 7.4)} = 26\,000$ $\phi\varepsilon_{366} \approx 5980$ [177] $\tau^* = 151$ μs (cyclization after photoisomerization) [106]	CINN-O-factor Xa Caged catalytic serine.		106, 177
	$\phi_{(trans\ to\ cis)} = 0.05$ $\lambda_{max} = 380$ nm $\varepsilon_{(366,\ pH\ 7.4)} = 18\,600$ $\phi\varepsilon_{366} \approx 740$ [177] $\tau^* = 287$ μs (cyclization after photoisomerization) [106]	CINN-O-thrombin Caged catalytic serine.		106, 177

Tab. 5.1.1 (continued)

Caging group	Caging group characteristics	Caged functionality and/or molecule; caging reagent	Comments	Refs.
Ph—C(=O)—C₆H₄—COXR X=O: 2-benzoyl-benzoate derivatives X=S: 2-benzoyl-benzenethioate derivatives		Photoprotection of primary and secondary alcohols (X=O) or thiol (X=S).	2-Benzoylbenzoate derivatives have been used to photoprotect alcohols ROH (X=O) and 2-benzoylbenzenethioate derivatives to protect thiols RSH (X=S) [178]. Substituted benzyl 2-benzoylbenzoate derivatives have been used for the modification of the catalytic serine of various proteases, the -XR group varying according to the specificity of the enzyme (see Tab. 5.1.2) [179]. For caged thiols, the photodeprotection yield is 60%, together with 20% formation of disulphide. For ROH, the yield was variable, but generally good.	178, 179

Groups that have actually been used to cage proteins are emphasized, although promising potential alternatives are also given. Reagents that are available for directly caging proteins are shown with emphasis on those that can be used to derivatize nucleophiles, especially the thiolate side-chains of Cys residues. Key: λ_{max}, maximum wavelength of absorption; ε_{xxx}, extinction coefficient at XXX nm in $M^{-1}\,cm^{-1}$; $\phi_{(xxx;\, pH\, Y)}$, quantum yield at XXX nm and pH Y. Unless otherwise stated, the product quantum yield (ϕ_P) is given, which is not always interchangeable with the quantum yield for uncaging (ϕ_u). In the text, we have often simply used ϕ for the purpose of general discussion; τ^*, lifetime of photolytic intermediate, as defined in the text. δ_u is the uncaging action cross-section, which is the product of the two-photon absorbance cross-section δ_a and the uncaging quantum yield ϕ_{u2}. Ideally, δ_u should exceed 0.1 Goeppert-Mayer (GM), where 1 GM is defined as $10^{-50}\,cm^4 \cdot s\, photon^{-1}$ [15].

The larger the value of E_{ud}, the better and the experience of many workers in the field of caged reagents suggests that $\phi_u\varepsilon_u$ should be $> 100\, M^{-1}\, cm^{-1}$, at least in the near-UV (350–400 nm), and even better when $> 1000\, M^{-1}\, cm^{-1}$. In addition, photolysis should be at the longest wavelengths possible (proteins and nucleic acids without prosthetic groups absorb little above 300 nm) and preferably in the absence of oxygen. Values of ϕ_u and ε_u for many caging groups that have or could be used with proteins have been compiled here (Tab. 5.1.1). Because these values have not been obtained at identical wavelengths, it can be difficult to compare $\phi_u\varepsilon_u$ values, but we have attempted to quote values at wavelengths that would be used experimentally.

The value of $\phi_u\varepsilon_u$ for a class of caging reagents can be manipulated by chemical substitution (Tab. 5.1.1). For example, in the case of 2-nitrobenzyl attached at the N atom of carbamoyl choline, the replacement of a methyl group with a carboxylate at the benzylic position increases ϕ from 0.25 to 0.8, resulting in an increase in $\phi_u\varepsilon_u$ (312 nm) from ∼300 to ∼1000 [10]. In another case, with a (2-nitrophenyl)ethyl group on the phenolic O atom of phenylephrine, the introduction of methoxy groups in positions 4 and 5 of the aromatic ring increases ε_{312} from 1300 to 4500 M^{-1} cm^{-1} (λ_{max} is increased from 260 to 340 nm), which with a quantum yield $\phi=0.13$ gives $\phi_u\varepsilon_u$ (312 nm) ~600 [11]. By comparison, the unsubstituted derivative with $\phi=0.05$ gives $\phi_u\varepsilon_u$ (312 nm) ~65. However, to the chagrin of the investigator, an improvement in ε_u often leads to a reduction in ϕ_u. In one such case, cysteine (Cys)- or thiophosphate-containing peptides were caged on the S atoms with 2-nitrobenzyl or 4,5-dimethoxy-2-nitrobenzyl and little was gained from substitution of the ring, which serves to increase ε_u [12]. Clearly, we have a great deal to learn about the mechanism(s) of photolysis of 2-nitrobenzyl reagents, which continue to be investigated [2, 13, 14].

A detailed study of substituted coumarins has been made by Furuta et al., who found $\phi_u\varepsilon_u$ (365 nm) values in the range of 100–630, with the highest value assigned to a tribrominated derivative [15]. By comparison, the 7-methoxycoumarins have $\phi_u\varepsilon_u$ (333 nm) values approaching 3000 [16]. Following their successful use of 4-hydroxyphenacyl (HPB) as a caging group [$\phi_u\varepsilon_u$(300 nm)=4400] [17], Givens et al. attempted to push the absorbance towards the visible by substitution of the aromatic ring with methoxy groups. The 3-methoxy and 3,5-dimethoxy derivatives have absorption tails above 400 nm, but the ϕ values are reduced to 0.03–0.04, which is accompanied by a change in the photolysis mechanism. In contrast to the underivatized 4-hydroxyphenacyl group, the formation of a phenylacetic acid from an initially released spiroketone (Fig. 5.1.3) is a minor pathway in the case of the methoxy reagents [18].

Additional examples can be found in Tab. 5.1.1. Further, while the determination of ε is a simple matter, the reader is cautioned to note that not all values of ϕ that make their way into tables in the literature (including Tab. 5.1.1) have been determined with equal rigor: some have been obtained by careful actinometry, while others have been obtained by comparison with other reagents for which the values of ϕ may in turn be of dubious origin.

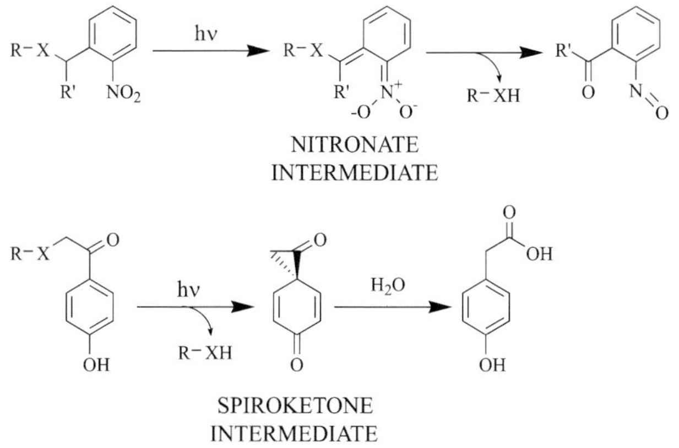

Fig. 5.1.3 Intermediates in the photolysis of two common caging groups: 2-nitrobenzyl (nitronate or *aci*-nitro intermediate) and 4-hydroxyphenacyl (spiroketone intermediate).

5.1.2.3 The Rate of Uncaging: Breakdown of Intermediates

The argument about the half-time for photolysis ($t_{1/2}$), given above, applies to cases in which the breakdown of the photogenerated intermediate is not rate limiting. In some applications, an intermediate might be long lived compared to the process under investigation. Such applications are unlikely to occur in cell biology, where the interest is usually in processes taking seconds to hours, but they can occur in biophysics, where for example there is renewed interest in examining protein folding on the submillisecond timescale using folding intermediates generated by flash photolysis [19–22].

Where they are available, values of the half-lives of caged reagents after the absorption of a photon, τ^*, have been tabulated (Tab. 5.1.1). The reader must be cautioned to take a careful look at how the measurements were made. Very often the appearance and decay of an intermediate in a flash photolysis experiment is followed by absorption spectroscopy and this does not necessarily indicate the rate of product release (e.g. recent work on 2-nitrobenzyl chemistry [13, 14]). Indeed, the identity of the intermediate is not always clear. Rapid-scan FTIR has a higher information content and might prove to be a method of choice for determining photochemical reaction sequences in the future [13, 14, 23–26]. In some cases, secondary intermediates are expected. For example, unstable carbamic acids are formed upon photolysis of molecules caged with the nitrobenzyloxycarbonyl group. By using a single-molecule approach, we recently examined the lifetimes of both the (presumed) nitronate (*aci*-nitro compound), generated initially, and the derived carbamic acid at the surface of a protein [27]. The nitronate had a half-life of ~ 2 s that was largely independent of pH, while the carbamic acid had a half-life of ~ 3 ms at pH 5.5 and ~ 5 s at pH 10. Clearly, these values are incompatible with, for example, rapid protein folding experiments.

The rates of breakdown of photolytic intermediates can vary hugely even within the same class of reagents. For example, in the case of 2-nitrobenzyl reagents (Fig. 5.1.3), the nitronate of the 1-(3,4-dimethoxy-6-nitrophenyl)ethyl ester of glycine breaks down with a unimolecular rate constant of $1\ \text{s}^{-1}$ [28], while the value for the nitronate of the corresponding *a*-carboxy-2-nitrobenzyl ester is $2 \times 10^5\ \text{s}^{-1}$ [29]. At present, the caging groups known to dissociate most rapidly after the absorption of a photon are the 4-hydroxyphenacyl group, $\tau^* \sim 1.2$ ns [17, 30] and the benzoins [20], which cleave in <1 ns. Photoinduced isomerizations can occur even more rapidly (see below).

5.1.2.4 The Effects of pH on Photolysis

Because there are often ionizable groups in caging reagents themselves or in photogenerated intermediates, both $t_{1/2}$ and τ^* can be affected by the prevailing pH. The two possibilities are not always clearly distinguished in the literature. Effects on $t_{1/2}$ arise because of effects of pH on ϕ and ε. The $t_{1/2}$ values are not meaningful in themselves because I_0 (the intensity of the photon source) can vary enormously.

The substituted 7-hydroxycoumarins of Furuta et al. provide an example of an effect of pH on $t_{1/2}$. Ionization of the coumarins led to a decrease in $t_{1/2}$ (365 nm), which was largely due to an increase in ε_u rather than ϕ_u [15]. In the case of 2-nitrobenzyl derivatives, pH can affect ϕ_u [3, 31]. In an example of an effect on τ^*, the rate of breakdown of the nitronate intermediate in the photolysis of 2-nitrobenzyl derivatives has been reported to be pH dependent over up to four orders of magnitude in certain cases, with faster breakdown at low pH [32]. In other cases, the pH dependence is weak [27, 29].

All these factors, ϕ_u, ε_u, τ^* and the effects of pH, will depend on the leaving group, which for proteins will often be the thiolate of a cysteinyl (Cys) residue (see below). Some measurements have been made on proteins derivatized on Cys or thiophosphate (Tab. 5.1.1), but more work is needed in this area. At present, the effects of pH should be considered to be unpredictable and the necessary exploratory experiments must be carried out for each new reagent caged protein.

5.1.2.5 Problems with the Released Caging Group

The released caging group, which is usually liberated in an altered form, must not interfere with the experiment that is underway. Therefore, the released group should be chemically inert and should not take part in further photochemistry, either by simply absorbing light and screening the remaining caged protein or by forming reactive intermediates. These issues are expected to be less of a problem in the case of cell biology experiments, where caged proteins are most often used at low concentrations (micromolar or less). In these cases, the photoproducts will be dilute and therefore do minimal damage, even if they are chemically reactive, and they will absorb little light, even if they have high ε values.

The most widely used caging reagent for proteins has been the 2-nitrobenzyl group (Tab. 5.1.1). Its main drawback is that it is released as a reactive nitrosoaldehyde or nitrosoketone, which has been shown to be deleterious to proteins by reaction with Cys residues in particular [25, 33, 34]. The nitroso compounds can be scavenged by the addition of a substantial concentration (>10 mM) of a thiol such as dithiothreitol (DTT) or 2-mercaptoethanol, prior to photolysis [34]. Where the caged protein is activated inside a living cell, the millimolar concentrations of intracellular thiols (glutathione, Cys, etc.) will serve to scavenge the reactive products. Semicarbazide at 10 µM has also been used to scavenge 2-nitrosobenzaldehyde, but its efficacy was not documented in detail [35].

Other photoproducts may be more benign. For example, 4-hydroxyphenacyl groups are released as a spiroketone, which rearranges to form 4-hydroxyphenylacetic acid, which is water-soluble, chemically inert and blue-shifted in absorbance compared to the caging reagent (Fig. 5.1.3) [17, 30]. A detailed comparison of this and other aspects of the nitrophenylethyl and hydroxyphenacyl caging groups has been made [25].

Several chromophores generate reactive oxygen species upon irradiation in the presence of oxygen. Therefore, where possible, uncaging should be carried out in an inert atmosphere, e.g. under nitrogen or argon. Alternatively, scavengers of the reactive species should be used. Thiols are good all round scavengers, but singlet oxygen is more effectively quenched with azide or carotenoids [36–38]. Many of the deleterious effects of short-wavelength light [8, 9] might be eliminated by two-photon uncaging at long wavelengths and this approach is under investigation [15, 39–41].

5.1.2.6 Photochemistry Peculiar to Proteins

While they have been treated here in the context of caged proteins, many of the phenomena described above also occur with small caged reagents. Certain phenomena that might affect the efficiency of photolysis are, however, peculiar to proteins (Fig. 5.1.4), e.g. the excited state of the caging group might be quenched by neighboring tryptophan. Conversely, it might be possible to uncage proteins by energy transfer to the caging group from tryptophan residues. This process has been used previously to activate photoaffinity reagents [42].

Other aspects of the microenvironment associated with caged residues in proteins include the local dielectric constant and pH. The dielectric constant can alter both the ground state absorption properties of a caging group, the photophysics after a photon is absorbed and the subsequent dark chemistry. The importance of pH has been noted above with reference to reactions in bulk solution. However, the "apparent pK_a" of a group in, say, the active site of an enzyme can be quite different from the bulk value, if the local environment differs. In other words, the protonation states of ground states and intermediates may differ significantly when a caged group is buried within a protein. Protonation and deprotonation rates may even be slowed to the extent that their kinetics dominate the deprotection process.

Steric hindrance to uncaging might also be of importance. For example, after a photon is absorbed, the caging group might not be able to rotate freely to attain the conformation that is required for bond rearrangement and cleavage (Fig. 5.1.4). From a practical viewpoint, these considerations indicate that it is not always possible to assess the properties of a caged protein from related solution photochemistry. Nevertheless, the circumstances suggest that, where possi-

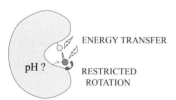

Fig. 5.1.4 The microenvironment at the site of caging in a protein can affect the photochemistry. For example, as depicted, the excited caging group might be quenched by energy transfer. Uncaging might be inhibited by restricted rotation. Local conditions such as pH and the dielectric constant might differ from bulk values.

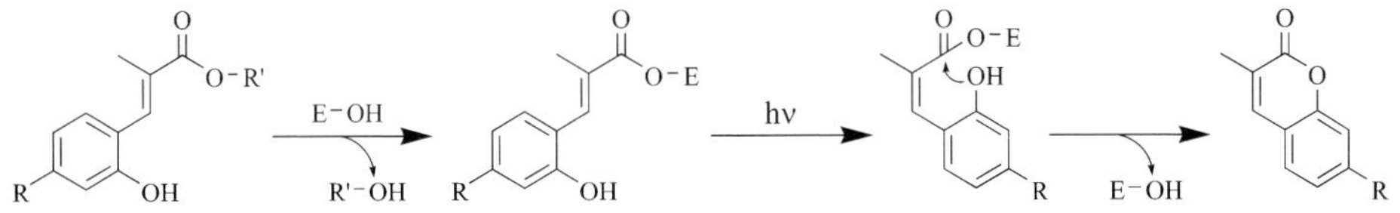

Fig. 5.1.5 Active site photochemistry of 2-hydroxycinnamoyl proteases. The enzyme (E-OH) is inhibited by acylation of the active site nucleophile, a serine side-chain. Upon photolysis, *trans–cis* isomerization occurs such that the phenolic hydroxyl of the caging group can attack the carbonyl of the acyl enzyme. The caging group then cyclizes and is released.

ble, a small mobile group should be used to cage a protein. Further, aromatic amino acids, which are potential quenchers, especially tryptophan, should be removed by mutagenesis from the vicinity of the site of modification. Access to solvent will also be important if the photochemistry that occurs in bulk solution is to be approximated. In some cases, the uncaging mechanism relies on active-site chemistry, as in the hydroxycinnamoyl-proteases of Porter et al. (for recent work, see [43–45]) (Fig. 5.1.5). For caged enzymes, the photoproducts should be released quickly from the active site and fail to rebind. This can be a tricky issue where a light-removable covalent inhibitor has been directed towards the active site. For example, Cys199 of the catalytic subunit of protein kinase A (PKA), situated at the entrance of the active site, was modified in preference to Cys343, with a substrate-like peptide linked to an *a*-bromo-2-nitrobenzyl group [46]. While the millimolar affinity of the peptide was sufficient to discriminate kinetically between the two Cys residues in the alkylation reaction (caging), it was, as desired, too weak for the peptide to remain in the active site after photolysis (uncaging).

Proteins are chiral and several caging reagents (Tab. 5.1.1) are attached through their own chiral centers forming diastereomers. Although the phenomenon has not been observed in practice, it is quite possible that the photochemistries of the two forms of caged protein differ.

5.1.2.7 Polypeptide Chain Cleavage

Occasionally, polypeptide chain cleavage has been noted during uncaging. For example, when the pore-forming protein *a*-hemolysin caged at Cys3 with 2-bromo-2-(nitrophenyl)acetic acid (BNPA) was irradiated in the near-UV, an apparently shortened fragment was observed upon sodium dodecylsulfate-polyacrylamide gel electrophoresis (SDS-PAGE) [47]. While other explanations are possible for the shift in electrophoretic mobility, such as intramolecular crosslinking, chain cleavage arising from scission at a site distant in the primary sequence is the most likely possibility. Less cleavage occurred when BNPA was attached at position 104 and in both cases the ratio of cleavage to deprotection was pH dependent [47]. In some cases, chain cleavage is intentional and desirable (see 5.1.3.10 and [48]).

5.1.2.8 Dominant Negative Effect

In the case of multisubunit proteins, "dominant negative" effects are a further consideration. Here, a small fraction of remaining caged protein or protein damaged during uncaging might remain in association with the properly uncaged protein and thereby reduce its activity. For example, in the case of caged a-hemolysin, which forms a heptameric pore, caging was about 60% complete as judged by SDS-PAGE, but only 10–15% of the pore-forming activity based on complete conversion was recovered [47].

5.1.3
Sites of Modification in Caged Proteins

5.1.3.1 Random Modification

Surprisingly, random modification has been used to cage proteins successfully [49]. For example, antibodies were treated with the oxycarbonylchloride of 1-(2-nitrophenyl)ethanol (NPE), which primarily reacts with the ε-amino groups of lysine (Lys) residues. When an Fc-specific anti-human IgG was "coated" with 30 NPE groups and the remaining Fc-binding fraction removed by affinity chromatography, a 15% recovery of IgG was obtained in which the specific Fc-binding activity was reduced to 1.5% of the original value. Upon irradiation with near-UV light, the specific binding activity increased to 29% of the original value. 3-Nitro-2-naphthalenemethanol has been used in much the same way [50]. For the unconjugated alcohol, $\phi_u = 0.63$ and $\varepsilon_u = 810 \text{ M}^{-1} \text{ cm}^{-1}$ ($\phi_u \varepsilon_u = 510$) in aqueous solution at 380 nm. In this case, the Protein A-binding activity of caged IgG increased from 18% to about 70% upon irradiation and then decreased upon further exposure. While simple in practice, the potential problems of random modification are clear. First, when the target protein has multiple functional domains, all of them will be affected by the caging reaction. Second, because multiple sites are likely to be modified by the caging reagent, uncaging will be a complicated process involving multiple photons, a lag phase and a complex mixture of intermediates with partly restored activity (Fig. 5.1.6). If the overall efficiency of removal of a caging group is E (the remaining positions being damaged in some manner), the final extent of uncaging will be $\sim E^n$, where n is the number of groups incorporated initially. However, the effect of multiple sites of caging on the final activity is actually less clear as each position at which caging takes place may have a different value of E and make a different contribution to the activity of the protein. On the positive side, a requirement for activation by multiple photons can aid in the spatial localization of uncaging (see 5.1.5.3).

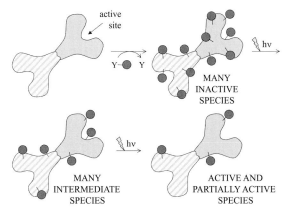

Fig. 5.1.6 Caging a protein by random modification. A two-domain protein is shown here. The protein is caged by "coating" with many protecting groups. Photolysis occurs through numerous partly deprotected intermediates that may or may not be active. In the case of a two-domain protein, intermediates may be formed in which one domain is active and the other is not.

5.1.3.2 **Active-site Directed Modification**

In an approach pioneered by Porter et al. [51], enzymes have been caged by making use of active-site chemistry. Various proteases have been inactivated as *trans*-2-hydroxycinnamoyl acyl enzymes by using ester substrates (for recent work, see [43–45]). Upon photolysis, the caging group undergoes *trans–cis* isomerization and the product then lactonizes in a dark reaction liberating the active site (Fig. 5.1.5). In a recent manifestation, the approach has been used to separate a specific enzyme from a mixture of proteases. In this case, the leaving group in the ester substrate confers enzyme specificity (it is an "inverse" substrate) and the acylating group contains an arm carrying biotin. The modified enzyme is extracted from the mixture with an avidin column and then released from the column by irradiation. Tagged acylating agents have been used for proteomic screening [52] and it is possible that photorelease might be useful in this context.

Additional active-site-directed reagents have been developed, e.g. the substrate-like peptide that reacts with a Cys residue at the entrance to the catalytic site of PKA [46]. However, there may not always be a need to tailor a reagent for an enzyme active site, as the latter often contain hyper-reactive residues associated with the catalytic activity. For example, the serine at the active site of a butyrylcholinesterase can be selectively reacted with the carbamylating reagent N-methyl-N-(2-nitrophenyl)carbamoyl chloride [53] and the catalytic Cys of a protein tyrosine phosphatase can be modified with various a-haloacetophenones, despite the presence of a total of four Cys residues in the protein [54].

It would seem likely that a modification strategy developed for one protein would be applicable to other members of its family, but this is not always so.

For example, N-methyl-N-(2-nitrophenyl)carbamoyl chloride efficiently inhibits both acetylcholinesterase and butyrylcholinesterase. However, only butyrylcholinesterase recovers activity efficiently upon irradiation [53].

5.1.3.3 The Use of Crosslinkers

A photocleavable, 2-nitrobenzyl, crosslinker was used to connect a ribosome-inactivating protein (PAP-S) to a monoclonal antibody or to the B-chain (lectin subunit) of ricin for delivery to cells [55]. Attachment to PAP-S was through a chloroformate and non-specific. A sulfhydryl group at the other end of the crosslinker was then deprotected for reaction with the monoclonal antibody, which had been premodified to carry maleimide groups. Alternatively, ricin B-chain was attached through another crosslinker carrying a maleimide. The conjugates were active towards cells in culture, but the activity was increased more than 20-fold in a plating efficiency assay after irradiation *in situ* at 350 nm, presumably because the liberated PAP-S has greater access to ribosomes (Fig. 5.1.7). This promising approach does not appear to have been pursued further.

There are many other potential applications of photocleavable crosslinkers for proteins. For example, they might be used to for the patterning of surfaces, including protein chips. One important possibility, for reagents carrying biotin on one arm [56], is in the purification of caged proteins. It has already been noted that it is often impossible to completely derivatize a protein with a caging re-

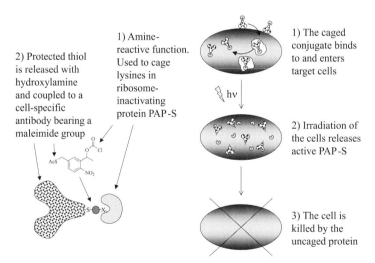

Fig. 5.1.7 Application of a photocleavable crosslinker. (Left) A small toxic protein (PAP-S) is coupled in two steps to a cell-specific antibody through a photocleavable crosslinker. (Right) The toxin–antibody conjugate recognizes target cells and is internalized, e.g. by endocytosis. Photocleavage of the crosslinker increases the toxicity of the internalized conjugate [55].

agent. With a biotinylated caging reagent, it would be possible to extract the derivatized fraction of the protein and this has been explored in the particular case of proteases by using active-site directed reagents [44]. The approach might be generalized by using Cys-directed biotinylated reagents to cage mutated proteins (see below).

5.1.3.4 Caging at Cys Residues

The deprotonated thiolate side-chains of Cys residues are by far the most nucleophilic groups in natural polypeptides. Therefore, under normal circumstances, a Cys residue will be selectively modified by an electrophilic caging reagent, even in the presence of hundreds of other reactive side-chains. "Normal circumstances" assumes that the Cys is not buried or that there is not a hyperreactive active-site nucleophile present in the target protein. The unperturbed pK_a value of a Cys residue in a polypeptide is 9.0–9.5.

The next issue then is to find or place a suitable Cys residue in the target protein (Fig. 5.1.8). In general, Cys residues are rarer than Lys residues and simply "coating" the protein with a Cys-reactive caging reagent is unlikely to work. Generally, a Cys residue in or close to a key functional site in the protein is required and can be introduced by mutagenesis. Occasionally, such a Cys residue will occur naturally; Cys residues are crucial active-site nucleophiles in several classes of enzyme including various proteases and phosphatases. For example, a protein tyrosine phosphatase has been caged with 4-hydroxyphenacyl bromide at the active-site Cys [54]. In other cases, a Cys residue in proximity to the active site may suffice and, where a structure is available, the residue is most readily identified by molecular graphics. For example, the catalytic subunit of PKA has been caged at Cys199, which lies near the entrance to the active site and in the so-called activation loop (PKA is activated by phosphorylation at Thr197). Care

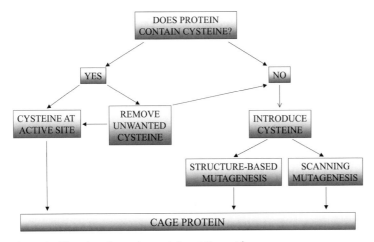

Fig. 5.1.8 Flow chart for caging proteins at Cys residues.

should be taken when caging enzymes at the periphery of the active site, because not all substrates may be affected equally by the covalent modification.

Where structural information is unavailable or unhelpful, cysteine-scanning mutagenesis can be a useful approach. Although it is tedious, the various spin-offs, e.g. mechanistic information and sites for the attachment of fluorescent probes, often make it worthwhile. Cys-scanning mutagenesis was performed on the pore-forming protein α-hemolysin (αHL) before the crystal structure had been determined and when little was known about the assembly mechanism. Wild-type αHL contains no Cys residues. Eighty-three mostly charged residues in the 293-residue polypeptide were changed individually to Cys, and the pore-forming activity of the protein was determined before and after modification at Cys with the bulky charged reagent 4-acetamido-4'-((iodoacetyl)amino)stilbene-2,2'-disulfonate, IASD [57]. The information gained in this way proved useful when it came to producing a caged αHL polypeptide [47]. Twenty-eight of 83 modified Cys mutants were substantially inactivated. Through various other negative criteria, including reduced activity before modification, incomplete inactivation by IASD, or a failure to be inactivated by a smaller reagent, the candidate positions were narrowed to four. One of these, Arg104, was investigated in detail and performed well when derivatized with the water-soluble caging reagent BNPA [47]. Subsequent structural information [58, 59] revealed that Arg-104 is in a region that undergoes a critical conformation change, including a 180° rotation at the ψ (C_α–C) angle of this residue.

In principle, proteins might be inactivated by modification at locations other than an active site. For example, it might be possible to use a caging group to prevent a protein–protein interaction, which could be reestablished upon photolysis. In practice, however, this may be more difficult than it appears. For example, a nitrobenzyl group located at the dimer interface in HIV protease did not prevent the interaction between the subunits [60]. In most cases, it is likely that a bulky reagent will be required for this purpose. For example, polymeric Cys-directed caging reagents might be developed [61] or biotinylated reagents [44, 56] might be used to block the interface after coupling to streptavidin.

An additional means of generating Cys residues at specific sites is by chemical ligation of polypeptide fragments, a methodology that is gaining ground for the total synthesis of proteins [62]. In this approach, two or more synthetic or recombinant fragments undergo chemical coupling. In general, the N-terminal fragment terminates in a reactive thioester at the C terminus and the C-terminal fragment contains an N-terminal Cys. For example, active channel proteins have been made in this way [63, 64]. After spontaneous ligation, the new polypeptide contains a Cys residue at the ligation site, which is available for chemical modification with a caging reagent.

In this section, the question of how to handle more than one Cys residue in the target protein has been neglected. As noted above, if several caged residues affect activity, the regeneration of activity upon photolysis will be a non-linear process. Because Cys residues are rare, this will not occur as often as it might with say non-specific modification at Lys. Again, because Cys is rare, it will often be possible to eliminate potentially offending residues by mutagenesis [65] (Fig. 5.1.8). In cases of extracellular proteins, Cys residues will often be present as disulfide bonds. The introduction of an additional Cys residue might then present a problem with disulfide scrambling, but often this can be circumvented by refolding in a redox buffer to encourage the formation of the natural disulfides and leave the new residue as a free sulfhydryl. Despite these issues, it is remarkable what can be achieved under potentially difficult circumstances. For instance, each 135 000 Da subunit of the β-galactosidase tetramer contains 16 Cys and 23 Met residues and yet it appears to be inactivated by the attachment of one 1-(4,5-dimethoxy-2-nitrophenyl)-2-nitroethene reagent per subunit [66]. Because the alkylation reaction is not reversed by 2-mercaptoethanol, the modification is unlikely to be at the active-site Met502. Upon irradiation, 89% of the enzyme activity was recovered.

5.1.3.5 Sites other than Cys for Site-specific Modification

Reactive side-chains other than Cys residues can be introduced into proteins for site-specific modification. For example, thiophosphate groups in peptides and proteins can be derivatized with various caging reagents. The thiophosphate group is an especially useful target because the activities of many cell-signaling proteins are controlled by phosphorylation. Thiophosphorylation can be carried out at serine and threonine by using the appropriate protein kinase and ATPγS [adenosine 5'-O-(3-thiotriphosphate)] [7]. In the case of tyrosine, the enzymatic phosphorylation conditions must be manipulated by the substitution of alternative divalent metal ions for magnesium [31]. Interestingly, the thiophosphate group is often a poor substrate for phosphatases, so proteins activated by thiophosphorylation rather than phosphorylation are expected to have longer lifetimes in vivo. A second useful feature of the thiophosphate group is that it can be alkylated at low pH in the presence of sulfhydryls [12]. The second pK_a value of a thiophosphate monoester is ~ 5.5 [67] and therefore at pH 4–5 the reactive monoanion will predominate, while Cys residues ($pK_a \sim 9.0$) will be protonated. With groups of similar nucleophilicity, selective modification can be obtained where $\Delta pK_a \sim 3$ or more.

In principle, the wide difference in reactivity between a thiophosphoryl and sulfhydryl group at low pH might allow two different groups to be placed on a protein, e.g. a caging group and a fluorescent probe. Therefore, it is worth considering additional residues that might react orthogonally. A prime candidate is selenocysteine (Sec), which when compared with Cys has a greatly reduced pK_a value (estimated to be 6.0–6.5 in a polypeptide chain [68]) and enhanced nucleophilicity [69, 70]. Sec can be introduced into proteins in vivo at TGA codons (although, in bacteria, the required cis signal disrupts the coding region) [71], by chemical ligation [70, 72] or, potentially, by unnatural amino acid mutagenesis. Interestingly, the relative rate of reaction of a selenophosphate over a thiophosphate monoester towards an alkyl iodide has been examined at pH 7.0 and found to be only ~ 3.5-fold higher [73].

More complex motifs can be derivatized even more selectively, as demonstrated by the work of Tsien et al. on tetracysteine (TC) sequences that can be reacted with bisarsenicals [74–76]. Reaction at these sites is sufficiently selective that proteins bearing them can be derivatized *in vivo*, although the generality of the approach for all types of cells has been questioned [77]. It is possible, for example, that caged TC motifs might be used to prevent protein–protein interactions in living cells. Other functionalities that might be useful for site-specific modification can be introduced by unnatural amino acid mutagenesis, and include azides [78, 79] and ketones [80].

5.1.3.6 Reagents for Caging by Protein Modification

It will be clear from the discussion above that the overwhelming majority of useful caging reagents for proteins are electrophilic molecules that modify nucleophilic amino acid side-chains (Tab. 5.1.1). Preferably, a reagent should be soluble in aqueous media. However, poorly soluble reagents can often be added from a concentrated solution in a water-miscible organic solvent, provided that the final concentration of organic solvent remains low (<5%; see, e.g. [65]). At higher concentrations of organic solvent many, but not all, proteins are irreversibly inactivated. The reagent and its linkage to the target protein should be stable in the storage buffer. For example, butyrylcholinesterase derivatized at the active site serine with the carbamylating reagent N-methyl-N-(2-nitrophenyl)carbamoyl chloride hydrolyzes over several days, but in this case decomposition is not a problem given the shorter timescale of the uncaging experiments [53]. The reactivity of the caging reagent should be tuned so that the desired residue on the protein is selectively modified. Often, the reaction rate can be controlled by manipulating the pH to adjust the protonation state of the reactive polypeptide side-chain. Excess reagent should be inactivated and removed so that it does not interfere with the biological system under study or simply screen photolysis. Typically, an electrophilic reagent is reacted with excess of a nucleophile, such as 2-mercaptoethanol, and removed by gel filtration or dialysis.

In principle, it would seem to be simple enough to choose a reagent (e.g. from Tab. 5.1.1) and modify a suitable Cys mutant of a target protein. In practice, there may be a need to test several reagents of differing size, shape and polarity to achieve a high extent of inactivation and the desired level of activation upon photolysis. For example, in the case of PKA, three nitrobenzyl reagents were tested for derivatizing Cys199: BNPA, 4,5-dimethoxy-2-nitrobenzyl bromide (DMNBB) and 2-nitrobenzyl bromide (NBB). Only one of them, NBB, gave an acceptable extent of inactivation (95–97%, after additional N-ethylmaleimide treatment) and reactivation (20- to 30-fold) [65]. The efficacy of NBB compared to the other reagents is not readily rationalized.

Site-specific reagents are not discussed individually here, because each case must be considered on its own merits. However, another potentially general class of modification should be noted. Lawrence et al. placed a Cys residue by mutagenesis into cofilin at the position of a Ser residue that is the site at which actin depolymerization activity is modulated by phosphorylation. The new Cys was modified with BNPA to yield a charged derivative that mimicked phosphoserine. This form of cofilin is inactive and is activated upon photolysis [81]. Because Cys is not a kinase substrate, the uncaged cofilin cannot be switched off by cellular kinases.

5.1.3.7 Determining the Extent of Protein Modification

As noted earlier, there are circumstances that require a high extent of inactivation when a protein is caged and others that require a high extent of reactivation upon irradiation. Often both are desirable. Unfortunately, inactivation is often incomplete, especially when less than 1% residual activity is sought. In the case of Cys residues, the possible reasons are varied. The residue in question may become oxidized during protein purification to a sulfenate (RSOH) or a disulfide (with say β-mercaptoethanol from a buffer). Sulfenates are inert towards most electrophilic reagents (but they will react with say 7-chloro-4-nitrobenzo-2-oxa-1,3-diazole (NBD-Cl)) [82]. Both sulfenates and disulfides can be converted back to Cys by treatment with a reducing agent such as DTT. Improved means for purifying the protein in the first place can help, and where small amounts of proteins are required preparation by *in vitro* translation may be preferable [27, 47]. We have found that the Cys residues in freshly translated protein are more likely to be fully modifiable than the same residues in protein that has been subjected to a relatively harsh purification procedure. Another possibility may be to modify a protein before purification. Other sources of incomplete caging at Cys include the presence of low-molecular-weight reactive impurities in the caging reagent that can modify but not inactivate a protein.

Although functional measurements are the most direct and best for detecting small fractions of residual activity, other approaches can be used to assess the fraction of a protein that is modified. Further, while the residual activity is easily measured for an enzyme, it is not always straightforward to determine in other systems. If sufficient material is available, the extent of modification can be determined by UV-vis absorption spectroscopy, based on the near-UV or visible absorption of the caging group [5, 83]. The most powerful approach is mass spectrometry, which also requires a relatively small amount of protein. A popular method is matrix assisted laser desorption ionization time-of-flight mass spectrometry (MALDI-MS) in which the protein is desorbed from an organic matrix by a laser pulse, which is sufficiently powerful to remove a fraction of the caging groups. The latter can be seen as a complication, in that it is difficult to quantify the extent of caging, or an advantage in that the spectrum of ions is characteristic of the caging reagent. The phenomenon has been well documented for 2-nitrobenzyl-protected Cys and thiophosphoryl peptides with a 337-nm laser [3, 12]. Besides simple removal of the caging group, a MH^+-16 peak is seen in the case of Cys peptides (loss of O from the NO_2 group?) and a MH^+-135 peak from caged thiophosphoryl peptides (loss of the 2-nitrobenzyl group with S/O exchange?). These peaks are not seen in electrospray ionization mass spectrometry (ESI-MS).

Another simple and clearly diagnostic approach is gel-shift electrophoresis [3, 35, 47], which is based on the fact that modified proteins may exhibit altered mobility upon SDS-PAGE that is greater than anticipated based on the increase in mass (and often in the opposite direction). For example, single-Cys mutants of α-hemolysin show distinct gel shifts after reaction with BNPA that permit quantification of the extent of modification [47]. Where the caging reagent does not produce a visible shift, an indirect approach is to show that the modified protein is no longer susceptible to a shift produced by a highly charged or high mass reagent, such as 4-acetamido-4′-[(iodoacetyl)amino]stilbene-2,2′-disulfonate [7, 47] or a sulfhydryl-directed poly(ethylene glycol) [84].

In a few cases, the crystal structure of a caged protein has been determined [53, 85]. These studies confirm the site and extent of caging (although small fractions of uncaged protein are not detected), and are important both in terms of the application of caged groups to study enzyme mechanisms and the examination of the mechanism of uncaging in an active site by time-resolved crystallography [85–87].

5.1.3.8 True Residual Activity

Of course, another possibility is that the fully and cleanly modified protein has true residual activity. When activity is blocked by steric hindrance or the prevention of a protein–protein interaction, rather than modification of a catalytic residue, the extent of inhibition might vary according to the bulk or charge of the photolabile group, even if the targeted residue is completely modified. When the caging group is attached at the periphery of the active site, it is even conceivable that a caged enzyme might have activity towards one class of substrates but not another. At the level of a few percent, it can be hard to distinguish between incomplete protein modification and a low level of activity after complete modification. However, after exhaustive investigation it was concluded that the catalytic subunit of PKA, caged with 4-hydroxyphenacyl bromide on thiophosphorylated Thr197, has ∼5% residual kinase activity towards a small peptide substrate [7]. In favorable cases, a systematic exploration of caging groups and locations may result in proteins with very low residual activity. For example, one of six caged RNase S peptides that were tested for their ability to complement RNase S yielded a complex (RNase S′) that was completely inactive. On irradiation, RNase activity equivalent to 36% of the native value was obtained [88]. The caged residue was adjacent to the catalytic His12.

5.1.3.9 Incomplete Uncaging

The photoregeneration of protein activity (uncaging) can also be incomplete and here even less is known. Very often details of photochemical side-reactions are poorly understood, even for model compounds of low mass, because the focus is usually on the major photolysis product. Additional details about the solution photolysis of small caged molecules can be found in this volume. Further, as indicated earlier, it is quite possible for photodeprotection to take a different course in the environment of a protein (Fig. 5.1.4). It might also simply be less efficient (reduced ϕ_p) and thereby allow competing side-reactions, such as photooxidation of the protein, to gain the upper hand. It would be very interesting to apply the power of mass spectrometry to this issue by determining the structure of protein byproducts formed during uncaging.

5.1.3.10 Means of Caging other than Chemical Modification

Caged proteins can be produced by means other than chemical modification of a protein or a mutant of it. Small proteins can now be produced by total synthesis or semisynthesis [62] and, as mentioned earlier, Cys residues generated at splice sites by chemical ligation might be modified by electrophilic reagents. Alternatively, unnatural photoactivatable amino acids might be incorporated into a polypeptide during chemical synthesis. Examples of small synthetic caged peptides include protein kinase inhibitors (caged Tyr [89]; caged Lys [90]), RNase S peptide (caged Asp, Glu, Lys, Gln [88]) and speract on a backbone NH [91]. The photomodulation of effectors and inhibitors is the topic of the review by Peng and Goeldner (Chapter 5.2). These approaches could be extended to entire synthetic or semisynthetic proteins. One interesting approach, which, however, cannot be generalized, was taken by Ueda et al. [92], who amidated Lys residues in phospholipase A2, removed three key N-terminal residues by Edman degradation and replaced them by covalently coupling tripeptides containing various unnatural amino acids at position 3. One of these amino acids, L-*p*-phenylazophenylalanine (PAP), yielded a *trans*-azo lipase with reduced catalytic activity that gained weak activity upon irradiation.

Unnatural amino acids can also be incorporated at specific sites during *in vitro* protein synthesis and this methodology has been used to make a range of caged proteins including enzymes [93] and ion channels [94]. Uncaging of *in vitro* incorporated amino acids has been used to initiate protein splicing [95], which might therefore also be used as a means to generate enzymatic activity. When the amino acid (2-nitrophenyl)glycine is introduced into proteins, intentional polypeptide chain cleavage occurs upon irradiation [48, 96]. In the case of ion channels, the proteins can be inactivated or specific modulatory regions, such as an N-terminal inactivation domain can be removed in situ. Further investigation of the generality and efficiency of chain cleavage by this means would be informative. Attempts to modulate protein–protein interactions through unnatural amino acid mutagenesis have been instructive. The p21^{ras} protein with the β-2-nitrobenzyl ester of Asp at position 38 was unable to act with the effector protein GAP. Residue 38 lies at the Ras–GAP interface. Upon photolysis, about $\sim$50% of the Ras activity was restored [35]. In contrast, when HIV protease was caged in the same way, at an active site Asp that lies at the dimer interface, the protein was inactivated but still able to dimerize [60]. All the examples of unnatural amino acid mutagenesis to produce caged proteins so far recorded have been based on 2-nitrobenzyl chemistry, but the approach is capable of handling a wide variety of amino acid side-chains [97, 98].

Advances are being made in applying the unnatural amino acid approach *in vivo*. Loaded tRNAs and mRNA can be injected into cells, most readily *Xenopus* oocytes [94, 99]. Recently, aminoacylated tRNA and mRNA (or DNA constructs) were loaded by electroporation into cells [100]. Bacteria containing mutant tRNA/aminoacyl-tRNA synthetase pairs have been prepared that permit the introduction of unnatural amino acids at amber sites *in vivo* [80, 101, 102]. It is only a matter of time before various classes of cells capable of incorporating caged amino acids into preselected proteins at specific sites are available. This is important because the need to rely on means such as microinjection to introduce caged proteins into cells and to the correct sites within them is a major deterrent to their more widespread use.

5.1.4
Applications of Caged Proteins

In this chapter, we have focussed on the means by which caged proteins with optimal properties can be prepared. The applications of such proteins *in vivo* are covered in Chapter 5.3. Here, for completeness, we have tabulated selected examples that have been examined for the most part *in vitro*, to illustrate the various classes of proteins that have been caged and to give emphasis to those cases where the protein has been used to generate new mechanistic or biophysical information (Tab. 5.1.2).

5.1.5
Future Prospects

5.1.5.1 Turning Proteins On *and* Off

Attempts to control reversibly the activity of proteins with light, like straightforward caging, go back many years [103–107]. Most commonly, proteins have been derivatized with azobenzenes that undergo *trans* to *cis* isomerization upon irradiation at short wavelengths, with reversal at longer wavelengths or by thermal relaxation [108]. With notable exceptions [92, 109], early attempts involved random modification with photoisomerizable reagents. Given the problems with simply caging proteins by this approach, it is not surprising that these efforts were only partly successful in that complete on–off switching was not achieved.

More recently systematic studies have been undertaken. The example of ribonuclease (RNase) is instructive [110]. The photoisomerizable amino acid PAP was incorporated into the RNase S peptide at six positions, 4, 7, 8, 10, 11 or 13, by chemical synthesis. Under the authors' conditions, the maximum extent of photoconversion of PAP in solution was from 96% *trans*/4% *cis* to 10% *trans*/90% *cis*, which places a limit on the maximum fold change in activity that might be achieved. In this case, the maximum feasible extent of switching was not translated into an effect on enzymatic activity because the effects of photoisomerization on kinetic constants were small and distributed among effects on S-peptide binding to RNase S, substrate binding and catalysis. In contrast, the apparently complete

Tab. 5.1.2 Caged proteins

Class	Example	Caged protein	Comments	Refs.
Enzymes	Protein kinase	Modification of Cys199 in the catalytic (C) subunit of PKA by NBB. Cys199 is located at the mouth of the active site. The other Cys in the C subunit, C343 is mutated to serine.	Various 2-nitrobenzyl derivatives (NBB, DMNBB and BNPA) are tested. The most effective is NBB, which gives 3–5% residual activity after further treatment with N-ethylmaleimide and 80–100% activity (a 20- to 30-fold increase) after irradiation, with $\phi_{(312;\ pH\ 6)}=0.84$ and $\phi_{(312;\ pH\ 8.5)}=0.14$	65
		Modification of Cys199 in the C subunit of PKA by a derivative of BNPA, in which the carboxylate is linked to a peptide that binds at the active site of the enzyme. The affinity of the peptide ($K_i=1.5$ mM) is sufficient to allow selective modification of Cys199 over Cys343, but low enough that the peptide dissociates after photolysis.	ESI mass spectrometry shows a single covalently bound CNB-peptide per protein. The C subunit shows 2% residual activity after modification and 50% of the original activity is regenerated upon irradiation. The caged PKA is microinjected into fibroblasts and morphological changes are observed upon photoactivation.	46
		Modification of a thiophosphorylated Thr197 residue in the catalytic (C) subunit of PKA with HPB. The C subunit of PKA is activated by phosphorylation at Thr197.	17-fold decrease in activity (5% residual activity) upon modification, and 15-fold increase of activity (85–90% uncaging) upon irradiation, with $\phi_{(312;\ pH\ 7.3)}=0.21$	7
	Protein phosphatase	Catalytic Cys of a tyrosine phosphatase caged by HPB.	HPB, 4-methoxyphenacyl bromide, 4-carboxymethylphenacyl bromide and 4-carboxymethylphenacyl chloride are tested. HPB proves to be the best irreversible inhibitor. It produces over 95% inactivation and 80% of the activity is recovered upon irradiation.	54

Tab. 5.1.2 (continued)

Class	Example	Caged protein	Comments	Refs.
	Proteases	p-Substituted o-hydroxy-α-methylcinnamoyl derivatives are used to acylate the catalytic serine of various proteases. The nature of the ester leaving group (-OR) is varied to achieve selectivity among chymotrypsin, thrombin and factor Xa. CINN-OR	Various studies have been carried out with the caged proteases [44, 51, 87, 180, 181]; including: time-resolved X-ray crystallography [85–87]; stimulation of localized thrombosis in the rabbit eye [139]; two-photon photolysis [182]; purification of proteases derivatized with a biotinylated arm, with an avidin column and release by photolysis [45]. More details are given in Tab. 5.1.1.	44, 45, 51, 85–87, 139, 177, 180–182
		p-Nitrophenyl 2-(2,5-dimethoxybenzoyl)benzoate ($R = NO_2$; $R' = OMe$) is used to cage the serine at the active site of chymotrypsin, making use of the catalytic mechanism.	Chymotrypsin is completely inactivated. Subsequent irradiation regenerates 70–80% of the activity. A biotinylated photolabile inhibitor enables the attachment of chymotrypsin to an avidin affinity column, with release by photolysis. p-Guanidinophenyl benzoylbenzoate ($R = NH(CNH)NH_2 \cdot HCl$; $R' = H$) is used in a similar way to cage thrombin. Acyl-thrombin does not regain any activity upon irradiation, but is rather shown to be photolabeled.	179
	HIV-1 protease	The active site of HIV-1 protease is at the interface of a homodimer. Asp-NB ester is incorporated at position 25 in the interface by the amber codon suppression method.	Surprisingly, dimer formation is not prevented, but the caged enzyme has minimal proteolytic activity. 97% of the activity is restored upon irradiation.	60

Tab. 5.1.2 (continued)

Class	Example	Caged protein	Comments	Refs.
	β-galacto-sidase	Modification of a Cys residue in β-galactosidase by 1-(4,5-dimethoxy-2-nitrophenyl)-2-nitroethene. MeO–C₆H₂(OMe)(NO₂)–CH=CH–NO₂	94% inhibition is obtained. Subsequent irradiation at 350 nm results in restoration of 83% of the enzymatic activity.	66
	Cholinesterases	The catalytic serine of butyrylcholinesterase is modified with N-methyl-N-(2-nitrophenyl)carbamoyl chloride (MNPCC). CH₃–N(C₆H₄-NO₂)–C(=O)–Cl	95% inactivation is achieved. Subsequent irradiation regenerates up to 100% activity. The specificity of the reagent for the active site is demonstrated by X-ray crystallography. Acetylcholinesterase can be efficiently inhibited by MNPCC, but photoregeneration is limited, 20–30% at most.	53
	Ribonuclease	Ribonuclease S′ (RNase S′) is composed of two fragments: S-peptide (residues 1–20) and S-protein (residues 21–124, also called RNase S). S-peptide complements the inactive S-protein to restore native-like structure and activity. 2-Nitrobenzyl caged amino acids are incorporated at different positions in the S-peptide by solid phase peptide synthesis.	Gln11 in the S-peptide stabilizes a phosphate ester of the RNA substrate. Caging of Gln11 leads to complete inhibition of the enzyme. Upon irradiation, 37% of the activity is recovered. S peptide caged at other positions shows a lesser or no effect on the ribonuclease activity. The authors report 20–30% photodamage of the control native enzyme.	88
	Lysozyme T4	Asp20 in T4 lysozyme is essential for catalytic activity. The active site Asp20 is replaced by Asp-NB ester using the amber-codon suppression method.	Irradiation of the inactivated NB-Asp20 lysozyme T4, at 315 nm restores 32% of the activity.	93

Tab. 5.1.2 (continued)

Class	Example	Caged protein	Comments	Refs.
Antibodies	Immunoglobulin	Random modification of amino groups of Lys residues in immunoglobulin with 1-(2-nitrophenyl)ethane chloroformate.	On average 30 NPE residues are bound per immunoglobulin protein. After removal of antibody that is not inactivated, the binding capacity of the modified protein in an ELISA falls to 1.5% of native IgG. On subsequent exposure to UV light as much as 29% of the binding capacity is regenerated. The unmodified control protein lost 30% of its activity upon irradiation for the same duration.	49
	Immunoglobulin	Lys residues in an antibody (Ab) that targets brain-derived neurotrophic factor (BDNF) are randomly modified with NVOC-Cl.	About 40% of the Lys residues of the BDNF Ab are modified, leading to a reduction of 80% in binding affinity. The derivatized antibody is effective when uncaged during an electrophysiological assay with hippocampal slices.	183
	Immunoglobulin	Random modification of Lys residues of immunoglobulin (IgG) by 3-nitro-2-naphthalyloxycarbonyl chloride (NMCC).	On average 22 residues are caged per protein, which decreases the activity to 18%. Upon irradiation, binding to Protein A is increased to 70–74% of the original value. $\lambda_{max} = 259$ nm; $\varepsilon_{259} = 7298$; $\varepsilon_{350} = 1289$; $\phi_{(380;\ pH\ 6.7)} = 0.83$.	50
	Ribosome-inactivating protein cross-linked with a targeting protein	Lys residues in the ribosome-inactivating protein PAP-S are modified with the chloroformate group of the cross-linker. The thiol is then deprotected and used to couple with an antibody carrying maleimide groups.	An average of 0.7 cross-linking groups are attached per PAP-S. Targeting proteins are either a monoclonal antibody, directed against the human transferrin receptor, or the B-chain of ricin; 85% of the protein–protein conjugate is cleaved upon irradiation.	55

Tab. 5.1.2 (continued)

Class	Example	Caged protein	Comments	Refs.
Cyto-skeleton	Colifin	Colifin is inactivated by phosphorylation at Ser-3. This residue is mutated into a Cys, which results in a constitutively active protein. Cys3 is then modified by BNPA. The negative charge of this reagent acts as a phosphate mimic and thereby inactivates colifin.	Caged colifin is microinjected into an adenocarcinoma cell line. Irradiation causes 80% uncaging of colifin, and the appearance of cleaved F-actin filaments.	81
	Thymosin β4 (Tβ4)	β Thymosins contain a Lys-rich segment, which binds G-actin and thereby regulates actin polymerization into F-actin. Tβ4 is inactivated by random modification with NVOC-Cl.	*In vitro*, caged Tβ4 shows no effect on the polymerization rate of actin. Once photolyzed, Tβ4 significantly inhibits actin polymerization. Caged Tβ4 is bead-loaded into keratocytes. Local photoactivation results in a regional locomotor response.	184
	Heavy meromyosin	Cys707 in heavy meromyosin is modified with DMNBB. [structure: benzene ring with Br, NO$_2$, MeO, OMe substituents]	The modification abolishes actin-activated ATPase activity. A 340–400-nm light pulse activates the movement of actin filaments with a velocity of 17–50% of that seen with native meromyosin.	185
	G-actin	Lys residues in G-actin are caged as carbamates and the oxirane group is subsequently used to cross-link the caged G-actin with thiolated fluorescent dextran. [structure: benzene ring with MeO, MeO, NO$_2$ substituents and oxirane/carbamate group with X leaving group]	Several cross-linking reagents with various X leaving groups for amine modification are made. Once the Lys-containing G-actin has been caged, it can be cross-linked to thiolated dextran by reaction with the oxirane group. The cross-linker is also used to cage the two aromatic amines of rhodamine 110, resulting in the non-fluorescent lactone form of the dye. The resulting reagent is coupled via one of the oxirane rings to Cys374 in G-actin, resulting in a protein that can both form F-actin and release rhodamine upon irradiation.	61

Tab. 5.1.2 (continued)

Class	Example	Caged protein	Comments	Refs.
		Lys61 is located at the actin–actin interface in the F-actin polymer. G-actin is modified with NVOC-Cl to allow photoregulation of the polymerization.	Three to five NVOC groups are attached per G-actin polypeptide, resulting in greatly reduced polymerization activity. Up to 95% polymerization into F-actin is induced upon irradiation.	83
Transcription factors	GAL4VP16	GAL4VP16 is randomly modified with NVOC-Cl.	Modification of GAL4VP16 with NVOC-Cl under basic conditions completely inhibits its DNA binding ability. More than 50% of the binding activity is recovered upon irradiation at 365 nm. GAL4VP16 modified at an average of 8 of 14 Lys residues with NVOC-Cl can be activated after injection into *Drosophila* embryos.	186, 187
Channels and pores	a-hemolysin	83 single Cys mutants of the a-hemolysin (aHL) monomer are screened and R104C is chosen to be modified by BNPA.	Only four of the unmodified mutants retain pore-forming activity similar to that of the wild-type and can be completely inhibited with sulfhydryl reagents. The R104C mutant, modified with a single CNB group, shows almost no residual activity when tested in a red cell lysis assay. Between 10 and 15% of the lytic activity of aHL-R104C is recovered upon uncaging. However, removal of the carboxynitrobenzyl group is 60% effective, as assessed by SDS-PAGE. Therefore, it is likely that one or a few inactive monomers prevent formation of the active heptameric pore.	47

Tab. 5.1.2 (continued)

Class	Example	Caged protein	Comments	Refs.
	Nicotinic acetylcholine receptor (nAChR)	Leu9′ in the M2 transmembrane segment of the γ subunit of nAChR is involved in channel gating. This residue is replaced by 2-nitrobenzyl tyrosine or 2-nitrobenzyl Cys by unnatural amino acid mutagenesis in *Xenopus* oocytes.	Cys-NB-9′ and Tyr-NB-9′ mutants both show an ACh-induced current before photolysis. Irradiation of Cys-NB-9′ causes a 50–150% current increase. For Tyr-NB-9′, the current increases up to 4-fold.	188
		Tyr93 and 198 in the α subunit of nAChR are highly conserved and thought to participate in agonist binding. They are replaced by 2-nitrobenzyl-tyrosine, by unnatural amino acid mutagenesis in *Xenopus* oocytes. The less crucial Tyr127 is caged the same way as a comparison.	The receptors containing Tyr-NB-93 and Tyr-NB-198 are barely affected by ACh until uncaged by irradiation. In the presence of ACh, photolysis produces an increase in current and subsequent desensitization. Up to 17% of the tyrosines are uncaged with a single 1-ms 300–350-nm flash.	142
	nAChR and K$^+$ channel	Irradiation of the unnatural amino-acid 2-nitrophenyl glycine (Npg) results in the cleavage of a polypeptide backbone. Npg is incorporated by unnatural amino acid mutagenesis at position 47 of the K$^+$ channel, just after the N-terminal inactivating "ball" domain. The nAChR contains a highly conserved disulfide loop, Cys128–Cys142, that is critical for protein folding and assembly. Npg is incorporated between the two Cys residues.	Both the K$^+$ channel and nAChR mutants are expressed in *Xenopus* oocytes. Irradiation of the K$^+$ channel results in a substantial reduction in its inactivation rate. Oocytes containing Npg-nAChR show a large loss of whole-cell current when irradiated, but the channels remain properly folded, as judged by α-bungarotoxin binding ability.	48

Tab. 5.1.2 (continued)

Class	Example	Caged protein	Comments	Refs.
	Kir2.1 channel	Tyr242 in the Kir2.1 potassium channel influences channel function, through its phosphorylation state. 2-Nitrobenzyl-caged tyrosine is introduced by unnatural amino acid mutagenesis in *Xenopus* oocytes.	Photolysis of the NB-Tyr242 mutant in the presence of tyrosine kinases results in a decrease of both current and cell capacitance, due to Kir2.1 endocytosis.	143
Splicing	Self-splicing proteins	Conserved nucleophilic residues (Cys, Ser or Thr) immediately following splice junctions in self-splicing proteins are essential to splicing. Ser1082 in the *T. litoralis* DNA polymerase precursor is replaced by NB-Ser by unnatural amino acid mutagenesis, abolishing splicing.	Irradiation of the truncated 56 kDa NB-Ser1082 DNA polymerase precursor, followed by incubation to allow splicing, results in excision of the expected 45-kDa intein as seen by Western blot analysis. However, attempts to detect the 10-kDa ligation product were unsuccessful, possibly because of the proteolytic sensitivity of the DNA polymerase in *Escherichia coli* extracts.	95
Protein–protein interactions	p21^{ras}	Asp38 in p21^{ras} forms an important interaction with Lys939 in p120-GAP (GTPase-activating protein). Asp38 is replaced by NB-Asp by the amber suppression technique.	NB-Asp38 p21^{ras} retains its intrinsic GTPase activity, but is unable to interact with p120-GAP until it has been uncaged by 355 nm irradiation. The GTPase activity is then increased 3.5-fold, up to about 50% of the activity of the wild-type control in the presence of GAP.	35
Caged fluorescent proteins	GFP	GFP contains a mixed population of neutral phenols (λ_{max}=397 nm) and phenolate chromophores (λ_{max}=475 nm). The T203H mutant (PA-GFP) is almost entirely in the neutral form. Irradiation at 413 nm causes conversion to the phenolate and an almost 100-fold increase in fluorescence. The activated form is stable for days.	PA-GFP fusion proteins can be activated *in vivo*.	122, 123

on–off switching of horseradish peroxidase has been accomplished [111]. Muranaka et al. placed PAP at 15 different positions in the peroxidase by using unnatural amino acid mutagenesis. Four of the mutants retained activity. In one of these (position 68), the *cis* form (UV activated) was more active than the *trans* form. Remarkably, for the protein substituted at position 179, the *cis* form was completely inactive and on–off switching was obtained, with 50% recovery in the first cycle utilizing reversal with visible light and no further loss of activity in the second cycle. Given, the observations with PAP in solution, it is surprising that no residual activity was seen. One possibility is that conversion to the *cis* form was more complete on the peroxidase, suggesting that the position of the photoequilibrium might vary with environment. There is precedent for such a possibility. To make a photomodulated DNA-binding protein, Caamaño et al. connected two 26-amino-acid peptides comprising GCN4 binding domains through C-terminal Cys residues by an azobenzene bridge [112]. Irradiation at 365 nm gave a mixture of 5% *trans*/95% *cis*. In the absence of DNA, this could be reversed by irradiation at 430 nm or by thermal relaxation. The *cis* form bound a target DNA sequence far more strongly than the *trans* form and, when bound to DNA, the *cis* form could not be switched back to the *trans* form by photolysis, demonstrating the high stability of the complex.

Under many circumstances, it seems likely that a photoisomerizable intramolecular crosslink would have a more profound effect on the structure of a protein than substitution with a photoisomerizable amino acid. This possibility has been explored by Woolley et al. with model peptides [113–115]. Recently, the group has devised a water-soluble azobenzene crosslinker containing two sulfonate groups for attachment at Cys residues [116]. Because it can be introduced without organic cosolvent, it should be especially useful for the direct modification of proteins.

The use of photoisomerizable reagents allows proteins to be activated by one wavelength of light and turned off by another. Alternatively, reversal can occur thermally. Therefore, it is possible to titrate the amount of activated protein in an experiment with a calibrated light pulse (assuming slow thermal reversal) or by continuous irradiation to produce a photo-steady state (assuming rapid thermal reversal). Additional means for turning proteins caged with protecting groups (rather than photoisomerizable reagents) into single-cycle reversible agents should be explored. For example, light activation as described earlier might be built into a temperature-sensitive mutant of a protein, so that activity can be turned on by light and off by a temperature shift. Temperature sensitivity can also be established by targeted chemical modification with polymers such as poly(*N*-isopropylacrylamide) that undergo a transition between an expanded and a condensed form [117]. Interestingly, the two ideas can be combined. Certain azopolymers undergo light-dependent expansion and condensation and, when attached at specific sites in a protein, they can mediate the control of ligand binding [118].

Additional constructs for on–off switching can be envisioned in which the protein must be modified at two sites. For example, it might be possible to activate a protein by uncaging with one wavelength of light and then inactivate it with another wavelength, e.g. by uncaging a sequence that targets the protein for proteolytic destruction in a cell or one that acts as an intramolecular inhibitor peptide, or by causing photolytic chain cleavage. Mutually exclusive modification at two sites might be achieved by using a nucleophile on the proteins that can be distinguished from Cys by virtue of its greater reactivity or lower pK_a, such as thiophosphate [12] or Sec [70]. The orthogonal photolysis required to carry out these possibilities has been demonstrated with model compounds [119, 120]. Two protecting groups, a dimethoxybenzoin (254 nm) and dimethoxynitrobenzyl (420 nm), were removed in either order with selectivities in the range of 5:1 to 10:1, thanks to an appropriate distribution of ϕ_u and ε_u values, and the lack of energy transfer between the short-lived excited state of dimethoxybenzoin and the dimethoxynitrobenzyl group. Incidentally, this result also suggests that selective activation should be possible in an experimental system containing two different caged proteins.

5.1.5.2 Genetically Encoded Caged Proteins and Caging within Cells

A major difficulty in using caged proteins within living cells is getting them into the cells in the first place, as addressed by Condeelis and Lawrence (Chapter 5.3). Techniques include microinjection, the use of Tat fusion proteins, electroporation, syringe loading and bead loading, all of which bring proteins in from the outside. In the case of fluorescent indicator proteins, rather than caged proteins, this issue has been addressed by using polypeptides that are completely genetically encoded, for the most part GFP fusion proteins [76]. It would be of great interest if caged proteins could also be genetically encoded, e.g. by using a fusion to a light-responsive signaling element such as photoactive yellow protein. In work along these lines, Shimizu-Sato et al. have produced a light-activatable gene expression system, which functions *in vivo* and is reversible [121] (Fig. 5.1.9). The system is based on the yeast two-hybrid model and utilizes the light-dependent phytochrome–PIF3 interaction to recruit a GAL4 DNA-binding domain to the desired promoter. In another case, the fluorescence of a GFP variant increases ∼100-fold when subjected to intense 400-nm radiation and can be thought of as a genetically encoded caged fluorescent marker [122, 123]. This protein is discussed further by Condeelis and Lawrence in Chapter 5.3. A second idea is to chemically modify a selected protein within a cell. Obviously, this cannot be done at a single Cys residue, because there are numerous endogenous proteins containing Cys residues. Nevertheless, it has been achieved by using genetically encoded TC motifs (TC tags) that couple selectively to small cell-permeant molecules containing two suitably positioned trivalent arsenic functionalities [76]. So far, caged proteins have not been made in this way, but the idea has been extended to chromophore-assisted light inactivation (CALI). In the original manifestation of CALI, a dye, malachite green, was attached to an antibody raised against a target protein and the conjugate was microinjected into target cells. Upon irradation, reactive oxygen species, perhaps hydroxyl radicals, were generated that inactivated the target, but spared proteins

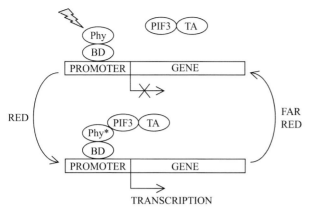

Fig. 5.1.9 A light-activated transcription factor. The system is based on the red light-induced binding of the plant photoreceptor phytochrome (Phy) to the protein PIF3 and reversal of the interaction by far red light. Phy is fused with a GAL4 DNA-binding domain (BD) and thus binds to the promoter of the target gene. Upon irradiation with red light, Phy recruits PIF3, which is fused with a transcription activation domain (TA). Upon irradiation with far red light, Phy dissociates from PIF3 and transcription ceases.

a short distance away [124–126]. Tsien et al. have shown that biarsenical derivatives of resorufin or fluorescein attached to a TC tag *in vivo* can mediate CALI, in this case most likely by the generation of singlet oxygen [127]. However, the degree of selectivity of biarsenicals in a cellular environment, in which natural Cys-rich proteins are present, has been questioned and further refinements of the technology can be expected [77].

5.1.5.3 Improved Spatial Control

A major advantage of the use of caged proteins is the spatiotemporal control of reagent release, and there is room for improvement in both the spatial and temporal aspects of control. By using two-photon uncaging with focussed light, photolysis within cells can be limited to a few cubic micrometers [15, 39] and new approaches to sharply focussed beams continue to be investigated [128]. In the two-photon approach, a laser beam is used to produce very high light intensity at the target site, allowing, for example, the excitation of near-UV transitions with two red photons. Because the absorption probability falls off with a quadratic dependence, the excitation volume is highly restricted. Further, although uncaged molecules may diffuse from the site of irradiation, they are rapidly diluted.

The use of a doubly caged glutamate that requires the sequential absorption of two photons for conversion to active glutamate also resulted in a substantial improvement in spatial resolution over singly caged glutamate when a focussed beam was used [129]. This "chemical two-photon uncaging" might also be applied to proteins. Local heating of biomolecules coupled to metal nanoparticles with RF magnetic fields is an interesting new approach for activation that has a spatial resolution of a few nanometers [130]. Spatial resolution might also be improved by placing tags on caged proteins that direct them to subcellular organelles, such as the nucleus or mitochondria. It is possible to screen for new tags by using approaches such as phage display of combinatorial libraries.

Spatial control is also important for patterning surfaces. For example, patterning at the micron level is desirable in the fabrication of protein chips. Surfaces can first be covered with protein, which is then released from specific sites by irradiation. Alternatively, protein can be attached to the desired sites by irradiation. In an example of photolytic removal, Ching and colleagues attached an 18-residue peptide to poly(4-vinylpyridine) via N-pyridinium linkages [131]. The linkages were cleaved by UV irradiation. Deposition brings us into the arena of photoaffinity labeling and related approaches. In a recent example, Holden and Cremer showed that dye-labeled proteins can be attached to a BSA-coated surface by irradiation, most likely through a singlet oxygen mechanism [132]. For example, an Alexa-IgG was coupled at 594 nm.

5.1.5.4 Temporal Control

In certain biophysical experiments, including time-resolved crystallography and *in vitro* protein folding, there is a need for very rapid, efficient uncaging. Interestingly, recent work in the area of protein folding has eschewed conventional caging reagents, such as nitrobenzyl groups, which are often released quite slowly (Tab. 5.1.1), for photoisomerizable molecules such as azobenzenes. It is known that photoinduced *trans–cis* isomerization in azobenzenes is very fast indeed, occurring on the subpicosecond timescale. The quantum yield is also high. Until recently, protein-folding studies had been limited to milliseconds by rapid mixing or dilution, or to microseconds by various "jump" techniques. New spectroscopic techniques are now being applied, many of which would benefit from rapid triggering of the folding process. Femtosecond time-resolved spectroscopy has been used to examine peptides cyclized with *cis*-azobenzene groups (Fig. 5.1.10). Photoisomerization, the subsequent dissipation of vibrational energy and the conformational relaxation of the *trans*-cyclic peptide were resolved [21]. Substantial conformational changes occurred on the 50-ps timescale, but additional changes in the polypeptide backbone were still occurring at 1 ns. The kinetics of helix unfolding in a 16-amino-acid peptide crosslinked with azobenzene have been studied on the nanosecond timescale by optical rotatory dispersion (ORD) [22]. In the *trans* form, the peptide favors a more helical structure (66% helix) and in the *cis* configuration the helical content is reduced. Thermal reisomerization takes minutes, so photochemistry could be studied by flash photolysis in the *trans* to *cis* direction. After a 7-ns laser pulse at 355 nm, the ORD signal at 230 nm was best fit to a single-exponential decay with a time constant of 55 ns. The folding and unfolding time constants for the *cis* peptide

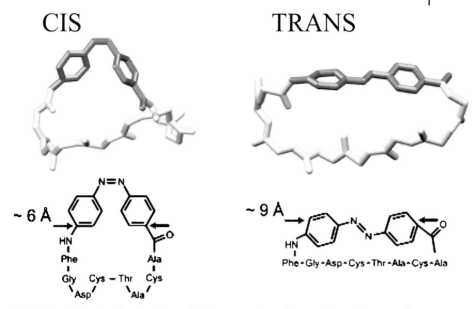

Fig. 5.1.10 Application of a photoisomerizable reagent. An azobenzene inserted into a cyclic peptide in the *cis* configuration serves as an ultrafast optical switch to study peptide conformational dynamics.

were estimated to be 330 and 66 ns, respectively. Alternative approaches in this area include the photolytic cleavage of crosslinks within conformationally constrained peptides, including aromatic disulfides (at 270 nm) [19], which are believed to cleave in <1 ps, and cyclic benzoins, which cleave in <1 ns [20].

In crystallography, an interesting way to observe intermediates consists of irradiating to completion a crystal of a caged protein at a low temperature (about 100–150 K) and then slowly increasing the temperature [133, 134]. In this way, caging groups that are inefficiently photolyzed can be used, which would not be the case with Laue diffraction, where uncaging with a single short light flash is generally employed [43].

5.1.5.5 Medical Applications

For many years, it has been realized that photoactivatable drugs might have advantages in certain circumstances [49, 135]. Indeed, the light-mediated effects of psoralens were among the earliest drug actions recorded. Proteins have been used to increase therapeutic specificity by carrying drugs such as photosensitizers into cells [136, 137] or onto bacteria [138], but there have been few developments of caged proteins as drugs. Examples include immunotoxins that have been shown to kill cells *in vitro* [55] (Fig. 5.1.7) and a caged hydroxycinnamoyl-thrombin that produces local thrombosis in living cornea after irradiation at 366 nm [139]. The latter might be useful for treating aberrant ocular neovascularization, a major cause of blindness. The infectivity of adenoviral vectors can be controlled by coating the viral particles with a water-soluble, photocleavable reagent [140]. The modified viral vectors possess little infectivity towards target cells. Exposure to 365-nm light induces a reversal of the neutralizing modification and the restoration of infectivity. The light-directed transduction of target cells by photoactivatable adenoviral vectors was demonstrated successfully both *in vitro* and *in vivo*. The modified vectors might find applications in gene therapy. Further advances in the medical application of caged proteins will require attention to a number of issues including the means of administration, binding or uptake by target cells and light penetration into tissues [135].

Acknowledgments

We thank Seong-Ho Shin and Maurice Goeldner for comments on the manuscript. S.L. was supported by a fellowship from the Fondation pour La Recherche Médicale and by the Welch Foundation (BE-1335). H.B. is the holder of a Royal Society–Wolfson Research Merit Award.

Note added in proof

In an excellent example of spatiotemporal control, the phosphocofilin mimick [81] has been microinjected into carcinoma cells and activated within a 3-μm diameter spot [189]. New examples of protein folding on a nanosecond timescale have appeared [190, 191]. As predicted, a caged amino acid has now been incorporated into proteins *in vivo* [192]. In related work *in vitro*, caged amino acids have been introduced at the active site at the dimer interface of an endonuclease by using a 4-base codon [193]. As in the case of HIV protease [60], caging and photoactivation do not affect dimerization. When PAP was used, a substantial increase in endonuclease activity was observed after irradiation [194]. The rate of activation was slower than photoisomerization in solution, suggesting that the azobenzene is constrained on the protein. Indeed, activation was not reversed by irradiation at long wavelengths. Finally, 2-nitrophenylglycine, which causes chain cleavage upon photolysis, has been introduced at the proteolytic processing site of caspase-3 [195]. A method for chemical peptide ligation that leaves a 2-nitrobenzylamide at the splice junction has been devised [196]. The ligation product is caged on the peptide backbone (cf. [91]). Chemical ligation has also been used to incorporate caged phosphoserines [197]. On the technical side, molecular antennae have been used to increase the efficiency of photolysis [198]. Improved sites for the attachment of biarsenicals have been developed [199].

References

1 G. Marriott (ed.), *Methods in Enzymology. Vol. 291. Caged Compounds*, Academic Press, New York, **1998**.
2 A. P. Pelliccioli, J. Wirz, *Photochem. Photobiol. Sci.* **2002**, *1*, 441–458.
3 H. Bayley, C.-Y. Chang, W. T. Miller, B. Niblack, P. Pan, *Methods Enzymol.* **1998**, *291*, 117–135.
4 K. Curley, D. S. Lawrence, *Curr. Opin. Chem. Biol.* **1999**, *3*, 84–88.
5 G. Marriott, P. Roy, K. Jacobson, *Methods Enzymol.* **2003**, *360*, 274–288.
6 E. J. Petersson, G. S. Brandt, N. M. Zacharias, D. A. Dougherty, H. A. Lester, *Methods Enzymol.* **2003**, *360*, 258–273.
7 K. Zou, S. Cheley, R. S. Givens, H. Bayley, *J. Am. Chem. Soc.* **2002**, *124*, 8220–8229.
8 J.-L. Ravanat, T. Douki, J. Cadet, *J. Photochem. Photobiol. B* **2001**, *63*, 88–102.
9 M. J. Davies, *Biochem. Biophys. Res. Commun.* **2003**, *305*, 761–770.
10 T. Milburn, N. Matsubara, A. P. Billington, J. B. Udgaonkar, J. W. Walker, B. K. Carpenter, W. W. Webb, J. Marque, W. Denk, J. A. McCray, G. P. Hess, *Biochemistry* **1989**, *28*, 49–55.
11 J. W. Walker, H. Martin, F. R. Schmitt, R. J. Barsotti, *Biochemistry* **1993**, *32*, 1338–1345.
12 P. Pan, H. Bayley, *FEBS Lett.* **1997**, *405*, 81–85.
13 J. E. T. Corrie, A. Barth, V. R. N. Munasinghe, D. R. Trentham, M. C. Hutter, *J. Am. Chem. Soc.* **2003**, *125*, 8546–8554.
14 Y. V. Il'ichev, M. A. Schwörer, J. Wirz, *J. Am. Chem. Soc.* **2004**, *126*, 4581–4595.
15 T. Furuta, S. S. H. Wang, J. L. Dantzker, T. M. Dore, W. J. Bybee, E. M. Callaway, W. Denk, R. Y. Tsien, *Proc. Natl Acad. Sci. USA* **1999**, *96*, 1193–1200.
16 B. Schade, V. Hagen, R. Schmidt, R. Herbrich, E. Krause, T. Eckardt, J. Bendig, *J. Org. Chem.* **1999**, *64*, 9109–9117.
17 R. S. Givens, C.-H. Park, *Tetrahedron Lett.* **1996**, *37*, 6259–6262.
18 P. G. Conrad, R. S. Givens, J. F. W. Weber, K. Kandler, *Org. Lett.* **2000**, *2*, 1545–1547.
19 H. S. M. Lu, M. Volk, Y. Kholodenko, E. Gooding, R. M. Hochstrasser, W. F. DeGrado, *J. Am. Chem. Soc.* **1997**, *119*, 7173–7180.
20 K. C. Hansen, R. S. Rock, R. W. Larsen, S. I. Chan, *J. Am. Chem. Soc.* **2000**, *122*, 11567–11568.
21 S. Spörlein, H. Carstens, H. Satzger, C. Renner, R. Behrendt, L. Moroder, P. Tavan, W. Zinth, J. Wachtveitl, *Proc. Natl Acad. Sci. USA* **2002**, *99*, 7998–8002.
22 E. Chen, J. R. Kumita, G. A. Woolley, D. S. Kliger, *J. Am. Chem. Soc.* **2003**, *125*, 12443–12449.
23 A. Barth, K. Hauser, W. Mäntele, J. E. T. Corrie, D. R. Trentham, *J. Am. Chem. Soc.* **1995**, *117*, 10311–10316.
24 X. Du, H. Frei, S. H. Kim, *J. Biol. Chem.* **2000**, *275*, 8492–8500.
25 X. Du, H. Frei, S. H. Kim, *Biopolymers* **2001**, *62*, 147–149.
26 Q. Cheng, M. G. Steinmetz, V. Jayaraman, *J. Am. Chem. Soc.* **2002**, *124*, 7676–7677.
27 T. Luchian, S.-H. Shin, H. Bayley, *Angew. Chem. Int. Ed.* **2003**, *42*, 1926–1929.
28 M. Wilcox, R. W. Viola, K. W. Johnson, A. P. Billington, B. K. Carpenter, J. A. McCray, A. P. Guzikowski, G. P. Hess, *J. Org. Chem.* **1990**, *55*, 1585–1589.
29 C. Grewer, J. Jäger, B. K. Carpenter, G. P. Hess, *Biochemistry* **2000**, *39*, 2063–2070.
30 C.-H. Park, R. S. Givens, *J. Am. Chem. Soc.* **1997**, *119*, 2453–2463.
31 K. Zou, W. T. Miller, R. S. Givens, H. Bayley, *Angew. Chem. Int. Ed. Engl.* **2001**, *40*, 3049–3051.
32 J. E. T. Corrie, D. R. Trentham, in: *Biological Applications of Photochemical Switches*, H. Morrison (ed.), Wiley, New York, **1993**, vol. 2, pp. 243–299.
33 P. Zuman, B. Shah, *Chem. Rev.* **1994**, *94*, 1621–1641.
34 A. Barth, J. E. T. Corrie, M. J. Gradwell, Y. Maeda, W. Mantele, T. Meier, D. R. Trentham, *J. Am. Chem. Soc.* **1997**, *119*, 4149–4159.

35 S. K. Pollitt, P. G. Schultz, Angew. Chem. Int. Ed. Engl. **1998**, *37*, 2104–2107.

36 J. E. González, R. Y. Tsien, Chem. Biol. **1997**, *4*, 269–277.

37 K. Briviba, H. Sies, Methods Enzymol. **2000**, *319*, 222–226.

38 S. Beutner, B. Bloedorn, T. Hoffmann, H. D. Martin, Methods Enzymol. **2000**, *319*, 226–241.

39 W. Denk, Proc. Natl Acad. Sci. USA **1994**, *91*, 6629–6633.

40 O. D. Fedoryak, T. M. Dore, Org. Lett. **2002**, *4*, 3419–3422.

41 M. Lu, O. D. Fedoryak, B. R. Moister, T. M. Dore, Org. Lett. **2003**, *5*, 2119–2122.

42 M. P. Goeldner, C. G. Hirth, Proc. Natl Acad. Sci. USA **1980**, *77*, 6439–6442.

43 B. L. Stoddard, B. E. Cohen, M. Brubaker, A. D. Mesecar, D. E. Koshland, Nat. Struct. Biol. **1998**, *5*, 891–897.

44 N. A. Porter, J. W. Thuring, H. Li, J. Am. Chem. Soc. **1999**, *121*, 7716–7717.

45 J. W. Thuring, H. Li, N. A. Porter, Biochemistry **2002**, *41*, 2002–2013.

46 K. Curley, D. S. Lawrence, J. Am. Chem. Soc. **1998**, *120*, 8573–8574.

47 C.-Y. Chang, B. Niblack, B. Walker, H. Bayley, Chem. Biol. **1995**, *2*, 391–400.

48 P. M. England, H. A. Lester, N. Davidson, D. A. Dougherty, Proc. Natl Acad. Sci. USA **1997**, *94*, 11025–11030.

49 C. H. Self, S. Thompson, Nat. Med. **1996**, *2*, 817–820.

50 A. K. Singh, P. K. Khade, Bioconjugate Chem. **2002**, *13*, 1286–1291.

51 A. D. Turner, S. V. Pizzo, G. W. Rozakis, N. A. Porter, J. Am. Chem. Soc. **1987**, *109*, 1274–1275.

52 G. C. Adam, J. Burbaum, J. W. Kozarich, M. P. Patricelli, B. F. Cravatt, J. Am. Chem. Soc. **2004**, *126*, 1363–1368.

53 S. Loudwig, Y. Nicolet, P. Masson, J. C. Fontecilla-Camps, S. Bon, F. Nachon, M. Goeldner, ChemBioChem **2003**, *4*, 762–767.

54 G. Arabaci, X.-C. Guo, K. D. Beebe, K. M. Coggeshall, D. Pei, J. Am. Chem. Soc. **1999**, *121*, 5085–5086.

55 V. S. Goldmacher, P. D. Senter, J. M. Lambert, W. A. Blättler, Bioconjugate Chem. **1992**, *3*, 104–107.

56 J. Olejnik, S. Sonar, E. Krzymañska-Olejnik, K. J. Rothschild, Proc. Natl Acad. Sci. USA **1995**, *92*, 7590–7594.

57 B. Walker, H. Bayley, J. Biol. Chem. **1995**, *270*, 23065–23071.

58 L. Song, M. R. Hobaugh, C. Shustak, S. Cheley, H. Bayley, J. E. Gouaux, Science **1996**, *274*, 1859–1865.

59 R. Olson, H. Nariya, K. Yokota, Y. Kamio, E. Gouaux, Nat. Struct. Biol. **1999**, *6*, 134–140.

60 G. F. Short III, M. Lodder, A. L. Laikhter, T. Arslan, S. M. Hecht, J. Am. Chem. Soc. **1999**, *121*, 478–479.

61 J. Ottl, D. Gabriel, G. Marriott, Bioconjugate Chem. **1998**, *9*, 143–151.

62 T. Muir, Annu. Rev. Biochem. **2003**, *72*, 249–289.

63 D. Clayton, G. Shapovalov, J. Maurer, J. A. Dougherty, H. A. Lester, G. G. Kochendoerfer, Proc. Natl Acad. Sci. USA **2004**, *101*, 4764–4769.

64 F. I. Valiyaveetil, M. Sekedat, T. W. Muir, R. MacKinnon, Angew. Chem. Int. Ed. **2004**, *43*, 2504–2507.

65 C.-Y. Chang, T. Fernandez, R. Panchal, H. Bayley, J. Am. Chem. Soc. **1998**, *120*, 7661–7662.

66 R. Golan, U. Zehavi, M. Naim, A. Patchornik, P. Smirnoff, Biochim. Biophys. Acta **1996**, *1293*, 238–242.

67 E. K. Jaffe, M. Cohn, Biochemistry **1978**, *17*, 652–657.

68 A. P. Arnold, K.-S. Tan, D. L. Rabenstein, Inorg. Chem. **1986**, *25*, 2433–2437.

69 R. G. Pearson, H. Sobel, J. Songstad, J. Am. Chem. Soc. **1968**, *90*, 319–326.

70 R. J. Hondal, B. L. Nilsson, R. T. Raines, J. Am. Chem. Soc. **2001**, *123*, 5140–5141.

71 K. E. Sandman, J. S. Benner, C. J. Noren, J. Am. Chem. Soc. **2000**, *122*, 960–961.

72 M. D. Gieselman, L. Xie, W. A. van der Donk, Org. Lett. **2001**, *3*, 1331–1334.

73 Y. Xu, E. T. Kool, J. Am. Chem. Soc. **2000**, *122*, 9040–9041.

74 B. A. Griffin, S. R. Adams, R. Y. Tsien, Science **1998**, *281*, 269–272.

75 S. R. Adams, R. E. Campbell, L. A. Gross, B. R. Martin, G. K. Walkup, Y. Yao, J. Llopis, R. Y. Tsien, J. Am. Chem. Soc. **2002**, *124*, 6063–6076.

76 J. Zhang, R. E. Campbell, A. Y. Ting, R. Y. Tsien, *Nat. Rev. Mol. Cell Biol.* **2002**, *3*, 906–918.
77 K. Stroffekova, C. Proenza, K. G. Beam, *Pflügers Arch. (Eur. J. Physiol.)* **2001**, *442*, 859–866.
78 H. Kolb, K. B. Sharpless, *Drug Disc. Today* **2003**, *8*, 1128–1137.
79 A. J. Link, D. A. Tirrell, *J. Am. Chem. Soc.* **2003**, *125*, 11164–11165.
80 Z. Zhang, B. A. Smith, L. Wang, A. Brock, C. Cho, P. G. Schultz, *Biochemistry* **2003**, *42*, 6735–6746.
81 M. Ghosh, I. Ichetovkin, X. Song, J. S. Condeelis, D. S. Lawrence, *J. Am. Chem. Soc.* **2002**, *124*, 2440–2441.
82 S. O. Kim, K. Merchant, R. Nudelman, W. F. Beyer, T. Keng, J. DeAngelo, A. Lausladen, J. S. Stamler, *Cell* **2002**, *109*, 383–396.
83 G. Marriott, *Biochemistry* **1994**, *33*, 9092–9097.
84 S. Howorka, L. Movileanu, X. Lu, M. Magnon, S. Cheley, O. Braha, H. Bayley, *J. Am. Chem. Soc.* **2000**, *122*, 2411–2416.
85 B. L. Stoddard, P. Koenigs, N. Porter, K. Petratos, G. A. Petsko, D. Ringe, *Proc. Natl Acad. Sci. USA* **1991**, *88*, 5503–5507.
86 B. L. Stoddard, J. Bruhnke, N. Porter, D. Ringe, G. A. Petsko, *Biochemistry* **1990**, *29*, 4871–4879.
87 B. L. Stoddard, J. Bruhnke, P. Koenigs, N. Porter, D. Ringe, G. A. Petsko, *Biochemistry* **1990**, *29*, 8042–8051.
88 T. Hiraoka, I. Hamachi, *Bioorg. Med. Chem. Lett.* **2003**, *13*, 13–15.
89 J. W. Walker, S. H. Gilbert, R. M. Drummond, M. Yamada, R. Sreekumar, R. E. Carraway, M. Ikebe, F. S. Fay, *Proc. Natl Acad. Sci. USA* **1998**, *95*, 1568–1573.
90 Y. Tatsu, Y. Shigeri, A. Ishida, I. Kameshita, H. Fujisawa, N. Yumoto, *Bioorg. Med. Chem. Lett.* **1999**, *9*, 1093–1096.
91 Y. Tatsu, T. Nishigaki, A. Darszon, N. Yumoto, *FEBS Lett.* **2002**, *525*, 20–24.
92 T. Ueda, K. Murayama, T. Yamamoto, S. Kimura, Y. Imanishi, *J. Chem. Soc. Perkin Trans. I* **1994**, 225–230.
93 D. Mendel, J. A. Ellman, P. G. Schultz, *J. Am. Chem. Soc.* **1991**, *113*, 2758–2760.
94 D. L. Beene, D. A. Dougherty, H. A. Lester, *Curr. Opin. Neurobiol.* **2003**, *13*, 264–270.
95 S. N. Cook, W. E. Jack, X. Xiong, L. E. Danley, J. A. Ellman, P. G. Schultz, C. J. Noren, *Angew. Chem. Int. Ed. Engl.* **1995**, *34*, 1629–1630.
96 P. M. England, Y. Zhang, D. A. Dougherty, H. A. Lester, *Cell* **1999**, *96*, 89–98.
97 T. Hohsaka, M. Sisido, *Curr. Opin. Chem. Biol.* **2002**, *6*, 809–815.
98 L. Wang, P. G. Schultz, *Chem. Commun.* **2002**, 1–11.
99 M. W. Nowak, P. C. Kearney, J. R. Sampson, M. E. Saks, C. G. Labarca, S. K. Silverman, W. Zhong, J. Thorson, J. N. Abelson, N. Davidson, P. G. Schultz, D. A. Dougherty, H. A. Lester, *Science* **1995**, *268*, 439–442.
100 S. L. Monahan, H. A. Lester, D. A. Dougherty, *Chem. Biol.* **2003**, *10*, 573–580.
101 L. Wang, A. Brock, B. Herberich, P. G. Schultz, *Science* **2001**, *292*, 498–500.
102 R. A. Mehl, J. C. Anderson, S. W. Santoro, L. Wang, A. B. Martin, D. S. King, D. M. Horn, P. G. Schultz, *J. Am. Chem. Soc.* **2003**, *125*, 935–939.
103 B. F. Erlanger, *Annu. Rev. Biochem.* **1976**, *45*, 267–283.
104 D. H. Hug, *Photochem. Photobiol. Rev.* **1978**, *3*, 1–33.
105 G. Montagnoli, S. Monti, L. Nannicini, M. P. Giovannitti, M. G. Ristori, *Photochem. Photobiol.* **1978**, *27*, 43–49.
106 N. A. Porter, J. D. Bruhnke, P. A. Koenigs, in: *Biological Applications of Photochemical Switches*, H. Morrison (ed.), Wiley, New York, **1993**, vol. 2, 197–241.
107 I. Willner, R. Shai, *Angew. Chem. Int. Ed. Engl.* **1996**, *35*, 367–385.
108 O. Pieroni, A. Fissi, N. Angelini, F. Lenci, *Acc. Chem. Res.* **2001**, *34*, 9–17.
109 H. A. Lester, M. E. Krouse, M. M. Nass, J. M. Nerbonne, N. H. Wassermann, B. F. Erlanger, *J. Gen. Physiol.* **1980**, *75*, 207–232.
110 D. A. James, D. C. Burns, G. A. Woolley, *Protein Eng.* **2001**, *14*, 983–991.

111 N. Muranaka, T. Hohsaka, M. Sisido, *FEBS Lett.* **2002**, *510*, 10–12.

112 A. M. Caamaño, M. E. Vázquez, J. Martínez-Costas, L. Castedo, J. L. Mascareñas, *Angew. Chem. Int. Ed.* **2000**, *39*, 3104–3107.

113 J. R. Kumita, O. S. Smart, G. A. Woolley, *Proc. Natl Acad. Sci. USA* **2000**, *97*, 3803–3808.

114 D. G. Flint, J. R. Kumita, O. S. Smart, G. A. Woolley, *Chem. Biol.* **2002**, *9*, 391–397.

115 J. R. Kumita, D. G. Flint, O. S. Smart, G. A. Woolley, *Protein Eng.* **2002**, *15*, 561–569.

116 Z. Zhang, D. C. Burns, J. R. Kumita, O. S. Smart, G. A. Woolley, *Bioconjugate Chem.* **2003**, *14*, 824–829.

117 T. Shimoboji, E. Larenas, T. Fowler, A. S. Hoffman, P. S. Stayton, *Bioconjugate Chem.* **2003**, *14*, 517–525.

118 T. Shimoboji, Z. L. Ding, P. S. Stayton, A. S. Hoffman, *Bioconjugate Chem.* **2002**, *13*, 915–919.

119 C. G. Bochet, *Angew. Chem. Int. Ed.* **2001**, *40*, 2071–2073.

120 A. Blanc, C. G. Bochet, *J. Org. Chem.* **2002**, *67*, 5567–5577.

121 S. Shimizu-Sato, E. Huq, J. M. Tepperman, P. H. Quail, *Nat. Biotechnol.* **2002**, *20*, 1041–1044.

122 G. H. Patterson, J. Lippincott-Schwartz, *Science* **2002**, *297*, 1873–1877.

123 G. H. Patterson, J. Lippincott-Schwartz, *Methods* **2004**, *32*, 445–450.

124 D. G. Jay, *Proc. Natl Acad. Sci. USA* **1988**, *85*, 5454–5458.

125 K. G. Linden, J. C. Liao, D. G. Jay, *Biophys. J.* **1992**, *61*, 956–962.

126 J. C. Liao, J. Roider, D. G. Jay, *Proc. Natl Acad. Sci. USA* **1994**, *91*, 2659–2663.

127 O. Tour, R. M. Meijer, D. A. Zacharias, S. R. Adams, R. Y. Tsien, *Nat. Biotechnol.* **2003**, *21*, 1505–1508.

128 R. Dorn, S. Quabis, G. Leuchs, *Phys. Rev. Lett.* **2003**, *91*, 233901-1–233901-4.

129 D. L. Pettit, S. S. H. Wang, K. R. Gee, G. J. Augustine, *Neuron* **1997**, *19*, 465–471.

130 K. Hamad-Schifferli, J. J. Schwartz, A. T. Santos, S. Zhang, J. M. Jacobson, *Nature* **2002**, *415*, 152–155.

131 J. Ching, K. I. Voivodov, T. W. Hutchens, *Bioconjugate Chem.* **1996**, *7*, 525–528.

132 M. A. Holden, P. S. Cremer, *J. Am. Chem. Soc.* **2003**, *125*, 8074–8075.

133 A. Specht, T. Ursby, M. Weik, L. Peng, J. Kroon, D. Bourgeois, M. Goeldner, *ChemBioChem* **2001**, *2*, 845–848.

134 T. Ursby, M. Weik, E. Fioravanti, M. Delarue, M. Goeldner, D. Bourgeois, *Acta Crystallogr. D* **2002**, *58*, 607–614.

135 H. Bayley, F. Gasparro, R. Edelson, *Trends Pharmacol. Sci.* **1987**, *8*, 138–143.

136 S. Yemul, C. Berger, A. Estabrook, S. Suarez, R. Edelson, H. Bayley, *Proc. Natl Acad. Sci. USA* **1987**, *84*, 246–250.

137 R. A. Firestone, *Bioconjugate Chem.* **1994**, *5*, 105–113.

138 F. Berthiaume, S. R. Reiken, M. Toner, R. G. Tompkins, M. L. Yarmush, *Biotechnology* **1994**, *12*, 703–706.

139 J. G. Arroyo, P. B. Jones, N. A. Porter, D. L. Hatchell, *Thromb. Haemostasis* **1997**, *78*, 791–793.

140 M. W. Pandori, D. A. Hobson, J. Olejnik., E. Krzymanska-Olejnik, K. J. Rothschild, A. A. Palmer, T. J. Phillips, T. Sano, *Chem. Biol.* **2002**, *9*, 567–573.

141 J. A. Barltrop, P. J. Plant, P. Schofield, *Chem. Commun.* **1966**, 822–823.

142 J. C. Miller, S. K. Silverman, P. M. England, D. A. Dougherty, H. A. Lester, *Neuron* **1998**, *20*, 619–624.

143 Y. Tong, G. S. Brandt, M. Li, G. Shapovalov, E. Slimko, A. Karschin, D. A. Dougherty, H. A. Lester, *J. Gen. Physiol.* **2001**, *117*, 103–118.

144 J. W. Walker, J. A. McCray, G. P. Hess, *Biochemistry* **1986**, *25*, 1799–1805.

145 A. Specht, M. Goeldner, *Angew. Chem. Int. Ed.* **2004**, *43*, 2008–2012.

146 J. M. Nerbonne, S. Richard, J. Nargeot, H. A. Lester, *Nature* **1984**, *310*, 74–76.

147 J. S. Wood, M. Koszelak, J. Liu, D. S. Lawrence, *J. Am. Chem. Soc.* **1998**, *120*, 7145–7146.

148 A. C. Pease, D. Solas, E. J. Sullivan, M. T. Cronin, C. P. Holmes, S. P. Fodor, *Proc. Natl Acad. Sci. USA* **1994**, *91*, 5022–5026.

149 J. Hebert, D. Gravel, *Can. J. Chem.* **1974**, *52*, 187–189.

150 D. Gravel, J. Hebert, D. Thoraval, *Can. J. Chem.* **1983**, *61*, 400–410.

151 M. J. Aurell, C. Boix, M. L. Ceita, C. Llopis, A. Tortajada, R. Mestres, *J. Chem. Res.* **1995**, *11*, 452–453.

152 L. Ceita, A. K. Maiti, R. Mestres, A. Tortajada, *J. Chem. Res.* **2001**, *10*, 403–413.

153 A. Blanc, C. G. Bochet, *J. Org. Chem.* **2003**, *68*, 1138–1141.

154 S. Walbert, W. Pfleiderer, U. E. Steiner, *Helv. Chim. Acta* **2001**, *84*, 1601–1611.

155 T. J. Albert, J. Norton, M. Ott, T. Richmond, K. Nuwaysir, E. F. Nuwaysir, K.-P. Stengele, R. D. Green, *Nucleic Acids Res.* **2003**, *31*, e35.

156 B. Amit, D. A. Ben-Efraim, A. Patchornik, *J. Am. Chem. Soc.* **1976**, *98*, 843–844.

157 G. Papageorgiou, J. E. T. Corrie, *Tetrahedron* **2000**, *56*, 8197–8205.

158 M. Matsuzaki, G. C. R. Ellis-Davies, T. Nemoto, Y. Miyashita, M. Lino, H. Kasai, *Nat. Neurosci.* **2001**, *4*, 1086–1092.

159 S. Loudwig, M. Goeldner, *Tetrahedron Lett.* **2001**, *42*, 7957–7959.

160 J. C. Sheehan, R. M. Wilson, *J. Am. Chem. Soc.* **1964**, *86*, 5277–5281.

161 J. C. Sheehan, R. M. Wilson, A. W. Oxford, *J. Am. Chem. Soc.* **1971**, *93*, 7222–7228.

162 R. S. Givens, B. Matuszewski, *J. Am. Chem. Soc.* **1984**, *106*, 6860–6861.

163 M. C. Pirrung, J. C. Bradley, *J. Org. Chem.* **1995**, *60*, 1116–1117.

164 M. C. Pirrung, C.-Y. Huang, *Tetrahedron Lett.* **1995**, *36*, 5883–5884.

165 J. F. Cameron, C. G. Willson, J. M. J. Frechet, *J. Am. Chem. Soc.* **1996**, *118*, 12925–12937.

166 C. S. Rajesh, R. S. Givens, J. Wirz, *J. Am. Chem. Soc.* **2000**, *122*, 611–618.

167 J. C. Sheehan, K. Umezawa, *J. Org. Chem.* **1973**, *38*, 3771–3774.

168 R. S. Givens, J. F. W. Weber, A. H. Jung, C.-H. Park, *Methods Enzymol.* **1998**, *291*, 1–29.

169 A. Specht, S. Loudwig, L. Peng, M. Goeldner, *Tetrahedron Lett.* **2002**, *43*, 8947–8950.

170 T. Furuta, M. Iwamura, *Methods Enzymol.* **1998**, *291*, 50–63.

171 T. Furuta, H. Torigai, M. Sugimoto, M. Iwamura, *J. Org. Chem.* **1995**, *60*, 3953–3956.

172 V. Hagen, J. Bendig, S. Frings, T. Eckardt, S. Helm, D. Reuter, U. B. Kaupp, *Angew. Chem. Int. Ed. Engl.* **2001**, *40*, 1045–1048.

173 D. Geissler, W. Kresse, B. Wiesner, J. Bendig, H. Kettenman, V. Hagen, *ChemBioChem* **2003**, *4*, 162–170.

174 T. Furuta, Y. Hirayama, M. Iwamura, *Org. Lett.* **2001**, *3*, 1809–1812.

175 A. Banerjee, C. Grewer, L. Ramakrishnan, J. Jager, A. Gameiro, H.-G. A. Breitinger, K. R. Gee, B. K. Carpenter, G. P. Hess, *J. Org. Chem.* **2003**, *68*, 8361–8367.

176 D. Ramesh, R. Wieboldt, L. Niu, B. K. Carpenter, G. P. Hess, *Proc. Natl Acad. Sci. USA* **1993**, *90*, 11074–11078.

177 P. M. Koenigs, B. C. Faust, N. A. Porter, *J. Am. Chem. Soc.* **1993**, *115*, 9371–9379.

178 P. B. Jones, M. P. Pollastri, N. A. Porter, *J. Org. Chem.* **1996**, *61*, 9455–9461.

179 P. B. Jones, N. A. Porter, *J. Am. Chem. Soc.* **1999**, *121*, 2753–2761.

180 A. D. Turner, S. V. Pizzo, G. Rozakis, N. A. Porter, *J. Am. Chem. Soc.* **1988**, *110*, 244–250.

181 N. A. Porter, J. D. Bruhnke, *Photochem. Photobiol.* **1990**, *51*, 37–43.

182 D. A. Pratt, D. F. Underwood, S. J. Rosenthal, N. A. Porter, *Abstr. 221st Am. Chem. Soc.* **2001**, ORGN-703.

183 A. H. Kossel, S. B. Cambridge, U. Wagner, T. Bonhoeffer, *Proc. Natl Acad. Sci. USA* **2001**, *98*, 14702–14707.

184 P. Roy, Z. Rajfur, D. Jones, G. Marriott, L. Loew, K. Jacobson, *J. Cell. Biol.* **2001**, *153*, 1035–1048.

185 G. Marriott, M. Heidecker, *Biochemistry* **1996**, *35*, 3170–3174.

186 J. Minden, R. Namba, J. Mergliano, S. Cambridge, *Sci. STKE.* **2000**, *62*, PL1.

187 S. B. Cambridge, R. L. Davis, J. S. Minden, *Science* **1997**, *277*, 825–828.

188 K. D. Philipson, J. P. Gallivan, G. S. Brandt, D. A. Dougherty, H. A. Lester, *Am. J. Physiol. Cell Physiol.* **2001**, *281*, 195–206.

189 M. Ghosh X. Song, G. Mouneimne, M. Sidani, D. S. Lawrence, J. S. Condeelis, *Science* **2004**, *304*, 743–746.

190 R.S. Rock, K.C. Hansen, R.W. Larsen, S.I. Chan, *Chem. Phys.* **2004**, *307*, 201–208.
191 R.P.Y. Chen, J.J.T. Huang, H.-L. Chen, H. Jan, M. Velusamy, C.-T. Lee, W. Fann, R.W. Larsen, S.I. Chan, *Proc. Natl. Acad. Sci. USA* **2004**, *101*, 7305–7310.
192 N. Wu, A. Deiters, T.A. Cropp, D. King, P.G. Schultz, *J. Am. Chem. Soc.* **2004**, *126*, 14306–14307.
193 M. Endo, K. Nakayama, T. Majima, *J. Org. Chem.* **2004**, *69*, 4292–4298.
194 K. Nakayama, M. Endo, T. Majima, *Chem.Commun.* **2004**, in press.
195 M. Endo, K. Nakayama, Y. Kaida, T. Majima, *Angew. Chem. Int. Ed.* **2004**, *43*, 5643–5645.
196 C. Marinzi, J. Offer, R. Longhi, P.E. Dawson, *Bioorg. Med. Chem.* **2004**, *12*, 2749–2757.
197 M.E. Hahn, T.W. Muir, *Angew. Chem. Int. Ed.* **2004**, *43*, 5800–5803.
198 G. Papageorgiou, D. Ogden, J.E.T. Corrie, *J. Org. Chem.* **2004**, *69*, 7228–7233.
199 S.R. Adams, R.E. Campbell, L.A. Gross, B.R. Martin, G.K. Walkup, Y. Yao, J. Llopis, R.Y. Tsien, *J. Am. Chem. Soc.* **2002**, *124*, 6063–6076.

5.2
Photochemical Enzyme Regulation using Caged Enzyme Modulators

Ling Peng and Maurice Goeldner

5.2.1
Introduction

Enzymes are important macromolecules that ensure numerous biological events such as cell metabolism, signal transduction, proliferation, differentiation, etc., and enzyme function is regulated by a number of different modulators, e.g. substrates, products, inhibitors and/or coenzymes. Thus, understanding the functioning of enzymes is important for us to study related biological events. Enzyme catalysis is a dynamic process that involves, successively, binding and breakdown of substrate with or without the help of a coenzyme, formation of product, as well as release of product. Enzymatic function can also be modulated with inhibitors that act at the active or a secondary-binding site. Therefore, photochemical enzyme regulation using caged enzyme modulators is of special interest, allowing control of the catalytic reaction. These caged modulators have a photoremovable protecting group which renders them biologically inactive either by transforming them to an inert chemical or by converting them to an enzyme inhibitor. Upon photoactivation, the caged compounds release the corresponding enzyme substrate, product, coenzyme or inhibitor and switch either on or off the enzymatic reaction. Thus, they are useful tools to photoregulate the enzyme activity and to undertake dynamic studies on their catalytic mechanism.

This chapter will cover different examples of photochemical enzyme regulation by caged enzyme modulators, including caged substrates, products, coenzymes or inhibitors. Our chapter will emphasize the functional regulation of the enzyme rather than the chemical and photochemical properties of some of these molecules, aspects that will be described in detail in other chapters of this volume. Also, this chapter will not include the use of irreversible inhibitors targeted for protein caging, a feature which will be described in separate chapters (see Chapters 5.1 and 5.3).

5.2.2
Photoregulation of ATPases

5.2.2.1 ATPases using Caged ATP Molecules

5.2.2.1.1 Na,K-ATPase
The first application of a caged compound in the photoregulation of enzyme activity was pioneered 1978 with the groundbreaking work of Kaplan et al. [1] using caged ATP (Scheme 5.2.1) to control the activity on the Na,K-ATPase. The caging group used was a 2-nitrobenzyl group, a frequently used protecting group in organic synthesis [2]. Both the P^3-2-nitrobenzyl ATP (1, NB-ATP) and the P^3-1-(2-nitrophenyl) ethyl ATP (2, NPE-ATP), upon irradiation at 340 nm, released ATP, with a better photolytic efficiency for the NPE-caged ATP (2) [3]. The mechanism and kinetics of the photochemical reaction of the caged ATP molecules have been examined in details [4]. Prior to irradiation, neither 1 nor 2 was a substrate nor an inhibitor of purified renal Na,K-ATPase. After photolysis, ATP was released from 2 and hydrolysed by Na,K-ATPase. The byproduct of the photoreaction, 2-nitroso acetophenone, produced inhibition of the enzymatic hydrolysis of ATP, which could be prevented by using scavengers such as glutathione or bisulfite in the irradiated enzyme solution. Further, caged ATP was incorporated into resealed human erythrocyte ghosts prepared from red blood cells depleted of internal energy stores [4]. While the Na,K pump was unable to use incorporated caged ATP molecules as a substrate, the ATP liberated by photolysis activated the pump as evidenced by measurements of K^+-dependent, ouabain-sensitive Na^+ efflux.

5.2.2.1.2 Transport ATPases and Carriers
A thorough analysis of the mechanism of ion transport across membranes using caged substrates for the chemical energy producing enzymes have been described in details by the groups of Fendler and Bamberg, and has been summarized recently [5]. These studies used the NPE-ATP (2) to photoregulate the different ATPases involved in ion transport processes: Na,K-ATPase, Ca-ATPase, Kdp-ATPase and H,K-ATPase, while a caged Ca^{2+} molecule was used to investigate the Na,Ca exchanger. More recently, these groups described in detail the mechanism of functioning of the Kdp system, a bacterial P-type ATPase of *Escherichia coli* that transports K^+ with high affinity [6]. Upon rapid release of ATP from 2, a transient current occurred in the absence of K^+, similar to the stationary current seen in the presence of K^+. Based on these studies, a kinetic model for the Kdp-ATPase similar to that of the Na,K-ATPase was proposed: (1) the K-independent

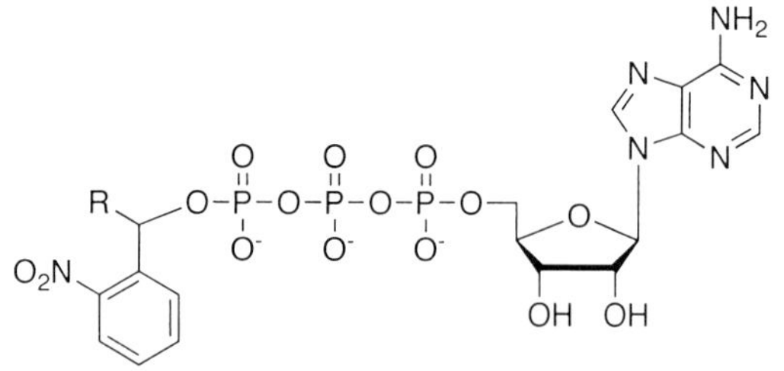

1: NB-ATP, R = H
2: NPE-ATP, R = CH₃

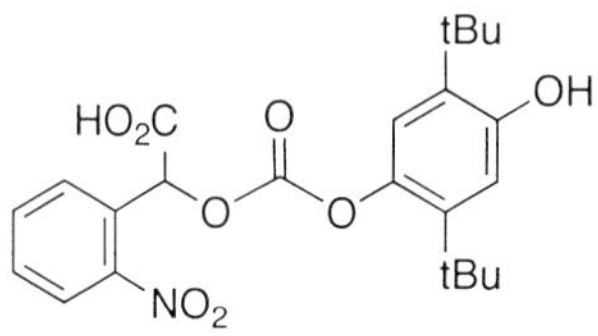

3: Nmoc-DBHQ

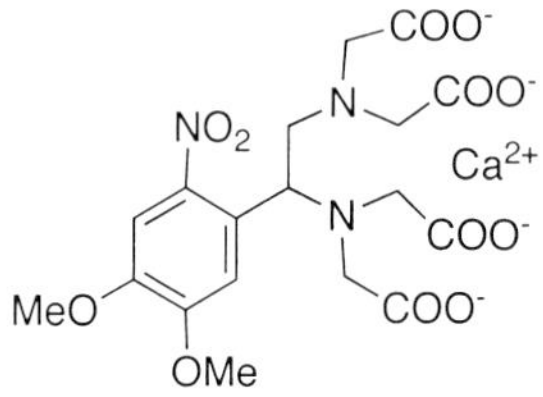

4: DM-N Ca

Scheme 5.2.1 Caged ATPase modulators: caged ATP (**1** and **2**); caged SERCA inhibitor (**3**); caged Ca^{2+}.

step is electrogenic and corresponds to the outward transport of a negative charge; and (2) the K-translocating step, probably also electrogenic, corresponds to transport of positive charge to the intracellular side of the protein [6].

5.2.2.1.3 Molecular Motors

Cell motility includes contraction of skeletal, cardiac and smooth muscles, and other motion. The motor proteins that are responsible for these essential functions are members of three superfamilies: kinesins, myosins and dyneins. They convert the chemical energy of ATP hydrolysis into mechanical work. The elementary mechanical steps like force generation should be related to chemical reactions such as ATP binding, and P_i and ADP release. Combining the measurement of force produced by kinesin or myosin with photolysis of caged compounds provides a powerful approach to study the mechano-chemical coupling of the molecular motors, which has been summarized recently by Dantzig et al. [7]. In order to investigate the relations between transient of force by single kinesin molecule and the elementary steps of the ATPase cycle [8], Higuchi et al. measured the time to force generation by kinesin after photorelease of ATP from **2**. The kinetics of force generation was consistent with a two-step reaction: ATP binding followed by force generation. The transient rate of force generation was close to the rate of the ATPase cycle in solution, suggesting that the rate-limiting step of the ATPase cycle was involved with the force generation.

5.2.2.2 Ca^{2+}-ATPase

The Ca^{2+}-ATPase of sarcoplasmic reticulum is a membrane protein which performs the active transport of Ca^{2+} from the cytoplasm of muscle cells into the sarcoplasmic reticulum, through hydrolysis of ATP [9]. It belongs to the P-type ATPase family and serves as a model for the study of these enzymes. A reaction scheme has been proposed for the Ca^{2+}-ATPase of sarcoplasmic reticulum, suggesting a change from a high Ca^{2+} affinity, ATP-phosphorylated enzyme state to a low Ca^{2+} affinity, P$_i$-phosphorylated state [9]. Details of the transport mechanism, especially the coupling between ATP hydrolysis and active Ca^{2+} transport, are unknown. Thus, different caged modulators, combined with time-resolved techniques, represent a powerful methodology to study the catalytic reaction of Ca^{2+}-ATPase.

5.2.2.2.1 Ca^{2+}-ATPase using Caged ATP

Mäntele et al. used caged ATP (**2**) to photoregulate the activity of the Ca^{2+}-ATPase and studied the subsequent reactions by Fourier transform IR (FTIR) spectroscopy [10]. Photolytic release of ATP from **2** started the catalytic cycle of Ca^{2+}-ATPase, permitting a sensitive probing of the molecular process such as ATP hydrolysis and formation of the phosphorylated enzyme during catalytic activity of Ca^{2+}-ATPase by FTIR spectroscopic studies. Their results suggested that the active Ca^{2+}-ATPase was the phosphorylated form with two Ca^{2+} ions. Further, Lewis and Thomas [11] undertook studies on microsecond rotational dynamics of Ca^{2+}-ATPase during enzymatic cycling initiated by photolysis of caged ATP **2**. The Ca^{2+}-ATPase was selectively labeled with an iodoacetamide spin label, which yielded electron paramagnetic resonance (EPR)-sensitive conformational changes of enzyme during ATP-induced enzymatic cycling. Their results indicated that substantial changes in protein mobility did not occur during the Ca^{2+}-ATPase cycle unless they occurred very early in the transient phase or at some later point in the cycle that was not significantly populated in the steady state. Further experiments [12] showed that transient changes in the amplitude of the restricted component associated with the pre-steady state of Ca^{2+} pumping were detected with 10-ms time resolution after an ATP jump produced by laser flash photolysis of **2**. More light would be shed to this question with caged ATP molecules displaying higher time resolution or with different caged compounds such as caged phosphate, caged calcium, etc.

5.2.2.2.2 Ca^{2+}-ATPase using Caged Inhibitor

The family of sarcoplasmic/endoplasmic reticulum Ca^{2+}-ATPases (SERCA) that transport Ca^{2+} into the sarcoplasmic reticulum and endoplasmic reticulum are important regulators of cytosolic free Ca^{2+} levels. The effect of these pumps on Ca^{2+} oscillations has been investigated by the use of a caged SERCA inhibitor: Nmoc-DBHQ (**3**, Scheme 5.2.1) [13]. DBHQ is a membrane-permeant, reversible inhibitor of SERCA. Photolysis of Nmoc-DBHQ is rapid and the subsequent release of DBHQ takes around 5 ms. The rate-limiting step in the release of the DBHQ inhibitor is the decarboxylation reaction of the photolytic carbonate intermediate. Nmoc-DBHQ was used for photomodulating the SERCA activity and for probing Ca^{2+} signaling dynamics such as $[Ca^{2+}]$ oscillations in fibroblasts, which were monitored through the Ca^{2+}-sensitive fluorescence of fluo-3 [13].

5.2.2.2.3 Ca^{2+}-ATPase using Caged Ca^{2+}

Using caged Ca^{2+}, DM-nitrophen Ca (4, Scheme 5.2.1), Troullier et al. [14] studied the calcium binding at the high-affinity transport sites of sarcoplasmic reticulum Ca^{2+}-ATPase through time-resolved FTIR spectroscopy. DM-nitrophen is a calcium chelator consisting of a photolabile group linked to an EDTA molecule, having a high affinity for calcium (5 nM at pH 7.1) [15]. Upon photolysis, the EDTA moiety of DM-nitrophen is cleaved into two parts and the affinity of the photoproducts for calcium (3 mM at pH 7.1) is dramatically reduced [15]. Thus, DM-nitrophen Ca was used to trigger the efficient release of calcium in the IR samples. FTIR analyses indicated that glutamic and/or aspartic side-chains were deprotonated upon calcium binding, while main-chain carbonyl groups could also play a role in calcium binding [15].

5.2.3
Photoregulation of GTPases: Ras Protein

The guanine nucleotide-binding protein Ras [16] plays a central role in cellular processes acting as a switch between an active GTP-bound and an inactive GDP-bound form. Such a switch mechanism is found in pathways involved in cell proliferation, signal transduction, differentiation, protein synthesis and protein transport. The intrinsic GTP hydrolysis rate by Ras is rather low and it is greatly enhanced by GTPase-activating proteins (GAPs). Therefore, understanding the detailed GTPase mechanism of Ras might provide a paradigm for many similar processes. Furthermore, the molecular GTPase mechanism of Ras is particularly relevant since oncogenic Ras mutants appear to be involved in 25–30% of all human tumors.

5.2.3.1 Time-resolved Crystallographic Studies with Caged GTP

Crystals of Ha-Ras p21 with NPE-GTP (5, Scheme 5.2.2) at the active site [17] have been used to study the conformational change of p21 on GTP hydrolysis. Upon photolysis, GTP was released from the caged compound and the structure of the short-lived p21–GTP complex was determined by Laue diffraction methods, although the structural resolution was low [18]. The binding of GTP in p21 was not significantly affected by the caging group, when the optically pure caged GTP (Scheme 5.2.2) was employed [19]. The resolution was, however, significantly improved when the crystals were frozen for monochromatic X-ray data collection [20]. The structural analyses showed local events around the γ-phosphate group of GTP played an important role in GTP hydrolysis.

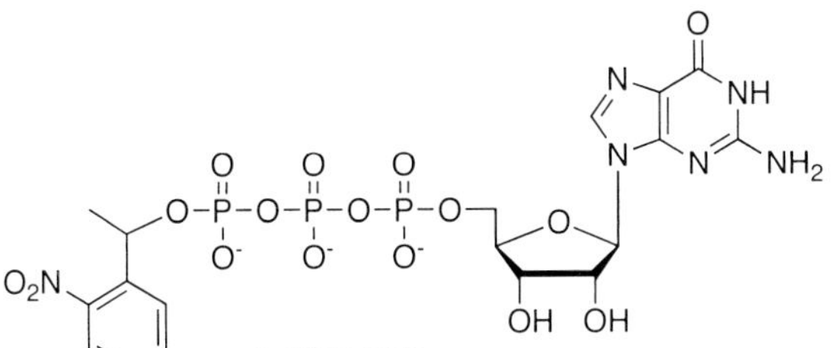

Scheme 5.2.2 Caged GTP.

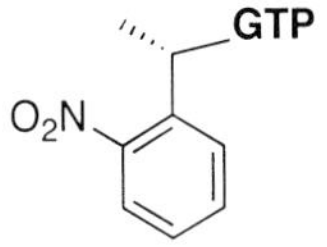

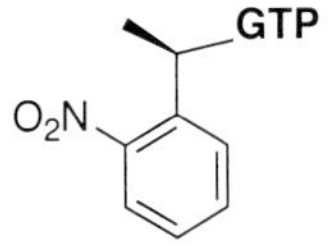

5.2.3.2 Time-resolved FTIR Studies with Caged GTP

The group of Gerwert has employed the technique of time-resolved FTIR difference spectroscopy (see Chapters 7.1 and 7.2) to study, in solution, the GTPase reaction of Ras, at atomic resolution, in a time-resolved mode [21]. The phosphate vibration was analyzed using site specifically ^{18}O-labeled caged GTP isotopomers [22]. One non-bridging oxygen per nucleotide was replaced for an ^{18}O isotope in the α, β or γ position of the phosphate chain. In photolysis experiments with free caged GTP, strong vibrational couplings were observed among all phosphate groups. The investigation of the photolysed [Ras·(NPE-GTP)] complex and the subsequent hydrolysis reaction of Ras·GTP showed that the phosphate vibrations are largely decoupled by interaction with the protein in contrast to free GTP. The unusual low frequency of the $\beta(PO_2^-)$ vibration of Ras-bound GTP, as compared to free GTP, indicated a large decrease in the P–O bond order. This revealed that the oxygen atoms of $\beta(PO_2^-)$ group interact much more strongly with the protein than the γ-oxygen atoms, leading to partial bond breakage or at least weakening the bond between the β/γ bridging oxygen, which preprograms the hydrolysis of GTP.

Using FTIR technique, Du et al. [23] found also stronger binding of the GDP moiety by the Ras mutant with higher activity, suggesting that the transition state is largely GDP-like. They analyzed the photolysis and hydrolysis FTIR spectra of the [β-non-bridging-^{18}O$_2$, α/β-non-bridging-^{18}O]GTP isotopomer to probe for positional isotope exchange. However, no positional isotope exchange was observed during GTP hydrolysis by Ras. Thus, they proposed a concerted mechanism with the transition state having a considerable amount of dissociative character.

Further, the mechanism of the Ras GTPase reaction, in the presence of catalytic amounts of GAP, was also probed by time-resolved FTIR using caged GTP (**5**) as a photolabile trigger [24]. This approach provided the complete GTPase reaction pathway at the atomic level with milliseconds time resolution. The addition of GAP reduced the measuring time by two orders of magnitude, which improved significantly the quality of the data [24]. The spectral data indicated that binding to Ras forced the flexible GTP molecule into a strained conformation and induced a specific charge distribution different from that in the unbound case. Binding of GAP to Ras·GTP shifted the negative charge from γ- to β-phosphate. Such a shift was already identified by FTIR because of Ras binding and was enhanced by GAP binding. Thus, the charge distribution of the GAP–Ras–GTP complex resembled dissociative-like transition state and was more like that in GDP. An intermediate was observed on the reaction pathway that appeared when the bond between β- and γ-phosphate was cleaved. In the intermediate, the released P_i was strongly bound to the protein and surprisingly showed bands typical of those seen for phosphorylated enzyme intermediates. The release of P_i from the protein complex is the rate-limiting step for the GAP-catalyzed reaction. These results provided a mechanistic picture that is different from the intrinsic GTPase reaction of Ras [25].

5.2.4
Photoregulation of Kinases

Kinases catalyze the transfer of a phosphate group from ATP to an acceptor including amino acids (serine, threonine, tyrosine and histidine), nucleotides (mono- and diphosphates) and glucides. This action can be reversed by phosphatases, and the activity of the kinases is closely regulated by intracellular signals such as the concentration of cAMP and of Ca^{2+}. Among the different protein kinases, PKA, activated by cAMP, and PKC, activated by Ca^{2+} and diacylglycerol (DAG), represent important regulatory enzymes for signal transduction pathways. Interestingly, as early as 1977, even earlier to the work reported by Kaplan et al. on caged ATP, Engels and Schlaeger [26] reported the synthesis of adenosine cyclic 3′,5′-phosphate-2-nitrobenzyl triester and their characterization as photolabile precursor of cAMP as demonstrated by the photochemical regulation of the activity of protein kinase. Caged cAMP with a 2-nitrobenzyl caging group was synthesized by treatment of the free acid of cAMP with diazo-2-nitrotoluene leading to two diastereomeric isomers of 2NB-caged cAMP, axial (**6**) and equatorial (**7**), respectively (Scheme 5.2.3). Clearly this pioneering work has opened the way to the use of caged regulators of signaling pathways [27].

5.2.4.1 cAMP-dependent Protein Kinase
cAMP is a second messenger which plays an important role in regulating a variety of cellular processes in eukaryotic organisms. Its intracellular level is regulated by two enzymes of opposite action, i.e. synthesis versus degradation, cata-

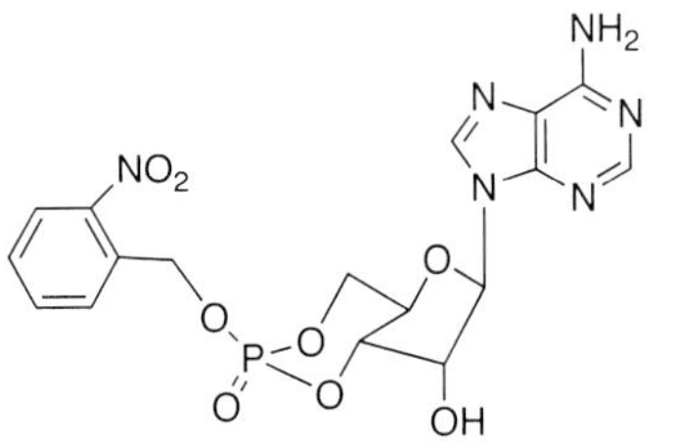

Scheme 5.2.3 Caged PKA modulators: caged cAMP (6–8); caged PKA peptide inhibitor (9).

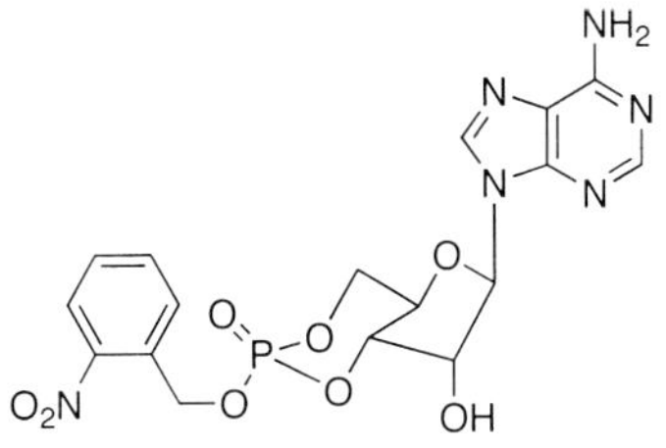

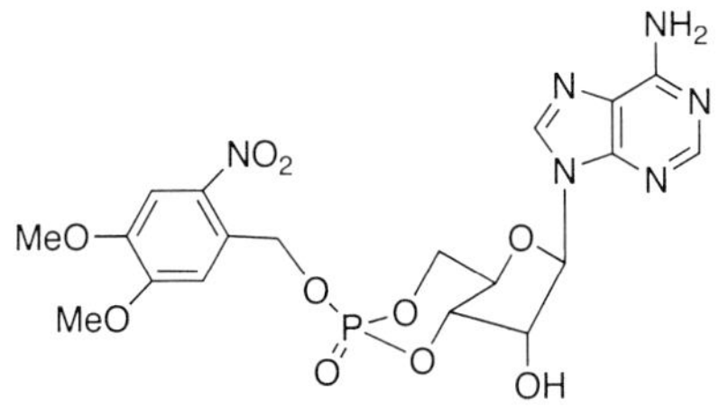

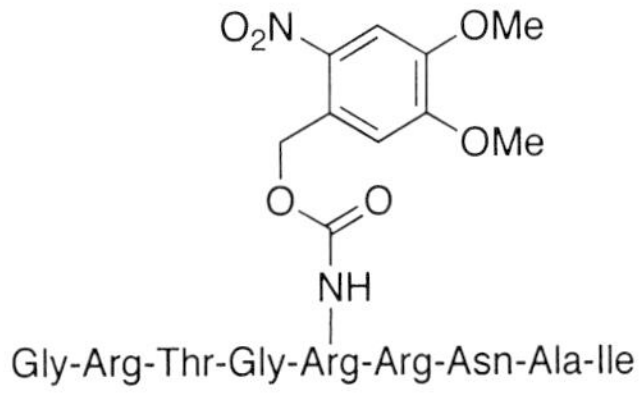

lyzed by adenylate cyclase and phosphodiesterase, respectively. cAMP activates specific cellular enzymes called cAMP-dependent protein kinases which catalyze the transfer of a phosphate group from ATP to a specific serine or threonine. Due to the cellular importance of cAMP, it is not surprising that this second messenger has been targeted for caging studies to allow a spatiotemporal control of its liberation in living cells is developed in detail in Chapter 4.1.

5.2.4.1.1 PKA using Caged cAMP

As already mentioned, the photochemical activation of PKA using caged cAMP molecules **6** or **7** was demonstrated *in vitro* on purified bovine brain cAMP-dependent protein kinase [26]. Another example assessed the role of cAMP and PKA in 11,12-epoxyeicosatrienoic acid-induced changes in endothelial cell coupling. This was demonstrated by a similar effect observed with forskolin (an adenylate cyclase activator), PKA activators or the photochemical release of cAMP from DMNB-cAMP (**8**, Scheme 5.2.3) on the enhanced endothelial cell coupling [28]. Finally, the cAMP-dependent regulation of intracellular Ca^{2+} involves activation of cAMP-dependent PK as suggested by phosphorylation of IP_3 receptor. Using DMNB-cAMP as well as other caged second messengers (Ca^{2+} and IP_3) it could be concluded that cAMP elevation inhibits IP_3-induced Ca^{2+} release in an intact cell and that this inhibition was mediated by cAMP-dependent PK [29].

5.2.4.1.2 PKA using Caged Inhibitor

Lawrence et al. described the synthesis and the photochemical properties of a caged peptide PKA inhibitor [30]. The incorporation of the NVOC-caging group on one arginine side-chain of the nonapeptide Gly-Arg-Thr-Gly-*Arg*-Arg-Asn-Ala-Ile (**9**, Scheme 5.2.3) generated a PKA competitive inhibitor of reduced affinity (20 µM for the caged inhibitor versus 420 nM for the uncaged inhibitor). The *in vivo* inhibitory activity of the nonapeptide on rat embryo fibroblasts, through PKA-dependent inhibition of morphological changes, was demonstrated by the light-induced changes in morphology observed in the presence of caged inhibitor on this cell line.

5.2.4.2 PKC

PKC is a family of enzymes that regulate several cellular proteins by phosphorylation. It is inactive in the cytoplasm until its primary activator DAG is released by phospholipase C.

5.2.4.2.1 PKC using Caged Activators

The synthesis and the photochemical properties of an α-carboxyl-2,4-dinitrobenzyl-1,2-dioctanoyl-*rac*-glycerol (**10**, Scheme 5.2.4) has been described and used to photochemically activate PKC purified from rat brain [31]. The caged DAG derivative was successfully incorporated in cell membranes. Illumination of electrically stimulated cardiac myocytes loaded with the caged derivative led to a complex cellular response including an initial enhancement of cell contraction followed by a loss of responsiveness. Both phases were inhibited by chelerythrine, a PKC antagonist.

The caging of fatty acids as potential PKC activators led to the development of a new caging group for carboxylic acids: 1-(2'-nitrophenyl)-2-ethanediol derivatives (NED-caged derivative) [32]. This new reagent was used to cage arachidonic acid (**11**, Scheme 5.2.4) and was developed mainly because of the hydrolytic instability of the corresponding CNB-caged derivative (**12**, Scheme 5.2.4). The group, besides its hydrolytic stability, displayed fairly good quantum yields (0.2). Activation of PKC could be clearly related to the photochemical release of arachidonic acid from **11**.

5.2.4.2.2 PKC using Caged Fluorescent Substrates

A caged fluorescent PKC substrate was conceived as a peptide derivative where the N-terminal serine side-chain was protected from phosphorylation by a DMNB-caging group and the N-terminal substituted by an NBD fluorescent moiety (**13**, Scheme 5.2.4) [33]. A 2.5-fold increase in fluorescence intensity was observed upon phosphorylation of the uncaged peptide [34]. This caged fluorescent PKC substrate constitutes a first example of a spatiotemporal sensor of PKC in living cells.

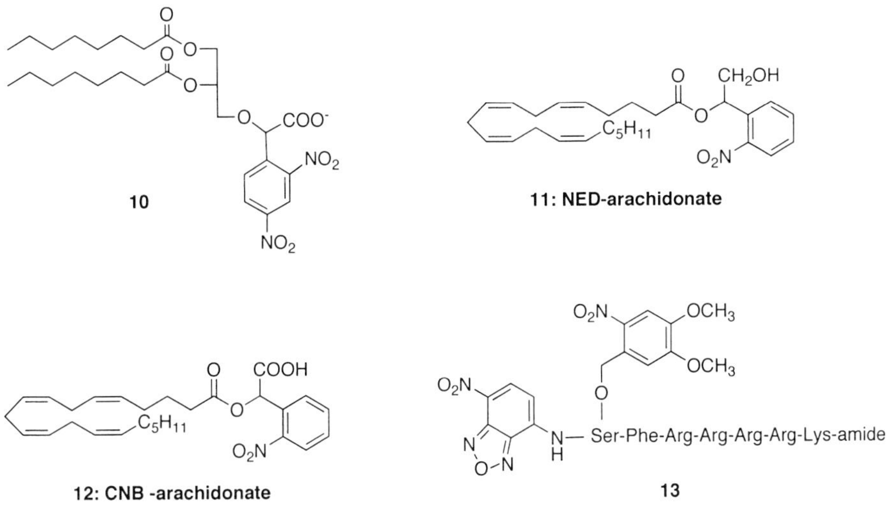

Scheme 5.2.4 Caged PKC modulators: caged DAG (**10**); caged arachidonic acid (**11** and **12**); caged fluorescent PKC peptide substrate (**13**).

5.2.4.3 Calmodulin-dependent Protein Kinases

5.2.4.3.1 Myosin Light Chain Kinase (MLCK) using Caged Peptide Inhibitors

To probe the function of calmodulin, caged peptide inhibitors of MLCK were synthesized and tested as potential blockers of eosinophil cell motility induced by calcium-bound calmodulin (CAM) binding to MLCK [35]. A 20-amino-acid peptide, representing the target sequence in MLCK and which binds tightly to CAM, has been modified as a CNB-caged Tyr derivative (**14**, Scheme 5.2.5). This caged peptide binds CAM 50 times less tightly than the parent peptide and is effectively inactive when assayed in smooth muscle cells known to be CAM dependent. This caged peptide was microinjected in eosinophil cells exhibiting normal motility and promptly blocked cell locomotion when exposed to light.

5.2.4.3.2 Calmodulin-dependent Protein Kinase II (CaMKII) using Caged Peptide Inhibitors

CaMKII is a second-messenger-responsive multifunctional protein kinase that plays important roles in controlling a variety of cellular functions including memory in response to an increase in intracellular Ca^{2+} [36]. A 13-amino-acid peptide (Lys-Lys-Ala-Leu-Arg-Arg-Gln-Glu-Ala-Val-Asp-Ala-Leu), designated as autocamitide-2-related inhibitory peptide (AIP), was found to inhibit CaMKII in a highly specific manner [37]. Studies showed the importance of the first two Lys residues in the inhibitory activity [38]. Therefore, caged AIPs (**15** and **16**, Scheme 5.2.5) were synthesized and characterized to elucidate the mechanism of CaMKII [39]. The caged AIP resulted in the decrease of the inhibitory activity

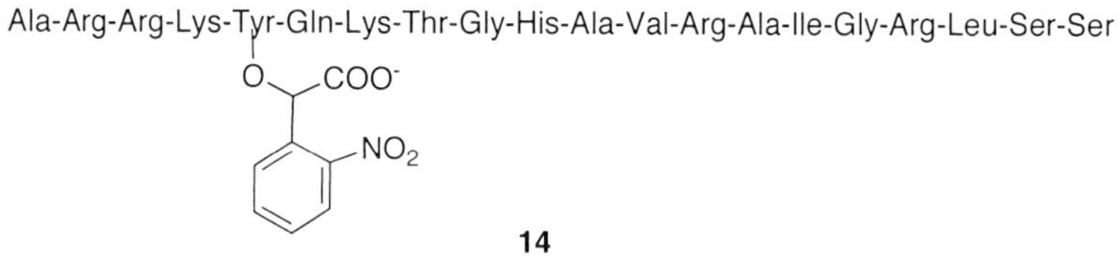

Ala-Arg-Arg-Lys-Tyr-Gln-Lys-Thr-Gly-His-Ala-Val-Arg-Ala-Ile-Gly-Arg-Leu-Ser-Ser

14

Lys-Lys-Ala-Leu-Arg-Arg-Gln-Glu-Ala-Val-Asp-Ala-Leu

15

Lys-Lys-Ala-Leu-Arg-Arg-Gln-Glu-Ala-Val-Asp-Ala-Leu

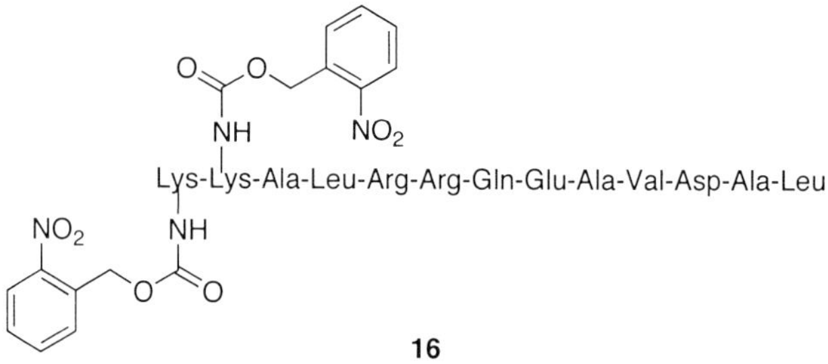

16

Scheme 5.2.5 Caged calmodulin-dependent PK peptide inhibitors (**14–16**).

on CaMKII. Upon irradiation, caged AIP regenerates the parent peptide AIP, thus restores the inhibitory activity as expected.

5.2.5
Photoregulation of Alcohol Dehydrogenases

Dehydrogenases are oxidoreductases which utilize different types of co-factors: nucleotide coenzymes (NAD$^+$, NADP$^+$), flavine coenzymes (FAD, FMN), quinone coenzymes (PQQ, TTQ) or metallic coenzymes. These oxido-reductions catalyze the transfer of different pairs of reactive species: hydride and proton, or one or two protons together with an equal number of electrons.

5.2.5.1 Isocitrate Dehydrogenase (IDH): Caged NAD and NADP

Pyridine nucleotides NAD and NADP are the most abundant coenzymes in eukaryotic cells, and serve as oxidative cofactors for many dehydrogenases, reductases and hydroxylases. They are major carriers of H$^+$ and e$^-$ in many important metabolic systems such as the glycolysis pathway, the tricarboxylic acid cycle, fatty acid synthesis and sterol synthesis.

A series of caged NADP and NAD analogs has been synthesized and investigated for potential use in time-resolved crystallography study of IDH, an

NADP-dependent enzyme of the tricarboxylic acid cycle which converts isocitrate to α-ketoglutarate [40]. Biochemically distinct coenzymes analogs have been synthesized and characterized (Scheme 5.2.6). The first, a "catalytically caged" NADP **17**, where the nicotinamide group is modified by a CNB-caging group, was designed to bind to IDH prior to photolysis, but is expected to become catalytically inactive. The second, an "affinity caged" NADP **18**, where a phosphate group modified by a DMNPE-caging group, was expected to reduce its affinity for IDH. The incorporation at the 2′-phosphate is expected to most significantly reduce the affinity of NADP for IDH.

The catalytic mechanism of IDH was investigated using these caged cofactors as well as a caged substrate, NPE-caged isocitrate (**19**, Scheme 5.2.6) [41]. This enzyme is fully reactive in the crystal and displays a kinetic mechanism similar to that in solution, with an overall turnover rate of about $60\,\mathrm{s}^{-1}$ at $22\,°\mathrm{C}$ and with a final dissociation of products as rate limiting step. Millisecond Laue structures of the enzyme/α-ketoglutarate/NADPH complex (the product complex) were obtained using these caged modulators [42], and the most well-refined structure was obtained from the experiment using the DMNPE-caged NADP molecule which displayed a better photochemical efficiency and the highest fragmentation rate ($13\,000\,\mathrm{s}^{-1}$).

5.2.5.2 Glucose-6-Phosphate Dehydrogenase (G6PDH): Caged Glucose 6-Phosphate and Caged NADP

The *in vivo* rate of G6PDH activity in sea urchin eggs has been investigated using radiolabeled (^{3}H or ^{14}C) NPE-caged phosphate esters of the glucose 6-phosphate substrate, NPE-glucose-6-phosphate (**20**, Scheme 5.2.6) [43]. Exposure to intense light flashes liberates the labeled glucose-6-phosphate within the cell, and the rates of formation of $^{14}CO_2$ and ^3H_2O after photolysis correlate with the *in vivo* rates of G6PDH activity. These results are discussed in terms of various hypotheses regarding the modulation of G6PDH activity by fertilization.

The catalytic activity of *Saccharomyces cerevisiae* G6PDH was also photoregulated using a NB-caged amide of NADP and allowing a 60% enzymatic reduction of photoreleased NADP to NADPH [44].

5.2.5.3 7α-Hydroxy Steroid Dehydrogenase (7α-HSDH): Caged Cholic Acid

Bile acids are essential to solubilizing and transporting dietary lipids. Recently, bile acids have been found to bind at orphan nuclear receptor, farnesoid X receptor, which suppresses transcription of the gene encoding cholesterol 7α-hydroxylase when bound to bile acids. The signaling mediated by such nuclear receptors is essential for the regulation of endogenous hormones as well as for metabolic elimination of xenobiotic chemicals. Thus, caged bile acids (**21** and **22**, Scheme 5.2.6) were developed to study the signaling mediated by steroid hormones [45]. Indeed, these caged compounds produced the parent bile acid upon photoirradia-

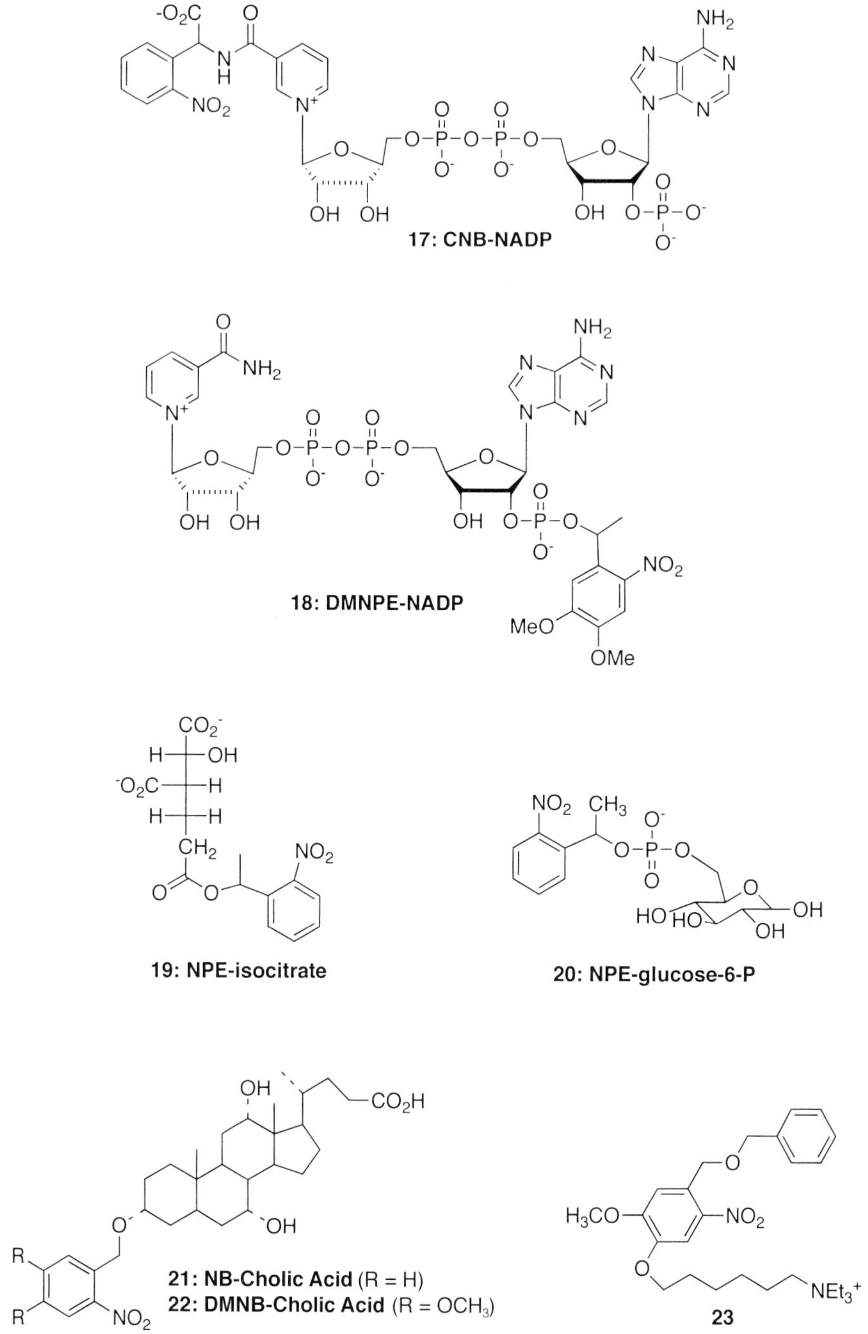

316 | *5 Photoregulation of Proteins*

Scheme 5.2.6 Caged dehydrogenase modulators: caged NADP (**17** and **18**); caged dehydrogenases substrates (**19–23**).

tion at 350 nm. The enzymatic activities of these compounds using 7α-HSDH, which oxidizes cholic acid to 7-oxo-cholic acid, were recovered after irradiation.

5.2.5.4 Caged Benzyl Alcohol
The catalytic oxidation of benzyl alcohol by horse liver alcohol dehydrogenase was investigated at sub zero temperature using a caged benzyl alcohol derivative (**23**, Scheme 5.2.6) [46].

5.2.6 Photoregulation of Cholinesterases

Acetylcholinesterase (AChE) is a serine hydrolase that terminates cholinergic transmission by rapid hydrolysis of the neurotransmitter acetylcholine, with a turnover number approaching $20\,000\,s^{-1}$ [47]. Butyrylcholinesterase (BuChE) [48], termed as such because of its faster hydrolysis of butyrylcholine over acetylcholine, besides having no identified endogenous substrate, plays an important role in detoxification by degrading esters like the muscle relaxant succinylcholine and cocaine. The description of the three-dimensional (3-D) structure of AChE [49] and of BuChE [50] as well as of several AChE-inhibitor complexes, has permitted a better understanding of structure-function relationships in the cholinesterases (ChEs). It has, however, also raised new questions concerning the traffic of substrate and products to and from the active site in view of the high turnover rate. To study these dynamic issues of ChEs, a fast and efficient triggering of the enzyme reaction is required, and, clearly, a temporally and spatially controlled photoregulation of enzyme activity represents the best adapted methodology and prompted the synthesis of several caged ChEs modulators (Scheme 5.2.7) [51].

5.2.6.1 Caged Choline Derivatives
As choline is the enzymatic product of ChEs, "caged" choline derivatives **24** and **25** (Scheme 5.2.7) were designed [52] to be complexed at the ChEs active site and to generate choline by subsequent photoactivation, allowing us to investigate the rapid clearance of choline from the active site. The arseno-choline derivatives **26** and **27** (Scheme 5.2.7) are heavy-atom analogs of caged choline, designed to track the path of choline exit by time-resolved crystallographic studies [53]. We have established the photoregulation of the ChE activities using probe **25**, by restoring the enzymatic activity after photolysis of a ChE/**25** complex [54].

The experiments in which we established the recovery of enzymic activity after flash laser photolysis of solutions of AChE or BuChE containing **25** (Fig. 5.2.1) suggested not only that the byproduct 2-nitrosoacetophenone did not have an inhibitory effect on the two enzymes in the experimental conditions employed, but that both 2-nitrosoacetophenone and choline, which are generated concomitantly within the active sites of the enzymes, were cleared from their active sites [54]. Thus, the experimental system can indeed serve as a paradigm for studying the

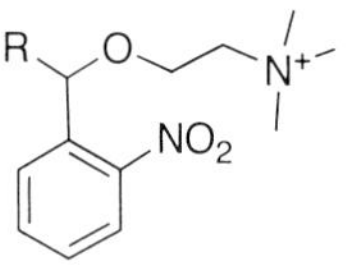

24: R = H
25: R = CH_3

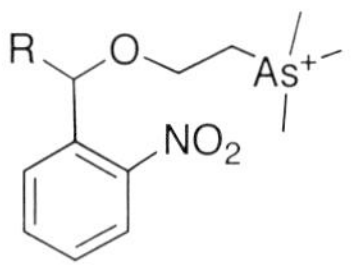

26: R = H
27: R = CH_3

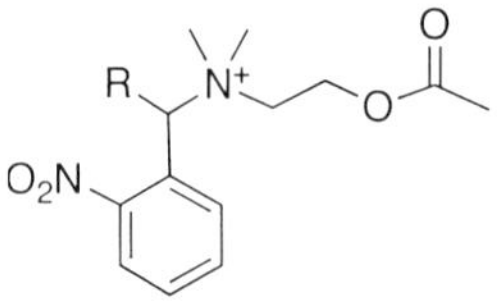

28: R = H
29: R = CH_3

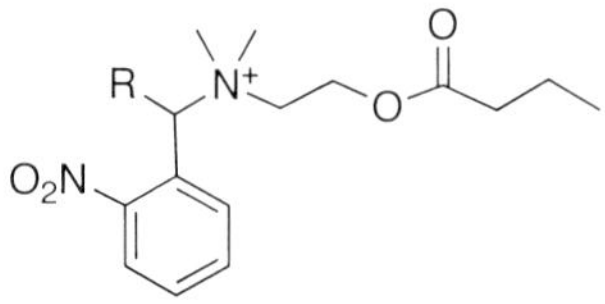

30: R = H
31: R = CH_3

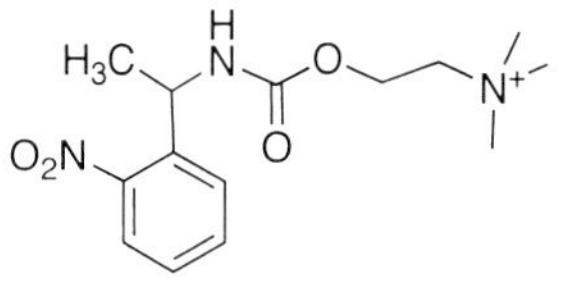

32

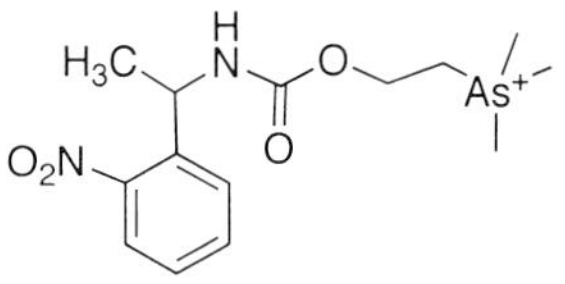

33

Scheme 5.2.7 Caged cholinesterase modulators: caged choline (**24** and **25**) and arsenocholine (**26** and **27**); caged nor-acetylcholine (**28** and **29**) and nor-butyrylcholine (**30** and **31**); caged carbamylcholine (**32**) and arsenocarbamylcholine (**33**).

exit of choline from the active site of AChE or BuChE by means of time-resolved crystallography. Recently, the 3-D structure of caged arseno-choline **27** complexed to AChE has been obtained [55] opening the way to anomalous diffraction analyses. These studies, however, will need to be reconsidered in the light of recent studies demonstrating [56] that photolytic release of alcohols from caged nitrobenzyl ether derivatives is much slower than initially claimed.

5.2.6.2 **Caged Cholinesterase Substrate**

Nor-acetylcholine is a close analog of the endogenous substrate of AChE. Hydrolysis of nor-acetylcholine and of acetylcholine is chemically identical: after the binary enzyme–substrate complex formation, successive acylation and deacylation steps follow as illustrated in Scheme 5.2.8. The advantage of caged nor-acetylcholine (**28** and **29**, Scheme 5.2.7) [57] is that it offers the possibility of studying the mechanism of the entire catalytic process shown in Scheme 5.2.8. Following the same strategy, caged nor-butyrylcholine derivatives (**30** and **31**, Scheme 5.2.7) [58] were designed for the studies on BuChE. Unfortunately, both

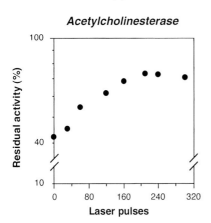

Fig. 5.2.1 Regeneration of enzymic activity after laser flash photolysis (351 nm) of (a) AChE in the presence of **25** (5×10^{-4} M in phosphate buffer, pH 6.5) and (b) BuChE in the presence of **25** (5.4×10^{-5} M in phosphate buffer, pH 7.2). (Reproduced with the permission of the American Chemical Society)

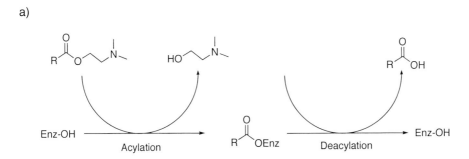

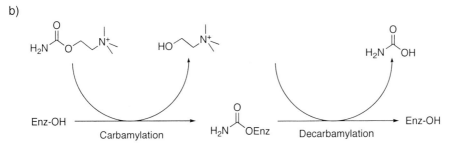

Scheme 5.2.8 Mode of action of different caged cholinesterase modulators: (a) caged nor-acetylcholine and (b) caged carbamylcholine.

28/29 and 30/31 are slowly hydrolysed by AChE and BuChE, respectively, impeding a further utilization of these probes.

Carbamylcholine serves as a slow substrate for AChE: rapid carbamylation with release of choline is followed by much slower decarbamylation which can be accelerated by dilution (Scheme 5.2.8). Thus, "caged" carbamylcholine and arseno-carbamylcholine, **32** [59] and **33** [53] (Scheme 5.2.7), can offer the possibility to study the mechanism of carbamylation of AChE as well as to investigate the enzymatic release of choline (Scheme 5.2.8). The photoregulation of the AChE activity could be demonstrated using **32**, by inactivating the AChE activity through photoinduced carbamylation after photolysis of the AChE/**32** complex [54]. In this experiment, even a single laser pulse caused time-dependent loss of enzymic activity to 60% of the control value, while 20 pulses decreased the enzyme activity by a further 15% (Fig. 5.2.2a). That this observed inactivation was due to photoinduced carbamylation was demonstrated by the fact that >90% of control activity could be regenerated in a progressive fashion upon subsequent dilution of carbamylated enzyme (Fig. 5.2.2b), similar to what was observed for AChE carbamylated directly by carbamylcholine. Such photoregulation reactions using caged carbamylcholine derivatives could not be applied to BuChE, since carbamylcholine has a very low affinity for BuChE and carbamylation of BuChE is a very slow process that requires in addition very high concentrations of carbamylcholine.

The obtained results show that **25** and **32**, or their quaternary arsonium analogs, are suitable for a photoregulation of the ChEs activities. They each release photochemically a different cholinergic modulator controlling the catalytic reactions at different steps. **25** photoreleases choline (the enzymatic product), while **32** photogenerates carbamylcholine (an AChE substrate), allowing the carbamylation of the enzyme with concomitant release of choline. Since these two compounds generate choline in two different ways, either by a direct photocleavage reaction (with **25**) or by enzymatic hydrolysis of a substrate generated by photocleavage (with **32**), they constitute complementary tools for time-resolved crystallographic studies envisaged [60]. To overcome the problem raised by the slow photolytic release of choline from the NPE-caged choline derivative **25**, a recently described alternative method combined cryophotolysis to kinetic crystallography [61], as illustrated in Chapter 7.3.

5.2.7
Miscellaneous Examples of Caged Enzyme Modulators

5.2.7.1 Caged Nitric Oxide Synthase (NOS) Inhibitor

The photo-control of the activity of NOS using a caged NOS inhibitor, Bhc-1400W (**34**, Scheme 5.2.9), has recently been established [62]. Incorporation of the Bhc-caging group induced about a 16-fold loss of affinity for NOS (IC_{50} ~1100 versus 70 nM for the uncaged inhibitor). Photochemical uncaging was possible either by UV irradiation or multiphoton excitation, although with a moderate first-order time constant (0.007 min^{-1} for multiphoton uncaging). This constitutes a first reported conjugation of a caging group with a large two-photon uncaging cross-section to a therapeutic agent.

(a)

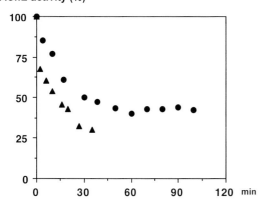

(b)

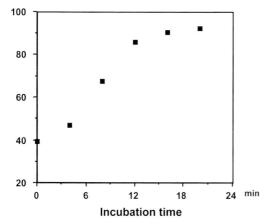

Fig. 5.2.2 (a) Time-dependent inactivation of AChE after laser flash photolysis (351 nm) of AChE in the presence of **32** (4.2×10^{-3} M) in phosphate buffer, pH 6.5): (circles): 1 pulse; (triangles): 20 pulses. (b) Regeneration of AChE activity after subsequent dilution. (Reproduced with the permission of the American Chemical Society)

5.2.7.2 Caged Substrate of Quinol-oxidizing Enzymes

Quinol oxidation is a common feature of respiration in mitochondria and aerobic bacteria. The quinol-oxidation enzymes utilize the energy released in electron transfer from quinols to electron acceptor-substrates to promote the translocation of protons across membranes. To investigate in detail the catalytic mecha-

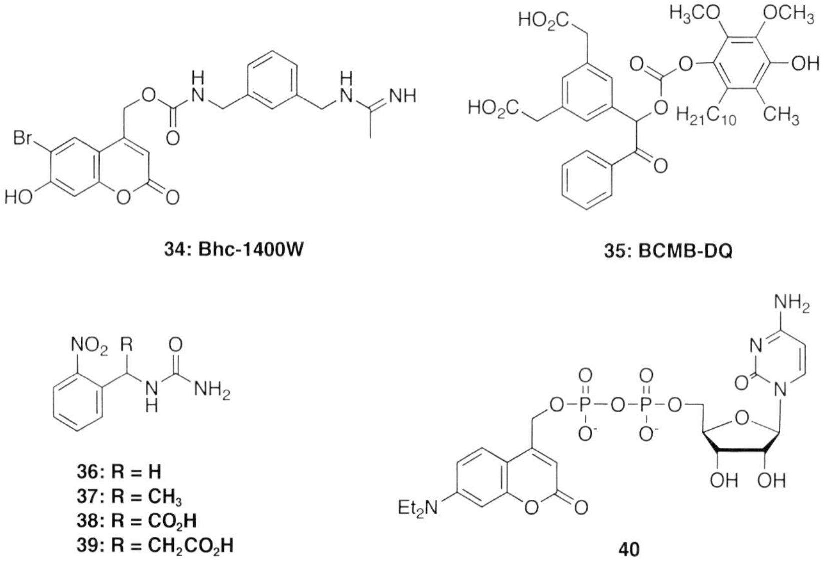

Scheme 5.2.9 Miscellaneous examples of caged enzyme modulators: caged nitric oxide synthase inhibitor (**34**); caged cytochrome bc_1 substrate (**35**); caged urease substrate (**36–39**) caged ribonuclease reductase substrate (**40**).

nisms of these enzymes (cytochrome bo_3 and cytochrome bc_1), a caged substrate (**35**, Scheme 5.2.9) was developed to overcome the use of stopped-flow mixing techniques [63]. A BCMB-caged decylubiquinol derivative, BCMB-DQ (**35**), was synthesized and tested for potential photoregulation of the quinol-oxidation enzymes. The electron input kinetics of cytochrome bc_1 could be investigated at a millisecond timescale corresponding to the rate-limiting decarboxylation of the photolytic DQ-carboxylate.

5.2.7.3 Caged Urease Substrate

A series of o-nitrobenzyl urea derivatives has been synthesized and tested for their photolytic properties (**36–39**, Scheme 5.2.9) [64]. Rates of photolysis were in the 10^4–10^5 s^{-1} range together with high quantum yields (up to 0.81). Irradiation of N-CNB-caged derivative (R=CO$_2$H) triggered the *Klebsiella aerogenes* urease activity, although only to a very moderate extent.

5.2.7.4 Caged Ribonuclease Reductase Substrate

A coumarin-caged cytidine 5′-diphosphate derivative was synthesized and tested for its photolytic properties (**40**, Scheme 5.2.9) [65]. The release of CDP from the caged derivative was ultra-fast (2×10^8 s^{-1}) while the moderate quantum yields (~ 0.03) are compensated for by the high absorptivity of the coumarin chromophore above 300 nm. The potential use of this derivative for photochemical triggering of Ribonuclease reductase (RNR) activity using this new caged substrate has not been reported yet.

5.2.8
Conclusions and Perspectives

As largely exemplified in this chapter, caged enzyme modulators are indeed useful tools to trigger enzymatic reactions or to control their inhibition. Such photoregulation represent a complementary method to the light-activated enzymes involving caging of protein residues. Photochemical enzyme regulation using caged modulators is of special interest, allowing a spatiotemporal control of the catalytic reaction. Combined with time-resolved studies such as FTIR and X-ray crystallography, it is possible to study, at the atomic level, the enzyme functions and mechanisms such as binding or breakdown of a substrate, with or without the help of a coenzyme, formation of product and release of product. Future endeavors will be directed towards the design of new caged enzyme modulators with caging groups displaying improved chemical and photochemical properties together with the development of improved time-resolved techniques with higher resolution for structure elucidation and functional study of enzymes.

References

1 J. H. KAPLAN, B. FORBUSH III, J. F. HOFFMAN, *Biochemistry* **1978**, *17*, 1929–1935.
2 (a) V. N. R. PILLAI, *Synthesis* **1980**, 1–26; (b) T. W. GREENE, P. G. M. WUTS, *Protective Groups in Organic Synthesis*, 3rd edn, Wiley, New York, **1999**.
3 A comparative analysis of the properties of the caged ATP molecules can be found in Chapter 1.4.
4 J. W. WALKER, G. P. REID, J. A. MCCRAY, D. R. TRENTHAM, *J. Am. Chem. Soc.* **1988**, *110*, 7170–7177.
5 K. FENDLER, K. HARTUNG, G. NAGEL, E. BAMBERG, *Methods Enzymol.* **1998**, *291*, 289–306.
6 K. FENDLER, S. DRÖSE, W. EPSTEIN, E. BAMBERG, K. ALTENDORF, *Biochemistry* **1999**, *38*, 1850–1856.
7 J. A. DANTZIG, H. HIGUCHI, Y. E. GOLDMAN, *Methods Enzymol.* **1998**, *291*, 307–348.
8 H. HIGUCHI, E. MUTO, Y. INOUE, T. YANAGIDA, *Proc. Natl Acad. Sci. USA* **1997**, *94*, 4395–4400.
9 C. TOYOSHIMA, H. NOMURA, Y. SUGITA, *FEBS Lett.* **2003**, *555*, 106–110.
10 A. BARTH, W. KREUZ, W. MÄNTELE, *FEBS Lett.* **1990**, *277*, 147–150.
11 S. LEWIS, D. D. THOMAS, *Biochemistry* **1991**, *30*, 8331–8339.
12 J. E. MAHANEY, J. P. FROEHLICH, D. D. THOMAS, *Biochemistry* **1995**, *34*, 4864–4879.
13 F. M. ROSSI, J. P. Y. KAO, *J. Biol. Chem.* **1997**, *272*, 3266–3271.
14 A. TROULLIER, K. GERWERT, Y. DUPONT, *Biophys. J.* **1996**, *71*, 2970–2983.
15 J. H. KAPLAN, G. C. R. ELLIS-DAVIS, *Proc. Natl Acad. Sci. USA* **1988**, *85*, 6571–6575.
16 K. KINBARA, L. E. GOLDFINGER, M. HANSEN, F. L. CHOU, M. H. GINSBERG, *Nat. Rev. Mol. Cell Biol.* **2003**, *4*, 767–776.
17 I. SCHLICHTING, G. RAPP, J. JOHN, A. WITTINGHOFER, E. F. PAI, R. S. GOODY, *Proc. Natl Acad. Sci. USA* **1989**, *86*, 7687–7690.
18 I. SCHLICHTING, S. C. ALMO, G. RAPP, K. WILSON, K. PETRATOS, A. LENTFER, A.

Wittinghofer, W. Kabsch, E. F. Pai, G. A. Petsko, R. S. Goody, *Nature* **1990**, *345*, 309–315.
19 A. J. Scheidig, S. M. Franken, J. E. T. Corrie, G. P. Reid, A. Wittinghofer, E. F. Pai, R. S. Goody, *J. Mol. Biol.* **1995**, *253*, 132–150.
20 A. J. Scheidig, C. Burmester, R. S. Goody, *Structure* **1999**, *7*, 1311–1324.
21 K. Gerwert, *Biol. Chem.* **1999**, *380*, 931–935.
22 V. Cepus, A. J. Scheidig, R. S. Goody, K. Gerwert, *Biochemistry* **1998**, *37*, 10263–10271.
23 X. Du, H. Frei, S.-H. Kim, *J. Biol. Chem.* **2000**, *275*, 8492–8500.
24 C. Allin, K. Gerwert, *Biochemistry* **2001**, *40*, 3037–3046.
25 C. Allin, M. R. Ahmadian, A. Wittinghofer, K. Gerwert, *Proc. Natl Acad. Sci. USA* **2001**, *98*, 7754–7759.
26 J. Engels, E.-J. Schlaeger, *J. Med. Chem.* **1977**, *20*, 907–911.
27 K. Curley, D. S. Lawrence, *Pharmacol. Ther.* **1999**, *82*, 347–354.
28 R. Popp, R. P. Brandes, G. Ott, R. Busse, I. Fleming, *Circ. Res.* **2002**, *90*, 800–806.
29 S. Tertyshnikova, A. Fein, *Proc. Natl Acad. Sci. USA* **1998**, *95*, 1613–1617.
30 J. S. Wood, M. Koszelac, J. Liu, D. S. Lawrence, *J. Am. Chem. Soc.* **1998**, *120*, 7145–7146.
31 R. Sreekumar, Y. Q. Pi, X. P. Huang, J. W. Walker, *Bioorg. Med. Chem. Lett.* **1997**, *7*, 341–346.
32 J. Xia, X. Huang, R. Sreekumar, J. W. Walker, *Bioorg. Med. Chem. Lett.* **1997**, *7*, 1243–1248.
33 W. F. Veldhuyzen, Q. Nguyen, G. McMaster, D. S. Lawrence, *J. Am. Chem. Soc.* **2003**, *125*, 13358–13359.
34 R.-H. Yeh, X. Yan, M. Cammer, A. R. Bresnick, D. S. Lawrence, *J. Biol. Chem.* **2002**, *277*, 11527–11532.
35 J. W. Walker, S. H. Gilbert, R. M. Drummond, M. Yamada, R. Sreekumar, R. E. Carraway, M. Ikebe, F. S. Fay, *Proc. Natl Acad. Sci. USA* **1998**, *95*, 1568–1573.
36 J. Lisman, R. C. Malenka, R. A. Nicoll, R. Malinow, *Science* **1997**, *276*, 2001–2002.
37 A. Ishida, I. Kameshita, S. Okuno, T. Kitani, H. Fujisawa, *Biochem. Biophys. Res. Commun.* **1995**, *212*, 806–812.
38 A. Ishida, Y. Shigeri, Y. Tatsu, K. Uegaki, I. Kameshita, S. Okuno, T. Kitani, N. Yumoto, H. Fujisawa, *FEBS Lett.* **1998**, *427*, 115–118.
39 Y. Tatsu, Y. Shigeri, A. Ishida, I. Kameshita, H. Fujisawa, N. Yumoto, *Bioorg. Med. Chem. Lett.* **1999**, *9*, 1093–1096.
40 B. E. Cohen, B. L. Stoddart, D. E. Koshland Jr, *Biochemistry* **1997**, *36*, 9035–9044.
41 M. J. Brubaker, D. H. Dyer, B. L. Stoddard, D. E. Koshland Jr, *Biochemistry* **1996**, *35*, 2854–2864.
42 B. L. Stoddard, B. E. Cohen, M. Brubaker, A. D. Mesecar, D. E. Koshland Jr, *Nat. Struct. Biol.* **1998**, *5*, 891–897.
43 R. R. Swezey, D. Epel, *Dev. Biol.* **1995**, *169*, 733–744.
44 C. P. Salerno, M. Resat, D. Magde, J. Kraut, *J. Am. Chem. Soc.* **1997**, *119*, 3403–3404.
45 Y. Hirayama, M. Iwamura, T. Furuta, *Bioorg. Med. Chem. Lett.* **2003**, *13*, 905–908.
46 S.-C. Tsai, J. P. Klinman, *Bioorg. Chem.* **2003**, *31*, 172–190.
47 J. Massoulié, L. Pezzementi, S. Bon, E. Krejci, F. M. Valette, *Prog. Neurobiol.* **1993**, *41*, 31–91.
48 O. Lockridge, *Pharmacol. Ther.* **1990**, *47*, 35–60.
49 J. L. Sussman, M. Harel, F. Frolow, C. Oefner, A. Goldman, L. Toker, I. Silman, *Science* **1991**, *253*, 872–879.
50 Y. Nicolet, O. Lockridge, P. Masson, J. C. Fontecilla-Camps, F. Nachon, *J. Biol. Chem.* **2003**, *42*, 41141–41147.
51 L. Peng, C. Colas, M. Goeldner, *Pure Appl. Chem.* **1997**, *69*, 755–759.
52 L. Peng, M. Goeldner, *J. Org. Chem.* **1996**, *61*, 185–191.
53 L. Peng, F. Nachon, J. Wirz, M. Goeldner, *Angew. Chem. Int. Ed.* **1998**, *37*, 2691–2693.
54 L. Peng, I. Silman, J. Sussman, M. Goeldner, *Biochemistry* **1996**, *35*, 10854–10861.
55 M. Weik, Personal communication.

56 J. E. T. Corrie, A. Barth, V. R. N. Munasinghe, D. R. Trentham, M. C. Hutter, J. Am. Chem. Soc. **2003**, *125*, 8546–8554.
57 L. Peng, J. Wirz, M. Goeldner, Angew. Chem. Int. Ed. **1997**, *36*, 398–400.
58 L. Peng, J. Wirz, M. Goeldner, Tetrahedron Lett. **1997**, *38*, 2961–2964.
59 J. W. Walker, J. A. McCray, G. P. Hess, Biochemistry **1986**, *25*, 1799.
60 L. Peng, M. Goeldner, Methods Enzymol. **1998**, *291*, 265–278.
61 A. Specht, T. Ursby, M. Weik, L. Peng, J. Kroon, D. Bourgeois, M. Goeldner, ChemBioChem **2001**, *2*, 845–848.
62 H. J. Montgomery, B. Perdicakis, D. Fishlock, G. A. Lajoie, E. Jervis, J. G. Guillemette, Bioorg. Med. Chem. **2002**, *10*, 1919–1927.
63 K. C. Hansen, B. E. Schultz, G. Wang, S. I. Chan, Biochim. Biophys. Acta **2000**, *1456*, 121–137.
64 R. Wieboldt, D. Ramesh, E. Jabri, P. A. Karplus, B. K. Carpenter, G. P. Hess, J. Org. Chem. **2002**, *67*, 8827–8831.
65 R. O. Schönleber, J. Bendig, V. Hagen, B. Giese, Bioorg. Med. Chem. **2002**, *10*, 97–101.

5.3
The Use of Caged Proteins in Cell-based Systems

John S. Condeelis and David S. Lawrence

5.3.1
Introduction

The extraordinary complexity of the protein-mediated biochemical events that drive cellular behavior presents enormous challenges with respect to correlating protein function with cellular activity. Although a wide variety of strategies, such as knock-outs, knock-ins and site-specific protein mutants, allows one to begin to address issues related to the biological role of specific proteins, these techniques are not without limitations. Knocking-out a specific gene, and the protein its codes for, in a living animal often has little or no effect *or* may prove to be embryonic lethal. The former may to due to biochemical compensation. However, either result precludes an assessment of the role of the protein in the adult animal. Cell-based studies using these strategies can likewise furnish less than satisfying results, particularly if the biological event under study is fast relative to protein expression (e.g. mitosis, long-term potentiation, motility, etc.). In this regard, light-activatable proteins represent an elegant approach to explore the biochemical basis of temporally and spatially sensitive cellular processes.

Although the use of caged proteins as tools to explore the biochemistry and molecular biology of the cell holds great promise, their actual application in a biological context has, thus far, been quite limited. The design, synthesis and characterization of caged proteins still represents a formidable task (see Chapter 5.1). The subsequent introduction of caged proteins into cells (i.e. via microinjection) is non-trivial. Finally, photouncaging the target protein in a temporally and/or spatially specific fashion requires expensive and highly sophisticated instrumentation. Nonetheless, the relatively few studies that have been reported to date clearly establish the potential utility of these reagents to correlate cellular behavior with protein function at a temporal, spatial, structural and mechanistic level of resolution that is unprecedented.

5.3.2
Cell Motility

5.3.2.1 β-Thymosin

The migration of a cell in response to an environmental stimulus involves a series of temporally and spatially regulated steps. These include lamellipod formation (due to actin polymerization which pushes the membrane outward), adhesion of the leading lamella to the extracellular matrix and the formation of focal contacts, development of contractile force, selective de-adhesion of focal contacts at the tail of the cell, and contraction causing tail retraction toward the leading lamella. These steps are finely choreographed requiring balanced actin polymerization transients at the appropriate time and place within the cell to afford a coherent response to the environmental stimulus. Proteins that control the actin polymer status must likewise be activated (and deactivated) with exquisite spatial and temporal precision. Unfortunately, the powerful tools of molecular biology (knock-ins, knock-outs, constitutively active analogs, dominant-negative analogs, etc.) lack the finesse required to address the role that a particular protein might play in a spatially and temporally regulated event.

Thymosin β4 (Tβ4) plays a key role in maintaining the intracellular pool of G-actin. The spontaneous nucleation of actin polymerization is impeded when large quantities of the latter are bound to Tβ4. By contrast, under stimulatory conditions when free barbed ends of actin filaments become available, Tβ4 will release free G-actin, which can then polymerize onto these free barbed ends [1]. Given the importance of F-actin transients in driving cell motility, Roy et al. investigated the role of Tβ4 in controlling directionality in locomoting fish scale keratocytes. Specifically, a caged Tβ4, which is unable to bind G-actin, should have little or no effect on the dynamics of cell motility [2]. However, the sudden light-driven generation of active Tβ4 in a highly localized region of the cell should bind to and therefore sequester G-actin within that region of the cell, thereby locally reducing the rate of actin polymerization and globally altering cell directionality.

Previous studies demonstrated that a short lysine rich segment (specifically lysine residues 18 and 19) in all β thymosins is responsible for actin coordination. Consequently, Tβ4 was treated with NVOC-Cl, a caging compound that covalently modifies amine groups. An *in vitro* actin polymerization assay was initially employed to assess the efficacy of Tβ4 caging and its subsequent photorelease. Acrylodan-actin, upon polymerization, displays a nearly 2-fold fluorescence enhancement upon polymerization and therefore serves as a useful assessment of the rate at which G-actin is converted to its polymeric counterpart. Caged Tβ4 (4-fold molar excess relative to G-actin) has no effect on polymerization rate. By contrast, both native Tβ4 and photoilluminated caged Tβ4 (i.e. uncaged Tβ4) dramatically reduces the rate of F-actin generation.

Intracellular studies using caged Tβ4 employed UV illumination times of 100 ms, thereby demonstrating exquisite temporal control over when Tβ4 is activated. However, spatial control over where the activated Tβ4 molecules are present upon photorelease is dependent upon the inherent biochemical properties of the protein under evaluation. Small molecules exhibit high intracellular diffusion rates, thereby precluding the ability to generate high local concentrations of the uncaged molecule within the context of a biological event. Tβ4 is a small protein ($\sim$ 5 kDa) and thus, one might expect, rapidly diffuses throughout the cytoplasm. Diffusion coefficients of FITC–dextran (10 kDa), FITC–Tβ4 and FITC–GST–Tβ4 were measured in fibroblasts using the FRAP bleaching technique. The diffusion coefficients are two orders of magnitude less than those measured in aqueous solution. Virtual Cell simulation software indicated that the G-actin–Tβ4 complex should persist within the local region of photoactivation for 5–10 s, long enough to reduce the local G-actin concentration and thereby perturb the directionality of cell motility.

Caged Tβ4 was bead loaded into keratocytes, a process that employs small glass beads (425–600 μm in diameter) that, upon sprinkling onto cells, generates temporary pores in the plasma membrane. A He:Cd laser was employed to generate illuminated spots 3 μm in diameter. Interestingly, local photouncaging of Tβ4 has no immediate effect on keratocyte locomotion. However, after approximately 1 min, the illuminated region becomes fixed to the substrate and the cell pivots around this site (Fig. 5.3.1). Ultimately, the illuminated region is released from the substrate and the cell resumes its normal migratory behavior.

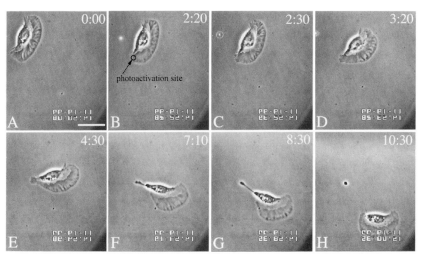

Fig. 5.3.1 Time-lapse images of a keratocyte injected with caged β-thymosin following photoillumination (frame B). (Reproduced from [2] by permission of The Rockefeller University Press)

A series of control experiments were performed, including irradiation alone (i.e. no caged Tβ4 present) and irradiation in the presence of caged FITC–dextran. Neither of these treatments resulted in the creation of a cellular pivot point. The 1-min lag time observed with caged Tβ4 and illumination is curious given the prediction that G-actin sequestration by photo-released Tβ4 should be complete by 10 s. The authors offer the possible explanation that the local disruption of the G-actin/F-actin equilibrium could serve as a signal that activates biochemical pathways that result in cell turning. The latter might be responsible for the observed temporal delay between the initial irradiation and the subsequent cellular response. Sequestration of G-actin *in vivo* might result in the disassembly of existing actin filaments as well as inhibition in growth of new filaments. This should produce a decrease in the propulsive forces in this region of the cell. Since the propulsive forces in other regions of the cell would not be affected, rotation of the cell about the region of illumination until the G-actin/F-actin equilibrium is reestablished would occur.

5.3.2.2 Cofilin

A number of proteins directly act upon F-actin and thereby influence the fashion by which the cell responds to environmental stimuli. Cofilin (ADF or actin depolymerization factor) is known to both sever and promote the depolymerization of F-actin. Cofilin is active in the non-phosphorylated (**1**) state and is rendered inactive upon phosphorylation (**2**) (Scheme 5.3.1), processes mediated by slingshot (a phosphatase) and LIM kinase, respectively. Agents that stimulate motility induce the dephosphorylation and therefore activation of cofilin. The severing activity of cofilin creates new F-actin barbed ends, which serve as initiation sites for polymerization. The depolymerization activity of cofilin makes G-actin available from older ADP filaments for ADP to ATP exchange and the po-

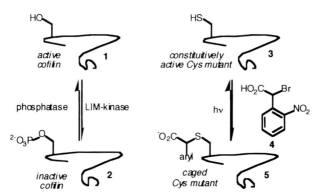

Scheme 5.3.1 Biochemical and photochemical regulation of cofilin and caged cofilin, respectively. (Reproduced from [4]. Copyright 2002 American Chemical Society)

lymerization of new ATP filaments. Consequently, cofilin's depolymerization activity is viewed as a prerequisite for enhanced actin polymerization [3].

Cofilin is a much larger protein than Tβ4 (18 versus 5 kDa) and contains multiple lysine residues. Although it is likely that NVOC-Cl treatment of cofilin will result in a modified protein that is no longer able to bind to F-actin, the modification of multiple residues requires their eventual removal. The latter carries a potential arithmetic penalty (i.e. Φ for each caged residue) that could ultimately result in vanishingly small quantities of the desired, completely uncaged, protein. Furthermore, some lysine residues may be more difficult to modify than others, thereby potentially resulting in the formation of a series of different cofilins with different properties. Given the established role of the Ser3 position in controlling cofilin activity, Ghosh et al. employed a different strategy to produce a caged cofilin [4].

A constitutively active cofilin was prepared via site-directed mutagenesis of Ser3 to a cysteine residue 3 (Scheme 5.3.1). Since Ser/Thr protein kinases do not catalyze the phosphorylation of cysteine residues in active site-directed peptides [5], it was expected that Cys3-cofilin (3) would fail to serve as a substrate for LIM kinase; an expectation that was experimentally validated [4]. Consequently, the severing/depolymerizing activity of this cofilin analog cannot be altered by endogenous LIM kinase. Cys3-cofilin was subsequently caged using the o-nitrobenzylbromide derivative 4. The latter contains a carboxylic acid moiety that, like the phosphate in 1, is ionized at physiological pH. Indeed, the caged cofilin 5 is unable to promote the depolymerization of cofilin, whereas cofilin depolymerizing activity is restored upon illumination. Mass spectrometry confirmed that the caged cofilin contains only a single caging group and that photolysis converts caged cofilin to its uncaged counterpart. In addition, a light microscope assay visually demonstrated that caged cofilin lacks F-actin severing activity whereas illuminated caged cofilin cleaves F-actin filaments into separate fragments (Fig. 5.3.2).

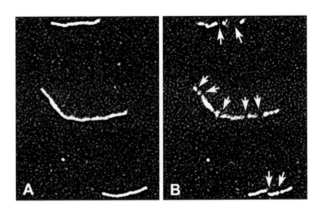

Fig. 5.3.2 (A) Rhodamine-labeled F-actin filaments (B) in the presence of irradiated (15 min) caged Cys3-cofilin. Cleavage sites along the F-actin filaments are marked with arrows. (Reproduced from [4]. Copyright 2002 American Chemical Society)

Caged cofilin **3** was microinjected into MTLn3 cells, a highly metastatic adenocarcinoma cell line [6]. A number of issues were addressed concerning the role of cofilin as a biochemical participant in cell motility. The first question to be probed concerned the known activities of cofilin (F-actin depolymerizing and severing) and their relationship to the depolymerization and/or polymerization of F-actin during motility. Global uncaging of cofilin **3** in MTLn3 cells generated new barbed ends and produced sizable enhancements in F-actin content ($\sim 40\%$). Furthermore, cells containing the photounleashed cofilin displayed an increase in instantaneous speed relative to their control counterparts. These observations are consistent with the notion that cofilin action on existing F-actin filaments is responsible for creating free barbed ends as new sites of polymerization, which in turn enhances F-actin content and produces higher rates of motility.

An early stage of cell motility is the formation of a lamellipod, which is generated via the protrusive force exerted by the polymerization of F-actin at the leading edge of the cell. Based on the results obtained in the global uncaging studies, Ghosh et al. reasoned that highly localized spatial uncaging should generate spatially restricted bursts of actin polymerization, which should produce lamellipods. Cells were locally irradiated in the cell cortex in a 3-µm diameter spot for 2 s and time-lapse imaged to assess the effect of the local photorelease of cofilin activity over the next several minutes. Localized illumination induced the formation of a protrusion at the site of irradiation. Furthermore, when a second site was illuminated in the cell, a protrusion formed at the new site. By contrast, non-microinjected cells treated in the above fashion failed to exhibit protrusions and, under longer photoillumination times, even displayed a slight retraction at the site of irradiation. Finally, the centroid of the cell was determined as a function of time in response to localized cofilin uncaging. The continuous photorelease of cofilin in a 3-µm spot determines the direction of migration (i.e. if the cell was photoilluminated in its southeast quadrant, the cell migrated in a southeasterly direction). These results indicated that cofilin serves as a key component of the steering wheel of the cell.

5.3.3
Electrical Conductance

5.3.3.1 Antibodies
Antibodies are arguably the most universally employed of all protein-based reagents. Consequently, the corresponding light-activatable derivatives would undoubtedly prove to be extremely useful in studying spatially and temporally sensitive phenomena. To this end, a generally applicable strategy for preparing caged antibodies could have a profound impact in modern biological research [7]. Unfortunately, although antibodies share a number of common structural features, their sheer size and complexity, not to mention the unique attributes of the individual antigen binding sites, suggests that structurally well-defined photoactivatable antibodies may be difficult to prepare.

Bonhoeffer et al. prepared a caged antibody that targets brain-derived neurotrophic factor (BDNF) and demonstrated for the first time that a light-activatable antibody could serve as a useful reagent in a biological context [8]. BDNF has been implicated as an essential protein component in the pathways that mediate long-term potentiation (LTP) in the brain. Both long-term increases (LTP) and decreases (depression; LTD) in synaptic strength are thought to play key roles in memory processing and storage. The locations, mediators and physiologic changes underlying synaptic plasticity remain areas of intense debate. In addition, the signaling pathways responsible for eliciting LTP and LTD must be activated in a temporally precise fashion. Until the study by Bonhoeffer, only caged small molecules (e.g. NO) have been useful in demonstrating the importance of specific time-dependent biochemical phenomena in synaptic plasticity.

Synaptic transmission at Schaffer collateral-CA1 synapses is dramatically enhanced upon the exogenous application of BDNF to hippocampal slices. The concern with this approach is that the amount of BDNF added significantly exceeds normal physiological concentrations and that the spatiotemporal dynamics alluded to above cannot be addressed. The use of BDNF antibodies allows one to work with existing endogenous levels of BDNF. As one would expect with function blocking antibodies, anti-BDNF interferes with BDNF signaling and thus results in a dramatic reduction in LTP. Unfortunately, the slow diffusion of antibodies into slice preparations provides limited control over both when and where BDNF blocking activity is observed. A caged anti-BDNF antibody was prepared by exposing the antibody to the lysine-modifying agent NVOC-Cl. The apparent BDNF-binding affinity is reduced by 80%. However, given the structural complexity of antibodies the precise interpretation of this result is not straightforward. Modification might be identical on all antibody molecules, thereby leading to the conclusion that binding affinity has been truly reduced by 80%. Alternatively, it is possible that the caged protein product is actually a mixture of differentially modified proteins that display a range of different affinities for BDNF. In either event, the authors were able to identify an optimal concentration for the caged antibody such that no biological effect was observed prior to photoirradiation. In addition, the BDNF antibodies were modified with NVOC-caged rhodamine green. These derivatives are not fluorescent until photoirradiated. Although these labeled antibodies were not employed as function blocking species they were used to assess diffusion rates, which proved to be slow on the experimental timescale. Specifically, the fluorescence intensity of an illuminated 800-μm diameter spot decreased only 38% after 1 h.

The caged function-blocking antibody was diffused, over a 2-h period, into hippocampal slices. As expected, the caged antibody has no effect on synaptic strength in the absence of irradiation. By contrast, upon photoillumination [short bursts (50–250 ms) at specific intervals (0.5–1 s); total illumination time: 12 s] starting 2 min prior to stimulation and ending 2 min after stimulation significantly reduced synaptic potentiation up to nearly 1 h following Schaffer collateral stimulation. A more narrow temporal photoirradiation window (starting only 7 s prior to stimulation; total illumination time: 700 ms) also produced an effect, but one that is less potentiated and of shorter duration (10–15 min) than the first experiment. Presumably this is due to the fact that less antibody has been uncaged, although the timing of the uncaging prior to stimulation could play a role as well. Nonetheless, these experiments do demonstrate that BDNF plays an important, almost immediate role in the biochemical pathways that elicit LTP.

5.3.3.2 Ion Channels

Dougherty, Lester and their colleagues have described the biosynthesis of several caged proteins in *Xenopus* oocytes [9–12]. The gene coding for the caged protein is synthesized containing a nonsense codon at the position of interest. A tRNA containing the appropriate nonsense anticodon and charged with a photolabile amino acid is also prepared. Both the gene and the tRNA are subsequently injected into *Xenopus* oocytes and suppression of the nonsense codon by the modified tRNA furnishes a caged protein. This strategy has been used to prepare caged derivatives of the nicotinic acetylcholine receptor, an inward rectifier potassium channel known as kir2.1 and the *Drosophila* Shaker K^+ channel. The nonsense suppression method, when applied in an intracellular setting, is especially useful for expressing properly localized membrane-embedded proteins.

Caged proteins are typically conceived as inert derivatives whose activity is unleashed upon photolysis. The Cal Tech group devised a variant of this concept by preparing a protein that, upon photolysis, losses its ability to perform a specific task [9]. The *Drosophila* Shaker B (ShB) K^+ channel opens in response to changes in membrane potential (i.e. the channel is "voltage-gated") and then rapidly closes within milliseconds. By contrast, a mutant K^+ channel missing the amino acid residues 6–46 does not inactivate. This observation, in conjunction with others, lead to the formulation of a model in which the N-terminus of the protein forms a ball-like structure that rapidly occludes the channel pore. The unnatural amino acid, o-nitrophenylglycine (Npg) is a photosensitive moiety that, upon illumination, results in the cleavage of the protein backbone (Scheme 5.3.2). Consequently, incorporation of Npg at the appropriate site in the *Drosophila* ShB K^+ channel furnishes a protein that can undergo a light-induced loss of its ability to rapidly block K^+ flow.

The Npg residue was inserted at position 47 of the K^+ channel. As expected, in the absence photoirradiation, the channel functioned normally by displaying activation and inactivation kinetics analogous to that exhibited by the wild-type protein. By contrast, irradiation of oocytes expressing the caged channel results

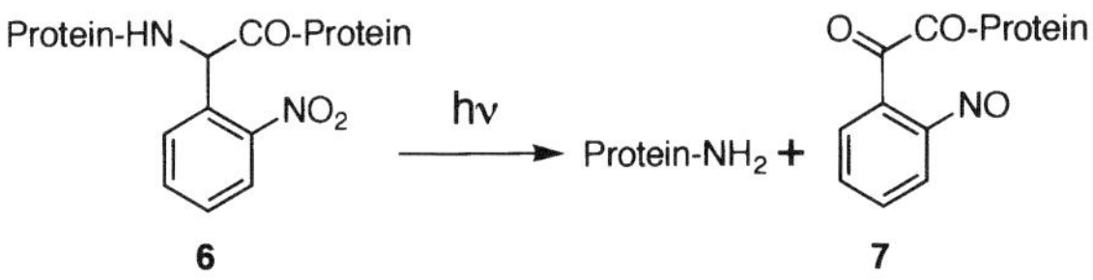

Scheme 5.3.2 Light-induced cleavage of an Npg-containing protein.

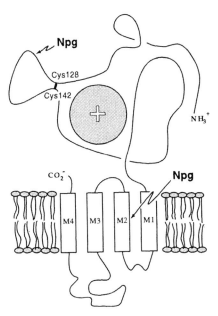

Fig. 5.3.3 Schematic of the membrane-embedded acetylcholine receptor highlighting the Cys128–Cys142 disulfide loop. (Reproduced from [9]. Copyright 1997 National Academy of Sciences, USA)

in a dramatic loss in the rate of channel inactivation. The latter is a consequence of the light-induced cleavage of the N-terminal pore-occluding "ball" from the rest of the protein channel.

A study analogous to the *Drosophila* ShB K$^+$ channel was performed with the nicotinic acetylcholine receptor, an ion channel that opens and closes in response to the acetylcholine ligand (i.e. a "ligand-gated" channel). The a-subunit of the muscle ($a_2\beta\gamma\delta$) receptor contains a highly conserved disulfide loop (Cys128–Cys142) near the N-terminus that appears to be critical for proper folding and assembly of functional protein (Fig. 5.3.3). Indeed, functional protein is not produced when either of the critical cysteine residues are replaced via site-directed mutagenesis. The Npg residue was inserted between the two cysteine residues. Light-directed amide bond cleavage should perturb only the conformational state of the disulfide loop since the overall structure of the a-subunit should remain intact (i.e. the N- and C-terminal components of the protein remain attached via the disulfide bond between Cys128 and Cys142). Oocytes expressing the Npg-containing a-subunit, in the absence of photoirradiation, display an acetylcholine-induced current similar to that exhibited by the wild-type receptor. However, there is a dramatic loss in whole cell current upon photolysis. Consequently, the disulfide loop is not only essential for receptor assembly, but it also plays a key role in receptor function.

The contribution of three different tyrosine residues of the a-subunit (a93, a127 and a198) to acetylcholine receptor function was probed by incorporating side-chain caged tyrosine analogs via the nonsense suppression strategy. Caged a93 and a198 mutants expressed in patch clamped oocytes were exposed to acetylcholine. Only a weak inward current was induced followed by the expected desensitization. Brief photolysis, which is sufficient to produce 5% uncaging, induces a rapid inward current and subsequent desensitization. Flash photolysis was repeated 70 times with the majority of measured current increase coming within the first 20 flashes. By contrast, the $a127$ mutant displays significant activity prior to photolysis and subsequent photouncaging increased currents by only 2-fold. These results suggest that the tyrosine residue at this position is less influenced by acetylcholine than its counterparts at $a93$ and $a198$. The role of a leucine residue in the transmembrane segment of the γ-subunit was likewise probed using both a caged tyrosine residue and a caged cysteine moiety [10]. This particular site had been previously implicated in gating the flow of ions through the channel. Acetylcholine-induced current was observed for the caged cysteine mutant and, to a lesser extent, for the caged tyrosine mutant, prior to uncaging. This suggests that the side-chain of these residues only partially blocks the flow of ions. These investigators found that the free tyrosine residue increases the binding affinity of acetylcholine for the pore, thereby leading to inhibition of ion flow. Indeed, photolysis of the caged tyrosine mutant resulted in a loss of current that could be transiently restored upon removal of acetylcholine from the solution.

The role of the tyrosine at position 242 in Kir2.1, an inwardly rectifying K^+ channel, was examined in a fashion analogous to that described for the acetylcholine receptor [11]. In this case, the phosphorylation status of the tyrosine moiety and/or its role in key protein–protein interactions is/are believed to influence channel function. Although the now conventional site-specific mutation approach (e.g. tyrosine-to-phenylalanine) allows one to probe the importance of particular residues in protein function, the light-activated strategy allows one to restore activity, and thereby confirm that the protein has been synthesized, folded properly and translocated to its proper intracellular site. Temporally sensitive measurements of Kir2.1 activity were performed as a function of photochemical uncaging. Based on previous studies, these investigators expected and subsequently found that flash uncaging of the tyrosine moiety results in a partial loss of K^+ currents in the presence of overexpressed Src tyrosine kinase. Unexpectedly, the light-induced chemical transformation also leads to massive endocytosis of the channel. The latter suggests that the tyrosine residue (or its phosphorylated counterpart) participates in a key protein–protein interaction that is a prerequisite for endocytosis. However, the Cal Tech team was unable to obtain any evidence that the free tyrosine moiety suffers phosphorylation, which implies that the Src-mediated down-regulation of Kir2.1 proceeds in a mechanistic fashion that does not directly involve the post-translational modification of Tyr242. Furthermore, although it is tempting to conclude that endocytosis is directly responsible for the observed loss of K^+ conducting activity, these investigators demonstrated that the light-induced decrease in channel activity occurs even if endocytosis is blocked.

5.3.4
Embryogenesis

In addition to the direct light-driven conversion of an inactive protein to its active counterpart, there has been a rise in the number of studies describing the photo-initiated expression of gene products. Caged DNA [13], mRNA [14], as well as small molecule activators [15, 16] of gene expression have been reported. However, the first study to describe the light-driven transcription of a gene was reported by Minden et al. who employed the polypeptide transcriptional activator GAL4VP16 [17, 18]. Treatment of the latter with NVOC-Cl furnished a GAL4VP16 derivative in which eight out of the 14 lysine residues were covalently modified. Although caged GAL4VP16 is still able to bind a 19-nt GAL4-containing consensus sequence *in vitro*, the modified polypeptide is no longer able to drive transcriptional activity in *Drosophila* embryos. The authors note that it is possible that this level of caging might preclude binding to chromatin but not free DNA. Alternatively, the caged GAL4VP16 might be unable to engage in key protein–protein interactions that are essential for transcriptional activity.

Given the time lag between photoillumination and protein expression, the light-initiated gene expression strategy will clearly not be useful for studying relative rapid biological events, such as LTP or cell motility. However, longer-term phenomena, such as embryogenesis or disease progression (e.g. metastasis) can certainly be probed with this approach. Minden et al. employed caged GAL4VP16 as a means to examine the fate of individual cells in the developing *Drosophila* embryo carrying an appropriate GAL4VP16-sensitive transgene. Light-initiated formation of free GAL4VP16 was used to drive either β-galactosidase or green fluorescent protein (GFP) expression in specific cells, which served as markers for determining cell fate.

Mitotic domains are clusters of cells that appear early in development. Cells within these domains are thought to give rise to cellular progeny restricted to a limited set of fates. Different mitotic domains are identified by their shape, location and time of appearance. For example, the photoactivated gene expression system was used to demonstrate that the embryonic brain develops from five distinct mitotic domains. These experiments were conducted by injecting *Drosophila* embryos with GAL4VP16 during the syncytial stage of development. Photoillumination for 3–4 s was subsequently performed on individual cells or groups of cells. Although a detailed discussion of the embryonic brain fate map is well beyond the scope of this review, the authors found that the five different mitotic domains ultimately furnish distinct cell populations within the brain. For example, mitotic domain B produces glial cells whereas mitotic domain 5 furnishes the tetrad of cell clusters that comprise the antennal sensory system. One of the exciting aspects of the cell labeling approach employed in these studies is the fact that the mitotic domain starting points furnish a global context for monitoring the changing morphology of the developing embryonic brain.

5.3.5
Protein Trafficking

Patterson and Lippincott-Schwartz recently described a novel GFP derivative whose fluorescence is increased 100-fold upon irradiation with 413-nm light [19]. This protein, like the gene-activation strategy described above, can be used to photolabel individual cells. In addition, protein–GFP conjugates can be prepared as well. The fluorophore within GFP is present as both the neutral phenol (397 nm absorbance; major) and anionic phenolate (475 nm absorbance; minor). Illumination of the 397-nm peak results in the conversion of the phenol to the phenolate and thereby leads to an approximately 3-fold increase in fluorescence upon long wavelength excitation. The authors sought to identify a GFP analog that exhibited little or no 475-nm peak. Such a GFP analog should exist almost completely in the phenol form. Subsequent photoconversion to the phenolate should result in an even more pronounced increase in long wavelength fluorescence. Previous studies had shown that the long wavelength absorbance is dramatically impaired upon replacement of threonine 203 with an isoleucine. Consequently, a series of GFP analogs were prepared in which the 203 position was mutated to a variety of different residues. The corresponding GFP histidine derivative displays a minimal absorbance peak at 475 nm and upon photoconversion by 413-nm light, this absorbance peak is dramatically enhanced. Living cells expressing this GFP analog display a greater than 60-fold increase in fluorescence following 1 s photoactivation (Fig. 5.3.4).

The authors evaluated the utility of the histidine 203 GFP mutant as a labeling agent by appending it to lpg120, a lysosomal membrane protein. A small region of COS 7 cells expressing the GFP mutant–lpg120 construct was photoirradiated for 1 s to effect conversion of the GFP to the fluorescent phenolate form. A fluorescent signal was observed in lysosomes outside of the photoirradiated region within 10 s. Essentially all lysosomes displayed a significant fluorescent signal within 20 min. These results indicate that the GFP mutant–lpg120 construct exhibits a significant level of exchange between lysosomes. Furthermore, this observed exchange is blocked in the presence of microtubule disrupting agent nocodazole, which indicates that the microtubule network is essential for the movement of GFP mutant–lpg120 between lysosomes.

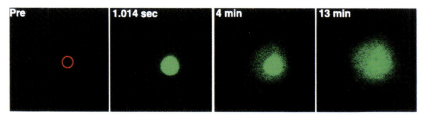

Fig. 5.3.4 A COS 7 cell expressing photoactivatable GFP prior ("pre") to and following exposure to 413-nm light. (Reproduced from *Science* **2003**, *300*, 87–91)

One of the obvious advantages associated with this GFP derivative is that it can be "switched on" within seconds, thereby allowing one to probe relatively rapid biological processes. Furthermore, the use of a cell line containing the GFP gene precludes the need to prepare, purify and microinject a caged protein.

5.3.6
Signal Transduction

Protein kinases play a critical role in mediating signal transduction pathways via the reversible phosphorylation of protein targets. These kinase pathways enable the cell to respond rapidly in a spatially sensitive fashion to extracellular and intracellular environmental changes. Although caged products of these pathways (caged cAMP, NO and Ca^{2+}) have found utility in addressing various aspects of the spatiotemporal dynamics of cell signaling, there have been very few studies devoted to the study of caged proteins as effectors of cell signaling. The groups of Bayley [20, 21] and Lawrence [22] have described the preparation of several different analogs of a caged cAMP-dependent protein kinase (PKA). PKA is a key participant in the pathways responsible for a variety of physiologically important processes (the regulation of glycogenolysis, cell motility, memory and learning, apoptosis, the cell cycle, etc.). As the name implies, the cAMP-dependent protein kinase is a direct downstream mediator of the second messenger cAMP. The enzyme is a tetrameric entity comprised of two non-covalently appended catalytic (C) subunits and a single dimeric regulatory (R) subunit. PKA is inactive in the holoenzyme form (R_2C_2), but upon association of the R subunits with cAMP, the C subunits dissociate from the complex and are now free to catalyze the phosphorylation of appropriate protein substrates. The relative intracellular concentration of the R and C subunits is essentially the same. Consequently, the free C subunit can be directly caged *in vitro* and subsequently employed in intracellular studies without fear that, upon introduc-

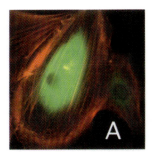 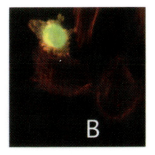

Fig. 5.3.5 Overlay images of rhodamine–phalloidin staining for F-actin (red) and FITC–IgG staining for microinjected cells (green) in (A) rat embryo fibroblasts microinjected with caged PKA, but not photochemically activated, and (B) rat embryo fibroblasts microinjected with caged PKA and photochemically activated. (Reproduced from [22]. Copyright 1998 American Chemical Society)

tion and photoactivation in the cell, its activity might be immediately downregulated by excess adventitious R subunit.

Fibroblasts are known to respond to cell permeable cAMP analogs by undergoing dramatic morphological changes. Prior to exposure, these cells exhibit stress fibers, display a relatively smooth surface, and assume a somewhat elongated shape. However, upon introduction of 8-(4-chlorophenylthio)-cAMP, the stress fibers of the fibroblasts disappear, the cell membrane ruffles, and the cells adopt a more rounded and less elongated form. Presumably this cellular behavior elicited by cAMP is mediated primarily (or completely) via PKA. This issue could be addressed by directly microinjecting active C subunit into target cells. However, it is possible that any subsequent morphological transformation might be due to the mere physical presence of free catalytic subunit and not necessarily to the activity of the enzyme. Furthermore, microinjection can have deleterious consequences that could prevent the cell from immediately exhibiting the expected response. The use of a caged protein not only allows the cell sufficient time to recover from microinjection, but subsequent photoactivation furnishes a starting point so that the response time from enzyme action to cellular behavior can be easily quantified. This key point is shown visually within the same microscopic field in Fig. 5.3.5. Cells microinjected with caged PKA were simultaneously co-injected with an inert protein that was used to subsequently identify those cells containing the modified PKA [22]. All cells were photolyzed and ultimately exposed to rhodamine-conjugated phalloidin, which binds to F-actin. Cells containing microinjected PKA (green color with red halo) have undergone a dramatic light-induced change in their appearance (loss of stress fibers, rounded shape and membrane ruffling), a process that can be directly linked to the unmasking of PKA activity (Fig. 5.3.5 B). By contrast, neighboring uninjected cells (solid red color), although exposed to photo-trigger, lack caged PKA and therefore remain in their non-PKA-activated state.

5.3.7
Bacterial Exotoxins

Bacterial defense mechanisms include the production of exotoxins, such as the pore-forming proteins α-hemolysin and streptolysin. These proteins have been the objects of a wide variety of studies that range from structural and mechanistic to applied. α-Hemolysin is secreted by *Staphylococcus aureus* as a monomer. However, it forms a heptameric pore upon insertion into target cells. Indeed, the pore-creating activity of this protein has found utility as an agent for the introduction of large molecules into cells.

Bayley et al. have created a light-activated analog of α-hemolysin and subsequently evaluated its activity as a hemolytic agent [23]. Previous extensive structure–function studies by the Bayley group resulted in the production of more than 80 different cysteine-containing hemolysin mutants (note that the wild-type protein lacks a free cysteine residue). Furthermore, approximately 20 of these mutants are inactivated upon covalent modification with an alkylating agent or Ellman's reagent. The latter suggested that modification with a photolabile substituent should likewise produce an inactive α-hemolysin. However, of the nearly 20 different mutants, only four (1) have their activity completely compromised by cysteine modification, (2) display activity, in the absence of modification, similar to that of the wild-type enzyme and (3) are able to bind to the cell surface even when covalently modified. The latter property was deemed to be essential since this would allow the corresponding caged protein to assume a pre-activated self-assembled state prior to photolysis.

Bayley et al. introduced the o-nitrobenzylbromide derivative 4 (Scheme 5.3.1) as a general water-soluble caging agent for the modification of cysteine residues in peptides and proteins. The covalently modified analog of α-hemolysin-Arg104Cys shows almost no residual activity. Approximately 80% mono-substitution occurs with the residual protein present being inactive non-cysteine-containing contaminants that are not photoactivatable. Subsequent photoirradiation restores approximately 60% of the free α-hemolysin-Arg104Cys as assessed by SDS-PAGE, but only 15% as determined by lytic activity toward red blood cells. This apparent discrepancy is explained by the fact that lytic activity requires the formation of a heptameric species. Any inactive monomers that are present likely compromise the pore-forming ability of the assemblage.

This study is significant in a number of different ways. First, the introduction of the water-soluble caging agent 4 has found utility in a number of subsequent studies. Second, although pore-forming proteins are useful as tools for introducing large molecules into cells, their application is decidedly nontrivial since exposure could compromise the integrity of the cell. The ability to turn on and subsequently rapidly turn off pore-forming activity could be extremely useful in this regard. Indeed, the Bayley group has identified α-hemolysin analogs that can be switched off by altering divalent metal ion concentration. Finally, the light-activatable α-hemolysin-Arg104Cys represents the first time that a caged protein had been used in cell-based experiments.

5.3.8
Summary and Conclusions

A small array of caged proteins has been prepared using a variety of techniques. Even at this early stage in development, the potential utility of these reagents is quite clear. Caged proteins have been used to evaluate spatially and temporally the pathways involved in cell motility, synaptic plasticity and electrical conductivity, embryogenesis, protein trafficking, cell signaling, and gene expression. These studies have demonstrated the role of specific amino acid residues in protein function, correlated temporally and spatially sensitive protein activity with cellular behavior, furnished a real-time view of protein trafficking, and identified the ultimate fate of specific cells in a developing organism. These initial studies suggest that caged proteins will serve as uniquely powerful reagents in the investigation of the biochemical basis of a wide variety of biological phenomena.

References

1 L. Cassimeris, D. Safer, V.T. Nachmias, S.H. Zigmond, *J. Cell. Biol.* **1992**, *119*, 1261–1270.
2 P. Roy, Z. Rajfur, D. Jones, G. Marriott, L. Loew, K. Jacobson, *J. Cell. Biol.* **2001**, *153*, 1035–1048.
3 J.R. Bamburg, *Annu. Rev. Cell. Dev. Biol.* **1999**, *15*, 185–230.
4 M. Ghosh, I. Ichetovkin, X. Song, J.S. Condeelis, D.S. Lawrence, *J. Am. Chem. Soc.* **2002**, *124*, 2440–2441.
5 M.F. Prorok, D.S. Lawrence, Unpublished results.
6 M. Ghosh, X. Song, G. Mouneimne, M. Sidani, D.S. Lawrence, J.S. Condeelis, *Science* **2004**, *304*, 743–746.
7 C.H. Self, S. Thompson, *Nat. Med.* **1996**, *2*, 817–820.
8 A.H. Kossel, S.B. Cambridge, U. Wagner, T. Bonhoeffer, *Proc. Natl Acad. Sci. USA* **2001**, *98*, 14702–14707.
9 P.M. England, H.A. Lester, N. Davidson, D.A. Dougherty, *Proc. Natl Acad. Sci. USA* **1997**, *94*, 11025–11030.
10 K.D. Philipson, J.P. Gallivan, G.S. Brandt, D.A. Dougherty, H.A. Lester, *Am. J. Physiol. Cell. Physiol.* **2001**, *281*, C195–C206.
11 Y. Tong, G.S. Brandt, M. Li, G. Shapovalov, E. Slimko, A. Karschin, D.A. Dougherty, H.A. Lester, *J. Gen. Physiol.* **2001**, *117*, 103–118.
12 J.C. Miller, S.K. Silverman, P.M. England, D.A. Dougherty, H.A. Lester, *Neuron* **1998**, *20*, 619–624.
13 W.T. Monroe, M.M. McQuain, M.S. Chang, J.S. Alexander, F.R. Haselton, *J. Biol. Chem.* **1999**, *274*, 20895–20900.
14 H. Ando, T. Furuta, R.Y. Tsien, H. Okamoto, *Nat. Genet.* **2001**, *28*, 317–325.
15 W. Lin, C. Albanese, R.G. Pestell, D.S. Lawrence, *Chem. Biol.* **2002**, *9*, 1347–1353.
16 F.F. Cruz, J.T. Koh, K.H. Link, *J. Am. Chem. Soc.* **2000**, *122*, 8777–8778.
17 J. Minden, R. Namba, J. Mergliano, S. Cambridge, *Sci. STKE* **2000**, *2000*, PL1.
18 S.B. Cambridge, R.L. Davis, J.S. Minden, *Science* **1997**, *277*, 825–828.
19 G.H. Patterson, J. Lippincott-Schwartz, *Science* **2002**, *297*, 1873–1877.
20 C. Chang, T. Fernandez, R. Panchal, H. Bayley, *J. Am. Chem. Soc.* **1998**, *120*, 7661–7662.
21 K. Zou, S. Cheley, R.S. Givens, H. Bayley, *J. Am. Chem. Soc.* **2002**, *124*, 8220–8229.
22 K. Curley, D.S. Lawrence, *J. Am. Chem. Soc.* **1998**, *120*, 8573–8574.
23 C.Y. Chang, B. Niblack, B. Walker, H. Bayley, *Chem. Biol.* **1995**, *2*, 391–400.

6
Photoremovable Protecting Groups in DNA Synthesis and Microarray Fabrication

Michael C. Pirrung and Vipul S. Rana

6.1
Introduction

Light and DNA are sometimes thought to be incompatible, with great concerns about the genetic mutations created by excessive sunlight exposure. However, when applied judiciously, light is an excellent reagent for promoting chemical changes in DNA molecules, especially when those changes must be restricted in time or space – one of the major themes of this volume.

6.2
Photoremovable Groups used in Conventional Nucleic Acid Synthesis

In nucleoside and nucleotide synthesis, the presence of multiple functionalities demands a selection of protecting groups that can be used simultaneously and removed selectively. Light can be exploited as one of the "orthogonal" reagents used to remove specific groups. Photochemical removal should generally occur in the 300–400-nm region. These wavelengths are long enough that the absorptions of the heterocyclic bases are weak, so that photochemical side-reactions of the bases should be minimal, yet short enough that concerns with unintended removal of the protecting group due to laboratory lighting are small, permitting handling of compounds without extraordinary precautions.

An early example of the use of a photoremovable protecting group in nucleoside chemistry (Scheme 6.1) comes from Patchornik [1], who originated and developed much of the chemistry of the prototypical photoremovable protecting groups, the *o*-nitrobenzyl groups. He prepared reagent **1** and used it in the phosphorylation of the 5'-hydroxyl of thymidine. Irradiation to produce **3** was performed in the presence of a polymer-bound acyl hydrazine, which serves to sequester a reactive and problematic byproduct of most nitrobenzyl photochemical deprotection reactions – the nitrosocarbonyl compound.

A serious concern in RNA synthesis is protection for the 2'-hydroxyl group of ribonucleosides, because the hydrolytic instability of RNA is a direct consequence of the participation of the 2'-hydroxyl in phosphodiester cleavage. The

Dynamic Studies in Biology. Edited by M. Goeldner, R. Givens
Copyright © 2005 WILEY-VCH Verlag GmbH & Co. KGaA, Weinheim
ISBN: 3-527-30783-4

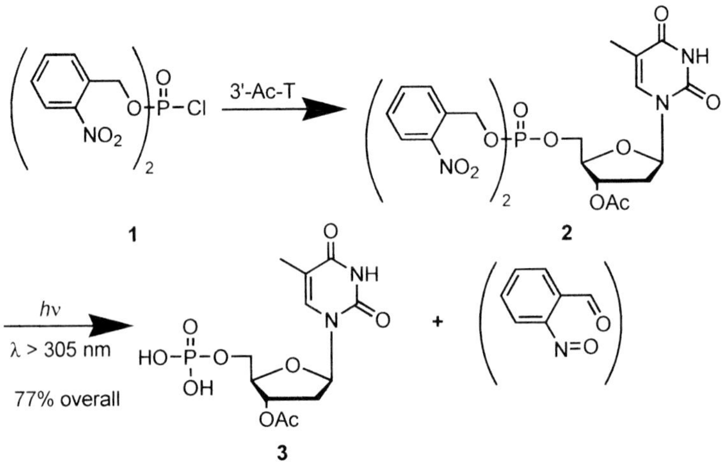

Scheme 6.1.1 Use of nitrobenzyl groups for protection in deoxyribonucleoside phosphorylation.

2′-hydroxyl must therefore remain protected throughout solid-phase synthesis and during all protecting group removal following the synthesis. The final step before use of the RNA must be removal of the 2′-protecting group. To prevent RNA degradation, the reaction conditions for 2′-protecting group deprotection must not be basic and likewise for manipulations following deprotection. Photochemistry offers an excellent solution to this challenge, since no deprotection reagents need be removed. Introduction of the nitrobenzyl group is performed by reaction of diazoalkane **4** with the native ribonucleosides [for U, C, A and I (inosine) bases, Scheme 6.2] [2] or by reaction of the nitrobenzyl bromide with the N-protected nucleoside under basic conditions (for the G base) [3]. These are not selective reactions, with the consequence that RNA building blocks like **5** cannot be obtained efficiently, but they can be used to form oligonucleotides. The photochemical removal of the nitrobenzyl groups from the building blocks (350-nm lamps in a Rayonet reactor) can be accomplished without affecting either purine or pyrimidine bases, but the removal of these protecting groups from synthetic RNAs with light above 280 nm is tolerated only by pyrimidine-containing sequences [4].

Because the formation of nitrobenzyl ethers of sugars and nucleosides has presented an ongoing synthetic problem, alternative protecting groups have been developed, the nitrobenzyloxymethyl ethers (Scheme 6.3) [5]. The o-nitrobenzyloxymethyl chloride **7** is more reactive than o-nitrobenzyl bromide and is used to alkylate a ribonucleoside stannylene derivative such as **6**. This reaction is somewhat selective for the 2′-OH. The resulting nucleosides **8** can be converted to 5′-O-DMTr-3′-O-β-cyanoethylphosphoramidites for use in synthesis of RNAs such as hammerhead ribozyme sequences. After removal from the support and deprotection, the 2′-protected oligoribonucleotides were exposed to

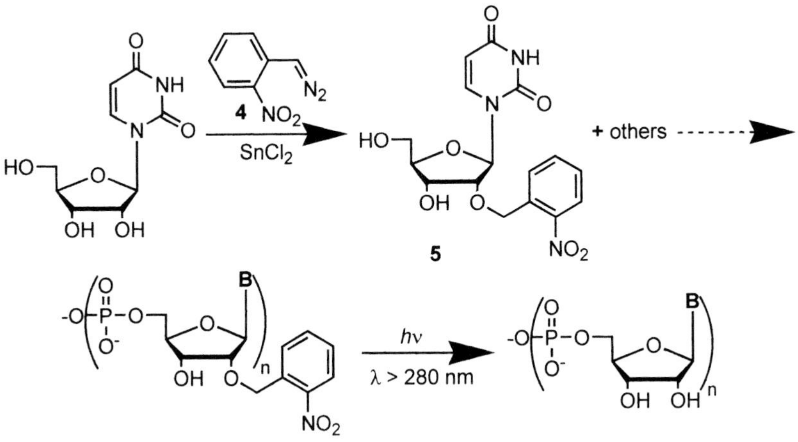

Scheme 6.1.2 Use of 2'-O-o-nitrobenzyl groups for protection in (poly)ribonucleotide synthesis.

"long-wavelength" UV light in 50% aq. tert-BuOH, pH 3.7, to remove the protecting group.

The deprotection of nitrobenzyloxymethyl groups presumably occurs in two distinct stages, with the photochemical stage generating a formaldehyde hemiacetal and an acid-catalyzed stage freeing the 2'-hydroxyl. The overall deprotection reaction is said to be significantly accelerated by acid, but the basis for this acceleration has not been explained. The pH dependence of the elementary steps in the photochemical deprotection of a simple nitrobenzyl ether has been studied; at high pH, the conversion of **9** to **10** is rate determining (Scheme 6.4) [6]. At lower pH, the release of XH is rate determining, with the minimum deprotection rate occurring at pH 3–6. Since the overall nitrobenzyloxymethyl deprotection reaction is fastest at pH 3.7, the rate-determining stage in its deprotection must be the hydrolysis of the formaldehyde hemiacetal intermediate. Interestingly, nitrobenzyloxymethyl groups also can be removed from synthetic RNA sequences fairly easily by fluoride ion (anhydrous TBAF/THF) [7]. One can speculate that this process is a base-promoted reaction that would apply equally well to simple nitrobenzyl ethers, so the sensitivity of both of these groups to fluoride must be kept in mind. Presumably, fluoride-promoted deprotection involves generation of the *aci*-nitro compound **11**, which is protonated to give **9** in a net tautomerization. This quinone methide-like intermediate is known to be formed in o-alkylnitrobenzene photochemistry [8] and is thought to be important in nitrobenzyl photochemical deprotection reactions. Evidently, the basicity of fluoride ion is adequate to promote the tautomerization reaction but insufficient to promote RNA hydrolysis.

The nitrobenzyloxymethyl method has been applied to the synthesis of unnatural enantiomers of RNAs [9] and a protocol has been developed to deal with the highly absorbing nitrosobenzaldehyde byproduct that filters the light being

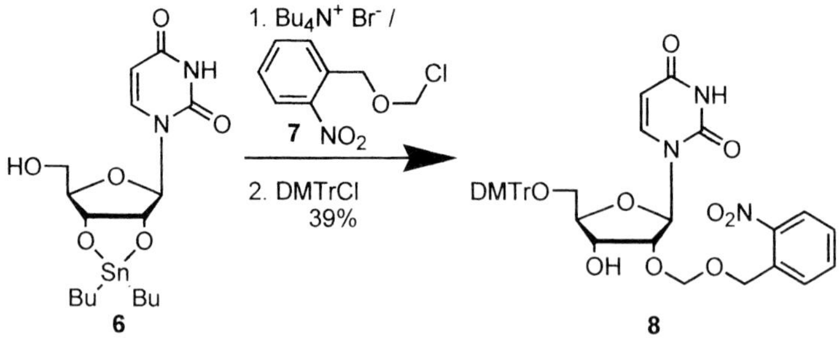

Scheme 6.1.3 Use of 2′-O-o-nitrobenzyloxymethyl groups for protection of ribonucleotides.

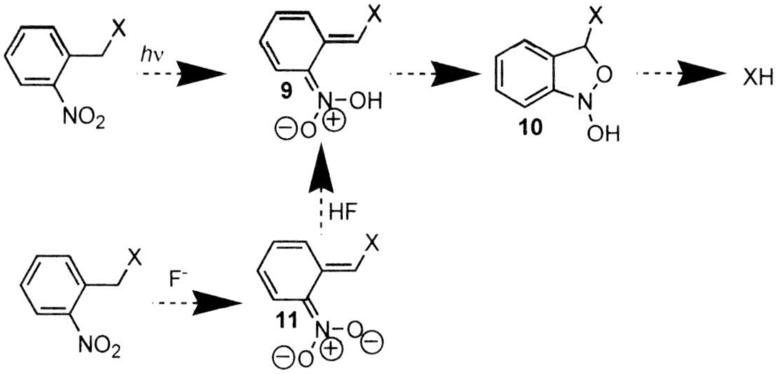

Scheme 6.1.4 o-Nitrobenzyl group deprotection under photochemical and fluoride conditions.

used for the final deprotection of the RNA. A continuous extraction with an organic/aqueous solvent system removes this aldehyde. The nitrobenzyloxymethyl method has been extended to the o-nitrophenethyloxymethyl group (Scheme 6.5) [10]. This group is introduced by protection of the 5′-O-DMTr, base-protected ribonucleosides with a chloromethyl ether (similar to **7**), providing **12**. The specific value of methyl substitution on the nitrobenzyl group is not explained, though it is stated that the o-nitrophenethyloxymethyl group is completely orthogonal to the (triisopropyl)siloxymethyl group that is also used as a 2′-O-protecting group in the same work; it of course is removed by fluoride ion (TBAF·3H$_2$O/THF). Perhaps stability to fluoride is conferred by the methyl substitution, which might be based on steric effects. Another benefit of methyl substitution at the benzylic position might be enhanced photochemical reactivity, which has been observed earlier (*vide infra*). It has been suggested that alkyl substitution stabilizes *aci*-nitro intermediates like **9**, favoring their subsequent cyclization. The idea that benzylic alkyl substitution stabilizes the *aci*-nitro intermediate runs counter to the observation of increased resistance to fluoride-cata-

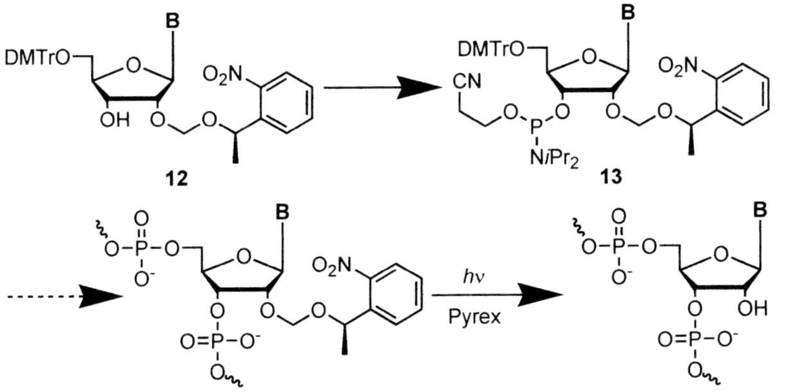

Scheme 6.1.5 o-Nitrophenethyloxymethyl can be used in "caged" RNA synthesis.

lyzed deprotection, however. Benzylic methyl substitution also leads to a nitrosoketone byproduct, which is surely less reactive than nitrosobenzaldehyde and therefore less problematic, although the nitrosoketone should have UV absorption similar to the aldehyde. After conversion of nucleosides **12** to the activated building blocks **13**, these materials were used with 2′-O-silyloxymethyl-protected building blocks in phosphoramidite-based RNA synthesis. The resulting sequences bear a photolabile 2′-protecting group at specific sites in the RNA that were of interest for their catalytic roles in ribozymes and thus constitute "caged" RNAs. The removal of the 2′-O-nitrophenethyloxymethyl groups from the ultimate RNA sequences is performed with Pyrex-filtered UV light. Despite some of the concerns with these acetal-type protecting groups, their decreased steric bulk as compared to the 2′-O-tert-butyldimethylsilyl group traditionally used to protect the 2′-hydroxyl in RNA synthesis offers the advantage of more rapid and complete phosphoramidite coupling. These methods may be competitive with the "2′-ACE" method for RNA synthesis, which uses an orthoester as a protecting group for the 2′-hydroxyl and a silyl group for the 5′-hydroxyl [11].

Nitrobenzyl groups have also been used to protect otherwise reactive sites in DNA during solid-phase synthesis. One such site is the highly base-labile deoxyribonolactone, which is a product of oxidative DNA damage. Sheppard has prepared the amidite building block **14** from deoxyribose. It can be incorporated into DNA strands using conventional automated methods (Scheme 6.6) [12]. In buffered aqueous solution, in the context of either single-stranded or double-stranded DNA, the nitrobenzyl group can be removed by irradiation. The resulting cyanohydrin spontaneously converts to lactone **15**. This methodology has enabled the characterization of deoxyribonolactone-containing DNA by mass spectrometry and permitted its subsequent reactions to be carefully studied.

An interesting facet of earlier work on nitrobenzyl protection for nucleosides and nucleotides is that much of this work was done with the unadorned nitrobenzyl group, and little concern has been expressed about the absorption prop-

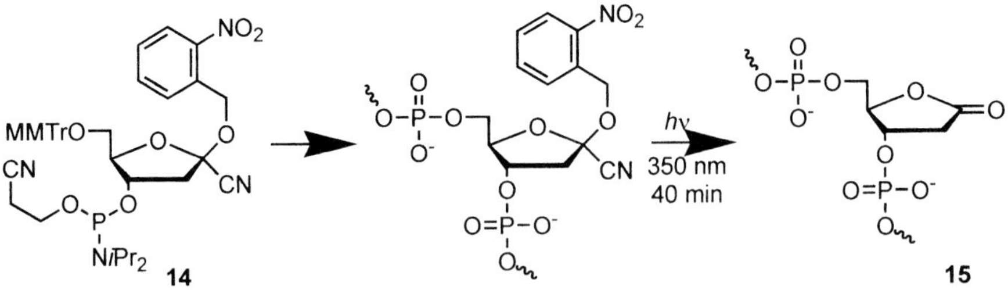

Scheme 6.1.6 o-Nitrobenzyl cyanohydrin glycosides are precursors to unstable deoxyribonolactones.

erties of the nitrobenzyl chromophore or the wavelength and rate at which deprotection occurs. Conversely, the deprotection wavelength has occupied a significant part of the research on superior protecting groups for DNA microarray production.

The 3′,5′-dimethoxybenzoin group (DMB, **16**) is another protecting group that has been used in oligonucleotide synthesis. It was first reported as a photosensitive protecting group for carboxylic acids by Sheehan et al. [13]. Deprotection of 3′,5′-DMB esters occurs with high quantum yield (0.64) and formation of the relatively inert 2-phenyl-5,7-dimethylbenzofuran byproduct **17** (Scheme 6.7). Under some circumstances, this compound undergoes [2+2] photocycloaddition to give a dimer. DMB protection and photochemical removal could be extended to phosphate esters and have been extensively used for applications in caged phosphates [14]. The DMB phosphate protecting group was subsequently adapted for a 5′→3′ inverse photochemical phosphotriester DNA synthesis method [15]. Nucleotide building blocks **18** were synthesized from nucleosides bearing (allyloxy)carbonyl [alloc (AOC)] N-protecting groups. Earlier studies (discussed below) showed that benzoyl N-protecting groups trigger DNA destruction during photochemical deprotection reactions. In contrast, the alloc groups present no difficulties during removal of the DMB phosphate group, which can be accomplished in essentially quantitative yield. The 3′-phosphotriesters **18** bearing conventional β-cyanoethyl and photolabile DMB protecting groups were prepared from 5′-silyl nucleosides via coupling of DMB-OH to the 3′-phosphoramidites and desilylation. Oligonucleotide synthesis begins with nucleotide **19** derivatized at its 5′-end. Irradiation deprotects the phosphotriester, generating a phosphodiester that is converted to its amine salt (Scheme 6.8). Such salts are conventionally used for coupling with nucleoside hydroxyl groups such as are present in building blocks **18**. In this example the reaction is promoted in solution by a sulfonyl halide in the presence of a nucleophilic catalyst to prepare the dinucleotide **B₁–B₂**. DMB photochemical deprotection and sulfonyl halide-promoted coupling constitute the coupling cycle for chain extension by one nucleotide. This method could be used to prepare short oligomers, including a sequence with adjacent T residues. Degradation and analysis of this product demonstrated that no T–T cyclobutane dimers had been generated during the photochemical

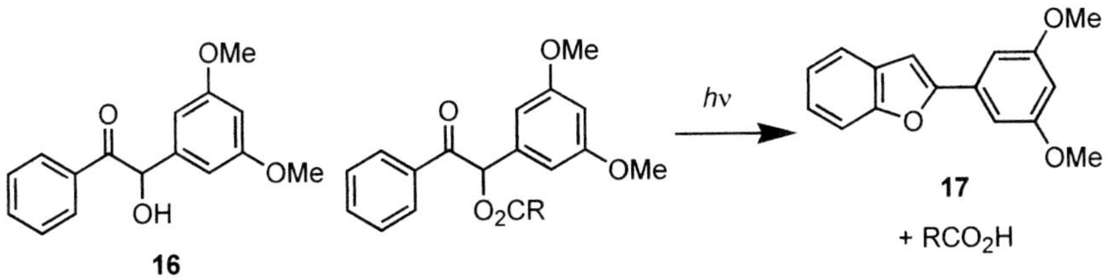

Scheme 6.1.7 Deprotection of DMB esters.

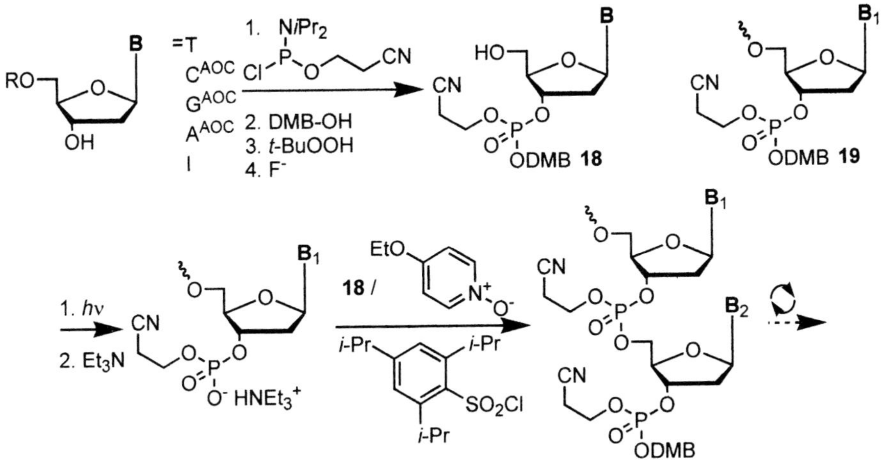

Scheme 6.1.8 Inverse DNA synthesis with 3'-DMB phosphate.

deprotection step. This work permitted a readily available and known protecting group for phosphate to be used for a light-based DNA synthesis, since protecting groups for nucleoside hydroxyls that were photoremovable with high efficiency had not then been developed. The relatively slow coupling of the β-cyanoethyl phosphodiester frustrated application of this method to light-directed synthesis (*vide infra*), however. The *o*-chlorophenyl group is more commonly used as the internucleotide protecting group in phosphotriester DNA synthesis, but resulted in side reactions when paired with the DMB group. (The β-ketone of the DMB group causes the difficulty. Nucleophilic addition to the phosphate is generally accelerated with electron-deficient ester groups like *o*-chlorophenyl. The same effect that promotes the coupling reaction promotes participation of the β-ketone of the DMB group, creating a cyclic phosphate intermediate that does not couple well with nucleoside 5'-hydroxyl groups.)

6.3
The Photolithographic Method for Microarray Fabrication

Some principles for the fabrication of molecular microarrays using light were delineated in 1991 [16]. This technology combines two fields, i.e. photolithography and solid-phase synthesis, that themselves had been highly developed and optimized. As disparate fields, their combination was not trivial, but the method is simple conceptually. The utility of microarrays, particularly DNA microarrays, has become sufficiently apparent that many other methods for microarray fabrication have been developed [17–19]. An essential feature of microarrays is that the identity of each compound on the array is based on its position. Microarrays prepared by *in situ* synthesis are thus examples of spatially addressable combinatorial libraries [20]. Whereas conventional solid-phase synthesis is conducted such that molecules can be released from the support for use following completion of the synthesis, it is essential that molecules prepared on a microarray remain attached to it during use in order to maintain their identity. This requirement means that the linkers used for conventional solid-phase synthesis must be replaced with linkers that are stable to all deprotection conditions. While this is rarely a challenging problem chemically, another requirement of light-directed synthesis has been quite challenging: the development of photoremovable protecting groups that have all of the desirable performance characteristics that are the focus of this review. The discovery of such groups has taken the better part of a decade.

The overall protocol for light-directed synthesis is given in Fig. 6.1. A pattern of illumination determines the regions of a surface that are activated for chemical coupling. Initial work used photolithographic masks, which can be thought of as analogous to stencils, to direct light to some areas and not others. In this example, light removes photolabile protecting groups from selected areas and then the entire surface is exposed to one member of a set of building blocks, each of which bears a photolabile protecting group. Building blocks react only at sites that were addressed by light in the preceding step. The substrate is illuminated in another area for deprotection and coupling with a second building block. The steps of (1) addressing by light and (2) coupling are repeated to build up various oligomers at different sites on the surface. The patterns of the light and the sequence of the building blocks used in each step define the ultimate sequences synthesized and their locations on the surface. While not shown in Fig. 6.1, light-directed synthesis permits the preparation of many more sequences than the number of steps in the synthesis (one definition of combinatorial synthesis) because it conducts reactions in parallel at many locations simultaneously.

As in all syntheses, the overall yield in a solid-phase synthesis is the product of the yields of the individual steps and, because solid-phase synthesis does not permit purification of synthetic intermediates, the purity of the synthesized product can be no greater than the overall yield. Premiums have thus been placed on developing highly reliable chemistries for solid-phase synthesis that can proceed in as close to quantitative yield as possible. As will be seen, microarray fabrication

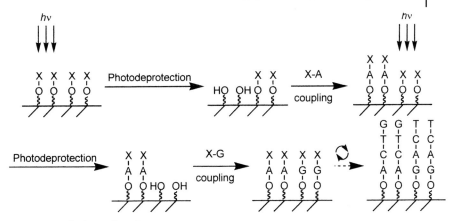

Fig. 6.1.1 Light-directed synthesis.

has even more demanding requirements. Conventional solid-phase synthesis has several modes of failure. One is incomplete deprotection, which will lead to molecules containing a deletion from the intended sequence if deprotection occurs in a subsequent cycle. These failures are generally addressed by being certain to carry deprotection reactions to completion, far beyond many half-times. Another failure is incomplete coupling, which will also lead to a deletion sequence if coupling occurs in a subsequent cycle. Unlike deprotection, which ought to be fairly independent of the sequence to be prepared, coupling can be quite dependent on the particular monomer being added and the monomer at the terminus of the growing oligomer chain. Coupling failures can be addressed in two ways. Carrying the reaction to completion can be attempted by repeating the coupling reaction ("double coupling"). "Capping" can be performed by irreversibly terminating the synthesis using a highly reactive reagent to create a product that is inert to further deprotection. Capping creates truncation sequences, which are generally more easily separated from full-length sequences than deletion sequences, but does nothing to affect the overall yield.

Like conventional solid-phase synthesis, in light-directed synthesis *intermediates* are not purified, but unlike solid-phase synthesis, in light-directed synthesis *products* cannot be purified before use. They must remain attached to the site at which they were prepared, in whatever state of purity they were generated. Capping is thus irrelevant to product purity in light-directed synthesis, but the choice of whether or not capping is used determines whether impurities are deletion or truncation sequences. Because the use of DNA arrays is based on hybridization, which is dependent on the length of the probe DNA that has been synthesized on the support, truncation sequences would usually be the more "desirable" impurity. However, the effects of truncation sequences on the performance of a DNA probe have been modeled experimentally, with the conclusion that, compared to a faithful probe, the presence of truncations reduces the ability of DNA probes to distinguish similar DNA sequences [21]. The desire therefore persists to faithfully generate intended sequences by taking deprotection and coupling steps as far toward completion as possible. However, another feature of light-directed synthesis, the necessity for deprotection to occur at some sites and not at others, limits the usefulness of the strategy of taking deprotection reactions to completion. If a deprotection reaction occurs at a site that was not intended for a coupling, an unintended monomer will be inserted into the intended sequence at that site. Such events, called "stowaways", are not a feature of conventional solid-phase synthesis. They occur because of imperfections in the ability to use light to limit reactions to specific, intended sites. Light may fall into unintended areas because of optical faults, but it can also cause deprotection in unintended areas simply because of the first-order kinetics of photochemical deprotection reactions [22]. These events lead to stowaways because the whole array is exposed to reagents in each coupling cycle. This feature is essential to gain the advantage of parallelism in the preparation of the array, yet it permits coupling to occur at unintended sites. One can easily see how stowaways limit the utility of the strategy of taking deprotection reactions to completion. Longer deprotection times at intended sites also cause greater deprotection at unintended sites, leading to stowaways. A detailed kinetic analysis of this phenomenon has recently been reported [22] and the consequences in one array fabrication technology have also been shown [23].

Considering all of these factors, the drive toward quantitative yields in deprotection and coupling reactions to generate the most faithful sequences has a greater imperative on microarrays than it does in conventional solid-phase synthesis. Yet, until very recently, attaining quantitative photochemical deprotection yields on surfaces has not been possible. The advent of the NPPoc group has made this possible. While there is no proof on this point, it seems likely that the deprotection reaction is the yield-limiting step in each coupling cycle. An advantage of microarray fabrication over conventional solid-phase synthesis is that the reaction sites bear such small molar quantities of functional groups that coupling reagents are used in huge molar excesses without prohibitive costs. Still, it is fair to say that the sequences that have been generated on extant microarrays have not been very pure. The utility to date of photolithographically generated DNA microarrays in genetic analysis is thus not related to the quality of the sequences thereon. Rather, it must reflect properties intrinsic to the way DNA sequences are used in hybridization and the considerable experience among molecular biologists in gaining valid data from DNA molecules that would not meet a criterion of "purity" to which most chemists would adhere.

6.3.1
Background – Polypeptide Microarrays

The initial demonstration of the preparation of a molecular microarray [16] generated peptides based on the (one-letter code) sequence YGGFL, the Leu-enkephalin pharmacophore. The amino acid building blocks **20** required for the synthesis are hydroxybenzotriazole active esters and bear the (nitroveratryloxy)-

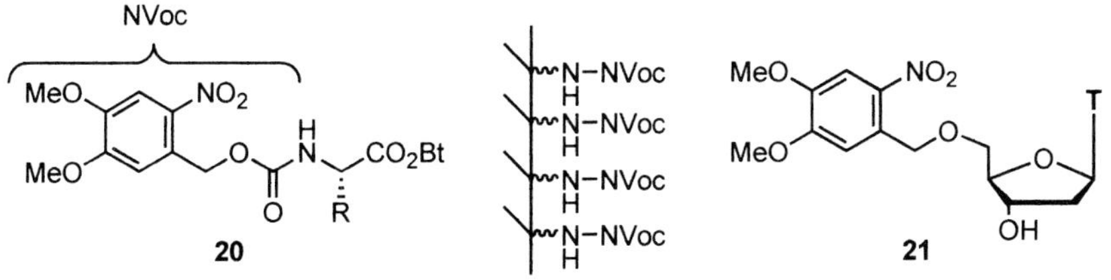

Fig. 6.1.2 Reagents and substrate for the initial demonstration of peptide and nucleotide bond formation on microarrays.

carbonyl (NVoc) protecting group, which had been developed for light-based peptide synthesis [24]. NVoc is preferred over a simple nitrobenzyl chromophore because of its red-shifted λ_{max} that permits deprotection to occur at wavelengths above 320 nm, at which tryptophan is not affected. Also required was a surface to which linker molecules were attached and which were additionally terminated with amines protected with NVoc groups (Fig. 6.2). Through repeated irradiation and coupling, sequences based on Leu-enkephalin were built up and their binding to an antibody that recognizes Leu-enkephalin was assessed by fluorescence microscopy. The domains in which these peptides were prepared were as small as 50 µm×50 µm. This work also reported the synthesis of a dinucleotide. The nitroveratryl-protected thymidine **21** was prepared by methods described above and attached to a glass surface by its 3′ end. Following patterned irradiation, it was coupled with a deoxycytidine phosphoramidite that was functionalized with a protected amine. After deprotection, the amine was labeled with fluorescein for detection of the areas in which the 5′-nitroveratryl group had been removed and phosphoramidite coupling had occurred. While this early work did not address the quantum efficiency or rate of photochemical deprotection, literature precedents predict that the efficiency of the deprotection of nitroveratryl ethers and carbamates would be low. NVoc-cyclohexylamine is deprotected with a quantum yield of 0.06, whereas nitroveratryl methyl ether is deprotected with a quantum yield of 0.09.

6.3.2
Early DNA Microarray Studies

The first DNA microarray of significant complexity generated by photolithography was reported in 1994 [25]. This work introduced the methylnitropiperonyl-oxycarbonyl (MeNPoc) photoremovable group, which via its cyclic ethers has a significantly greater absorptivity at longer wavelengths than unsubstituted nitrobenzyl groups. The fact that it is a mixed carbonate makes for far easier introduction than the nitrobenzyl ethers discussed above and the carbonate has not proved unstable or to interfere with phosphoramidite-based DNA coupling chemistries. The MeNPoc group was added to the 5′-hydroxyls of N-protected nucleosides to give derivatives **22** (Fig. 6.3). The rates of deprotection of these

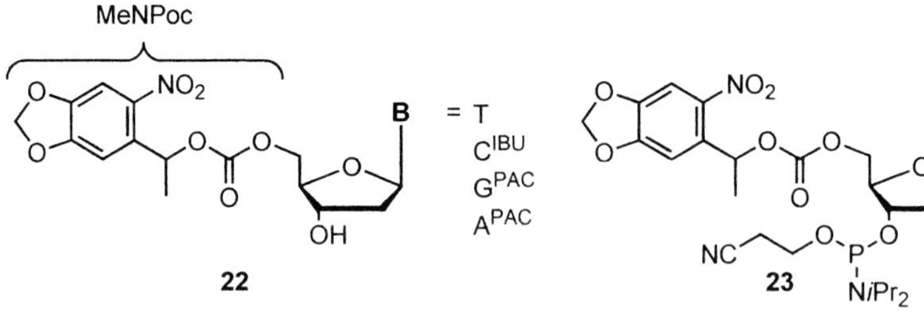

[Structures 22 and 23 at top]

Fig. 6.1.3 5′-MeNPoc nucleosides and phosphoramidite building blocks.

compounds in short pathlength dioxane solutions with 365 nm light of known intensity were determined, giving half-times of 18–31 s. Based on these rates, their irradiation time for deprotection was set at nine times the half-time of the slowest group, or 4.5 min. Validation of DNA synthesis with these building blocks was first performed on controlled-pore glass beads, which gave coupling efficiencies for amidites **23** of 95–100%. The coupling efficiencies on glass slide substrates were determined as shown in Scheme 6.9. A building block B_1 was attached to the substrate and a section was deprotected and coupled with a second building block B_2. Another section was deprotected and coupled with a fluorescein phosphoramidite **24** (Fig. 6.4), which can react with newly deprotected hydroxyl groups as well as any hydroxyl groups remaining from incomplete coupling with B_2. After deprotection of the nucleotides, the surface is imaged and the ratio of fluorescence in the first and second sections provides a measure of the efficiency in coupling of B_2 to B_1, which was 85–98%. This work represents the initial description of a general protocol for the determination of "yields" of reactions on a surface, wherein fluorescent labeling of free hydroxyl groups and imaging determines the number of hydroxyl groups in different sections of the substrate and thereby provides relative yields. Note that because these cycle efficiencies are relative, they do not provide information concerning overall yield or imputed purity of the sequences. Consider a situation in which the first reaction is taken only to 50% conversion in a given region with a given irradiation time. Using that same irradiation time in a subsequent irradiation in

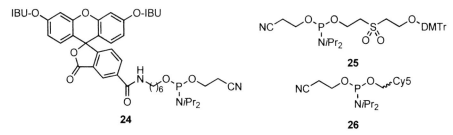

Fig. 6.1.4 Reagents for fluorescent labeling of hydroxyl groups on microarray surfaces (Cy5 = a cyanine dye)

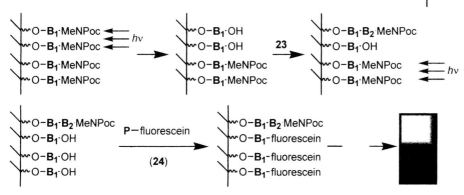

Scheme 6.1.9 Determination of surface coupling efficiency of MeNPoc amidites.

an adjacent region should deprotect the same 50% of hydroxyl groups and give the same fluorescence signal in a labeling reaction. The relative yield is 100%, but the absolute yield in each region is still 50%, with the result that coupling reactions at such sites will create mixtures of sequences.

6.3.3
Methods to Assess Chemical Reactions on DNA Microarrays

The foregoing has emphasized the importance of knowing the extent of deprotection on a glass substrate. Demonstrating quantitative deprotection and coupling for light-directed synthesis on surfaces is much more challenging than it is for conventional solid-phase synthesis, where the cleave-and-characterize approach always works. The amounts of material at individual sites on a microarray are so small that such direct approaches fail. Much of the progress in methods for assessing chemical reactions on a chip surface has been made by McGall [26]. The rate of deprotection on a substrate of a family of nucleosides bearing a specific photoremovable group can be determined as follows. Irradiation in stripes and coupling of individual building blocks creates substrate **27** (Scheme 6.10). Irradiation in a stripe orthogonal to the first for a specified time followed by translation of the stripe and irradiation for longer times provides a set of time points for the extent of alcohol release. The resulting alcohols are coupled with a fluorescein phosphoramidite and the fluorescence in each stripe is imaged. The fluorescence intensity in each stripe is plotted versus time to give a reaction progress curve from which rate constants can be determined.

Similar methods can be used to study the cycle yields in microarray preparation. In this example, a surface is derivatized with a nucleoside phosphoramidite bearing a MeNPoc group (Scheme 6.11). A vertically striped photolithographic mask is used in a first irradiation, and the phosphoramidite is coupled to exposed sites. The mask is then translated horizontally by the fraction of its width corresponding to the inverse of the number of coupling cycles over which yields are to be determined. That is, if the yield over 12 coupling cycles is de-

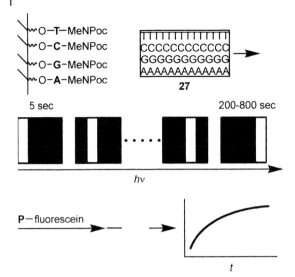

Scheme 6.1.10 Determination of deprotection rates for amidites **23** on a glass substrate **27** (black areas are masked from light).

sired, the mask should be translated by 1/12th of its width. Irradiation and coupling are iterated. Due to overlap of the areas of irradiation at each of the steps, this protocol leads to the synthesis of dodecanucleotides in the right-most region, reflecting 12 coupling steps. Regions to the left reflect monotonically decreasing numbers of coupling cycles (the mirror image of this arrangement is also produced to the right). The array is uniformly irradiated to free 5'-hydroxyl groups in all regions, which are coupled with fluorescein phosphoramidite. Fluorescence imaging permits the yield for each cycle to be determined based on the relative fluorescence intensity of adjacent regions, whose oligonucleotides differ in length by one base.

A somewhat related concept was developed by Beier and Hoheisel in the validation of one of the newer protecting groups, NPPoc [27]. Following the synthesis of a target oligonucleotide array, a final coupling is performed with the phosphoramidite **25** and then the dye-phosphoramidite **26** (Scheme 6.12). Provided that all failure sequences are capped, this reaction should label only full-length sequences. Imaging of substrate **28** can establish that each of the probe sites on the array bears equivalent fluorescence intensity. However, since the array is generally used with fluorescent target DNA, it is necessary that this fluorescent group be removed from the probes. This is facilitated by the linker in **25**, which is readily cleaved by β-elimination under the same ammonia conditions that are used to remove all of the exocyclic amine [phenoxyacetyl (PAC)] and internucleotide phosphate (β-cyanoethyl) protecting groups in the synthesized sequences. This method enables sites with synthesis failures to be identified be-

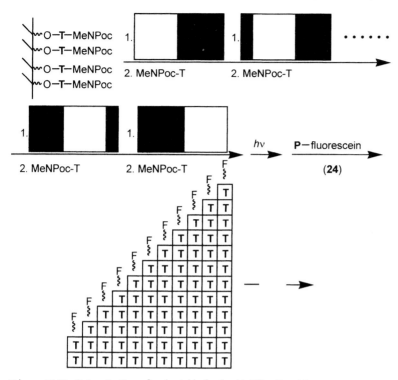

Scheme 6.1.11 Determination of cycle yields for the MeNPoc-T amidite.

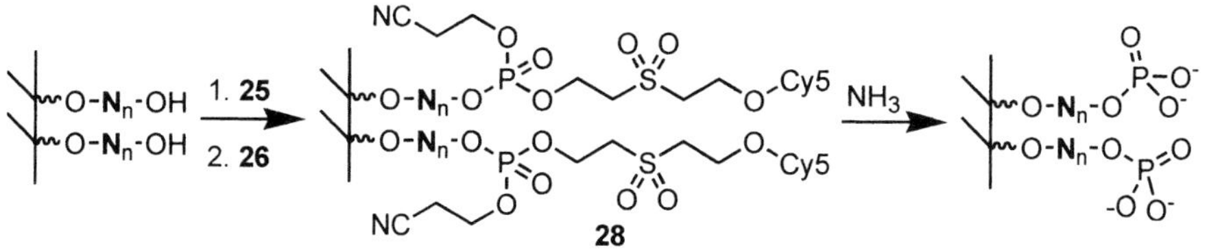

Scheme 6.1.12 Determination of oligonucleotide quality on a microarray (N_n = oligonucleotide; Cy5 = a cyanine fluorescent dye).

fore the array is used. It also creates DNA terminated with a phosphate group, which is not an issue in hybridization assays but might be a problem in advanced methods for microarray use involving enzymatic manipulations at the terminus of the DNA [28]. The presence of this phosphate could prevent enzymatic processing.

6.3.4
Protecting Groups

A wide variety of photochemically removable protecting groups are certainly known and quite a few are suitable for nucleoside hydroxyl groups. An ongoing concern has been the wavelength of deprotection, which certainly must be above 300 nm to avoid creation of photochemical side products such as T–T dimers. Nevertheless, the photochemical reactivity of N-protected bases during synthesis has proved to be a much larger problem than T–T dimer formation. The absorption maximum of the photoremovable protecting group can be shifted into the near UV with substituents, and the mercury lines at 365, 400 and 435 nm can be used for deprotection. A balance must be struck between red-shifting groups that enhance absorption at longer wavelength but correspondingly decrease the energy of the excited state, which may decrease its reactivity and therefore the product quantum yield. While earlier work on photochemically removable protecting groups has not investigated sensitization in much detail, the use of sensitizers has also been fruitful in promoting deprotection reactions at wavelengths not injurious to DNA.

A useful measure of the overall photochemical reactivity of a group at a particular wavelength is the product of its extinction coefficient and quantum yield [29]. Such data are given for several of the groups discussed in this work in Tab. 6.1. For dilute solutions at low absorbance, in principle there should be a relationship between this reactivity parameter, the light intensity, and the rate of deprotection, such as is expressed in Eqn. (1) [22]. During the preparation of this review, attempts to generate a linear correlation between rate and reported irradiation power (scaled by $\varepsilon\Phi$) for all of the different protecting groups presented here were unsuccessful. This is likely due to the fact that data were reported from different laboratories using different experimental set-ups. However, within a single study, this relationship appears to be obeyed.

$$K = N_a I \varepsilon \Phi \tag{1}$$

where Φ is the quantum yield, ε is the extinction coefficient at the wavelength of irradiation, I is the irradiation intensity and N_a is Avogadro's number.

Tab. 6.1.1 Photophysical properties of 5'-O-protected thymidines

	ε_{365} (M^{-1} cm^{-1})	Φ	$\varepsilon\Phi$
NPEoc	179	0.033	5.90
NVoc	5100	0.0013	6.63
NPPoc	294	0.35	102.9
DMBoc	22	0.64	14
MeNPoc	3000	0.1	300

6.3.4.1 Nitrobenzyl Groups

The MeNPoc group developed by Holmes [30] enabled a great deal of progress in DNA microarray fabrication. The work of McGall [26] at Affymetrix established that deprotection of MeNPoc groups on nucleosides can be achieved using near-UV wavelengths (λ_{max} = 345 nm, $\Phi \sim 0.05$). Some simple light-filtering studies established that deprotection using an arc lamp source is due almost exclusively to the mercury line at 365 nm. The reaction rates are linearly dependent on power in the range of 5–50 mW cm^{-2}, rapid (less than 1 min to complete at 50 mW cm^{-2}) and independent of either the nucleotide base or oligomer length, making it convenient to use a single exposure time during an entire microarray fabrication. Deprotection does not require any special catalysts or co-reactants and is most efficient in the presence of non-polar solvents or in the absence of solvent ("dry" irradiation). Irradiation without solvent offers attractions during manufacturing, so in fact DNA microarrays marketed by Affymetrix are prepared in this way. The building blocks for MeNPoc microarray fabrication are prepared via reaction of nucleosides bearing "easy off" N-protecting groups with MeNPoc-Cl (Scheme 6.13). The choice of N-protecting groups is important. Benzoyl is commonly used in conventional solid-phase DNA synthesis, but can cause destruction of the DNA when present during photochemical synthesis, particularly on C. Bases bearing the PAC group are not photochemically reactive at 365 nm and, because of the inductive effect of the alkoxy group, PAC can be removed under mild conditions that do not affect the linkage of the DNA to the array. Isobutyroyl (IBU) can be used in a similar way, primarily on C.

The main problem with the MeNPoc group is its cycle yields. Coupling cycles using MeNPoc building blocks for preparation of homopolymers (dodecamers) on a microarray have stepwise efficiencies of 81–94%, which are believed to be primarily determined by the photochemical deprotection step. The purines, particularly G, are the worst performers. These cycle yields proved refractory to extensive modification of reaction conditions. In contrast to the situation on the array, not only could compounds **22** be deprotected in solution in quantitative yield, but compound **29** (Fig. 6.5), a representative model for the chain end of a growing oligonucleotide, could be also. Attempting to mimic the close spacing of chain termini on an array, compound **30** was prepared, but it could also be deprotected in quantitative yield in solution. Thus, the reasons for the less-than-quantitative yields for MeNPoc deprotection on the array surface remain obscure.

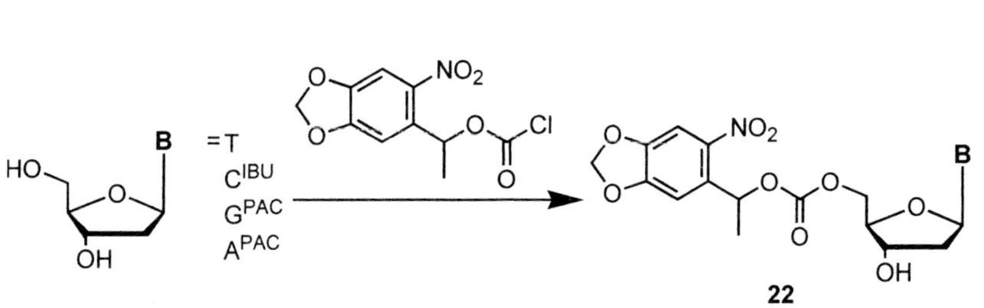

Scheme 6.1.13 Preparation of MeNPoc nucleoside building blocks.

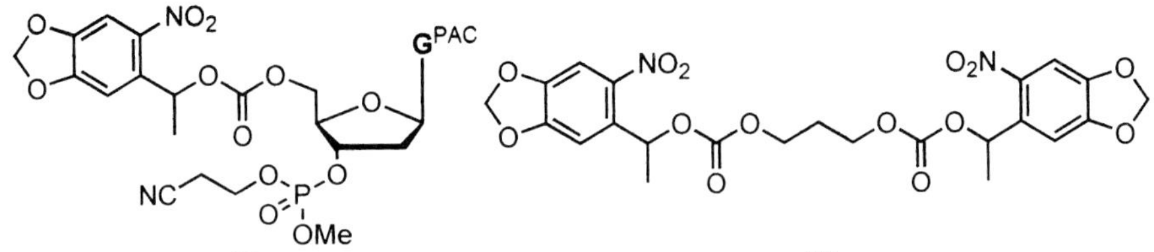

Fig. 6.1.5 Model compounds for investigation of photochemical deprotection reactions of MeNPoc.

6.3.4.2 Dimethoxybenzoin Carbonate Groups

The photochemical generation of radicals from benzoin ethers by α-cleavage [31] is distinct from the deprotection photochemistry of benzoin esters. Thus, it would not be possible to use 3′,5′-DMB for the direct protection of a nucleoside 5′-hydroxyl as an ether. Adaptation of dimethoxybenzoin was required. A carbonate group serves as a linker that makes the photochemical step in an alcohol deprotection completely analogous to the photochemical deprotection of esters. The resulting nucleoside hemi-carbonate can readily decarboxylate to give the alcohol. The preparation of DMB carbonates of nucleosides is performed with reagent **31**, which is prepared from the benzoin and bis-methylated carbonyldiimidazole [32]. Comparison of the rates of photochemical deprotection of DMB carbonates to the rate of dimethoxybenzoin acetate suggests they have similar quantum yields, ∼0.65. With a variety of alcohols, including N-protected nucleosides, the isolated yield of deprotected alcohol is essentially quantitative, making DMB carbonates attractive alternatives to the MeNPoc group for photochemical DNA synthesis. Moreover, the dimethoxybenzofuran **17** (BF) is also produced in high yield. Its significant UV absorption and fluorescence potentially offer a method for directly assessing the cycle yield, similar to the way dimethoxytrityl cation is used to measure cycle yields in conventional DNA synthesis.

The use of DMB carbonate in phosphoramidite-based solid-phase DNA synthesis was first examined on controlled-pore glass [33]. Compounds **32** were attached to the support and loading was evaluated by photochemical BF release. The perplexing observation was made that the amount of BF measured in the effluent was about half the actual amount of DMB carbonate deprotected (determined by other measures). Several reasons for this observation were investigated without conclusion, and solutions to it were investigated without improvement. Although relative cycle yields could still be evaluated by BF release, a more reliable method to evaluate absolute efficiency in photochemical DMB carbonate DNA synthesis was product release and quantitative HPLC analysis. Compounds **32** were converted to their β-cyanoethyl phosphoramidites for this purpose. These studies showed that DMB carbonate and MeNPoc give quite comparable cycle yields in syntheses of DNA up to decanucleotides, but that they provide only about 80% cycle yield compared to a conventional DMTr synthesis. This work again demonstrated that T–T dimer formation is not a concern, but that benzoyl cytidine is photochemically reactive even under long-wavelength irradiation.

Evaluation of the DMBoc chemistry on glass substrates used methods described above [34]. Nucleoside building blocks **32** with "easy off" N-protecting groups (PAC for A and G, IBU for C) were prepared (Scheme 6.14). The half-times for deprotection of these protected nucleosides attached to a glass slide contacted with dioxane were 5.5–12 s at 310 nm (5.0 mW cm^{-2}) and 5.6–17 s at 365 nm (44.5 mW cm^{-2}). These data provide a good example of the importance of both absorptivity and quantum yield in photochemical deprotection. The DMB absorption is much greater at 310 nm than at 365 nm, but the quantum yield is higher at the longer wavelength [35] and the 10-fold greater light intensity at 365 nm makes the rates comparable. These values can also be compared to MeNPoc, which are 10–13 s (27.5 mW cm^{-2}). The dependence of rate on solvent for deprotection of the pyrimidines was modest, spanning only a 2-fold range, with the fastest reactions occurring in the absence of solvent. The coupling efficiencies of these amidites in the synthesis of homododecanucleotides were: T, 91–98%; C, 82–95%; G, 79–92%; A, 74–84%. Comparison with MeNPoc-protected nucleosides shows 92–94% average stepwise yield. The DMBoc group performs better than would have been predicted based on the results from experiments on controlled-pore glass. Thus, DMBoc and MeNPoc protecting groups show very similar performance characteristics in DNA array fabrication.

It is interesting that two different protecting groups with different mechanisms of photochemical deprotection give quite comparable (and far inferior, compared to DMTr!) outcomes in DNA synthesis on glass slides. Unfortunately, no direct evidence was obtained in the study of either of these groups that would suggest how cycle yields could be improved. The dependence of deprotection on the nucleotide to which the protecting group is attached is also perplexing. Differential quenching mechanisms must always be considered, but no evidence supporting this as a problem could be gained in spectroscopic studies. Empirical means would be needed to discover a novel photochemical DNA synthesis protocol and appropriate groups that would permit quantitative cycle yields to be obtained.

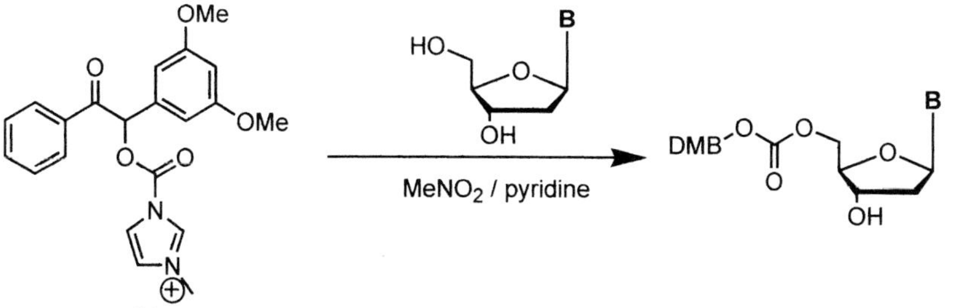

Scheme 6.1.14 Synthesis of nucleoside 5'-DMB carbonates.

6.3.4.3 (Nitrophenylpropyloxy)carbonyl Groups

The discovery was made by Pfleiderer et al. [36] that the family of nitrophenylethoxycarbonyl groups can be readily removed photochemically (Scheme 6.15). These compounds follow a reaction pathway different from nitrobenzyl groups, undergoing a β-elimination rather than an internal redox reaction (though this type of product has been observed in a few instances). A variety of substitution patterns in the basic skeleton were investigated in initial work, and it was found that the (nitrophenylpropyloxy)carbonyl group (NPPoc) 33 has the most rapid reaction under 365 nm irradiation (half-time 40 s, X = 5′-O-thymidine). For example, the half-time of (nitrophenylethoxy)carbonyl is 150 s, demonstrating the beneficial effect of benzylic alkyl substitution. NPPoc is also significantly faster than MeNPoc, which has a half-time of 150 s under these conditions. The NPPoc derivatives of nucleosides were generated by reaction of the chloroformate (33, X = Cl) with the 5′-OH.

Development of the NPPoc group represented a breakthrough in photochemical DNA array fabrication (*vide infra*) and prompted much further investigation concerning groups in this class. The mechanism of the photochemical cleavage process has been studied using nanosecond laser flash photolysis and stationary illumination techniques, and the reaction has been shown to proceed via the triplet state. The quantum yield for NPPoc thymidine is 0.40, about 4-fold higher than MeNPoc and about 10-fold higher than for nitrophenylethoxycarbonyl [37]. The effect of alkyl substitution on the quantum yield was explained by stabilization of the *aci*-nitro intermediate, providing greater opportunity for the subsequent elimination. A truly heroic number of substituted NPPoc variants were prepared and used to protect the 5′-hydroxyl of thymidine [38]. Examination of their deprotection reactions in dilute MeOH/H_2O solution with irradiation at 365 nm allowed comparisons to NPPoc, which in this study exhibited a 48 s half-time and gave thymidine in 88% yield. Good performance was seen with a variety of 5-aryl groups, with 5-phenoxyphenyl-NPPoc (34; Fig. 6.6) emerging as a superior group (16 s half-time and 98% yield). Other interesting observations included the beneficial effect of benzylic methylation (35 versus 36) and the positive effect of Cl or Br at the 5-position (38 and 32 s half-times and 95% yields). Enhanced intersystem crossing may be responsible for the latter observation. Other workers have reported an NPPoc group with red-shifting methylenedioxy substitution (37) which has a significantly enhanced 365 nm absorption, similar to MeNPoc, and which also undergoes deprotection in high yield [39].

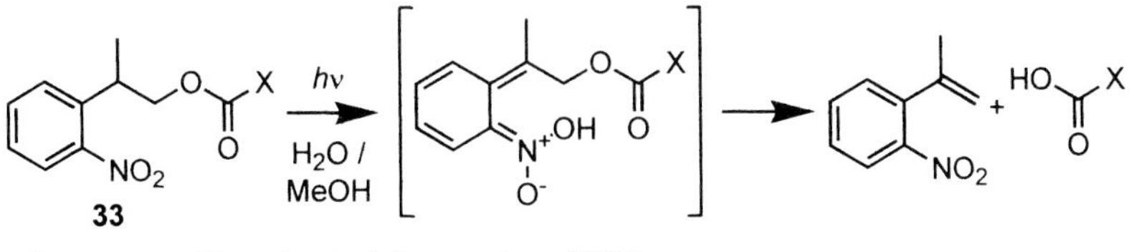

Scheme 6.1.15 Photochemical deprotection of NPPoc groups.

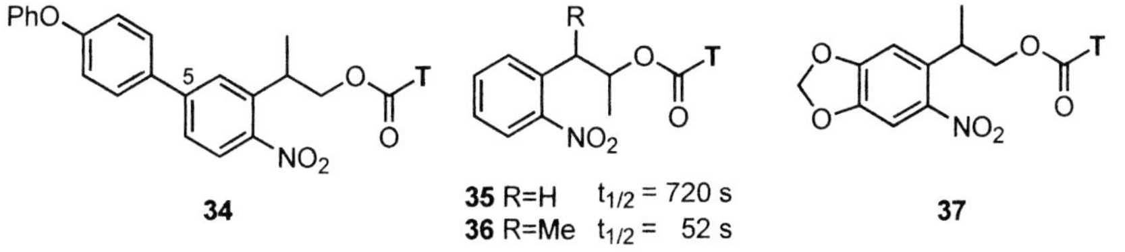

Fig. 6.1.6 Substituted NPPoc groups.

34

35 R=H $t_{1/2}$ = 720 s
36 R=Me $t_{1/2}$ = 52 s

37

The NPPoc group was adopted by Beier and Hoheisel for DNA array fabrication. They prepared nucleoside building blocks **38** [27] (Fig. 6.7), which had also been achieved by Pfleiderer [36]. Deprotection on the array was inefficient using the conditions of Pfleiderer, but 365 nm irradiation in 50 mM diisopropylethylamine/acetonitrile or piperidine/acetonitrile at a power of 25 mW cm^{-2} gives clean deprotection within 5 min. The kinetics of deprotection were assessed using the methods described above and showed 50–60 s half-times. The NPPoc group was also compared directly to MeNPoc (irradiated under dry conditions) on a glass slide substrate. In homopolymer synthesis, NPPoc showed 12% higher cycle yields on average than MeNPoc. Using the reversible fluorescent staining method described above for measuring the quality of probe sites, high overall yields of oligonucleotides when using NPPoc protection were established.

Several applications of DNA arrays require surface-bound oligonucleotides that present free 3′-hydroxyl groups. These oligonucleotides therefore must be attached via their 5′ ends and synthesized from the 5′ end toward the 3′ end. This requires swapping the positions of the NPPoc group and the phosphoramidite on the nucleoside (**39**; Fig. 6.8). The preparation of appropriately derivatized pyrimidines and their use in solid-phase, light-based DNA synthesis with quantitative cycle yields was reported in 2001 [40]. The preparation of a similar building block set including the purines and its use in 5′ → 3′ DNA array fabrication was reported in 2003 [41].

A recent study has shown that the removal of NPPoc groups can be sensitized by thioxanthone [42]. Its significantly higher 365 nm absorption (ε = 4918 cm^{-1} M^{-1} compared 225 cm^{-1} M^{-1} for NPPoc) followed by energy transfer provides to a 2- to 3-fold enhancement in deprotection rate on the microarray. Another advance emerging from sensitization is that the deprotection reaction on a glass substrate

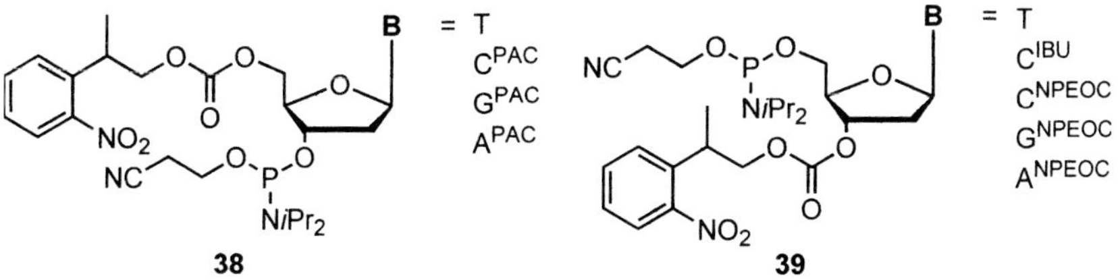

B = T
C^{PAC}
G^{PAC}
A^{PAC}

B = T
C^{IBU}
C^{NPEOC}
G^{NPEOC}
A^{NPEOC}

38

39

Fig. 6.1.7 NPPoc phosphoramidites for light-directed DNA synthesis.

proceeds nearly linearly with time to completion, rather than the asymptotic approach to completion observed with direct irradiation. Photophysical studies have established that sensitization is a diffusion-controlled triplet–triplet energy transfer that is zero order in sensitizer when conducted on an array surface. Sensitization thus appears to provide a useful tool to enhance the contrast between intended and unintended areas in DNA microarray fabrication.

6.3.4.4 Other Photoremovable Protecting Groups

The drive to improve DNA microarray fabrication with superior photochemically removable groups has certainly been significant. An additional technique that has been developed to measure the quality of DNA produced by light-directed synthesis is HPLC analysis of fluorescent oligonucleotides cleaved from the chip (Scheme 6.16). Only one sequence can be made and analyzed per chip using this assay, but it is a direct indicator of the quality of the DNA that is actually prepared on the chip surface. One group that has been investigated by Affymetrix scientists using this method is the nitronaphthylethoxycarbonyl (NNEoc) group (**40**) [43]. Presumably, deprotection of this group involves a process similar to nitrobenzyl groups, internal oxygen-atom transfer and decomposition of the resulting hemiacetal. While NNEoc provides 96% stepwise coupling yields, it also requires immersion of the chip surface in a basic DMSO solution during irradiation. Irradiation under solvent-free conditions during DNA microarray manufacture offers Affymetrix advantages in the optical set-up for photolithography (front irradiation versus irradiation through the back of the array). Their

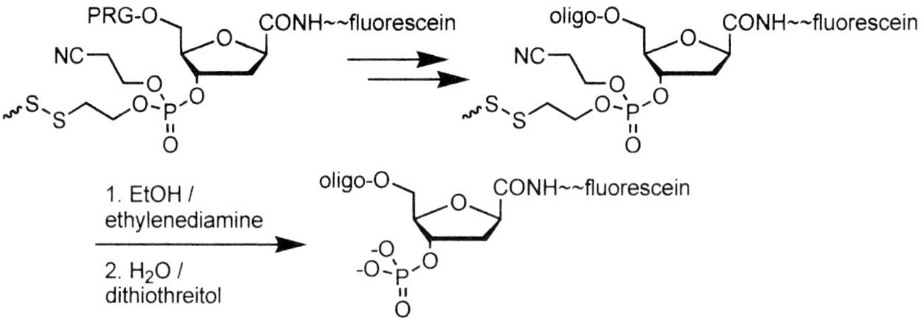

Scheme 6.1.16 Cleave-and-characterize assay for light-directed DNA synthesis

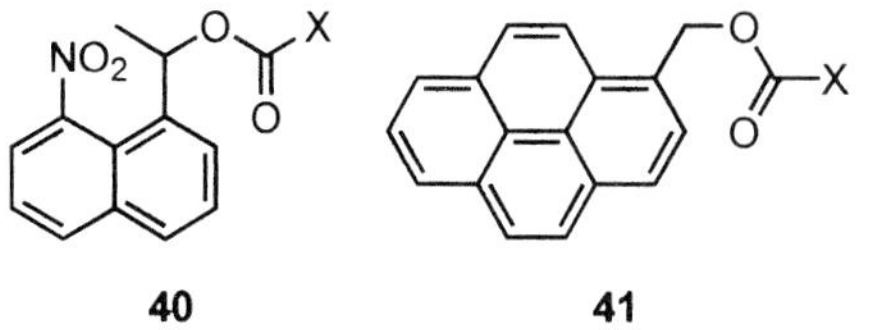

Fig. 6.1.8 Alternative hydroxyl protecting groups for light-directed DNA synthesis.

preference for "dry" irradiation has dissuaded them from adopting such groups despite their superior chemical performance. A similar situation presumably pertains to the (pyrenylmethyloxy)carbonyl (PYMoc) group (41) [44], which gives 96% stepwise coupling yields when the chip surface is immersed in MeOH during irradiation. The known photosolvolysis of (pyrenyl)methyl esters [45], which requires a polar and nucleophilic solvent like MeOH, is presumably the inspiration for this reaction.

6.3.4.5 Photoacids with DMTr Groups

An alternative to the photochemically removable protecting group strategy discussed so far is generation of an acid that cleaves a conventional acid-sensitive group, such as a Boc or DMTr group. This approach is inspired by photoacid-based, chemically amplified resist technologies that have been extensively developed for semiconductor photolithography [46]. Typically, the photoacid precursor is an aryl sulfonium or iodonium salt. Since the deprotection of DMTr groups is catalytic in acid, which is diffusible, this idea may at first seem questionable, as photoacid might cause indiscriminate deprotection. However, diffusion can be limited by polymeric layers [47] or by physical segregation of oligonucleotide probe sites with non-wetting surface coatings [48, 49]. These pre-defined sites or "wells" are created with feature sizes as small as 150 μm using conventional photolithographic processing [50]. Then, light is directed using a spatial light modulator, such as the digital micromirror device (vide infra), to wells intended for deprotection. Because these devices are manufactured for use in the visible region of the spectrum (as in video projection), they transmit light of $\lambda < 400$ nm poorly. For activation of onium salts, which absorb at shorter wavelengths, a sensitizer such as thioxanthone must be used (Scheme 6.17). Use of the 405-nm Hg line at 30 mW cm^{-2} results in deprotection times below 15 s. Use of a neutralizing, pyridine wash solution following deprotection prevents adventitious deprotection of unirradiated wells adjacent to irradiated wells. This method does offer an attraction that many modified DMTr/phosphoramidite building blocks for DNA synthesis are commercially available, increasing the diversity of the molecules that can be made with light-directed synthesis. This convenience

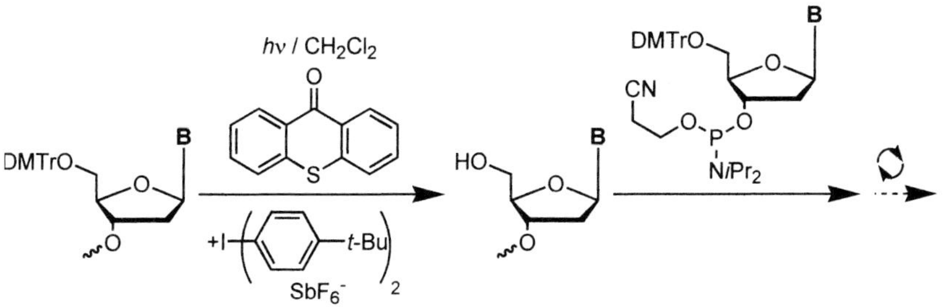

Scheme 6.1.17 Photoacid-based deprotection for solid-phase DNA synthesis.

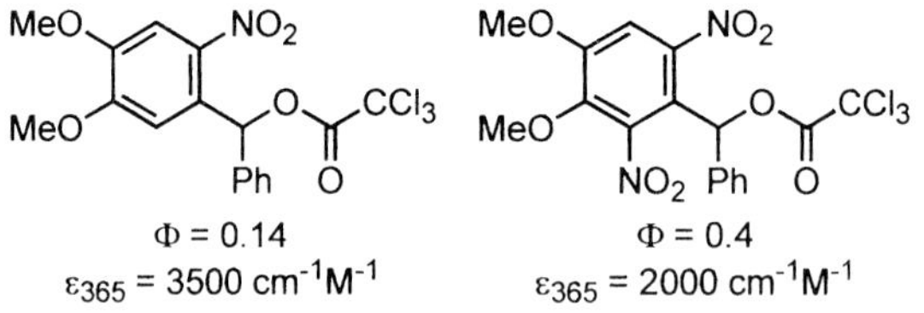

Fig. 6.1.9 Photochemical precursors to TCA.

comes at the cost of the number of sequences that can be prepared and the dimensions at which they are made. The well sizes in this method are at least 10 times larger than probe sites in other DNA microarrays generated by light-directed synthesis, meaning that a 100- or 1000-fold fewer sequences can be prepared. For many applications in molecular biology, this would be a serious limitation.

Another concern with photogenerated acid deprotection is the possibility of acid-based side reactions (such as depurination) that can result from their high acidity ($HSbF_6$ $pK_a < -12$). Perhaps buffering of the solution should be considered. Another solution to the high acidity of conventional photoacids is photogeneration of the acid already proven for DMTr deprotection in oligonucleotide synthesis, trichloroacetic acid [51]. An obvious light-based precursor to trichloroacetic acid (TCA) is a nitrobenzyl ester. Examination of a variety of nitrobenzyl trichloroacetates revealed two compounds with the most desirable performance features (Fig. 6.9). These reagents were shown to provide cycle yields comparable to conventional 3% TCA deprotection using a DNA synthesizer modified as earlier described [40] for light-based deprotection. These esters have high absorptivity and good quantum yields, and half-times for DMTr deprotection promoted by the acid generated from their irradiation in thin films are below 20 s at a power of 25 mW cm^{-2}.

6.4
The Future

While it is always dangerous to predict the direction of science, at least some of the recent trends in microarrays should persist. One breakthrough has been sensitization of photochemical deprotection and, particularly, the kinetic advantages it offers. A significant challenge in developing protecting groups has been finding molecules that have both the absorption properties required and the excited state chemical reactivity needed. Sensitization allows these two tasks to be carried out by different molecules, each of which can be optimized for its specialized function. In addition, at least two other subjects are likely to be on the microarray horizon.

6.4.1
Multiphoton Sensitive Protecting Groups

One strategy to increase the fidelity in DNA chip fabrication by decreasing edge effects is the use of multiphoton processes in the photochemical deprotection step. The rate of photochemical deprotection, which is directly proportional to the % of maximal yield, for a one-photon group depends linearly on light intensity. The rate/yield for a two-photon group depends quadratically on the light intensity. Consider a chip (Fig. 6.10) being irradiated with light of flux 1 through a mask with an optical density of 100 (arbitrary values; the relative ratios to be calculated are independent of them). We analyze the deprotection rate at three positions on the chip. Position a is in a masked zone, position b is at the edge of masked and unmasked zones, and in this analysis we assume it receives a light flux half that of the unmasked zone at position c. For one-photon deprotection, the rates in each zone are linearly proportional to the flux of light received. The contrast ratio between the unmasked and masked zones is the ratio of their rates, k_c/k_a, and the contrast ratio for the unmasked zone and the edge is k_c/k_b. Calculation shows that the intermediate light flux at the edge (position b) decreases the lateral definition of the feature. For two-photon deprotection, the rates in each zone are the *square* of the light flux received. The contrast ratio between the unmasked and masked zones (k_c/k_a) is enhanced 100-fold in this example. Further, the contrast ratio between the unmasked zone and the edge is increased 2-fold, enhancing lateral feature definition. This effect is basically the difference between a linear function and a parabolic function. The benefits of multiphoton photochemistry described here would also apply to the new methods for maskless microarray fabrication using micromirror arrays described below.

6.4.2
Direct Write Microarray Fabrication Systems

In the development of the original concepts around light-directed synthesis, Read [52] proposed using tightly focused lasers to deprotect each synthesis site. Because this would be a serial process, requiring scanning of the laser to each site, it could result in protracted deprotection times for each synthesis cycle, de-

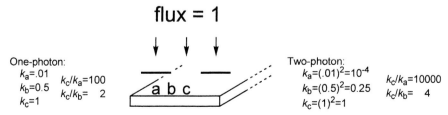

Fig. 6.1.10 Comparison of deprotection rate and contrast ratio with one- and two-photon deprotection.

pending on the dwell time at each site. The attraction of using photolithographic masks to direct the light was the ability to address multiple sites simultaneously, permitting parallelism in the deprotection reaction as well as in the coupling reaction. The drawback of masks is that they are rather static entities, and redesign of an array can require fabrication of a new set of masks. Such a redesign might be prompted by the discovery that some sequences in the original array give poor performance in their designed use, or by the introduction of new ideas concerning best practices for array fabrication, or by the natural evolution of the genomic knowledge base on which DNA array sequences are based. As new and better genomic studies are conducted, sequences that were originally deposited into databanks will be revised, requiring revisions in the microarray sequences that interrogate those genes. The main point, though, is that methods for directing light that are more versatile could offer certain attractive features not available when using masks. Several "on-the-fly" light-direction methods have been developed for light-directed synthesis, and more can be anticipated in the future.

An example of a method that offers versatility in microarray fabrication is maskless array synthesis using micromirror chips to direct the light [53, 54]. Because the pattern of light is computer-directed, DNA microarrays fabricated with this technique can be readily changed. The image of the micromirror array, with some mirrors switched on and some mirrors switched off, is projected onto the chip that is being written, so this method maintains the parallelism in deprotection that is characteristic of mask-based light-directed synthesis. Optoelectronic devices related to micromirror chips, such as liquid crystal spatial light modulators, might be used in similar ways. Future developments in optoelectronics that cannot be anticipated could be applied similarly. These methods, which might in some ways be considered "direct write" techniques, are currently under investigation in two development-stage companies.

Laser direct-write methods are gaining interest in conventional semiconductor photolithography [55, 56]. By scanning a laser beam across the work-piece, they direct light for the removal of resist materials. Many of the laser-promoted chemical reactions used are thermal rather than photochemical processes and the laser serves mainly to heat the site of reaction. The exponential dependence of reaction rate on temperature, as expressed in the Arrhenius relationship, permits these methods to produce features that are smaller (or larger) than the laser beam size. The relationship between beam size and feature size is dependent not only on the total energy delivered to the feature but the rate at which it is delivered and the thermal conductivity of the work-piece and the medium. Smaller features are produced when pulsing or rapidly scanning the laser or reducing its power. Because thermal reactions exhibit the same non-linear dependence of chemistry on light intensity that was discussed in Section 6.4.1, adapting the laser direct-write strategy for photolithography to light-directed synthesis using thermally based deprotection may be desirable for microarray fabrication with high spatial resolution. This approach will require protecting groups that are thermally removable under conditions that do no damage to the oligomers to be prepared. While there are certainly many chemical reactions that require heat, only a few deprotection reactions have been specifically reported to require only heat [57]. Advances in protecting groups that can be removed with thermal, directed energy will be required if laser direct-write microarray fabrication is to become practical. This strategy can open opportunities for development of new chemistry, although it may not be photochemistry.

Acknowledgments

We are grateful to several authors for providing reprints and preprints of their work in this field.

References

1 RUBINSTEIN, M., AMIT, B., PATCHORNIK, A. *Tetrahedron Lett.* **1975**, *17*, 1445.
2 BARTHOLOMEW, D.G., BROOM, A.D. *J. Chem. Soc. Chem. Commun.* **1975**, 38.
3 OHTSUKA, C., TANAKA, S., IKEHARA, M. *Synthesis* **1977**, 453–454.
4 OHTSUKA, C., TANAKA, S., IKEHARA, M. *Nucleic Acids Res.* **1974**, *1*, 1351.
5 SCHWARTZ, M.E., BREAKER, R.R., ASTERIADIS, G.T., deBEAR, J.S., GOUGH, G.R. *Bioorg. Med. Chem. Lett.* **1992**, *2*, 1019.
6 (a) SCHWÖRER, M., WIRZ, J. *Helv. Chim. Acta* **2001**, *84*, 1441; (b) IL'ICHEV, Y.V., SCHWÖRER, M., WIRZ, J. *J. Am. Chem. Soc.* **2004**, *126*, 4581.
7 GOUGH, G.R., MILLER, T.J., MANTICK, N.A. *Tetrahedron Lett.* **1996**, *37*, 981.
8 YIP, R.W., SHARMA, D.K., GIASSON, R., GRAVEL, D. *J. Phys. Chem.* **1985**, *89*, 5328.
9 PITSCH, S. *Helv. Chim. Acta* **1997**, *80*, 2286.
10 PITSCH, S., WEISS, P.A., WU, X., ACKERMANN, D., HONEGGER, T. *Helv. Chim. Acta* **1999**, *82*, 1753.
11 SCARINGE, S.A. *Methods Enzymol.* **2000**, *317*, 3.
12 LENOX, H.J., McCOY, C.P., SHEPPARD, T.L. *Org. Lett.* **2001**, *3*, 2415.
13 SHEEHAN, J.C., WILSON R.M., OXFORD, A.W. *J. Am. Chem. Soc.* **1971**, *93*, 7222.
14 (a) CORRIE, J.E.T., TRENTHAM, D.R. *J. Chem. Soc. Perkin Trans. 1* **1992**, 2409; (b) GIVENS, R.S., KEUPER, L.W. *Chem. Rev.* **1993**, *93*, 55.
15 PIRRUNG, M.C., FALLON, L., LEVER, D.C., SHUEY, S.W. *J. Org. Chem.* **1996**, *61*, 2129.
16 FODOR, S.P.A., READ, J.L., PIRRUNG, M.C., STRYER, L., LIU, A.T., SOLAS, D. *Science* **1991**, *251*, 767.
17 KODADEK, T. *Chem. Biol.* **2001**, *8*, 105.
18 PIRRUNG, M.C. *Angew. Chem. Int. Ed.* **2002**, *41*, 1276.
19 WILSON, D.S., NOCK, S. *Angew. Chem. Int. Ed.* **2003**, *42*, 494.
20 PIRRUNG, M.C. *Chem. Rev.* **1997**, *97*, 473.
21 JOBS, M., FREDRIKSSON, S., BROOKES, A.J., LANDEGREN, U. *Anal. Chem.* **2002**, *74*, 199.
22 PIRRUNG, M.C., PIEPER, W., KALIAPPAN, K.P., DHANANJEYAN, M.R. *Proc. Natl. Acad. Sci. USA* **2003**, *100*, 12548.
23 GARLAND, P.B., SERAFINOWSKI, P.J. *Nucleic Acids Res.* **2002**, *30*, e99.
24 PATCHORNIK, A., AMIT, B., WOODWARD, R.B. *J. Am. Chem. Soc.* **1970**, *92*, 6333.
25 PEASE, A.C., SOLAS, D., SULLIVAN, E.J., CRONIN, M.T., HOLMES, C.P., FODOR, S.P.A. *Proc. Natl. Acad. Sci. USA* **1994**, *91*, 5022.
26 McGALL, G.H., BARONE, A.D., DIGGELMANN, M., FODOR, S.P.A., GENTALEN, E., NGO, N. *J. Am. Chem. Soc.* **1997**, *119*, 5081.
27 BEIER, M., HOHEISEL, J.D. *Nucleic Acids Res.* **2000**, *28*, e11.
28 TILLIB, S.V., MIRZABEKOV, A.D. *Curr. Opin. Biotechnol.* **2001**, *12*, 53.

29 Turro, N.J. *J. Chem. Ed.* **1967**, *44*, 536.
30 Holmes, C.P., Kiangsoontra, B. In *Peptides. Chemistry, Structure and Biology. Proc. 14th Am. Peptide Symp.*, Hodges, R.S., Smith, J.A. (eds), ESCOM, Leiden, **1994**, p. 110.
31 Pappas, S.P. *Progr. Org. Coat.* **1974**, *2*, 333.
32 Pirrung, M.C., Bradley, J.-C. *J. Org. Chem.* **1995**, *60*, 1116.
33 Pirrung, M.C., Bradley, J.-C. *J. Org. Chem.* **1995**, *60*, 6270.
34 Pirrung, M.C., Fallon, L., McGall, G. *J. Org. Chem.* **1998**, *63*, 241.
35 Cameron, J.F., Willson, C.G., Fréchet, J.M.J. *J. Chem. Soc. Chem. Commun.* **1995**, 923.
36 Giegrich, H., Eisele-Bühler, S., Hermann, C., Kvassyuk, E., Charubala, R., Pfleiderer, W. *Nucleosides Nucleotides* **1998**, *17*, 1987.
37 Walbert, S., Pfleiderer, W., Steiner, U. *Helv. Chim. Acta* **2001**, *84*, 1601.
38 Bühler, S., Lajoga, I., Giegrich, H., Stengele, K.-P., Pfleiderer, W. *Helv. Chim. Acta* **2004**, *87*, 620.
39 Berroy, P., Viriot, M.L., Carre, M.C. *Sensors Actuators B: Chemical* **2001**, *74*, 186.
40 Pirrung, M.C., Wang, L., Montague-Smith, M.P. *Org. Lett.* **2001**, *3*, 1105.
41 Albert, T.J., Norton, J., Ott, M., Richmond, T., Nuwaysir, K., Nuwaysir, E.F., Stengele, K.-P., Green, R.D. *Nucleic Acids Res.* **2003**, *31*, e35.
42 (a) Wöll, D., Walbert, S., Stengele, K.P., Green, R., Albert, T., Pfleiderer, W., Steiner, U.E. *Nucleosides Nucleotides Nucleic Acids* **2003**, *22*, 1395; (b) Wöll, D., Walbert, S., Stengele, K.P., Albert, T., Richmond, T., Norton, J., Singer, M., Green, R., Pfleiderer, W., Steiner, U.E. *Helv. Chim. Acta* **2004**, *87*, 28.
43 Barone, A.D., Beecher, J.E., Bury, P.A., Chen, C., Doede, T., Fidanza, J.A., McGall, G.H. *Nucleosides Nucleotides Nucleic Acids* **2001**, *20*, 525.
44 McGall, G.H. in *Biochip Arrays and Integrated Devices for Clinical Diagnostics*, Hori, W. (ed.), IBC Library Series, Southboro, MA, **1997**, p. 2.1.
45 Furuta, T., Torigai, H., Osawa, T., Iwamura, M. *Chem. Lett.* **1993**, 1179.
46 MacDonald, S.A., Willson, C.G., Frechet, J.M.J. *Acc. Chem. Res.* **1994**, *27*, 151.
47 (a) McGall, G., Labadie, J., Brock, P., Wallraff, G., Nguyen, T., Hinsberg, W. *Proc. Natl Acad. Sci. USA* **1996**, *93*, 13555; (b) Wallraff, G., Labadie, J., Brock, P., Nguyen, T., Huynh, T. Hinsberg, W., McGall, G. *Chemtech* **1997**, 22.
48 LeProust, E., Pellois, J.P., Yu, P., Zhang, H., Gao, X., Srivannavit, O., Gulari, E., Zhou, X. *J. Comb. Chem.* **2000**, *2*, 349.
49 Gao, X., LeProust, E., Zhang, H., Srivannavit, O., Gulari, E., Yu, P., Nishiguchi, C., Xiang, Q., Zhou, X. *Nucleic Acids Res.* **2001**, *29*, 4744.
50 Butler, J.H., Cronin, M., Anderson, K.M., Biddison, G.M., Chatelain, F., Cummer, M., Davi, D.J., Fisher, L., Frauendorf, A.W., Frueh, F.W., Gjerstad, C., Harper, T.F., Kernahan, S.D., Long, D., Pho, Q.M., Walker, J.A., Brennan, T.M. *J. Am. Chem. Soc.* **2001**, *123*, 8887.
51 Serafinowski, P.J., Garland, P.B. *J. Am. Chem. Soc.* **2003**, *125*, 962.
52 Pirrung, M.C., Read, J.L. US Patent Appl., USSN 07/362901, June 7, **1989**.
53 Singh-Gasson, S., Green, R.D., Yue, Y., Nelson, C., Blattner, F., Sussman, M.R., Cerrina, F. *Nat. Biotechnol.* **1999**, *17*, 974.
54 Guimil, R., Beier, M., Scheffler, M., Rebscher, H., Funk, J., Wixmerten, A., Baum, M., Hermann, C., Tahedl, H., Moschel, E., Obermeier, F., Sommer, I., Buchner, D., Viehweger, R., Burgmaier, J., Stahler, C.F., Muller, M., Stahler, P.F. *Nucleosides Nucleotides Nucleic Acids* **2003**, *22*, 1721.
55 Kunz, R.R., Horn, M.W., Bloomstein T.M., Ehrlich, D.J. *Appl. Surf. Sci.* **1994**, *79/80*, 12.
56 Ashby, C.I.H. *Thin Solid Films* **1992**, *218*, 252.
57 Ahn, K.-D., Koo, D.-I., Willson, C.G. *Polymer* **1995**, *36*, 2621.

7
Analytical Time-resolved Studies using Photochemical Triggering Methods

7.1
Time-resolved IR Spectroscopy with Caged Compounds: An Introduction

Andreas Barth

7.1.1
IR Spectroscopy

7.1.1.1 Introduction

IR spectroscopy is one of the classical methods for structure determination of small molecules. This standing is due to its sensitivity to the chemical composition and architecture of molecules. The wealth of information encoded in the IR spectrum has also been exploited in time-resolved biophysical studies of protein reactions in the last 25 years and caged compounds entered this field as a tool to trigger protein reactions 15 years ago [1]. This chapter introduces time-resolved IR spectroscopy and the application of IR spectroscopy to the study of caged compound photolysis. The following chapter [2] describes the use of caged compounds in IR spectroscopic experiments which further the molecular understanding of protein reactions.

IR radiation was discovered in 1800 by the astronomer and musician F. W. Herschel, who passed sunlight through a prism that dispersed the light into its spectral components. He measured the temperature in dependence of the wavelength and found that the maximum of the temperature curve was outside the visible spectrum, beyond the red part of the visible spectrum, giving evidence for IR radiation.

More molecular information than in the near-IR spectral region discovered by Herschel is contained in the mid-IR spectral region extending from 2.5 to 50 µm. In this region, IR light can be absorbed by a molecular vibration when the frequencies of light and vibration coincide. The frequency of a vibration depends on the *strength of the vibrating bonds* and is thus influenced by intra- and intermolecular effects. The stronger the vibrating bonds, the higher the frequency. The vibrational frequency also depends on the *masses of the vibrating atoms*, which can be exploited in isotopic substitution experiments to assign an

Dynamic Studies in Biology. Edited by M. Goeldner, R. Givens
Copyright © 2005 WILEY-VCH Verlag GmbH & Co. KGaA, Weinheim
ISBN: 3-527-30783-4

absorbance band to a specific group of a molecule. Apart from its sensitivity to electron density and vibrating masses, IR spectroscopy also provides insight into *molecular geometry*. This is because vibrations of similar frequency can couple which depends on the molecular structure. Coupling is detected in the IR spectrum because the frequency of the coupled vibration usually differs from that of the isolated vibrations.

The sensitivity of IR spectroscopy to molecular structure has widely been used to deduce the structure of small molecules from the spectrum (see, e.g. [3, 4]). It is also valuable to elucidate molecular reaction mechanisms: when the molecular structure is modified in a reaction, the vibrations are affected and as a consequence the IR spectrum is altered – changes in IR absorption can then be followed that reflect the progress of the reaction, such as formation/decay of intermediates and formation of the final products.

All atoms of a molecule are in motion and it might therefore seem futile to extract molecular information on a particular subset of atoms from vibrations. However, it is often the case that only a few atoms are involved in a particular vibration and some vibrations are essentially localized on a particular bond. An example of a localized vibration is a double bond connected with single bonds to the rest of the molecule. The double-bond vibration is virtually independent from the rest of the molecule because its vibrational frequency is much higher than that of adjacent bonds, resulting in a weak coupling to the rest of the molecule. Another example for localized vibrations are parts of a molecule that are separated by a heavy atom. Because of its large mass, a heavy atom hardly moves and acts like a soft cushion, isolating vibrations on different sides of the heavy atom. Vibrations that are rather independent from the rest of the molecule are called *group vibrations*.

The probability of IR absorption depends on the change in dipole moment during the vibration. As a rule of thumb this can be correlated to the polarity of the vibrating bonds: the higher the polarity, the higher the probability of absorption. Because of this, $C=O$, $C-O$, $N-O$ and $P-O$ bonds are particularly good IR absorbers, whereas $C-C$ bonds absorb poorly. The absorption index (extinction coefficient) of molecular vibrations for IR radiation is typically a factor of 100 lower than that of electronic transitions in the UV-vis spectral range. This means that IR spectroscopy can be a struggle against noise, in particular when molecular groups at low concentrations (less than millimolar) are investigated.

The IR spectrum is usually not plotted against the wavelength, but against the *wavenumber* (in units of cm^{-1}) which is the inverse of the wavelength. The wavenumber of an IR absorption maximum is proportional to the vibrational frequency and to the vibrational energy (in most relevant cases). Therefore wavenumber and frequency are often used synonymously. The mid-IR spectral range extending from 2.5 to 50 µm corresponds to 4000 to 200 cm^{-1}.

7.1.1.2 Information that can be Derived from the IR Spectrum

The IR spectrum is sensitive to a number of factors which can be exploited to obtain molecular information on the absorbing molecules and their environment:

- The *chemical structure* of a molecule is the dominating effect that determines vibrational frequencies via the strengths of the vibrating bonds and the masses of the vibrating atoms.
- The *conformation* of a molecule affects the coupling of vibrations because this coupling depends on details of the molecular geometry, like bond angles. Coupling alters vibrational frequencies, so can be detected in the IR spectrum and provides insight into the three-dimensional structure of molecules.
- *Environmental effects* like hydrogen bonding and the electric field produced by a structured environment modify the electron density distribution of a given molecule and will thus reflect in the IR spectrum.
- *Conformational freedom* affects the width of IR bands. IR spectroscopy provides a snapshot of the sample conformer population due to its short characteristic time scale on the order of 10^{-13} s. As the band position for each conformer is usually slightly different, heterogeneous band broadening is the consequence. Flexible structures will thus give broader bands than rigid structures and the bandwidth is a measure of conformational freedom.

The sensitivity of the IR spectrum to many different factors is an advantage and a drawback at the same time: at first an IR spectrum may seem hopelessly crowded with many overlapping bands; however, with increased understanding, the spectrum enables very precise control of the sample since it allows one to monitor a variety of sample parameters in addition to the molecular events that are of prime interest. Examples are protein concentration, enzyme activity, progress of caged compound photolysis and its yield, as well as buffer protonation or deprotonation.

7.1.1.3 Instrumentation

7.1.1.3.1 Fourier Transform IR (FTIR) Spectroscopy

Modern IR spectrometers are usually *FTIR spectrometers*. The heart of an FTIR spectrometer is an interferometer, like the Michelson interferometer shown in Fig. 7.1.1. It has a fixed and a movable mirror. Light emitted from the light source is split by the beam splitter: about half of it is reflected towards the fixed mirror (black arrow in Fig. 7.1.1), the other half of the initial light intensity passes the beam splitter; 50% of the light reflected from the fixed mirror passes the beamsplitter on its second encounter to reach the detector (grey arrow in Fig. 7.1.1), 50% of the light reflected by the movable mirror is reflected by the beam splitter and reaches the detector. When the two beams recombine, they interfere with one another and there will be constructive or destructive interference depending on the optical path difference between the two paths. The in-

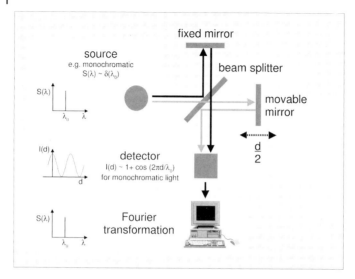

Fig. 7.1.1 Scheme of a FTIR spectrometer.

strument measures the light intensity relative to the position of the movable mirror and this is called an *interferogram*. It turns out that the interferogram is the Fourier transform of the spectrum. A second Fourier transform performed by a computer converts the measured data back into a spectrum. In total, a Fourier transform spectrometer does two Fourier transformations – one performed by the interferometer, one by the computer.

That the interferometer produces the Fourier transform of the spectrum is best seen when a monochromatic source is considered. Depending on the position of the movable mirror, one obtains constructive or destructive interference at the detector and the detector signal varies in a cosine function with the mirror position. This cosine function and the function describing the monochromatic spectrum are related by a Fourier transformation.

The main advantage of Fourier transform spectrometers is the high light intensity at the detector and in consequence the high signal-to-noise ratio. Therefore a complete spectrum can be recorded in as few as 10 ms.

7.1.1.3.2 Dispersive IR Spectrometers

Dispersive IR spectrometers are no longer used for routine measurements, but are still valuable for special applications like time-resolved measurements. Briefly, a dispersive IR spectrometer consists of a light source, optics (often mirrors) that focus the light on the sample and then on the entrance slit of a monochromator, a monochromator, another mirror to focus the light on a detector, and a detector. The positioning of the monochromator behind the sample avoids the detection of heat radiation from the sample. The disadvantage of these instruments is that the incident light is split into its spectral components. Therefore, the light intensity reaching the detector is low, which results in a relatively poor signal-to-noise ratio.

7.1.1.3.3 IR Samples

One drawback of IR spectroscopy of aqueous solutions is the strong absorbance of water in the mid-IR spectral region (near 1640 cm^{-1}) [5]. This demands a short path length for aqueous samples, which is typically around 5 µm, and in turn relatively high concentrations, of typically 0.1–1 mM for proteins and 1–100 mM for small molecules. Using ^{2}H$_2$O, the pathlength can be increased to 50 µm and the concentration lowered because the water band is downshifted to 1200 cm^{-1}.

IR cuvettes cannot be made of glass since glass is not transparent in the mid-IR spectral region. BaF$_2$ and CaF$_2$ are often the choice, when excitation with UV light is necessary. A simple demountable IR cuvette consists either of two plane windows separated by a spacer that defines the pathlength or of one plane window and a window with a trough. Recently a number of flow cells have been developed that enable mixing experiments [6–9].

7.1.1.4 Time-resolved IR Spectroscopy

7.1.1.4.1 Overview

Three time-resolved IR techniques have been applied to research on or with caged compounds and will be discussed here: the rapid scan technique, the step scan technique and single-wavelength measurements. The former two are performed with FTIR spectrometers, while the latter is done on dispersive instruments. Not discussed are the pump-probe technique and stroboscopic FTIR spectroscopy. For reviews on time-resolved IR spectroscopy and biological or photochemical applications thereof, see [10–21].

7.1.1.4.2 The Rapid Scan Technique

In the rapid scan technique the movable interferometer mirror is moved at a maximum speed of 10 cm s^{-1}. From one complete forward and backward movement of the mirror, up to four spectra can be obtained which results in a maximum time resolution of about 10 ms at 12 cm^{-1} optical resolution. A single experiment can yield a full series of time-resolved spectra.

7.1.1.4.3 The Step Scan Technique

Data recording in an FTIR spectrometer is not continuous, but done at discrete positions of the movable mirror. In the step scan technique the movement of the movable mirror is stopped at these positions, a time-resolved experiment performed and the time course of intensity at the detector recorded. Then the mirror is moved to the next position and the experiment is repeated. One obtains a series of time-resolved intensity measurements at the different mirror positions. When intensity data of all mirror positions at a given time are combined, one obtains the interferogram at that time which can be transformed into a spectrum. Accordingly, all data are reshuffled to obtain a time-resolved series of interferograms which is then transformed into a series of time-resolved spectra. The time resolution of this technique is limited by the response time of detector and electronics and can be of the order of several nanoseconds. A requirement for the step scan technique is that the experiment can be accurately reproduced at least several hundred times, since kinetic traces at typically about 600 mirror positions have to be sampled (at 4 cm^{-1} optical resolution, see discussion in [22]). This requirement is either met by cyclic or reversible systems where the reaction of interest can be repeated many times with the same sample, by repeating the experiment at different small sample spots for each interferometer position, or by using a flow cell to conveniently refill the IR cuvette with fresh material. Because photolysis of caged compounds is an intrinsically irreversible process only the latter two options have been applied to their study [23, 24, 72].

7.1.1.4.4 Single-wavelength Measurements

Single-wavelength measurements are done on dispersive instruments. The monochromator is set to a wavenumber of interest and the intensity change at the detector during the time-resolved experiment is recorded. The advantage of this method is that a kinetic trace can be obtained from a single experiment. However, the relatively low signal to noise ratio of dispersive instruments has the consequence that only relatively large absorbance changes ($\Delta A > 0.002$–0.01) can be resolved in a one-shot experiment. The lower limit of sensitivity applies when the IR band of interest is broad (20 cm^{-1}) which allows one to increase the spectral width that is detected. The signal to noise ratio can be improved considerably by using IR lasers (reviewed in [18–20, 25]) or by repeating the experiment many times (e.g. [26]). The time resolution is comparable to that of the step scan technique.

7.1.2
IR Spectroscopy of Caged Compounds

7.1.2.1 Introduction

This section describes the application of IR spectroscopy to the research on caged compounds. The first part introduces IR difference spectroscopy of caged compounds using the example of the archetype of caged compounds – the 1-(2-nitrophenyl)ethyl derivative of ATP (NPE-caged ATP) [27]. The spectra obtained upon photolysis of NPE-caged ATP and of side-reactions of the photolysis by-product are described, and a detailed band discussion is given in Tabs. 7.1.1–7.1.7. Finally, the use of IR spectroscopy to measure the rate of product release and to elucidate the molecular mechanism of photolysis is discussed.

As early as 1933, IR spectroscopy was used to study photochemical reactions of nitro compounds [28]. While the IR experiments in this study were confined to the educt of the photoreaction, 2-nitrobenzaldehyde, and to the spectral region above 2000 cm^{-1}, later studies in the 1960s [29–32] explored irradiation-induced absorbance changes in the mid-IR spectral region down to 700 cm^{-1}. These studies already describe the disappearance of IR absorption of the nitro group and the appearance of that of a keto group upon photolysis [30–32]. Furthermore, hydrogen abstraction upon irradiation of 2-nitrotoluene has been demonstrated by the formation of a C–^{2}H bond in ^{2}H$_2$O [29]. The number of publications on time-resolved IR spectroscopy of photochemical reactions is considerable and some selected reviews [18–21, 33] will provide a first guide for the interested reader. IR studies of caged compounds have been reviewed before [21, 34].

7.1.2.2 The IR Absorbance Spectrum of NPE-caged ATP

Fig. 7.1.2 shows the absorbance spectrum of NPE-caged ATP in the mid-IR spectral region. Labeled bands originate from the nitrobenzyl moiety and the phosphate groups. Most prominent are bands of the nitro group (at 1525 and 1345 cm^{-1}) and of the phosphate groups (bands below 1270 cm^{-1}). For a detailed band assignment see Tab. 7.1.1. Non-labeled bands originate from the ribose and adenine moieties of ATP.

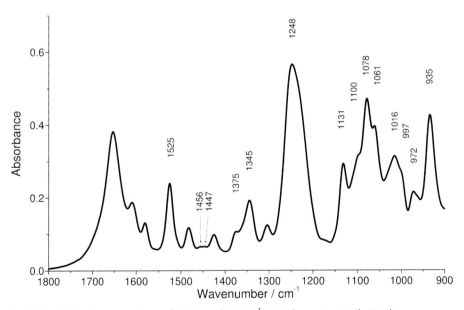

Fig. 7.1.2 IR absorbance spectrum of NPE-caged ATP in ^{1}H$_2$O. The water contribution has been subtracted.

7.1.2.3 Difference Spectroscopy

IR spectroscopy is at its best when the molecular mechanism of reactions is investigated. Here, two of its advantages are combined: the wealth of molecular information provided and the high time resolution. These studies usually monitor only those changes in IR absorption that are associated with the reaction, which are displayed as an IR difference spectrum. For time-resolved measurements and highest sensitivity, the reaction of interest has to be initiated directly in the cuvette.

Fig. 7.1.3 illustrates how a typical reaction-induced difference spectrum is created at the example of NPE-caged ATP photolysis. First, a spectrum is recorded that characterizes the absorbance of caged ATP. Then, photolysis is triggered by an intense light flash, which produces reaction intermediates and final products. This is monitored by recording time-resolved spectra. Note that there is usually no obvious difference between the absorbance spectra recorded before and after the flash, as indicated by the identical spectra shown on the two top

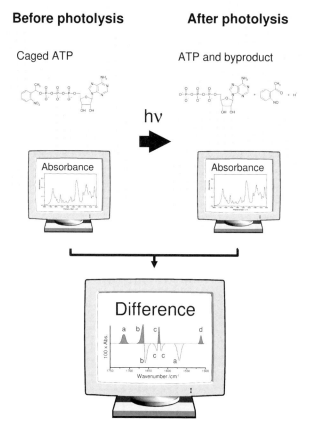

Fig. 7.1.3 The principle of IR difference spectroscopy. See text for further explanations.

screens in Fig. 7.1.3. However, when the spectrum recorded before the flash is subtracted from the spectra after the flash, small absorbance changes caused by photolysis are revealed clearly. An idealized difference spectrum calculated in that way is shown on the bottom screen of Fig. 7.1.3. Bands in a difference spectrum originate only from those molecular groups that are affected by the reaction. All "passive" groups are invisible in the difference spectrum, which therefore exhibits molecular details of the reaction mechanism despite a large background absorption. In the idealized difference spectrum in Fig. 7.1.3, negative bands are characteristic of the educt caged ATP, while positive bands reflect photolysis intermediate(s) or photolysis products in spectra recorded during the reaction or after the reaction, respectively. For simplicity, intermediates will not be considered in the following discussion. However, the principles discussed below for product bands apply as well for bands of intermediates.

Difference bands arise from several sources and four examples are given in Fig. 7.1.3:

(a) Chemical reactions transform molecular groups from educt groups to product groups which usually have different IR absorption. In the difference spectrum of the reaction, disappearing educt groups produce negative bands, while product groups give rise to positive bands. The bands of the appearing and disappearing groups may be widely separated in the spectrum. In Fig. 7.1.3 this is illustrated with the two difference bands marked "a".

(b) Alternatively, a vibration may experience a shift in frequency, due to a conformational or environmental change that alters the electron density of the vibrating bonds or the coupling with other vibrations. This band shift leads to a pair of signals, composed of a negative and a positive band, which are close together. An example is shown in Fig. 7.1.3 for the two bands marked "b". In this example, the educt band is found at lower wavenumber than the product band.

(c) A difference band with side lobes of opposite sign is produced when the width of a band changes in the reaction. If a decrease of bandwidth is considered, the intensity will decrease on the sides of the band but will increase at the center (if the absorption index remains constant) leading to a positive band with negative side lobes. This case is shown in Fig. 7.1.3 for bands marked "c". As the bandwidth is a measure of conformational flexibility, the decrease of bandwidth shown indicates a more rigid structure of the product as compared to the educt.

(d) Only one band is observed when the reaction results in a change of the absorption index of a vibrational mode, e.g. because of a polarity change of the vibrating bond(s). A minimum (maximum) in the difference spectrum then indicates a reduced (increased) absorption of the product as compared to the educt. This case is illustrated with the band marked "d".

7.1.2.4 How to Interpret IR Spectra

As shown in Fig. 7.1.2, an IR spectrum is usually rather complicated because it consists of the absorbance bands of many of the molecular groups that make up a molecule. Naturally, the larger the molecule is, the more complicated is its absorbance spectrum. The number of bands in a difference spectrum is fewer, since it reports only those groups that are modified or change their environment in a reaction. Nevertheless, difference spectra are usually complicated and a straightforward assignment of bands to molecular groups is difficult. A first "educated guess" will be to assign bands to group vibrations (see Section 7.1.1.1). Group frequencies are listed (e.g. [3, 4]) and most straightforward is the assignment to strongly absorbing groups like carbonyl, carboxyl, nitro or phosphate groups. If the molecule is simple, this approach can lead to a reasonable assignment of IR bands. However, in many cases further experiments are necessary and have been applied to the study of caged compounds, like recording of spectra of model substances that lack certain molecular groups of the parent molecule (mutants for proteins) [35], recording of spectra in ^2H_2O which deuterates all acidic groups and isotopic labeling of specific atoms of a molecule [34, 36, 37]. The latter shifts IR bands because of the mass effect on the vibrational frequency and provides a definitive assignment of the shifted bands. Experimental assignments can then be complemented by *ab initio* calculations of vibrational frequencies [38, 72] which can give further insight into the conformation and electron density distribution of the molecule studied.

7.1.2.5 The Difference Spectrum of NPE-caged ATP Photolysis

Scheme 7.1.1 shows a simplified scheme of NPE-caged ATP photolysis: photolysis proceeds via an *aci*-nitro intermediate to the photolysis products 2-nitrosoacetophenone and ATP [39]. Fig. 7.1.4 presents difference spectra recorded at two different times during NPE-caged ATP photolysis: when the *aci*-nitro intermediate is present (thin line) and after ATP release (bold line). They are named "*aci-nitro formation spectrum*" and "*product formation spectrum*" in the following. Positive bands are characteristic of the *aci*-nitro intermediate in the "thin line" spectrum and of ATP plus the nitrosoacetophenone byproduct in the "bold line" spectrum. The reference spectrum for both difference spectra was recorded be-

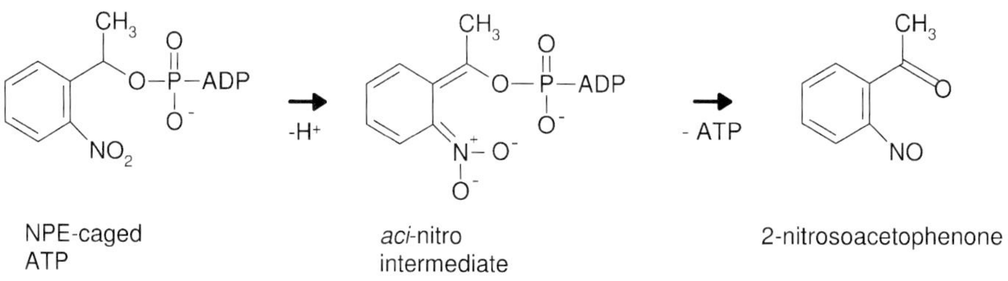

Scheme 7.1.1

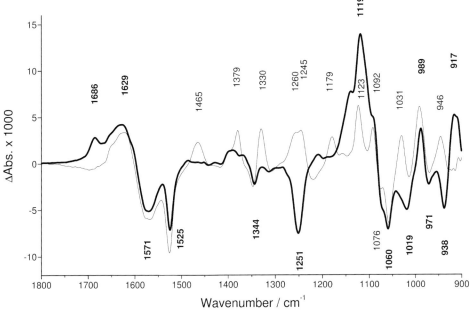

Fig. 7.1.4 IR difference spectra of NPE-caged ATP photolysis at 1 °C and pH 8.5, 90 mM NPE-caged ATP: *aci*-nitro formation spectrum (thin line, normal labels, obtained 1–60 ms after the photolysis flash) and product formation spectrum (bold line, bold labels, obtained 2.7–22.5 s after the photolysis flash) [36].

fore photolysis, negative bands are therefore characteristic of NPE-caged ATP. *Photolysis spectra* will be the general name here for all difference spectra calculated from a spectrum recorded at any time after the photolysis flash and a reference spectrum recorded before photolysis.

IR absorption bands of NPE-caged ATP, the *aci*-nitro intermediate, 2-nitrosoacetophenone and phosphate bands of ATP are discussed in detail in Tabs. 7.1.1–7.1.4 and in [34, 36, 37]. Note that the spectral positions of bands in difference spectra and in absorbance spectra often differ slightly because of different overlap by other bands. Here it is worth mentioning three bands of the *aci*-nitro formation spectrum and the product formation spectrum which illustrate that time-resolved IR spectroscopy monitors several chemical groups in a single experiment. A negative band near 1525 cm^{-1} appears in the *aci*-nitro formation spectrum and in the product formation spectrum. It has been assigned to the antisymmetric stretching vibration of the nitro group from its spectral position [1] and from its shift upon ^{15}N labeling [37]. Obviously, this group disappears already upon formation of the *aci*-nitro intermediate in line with the established photolysis reaction scheme. In contrast, there are two large bands in the phosphate region of the spectrum that appear only in the product formation spectrum: the negative band at 1251 cm^{-1} and the broad positive band centered at 1119 cm^{-1}. From their shifts upon ^{18}O labeling [34, 36, 37] they have been assigned to the antisymmetric stretching vibrations of NPE-caged ATP's PO_2^- groups and ATP's γ-PO_3^{2-} group, respectively. These bands appear in the difference spectrum because the bond strengths of mainly the γ-phosphate are altered upon ATP formation at pH > 7 where the γ-phosphate is unprotonated [1]: this group is a PO_2^- group in NPE-caged ATP, but a PO_3^{2-} group in ATP with the former having stronger non-bridging P–O bonds. The negative band appears because the absorbance of the γ-PO_2^- group of NPE-caged ATP is lost upon ATP release and the absorbance of the two remaining PO_2^- groups is affected by this. The positive band appears because the γ-PO_3^{2-} group is formed when ATP is released. The possibility to observe the γ-PO_3^{2-} group of ATP implies a direct way to measure the product release rate as discussed below. In addition to bands from caged ATP and its photolysis products, the spectra show two bands at 1629 and 1571 cm^{-1} which are due to protonation of the bicine buffer upon *aci*-nitro formation.

Absorbance bands of the protonated form of the *aci*-nitro intermediate – the nitronic acid – have been obtained for 2-nitrobenzyl methyl ether (see Scheme 7.1.2) in argon and nitrogen matrixes [40] and in C^2H_3CN [72]. They have been assigned with the aid of density functional theory calculations [38, 72]. These spectra are substantially different from those of the *aci*-nitro anion of a closely related 2-nitrobenzyl ether (Scheme 7.1.2) [41] in aqueous solution which resemble those shown in Fig. 7.1.4. Particularly striking is the difference in the spectral region of double bond stretching vibrations. The highest frequency band of the *aci*-nitro anion is found near 1465 cm^{-1} for both, the 2-nitrobenzyl-ether (2-nitrobenzyloxy)acetic acid [41] and NPE-caged ATP [36], whereas up to three bands are observed above 1500 cm^{-1} for nitronic acid [40, 72]. As an example, the C=N stretching vibration of nitronic acid is predicted by density functional theory calculations between 1620 and 1650 cm^{-1} [38, 72], whereas it contributes to the 1465 cm^{-1} band of *aci*-nitro anions. The latter assignment is based on a band shift observed upon ^{15}N labeling [37]. The lower frequency of double bond vibrations of the *aci*-nitro anion indicates that they are less localized on individual bonds than in nitronic acid. As a consequence, the electron density seems to be more evenly distributed in the *aci*-nitro anion leading to highly delocalized vibrations with vibrational coupling between groups quite far apart in the structure [37].

Product formation spectra of the following caged compounds have been described or published: NPE-caged ATP [1, 34, 36, 37, 42, 43], P^3-[2-(4-hydroxyphe-

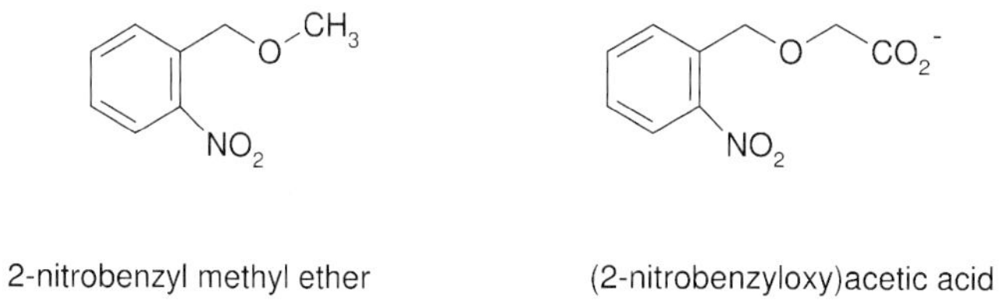

2-nitrobenzyl methyl ether   (2-nitrobenzyloxy)acetic acid

Scheme 7.1.2

nyl)-2-oxo]ethyl ATP (pHP-caged ATP, see Scheme 7.1.4) [44], NPE-caged GTP [34, 45], NPE-caged β,γ-imido ATP (AMP-PNP) [46], NPE-caged ADP [42, 43, 46–50], NPE-caged phosphate [34, 48], N-[1-(2-nitrophenyl)ethyl]carbamoylcholine and N-(α-carboxy-2-nitrobenzyl)carbamoylcholine (caged carbamoylcholines) [51], 1-(2-nitrophenyl)ethyl sulfate (caged sulfate and caged proton) [52], 1,2-amino-5-[1-hydroxy-1-(2-nitro-4,5-methylene dioxyphenyl)methyl]phenoxyl-2-(2'-amino-5'-methylphenoxy)ethane-N,N,N'',N'-tetraacet ic acid (Nitr-5) [53] and DM-nitrophen [34, 54, 55] (two caged calciums), an acetate derivative of an α-keto amide (caged acetate) [56], γ-(α-carboxy-2-nitrobenzyl)-L-glutamic acid (caged glutamate) [24, 35], 1-[5-(dihydroxyphosphoryloxy)pentanoyl]indoline-5-acetate (a caged carboxylate) [57], (2-nitrobenzyloxy)acetic acid and methyl 2-[1-(2-nitrophenyl)ethoxy]ethyl phosphate (two caged alcohols) [41], and (μ-peroxo)(μ-hydroxo)-bis[bis(bipyridyl)cobalt(III)] complex (caged oxygen) [58].

7.1.2.6 Side-reactions of the Photolysis Byproduct

Two side-reactions of the nitrosoacetophenone byproduct have been investigated by IR spectroscopy: (1) dimerization of the nitroso group and (2) its reduction by thiol reagents like dithiothreitol (DTT) [37]. The former reaction is discussed in this chapter, the latter elsewhere in this volume [2, 73]. For molecules involved in both reactions, the IR bands observed are listed in Tabs. 7.1.3 and 7.1.5–7.1.7 (see Scheme 7.1.3 for intermediates and product of the reaction with DTT).

At the high NPE-caged compound concentrations used in IR spectroscopy (typically several millimolar or more), the nitroso group of the photolysis byproduct forms cis- and trans-dimers. As shown in Fig. 7.1.5 (a), a band characteristic of the nitroso monomer decays at 1501 cm^{-1} (half-time $\sim$0.2 s at p^{2}H 6 and 35 °C, half-time $\sim$0.8 s at p^{2}H 6 and 1 °C) and at 1383 cm^{-1} a band of the cis-dimer rises. At 1426 cm^{-1}, a monomer and a cis-dimer band overlap such that the spectrum is not significantly altered in that region. These two bands are only revealed upon ^{15}N labeling of the nitroso group which shifts them to different extents. The absorbance changes upon dimerization to the cis-nitroso dimer are best revealed when the difference of the two spectra shown in Fig. 7.1.5 (a) is calculated (absorbance of cis-dimer minus absorbance of NO monomer). The resulting spectrum is shown as full line in Fig. 7.1.5 (b).

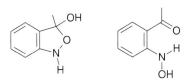

2-hydroxylaminoacetophenone
closed form   open form

3-methylanthranil

Scheme 7.1.3

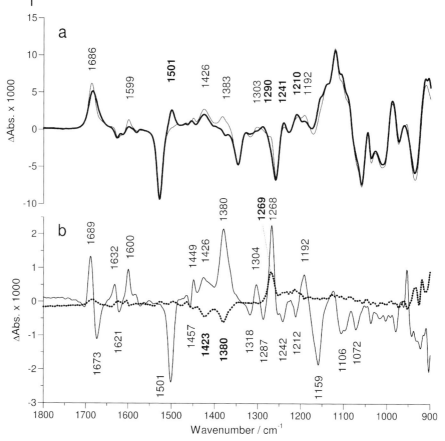

Fig. 7.1.5 IR difference spectra of NPE-caged ATP photolysis showing the dimerization of nitrosoacetophenone. (a) Photolysis spectra obtained at p^{2}H 6.0, 1 °C and 90 mM NPE-caged ATP in ^{2}H$_2$O. Bold line, bold labels: recorded in the presence of the nitroso monomer (10–320 ms after the photolysis flash, absorbance of nitroso monomer and ATP minus absorbance of NPE-caged ATP). Thin line, normal labels: recorded in presence of the nitroso cis-dimer (23–100 s after the photolysis flash, absorbance of nitroso cis-dimer and ATP minus absorbance of NPE-caged ATP). (b) Full line: difference spectrum of nitroso dimerization calculated by subtracting the spectra shown in (a) (absorbance of nitroso cis-dimer minus absorbance of nitroso monomer). Dotted line: difference spectrum of conversion of the nitroso cis-dimer to the trans-dimer obtained at p^{2}H 6.0, 35 °C and 90 mM NPE-caged ATP in ^{2}H$_2$O by subtracting a spectrum obtained 23–42 s after the photolysis flash from one obtained 1.0–3.6 s after the flash.

At 1 °C and p^{2}H 6.0, the cis-dimer is stable for several minutes. At the higher temperature of 35 °C, the cis-dimer partly transforms to the trans-dimer within one minute. The difference spectrum of that reaction (absorbance of trans-dimer minus absorbance of cis-dimer) is shown as dotted line spectrum in Fig. 7.1.5 (b). The reaction results in the decay of cis-dimer bands at 1423 and 1380 cm^{-1}, and the evolution of a trans-dimer band at 1269 cm^{-1}. Since the cis-dimer also shows a band in this region (at 1268 cm^{-1}) the two dimers are best distinguished upon ^{15}N labeling which shifts the *trans*-dimer band, but not the *cis*-dimer band, because only the corresponding *trans*-dimer vibration involves the nitrogen atom to a significant extent. When an NPE-caged compound solution is flashed several times, the nitroso dimers already formed in previous flashes transiently revert to monomers.

7.1.2.7 Measuring Product Release Rates with IR Spectroscopy

The product release rate of NPE-caged compounds has been usually inferred from the decay rate of the *aci*-nitro intermediate which can be measured at 406 nm [39] or assessed by indirect means [59]. However, equating the *aci*-nitro decay with product release is not always justified and examples are given in the following section. Therefore, a direct measurement of product release is preferable and this can be done with IR spectroscopy. We have seen above that the IR difference spectrum of caged compound photolysis contains bands which are characteristic of the released product. Monitoring these bands in a time-resolved experiment enables a direct measurement of the rate of product release. This approach was first validated with NPE-caged ATP [36] and has confirmed for this compound that *aci*-nitro decay is the rate-limiting step in product release [37]. Fig. 7.1.6 shows the kinetics

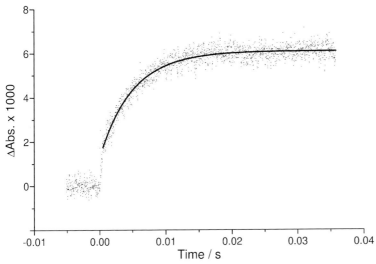

Fig. 7.1.6 Kinetics of ATP release from NPE-caged ATP at pH 7 and 22 °C [36] obtained by monitoring the formation of the IR difference band around 1119 cm^{-1} in the product formation spectrum (bold line in Fig. 7.1.4). The kinetics can be described by a small step change followed by an exponential increase. The step change is due to mainly the 1123 cm^{-1} band of the *aci*-nitro intermediate whereas the exponential increase originates from the appearance of the 1119 cm^{-1} band in the product formation spectrum. This band is assigned to the $v_{as}(PO_3^{2-})$ vibration of the γ-phosphate of ATP and its amplitude proportional to the concentration of released ATP.

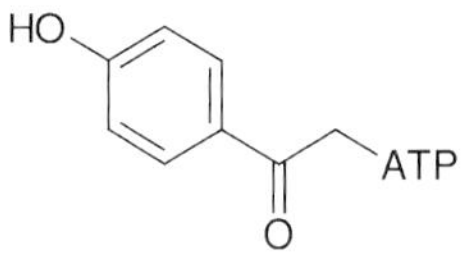

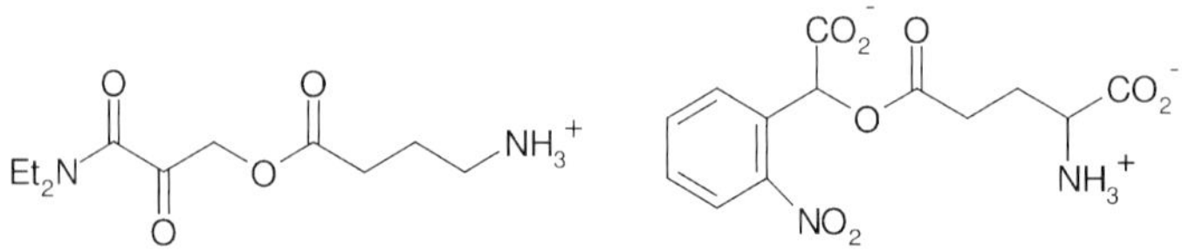

Scheme 7.1.4

of formation of the γ-PO$_3^{2-}$ group of ATP which proceeds with a time constant of 5 ms at 22 °C and pH 7 [36] (see legend for additional information). Another example is shown in Fig. 7.1.7 which monitors the same band of P^3-[2-(4-hydroxyphenyl)-2-oxo]ethyl ATP (pHP-caged ATP, see Scheme 7.1.4) [44]. In this example, ATP release is equal or faster than the 1-µs time resolution of the experiment (24 °C, pH 7). The above measurements were done with a dispersive spectrometer at a single wavenumber. Using the rapid scan technique, release of γ-aminobutyrate (GABA) from a caged precursor (see Scheme 7.1.4) was found to proceed with a time constant of 30 ms [56]. Combining a flow cell and the step scan technique, the time constant of release of glutamate from γ-(α-carboxy-2-nitrobenzyl)glutamate (see Scheme 7.1.4) was determined to be 23 µs [24].

7.1.2.8 IR Spectroscopy for Mechanistic Studies

Because of its sensitivity to molecular structure, IR spectroscopy is a powerful tool to elucidate the photolysis mechanism of caged compounds. It has been valuable to prove the release of a presumed photoproduct [35, 60], to elucidate the mechanism of the photolysis byproduct reaction with DTT [37] and to confirm an aspect of the photolysis mechanism of NPE-caged ATP which had been established before based on less direct experiments [59]: *aci*-nitro decay was found to be the rate-limiting step of ATP release [37].

Examples are given below where *aci*-nitro decay cannot be equated with the formation of the final photolysis product and where product formation is considerably slower than *aci*-nitro decay. They illustrate how IR spectroscopy can give striking insight into the photolysis mechanism of caged compounds: the time-resolved spectra are like a movie of the molecular rearrangements upon photolysis. The first example given is the photolysis of a caged alcohol – methyl 2-[1-(2-nitrophenyl)ethoxy]ethyl phosphate (see Scheme 7.1.5) and related ethers

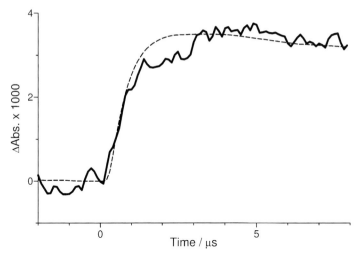

Fig. 7.1.7 Kinetics of ATP release from P^3-[2-(4-hydroxyphenyl)-2-oxo]ethyl ATP (pHP-caged ATP) at pH 7 and 24 °C [44]. Full line: signal of the IR difference band at 1140 cm^{-1} which originates from the γ-phosphate $\nu_{as}(PO_3^{2-})$ vibration. Note that the center of the broad PO_3^{2-} band is different for pHP-caged ATP (1140 cm^{-1}) compared to NPE-caged ATP (1119 cm^{-1}, see Fig. 7.1.4) because of overlap of negative caged ATP bands which are different for pHP-caged ATP and NPE-caged ATP. Dashed line: signal of the 32nd order of an excimer laser stray light signal measured at 32 × 308 nm and normalized to the amplitude of the 1140-cm^{-1} signal. The rise time of the laser signal is limited by the time constants of detector and electronics and indicates that the time resolution of the instrument is 1 µs. The comparison of the two signals shows that the increase of the 1140-cm^{-1} signal is limited by the time resolution of the instrument. The time constant of ATP release from pHP-caged ATP is therefore 1 µs or less.

[41]. Selected spectra of a time-resolved series of spectra are shown in Fig. 7.1.8. In the experiment, the *aci*-nitro intermediate decays too fast to be monitored with the rapid-scan technique. Instead, the first spectrum monitors a species having a nitroso monomer band at 1497 cm^{-1}. This species is clearly an intermediate in the reaction since the difference spectrum changes as time evolves. The final product shows the expected carbonyl band at 1688 cm^{-1} for the nitrosoketone byproduct, the nitroso monomer band has decayed and instead bands

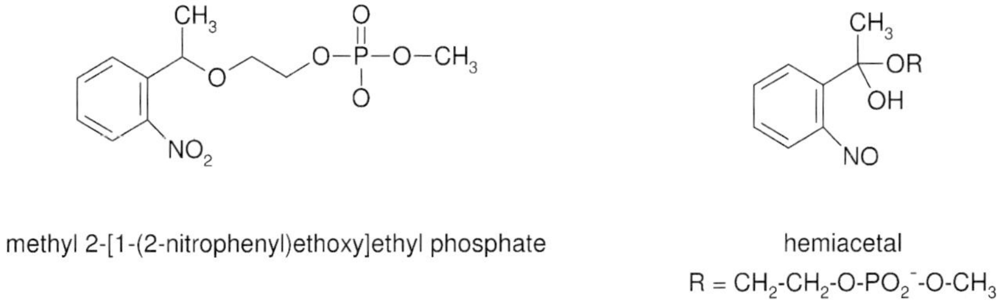

methyl 2-[1-(2-nitrophenyl)ethoxy]ethyl phosphate

hemiacetal
R = CH_2-CH_2-O-PO_2^--O-CH_3

Scheme 7.1.5

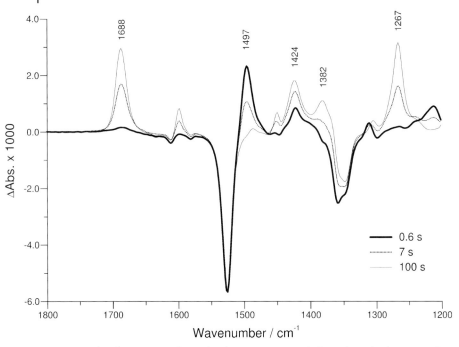

Fig. 7.1.8 Photolysis spectra obtained with methyl 2-[1-(2-nitrophenyl)ethoxy]ethyl phosphate at pH 7.0 and 1 °C. The average time after the photolysis flash at which spectra were recorded is indicated. The spectra demonstrate that the nitroso group (band at 1497 cm^{-1}) forms before the keto group (1688 cm^{-1}) [41].

characteristic of nitroso dimers are present at 1424, 1382 and 1267 cm^{-1}. This observation has lead us to suggest a photolysis reaction mechanism where *aci*-nitro decay results in formation of a hemiacetal (see Scheme 7.1.5) which then breaks down to the final products alcohol and nitrosoketone. Further experiments in the UV-vis spectral range that monitor nitroso formation at 740 nm, pH transients and *aci*-nitro decay at 406 nm showed that this pathway represents approximately 20% of the reaction flux. The major pathway accounting for 80% of the reaction flux involves rapid formation (time constant < 10 ms, 2 °C, pH 8.5) of the hemiacetal without formation of an *aci*-nitro anion.

A hemiacetal intermediate has also been observed for a nitrobenzyl based caged alcohol (2-nitrobenzyl methyl ether, see Scheme 7.1.2) [72]: step scan IR spectroscopy with millisecond time resolution of this compound in C^{2}H$_3$CN has detected three intermediates upon photolysis: nitronic acid, a bicyclic intermediate characterized by only one strong IR band at 1080 cm^{-1} assigned to the O–CH$_3$ stretching vibration and the hemiacetal intermediate. Based on these IR findings, time-resolved UV-vis spectroscopy detected the intermediates also in aqueous solutions, the cyclic intermediate below pH 6 and the hemiacetal intermediate below pH 8. The mechanistic aspects of these findings is discussed elsewhere in this volume [73].

7.1.3
Summary and Conclusions

This review summarizes briefly different time-resolved techniques of IR spectroscopy and discusses their application to the research on caged compounds. The IR absorbance spectrum of the 1-(2-nitrophenyl)ethyl derivative of ATP (NPE-caged ATP) and its photolysis-induced IR difference spectra are presented, and a detailed band discussion given. Phosphate bands of ATP are also reviewed. For the research on caged compounds, IR spectroscopy is valuable to determine directly the rate of product release and to elucidate the molecular mechanism of photolysis. The latter point is illustrated at the example of photolysis of caged alcohols where IR spectroscopy has detected a hemiacetal intermediate implying that product formation is considerably slower than decay of the *aci*-nitro intermediate.

We can conclude that high time resolution and sensitivity to molecular structure and environment make IR spectroscopy a prime method for the study of molecular reactions. Caged compounds enter this field in two aspects: (1) as objects of study in their own right and (2) as biophysical tools to initiate a reaction of interest [2, 61]. Regarding the former aspect, IR spectroscopy is particularly useful to confirm an expected photolysis product, to determine directly the product release rate and to elucidate the photolysis mechanism.

Acknowledgments

I am grateful to John Corrie (National Institute for Medical Research, London) for excellent collaboration as well as to Werner Mäntele (Johann Wolfgang Goethe-Universität, Frankfurt am Main) for his support and the provision of equipment. J. Wirz (Universität Basel) is acknowledged for sending me his manuscript prior to publication. Our work on caged compounds was supported by a Royal Society European Science Exchange Program grant, by an ARC grant from the Deutscher Akademischer Austauschdienst and the British Council, and by a fellowship from the Ciba-Geigy Foundation (to Y. Maeda).

Tab. 7.1.1 Caged ATP bands in the spectral range from 1800 to 900 cm^{-1}

Band position (cm^{-1})	Assignment	Reference for assignment	Comments
1616	v(CC) of benzene ring	3	weak, observed in the spectrum of NPE-caged methyl phosphate (data not shown) and at similar position for 2-nitrotoluene (data not shown), overlayed by stronger adenine absorption in the spectrum of NPE-caged ATP
1580	v(CC) of benzene ring	3	weak, see comment of 1616 cm^{-1} band
1525 a, i, p	v_{as}(NO$_2$)	1, 37	shifts −26 cm^{-1} upon ^{15}N labeling of the nitro group [37]
1482	v(CC) of benzene ring	3	very weak, see comment of 1616 cm^{-1} band
1456 a	v_{as}(CH$_3$)	3	very weak
1447 a	v(CC) of benzene ring	3	very weak, shifts −3 cm^{-1} upon ^{15}N labeling
1375 a	v_s(CH$_3$)	3	
1345 a, i, p	v_s(NO$_2$)	1, 37	shifts −18 cm^{-1} upon ^{15}N labeling of the nitro group [37]
1302			weak, see comment of 1616 cm^{-1} band
1272			very weak, see comment of 1616 cm^{-1} band
1248, a, p	v_{as}(PO$_2^-$) of α-, β- and γ-phosphate	34, 36	shifts −6 cm^{-1} for γ-^{18}O$_3$ labeling, −14 cm^{-1} for $\beta,\beta\gamma$-^{18}O$_3$ labeling, −3 cm^{-1} for ^{13}C labeling of the benzylic carbon (^{13}C–O–P$_\gamma$) and −3 cm^{-1} for ^{15}N labeling
1225 (a)	v_a (PO$_2^-$) with a strong contribution from γ- and β-phosphate	36, 37	canceled in the photolysis spectrum by an ATP band, shifts −41 cm^{-1} upon γ-^{18}O$_3$ labeling [36] and −27 cm^{-1} upon $\beta,\beta\gamma$-^{18}O$_3$ labeling
∼1150 (a)	delocalized v(PO$_2^-$)		weak in the absorbance spectrum, shifts at least −10 cm^{-1} upon γ-^{18}O$_3$ labeling and approximately −14 cm^{-1} upon $\beta,\beta\gamma$-^{18}O$_3$ labeling
1131, a, (p)	delocalized v_s(PO$_2^-$)	36	causes a shoulder in an overlapped positive band in the product formation spectrum; band shifts −8 cm^{-1} upon γ-^{18}O$_3$ labeling, −16 cm^{-1} for $\beta,\beta\gamma$-^{18}O$_3$ labeling and −2 cm^{-1} for ^{13}C labeling of the benzylic carbon (^{13}C–O–P$_\gamma$)

Tab. 7.1.1 (continued)

Band position (cm^{-1})	Assignment	Reference for assignment	Comments
~1100, (a), (p)	delocalized $\nu_s(PO_2^-)$ coupled to the nitrobenzyl moiety	36	overlapped by a positive band in the product formation spectrum; band shifts -14 cm^{-1} upon γ-$^{18}O_3$ labeling, -5 cm^{-1} for $\beta,\beta\gamma$-$^{18}O_3$ labeling, -6 cm^{-1} for ^{13}C labeling of the benzylic carbon (^{13}C–O–P$_\gamma$) and about -2 cm^{-1} for ^{15}N labeling
1078, a, i, (p)	$\nu_s(PO_2^-)$ with a strong contribution from the β-phosphate	36	shifts upon γ-$^{18}O_3$ labeling are not obvious in the absorbance spectrum. According to second derivative spectra, part of the band seems to shift -12 cm^{-1}. In line with that, the band in the *aci*-nitro formation spectrum shifts -8 cm^{-1}. Upon $\beta,\beta\gamma$-$^{18}O_3$ labeling, part of the intensity shifts -28 cm^{-1} in the absorbance spectrum (no shift in the *aci*-nitro formation spectrum, but intensity loss in the product formation spectrum). Band shifts -3 cm^{-1} for ^{13}C labeling of the benzylic carbon (^{13}C–O–P$_\gamma$) and -2 cm^{-1} for ^{15}N labeling
1061, a, i, p	$\nu_s(PO_2^-)$ with a strong contribution from the γ-phosphate	36	shifts -23 cm^{-1} upon γ-$^{18}O_3$ labeling [36], possibly part of this band shifts -13 cm^{-1} for $\beta,\beta\gamma$-$^{18}O_3$ labeling [37] (no shift in the *aci*-nitro formation spectrum, but intensity loss in the product formation spectrum), shifts -5 cm^{-1} for ^{13}C labeling of the benzylic carbon (^{13}C–O–P$_\gamma$)
1016, a, i, p	backbone $\nu(CO)$ and $\nu(PO)$	36	shifts -13 cm^{-1} upon C–^{18}O–P$_\gamma$ labeling [36]. The same shift is observed for γ-$^{18}O_3$ labeling in *aci*-nitro formation and product formation spectra, but is not obvious in absorbance spectra. Band shifts -2 cm^{-1} and looses intensity for $\beta,\beta\gamma$-$^{18}O_3$ labeling in absorbance spectra, but no significant change is observed in *aci*-nitro formation spectra and a slight upshift under intensity loss in product formation spectra; band shifts -6 cm^{-1} for ^{13}C labeling of the benzylic carbon (^{13}C–O–P$_\gamma$)
997, (a)	possibly backbone $\nu(CO)$ and $\nu(PO)$ close to α-phosphate		no clear shift upon γ-$^{18}O_3$ labeling, shifts -3 cm^{-1} and intensifies upon $\beta,\beta\gamma$-$^{18}O_3$ labeling

Tab. 7.1.1 (continued)

Band position (cm^{-1})	Assignment	Reference for assignment	Comments
972, a, i, p	backbone v(CO) and v(PO) and non-bridging (PO) with a strong contribution from the γ-phosphate	36	shifts −12, −20, −3 and −7 cm^{-1} upon C−^{18}O−P$_\gamma$, γ-^{18}O$_3$, $\beta,\beta\gamma$-^{18}O$_3$ and ^{13}C−O−P$_\gamma$ labeling, respectively
935, a, p	backbone v(CO), backbone v(PO), and non-bridging v(PO) involving also the nitrobenzyl moiety	36	shifts −7, −12 and −16 cm^{-1} upon C−^{18}O−P$_\gamma$, γ-^{18}O$_3$ and $\beta,\beta\gamma$-^{18}O$_3$ labeling, respectively [36, 37], −4 cm^{-1} for ^{13}C labeling of the benzylic carbon (^{13}C−O−P$_\gamma$) and −3 cm^{-1} for ^{15}N labeling

The wavenumbers and shifts upon isotopic substitution given are from absorbance spectra (Fig. 7.1.2 and unpublished) in ^{1}H$_2$O and might differ from those found in the photolysis difference spectra by a few wavenumbers [1, 34, 36, 37]. Note that the α- and γ-labeled compounds used by Cepus et al. [34, 45] seem to have been a mixture of both isotopomers [62]. Only those bands are listed that can be assigned to either the nitrophenyl moiety or the phosphate groups of NPE-caged ATP. Band shifts of less than 2 cm^{-1} are not reported. For band amplitudes in the *aci*-nitro formation spectrum and the product formation spectrum, see Fig. 7.1.4.

Abbreviations: a: observable in the caged ATP absorbance spectrum as peak (no brackets) or shoulder (in brackets); i: observable in the *aci*-nitro formation spectrum (caged ATP → *aci*-nitro intermediate) as peak (no brackets) or shoulder (in brackets); p: observable in the product formation spectrum (caged ATP → ATP and byproduct) as peak (no brackets) or shoulder (in brackets). v: stretching vibration, v_{as}: antisymmetric or asymmetric stretching vibration, v_s: symmetric stretching vibration.

Tab. 7.1.2 Aci-nitro intermediate bands in the spectral range from 1800 to 900 cm^{-1}

Band position (cm^{-1})	Assignment	Reference for assignment	Comments
1465	delocalized mode involving $v(C=N)$	37	shifts -14 cm^{-1} upon ^{15}N labeling [37]
1379		37	shifts -2 cm^{-1} upon ^{13}C labeling of the benzylic carbon (^{13}C–O–P$_\gamma$) [37]
1330	delocalized mode involving $v_{as}(NO_2^-)$	37, 63	shifts -12 cm^{-1} upon ^{15}N labeling [37] and -4 cm^{-1} upon ^{13}C labeling of the benzylic carbon (^{13}C–O–P$_\gamma$) [37]
~1260	$v_{as}(PO_2^-)$ with contributions from β- and γ-phosphate	36, 37	reduced upon γ-^{18}O$_3$ labeling [36], shifts approximately -15 cm^{-1} upon $\beta,\beta\gamma$-^{18}O$_3$ labeling [37]
1245	$v(C=C-O-P_\gamma)$	36	shifts -11 cm^{-1} upon ^{13}C labeling of the benzylic carbon (^{13}C–O–P$_\gamma$) [37], shifts -3 cm^{-1} upon C–^{18}O–P$_\gamma$ and γ-^{18}O$_3$ labeling [36]
1179	delocalized mode involving $v_s(NO_2^-)$	63	shifts -15 cm^{-1} upon ^{15}N labeling [37]
1123	delocalized vibration involving $v_s(PO_2^-)$ with contributions from $v(P_\gamma O)$ and nitrobenzyl moiety	36	shifts -2 and -4 cm^{-1} upon C–^{18}O–P$_\gamma$ and γ-^{18}O$_3$ labeling [36], respectively, splits upon $\beta,\beta\gamma$-^{18}O$_3$ labeling [37], shifts -4 cm^{-1} upon ^{13}C labeling of the benzylic carbon (^{13}C–O–P$_\gamma$) [37], and -2 cm^{-1} upon ^{15}N labeling [37]
1092	$v_s(PO_2^-)$ involving $v(P_\gamma O)$	36	less intense upon γ-^{18}O$_3$ labeling [36], shifts -2 cm^{-1} upon ^{13}C labeling of the benzylic carbon (^{13}C–O–P$_\gamma$) [37]
1031	delocalized mode involving $v_s(\gamma\text{-}PO_2^-)$		shifts -11 cm^{-1} upon γ-^{18}O$_3$ labeling [36], shifts -3 cm^{-1} upon ^{15}N labeling [37]
992	backbone $v(CO)$ and $v(PO)$, $v_s(PO_2^-)$	36	shifts -10 and -13 cm^{-1} upon C–^{18}O–P$_\gamma$ and γ-^{18}O$_3$ labeling [36], less intense upon $\beta,\beta\gamma$-^{18}O$_3$ labeling, shifts -7 cm^{-1} upon ^{13}C labeling of the benzylic carbon (^{13}C–O–P$_\gamma$) [37]
946	vibration involving $v(COP_\gamma)$	36	shifts -10 cm^{-1} upon C–^{18}O–P$_\gamma$ and γ-^{18}O$_3$ labeling [36]

Wavenumbers refer to peak positions in the aci-nitro formation spectrum in ^{1}H$_2$O [36, 37]. For band amplitudes, see Fig. 7.1.4. Band shifts of less than 2 cm^{-1} are not reported. For abbreviations, see Tab. 7.1.1.

Tab. 7.1.3 Nitrosoacetophenone bands in the spectral range from 1800 to 1150 cm^{-1}

Band position (cm^{-1})	Assignment	Reference for assignment	Comments
1686	ν(C=O) of NO dimer	37, 64	shifts –37 cm^{-1} upon ^{13}C labeling of the benzylic carbon (^{13}C–O–P$_7$) [37], in ^{1}H$_2$O at 1687 cm^{-1}, no shift upon dimerization
1684	ν(C=O) of NO monomer	37, 64	shifts –37 cm^{-1} upon ^{13}C labeling of the benzylic carbon (^{13}C–O–P$_7$) [37], in ^{1}H$_2$O at 1687 cm^{-1}
1632	NO *cis*-dimer		band position from NO *cis*-dimer minus monomer difference spectrum (Fig. 7.1.5b), at 1647 cm^{-1} (shoulder) in a 2-nitrosoacetophenone *trans*-dimer powder absorbance spectrum (not shown)
1599	ν(CC) of benzene ring of NO dimer	3	at 1600 cm^{-1} in a 2-nitrosoacetophenone *trans*-dimer powder absorbance spectrum
1584	ν(CC) of benzene ring of NO *cis*-dimer	3	band position from NO *cis*-dimer minus monomer difference spectrum (Fig. 7.1.5b), at 1574 cm^{-1} in a 2-nitrosoacetophenone *trans*-dimer powder absorbance spectrum
1501	ν(NO) of NO monomer	37, 65	shifts –18 cm^{-1} upon ^{15}N labeling [37]
1456	ν(CC) of benzene ring of NO monomer	37	shifts –4 cm^{-1} upon ^{15}N labeling
1451	ν(CC) of benzene ring of NO *cis*-dimer	37	at 1442 cm^{-1} in a 2-nitrosoacetophenone *trans*-dimer powder absorbance spectrum
1426	delocalized vibration involving ν(NO) and/or ν(CN) of NO monomer	37	shifts –10 cm^{-1} upon ^{15}N labeling [37]
1426	ν(NO), ν(NN) and/or ν(CN) of NO *cis*-dimer	37, 65	shifts –32 cm^{-1} upon ^{15}N labeling [37], at 1424 cm^{-1} in a 2-nitrosoacetophenone *trans*-dimer powder absorbance spectrum
1380	ν(NO), ν(NN) and/or ν(CN) of NO *cis*-dimer	37, 65	band position from NO *cis*-dimer minus monomer difference spectrum (Fig. 7.1.5b), shifts –34 cm^{-1} upon ^{15}N labeling [37]
1318?	NO monomer		unclear whether this band is an NO monomer band, since it is only seen in the NO *cis*-dimer minus monomer difference spectrum (Fig. 7.1.5b), there is no obvious band in the photolysis spectrum and no corresponding band in a 2-nitrosoacetophenone *trans*-dimer powder absorbance spectrum

Tab. 7.1.3 (continued)

Band position (cm^{-1})	Assignment	Reference for assignment	Comments
1303	NO cis-dimer		at 1300 cm^{-1} in a 2-nitrosoacetophenone trans-dimer powder absorbance spectrum
1290	NO monomer		shifts –4 cm^{-1} upon ^{15}N labeling
1269	delocalized vibration of NO trans-dimer with considerable contributions from $v(NN)$ and $v(NO)$	37, 65	band position from NO trans-dimer minus cis-dimer difference spectrum (Fig. 7.1.5 b), shifts –26 cm^{-1} upon ^{15}N labeling [37]
1268	NO cis-dimer		band position from the NO cis-dimer minus monomer difference spectrum, shifts –18 cm^{-1} upon ^{13}C labeling of the benzylic carbon (^{13}C–O–P_γ)
1241	NO monomer and dimer		more intense for the monomer
1210	NO monomer		
1207	NO cis-dimer		
1193	NO monomer		
1192	delocalized vibration involving $v(NN)$, v and/or $v(CN)$ of NO cis-dimer		shifts –8 cm^{-1} upon ^{15}N labeling
1159	NO monomer		band position from NO cis-dimer minus monomer difference spectrum (Fig. 7.1.5b), shifts –2 cm^{-1} upon ^{15}N labeling

Wavenumbers refer to peak positions in product formation spectra at p^2H 6.0, 1 or 35 °C in ^2H_2O [37] or, when indicated, to nitroso monomer minus nitroso cis-dimer difference spectra at p^2H 6.0 and 1 °C in ^2H_2O or to nitroso trans-dimer minus cis-dimer difference spectra at p^2H 6.0 and 35 °C in ^2H_2O (Fig. 7.1.5). If not stated otherwise, no significant band shifts have been observed in ^1H_2O. Band shifts of less than 2 cm^{-1} are not reported. For band amplitudes, see Fig. 7.1.5. For abbreviations see Tab. 7.1.1.

Tab. 7.1.4 ATP bands in the spectral region of phosphate absorption from 1250 to 900 cm^{-1}

Band position (cm^{-1})	Assignment	Reference for assignment	Comments
1234	$v_{as}(\alpha\text{-}PO_2^-)$ and $v_{as}(\beta\text{-}PO_2^-)$	36, 66, 67	up to -4 cm^{-1} shift upon $\beta\gamma,\gamma\text{-}^{18}O_4$ or $\gamma\text{-}^{18}O_3$ labeling [66, 67]; part of the band shifts approximately -40 cm^{-1} upon either $\alpha\text{-}^{18}O_2$, $\alpha\beta,\beta\text{-}^{18}O_3$ or $\beta,\beta\gamma\text{-}^{18}O_3$ labeling, the remaining band shifts about -10 cm^{-1} [66, 67]
1216 sh	$v_{as}(PO_2^-)$ with a strong contribution from the β-phosphate	37	shifts approximately -40 cm^{-1} upon $\beta,\beta\gamma\text{-}^{18}O_3$ labeling [37]; the band position of the $\beta,\beta\gamma,\text{-}^{18}O_3$ isotopomer was observed in the product formation spectrum upon photolysis of labeled NPE-caged ATP [37] and is similar to a band observed in the absorbance spectrum of labeled GTP [67] which was not discussed in the reference; the band position for the unlabeled compound was inferred from calculating the difference between a product formation spectrum obtained with unlabeled and labeled ATP; this difference spectrum shows only the effects of isotopic substitution and has a clear shoulder near 1215 cm^{-1} (not shown) where also the absorbance spectrum of unlabeled ATP has a shoulder
~1120 broad	$v_{as}(PO_3^{2-})$ coupled to out of phase $v_s(PO_2^-)$ and possibly $v(P_\beta OP_\gamma)$	66, 67	$v_{as}(PO_3^{2-})$ weakly coupled to $v_s(\alpha\text{-}PO_2^-)$, strongly coupled to $v_s(\beta\text{-}PO_2^-)$ [66], shifts -3, -10 and -17 cm^{-1} upon $\alpha\text{-}^{18}O_2$, $\alpha\beta,\beta\text{-}^{18}O_3$ and $\beta,\beta\gamma\text{-}^{18}O_3$ labeling, respectively [66, 67], shifts -27 cm^{-1} upon $\gamma\text{-}^{18}O_3$ labeling [67] and -34 cm^{-1} upon $\beta\gamma,\gamma\text{-}^{18}O_4$ labeling [66]; the additional shifts caused by labeling of the $\beta\gamma$-oxygen seems to indicate a contribution by a P–O–P stretching vibration
1116	$v_s(PO_2^-)$ in phase	66, 67	does not shift significantly upon $\gamma\text{-}^{18}O_3$ [67] and $\beta\gamma,\gamma\text{-}^{18}O_4$ labeling [66], shifts -7 upon $\alpha\text{-}^{18}O_2$ labeling [66], shifts -13 and -14 cm^{-1} upon $\alpha\beta,\beta\text{-}^{18}O_3$ [66] and $\beta,\beta\gamma\text{-}^{18}O_3$ labeling [67], respectively; the assignment in the references is from Raman spectra

Tab. 7.1.4 (continued)

Band position (cm^{-1})	Assignment	Reference for assignment	Comments
1087 sh	$v_s(PO_2^-)$ out of phase and possibly $v(P_\beta OP_\gamma)$	66, 67	no shift upon γ-$^{18}O_3$ or $\beta\gamma,\gamma$-$^{18}O_4$ labeling [66, 67], shifts approximately -22 cm^{-1} upon $\alpha\beta,\beta$-$^{18}O_3$ labeling [66] and -27 cm^{-1} upon $\beta,\beta\gamma$-$^{18}O_3$ labeling [67]; the additional shifts caused by labeling of the $\beta\gamma$-oxygen seems to indicate a contribution by a P–O–P vibration
1030			weak in the absorbance spectrum
990	$v_s(PO_3^{2-})$, $v(POP)$ with a considerable contribution of $v(P_\alpha OP_\beta)$	36, 66	shifts -10 to -16 cm^{-1} upon γ-$^{18}O_3$ labeling [36, 67]; upon $\beta\gamma,\gamma$-$^{18}O_4$ labeling two bands are observed at 993 and 965 cm^{-1} [66], possibly because backbone and non-bridging vibrations are decoupled in this isotopomer; shifts up to -3 cm^{-1} upon $\beta,\beta\gamma$-$^{18}O_3$ labeling [37, 67] and -14 cm^{-1} upon $\alpha\beta,\beta$-$^{18}O_3$ labeling [66]
917	$v_s(PO_3^{2-})$, $v(POP)$, $v_s(\beta$-$PO_2^-)$, $v_s(\alpha$-$PO_2^-)$	36, 37, 66	shifts -15 to -22 cm^{-1} upon γ-$^{18}O_3$ labeling [36, 67], -26 cm^{-1} upon $\beta\gamma,\gamma$-$^{18}O_4$ labeling [66], more than -15 cm^{-1} upon $\beta,\beta\gamma$-$^{18}O_3$ labeling [37], -16 cm^{-1} upon $\alpha\beta,\beta$-$^{18}O_3$ labeling [66] and -3 cm^{-1} upon α-$^{18}O_2$ labeling [66]

Band positions refer to our absorbance spectrum of ATP in water (not shown). Similar spectra can be found in the literature (see, e.g. [66–68]). The wavenumbers and shifts upon isotopic substitution in photolysis difference spectra might differ by a few wavenumbers [1, 34, 36, 37]. For band amplitudes in the product formation spectrum see Fig. 7.1.4. Band shifts of less than 2 cm^{-1} are not reported.

The ATP spectrum is sensitive to the binding of cations [66–69] or to a protein [45, 62, 70, 71]. Some of the spectra cited above were obtained with GTP instead of ATP discussed here. However, the different base is not expected to alter the absorption of the phosphate groups and this is supported by similar absorbance spectra published for GTP [67] and ATP [66, 68], and the identical photolysis spectra of NPE-caged ATP and NPE-caged GTP in the spectral range from 1300 to 950 cm^{-1} [34] which are sensitive to the absorption of all three phosphate groups (see Tab. 7.1.1). Note that the α- and γ-labeled compounds used by Cepus et al. [34, 45] seem to have been a mixture of both isotopomers [62]. For abbreviations, see Tab. 7.1.1.

Tab. 7.1.5 2-hydroxylaminoacetophenone (closed form) bands in the spectral range from 1800 to 1250 cm^{-1}

Band position (cm^{-1})	Comments
1481	weak, shifts –2 cm^{-1} upon ^{13}C labeling of the benzylic carbon (^{13}C–O–P$_{?}$)
1465	weak
1415	weak
1388	weak
1334	weak, band position from a difference spectrum between the open and the closed form of 2-hydroxylaminoacetophenone, observed at 1329 cm^{-1} in the product formation spectrum, shifts approximately –9 cm^{-1} upon ^{13}C labeling of the benzylic carbon (^{13}C–O–P$_{?}$)

Band positions are from photolysis spectra recorded within the first 60 ms after photolysis at 35 °C in ^{2}H$_2$O at p^{2}H 6.0 [37]. Band shifts of less than 2 cm^{-1} are not reported. Band intensities are judged relative to the strongest bands in the photolysis spectrum.

Tab. 7.1.6 2-Hydroxylaminoacetophenone (open form) bands in the spectral range from 1800 to 1200 cm^{-1}

Band position (cm^{-1})	Assignment	Reference for assignment	Comments
1685	v(C=O)		weak, shifts –30 to –40 cm^{-1} upon ^{13}C labeling of the benzylic carbon (^{13}C–O–P$_{?}$)
1641	v(C=O), internally hydrogen bonded	37	medium, shifts –32 cm^{-1} upon ^{13}C labeling of the benzylic carbon (^{13}C–O–P$_{?}$)
1605	v(CC) of benzene ring	3	weak
1562	v(CC) of benzene ring	3	weak, shifts –4 cm^{-1} upon ^{13}C labeling of the benzylic carbon (^{13}C–O–P$_{?}$), shifts –2 cm^{-1} upon ^{15}N labeling
1475	v(CC) of benzene ring	3	weak, shifts –3 cm^{-1} upon ^{15}N labeling
1451	v(CC) of benzene ring	3	weak
1415			weak
1368			weak, band position from a difference spectrum between the open form of 2-hydroxylaminoacetophenone and 3-methylanthranil, shifts –3 cm^{-1} upon ^{13}C labeling of the benzylic carbon (^{13}C–O–P$_{?}$)
1256	delocalized vibration involving v(CC) of carbonyl carbon		medium, band position from a difference spectrum between the open form of 2-hydroxylaminoacetophenone and 3-methylanthranil, shifts –12 cm^{-1} upon ^{13}C labeling of the benzylic carbon (^{13}C–O–P$_{?}$)

Band positions are from photolysis spectra recorded 0.3–3.6 s after photolysis at 35 °C in ^{2}H$_2$O at p^{2}H 6.0 [37]. Band shifts of less than 2 cm^{-1} are not reported. Band intensities are judged relative to the strongest bands in the photolysis spectrum. For abbreviations, see Tab. 7.1.1.

Tab. 7.1.7 3-Methylanthranil bands in the spectral range from 1800 to 1200 cm^{-1}

Band position (cm^{-1})	Assignment	Comments
1645	delocalized vibration with a considerable contribution from ν(C=C) of the heterocyclic ring	medium, shifts –6 cm^{-1} upon ^{13}C labeling of the benzylic carbon (^{13}C–O–P$_\gamma$)
1566	delocalized vibration with a considerable contribution from ν(C=C) of the heterocyclic ring	weak, shifts –9 cm^{-1} upon ^{13}C labeling of the benzylic carbon (^{13}C–O–P$_\gamma$)
1526	delocalized vibration involving the heterocyclic ring	medium, band position from a difference spectrum between the open form of 2-hydroxylaminoacetophenone and 3-methylanthranil, shifts –2 cm^{-1} upon ^{13}C labeling of the benzylic carbon (^{13}C–O–P$_\gamma$)
1466	delocalized vibration with a considerable contribution from the heterocyclic ring	medium, shifts –6 cm^{-1} upon ^{13}C labeling of the benzylic carbon (^{13}C–O–P$_\gamma$), shifts –4 cm^{-1} upon ^{15}N labeling
1442	delocalized vibration with a considerable contribution from the heterocyclic ring	weak, shifts –6 cm^{-1} upon ^{13}C labeling of the benzylic carbon (^{13}C–O–P$_\gamma$), shifts apparently –3 cm^{-1} upon ^{15}N labeling in the photolysis spectrum but not in the difference spectrum between the open form of 2-hydroxylaminoacetophenone and 3-methylanthranil; the ^{15}N-induced shift in the photolysis spectrum seems therefore to be due to shifts of overlapping bands and not to a shift of the 1442 cm^{-1} band itself
1415		weak
1403	delocalized vibration involving the heterocyclic ring	weak, band position from a difference spectrum between the open form of 2-hydroxylaminoacetophenone and 3-methylanthranil, shifts –3 cm^{-1} upon ^{13}C labeling of the benzylic carbon (^{13}C–O–P$_\gamma$)
1377		seen in the absorbance spectrum of methylanthranil, gives rise to a shoulder in the photolysis spectrum
1222		weak, band position from a difference spectrum between the open form of 2-hydroxylaminoacetophenone and 3-methylanthranil, reduced upon ^{13}C labeling of the benzylic carbon (^{13}C–O–P$_\gamma$)

Band positions and intensities are from photolysis spectra recorded 120–140 s after photolysis at 35 °C in ^{2}H$_2$O at p^{2}H 6.0 [37] unless stated otherwise. Band shifts of less than 2 cm^{-1} are not reported. Band intensities are judged relative to the strongest bands in the photolysis spectrum. For abbreviations see Tab. 7.1.1.

References

1. A. Barth, W. Mäntele, W. Kreutz, *FEBS Lett.* **1990**, *277*, 147–150.
2. V. Jayaraman, in *Dynamic Studies in Biology: Phototriggers, Photoswitches and Caged Biomolecules*, M. Goeldner, R. S. Givens (eds), Wiley-VCH, Weinheim, **2004**, chapter 7.2 (this volume).
3. N. B. Colthup, L. H. Daly, S. E. Wiberley, *Introduction to Infrared and Raman Spectroscopy*, 2nd edn, Academic Press, New York, **1975**.
4. L. J. Bellamy, *The Infrared Spectra of Complex Molecules*, 3rd edn, vol. 1, Chapman & Hall, London, **1975**.
5. S. Y. Venyaminov, F. G. Prendergast, *Anal. Biochem.* **1997**, *248*, 234–245.
6. P. Hinsmann, M. Haberkorn, J. Frank, P. Svasek, M. Harasek, B. Lendl, *Appl. Spectrosc.* **2001**, *55*, 241–251.
7. A. J. White, K. Drabble, C. W. Wharton, *Biochem. J.* **1995**, *306*, 843–849.
8. R. Masuch, D. A. Moss, in *Spectroscopy of Biological Molecules: New Directions*, J. Greve, G. J. Puppels, C. Otto (eds), Kluwer, Dordrecht, **1999**, pp. 689–690.
9. E. Kauffmann, N. C. Darnton, R. H. Austin, C. Batt, K. Gerwert, *Proc. Natl Acad. Sci. USA* **2001**, *98*, 6646–6649.
10. W. Mäntele, in *Biophysical Techniques in Photosynthesis*, J. Amesz, A. J. Hoff (eds), Kluwer, Dordrecht, **1996**, pp. 137–160.
11. K. Gerwert, *Curr. Opin. Struct. Biol.* **1993**, *3*, 769–773.
12. F. Siebert, in *Infrared Spectroscopy of Biomolecules*, H. H. Mantsch, D. Chapman (eds), Wiley-Liss, New York, **1996**, pp. 83–106.
13. B. R. Cowen, R. M. Hochstrasser, in *Infrared Spectroscopy of Biomolecules*, H. H. Mantsch, D. Chapman (eds), Wiley-Liss, New York, **1996**, pp. 107–129.
14. R. M. Slayton, P. A. Anfinrud, *Curr. Opin. Struct. Biol.* **1997**, *7*, 717–721.
15. M. Tasumi, *J. Mol. Struct.* **1993**, *292*, 289–293.
16. H. Georg, K. Hauser, C. Rödig, O. Weidlich, F. Siebert, *NATO ASI Series, Series E: Appl. Sci.* **1997**, *342*, 243–261.
17. H. Frei, *AIP Conf. Proc.* **1998**, *430*, 28–39.
18. E. Hirota, K. Kawaguchi, *Annu. Rev. Phys. Chem.* **1985**, *36*, 53–76.
19. M. W. George, M. Poliakoff, J. J. Turner, *Analyst* **1994**, *119*, 551–560.
20. K. McFarlane, B. Lee, J. Bridgewater, P. C. Ford, *J. Organomet. Chem.* **1998**, *554*, 49–61.
21. J. P. Toscano, *Adv. Photochem.* **2001**, *26*, 41–91.
22. W. Uhmann, A. Becker, C. Taran, F. Siebert, *Appl. Spectrosc.* **1991**, *45*, 390–397.
23. R. Rammelsberg, S. Boulas, H. Chorongiewski, K. Gerwert, *Vibrational Spectrosc.* **1999**, *19*, 143–149.
24. Q. Cheng, M. G. Steinmetz, V. Jayaraman, *J. Am. Chem. Soc.* **2002**, *124*, 7676–7677.
25. W. Mäntele, R. Hienerwadel, F. Lenz, W. J. Riedel, R. Grisar, M. Tacke, *Spectrosc. Int.* **1990**, *2*, 29–35.
26. T. Yuzawa, C. Kato, M. W. George, H. Hamaguchi, *Appl. Spectrosc.* **1994**, *48*, 684–690.
27. J. H. Kaplan, B. Forbush, J. F. Hoffman, *Biochemistry* **1978**, *17*, 1929–1935.
28. K. C. Zimmer, *Z. Physik. Chem.* **1933**, *B23*, 239–255.
29. H. Morrison, B. H. Migdalof, *J. Org. Chem.* **1965**, *30*, 3996.
30. J. N. Pitts, J. K. S. Wan, E. A. Schuck, *J. Am. Chem. Soc.* **1964**, *86*, 3606–3610.
31. A. V. Uvarov, S. V. Yakubovich, *Lakokras. Mater. Ikh Primen.* **1961**, *6*, 49–52 (*Chem. Abstr.* **1962**, *56*, 13131b).
32. S. Mager, M. Ionescu, *Rev. Roum. Chim.* **1966**, *11*, 533–539.
33. G. C. Pimentel, *Pure Appl. Chem.* **1965**, *11*, 563–569.
34. V. Cepus, C. Ulbrich, C. Allin, A. Troullier, K. Gerwert, *Methods Enzymol.* **1998**, *291*, 223–245.
35. V. Jayaraman, S. Thiran, D. R. Madden, *FEBS Lett.* **2000**, *475*, 278–282.
36. A. Barth, K. Hauser, W. Mäntele, J. E. T. Corrie, D. R. Trentham, *J. Am. Chem. Soc.* **1995**, *117*, 10311–10316.
37. A. Barth, J. E. T. Corrie, M. J. Gradwell, Y. Maeda, W. Mäntele, T. Meier, D. R. Trentham, *J. Am. Chem. Soc.* **1997**, *119*, 4149–4159.

38 I. R. Dunkin, J. Gebicki, M. Kiszka, D. Sanín-Leira, *J. Chem. Soc. Perkin Trans. 2*, **2001**, 1414–1425.

39 J. W. Walker, G. P. Reid, J. A. McCray, D. R. Trentham, *J. Am. Chem. Soc.* **1988**, *110*, 7170–7177.

40 I. R. Dunkin, J. Gebicki, M. Kiszka, D. Sanín-Leira, *Spectrochim. Acta A* **1997**, *53*, 2553–2557.

41 J. E. T. Corrie, A. Barth, V. R. N. Munasinghe, D. R. Trentham, M. C. Hutter, *J. Am. Chem. Soc.* **2003**, *125*, 8546–8554.

42 F. Von Germar, A. Galán, O. Llorca, J. L. Carrascosa, J. M. Valpuesta, W. Mäntele, A. Muga, *J. Biol. Chem.* **1999**, *274*, 5508–5513.

43 C. Raimbault, F. Besson, R. Buchet, *Eur. J. Biochem.* **1997**, *244*, 343–351.

44 S. Geibel, A. Barth, S. Amslinger, A. H. Jung, C. Burzik, R. J. Clarke, R. S. Givens, K. Fendler, *Biophys. J.* **2000**, *79*, 1346–1357.

45 V. Cepus, A. J. Scheidig, R. S. Goody, K. Gerwert, *Biochemistry* **1998**, *37*, 10263–10271.

46 F. Von Germar, A. Barth, W. Mäntele, *Biophys. J.* **2000**, *78*, 1531–1540.

47 A. Barth, W. Kreutz, W. Mäntele, *Biochim. Biophys. Acta* **1994**, *1194*, 75–91.

48 C. Raimbault, R. Buchet, C. Vial, *Eur. J. Biochem.* **1996**, *240*, 134–142.

49 T. Granjon, M.-J. Vacheron, C. Vial, R. Buchet, *Biochemistry* **2001**, *40*, 2988–2994.

50 C. Raimbault, E. Clottes, C. Leydier, C. Vial, R. Buchet, *Eur. J. Biochem.* **1997**, *247*, 1197–1208.

51 U. Görne-Tschelnokow, F. Hucho, D. Naumann, A. Barth, W. Mäntele, *FEBS Lett.* **1992**, *309*, 213–217.

52 A. Barth, J. E. T. Corrie, *Biophys. J.* **2002**, *83*, 2864–2871.

53 R. Buchet, I. Jona, A. Martonosi, *Biochim. Biophys. Acta* **1991**, *1069*, 209–217.

54 H. Georg, A. Barth, W. Kreutz, F. Siebert, W. Mäntele, *Biochim. Biophys. Acta* **1994**, *1188*, 139–150.

55 A. Troullier, K. Gerwert, Y. Dupont, *Biophys. J.* **1996**, *71*, 2970–2983.

56 C. Ma, M. G. Steinmetz, Q. Cheng, V. Jayaraman, *Org. Lett.* **2003**, *5*, 71–74.

57 J. E. T. Corrie, A. Barth, G. Papageorgiou, *J. Labeled Cpd Radiopharm.* **2001**, *44*, 619–626.

58 C. Ludovici, R. Fröhlich, K. Vogt, B. Mamat, M. Lübben, *Eur. J. Biochem.* **2002**, *269*, 2630–2637.

59 J. A. McCray, L. Herbette, T. Kihara, D. R. Trentham, *Proc. Natl Acad. Sci. USA* **1980**, *77*, 7237–7241.

60 J. F. Cameron, J. M. J. Fréchet, *J. Am. Chem. Soc.* **1991**, *113*, 4303–4313.

61 A. Barth, C. Zscherp, *FEBS Lett.* **2000**, *477*, 151–156.

62 C. Allin, K. Gerwert, *Biochemistry* **2001**, *40*, 3037–3046.

63 H. Feuer, C. Savides, C. N. R. Rao, *Spectrochim. Acta* **1963**, *19*, 431–434.

64 A. Barth, W. Mäntele, W. Kreutz, *Biochim. Biophys. Acta* **1991**, *1057*, 115–123.

65 C. N. R. Rao, K. R. Bashkar, *The Chemistry of the Nitro and Nitroso Groups. Part 1*, H. Feuer (ed.), Wiley, New York, **1970**, pp. 137–161.

66 H. Takeuchi, H. Murata, I. Harada, *J. Am. Chem. Soc.* **1988**, *110*, 392–397.

67 J. H. Wang, D. G. Xiao, H. Deng, R. Callender, M. R. Webb, *Biospectroscopy* **1998**, *4*, 219–227.

68 H. Brintzinger, *Biochim. Biophys. Acta* **1963**, *77*, 343–345.

69 H. Brintzinger, *Helv. Chim. Acta* **1965**, *48*, 47–54.

70 X. Du, H. Frei, S.-H. Kim, *J. Biol. Chem.* **2000**, *275*, 8492–8500.

71 H. Cheng, S. Sukal, H. Deng, T. S. Leyh, R. Callender, *Biochemistry* **2001**, *40*, 4035–4043.

72 Y. V. Il'ichev, M. A. Schwörer, J. Wirz, *J. Am. Chem. Soc.* **2004**, in press.

73 J. E. T. Corrie, in *Dynamic Studies in Biology: Phototriggers, Photoswitches and Caged Biomolecules*, M. Goeldner, R. S. Givens (eds), Wiley-VCH, Weinheim, **2004**, chapter 1.1 (this volume).

7.2
IR Spectroscopy with Caged Compounds: Selected Applications

Vasanthi Jayaraman

7.2.1
Introduction

Molecular recognition and specific interactions between substrates or ligands and proteins play an important role in various biological processes. In order to gain a complete understanding of the biological reaction or protein function, it is thus essential to characterize these specific interactions and follow changes in these interactions as the protein carries out its function. Spectroscopic investigations, such as Fourier transform IR (FTIR) spectroscopy, are the method of choice for such investigations as they possess the required sensitivity to probe these specific interactions in detail and provide molecular-level insight into the interactions [1–5]. More importantly, when combined with the ability to initiate the reaction using inert precursors that photolytically release the substrates or ligands (caged compounds), FTIR spectroscopy has the ability to investigate the changes in the protein in a time-resolved manner [6–8].

FTIR spectroscopy probes the IR-active vibrational modes of the chemical moieties being studied. Hence, when studying macromolecules, such as proteins, this method is limited by the high degree of spectral congestion due the vibrations from all the constituents of the macromolecule. This can be overcome by studying the changes in the vibrational modes due to the reaction being investigated, by obtaining a difference spectrum between the product or intermediate and the initial state. In doing this, only specific molecular changes that are critical to the reaction or function of the protein are reported in the difference spectrum. On the other hand, the vibrations from the bulk of the protein, which do not undergo any change, are not reported. Since only a small fraction of the protein undergoes changes during the process being investigated, these differences in the IR absorbance intensities are typically small, usually less than 1% of parent absorption. As a consequence, obtaining a difference spectrum with two different samples such as product and reactants, with errors in concentration and pathlength being important, can introduce debilitating artifacts preventing any useful interpretation of the data. Instead it is more accurate to initiate the biological reaction in the sample holder and obtain a difference spectrum on the same sample before and after the reaction. Additionally, initiating the reaction in the sample holder allows for a time-resolved investigation of the reaction by obtaining a difference spectrum at various times during the reaction.

A convenient method to initiate the reaction in the sample holder is to photolytically release the substrate/ligand from a "caged" substrate/ligand that is biologically inert prior to photolysis. In some cases, such as caged glutamate, this method has the added advantage that it allows for investigations in the submillisecond timescale [9], while classical mixing devices where the substrate/ligand

and protein solutions are mixed to initiate the ligand binding process typically have a millisecond time resolution. This chapter discusses the application of the combined use of caged compounds and FTIR spectroscopy in three biological systems: glutamate binding to the ligand binding segment of glutamate receptor, GTP hydrolysis by Ras and ATP hydrolysis by ATPases. These investigations clearly indicate the advantages of using caged compounds and FTIR spectroscopy to not only obtain detailed insights into the strength of the interactions at specific chemical moieties of the ligand and protein, but in also providing structural insights into the mechanism of the process being studied such as enzyme catalysis in the case of GTPase and ATPase.

7.2.2
Selected Examples

7.2.2.1 Allosteric Mechanism of Activation of the Glutamate Receptor

Communication between nerve cells serves as the basis of all brain activity and one of the fundamental steps involved in signal transmission between the nerve cells is the conversion of a "chemical" signal liberated at the end of one nerve cell into an "electrical" signal at the second nerve cell. This step is mediated by a class of membrane-bound proteins known as neurotransmitter receptors. Glutamate receptors belong to this family of proteins and are the main excitatory receptors in the central nervous system [10, 11]. Glutamate binding to the receptor triggers the formation of transmembrane channels permitting cations to flow down their electrochemical gradients across the postsynaptic membrane, depolarizing it and thereby stimulating the receiving cell. One of the central questions in understanding the function of this class of proteins is how agonist (glutamate) binding to the receptor leads to the sequence of conformational changes that regulate ion flow. An important breakthrough in addressing this question was the expression of the ligand-binding domain as a soluble fusion protein (S1S2) for the a-amino-3-hydroxy-5-methylisoxazole-4-propionic acid (AMPA) receptor subtype of the glutamate receptors [12, 13]. This large-scale expression of the S1S2 protein has lead to the relatively rapid determination of its structure in the unligated and various ligated states [14–17]. These X-ray structures act as an excellent foundation for detailed investigations of the structural dynamics of the ligand-binding process using a powerful spectroscopic probe such as FTIR that is capable of high-resolution structural definition and adequate temporal resolution. FTIR spectroscopy is particularly useful for studying S1S2 since the ligand glutamate of the glutamate receptor has characteristic vibrational modes in the 1400–1800 cm^{-1} frequency region [18, 19], which allows the characterization of the ligand environment in the protein. Additionally, the amide I vibration of the S1S2 protein backbone is an excellent probe for investigating the changes in the secondary structure and hence overall conformational changes in the protein. The use of these modes in gaining insight into the ligand controlled receptor function in the glutamate receptor and the possibility for future time-resolved FTIR spectroscopy investigations using these modes as markers for determining the structural changes during the ligand binding process is discussed here.

Since the differences in the FTIR spectral features due to glutamate binding to the S1S2 protein are small, the best method to obtain a difference FTIR spectrum between the apo and glutamate bound forms is to photolytically release glutamate from caged glutamate in the sample holder. This eliminates the need to change the samples, which as discussed in the Introduction introduces changes in pathlength and concentration, hence making it difficult to subtract the two data sets. The difference spectrum thus obtained by subtracting the spectrum of the S1S2 protein in the presence of caged glutamate after and before photolysis is shown in Fig. 7.2.1 (trace A) [19]. This difference spectrum reflects the features due to bound glutamate as well as changes in the S1S2 protein induced by glutamate binding. The bands due to bound glutamate have been identified using isotopically labeled glutamate. Isotopic labeling selectively shifts the vibrational band from the moiety labeled hence identifying the modes from that moiety. These labeling studies indicate that in the 1500–1800 cm^{-1} frequency region the ligand glutamate has two vibrational modes arising from the two, a and γ, carboxylate moieties in the spectral region. Since the frequencies of the vibrational modes of a specific moiety are a signature of its electronic environment, the carboxylate vibrational modes of the ligand glutamate provide insight into the strength of the non-covalent interactions between the carboxylate moieties of glutamate and the protein. Specifically, in the case of the asymmetric carboxylate vibrations, the shift in the frequency can be directly related to the change in enthalpy associated with the change in the non-covalent interactions at this moiety [20]. Therefore, by comparing the frequencies of the carboxylate vibrations of glutamate in the protein versus in the free state it is evident that the non-covalent interactions at the a-carboxylate of the glutamate have favorable ($\Delta H \sim 1.3$ kcal) interactions in the protein relative to the free state of the ligand, while the γ-carboxylate of the ligand glutamate is in an unfavorable environment relative to that in D$_2$O ($\Delta H \sim 2.3$ kcal). These results,

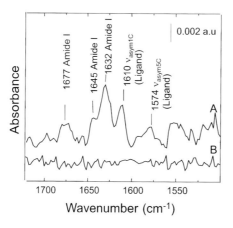

Fig. 7.2.1 (A) Difference FTIR spectrum between S1S2 protein in D$_2$O in the presence of caged glutamate after and before photolysis. Features due to excess caged glutamate and photolysis products have been subtracted. (B) FTIR spectrum generated by subtracting the spectrum of S1S2 protein in the presence of caged glutamate in D$_2$O with the spectrum of S1S2 and the spectrum of caged glutamate. (Adapted from [19] with permission from the Federation of the European Biochemical Societies)

when taken in the context of the X-ray structures of the glutamate-bound form of S1S2 protein, allowed for the identification of the specific residues that interact with the α- and γ-carboxylates in the dynamic state of the protein. For instance, in the case of the γ-carboxylate, these studies indicated that the negatively charged carboxylate of a side-chain glutamate residue of the protein leads to the unfavorable environment and overrides the favorable interactions identified by the X-ray structures.

In addition to the ligand modes the positive features at 1677, 1645 and 1632 cm^{-1}, in difference spectrum between the glutamate-bound and unligated forms of the protein (Fig. 7.2.1, trace A) are characteristic of amide I modes arising from turns, irregular β-sheet/solvent exposed helices and β-sheet secondary structures of the protein respectively [21–23]. The positive bands at these frequencies indicate that there is a modest increase in the content of all three ordered secondary structure elements in the protein when it binds to glutamate, suggesting that the protein is more ordered upon binding the ligand. These results are consistent with the NMR structures that indicate a decrease in the dynamics of the bilobed S1S2 protein upon binding glutamate [24].

The identification of the protein and ligand vibrational modes in the S1S2 protein as well as the establishment that caged glutamate is inert prior to photolysis sets the foundation for future time-resolved investigations, where combining time-resolved vibrational spectroscopy with the photolytic release from caged glutamate should make it possible to map the allosteric pathway, i.e. the steps involved in the ligand-binding process, in the S1S2 protein.

Another important aspect of these FTIR investigations is the ability to directly probe the inertness of the caged glutamate prior to photolysis. The biological inertness is typically investigated by probing the absence of functional changes in the protein due to the presence of the caged compound. However, it is possible for the caged compound to be docked in the ligand binding site or some other site on the protein and not produce any functional consequences. Such a docking could lead to altered "apparent" kinetics, specifically affecting the ligand binding steps. In the FTIR investigations such an interaction between the caged compound and the protein can be studied by obtaining the spectrum of protein in the presence of caged compound and subtracting the features due to free protein and due to free caged compound. A featureless spectrum as shown in the case of caged glutamate in the presence of S1S2 protein (shown in Fig. 7.2.1, trace B) [20], indicates that caged compound does not interact with the protein and hence is biologically inert.

7.2.2.2 GTP Hydrolysis by Ras: Nature of the Intermediate

Ras, a guanine nucleotide-binding protein, belongs to a family of G-proteins that play a central role in a number of cell signaling pathways [25–27]. The signal is transmitted by the active GTP-bound form of the protein, while hydrolysis of bound GTP to bound GDP leads to inactivation of the protein, which turns the signal off. The signals transmitted by these proteins control a number of cellular pathways such as cell proliferation, metabolism and differentiation [27]. For this reason, the mechanism of GTP hydrolysis catalyzed by this class of proteins has been the subject of extensive investigations.

In order to gain an understanding of the mechanism of enzyme catalysis, it is essential to determine the nature of the transition state and how this transition state (stabilized by the enzyme) is different from the transition state of an uncatalyzed reaction. In the case of GTP hydrolysis, a continuum of potential transition states exist which range from "dissociative" to "associative" depending on the extent of bond cleavage and bond formation (shown in Fig. 7.2.2) [28]. A dissociative transition state is one in which the bond to the leaving group is fully or nearly cleaved and the bond to the incoming nucleophile is absent or barely formed. An extreme case of this would result in the formation of a metaphosphate group. In the case of GTP, this leads to a decrease in charge at the γ-phosphate group in the transition state. In contrast, the associative mechanism has a trigonal bipyramidal transition state, with a large amount of bond formation with the incoming nucleophile and very little bond cleavage with the leaving group. In the case of GTP, this would lead to a net increase in charge at the γ-phosphate group.

The solution based uncatalyzed phosphate ester hydrolysis has been extensively characterized since the 1950s and these studies suggest a dissociative metaphosphate-like transition state [29]. Furthermore, linear free energy relationships for the uncatalyzed hydrolysis of GTP are also consistent with a dissociative transition state [30]. In spite of all the evidence for a dissociative transition state in the uncatalyzed reaction, the associative pathway cannot be ruled out in the enzyme-catalyzed reaction. Specifically, the presence of large number of positively charged residues and metal ions observed in the catalytic site of the end-state X-ray structures of Ras, if involved in interactions with the negative charge at the γ-phosphate, could stabilize an associative transition state.

Crystal structures of the active and inactive states of Ras have provided a detailed picture of the end states in the GTP hydrolysis reaction, and possible

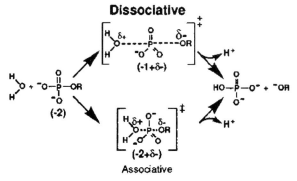

Fig. 7.2.2 Pathways for the phosphoryl transfer reactions. (Reproduced with permission from [28]. Copyright 1996 National Academy of Sciences, USA)

mechanisms have been developed based on these end state structures [31, 32]. However, in order for one to gain a deeper understanding of the reaction pathway it is essential to follow the structural changes during the reaction pathway and use a solution-based assay that is not limited by crystallographic constraints. Time-resolved vibrational spectroscopy is ideal for such an investigation and hence has been the method of choice by a number of investigators studying this enzyme.

An important requirement for such a time-resolved investigation, as described in detail in the Introduction, is a method for rapid initiation of the reaction. As discussed earlier, a convenient method to initiate the GTP hydrolysis reaction is to photolytically release the substrate (GTP) from caged GTP. Nitrophenylacyl derivatives and hydroxyphenylacyl derivatives of GTP have been proven to be excellent caged compounds for GTP for investigating Ras hydrolysis [33–36]. Both these compounds interact with the enzyme prior to photolysis, but do not lead to the activation of the enzyme prior to photolysis, and liberate GTP upon irradiation with UV light. By coupling the use of caged GTP and vibrational spectroscopy, the mechanism of GTP hydrolysis by Ras has been investigated in the presence and absence of GTPase-activating protein, and these studies are summarized below.

A detailed study of the environments and interactions of the α-, β- and γ-phosphate of GTP when bound to Ras has been possible by studying the asymmetric and symmetric vibrations of these phosphate groups [33–36]. This required the specific identification of these modes using isotopic ^{18}O labeling at these specific sites. These studies reveal that there is an unusual large downshift in the asymmetric vibrational mode of the β-phosphate group of GTP (given in Fig. 7.2.3, reproduced from [33]) upon binding to Ras relative to its free form, revealing strong hydrogen bonding and electrostatic interaction at this group in the active site of Ras. This observation would be consistent with an increase in the negative charge at the non-bridging β-phosphate oxygens. Interestingly, the asymmetric vibrational mode of the γ-phosphate group is upshifted suggesting a decrease in the negative charge at the oxygens of this phosphate (tabulated in Fig. 7.2.3). This pattern of shifts in the charges provides clear evidence for a dissociative transition state (Fig. 7.2.2). An important finding is that GTP binding to Ras in the presence of GTPase-activating protein causes an even larger shift in the asymmetric vibrational mode of the β-phosphate group consistent with a further accumulation of negative charge on the β-phosphate oxygens [33]. This shift is the only change observed due to the presence of GTPase-activating protein, suggesting that the charge transfer to the β-phosphate oxygens plays a critical role in reducing the activation energy for GTP hydrolysis and is an important pathway through which this reaction is catalyzed. Further evidence that the hydrolysis is a concerted mechanism involving a charge transfer and not the formation of a discrete metaphosphate intermediate, the extreme case scenario for the dissociative mechanism, is obtained from the positional isotope exchange investigations (Fig. 7.2.4) [36]. In these investigations, the [β-^{18}O$_3$]GTP was used and the extent of scrambling at this position

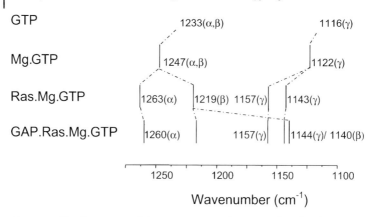

Fig. 7.2.3 The frequencies of the asymmetric vibrational modes of the α-, β- and γ-phosphates of GTP in the free form, and when bound to Mg, Ras and Ras–GAP complex. (Reproduced with permission from [33]. Copyright 2001 National Academy of Sciences, USA)

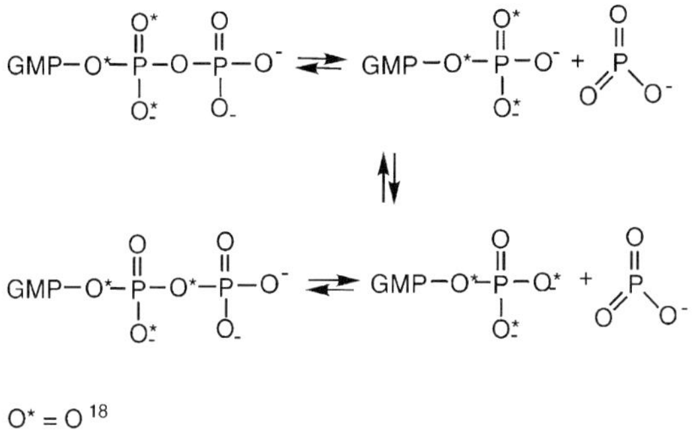

Fig. 7.2.4 Scrambling in the isotopic labeling of the O atoms at the -position of GTP due to the formation of metaphosphate. (Reproduced with permission from [36])

monitored. The formation of the metaphosphate intermediate would lead to the conversion of the isotopically labeled non-bridging oxygen to a bridging position and *vice versa* which would lead to asymmetric stretches at both ^{18}O and ^{16}O position for the β-phosphate. However, this was not observed, ruling out the formation of the metaphosphate.

The assignments of the vibrational modes from α-, β- and γ-phosphate of GTP not only allowed for a detailed investigation of their environments in the enzyme, but more importantly allowed for a detailed characterization of the changes at these positions during the hydrolysis when studied in the time-resolved mode. The time-resolved investigation of the Ras catalyzed hydrolysis of GTP in the presence of GTPase-activating protein, for instance, revealed three distinct steps. The first was the photolysis of the caged GTP to release GTP, followed by cleavage of the γ-phosphate accompanied by the appearance of the protein-bound phosphate and a final step involving the release of the phosphate from the protein into the bulk. These studies provide the first direct evidence that the release of phosphate is the rate-limiting step in Ras-catalyzed hydrolysis of GTP in the presence of GTPase-activating protein.

7.2.2.3 Active Transport of Ca^{2+} by Ca-ATPase

Muscle relaxation is mediated by the removal of Ca^{2+} from the cytoplasm of the muscle cells into the sarcoplasmic reticulum by Ca-ATPase. The enzyme, Ca-ATPase, actively transports two Ca^{2+} ions by using the energy derived from the hydrolysis of one ATP molecule and also catalyzes the reverse process, i.e. the synthesis of ATP using the energy derived from the Ca^{2+} gradient. The initial studies in the 1970s suggested a two state model (E1 and E2) for the Ca^{2+} transport by ATPase, according to which the translocation of Ca^{2+} ions occurred by a rate-limiting conformational change from an ADP-sensitive phosphoenzyme (E1-P-Ca$_2$) which is formed from ATP and cytoplasmic Ca^{2+}, to an ADP-insensitive phosphoenzyme (E2-P-Ca$_2$), from which Ca^{2+} ions dissociate to the lumen of the sarcoplasmic reticulum (Fig. 7.2.5) [37]. However, kinetic investigations monitoring the formation of the phosphoenzyme from [^{32}P]phosphate could not detect the presence of two Ca^{2+}-bound phosphoenzyme intermediates in the reaction pathway and the single intermediate observed could be best attributed to E1-P-Ca$_2$ [38]. Since time-resolved FTIR spectroscopy can monitor the changes at the substrate and the enzyme, and hence provide detailed insights into kinetics and structures of the intermediates, it has been used to further test the two-state model. In these investigations, the changes in the vibrational spectrum of sarcoplasmic reticulum Ca-ATPase were followed during the course of its catalytic cycle by photolytically releasing ATP from caged ATP [39–42]. Specifically, these investigations could monitor changes of specific residues at the catalytic sites, such as the phosphate modes of ATP, ADP and bound phosphate, modes that could be assigned to Ca^{2+} release, as well as global conformational changes in the enzyme through the polypeptide backbone vibrations; hence, providing the ability to discriminate between reactions at specific sites of the substrate and enzyme and the resulting or preceding overall conformation changes in the

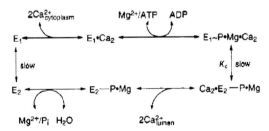

Fig. 7.2.5 Mechanism of Ca^{2+} transport by Ca-ATPase. (Reproduced with permission from [38]. Copyright 1995 American Chemical Society)

enzyme. The evolution of the spectral features in these time-resolved FTIR investigations clearly indicated two kinetic steps following ATP binding to the enzyme, and based on the vibrational modes contributing to these phases the steps could be identified as those leading to the formation of E1-P-Ca_2 and E2P. There was no evidence for the presence of a long-lived E2-P-Ca_2, consistent with the radioactive labeling studies by Jencks et al. These studies have lead to the modification of the mechanism of Ca-ATPase with the elimination of the E2-P-Ca_2 state.

Aside from the use of FTIR spectroscopy in combination with caged compounds to probe the mechanism and detailed interactions of the substrate and protein/enzyme, the sensitivity of this method to probe for small structural changes also allows for structure–activity studies, which provide the basis for the design of future generation drugs that have a high specificity for the enzyme/protein. For such investigations, caged compounds can be generated for the various substrates being studied. For instance, in the case of Ca-ATPase, caged derivatives of 2-deoxy-ATP, 3-deoxy-ATP and inosine 5′-triphosphate (Fig. 7.2.6), were synthesized and used to liberate the various substrates, in order to study their properties and compare it to those of ATP [43]. These studies were able to detect the effect of alterations at each of the sites in the substrate on specific steps in the catalysis reaction and hence identify the role of each of these moieties in the enzyme catalysis. For instance, modifications at the 3-OH and amine group of the adenine ring produced more pronounced effects than changes at the 2-OH position in the extent of binding-induced conformational change of the enzyme. However, the transfer of the phosphate group from the substrate to the protein was only detected for 2-deoxy-ATP and inosine 5′-triphosphate, but not for 3-deoxy-ATP. One of the drawbacks of such a structure–activity study is the necessity to synthesize the corresponding caged compounds. This has been overcome by a strategy recently employed by Barth et al., in which the conformational changes induced by 2′,3′-O-(2,4,6-trinitrophenyl)adenosine 5′-monophosphate (TNPAMP) binding to Ca-ATPase was monitored through a competitive experiment. In these experiments the TNP-AMP initially

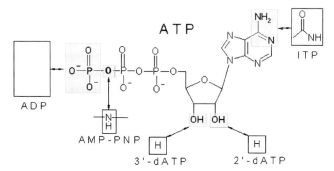

Fig. 7.2.6 ATP and analogs of ATP used for structure–activity investigations. (Reproduced with permission from [43])

bound to the enzyme was displaced by the competitive substrate β-γ-iminoadenosine 5′-triphosphate (AMPPNP) [44]. The difference spectrum obtained before and after displacement was then compared to the difference spectrum obtained between the apo and AMPPNP bound forms of the enzyme, to identify changes induced due to TNPAMP binding. These studies indicated that TNPAMP binding led to a conformation similar to that the E1Ca$_2$ state and not E1ATPCa$_2$ state. The identification of the specific conformation of the protein in the presence of TNPAMP as investigated in the above study is important because the TNP analogs of adenosine phosphates are fluorescent and have been used extensively to study the Ca-ATPase catalysis reaction.

7.2.3
Conclusions

The examples discussed above demonstrate the combined use of caged compounds and IR spectroscopy in obtaining detailed insights into protein function and enzyme catalysis. FTIR spectroscopy has the sensitivity to probe changes at individual chemical moieties and the capability to probe large-scale conformational changes such as changes in secondary structure of the proteins, while the use of caged compounds provides time resolution, accuracy and convenience for such investigations.

Acknowledgments

I would like to thank Dr. Ramanan Krishnamoorti and Dr. Andreas Barth for reading the manuscript and providing helpful comments. This work was supported by National Science Foundation grant NSF-0096635 and a grant from the American Heart Association of Texas (V.J.).

References

1 F. SIEBERT, *Methods Enzymol.* **1995**, *246*, 501–526.
2 M. JACKSON, H.H. MANTSCH, *Crit. Rev. Biochem. Mol. Biol.* **1995**, *30*, 95–120.
3 A. BARTH, C. ZSCHERP, *FEBS Lett.* **2000**, *477*, 151–156.
4 C. JUNG, *J. Mol. Recognit.* **2000**, *13*, 325–351.
5 A. BARTH, C. ZSCHERP, *Q. Rev. Biophys.* **2002**, *35*, 369–430.
6 A. BARTH, et al., *J. Biol. Chem.* **1996**, *271*, 30637–30646.
7 K. GERWERT, *Biol. Chem.* **1999**, *380*, 931–935.
8 S. GEIBEL, et al., *Biophys. J.* **2000**, *79*, 1346–1357.
9 Q. CHENG, M.G. STEINMETZ, V. JAYARAMAN, *J. Am. Chem. Soc.* **2002**, *124*, 7676–7677.
10 B. SOMMER, P.H. SEEBURG, *Trends Neurosci.* **1992**, *13*, 291–296.
11 M. HOLLMANN, S. HEINEMANN, *Annu. Rev. Neurosci.* **1994**, *17*, 31–108.
12 A. KUUSINEN, M. ARVOLA, K. KEINÄNEN, *EMBO J.* **1995**, *14*, 6327–6332.
13 G.-Q. CHEN, E. GOUAUX, *Proc. Natl Acad. Sci. USA* **1997**, *94*, 13431–13436.
14 N. ARMSTRONG, et al., *Nature* **1998**, *395*, 913–917.

15 N. Armstrong, E. Gouaux, *Neuron* **2000**, *28*, 165–181.
16 A. Hogner, et al., *J. Mol. Biol.* **2002**, *322*, 93–109.
17 Y. Sun, et al., *Nature* **2002**, *417*, 245–253.
18 T. G. Spiro, *Biological Applications of Raman Spectroscopy*, Wiley, New York, **1987**.
19 V. Jayaraman, S. Thiran, D. R. Madden, *FEBS Lett.* **2000**, *475*, 278–282.
20 Q. Cheng, et al., *Biochemistry* **2002**, *41*, 1602–1608.
21 J. L. Koenig, D. L. Tabb, in *Analytical Applications of FT-IR to Molecular, Biological Systems*, J. R. Durig (ed.), Kluwer, Dordrecht, **1980**.
22 D. M. Byler, H. Susi, *Biopolymers* **1986**, *25*, 469–487.
23 D. Chapman, M. Jackson, P. I. Harris, *Biochem. Soc. Trans.* **1989**, *17*, 617–619.
24 R. L. McFeeters, R. E. Oswald, *Biochemistry* **2002**, *41*, 10472–10481.
25 M. Barbacid, *Annu. Rev. Biochem.* **1987**, *56*, 779–827.
26 H. R. Bourne, D. A. Sanders, F. McCormick, *Nature* **1991**, *349*, 117–127.
27 S. R. Sprang, *Annu. Rev. Biochem.* **1997**, *66*, 639–678.
28 K. A. Maegley, S. J. Admiraal, D. Herschlag, *Proc. Natl Acad. Sci. USA* **1996**, *93*, 8160–8166.
29 J. Kumamoto, F. H. Westheimer, *J. Am. Chem. Soc.* **1955**, *77*, 2515–2518.
30 S. J. Admiraal, D. Herschlag, *Chem. Biol.* **1995**, *2*, 729–739.
31 E. F. Pai, et al., *EMBO J.* **1990**, *9*, 2351–2359.
32 L. A. Tong, et al., *J. Mol. Biol.* **1991**, *217*, 503–516.
33 C. Allin, et al., *Proc. Natl Acad. Sci. USA* **2001**, *98*, 7754–7759.
34 C. Allin, K. Gerwert, *Biochemistry* **2001**, *40*, 3037–3046.
35 V. Cepus, et al., *Biochemistry* **1998**, *37*, 10263–10271.
36 X. L. Du, H. Frei, S. H. Kim, *J. Biol. Chem.* **2000**, *275*, 8492–8500.
37 L. de Meis, A. L. Vianna, *Annu. Rev. Biochem.* **1979**, *48*, 275–292.
38 J. Myung, W. P. Jencks, *Biochemistry* **1995**, *34*, 3077–3083.
39 A. Barth, W. Kreutz, W. Mantele, *FEBS Lett.* **1990**, *277*, 147–150.
40 A. Barth, et al., *J. Biol. Chem.* **1996**, *271*, 30637–30646.
41 A. Barth, W. Mantele, *Biophys. J.* **1998**, *75*, 538–544.
42 F. von Germar, A. Barth, W. Mantele, *Biophys. J.* **2000**, *78*, 1531–1540.
43 M. Liu, A. Barth, *J. Biol. Chem.* **2003**, *278*, 10112–10118.
44 M. Liu, A. Barth, *Biophys. J.* **2003**, *85*, 3262–3267.

7.3
New Perspectives in Kinetic Protein Crystallography using Caged Compounds

Dominique Bourgeois and Martin Weik

7.3.1
Introduction: Principles of Kinetic Crystallography

7.3.1.1 Protein Activity and Conformational Changes in Crystals

It is quite astonishing that such complex and highly flexible objects as biological macromolecules can arrange themselves in a perfectly ordered fashion to form crystals. Nevertheless, more than 18 000 protein structures have now been determined to near-atomic resolution by X-ray crystallography and this number is just a beginning: with the advent of high-throughput structural proteomics, a world-wide initiative aimed at systematic structural investigations of proteins expressed by given genomes, an explosion of new structures will undoubtedly ap-

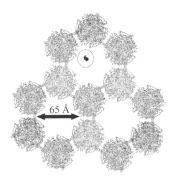

Fig. 7.3.1 Representation of the solvent channels within a trigonal crystal of acetylcholinesterase, viewed down the c-axis. The protein molecules are bathed within solvent and small molecules such as substrate or cofactors (an NPE-Ch molecule is shown encircled) may diffuse quite freely through the channels whose diameter attains 65 Å in this case. Structural data are from PDB entry 1EA5.

pear in the near future. However, despite the considerable efforts behind this evolution, our knowledge of static X-ray structures remains insufficient to understand how proteins function. What characterizes life is motion, and biological molecules constantly breathe around their average shape, wandering in a conformational landscape at the basis of activity. Fortunately, watching crystalline proteins in action is not an impossible task. Indeed, macromolecular crystals are fragile samples maintained by weak contact forces, of the level of a single hydrogen bond (~ 10 kcal mol^{-1}). Therefore, individual molecules in the crystal remain flexible, with only few surface residues interacting with neighboring molecules and most of the others being bathed in solvent (Fig. 7.3.1). Indeed, the presence of solvent channels, sometimes making up to 90% of the crystal volume content, is a characteristic feature of protein crystals. Under these conditions, the observation that many enzymes are fully active in the crystalline state is not so surprising. Following conformational changes by X-ray crystallography is nevertheless a considerable challenge. In some cases, the catalysis driving force may lead to crystal disorder or even rupture, ruining the diffraction power of the sample. In other cases, access to the protein active site by a substrate may be restricted by unfortunate crystal contacts or by the competitive presence of some constituent indispensable to crystallization and present in large concentration in the medium. Additionally, the pH, dielectric constant and solvent viscosity generally deviate from their optimal values in physiological conditions, which often slows down catalysis and may alter its mechanism or even abolish it. Last, but not least, crystallography yields an average structure from an ensemble of typically 10^{13}–10^{15} individual molecules and a significant portion, if not all, of these must "walk in pace" for a coherent structural picture of catalysis to be delivered. The emerging field of "kinetic crystallography" [1–3], aimed at solving the structures of transient species along the reaction pathway, is therefore paved with a number of exciting challenges. One of the most critical issues concerns the question of homogeneous and synchronous triggering of a reaction in a protein crystal [4]. Caged compounds – photolabile precursors of substrates, cofactors, amino acids or even protons – represent highly effective tools to address this question and their use in the context of kinetic crystallography has already produced results of outstanding biological significance [5–10].

Properly handling them for such experiments, however, is delicate, and it is the purpose of this chapter to highlight both their potential and limitations in this field of research.

7.3.1.2 Timescales and Activation States

Enzymes catalyze biological reactions over a broad range of timescales, extending from nanoseconds to seconds or more. In general, rapid reactions involve subtle, small conformational changes, e.g. side-chain motions, associated with small activation barriers, whereas slower reactions involve larger changes, e.g. helix or loop motions, associated with high barriers. Larger motions are not necessarily easier to observe crystallographically because the reduced requirement for high-resolution data may be compromised by the generation of stronger disorder in the crystal during catalysis.

When designing a reaction initiation strategy in a kinetic crystallography experiment, the sole knowledge of the overall reaction rate as measured in solution is largely insufficient. First, rates must be considered in the crystalline state, which in general slows down turnover [1], but in certain circumstances may accelerate it (e.g. due to a short-cut pathway imposed by lattice forces [11]). Second, what counts is the lifetime of the particular intermediate targeted in the study, which can be considerably shorter than the overall reaction cycle. As an example, the lifetime of the K intermediate in the reaction cycle of bacteriorhodopsin, a light-driven proton pump, is a few microseconds [12], whereas the overall cycle takes place in milliseconds. Third, for a significant population of an intermediate to build-up, the rate of formation of the latter must be greater than its rate of decay. These rates (taking the form $k = v \exp{-E_a/kT}$) depend on activation energy barriers E_a, as well as on frequency factors v, which can be understood as how often per second the molecule attempts to jump over the activation barrier. As a consequence, extensive kinetic studies are generally required, both in the crystal and in solution, before embarking on non-interpretable diffraction experiments.

Once timescales are approximately known, one must decide on a suitable reaction triggering method, activation profile and data collection strategy. It is useful to distinguish between the triggering method, which brings together the ingredients (protein, ligand, substrate and/or cofactor) necessary to start a chemical reaction, and the activation profile, which controls how fast or how far the reaction may proceed by acting on physicochemical parameters such as temperature, pressure, viscosity or pH. Accumulation of a transient species in the crystal will, in principle, be possible whenever reaction triggering is rapid enough to ensure synchrony amongst the molecules of the sample in the chosen activation state. Suppose, for example, that the activation state is first lowered by reducing the temperature to the point where no activity may occur. Under such conditions, reaction triggering, e.g. delivering the substrate to an enzyme, may be as slow as practical considerations permit and accumulation of the intermediate will still be possible, depending on subsequent, adequate temperature elevation. On the contrary, if the activation state of the sample is kept high, optimal turnover occurs immediately upon triggering, which therefore must be rapid enough to keep synchrony. The same considerations hold true for diffraction data collection. If the biological system is brought to a low activation state (e.g. by cryo-cooling), data collection may be slow and single-wavelength X-rays may be used with standard protocols. In a high activation state, transient species must be caught "on the fly" and rapid data collection must be used, such as the Laue technique making use of polychromatic X-rays.

Although most kinetic protein crystallographic experiments to date are based on single-turnover strategies, the possibility to rely on a steady-state build-up of an intermediate, where substrates and products are continuously delivered to and withdrawn from the sample with a flow-cell apparatus, has been used successfully [13, 14].

The recent evolution of protein crystallography shows a global standardization, and even automation, towards data collection at cryo-temperatures (typically 100 K or less, i.e. in a low activation state) using samples mounted in nylon cryo-loops [15, 16]. This is mostly due to the extreme brilliance of recent synchrotron sources, resulting in rapid radiation damage, which is minimized [17], but not abolished, at 100 K [17–20]. Monochromatic data collection at close to ambient temperature from crystals mounted in capillaries remains generally only possible with laboratory X-ray sources and provides high-resolution density maps with only a small number of friendly crystalline proteins. In parallel, the availability of Laue beamlines is scarce, because of the high level of instrumental sophistication required combined with the low number of favorable samples. As a consequence, kinetic crystallography nowadays receives practical constraints from the structural proteomics world. The dedicated methodology must adapt and we have to consider the use of caged compounds in this new context.

7.3.1.3 Accumulating Unstable Species in the Crystal

In principle, the easiest way of triggering catalysis in a crystalline enzyme consists of letting the substrate or a cofactor rapidly diffuse into the crystal by soaking the latter into a proper solution. However, diffusion throughout the narrow crystal channels is often exceedingly slow and may require up to hours or even days, becoming rate limiting [21–23] and preventing the synchronous build-up of intermediate states if the enzyme is in a high activation state. In extreme cases, a reactant may not diffuse at all, being turned over into product by the first layers of macromolecules crossed in the crystal. Despite these limitations, a number of intermediates could be captured using this simple methodology [24–26]. Slow diffusion could be tolerated if the protein could be kept in a low activation state in a way that does not interfere with the diffusion process. For example, lowering the temperature below solvent crystallization or embedding the protein in a glassy medium would arrest diffusion and cannot be envisaged. However, a combination with a pH jump is possible [27, 28] or with engineered proteins that are mutated to block the reaction at a certain step [13, 29]. The other triggering protocol most largely used in protein crystallography relies on direct light activation of built-in photosensitive moieties. These can be natural chromophores such as the retinal of bacteriorhodopsin [30] or the 4-hydroxy cinamoic acid from photoactive yellow protein [31], or photosensitive coordination bonds such as the iron–carbonmonoxy bonds in heme containing proteins [32]. The advantages of this strategy are immense: (1) proper choice of the excitation wavelength and light intensity often allows homogeneous activation throughout the sample, (2) photoactivation is extremely fast, taking place on the timescale of Franck-Condon absorption ($\sim 10^{-18}$ s), allowing synchronization in real-time-resolved experiments at the level of the pulse duration of the used light source, and (3) photoactivation occurs at cryo-temperatures, while protein activity is partially or completely stopped. Using this latter strategy, the early steps of extremely fast reactions such as the photocycle of bacteriorhodopsin [12, 33] or PYP [34–36] and the relaxation of myoglobin [37–40] or hemoglobin [41] after CO photolysis could be studied at high resolution using gentle light sources combined with cryo-crystallography. Unfortunately, natural photosensitivity is scarce in nature and other widely applicable triggering strategies are necessary. Recently, the observation of radiation damage caused by X-ray-induced photoelectrons has been turned into an elegant strategy to trigger photoreduction in redox crystalline enzymes at low temperature [42, 43]. For the huge number of non-redox proteins that do not contain a naturally photoactivable group, caged compounds represent an elegant way to trigger catalysis by UV-vis light.

7.3.2
Real-time Crystallography and the Laue Technique

The main reason for using caged compounds in biological sciences is the rapidity with which they can be photolysed (typically in the millisecond timescale). This is because in practically all fields of research using these compounds [44], this property is a major issue, the biological system being kept in its physiological, high activation state and fast synchronization being essential. In this context, caged compounds were first envisaged in crystallography for their use in real-time-resolved experiments, where intermediate states are caught "on the fly" as they transiently build-up in the crystal. This was at a time where these types of experiments were largely put forward as one of the most elegant applications of high-brilliance synchrotron machines in protein science. Indeed, synchrotrons deliver intense X-rays over a broad bandpass of wavelengths, allowing us to reduce the exposure time necessary to record a diffraction pattern down to well below typical durations of single-turnover biological reactions. This is achieved by employing the Laue technique, where the white beam is allowed to diffract at once on a non-rotating, static crystal. With the advent of third-generation facilities such as the ESRF, Spring-8 or APS, exposure times can be as short as 100 ps [45] – a time much shorter than the photofragmentation time of most known caged compounds. It was therefore anticipated that real movies of catalysis could be obtained where the time resolution would be as high as permitted by the photoactivation process (down to a few tens of microseconds). However, it is striking to observe that amongst all publications that have used Laue diffraction in combination with photolysis of caged compounds, only one of them achieved a time resolution down to the millisecond [8]. For all the others [5–7], the time resolution was rather in the second or even minute timescale. Two reasons explain this fact. The first reason is that the collection of a single Laue diffraction image, in the vast majority of cases [46], is largely insufficient to completely sample the crystal reciprocal space. Therefore, several images are needed to reconstruct interpretable electron density maps. Since photolysis of caged compounds is a non-repeatable process, the time resolution of the experiment is given by the total experimental time necessary to collect a sufficient number of images if a single crystal is employed. The latter comprises delays to rotate the sample and readout the detector between X-ray exposures. Even with the fastest CCD detectors currently available, the readout time cannot be reduced to below ~ 100 ms and the total experimental time in practice reaches the order of several seconds. This limitation can, in principle, be alleviated if a series of crystals are employed, each of them being exposed once at complementary orientations. Such a strategy, however, is experimentally very demanding, requiring the preparation of many high-quality, pre-oriented samples. Furthermore, it is based on proper scaling between the different frames – an extremely difficult task due to crystal-to-crystal variations in photolysing efficiency and diffraction quality.

7.3.2.1 A Favorable Case: Isocitrate Dehydrogenase

The problem addressed above was overcome successfully by Stoddard et al. [8] in studies of isocitrate dehydrogenase. This enzyme uses NAD(P) to convert isocitrate into a-ketoglutarate and carbon dioxide. The rate-limiting product complex, with a life-time of 40–50 ms under the experimental conditions, could be accumulated in the crystal by using the following strategy: several crystals were soaked into free isocitrate and caged (DMNPE)-NADP ($P^{2'}$-[1-(4,5-dimethoxy-2-nitrophenyl)-ethyl] NADP) (Fig. 7.3.2). In the dark, isocitrate bound to the enzyme active site, while caged NADP remained in the solvent channels because the cage lowered dramatically the affinity of the cofactor for the enzyme. A single 0.5-ms xenon light pulse was then sufficient to release a sufficient number of NADP molecules to completely saturate the entire population of enzyme molecules in each crystal despite crystal-to-crystal variations in photolysis yield. This was possible because the concentration of enzyme molecules in the crystals was rather low (5 mM) and because NADP has an affinity for the enzyme in the micromolar range, so that a 10-fold excess of caged cofactor could be used (50 mM) without compromising the possibility of a uniform photolysis due to high optical density. In such conditions, several crystals could be used, each of them providing partial data sets with millisecond time resolution that could be scaled together successfully because even partial photolysis (20–50%) led to complete saturation of the active sites in each sample.

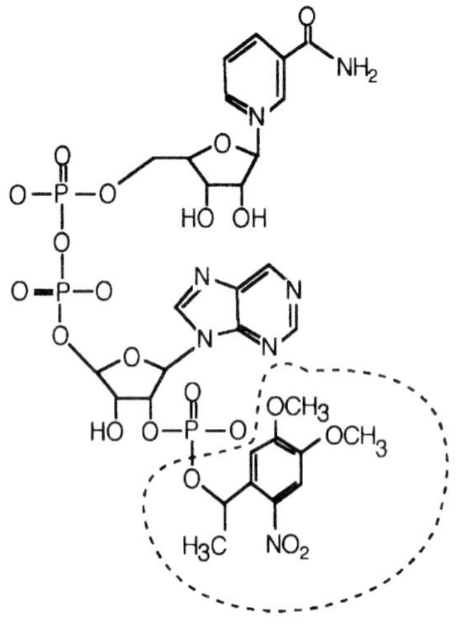

Fig. 7.3.2 Representation of caged NADP. The DMNPE blocking group is outlined by a dashed border. (Reproduced with permission from [8])

The studies of Stoddard et al. [8] could possibly be tackled by two other methods allowing the use of a unique crystal. The first would consist in using a flow cell to regenerate the crystal content in between two X-ray exposures by flowing in fresh isocitrate and caged NADP, while flushing out the products liberated by the last round of reaction triggering. The second, still hypothetical as of today, would rely on the ongoing development of so-called pixel detectors, which are able to continuously readout pixels as they are hit by X-ray photons at a very high rate, thus considerably decreasing the overall data collection time of a Laue experiment provided the crystal can be rotated sufficiently quickly [47].

The advantage given here by the so-called "affinity caging" might in general be offset by the concomitant decrease in time resolution, which is limited by the time it takes for the released co-factor to diffuse and bind into the active site. The diffusion time is small because the distance to travel is tiny, of the order of the unit-cell length. With a conservative diffusion coefficient of $D = 10^{-9}$ cm^2 s^{-1} (representing a diffusion time of a day through a 0.2-mm thick crystal [48]), the mean distance traveled within a millisecond is given by $d = \sqrt{(2Dt)} = 140$ Å, that is about the size of a unit cell. The binding time also depends on the bimolecular association (and dissociation) rate constant, and reactant concentrations. In the case of isocitrate dehydrogenase, this penalty had no consequence, considering the high association rate constant (10^5–10^6 M^{-1} s^{-1}) and high reactant concentrations (5 and 10 mM).

The experiment by Stoddard et al. [8] was also successful for another reason, i.e. the radiation hardness of the biological system with respect to intense and short UV-vis radiation, of the order of 5 mJ (in 0.5 ms) delivered onto the sample in the case of isocitrate dehydrogenase. This value can be compared to the theoretical energy needed to photolyse a caged compound in a crystal, expressed by:

$$E_{\text{delivered}}[\text{mJ}] = 12 \times 10^5 \times d[\text{mm}] \times c[\text{mol l}^{-1}] \times S[\text{mm}^2] / \{1 - 10^{-\text{OD}}\} \times \lambda[\text{nm}] \times \phi\}$$

where d is the crystal thickness, c is the concentration of caged compound that needs to be photolysed in the crystal, S is the photolysing beam section, OD is the optical density at the photolysing wavelength λ and ϕ is the photolysis yield. For crystals of 0.1 mm thickness, a concentration of caged compound of 10 mM, an optical density of 0.1, a mean wavelength of 450 nm, a yield of 0.2 and a beam cross-section of 2 mm², we find $E_{\text{delivered}} = 13$ mJ. This energy is commonly achievable by pulsed flash lamps. Provided good care is taken in alignment and focusing procedures, the energy actually deposited within the crystal is expressed by:

$$E_{\text{deposited}}[\text{mJ}] = 12 \times 10^5 \times V[\text{mm}^3] \times c[\text{mol}\, l^{-1}]/(\lambda[\text{nm}] \times \phi)$$

with V the crystal volume. In the case above, the energy deposited into a $1 \times 1 \times 0.1$-mm³ crystal amounts to 1.3 mJ. The adiabatic temperature rise resulting from such energy deposition can be calculated as:

$$\Delta T[\text{K}] = E_{\text{deposited}}[\text{J}]/(m[\text{mol}] \times C[\text{J K}^{-1}\text{mol}^{-1}])$$

with m the number of moles of matter (taken as water) in the crystal and C the molar heat capacity of the crystal (taken as water, i.e. 75.3 J K⁻¹ mol⁻¹). The temperature rise amounts to about 30 K, a very substantial increase. Unfortunately, in addition to this isotropic temperature elevation, thermal gradients will unavoidably set in throughout the crystal, which may often result in strong crystal disorder and significant loss of X-ray diffraction.

In the case of isocitrate dehydrogenase, the employed caged compound could be efficiently photolysed in a single pulse at around 450 nm. At this wavelength, no competitive absorption by other species in the sample occurred and, again, the situation was favorable. In general, the absorption patterns of common caged compounds like NPE compounds tend to overlap significantly with the ones of aromatic amino acids like tryptophans or possibly with the ones of other chromophores present in the sample (e.g. NADH). As a consequence, the choice of a proper excitation wavelength is difficult. If a wavelength close to the absorption maximum of the caged compound is chosen, the overall optical density of the sample will be high (at some point, light will only penetrate a fraction of the crystal volume), resulting in low overall photolysis and in a strong thermal gradient within the sample inducing transient or irreversible damage. On the other hand, if a wavelength remote from the absorption maximum is chosen, the photolysis yield will be lower than optimal and a very intense light source (that may not easily be available) will be required together with a sacrifice in time resolution. The situation may even be more complicated if the build-up of photolytic products results in further competitive absorption and acts as a shield against the photolysing beam. For these reasons, a sufficient level of photolysis generally requires two or more gentle flashes, with the crystal being ideally rotated in between the flashes for a most uniform photoactivation. Otherwise, the photolysis level is too low or crystal damage is too high. In practice, therefore, the achievable time resolution is also limited by the total photolysing time, which can be as long as several seconds at room temperature when employing a flash lamp, possibly being reduced to the 100-ms timescale if using a continuous light source or laser connected to a fast shutter.

Finally, isocitrate dehydrogenase crystals suffered only moderately from a further complication linked to the Laue technique. The geometry of Laue diffraction renders this technique extremely sensitive to crystalline disorder. Therefore, crystals amenable to Laue diffraction must not only grow with low mosaicity (≪0.5), but also must not become mosaic as a result of either X-ray radiation damage, photolysis damage or transient constraints on the lattice resulting from the genuine conformational changes that are under study. In addition, the possibility of subtle structural modifications being induced by the very intense X-ray beam on the crystal kept at close to room temperature must be carefully considered. The likelihood of such artifacts is indeed quite high, since at room temperature, primary radiation damage (inelastic interaction of a X-ray photon with atoms in the crystal) is accompanied by extensive secondary damage (caused by diffusing secondary radicals that are created as a consequence of primary events).

Overall, the number of macromolecular crystals fulfilling the above-mentioned requirements is unfortunately low, explaining for a large part the small number of successful studies that have employed the Laue technique. Only a very small number of favorable biological systems like isocitrate dehydrogenase are amenable to Laue diffraction and to single-flash photolysis, which are both required to reach a time resolution of a few milliseconds. In the majority of cases, successful photolysis takes of the order of seconds or reproducible kinetics may not be achieved and only the study of intermediates of long lifetimes may be envisaged. Such lifetimes are greater than the time it takes to flash-cool a protein crystal (<1 s) [49]. Although the technique of flash-cooling introduces its own difficulties, we will see in the next section that trapping intermediates in this way and collecting the data with monochromatic radiation rather than with Laue diffraction may be the best compromise. As a consequence, white beam stations at third-generation synchrotron sources nowadays favor a few number of studies on naturally photosensitive biological systems, requiring extremely high time-resolution (nanoseconds or even picoseconds [39, 40]) and receive very few requests for caged-compound-based Laue studies.

7.3.2.2 The Case of Ha-*ras* p21: From Laue to Monochromatic Diffraction and Further Difficulties with Caged Compounds in Kinetic Crystallography

A representative case reflecting the evolution of the use of caged compounds in kinetic crystallography concerns the study of the enzyme p21^{ras}, a small guanine binding protein involved in a signal transduction pathway similar to those involving G-proteins. The protein switches from a GTP-bound, active state to a GDP-bound, inactive state. The switch in signaling state involves conformational changes associated to the hydrolysis of GTP to GDP, that occur on a timescale of several minutes. Photolysis of NPE-GTP (Fig. 7.3.3) in crystals of

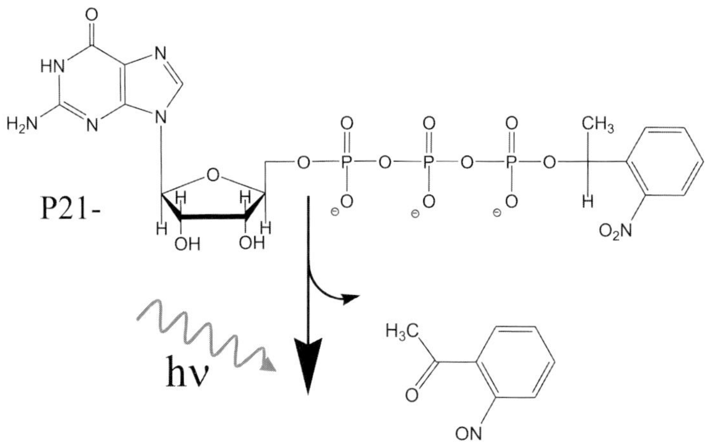

Fig. 7.3.3 Reaction scheme for time-resolved and kinetic crystallography on Ras p21. Ras p21 in complex with GTP is unstable and cannot be crystallized. Caged GTP is a biochemical inert precursor of GTP and the 1:1 complex with Ras p21 is stable and can be crystallized. Illumination with light cleaves off the protection group and converts the crystal of Ras p21-caged GTP into a crystal of Ras p21–GTP. In time-resolved crystallography the data collection is at room temperature and has to be finished within 2 min after complete photolysis of the crystal (the Laue X-ray diffraction method is used). In kinetic crystallography the GTP state is stabilized by freeze-trapping and data collection is performed at cryogenic temperatures. Courtesy of Axel Scheidig.

p21^{ras} was first used in combination with Laue diffraction to observe the relatively short-lived GTP complex [6]. Ten flashes with a xenon lamp were delivered in about 1 min to activate crystals and Laue data collection took 5 min. However, it was realized over the years that a number of pitfalls compromised the data quality. Considerable improvement resulted from several modifications in the experimental strategy. Firstly, the use of gentle photolysis using monochromatic light at a wavelength (313±2 nm) remote from absorption bands of aromatic amino acids and at moderately low temperature (2 °C) enabled the preservation of the diffraction quality of the crystals, and sometimes even improved it slightly. Complete photolysis could be achieved in this way within 2–3 min. During exposure, continuous rotation of crystals chosen for their relatively small size (about 150 µm in the largest dimension) allowed uniform photolysis throughout the crystal volume, while eliminating disorder that could otherwise result from thermal gradients. Second, the crystal quality was improved by using a pure diastereoisomeric form of caged GTP instead of a racemic mixture of this chiral molecule as employed in the first studies [50, 51]. The use of the pure R-diastereoisomer yielded clear electron density maps consistent with the conformation of the nucleotide precursor being similar to that of the non-hydrolysable form GppNHp, at variance with the early observation of an abnormal binding when using the R/S racemic mixture. In fact, cocrystallization with either form of the caged precursor gave crystals of different morphology and diffraction power, showing that interference may occur between the precursor binding mode and the lattice packing forces. Interestingly, non-chiral caged GTP molecules such as those based on 2-nitrobenzyl or bis(2-nitrophenyl)-methyl caging groups gave crystals of lower diffraction quality, emphasizing that subtle chemical modifications of the caged molecules may drastically influence crystalline order in a rather unpredictable way. In these studies, "catalytic caging" was used, i.e. the caged compound displayed high affinity for the protein and did bind to the active site prior to photolysis (as opposed to "affinity caging" where the caged precursor is prevented from binding, as in the case of isocitrate dehydrogenase). The results on p21^{ras} showed that where "catalytic caging" takes place, the presence of the cage may modify the crystallization conditions, alter the packing constraints and modify the configuration of the active site due to interactions between the cage and various amino acids. Whereas this is acceptable for the unphotolysed structure, which is not expected to be of physiological relevance, this may lead to trouble if, upon photolysis, the released, highly reactive cage (orthonitroacetophenone in the case of NPE compounds) stays in the active site and continues to interact with it. To minimize this potential problem, it has been suggested to use reducing agents such as dithiothreitol (DTT) in high concentration in the crystallization medium so as to scavenge the released cage [44]. This concern seems to be of less importance in the case of affinity caging, when the cage has no affinity for the active site. However, DTT should always be used to avoid residual interactions with other parts of the protein.

Third, instead of relying on Laue diffraction to catch the transient GTP-bound complex, this unstable species was trapped by rapid flash-cooling and structural data were obtained by the standard monochromatic data collection method [9]. Using this strategy, the overall time resolution of the experiment, which previously was limited by the Laue data collection time (several minutes), was improved, since flash-cooling a protein crystal can be achieved in <1 s. Also, the likelihood of residual radiolysis of the caged compound by the white X-ray beam, resulting in further degradation or "blurring" of the time resolution, was much reduced (see below). Furthermore, diffraction data of higher quality and higher completeness could be collected on smaller crystals.

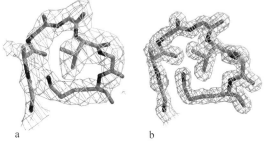

Fig. 7.3.4 Comparison of the electron density distribution around the crucial catalytic loop (P-loop) of Ras p21, obtained with Laue (a) and monochromatic (b) diffraction. A clear improvement is obtained with monochromatic data collection. Courtesy of Axel Scheidig.

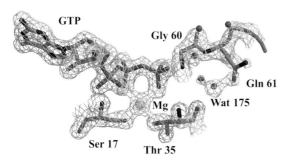

Fig. 7.3.5 Freeze-trapped structure of Ras p21–GTP. Electron density distribution around the nucleotide GTP and the catalytic loop of Ras p21. Displayed are the nucleotide, the magnesium(II) ion (together with the coordinating amino acids Ser17 and Thr35), the residues Gly60 and Gln61, and four water molecules close to the γ-phosphate. Note the quality of the electron density, which is high enough to clearly identify the binding mode of GTP. Courtesy of Axel Scheidig.

Overall, the gain in data quality jumped from 2.8 Å with 60% completeness to 1.6 Å with >90% completeness using gentle photolysis at 2 °C, small crystals, pure diastereoisomer, and monochromatic data collection (Fig. 7.3.4). This drastic improvement allowed the authors to draw firm conclusions on the hydrolysis mechanism, particularly concerning the role of the catalytic loop, especially Gln61, and of water molecules positioned around the GTP γ-phosphate group (Fig. 7.3.5).

7.3.3
Trapping of Intermediate States

Progress made in the case of p21^{ras} is in line with a trend showing that the structures of transients are often more easily visualized by crystallography using trapping methods rather than "real-time-resolved" methods based on fast data collection techniques such as the Laue technique. However, capturing intermediate states in rapid enzymes, with turnover $\gg 1$ s^{-1}, still represents a considerable challenge. Indeed, the application of the experimental protocol applied to, for example, p21^{ras} may not be envisaged. On the other hand, a number of trapping studies have been performed on extremely fast proteins displaying a built-in photosensitivity [12, 33, 35, 41, 52, 53]. These studies have allowed researchers to capture intermediates with lifetimes as short as nanoseconds by initiating the photoreaction at cryo-temperatures. At such temperatures, the reaction cycle may not proceed entirely and is stopped whenever an enthalpy barrier cannot be crossed due to the lack of thermal energy or when motions throughout the conformational landscape of the protein are arrested below critical temperatures. The low-temperature intermediates trapped in this way, although they may not thoroughly resemble their physiological counterparts, have nevertheless provided considerable insight

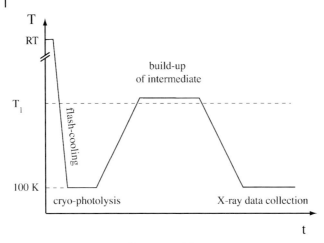

Fig. 7.3.6 Temperature profile proposed for a trapping experiment based on cryo-photolysis.

into the structure–function relationships of the biological molecules studied in this way. A similar approach may in principle be used in the case of very fast crystalline enzymes, where the built-in photosensitivity is replaced by the exogenous photosensitivity of a caged compound introduced into the crystal [10]. The following strategy, based on a combination of cryo-photolysis of caged compounds and transient temperature increase, can be envisaged (Fig. 7.3.6). The caged compound is first soaked into the crystal at room temperature and either binds specifically to the active site (case of "catalytic caging") or remains in the solvent part of the crystal (case of "affinity caging"). The crystal is then flash-cooled to 100 K and illuminated with UV light. At such a low temperature, neither the photolysis products nor the protein itself are anticipated to move significantly. Only upon transiently raising the temperature to above a critical value (T_1 in Fig. 7.3.6) would the protein flexibility and solvent viscosity be such that the un-caged substrate can be expected to reach the active site and/or to be turned over partially by the enzyme. As a consequence, an intermediate state may accumulate above T_1 that can be trapped by cooling the crystal back to 100 K, so that its structure can be solved by conventional monochromatic X-ray crystallography. Two critical questions arise from such a strategy. First, is it possible to cryo-photolyse a caged compound? Second, which temperature profile should be chosen? These issues are addressed in the following sections.

7.3.3.1 Cryo-photolysis of NPE-caged Compounds

It is well known that photolysis of caged compounds results from complex photofragmentation reactions involving rate limiting dark reactions [44]. In contrast to the photoexcitation of built-in chromophores that immediately leads to a response of biological relevance, the decomposition of a caged compound into its biologically active form is comparatively much slower and occurs at a rate that decreases as the temperature is lowered. At cryo-temperatures around 100 K, it would be expected that the active compound may not be released at all. For example, the activation enthalpy of NPE-ATP has been measured to be ~ 55 kJ mol^{-1} at pH 6.0–8.0, at ambient temperature [54]. If we (naively) extrapolate the Arrhenius behavior of the photolytic process as measured in the range 280–300 K down to cryo-temperatures, a rate constant of 10^{-16} s^{-1} at 100 K is predicted, meaning that photofragmentation may simply not occur at such a temperature.

In practice, however, preliminary experiments by absorption microspectrophotometry [55] have suggested that the photolysis of several caged compounds was possible at temperatures as low as 100 K [56]. In these experiments, nano-volumes of solutions of NPE ether (NPE-choline) or phosphate ester (NPE-TMP, NPE-ATP) compounds are flash-cooled and subjected to photolysis at 100 K by a xenon flash lamp or a UV laser delivering light at 355 nm. Typical absorption bands at 290 and 320 nm generally assigned to the nitrosoketone photoproduct appear on the minute timescale (Fig. 7.3.7a and b). In agreement with these spectrophotometric results, HPLC analysis has shown that quantitative uncaging of NPE-choline and NPE-TMP is possible by illumination at 100 K followed by subsequent excursion to room temperature [56]. Furthermore, a crystallographic experiment revealed directly that biologically active compounds could be released by cryo-photolysis followed by transient warming to room temperature. NPE-caged ATP in flash-cooled crystals of *Mycobacterium tuberculosis* thymidylate kinase was successfully photolysed at 100 K as assessed by the structural observation of ATP-dependent enzymatic conversion of TMP to TDP after temporarily warming the photolysed crystals to room temperature [10].

To account for the observation of photofragmentation reactions at cryo-temperatures, it was suggested that the absorption of UV photons could generate heat locally that may serve to carry forward the fragmentation reaction without affecting significantly the overall temperature of the sample [10, 56]. Alternatively, a possible photosensitivity of one or several intermediates on the photofragmentation pathway was hypothesized. Overall, the mechanism of cryo-photolysis was not fully elucidated.

A recent article by Corrie et al. [57] made us realize that the experiments carried out so far could not demonstrate unambiguously the effective release of active compounds at ~ 100 K. In that article, it was shown that the rate-limiting step in the photofragmentation pathway of NPE ethers is the decomposition of a hemiacetal species whose absorption spectrum closely resembles the one of the nitrosoketone product. Therefore, at least in the case of NPE-choline, successful photolysis at 100 K cannot be concluded from the appearance of absorption bands at 290–320 nm. It is possible that only the hemiacetal species formed and that free choline is not released at 100 K. In contrast, in the case of NPE-phosphate esters, the fragmentation reaction is expected to be rate limited by decomposition of the well-established *aci*-nitro intermediate and microspectrophotometric results are probably more conclusive.

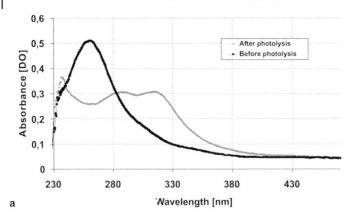

a

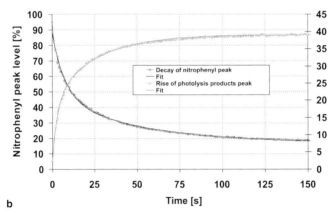

b

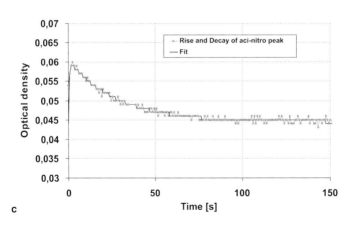

c

In total, a clear conclusion obtained so far is that cryo-photolysis at ~100 K followed by transient warming to room temperature leads to effective photofragmentation. The amount of uncaging is partial, increases with the photolysing temperature and depends on the type of caged compound [56].

Although a complete elucidation of the mechanism of photofragmentation at cryo-temperatures will clearly require further investigations, recent experiments were carried out (Fig. 7.3.7) that reveal new interesting features, some of which are summarized below. In these experiments, a 80-μW laser beam at 355 nm focused on a 50-μm diameter spot was directed on an amorphous solution containing 15 mM NPE-arseno-choline (NPE-AsCh), 50 mM MES, pH 6.0 and ~25% glycerol. In NPE-AsCh, the nitrogen atom of the ammonium group is replaced by arsenic. This modification does not modify the photofragmentation pathway as compared to normal NPE-choline, but has the advantage in crystallographic experiments that it provides a strong and characteristic anomalous X-ray diffraction signal allowing to readily position this atom in electron density maps [58]. A quantitative evaluation of the reaction kinetics under continuous laser illumination shows the disappearance of ~80% of the initial NPE compound in about 100 s (Fig. 7.3.7b). Therefore, photolysis of the initial molecules is nearly complete on this timescale. However, only ~40% of these molecules are transformed into species absorbing at 290–320 nm. These species may either consist of the nitrosoketone final product or a hemiacetal still bound to the molecule to be uncaged [57]. The other 40% photolysed molecules probably end-up in species that are spectroscopically silent in the range 260–600 nm. Prolonged exposure to the laser beam does not modify these amounts substantially, suggesting that an equilibrium is reached. The decay of the initial NPE compound and the build-up of the 290–320-nm (Fig. 7.3.7b) species both display biexponential behaviors. In each case, a fast component ($t_{1/2}<8.0$ s) coexists with a slower component ($t_{1/2}>25$ s), consistent with the idea that upon flash-cooling at least two different photofragmentation pathways exist. Interestingly, fitting with a simple kinetic model reveals that the *aci*-nitro intermediate (monitored at 400 nm) builds up rapidly (<1 s) and decays with a half-time of ~28 s. This

Fig. 7.3.7 (a) Absorption spectrum of NPE-AsCh before and after photolysis at 103 K with 355-nm light. (b) Kinetics of disappearance of the initial nitro-phenyl compound [black curve in (a)] and build up of the photolysis product (putative nitroso or hemiacetal compound) [gray curve in (a)]. The relative contribution of these species were obtained by fitting the absorption spectra with a linear combination of reference spectra corresponding to 0% and (estimated) 100% photolysis (at room temperature). This procedure allowed us to estimate the total photolysis yield as a function of time. The experimental data (in black) are properly fitted with a biexponential kinetic scheme (in gray). (c) Kinetics of appearance and disappearance of the *aci*-nitro intermediate, monitored by the absorbance at 400 nm. In this case, no deconvolution was necessary and the raw data could be used. The experimental data (in black) are properly fitted with a simple kinetic scheme taking into account the conditions of steady state illumination (in gray), of type $C(t) = k_2 N_0/(k_2 - k_1) \times (\exp{-k_1 t} - \exp{-k_2 t})$, where C is the concentration of the *aci*-nitro intermediate as a function of time, N_0 is a constant, and k_1 and k_2 are the decay and build-up time constants of the intermediate, respectively.

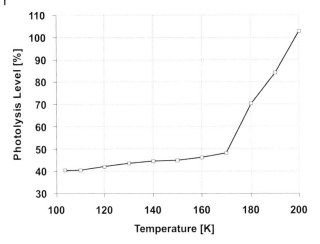

Fig. 7.3.8 Evolution of the photolysis product formation upon temperature increase in the dark after illumination at 100 K by UV light at 355 nm for 400 s, as reported by the increase in optical density monitored at 320 nm.

finding indicates that the *aci*-nitro intermediate is involved only in the slower reaction pathway leading to the formation of the 290–320-nm species, not in the faster one, in agreement with the conclusions of Corrie et al. [57].

If, subsequently to laser illumination at 355 nm at 100 K, the temperature is elevated (at a rate of ∼ 5 K min^{-1}), in the dark, the absorbencies at 290–320 nm increase. They increase slowly below 170 K and then rapidly above this temperature, reaching a level suggestive of full conversion to the 290–320-nm species at around 200 K (Fig. 7.3.8; above this temperature, rapid ice formation takes place and obscures further measurements). This observation means that, as the temperature is elevated, cross reactions between the above mentioned reaction pathways may occur, finally leading to the production of the 290–320-nm species, which is the most thermodynamically favored species at high temperatures. It is tempting to correlate the change in the rate of formation of the 290–320-nm species at ∼ 170 K with the occurrence of a glass transition in the sample, resulting in a greatly enhanced rotational and translational freedom of the molecules [59].

Taken together, our results show that photolysis at 100 K followed by transient warming above a critical temperature (T_1 in Fig. 7.3.6) leads to complete liberation of photolysis products absorbing at 290–320 nm. In the case of NPE compounds, assuming that these products correspond to final nitroso species (which remains to be demonstrated for NPE-ether compounds [57]), a complete release of the active molecule is expected.

In other words, and as substantiated below, the photofragmentation process is in general not expected to be rate limiting for subsequent catalysis in a protein crystal at cryo-temperatures, opening the possibility to efficiently trap intermediate states.

7.3.3.2 Temperature-dependent Enzymatic Activity and Dynamical Transitions

Enzymatic turnover often ceases below a certain temperature called the dynamical transition temperature. This transition (generally within the range 150–250 K) marks the onset of an-harmonic motions of protein atoms occurring on a timescale from picoseconds to 100 ns as evidenced by, for example, neutron [60, 61] or Mössbauer spectroscopy [62]. A correlation between the dynamical transition and the onset of biological activity has been established for several proteins, such as bacteriorhodopsin [61], ribonuclease A [63, 64] or myoglobin [52, 65], but in other cases this correlation was not observed [66] and the issue remains controversial. Whether or not the onset of biological activity coincides with the dynamical transition might depend on the type of protein motions that are monitored, which are either determinant or not for biological function. Although the dynamical transition is still a poorly understood process, especially in the case of crystalline proteins, there are further indications suggesting that this transition might be coupled to a solvent glass transition occurring within the crystal channels surrounding the protein molecules [67, 68]. Above the glass transition, the solvent molecules gain rotational and translational freedom, meaning that some degree of liquid state is reached and diffusion may occur.

The determination of the proper temperature T_1 (Fig. 7.3.6) to be used experimentally is tricky, because transitions are difficult to detect and because experimental caveats also come into play, such as ice formation within the crystal that might slowly develop as the glass transition is passed and that may irreversibly damage the crystal. These questions are currently being investigated and some of them might be answered by techniques such as fluorescence microspectrophotometry, whose sensitivity allows us to probe subtle changes in protein dynamics as a function of temperature within a single crystal [69].

Finally, our current view is that, if proper conditions are found, the combination of cryo-photolysis at a temperature below the dynamical transition followed by transient warming above this temperature should allow a substantial amount of caged compound to be effectively released. Moreover, in the case of "affinity caging", a released substrate, for example, should be able to diffuse from the solvent channels to the nearest active site centre, typically 100 Å away, within a few minutes. Furthermore, partial catalysis may take place and it is possible that an enzymatic intermediate will accumulate in the crystal.

7.3.3.3 Potential Applications of Cryo-photolysis Combined with Temperature-controlled Protein Crystallography

The strategy described above and summarized in Fig. 7.3.6 may enable the effective accumulation and trapping of an intermediate state of biological interest within the crystal, in order to solve its three-dimensional structure. This strategy appears especially interesting in the case of rapid enzymes, for which any other known trapping method is prone to failure due to suboptimal synchrony between the molecules of the crystal. In fact, cryo-photolysis below the dynamical transition temperature simply eliminates the requirement for synchrony. As a consequence, this technique may also widen the panel of caged compounds adequate for kinetic crystallography experiments. Indeed, caged compounds of low quantum yield may be used [44], since, within practical limits, the rate of photolysis is not an issue as long as protein activity is arrested below the dynamical transition. Enzymes for which there are no compounds of high quantum yield available could therefore be studied. For example, caged dioxygen [70], displaying a quantum yield of 0.04 at physiological pH, could serve to study the mechanisms of dioxygenases and preliminary experiments have shown that cryo-photolysis of this compound is successful (Grummit, personal communication). Caged protons [71] could also be used to trigger pH-dependent catalytic activity at low temperature.

7.3.3.4 Cryo-photolysis Wavelength and Cryo-radiolysis of NPE-caged Compounds

The possibility to use gentle (low-power) light sources is a key advantage of cryo-photolysis, due to the absence of requirement for a high photolysis rate. Cryo-photolysis can also be envisaged by two-photon absorption using light in the range 640–700 nm [72]. However, the choice of the photolysing wavelength leading to optimal cryo-photolysis is always a delicate problem. For example, in the case of amorphous solutions of NPE-AsCh (see above), we observed that illumination at 266 nm leads to a higher photolysis yield than at 355 nm, independently of the illumination time (not shown). This finding is probably related to the wavelength-dependent photosensitivity of various species along the fragmentation pathways. However, illuminating a protein crystal at 266 nm leads to competitive absorption by the aromatic residues, which is expected to offset completely the advantage of a higher photolysis yield. The high optical density of the crystal at this wavelength will result in significant temperature elevation of the sample, thermal gradients, UV damage and even a lack of available photons to interact with the caged molecules located in the crystal centre. An adequate solution may consist in chemically modifying the caged compound so as to tune its absorption spectrum away from the 280-nm region, which often can be achieved by incorporating two met-oxy groups on the aromatic group [8].

A serious pitfall complicates the usage of caged compounds in protein crystallography. Experiments by absorption microspectrophotometry have suggested that NPE-caged compounds can be cleaved at cryo-temperatures by employing a highly intense X-ray beam rather than visible light [10]. In the case of NPE-TMP soaked into crystals of *M. tuberculosis* thymidylate kinase, spectral changes induced by X-irradiation after collection of a full crystallographic data set at 100 K on a undulator beamline (ID9 at the ESRF, Grenoble) were comparable to those induced by cryo-photolysis with 355-nm laser light at the same temperature. Recently online microspectrophotometry on beamline ID14-4 at the ESRF (R. Ravelli et al., in preparation) enabled us to monitor the spectral changes induced by X-ray irradiation in real-time. Thin films of amorphous NPE-AsCh solutions, as described above, were flash-cooled to 100 K using a cryo-loop, and exposed to monochromatic X-rays ($\lambda = 0.94$ Å; $\sim 5 \times 10^{11}$ photons s^{-1} through slits of dimensions

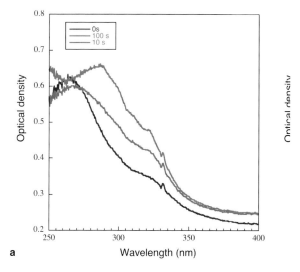

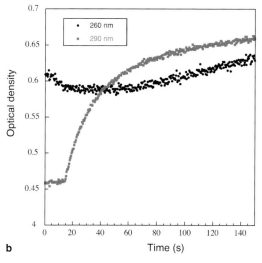

Fig. 7.3.9 (a) Absorption spectrum of NPE-AsCh before and after 10 and 100 s of radiolysis at 100 K with X-rays at 0.94 Å. (b) Kinetics of disappearance of the initial nitrophenyl compound (260 nm) and build up of the photolysis product (putative nitroso or hemiacetal compound, 290 nm peak). In this case, no deconvolution was attempted and the results are only qualitative. Already before X-ray irradiation a peak at 290–320 nm is visible, originating from photolysis products that formed under the influence of the measuring light (deuterium lamp) during alignment of the sample. A flash-cooled control solution, containing 50 mM MES, pH 6.0, and ~25% glycerol but no caged compound, did not show any of the features observed in this figure upon X-ray irradiation (data not shown).

100×100 µm²). Absorbance spectra were recorded every ∼0.4 s during X-ray exposure. Figure 7.3.9(a) shows spectra before, after 10 s and after 100 s of X-ray irradiation. After 10 s of exposure, the peak height at 320 nm has increased and the peak at 290 nm is clearly visible. The spectrum after 100 s of exposure resembles the one after cryo-photolysis with a 355-nm laser, although the ratio of the absorbances at 290 and 320 nm is largely different, suggestive of different fragmentation mechanisms (compare with Fig. 7.3.7). The time course of the spectral changes at 260 and 290 nm is shown in Fig. 7.3.9(b). Up to a time point of 15 s, corresponding to the onset of X-ray irradiation, the absorbance at 290 nm slowly increases and the one at 260 nm decreases, due to significant photolysis by the measuring light. At time points longer than 15 s, radiolysis is evident, leading to a rapid increase of absorbance at 290–320 nm. The absorbance at 260 nm first decreases and then increases due to a competition between the vanishing peak at 260 nm and the rising peak at 290 nm.

Synchrotron X-ray irradiation not only leads to changes in the spectrum of caged compounds, but can also alter them structurally. Electron densities for un-illuminated NPE-AsCh in the active site of trigonal crystals of *Torpedo californica* acetylcholinesterase [73] are shown for data collected at 100 K on a laboratory X-ray source and on beamline ID14-EH2 at the ESRF (Fig. 7.3.10a and b). Clearly, irradiation with highly intense synchrotron radiation leads to a loss of definition of the caged compound, whereas radiation from a laboratory X-ray source necessary to collect a full data set preserves its full definition. It is not clear yet how spectroscopic changes induced by radiolysis (Fig. 7.3.9) relate to the observed structural damage (Fig. 7.3.10) and experiments will be needed to clarify this issue. However, we observed that it is possible to preserve full definition of the caged compound even on intense synchrotron beamlines (data not shown) if the X-ray dose necessary to collect a full data set is carefully minimized. The requirements for suitable X-ray data collection in combination with the strategy of Fig. 7.3.6 are summarized below.

7.3.3.5 X-ray Data Collection Strategy

Proper X-ray data collection protocols are needed to establish unambiguously the occurrence of a catalytically induced conformational change. In general, structural motions are best detected by computing difference Fourier electron density maps between a "trapped" state and a reference "resting" state. However, the problem of how to collect the corresponding diffraction data is far from obvious. The two data sets should be collected at the same temperature, typically 100 K, where secondary X-ray radiation damage is minimized. Searching for the best isomorphism between data sets would indicate to use a unique crystal for the collection of both sets, one before warming to T_1 and one after (Fig. 7.3.6). Unfortunately, a temperature profile transiently driving the crystal above the critical glass and/or dynamical transition temperatures will likely result in diffusion throughout the crystal of radicals formed during the first data collection, leading to disastrous damage apparent in the second data set that will alter the

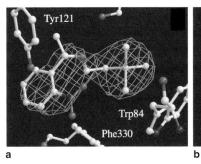 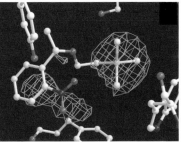

a **b**

Fig. 7.3.10 Density maps of NPE-Ch bound to the active site of *T. californica* acetylcholinesterase. Diffraction data were collected at 100 K on a laboratory X-ray source (a) and on the undulator beamline ID14-EH2 at the ESRF (b). A model of the caged compound refined from the data collected on the laboratory X-ray source is shown (unpublished result, obtained in collaboration with M. Goeldner, L. Peng, P. Gros, I. Silman and J. Sussman). Clearly the intense X-rays from a synchrotron source induced a structural alteration of the caged compound.

structure, if not ruin the diffraction quality. Separating radiation-induced structural changes from genuine conformational changes might be feasible to a certain extent if the specific alterations expected from the former are known, but this will remain a delicate task. For the same reason, deliberately using X-rays to cryo-radiolyse a caged substrate or cofactor would introduce the risk of damaging the entire crystal at the same time. It seems, therefore, preferable to use two different crystals – one serving as a reference, and the other containing the trapped intermediate generated by UV photolysis and transient temperature rise. However, the following difficulties appear: (1) the reference data set may already reveal a modified caged compound, due to X-ray radiolysis, (2) the caged compound may have different occupancies in different crystals, (3) non-isomorphism between crystals may significantly deteriorate the quality of experimental difference electron density maps (most informative because least biased by modeling errors) and (4) many crystals need to be tediously prepared if the diffraction quality tends to vary randomly. Additional control experiments are necessary that address the structural consequences of either the sole application of a temperature profile on a non cryo-photolysed crystal or the sole UV illumination at 100 K.

7.3.4
Conclusion

In conclusion, kinetic protein crystallography is a challenging field of research standing at the forefront of protein crystallography. Its development, which will certainly receive great attention from the structural proteomics world in the future, is still in its infancy and will still require considerable efforts. It has now been realized that real-time-resolved crystallography employing the Laue method, although an extremely powerful and elegant technique, is only adequate for a limited number of proteins displaying built-in photosensitivity and may not be easily used in combination with caged compounds. New perspectives are directed towards trapping strategies, where caged compounds still offer considerable interest, especially through their ability to be photolysed at cryo-temperatures, as recently demonstrated [10, 56]. This possibility, combined with the use of physicochemical properties of biological macromolecules such as temperature-dependent activity and dynamical transitions, adds an additional tool to the panel of methods based on cryo-crystallography [3, 74] to trap intermediate states in protein crystals. In particular, it opens the way to the study of structural changes in rapid enzymes such as cholinesterases, a field that was hitherto reserved for a limited number of proteins displaying endogenous photosensitivity. A wide range of caged compounds can benefit from the technique, including those that are moderately photosensitive, and therefore a wider range of enzymes may in principle be studied.

Acknowledgments

We acknowledge M. Goeldner, A. Specht, I. Silman, J. Sussman and P. Gros for fruitful collaborations and numerous discussions. T. Ursby is acknowledged for his essential contribution to cryo-photolysis studies and the development of absorption microspectrophotometry. We thank R. Ravelli and J. Colletier for their help in cryo-radiolysis experiments.

References

1. B.L. Stoddard, *Methods* **2001**, *24*, 125–138.
2. G.A. Petsko, D. Ringe, *Curr. Opin. Chem. Biol.* **2000**, *4*, 89–94.
3. I. Schlichting, *Acc. Chem. Res.* **2000**, *33*, 532–538.
4. I. Schlichting, R.S. Goody, *Methods Enzymol.* **1997**, *277*, 467–490.
5. E.M. Duke, S. Wakatsuki, A. Hadfield, L.N. Johnson, *Protein Sci.* **1994**, *3*, 1178–1196.
6. I. Schlichting, S.C. Almo, G. Rapp, K. Wilson, K. Petratos, A. Lentfer, A. Wittinghofer, W. Kabsch, E.F. Pai, G.A. Petsko, R.S. Goody, *Nature* **1990**, *345*, 309–315.
7. B.L. Stoddard, P. Koenigs, N. Porter, K. Petratos, G.A. Petsko, D. Ringe, *Proc. Natl Acad. Sci. USA* **1991**, *88*, 5503–5507.
8. B.L. Stoddard, B.E. Cohen, M. Brubaker, A.D. Mesecar, D.E. Koshland, Jr, *Nat. Struct. Biol.* **1998**, *5*, 891–897.
9. A.J. Scheidig, C. Burmester, R.S. Goody, *Struct. Fold. Des.* **1999**, *7*, 1311–1324.
10. T. Ursby, M. Weik, E. Fioravanti, M. Delarue, M. Goeldner, D. Bourgeois, *Acta Crystallogr. D* **2002**, *58*, 607–614.
11. D.M. van Aalten, W. Crielaard, K.J. Hellingwerf, L. Joshua-Tor, *Protein Sci.* **2000**, *9*, 64–72.
12. K. Edman, P. Nollert, A. Royant, H. Belrhali, E. Pebay-Peyroula, J. Hajdu, R. Neutze, E.M. Landau, *Nature* **1999**, *401*, 822–826.
13. J.M. Bolduc, D.H. Dyer, W.G. Scott, P. Singer, R.M. Sweet, D.E. Koshland, Jr, B.L. Stoddard, *Science* **1995**, *268*, 1312–1318.

14 T. R. Schneider, E. Gerhardt, M. Lee, P. H. Liang, K. S. Anderson, I. Schlichting, *Biochemistry* **1998**, *37*, 5394–5406.

15 E. Garman, T. Schneider, *J. Appl. Crystallogr.* **1997**, *27*, 211–237.

16 T. Y. Teng, *J. Appl. Crystallogr.* **1990**, *23*, 387–391.

17 M. Weik, R. B. Ravelli, I. Silman, J. L. Sussman, P. Gros, J. Kroon, *Protein Sci.* **2001**, *10*, 1953–1961.

18 R. B. Ravelli, S. M. McSweeney, *Structure Fold. Des.* **2000**, *8*, 315–328.

19 W. P. Burmeister, *Acta Crystallogr. D* **2000**, *56*, 328–341.

20 M. Weik, R. B. Ravelli, G. Kryger, S. McSweeney, M. L. Raves, M. Harel, P. Gros, I. Silman, J. Kroon, J. L. Sussman, *Proc. Natl Acad. Sci. USA* **2000**, *97*, 623–628.

21 M. W. Makinen, A. L. Fink, *Annu. Rev. Biophys. Bioeng.* **1977**, *6*, 301–343.

22 B. L. Stoddard, G. K. Farber, *Structure* **1995**, *3*, 991–996.

23 P. O'Hara, P. Goodwin, B. L. Stoddard, *J. Appl. Crystallogr.* **1995**, *28*, 829–833.

24 W. G. Scott, J. B. Murray, J. R. Arnold, B. L. Stoddard, A. Klug, *Science* **1996**, *274*, 2065–2069.

25 P. Gouet, H. M. Jouve, P. A. Williams, I. Andersson, P. Andreoletti, L. Nussaume, J. Hajdu, *Nat. Struct. Biol.* **1996**, *3*, 951–956.

26 H. Kack, K. J. Gibson, Y. Lindqvist, G. Schneider, *Proc. Natl Acad. Sci. USA* **1998**, *95*, 5495–5500.

27 R. C. Wilmouth, K. Edman, R. Neutze, P. A. Wright, I. J. Clifton, T. R. Schneider, C. J. Schofield, J. Hajdu, *Nat. Struct. Biol.* **2001**, *8*, 689–694.

28 P. T. Singer, A. Smalas, R. P. Carty, W. F. Mangel, R. M. Sweet, *Science* **1993**, *259*, 669–673.

29 A. D. Pannifer, A. J. Flint, N. K. Tonks, D. Barford, *J. Biol. Chem.* **1998**, *273*, 10454–10462.

30 D. Oesterhelt, W. Stoeckenius, *Nat. New Biol.* **1971**, *233*, 149–152.

31 W. D. Hoff, P. Dux, K. Hard, B. Devreese, I. M. Nugteren-Roodzant, W. Crielaard, R. Boelens, R. Kaptein, J. van Beeumen, K. J. Hellingwerf, *Biochemistry* **1994**, *33*, 13959–13962.

32 R. H. Austin, K. W. Beeson, L. Eisenstein, H. Frauenfelder, I. C. Gunsalus, *Biochemistry* **1975**, *14*, 5355–5373.

33 A. Royant, K. Edman, T. Ursby, E. Pebay-Peyroula, E. M. Landau, R. Neutze, *Nature* **2000**, *406*, 645–648.

34 U. K. Genick, G. E. Borgstahl, K. Ng, Z. Ren, C. Pradervand, P. M. Burke, V. Srajer, T. Y. Teng, W. Schildkamp, D. E. McRee, K. Moffat, E. D. Getzoff, *Science* **1997**, *275*, 1471–1475.

35 U. K. Genick, S. M. Soltis, P. Kuhn, I. L. Canestrelli, E. D. Getzoff, *Nature* **1998**, *392*, 206–209.

36 B. Perman, V. Srajer, Z. Ren, T. Teng, C. Pradervand, T. Ursby, D. Bourgeois, F. Schotte, M. Wulff, R. Kort, K. Hellingwerf, K. Moffat, *Science* **1998**, *279*, 1946–1950.

37 I. Schlichting, J. Berendzen, G. N. Phillips, Jr, R. M. Sweet, *Nature* **1994**, *371*, 808–812.

38 I. Schlichting, K. Chu, *Curr. Opin. Struct. Biol.* **2000**, *10*, 744–752.

39 D. Bourgeois, B. Vallone, F. Schotte, A. Arcovito, A. E. Miele, G. Sciara, M. Wulff, P. Anfinrud, M. Brunori, *Proc. Natl Acad. Sci. USA* **2003**, *100*, 8704–8709.

40 F. Schotte, M. Lim, T. A. Jackson, A. V. Smirnov, J. Soman, J. S. Olson, G. N. Phillips, Jr, M. Wulff, P. A. Anfinrud, *Science* **2003**, *300*, 1944–1947.

41 S. Adachi, S. Y. Park, J. R. Tame, Y. Shiro, N. Shibayama, *Proc. Natl Acad. Sci. USA* **2003**, *100*, 7039–7044.

42 I. Schlichting, J. Berendzen, K. Chu, A. M. Stock, S. A. Maves, D. E. Benson, R. M. Sweet, D. Ringe, G. A. Petsko, S. G. Sligar, *Science* **2000**, *287*, 1615–1622.

43 G. I. Berglund, G. H. Carlsson, A. T. Smith, H. Szoke, A. Henriksen, J. Hajdu, *Nature* **2002**, *417*, 463–468.

44 J. A. McCray, D. R. Trentham, *Annu. Rev. Biophys. Biophys. Chem.* **1989**, *18*, 239–270.

45 D. Bourgeois, T. Ursby, M. Wulff, C. Pradervand, A. Legrand, W. Schildkamp, S. Labouré, V. Srajer, T. Y. Teng, M. Roth, K. Moffat, *J. Synchrotron Radiat.* **1996**, *3*, 65–74.

46 I. J. Clifton, M. Elder, J. Hajdu, *J. Appl. Crystallogr.* **1991**, *24*, 267–277.

47 J. Hajdu, *Annu. Rev. Biophys. Biomol. Struct.* **1993**, *22*, 467–498.

48 A. L. Fink, G. A. Petsko, *Adv. Enzymol. Relat. Areas. Mol. Biol.* **1981**, *52*, 177–246.

49 E. H. Snell, R. A. Judge, M. Larson, M. J. van der Woerd, *J. Synchrotron Radiat.* **2002**, *9*, 361–367.

50 A. J. Scheidig, A. Sanchez-Llorente, A. Lautwein, E. F. Pai, J. E. Corrie, G. P. Reid, A. Wittinghofer, R. S. Goody, *Acta Crystallogr. D* **1994**, *50*, 512–520.

51 A. J. Scheidig, S. M. Franken, J. E. Corrie, G. P. Reid, A. Wittinghofer, E. F. Pai, R. S. Goody, *J. Mol. Biol.* **1995**, *253*, 132–150.

52 A. Ostermann, R. Waschipky, F. G. Parak, G. U. Nienhaus, *Nature* **2000**, *404*, 205–208.

53 M. Brunori, B. Vallone, F. Cutruzzola, C. Travaglini-Allocatelli, J. Berendzen, K. Chu, R. M. Sweet, I. Schlichting, *Proc. Natl Acad. Sci. USA* **2000**, *97*, 2058–2063.

54 K. Barabas, L. Keszthelyi, *Acta Biochim. Biophys. Acad. Sci. Hung.* **1984**, *19*, 305–309.

55 D. Bourgeois, X. Vernede, V. Adam, E. Fioravanti, T. Ursby, *J. Appl. Crystallogr.* **2002**, *35*, 319–326.

56 A. Specht, T. Ursby, M. Weik, L. Peng, J. Kroon, D. Bourgeois, M. Goeldner, *ChemBioChem* **2001**, *2*, 845–848.

57 J. E. Corrie, A. Barth, V. R. Munasinghe, D. R. Trentham, M. C. Hutter, *J. Am. Chem. Soc.* **2003**, *125*, 8546–8554.

58 L. Peng, F. Nachon, J. Wirz, M. Goeldner, *Angew. Chem. Int. Ed.* **1998**, *37*, 2691–2693.

59 M. Fisher, J. P. Devlin, *J. Phys. Chem.* **1995**, *99*, 11584–11590.

60 W. Doster, S. Cusack, W. Petry, *Nature* **1989**, *337*, 754–756.

61 M. Ferrand, A. J. Dianoux, W. Petry, G. Zaccai, *Proc. Natl Acad. Sci. USA* **1993**, *90*, 9668–9672.

62 F. Parak, E. W. Knapp, D. Kucheida, *J. Mol. Biol.* **1982**, *161*, 177–194.

63 R. F. Tilton, Jr, J. C. Dewan, G. A. Petsko, *Biochemistry* **1992**, *31*, 2469–2481.

64 B. F. Rasmussen, A. M. Stock, D. Ringe, G. A. Petsko, *Nature* **1992**, *357*, 423–424.

65 H. Lichtenegger, W. Doster, T. Kleinert, A. Birk, B. Sepiol, G. Vogl, *Biophys. J.* **1999**, *76*, 414–422.

66 R. M. Daniel, J. C. Smith, M. Ferrand, S. Hery, R. Dunn, J. L. Finney, *Biophys. J.* **1998**, *75*, 2504–2507.

67 P. W. Fenimore, H. Frauenfelder, B. H. McMahon, F. G. Parak, *Proc. Natl Acad. Sci. USA* **2002**, *99*, 16047–16051.

68 M. Weik, *Eur. Phys. J. E* **2003**, *12*, 153–158.

69 M. Weik, X. Vernede, R. Royant, D. Bourgeois, *Biophys. J.* **2004**, *86*, 3176–3185.

70 R. MacArthur, A. Sucheta, F. F. Chong, O. Einarsdottir, *Proc. Natl Acad. Sci. USA* **1995**, *92*, 8105–8109.

71 S. Khan, F. Castellano, J. L. Spudich, J. A. McCray, R. S. Goody, G. P. Reid, D. R. Trentham, *Biophys. J.* **1993**, *65*, 2368–2382.

72 W. Denk, J. H. Strickler, W. W. Webb, *Science* **1990**, *248*, 73–76.

73 J. L. Sussman, M. Harel, F. Frolow, C. Oefner, A. Goldman, L. Toker, I. Silman, *Science* **1991**, *253*, 872–879.

74 K. Moffat, R. Henderson, *Curr. Opin. Struct. Biol.* **1995**, *5*, 656–663.

8
Multiphoton Phototriggers for Exploring Cell Physiology

Timothy M. Dore

8.1
Introduction and History

Most activation protocols for phototriggers, photoswitches and caged compounds for biological use employ UV light, which can be damaging to biological samples and lacks three-dimensional (3-D) spatial selectivity of the excitation. A relatively new aspect in this field is the use of a non-linear optical process, called multiphoton excitation (MPE), to activate triggers and switches and uncage biological effectors. MPE is the general term for excitation by two or more photons, and when it is used the photochemistry is temporally *and spatially* controlled, potentially in a volume as small as a single synapse of a neuron or other subcellular structures. This enables the non-invasive exploration of local biochemical responses to stimuli in complex tissues. MPE of phototriggers, photoswitches and caged compounds provides an excellent method for controlling the temporal and spatial release of biological effectors. It has the potential to become a powerful technique for probing biological function in real-time and on living tissue.

In 1931, Maria Göppert-Mayer predicted that an atom could become excited through the simultaneous absorption of two quanta of light, whose sum of frequencies equaled the frequency required for excitation of the atom [1, 2]. The invention of the laser enabled Kaiser and Garrett to demonstrate the two-photon excitation (2PE)-induced fluorescence of $CaF_2:Eu^{3+}$ in 1961 [3]. Three years later, Singh and Bradley used three-photon excitation (3PE) to generate fluorescence from naphthalene crystals [4]. MPE was mainly used for spectroscopic purposes until 1990, when Denk et al. introduced the concept of 2PE-induced fluorescence for microscopy and spatially resolved photochemistry [5]. There are several excellent reviews on the use of MPE in biological microscopy [6–13]. Subsequently, Denk executed the release of a biological effector by 2PE in cell culture, using a caged version of a neurotransmitter [14]. Since then, a number of advances in multiphoton imaging and uncaging have been made. This chapter will focus on the two-photon-mediated release of biological effectors, and the use of this technology in biology and medicine.

Dynamic Studies in Biology. Edited by M. Goeldner, R. Givens
Copyright © 2005 WILEY-VCH Verlag GmbH & Co. KGaA, Weinheim
ISBN: 3-527-30783-4

8.2
Theory

In 2PE, a single chromophore absorbs two non-resonant photons of the same wavelength nearly simultaneously (within 1 fs) to produce an electronically excited state, which can then undergo photophysical or photochemical processes, such as fluorescence, intersystem crossing to the triplet state and phosphorescence, or chemical reaction. The two photons exploit the virtual excited state, delivering the same energy as a single photon of half the wavelength. In a 2PE-mediated photolysis, such as that used to release a biological effector molecule from a caged compound, after the excitation, the mechanistic steps are no different than if the process was initiated through conventional one-photon excitation. Only the process of excitation is different. High light intensities, such as those obtained from pulsed lasers, are required to generate an appreciable amount of 2PE, because the probability of simultaneous two-photon absorption depends on the square of the intensity of light. The result is that excitation occurs only at the focus of the laser beam, the focal volume, which can be as small as $\sim 1~\mu m^3$ for a tightly focused laser, and corresponds to the volume of a bacteria cell. Outside of this space, no excitation occurs, in contrast to a single-photon process, which lacks 3-D selectivity of excitation because all of the photons, not just the ones in the focus of the laser, have sufficient energy to excite the chromophore (Fig. 8.1). Because excitation is proportional to the square of the intensity rather than having a linear dependence as in one-photon excitation (1PE), it is called a non-linear process, which differs from sequential single-photon events that have also been referred to as "two-photon" processes [15–21]. These excitations do not follow the squared dependence on incident light power. Different quantum mechanical selection rules apply to 2PE, and large differences between 1PE and 2PE spectra can exist [7, 8]. The wavelength at which a chromophore is most sensitive to 2PE is not always twice the single-photon absorbance maximum, but this is a good qualitative prediction of where 2PE of a chromophore will occur [6].

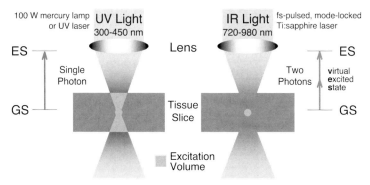

Fig. 8.1 1PE versus 2PE of a chromophore.

To excite common organic or biological chromophores, the wavelength of light used is in the near-IR (700–1100 nm), as this corresponds to roughly twice their absorbance wavelength. Cells and tissues are relatively transparent to these wavelengths, as there are few endogenous chromophores in this range. Tissues lacking hemoglobin, melanin or chlorophyll have only water as the principal IR absorber, but its absorption coefficient at these wavelengths is very low. Consequently, out-of-focus absorption does not occur and the longer wavelengths of the near-IR region reduce light scattering, contributing to increased depth penetration of the excitation light. Additionally, out-of-focus photodamage to the tissue is minimized, because of the low probability of a 2PE event occurring outside the focal volume of the beam.

Pulsed and mode-locked lasers make 2PE activation of phototriggers, photoswitches and caged compounds practical, because they increase the probability that two photons will simultaneously interact with a chromophore, but keep the average power relatively low [7, 8, 13]. Several different types of lasers have been used for 2PE. The first two-photon microscope used a colliding-pulse mode-locked (CPM) laser, which gave 15 mW average power of 100-fs pulses at 100 MHz of 630-nm light [5]. Tunable (550–700 nm) 100–400-fs pulses at 76 MHz and 90–150 mW are generated from dye lasers that are pumped by a mode-locked argon-ion or a frequency doubled Nd:YAG laser. (A "mode-locked" laser is one in which only a certain set of frequencies, or modes, propagate in the laser cavity. Locking results from destructive interference of the propagating modes everywhere in the cavity except at one point, where there is constructive interference, creating a single circulating pulse.) Nd:YAG lasers can provide 100-ps pulses at 70 MHz of 1060-nm light. All of these lasers lack tunability over a wider range of wavelengths, limiting the number of usable chromophores. The light source of choice for 2PE is the femtosecond-pulsed, mode-locked, titanium:sapphire (Ti:S) laser pumped by a 532-nm continuous wave $Nd:YVO_4$ or argon-ion laser. These commercially available [22, 23], easy-to-use, turnkey systems offer tuning ranges from 700 to 1000 nm and pulses of less than 100 fs at a repetition rate of 76 MHz. Depending on the strength of the pump laser and the wavelength, average powers of 650 mW to 1.3 W are possible. Ti:S lasers designed specifically for MPE microscopy have become available [22, 23], and boast tuning ranges of 720–980 nm, <140-fs pulses, 90-MHz repetition rates and average power of up to 1000 mW. They are even easier to use, because the Ti:S and the pump laser are contained in a single sealed box, and tuning of the wavelength and optimization of power output and pulse width are automated. One no longer needs to be a laser expert to conduct MPE experiments.

Experiments on cells or tissue using 2PE are typically conducted on a multiphoton microscope, which is essentially a laser-scanning confocal microscope without the confocal aperture, and modified with mirrors for reflecting near-IR and a few peripherals specific for MPE microscopy [6, 10, 13]. Commercial microscopes are available from Bio-Rad, Zeiss and Leica, but many researchers have built custom systems from confocal microscopes. The beam intensity is controlled by neutral density filters and an electro-optic modulator (EOM or Pockels cell), which can shutter the beam. The z-axis excitation is controlled by the focus of a conventional high numerical aperture (NA) objective. A beam scanner controls the position of the focus in the xy-plane. A non-resonant point scanner enables variable scan speeds and the ability to park the laser beam at any position in the xy-plane. Resonant galvanometers provide video-rate scanning, but only at a fixed frequency.

8.3
The Two-photon Action Cross-section, δ_u

The sensitivity of a phototrigger, photoswitch or caged compound to 2PE-induced photolysis is quantified by its two-photon action cross-section, δ_u, given in Göppert-Mayer (GM, 10^{-50} cm^4·s photon^{-1}), which is equal to the chromophore's two-photon absorbance cross-section, δ_a, times the quantum yield for the subsequent chemical reaction, Q_R. It represents the probability that a two-photon absorption and subsequent reaction will occur, and it is analogous to the molar absorptivity (ε) of a compound in UV-vis spectroscopy, which is a measure of the probability that an absorption will occur at a specified wavelength. To be biologically useful, δ_u should exceed 0.1 GM [24].

For measuring δ_u of chromophores that release a fluorescent molecule or one that can be chelated by a fluorescent indicator dye, Webb et al. have developed a method using a two-photon microscope (Fig. 8.2) [25]. A brief (8.5 µs), high-intensity pulse train from the laser releases a chelatable species, which is rapidly bound to the fluorescent indicator dye. The same laser beam, which is attenuated immediately after the photolysis pulse by a Pockels cell, is used to excite the dye before it diffuses out of the focal volume of the beam. The amplitude of

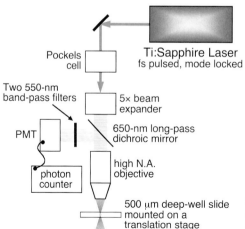

Fig. 8.2 Schematic for measuring two-photon action cross-sections with a microscope.

the signal measured is proportional to δ_u, and it will decay as the indicator-bound species diffuses out of the excitation volume. An estimation of the diffusion rates for each species and a kinetic model for the action of the caging group and fluorescent dye are needed to determine the concentrations of the released and cage-bound species in the sample chamber. Error in the δ_u measurement arises from uncertainty in the calculation of the size of the illuminated volume and the determination of the efficiency of the detector. These errors are not meaningful when comparing the relative δ_u values of two caged compounds on the same instrument, but the absolute values of δ_u obtained are correct to within a factor of 2.

Tsien et al. [24] developed a method that relates δ_u to the two-photon absorbance cross-section (δ_{aF}) and fluorescence quantum yield (Q_{fF}) of fluorescein, which are known [26, 27], by the following equation:

$$\delta_u = (N_p \phi Q_{fF} \delta_{aF} C_F)/(\langle F(t) \rangle C_S)$$

where ϕ is the collection efficiency of the detector, C_F is the concentration of fluorescein, C_S is the initial concentration of caged substrate and $\langle F(t) \rangle$ is the time averaged fluorescent photon flux from the fluorescein standard. Using a simple apparatus (Fig. 8.3), samples are placed into a cuvette with $1 \times 1 \times 10$ mm illuminated dimensions (effective filling volume of 20 µL) and a side window to measure fluorescence. The laser is focused on the center of the sample with a 25-mm focal length lens and the fluorescence output from a 20-µL fluorescein reference sample (pH ~ 11) is measured by the radiometer to determine $\langle F(t) \rangle$. Then, 20-µL aliquots of the analyte sample are irradiated successively with the laser and HPLC is used to measure the rate of disappearance of starting material, appearance of products or both. The value of N_p is calculated from the initial rate of photolysis. Similar to the first method, absolute values of δ_u are probably reliable within a factor of 2, but relative values measured on the same instrument are more accurate. Additionally, the value of δ_u depends on the accuracy of the assumed values of fluorescein fluorescence quantum efficiency and two-photon absorbance cross-section. In a variant of this method, one can compare

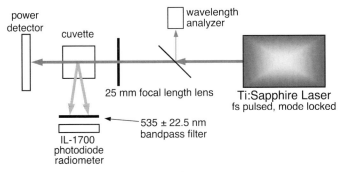

Fig. 8.3 Schematic for measuring two-photon action cross-sections in a cuvette.

the amount of photoproducts formed from two-photon photolysis of a caged compound with that of one with known δ_u [28].

8.4
Chromophores for Two-photon Release of Small Organic Ligands or Metal Ions

Molecules with large two-photon absorbance cross-sections (δ_a) typically possess extended conjugation, polarizability and high extinction coefficients for π–π^* transitions [29, 30]. The design and synthesis of new materials with large δ_a is an active area of research [31–38]. Nevertheless, simply having the capacity to absorb two non-resonant photons simultaneously is not sufficient for a chromophore to act as a good two-photon phototrigger or photoswitch. The excitation process has to initiate a photochemical reaction with good efficiency. For caged compounds, the 2PE must lead to the release of the attached effector molecule and, to be useful biologically, δ_u should be > 0.1 GM [24]. Several research groups have reported chromophores with adequate δ_u for biological use, but this is not the only consideration, as the chromophore must also meet the criteria for being a good photoremovable protecting group for use in a biological system. Briefly, it should be biocompatible and release the biological effector in high yield with kinetics more rapid than the physiological phenomena under study, and it should not interfere with the method used to measure the physiological output. The caged conjugate must not have activity prior to irradiation and, ideally, it will be water-soluble at the high ionic strengths of physiological buffers.

To achieve localized release of biological effectors from caged compounds, one must pay careful attention to the mechanism of the photoreaction, and understand the kinetics of intermediate and product formation, not just the excitation step. A photochemical reaction has a number of "dark" steps along the pathway toward the final products (active messenger and caging remnant) that typically have slower kinetics than the excitation process, which occurs on the 10^{-15}-s timescale. Light from the laser is delivered in 80–200-fs pulses with ~10-ns intervals in between, meaning that, depending on the photolysis efficiency, some time passes before sufficient quantities of released effector are produced in the excitation volume. All of the intermediates in the reaction are subject to exchange with the bulk solution through diffusion on timescales of $\tau = 113$–900 µs, depending on the diffusion coefficients of the species involved. Therefore, in order to minimize spread of the release volume, photochemical reactions must proceed faster than the rates of diffusion. The timescale of diffusion also determines the optimum length of irradiation, which should match τ; longer exposure does not increase local concentration of the effector, but contributes to its spread beyond the volume of excitation [39].

Out-of-focus photodamage to cells is negligible with 2PE microscopy techniques, because biological preparations generally do not absorb light in the near-IR. Nevertheless, the high intensity lasers required for 2PE mean that photodamage to the tissues within the focal volume can be a problem. Studies of two-photon microscopy of cells in culture revealed that at 700–1000 nm wavelengths, damage from 1PE and 3PE is negligible, but destruction from 2PE processes could be of concern at high excitation doses [40, 41]. At powers >7 mW, König et al. [42] found that reactive oxygen species (ROS) were generated using 170-fs pulses of 800-nm light. This led to "membrane barrier dysfunction", damage to the nucleus, DNA cleavage and cell death of PtK2 cells. Generation of ROS would be detrimental to studies using caged compounds, but controlled tissue destruction has some utility. Tirlapur and König used 2PE to make localized perforations in the membranes of CHO and PtK2 cells in order to target the transfection of green fluorescent protein (GFP) to individual cells in culture [43]. Ogden et al. [44] determined that photodamage to synaptic junctions of the snake neuromuscular junction occurred with 5-ms laser flashes of 640-nm light from a 100–400-fs pulsed dye laser at average powers >5 mW. The investigators studied the fluorophore release from the o-1-(2-nitrophenyl)ethyl ether of 8-hydroxypyrene-1,3,6-trisulfonate (NPE-HPTS, see Fig. 8.9) to show that in order to achieve sufficient levels of effector release, average laser powers of around 72 mW, well-above the threshold for tissue damage, were required.

As the preceding example illustrates, 2-nitrobenzyls, the most widely used photoremovable protecting groups, are not very sensitive to two-photon excitation (Tab. 8.1). The 4,5-dimethoxy-2-nitrobenzyl group (DMNB, Fig. 8.4) has $\delta_u = 0.03$ GM [24], below the biologically useful threshold (Tab. 8.1). The (α-carboxy)-2-nitrobenzyl group (CNB) has also been used [14], but its 2PE wavelength of 640 nm is outside the tuning range of most Ti:S lasers. The δ_u of CNB has not been explicitly measured, but the fact that its mechanism of action is the same as the DMNB group, and it requires high concentrations of CNB-caged compounds and high laser power to uncage it, suggests that its cross-section is probably similar to the value for DMNB. The δ_u of the 1-(2-nitrophenyl)ethyl group (NPE) is similar to DMNB in the near-IR [44], but it is several orders of

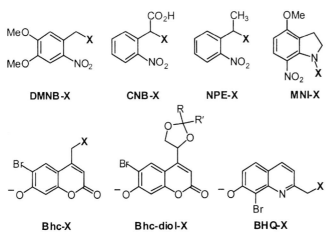

Fig. 8.4 Chromophores for caged compounds with sensitivity to 2PE.

Tab. 8.1 Summary data for 2PE-sensitive chromophores for caged compounds

Chromophore	λ_{max} (nm)	ε (M^{-1} cm^{-1})	$Q_u^{a)}$ (mol ein^{-1})	$\delta_u(\lambda)$ [GM (nm)]$^{b)}$	Functional groups protected	Release rate (s^{-1})
DMNB [24, 60]	346	6100	0.005	0.03 (740)	carboxylates, phosphates, alcohols and amines	10^2
NPE [44, 45, 60, 61]	265	4200	0.52	0.019 (640); 4.0 (460)	carboxylates, phosphates, alcohols and amines	10^1
CNB [62–64]	262	5100	0.14	NA$^{c)}$	carboxylates, phosphates, alcohols and amines	10^4
Bhc [24, 47, 48, 50, 52]	370	15000	0.037	0.72 (740)	carboxylates, phosphates, diols, amines (as carbamates), and alcohols and phenols (as carbonates)	$>10^6$
Bhc-diol [51]	370	18500	0.057	0.90 (740)	ketones and aldehydes	NA$^{c)}$
BHQ [54]	369	2600	0.29	0.59 (740)	carboxylates, phosphates and diols	NA$^{c)}$
MNI [28, 57, 58]	347	4330	0.085	0.06 (720)	carboxylates	$\sim 10^6$

a) Q_u = quantum yield for uncaging.
b) GM = Göppert-Mayer (10^{-50} cm$^4 \cdot$ s photon^{-1}).
c) Data not available.

magnitude higher at shorter wavelengths [45]. Use of these types of chromophores is likely to lead to cell or tissue damage; nevertheless, some researchers have used them with some success.

Furuta et al. introduced the 6-bromo-7-hydroxycoumarin-4-ylmethyl (Bhc, Fig. 8.4) chromophore as a multiphoton-sensitive caging agent for neurotransmitters [24]. It has the highest two-photon action cross-section of the known 2PE-based protecting groups (Tab. 8.1), and it is capable of protecting carboxylates [24], phosphates [46, 47], diols [48], amines (as the carbamate) [24, 49], and alcohols and phenols (as the carbonate) [50]. A variant called Bhc-diol [51] releases aldehydes and ketones to complete the coverage of the most biologically relevant functional groups. The synthesis of Bhc is straightforward, involving known chemistry of coumarins. It offers moderate water solubility, good cell permeability (depending on the substrate) and rapid release of the substrates. Time-resolved flash photolysis experiments on Bhc-protected acetate (Bhc-OAc) in 10 mM KMOPS or phosphate buffer at pH 7.2 revealed that photolysis occurs rapidly on the submicrosecond timescale (time constant, $\tau = 0.65$ µs) [52]. Photolysis rates were faster at higher pH and slower at lower MOPS or phosphate buffer concentrations.

8.4 Chromophores for Two-photon Release of Small Organic Ligands or Metal Ions

The mechanism of ligand release by Bhc likely proceeds through a singlet state and an S_N1 mechanism, i.e. a solvent-assisted photoheterolysis (Scheme 8.1) [53]. At physiological pH, Bhc exists as the phenolate, which increases its λ_{max} and molar absorptivity. ^{18}O-labeling studies suggest that the oxygen attached at C-11 of Bhc-OH comes from the solvent and the ligand **X** must be a good leaving group. Alcohols and amines cannot be released directly from Bhc. Instead, they are attached through a carbamate or carbonate linkage. Upon photolysis, the Bhc releases the corresponding carbamate or carbonate, which releases carbon dioxide in a thermolytic hydrolysis step. The timescale for the dark reaction is on the order of several milliseconds, so this poses problems for maintaining tight control over the 3-D localization of ligand release. The unactivated ligand has time to diffuse away from the excitation volume before it is fully activated.

8-Bromo-7-hydroxyquinoline (BHQ) releases carboxylates [54], but it has not yet been used in a biological system. It has good sensitivity to 2PE, but its cross-section is not quite as high as the value for Bhc. Its mechanism of photolysis is probably similar to that of Bhc; therefore, future work will likely show that BHQ can also photorelease phosphates, diols, ketones and aldehydes, amines (as the carbamate), and alcohols (as the carbonate). BHQ and derivatives of it are synthetically accessible through standard quinoline chemistry. BHQ has several advantages over other protecting groups: the quantum efficiency of single-photon uncaging is much larger, it is highly soluble in aqueous buffers and it exhibits low levels of fluorescence. This latter property will facilitate BHQ's simultaneous use with fluorescent indicators to measure the biological response to the release of an effector.

Invented by Papageorgiou and Corrie, 4-methoxy-7-nitroindoline (MNI) [55–57] releases carboxylates through 2PE [28], albeit with a cross-section that is less than 0.1 GM. The low sensitivity to TPE might limit the usefulness of MNI for applications requiring large increases in local concentrations of ligand. MNI-

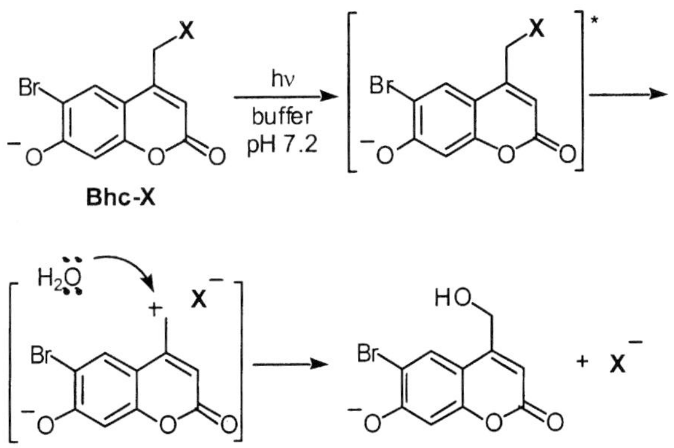

Scheme 8.1 Proposed mechanism of Bhc photolysis in aqueous solution.

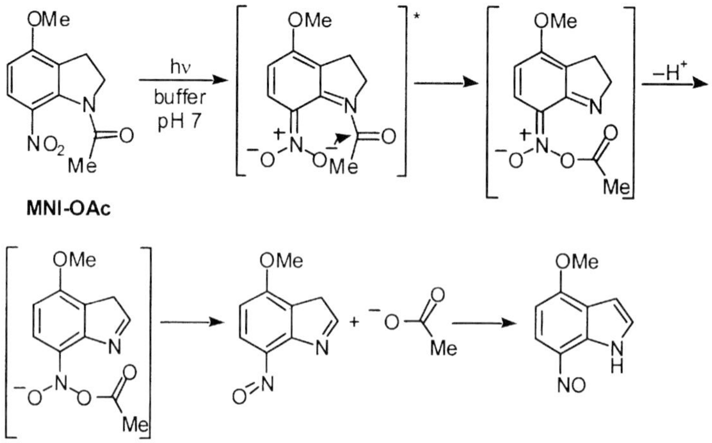

Scheme 8.2 Proposed mechanism of MNI photolysis in aqueous solution.

caged compounds are synthesized from 4-methoxyindole in three steps: (1) reduction of the indole to indoline, (2) acylation and (3) nitration. The sequence produces a 1:1 mixture of 5- and 7-nitro isomers that requires separation. Based on studies of related 1-acyl-7-nitroindolines [58, 59], the photolysis of MNI likely proceeds through a triplet excited state and involves an intramolecular attack of one of the nitro group's oxygens on the carbonyl (Scheme 8.2). The kinetics of release are on the microsecond timescale, but protection by MNI is limited to carboxylates.

8.5
Applications

A number of applications for two-photon phototriggers, photoswitches and caged compounds have emerged. Caged conjugates of neurotransmitters, signal transduction mediators, fluorophores, photodynamic therapy (PDT) compounds and drugs have been synthesized, and used to investigate the local biochemical effects of their action.

8.5.1
Neurotransmitters

The 3-D spatial selectivity of 2PE opens the door for this technology to enable new discoveries in neuroscience. The ability to simulate the release of neurotransmitters from vesicles in the pre-synaptic neuron would advance the understanding of neurotransmitter action on their receptors. Detailed 3-D maps of neurotransmitter receptors could be generated and used to study the function of neural networks. Toward this end, Denk achieved the first 2PE-mediated release of a caged compound in a biological system [14]. He recorded whole-cell current in conjunction with the release of the nicotinic acetylcholine receptor agonist, carbamoylcholine, from a CNB group (Fig. 8.5) to determine the distribution of ligand-gated ion channels in cultured BC3H1 cells.

Tsien et al. used 2PE to release glutamate in brain tissue slices (400 µm) from the hippocampus and cortex of rats [24]. The beam from a Ti:S laser (780 nm, 100-fs pulses) was raster scanned across neuron cell bodies and dendrites, and inward currents (measured by whole-cell patch clamping), non-linearly dependent on average beam power (0–75 mW), were evoked in the presence of Bhc-Glu (δ_u=0.95 GM at 740 nm, Fig. 8.5). Capitalizing on the ability of 2PE to release glutamate only in the focal plane, glutamate-evoked currents in the dendrites were mapped at 4-µm depth intervals. The electrophysiological responses of the neurons to released glutamate varied as a function of the focal plane and position of the scanning laser. Higher resolution mapping or greater precision of glutamate release is probably not possible with Bhc-Glu because of the slow kinetics of glutamate formation (several milliseconds) from the carbamic acid intermediate, which is released from Bhc on the submicrosecond timescale. Diffusion of the released intermediate away from the focal volume during the dark reaction is significant, thus limiting the 3-D spatial resolution of uncaging. Similarly, Bhc-GABA (δ_u=0.61 GM at 740 nm) has been released on Purkinje cells in slices of the cerebellum, but a non-GABAergic response to the caged neurotransmitter makes this compound less useful [65].

In order to investigate functional glutamate receptors at the level of the individual synapse, Kasai et al. developed MNI-Glu [28]. Experimenting with hippocampal neurons in culture and CA1 pyramidal neurons in slices, the researchers used 2PE-induced release of glutamate to mimic the spatiotemporal resolution exhibited by pre-synaptic terminals. The point spread of excitation in their setup was 0.29 µm laterally and 0.89 µm axially with 50-µs pulses of 720-nm light (7 mW), thus enabling a spatial resolution of receptor activation of 0.45–0.6 µm laterally and 1.1–1.4 µm axially. For comparison, the spatial resolution of glutamate release from pre-synaptic vesicles is ~0.4 µm in ~0.1 ms and the channel response time is 1–2 ms, so this technique adequately mimics the natural release of neurotransmitter. The researchers found a correlation between dendritic spine shape and the excitatory post-synaptic current (EPSC) observed

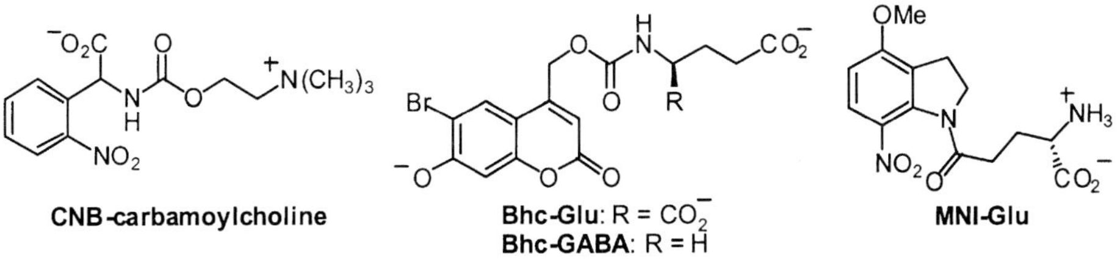

CNB-carbamoylcholine

Bhc-Glu: R = CO_2^-
Bhc-GABA: R = H

MNI-Glu

Fig. 8.5 Caged neurotransmitters.

with 2PE of MNI-Glu. Thin spines had little or no sensitivity to glutamate, while thick, mushroom-shaped spines had a large sensitivity. The research suggests that the expression of AMPA-sensitive glutamate receptors is independently controlled at individual dendritic spines, a finding that supports most models of learning in neural networks.

Smith et al. used MNI-Glu to examine how synapse location influences the input from Schaffer collateral axons onto excitatory synapses of CA1 pyramidal neurons in rats [66]. The experimental techniques used were similar to those used by Kasai [28]. Using 2-ms pulses of 720-nm light at 5 mW average power, glutamate was released onto single isolated dendritic spines in hippocampal slices. The spine head volume and the evoked current correlated with spine size, confirming earlier results, but the AMPA-receptor-mediated current evoked from proximal synapses was lower (42 ± 3 pA) than that from distal ones (109 ± 8 pA). Further analysis revealed that spines distal from the soma of the CA1 neuron had twice the number of AMPA-sensitive glutamate receptors than the proximal spines had. Synapses made with Schaffer collateral axons function in a location-independent manner, which increases the computational ability of CA1 pyramidal neurons. From the preceding examples, it is clear that 2PE-mediated release of neurotransmitters will play an increasing role in the continued exploration of molecular mechanisms of neural function.

8.5.2
Signal Transduction Molecules

Cyclic nucleotide monophosphates (cNMP), such as cyclic adenosine monophosphate (cAMP) and cyclic guanosine monophosphate (cGMP), are intracellular second messengers that regulate many cellular functions, including growth, differentiation, motility, and gene expression. Furuta et al. have shown that Bhc-cAMP and Bhc-cGMP (Fig. 8.6) are efficiently photolyzed by 2PE ($\delta_u = 1.68$–2.28 GM at 740 nm) [24, 47]. These molecules are not membrane permeant, but acetylating the phenol of Bhc solves this problem. After migration across the

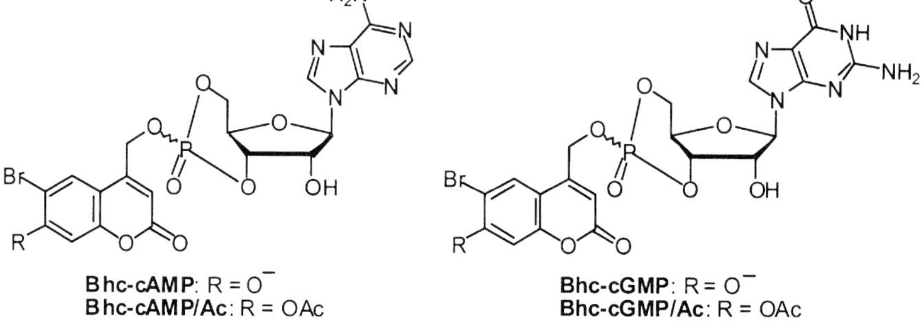

Bhc-cAMP: R = O$^-$
Bhc-cAMP/Ac: R = OAc

Bhc-cGMP: R = O$^-$
Bhc-cGMP/Ac: R = OAc

Fig. 8.6 Caged cAMP and cGMP.

cell membrane, non-specific esterases hydrolyze the acetate, revealing the active chromophore. The biological usefulness of Bhc-caged cNMPs has been demonstrated with 1PE, but not yet with 2PE.

Calcium ions are important physiological second messengers, and 2PE of caged calcium would provide a means to study the effects of localized jumps in Ca^{2+} concentration and the kinetics of cellular Ca^{2+} buffering. Spontaneous microscopic increases in Ca^{2+} concentration, known as "Ca^{2+} sparks", play a role in muscle contraction. Three caging groups have been shown to release Ca^{2+} upon 2PE: DM-nitrophen [25, 67–69], DMNPE-4 [70, 71] and azid-1 [25] (Fig. 8.7 and Tab. 8.2).

DM-nitrophen ($\delta_u = 0.013$ GM at 730 nm) [25] was first used by Lipp and Niggli [67, 68] to study Ca^{2+}-induced Ca^{2+} release (CICR) from the sarcoplasmic reticulum, which activates muscle contraction. The researchers focused 705-nm light from a 75-fs pulsed Ti:S laser for 50 ms inside a guinea-pig ventricular myocyte. A scanning confocal microscope was used to image the Ca^{2+} with the indicator fluo-3. Setting the average laser power to 40 mW, DM-nitrophen excitation resulted in a small response similar to a natural Ca^{2+} spark in amplitude and size, while increasing the laser power to 80 mW produced a Ca^{2+} wave that propagated through the cell. The researchers identified Ca^{2+} quarks, bursts of Ca^{2+} 20–40 times smaller than the Ca^{2+} spark that triggers CICR, that were observed after 2PE of DM-nitrophen with 30 mW average power. This technique was then used to investigate the control mechanism responsible for limiting the amplification of Ca^{2+} signaling and terminate Ca^{2+} release from the sarcoplasmic reticulum [75]. Photochemical triggering of CICR is independent of the L-type Ca^{2+} channels, so the spatial aspects of ryanodine receptor (RyR) activation by Ca^{2+} could be explored with 2PE-mediated release of Ca^{2+} from DM-nitrophen and compared to global release from 1PE. The researchers observed a marked difference in the way the myocyte behaved, depending on the excitation method. They concluded that the total amount of Ca^{2+} in the sarcoplasmic reticulum plays an essential function in terminating Ca^{2+} release, but other mechanisms underlie Ca^{2+} release termination during Ca^{2+} sparks.

Some caution should be used in interpreting the results of these experiments. As Soeller and Cannell point out [69], the possibility of cell damage occurring

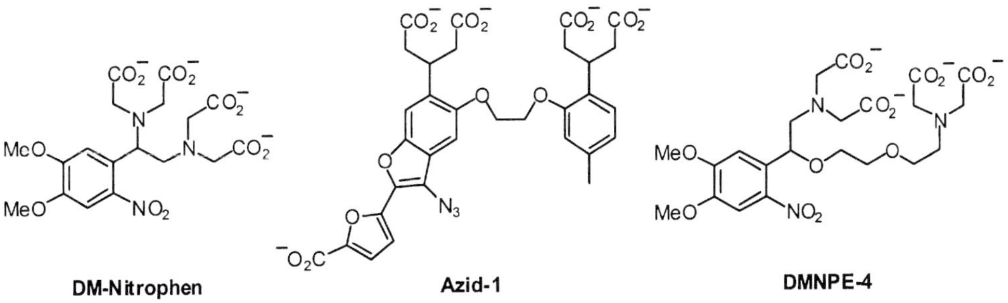

Fig. 8.7 Cages for calcium.

Tab. 8.2 Summary data for 2PE-sensitive chromophores for caged calcium

Chromophore	λ_{max} (nm)	ε (M^{-1} cm^{-1})	$Q_u^{a)}$ (mol ein^{-1})	$\delta_u(\lambda)$ [GM (nm)]$^{b)}$	Pre-hv Ca^{2+} K_d (μM)	Post-hv Ca^{2+} K_d (μM)
DM-nitrophen [25, 72, 73]	350	4330	0.18	0.013 (730)	0.005	3000
DMPE-4 [70]	347	5140	0.20	NA$^{c)}$ (700)	0.048	1000
Azid-1 [25, 74]	342	33000	1.00	1.4 (700)	0.23	120

a) Q_u = quantum yield for uncaging Ca^{2+}.
b) GM = Göppert-Mayer (10^{-50} cm$^4 \cdot$ s photon^{-1}).
c) Data not available.

from the 2PE cannot be ruled out. Release of Ca^{2+} could be coming from the rupture of mitochondria, endoplasmic reticulum (ER) or sarcoplasmic reticulum membranes. They examined the spatial and temporal spread of Ca^{2+} release from DM-nitrophen using fluo-3 fluorescence to detect Ca^{2+}. The spatial spread of Ca^{2+} release was wider than the photolysis spot. At 54 and 18 mW average laser power (10-ms flashes, 730 nm), significant bleaching of the fluo-3 occurred, which was mitigated when the power was reduced to 8 mW. The authors contend that cell damage is of concern when the power levels reach the point where photobleaching of indicators occurs. Using intact rat ventricular myocytes, Soeller and Cannell also explored the use of 2PE and caged calcium to simulate Ca^{2+} sparks, and found that the artificially generated sparks had spatio-temporal properties that were nearly identical to naturally occurring sparks. No obvious cell damage was observed with 20-ms pulses and 4 mW average laser power. In a more recent study, Soeller et al. used DM-nitrophen to test an algorithm to reconstruct the Ca^{2+} release flux underlying the fluorescence signal [71]. On a millisecond timescale and a spatial resolution of 1 µm, the authors obtained a reasonable description of the time course of Ca^{2+} release. Applying the algorithm to sparks observed in a rat ventricular myocyte showed that around 15 RyRs release approximately 200 000 Ca^{2+} ions in a Ca^{2+} spark.

Heart muscle tissues exhibit highly coordinated intercellular signaling to carry out organized functions that are coordinated by pacemaker cells and mediated by cytosolic Ca^{2+}, but organized intercellular signaling also exists in epithelial cells. To examine this phenomenon, Nathanson et al. [76] used the two-photon-mediated release of Ca^{2+} from DM-nitrophen with 730-nm light and 2–4 mW of average power at the focal plane. In SkHep1 cell culture, the researchers found that Ca^{2+} does not act as an intercellular messenger, or as a global intracellular messenger, in cells without activated inositol-1,4,5-triphosphate receptors (InsP$_3$Rs). Further, they showed that gap junctions were required for synchronizing intercellular Ca^{2+} signals and that cells with increased expression of the hormone vasopressin, which activates InsP$_3$ receptors in the presence of Ca^{2+}, could act as pacemakers for nearby epithelial cells.

Subcellular Ca^{2+} release through 2PE of DM-nitrophen has provided insight into the interplay between the two main Ca^{2+}-release channels, RyRs and InsP$_3$Rs, in generating Ca^{2+} waves. In polarized epithelia, such as pancreatic acinar cells, the RyRs and InsP$_3$Rs are located in apical and basolateral subcellular regions, respectively. Ca^{2+} was released by 2PE of DM-nitrophen in each of these regions of the cell and the resulting Ca^{2+} wave was monitored by confocal microscopy of fluo-3. The research showed that the activity of the two receptors is coordinated by the subcellular release of Ca^{2+}; apical release of Ca^{2+} from InsP$_3$Rs causes further release of Ca^{2+} from the basolateral RyRs [77].

Azid-1 has a significantly higher sensitivity to 2PE [25] and it has been used with 1PE to show a link between Ca^{2+} and synaptic plasticity in Purkinje cells of the cerebellum [74]. Complete photolysis of azid-1 within the 2PE focal volume requires 10-μs pulse of 700-nm light at 7 mW average power. To achieve the same effect with DM-nitrophen requires 74 mW of power, which is above the threshold for potential cellular photodamage. One drawback to azid-1 is that it undergoes a 500-fold increase in Ca^{2+} dissociation constant (K_d) upon photolysis, which is small compared to the 600 000-fold increase achieved for DM-nitrophen. Azid-1 has low affinity for Mg^{2+} ($K_d=7.6$ mM).

Niggli et al. report [70] that DMNPE-4 exhibits good 2PE at 705 nm, but give no explicit measurement of δ_u for this Ca^{2+} chelator. Upon photolysis, it changes its affinity for Ca^{2+} by 21 000-fold and has the advantage that it has a lower Mg^{2+} affinity than DM-nitrophen ($K_d=7$ mM and 2.5 μM, respectively). Mg^{2+} buffering by the Ca^{2+} chelator can be problematic in some biological systems, limiting the amount of Ca^{2+} liberated. In droplets, DMNPE-4 released Ca^{2+} upon 2PE in the presence of 10 mM Mg^{2+}, whereas with DM-nitrophen no Ca^{2+} liberation was observed. In guinea-pig ventricular myocytes, 30-ms pulses of 50-mW laser light produced cellular Ca^{2+} signals that resemble Ca^{2+} sparks.

DMNPE-4 has been applied to the study of Ca^{2+} movements within microscopic intracellular compartments like the nucleus. Soeller et al. [71] probed nuclear Ca^{2+} diffusion by injecting DMNPE-4 into mouse oocytes, photolyzing with 60-ms flashes of 700-nm light from a mode-locked Ti:S laser at "powers that were significantly below intensities that caused visible photodamage" and observing the diffusion of Ca^{2+} with fluo-3. Interestingly, the fluorescence from the indicator spread half as quickly in the nucleolus, indicating a slower rate of Ca^{2+} diffusion than in the cytoplasm.

In the nucleus, Ca^{2+} signals impact gene transcription and cell growth, requiring intricate machinery for the regulation of local concentrations of this important second messenger. Nathanson et al. [45] identified a structure analogous to the ER, the nucleoplasmic reticulum (NR), where Ca^{2+} is stored and InsP$_3$ receptors exist. They investigated the effects of InsP$_3$ on subnuclear Ca^{2+} release in SKHep1 by 2PE of NPE-caged InsP$_3$ (Fig. 8.8), which was injected into the cells. Light from a Ti:S laser tuned to 920 nm was frequency doubled with a 0.5-mm LBO crystal to produce 460-nm light. At this wavelength, δ_u of NPE is 4 GM, the highest uncaging action cross-section reported to date. Release of InsP$_3$ 1 μm from the NR structures generated increases in Ca^{2+} concentrations

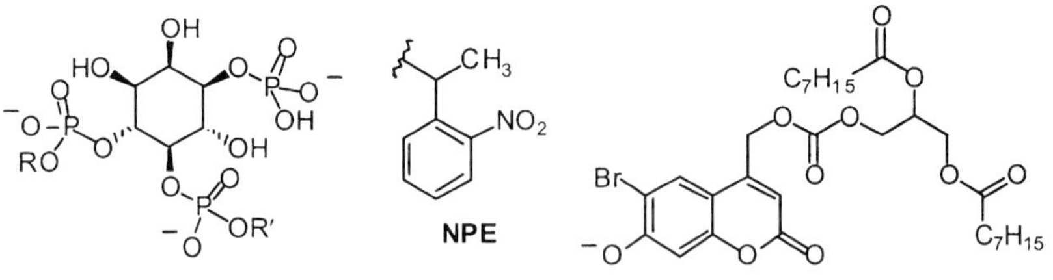

Fig. 8.8 Caged second messengers.

originating from the NR, as measured by simultaneous confocal imaging of fluo-4. In the absence of NPE-InsP$_3$, irradiation at 460 nm had no effect on Ca^{2+} levels. The researchers also report that cells could "tolerate up to several seconds of pulsed light at 460 nm without...evidence of damage". From these studies, they conclude that the InsP$_3$ receptors in the NR gate Ca^{2+} to generate local signals within the nucleus. As an extension of this work, Nathanson et al. [45] visualized the effects of nuclear and cytosolic Ca^{2+} release on the distribution of protein kinase C (PKC) in the cell. DM-nitrophen-bound Ca^{2+} was released by 2PE in the nucleus, inducing PKC-γ in the nucleus to translocate to the nuclear envelope without affecting cytosolic PKC-γ. Conversely, localized Ca^{2+} release in the cytoplasm caused translocation of cytoplasmic PKC-γ, but did not affect nuclear PKC-γ.

Using a Bhc-caged version of a diacylglycerol (DAG), dioctanoylglycerol (diC$_8$), Walker et al. investigated how the location of PKC activation by DAG determines whether cardiac function is stimulated or inhibited [78]. Two-photon photolysis (780 nm, 20–25 mW) of Bhc-caged diC$_8$ (Bhc-diC$_8$, Fig. 8.8) in different subdomains of rat cardiac myocytes yielded different physiological responses, presumably because different subsets of PKC were activated. Noteworthy in this study is that Bhc-diC$_8$ could be used at 4-fold lower concentration than a nitrobenzyl ether caged diC$_8$, owing to the greater sensitivity of Bhc to 1PE and 2PE.

8.5.3
Fluorophores

One technique to trace the lineage of cells through cell divisions is to use a fluorescent dye conjugated to a dextran to keep the dye confined to daughter cells arising from the parent. Fluorescence microscopy is then used to follow the cells through cell division. A limitation to this technique is that in order to be marked, the cell must be large enough to tolerate microinjection, yet this is not always the case. This problem can be overcome if a caged dye–dextran conjugate (i.e. non-fluorescent) is injected into a relatively large cell at an early stage of development (such as the single-cell stage), followed by light activation of the fluorophore. Uncaging the dye through 2PE enables smaller cells to be traced through development without the risk of damage through microinjection or UV light if 1PE is used to activate the dye. Summers et al. [79] developed this technique to study the relationship between the orientation of the first cleavage furrow of a two-cell stage sea urchin embryo (*Lytechinus variegatus*) and the axes of bilateral symmetry. Sea urchin embryos are an important model system for studying many aspects of developmental biology. DMNB–fluorescein–dextran (Fig. 8.9) was microinjected into a single-celled zygote and activated with a 30-s flash of 700-nm light from a Ti:S laser in one of the two blastomeres at the two-cell stage. They compared the fluorescent labeling results with an established method and found better survival among the embryos with the 2PE technique.

In a later study on sea urchin, Piston et al. [80] investigated epithelial cell involution during archenteron and gastrointestinal tract development. Single-cell embryos were injected with DMNB–fluorescein–dextran, and small groups of cells were marked by exposure of undisclosed duration to 700-nm light of undisclosed average power from a Ti:S laser. Recruitment of endothelial cells into the archenterons occurred during primary and secondary invagination, while further involution after the secondary invagination formed the larval intestine. These results, which are inconsistent with previous sea urchin gastrulation models, demonstrate the potential power of 2PE of a caged fluorophore to open new views to physiological function.

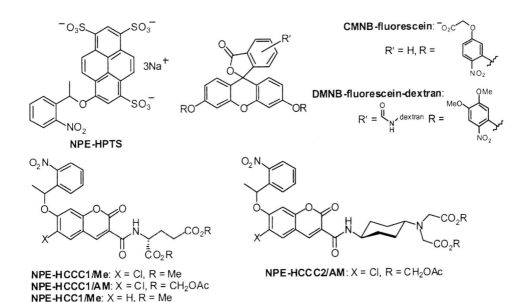

Fig. 8.9 Caged fluorophores.

Diffusional exchange of materials between the spine heads and dendritic shafts of neurons occurs through a narrow neck that connects the head to the shaft. Understanding how materials flow between the spines and the shaft is important for understanding synaptic plasticity. To investigate diffusional coupling of dendritic spines to the shafts, Svoboda et al. [81] used the photorelease of fluorescein–dextran from DMNB–fluorescein–dextran to create a burst of fluorescence in a single spine head. They monitored the diffusion of fluorescence inside the spine head to measure the timescale of diffusional equilibration and demonstrated that second messengers could be confined to the spine, which might be important for the induction of synapse-specific plasticity.

To study the role of molecular mobility in intracellular communication, Soeller et al. [71, 82] used CMNB–fluorescein (Fig. 8.9) to generate a point source of fluorescence to probe local diffusion in rat lenses. In mammalian lenses, membrane channels called gap junctions allow low-molecular-weight molecules to travel between cells and enable intercellular communication, which is important for maintaining the transparency of the lens. Observing dye transfer is one method of studying gap junction function, but loading can be a problem unless a dye activated by 2PE is used. The researchers released fluorescein inside fiber cells (750 nm, < 30 mW average power), and using confocal microscopy, they monitored the diffusion of the dye to neighboring cells. They found variations in intercellular coupling that depended upon whether the cells were located in the periphery or deep inside the lens. The density and orientation of gap junctions probably play a role in the diffusion of materials between cells. In another application of 2PE-mediated photolysis of CMNB–fluorescein, Baumgartner and Montrose showed that the diffusion rates at the gastric surface were the same as those in free solution [83]. These experiments were part of a larger study on the role of mucus, the unstirred layer and acid/alkali secretion in regulating the pH of gastric surfaces.

Fluorescein dyes photobleach relatively easily, and loading high concentrations of caged fluorescein can be a problem. To overcome these problems, Li et al. [35] devised a series of NPE-caged coumarins: NPE-HCCC1/Me, NPE-HCCC1/AM and NPE-HCC1/Me (Fig. 8.9). The coumarin is less susceptible to photobleaching and it acts as an "antenna" for "substrate-assisted photolysis" of the NPE moiety. As a result, NPE-HCCC1/Me and NPE-HCC1/Me exhibited δ_u values of 0.37 and 0.68 GM at 740 nm, respectively, which are significantly higher than other NPE-caged conjugates. High concentrations of cell-permeant versions, NPE-HCCC1/AM and NPE-HCCC2/AM, were loaded into a variety of cell lines, but the researchers did not demonstrate fluorophore release by 2PE in cell culture. Nevertheless, this strategy for creating caged compounds sensitive to 2PE holds promise.

8.5.4
PDT and Gene Inactivation

One of the challenges of developing PDT as a means of treating disease is depth penetration of the light that produces the ROS, typically singlet oxygen. The use of 2PE to excite a singlet oxygen sensitizer is an appealing application of the technology because of the greater depth penetration into biological tissues achieved with IR wavelengths. Ogilby et al. have created chromophores, based on difuronaphthalene and distyryl benzene, for generating singlet oxygen by 2PE [84]. Irradiating singlet oxygen generators **1a–e** (Fig. 8.10) with 590–680-nm light and **2** with 780–880-nm light, produced from a nanosecond-pulsed Nd:YAG-pumped optical parametric oscillator (OPO), initiated the production of singlet oxygen (observed by detecting the oxygen phosphorescence at 1270 nm). For use on biological systems, a femtosecond-pulsed laser is more desirable and sensitizer **2** produced singlet oxygen in an oxygen-saturated solution of toluene upon 2PE with a femtosecond-pulsed Ti:S laser tuned to 802 nm. An *ab initio* computational study of these compounds modeled δ_a in an effort to develop a predictive tool for the design of 2PE-sensitive singlet oxygen sensitizers [85]. Many obstacles must be cleared before 2PE-induced singlet oxygen production in a biological system becomes a reality with this system.

Psoralens in combination with UV irradiation have been used in the treatment of a variety of skin disorders. The psoralens intercalate with DNA and, upon exposure to light, inflict damage to the DNA. In addition to indiscriminant tissue destruction, this method suffers from poor penetration to deeper levels of the dermis. Using 2PE of the chromophore would increase the depth of penetration, improve localization of the photodynamic effect and minimize absorption from other chromophores in the skin. Wachter et al. [86, 87] evaluated 8-methoxypsoralen (8-MOP, Fig. 8.11) and 4′-aminomethyl-4,5′,8-trimethylpsoralen (AMT) for their 2PE properties, finding that both compounds could be excited by 730-nm light. An Ames II test, which detects mutations in a histidine auxotrophic strain of *Salmo*-

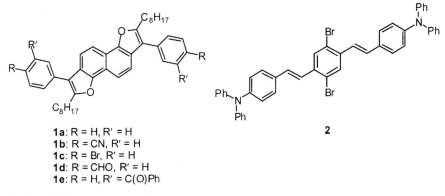

Fig. 8.10 Singlet oxygen generators.

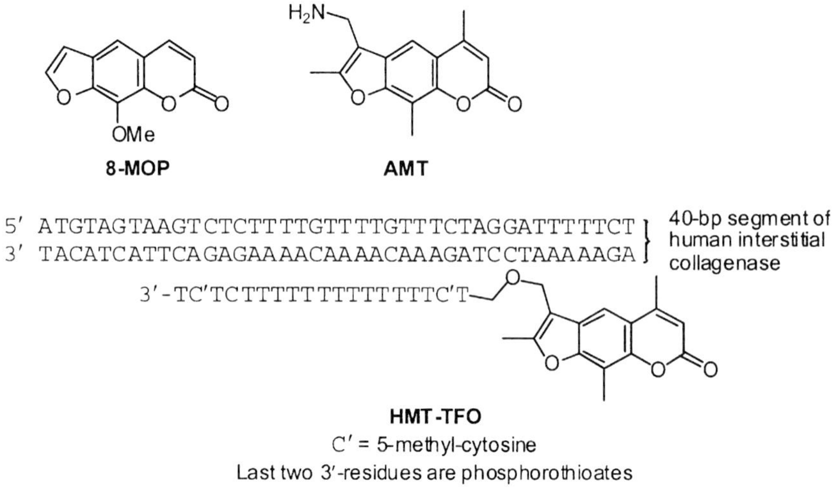

8-MOP

AMT

5' ATGTAGTAAGTCTCTTTTGTTTTGTTTCTAGGATTTTTCT
3' TACATCATTCAGAGAAAACAAAACAAAGATCCTAAAAAGA } 40-bp segment of human interstitial collagenase

3'-TC'TCTTTTTTTTTTTTC'T-

HMT-TFO

C' = 5-methyl-cytosine
Last two 3'-residues are phosphorothioates

Fig. 8.11 Psoralens for 2PE-mediated photodynamic therapy.

nella typhimurium, showed the cytotoxic effect of exposing the cells to 730-nm light (500 mW) in the presence of AMT. Experiments conducted in the absence of AMT imply that the laser irradiation is non-toxic to the cells.

Oh et al. have used 2PE of 4'-hydroxymethyl-4,5',8-trimethylpsoralen [88], which is linked to a triple helix-forming oligonucleotide (HMT-TFO, Fig. 8.11), to bind a targeted DNA sequence and generate sequence-specific DNA damage [89]. The TFO targets HMT to a specific DNA sequence and the 2PE of HMT localizes the DNA damage to the desired volume. The authors observed HMT-TFO-mediated sequence-specific formation of HMT-DNA monoadducts (no cross-linking of DNA strands was observed) in solution after irradiation for 13 h with 760-nm light from a Ti:S laser seeded regenerative amplifier (140–200 fs pulse width, 450–700 mW average power). When the target DNA and HMT-TSO were immobilized in a polyacrylamide gel, 2PE enabled 3-D control of monoadduct formation. Treatment of human dermal fibroblasts with a tritiated version of the psoralen–TFO conjugate, [^{3}H]HMT–TFO, and 2PE showed significantly more radiation counts associated with the cells' DNA than untreated or unirradiated controls, indicating that [^{3}H]HMT–DNA monoadducts had formed. Importantly, the cells were metabolically active and morphologically normal after treatment, indicating that the treatment was not toxic. Using a regenerative amplifier with a repetition rate of 1 kHz, as opposed to ~80 MHz for a Ti:S laser probably helped prevent non-specific thermal damage to the cells. These experiments provide proof-of-principle that 2PE can be used to target photochemical DNA damage and show that it might be useful for photodynamic therapy applications in optically accessible tissues such as skin.

The selective inactivation of genes would provide a useful means of probing genetic function. Berns et al. [90] have used 2PE-mediated inactivation of the ri-

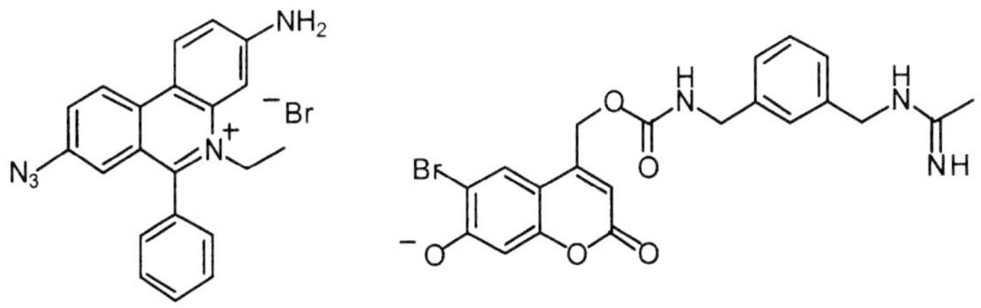

Ethidium Monoazide Bromide   **Bhc-1400W**

Fig. 8.12 A mediator of two-photon gene inactivation and a caged drug.

bosomal (nucleolar organizer) genes in PtK2 cells. Of importance is that 100% of the cells undergoing the laser-based protocol survived, in contrast to earlier studies with UV lasers and nanosecond-pulsed and high-energy 2PE lasers. Prophase PtK2 cells in culture were treated with ethidium monoazide bromide (phenanthridium, Fig. 8.12), which intercalates into the cell's DNA and, when exposed to light, forms a covalent bond to the DNA. The cells were exposed for 30–60 s of irradiation from a pulsed (100-ps) laser beam (1060 nm) focused on the nucleolus and attached chromosomes. Monitoring through mitosis revealed a decrease in the number of nucleoli in the daughter cells, indicating that the nucleolar genes had been inactivated.

8.5.5
Drug Delivery

Controlling the timing and location of drug release *in vivo* would represent a leap forward in improving the selectivity of drug action. A Bhc-caged version of a nitric oxide synthase (NOS) inhibitor, 1400W, has been used as a vehicle to regulate the generation of NO [49]. Hypoxia, ischemia, or immunostimulation initiates unregulated NO production by inducible NOS (iNOS), leading to local tissue damage. Inhibiting iNOS only in selected tissues and locations after a stroke or other pathological process involving NO overproduction could potentially enable better treatment programs. Using Bhc-1400W (Fig. 8.12), Guillemette et al. showed that active drug released with 2PE (740 nm) effectively inhibited iNOS in a cuvette. While no *in vitro* studies on the photorelease of 1400W were conducted, 80% of PC12 cells remained viable after treatment with 100 µM Bhc-1400W. More experiments are required to show that 2PE of a caged inhibitor can be used to regulate the antagonist in a living organism, but this approach to drug delivery appears promising.

8.6
Conclusion

The unique capability of 2PE to localize the triggering and release of biological effectors has made an impact on many areas of biology. Compared to UV-mediated uncaging, it provides a less invasive and less destructive method of delivering biological effectors, such as neurotransmitters, second messengers and drugs. The transparency of tissues to near-IR light also enables the manipulation of physiological processes deeper in biological preparations than is possible with UV light. The development of this technology is in its early stages and, as the examples show, exploiting this technology to further understand local physiological processes will require the development of new chromophores that are more sensitive to 2PE and possess uncaging kinetics faster than diffusion. Advances in laser and microscope technology have begun to lower the barriers to more widespread implementation of two-photon microscopy. Indeed, the new Ti:S lasers do not require sophisticated expertise to operate, and commercially available microscope systems are becoming more common. As a result, multiphoton phototriggers will play an increasing role in the exploration of cell physiology.

Acknowledgments

I would like to thank Richard K. Hill and Mark Farmer for helpful discussions, and the National Science Foundation for financial support through a CAREER award (CHE-0349059).

References

1 GÖPPERT-MAYER, M. *Ann. Phys. (Berlin)* **1931**, *9*, 273–294.
2 GÖPPERT, M. *Naturwissenschaften* **1929**, *17*, 932.
3 KAISER, W., GARRETT, C.G.B. *Phys. Rev. Lett.* **1961**, *7*, 229–231.
4 SINGH, S., BRADLEY, L.T. *Phys. Rev. Lett.* **1964**, *12*, 612–614.
5 DENK, W., STRICKLER, J.H., WEBB, W.W. *Science* **1990**, *248*, 73–76.
6 DENK, W., PISTON, D.W., WEBB, W.W. Two-photon molecular excitation in laser scanning microscopy, in *Handbook of Biological Confocal Microscopy*, PAWLEY, J.B. (ed.), Plenum Press, New York, **1995**, pp. 445–458.
7 XU, C., ZIPFEL, W., SHEAR, J.B., WILLIAMS, R.M., WEBB, W.W. *Proc. Natl Acad. Sci. USA* **1996**, *93*, 10763–10768.
8 XU, C., WEBB, W.W. Multiphoton excitation of molecular fluorophores and nonlinear laser microscopy, in *Topics in Fluorescence Spectroscopy. Vol. 5. Nonlinear and Two-photon-induced Fluorescence*, LAKOWICZ, J. (ed.), Plenum Press, New York, **1997**, pp. 471–540.
9 DENK, W., SVOBODA, K. *Neuron* **1997**, *18*, 351–357.
10 KÖNIG, K. *J. Microsc.* **2000**, *200*, 83–104.
11 WILLIAMS, R.M., ZIPFEL, W.R., WEBB, W.W. *Curr. Opin. Chem. Biol.* **2001**, *5*, 603–608.

12 Helmchen, F., Denk, W. *Curr. Opin. Neurobiol.* **2002**, *12*, 593–601.
13 Zipfel, W. R., Williams, R. M., Webb, W. W. *Nat. Biotechnol.* **2003**, *21*, 1369–1377.
14 Denk, W. *Proc. Natl Acad. Sci. USA* **1994**, *91*, 6629–6633.
15 Uchida, M., Irie, M. *J. Am. Chem. Soc.* **1993**, *115*, 6442–6443.
16 Perez-Prieto, J., Miranda, M. A., Garcia, H., Konya, K., Scaiano, J. C. *J. Org. Chem.* **1996**, *61*, 3773–3777.
17 Miranda, M. A., Perez-Prieto, J., Font-Sanchis, E., Konya, K., Scaiano, J. C. *J. Org. Chem.* **1997**, *62*, 5713–5719.
18 Pettit, D. L., Wang, S. S. H., Gee, K. R., Augustine, G. J. *Neuron* **1997**, *19*, 465–471.
19 Qiao, L., Kozikowski, A. P., Olivera, A., Spiegel, S. *Bioorg. Med. Chem. Lett.* **1998**, *8*, 711–714.
20 Wang, S. S. H., Khiroug, L., Augustine, G. J. *Proc. Natl Acad. Sci. USA* **2000**, *97*, 8635–8640.
21 Pirrung, M. C., Pieper, W. H., Kaliappan, K. P., Dhananjeyan, M. R. *Proc. Natl Acad. Sci. USA* **2003**, *100*, 12548–12553.
22 Coherent, Inc. http://www.coherentinc.com.
23 Spectra-Physics. http://www.splasers.com.
24 Furuta, T., Wang, S. S. H., Dantzker, J. L., Dore, T. M., Bybee, W. J., Callaway, E. M., Denk, W., Tsien, R. Y. *Proc. Natl Acad. Sci. USA* **1999**, *96*, 1193–1200.
25 Brown, E. B., Shear, J. B., Adams, S. R., Tsien, R. Y., Webb, W. W. *Biophys. J.* **1999**, *76*, 489–499.
26 Xu, C., Guild, J., Webb, W. W., Denk, W. *Optics Lett.* **1995**, *20*, 2372–2374.
27 Xu, C., Webb, W. W. *J. Opt. Soc. Am. B* **1996**, *13*, 481–491.
28 Matsuzaki, M., Ellis-Davies, G. C. R., Nemoto, T., Miyashita, Y., Iino, M., Kasai, H. *Nat. Neurosci.* **2001**, *4*, 1086–1092.
29 Albota, M., Beljonne, D., Bredas, J.-L., Ehrlich, J. E., Fu, J.-Y., Heikal, A. A., Hess, S. E., Kogej, T., Levin, M. D., Marder, S. R., McCord-Maughon, D., Perry, J. W., Rockel, H., Rumi, M., Subramaniam, G., Webb, W. W., Wu, X.-L., Xu, C. *Science* **1998**, *281*, 1653–1656.
30 Reinhardt, B. A., Brott, L. L., Clarson, S. J., Dillard, A. G., Bhatt, J. C., Kannan, R., Yuan, L., He, G. S., Prasad, P. N. *Chem. Mater.* **1998**, *10*, 1863–1874.
31 Belfield, K. D., Schafer, K. J., Liu, Y., Liu, J., Ren, X., Van Stryland, E. W. *J. Phys. Org. Chem.* **2000**, *13*, 837–849.
32 Pati, S. K., Marks, T. J., Ratner, M. A. *J. Am. Chem. Soc.* **2001**, *123*, 7287–7291.
33 Brousmiche, D. W., Serin, J. M., Frechet, J. M. J., He, G. S., Lin, T.-C., Chung, S. J., Prasad, P. N. *J. Am. Chem. Soc.* **2003**, *125*, 1448–1449.
34 Chung, S. J., Lin, T. C., Kim, K. S., He, G. S., Swiatkiewicz, J., Prasad, P. N., Baker, G. A., Bright, F. V. *Chem. Mater.* **2001**, *13*, 4071–4076.
35 Zhao, Y., Zheng, Q., Dakin, K., Xu, K., Martinez, M. L., Li, W.-H. *J. Am. Chem. Soc.* **2004**, *126*, 4653–4663.
36 Wang, X., Zhou, Y., Zhou, G., Jiang, W., Jiang, M. *Bull. Chem. Soc. Jpn.* **2002**, *75*, 1847–1854.
37 Wang, X., Zhou, G., Wang, D., Wang, C., Fang, Q., Jiang, M. *Bull. Chem. Soc. Jpn.* **2001**, *74*, 1977–1982.
38 Zhou, X., Ren, A.-M., Feng, J.-K., Liu, X.-J., Zhang, J., Liu, J. *Phys. Chem. Chem. Phys.* **2002**, *4*, 4346–4352.
39 Kiskin, N. I., Ogden, D. *Eur. Biophys. J.* **2002**, *30*, 571–587.
40 Koester, H. J., Baur, D., Uhl, R., Hell, S. W. *Biophys. J.* **1999**, *77*, 2226–2236.
41 Hopt, A., Neher, E. *Biophys. J.* **2001**, *80*, 2029–2036.
42 Tirlapur, U. K., Koenig, K., Peuckert, C., Krieg, R., Halbhuber, K.-J. *Exp. Cell Res.* **2001**, *263*, 88–97.
43 Tirlapur, U. K., Koenig, K. *Nature* **2002**, *418*, 290–291.
44 Kiskin, N. I., Chillingworth, R., McCray, J. A., Piston, D., Ogden, D. *Eur. BioPhys. J.* **2002**, *30*, 588–604.
45 Echevarria, W., Leite, M. F., Guerra, M. T., Zipfel, W. R., Nathanson, M. H. *Nat. Cell Biol.* **2003**, *5*, 440–446.
46 Ando, H., Furuta, T., Tsien, R. Y., Okamoto, H. *Nat. Genet.* **2001**, *28*, 317–325.

47 Furuta, T., Takeuchi, H., Isozaki, M., Takahashi, Y., Kanehara, M., Sugimoto, M., Watanabe, T., Noguchi, K., Dore, T. M., Kurahashi, T., Iwamura, M., Tsien, R. Y. *ChemBioChem* **2004**, *5*, 1119–1128.

48 Lin, W., Lawrence, D. S. *J. Org. Chem.* **2002**, *67*, 2723–2726.

49 Montgomery, H. J., Perdicakis, B., Fishlock, D., Lajoie, G. A., Jervis, E., Guillemette, J. G. *Bioorg. Med. Chem.* **2002**, *10*, 1919–1927.

50 Suzuki, A. Z., Watanabe, T., Kawamoto, M., Nishiyama, K., Yamashita, H., Ishii, M., Iwamura, M., Furuta, T. *Org. Lett.* **2003**, *5*, 4867–4870.

51 Lu, M., Fedoryak, O. D., Moister, B. R., Dore, T. M. *Org. Lett.* **2003**, *5*, 2119–2122.

52 Magde, D., Tsien, R. Y., Dore, T. M. Unpublished results. University of California, San Diego, CA **1999**.

53 Schade, B., Hagen, V., Schmidt, R., Herbich, R., Krause, E., Eckardt, T., Bendig, J. *J. Org. Chem.* **1999**, *64*, 9109–9117.

54 Fedoryak, O. D., Dore, T. M. *Org. Lett.* **2002**, *4*, 3419–3422.

55 Papageorgiou, G., Corrie, J. E. T. *Synth. Commun.* **2002**, *32*, 1571–1577.

56 Canepari, M., Nelson, L., Papageorgiou, G., Corrie, J. E. T., Ogden, D. *J. Neurosci. Methods* **2001**, *112*, 29–42.

57 Papageorgiou, G., Corrie, J. E. T. *Tetrahedron* **2000**, *56*, 8197–8205.

58 Morrison, J., Wan, P., Corrie, J. E. T., Papageorgiou, G. *Photochem. Photobiol. Sci.* **2002**, *1*, 960–969.

59 Papageorgiou, G., Ogden, D. C., Barth, A., Corrie, J. E. T. *J. Am. Chem. Soc.* **1999**, *121*, 6503–6504.

60 Corrie, J. E. T., Trentham, D. R. Caged nucleotides and neurotransmitters, in *Biological Applications of Photochemical Switches*, Morrison, H. (ed.), Wiley, New York, **1993**, pp. 243–305.

61 Kaplan, J. H., Forbush, B., III, Hoffman, J. F. *Biochemistry* **1978**, *17*, 1929–1935.

62 Billington, A. P., Walstrom, K. M., Ramesh, D., Guzikowski, A. P., Carpenter, B. K., Hess, G. P. *Biochemistry* **1992**, *31*, 5500–5507.

63 Milburn, T., Matsubara, N., Billington, A. P., Udgaonkar, J. B., Walker, J. W., Carpenter, B. K., Webb, W. W., Marque, J., Denk, W., McCray, J. A., Hess, G. P. *Biochemistry* **1989**, *28*, 49–55.

64 Wieboldt, R., Gee, K. R., Niu, L., Ramesh, D., Carpenter, B. K., Hess, G. P. *Proc. Natl Acad. Sci. USA* **1994**, *91*, 8752–8756.

65 Kovalchuk, Y., Lev-Ram, V., Dore, T. M., Konnerth, A., Tsien, R. Y. Unpublished results. University of California, San Diego, CA, **2000**.

66 Smith, M. A., Ellis-Davies, G. C. R., Magee, J. C. *J. Physiol. (Cambridge)* **2003**, *548*, 245–258.

67 Lipp, P., Niggli, E. *J. Physiol. (Cambridge)* **1998**, *508*, 801–809.

68 Lipp, P., Niggli, E. *J. Physiol. (Cambridge)* **1998**, *510*, 987.

69 Soeller, C., Cannell, M. B. *Microsc. Res. Tech.* **1999**, *47*, 182–195.

70 DelPrincipe, F., Egger, M., Ellis-Davies, G. C. R., Niggli, E. *Cell Calcium* **1999**, *25*, 85–91.

71 Soeller, C., Jacobs, M. D., Donaldson, P. J., Cannell, M. B., Jones, K. T., Ellis-Davies, G. C. R. *J. Biomed. Opt.* **2003**, *8*, 418–427.

72 Kaplan, J. H., Ellis-Davies, G. C. R. *Proc. Natl Acad. Sci. USA* **1988**, *85*, 6571–6575.

73 Ellis-Davies, G. C. R., Kaplan, J. H. *J. Org. Chem.* **1988**, *53*, 1966–1969.

74 Adams, S. R., Lev-Ram, V., Tsien, R. Y. *Chem. Biol.* **1997**, *4*, 867–878.

75 DelPrincipe, F., Egger, M., Niggli, E. *Nat. Cell Biol.* **1999**, *1*, 323–329.

76 Leite, M. F., Hirata, K., Pusl, T., Burgstahler, A. D., Okazaki, K., Ortega, J. M., Goes, A. M., Prado, M. A. M., Spray, D. C., Nathanson, M. H. *J. Biol. Chem.* **2002**, *277*, 16313–16323.

77 Leite, M. F., Burgstahler, A. D., Nathanson, M. H. *Gastroenterology* **2002**, *122*, 415–427.

78 Robu, V. G., Pfeiffer, E. S., Robia, S. L., Balijepalli, R. C., Pi, Y., Kamp, T. J., Walker, J. W. *J. Biol. Chem.* **2003**, *278*, 48154–48161.

79 Summers, R. G., Piston, D. W., Harris, K. M., Morrill, J. B. *Dev. Biol.* **1996**, *175*, 177–183.
80 Piston, D. W., Summers, R. G., Knobel, S. M., Morrill, J. B. *Microsc. Microanal.* **1998**, *4*, 404–414.
81 Svoboda, K., Tank, D. W., Denk, W. *Science* **1996**, *272*, 716–719.
82 Cannell, M. B., Jacobs, M. D., Donaldson, P. J., Soeller, C. *Microsc. Res. Tech.* **2004**, *63*, 50–57.
83 Baumgartner, H. K., Montrose, M. H. *Gastroenterology* **2004**, *126*, 774–783.
84 Poulsen, T. D., Frederiksen, P. K., Jorgensen, M., Mikkelsen, K. V., Ogilby, P. R. *J. Phys. Chem. A* **2001**, *105*, 11488–11495.
85 Frederiksen, P. K., Jorgensen, M., Ogilby, P. R. *J. Am. Chem. Soc.* **2001**, *123*, 1215–1221.
86 Fisher, W. G., Partridge, W. P., Jr., Dees, C., Wachter, E. A. *Photochem. Photobiol.* **1997**, *66*, 141–155.
87 Wachter, E. A., Partridge, W. P., Fisher, W. G., Dees, H. C., Petersen, M. G. *Proc. SPIE Int. Soc. Opt. Eng.* **1998**, *3269*, 68–75.
88 Oh, D. H., Stanley, R. J., Lin, M., Hoeffler, W. K., Boxer, S. G., Berns, M. W., Bauer, E. A. *Photochem. Photobiol.* **1997**, *65*, 91–95.
89 Oh, D. H., King, B. A., Boxer, S. G., Hanawalt, P. C. *Proc. Natl Acad. Sci. USA* **2001**, *98*, 11271–11276.
90 Berns, M. W., Wang, Z., Dunn, A., Wallace, V., Venugopalan, V. *Proc. Natl Acad. Sci. USA* **2000**, *97*, 9504–9507.

9
New Challenges

9.1
Laser-induced T-Jump Method:
A Non-conventional Photoreleasing Approach to Study Protein Folding

Yongjin Zhu, Ting Wang, and Feng Gai

9.1.1
Introduction

Protein folding is one of the fundamental problems in contemporary structural biology and has been studied extensively [1–9]. The goal is to provide a quantitative and predictive understanding of factors that control or determine folding pathways, the stable structures, as well as thermally and kinetically accessible conformational substates.

Protein folding can be viewed as a self-assembly process at the molecular level during which an ensemble of largely disorganized and unfolded conformations evolves toward a highly compact native structure with greatly reduced conformational entropy. Although a quantitative and predictive understanding of the folding mechanism is still elusive, the study of protein-folding kinetics on the millisecond timescale has reached an advanced stage. Using the techniques of site-directed mutagenesis in conjunction with stopped-flow kinetics and hydrogen-deuterium exchange, it has been possible to define the sequence of events involved in folding on the millisecond timescale. However, the kinetics of folding of individual segments of secondary structures or small folding motifs occurs on a more rapid timescale [10–12] and thereby has been much less extensively studied. This is partly due to the fact that studying the early protein-folding events requires initiation methods that can induce the folding/unfolding process on the submillisecond timescale.

Recently, a number of approaches that are capable of triggering a folding/unfolding event on the submillisecond timescale, including rapid mixing [13, 14], phototriggering [15–20], pH-jump [21, 22], pressure-jump [23–26] and laser-induced temperature-jump (T-jump) [27–29] techniques, have been introduced to study a variety of conformational processes in protein folding. Of particular interest are those techniques that involve the use of photo or photochemical triggers. Existing methods include those that can induce conformational changes

Dynamic Studies in Biology. Edited by M. Goeldner, R. Givens
Copyright © 2005 WILEY-VCH Verlag GmbH & Co. KGaA, Weinheim
ISBN: 3-527-30783-4

by (1) photodissociation of a bound ligand [20], (2) photo-induced electron transfer [16], (3) photo-induced dipole moment change [19], (4) photodissociation of aromatic disulfides [15], (5) photodissociation of benzoinyl cages [17] and (6) photoisomerization of azobenzene derivatives [18]. Although these photo or photochemical methods represent significant advances in the application of photoreleasing approaches to the study of protein folding and conformational dynamics, a comprehensive review of all the existing methods is beyond the scope of this chapter. Below, we only briefly discuss some of these methods. The method of photodissociation of a bound ligand is limited to the study of mostly hemoproteins, such as cytochrome c [20]. Similarly, the photo-induced electron transfer method can only be applied to redox-active proteins whose conformations are sensitive to the oxidation state of the bound cofactor under appropriate conditions [16, 30]. The time resolution of this technique is determined by the rate of the electron transfer from the photo-oxidized electron donor to the cofactor in the protein, while the length of the observation window is determined by the rate of the back-electron transfer process. Studies have shown that the time resolution of this method can be increased by covalently attaching the donor to the protein, whereas reductive quenching of the oxidized electron donor can enlarge the observation time window [30].

Amide derivatives of tris-bipyridyl complexes [31] such as $(Rub_2m\text{-}OH)^{2+}$ (where Ru = Ru(II), b = 2,2' bipyridine and m-OH = 4'-methyl-2,2'-bipyridyl-4'-carboxylic acid) can also be used to trigger protein folding. Upon irradiation with visible light, this ruthenium complex undergoes a rapid metal-to-ligand charge transfer (MLCT) at the excited state, which generates a photo-induced dipole moment change of ~ 5–9 Debye [31]. By attaching such a ruthenium complex to the N-terminus of an alanine-based peptide, Huang et al. have shown that upon photo-excitation with a nanosecond laser pulse, the favorable interaction between the increased dipole moment of the ruthenium complex and the helix macro-dipole induces an overall coil–helix transition [19].

Other photochemical triggering methods generally involve the removal of a designed conformational constraint by employing a laser pulse. For example, an aryl disulfide bond has been used to constrain an alanine-based peptide to be distorted from its equilibrium form, and after photolysis the dissociation of the disulfide bond generates two thiyl radicals and the peptide then commences to fold into an a-helical conformation [32]. The time resolution of this scheme can potentially reach the subpicosecond timescale (the rate of photodissociation of the disulfide bond) and this method has been used to study very early events in peptide conformational dynamics [15].

Similarly, a small loop formed between an internal cysteine residue and a 3'-(carboxymethoxy)benzoin (CMB) molecule, which is covalently attached to the N-terminus of a polypeptide as a linker, has also been used as a conformational constraint to alter the unfolded state of proteins. For example, this approach has been applied to the folding study of a mutant of the small a-helical villin headpiece subdomain (i.e. cVHP-34 M12C) [17]. Upon illumination with a 355-nm laser pulse, this linker molecule is irreversibly cleaved; the loop is therefore broken to yield the "linear" protein that is free to refold [17]. The time resolution of this method is determined by the rate of photolysis of the linker, which is faster than 1 ns [17].

The reversible photoisomerization of azobenzene, linking two cysteine residues in a designed peptide, has also been demonstrated to have the capability to control the helix content in steady state [18]. The azobenzene crosslinker, which was designed to work as a switch, undergoes *cis/trans* photoisomerization and reversibly disrupts the helix hydrogen bonding due to distance change of its two ends. Simple azobenzene compounds are photoisomerized on the picosecond timescale and therefore suitable for studying the early events in protein folding as well [33].

However, it is worth pointing out that while these techniques are applicable for triggering conformational processes on the nanosecond or even picosecond timescale, they also suffer from several limitations. For example, light-initiated ligand dissociation and electron transfer-initiated folding are limited to hemoproteins, whereas photodissociation of benzoinyl cages and photoisomerization of azobenzene derivatives are not readily reversible, which precludes signal averaging of a single sample, and studies with azobenzene and benzoinyl cages as well as disulfide derivatives have involved the preparation of cyclic structures that present highly restrained conformational states.

Among those fast folding initiation methods, the *T*-jump method seems to be the most versatile because it does not require the additional chemical modification of the polypeptide of interest, even though it is limited to conditions where the protein undergoes a net unfolding reaction or the folding of proteins that undergo cold denaturation. Nevertheless, the *T*-jump method has contributed greatly to our understanding of the primary processes in protein folding, such as the helix–coil transition, β-hairpin formation and hydrophobic collapse. Here we focus on the discussion of *T*-jump-coupled IR method as well as its application to the study of several important processes in protein folding. For interested readers, we would also like to direct their attention to other reviews on similar topics [10, 34–36].

9.1.2
Laser-induced *T*-jump IR Technique

Proteins can be denatured chemically or thermally. The *T*-jump technique simply relies on the fact that the stability of a protein, or its free energy, depends on temperature. In aqueous solution, proteins usually have a temperature range over which their native structures are stable. Temperatures falling out of this range will cause the native structures to unfold; either increasing or decreasing temperature, therefore, can achieve thermal denaturation. For many proteins, especially some small ones, the denaturation is reversible; they regain their native conformations spontaneously if the denaturation condition is removed. For example, some proteins spontaneously refold if cooled after thermal denaturation. Such reversibility is important to folding studies because most probing techniques require repetitive measurements.

Depending upon (1) the thermal properties of the protein of interest, (2) the initial temperature and (3) the magnitude of the T-jump, the final state in a T-jump experiment may correspond to an equilibrium position that favors either folded or unfolded conformation, thus allowing the study of both refolding and unfolding. Strictly speaking, the T-jump method is essentially a relaxation approach, and both refolding and unfolding may be studied no matter in which direction the equilibrium shifts. In other words, the relaxation kinetics to the new equilibrium position contain contributions from both folding and unfolding kinetics. For example, in a simple two-state system, e.g. U $\Longleftrightarrow$ N (U and N represent the unfolded and the native states, respectively), the observed rate constant is exactly the sum of the folding rate constant (k_f) and the unfolding rate constant (k_u). The unfolding and folding rate constants can be obtained individually if the equilibrium constant of this two-state reaction at the final temperature is known since $K_{eq} = k_f/k_u$.

The first fast T-jump method, pioneered by Eigen and De Maeyer, utilized rapid capacitance discharge induced Joule heating to increase the temperature of a conducting solution on the microsecond timescale [37]. To achieve an even faster T-jump, pulsed lasers are usually used to provide a burst of energy to heat up the sample solution within a very short period of time, e.g. nanoseconds. In aqueous solution, this can be achieved by exciting water's near-IR overtone transitions in the vicinity of either 1.5 or 2 μm [38]. The absorbed laser energy is quickly dissipated and equilibrated within the laser interaction volume, leading to a T-jump whose amplitude depends on the pump volume and the pulse energy, while the dead time is typically determined by the pump pulse width since vibrational relaxation takes place on the picosecond timescale [39]. In addition, the useful time window of this technique in which the temperature remains approximately constant is governed by the time it takes for the heat to diffuse out of the laser interaction volume, ranging from a few to tens of milliseconds, depending on the setup.

Fast T-jump initiation can be combined with a variety of time-resolved spectroscopic methods, such as fluorescence [40], circular dichroism [41–43] and vibrational spectroscopy [44–48], to study protein-folding dynamics. IR spectroscopy has been proven to be versatile in protein conformational studies. For example, both the vibrations from the backbone amide groups and amino acid side-chains have been extensively used to study protein conformational changes [47, 48]. The backbone amide vibrations are valuable markers of protein secondary structures; in contrast, the amino acid side-chain vibrations are suitable reporters of local environment [48, 49]. Among the eight amide vibrational modes, the amide I/I' band [44, 50], which arises mostly from the amide C=O stretching vibration, is the most intense and best characterized [44, 50]. The amide I/I' band of proteins or peptides has long been used as a global conformational reporter. The exact shape and position of the amide I/I' band of a protein are determined by many interactions, including through-bond coupling, hydrogen bonding, transition dipole coupling (TDC), etc. [47, 48].

IR studies of proteins in aqueous solution are complicated by the fact that the H-O-H bending vibrational mode of H_2O absorbs very strongly, which overlaps

9.1 Laser-induced T-Jump Method: A Non-conventional Photoreleasing Approach

with the amide I band. D_2O, on the other hand, has a relatively low absorbance between 1400 and 1800 cm^{-1} and is therefore commonly used in IR studies [47] (in D_2O the amide I band is referred as amide I'). In a T-jump IR experiment, however, the T-jump pulses still induce a change in the D_2O absorbance. This solvent signal should be appropriately measured and subtracted from that measured with the protein sample.

Moreover, the amide bands of proteins are broadened inhomogeneously and homogeneously. As a result, it is often difficult to quantitatively interpret amide I' bands due to the fact that they are almost invariably congested with components arising from different structural ensembles. To obtain site-specific structural information, a common technique is to employ isotope editing, where several amide $^{12}C=Os$ are substituted with $^{13}C=Os$. Since vibrational transitions are sensitive to isotopic substitution, the amide I' absorbance of the ^{13}C-labeled carbonyls shifts to lower frequency, therefore (sometimes) permitting site-specific studies [51].

The T-jump IR apparatus used in our studies has been described in details previously [28] (Fig. 9.1.1). Briefly, the 1.9-µm T-jump pulse (3 ns and 10 Hz) is generated via Raman shifting the fundamental output of an Nd:YAG laser in a Raman cell that is pressurized with a mixture of H_2 and Ar to 750 p.s.i. The presence of an inert gas, such as Ar, greatly reduces the thermal effects and therefore stabilizes the 1.9-µm output. A stable pump source is rather critical to T-jump coupled measurements because the observed kinetics are usually temperature dependent. A CW lead salts IR diode laser serves as the probe which is tunable from 1550 to 1800 cm^{-1}. Transient absorbance changes of the probe induced by the T-jump pulses are detected by a 50-MHz MCT detector. Digitization of the signal is accomplished by a digital oscilloscope. To ensure a uniform

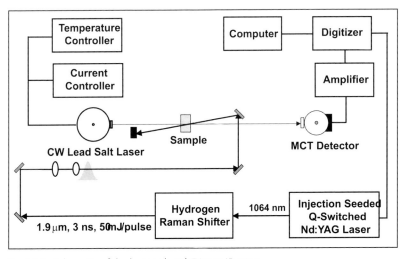

Fig. 9.1.1 Schematic of the laser-induced T-jump IR setup.

T-jump distribution within the laser interaction volume, a thin optical path length, typically 52 μm, is used. Furthermore, to provide information for both background subtraction and T-jump amplitude calibration in the time-resolved measurements, a sample cell with dual compartments is used to measure the T-jump induced signals of both the sample and reference under the same conditions. Calibration of the T-jump amplitude is done by using the T-jump induced absorbance change of the buffer (D_2O) at the probing frequency v, $\Delta A(\Delta T, v)$ and the following equation: $\Delta A(\Delta T, v) = a(v)*\Delta T + b(v)*\Delta T^2$, where ΔA is the absorbance change, $\Delta T = T_f - T_i$ is the T-jump amplitude; T_f and T_i correspond to the final and initial temperatures, respectively; and $a(v)$ and $b(v)$ are constants that are determined by the steady-state FTIR spectra of D_2O measured at different temperatures.

9.1.3
Helix–Coil Transition Kinetics

Helices are ubiquitous in proteins. As such, the helix–coil transition has been studied extensively. Seminal works of Schellman, Zimm and Bragg, Lifson and Roig and Poland and Sheraga form the basis of our understanding of the thermodynamics of monomeric helices [52–55]. The classical helix–coil transition theory describes the mechanism of helix formation as sequential events, starting from a so-called nucleation step where the first helical hydrogen bond, formed between the amide carbonyl of residue i and the amide hydrogen of residue $i + 4$, is generated. The subsequent steps involve the elongation of the pre-existing helical structures by adding an extra hydrogen bond at the end. Because the nucleation step is entropically unfavorable, it has been suggested that it encounters the largest free energy barrier during the course of helix formation. Although the helix–coil transition represents the simplest scenario in protein folding, only recently were its kinetics studied in detail by rapid T-jump methods [26, 27, 56–58].

Recent studies on the kinetics of the helix–coil transition using either computer simulation [59–64] or laser-induced T-jump method [26–28, 56–58] provide new insights into our understanding of the kinetic aspects of helix formation. Williams et al. [26] measured a relaxation time of 160 ns for the Fs peptide [sequence: $(A)_5$-$(AAARA)_3$-A-NH_2] following a T-jump from 9.3 to 27.4 °C, by monitoring the amide I' IR absorbance of the peptide backbone. This result establishes the fact that the helix–coil transition is a submicrosecond event. Using a fluorescent probe (MABA) attached at the N-terminus of the Fs peptide, however, Thompson et al. [27] observed a much faster relaxation process (~ 20 ns near the thermal melting temperature, 303 K). They attributed this relaxation to rapid fraying of the helix ends in response to the T-jump. Using a "kinetic zipper" model, they have also calculated the decay of the average helix content and their results indicate that a slower rate should account for the transition between the helix-containing and non-helix-containing structural ensembles, due to the energy barrier associated with the nucleation process. Subsequently, these authors [56] reported the observation of a slower relaxation process (220 ns at 300 K) in a 21-residue helical peptide, Ac-WAAAH$^+$(AAAR$^+$A)$_3$A-NH$_2$, with a tryptophan residue in position 1 to serve as the fluorescent probe. This relaxation is temperature-dependent and has an apparent activation energy of $\sim$ 8 kcal mol^{-1}. Using UV resonance Raman as a probe, Lednev et al. [57, 58] also observed a single exponential relaxation process that is weakly temperature dependent for the Fs peptide, following a 3-ns T-jump pulse [65]. Using TFE as a cosolvent, Dyer et al. have been able to study the folding kinetics of a helical peptide from its cold denatured states [66].

Using the T-jump IR method, we have also studied the temperature dependence of the helix–coil transition rate of an alanine-based 21-residue helical peptide and observed a relatively large activation energy for the relaxation, $\sim$ 15.5 kcal mol^{-1} [67]. Presumably, part of the observed energetic barrier comes from frictions exerted by solvent molecules, as indicated in the study of Eaton et al. [68]. Furthermore, we have studied [28] in detail the helix–coil transition kinetics of another alanine-based peptide using an isotope-editing strategy introduced by Decatur et al. [51]. The labeled peptides contain a block of amide ^{13}C=Os as indicated by underlines:

NL peptide: Ac-YGSPE<u>AAA</u>KA$_4$KA$_4$r-CONH$_2$
ML peptide: Ac-YGSPEA$_3$K<u>AAAA</u>KA$_4$r-CONH$_2$
CL peptide: Ac-YGSPEA$_3$KA$_4$K<u>AAAA</u>r-CONH$_2$

where r represents D-arginine. As shown (Fig. 9.1.2), the red-shifted amide I′ absorbance of the ^{13}C-labeled carbonyls indeed provides site-specific structural information. In addition, we have also demonstrated [69] that the T-jump-induced relaxations exhibit non-exponential behaviors, and are sensitive to both initial and final temperatures (Fig. 9.1.3). The non-exponential folding behavior of helices has also been observed in computer simulations [70, 71]. Moreover, these results provide strong evidence supporting a picture that the nucleation process is fast, on the nanosecond or subnanosecond timescale, and the helix formation process can be described by a diffusion search within the coil states, as suggested by theoretical studies and molecular dynamics (MD) simulations [63, 70, 72].

Although much less studied, sequence variation is another factor that would have strong effects on the kinetics of helix formation. For example, it has long been recognized that helix-capping sequences provide very significant thermodynamic stability to the a-helix [73–78], but their role in the kinetics of folding is not yet known. The classical view would be that the initiation of a single turn of helix occurs in a more or less sequence-independent manner [79]; hence the role of end-caps is to lock in pre-existing helices. However, it is also possible that capping sequences can effectively serve as helix initiators. These two opposing views lead to different experimental results. If the capping is not involved in initiation, the rate of helix initiation should not be much affected for peptides with different end-caps. On the other hand, if capping interactions are able to initiate helix formation then the rates of initiation of the helices should be very profoundly affected. The peptide length is another factor that may also pro-

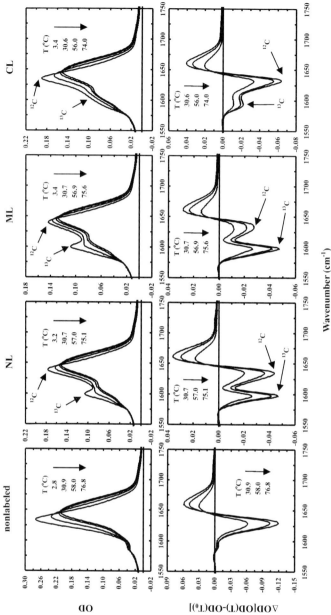

Fig. 9.1.2 Temperature-dependent equilibrium (top traces) and difference (bottom traces) FTIR spectra of the labeled and non-labeled peptides, as indicated in the plot. Difference spectra were generated by subtracting the spectrum collected at the lowest temperature from the spectra collected at higher temperatures. The bands at ~ 1600 and ~ 1636 cm^{-1} are assigned to the amide I' absorbance of the ^{13}C-labeled and non-labeled residues, respectively. (From [69])

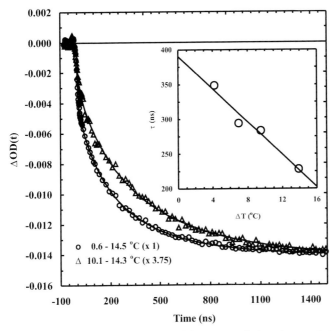

Fig. 9.1.3 T-jump induced relaxations of the ML peptide probed at 1600 cm^{-1}. The T-jump is from different initial temperatures to the same final temperature, as indicated in the plot. The smooth lines are fits to a stretched exponential function. Note that the signals have been scaled to reflect the difference between relaxations corresponding to different T-jump amplitudes. Inset: T-jump amplitude-dependent relaxation time constants. The solid line is the linear regression to the data points. (From [69])

foundly affect the kinetics of the helix–coil transition, as indicated by a number of theoretical studies [59, 80]. If the kinetically and thermodynamically difficult step in the process of the coil–helix transition is initiation of a single turn of helix, as one proceeds in peptide length from short to long the rate of helix formation should not be much affected. Only a small increase would be observed due to a linear, length-dependent increase in the probability of initiation as the peptide is lengthened.

To understand how peptide sequence affects the kinetics of the helix–coil transition, we have studied [81] a series of alanine-based peptides of different lengths, with either excellent or poor end-capping groups:

AKA$_n$ peptides: YGAKAAAA(KAAAA)$_n$G, $n=2–5$
SPE$_n$ peptides: YGSPEAAA(KAAAA)$_n$r, $n=2–5$

Here, r also represents D-arginine and is one of the best C-caps [82] and the Ser-Pro-Glu tripeptide (SPE) serves as a helix-stabilizing N-cap. Our results show definitively that good end-capping sequences can increase the rate of helix formation, whereas the peptide length has a rather complicated effect on the kinetics of the helix–coil transition (Fig. 9.1.4).

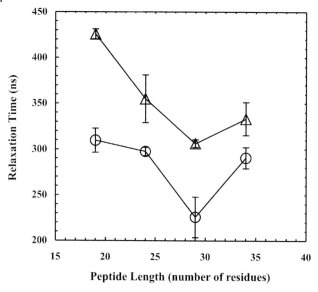

Fig. 9.1.4 Relaxation times of AKA$_n$ (△) and SPE$_n$ (○) peptides following a T-jump from 1 to 11 °C. (From [81])

We found that the relaxation rates for peptides with efficient end-caps are more rapid than those of the corresponding peptides without good end-caps. These results indicate that efficient end-capping sequences cannot only stabilize pre-existing helices, but also promote helix formation through initiation. Furthermore, we found that the relaxation times of these peptides following a T-jump of 1–11 °C show rather complex behaviors as a function of the peptide length, in disagreement with theoretical predictions. These results are not readily explained by theories in which alanine is taken to have a single helical propensity. However, if one were to assume that the propagation constant, s, depends on chain length [83, 84], the mean first-passage times of the coil–helix transition obtained, based on the model developed by Straub et al. [80], show similar dependence on the peptide length as those observed experimentally [81].

9.1.4
The Folding Mechanisms of β-Hairpins

Short peptides that fold into well-defined structures in aqueous solution provide ideal model systems for the study of fundamental questions in protein folding in detail [85]. Of particular interest are β-hairpins. With two short antiparallel β-strands that are connected by a turn or loop, β-hairpin structure can be viewed as the smallest β-sheet unit. Despite their small size (normally 12–16 residues), however, many β-hairpins possess properties that are typical to globular proteins. For example, the most extensively studied 16-residue β-hairpin, which is derived from the C-terminal fragment (41–56) of protein GB1, contains a hydrophobic "core" and also shows cooperative thermal folding-unfolding transition [86–88]. Therefore, the β-hairpin motif has been increasingly used to probe factors that govern the conformational stability as well as the folding mechanism of β-sheets [89–91]. Moreover, there is increasing evidence suggesting that a β-hairpin or β-turn could act as the nucleation site in the early stages of folding [92–96]. Thus, understanding how β-hairpins fold can also shed light on the mechanism of protein folding in general. Theoretical and computational studies [87, 97–100] have indeed revealed rich features associated with the folding energetics and mechanisms of β-hairpins. Several factors, including the rigidity of the turn and the relative position of the hydrophobic cluster, may have rather complex effects on the rate of β-hairpin folding.

The first experimental study of β-hairpin folding kinetics in the microsecond timescale was carried out by Muñoz et al. [86], who measured the folding time of the 16-residue β-hairpin derived from protein GB1 using a T-jump fluorescence technique. Their results indicate that this β-hairpin folds in a two-state manner with a folding time constant of 6 μs at 297 K. Subsequently, these authors developed a statistical zipper model [98] to explain the apparent two-state folding behavior of this β-hairpin. This model essentially suggests that the most probable first step in the formation of a β-hairpin begins with the formation of the β-turn, which is followed by the formation of interstrand hydrogen bonds. The transition state can only be stabilized when hydrophobic contacts among side-chains start to form.

We have studied the thermodynamics and folding kinetics of a different β-hairpin peptide with 15 residues using the T-jump IR method [101]. This peptide was originally described by Jiménez et al. [102], who have shown by NMR spectroscopy that this peptide, which has a sequence of SESYINPDGTWTVTE, folds into a β-hairpin structure in aqueous solution with a type I+G1 β-bulge turn involving residues Asn6, Pro7, Asp8 and Gly9 [102, 103]. Furthermore, not only does this peptide yield appreciable β-hairpin population at room temperature [102], but it can also reach millimolar concentration in aqueous solution without forming detectable aggregates. The latter is important for IR studies because IR techniques normally require a relatively high peptide concentration.

Equilibrium IR measurements indicate that the thermal unfolding of this β-hairpin is fairly broad. However, it can be described by a two-state transition with a thermal melting temperature of $\sim 29\,°C$ [101]. Time-resolved IR measurements following a T-jump, probed at 1634 cm^{-1}, indicate that the folding of this β-hairpin follows first-order kinetics and is surprisingly fast. At 300 K, the folding time is approximately 0.8 μs, which is only 2–3 times slower than that of α-helix formation (Fig. 9.1.5). Additionally, the energetic barrier for folding is small (~ 2 kcal mol^{-1}). These results, in conjunction with results from other studies, support a view that the details of native contacts play a dominant role in the kinetics of β-hairpin folding. Moreover, the small enthalpic barrier observed for folding of this peptide as well as other β-hairpins also indicate that folding free energy barrier is dominated by entropic contribution [101]. This is also consistent with the picture that the search for native structure within an

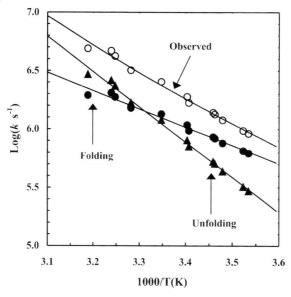

Fig. 9.1.5 Arrhenius plots of the observed (○), folding (●) and unfolding (▲) rate constants of β-hairpin SESYINPDGTWTVTE. The extrapolated folding time at 300 K is ~ 0.76 μs. (From [101])

ensemble of collapsed conformations dominates the rate of β-hairpin formation, as indicated by computer simulations [99, 104]. A recent study by Dyer et al. [105] on the folding kinetics of a series of cyclic β-hairpin peptides also supports such a folding mechanism.

9.1.5
Ultrafast Folding of Helical Bundles

Many simple, single-domain proteins fold via first order kinetics as the result of a single free energy barrier. Similar to gas phase chemical dynamics, the folding of two-state proteins may be described by transition state theory. Recently, there has been growing interest in studying/identifying proteins that are capable of folding on ultrafast timescales, e.g. microseconds [106–112]. The goal is to determine the speed limit of protein folding and also provide model systems for computer simulations [113].

9.1.5.1 Folding Kinetics of 1prb$_{7-53}$

1prb$_{7-53}$, the GA module of an albumin-binding domain [114], adopts a simple three-helix bundle structure in solution. Because of its small size (47 residues) and simple topology, 1prb$_{7-53}$ is considered to be a good candidate for exploring fast folding. In fact, a recent simulation study carried out by Takada [115], who employed a coarse-grained protein model that takes into account solvent effects, suggests that this protein can fold within ~ 1 μs from its unfolded state. Our T-jump IR experiments confirmed the fast folding nature of 1prb$_{7-53}$ and revealed that its minimum folding time constant is ~ 6 μs [116]. This property makes 1prb$_{7-53}$ one of the fastest folding proteins known to date. The folding of 1prb$_{7-53}$ brings side-chains into tertiary contacts that are separated by up to 30 residues in sequence. Two residues separated by this distance in sequence would be expected to collide on the 0.3–1 μs timescale [117], yet one of its double mutants, i.e. K5I/K39V, achieves the fully folded state on the same timescale. This suggests that the folding energy landscape of K5I/K39V is probably very smooth and the free energy barrier for folding is small [118].

9.1.5.2 Folding Kinetics of a_3D

a_3D, a *de novo* designed three-helix bundle protein with 73 residues [119], is another excellent system for probing the maximal rate of folding in a small globular protein. It was designed to be stabilized only by the packing of hydrophobic side-chains and it lacks buried polar residues or structured loops that might introduce significant kinetic barriers to folding [120]. This protein adopts a well-defined tertiary structure in which most of its solvent-inaccessible side-chains adopt predominantly a single conformation [121]. Also, the observed rates of amide hydrogen–deuterium exchange and the thermodynamic parameters for folding are within the range expected for a protein of its size, indicating that its core is relatively well packed [122]. We used T-jump IR as well as an ultrafast fluorescence mixing technique [14] to monitor the folding of this protein and found that both methods reveal a single-exponential process consistent with a minimal folding time of 3.2±1.2 μs (at ~ 50 °C) (Fig. 9.1.6). We also performed MD simulations of unfolding [123–127], which, combined with IR spectroscopy, provide a molecular rationale for the rapid, single exponential folding of this protein. a_3D shows a significant bias towards local helical structure in the thermally denatured state and the MD-simulated transition state (TS) ensemble is highly heterogeneous and dynamic, allowing access to the TS via multiple pathways [112].

Taken together, these studies unambiguously demonstrate that a protein can fold into a fully native conformation on the 1–5-μs timescale. Furthermore, the single exponential nature of the relaxation indicates that the pre-factor for TS folding models is probably $\geq 1\ \mu s^{-1}$ for a protein of similar size and topology [128–130].

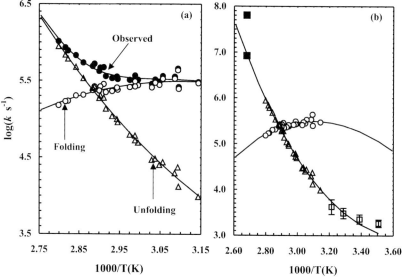

Fig. 9.1.6 (a) Arrhenius plots of the observed (●), folding (○) and unfolding (△) rate constants of $\alpha_3 D$. Lines are fits to the Eyring equation, i.e. $\ln(k) = \ln(D) - \Delta G^{\neq}/RT$, where D is a constant and $\Delta G^{\neq}$ is the free energy of activation that is a function of temperature, as described by the following function: $\Delta G^{\neq} = \Delta H^{\neq}(T_m) + \Delta C_p^{\neq}(T - T_m) - T[\Delta S^{\neq}(T_m) + \Delta C_p^{\neq} \ln(T/T_m)]$, where T_m is the thermal melting temperature, 73.2 °C. Global fitting of the folding and unfolding kinetics yields the following thermodynamic parameters of activation when $D = 10^6$ s^{-1} is used: (1) for folding, $\Delta H^{\neq}(T_m) = -6.1$ kcal mol^{-1}, $\Delta S^{\neq}(T_m) = -20.5$ cal mol^{-1} K^{-1}, $\Delta C_p^{\neq} = -255$ cal mol^{-1} K^{-1}; (2) for unfolding, $\Delta H^{\neq}(T_m) = 29.8$ kcal mol^{-1}, $\Delta S^{\neq}(T_m) = 83.1$ cal mol^{-1} K^{-1}, $\Delta C_p^{\neq} = 392$ cal mol^{-1} K^{-1}. (b) Same as (a) but with an extended x-axis. For comparison, unfolding rate constants (□) measured in mixing experiments are also shown. These data show that both folding and unfolding rates exhibit non-linear temperature dependence. The folding time at 25 °C is predicted to be ~ 4.8 µs according to the fit in (a) and the unfolding time at 100 °C is predicted to be ~ 100 ns. The times required to reach the transition state in two independent simulations at 100 °C are shown as filled squares (■). (From [112].)

9.1.6
Summary and Conclusions

The problem of protein folding is difficult to study. Relevant folding or conformational processes can take place on very different timescales. Using the stopped-flow initiation technique in conjunction with site-directed mutagenesis, hydrogen–deuterium exchange and spectroscopy, it has been possible to understand the sequence of events involved in folding/unfolding on slow timescales (e.g. milliseconds). To study the early folding events, however, fast initiation methods are required. Recently, many elegant photo and photochemical triggering methods have been developed to serve this purpose. In the current chapter, we have focused on the discussion of the laser-induced T-jump IR method and its application in several folding studies. Although the past few years have seen great advances in our understanding of some fundamental aspects of the dynamics of protein folding, many questions remained unanswered. For example, the nucleation step in helix formation and the turn formation in β-hairpin folding have not been observed directly. This is due partly to the lack of appropriate initiation methods. Further developments of new phototriggering methods should make the study of these problems possible.

Acknowledgments

We wish to thank our colleagues in the Gai research group for their contributions to this work, particularly Cheng-Yen Huang, Yao Xu and Rolando Oyola. We also wish to thank our collaborators for their contributions, particularly Drs. W. F. DeGrado, H. Roder, J. G. Saven and Z. Getahun.

References

1 H. A. SCHERAGA, *Chem. Rev.* **1971**, *71*, 195–217.
2 C. B. ANFINSEN, H. A. SCHERAGA, *Adv. Protein Chem.* **1975**, *29*, 205–300.
3 R. L. JERNIGAN, *Curr. Opin. Struct. Biol.* **1992**, *2*, 248–256.
4 P. G. WOLYNES, Z. LUTHEY-SCHULTEN, J. N. ONUCHIC, *Chem. Biol.* **1996**, *3*, 425–432.
5 C. M. DOBSON, A. SALI, M. KARPLUS, *Angew. Chem. Int. Ed.* **1998**, *37*, 869–893.
6 C. M. DOBSON, M. KARPLUS, *Curr. Opin. Struct. Biol.* **1999**, *9*, 92–101.
7 B. HONIG, *J. Mol. Biol.* **1999**, *293*, 283–293.
8 R. L. BALDWIN, *Nat. Struct. Biol.* **1999**, *6*, 814–817.
9 J. W. H. SCHYMKOWITZ, F. ROUSSEAU, L. SERRANO, *Proc. Natl Acad. Sci. USA* **2002**, *99*, 15846–15848.
10 E. A. EATON, V. MUÑOZ, P. A. THOMPSON, E. R. HENRY, J. HOFRICHTER, *Acc. Chem. Res.* **1998**, *31*, 745–753.
11 O. BIERI, T. KIEFHABER, *Biol. Chem.* **1999**, *380*, 923–929.
12 E. A. EATON, V. MUÑOZ, S. J. HAGEN, G. S. JAS, L. J. LAPIDUS, E. R. HENRY, *Annu. Rev. Biophys. Biomol. Struct.* **2000**, *29*, 327–359.
13 C. K. CHAN, Y. HU, S. TAKAHASHI, D. L. ROUSSEAU, E. A. EATON, J. HOFRICHTER, *Proc. Natl Acad. Sci. USA* **1997**, *94*, 1779–1784.
14 M. C. R. SHASTRY, S. D. LUCK, H. RODER, *Biophys. J.* **1998**, *74*, 2714–2721.
15 M. VOLK, Y. KHOLODENKO, H. S. M. LU, E. A. GOODING, W. F. DEGRADO, R. M. HOCHSTRASSER, *J. Phys. Chem.* **1997**, *101*, 8607–8616.
16 P. WITTUNG-STAFSHEDE, J. C. LEE, J. R. WINKLER, H. B. GRAY, *Proc. Natl Acad. Sci. USA* **1999**, *96*, 6587–6590.
17 K. C. HANSEN, R. S. ROCK, R. W. LARSEN, S. I. CHAN, *J. Am. Chem. Soc.* **2000**, *122*, 11567–11568.
18 J. R. KUMITA, O. S. SMART, G. A. WOOLLEY, *Proc. Natl Acad. Sci. USA* **2000**, *97*, 3803–3808.
19 C.-Y. HUANG, S. HE, W. F. DEGRADO, D. G. MCCAFFERTY, F. GAI, *J. Am. Chem. Soc.* **2002**, *124*, 12674–12675.
20 C. M. JONES, E. R. HENRY, Y. HU, C. K. CHAN, S. D. LUCK, A. BHUYAN, H. RODER, J. HOFRICHTER, E. A. EATON, *Proc. Natl Acad. Sci. USA* **1993**, *90*, 11860–11864.
21 S. ABBRUZZETTI, E. CREMA, L. MASINO, A. VECLI, C. VIAPPIANI, J. R. SMALL, L. J. LIBERTINI, E. W. SMALL, *Biophys. J.* **2000**, *78*, 405–415.
22 A. BARTH, J. E. T. CORRIE, *Biophys. J.* **2002**, *83*, 2864–2871.

23 G. Desai, G. Panick, M. Zein, R. Winter, C. A. Royer, *J. Mol. Biol.* **1999**, *288*, 461–475.

24 J. Woenckhaus, R. Kohling, P. Thiyagarajan, K. C. Littrell, S. Seifert, C. A. Royer, R. Winter, *Biophys. J.* **2001**, *80*, 1518–1523.

25 R. Kitahara, C. Royer, H. Yamada, M. Boyer, J.-L. Saldana, K. Akasaka, C. Roumestand, *J. Mol. Biol.* **2002**, *320*, 609–628.

26 H. Herberhold, R. Winter, *Biochemistry* **2002**, *41*, 2396–2401.

27 S. Williams, T. P. Causgrove, R. Gilmanshin, K. S. Fang, R. H. Callender, W. H. Woodruff, R. B. Dyer, *Biochemistry* **1996**, *35*, 691–697.

28 P. A. Thompson, E. A. Eaton, J. Hofrichter, *Biochemistry* **1997**, *36*, 9200–9210.

29 C.-Y. Huang, J. W. Klemke, Z. Getahun, W. F. DeGrado, F. Gai, *J. Am. Chem. Soc.* **2001**, *123*, 9235–9238.

30 T. Pascher, J. P. Chesick, J. R. Winkler, H. B. Gray, *Science* **1996**, *271*, 1558–1560.

31 D. H. Oh, S. G. Boxer, *J. Am. Chem. Soc.* **1989**, *111*, 1130–1131.

32 H. S. M. Lu, M. Volk, Y. Kholodenko, E. Gooding, R. M. Hochstrasser, W. F. DeGrado, *J. Am. Chem. Soc.* **1997**, *119*, 7173–7180.

33 J. Bredenbeck, J. Helbing, A. Sieg, T. Schrader, W. Zinth, R. Renner, R. Behrendt, L. Moroder, J. Wachtveitl, P. Hamm, *Proc. Natl Acad. Sci. USA* **2003**, *100*, 6452–6457.

34 J. Hofrichter, *Methods Mol. Biol.* **2001**, *168*, 159–191.

35 R. Callender, R. B. Dyer, *Curr. Opin. Struct. Biol.* **2002**, *12*, 628–633.

36 M. Gruebele, J. Sabelko, R. Ballew, J. Ervin, *Acc. Chem. Res.* **1998**, *31*, 699–707.

37 M. Eigen, L. D. De Maeyer, in *Techniques of Organic Chemistry*, S. L. Friess, E. S. Lewis, A. Weissberger (eds), Interscience, New York, **1963**.

38 K. F. Palmer, D. Williams, *J. Opt. Soc. Am.* **1974**, *64*, 1107–1110.

39 J. C. Owrutsky, D. Raftery, R. M. Hochstrasser, *Annu. Rev. Phys. Chem.* **1994**, *45*, 519–555.

40 J. M. Beechem, L. Brand, *Annu. Rev. Biochem.* **1985**, *54*, 43–71.

41 A. J. Adler, N. J. Greenfield, G. D. Fasman, *Methods Enzymol.* **1973**, *27*, 675–735.

42 C. F. Zhang, J. W. Lewis, R. Cerpa, I. D. Kuntz, D. S. Kliger, *J. Phys. Chem.* **1993**, *97*, 5499–5505.

43 Y. X. Wen, E. Chen, J. W. Lewis, D. S. Kliger, *Rev. Sci. Instrum.* **1996**, *67*, 3010–3016.

44 S. Krimm, J. Bandekar, *Adv. Protein Chem.* **1986**, *38*, 181–364.

45 T. Kitagawa, S. Hirota, in *Handbook of Vibrational Spectroscopy*, J. M. Chalmers, P. R. Griffiths (eds), Wiley, New York, **2002**.

46 T. G. Spiro, R. S. Czernuszewicz, *Methods Enzymol.* **1995**, *246*, 416–460.

47 H. Fabian, W. Mantele, in *Handbook of Vibrational Spectroscopy*, J. M. Chalmers, P. R. Griffiths (eds), Wiley, New York, **2002**.

48 A. Barth, C. Zscherp, *Q. Rev. Biophys.* **2002**, *35*, 369–430.

49 Z. Getahun, C.-Y. Huang, T. Wang, B. De Leon, W. F. DeGrado, F. Gai, *J. Am. Chem. Soc.* **2003**, *125*, 405–411.

50 H. Susi, *Methods Enzymol.* **1972**, *26*, 455–472.

51 S. M. Decatur, J. Antonic, *J. Am. Chem. Soc.* **1999**, *121*, 11914–11915.

52 J. A. Schellman, *J. Phys. Chem.* **1958**, *62*, 1485–1494.

53 B. H. Zimm, J. K. Bragg, *J. Chem. Phys.* **1959**, *31*, 526–535.

54 S. Lifson, A. Roig, *J. Chem. Phys.* **1961**, *34*, 1963–1974.

55 D. Poland, H. A. Scheraga, *Theory of Helix–Coil Transition in Biopolymers*, Academic Press, New York, **1970**.

56 P. A. Thompson, V. Muñoz, G. S. Jas, E. R. Henry, E. A. Eaton, J. Hofrichter, *J. Phys. Chem.* **2000**, *104*, 378–389.

57 I. K. Lednev, A. S. Karnoup, M. C. Sparrow, S. A. Asher, *J. Am. Chem. Soc.* **1999**, *121*, 4076–4077.

58 I. K. Lednev, A. S. Karnoup, M. C. Sparrow, S. A. Asher, *J. Am. Chem. Soc.* **1999**, *121*, 8074–8086.

59 C. L. Brooks, *J. Phys. Chem.* **1996**, *100*, 2546–2549.

60 M. Schaefer, C. Bartels, M. Karplus, *J. Mol. Biol.* **1998**, *284*, 835–848.

61 Y. Duan, P. A. Kollman, *Science* **1998**, *282*, 740–744.

62 S. Takada, Z. Luthey-Schulten, P. G. Wolynes, *J. Chem. Phys.* **1999**, *110*, 11616–11629.
63 G. Hummer, A. E. Garcia, S. Grade, *Proteins* **2001**, *42*, 77–84.
64 J. Schimada, E. L. Kussell, E. I. Shakhnovich, *J. Mol. Biol.* **2001**, *308*, 79–95.
65 I. Nishii, M. Kataoka, F. Tokunaga, Y. Goto, *Biochemistry* **1994**, *33*, 4903–4909.
66 J. H. Werner, R. B. Dyer, R. M. Fesinmeyer, N. H. Andersen, *J. Phys. Chem.* **2002**, *106*, 487–494.
67 C.-Y. Huang, Z. Getahun, T. Wang, W. F. DeGrado, F. Gai, *J. Am. Chem. Soc.* **2001**, *123*, 12111–12112.
68 G. S. Jas, E. A. Eaton, J. Hofrichter, *J. Phys. Chem.* **2001**, *105*, 261–272.
69 C.-Y. Huang, Z. Getahun, Y. Zhu, J. W. Klemke, W. F. DeGrado, F. Gai, *Proc. Natl Acad. Sci. USA* **2002**, *99*, 2788–2793.
70 G. Hummer, A. E. Garcia, S. Grade, *Phys. Rev. Lett.* **2000**, *85*, 2637–2640.
71 S. Elmer, V. S. Pande, *J. Phys. Chem.* **2001**, *105*, 482–485.
72 J. D. Bryngelson, J. N. Onuchic, P. G. Wolynes, *Proteins* **1995**, *21*, 167–195.
73 L. G. Presta, G. D. Rose, *Science* **1988**, *240*, 1632–1641.
74 J. S. Richardson, D. C. Richardson, *Science* **1988**, *240*, 1648–1652.
75 L. Regan, *Proc. Natl Acad. Sci. USA* **1993**, *90*, 10907–10908.
76 R. Aurora, T. P. Creamer, R. Srinivasan, G. D. Rose, *J. Biol. Chem.* **1997**, *272*, 1413–1416.
77 R. Aurora, G. D. Rose, *Protein Sci.* **1998**, *7*, 21–38.
78 S. Dasgupta, J. A. Bell, *Int. J. Peptide Protein Res.* **1993**, *41*, 499–511.
79 J. M. Scholtz, R. L. Baldwin, *Peptides: Synthesis, Structures and Applications*, Academic Press, New York, **1995**.
80 N.-V. Buchete, J. E. Straub, *J. Phys. Chem.* **2001**, *105*, 6684–6697.
81 T. Wang, Y. Zhu, Z. Getahun, D. Du, C.-Y. Huang, W. F. DeGrado, F. Gai, *J. Phys. Chem.* **2004**, *108*, 15301–15310.
82 J. P. Schneider, W. F. DeGrado, *J. Am. Chem. Soc.* **1998**, *120*, 2764–2767.
83 R. J. Kennedy, K.-T. Tsang, D. S. Kemp, *J. Am. Chem. Soc.* **2002**, *124*, 934–944.
84 Y. Z. Ohkubo, C. L. Brooks, *Proc. Natl Acad. Sci. USA* **2003**, *100*, 13916–13921.
85 N. Ferguson, A. R. Fersht, *Curr. Opin. Struct. Biol.* **2003**, *13*, 75–81.
86 V. Muñoz, P. A. Thompson, J. Hofrichter, E. A. Eaton, *Nature* **1997**, *390*, 196–199.
87 V. S. Pande, D. S. Rokhsar, *Proc. Natl Acad. Sci. USA* **1999**, *96*, 9062–9067.
88 S. Honda, N. Kobayashi, E. Munekata, *J. Mol. Biol.* **2000**, *295*, 269–278.
89 L. Serrano, *Adv. Protein Chem.* **2000**, *53*, 49–85.
90 M. S. Searle, *J. Chem. Soc., Perkin Trans.* **2001**, *2*, 1011–1020.
91 J. F. Espinosa, F. A. Syud, S. H. Gellman, *Protein Sci.* **2002**, *11*, 1492–1505.
92 Z. Y. Guo, D. Thirumalai, *Biopolymers* **1995**, *36*, 83–102.
93 M. Gruebele, P. G. Wolynes, *Nat. Struct. Biol.* **1998**, *5*, 662–665.
94 V. P. Grantcharova, D. S. Riddle, J. V. Santiago, D. Baker, *Nat. Struct. Biol.* **1998**, *5*, 714–720.
95 J. C. Martinez, M. T. Pisabarro, L. Serrano, *Nat. Struct. Biol.* **1998**, *5*, 721–729.
96 W. F. Walkenhorst, J. A. Edwards, J. L. Markley, H. Roder, *Protein Sci.* **2002**, *11*, 82–91.
97 A. M. J. J. Bonvin, W. F. van Gunsteren, *J. Mol. Biol.* **2000**, *296*, 255–268.
98 V. Muñoz, E. R. Henry, J. Hofrichter, E. A. Eaton, *Proc. Natl Acad. Sci. USA* **1998**, *95*, 5872–5879.
99 A. R. Dinner, T. Lazaridis, M. Karplus, *Proc. Natl Acad. Sci. USA* **1999**, *96*, 9068–9073.
100 D. K. Klimov, D. Thirumalai, *Proc. Natl Acad. Sci. USA* **2000**, *97*, 2544–2549.
101 Y. Xu, R. Oyola, F. Gai, *J. Am. Chem. Soc.* **2003**, *125*, 15388–15394.
102 C. M. Santiveri, J. Santoro, M. Rico, M. A. Jiménez, *J. Am. Chem. Soc.* **2002**, *124*, 14903–14909.
103 E. de Alba, M. A. Jiménez, M. Rico, J. L. Nieto, *Folding Des.* **1996**, *1*, 133–144.
104 C. D. Snow, L. Qiu, D. Du, F. Gai, S. J. Hagen, V. S. Pande, *Proc. Natl Acad. Sci. USA* **2004**, *101*, 4077–4082.
105 S. J. Maness, S. Franzen, A. C. Gibbs, T. P. Causgrove, R. B. Dyer, *Biophys. J.* **2003**, *84*, 3874–3882.
106 U. Mayor, N. R. Guydosh, C. M. Johnson, J. G. Grossmann, S. Sato, G. S. Jas, S. M. Freund, D. O. V. Alonso, V.

Daggett, A. R. Fersht, *Nature* **2003**, *421*, 863–867.
107 W. Y. Yang, M. Gruebele, *Nature* **2003**, *423*, 193–197.
108 B. Gillespie, D. M. Vu, P. S. Shah, S. A. Marshall, R. B. Dyer, S. L. Mayo, K. W. Plaxco, *J. Mol. Biol.* **2003**, *330*, 813–819.
109 L. Qiu, S. A. Pabit, A. E. Roitberg, S. J. Hagen, *J. Am. Chem. Soc.* **2002**, *124*, 12952–12953.
110 J. W. Neidigh, R. M. Fesinmeyer, N. H. Andersen, *Nat. Struct. Biol.* **2002**, *9*, 425–430.
111 J. Kubelka, E. A. Eaton, J. Hofrichter, *J. Mol. Biol.* **2003**, *329*, 625–630.
112 Y. Zhu, D. O. V. Alonso, K. Maki, C.-Y. Huang, S. J. Lahr, V. Daggett, H. Roder, W. F. DeGrado, F. Gai, *Proc. Natl Acad. Sci. USA* **2003**, *100*, 15486–15491.
113 A. E. Garcia, J. N. Onuchic, *Proc. Natl Acad. Sci. USA* **2003**, *100*, 13898–13903.
114 M. U. Johansson, M. de Château, L. Björck, S. Forsén, T. Drakenberg, M. Wikström, *FEBS Lett.* **1995**, *374*, 257–261.
115 S. Takada, *Proteins* **2001**, *42*, 85–98.
116 Y. Zhu, X. Fu, T. Wang, A. Tamura, S. Takada, J. G. Saven, F. Gai, *Chem. Phys.* **2004**, *307*, 99–109.
117 S. J. Hagen, J. Hofrichter, A. Szabo, E. A. Eaton, *Proc. Natl Acad. Sci. USA* **1996**, *93*, 11615–11617.
118 T. Wang, Y. Zhu, F. Gai, *J. Phys. Chem.* **2004**, *108*, 3694–3697.
119 J. W. Bryson, J. R. Desjarlais, T. M. Handel, W. F. DeGrado, *Protein Sci.* **1998**, *7*, 1404–1414.
120 C. D. Waldburger, T. Jonsson, R. T. Sauer, *Proc. Natl Acad. Sci. USA* **1996**, *93*, 2629–2634.
121 S. T. R. Walsh, H. Cheng, J. W. Bryson, H. Roder, W. F. DeGrado, *Proc. Natl Acad. Sci. USA* **1999**, *96*, 5486–5491.
122 S. T. R. Walsh, V. I. Sukharev, S. F. Betz, N. L. Vekshin, W. F. DeGrado, *J. Mol. Biol.* **2001**, *305*, 361–373.
123 M. Levitt, *ENCAD, Computer Program, Energy Calculations and Dynamics.* Molecular Applications Group, Palo Alto, CA and Yeda, Rehovot, Israel, **1990**.
124 M. Levitt, M. Hirshberg, R. Sharon, V. Daggett, *Comput. Phys. Commun.* **1995**, *91*, 215–231.
125 M. Levitt, M. Hirshberg, R. Sharon, K. E. Laidig, V. Daggett, *J. Phys. Chem.* **1997**, *101*, 5051–5061.
126 G. S. Kell, *J. Chem. Eng. Data* **1967**, *12*, 66–69.
127 A. Li, V. Daggett, *Proc. Natl Acad. Sci. USA* **1994**, *91*, 10430–10434.
128 J. J. Portman, S. Takada, P. G. Wolynes, *J. Chem. Phys.* **2001**, *114*, 5082–5096.
129 I. J. Chang, J. C. Lee, J. R. Winkler, H. B. Gray, *Proc. Natl Acad. Sci. USA* **2003**, *100*, 3838–3840.
130 M. S. Li, D. K. Klimov, D. Thirumalai, *Polymer* **2004**, *45*, 573–579.

9.2
Early Kinetic Events in Protein Folding:
The Development and Applications of Caged Peptides

Sunney I. Chan, Joseph J.-T. Huang, Randy W. Larsen, Ronald S. Rock, and Kirk C. Hansen

9.2.1
Introduction

Studies of the early kinetics events in protein folding provide a means to explore the free energy landscape of a protein. However, these processes are difficult to follow. First, there is the inherent difficulty of initiating the refolding of a protein from a well-defined state due to the extremely high density of energy states associated with the unfolded protein. Second, the protein-folding process from the "random coil" is essentially an enthalpic collapse from a complex manifold of states toward the relatively well-defined native state. Clearly the process must follow a well-defined set of trajectories with varying sequences of energy dephasing and energy dissipation steps specified by the width and depth of the "protein-folding funnel" as well as the roughness of the free energy landscape associated with the protein, if kinetic competence is to be attained [1, 2]. Moreover, the timescales associated with these energy dephasing and energy dissipation events are rapid. Accordingly, the overall pathway is extremely difficult to define kinetically, except, perhaps, in the special case of a protein with a very smooth "protein-folding funnel" [3].

Experimentally, the traditional stopped-flow method has been used to trace the protein-folding reaction. However, these experiments are limited in several fundamental aspects, including the long dead time of mixing, and the need to use denaturants to slow down the kinetics and to trigger the process. In order to surmount the limitations related to long mixing times, new approaches have been established over the years based on temperature jump (T-jump) [4], pH jump [5], pressure jump [6], flash photolysis of heme ligands [7], photoreduction of metalloproteins [8] and the photolysis of engineered disulfides [9]. Although these techniques have greatly moved back the window of observation to shorter times to allow observation of earlier protein-folding events, most of these experiments involved perturbation of the protein away from equilibrium and measuring the time course of the return to equilibrium. The extent of the perturbation needed to be determined in each case, often not a simple exercise due to the complexity of protein systems. Moreover, in many situations, the studies must still be done in the presence of denaturant, which might compromise the sequence of protein-folding events that the experiment is intended to probe. While it is apparent that the presence of denaturant squashes the protein-folding funnel, the details of how the thermodynamics and the kinetics of individual protein-folding steps or events are affected by the denaturant remain to be addressed. What is known, however, is that denaturant shifts the overall equilibrium between the denatured and native states of the protein, and slows down the overall kinetics in any case.

9.2.2
An Ultrafast Photochemical Triggering System

To simplify the study of the rapid kinetic events inherent in protein-folding processes, a trigger event that is rapid, irreversible and results in a large change in stability ($\Delta\Delta G^0$) is required. In the approach described in this chapter, the picosecond flash photolysis of small organic chromophores, known as cages, attached to protein functional groups is employed. Instead of manipulating the external conditions, such as denaturant concentration, pH or temperature, the covalent structure of the protein is altered in the triggering event.

The ideal conformational trigger for protein folding and unfolding studies would have the following properties. First, the trigger must involve a large stability change, i.e. a large $\Delta\Delta G^0$. Such a large change in stability allows a wide variety of conditions to be examined, especially strongly native, zero-denaturant conditions. Second, the trigger event must be rapid. In protein folding, some conformational changes might occur on timescales as rapid as 100 ps, depending on the system. As a result, the ideal trigger event would be complete in less than this time. Third, in order to be of general applicability, the trigger must not rely on protein prosthetic groups, such as hemes or metal ions that are not present in all proteins. Hemes, in particular, are to be avoided, since a heme is a significant fraction of the protein mass and may be expected to contribute interactions that would dominate the conformational change. Finally, and perhaps most importantly, the trigger must be irreversible. All current rapid conformational triggers (with the exception of capillary mixing) suffer from reversibility to a greater or lesser extent. Protein conformational changes can be enormously complex, with events spanning timescales from picoseconds to seconds. It is simply too difficult to construct an accurate kinetic framework by combining data from different triggering methods, each with its own observation window, when a single, rapid, irreversible trigger would provide the same information. Some have claimed that reversibility is a desired property of a conformational trigger, in that signal averaging from multiple triggering events is allowed [10]. This would be true, if the reverse reaction occurred on the timescale of seconds or more, after all relevant conformational changes were complete. However, with the current systems, timescales spanning at most three orders of magnitude may be studied, due to reverse reactions that are too rapid. On the other hand, with an irreversible trigger, signal averaging may always be achieved with additional peptide material.

One triggering technique that can meet all of the above requirements is the flash photolysis of a class of organic protecting groups known as cages. Irradiation of a cage chromophore leads to specific irreversible bond cleavage. Carefully placed, a cage group will block the reaction of interest until a photolysis pulse is applied. A number of cages, with a variety of photochemical properties, are candidates for protein conformational triggers [11, 12]. One that is particu-

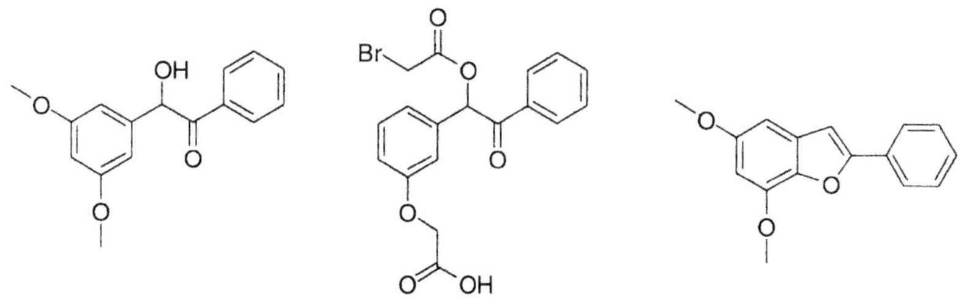

Scheme 9.2.1

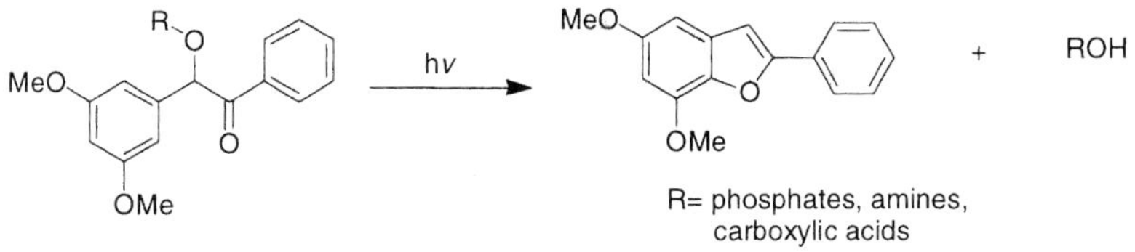

Fig. 9.2.1 Photochemistry of benzoin.

larly suited for the applications examined here is 3′,5′-dimethoxybenzoin (DMB, **1**), first studied in depth by Sheehan et al. [13, 14] (Scheme 9.2.1).

DMB forms carboxylate and phosphate esters that may be cleaved to form the free acid when irradiated at 300–360 nm. The sole photoproducts are 5,7-dimethoxy-2-phenyl benzofuran (**3**) and the parent acid (Fig. 9.2.1). Triplet quenchers such as piperylene or naphthalene cannot quench the photolysis of DMB, while unsubstituted benzoinyl esters which photolyse to yield the same products are quenched. Stern-Volmer analysis suggests that the photolysis rate is in excess of 10^{10} s^{-1} [14] and may be as high as 10^{12} s^{-1} [15]: The quantum yield of photolysis is high (0.64) [15–17], allowing short photolysis laser pulses, as multiple excitation of the DMB chromophore is not required. Details of the photochemistry have been described elsewhere [15]. Finally, the phenylbenzofuran photoproduct is inert, so further reactions with polypeptide functional groups are not expected. These characteristics of the benzoin system make it the cage of choice for the present application.

9.2.3
Caging Strategies and the Development of a Photolabile Linker

Three methods for denaturing a protein can be envisioned using photolabile protecting groups (Fig. 9.2.2). First, the photolabile protecting group could be attached to a side-chain residue to create steric bulk. It could also be used to cyclize the peptide head-to-tail or head-to-side-chain. In the cyclization methods, the photolabile protecting group must also act as a linker.

Fig. 9.2.2 Strategies of "caging" a polypeptide to denature a protein. (Top) Head-to-tail cyclization. (Middle) Side-chain protection. (Bottom) Head-to-side-chain cyclization.

The simplest version of this technology is the side-chain "cage". In the side-chain protection method, it is believed that if the "cage" is placed in the core of the protein, the steric bulk of the group will cause unfolding of the polypeptide. For example, DMB can be selectively attached to the side-chain of an aspartic acid (Asp) or a glutamic acid (Glu) as an ester and incorporated into the polypeptide by solid-phase peptide synthesis. The existence of the cage interferes with hydrogen bonding, ionic interactions, as well as hydrophobic interactions, and the disruption often unfolds the polypeptide. When this ester of DMB is photolysed by a laser pulse, the benzoin is rapidly released on the picosecond timescale, generating the parent acid on the peptide-chain and the inert photoproduct 5,7-dimethoxy-2-phenylbenzofuran (Fig. 9.2.3). The peptide then starts to fold toward its native equilibrium structure (see below, Fig. 9.2.5a).

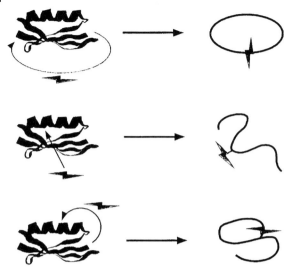

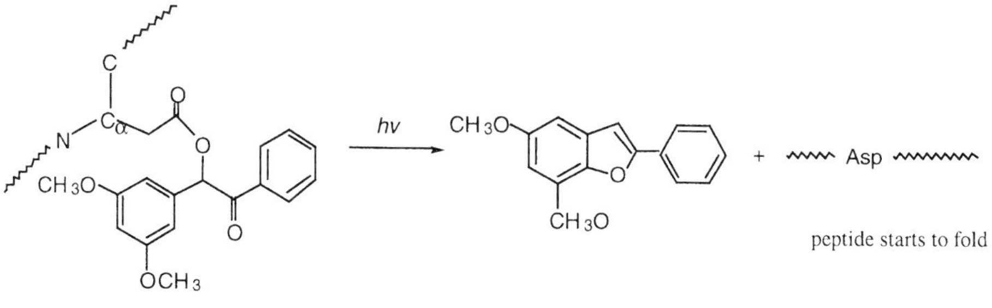

Fig. 9.2.3 Photolysis scheme for a peptide containing Asp(DMB).

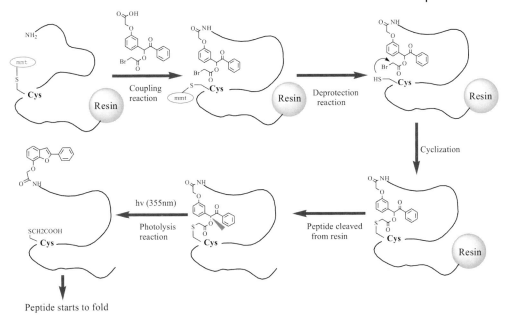

Fig. 9.2.4 Reaction scheme for the head-to-side-chain strategy.

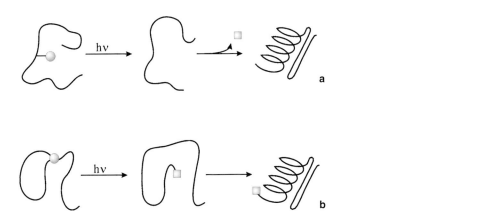

Fig. 9.2.5 (a) The side-chain caging scheme. (b) The head-to-side-chain cyclization scheme.

DMB is the photolabile protecting group of choice to cage the side-chains of Asp and Glu. The technology is simple and easy to apply. Asp and Glu that are buried or involved in salt bridges are obvious candidates. Given that Asp and Glu are usually found at the protein surface, the choice of the Asp or Glu needs to be judiciously made. A side-chain cage would not be expected to unfold the protein if the protecting group is attached to an Asp or Glu that is located on the surface of the three-dimensional protein fold in the native structure.

The cyclization strategy is more likely to disrupting the protein structure compared to the side-chain cage. However, the head-to-tail cyclization has proven to be difficult to develop. For these reasons, the head-to-side-chain cyclization strategy has inherent advantages. To accomplish the head-to-side-chain cyclization, we have developed the benzoin derivative bromoacetyl-carboxy-methoxybenzoin (BrAc-CMB, **2**) to cyclize a peptide from the N-terminus (head) to the side-chain (Fig. 9.2.4). BrAc-CMB has the ability to couple with the N-terminal group at one end and link to the thio group of a Cys residue on the peptide. The head-to-side-chain cyclization strategy is shown in Fig. 9.2.5 b. Under UV photolysis, the linker on the caged peptide will break and immediately initiate the folding of the peptide.

9.2.4
Synthesis of Cages and Caging of a Protein/Peptide

The bottleneck in the development and applications of this kind of technology is, of course, the synthesis of the "cage", or the photoactive protecting group, and the design and preparation of the caged peptides. Accordingly, we outline below the synthesis of the three cages that we have developed to date.

9.2.4.1 DMB, Fmoc-Asp(DMB)-OH and Fmoc-Glu(DMB)-OH

Synthetic schemes for DMB, Fmoc-Asp(DMB)-OH and Fmoc-Glu(DMB)-OH are outlined in Scheme 9.2.2.

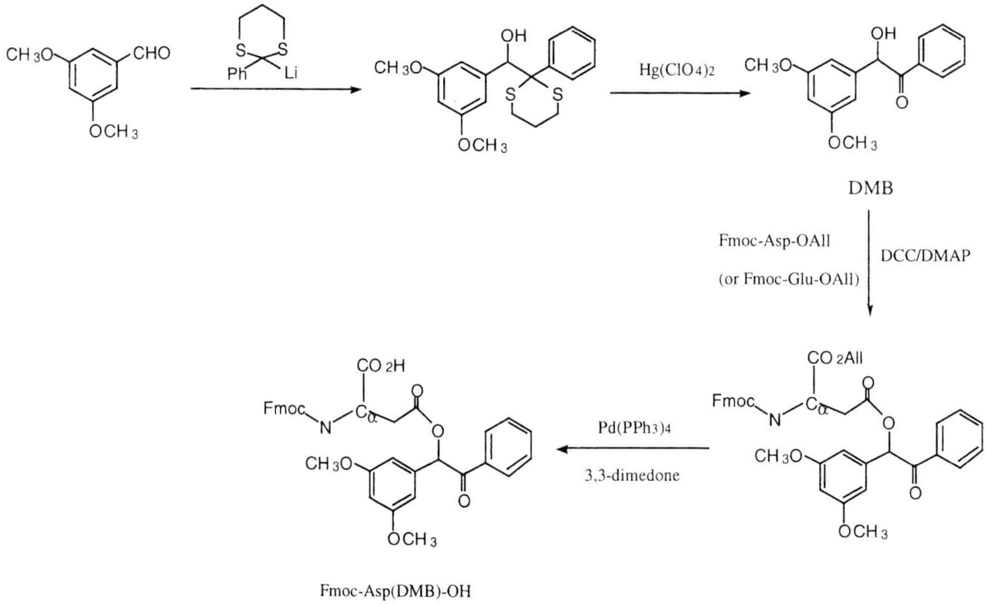

Scheme 9.2.2 Synthesis of Fmoc-Asp(DMB)-OH.

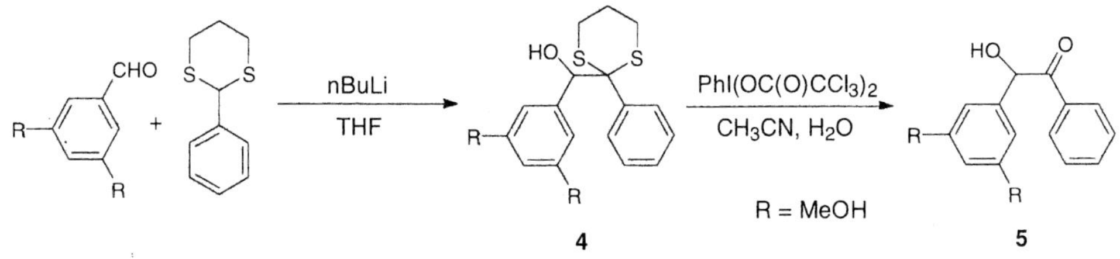

Scheme 9.2.3 Synthesis of DMB.

Although much effort has been put on the synthesis of acyloins, many of these methods have proven to be inefficient, in particular 3′,5′-DMB. Second, due to the high quantum yield of 3′,5′-DMB, photolysis may occur in standard laboratory light, retarding subsequent synthesis. Consequently, our synthesis of benzoins is based on the Corey-Seebach dithiane addition method (Scheme 9.2.3) [16]. The dithiane adduct 4 serves as a very convenient synthon for the introduction of the 3′,5′-DMB protecting group [16] into its intended target, as the complex will remain photochemically stable until the dithiane protecting group is removed. Without this technical advance, it seems unlikely that we could have made the progress that we did in the development of our strategy for the caging of peptides.

9.2.4.2 3′,5′-Bis(carboxymethoxy)benzoin (BCMB)

In addition to 3′,5′-DMB, we have developed the synthesis of the analogous cage BCMB, which is more water soluble, by introducing the appropriate charged functionalities into the DMB group. The synthesis of BCMB protection group starts with 3,5-dihydroxy-benzaldehyde, where the hydroxyls are protected as t-butyldimethylsilyl (TBDMS) ethers (Scheme 9.2.4). The benzaldehyde is condensed with phenyl dithiane lithium anion, and the TBDMS protected hydroxyls are then unmasked and alkylated with t-butylbromoacetate in a single step. After hydrolysis of the dithiane, the BCMB is generated by acylation with acetic anhydride followed by hydrolysis of the 3′ and 5′ esters.

9.2.4.3 BrAc-CMB

The synthesis of BrAc-CMB is shown in Scheme 9.2.5. The linker was synthesized from the 3-hydroxybenzaldehyde following the similar method for benzoinyl linker in six steps. After four steps, the intermediate CMB-OtBu was obtained. The BrAc-CMB was acquired by dicyclohexylcarbodiimide (DCC)-activated esterification of CMB-OtBu with bromoacetic acid, followed by cleavage of the tBu group with 50% TFA.

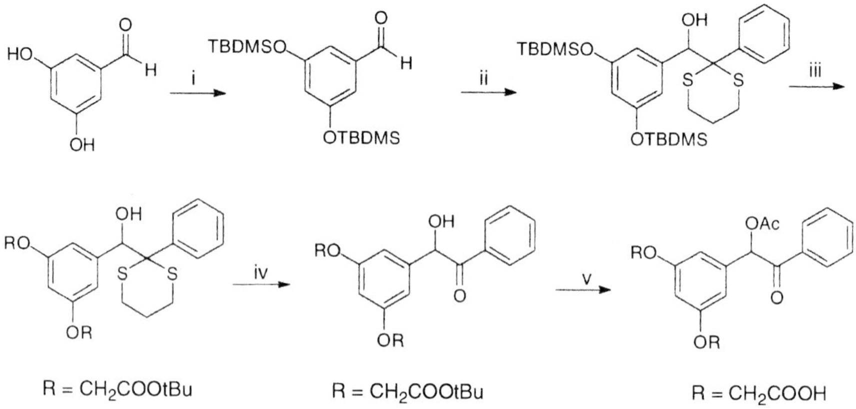

Reagents: i: TBDMSCl, Et$_3$N. ii: Phenyl dithiane lithium anion, H$_2$O. iii: BrCH$_2$COOtBu, TBAF. iv: Hg(ClO$_4$)$_2$, H$_2$O. v: a) Ac$_2$O, DMAP; b) TFA

Scheme 9.2.4 Synthesis of BCMB.

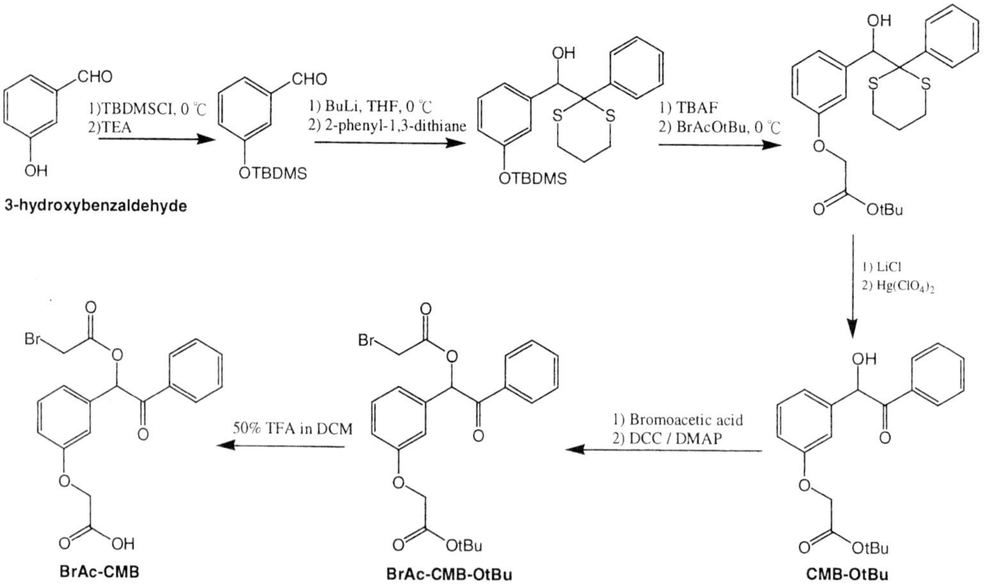

Scheme 9.2.5 Synthesis of BrAc-CMB.

9.2.4.4 Protein Head-to-Side-chain Cyclization by BrAc-CMB

The head-to-side-chain cyclization strategy is shown in Scheme 9.2.6. BrAc-CMB is added to the N-terminus of a cysteine-containing peptide by solid-phase peptide synthesis. The thiol group is protected by an acid-labile protecting group, such as methoxytrityl, which could be deprotected by dilute acid leaving

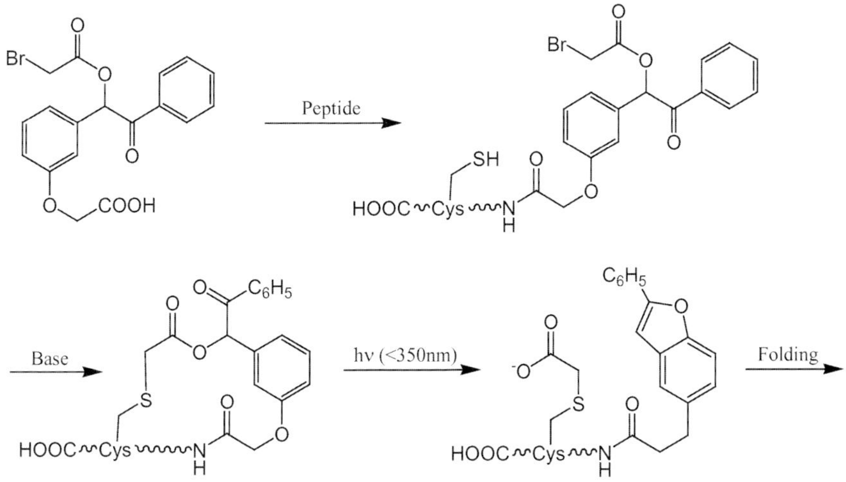

Scheme 9.2.6 General synthesis strategy for head-to-side-chain cyclization of a protein.

other protecting groups intact. Nucleophilic attack of the cysteine thiol on the bromoacetyl cyclizes the peptide to give the carboxymethylcysteine linkage. Upon irradiation at 355 nm, the caged compound is released and the peptide containing the thioether analog of glutamate residue folds. The quantum yield (Φ) of the photocleavage of the BrAc-CMB has recently been determined to be 0.72, which is certainly adequate to serve as an efficient photolabile linker.

9.2.5
Monitoring the Refolding Kinetics of Protein Structural Motifs by Laser Flash Photolysis of "Caged Peptides"

Although photoactive protecting groups are now in hand for the caging of peptides and proteins for ultrafast triggering of folding or unfolding events, at least for some applications, we must still couple the systems under observation to a laser system for rapid flash photolysis to trigger the folding or unfolding process and a sufficiently sensitive method or device to monitor the events in real-time over several orders in time. In the work to date, we have appealed to two related methods: photoacoustic calorimetry (PAC) and photothermal beam deflection (PBD). In principle, other methods are possible for the monitoring of the early kinetic events, including Fourier transform IR spectroscopy (FTIR), circular dichroism (CD) and fluorescence resonance energy transfer (FRET), and we are in the process of developing them for use in our experiments. In any case, space will not allow us to go into any of these methods in great detail.

Aside from the timescales of the kinetic events, PAC and PBD provide estimates of the volume changes and enthalpies associated with the refolding of the destabilized peptide or protein. Since the refolding reactions begin from a

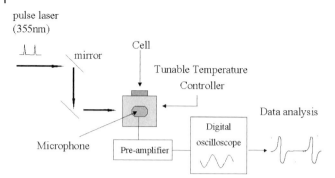

Fig. 9.2.6 Schematic setup of a PAC apparatus.

single conformation, as opposed to refolding reactions that begin from a random coil ensemble, the interpretation of the kinetic data is relatively straightforward. Both techniques involve photoexcitation of a molecular system of interest in which excess energy is dissipated by vibrational relaxation to the ground state. During this process, the excitation energy is converted to thermal motion of the surrounding solvent molecules. This photothermal heating of the solvent (such as water) causes a rapid volume expansion that changes the refractive index of the solution and develops an acoustic wave [18, 19]. Additionally, volume changes in the molecular system of interest induced by a photoinitiated reaction also contribute to the refractive index change and acoustic wave. In PAC, the volume expansion is monitored by detecting the accompanying acoustic wave; in PBD, changes in the refractive index are followed optically by a laser probe beam. Generally, the PAC could measure enthalpy/volume changes from 30 ns to 5 µs and the PBD system can resolve volume/enthalpy changes with lifetimes ranging from roughly 1 µs to 50 ms. Schematics of the experimental setups for these measurements are depicted in Figs. 9.2.6 and 9.2.7.

In the PAC experiment, the transient signals(s) are sampled by a microphone mounted to the sidewall of the cuvette, whereas in the PBD, the deflected

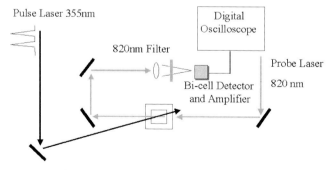

Fig. 9.2.7 Schematic setup of a PBD apparatus.

He:Ne laser beam is monitored by an optical detector. In each case, the signal amplitudes (S) are fitted to the equation:

$$S = a_0 + \Sigma_i a_i(1 - \exp[-t/\tau_i])$$

where a_i denotes the amplitude for each intermediate step in the PAC/PBD experiment, and τ_i is the corresponding lifetime. In both PAC and PBD, the enthalpy of the reaction is obtained from the PAC/PBD amplitude, where $\Phi \cdot \Delta H = E_{hv} - Q$ for a_0, and for the subsequent a_is, $\Delta H = -Q$. These amplitudes are a function of the heat released Q and the volume change ΔV associated with each step, and the two parameters must be determined from analysis of temperature data.

9.2.6
Early Kinetic Events in Protein Folding

As our research has been directed toward understanding the early kinetic events during protein folding, we have focused on the applications of the caging strategies that have been developed towards the real time observation of the folding and unfolding kinetics of simple structural motifs (α-helices, β-hairpins and β-sheets) commonly found in proteins [20–22]. Three specific examples will be summarized in this review.

9.2.6.1 Side-chain Caging of the GCN4-p1 Leucine Zipper

Rock et al. [20] have applied the side-chain caging strategy to study the kinetic events associated with the unfolding of the GCN4-p1 leucine zipper. GCN4-p1 is a 33-residue peptide derived from the bZIP class of yeast transcription factors [23–26]. It is a homodimeric coiled-coil in solution. While the coiled-coil interface consists of primarily hydrophobic residues, it contains a single Asn at position 16 in each strand, and together they form a buried interstrand hydrogen bond to stabilize the coiled-coil. Degrado et al. have examined Asn16 in GCN4-p1 as a potential site for a designed, buried salt bridge [27]. In their proposed design, Asn16 in one strand is mutated to Asp16 [GCN4-p1(N16D)], while the Asn16 in the neighboring strand is converted to diaminopropionic acid (Dsp) or diaminobutryric acid (Dab). Homodimers do not form in the case of the latter peptides due to charge repulsion.

Rock et al. [20] used the analogous dimer of the mutant peptide GCN4-p1 (N16D) and caged the aspartate side-chain introduced at residue 16 by DMB. GCN4-p1 [N16D(DMB)] remained a homodimer and CD measurements on the peptide indicated that the dimer is in fact quite stable. However, cleavage of the DMB cage by laser photolysis unmasked the unfavorable charge–charge repulsion between the two opposing negatively charged Asp carboxylates, and the coiled-coil dissociated and unfolded (Fig. 9.2.8). Rock et al. [20] monitored the kinetic events following the photoinitiated rapid unfolding process by PAC and PBD.

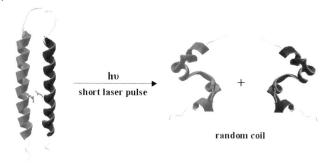

Fig. 9.2.8 Photolysis of caged GCN4-p1.

Photolysis of Ac-DMB as a control revealed a volume contraction but no kinetic processes that could be resolved by either PAC or PBD, suggesting that cage photolysis, loss of the acetyl group, and all subsequent rearrangements of the solvation sphere occurred on the nanosecond or faster timescale. For GCN4-p1 [N16D(DMB)], PAC revealed an instantaneous volume contraction assigned to the photolysis event, followed by two kinetic phases with time constants around 300 ns and 710 ns. These events corresponded to a volume expansion and contraction, respectively. PBD revealed a fast volume contraction of -7.6 mL mol^{-1}, on a timescale faster than could be resolved by the setup, encompassing the two kinetic processes detected by PAC. This fast phase was followed by a single additional volume expansion of 3.0 mL mol^{-1}, occurring at 5.5 μs (Fig. 9.2.9). No processes on longer timescales were detected. These results are consistent with an initial electrostrictive contraction of the solvent upon generation of the free carboxylate at residue 16 that occurs upon photolysis (nanosecond timescale or faster), followed by a pair of intermediates involving exposure and reburial of hydrophobic surfaces. Such a reburial of hydrophobic surface area may be expected under the strongly native solvent conditions of this experiment. These volume changes culminate in an overall expansion from exposure of the core residues of GCN4-p1 upon complete unfolding.

9.2.6.2 Head-to-Side-chain Cyclization of β-Sheets

The most prominent, stable, protein secondary structures are the α-helices and β-structures. Information regarding the mechanism and time scales of formation of these structural elements can provide a better insight into the early events in the protein-folding process.

One of our target peptides is the three-stranded β-sheet first reported by Gellman et al. [28]. This peptide is denoted the 20mer and the sequence of the peptide is shown: VFITS^{D}PGKTYTEV^{D}PGOKILQ. Studies from this laboratory has revealed that mutating ^{D}P to D will change the turn type from a four-residue turn to a five-residue turn and change the side-chain pairing between the first

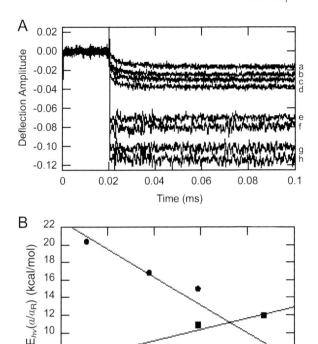

Fig. 9.2.9 (A) Overlay of the PBD traces of GCN4-p1 [N16D(DMB)] subsequent to photolysis and the reference compound (R). Sample: (a) 13, (b) 20, (c) 26 and (d) 35 °C; reference: (e) 13, (f) 20, (g) 26 and (h) 35 °C. (B) Plots of $(a_1/a_R)E_{hv}$ versus $pC_p/(dn/dT)$ for a_0 (circles) and a_1 (squares), where ρ is the density of the buffer, C_p is the heat capacity and dn/dT is the temperature gradient of the refractive index. The slopes of these plots are proportional to ΔV and the intercepts to the amount of heat released to the solvent.

two strands. The resulting frame-shift is for accommodating the disfavored turn sequence T-S-D-G-K. The mutation not only alters the strand register, but also decreases the structural stability of the P6D mutant peptide [29].

Chen et al. [21] have recently applied the head-to-side-chain caging strategy to cyclize the β-sheet peptides with these two different turns and monitored the folding process using laser flash photolysis combined with photoacoustic calorimetry (Fig. 9.2.10). These experiments revealed a refolding time constant for the peptide with the "S-^{D}P-G-K" turn (c-E12C) of around 40 ns at 20 °C. Interestingly, one additional phase was observed for the peptide with the "T-S-D-G-K" turn (c-P6DE12C) following cleavage of the photolabile linker. The time constant associated with this component was about 150 ns. Apparently, the peptide with the T-S-D-G-K turn needed a longer time to return to its equilibrium state

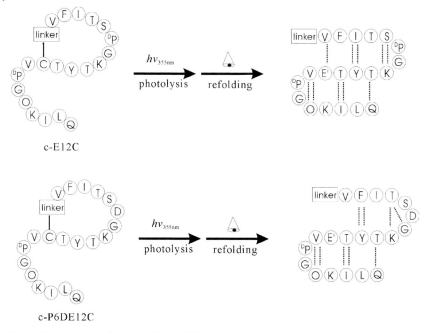

Fig. 9.2.10 Experimental design of the refolding of c-E12C and c-P6DE12C.

than the peptide with the more favored S-^{D}P-G-K turn. Thus, the turn region in the β-sheet plays a key role in guiding the subsequent assembly of the structure during the early stages of the folding process. In any case, the kinetics of the β-sheet conformational rearrangements are very rapid, and it is clear that the caging strategy that we have developed is capable of following elementary folding processes occurring on the nanosecond timescale.

9.2.6.3 Head-to-Side-chain Caging of the α-Helical Villin Headpiece Subdomain

As an extension of the caged peptide strategy, Hansen et al. [22] have targeted a more complex system, the α-helical villin headpiece subdomain. The F-actin binding protein villin contains an autonomously folding 35-residue subdomain (HP-35) at its extreme C terminus. The HP-35 subdomain is the shortest autonomously folded sequence forming a compact primarily α-helical structure; accordingly, it is a perfect model for protein-folding study.

In the work of Hansen et al., residue Met 12 was chosen for mutation to Cys due to its close proximity in sequence to the N terminus and its apparent high solvent exposure (Fig. 9.2.11) [22]. However, using this caging strategy, the local structure of only one of the three helices could be destroyed. It was possible to synthesize the caged villin headpiece subdomain (c-VHP) in high yields. The UV absorption spectra of c-VHP before and after irradiation at 352 nm revealed a decrease in absorption around 255 nm with simultaneous increase around

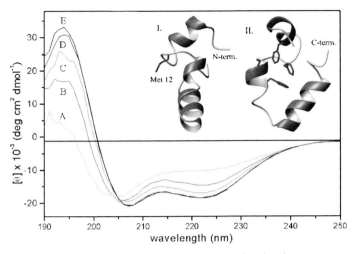

Fig. 9.2.11 The steady-state CD spectrum of c-VHP under photolysis.

285 nm, indicating that the linker was photolysed after UV irradiation. Circular dichroism data showed that there was a secondary structure difference of the c-VHP before and after photolysis (Fig. 9.2.11). Deconvolution of the experimental PAC data revealed two kinetic phases with time constants of approximately 100 and 400 ns.

9.2.7
Summary and Conclusions

We have developed two photoactive peptide-protecting strategies, one to cage a side-chain of a peptide or protein and the other to cyclize the N-terminus of a peptide to a side-chain, to facilitate the observation of the early kinetic events in protein folding and unfolding. Due to the rapid photochemistry of cleavage associated with the photoactive protecting group ($>10^{10}$ s^{-1}) and the high efficiency of photocleavage (quantum yield of ~0.70), these linkers provide an efficient and ultrafast photochemical triggering system to initiate protein folding and unfolding so that the early kinetic events could be followed. In principle, conformational rearrangements or other elementary processes as early as a few nanoseconds could be studied. To date, we have applied this strategy to study the early kinetic events in the refolding or unfolding of three systems. Although we have so far tracked these events by photoacoustic calorimetry and photothermal beam deflection, other methods of observation are possible, in principle, including CD, FTIR, FRET, etc. The method is particularly useful to detect local sequence effects on the early kinetic events of protein folding under ambient conditions without resorting to temperature, pH or pressure jumps, and/or the addition of denaturants to the system. While a number of technical difficulties remain to be overcome, including the low solubility and yield of the caged peptides, it is clear that the overall strategy works and will allow us to move the protein-folding field to a new phase.

References

1 WOLYNES, P.G., ONUCHIC, J.N., THIRUMALAI, D. *Science* **1995**, *267*, 1619–1625.
2 LEOPOLD, P., ONUCHIC, J.N. *Proc. Natl Acad. Sci. USA* **1992**, *89*, 8721–8725.
3 GRUEBELE, M. *Curr. Opin. Struct. Biol.* **2002**, *12*, 161–168.
4 BALLEW, R.M., SABELKO, J., GRUEBELE, M. *Proc. Natl Acad. Sci. USA* **1996**, *93*, 5759–5764.
5 ABBRUZZETTI, S., CREMA, E., MASINO, L., VECLI, A., VIAPPIANI, C., SMALL, J.R., LIBERTINI, L.J., SMALL, E.W. *Biophys. J.* **2000**, *78*, 405–415.
6 JACOB, M., HOLTERMANN, G., PERL, D., REINSTEIN, J., SCHINDLER, T., GEEVES, M.A., SCHMID, F.X. *Biochemistry* **1999**, *38*, 2882–2891.
7 JONES, C.M., HENRY, E.R., HU, Y., CHAN, C.K., LUCK, S.D., BHUYAN, A., RODER, H., HOFRICHTER, J., EATON, W.A. *Proc. Natl Acad. Sci. USA* **1993**, *90*, 11860–11864.
8 PASCHER, T., CHESICK, J.P., WINKLER, J.R., GRAY, H.B. *Science* **1996**, *271*, 1558–1560.
9 RODER, H., SHASTRY, M.R. *Curr. Opin. Struct. Biol.* **1999**, *9*, 620–626.
10 LU, H.S.M., VOLK, M., KHOLODENKO, Y., GOODING, E., HOCHSTRASSER, R.M., DEGRADO, W.F. *J. Am. Chem. Soc.* **1997**, *119*, 7173–7180.
11 GURNEY, A.M., LESTER, H.A. *Physiol. Rev.* **1987**, *67*, 583–617.
12 ADAMS, S.R., TSIEN, R.Y. *Annu. Rev. Physiol.* **1993**, *55*, 755–784.
13 SHEEHAN, J.C., WILSON, R.M. *J. Am. Chem. Soc.* **1964**, *86*, 5277–5281.
14 SHEEHAN, J.C., WILSON, R.M., OXFORD, A.W. *J. Am. Chem. Soc.* **1971**, *93*, 7222–7228.
15 ROCK, R.S., CHAN, S.I. *J. Am. Chem. Soc.* **1998**, *120*, 10766–10767.
16 STOWELL, M.H.B., ROCK, R.S., REES, D.C., CHAN, S.I. *Tetrahedron Lett.* **1996**, *37*, 307–310.
17 ROCK, R.S., CHAN, S.I. *J. Org. Chem.* **1996**, *61*, 1526–1529.
18 BRASLAVSKY, S.E., HEIBEL, G.E. *Chem. Rev.* **1992**, *92*, 1381–1410.
19 YEH, S.R., FALVEY, D.E. *J. Photochem. Photobiol. A Chem.* **1995**, *87*, 13–21.
20 ROCK, R.S., HANSEN, K.C., LARSEN, R.W., CHAN, S.I. *J. Chem. Phys.* **2004**, *307(2, 3)*, 201–208.
21 CHEN, R.P.-Y., HUANG, J.J.-T., CHEN, H.L., JAN, H., VELUSAMY, M., LEE, C.T., FANN, W., LARSEN, R.W., CHAN, S.I. *Proc. Natl Acad. Sci. USA* **2004**, *101*, 7305–7310.
22 HANSEN, K.C., ROCK, R.S., LARSEN, R.W., CHAN, S.I. *J. Am. Chem. Soc.* **2000**, *122*, 11567–11568.
23 O'SHEA, E.K., RUTKOWSKI, R., KIM, P.S. *Science* **1989**, *243*, 538–542.
24 RASMUSSEN, R., BENVEGNU, D., O'SHEA, E.K., KIM, P.S., ALBER, T. *Proc. Natl Acad. Sci. USA* **1991**, *88*, 561–564.
25 OAS, T.G., MCINTOSH, L.P., O'SHEA, E.K., DAHLQUIST, F.W., KIM, P.S. *Biochemistry* **1990**, *29*, 2891–2894.
26 O'SHEA, E.K., KLEMM, J.D., KIM, P.S., ALBER, T. *Science* **1991**, *254*, 593–544.
27 SCHNEIDER, J.P., LEAR, J.D., DEGRADO, W.F. *J. Am. Chem. Soc.* **1997**, *119*, 5742–5743.
28 SCHENCK, H.L., GELLMAN, S.H. *J. Am. Chem. Soc.* **1998**, *120*, 4869–4870.
29 CHEN, P.Y., LIN, C.K., LEE, C.T., JAN, H., CHAN, S.I. *Protein Sci.* **2001**, *10*, 1794–1800.

9.3
Photocontrol of RNA Processing

Steven G. Chaulk, Oliver A. Kent, and Andrew M. MacMillan

9.3.1
Introduction

In addition to being a carrier of genetic information, RNA is involved in the regulation of both transcription and translation, plays a key role in the splicing of pre-mRNA in eukaryotes, and is responsible for catalysis of amide bond formation at the heart of the ribosome [1, 2]. The study of many RNA systems is complicated by the dynamic nature of RNA secondary and tertiary structures, the transient nature of RNA–RNA or RNA–protein complexes, and in some cases the chemical reactivity of the RNA itself. One useful approach to the study of discrete RNA (and DNA) structures has been to limit available conformations through site-specific intra-strand crosslinking [3–7]. A complementary approach which allows the isolation of specific RNA structures or complexes involves the transient blocking or "caging" of RNA functional groups involved in the transition between two different states. The caging of cofactors or reactive substrates with photolabile groups has proven useful in a wide variety of investigations ranging from mechanistic enzymology to cell biology [8–10]. Caging of specific functionalities within proteins has been used to control reactivity in these systems, e.g. Noren et al. caged a specific serine residue in *Thermococcus litoralis* Vent polymerase for studies on the mechanism of protein splicing [11].

In the context of an RNA molecule, a caging approach might be used to block either chemical reactivity of the RNA or formation of secondary or tertiary structure. The caged RNA system can be studied both before and after photolysis, thus permitting characterization of the two states and the transition between them. The first application of the caging approach to studies of RNA structure and function has involved blocking the chemical reactivity associated with an RNA functionality, the 2′-hydroxyl group, since specific RNA 2′-hydroxyls act as nucleophiles in a number of biologically important transesterifications [12–17]. Caging of a single 2′-hydroxyl in a short synthetic oligonucleotide blocks the cleavage reaction catalyzed by the hammerhead ribozyme; cleavage is initiated by photolysis of the ribozyme–substrate complex [18]. This approach has also been extended to the caging of the branch adenosine in a full-length pre-mRNA for studies of RNA processing by the mammalian spliceosome [19]. Caging effectively isolates spliceosome assembly from catalysis of the splicing transesterifications permitting a closer examination of the mechanisms of each.

9.3.2
Caging the Hammerhead Ribozyme

The hammerhead system was chosen for preliminary studies because it has been the subject of intense investigation culminating in high-resolution X-ray structures of both model and native hammerhead–substrate complexes [20–22]. The hammerhead ribozyme is a site-specific RNA endonuclease derived from self-cleaving plant viroid and satellite RNAs [12]. The ribozyme–substrate complex contains three base-paired stems and a central core region of 13 conserved nucleotides which form the catalytic site (Fig. 9.3.1 A). Cleavage of the substrate RNA (RNA 1 in Fig. 9.3.1 A) is magnesium dependent and occurs through intramolecular attack of a 2′-hydroxyl on the adjacent phosphodiester (Fig. 9.3.1 B). As part of efforts to understand the hammerhead system, many studies have involved the use of altered reaction conditions or mutant ribozymes or substrates.

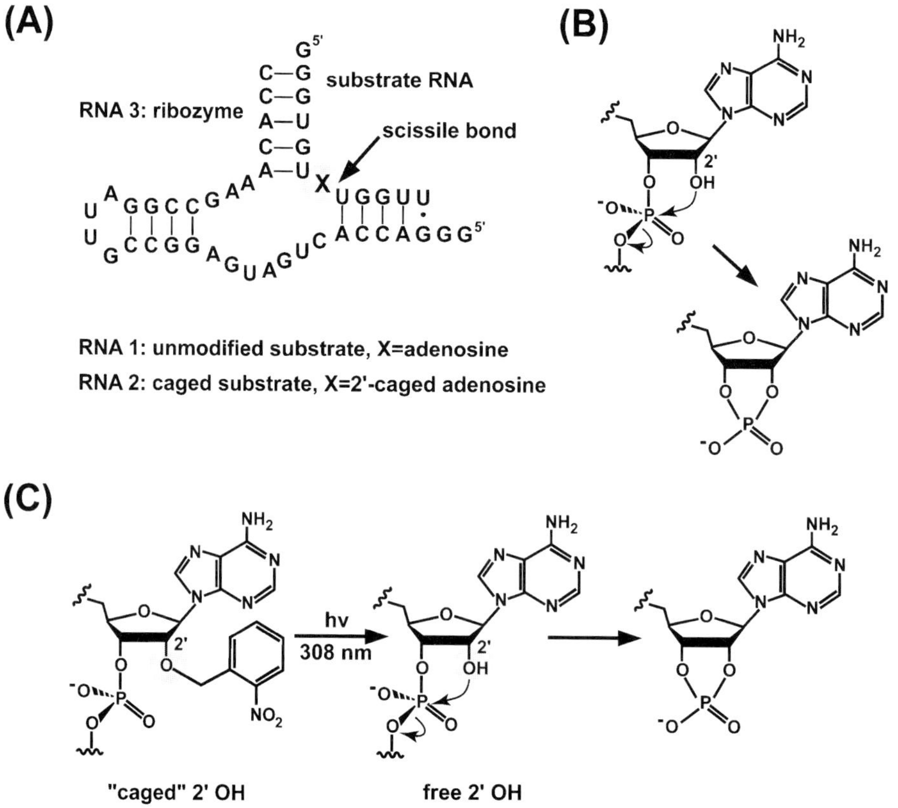

Fig. 9.3.1 (A) Secondary structure of a canonical hammerhead–ribozyme-substrate complex showing the scissile phosphodiester linkage. (B) Intramolecular transesterification catalyzed by the hammerhead ribozyme. (C) Photolysis of a 2′-caged hammerhead substrate to initiate the hammerhead cleavage reaction.

For example, since the ribozyme-mediated cleavage is dependent on the presence of a 2'-hydroxyl 5' to the site of cleavage, replacement of this functionality with hydrogen or OMe substituents has allowed the study of catalytically inactive ribozyme–substrate complexes [20, 21]. Modification of this single 2'-hydroxyl of the hammerhead with a photodissociable group would represent a caging of this reactive functionality, thus permitting photocontrol of the hammerhead mediated reaction (Fig. 9.3.1 C).

For initial studies on the hammerhead system, the 2-nitrobenzyl group was chosen as the caging functionality because its photochemistry has been well characterized and because it can be removed, by irradiation at 308 nm, under conditions which will not damage nucleic acids or proteins. The residue 5' to the cleavage site in the hammerhead substrate can be a C, U or A residue. Caged adenosine was chosen for incorporation into a hammerhead substrate because of applications of this monomer to studies of Group II self-splicing introns and pre-mRNA splicing. The nucleoside 2-nitrobenzyladenosine was synthesized [23–25] and its reactivity studied under a variety of conditions. Upon photolysis at 308 nm with an excimer laser, the 2-nitrobenzyladenosine was quickly, cleanly and quantitatively converted to adenosine as monitored by C_{18} reverse-phase high-performance liquid chromatography (HPLC) with the yield of uncaged adenosine independent of pH in the range of 6–8. Next, the compatibility of the 2-nitrobenzyl modification with standard RNA synthesis conditions was assessed. While the 2-nitrobenzyl ether was stable to most conditions of RNA synthesis, it was partially removed by fluoride under the conditions used in deprotection of 2'-silyl protected RNAs (~20% cleavage observed). Thus the 2'-*tert*-butyldimethylsilyl (TBDMS) group commonly employed in RNA synthesis was incompatible with the introduction of the caged monomer into RNA. As an alternative, the acid labile [1-(2-fluorophenyl)-4-methoxypiperidin-1-yl (Fpmp)] group was used as a 2'-protecting functionality [26]. The 2'-Fpmp group is quantitatively removed upon treatment with mild aqueous acid over 30–40 h, conditions under which the nitrobenzyl ether is stable, with RNA synthesis yields comparable with those obtained using the 2'-TBDMS group. In order to incorporate the caged monomer into RNA, the 2'-modified adenosine was elaborated into a 5'-dimethoxytrityl-3'-phosphoramidite by standard procedures. This monomer was used, along with 2'-Fpmp phosphoramidites, in the solid phase synthesis of the RNA **2**: 5'-GGGUGUA*UGGUU-3' (Fig. 1A; a modified version of a well-characterized hammerhead substrate in which **A*** represents a caged nucleotide 5' to the cleavage site [27]). Deprotection of the crude product with methanolic ammonia followed by aqueous acid afforded an RNA in which a single residue was modified at the 2' position with the desired caging functionality. The unmodified control oligonucleotide substrate **1**: 5'-GGGUGUAUG-GUU-3' was synthesized by standard solid phase synthesis procedures and the ribozyme **3** was synthesized by T7 transcription from a synthetic DNA template [28]. HPLC analysis of the purified caged RNA, **2**, following enzymatic digestion with snake venom phosphodiesterase and calf intestinal alkaline phosphatase, showed the presence of a single nitrobenzyl modified adenosine residue. Upon photolysis of **2** at 308 nm, the caged adenosine was cleanly and quantitatively converted to adenosine with no changes in the composition of the oligonucleotide as monitored by denaturing polyacrylamide gel electrophoresis (PAGE) and HPLC of the digested RNA.

As expected, incubation of the control hammerhead substrate **1** with catalytic amounts of ribozyme **3** led to site-specific cleavage of **1** (Fig. 9.3.2A, lane 2) with kinetics similar to those reported for related substrates [27]. Under identical conditions, the caged RNA substrate **2** was not cleaved (Fig. 9.3.2A, lane 5). Following photolysis, however, the uncaged RNA was site-specifically cleaved in a magnesium-dependent reaction to the same extent as the unmodified substrate **1** (Fig. 9.3.2A, lane 6; typically, 70–80% cleavage of either unmodified or uncaged substrate was observed). In order to further characterize the caged RNA substrate **2**, it was compared with the unmodified RNA substrate **1** in the presence of saturating ribozyme. Under these conditions, following photolysis, the uncaged substrate was cleaved as quickly as the unmodified RNA substrate ($k_{obs} \approx 0.2$ min^{-1} for RNAs **1** and **2**; Fig. 9.3.2B). In order to measure the effect, if any, of the caging functionality on the stability of the ribozyme–substrate complex, equilibrium binding studies measuring the dissociation constants for interaction of the ribozyme **3** with the caged RNA **2** and a modified version of RNA **1** containing 2′-deoxy-adenosine 5′ to the cleavage site were performed [29]. The K_ds for both of the ribozyme complexes were measured to be ~200 nM suggesting that the caging functionality does not appreciably disrupt the ribozyme–substrate complex. Together, these results demonstrate that the reactive 2′-hydroxyl functionality in a hammerhead substrate can be caged and that an efficient and accurate hammerhead reaction is initiated upon photolysis of the caged substrate. Furthermore, neither the photolysis conditions nor the nitroso-aldehyde product of photolysis adversely affect the course of the reaction even under saturating conditions. Finally, the presence of the caging functionality does not disrupt the ribozyme–substrate complex as evidenced by equilibrium binding studies.

9.3.3
Splicing

9.3.3.1 Photocontrol of Splicing using Caged pre-mRNAs

Pre-mRNAs in eukaryotes are characterized by a split-gene structure in which coding exon sequences are separated by non-coding intron sequences [30]. The process by which the introns are excised from the pre-mRNA and the exons are joined together is known as pre-mRNA splicing, and is catalyzed by the spliceosome – a biochemical machine containing both protein and RNA components [31]. The spliceosome includes the U1, U2 and U4/U5/U6 small nuclear ribonucleoprotein particles (snRNPs), each containing a unique RNA and associated proteins. The functional spliceosome is not formed independent of its substrate; instead the active splicing catalyst assembles in an ATP dependent, stepwise, ordered fashion on pre-mRNA substrates through the A, B and C complexes

(A)

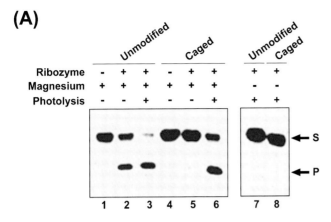

(B)

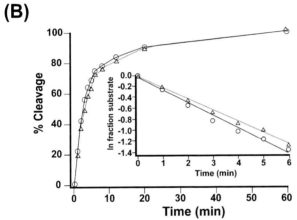

Fig. 9.3.2 (A) Denaturing PAGE analysis of ^{32}P-end-labeled unmodified (RNA 1; S) and caged (RNA 2; S) substrates and 7-nt product (P) of ribozyme-mediated cleavage under conditions of catalytic ribozyme. Lane 1, unmodified substrate; lane 2, unmodified substrate and ribozyme, $t=120$ min; lane 3, unmodified substrate and ribozyme, photolysed, $t=120$ min; lane 4, caged substrate; lane 5, caged substrate and ribozyme, $t=120$ min; lane 6, caged substrate and ribozyme, photolysed, $t=120$ min; lane 7, unmodified substrate and ribozyme in the absence of magnesium, $t=120$ min; lane 8, caged substrate and ribozyme, photolysed, in the absence of magnesium, $t=120$ min. (B) Kinetics of hammerhead catalyzed cleavage of RNA 1 and RNA 2, following photolysis with an excimer laser, under conditions of saturating ribozyme: normalized percent cleavage as a function of time for RNA 1 (△) and RNA 2 (○); plots of ln fraction substrate present versus time give k_{obs} values of 0.20 ± 0.02 min^{-1} for RNA 1 and 0.22 ± 0.02 min^{-1} for RNA 2.

(Fig. 9.3.3 A) which may be visualized by native gel electrophoresis [32]. It would clearly be useful to separate spliceosome assembly from the subsequent chemical steps of pre-mRNA splicing such that the requirements for each could be studied independently.

The chemistry of pre-mRNA splicing involves two sequential transesterification reactions. During the course of spliceosome assembly, a specific adenosine residue in the branch region of the intron becomes bulged from an RNA duplex formed between the pre-mRNA and the U2 snRNA. The 2'-hydroxyl of this residue carries out a nucleophilic attack at the 5' splice site to generate a free 5' exon and a cyclic intermediate containing a 2'–5' phosphodiester branch [33]. Attack of the free 5' exon at the 3' splice site then yields ligated exons and a cyclic intron product (Fig. 9.3.3 B). Removal of the nucleophilic hydroxyl of the branch adenosine by placement of a 2'-deoxy residue at this position prevents branch formation at this position [34].

The demonstration in the hammerhead system that the reactivity of a unique RNA 2'-hydroxyl can be transiently blocked by the presence of a photolabile nitrobenzyl ether at this position suggested that a similar strategy could be applied to study of the splicing reaction. In order to prepare a caged pre-mRNA, a 10-base oligomer was chemically synthesized in which the branch adenosine was modified at the 2' position as an *ortho*-nitrobenzyl ether and all other 2' positions were protected the Fpmp group. Following standard work-up, the purified caged oligomer was ligated to products of T7 RNA polymerase transcription representing the 5' and 3' portions of the PIP85.B pre-mRNA substrate in the presence of a DNA-bridging oligonucleotide to yield a full-length pre-mRNA with caged branch nucleotide (Fig. 9.3.4 A) [34–36]. The resulting RNA contained a single ^{32}P radiolabel that was 2 nt 3' of the caged nucleotide (Fig. 9.3.4 A) facilitating analysis by denaturing PAGE.

The reactivity of the caged RNA was examined by carrying out photolysis experiments using a xenon arc lamp as a light source followed by RNase A and RNase T1 digestion, and thin-layer chromatography (TLC) analysis of the resulting RNA fragments. Although control RNAs were unaffected by photolysis, the caging group was 90% removed upon irradiation to yield the free 2'-hydroxyl (Fig. 9.3.4 B).

The behavior of the caged pre-mRNA under splicing conditions was tested by incubating it in HeLa nuclear extract in the presence of ATP and Mg^{2+} – both required cofactors for pre-mRNA splicing. Products associated with the first and second transesterification reactions of splicing are typically detected after an approximately 20-min lag time required for the assembly of the spliceosome. However, in the case of the caged pre-mRNA, a negligible amount of splicing product was observed (~3% due to some uncaging during handling; Fig. 9.3.5 A). The formation of spliceosomes on the caged pre-mRNA was confirmed by non-denaturing PAGE (Fig. 9.3.5 B) [32]. Thus, while the caging group blocks the first chemical step of splicing, the caged pre-mRNA is still able to direct formation of the spliceosome.

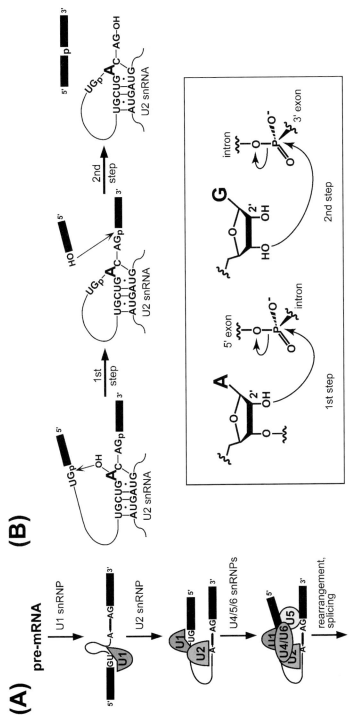

Fig. 9.3.3 (A) Assembly of individual snRNP particles on a pre-mRNA to form the active spliceosome showing conserved 5′ splice site (GU), branch site (A) and 3′ splice site (AG) sequences on the pre-mRNA. (B) Sequential transesterification reactions catalyzed by the spliceosome. In the first step, the 2′-hydroxyl of the branch adenosine within the intron carries out a nucleophilic displacement at the 5′ splice site generating a free 5′ exon and a cyclic (lariat) intermediate. Attack of the free 5′ exon at the 3′ splice site then yields ligated exons and a lariat intron product.

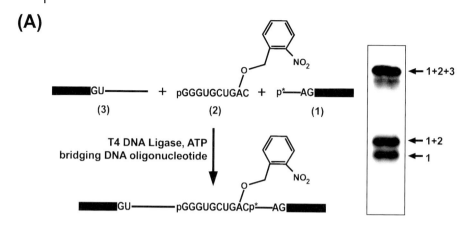

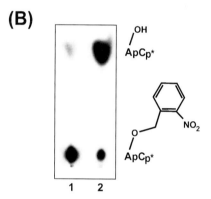

Fig. 9.3.4 (A) Synthesis of pre-mRNAs containing a caged 2′-hydroxyl at the branch position. A synthetic oligomer containing a single modified adenosine residue was ligated to T7 RNA polymerase transcripts to yield full length caged pre-mRNA containing a single ^{32}P radio-label (**p***), right: denaturing PAGE purification of ligation; (B) TLC analysis of photolysis of caged pre-mRNA. Lane 1: unphotolysed caged pre-mRNA digested with RNase A and RNase T1; lane 2: photolysed uncaged pre-mRNA digested with RNase A and RNase T1.

In order to determine whether the spliceosomes formed on the caged substrates were catalytically competent, caged pre-mRNA was incubated in HeLa extract under splicing conditions, the resulting complexes were photolysed, aliquots were removed over time and the RNA analyzed by denaturing PAGE. These experiments showed that uncaging of the pre-mRNA occurred cleanly within the spliceosome and furthermore that the two chemical steps of splicing proceeded without a delay, consistent with the fact that the caged RNA is sequestered within a fully formed spliceosome (Fig. 9.3.5 C).

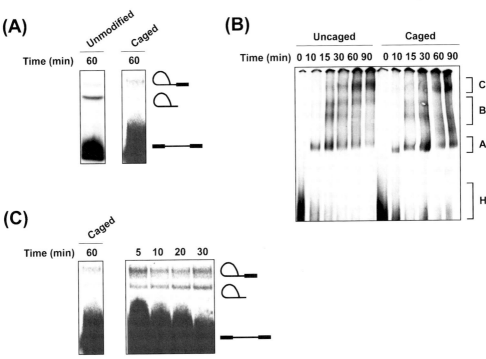

Fig. 9.3.5 (A) Splicing analysis of unmodified and caged pre-mRNAs. Unmodified and caged ^{32}P-labeled pre-mRNAs were incubated under splicing conditions in HeLa cell nuclear extract for 60 min. Left: unmodified pre-mRNA incubated for 60 min under splicing conditions showing production of lariat intermediate (first chemical step) and lariat intron (second chemical step) products. Right: caged pre-mRNA incubated for 60 min under splicing conditions showing no splicing products. (B) Spliceosome formation on uncaged and caged pre-mRNAs. Uncaged and caged ^{32}P-labeled pre-mRNAs were incubated under splicing conditions in HeLa cell nuclear extract for the indicated times to allow formation of the A, B and C (spliceosome) complexes, and analyzed by native gel electrophoresis (4%; 80:1; 50 mM Tris-glycine). (C) Uncaging of pre-mRNA in HeLa cell nuclear extract; denaturing PAGE analysis of ^{32}P-labeled RNA. Left: caged pre-mRNA incubated for 60 min under splicing conditions; right: time course of splicing following photolysis of pre-mRNA–spliceosome complexes showing production of lariat intermediate (first chemical step) and lariat intron (second chemical step) products.

9.3.3.2
ATP Requirements for Splicing

The ATP requirements of splicing were examined by incubating caged pre-mRNA in HeLa extract in the presence of ATP to allow the ATP-dependent formation of the spliceosome. Following a 60-min incubation, the reaction mixture was depleted of ATP by the addition of glucose and the enzyme hexokinase [37]. The reactions were irradiated to uncage the adenosine nucleophile and products were analyzed by denaturing PAGE. Although the first step of splicing was ob-

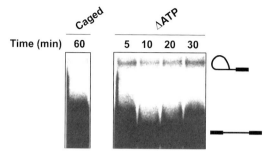

Fig. 9.3.6 ATP requirement following assembly of the spliceosome; denaturing PAGE analysis of ^{32}P-labeled RNA. Caged pre-mRNA was incubated in HeLa extract under splicing conditions to form pre-mRNA–spliceosome complexes after which ATP was depleted (ΔATP) from the reaction by the addition of glucose and hexokinase. Left: caged pre-mRNA incubated for 60 min under splicing conditions; right: time course of splicing following ATP depletion and photolysis of pre-mRNA–spliceosome complexes showing production of lariat intermediate (first chemical step), but no lariat intron (second chemical step) products.

served under these conditions, depletion of ATP had the effect of blocking the second step (Fig. 9.3.6). Addition of excess ATP resulted in complete recovery of the second step of splicing in these reactions (not shown).

Although ATP is required for spliceosome assembly, depletion of ATP from reactions containing fully formed spliceosomes did not prevent first step chemistry. This suggests that the caged pre-mRNA was poised to perform the first transesterification step and further indicates that ATP hydrolysis is in no way coupled to the nucleophilic attack at the 5′ splice site. ATP depletion did prevent the second step of splicing which is in agreement with both genetic studies in yeast [38] and also the ATP requirement for second step chemistry in a *trans*-splicing system [39]. This requirement most likely reflects the ATP requirement of RNA helicases involved in the rearrangement of RNA structure for the second step [31].

9.3.3.3 Phosphatase Activity and Spliceosome Assembly

Lamond et al. have performed a series of studies suggesting that protein phosphorylation and dephosphorylation are critical for both spliceosome assembly and the splicing process itself. Addition of the protein phosphatase PP1 to HeLa extract blocked the formation of E complex on a pre-mRNA during the course of spliceosome assembly [40]. However, if phosphatase activity was blocked by the addition of tautomycin, okadaic acid or microcystin LR spliceosome assembly proceeded through to C complex, as assayed by native gel electrophoresis, but the transesterifications themselves were blocked [41]. Specific inhibition of PP2A using low concentrations of okadaic acid resulted in normal first-step chemistry, but blocked the second step of splicing. Together, these results suggest the requirement of a PP1-specific dephosphorylation immediately prior to

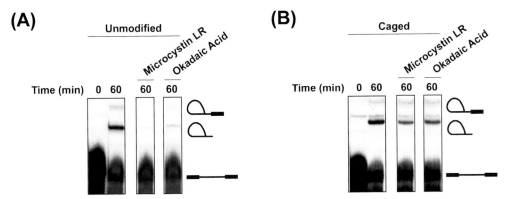

Fig. 9.3.7 Phosphatase requirements of the chemical steps of splicing using ^{32}P-labeled pre-mRNA. (A) Generation of spliced products (left) from unmodified pre-mRNA is blocked by the addition of either microcystin LR or okadaic acid (right). (B) Neither microcystin LR nor okadaic acid block the generation of spliced products from a photolysed caged spliceosome.

the first step of splicing and a PP2-catalyzed dephosphorylation between the two steps of splicing.

In a preliminary series of experiments, the phosphatase requirements for pre-mRNA splicing have been examined using caged pre-mRNA. First, it was confirmed that pre-incubation of HeLa nuclear extract with the phosphatase inhibitors microcystin or okadaic acid inhibited the splicing of an unmodified pre-mRNA (Fig. 9.3.7 A). In a second set of experiments, caged pre-mRNA was incubated in HeLa extract to allow spliceosome assembly to occur, high concentrations of microcystin or okadaic acid were added and the reactions were photolysed to remove the caging group. No inhibition of either step of splicing was observed (Fig. 9.3.7 B). Taken together with the observations of Lamond, these results indicate that a phosphatase activity for pre-mRNA splicing is active within the fully assembled spliceosome before the first step of splicing, but is not directly coupled to that step. Second, if a phosphatase activity is required for the second step of splicing, the relevant dephosphorylation occurs before the *first* step of splicing.

9.3.4
Future Prospects

Caging of the 2′-hydroxyl in an RNA can be used to control RNA processing by the hammerhead ribozyme and the catalytic center of the spliceosome. This methodology could be applied to studies of other systems in which a 2′-hydroxyl is the reactive functionality including the Group II intron, the hepatitis δ, hairpin and VS ribozymes. Finally, the hydrogen bond donor/acceptors of the nucleic acid bases are the principal determinants governing most nucleic acid–nucleic acid and protein–nucleic acid interactions through both Watson-Crick and non-Watson-Crick hydrogen bonding patterns [42]. Caging of these base functionalities will allow the disruption of specific recognition events allowing characterization of the formation of and transition between different nucleic acid complexes.

9.3.5
Experimental

9.3.5.1 **Caged Hammerhead Ribozyme**

9.3.5.1.1 **Synthesis of 2′-O-(2-nitrobenzyl)adenosine Phosphoramidite**
Chemical reagents were purchased from Aldrich Chemical Company. Silica gel (0.03–0.07 mm) for flash chromatography was purchased from Acros. Solvents were distilled as follows: tetrahydrofuran from Na/benzophenone; pyridine, dichloromethane and triethylamine from CaH_2. ^{1}H-NMR spectra were collected on a Varian Gemini-200 instrument (200 MHz) and ^{31}P-NMR on a Varian Gemini-300 instrument (300 MHz). All FAB-HRMS analyses were performed on a VG ZAB-SE machine (Medical Science Mass Spectrometry Facility, University of Toronto). HPLC was performed using a Waters 501 HPLC pump and an analytical C_{18} column (3.9–300 mm; Bondapak) with detection at 254 nm on a Waters 486 detector.

9.3.5.1.2 **2′-O-(2-nitrobenzyl)adenosine**
Adenosine (1.0 g, 3.75 mmol; evaporated 3 times from dry pyridine) was dissolved in 34 mL of hot DMF. To this solution was added NaH (225 mg, 60% in oil, washed 3 times with hexanes) as a suspension in 4 mL of DMF. The resulting solution was stirred at 0 °C for 45 min after which 2-nitrobenzylbromide (1.21 g, 5.6 mmol) in 2 mL of DMF was added. The reaction mixture was then stirred at room temperature under nitrogen for 5 h after which it was poured into 375 mL of ice-cold H_2O and stirred overnight. The resulting yellow precipitate was collected by vacuum filtration and concentrated *in vacuo*. The ^{1}H-NMR (200 MHz) of the crude product was consistent with the literature [23]. The yield was assumed to be 100% for the next step in the preparation.

9.3.5.1.3 **2′-O-(2-nitrobenzyl)-N⁶-benzoyladenosine**
Trimethylsilylchloride (3.8 mL, 30 mmol) was added to a suspension of 2′-O-(2-nitrobenzyl)adenosine (1.5 g, 3.73 mmol) in 20 mL pyridine under nitrogen. The mixture was then stirred for 30 min at room temperature after which time benzoyl chloride (2.4 mL, 20.6 mmol) was added and the reaction stirred for another 2.5 h. The resulting mixture was then cooled to 0 °C, H_2O (4 mL) was added, the reaction was stirred for 5 min, NH_4OH (33%, 8 mL) was added and

stirring continued for an additional 30 min. The mixture was concentrated *in vacuo* and subjected to silica gel flash chromatography in 5% MeOH/CH$_2$Cl$_2$ (R_f=0.3) to yield 647 mg of 2‑O‑(2‑nitrobenzoyl)‑N^6‑benzoyladenosine (43% from adenosine).

^{1}H‑NMR (CDCl$_3$, 200 MHz): δ (p.p.m.) 8.72 (s, 1H, H8), 8.12 (s, 1H, H2), 8.04 (d, 2H, NO$_2$‑ArH), 7.85–7.82 (m, 1H, NO$_2$‑ArH), 7.61–7.28 (m, 6H, ArH), 6.00 (d, 1H, 1′H), 5.09–4.91 (m, 2H, methylene), 4.71 (s, 1H, 3′H), 4.65 (d, 1H, 2′H), 4.21 (s, 1H, 4′H), 3.87 (dd, 2H, 5′H). The 2′ position of the nitrobenzyl group was confirmed by an NOED experiment. FAB-HRMS: calculated for C$_{24}$H$_{22}$N$_6$O$_7$ (MH$^+$), 507.1628; observed, 507.1612.

9.3.5.1.4 5′-O-(4,4′-dimethoxytrityl)-2′-O-(2-nitrobenzyl)-N^6-benzoyladenosine-3′-O-(2-cyanoethyl-N,N-diisopropylamino) Phosphoramidite

5′-O-(4,4′-dimethoxytrityl)-2′-O-(2-nitrobenzyl)-N^6-benzoyl-adenosine (520 mg, 0.6 mmol) was dissolved in dry distilled THF (3 mL) under nitrogen in flame-dried glassware. Then Et(iPr)$_2$N (0.71 mL, 4 mmol) and 2-cyanoethyl-N,N-diisopropylaminochloro-phosphite (0.29 mL, 1.3 mmol) were added and the reaction was allowed to proceed for 6 h. The resulting mixture was poured into EtOAc (200 mL), washed 3 times with 5% NaHCO$_3$ (100 mL), dried over MgSO$_4$, concentrated *in vacuo* and subjected to silica gel flash chromatography in 80% EtOAc/hexanes (R_f=0.5–0.6, both diastereomers) to yield 622 mg (92%) of 5′-O-(4,4′-dimethoxy-trityl)-2′-O-(2-nitrobenzyl)-N^6-benzoyladenosine-3′-O-(2-cyano-ethyl-N,N-diisopropylamino) phosphoramidite.

^{1}H‑NMR (CDCl$_3$, 200 MHz): δ (p.p.m.) 9.55 (bs, 2H, NH$_i$, NH$_{ii}$), 8.61 (d, 2H, H8$_i$, H8$_{ii}$), 8.23 (d, 2H, NO$_2$-ArH$_i$, NO$_2$-ArH$_{ii}$), 7.94 (s, 2H, 2H$_i$, 2H$_{ii}$), 7.94 (d, 4H, NO$_2$-ArH$_i$ NO$_2$-ArH$_{ii}$), 7.64 (d, 2H, NO$_2$-ArH$_i$, NO$_2$-ArH$_{ii}$), 7.51–7.12 (m, 30H, ArH$_i$, ArH$_{ii}$), 6.80–6.74 (m, 8H, DMT-ArH$_i$, DMT-ArH$_{ii}$), 6.26 (m, 2H, 1′H$_i$, 1′H$_{ii}$), 5.26–4.90 [m, 4H, methylene(i), methylene(ii)], 4.69 (m, 2H, 3′H$_i$, 3′H$_{ii}$), 4.46 (m, 2H, 2′H$_i$, 2′H$_{ii}$), 4.39 (m, 2H, 4′H$_i$, 4′H$_{ii}$), 4.09–3.99 (m, 4H, 5′H$_i$, 5′H$_{ii}$), 3.78–3.31 [m, 8H, ethylene, cyanoethyl$_{(i)}$, cyanoethyl$_{(ii)}$], 3.70 [d, 12H, methoxy$_{(i)}$, methoxy$_{(ii)}$], 2.49–2.33 [m, 4H, methine, iPr$_{(i)}$, iPr$_{(ii)}$], 1.29–0.91 [m, 24H, methyl, iPr$_{(i)}$, iPr$_{(ii)}$].

^{31}P-NMR (CDCl$_3$, 300 MHz): δ, 151.27 p.p.m., δ, 150.94 p.p.m. (85% H$_3$PO$_4$ in H$_2$O as external standard). FAB-HRMS: calculated for C$_{54}$H$_{57}$N$_8$O$_{10}$P (MH$^+$), 1009.4013; observed, 1009.4019.

9.3.5.1.5 Nucleoside Model Studies

A 200-μl solution of 2′-O-(2-nitrobenzyl)-adenosine (32 μM) in 20 mM Tris, pH 8.0 in a Pyrex reaction vessel was irradiated with 100 pulses (308 nm; 300 mJ pulse^{-1}; full beam: 1×3 cm) from a Lambda Physik EMG 201 MSC excimer laser. The reaction mixture was directly injected onto a C$_{18}$ reverse-phase HPLC column (Waters) and was analyzed using an elution gradient from

0.05 M TEAA (triethylammonium acetate) to 1:1 0.1 M triethylammonium acetate/acetonitrile (16 min; 1 mL/min). The chromatogram was monitored at 254 nm, peaks were identified by comparison with authentic standards and conversion yields determined by integration of the peaks. Irradiation at 300 mJ pulse^{-1} resulted in 100% conversion, while irradiation at 45 mJ pulse^{-1} resulted in 80–85% conversion.

9.3.5.1.6 T7 Transcription

The hammerhead ribozyme, **3**: 5'-GGGACCACUGAUGAGGCCGUUAGGCC-GAAACACC-3', was synthesized by *in vitro* transcription with T7 RNA polymerase from synthetic DNA templates. Transcription reactions (1 mL) contained 40 mM Tris, pH 7.5, 5 mM DTT, 1 mM spermidine, 3 mM nucleoside triphosphates (Pharmacia), 0.01% Triton X-100, 0.2 mM DNA template (from a template stock solution which contained 10 mM Tris, pH 7.5, 10 mM $MgCl_2$ and 2 mM DNA template containing the T7 promoter region), 2 U T7 RNA polymerase (Promega) and 20 mM $MgCl_2$. The transcriptions were performed at 37°C for 3 h after which time the reaction mixture was extracted with phenol/chloroform/isoamyl alcohol and chloroform and then ethanol precipitated. The crude product was then purified by denaturing 20% (19:1) PAGE.

9.3.5.1.7 Automated Synthesis

All oligonucleotide synthesis was carried out on an Applied Biosystems 392 DNA/RNA Synthesizer. DNA phosphoramidites and 2'-TBDMS-protected RNA phosphoramidites, and 500-Å controlled-pore glass resin for the synthesis of DNA and unmodified RNA substrate **1** were from CPG. The 2'-Fpmp RNA phosphoramidites and controlled-pore glass resin used in the synthesis of caged RNA **2** were from Cruachem. Standard DNA and RNA synthesis cycles were used in all cases except that the standard 1-µmol RNA synthesis cycle was modified to give a coupling time of 15 min. All phosphoramidites were 0.1 M in acetonitrile. The unmodified hammerhead substrate RNA **1**: 5'-GGGUGUAUG-GUU-3' was synthesized using phosphoramidites with 2'-O-tert-butyldimethylsilyl protecting groups on a 1-µmol scale. The RNA was cleaved from the resin and deprotected by treatment with saturated NH_3/MeOH (1 mL) at 55°C for 20 h. The supernatant was recovered and lyophilized to dryness. Deprotection of the 2'-hydroxyls was carried out by treatment with tetrabutylammonium fluoride (600 µL, 0.1 M in THF) at room temperature for 24 h. The oligonucleotide solution was added to 100 µL of 1 M triethylammonium bicarbonate, loaded onto a C_{18} cartridge (Waters/Millipore), washed with 20 mL of 20 mM triethylammonium bicarbonate, eluted with 30% CH_3CN/0.1 M triethylammonium bicarbonate and lyophilized to dryness. The resulting crude product was then purified by denaturing 20% (19:1) PAGE. The modified hammerhead substrate RNA **2**: 5'-GGGUG-UA*UGGUU-3' (A* represents 2' caged adenosine) was synthesized using the 2'-O-(2-nitrobenzyl)adenosine phosphoramidite and 2'-Fpmp phosphoramidites on a 1-μmol scale. The RNA was cleaved from the resin and deprotected by treatment with saturated NH_3/MeOH (1 mL) at 55 °C for 20 h. The supernatant was recovered and lyophilized to dryness. The 2′-Fpmp groups were removed by treatment with 500 μL NaOAc (pH 3.25) for 40 h at room temperature after which the mixture was neutralized with 500 μL Tris buffer (3.15 M, pH 9). The mixture was ethanol precipitated, the residue lyophilized to dryness and the crude product purified by denaturing 20% (19:1) PAGE. The recovered RNA consisted of an 85:15 mixture of RNAs **2** and **1** and was subjected to C_{18} reverse-phase HPLC to yield pure RNA **2** (gradient elution from 9:1 0.05 M triethylammonium acetate/acetonitrile to 7:3 0.05 M triethylammonium acetate/acetonitrile over 25 min; 1 mL/min). Yields of unmodified RNA **1** and modified RNA **2** were 20% as determined by UV absorption at 260 nm.

9.3.5.1.8 Nucleoside Composition Analysis

Enzymatic RNA digests (37 °C, 8 h) of modified and unmodified RNAs were performed in 60 μL reactions containing RNA substrate (5 nmol), 0.2 mM $ZnCl_2$, 16 mM $MgCl_2$, 250 mM Tris, pH 6.0, 0.2 U snake venom phosphodiesterase (Pharmacia) and 4 U calf-intestinal alkaline phosphatase (Boehringer-Mannheim). Following digestion, samples were injected onto a reverse-phase C_{18} HPLC column (Waters) with a gradient elution from 0.05 M triethylammonium acetate to 1:1 0.1 M triethylammonium acetate/acetonitrile (16 min; 1 mL/min). Peaks corresponding to U, G, A and the modified nucleoside 2′-O-(2-nitrobenzyl)adenosine were identified by co-injection of nucleoside standards with the chromatogram monitored at 254 nm.

9.3.5.1.9 RNA Photolysis

RNAs (5 nmol) were photolysed in a Pyrex reaction vessel essentially as described above (308 nm; 140 pulses; 250 mJ pulse^{-1}), enzymatically digested and analyzed by reverse-phase HPLC as described above. Hammerhead·substrate solutions were cooled to 0 °C before flashing and were immediately transferred to a 30 °C bath following photolysis.

9.3.5.1.10 Catalytic Ribozyme Cleavage Reaction

RNA was labeled in reactions (20 μL) containing: 10 pmol RNA, 70 mM Tris pH 7.6, 10 mM $MgCl_2$, 5 mM DTT, 1 U T4 polynucleotide kinase (New England Biolabs) and 10 μL [γ-^{32}P]ATP (3000 Ci/mmol, 5 mCi/mL, NEN). Reactions were carried out at 37 °C for 10 min, extracted with phenol/chloroform/isoamyl alcohol and chloroform, and then ethanol precipitated. Excess substrate cleavage reactions contained 40 nM ribozyme **3**, 100 nM substrate (RNA **1** or **2**), 50 mM Tris pH 6.0 and 10 mM $MgCl_2$ (or 10 mM EDTA for control reactions) in a reaction volume of 20 μL. Ribozyme and substrate solutions in reaction buffer were heated to 90–95 °C for 1 min and then allowed to cool to 22 °C over 15 min after which the substrate solution was brought to 10 mM in MgCl$_2$ (or 10 mM in EDTA). The cleavage reaction was initiated by adding ribozyme to the substrate and then flashed, if required, as described above (308 nm; 140 pulses; 250 mJ pulse^{-1}). After 2 h, the reaction was stopped by the addition of 20 µL of stop solution (0.05% bromophenol blue, 0.05% xylene cyanol, 8 M urea and 100 mM EDTA) and heated to 90–95 °C for 1 min before loading onto 20% (19:1) polyacrylamide denaturing sequencing gels for quantification. Gels were quantified using a Molecular Dynamics PhosphorImager and ImageQuant software version 3.22.

9.3.5.1.11 Saturating Ribozyme Kinetics

Hammerhead cleavage reactions were performed under saturating ribozyme conditions at pH 7.5. The cleavage reactions contained 7500 nM ribozyme 3, 10 nM substrate (RNA **1** or **2**), 50 mM Tris, pH 7.5 and 10 mM MgCl$_2$ in a total volume of 100 µL. Separate solutions of ribozyme and substrate were prepared, heated to 90–95 °C for 1 min and allowed to cool over 5 min to 30 °C. Then both solutions were brought to 10 mM in MgCl$_2$ and 50 mM in Tris, pH 7.5, and combined to initiate the reaction. Reactions containing the modified hammerhead substrate were photolysed as described above (308 nm; 140 pulses; 250 mJ pulse^{-1}). Aliquots, from 0 to 60 min, taken from the reactions were combined with an equal volume of stop solution (0.05% bromophenol blue, 0.05% xylene cyanol, 8 M urea and 100 mM EDTA) and heated to 90–95 °C for 1 min before loading onto 20% (19:1) polyacrylamide denaturing sequencing gels for quantification. Gels were quantified using a Molecular Dynamics PhosphorImager and ImageQuant software version 3.22.

9.3.5.1.12 Measurement of Equilibrium Dissociation Constants for Ribozyme–RNA Complexes

Equilibrium dissociation constants (K_d values) for ribozyme–RNA complexes were determined as reported elsewhere [29]. Binding experiments were carried out under native conditions (50 mM Tris, pH 7.5, 10 mM MgCl$_2$, 5% sucrose, 0.02% bromophenol blue, 0.02% xylene cyanol) comparing the caged RNA **2** with an RNA containing a single deoxy-adenosine residue 5′ to the cleavage site. Binding reactions were allowed to equilibrate for 15 h at room temperature and resolved on 15% native polyacrylamide gels (19:1; 13–22–0.15 cm) in 50 mM Tris acetate, pH 7.5, 10 mM magnesium acetate buffer (pre-run for 2 h at 6 W followed by buffer exchange and then run at 6 W for 6 h at 4 °C). Gels were quantified using a Molecular Dynamics PhosphorImager and ImageQuant software version 3.22. K_d values were determined from Scatchard analysis of the binding data.

9.3.5.2 Caged Pre-mRNA

9.3.5.2.1 Preparation of Caged Pre-mRNA

A 10-base oligoribonucleotide containing a single caged adenosine was synthesized and purified as described above. Full length caged pre-mRNA was synthesized by incubating 5′-phosphorylated synthetic branch oligomer (300 pmol) with upstream (300 pmol) and 5′ ^{32}P-labeled downstream (100 pmol) T7 RNA transcription products representing the 5′ and 3′ portions of the PIP85.B pre-mRNA in the presence of a bridging DNA oligonucleotide and 40 U T4 DNA ligase (60 mM Tris, pH 7.8, 20 mM MgCl$_2$, 36 U ribonuclease inhibitor, 1.2 mM ATP, 2.4% PVP-40, 5 mM DTT) at 30 °C for 3 h. Ligations were purified directly by 15% PAGE. The products were visualized by autoradiography, extracted, dissolved in double-distilled water and stored at –20 °C.

9.3.5.2.2 TLC Analysis of Pre-mRNA

Caged pre-mRNAs containing a single ^{32}P label two nucleotides 3′ to the modification were digested with 2 U each of RNase T1 and RNase A (20 µL reaction volume, 10 mM Tris, pH 7) both before and after photolysis (irradiation at a distance of 1 cm with a 1000 W Oriel Xenon arc lamp for 4 s in 5-mm Pyrex reaction vessels). Reactions were then concentrated and loaded onto a cellulose PEI TLC plate (4×10 cm, J.T. Baker) and eluted for 1 h in 79:19:1 saturated (NH$_4$)$_2$SO$_4$/1 M NH$_4$OAc/isopropanol. The air-dried TLC plate was exposed to a Molecular Dynamics phosphor screen and then scanned using a Molecular Dynamics Storm 860 PhosphorImager.

9.3.5.2.3 Splicing and Photolysis of Caged Pre-mRNA

RNAs (50–100×10^3 c.p.m.) were incubated in 10 µL reactions containing 40% HeLa nuclear extract, 2 mM MgCl$_2$, 60 mM KCl, 1 mM ATP, 5 mM creatine phosphate, 4 U ribonuclease inhibitor and 4 µg tRNA. Following a 60-min incubation period at 30 °C, splicing complexes were photolysed as described above. ATP depletions were effected by adding glucose to a final concentration of 7 mM, adding 0.3 U of hexokinase and incubating at 37 °C for 10 min. Phosphatase inhibition was effected by the addition to splicing extracts of either Microcystin LR (Sigma) or okadaic acid (Sigma) to a final concentration of 11 µM. During the course of splicing, aliquots were removed at various times, the reactions were quenched by extraction with phenol/chloroform/isoamyl alcohol, ethanol precipitated and then subjected to denaturing PAGE (15%, 29:1). Dried gels were exposed to a Molecular Dynamics phosphor screen which was then scanned using a Molecular Dynamics Storm 860 PhosphorImager.

Acknowledgments

This work was supported by the Natural Sciences and Engineering Research Council of Canada (NSERC), by a grant from the Connaught Fund of the University of Toronto, and by a Research Opportunity Award (Research Corporation). AMM acknowledges support from the Natural Sciences and Engineering Research Council of Canada (NSERC) and the Alberta Heritage Foundation for Medical Research (AHFMR). OAK is supported by an AHFMR graduate fellowship.

References

1 R. F. Gesteland, J. F. Atkins (eds), *The RNA World*. Cold Spring Harbor Laboratory Press, Cold Spring Harbor, NY, **1992**.
2 M. V. Rodnina, W. Wintermeyer, *Curr. Opin. Struct. Biol.* **2003**, *13*, 334–340.
3 A. E. Ferentz, G. L. Verdine, *J. Am. Chem. Soc.* **1991**, *113*, 4000–4002.
4 S. A. Wolfe, G. L. Verdine, *J. Am. Chem. Soc.* **1993**, *115*, 12585–12586.
5 C. R. Allerson, G. L. Verdine, *Chem. Biol.* **1995**, *2*, 667–675.
6 J. T. Goodwin, G. D. Glick, *Tetrahedron Lett.* **1994**, *35*, 1647–1649.
7 S. T Sigurdsson, T. Tuschl, F. Eckstein, *RNA* **1995**, *1*, 575–583.
8 J. A. McCray, D. R. Trentham, *Annu. Rev. Biophys. Biophys. Chem.* **1989**, *18*, 270–270.
9 S. R. Adams, R. H. Symons, *Annu. Rev. Biochem.* **1992**, *61*, 641–671.
10 R. Y. Tsien, *Annu. Rev. Physiol.* **1993**, *55*, 755–784.
11 S. N. Cook, W. E. Jack, X. Xiong, L. E. Danley, J. A. Ellman, P. G. Schultz, C. J. Noren, *Angew. Chem. Int. Ed.* **1995**, *34*, 1629–1630.
12 O. C. Uhlenbeck, *Nature* **1987**, *328*, 596–600.
13 A. D. Branch, H. D. Robertson, *Proc. Natl. Acad. Sci. USA* **1991**, *88*, 10163–10167.
14 J. M. Buzayan, W. L. Gerlach, G. Bruening, *Nature* **1986**, *323*, 349–353.
15 H. C. Guo, R. A. Collins, *EMBO J.* **1995**, *14*, 368–376.
16 C. L. Peebles, P. S. Perlman, K. L. Mecklenburg, M. L. Petrillo, J. H. Tabor, K. A. Jarrell, H. L. Cheng, *Cell* **1986**, *44*, 213–223.
17 M. M. Konarska, P. J. Grabowski, R. A. Padgett, P. A. Sharp, *Nature* **1985**, *313*, 552–557.
18 S. G. Chaulk, A. M. MacMillan, *Nucleic Acids Res.* **1998**, *26*, 3173–3178.
19 S. G. Chaulk, A. M. MacMillan, *Angew. Chem. Int. Ed.* **2001**, *40*, 2149–2152.
20 H. W. Pley, K. M. Flaherty, D. B. McKay, *Nature* **1994**, *372*, 68–74.
21 W. G. Scott, J. T. Finch, A. Klug, *Cell* **1995**, *81*, 991–1002.
22 W. G. Scott, J. B. Murray, J. R. P. Arnold, B. L. Stoddard, A. Klug, *Science* **1996**, *274*, 2065–2069.
23 E. Ohtsuka, S. Tanaka, M. Ikehara, *Chem. Pharm. Bull.* **1977**, *25*, 949–959.
24 D. G. Bartholomew, A. D. Broom, *J. Chem. Soc. Chem. Commun.* **1975**, 38.
25 E. Ohtsuka, T. Wakabayashi, S. Tanaka, T. Tanaka, K. Oshie, A. Hasegawa, M. Ikehara, *Chem. Pharm. Bull.* **1981**, *29*, 318–324.
26 D. C. Capaldi, C. B. Reese, *Nucleic Acids Res.* **1994**, *22*, 2209–2216.
27 D. M. Williams, W. A. Pieken, F. Eckstein, *Proc. Natl Acad. Sci. USA* **1992**, *89*, 918–921.
28 J. F. Milligan, O. C. Uhlenbeck, *Methods Enzymol.* **1989**, *180*, 51–62.
29 M. J. Fedor, O. C. Uhlenbeck, *Biochemistry* **1992**, *31*, 12042–12054.
30 P. A. Sharp, *Angew. Chem. Int. Ed.* **1994**, *33*, 1229–1240.
31 J. P. Staley, C. Guthrie, *Cell* **1998**, *92*, 315–326.
32 M. M. Konarska, P. A. Sharp, *Cell* **1987**, *49*, 763–774.

33 M. M. Konarska, P. J. Grabowski, P. A. Padgett, P. A. Sharp, *Nature* **1985**, *313*, 552–557.
34 C. C. Query, M. J. Moore, P. A. Sharp, *Genes Dev.* **1994**, *8*, 587–597.
35 M. J. Moore, P. A. Sharp, *Science* **1992**, *256*, 992–997.
36 A. M. MacMillan, C. C. Query, C. R. Allerson, S. Chen, G. L. Verdine, P. A. Sharp, *Genes Dev.* **1994**, *8*, 3008–3020.
37 L. A. Lindsey, A. J. Crow, M. A. Garcia-Blanco, *J. Biol. Chem.* **1995**, *270*, 13415–13421.
38 B. Schwer, C. Guthrie, *EMBO J.* **1992**, *11*, 5033–5039.
39 K. Anderson, M. J. Moore, *Science* **1997**, *276*, 1712–1716.
40 J. E. Mermoud, P. T. Cohen, A. I. Lamond, *EMBO J.* **1994**, *13* 5679–5688.
41 J. E. Mermoud, P. Cohen, A. I. Lamond, *Nucleic Acids Res.* **1992**, *20*, 5263–5269.
42 W. Saenger, *Principles of Nucleic Acid Structure*, Springer, New York, **1984**.

9.4
Light Reversible Suppression of DNA Bioactivity with Cage Compounds

W. Todd Monroe and Frederick R. Haselton

9.4.1
Introduction

In this chapter we examine progress in photochemical control of DNA-dependent processes such as transcription, translation, DNA–DNA hybridization and DNA–protein interactions. Both *in vitro* and *in vivo* examples are described. Emphasis is placed on the functional regulation of these processes and their utility in biology, rather than on the chemical synthesis and photolysis of the caged complexes, which is covered in greater detail in other chapters of this book. As in other chapters, the basis of these efforts is the stable covalent modification of DNA structure by photocleavable cage structures. The presence of the cage structures suppresses or blocks the native DNA bioactivity of interest, but upon exposure to light native bioactivity is restored.

The native bioactivities of DNA are manifold and it would be impossible to attempt to describe all of the potential uses for caged DNA. Some examples are sketched in Fig. 9.4.1. In this chapter we focus on demonstrating several applications, generally presented in order of increasing complexity. Reports documenting reversible changes in DNA structure are followed by examples of reversible blockade of restriction enzyme digestion, *in vitro* translation, *in vitro* hybridization, cellular blockade of expression by exogenous plasmids, blockade of antisense knock down in cells and, finally, examples of blockade of expression *in vivo*.

To date, there have been few publications in this area. Our efforts to cage DNA with 1-(4,5-dimethoxy-2-nitrophenyl)diazoethane (DMNPE) were the first demonstration of photoactivated gene expression with cage compounds [1, 2]. A

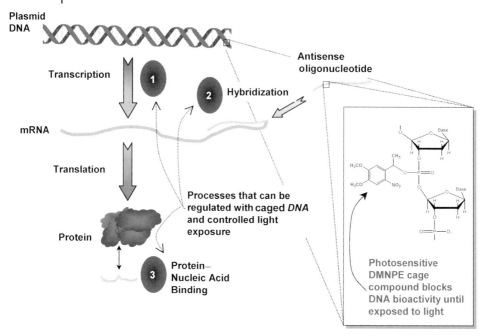

Fig. 9.4.1 Examples of DNA-dependent bioactivities that can be controlled with cage compounds.

second group has carried out some of the same studies using the caging group 6-bromo-7-hydroxycoumarin-4-ylmethyl (Bhc) and extended some of our findings in an attempt to develop a conditional expression strategy of caged microinjected mRNA in zebrafish embryos [3–5]. Both of these caging strategies are characterized by random attachment of cage groups during the caging reaction. Other groups have developed directed caging strategies to examine additional DNA processes, which are also discussed [6, 7].

9.4.2
Evidence for DNA Caging

We adapted Trentham et al.'s published synthesis of DMNPE-caged ATP to cage plasmid DNA [8]. The reaction protocol for this random attachment of DMNPE to plasmid involves activation of the hydrazone 4,5-dimethoxy-2-nitroacetophenone with manganese(IV) oxide by gentle agitation in dimethylsulfoxide (DMSO) at room temperature for 20 min. The oxidation reaction is rapid and can be monitored by a change in the solution from an amber color to a deep red tint when activated. To remove the oxidizing agent from the activated cage before addition to target DNA, manganese oxide can be filtered with diatomaceous earth supported by glass wool in a syringe. The activated cage filtrate (150 µL) is then added to 150 µg of plasmid DNA in 300 µL of Bis–Tris buffer

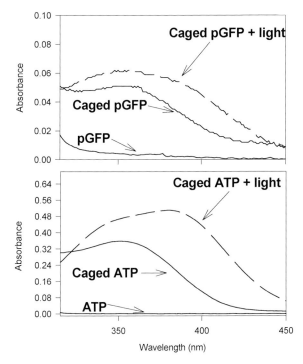

Fig. 9.4.2 Light-induced changes in absorbance of DMNPE-caged ATP and DMNPE-caged pGFP [1].

(10 mM Bis–Tris–Cl, pH 5.5) and gently agitated for 24 h at 4 °C. Caged DNA is then extracted twice with an equal volume of chloroform, followed by ethanol precipitation and stored in water or TE buffer at 4 °C. Similar methods have been described by others [4].

We are unaware of any published reports that provide direct evidence for the covalent attachment of cage groups to DNA, although it has been speculated to couple at one of the non-bridging oxygens on the phosphate backbone [1, 4]. This finding is in agreement with a structural study that confirmed dinucleotides caged with DMNPE showed adduction at the bridging phosphate [9]. Most have sought indirect evidence by measuring changes in light absorption or in gel behavior. For example, Fig. 9.4.2 illustrates the changes in light absorption observed with caged plasmid DNA. Similar results have been reported by others using Bhc caging [4]. Reversible changes in gel migration patterns have also been reported. An example of these shifts is shown in Fig. 9.4.3. Similar shifts have been reported for caged mRNA [4].

Because of the large extinction coefficients of the nitrobenzyl cage compounds, the degree of cage adduction can be estimated using spectrophotometric analysis. Fig. 9.4.2 contains absorbance spectra comparing DMNPE-caged pGFP plasmid and DMNPE-caged ATP. The solid curves are absorbance spectra before caged ma-

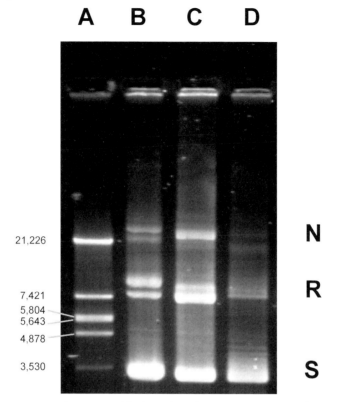

Fig. 9.4.3 Caging reversibly alters plasmid structure. DNA gel of caged and caged-light-exposed samples of GFP plasmids. The markers in lane (A) are from an EcoRI digest of λ DNA. Lane (B) is caged plasmid. Lane (C) contains caged-light-exposed plasmid exposed to 20 min of 365-nm light. Lane (D) contains plasmid that was exposed to the caging reaction conditions but without the caging compound. Typical plasmid conformation bands are observed; the bands with greatest mobility correspond to multiple conformations observed with supercoiled bands (S), the least mobile bands correspond to the nicked conformations (N) and the bands that appear between these two groups are presumed to be relaxed (R) [1].

terials were exposed to light, and the dashed curves are spectra after exposure to 20 min of 365-nm light. Similar shifts in absorbance at 355 nm are observed for DMNPE-caged pGFP and commercially available DMNPE-caged ATP (Molecular Probes, Eugene, OR). After exposure to 365-nm light, increases in absorbance at 390 nm are observed for both compounds as a result of photolysis of the caging group.

Light changes characteristic of DMNPE-caged ATP can be used to estimate the degree of DMNPE caging of DNA [1]. The extinction coefficient ($\varepsilon_\lambda = 355$) of the bound caging DMNPE group was calculated from the local absorbance peak at 355 nm before exposures to 365-nm light of commercially available DMNPE-caged ATP (Molecular Probes). The extinction coefficient for DMNPE ($\varepsilon_\lambda = 355$) was estimated to be 5340 M^{-1} cm^{-1} [1] and used to determine the molar concentration of DMNPE in the caged plasmid samples. Based on the extinction coefficient of attached DMNPE at 355 nm, the number of caging groups per plasmid for the caged plasmid shown in Fig. 9.4.2 is ~270, which is 2.7% of the ~10 000 available phosphate sites on the DNA backbone [1].

9.4.3
Reversible Blockade of Restriction Enzyme Activity

We have found that caging of plasmid DNA can interfere with some DNA–protein interactions. For example, caging plasmids with DMNPE reversibly blocks cleavage by restriction enzymes. Plasmids caged as described above were digested with *Bam*HI (NEB, Beverly, MA), a restriction endonuclease that cuts at one site on the pGreenLantern-1 (pGFP) plasmid. Native, caged and caged-light-exposed pGFP were digested with *Bam*HI (0.02 µg µL^{-1}) for 24 h at 37 °C and the products run on 1% agarose gels. Figure 9.4.4 shows a gel of the *Bam*HI digestion products. Banding patterns show reduced cleavage for caged plasmid, as there is a distinguishable band corresponding to the supercoiled conformation that remains uncut (lane D). Caged plasmids exposed to 365-nm light (5 J cm^{-2}) show increased cleavage by *Bam*HI, as there is decreased intensity in the supercoiled band, similar to the pattern seen with digested native pGFP (lanes E and B). These results indicate that the presence of the DMNPE group at or near a restriction enzyme's recognition site on plasmid DNA may block its complete cleavage. Blockade of DNA–protein interactions by DMNPE adduction would be expected, as the attachment of single methyl groups and other small compounds has been shown to effect histone–DNA interactions in gene expression (reviewed in [10]). Similar to methylation-based diagnostic techniques, this information could lend information about the changes in DNA structure produced by the covalent attachment of DMNPE.

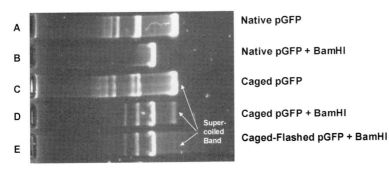

Fig. 9.4.4 Agarose gel of *Bam*HI-digested native, caged and caged-light-exposed pGFP plasmids [2].

9.4.4
Reversible Blockade of *In Vitro* Transcription

In vitro, caging interferes with DNA transcription. mRNA produced using an *in vitro* transcription kit from native (non-caged), DMNPE-caged, and caged-flashed (light-exposed) plasmids were compared on an agarose gel (Fig. 9.4.5). DMNPE-caged pGFP plasmids are not transcribed *in vitro* (D). Caged-light-exposed plasmids produce an mRNA band (E) similar to that produced by native plasmids (B), and caging reaction controls (C). In Fig. 9.4.5, the intensities for the native template and the template subjected to the extraction and isolation procedures of the caging reaction were within 10% of each other (103 264 and 114 830 counts, respectively). Average background (29 731) was subtracted and these values were averaged and used as a standard for comparison. The caged template and caged-light-exposed template samples showed intensities of 35 522 and 59 311 counts. These values were halved to account for the increased (×2) loading of the caged and light-exposed lanes, resulting in mRNA production estimates of 3.6% for caged and 19% for caged-light-exposed of the average native template.

Similar to our observations with DNA, caged and biologically inactive mRNA can also be restored to its native bioactive state with light [4]. mRNA generated from an *in vitro* transcription kit was randomly caged with Bhc, a modification

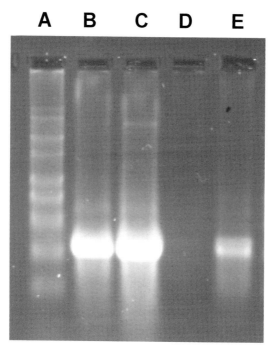

Fig. 9.4.5 Agarose–formaldehyde gel (1.5%) of mRNA from *in vitro* transcription reaction of caged and caged-light-exposed plasmids [1].

of Tsien's bromocoumarin cage [11]. Bhc reportedly has better quantum efficiencies than DMNPE, which would require less light for complete photocleavage after delivery to cells. Okamoto et al. showed complete blockage of *in vitro* translation with approximately 3% caging of supposed phosphate backbone moieties on mRNA. Exposure to low levels of 355–365-nm light (100 mJ cm^{-2}) yielded recovery of translational activity [4].

9.4.5
Reversible Blockade of Hybridization

Caging oligonucleotides also appears to reversible block the binding of a DNA strand with its complement. We used a molecular beacon to demonstrate reversible blockade of hybridization as illustrated in Fig. 9.4.6. Samples of 2 μg of a caged 20mer "Probe" phosphorothioated DNA oligonucleotide and 2 μg of a 30mer "Target" complementary Molecular Beacon oligonucleotide (Biosearchtech, Novato, CA) were denatured at 90 °C for 1 min and allowed to incubate for 2 h at 25 °C in 100 mM NaCl, 1 mM EDTA. Immediately before denaturing, the sample was halved and one half exposed to light while the other half was kept in the dark. The 30mer molecular beacon contains a 5'-FAM Fluorophore, a 3'-quenching group, and five complementary bases on the 3' and 5' ends, so that unless hybridized to a complementary strand, the fluorescence of the beacon in its hairpin configuration is quenched. The basic scheme of this design is shown in Fig. 9.4.6. The hybridization reaction was then assayed by non-denaturing gel electrophoresis (Fig. 9.4.7) and fluorimetry (Fig. 9.4.8). For electrophoresis, 200 ng of the hybridization mixture was run in 15% non-denaturing polyacrylamide, Tris-borate buffer, for 90 min. Gels were stained for 30 min with Sybr-Gold Nucleic Acid stain (Molecular Probes). For solution fluorescence quantitation, 200 ng of the hybridization mixture was diluted in 100 μL Tris-acetate buffer, excited at 488 nm and emitted fluorescence quantified at 512 nm in a Perkin Elmer LS50B luminescence spectrophotometer.

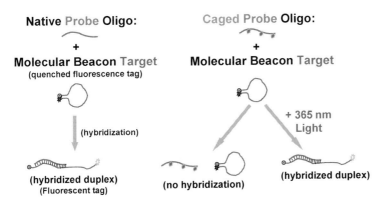

Fig. 9.4.6 Hybridization assay of caged probe and complementary molecular beacon [2, 12].

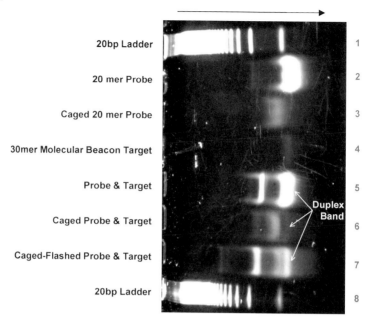

Fig. 9.4.7 Non-denaturing gel of hybridization products from caged and caged-flashed samples of hybridized oligonucleotides [2, 12].

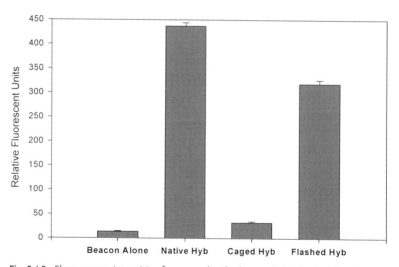

Fig. 9.4.8 Fluorescence intensities from a molecular beacon hybridized with native, caged or caged-light-exposed ODNs [2, 12].

Fluorescence emissions from a complementary molecular beacon indicate differing amounts of hybridization for caged and native ODNs (Fig. 9.4.8). Relative fluorescence units from a hybridization mixture of caged ODNs and complementary beacon are similar to levels from molecular beacon alone. When exposed to 5.6 J cm^{-2} of 365 nm light prior to hybridization, fluorescence emission increases, indicating a restoration of hybridization activity. The level of hybridization activity between the caged and caged-light-exposed ODN-beacon hybridizations increases from 5 to 72% of native ODN activity when the background of native probe alone is subtracted.

Non-denaturing polyacrylamide electrophoresis shows shifts in hybridized oligonucleotides. Note the dramatic shifts in the hybridized duplex band intensity between caged and caged-flashed samples (lanes 6 and 7). The caged 20mer does not completely hybridize with its complementary sequence on the 30mer molecular beacon. However, the half of the sample that was exposed to light shows more oligonucleotide in the duplex form, resembling that of the native probe hybridization. The native oligonucleotide hybridization banding pattern is shown for comparison (lane 5). These results indicate that the presence of the cage compound on the 20mer probe oligonucleotide can reversibly alter its hybridization with a complementary target.

9.4.6
Reversible Inhibition of Antisense Activity

We used cell-surface expression of ICAM-1 in combination with an antisense inhibitor of ICAM-1 expression to demonstrate reversible blockade of antisense activity. The scheme is illustrated in Fig. 9.4.9. This model of antisense activity was developed using ISIS 2302, a 20mer phosphorothioate oligonucleotide designed by ISIS Pharmaceuticals to block expression of intracellular adhesion molecule-1 (ICAM-1) in cultured HeLa cells [13]. ISIS 2302 is a 20mer phosphorothioate (PS) ODN engineered to block the cytokine-induced expression of ICAM-1. This system is particularly appropriate for study in antisense blockade assays, because ICAM-1 expression induced by treatment of interferon (IFN)-γ in HeLa cell cultures is dependent on *de novo* mRNA synthesis and translation of the protein, and is reversible [14].

In these studies, HeLa cells were liposome-transfected with 0.75 µg of caged, caged-light-exposed or native ISIS 2302 oligonucleotide. Oligonucleotides were complexed with GenePorter lipid transfection reagent (Gene Therapy Systems, San Diego, CA) 30 min prior to application. Following a 4-h transfection period, cells were treated with 0.5 µM IFN-γ for 20 min to stimulate inducible ICAM-1 expression. After 24 h, cells were harvested, and labeled with a FITC-conjugated anti-ICAM-1 antibody (Sigma, St Louis, MO). Expression of ICAM-1 was then quantified by flow cytometry and expressed as a percent of the cells receiving no antisense:

% Control activity=[(Sample $-$ Negative control)/(IFN-γ $-$ Negative control)]$\times$100

ICAM-1 induction in HeLa cells:

Antisense blockade of ICAM-1 expression:

Control of antisense activity:

Fig. 9.4.9 *In vitro* antisense assay for ICAM-1 in HeLa cells [2].

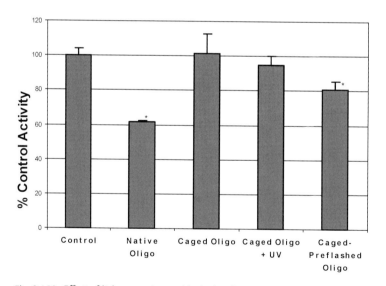

Fig. 9.4.10 Effect of light on antisense blockade of ICAM-1 expression in HeLa cells [2].

where "Sample" is the expression level of the sample being calculated, "Negative control" is the ICAM-1 expression level of cells that are not stimulated with IFN-γ and "IFN-γ" is the ICAM-1 expression level of cells that have been stimulated with IFN-γ.

As shown in Fig. 9.4.10, we found that the expression level of ICAM-1 in cells treated with native ISIS 2302 was 62±11% of untreated samples ($n=8$, mean±SD). Without exposure to light (i.e. with 0 J cm^{-2} of 365-nm light) ICAM-1 expression in samples transfected with caged ISIS 2302 is equal to that of untreated samples (101±5%), indicating that caged oligos have no antisense activity. After exposure to 0.25 mJ cm^{-2} of light, the activity of caged antisense increases, as the ICAM-1 expression is 95±10% of control. Cultures transfected with caged ISIS 2302 exposed to light before transfection showed ICAM-1 levels of 80±7% of untreated samples, indicating a further restoration of antisense activity. Asterisks indicate significant difference from ICAM-1 expression in untreated samples ($p<0.05$, Bonferroni's t-test).

9.4.7
Reversible Blockade of GFP Expression in HeLa Cells

Caged plasmids introduced into cells using liposomes can be induced to express GFP (Fig. 9.4.11). Without exposure to light (i.e. with 0 J cm^{-2} of 365-nm light, the fraction of HeLa cells that express caged pGFP (solid bar) is about one-fourth of the level

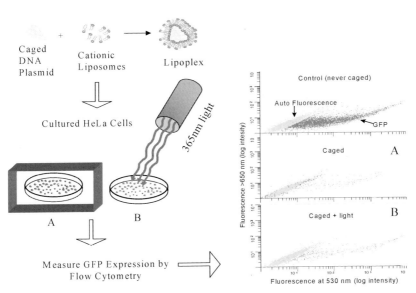

Fig. 9.4.11 In vitro GFP assay in HeLa cells. Using liposomal techniques, cells are transfected with caged plasmids, and selected cultures (A) are held in the dark, while others (B) are exposed to light. Expression of caged (A) versus caged+light (B) cultures are measured using flow cytometry [1].

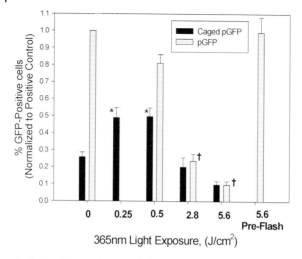

Fig. 9.4.12 Effects of 365-nm light dosage on pGFP and caged-pGFP expression [1].

of expression of non-caged material (crosshatched bars, Fig. 9.4.12). The expression levels of native non-caged pGFP was $43 \pm 4.3\%$ of cells, and seemed to vary from culture to culture ($n=7$, mean $\pm$ SEM). To reduce variability among experiments, percentages of expression were normalized to this group. After exposure to 0.25 or 0.5 J cm^{-2} of light, expression of the caged material increases to 50% of control. The asterisks above these bars of light indicate significant difference from expression of caged plasmids that received no light exposure ($p<0.05$, Bonferroni's t-test). Beyond 0.5 J cm^{-2}, increased light dosages caused a decrease in expression levels of caged pGFP. Cultures transfected with caged pGFP and treated with 2.8 and 5.6 J cm^{-2} of light showed expression levels of 20 and 10%, respectively. This decrease in expression with increasing light dosage is also observed for non-caged pGFP plasmids. GFP expression levels decreased with increasing post-transfection light exposure, from a normalized 100% with no light exposure, to 81, 24 and 10% with light exposures of 0.5, 2.6 or 5.6 J cm^{-2}, respectively. Cultures that were exposed to 2.8 or 5.6 J cm^{-2} of light after transfection showed significantly lower levels of expression than those that received no light (denoted by crosses, $p<0.05$, Bonferroni's t-test). This decrease is evidently not an effect of light on plasmid since pGFP plasmids exposed to the highest dose of light before transfection (bar labeled "Pre-Flash") express at levels equal to control plasmids that received no light.

9.4.8
Reversible Blockade of Plasmid Expression in Cultured Corneas

In an initial feasibility test, we utilized gene gun delivery of caged plasmids into a cultured cornea system to demonstrate reversible blockade of plasmid gene expression in tissues. Corneas from New Zealand white albino rabbits were placed

Fig. 9.4.13 Expression of DMNPE-caged β-galactosidase in corneal tissues [15].

in an organ culture medium for 5 days prior to gene transfer. LacZ with CMV promoter was used as a reporter gene. An Accell Pulse Gun (Auragen, Middleton, WI) was used to deliver gold particles with either DMNPE caged or native plasmids to the cultured corneas with blasts of helium at 250 p.s.i. One of the transfected corneas was exposed to 355-nm light for 15 min. The corneas were stained for LacZ expression 24 h after transfection.

The two top panels of Fig. 9.4.13 show the low level of expression of β-galactosidase (small dark dots) when transfected with caged pSV β-galactosidase plasmid (2 mags). As shown in the lower panels, after the tissue containing caged pSV β-galactosidase plasmid was exposed to 365-nm light, the expression of β-galactosidase protein increased dramatically. Highest LacZ expression was seen in the corneas transfected with control plasmids (not shown) and lowest expression was seen in corneas with caged plasmid without 365-nm light exposure (upper panels). Although light exposure did not achieve the level of expression seen with plasmids that were not caged, the increase in expression with light treatment (lower panels) compared to untreated caged plasmids (upper panels) is substantial. This suggests that expression of caged plasmid can be activated by light following gene gun delivery to the corneal epithelial layer and it is possible to control the expression of DNA after delivery to a tissue.

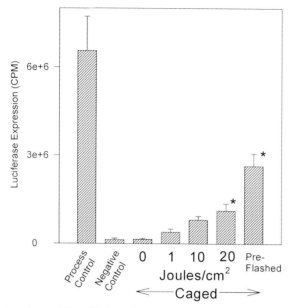

Fig. 9.4.14 Effect of light on luciferase expression in rat skin [1].

9.4.9
Reversible Blockade of Plasmid Expression *In Vivo*

In addition to controlling the kinetics of a particular molecular target, caging affords the ability to restrict re-activation to a localized area of interest to achieve targeted expression. In a design similar to our corneal experiments, we used a gene gun delivery system to deliver plasmids to the skin of rats (Fig. 9.4.14). We found that luciferase expression of skin sites transfected with caged plasmid is equal to levels in untransfected skin. Exposure of skin sites transfected with caged plasmids to increasing amounts of 355-nm laser light increases expression to 6 ± 3, 12 ± 4 and $17\pm6\%$ of control, respectively. Right bar indicates highest expression ($40\pm12\%$ of control) from skin sites transfected with caged plasmids exposed to light before delivery. The asterisks indicate difference from no light exposure by two-way repeated measures ANOVA ($n=4$; mean±SEM).

9.4.10
Reversible Blockade of mRNA for Expression Control *In Vivo*

Caged mRNA has also been microinjected in zebrafish embryos to induce expression of certain genes and study the effect of their expression on developmental patterns [4]. This study took an important next step and microinjected caged mRNA into zebrafish embryos at the single-cell stage. Caged mRNA was not degraded and spread throughout the developing embryo with subsequent cell divisions. Photo-mediated activation of *eng2a* (which encodes the transcription factor

Engrailed2a) in the head region during early development caused a severe reduction in the size of the eye and other effects demonstrating a powerful tool to study the spatial effects of gene expression in development (see Fig. 2 from [4]).

9.4.11
Caged Hormones as Spatial Regulators of Gene Expression

A related strategy for controlling gene expression is based on caging a signaling molecule. Koh et al. improved upon hormonal inducing strategies by caging estradiol, a hormone analog that triggers gene expression in cells transfected with the estrogen receptor transcriptional regulator [16]. 4,5-Dimethoxy-2-nitrobenzyl-bromide was used to alkylate the 1-hydroxyl group of estradiol. In transiently transfected HEK293 cell cultures, caged estradiol did not activate the estrogen receptor, which was a promoter for expression of luciferase. Exposure of cultures to 365-nm light released active estradiol intracellularly, inducing measurable luciferase expression to 85% of native (non-caged) estradiol levels. Another hormone-based inducing system, β-ecdysone, has since been used to demonstrate photoactivated expression in cell cultures. Lawrence et al. prepared a racemic mixture of DMNPE-caged ecdysone, an insect molting hormone, to phototrigger retinoid receptor binding and activate luciferase expression. Caging combined with focused illumination of transiently transfected 293T cells [17]. These approaches can be applied to other systems with any target gene of interest with integrated hormone-inducible transcriptional activators.

9.4.12
Concerns for Caging to Control Gene Expression

One important factor in evaluating the effectiveness of a strategy for targeting gene expression is its basal or un-induced expression. Ideally, un-induced levels of expression should be non-existent. However, many of the inducible systems have "leakage" – the expression of the therapeutic gene is measurable even without application of the chemical stimuli. Thus, the ability of the attached caging compound on the DNA to block its transcription or hybridization must be characterized. If caged DNA has activity before light exposure, the targeting efficiency of this strategy will be greatly reduced. It is important to determine the number of cage groups necessary to block transcription or hybridization of DNA. Methylation has been found to have a strong silencing effect on gene expression *in vivo* [18, 19]. The addition of a single methyl group to the cytosine of plasmids is apparently sufficient to block their transcription. Similarly, it is hypothesized that the attachment of the somewhat larger DMNPE (the cage compound used in these studies) also inactivates DNA by blocking transcription. Hybridization of an ODN to a complementary target may require relatively few attached cage groups as well. Antisense ODNs are designed to inhibit expression of a specific mRNA – this specificity arises from their ability to discriminate between targets differing in only 1 or 2 bases in the complementary region [20]. Even a single base point mutation can lead to a 400-fold decrease in binding kinetics [21]. Two mismatches out of 15 complementary bases can lead to a complete loss of antisense activity [22].

Another important aspect of this strategy is the bioactivity of the released caging group inside the target cell. Published data on the biohazards of benzaldehyde compounds released from caging groups is scarce. Of the few studies that mention effects, the findings are not definitive. In Walker's report on the synthesis of nitrobenzyl-caged nucleotides, there is mention of the reactivity of released 2-nitrosoacetophenone being less reactive than previous caging groups that release 2-nitrosobenzaldehyde, but this reactivity is not quantified [8]. In a more recent report, it is mentioned that 2-nitrosoacetophenone is reactive and forms covalent adducts with reactive sulfhydryls, such as cysteine residues on proteins [23]. However, control experiments run by Patton et al. indicate no significant morphological changes or protein modifications in cultured endothelium as a result of treatment with 2-nitroacetophenone [24]. Ishihara et al. also report no cytotoxicity of the released photoproduct in their biological preparation [25]. There is a clear need for more investigations into this area before the caging strategy is deemed feasible. A more in-depth discussion of this topic is contained in chapters (Chapters 1.1 and 2).

Similar to biological effects of the released cage, another important factor for consideration is the potential for the light required to produce photorelease to also damage target tissues. As described above, the wavelengths of light that can effectively cleave nitrobenzyl caging groups are longer than UV and do not share its deleterious effects [26]. The caging group used in these studies, DMNPE, can be photocleaved with light ranging from 345 to 365 nm [25]. Light from the sources used in these studies (355 and 365 nm) falls into the UVA classification and is thought to be far less damaging to cells than UVB light [27, 28]. The maximum light dosage used to uncage in these experiments is less than half of the UVA exposure that might be received in a single visit to a tanning parlor [27]. Our preliminary studies on cellular effects of photoactivation indicate that the light sources cited here do not trigger apoptosis or necrosis in cell cultures. HeLa cells were seeded in 35-mm dishes and exposed to 5.6 J cm^{-2} of 365-nm light as in the other culture assays described. Selected cells were incubated in 1 µM staurosporine for 12 h as a positive control. Twenty-four hours later, cells were harvested and stained with Molecular Probes' Apoptosis Assay kit 2 (Eugene, OR) containing propidium iodide and Alexa 488-labeled Annexin V. Cells were then analyzed with a FACSCalibur (Becton-Dickinson, Mountain View, CA) flow cytometer. HeLa cells exposed to the highest dose of 365-nm light (5.6 J cm^{-2}) used for photoactivation in our previous studies do not show signs of induced apoptosis or necrosis. Cells that were exposed to 5.6 J cm^{-2} show similar fluorescence to viable cells that received no treatment (Fig. 9.4.15, top two panels). Cells that were treated with staurosporine, a known apoptosis-inducing agent, do show signs of both apoptosis, and necrosis, as seen in the lower panel of Fig. 9.4.15. However, little is known about the specific inhibitory effects of 365-nm light on gene expression and cellular damage. If there are biological effects associated with this treatment, acceptable

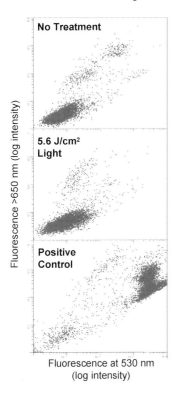

Fig. 9.4.15 Effects of 365-nm light on cell cultures.

levels of light must be determined such that they minimize cellular damage, yet still release the caging group from the delivered transgene.

9.4.13
Examples of Site-specific Caging of DNA and RNA

Unlike the randomly caged mRNA and DNA examples cited above, several groups have followed an incorporation strategy to build caged DNA and RNA structures with cage groups placed at precise locations. Several groups have reported the caging of single nucleotides and insertion into strands of DNA or RNA to enable the temporal control of hybridization and ribozyme reactions. Nitrobenzyl caging groups have been used to trigger single- and double-strand breaks in double-stranded DNA [29]. A caged building block was incorporated into the stem portion of a hairpin oligonucleotide, such that upon irradiation, the hairpin would open up and be available for hybridization. This system can act as a tool for measurement of the kinetics of DNA probe hybridization. A similar development in the use of caged nucleotides has been to temporally control ribozyme and spliceosome reactions by including caged adenosine within synthesized RNA oligonucleotides [30, 31]. A nuclear localization signal peptide caged at lysines was recently used to study the kinetics of nuclear import of proteins [32]. Additionally, backbone caging of a receptor-binding peptide has demonstrated control over peptide-receptor binding [33]. These later studies demonstrate the site-specific caging of a larger molecule by using caged building blocks in standard peptide synthesizers.

9.4.14
Future Directions and Conclusions

One possible technique that could be employed to photoactivate caged DNA more effectively *in vivo* is two-photon excitation. Two-photon excitation (2PE) employs light at a wavelength that is twice the value for single-photon uncaging wavelength. The principle of the technique is that two photons of lower energy (e.g. 700 nm) arriving simultaneously (10^{-18} s apart) deliver an additive effect of transferred energy, which is equivalent to a single 350-nm photon impacting the target [34]. If high enough intensities of light are used, then the probability of two photons impacting the molecule simultaneously increases. Pulsed lasers focused to a specified depth within a tissue create a focal region with photon density high enough to induce a two-photon event. There are advantages to using 2PE excitation that make it attractive for releasing caged molecules. First, 700-nm light has lower energy than 350-nm light and causes less photodamage to cells and tissues [34]. Studies comparing cellular responses to pulsed irradiation from both 365- and 730-nm sources find that significantly less oxidative stress results from the two-photon source (730 nm) if the peak pulse power is kept below a threshold level [36]. The longer wavelength also allows for deeper penetration, since less absorption and scattering occurs. This allows for targeting at greater depths within tissues. Second, 2PE is localized to the focal point of the laser beam, which can be manipulated to control the depth of photoactivation within a tissue target. The use of 2PE to photoactivate caged compounds was demonstrated by Denk et al., who used DMNPE-caged ATP as an example of the technique's feasibility [37].

Perhaps the greatest need in this area of research is the production of a caged phosphoramidite which could be used to synthesize specific DNA structures. The site-specific caging of only select nucleobases or backbone moieties would facilitate these studies, such that a minimal number of cages can be attached to sufficiently block bioactivity. This would require less light to photocleave and restore bioactivity, and would also allow cage groups to be attached at the most critical sites for blockade, depending on the intended cage-induced interference (e.g. at the transcription start site of plasmid DNA). The current DNA and RNA caging reactions for gene control that have been published to date are batch-type process that yields a random attachment of the cage to DNA, and varying degree of cage attachment. A site-directed caging strategy employing a caged building block (e.g. a caged phosphoramidite for incorporation into a standard DNA synthesizer) will be the next step in improving the overall strategy.

Caging methods specifically inactivating a particular nucleic acid dependent function might have utility as a new tool to further understand and control these bioactivities. Important aspects of this evaluation will be the extent of the blockade and the efficiency of the restoration of bioactivity. There is a wide array of nucleic acid interaction assays that might be employed to further explore these bioactivities.

Acknowledgments

This work would not have been possible without the continued generous support of Kyle Gee at Molecular Probes, Inc. The work was supported in part by NIH grant EY13451, the National Science Foundation under grant EPS-0346411, and the State of Louisiana Board of Regents Support Fund.

References

1 MONROE, W. T., et al., *J. Biol. Chem.* **1999**, *274*, 20895–20900.
2 MONROE, W. T., Targeting DNA expression and hybridization with light: application to plasmid transfections and antisense oligonucleotides, *PhD Thesis*, Biomedical Engineering, Vanderbilt University, **2001**.
3 OKAMOTO, H., Y. HIRATE, H. ANDO, *Front. Biosci.* **2004**, *9*, 93–99.
4 ANDO, H., et al., *Nat. Genet.* **2001**, *28*, 317–325.
5 ANDO, H., H. OKAMOTO, *Methods Cell Sci.* **2003**, *25*, 25–31.
6 ORDOUKHANIAN, P., J. S. TAYLOR, *Bioconjugate Chem.* **2000**, *11*, 94–103.
7 ZHANG, K., J. S. TAYLOR, *Biochemistry* **2001**, *40*, 153–159.
8 WALKER, C. J., et al., *J. Am. Chem. Soc.* **1988**, *110*, 7170–7177.
9 ABRAMOVA, T. V., et al., *Bioorg. Khim.* **2000**, *26*, 197–205.
10 ATTWOOD, J. T., R. L. YUNG, B. C. RICHARDSON, *Cell Mol. Life Sci.* **2002**, *59*, 241–257.
11 FURUTA, T., et al., *Proc. Natl Acad. Sci. USA* **1999**, *96*, 1193–1200.
12 GHOSN, B., et al., *Photochem. Photobiol.* **2004**, in review.
13 BENNETT, C. F., et al., *J. Immunol.* **1994**, *152*, 3530–3540.
14 CHIANG, M. Y., et al., *J. Biol. Chem.* **1991**, *266*, 18162–18171.
15 HASELTON, F. R., W.-C. TSENG, M. S. CHANG, Light activated protein expression using caged transfected plasmid II: delivery by gene gun to organ cultured corneas, presented at in *ARVO Meeting: Investigative Ophthalmology and Visual Sciences*, Fort Lauderdale, FL, **1997**.
16 CRUZ, F. G., J. T. KOH, K. H. LINK, *J. Am. Chem. Soc.* **2000**, *122*, 8777–8778.
17 LIN, W., et al., *Chem. Biol.* **2002**, *9*, 1347–1353.
18 EDEN, S., H. CEDAR, *Curr. Opin. Genet. Dev.* **1994**, *4*, 255–259.
19 RAMSAHOYE, B. H., C. S. DAVIES, K. I. MILLS, *Blood Rev.* **1996**, *10*, 249–261.
20 MIRABELLI, C. K., et al., *Anticancer Drug Des.* **1991**, *6*, 647–661.
21 MONIA, B. P., et al., *J. Biol. Chem.* **1992**, *267*, 19954–19962.
22 HOLT, J. T., R. L. REDNER, A. W. NIENHUIS, *Mol. Cell Biol.* **1988**, *8*, 963–973.
23 DANTZIG, J. A., H. HIGUCHI, Y. E. GOLDMAN, *Methods Enzymol.* **1998**, *291*, 307–348.
24 PATTON, W. F., et al., *Anal. Biochem.* **1991**, *196*, 31–38.
25 ISHIHARA, A., et al., *Biotechniques* **1997**, *23*, 268–274.
26 McCRAY, J. A., D. R. TRENTHAM, *Annu. Rev. Biophys. Biophys. Chem.* **1989**, *18*, 239–270.
27 ROBERT, C., et al., *J. Invest. Dermatol.* **1996**, *106*, 721–728.
28 SCHINDL, A., et al., *J. Photochem. Photobiol. B* **1998**, *44*, 97–106.
29 ORDOUKHANIAN, P., Caged DNA, *PhD Thesis*, Chemistry, Washington University, **1997**.
30 CHAULK, S. G., A. M. MACMILLAN, *Nucleic Acids Res.* **1998**, *26*, 3173–3178.
31 CHAULK, S. G., A. M. MACMILLAN, *Angew. Chem. Int. Ed.* **2001**, *40*, 2149–2152.
32 WATAI, Y., et al., *FEBS Lett.* **2001**, *488*, 39–44.
33 TATSU, Y., et al., *FEBS Lett.* **2002**, *525*, 20–24.
34 PISTON, D. W., *Trends Cell Biol.* **1999**, *9*, 66–69.
35 DENK, W., D. PISTON, W. W. WEBB, Two-photon molecular excitation in laser-scanning microscopy, in *Handbook of Biological Confocal Microscopy*, J. B. PAWLEY (ed.), Plenum Press, New York, **1995**.
36 KOENIG, K., et al., *J. Microsc.* **1996**, *183*, 197–204.
37 DENK, W., J. H. STRICKLER, W. W. WEBB, *Science* **1990**, *248*, 73–76.

9.5
Photoactivated Gene Expression through Small Molecule Inducers

Sidney B. Cambridge

9.5.1
Introduction

Regulated expression of genes is central to long-term biological processes such as cell fate specification during development, synaptic plasticity in the nervous system and wound healing induced migration of fibroblasts. To unravel the complex gene expression patterns underlying these processes, researchers often missexpress relevant genes in a spatially and/or temporally aberrant manner. If, in turn, the missexpression affects the phenotype, then the function of the gene may be deduced based on the experimental context. Missexpression or conditional gene expression has recently become a popular research tool with the advent of small molecule inducers which can activate the expression of transgenes in many different systems. The most prominent small molecule inducers are analogs of tetracycline [1], ecdysone [2] and estradiol [3], but also other molecules such as dexamethasone [4] have been employed. These inducers are cell membrane-permeant and mostly biologically inert except for their binding to one particular transcriptional activator that is not normally found in the model system being studied. Following formation of an inducer–activator complex, a specific DNA-binding site is targeted and transcription of downstream genes will be initiated (Fig. 9.5.1). Such a two-component system (transcriptional activator and activator-dependent transgene) affords good spatial and temporal control over conditional gene expression. Temporally, gene expression is simply controlled by the addition of the inducer to the system. Spatial control is achieved by expressing the transcriptional activator only in a subset of cells, e.g. by driving its expression through a well-characterized endogenous promoter. However, despite the potential to activate transgenes in a spatially defined manner, many research applications often require an even better spatiotemporal resolution including the possibility to manipulate single cells.

To improve the resolution of conditional gene expression, several groups have recently synthesized caged analogs of some of the small molecule inducers. The idea being the induction of transgenes through irradiation with UV light, i.e. photoactivated gene expression. This will allow researchers to express transgenes at will simply by modulation of the irradiation beam. Our group was the first to demonstrate the concept of photoactivated gene expression (also Chapter 5.3) by injection of a caged transcription factor into syncytial *Drosophila* embryos prior to formation of cell membranes [5]. We could show that photoactivated gene expression allows transgene induction in single cells and that this technique was also suitable to drive transgene expression in *Xenopus* tadpoles. However, transgene expression depended on photoactivation of a caged protein rather than a small diffusible molecule. This required injection and thus pre-

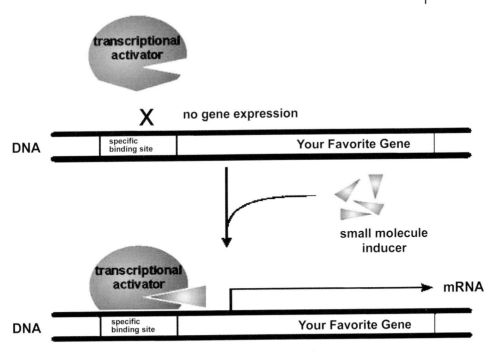

Fig. 9.5.1 Conditional gene expression based on a small molecule inducer.

cluded application in dense three-dimensional tissue. Therefore, caged small molecule inducers will elevate photoactivated gene expression to a new level so that it is applicable in many different biological systems. In this chapter, I will describe the use of three different caged inducers in the context of their respective conditional gene expression paradigm.

9.5.2
Tamoxifen and Estrogen Receptor (ER)

Induction of transgenes via tamoxifen and a modified ER has become increasingly popular in recent years. The system is based on a mutated form of the ER that only binds to tamoxifen and hydroxytamoxifen, but not to endogenous estradiol [6]. In the absence of ligand, the nuclear estrogen receptor is inactive due to complexation with heat-shock proteins. Binding of tamoxifen induces a conformational change that releases the blockade of heat-shock proteins and increases the affinity for DNA estrogen response elements (ERE). Transgenes under control of the ERE can thus be induced upon administration of tamoxifen. However, tamoxifen is now mainly employed to regulate the activity of Cre recombinase. Cre recombinase is a bacteriophage P1 protein that can excise genomic DNA flanked by loxP recognition sites [7]. This offers the exciting possibility to modify genes *in vivo* such as "knocking-out" transgenes flanked by loxP sites. Fusion of Cre recombinase with the ER produces an inducible knockout paradigm that is particularly useful for studying embryonic lethal genes in a tissue-specific manner. The efficiency of inducible site-specific recombination appears to depend mainly on the levels of the Cre–ER fusion protein present in the respective tissue [6]. Levels of recombination reached almost 50% in tissues such as the kidney were strong expression was achieved, but only 10% efficiency in low expressing regions such as the brain. Given the further improvements of the ER construct [8], tamoxifen inducible control of gene expression will continue to be a powerful research tool.

Caging of hydroxytamoxifen was recently accomplished by Shi and Koh [9]. Using standard photosensitive protection chemistry, the tamoxifen analog was alkylated with 4,5-dimethoxy-2-nitrobenzylbromide to yield NB-Htam (nitrobenzyl caged hydroxytamoxifen). To test for photoactivated gene expression, the authors used HeLa cells which were transiently transfected with two plasmids, one expressing either ERα or ERβ and one being a reporter plasmid with an ER-dependent luciferase construct. Hydroxytamoxifen is an agonist for ERβ, but a partial antagonist for ERα as it antagonizes the effect of estradiol. This can be particularly useful for regulating repression and induction of different transgenes simultaneously.

Administration of NB-Htam to ERβ transfected HeLa cells followed by long-wavelength UV irradiation (>345 nm) led to a marked increase in luciferase expression after 24 h compared to unirradiated controls. Similarly, photoactivation of ERα transfected cells significantly reduced estradiol-dependent expression of the reporter gene. This clearly demonstrated that irradiation generated biologically active hydroxytamoxifen. Unfortunately, however, the authors did not use NB-Htam to spatially restrict expression in a subset of cells. Rather, NB-Htam was photoactivated in the entire cell medium so that the released hydroxytamoxifen was globally present. Shi and Koh noted that tamoxifen analogs presumably cross cell membranes quickly, thereby decreasing the possibility of intracellularly released ligand to bind to the ER. Therefore, it may not be possible to use NB-Htam for photoactivation in a disperse group of cells. Nevertheless, in light of the continued success of the tamoxifen paradigm, having a light-inducible version of this system will be valuable research tool.

9.5.3
Ecdysone

Another system for inducible conditional gene expression is the ecdysone system developed by Evans et al. Ecdysone is an insect molting steroid that triggers metamorphosis in *Drosophila* and other insects. In heterologous cells, coexpression of a modified ecdysone receptor (EcR) and its mammalian counterpart, the retinoic acid receptor (RXR), will allow ecdysone-dependent expression of transgenes under control of the respective response elements [2]. Recently, more potent analogs of ecdysone were identified and one of these ligands, ponasteroneA, improved inducibility several fold. Furthermore, the simultaneous addition of retinoic acid receptor ligands also improved the levels of transgene expression 4- to 5-fold [10].

Despite the suitability of this system for conditional gene expression, it has enjoyed only limited use so far. One of the reasons may be that the small molecule inducers such as ponasteroneA are rather expensive and are not easily purchased in sufficient quantities for chemical modification. The group of D. Lawrence recently synthesized caged β-ecdysone instead of the more potent ponasteroneA for that very same reason. Because standard alkylation procedures failed, the authors developed an elegant scheme to first synthesize a tin-based acetal intermediate which then very efficiently (90%) alkylated with 4,5-dimethoxy-2-nitrobenzylbromide. This procedure yielded a monoalkylated β-ecdysone with high regioselectivity [11].

The maximal photoconversion of the caged analog in a cell-free assay was 60%, and not surprisingly, photoactivated gene expression also reached about 60% of the unmodified β-ecdysone. Gene expression was assessed by transient cotransfection of EcR/RXR and a luciferase reporter construct. Unmodified β-ecdysone increased induction of luciferase activity 88-fold, caged β-ecdysone 6.5-fold and irradiated caged β-ecdysone 50-fold. The small, but detectable, gene expression activity of caged β-ecdysone may be due to residual binding affinity to the receptors or due to instability of the caged compound.

A crucial issue of photoactivated gene expression paradigms is whether the caged inducer can diffuse through the plasma membrane. This is particularly important for single-cell photoactivation because this kind of spatial resolution is probably only attained if the active inducer can be photoactivated inside the cells. To test if the caged compound can penetrate the plasma membrane, the cells were first incubated with the analog for 16 h and then were washed with ecdysone-free medium. Irradiation of the washed cells induced marked luciferase expression, clearly demonstrating that the concentration of caged compound inside the cells was sufficient for photoactivated gene expression. Moreover, diffusion of active β-ecdysone out of the cells did not appear to be limiting for robust gene expression to occur. This exciting finding was further supported by local luciferase expression after local irradiation of a subset of cells. Fig. 9.5.2(a) shows transfected cells that were exposed to unmodified β-ecdysone and immunostained for luciferase expression. In Fig. 9.5.2(b), transfected cells were exposed to caged ecdysone, washed, irradiated and immunostained for luciferase expression. Note that the expression was exclusively localized within the irradiation spot, while no staining was seen in the unirradiated control area (Fig. 9.5.2c). This suggests that the caged ecdysone system has the potential for high-resolution, spatially restricted photoactivated gene expression. Future applications based on caged ecdysone therefore hold great promise for conditional gene expression.

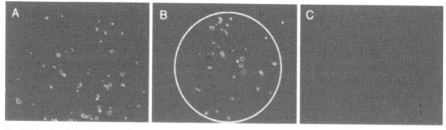

Fig. 9.5.2 Spatially discrete photoactivated expression of luciferase. (A) Control: unmodified β-ecdysone. (B) Caged ecdysone, irradiation within circular area. (C) Same sample as in (B), unirradiated region. (Reprinted from [11] with permission from Elsevier)

9.5.4
Tetracycline

The tetracycline (Tet)-system is by far the most prominent system currently used for conditional gene expression through small molecule inducers. The Tet-system was developed by Bujard et al. more than a decade ago [12]. Two different versions exist: the Tet-on system (induction upon tetracycline administration) and the Tet-off system (induction upon tetracycline removal), while only the Tet-on version is suitable for photoactivated gene expression. The Tet-on version is based on a mutated and codon optimized form of the bacterial Tet repressor fused to a transcriptional activator domain such as the VP16 domain from the herpes simplex virus. Originally, the Tet-system was quite 'leaky' as transgenes were found to be expressed even in the uninduced state, but PCR-based mutagenizing screens produced improved transactivators with virtually no background activity [13]. Doxycycline is the most potent tetracycline analog used with this system and thanks to its biocompatibility, it is commonly utilized for *in vivo* applications. In living animals, the Tet-system allows robust transgene expression in every tissue, although there are some strong limitations in the brain that are currently not well understood. Nevertheless, the Tet-system appears to be the method of choice for induction of transgenes in disperse cells, tissue culture and intact animals. Due to the popularity of the Tet-system, a large number of constructs and transgenic animals exist which provide a rich 'infrastructure' for photoactivated gene expression to be based upon.

We recently succeeded in alkylating doxycycline with the standard 4,5-dimethoxy-2-nitrobenzyl photosensitive protection moiety (Cambridge et al., unpublished results). Caged doxycycline is highly water-soluble so that 1000×–10000× stocks can easily be made in water. This is particularly relevant for applications in sensitive tissue such as the nervous system which is adversely affected by most vehicle solutions, including alcohols and organic solvents. Importantly, the caged analog is able to quickly cross the plasma membrane as determined by isothermic titration calorimetry (Sandro Keller, unpublished results).

The transcriptional activity of caged versus photoactivated doxycycline was compared in three different model systems (unpublished results). These in-

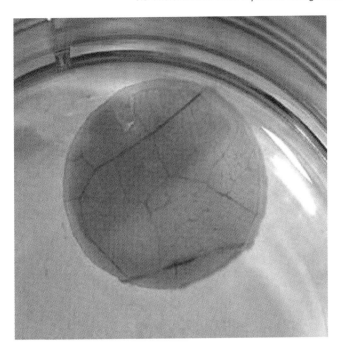

Fig. 9.5.3 Doxycycline-dependent photoactivated gene expression in a transgenic tobacco leaf. Expression of the GUS reporter gene is only detected in the irradiated half, but not in the unirradiated control area.

cluded CHO cells with Tet-dependent enhanced green fluorescent protein (EGFP) expression, transgenic tobacco leafs with Tet-dependent expression of a GUS (β-glucuronidase) reporter gene and organotypic hippocampal slice cultures of wild-type mouse brain transfected with two adenoviruses, one constitutively expressing the transactivator and one containing a Tet-dependent EGFP construct. In all cases, following administration of caged doxycycline and subsequent photoactivation, spatially restricted transgene expression was detected in irradiated areas but not in the unirradiated control area of the same sample. In Fig. 9.5.3, strong expression of GUS is detected in the irradiated half of the transgenic tobacco leaf and only background staining is present in the unirradiated other half. Note the sharp boundary between the two areas. This clearly demonstrates that conditional gene expression based on photoactivation of small molecule inducers can be used in thick three-dimensional tissue with excellent spatiotemporal control. We are currently synthesizing a two-photon-sensitive caged doxycycline analog to be used for high-resolution uncaging *in vitro* and possibly *in vivo*. In combination with two-photon microscopy, this will allow one to do single cell manipulations deep down in strongly scattering tissue of live animals. In the near future, interdisciplinary research using multiphoton photoactivation of caged compounds *in vitro* and *in vivo* will continue to increase.

This will certainly aid the development of a complex experimental paradigm such as *in vivo* photoactivated gene expression.

9.5.5
Summary and Conclusions

The establishment of a photoactivated gene expression system as a routine and reliable research tool will have an enormous impact on research in whole tissues of transgenic organisms. The first steps towards this direction have been accomplished with the synthesis of caged small molecule inducers. Several aspects including biocompatibility, efficiency and cost of synthesis, membrane permeability and solubility of the caged compound, as well as 'infrastructure' of the conditional gene expression paradigm will determine the success of photoactivated gene expression based on any particular caged inducer. In light of the tremendous need for such a technique, this research area may advance so far that soon there will be several different photoactivated gene expression paradigms to be used alone or in combination for missexpression of transgenes with high spatiotemporal control.

Acknowledgments

I thank Albrecht Kossel, Christian Lohmann and Jonathan Minden for critically reading the manuscript. Fig. 9.5.3 was produced in the lab of Christiane Gatz (Göttingen/Germany) with the help of Ronald Scholz.

References

1 M. Gossen, A.L. Bonin, S. Freundlieb, H. Bujard, *Curr. Opin. Biotechnol.* 1994, 5, 516.
2 D. No, T.P. Yao, R.M. Evans, *Proc. Natl Acad. Sci. USA* 1996, 93, 3346.
3 R. Feil, J. Brocard, B. Mascrez, M. Lemeur, D. Metzger, P. Chambon, *Proc. Natl Acad. Sci. USA* 1996, 93, 10887.
4 D.E. Jaalouk, N. Eliopoulos, C. Couture, S. Mader, J. Galipeau, *Hum. Gene. Ther.* 2000, 11, 1837.
5 S.B. Cambridge, R.L. Davis, J.S. Minden, *Science* 1997, 277, 825.
6 P.S. Danielian, R. White, S.A. Hoare, S.E. Fawell, M.G. Parker, *Mol. Endocrinol.* 1993, 7, 232.
7 B. Sauer, N. Henderson, *Proc. Natl Acad. Sci. USA* 1998, 85, 5166.
8 R. Feil, J. Wagner, D. Metzger, P. Chambon, *Biochem. Biophys. Res. Commun.* 1997, 237, 752.
9 Y. Shi, J.T. Koh, *ChemBioChem* 2004, 5, 788.
10 E. Saez, M.C. Nelson, B. Eshelman, E. Banayo, A. Koder, G.J. Cho, R.M. Evans, *Proc. Natl Acad. Sci. USA* 2000, 97, 14512.
11 W. Lin, C. Albanese, R.G. Pestell, D.S. Lawrence, *Chem. Biol.* 2002, 9, 1347.
12 M. Gossen, H. Bujard, *Proc. Natl Acad. Sci. USA* 1992, 89, 5547.
13 S. Urlinger, U. Baron, M. Thellmann, M.T. Hasan, H. Bujard, W. Hillen, *Proc. Natl Acad. Sci. USA* 2000, 97, 7963.

Subject Index

a

absorption spectrum   425
abstraction rate constants   98
acetophenone chromophore   56
acetylcholinesterase   317 f
acetycholine receptor   333
AChE (acetylcholinesterase)   317, 411
– trigonal crystal of   411
aci-nitro   97, 99, 100, 102, 126
aci-nitro anion   6
aci-nitro decay rate   7
aci-nitro formation spectrum   388
aci-nitro intermediate   99 f, 143
– decay of   100
aci-nitro intermediate band   378
ACM (7-(acyloxy)coumarin-4-yl]methyl)   160
acrylodan-actin   326
active-site-directed reagent   275
actin depolymerization factor   328
actin polymerization assay   326
activation loop   277
1-acyl-5-bromo-7-nitroindoline   2
1-acyl-nitroindoline   3
– photolysis reaction   3
1-acyl-7-nitroindoline derivative   262
1-acyl-7-nitroindoline   2, 21
– photochemical cleavage   21
– photocleavage   2
acyl transfer   142
– photoremovable scaffold   142
adenosine   506
adenosine-5'-triphosphate caged molecule   79
adenosine-5'-triphosphate derivative   78, 83, 92
– chemical and photochemical properties   83
– synthesis of the P^3-caged   78

adenosine-5'-triphosphate   78, 80 ff
– caged   78, 80
– coupling of caged monophosphate to ADP   81
– diazo-coupling   82
adiabatic/diabatic decay process   113
adiabatic ESIPT process   112
adiabatic ionization   65
adiabatic proton transfer   64
ADP (adenosine-5'-diphosphate)   36, 78, 309, 310, 311
– filament   328
– ribose   4
affinity caging   416, 419
AIP   147
Ala-Ala   58, 59, 61, 68, 69
alcohol   16, 45, 50
– caged   16
– dehydrogenase   314
– glucose 6-phosphate dehydrogenase   314, 315
– horse liver alcohol dehydrogenase   317
– 7α-hydroxy steroid dehydrogenase   317
– isocitrate dehydrogenase   314
– photolysis of Bhcmoc-caged   45
– precursor molecule   50
aldehyde   46
– photolysis of Bhc-diol-protected   46
aliphatic ether   45
alkyl nitrite   189
– photochemistry   189
amide   14
– caged   14
amidite   354 f
– determination of cycle yield   355
– determination of deprotection rate   354
amine   14, 43, 50, 136
– caged   14

- coumarine caged 43
- photocleavable protection 136
- precursor molecule with 50
- protection 136

α-amino-3-hydroxy-5-methylisoxazole-4-propionic acid 401

amino acid residue 136, 151
- photoaffinity of 151

AMP (adenosine-5′-monophosphate) 36, 78

D-aminophosphonovalerate 66

amyloidogenic peptide 13

2-anthraquinonlylmethyl 165

antibody 288, 330

antisense activity 521
- reversible inhibition 521

antisense assay 522
- in vitro 522

antisense blockade 522
- effect of light on 522

aqueous solubility 77, 80, 81, 83, 89, 90, 92

AQMOC 266

arachidonic acid 14
- caged 14

aromatic nitro compound 190

2-arylbenzofurans 102

arylmethyl 96, 125

arylmethyl derivative 122

astrocytes 59, 68, 67, 68

astrocytic circuit 237

ATP (adenosine-5′-triphosphate) 1, 36, 76, 78, 80, 82, 84, 88, 91, 92, 95, 101, 304–307, 309, 375, 381
- binding 73, 74
- in the spectra of phosphate absorption 381

ATPases 304, 306
- Ca^{2+}-ATPase 305–307
- H^+,K^+-ATPase 305
- Kdp-ATPase 305
- Na^+,K^+-ATPases 304
- transport ATPase 305

azid-1 449

azlactone formation 107

azopolymer 293

b

bacterial exotoxin 338

bacteriorhodopsin 413, 426

bathochromic shift 111, 124

BAY K 8644 190
- structure of 190

BCMB (3′,5′bis-(carboxymethoxy)benzoin) 462, 485
- acetate 105

BCMCNB (4-[N,N-bis(carboxymethyl)carbamoyl]-2-nitrobenzyl ester) 156

BCMNB (4,5-bis-d(carboxymethoxy)-2-nitrobenzyl ester) 156

BDNF (brain-derived neurotrophic factor) 331

beam splitter 371

benzisoxazole 100
- intermediate 100

benzisoxazoline 6

benzo-α-pyrone 29

benzoin 96, 102, 262, 481
- photochemistry 481
- photoremovable protecting group 102

benzoin acetate 104
- effect of methoxy substitution 104
- photochemistry 104

benzoin carbamate synthesis 108
- competitive by-product formation 108

benzoin derivative 102
- as protecting group in biologically active substrate 102

benzoin diethyl phosphate 106

benzoin phosphate 106, 108
- photosolvolysis 108

benzoin 102, 262, 481
- photochemistry 481
- photoremovable protecting group 102

benzophenone 152

2-benzoylbenzoate derivative 268

benzyl 96, 125

benzyl alcohol 317

benzyl carboxylate 122
- photolysis of 122

benzyl cation 107

benzyloxycarbonyl group 135

Bhc (6-bromo-7-hydroxycoumarin-4-ylmethyl) 4, 125
- diol 266
- methanol 31

BHQ (8-bromo-7-hydroxyquinoline) 266, 443

Biological switch 73

biomolecule 1
- caging 1

biotin 16, 276

biotin/avidin 16

biradical intermediate 105

Subject Index

bis(2,2'-bipyridine)nitrosylruthenium(II) derivative 187
bis-2-chlorophenylacetylene 191
biscyanoethyl phosphoramidite 78
BNN (bis-*N*-nitroso-*p*-phenylenediamine) 198
– derivative 199
– intermediates involved in the photochemistry of 199
BNPA (2-bromo-2(nitrophenyl)acetic acid) 273
bold line spectrum 388
Bonferroni's *t*-test 524
bovine serum albumin 180
BrAc-CMB 485
bradykinin 58, 59, 61, 62, 66, 70
– caged 59, 61, 62, 66, 67, 68
bromocoumarin cage 519
2-bromo-4'-hydroxyacetophenone 60
6-bromo-7-hydroxycoumarin-4yl methylacetal 46
– photolysis of 46
6-bromo-7-methoxymethoxycoumarin-4-yl)methyl chloroformate 50
8-bromo-7-hydroxyquinoline (BHQ) 443
1-bromo-1-(2-nitrophenyl)-ethane 18
4-bromoresorcinol 48
BuChE (butyrylcholinesterase) 275 f, 317–319
2'-*O*-*tert*-butyldimethylsilyl group 345

c

Ca^{2+} 306, 307, 309, 311, 313
Ca^{2+}-ATPase 306 f, 406
– active transport of Ca^{2+} by 406
– using caged ATP 306
– using caged Ca^{2+} 307
– using caged inhibitor 307
cage release 95
caged arsenocarbamylcholine 317, 318
caged arsenocholine 317, 318
caged ATP 4, 76, 78, 80, 304, 306, 310
– NB-ATP 304, 305
– NPE-ATP 304, 305
caged ATP derivatives 79
– CNB-caged ATP 78, 79, 80, 83
– Desyl-caged ATP 78, 79, 82, 83
– DMACM-caged ATP 78, 79, 80, 82, 83, 84
– DMB-caged ATP 78, 79, 80, 82, 83
– DMNPE-caged ATP 78, 79, 81, 83
– DNP-caged 78, 79, 83
– NB-caged ATP 78, 79, 83

– NPE-caged ATP 78, 79, 80, 83, 84
– *p*-HP-caged ATP 78, 79, 82, 83, 84
caged biomolecules 76, 77, 92
caged Ca^{2+} 305, 307
– DM-nitrophen Ca 307
caged cAMP 310, 311
– DMNB-CAMP 310, 311
– NB-CAMP 310
caged choline 317
– NPE-choline 321
caged cholinesterase substrate 318
caged cytochrome *bc*1 substrate 322, 323
caged DAG 311, 312
caged glucose 6-phosphate 314
– NPE-glucose 6-phosphae 314, 315
caged glutamate 58, 66, 67, 76, 84, 91, 92
caged glutamate derivatives: 85, 86
– γ-ANB 85
– *N*-BHC 86, 91
– γ-BhcMe 86
– bis-α,γ-CNB- 85
– α-CNB- 85
– γ-CNB- 85–91
– *N*-CNB- 85, 91
– α-*O*-Desyl 85, 91
– γ-*O*-Desyl 85
– *N*-Desyl 85
– α-DMNB 85
– α-DMNPE- 85
– α-DMNPT- 85
– γ-DMNPP- 86
– γ-*p*-HP 85, 91
– γ-*p*-HPDM 85
– γ-*p*-HPM 85
– γ-MNI- 86, 91
– *N*-NB- 85, 91
– γ-NI- 86, 91
– *N*-Nmoc- 85, 91
– γ-NPE- 85
– *N*-NPE- 85, 91
– *N*-NPEOC- 85, 91
caged GTP 308, 309
– NPE-GTP 308
– ras 72
caged inhibitor 307, 311
caged isocitrate 315
– NPE-isocitrate 315
caged modulators 304, 305, 307, 317, 320, 322
caged NAD 314
caged NADP 314
CNB-NADP 315
– DMNPE-NADP 315

- caged pre-mRNA 511
- NB-NADP 317
caged nitric oxide synthase inhibitor 321
caged nucleotides 70
caged peptide inhibitor 312, 313
- CNB-caged peptide 312, 313
- NB-caged peptide 313
caged PKC activator 311
- CNB-arachidonate 312
- NDE-arachidonate 311, 312
caged proteins 325 f
caged ribonuclease reductase substrate 322, 323
cage urease substrate 322, 323
cages 95
caging agent 46
caging group 76, 257, 271
- for protein 257
- problem with the released 271
- property for 76
calcium signal 67
calcium 19, 447
- cage for 447
- caged 19
CALI (chromophore-assisted light inactivation) 294
CAM (calcium-bound calmodulin) 147, 312
cAMP 78, 309 f
- dependent protein kinase 310, 311
CaMKII (calmodulin-dependent protein kinase II) 313
- using caged peptide inhibitor 313
cAMP (adenosine 3′,5′-cyclic monophosphate) 73, 95, 155, 446
- permeability 174
capping 349
carbamate-caged amine 15
- photolysis of 15
carbamic acid 43
carbamoylcholine 211, 213, 218
carbamylcholine 257, 318
- caged 318
carbonyl compound 46
carbonyl diimidazole 62
carboxyamide 138
- protection 138
carboxyl group 137
- protection 137
carboxylates 13, 41, 50, 76, 80, 84, 91
- caged 13
- coumarine caged 41

- precursor molecule with 50
catalytic caging 419
catalytic mechanism 304, 319
CCD detector 414
CDI (diimidazole) 62
CDP (cytidine-5′-diphosphate) 36
cell flow technique 209
cell lineage measurement 20
cell motility 326
cell physiology 435
cell signalling 71
- protein 73
cellular activity 155
cellular pivot point 328
cerebellar purkinje cell 245
CFTR channel 175
cGMP (cyclic guanosine monophosphate) 78, 155, 446
- signaling pathway 173
- in sperm 173
charge (qi) density 114
charge transfer intramolecular exciplex 105
chemical two-photon uncaging 296
chemotactic peptide 149
chemotaxis 173
chloroformate precursor 50
2-chloro-4′-hydroxyacetophenone 110
2-chlorotrityl resin 61
4-coumarinylmethyl 12
cholic acid 315
choline derivatives 317
cholinesterase 287, 317, 318, 319
- modulator 317, 319
- substrate 318
chromium-based photosensitive precursor 188
chromophore 440, 441
- for caged compound 441
- two-photon release 440
3,4-*bis*-(2′-chlorophenyl)furoxan 191
cinnamate ester derivative 267
circular dichroism 464, 487
CMB (3′-(carboxymethyl)benzoin) 462
CNB (α-carboxy-2-nitrobenzyl) 6, 258
- caged compound 10
- 2-carboxy-2-nitrobenzyl 258
- protecting group 213
cNMP (cyclic nucleotide monophosphate) 446
cocaine inhibition 223
coenzyme 303, 314, 322
cofilin 289, 328

- biochemical and photochemical regulation of 328
n,$\pi*$/π,$\pi*$ configurations 111
confocal imaging 67
conformation 371
- by IR spectroscopy 371
conformational freedom 371
- by IR spectroscopy 371
conjugated base 113
continuous irradiation 255, 293
conversion (%) 69
Corey-Seebach dithiane addition method 485
correction equation 209
coumarin phototrigger 32, 49
- spectroscopic and photochemical properties 32
coumarins 29, 33 ff, 47, 122
- 7-dialkylamino-substituted 35
- 6-bromo-7-alkoxy-substituted 34
- 6,7-dialkoxy-substituted 34
- 7-alkoxy-substituted 33
- structure of commercially available 47
coumarin-4-yl-methanol 29, 31
- UV/Vis spectra of 31
coumarin-4-ylmethoxycarbonyl 43
coumarin-4-ylmethyl carbonate 44
coumarin-4-yl-methyl derivative 160 ff
- photolysis of the diasteromer 162
- photophysical property 162
- structure of the axial and equatorial isomer 161
coumarin-4-ylmethyl phototrigger 29 f
- structure and acronym of 30
coumarin-caged alcohol 45
coumarin-caged amine 43
coumarin-caged carboxylate 41
coumarin-caged compound 42, 47, 51
- biological application of 51
- precursor molecule for synthesis 47
- spectroscopic and photochemical properties 42
coumarin-caged phenol 45
coumarin-caged phosphate 37 f
- spectroscopic and photochemical properties 37
coumarin-caged sulfate 41
coumarinyl amine 44
- mechanism of photolysis 44
coumarinylmethyl carbocation 162
coumaryl (cou) 96
coumaryl (cou) derivative 122–124, 126
crosslinker 276

- photocleavable 276
cryo-crystallography 413
cryo-photolysis 422, 428
- of NPE-caged compound 422
- trapping 422
- wavelength 428
cryo-radiolysis 428
- of NPE-caged compound 428
crystal 410
- conformational change in 410
crystallography 297
cupferron 198
- production of NO from 198
6-cyano-7-nitroquinoxaline-2,3-dione 66
cyclic nucleotide 11, 36, 155
- caged 11
cyclic nucleotide-gated (CNG) channel 168
cyclic peptide 142
- synthesis 142
cyclization 107
cyclopropanone intermediate 110, 113
cys scanning mutagenesis 278
cysteines 69, 72
- phosphorylated 72
cytochrome substrate 321
cytoskeleton 289

d

DAG 310, 311
DANP 266
DEACM-OH (7-diethylaminocoumarin-4-yl)methanol) 31
decay kinetics 8
degree of cage adduction 515
dehydrogenase 314, 315
- modulator 314
denaturing page 502
deoxyribonolactone 346
deoxyribonucleoside phosphorylation 342
3′-thio-2′-deoxythymidine phosphate 263
deprotection yield (see also conversion) 73
deprotonation 114
desyl 165
desyl group 102
Dewar mechanism 115
DFT calculation 107, 113
diacylglycerol 17, 450
- caged 17
1-(N,N-dialkylamino)diazen-1-ium-1,2-diolate 193
- release of NO from 193
diaminopropionic acid 489
diastereomer 164

– axial and equatorial 164
diazeniumdiolate 193, 195f, 198
– carbon-based 198
– dialcylamino-based 193
– O^2-naphthylmethyl- and O^2-naphthylallyl-substituted 196
– photochemistry of *meta*-substituted O^2-benzyl-substituted 195
– photosensitive precursor to 193
diazeniumdiolate derivative 194
– expected photochemistry of O^2-2-nitrobenzyl 194
– photochemistry of O^2-benzyl 194
diazirine 153
diazoalkane cage compound 515
dibenzyl tetramethylammonium phosphate 62
di-*t*-butylnitroxide 192
dicyclohexylcarbodiimide 139
dielectric constant 272
7-diethylamino coumaryl cage 123
difference FTIR spectrum 402
– S1S2 protein the presence of caged glutamate 402
difference spectroscopy 385
difference spectrum 388
– of NPE-caged ATP photolysis 388
diisopropyl ethylamine 86
2,5-di-*t*-butylhydroquinone 17
2–4′-dihydroxyacetophenone 64
3′,5′-dimethoxybenzoin group 346
3′,5′-dimethoxybenzoin phosphate 80
dimethoxybenzoin 294
dimethoxybenzoin carbonate group 358
– in microarray fabrication 358
m,m′-dimethoxybenzoin acetate 105
dimethoxybenzoin derivative 15
m,m′-dimethoxybenzyl acetate 119
m,m′-dimethxybenzyl chromophore 118
o-4,5-dimethoxybenzyl-serine 138
3,4-dimethoxycarbonyl protection 136
3,5-dimethoxycarbonyl phenylglycine 137
4,5-dimethoxy-2-nitrobenzylbromide 535
1-(4,5-dimethoxy-2-nitrophenyl)diazoethane 4, 513
1-(4,5-dimethoxy-2-nitrophenyl)-2-nitroethene 260
2-(4,5-dimethoxy-2-nitrophenyl)propanol 91
5,7-dimethoxy-2-phenylbenzofuran 104, 482
dinitrosyl molybdenum complex 189
– photochemistry of 189

dioctanoylglycerol 450
1,3-dioxane 45
dioxaphosphole 80
dipeptide 58
– pHp-protected 58
dipivaloyl peroxide 122
– thermal decomposition 122
directed modification 275
dispersive IR spectrometer 372
dithiane „safety-catch" 141
dithiol 9
DMACM 264
DMCM-OH (6,7-dimethoxycoumarin-4-yl)methanol) 31
DMNB (4,5-dimethoxy-2-nitrobenzyl) 5, 259
– ester 259
DMNB (4,5-dimethoxy-3-nitrobenzyl) 156
DMNBB (4,5-dimethoxy-2-nitrobenzyl bromide) 280
DMNB-OH (4,5-dimethoxy-2-nitrobenzyl alcohol) 31
DMNPE (1-(4,5-dimethoxy-2-nitrophenyl)ethyl) 156
DM-nitrophen 19
DMNPT (1-(4,5-dimethoxy-2-nitrophenyl)-2,2,2-trifluoroethyl) 259
– ester 259
DNA caging 514
DNA microarray 353
– chemical reaction on 353
DNA transcription 518
DNA-protein interaction 517
dominant negative effect 274
dorsal root ganglion cell 172
double coupling 349
doxycycline 536
drug delivery 455
DTT (dithiothreitol) 9, 271, 420
dynamical transition 426
dynein 305

e

ecdysone 534
– receptor 534
β-ecdysone 527, 535
efficiency 96
efficiency of release (see also quantum efficiency) 74
electrical conductance 330
electron transfer 64
electrophysiological technique 207
embryogenesis 335

endothelial-derived nitric oxide
  synthase 179
Engels 95
enhanced green fluorescent protein 537
environmental effect 371
– by IR spectroscopy 371
enzymatic activity 303, 304, 317, 318, 319, 426
– temperature-dependent 426
enzymatic function 303
enzyme 273, 285, 303
enzyme inhibitor 303, 304, 321
enzyme 303, 319, 322, 321
– caged 322
enzyme photoswitches 72
enzyme substrate 303, 318, 319, 321, 322
ependymal cell 175
– ciliary beating of 175
epilepsy 223
EPR (electron paramagnetic resonance)
  spectroscopy 306
EPR spectrometer 8
equilibrium constant 218
– of channel-opening mechanism 218
equilibrium dissociation constant 510
ER (estrogen receptor) 533
ERE (estrogen response element) 533
ESPIT (excited-state intromolecular proton
  transfer) process 112
excitatory post-synaptic current 445
excited singlet and triplet states 98, 103
excited singlet of 2-NB 98
excited triplet state reactivity 113
excited-state pKa's 65
extinction coefficient 159, 163, 254, 516

f

F-actin 328, 329
F-actin filament 329
– cleavage site 329
– rhodamine-labeled 329
FACS calibur flow cytometer 528
Favorskii cyclopropanone ring formation 109
Favorskii process 110
Favorskii rearrangement 110, 115
– mechanism of ground state 115
Favorskii-like rearrangement 109, 110, 113, 115, 116, 126
Fe(II)-proline-dithiocarbamate 192
flash photolysis 169, 185, 194
fluorescein 20, 439
fluorescein dye 452

fluorescence 170, 165, 464
– efficiency 123
– measurement 172
– microscopy 450
– resonance energy transfer 487
– spectra 165
– superposition of 170
fluorescence quantum yield 40, 164, 439
– of coumarinylmethyl-caged cNMP 164
fluorescent labeling agent 29
fluorescent PKC substrate 312
fluorescent property 40
fluorescent protein 292
– caged 292
fluorogenic enzyme substrate 29
fluorophore 20, 450 f
– caged 20, 451
flutamide 190
Fmoc-Asp(DMB)-OH 484
Fmoc-Glu(DMB)-OH 484
focal uncaging 237
folding kinetics 471 ff
– of a_3D 473
– of 1prb$_{7-53}$ 472
FTIR (Fourier transform IR spectroscopy) 306, 307, 308, 309, 322, 371, 399, 487
– spectrometer 372
Franck-Condon absorption 413
fragmentation kinetics 82, 84, 91
FRAP bleaching technique 327
FTIR spectra 468
– labeled and nonlabeled peptide 468
furoxan 191 f
– isomerization of 3,4-disubstituted 191
– release of NO from 3,4-disubstituted 192
– structure 191

g

GABA (γ-aminobutyric acid) 4, 22, 58, 60, 211
– nitroindoline-caged 4
G-actin 326
galactose 266
β-galactosidase 278, 287, 525
– tetramer 278
gap junction 174
GDP 307, 309
gene expression 527, 532 f
– conditional 533
– photoactivated 532
gene inactivation 453
general acid catalytic decay 99

general acid catalyzed breakdown 101
genetically encoded caged protein 294
β-glucuronidase 537
glucose-6-phosphate 314
glucose-6-phosphate dehydrogenase (G6PDH) 314
– caged glucose 6-phosphate and caged NADP 314
glutamate 58, 66, 67, 76, 84, 91, 92, 211, 235, 295
– caged 66, 84
– caging 76
– double caged 235, 295
– property of the caged 91
– single-caged 235
glutamate binding 401
glutamate derivative 86, 89, 92
– chemical and photochemical properties 89
– synthesis of the caged 86
glutamate precursor 86
glutamate receptor antagonist 66
glutamate receptor density 241
glutamate receptors 66, 67, 68, 400
– activation of 400
L-glutamate 22 f
glutamide 85
glutathione 9, 69, 70
glycine 4, 22, 211
– nitroindoline-caged 4
glycine receptor 219
Griess assay 192
group vibration 370
Growth cone guidance 174
GRPase-activating protein 405
Grunewald-Winstein value 115
GTP 307, 308, 309
GTPases 307
GTP hydrolysis 307–309, 403
guanidine 139
– protection 139
p-guanidinophenyl benzoylbenzonate 286
guanosine 3′,5′-cyclic monophosphate (cGMP) 155

h

Hagen and Bendig's mechanism 124
β-hairpin 470
– folding 471
half-time for activation 254
α-haloacetophenone 275
hammerhead ribozyme 496, 506
– caged 506

hammerhead substrate 498
Ha-Ras p21 308
HCN (hyperpolarization-activated, cyclic nucleotide-gated) channel 172
HCM (7-hydroxycoumarin-4-yl)methyl-) 160
HDMP 264
head-to-side-chain caging 492
– of α-helical villin headpiece subdomain 492
head-to-side-chain cyclization 482, 483, 490
– β-sheet 490
head-to-side-chain strategy 483
head-to-tail cyclization 482
helix-coil transition kinetics 466
hemiacetal 97, 99, 100, 101, 395
hemiacetal decay rate 7
hemiacetal intermediate 97
hemiacetal/ketal 99, 101
hemiketal 101
α-hemolysin 290
HEPES buffer solution 63
heptapeptide kemptide 257
heterolysis 105, 109, 113
heterolytic cleavage 64, 109
heterolytic dissociation process 119
high-performance liquid chromatography 497
holoenzyme 337
homolysis 105
– of pHP esters 58
– of phosphotyrosine 72
homolytic cleavage 43, 64
homolytic dissociation process 119
hormones 527
– caged, as spatial regulator 527
HPLC chromatography 134
hybridization assay 519
hydrazone 4,5-dimethoxy-2-nitroacetophenone 514
hydrogen abstraction 98
hydrogen atom abstraction rate constants 98
hydrogen atom transfer 98
hydrogen-bond shift 111
hydrolytic stability 77, 80, 82, 83, 89, 90
p-hydroxyacetophenone 59, 60, 61, 65
2-hydroxycinnamoyl protease 273
hydroxycinnamoylthrombin 297
7-hydroxycoumarine 20
– NPE-caged 20
hydroxyl group 138

– protection of 138
2-hydroxylaminoacetophenone band 383
2-hydroxylaminoacetophenone 391
– closed form 391
– open form 391
4-(hydroxymethyl)coumarin 162
p-hydroxyphenacyl 55, 56, 57, 59, 64, 96, 109, 114
– substrate-caged 59
p-hydroxyphenacyl ATP 62, 71
– release of ATP from 71
– synthesis of 62
p-hydroxyphenacyl ATP (PHP ATP) 55, 56, 58, 62, 70, 71
p-hydroxyphenacyl ATP (PHP Br) 60, 61, 62, 63, 69, 73
p-hydroxyphenacyl bradykinin 62
– synthesis of 62
4-hydroxyphenacyl bromide 263, 282
p-hydroxyphenacyl caged substrate 60 f
– chemical protection 61
– photoprotected 61
– synthesis of 60
p-hydroxyphenacyl chromophore 59, 110
– mechanism for photorelease of substrate from 110
p-hydroxyphenacyl derivative 65, 117
– ground excited state 65
– physical and photochemical properties of 117
– quantum efficiency for release of GABA 65
p-hyroxyphenacyl diethyl phosphate 57
p-hydroxyphenacyl GABA derivative 60
– spectral and photochemical properties 60
p-hydroxyphenacyl group 64, 69, 74
– mechanism for photorelease of substrate from 64
– photochemistry 69
– photoremovable protecting group 55, 74
4-hydroxyphenacyl phosphate 80
p-hydroxyphenacyl triplet state 114
– ipso position on 114
p-hydroxyphenacyl-Ala-Ala 61, 69
– quantum efficiencies for 69
– synthesis of 61
p-hydroxyphenyl acetic acid 56 f
hydroxyphenyl phospate 19
– NPE-caged 19
8-hydroxypyrene-1,3,6-trisulfonate 20, 441
hydroxytamoxifen 534
7a-hydroxy steroid dehydrogenase 317

i

IASD (4-acetamido-4'-(iodoacetyl)amino)stilbene-2,2'-disulfonate 278
imidazole 139
– protection 139
immunoglobulin 288
infrared absorbance 71
inhibitor, caged 307, 311
inositol phosphate 18
intercellular signaling 67, 68
interferon 521
intermediate 270
– breakdown of 270
interneuron 236
– adapting 236
– fast-spiking 236
intracellular transport 70
intracellular signal 309
inverse nucleic acid synthesis 347
– with 3'-DMB phosphate 347
ion channels 332
ion-derived pathway 121
ipso carbon 113
IR absorbance spectrum 380
– of NPE-caged ATP 380
IR spectroscopy 386
– principle of 386
IR radiation 369
IR sample 373
IR spectra 387
IR spectroscopy 254, 371, 374, 390, 394, 399
– for mechanistic study 394
– information from 371
– measuring product release rate with 391
– of caged compound 374
– side-reaction investigation 390
– with caged compounds 399
IR spectrum 371
iron(III) cupferronate 186
– photochemistry of 186
iron-based photosensitive precursor 186
irradiation 276
– wavelength 77, 80, 82, 83, 84, 89, 90, 91, 92
isocitrate dehydrogenase 314, 415 ff
– caged NAD and NADP 314
isoxazoles 100, 102

k

kainate 211
Kaplan 95
keratocyte 327

ketone 46
- photolysis of Bhc-diol-protected 46
kinase 309
- photoregulation 309
kinesin 305, 306
kinetic crystallography 411, 418
kinetic protein crystallography 410
- using caged compound 410
kinetic zipper model 466
Knoevenagel condensation 48

l

laser dye 29
laser flash photolysis 64, 107, 199, 318, 320
- triplet-sensitized 199
laser scanning photostimulation 234 f
laser-induced T-jump 461, 463, 465, 466
- IR setup 465
- IR technique 463
- kinetic of the helix-coil transition using 466
- method 461, 466
lateral superior olive 237
Laue diffraction 414, 418
Laue technique 414
leaving group 101
- role of 101
lectin subunit 276
leuenkephalin pharmacophore 350
lifetime 105
ligand-gated channel 333
ligation 278
- of polypeptide fragment 278
light reversible suppression 513
light source 33
light-activated transcription factor 295
light-directed DNA synthesis 361 f
- cleave-and-characterize assay for 362
- hydroxyl protecting group for 362
- NPPoc phosphoramidite for 361
light-directed synthesis 349
light-induced cleavage 332
light-initiated gene expression 335
LIM kinase 328
limiting rates 95
linsidomine 192
- release of NO from a caged form of 192
Loftfield mechanism 115
long peptide 141
- formation 141
long-wavelength absorption maximum 159, 163

low-temperature phosphorescence emission 112
LTD (long-term depression) 174, 243 f, 331
- glutamate response in hippocampal neuron 244
- postsynaptic 244
LTP (long-term potentiation) 243, 331
luciferase expression 336, 526
lysosome 287, 336

m

manganese dioxide 48
Marcus correlation 121
MS (mass spectrometry) 281
- MALDI (matrix-assisted laser desorption ionization) 135
MCM-OH (7-methoxycoumarin-4-yl)methanol 31
medial nucleus of the trapezoid body 237
membrane barrier dysfunction 441
membrane protein 306
2-mercaptoethanol (β-mercaptoethanol) 271, 280, 281
meromyosin 289
meta effect 119–121
methionine-caged protein 13
7-methoxycoumarin-4yl)methanol-caged phosphate 40
- mechanism of photolysis 40
7-methoxycoumarin-4-yl)methanol (MCM-OH) 31
7-methoxycoumarin-4-yl)methyl (MCM) 160
4-methoxy-7-nitroindoline cage 4
4-methoxy-7-nitroindoline (MNI) 443
p-methoxyphenylacetate 57
4-methoxy-2-phenyl-benzofuran 104
7-methoxy-2-phenyl benzofuran 104
5,7-methoxy-2-phenylbenzofuran 104
5-methoxy-2-phenyl-benzofuran 104
methyl 2-[1-(2-nitrophenyl)ethoxy]ethyl phosphate 395
3-methylanthranil 9, 391
- bands 384
4-methylcoumarin 48
4-methylesculetin 47
4-methylumbelliferone 47
Michael acceptor 13
microarray 350 f, 353, 355
- chemical reaction on DNA 353
- determination of oligonucleotide quality on 355

Subject Index | 549

- peptide and nucleotide bond formation on 351
- preparation of a molecular 350

microarray fabrication 348, 365
- direct write 365

microarray study 351
microarray surface 352
- fluorescent labeling of hydroxyl group on 352

microcystin 505
micromirror chip 366
microsecond-to-millisecond time region 205
millisecond Laue structure 314
mitotic domain 335
MNI, see also 4-methoxy-7-nitroindoline 443
MNPCC (N-methyl-N-(2-nitrobenzyl)carbamoyl chloride) 262, 275 f, 287
molecular beacon 520
- oligonucleotide 519

molecular dynamic simulation 467
molecular geometry 370
molybdenum-based photosensitive precursor 189
monoalkyl carbonate 17
- formed on photolysis 17

monochromatic diffraction 418
monochromator 372
monothiol 9
Mössbauer spectroscopy 426
motors 305, 306
MPE microscopy 437
MS (mass spectrometry) 135, 281
- ESI (electrospray ionization) 281

multiphoton phototrigger 435
multiphoton excitation (MPE) 435
multiphoton sensitive protecting group 365
- in DNA chip fabrication 365

muscle nicotinic acetylcholine receptor 210
1-D-myo-inositol 1,4-bisphosphate 5-phosphorothioate 18
myosin 305, 306
myosin light chain kinase (MLCK) 147, 312
- using caged peptide inhibitor 312

n

N-acetylcysteine 9, 180
N-benzyloxycarbamoyl glycine 118

N-α-Boc-N-im-nitrobenzyl-histidine 139
N,N-dimethylnitrosamine 179
N,N'-ethylenebis(salicylideneiminato) dianion 187
N-hydroxybenzisocazoline 6
N-methyl-D-asparate 213
N-2-nitrobenzyloxycarbonyl group 135
N,N'1,2-phenylenebis(salicylideneiminato) dianion 187
N-terminal amino group 132
- permanent 132
- temporary 132

NAD (nicotinaminde adenine dinucleotide) 4
nanosecond/picosecond laser flash photolysis 103, 107
nanosecond/picosecond TR resonance Raman spectroscopy 113
nanosecond laser flash 99
- photolysis 103

naphthalene 107
- crystal 435

naphthylallyl derivative 197
2-naphthylmethyl 165
NBD-Cl (7-chloro-4-nitro-benzo-2-oxa-1,3-diazole) 281
2-NB triplet states 99
NCM 264
neocupferron 198
- production of NO from 198

neuronal connection 232 f, 235
- functional mapping 233
- mapping with caged glutamate 235

neuronal integration 239
- probing 239

neuronal long-term depression 174
neurotransmitter 66, 205, 208, 212, 444 f
- caged 208, 445
- caged derivative 66
- photochemical property of caged 212
- photochemical release of 205
- release 66

neurotransmitter receptor 205
- membrane-bound 205
- complex 208

nicotinic acetylcholine receptor 218, 291
nitric oxide 178 f, 182, 189
- caged 178
- inorganic photosensitive precursor 182
- organic photosensitive precursor to 189
- photosensitive precursor to 179

nitric oxide synthase (NOS) inhibitor 320
o-nitroacetophenone 420

9-nitroanthracene   190
– photochemistry of   190
2-nitrobenzaldehyde   2, 19
– photochemical isomerization   2
– photomerization   19
o-nitrobenzaldehyde   88
6-nitrobenzo[a]pyrene   190
– structure of   190
5-thio-2-nitrobenzoic acid (TNB)-thiol agarose   73
2-nitrobenzyl   5, 96, 126
– major photoremovable protecting group   96
2-nitrobenzyl bromide   280
o-nitrobenzyl bromide   342
2-nitrobenzyl cage   5 f
– mechanistic aspect of photocleavage   6
nitrobenzyl caged methyl phosphate   7
nitrobenzyl caged cNMP   158
– photolysis of the diastereomer of   158
nitrobenzyl caged glucose   16
2-nitrobenzyl chromophore   98
– reactive excited state   98
– UV absorption   98
o-nitrobenzyl cyanohydrin glycoside   346
o-nitrobenzyl derivative   156
2-nitrobenzyl derivative   1
2-nitrobenzyl ester   159
– irradiation   156
– properties   159
2-nitrobenzyl group   142 f
– bound to peptide   143
– in microarray fabrication   357
– photolysis   142
– photorelease of a substrate   97
2-nitrobenzyl methyl ether   390
2-nitrobenzyl phoshate   2
2-nitrobenzyl photochemistry   8
2′-(2-nitrobenzyl)adenosine   506
2′-O-(2-nitrobenzyl)adenosine phosphoramidite   506
2′-O-(2-nitrobenzyl)-N^6-benzoyladenosine   506
2-nitrobenzylated peptide   132
nitrobenzylglycine   15
2-nitrobenzyloxyacetic acid   390
o-nitrobenzyloxymethyl chloride   342
nitrobenzyloxymethyl deprotection   343
nitroindoline   21
7-nitroindoline cage   21
7-nitroindoline caged compound   22
7-nitroindoline derivative   1
nitroindoline photolysis   23

nitroindolinyl carbamate EDTA derivative   23
nitronate   270
– intermediate   271
E-nitronic acid   6
Z-nitronic acid   6
3-nitro-2-naphthalenemethanol   274
1-(2-nitro-4,5-methylenedioxyphenyl)-ethyl-1-oxycarbonyl chloride   261
nitronaphthylethoxycarbonyl group   362
– in microarray fabrication   362
o-nitrophenethyloxymethyl group   344
1-(2′-nitrophenyl)-2-ethanediol   311
4-nitrophenylalanine   152
1-(2-nitrophenyl)diazoethane   4
p-nitrophenyl 2-(2,5-dimethylbenzoyl)benzoate   286
1-(2-nitrophenyl)ethanol   274
1-(2-nitrophenyl)ethyl   2, 5, 156
1-(2-nitrophenyl)ethyl ATP   101, 102
1-(2-nitrophenyl)ethyl caged ATP   10
1-(2-nitropheny)ethyl caged choline   16
1-(2-nitrophenyl)ethyl caged methyl phosphate   7
o-1-(2-nitrophenyl)ethyl ether   441
1-(2-nitrophenyl)ethyl methyl ether   101
– photocleavage of   10
1-(2-nitrophenyl)ethyl phosphate   80
1-(2-nitropheny)ethyl phosphoramidite reagent   13
2-nitrophenylglycine   138, 283, 291
1-(o-nitrophenyl-[2,2,2]-trifluoro)ethanol   91
(nitrophenylpropyloxy)carbonyl group   360 f
– in microarray farbrication   360
– photochemical deprotection of   360
– substituted   361
nitrosamine   181
– photochemistry of   181
nitroso derivatives   304, 317
nitrosoacetophenone   8 f, 80, 97, 317
– reaction of   9
– by-product of the NB cage   9
nitrosoacetophenone band   379
2-nitrosobenzaldehyde   101
nitrosoindole   21
– by-product from photolysis   21
nitrosopyruvate   10
– NB cage by-product   10
nitrosyl chloride   186
nitrosyl derivatives   80, 91, 186
– of ruthenium   186
nitrosyl ruthenium complex   188
– photochemistry of a pyrazine-based   188

– pyrazine-bridged 188
NMCC (3-nitro-2-naphthalylocycarbonyl chloride) 261, 288
non-denaturing gel 520
nonsense codon suppression technique 149
– caged peptide by 149
nor-butyrylcholine derivative 318
NOS (nitric oxide synthese) 320
NPE (1-(2-nitrophenyl)ethyl) 5, 258
– ester 258
Npg (o-nitrophenylglycine) 332
NPPOC 261
NPY (neuropeptide Y) 145
nuclear localization signal 147
nucleic acid synthesis 341, 345
– o-nitrophenethyloxymethyl in caged 345
– photoremovable group used in 341
nucleoside building block 357
nucleoside composition analysis 509
nucleoside 5'-DMB carbonate 359
– synthesis of 359
nucleoside model study 507
nucleotide 11, 70
– caged 70
– nitrobenzyl-caged 11
nucleotide release 70
NVOC (2-nitroveratryloxycarbonyl group) 135, 350 f
NVOC-Cl (6-nitroveratryloxycarbonyl chloride) 260
NVOC-OH (4,5-dimethoxy-2-nitrobenzyl alcohol) 31
β-$^{18}O_3$ labeled GTP 72

o

okadaic acid block 505
oligonucleotide synthesis 508
oligonucleotides 12
– photomanipulated 12
oligopeptide 58, 59, 61, 62
– pHp-protected 58
on-the-fly light-direction method 366
ORD (optical rotatory dispersion) 296
organic nitrite 189
oxazole 97
oxetane 105
oxidoreductase 314
oxyallyl 64
oxyallyl intermediate 113
oxyallyl triplet 113
oxyallyl zwitterions 115

p

PACAP (pituitary adenylate cyclase-activating polypeptide) 175
PaPy$_3$ 186
– structure of 186
PBN (phenyl-N-tert-butyl nitrone) 192
Pechmann reaction 48
pentacyanonitrosylferrate 182
peptide 12 f, 131, 143 ff, 151
– building block for caged 145
– design of caged 144
– nitrobenzyl caged 12
– pentacationic modified 13
– photoaffinity 151
– photochromic 151
– photolysis of 131
– photolysis of 2-nitrobenzyl group to 143
– synthesis of caged 144
peptide bond formation 141
peptide fragment 133, 141
– formation of 133
– photocleavable linker within 141
peptide fragment elongation 134
– outline of 134
peptide inhibitor 312
peptide library 141
peptide PKA inhibitor 311
peptide release 68
peptide synthesis 134 f, 140
– amino protection 135
– photocleavable protecting group in 135
– photocleavable support for 140
– purification and identification 134
pH-dependent rate constants 101
phenacyl (pHP) 109
– photoremovable protecting group 109
phenacyl chloride 116
– photochemical study of substituted 116
phenacyl phosphate 57
– p-substituted 57
phenol 16, 45, 50
– caged 16
– photolysis of Bhcmoc-caged 45
– precursor molecule with 50
phenolic carbonate 17
– CNB-caged 17
L-p-phenylazophenylalanine 283
2-phenylbenzofuran 104–107, 117
2-phenyl-5,7-dimethoxybenzofuran 103, 104
2-phenyl-5,7-dimethylbenzofuran 346
phenylephrine 17, 257 f
3'-phosphoramidite 346

phenyl-N-tert-butyl nitrone (PBN)   192 f
– photochemistry   193
phosphatase activity   504
phosphates   7, 18, 36, 49, 63, 72, 76, 78
– caged   18
– coumarine caged   36
– deca rate   7
– precursor molecule with   49
phosphopeptide   18, 149
– building bock for caged   149
– caged   18
phosphoprotein   148
– caged   148
phosphoramide mustard   18
phosphoramidite   18, 507
– building block   352
phosphorescence emission spectra   111
3′-phosphotriester   346
phosphorylated cysteinyl group   72
phosphotriester   148
phosphotyrosine   72
photoacid   363
– with DMTr group   363
– precursor   363
photoacoustic calorimetry   487 f
– schematic setup   488
photoacivated release   95
photoaffinity peptide   151 f
photochemical amide synthesis   21
photochemical decarboxylation   15
photochemical efficiency   315
photochemical enzyme regulation   303
photochemical Favorskii process   110
photochemical quantum yield   159, 163 f
– of coumarinylmethyl-caged derivative   164
photochemical triggering   76
photochemical uncaging   320
photochemistry   3
photochromic peptide   151
photocleavable linker   139
photocleavable protecting group   212
– for biologically active compounds   212
photocleavage   15
– kinetics   13
photoconversion   3
photodeprotection   3, 95
photodynamic therapy   178, 444 f
– psoralen for 2PE-mediated   454
photoefficiency   22
photoheterolysis   102
photoinduced intramolecular redox reaction   97

photoredox hydrogen abstraction   98
photorelease sequence adiabatic   110
photoremovable protecting group   55, 56, 74, 95, 96, 102, 110, 341
– in DNA synthesis   341
photo-Favorskii rearrangement   56, 64
– triplet   64
photofragmentation   91, 423
– at cryo-temperature   423
photoreleased PsT197Ca   73
photoinduced homolytic cleavage   180
photoinduced isomerization   270
photoirradiation   136
– of protected amine   136
photoisomerizable intramolecular crossling   293
photoisomerization   296
photolabile derivative   211
– of neurotransmitter   211
photolabile linker   481
– development of   481
photolithographic mask   348
photolithographic processing   363
photolithography   348
photolysis   2, 3, 7, 270
– caged amide   7
– caged-compound   2
– effect of pH on   270
photolysis quantum yield   41
photolysis spectra   388
photolytic chain cleavage   294
photolytic cleavage   232
– of caged glutamate   232
photolytic efficiency   41, 77, 80, 82, 89, 90, 91, 92
photolytic quantum yield   44
photolytic rate constants   77, 80, 82, 83, 89, 90, 91
photomethanolysis   116
– substituted phenacyl chloride   116
photorearrangement   112
– pHP   112
photoregulation   304, 307, 309, 314, 317, 318, 319, 322
– of alcohol dehydrogenase   314
– of ATPase   304
– of cholinesterase   317
– of GTPase   307
– of kinase   309
photorelaxation   179
photorelease   110
photoremovable protecting group   1, 55, 95, 304

- for caging bioactive substrates 55
- for caging biomolecule 1
photoremovable scaffold 141
photoresponsive melittin 151
photosensitivity 22, 41
photosensivitve precursor 188 f
- chromium-based 188
- molybdenum-based organic 189
photosolvolysis 118, 123
- quantum efficiency 123
photoswitch 435
photothermal beam deflection 487 f
- schematic setup 488
phototriggers 29, 33, 57, 95, 156, 166, 435
- for cNMP 156
- selection of 166
pHP (p-hydroxyphenacyl) 57
pHP Ala-Ala 61, 68, 69
pHP Bradykinin (pHP Bk) 59, 61, 62, 66, 67, 68
pHP excited singlet 111
pHP GTP 71, 72
pHP(thio)-protected enzyme 73
pHP triplet states 114
pHP-cated SHP-1, SHP-1 (ΔSH2) enzymes 73
pHP-PsT197Ca 63, 74
pH-rate profiles 99–102, 126
phytochrome 295
Pincock mechanism 122
^{3}pKa 114
pKa of ^{3}PHA 65
PKA (protein kinase A) 73 f, 273, 309, 310
- cAMP-dependent PK 310, 311
- calmodluin-dependent PK 312, 313
- modulator 310
- peptide inhibitor 312, 313
- structure of the C subunit of 74
- using caged cAMP 311
- using caged inhibitor 311
PKC 311 f
- using caged activator 311
- using caged fluorescent substrate 312
pockel cell 438
polyacrylamide electrophoresis 521
poly(ethylene glycol) 282
- sulfhydryl-directed 282
poly(N-isopropylacrylamide) 293
polypeptide chain cleavage 273
polypeptide microarray 350
polypeptide side-chain 280
poly-ribonucleotide synthesis 343

- 2'-O-o-nitrobenzyl group for protection in 343
postsynaptic effect 239
precursor molecule 46, 48
product formation spectrum 388
properties of caged-ATP derivatives 83
properties of caged-glutamate derivatives 89, 90
protease 286
protecting group 95, 133, 356
- chemical stability 133
- in microarray fabrication 356
- light-activated removal of 95
protection-deprotection chemistry 49
protein 12, 253 f, 256, 272, 274, 277, 284, 295 ff, 305, 325, 486
- at cys residue 277
- caged, in cell-based system 325
- caging group for 256
- coating 277
- head-to-side-chain cyclization by BrAc-CMB 486
- light-activated protein 253
- medical application 297
- motor 305
- nitrobenzyl caged 12
- photochemistry peculiar to 272
- photoregulation of 253
- procedure for caging and uncaging 254
- site of modification in caged 274
- spatial control 295
- temporal control 296
- turning on and off 284
protein activity 282, 410
- photoregeneration of 282
protein crystallography 412
protein folding 461, 479, 489
- kinetic event in 489
protein kinase 285, 310, 312
- A (PKA) 63, 73
- calmodulin-dependent 312
- cAMP-dependent 310
protein modification 280 f
- determining the extent of 281
- reagent for caging by 280
protein phosphatase 285
protein splicing 283
protein trafficking 336
protein-folding funnel 479
protein-protein interaction 278, 282, 292, 334
proton caged 19
proton inventory 118

psoralen   453
PTP (see also protein tyrosine phosphatase)   72
– inhibitors   72
Purkinje neuron   174
pyrenyl-methyl ester   363
pyridine nucleotide   314

q

Q-rhodamine   20
quantum efficiencies   59, 60, 65, 69, 77, 80, 81–83, 89, 90, 91, 95, 107, 112, 117, 120 f, 124
– acylmethyl derivative   121
– for a series of MCou substrate   124
– for mono-substituted benzyl acetate   121
– for ortho and meta derivative   120
– ortho/meta effect   121
quantum yield   254
quenching   105, 107
quinol-oxidizing enzyme   321
p-quinone methide pQM (intermediate)   111, 112
– caged substrate of   320

r

radical clock   121
radical-derived pathway   121
Raman spectroscopy   113
– TR resonance   113
random modification   274
rapid metal-to-ligand charge transfer   462
rapid scan technique   373
rapid singlet-triplet (ST) intersystem crossing   64
rapid-scan FTIR spectroscopy   9
ras protein   71, 307, 309
– of GTPase   307
ras-bound GTP   71
rate constant   40, 124, 218
– for a series of MCou substrate   124
– of channel-opening mechanism   218
rate determining cyclization   102, 107
rate determination step   99
rate limiting microscopic rate constants   99
rate limiting step   96
rate of uncaging   254, 270
Rayonet reactor   342
reactions efficiencies   110
reactive exciplex   105
reactive oxygen species   441
release rate constants   99
release rates   107

real-time crystallography   414
restriction enzyme activity   517
– reversible blockade of   517
retinoic acid receptor   534
reversible blockade   517 f, 519, 523 f, 525 f
– of GFP expression in HeLa cell   523
– of hybridization   519
– of in vitro transcription   518
– of mRNA   526
– of plasmid expression   524, 525
– of restriction enzyme   517
reversible inhibitors   72
rhodamine-phalloidin staining   337
ribonuclease reductase substrate   322 f
– caged   322
ribonuclease   147, 287, 426
ribonucleoside stannylene   342
ribonucleotide   344
ribosome-in-activating protein   276
ribozyme cleavage   509
ribozyme kinetic   510
ribozyme RNA complexe   510
ring closure   107
RNA photolysis   509
RNA processing   495
– photocontrol of   495
Roussin's black salt   184
Roussin's red ester   186
– photochemistry of   186
Roussin's red salt   184 f
– photochemistry of   185
– structure of   185
RS-20   147
ruthenium salen   187
ruthenium-based photosensitive precursor   186

s

(s)-α-methyl-4-carboxyphenylglycine (MCPG)   67
salophen nitrosyl   187
Sandros equation   199
Schaffer collateral-CA1 synapse   331
scission   273
sDD triplet   113
SDS-PAGE (sodium dodecylsulfate-polyacryl-amide gel electrophoresis)   273
second messenger   178, 310, 450
– photochemical release   178
Sheehan   56, 102, 103, 105
selenium dioxide   48
selenocysteine   278
self-assembled monolayer   191

semisynthetic protein   283
serotonin   175, 211
side-chain cage   482
side-chain caging   489
– of the GCN4-p1 leucine zipper   489
– scheme   483
side-chain protection   482
signal transducer   67
signal transduction   66, 337
– molecule   446
signalling pathway   71
silver oxide   49, 160
5′-silyl nucleoside   346
single-wavelength measurement   374
singlet excited state   98, 111, 105
singlet intramolecular exciplex mechanisms   103
singlet lifetime (τ_s) of 2-NB   98
singlet oxygen generator   453
singlet redox reaction   98
singlet state reactivity   103
singlet-triplet crossing   98, 111
– rate constant   98
site modification   275, 278
– in caged protein   275
– site other than cys for specific   278
site-specific caging   529
– of DNA and RNA   529
small molecule inducer   532
small nuclear ribonucleoprotein particle   498
S-nitrosoacetylpenicillamine   181
S-nitrosoalbumin   179
S-nitrosodithiothreitol   181
S-nitrosoglutathione   179 f
– photochemistry of   180
S-nitrosothiol   180
sodium nitroprusside (SNP)   182 ff
– equilibrium   184
– photochemistry of   183
– photochemistry of a caged form of   183
– primary photochemical reaction of   182
solid-phase DNA synthesis   363
– photoacid-based deprotection for   363
solid-phase peptide synthesis   131 f, 146
– caged peptide   146
– coupling technique in   133
– protecting group in   132
solution phase synthesis   149
– caged peptide by   149
solvent glass transition   427
solvent isotope effect   111, 117
spatiotemporal control   295, 310

S-peptide   147
sperm-activating peptide   148
S-P-G-K   491
sphingosine phosphate   18
spin (a_i) density   114
spirodienedione   64, 109, 110, 112, 113, 116
spirodienedione intermediate   70, 110
spiroketone intermediate   269
spliceosome assembly   500, 504
splicing   292, 498, 503, 511
– ATP requirement for   503
– of caged pre-mRNA   511
spontaneous hydrolysis rate   14
– for CNB ester   14
step scan technique   373
Stern-Volmer analysis   481
Stern-Volmer quenching technique   117
stowaway   350
strand break   12
streptozocin   181
– structure of   181
substrate-assisted photolysis   452
sulfanate   281
sulfate   41
– coumarine caged   41
sulforhodamine   215
synaptic plasticity   243
synchrotron   414
– X-ray irradiation   430
synthesis   60, 61, 62, 74
synthesis of caged-ATP derivatives   78, 81, 82
– of CNB-caged ATP   78, 81
– of Desyl-caged ATP   80, 81
– of DMACM-caged ATP   80, 81, 82
– of DMB-caged ATP   78, 81
– of DMNPE-caged ATP   78, 82
– of DNP-caged ATP   78, 81
– of NB-caged ATP   78, 81
– of NPE-caged ATP   78, 80, 81, 82, 84
– of p-HP-caged ATP   78, 82, 83, 84
syntheses of caged-glutamate derivatives   86 f
– N-Bhc-   86–88
– γ-BhcMe-   86, 87
– bis-α,γ-CNB-   87
– α-CNB-   86, 87
– γ-CNB-   86, 87
– N-CNB-   86, 87
– α-O-Desyl-   86, 87
– γ-O-Desyl-   86, 87
– N-Desyl-   86, 87, 88

Subject Index

- α-DMNB- 87, 88
- α-DMNPE- 87, 88
- α-DMNPT- 87, 88, 91
- γ-DMNPP- 86, 87, 88, 91
- γ-p-HP- 86, 87
- γ-p-HPDM- 86, 87
- γ-p-HPM- 86, 87
- γ-*MNI*- *86, 87, 88*
- N-NB- 87, 88
- γ-NI- 86, 87, 88
- N-Nmoc- 87, 88
- γ-NPE- 86, 87, 88
- N-NPE- 86, 87, 88
- N-NPEOC- 87, 88

t

T7 transcription 508
tagged acylating agent 275
tamoxifen 533
target molecule 36
target oligonucleotide array 354
TBHCM 264
t-butyl 2-nitromandelate 78
temperature-controlled protein crystallography 427
tet-off system 536
tet-on system 536
1,4,8,11-tetraazacyclotetradecane 188
tetracycline 536
tetracysteine (TC) sequence 280
tetrodotoxin 66
thermal reisomerization 296
thin line spectrum 388
thin-layer chromatography (TLC) 500, 511
– analysis of pre-mRNA 511
thiophosphoryl peptide 281
thiophosphorylation 278
thiophosphotyrosine 257
thioxo peptide 150
threonine 263
thymidine 356
– photophysical properties 356
thymosin 289
β-thymosin 326
time-resolved crystallographic study 308
– with caged GTP 308
time-resolved crystallography 308, 314, 317
– with caged GTP 308
time-resolved Fourier transform infrared (TR-FTIR) spectroscopy 71
time-resolved FTIR difference spectroscopy 308
– with caged GTP 308

time-resolved infrared (TRIR) spectra 100
– of aci-nitro decay 100
time-resolved infrared (TRIR) spectroscopy 99, 100, 194
time-resolved IR spectroscopy 369, 373
– with caged compound 369
time-resolved laser flash photolysis 117
time-resolved picosecond pump probe 99, 101
time-resolved resonance raman spectroscopy 113
time-resolved studies 306, 308, 317–320, 322
time-resolved X-ray diffraction 254
T-jump induced relaxation 469
– of the ML peptide 469
tosylhydrazone formation 48
target protein 288
trans-cis isomerization 275
transcription factor 290
transfer reaction 403
– pathway for the phosphoryl 403
transgene 532
transient absorbance 465
transient absorption spectrum 107
transient kinetic investigation 212
– of neurotransmitter receptor 212
transient kinetic technique 207
transient species 198
transition dipole coupling 464
transmembrane channel 400
transmembrane voltage 208
transport ATPase 305
trapping 421
– experiment 422
trichloroacetic acid (TCA) 364
– photochemical precursor to 364
triethylamine 139
trifluoroacetic acid 132
trifluoroacetoxy)iodobenzene 141
triggering technique 480
triisopropyl)siloxymethyl group 344
3,3'-bis(trimethyl aminomethyl) azobenzene 212
trimethylsilychloride 506
π,π^* triplet 113
triplet benzion cation $^3BC^+$ 107
triplet cation 107
– ($^3BC^+$)
triplet excited state 98, 111
triplet intramolecular hydrogen atom transfer 99
triplet lifetime 105

triplet oxyallyl   113
triplet oxyallyl 1,3-biradical intermediate   113
triplet pathway   3, 10, 107
triplet phenol   65
– role of   65
triplet state   105
– pathway   103
– reactivity   103
TR-FTIR   71, 102
TRIR difference spectra   101
TR-LFP studies   112
true residual activity   282
tryptophan residue   272
T-S-D-G-K turn   491
Two-photon   84, 89, 90, 92
two-photon absorbance cross-section   440
two-photon action cross-section   438 f
– schematic for measuring   439
two-photon excitation (2PE)-induced fluorescence   435
two-photon excitation   442
– sensitive chromophore for caged compound   442
– sensitive chromophore for caged calcium   448
two-photon gene inactivation   455
two-photon microscopy   240, 438
two-photon uncaging   234, 320
tyrosine phosphatase   275
tyrosine phosphorylation   72
tyrosine residue   333
tyrosine   72
– phosphorylated   72

u

ultrafast folding   472
ultrafast photochemical triggering system   480
uncaging action cross section   77, 80, 83, 89, 90
uncaging glutamate   235
uncaging rate constant   159
uncaging   282

– incomplete   282
unnatural amino acid mutagenesis   278, 283
unnatural photoactivatable amino acid   283
unquenchable triplet   105
urease substrate   322 f
– caged   322
UV photolysis   192
UV-vis absorption spectroscopy   281
UV-vis radiation   416
UV-vis spectra   98
– caged cyclic nucleotide   158
– of 2-NB methyl ether   98
– 2-nitrosobenzaldehyde   98
UV-vis spectroscopy   438

v

vibrating atom   369
– mass of   369
vibrating bond   369
– strength of   369
vibrational mode   405
– of the α, β and γ-phosphates of GTP   405
vibrational spectroscopy   464
villin headpiece subdomain   492
virtual cell simulation   327
voltage-gated channel   332

w

Wan's mechanism for benzoin   104
– photoheterolysis   104
wavelength   370
wavenumber   370
Wirz and Givens mechanism   108

x

X-ray crystallography   410
X-ray data collection strategy   430

z

zinc phthalocyanine   180
zwitterion   115
zwitterionic oxyallyl intermediate   115

DIE GESAMTE
BAUSCHREINEREI

EINSCHLIESSLICH

DER HOLZTREPPEN, DER GLASERARBEITEN UND DER BESCHLÄGE

HERAUSGEGEBEN
VON
THEODOR KRAUTH
ARCHITEKT, GROSSH. PROFESSOR UND REGIERUNGSRAT IN KARLSRUHE.

VIERTE, DURCHGESEHENE UND VERMEHRTE AUFLAGE

MIT 82 VOLLTAFELN UND 393 WEITEREN FIGUREN IM TEXT

ERSTER BAND: TEXT

LEIPZIG
VERLAG VON E. A. SEEMANN
1899.

Über die Schreinerarbeiten aus der zweiten Hälfte des 19. Jahrhunderts gibt es nur wenige gedruckte Quellen; die umfassendste und informativste ist sicherlich das hier im Reprint vorgelegte Schreinerbuch I „Die Bauschreinerei" verbunden mit dem bereits 1980 erschienenen Nachdruck des Schreinerbuches II „Die Möbelschreinerei". Beide Bände bilden zusammen ein Gesamtwerk, das infolge seiner Abbildungsfülle verbunden mit hoher Detailgenauigkeit und klaren Formulierungen den Fachleuten wie Laien umfassende Informationen über die Vielfalt tischlermäßig gefertigter Holzarbeiten bietet.

Obwohl in mehreren Auflagen erschienen, gehört es heute zu den gesuchten Fachbüchern und Nachschlagewerken für einen breiten Interessentenkreis in der holzverarbeitenden Industrie, dem Handwerk und allen kunstgewerblich orientierten Kreisen.

Die unwesentliche Verkleinerung des Reprints nimmt dem zweibändigen – hier zu einem Band zusammengefaßten – Werk nichts von seinem ursprünglichen Charakter.

Das Schreinerbuch I
Die gesamte Bauschreinerei
Bestellnummer 111

Das Schreinerbuch II
Die gesamte Möbelschreinerei
Bestellnummer 110

Das Originalwerk in der 4. Auflage von 1899 hat das Buchformat von 22 × 28,5 cm. Es besteht aus den Teilen „Die Möbelschreinerei" (2 Bände) sowie „Die Bauschreinerei" (2 Bände) und befindet sich im Besitz der Universitätsbibliothek Hannover.

Herausgegeben 1981 von der Edition »libri rari« Th. Schäfer GmbH, Hannover
Gesamtherstellung Th. Schäfer Druckerei GmbH, Hannover

DAS SCHREINERBUCH

VON

THEODOR KRAUTH UND FRANZ SALES MEYER

I.

DIE BAUSCHREINEREI

I. BAND: TEXT.

VORWORT ZUR ERSTEN AUFLAGE.

Der vorliegende Band des Schreinerbuches, welches sein Entstehen der Anregung des Herrn Verlegers verdankt, stellt im allgemeinen den Unterrichtsstoff der Baukonstruktionslehre dar, wie sie in der V. Klasse der Hochbau-Abteilung der Grossherzogl. Baugewerkeschule in Karlsruhe behandelt wird.

Der Verfasser hat diesen Unterricht seit einigen Jahren zu erteilen, und es ist ihm hierbei als ein schwerer Missstand erschienen, kein passendes Hilfsbuch für den Schüler zu besitzen. Das gesprochene Wort des Vortrages allein genügt nicht zum völligen Verständnis und insbesondere nicht zur dauernden Erhaltung desselben. Zweifellos ist ein geeignetes Handbuch zum Nachlesen und zum Selbststudium ein besseres Ergänzungsmittel, als dasjenige des zeitraubenden Diktierens und Nachschreibens.

Wenn dieser Umstand für die Abfassung des vorliegenden Buches in erster Linie massgebend war, so sollte es doch anderseits so gehalten sein, dass es sowohl dem praktischen Geschäftsmann als auch dem entwerfenden Techniker von Nutzen sei.

Ob dies gelungen, muss der Erfolg lehren. Wenn aber, dann ist dies zum grossen Teil der Art und Weise zu verdanken, in welcher der betreffende Unterricht an der Karlsruher Baugewerkeschule beim Eintritt des Verfassers in deren Lehrkörper bereits erteilt wurde und ihm als Vorbild diente. Die vorgefundene Lehrmethode, mit welcher er sich völlig einverstanden erklären konnte und zu welcher vornehmlich die für den Anschauungsunterricht so wichtigen parallelperspektivischen Erläuterungsskizzen gehören, war aber das Ergebnis der eingehenden Versuche und Studien seines Vorgängers, des jetzigen Direktors; sie war dem Anfänger im Lehrfache eine willkommene Einführung von höchstem Werte.

Der Verfasser glaubt daher, seinen Dank für diese Unterstützung, sowie für die im Laufe der Jahre ihm erteilten Ratschläge nicht besser zum Ausdruck bringen zu können, als indem er diese seine Arbeit

Herrn **Philipp Kircher**
Direktor der Grossherzogl. Baugewerkeschule in Karlsruhe

in dankbarer Hochachtung ergebenst widmet.

Karlsruhe, 1890.

Der Verfasser.

VORWORT ZUR VIERTEN AUFLAGE.

Anlässlich der verschiedenen Neuauflagen ist das Buch jeweils im ganzen durchgesehen, an einzelnen Stellen gekürzt, an anderen erweitert worden. Während die erste Auflage 64 Tafeln und 328 Textfiguren aufzuweisen hatte, ist das Illustrationsmaterial nun derart vermehrt, dass die vorliegende Ausgabe 82 Tafeln und 393 Figuren im Textband umfasst. Als Ergänzung zur Bauschreinerei sind in demselben Verlage erschienen:

Krauth, Th., Hausthüren und Glasabschlüsse. I. Einflügelige Hausthüren. 30 Tafeln. II. Glasabschlüsse. 30 Tafeln. Preis jedes Heftes broschiert 3 Mark.

Karlsruhe, Mai 1899.

INHALT.

	Seite
I. Das Material	1
1. Die Holzarten	1
2. Die Holzsorten und Schnittwaren	3
3. Der Aufbau des Holzes	5
4. Die Kennzeichen gesunder und schadhafter Bäume	6
5. Die Fällzeit des Holzes	7
6. Das Schneiden des Holzes	7
7. Das Trocknen des Holzes	8
a) Das Schwinden	10
b) Das Quellen	13
8. Die Zerstörung des Holzes (Trockenfäule, nasse Fäule, Hausschwamm)	14
II. Die Werkzeuge	17
1. Werkzeuge, welche jedem Arbeiter vom Meister gestellt werden	18

1. Hobelbank. — 2. Schlichthobel. — 3. Doppelhobel. — 4. Putzhobel. — 5. Rauhbank. — 6. Schrupphobel. — 7. Zahnhobel. — 8. Simshobel. — 9. Faustsäge. — 10. Handsäge. — 11. Absetzsäge. — 12. Stecheisen. — 13. Zirkel. — 14. Raspel und Feile. — 16. Sägefeile. — 17. Winkel. — 18. Winkelmass. — 19. Hammer. — 20. Klöpfel. — 21. Kropflade. — 22. Schraubzwinge. — 23. Oelgefäss.

	Seite
2. Allgemeine Werkzeuge, welche sämtlichen Arbeitern gemeinsam zur Verfügung stehen	25

24. Nuthobel. — 25. Grathobel. — 26. Kurvenhobel. — 27. Grundhobel. — 28. Falzhobel. — 29. Plattbank. — 30. Kehl-, Rundstab- und Gesimshobel. — 31. Schweifsäge. — 32. Fuchsschwanz. — 33. Lochsäge. — 34. Gratsäge. — 35. Feilkluppe. — 36. Gehrmass. — 37. Schmiege. — 38. Winkelstosslade. — 39. Gehrungsstoss- und Schneidelade. — 40. Lochbeitel. — 41. Fischbandeinstemmeisen. — 42. Wasserwage und Bleiwage. — 43. Richtscheit. — 44. Setzlatte. — 45. Dächsel. — 46. Handbeil. — 47. Schraubenschlüssel — 48. Schraubknechte. — 49. Schraubstock. — 50. Schleifstein. — 51. Bandsäge mit Hand- und Fussbetrieb. — 52. Bandsäge mit Kreissäge. — 53. Bandsäge mit Fräsmaschine. — 54. Bandsäge mit Decoupiersäge. — 55. Bohrmaschine. — 56. Bohr- und Stemmmaschine für Handbetrieb. — 57. Universalmaschine.

	Seite
3. Werkzeuge, welche sich jeder Arbeiter selbst beschafft	34

58. Zentrumbohrer. — 59. Amerikaner Bohrer. — 60. Nagelbohrer. — 61. Spitzbohrer. — 62. Ausreiber. — 63. Geissfuss. — 64. Stechzeug. — 65. Schnitzer. — 66. Versenker. — 67. Schraubenzieher. — 68. Schränkeisen. — 69. Ziehklinge. — 70. Abziehstein. — 71. Beisszange. — 72. Massstab.

	Seite
III. Die Verbindungen der Hölzer	36
1. Holzverbindungen nach der Breite	36
2. Das Stemmen (Gestemmte Arbeit)	39
3. Holzverbindungen nach der Länge	47
4. Eckverbindungen	47
5. Hilfsmittel zur Verbindung (Nägel, Schrauben, Drahtstifte und Leim)	51
IV. Die Fussböden	57
Allgemeines	57
1. Blindboden	61
2. Rauher Dielenboden	61
3. Gehobelter Dielenboden	61
4. Tafelfussboden	61
5. Friesboden	63
6. Riemenboden (Schiffboden)	64
7. Fischgrat-, Kapuziner- oder Stabfussboden	65
8. Desgleichen in Asphalt	66
9. Tafelparketten	67
V. Lambris und Täfelungen	69
1. Glatte Lambris (Sockelleisten, Fussockel, Sockel mit Fussleiste, Sockel mit Fuss- und Deckleiste)	69
2. Gestemmte Lambris (Brüstungslambris und Vertäfelungen)	72
VI. Thüren und Thore	86
Allgemeines	86
1. Einfache Thüren, a) Lattenthüre, — b) Riementhüre, — Stumpf verleimte Thüre	88
2. Verdoppelte Thüren	89

Inhalt.

	Seite
3. Gestemmte Thüren	91
a) Einflügelige Thüre	93
b) Zweiflügelige Thüre	100
c) Schiebthüre	108
4. Verglaste Thüren, Glasthüren	110
a) Balkon- und Verandathüre	111
b) Glasabschluss	112
c) Wartesaal- und Vorplatzthüre	114
d) Pendelthüre	114
e) Windfang	116
5. Hausthüren	118
a) Ein-, zwei- und dreiflügelige Thüre	118
b) Thor, Hausthor oder Thorweg	132
c) Magazin- und Scheunenthor	135
d) Hofeinfriedigungsthor	137
6. Verschiedene Thüren zu bestimmten Zwecken	139
Polsterthüre — Schalterthüre. — Abortthüre. — Kellerthüre. — Treppenthürchen. — Notthüre. — Fallthüre. — Rollenthüre. — Dachaussteigethüre. — Gartenthüre. — Barrière.	

VII. Die Fenster 143
Allgemeines	143
1. Die Bildung der Futterrahmen	145
2. Die Bildung der Fensterflügel	148
3. Das einfache Fenster	153
4. Das Doppelfenster	155
a) Das Vorfenster oder Winterfenster. — b) Das Kastenfenster. — c) Das Blumenfenster.	
5. Das Klappfenster	158
6. Das Drehfenster	159
7. Das Schiebfenster	159
8. Fenster in Fachwerkswänden	163
9. Das Schau-, Auslage- oder Ladenfenster	163
10. Das Glas	165
Tafelglas. — Spiegelglas. — Rohglas. — Kathedralglas. — Farbiges Hüttenglas. — Ueberfangglas. — Antikglas. — Mattglas. — Geätztes Glas. — Musselinglas. — Butzenscheiben.	
11. Das Verglasen	170

VIII. Fensterläden 171
A. Aeussere Läden	171
1. Klappläden. — 2. Rollläden und Rolljalousien. — 3. Schiebläden. — 4. Zugjalousien.	
B. Innere Läden	180

IX. Holzdecken 181
Allgemeines	181
1. Die Balkendecke	183
2. Die Kassettendecke	183
3. Die Felderdecke	184

X. Holztreppen 188
Allgemeines	188
1. Die Führung der Treppen, a) die gewöhnliche, b) die gemischte, c) die gewundene Treppe	189
2. Treppenarme und Treppenbenennung	190
3. Die Konstruktion der Treppen, a) die eingeschobene, b) die gestemmte, c) die aufgesattelte Treppe	191
4. Das Treppengeländer	198

XI. Abortsitze 201

XII. Die Beschläge 203
1. Beschläge zur Befestigung und Verbindung einzelner Teile	203
a) Steinschraube, — b) Bankeisen — c) Eckwinkel.	
2. Beschläge zur Bewegung einzelner Konstruktionsteile	205
a) Lang- und Kurzband, — b) Schippenband, — c) Winkelband, — d) Kreuzband, — e) Fischband, — f) Aufsatzband, — g) Paumelleband, — h) Scharnierband, — i) Zapfenband.	
3. Beschläge zum Festhalten einzelner Konstruktionsteile in bestimmten Lagen und zum Verschliessen	209
a) Riegel (Schiebriegel und Kantenriegel)	209
b) Vorreiber (einfacher und doppelter)	210
c) Ruderverschluss	210
d) Baskülenverschluss	210
e) Schwengelverschluss	214
f) Espagnolettstangenverschluss	215
g) Aufstellvorrichtungen für obere Fensterflügel	216
h) Festhaltungen für geöffnete Flügel (für Thüren und Thore, für Fenster und Läden)	218
i) Zuwerfungen für Thüren und Thore	219
k) Windfang- und Pendelthürbeschläge	224
l) Schlösser	226
(Fallenschloss. — Riegelschloss. — Eintouriges Kastenschloss mit hebender Falle. — Zweitouriges Kastenschloss ohne Ueberbau mit hebender Falle. — Ueberbautes zweitouriges Kastenschloss mit hebender Falle und Nachtriegel. — Einsteckschloss mit hebender und schiessender Falle. — Einsteckschloss für Schiebthüren. — Chubbschloss. — Stangenschloss. — Baskülenschloss.)	
m) Thürsicherungen anderer Art	237
(Selbständige Nachtriegel. — Thürsperrkette. — Schlosssicherung von Schubert & Werth.)	

I. DAS MATERIAL.

1. Die Holzarten. — 2. Die Holzsorten und Schnittwaren. — 3. Der Aufbau des Holzes. — 4. Die Kennzeichen gesunder und schadhafter Bäume. — 5. Die Fällzeit des Holzes. — 6. Das Schneiden des Holzes. — 7. Das Trocknen des Holzes. Das Schwinden und Quellen. — 8. Die Zerstörung des Holzes.

1. Die Holzarten.

Die Zahl der bei der Herstellung der gewöhnlichen Bauschreinerarbeiten in Betracht kommenden Holzarten ist eine geringe. Am häufigsten verwendet sind, schon des billigen Preises wegen, diejenigen unserer einheimischen Waldbäume: Tanne, Fichte, Lärche, Kiefer, Eiche und Buche. Weniger allgemeine Verwendung finden folgende Hölzer: Esche, Nussbaum, Pappel, Linde, Ahorn, Akazie und das amerikanische Pitchpine, obwohl sie für bestimmte Zwecke sehr wohl geeignet sind.

Die ausländischen, besonders ihrer Farbe wegen geschätzten Hölzer, wie Mahagoni, Palisander, Amarant und Ebenholz, finden nur bei der Parkettfabrikation und für ganz feine Ausstattungen von Innenräumen, hauptsächlich aber in der Möbelschreinerei Verwendung.

Tanne, Edeltanne, Weisstanne (Abies pectinata *D. C.*)

Von Farbe lichtgelb und glänzend. Leicht und elastisch, mit harten deutlichen Jahresringen, im übrigen weich und harzfrei; sehr gut zu bearbeiten und zu spalten. Der Splint ist breit, das Kernholz fehlt. Vorzugsweise in Süddeutschland verwendet.

Fichte, Rottanne (Abies excelsa *D. C.* — Picea excelsa *Lk.*)

Farbe gelblich-weiss bis rötlich-weiss. Das Holz ist zart und seiner weichen Jahresringe wegen zu Blindholz geeignet. Es ist gut zu hobeln und zu spalten. Das eigentliche Kernholz fehlt. Leichter Harzgeruch.

Lärche (Larix europaea *D. C.*)

Farbe des Kernholzes rotbraun, des Splintes hellgelb und glänzend. Grossfaserig und spröde. Leichter Harzgehalt. Das Holz lässt sich gut verarbeiten, auch gut schnitzen.

Kiefer, Forle, Föhre (Pinus silvestris *L.*).

Von Farbe im Splint geblich-rötlich, im Herbstholz braunrot, im Kern gelbrot. Harzreich, leicht, weich, grob, dauerhaft und weniger gut zu bearbeiten. Besonders für Arbeiten ins Freie geeignet.

Eiche, a) Sommereiche, Stieleiche (Quercus pedunculata *Ehrh.*).

b) Wintereiche, Steineiche (Quercus sessiliflora *Salisb.*).

Das Holz hat eine eigentümliche gelbe oder hellbraune Farbe in verschiedenen Abtönungen, glänzende, breite Markstrahlen und charakteristische Spiegel, breite, überall gleichmässig gebaute

Jahresringe und kräftigen, frischen Geruch. Es ist hart, schwer, zähe, dauerhaft, langfaserig und gerbsäurehaltig. Der Splint, gelblich und scharf getrennt vom Reifholz, muss unbedingt entfernt werden, da er dem Wurmfrass besonders unterworfen ist. Das Eichenholz lässt sich sehr sauber verarbeiten. Verwendet wird es seines hohen Preises wegen heute nicht mehr so allgemein wie ehedem, immerhin aber noch zu allen Arbeiten, welche den Unbilden der Witterung ausgesetzt sind, wie Thüren und Fenster, oder welche stark benützt werden, wie Fussböden etc.

Buche, Rotbuche (Fagus silvatica *L.*).

Farbe rötlich. Das Holz ist hart und lässt sich gut bearbeiten. Der Uebergang vom Splint zum Reifholz ist fast nicht bemerkbar. Die Jahresringe sind deutlich abgegrenzt durch das dunkler gefärbte Herbstholz. Die Markstrahlen erscheinen auf Querschnitten lichter, auf Längsschnitten dunkler als das umgebende Holz. Bis vor kurzem wurde das Buchenholz in der Bauschreinerei kaum verwendet, weil es ständig »arbeitet«. Erst der Neuzeit ist es zum Teil gelungen, diesem Nachteil durch Behandlung mit Dampf entgegen zu wirken, so dass man es jetzt des öfteren, und zwar zu Fussböden (Stab-, Kapuziner- oder Fischgratböden) benützt. Zu Werkzeugheften (für Stech- und Lochbeitel) ist es ungeeignet, da es in der Hand »brennt«.

Esche (Fraxinus excelsior *L.*).

Farbe des breiten Splintes weiss und durch das Reifholz allmählich in den bräunlichen Kern übergehend; die Jahresringe sind scharf abgegrenzt. Das Holz ist hart, zäh und elastisch, grobfaserig, geflammt, oft schön gemasert und wird daher häufig zu dekorativen Zwecken, für Füllungs-Einlagen und dergleichen verwendet.

Nussbaum (Juglans regia *L.*).

Farbe graubraun, Splint grauweisslich, Kern in verschiedenen Abstufungen braun bis rotbraun. Das Holz ist zart und schlicht, mittelhart und lässt sich vorzüglich verarbeiten, insbesondere gut polieren. Seine geflammte und oft schön gemaserte Zeichnung kommt gerade bei dieser letztgenannten Art der Behandlung gut zur Geltung. Verwendet wird es in der Bauschreinerei hauptsächlich zu feinen Täfelungen, Thüren etc. im Innern und zwar meist in Form von auf Blindholz aufgelegten Furnieren. Der Splint ist sehr oft unbrauchbar.

Pappel, a) Chaussee-, Pyramiden- oder italienische Pappel (Populus pyramidalis *Roz.*).

b) Silberpappel (Populus alba *L.*).

c) Kanadische oder Wald-Pappel (Populus canadensis *Mnch.*).

Farbe weiss. Das Holz ist sehr weich, leicht, schwammig und zart; seine Jahresringe sind kaum bemerkbar, weshalb es sich sehr gut als Blindholz eignet. Wird gern von Würmern angebohrt.

Linde, a) Sommerlinde, grossblätterige Linde (Tilia grandifolia *Ehrh.*)

b) Winterlinde, kleinblätterige Linde (Tilia parvifolia *Ehrh.*).

Die Linde hat breiten, weissen Splint, rötlich-weisses Holz mit deutlichen Jahresringen und Markstrahlen. Das Holz ist weich, aber doch wesentlich härter als Pappel, mit dem es die Eigenschaft gemein hat, sich gut als Blindholz verwenden zu lassen. Es eignet sich besonders gut zu Schnitzereien.

Ahorn, a) Bergahorn, stumpfblättriger Ahorn (Acer Pseudoplatanus *L.*).

b) Spitzahorn, spitzblättriger Ahorn (Acer platanoides *L.*).

Farbe weiss bis gelblich-weiss. Das Holz ist hart und spröde, lässt sich gut verarbeiten; besonders gut polieren.

Akazie (Robinia Pseudacacia *L.*).

Das Holz ist zäh, hart, elastisch und dauerhaft. Splint schmal, geblich, unbrauchbar. Kern gelbbraun. Splint und Kern oft auch grünlich. Schwer spaltbar.

Pitchpine, amerikanische Terpentin-Kiefer (Pinus australis *Mich.*).

Farbe gelb mit dunklen, rotbraunen Jahresringen. Ebenso harzreich wie unsere einheimische Kiefer und mindestens ebenso schwer zu bearbeiten wie diese. Wird Pitchpine in Verbindung mit dem einheimischen Kiefernholz verwendet, so lassen sich infolge der verschiedenartigen Färbung beider Hölzer bei gestemmten Arbeiten im Innern unserer Wohnräume schöne Effekte erzielen. Der Preis des Pitchpine-Holzes stellt sich bei uns, d. h. in Süddeutschland, ca. 20 bis 30 % höher als der des einheimischen Kiefernholzes.

Mahagoni (Swietenia Mahagoni *L.*).

Das Holz wird aus Amerika bezogen. Farbe gleichmässig gelb-rot; dieselbe dunkelt jedoch nach bis zu braunrot. Das schlichte Holz ist weich, hat feine Jahresringe, bleibt gut stehen und lässt sich gut bearbeiten.

Palisander (Jacaranda brasiliana *Pers.*).

Süd-Amerika. Farbe dunkelrotbraun und schwarz, geflammt, mit hellrötlichen Linien durchzogen. Das Holz ist hart, schwer, fest und spröde und liefert saubere Arbeit. Der feine, würzige Geruch beim Sägen desselben ist bekannt. Verwendet wird es in der Bauschreinerei hauptsächlich zu Einlagen, zu Adern in Parketttafeln.

Amarant (Copaifera bracteata *Benth.*).

Süd-Amerika. Farbe violett, bis ins Bräunliche übergehend mit hellroten Linien. Das Holz ist hart und schwer und wird ebenfalls in der Parkettfabrikation verwendet.

Ebenholz (Diospyros Ebenum *Retz.* und andere Bäume).

Ostindien, Afrika etc. Kern schwarz, Splint weiss. Es ist eines der härtesten, schwersten und sprödesten Hölzer und wird in der Bauschreinerei meist nur wie Palisander und Amarant benützt.

Kayoe-Bessie (Baum unbestimmt).

Celebes, Sunda-Inseln. Von warmbrauner Farbe; im Splint gelb. Härter wie Eiche, harzreich, in Wasser und Hitze gut stehend. Ab Amsterdam 90 Mark pro cbm. Empfehlenswertes Holz.

2. Die Holzsorten und Schnittwaren.

Die ausländischen Hölzer kommen vielfach als Blöcke oder Stämme in den Handel und auch in dieser Form wohl in die Schreinereien. Die einheimischen Hölzer dagegen werden nach den vorbereitenden Arbeiten im Walde auf die Sägemühlen geschafft und dort in der bekannten Weise zu Schnittwaren zerlegt, und erst in solcher Gestalt gelangen sie in den Handel und in die Werkstätte. Wird der Stamm in seiner Längsrichtung durch parallele Schnitte zerlegt, so entstehen Bohlen, Dielen und Bretter. Werden diese nochmals durch Schnitte in der Längsrichtung, aber senkrecht zur Ebene der ersten Schnitte in Streifen zerlegt, so entstehen Latten und Rahmenschenkel. Die Länge der gewöhnlichen Schnittware beträgt 4,5 bis 4,6 m.

Die Bezeichnung ist in Deutschland keine einheitliche. Unter Bohlen versteht man allgemein die Ware von 50 bis 100 mm Stärke. Was dünner ist, geht in Norddeutschland als Dielen; in Süddeutschland macht man eine weitere Unterscheidung und heisst so die Stärken von 30 bis 50 mm, während die geringeren Stärken Bretter genannt werden, wofür am Rhein wieder der Ausdruck Bord üblich ist. Die Stärken von 12 bis 20 mm heissen schwache Bretter oder Bord. Schleifdielen sind 36 mm stark, 29 cm breit und 4,56 m lang; sie gehen besonders nach Holland. Sattelbretter oder Bettseiten sind 30 mm stark; der Name stammt von ihrer Verwendung bei der Herstellung einfacher Bettladen. Schwarten sind die beiderseitigen Abfälle beim Zerlegen des Stammes in Dielen; sie sind einerseits rindig oder rindschälig und zeigen anderseits eine Schnittfläche.

Die Rahmenschenkel sind gewissermassen schwaches Balkenholz. Sie haben rechteckigen oder quadratischen Querschnitt; das letztere ist die Regel und die gewöhnlichen Stärken halten sich zwischen 4 und 12 cm.

Die Latten haben für gewöhnlich die doppelte Dicke zur Breite und der gangbare Querschnitt misst 24 auf 48 oder 25 auf 50 mm. Doppellatten sind 35 auf 50 mm stark und die Deck- oder Fugenlatten 15 auf 50 mm.

Rahmenschenkel und Latten sind allseitig beschnitten, Bohlen, Dielen und Bretter zunächst nur zweiseitig. Werden sie in diesem Zustande verkauft, was in Baden und der Schweiz noch üblich ist, so heisst die Ware ungesäumt. Anderwärts wird in der Regel nur gesäumte Ware verkauft; die Baumkanten zu beiden Seiten sind abgetrennt und die Dielen haben Rechtecksform und durchweg gleiche Breite.

Der Preis der Schnittwaren richtet sich nach Stärke und Qualität. Geringere Stärken sind im Verhältnis teurer wegen dem vermehrten Schneidelohn und weil mehr Holz in die Sägespäne geht.

Der Qualität nach unterscheidet man in Süddeutschland:

    a) reine, ganz reine Dielen oder Bord; sie dürfen keine Aeste haben, das Holz muss schön weiss, schön langfaserig, »schlicht« und »sauber« sein;

    b) halbreine Dielen oder Bord heissen solche von zartem Holz mit wenigen, gut verwachsenen Aestchen;

    c) ordinäre Dielen, d. s. solche, die lose, schwarze, etwas grössere Aeste zeigen; Kleinhändler wählen die besten dieser Sorte aus und verkaufen sie als sogen. halbgeschlachte Dielen;

    d) mit Brennbord bezeichnet man grobästige, faulstellige, zerrissene Bord, ohne Fehlen eines Teiles derselben; sie sind zum Verschalen und zur Anfertigung von Kisten geeignet. Der verbleibende Rest heisst

    e) Ausschuss.

Dielen oder Bretter irgend welcher Gattung und Holzart, welche blaue oder schwarze Flecken zeigen, sind, des unschönen Aussehens und ihrer beschränkten Verwendung wegen, minderwertig.

Bretter und Dielen werden je nach der örtlichen Gepflogenheit nach dem Kubikmass, nach dem Quadratmass oder nach dem Stück verkauft.

Die von den Bauschreinern am meisten bezogenen Holzsorten sind Bretter und Dielen von

Länge	Stärke	14,5	17	19	21,5	24	26,5	29	31,5	34 cm Breite
4,50 m	12 mm	»	»	»	»	»	»	»	»	»
» » »	15 »	»	»	»	»	»	»	»	»	»
» » »	18 »	»	»	»	»	»	»	»	»	»
» » »	20 »	»	»	»	»	»	»	»	»	»
» » »	24 »	»	»	»	»	»	»	»	»	»
» » »	30 »	—	—	—	—	—	»	»	»	»
» » »	36 »	—	—	—	—	—	»	»	»	»
» » »	48 »	—	—	—	—	—	»	»	»	»

sowie die Rahmenschenkel von 4,50 m Länge und 4×5, 6×6, 7×7, 7×9, 9×9, 9×12 cm Stärke und Latten von gleicher Länge und 24×48 mm Stärke.

3. Der Aufbau des Holzes.

Unter Holz verstehen wir im allgemeinen die Hauptmasse der Stämme, Aeste und Wurzeln unserer Bäume. Dasselbe ist von der Rinde überzogen, welche für die Bauschreinerei wertlos ist.

Schneiden wir nach Entfernen derselben einen Baumstamm senkrecht zur Holzfaser durch (Fig. 1), so erblicken wir das sogen. Hirnholz, und zwar in verschiedenen Reifestadien. Während die äussere, als konzentrischer Ring in verschiedener Stärke auftretende Schicht das jüngste und unreifste Holz, der sogen. Splint, ist, folgt nach innen das dichtere Reifholz bis zum dichtesten, festesten und härtesten Teil, dem sogen. Kernholz, auch das Herz genannt. Aus dieser Wahrnehmung ersehen wir, dass das Wachsen des Stammes aussen vor sich geht und die Verstärkung desselben am äusseren Umfang in Form von konzentrischen Schichten statthat. Die Verstärkung erfolgt bei den einheimischen Bäumen von Frühjahr bis Herbst, während im Winter Ruhe eintritt; das ungleiche Wachstum bedingt die Jahresringe. Aus der Anzahl der Jahresringe, welche bei den einzelnen Holzarten mehr oder minder deutlich auftreten, ist das Alter des Baumes zu ersehen. Rasch wachsende, weiche Hölzer, z. B. Tanne, Fichte und Forle, zeigen deutliche Jahresringe, härtere, wie Eiche, Ahorn, weniger deutliche.

Der Splint, die äusserste, noch nicht ausgereifte, wesentlich hellere und meist scharf abgegrenzte Schicht, welche sich nach und nach in Holz verwandelt, ist weich und sehr saftreich und kann daher, der Gefahr des Wurmfrasses wegen (die Würmer bohren dem Safte nach), bei verschiedenen Hölzern, wie Eiche, Akazie, Nussbaum nicht verwendet werden. Bei Tanne, Fichte etc. wird derselbe belassen, doch sind die daraus geschnittenen Dielen minderwertig und, ihrer Weichheit wegen, zu verschiedenen Arbeiten, wie Fussböden etc., unbrauchbar. Nach dem Reifholz, dem ausgereiften Holz, kommt das Kernholz, welches das Mark, eine weiche, schwammige, oft korkartige Masse in sich einschliesst, die bei den meisten Nutzhölzern nach und nach ganz verschwindet. Das Kernholz ist das innerste, dichteste und härteste Holz; aber gerade dieser scheinbare Vorzug beschränkt oft

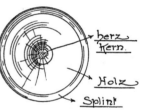

Fig. 1.

Querschnitt eines Baumes.

seine Verwendung, da es ganz wesentlich anders arbeitet, als Splint und Reifholz. Man unterscheidet übrigens zwischen Kernbäumen, Splintbäumen, Reifholzbäumen und Kern-Reifholzbäumen. Nur die letzteren zeigen in ausgesprochener Weise alle drei Entwickelungsformen. Bei den Kernbäumen geht der Splint unmittelbar in das Kernholz über; bei den Reifholzbäumen fehlt der eigentliche Kern und bei den Splintbäumen fehlen Reifholz und Kern.

Splint, Kernholz und der zwischen beiden liegende, meist bedeutendste Teil, das eigentliche Reifholz, bestehen bei allen Baumarten, auch bei ganz gerade gewachsenen Bäumen, aus leicht schraubenförmig gewundenen Faserbündeln, welche im Wasser unlöslich sind. Diese Fasern umschliessen längliche Zellen, in welchen sich der Saft, die Nahrung des Baumes, befindet. Nach dem Fällen des Holzes verdunsten dessen wässerige Bestandteile zum grössten Teil, während die festen in den Zellen zurückbleiben.

Sind die Holzfasern gerade und parallellaufend, so heisst das Holz schlicht; sind sie krumm, gebogen, gewellt und verschlungen, so heisst das Holz geflammt oder gemasert. Schöne Masern sind zu dekorativen Zwecken sehr geschätzt, z. B. bei Nussbaum, Ahorn, Esche etc. Sind die Fasern zart gebildet, so nennt man das Holz fein (z. B. Nussbaum), während es andernfalls grobfaserig genannt wird (Esche).

Bezüglich ihrer Dichtigkeit unterscheiden wir:

a) harte Hölzer, deren Gefüge gleichmässig dicht ist, bei welchen die Jahresringe deshalb schwer von einander zu unterscheiden sind und deren Gewicht naturgemäss am bedeutendsten ist; die Farbe derselben ist meist eine dunkle, obwohl auch Ausnahmen, wie Ahorn, vorkommen. Diese Hölzer schwinden weniger als weiche, werfen sich aber auch gern; zu den harten Hölzern gehören: Eiche, Esche, Ahorn, Buche, Ebenholz, Palisander, Amarant u. a.;

b) **mittelharte Hölzer** wie Nussbaum und einige Mahagoniarten;

c) **weiche Hölzer**; dieselben haben lockeres, schwammiges Gefüge mit mehr oder minder sichtbaren Jahresringen, helle Farbe und leichtes Gewicht; sie schwinden am meisten, bleiben aber nach richtiger Trocknung auch gut stehen und eignen sich daher vorzugsweise zu Blindholz; hierher gehören Tanne, Fichte, Lärche, Pappel, Linde etc.

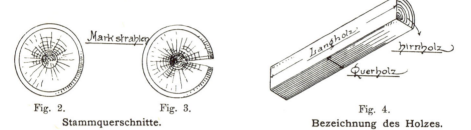

Fig. 2.   Fig. 3.   Fig. 4.
Stammquerschnitte.   Bezeichnung des Holzes.

Von dem Zentrum, dem Mark, ziehen sich, radial nach dem Umfang zu, die sog. Markstrahlen, nach welchen das Holz leicht zu spalten ist und nach deren Richtung es beim Trocknen und Schwinden gern aufreisst (Fig. 2 und 3). Dieselben heissen, sobald sie an der Oberfläche in ihrer Breitseite zu Tage treten, Spiegel und sind als solche ein charakteristisches Erkennungszeichen für viele Hölzer, wie z. B. Eichen und Buchen. Die Spaltfläche nach den Markstrahlen ist nicht eben, sondern windschief, eine leichte Schraubenfläche bildend.

Mit Langholz bezeichnen wir das parallel der Fasernrichtung ziehende Holz; Querholz heisst die quer zur Holzfaser ziehende Richtung, während, wie bereits erwähnt, als Hirnholz die senkrecht zur Holzfaser geschnittene Fläche gilt (Fig. 4). Da, abgesehen vom Hirnholz, die beiden übrigen Flächen das gleiche Holz aufweisen, so ist nur die Richtung massgebend und von Querholz wird man in Bezug auf ein bestimmtes Stück nur reden können, wenn es in der Richtung der Faser kürzer ist, als quer zu derselben.

4. Die Kennzeichen gesunder und schadhafter Bäume.

Der Baum soll einen schönen, schlanken, gerade gewachsenen Stamm mit glatter, unbeschädigter Rinde und eine volle Laubkrone haben. Er soll ohne Auswüchse, Frostrisse, abgebrochene Aststümpfe sein und keine faulen Stellen zeigen. Bäume mit fehlerhaftem Wuchs, krummer oder verdrehter Richtung des Stammes, beschädigter Rinde, abgebrochener Krone, Blitzschlag-, Frost- und Windrissen liefern kein gutes Holz. Bäume, auf sumpfigem, feuchtem Boden gewachsen, ergeben weiches, schwammiges Holz, während Bäume auf magerem Boden ein dichteres Holz liefern. Deswegen ist auf dem Berge gewachsenes Holz dem im Thal gewonnenen vorzuziehen (Bergholz — Thalholz). Im geschlossenen Bestand aufgewachsene Bäume geben das beste Holz. Frei oder am Waldrand stehende Bäume sind meistens krumm und drehwüchsig und auch anderen Schädigungen leichter zugänglich. Die Kernfäule, welche nicht immer äusserlich wahrzunehmen ist, kann höchstens durch Anbohren des Stammes bezw. Untersuchung der Bohrspäne festgestellt werden.

5. Die Fällzeit des Holzes.

Die zweckmässigste Fällzeit ist unstreitig die, in welcher die Saftthätigkeit im Baum am geringsten ist, also der Winter. Für diese Fällzeit sprechen ferner die geringeren Tagelöhne und der Umstand, dass während des Frostes auch solche Waldwege befahren werden können, die zu anderen Zeiten sich als unfährbar erweisen. Ueberall wo es sich ausführen lässt, wird man daher bei der Winterfällung bleiben; unbedingt beizuhalten ist sie, wo es sich um **Kiefer** und **Buche** handelt, da beide im Sommer gefällt, blau bezw. schwarz werden.

Anders verhält es sich im Hochgebirge, wo monatelang tiefer Schnee liegt, welcher die Fällung unmöglich macht. Hier ist man auf die Sommerfällung angewiesen. Der durch seine hochinteressanten Schriften über die Bauhölzer etc. bekannte Gelehrte, Dr. Hartig in München, hat in seinem Buch »Der echte Hausschwamm« (Berlin, Jul. Springer, 1885) den Beweis erbracht, dass das Sommerholz unter sonst gleichen Verhältnissen nicht empfänglicher für Schwammbildung ist, als das im Winter gefällte, und die Erfahrung lehrt, dass im Sommer gefälltes und gut getrocknetes Holz widerstandsfähiger und dauerhafter ist als Holz der Winterfällung mit ungenügender Trocknung. In Süddeutschland sind die Hauptbezugsquellen das bayrische Hochgebirge und der badische Schwarzwald; das meiste von da bezogene Holz ist Sommerholz, das niemand als schlecht bezeichnen wird, wenn es sonst fehlerfrei ist. Zugegeben muss aber werden, dass Sommerholz beim Trocknen stärker reisst als Winterholz und sich infolgedessen nicht für alle Arbeiten verwenden lässt.

---

6. Das Schneiden des Holzes.

Nach dem Fällen muss der Stamm, um ihn vor Wurmfrass oder Stockung zu schützen, entweder **ganz entrindet** oder doch **geringelt, gereppelt** werden. Ersteres Verfahren bewirkt ein rascheres Austrocknen und schützt mehr vor dem Wurmfrass, befördert aber auch das Reissen des Stammes; man zieht daher vielfach die zweite Art, das Reppeln, vor, nach welchem die Rinde mit dem Beil nur stück- oder stellenweise enfernt wird, um die Austrocknung etwas zu mässigen. Baldiges Schneiden des Holzes ist wünschenswert, doch kann Eiche 1—1½ Jahr, Kiefer (Winterfällung) bis Mai und Tanne bis Herbst ungeschnitten liegen bleiben ohne Schaden zu nehmen; bei längerem Liegen vor dem Schnitte wird Weichholz fleckig.

Um das Reissen wertvoller Stämme, wie Eichen, etwas zu verhindern, verklebt man die Hirnholzflächen mit Papier oder bestreicht sie mit Lehm

---

Das Schneiden des Holzes in Dielen und Bretter erfolgt auf verschiedene Weise. Im allgemeinen gilt für »kurante« Waren alles, was erzielt wird, wenn man den Stamm, ohne Rücksicht auf die Jahresringe, durch parallele Schnitte zerteilt (Fig. 5), wobei zwei Schwarten a und b abfallen, oder, wenn man von dem Stamm auf zwei Seiten die Schwarten f und g abschneidet, den Klotz dann stürzt und den Rest zerteilt wie vorher (Fig. 6 und 7). Im ersteren Fall erhalten wir ungesäumte Ware und zwei Schwarten, im letzteren gesäumte Ware und vier Schwarten. Die mittleren Dielen heissen **Herzdielen**.

Ist der Stamm stärker, so werden ausser den Schwarten noch zwei Dielen abgetrennt, worauf der Klotz gestürzt und, wie in beiden vorhergehenden Fällen, in Dielen und Bretter geschnitten wird (Fig. 8). Aus dem Rest schneidet man Latten.

Für das Holz der in neuerer Zeit so ausserordentlich beliebten Riemen- oder Schiffböden hat man eine andere Schneidart eingeführt. Dieselbe bezweckt, Riemen mit sogen. senkrechten Jahresringen zu gewinnen, welche sich als Bodenbelag weit besser bewähren, als Dielen gewöhnlicher Schneidart. Diese Riemen sind, entsprechend dem vermehrten Arbeitsaufwand, teurer im Preise, was sich aber ausgleicht durch die grössere Dauerhaftigkeit des Bodens. Sie werden meist nach Fig. 9 oder 10 geschnitten, seltener dagegen nach der besseren, aber wieder teureren Art, die Fig. 11 darstellt. Die Breite derselben ist ca. 12—15 cm, ihre Dicke meist so, dass sie,

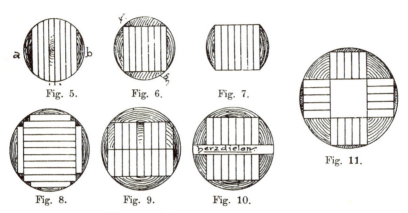

Die verschiedenen Holzschneidearten.

einseitig gehobelt, noch 30 mm stark sind. Die meisten der grösseren neuzeitlichen Sägewerke sind auch mit Holzbearbeitungsmaschinen versehen, so dass sie die Riemen vollständig fertig zum Verlegen im Bau, also gehobelt, gefälzt, genutet oder gespundet, je nach Wunsch, zu annehmbaren Preisen zu liefern im stande sind. Das ist ein Vorzug, welcher dem Meister, der rasch zu liefern gezwungen ist, sehr zu gute kommt. Auch fertigen sie für andere Zwecke Profilleisten (z. B. für Thürverkleidungen) und hobeln Profile an Fusssockel, Sockelleisten, Verschalbretter etc. (vergl. Fig. 12 und 13).

---

7. Das Trocknen des Holzes.

Nach dem Schneiden muss das Holz, soll es nicht stockig werden, sofort zum Trocknen aufgesetzt werden. Von dieser Behandlung hängt ungemein viel für die spätere Qualität der Ware ab. Das schönste Holz kann bei nachlässiger Behandlung minderwertig werden, ja verderben. Vor allem ist es nötig, die Schnittware vor den Unbilden der Witterung, vor Regen und Sonne und vor zu starker Zugluft zu schützen. Mit Latten zugeschlagene Holzschuppen eignen sich am besten zur Aufbewahrung. Die Ware muss auf ein solides, genau mit der Setzlatte hergestelltes Unterlager aufgelegt werden, um zu verhindern, dass sie krumm und windschief oder gar vom Boden her feucht werde. Man bezeichnet dieses Aufsetzen als »Aufholzen«. Erfahrungsgemäss genügen auf Dielenlänge (4,50 bis 4,60 m) vier Unterlager. Ist der erste Dielen gelegt, so werden an den beiden Enden desselben, und zwar bündig damit zwei, sowie in gleichen Entfernungen von einander zwei weitere Lattenstücke aufgelegt; sodann wird der zweite Dielen aufgebracht, welcher nun durch eine Luftschicht in der Stärke der Zwischenlatten von dem darunter liegenden Dielen getrennt ist. Das Bündiglegen der Hölzer an den Enden bezweckt, das Reissen der am meisten gefährdeten Dielenenden zu verhindern, indem auf solche Art die Luft an diesen Stellen von oben und unten abgehalten wird; jedoch ist es nötig, die Ware nach Jahresfrist umzusetzen, damit auch die von den Latten bedeckten Stellen austrocknen können. Man kann die Dielen auch der Höhe nach stellen, wodurch das Trocknen beschleunigt wird. Dabei ist natürlich Vorkehrung zu treffen, dass der Boden, auf den die Dielen gestellt werden, absolut trocken ist; ferner dass dieselben in gehöriger Entfernung von einander zu stehen kommen und stehen bleiben, und schliesslich, dass die Dielen sich nicht verziehen, nicht krumm oder windschief werden können.

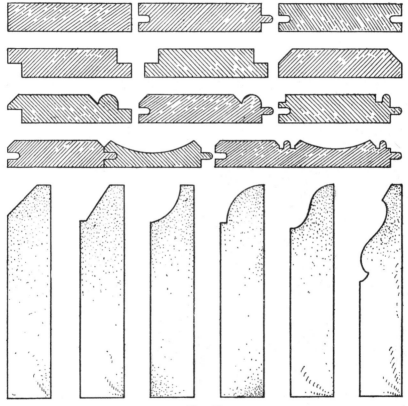

Fig. 12.
Profile verschiedener Holzfriese, Fusssockel etc. des Hobelwerkes von Th. Berger in Holzkirchen, Oberbayern.

Bleibt geschnittene Ware unaufgeholzt längere Zeit sitzen, so verliert sie ihre schöne helle und gesunde Farbe, sie wird äusserlich blau und schwarz, die Holzfaser verliert ihre Elastizität, und die relative Festigkeit des Holzes wird so gering, dass man ein starkes Stück Langholz ohne besondere Kraft beim Aufschlagen auf eine harte Kante zerbrechen kann — das Holz ist stockig geworden.

In der Möbelschreinerei verwirft man dieses Fehlers wegen solches Holz nicht, man benützt es sogar für gewisse Arbeiten nicht ungern, da es ruhig und gut stehen bleibt; man macht dabei keine Ansprüche an seine absolute und relative Festigkeit. Für die Bauschreinerei hat stockig gewordenes Holz keinen Wert. Man erkennt derartiges Holz, ausser an den bläulichen und schwärzlichen Flecken, besonders bei der Verarbeitung, beim Hobeln, da der von solchem Holz erzielte Hobelspan nicht ganz bleibt und sich aufrollt, sondern in kleine Stückchen zerbricht.

Wie schon erwähnt, schliessen die Zellen der Holzfaser eine Flüssigkeit, den Saft, ein. Dieser Saft, welcher die Nahrung des lebenden Baumes darstellt, und ohne welchen er nicht weiterwachsen kann, ist aber dem gefällten Baum nicht mehr zuträglich. Ja er kann demselben sogar verderblich werden, da er leicht in Gärung und Verwesung übergeht, dadurch das Holz in Mitleidenschaft zieht und überdies den Wurmfrass sehr begünstigt. Doch die Natur hat auch hier das Gegenmittel sofort zur Hand: die atmosphärische Luft. Dieselbe trocknet den grössten Teil der wässerigen Bestandteile des Saftes aus, wodurch die Fäulnis verhindert und dem Wurmfrass wirksam begegnet wird. Hiermit ist aber eine Verringerung des Volumens verbunden, das Holz schwindet.

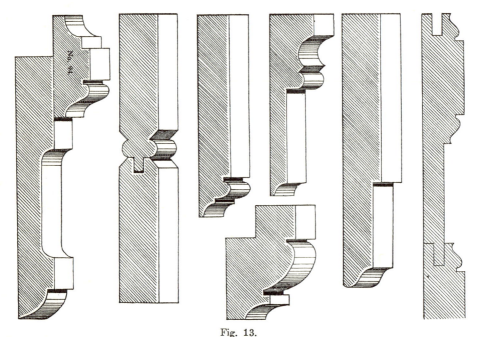

Fig. 13.
Profile verschiedener Holzfriese, Fusssockel etc. des Hobel- und Sägewerks von H. Fuchs Söhne in Karlsruhe. $^2/_3$ der nat. Gr.

a) Das Schwinden.

Dasselbe tritt umsomehr ein, je saftreicher das Holz und je grösser die Wärme ist, welcher dasselbe ausgesetzt wird. Weiches Holz schwindet mehr als hartes, Splint mehr als Reifholz, und dieses wieder mehr als das Kernholz.

Nach der Fasernrichtung (Langholz) ist das Schwinden so gering ($^1/_{10}\,^0/_0$), dass wir es bei den Schreinerarbeiten füglich unberücksichtigt lassen können, dagegen ist es quer zur Faser (Querholz) sehr bedeutend, doch keineswegs gleichmässig (3 bis $10\,^0/_0$). So schwindet z. B. ein dem äusseren Teil des Stammes entnommenes Brett a (Fig. 16) mehr als ein der Mitte desselben entnommenes b, desgleichen schwindet das Brett a mehr auf der Seite x als auf Seite y, es wird

7. Das Trocknen des Holzes.

hohl. Das Brett b vereinigt dreierlei Holz in sich (Splint, Reifholz, Herz) und ist demgemäss dreierlei Schwindmassen unterworfen. Es wird an den Kanten dünner als in der Mitte, und ebenso schwindet es von aussen herein mehr als in der Nähe des Kerns. Die verschiedenen Holzarten schwinden nicht in gleichem Masse unter sonst gleichen Bedingungen. Von den zu Beginn des Buches aufgeführten Hölzern schwindet durchschnittlich am wenigsten das Fichtenholz, am meisten das Akazienholz. Die ungefähre Reihenfolge ist: Fichte — Lärche — Kiefer — Tanne — Eiche — Nussbaum — Pappel — Esche — Ahorn — Buche — Linde — Akazie (nach Nördlinger). Werden die Jahresringe des Holzes schräg durchschnitten, was ja bei der schraubenartigen Form der Fasern sehr oft, ja fast immer eintritt, so wird das Brett auch noch der Länge nach seine Form zu verändern, zum mindesten sich zu drehen suchen; es wird — wenn auch nicht in hohem Masse, so doch etwas — **windschief**.

Auf die Verschiedenartigkeit des Schwindmasses von Splint und Kern ist auch das Aufreissen entrindeter saftreicher Stämme zurückzuführen. Splint trocknet sehr rasch und bedeutend ein, so dass — da der Kern sich ziemlich gleich bleibt — die Peripherie des Holzes schliesslich platzt und Risse bekommt (Fig. 14).

  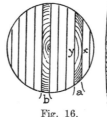

Fig. 14.   Fig. 15.   Fig. 16.   Fig. 17.   Fig. 18.
Reissen des Holzes.   **Schwinden des Holzes.**   **Halbiertes und gevierteltes Holz.**

Entfernt man das Herz des Stammes, etwa durch Ausbohren, so bleibt der Rest ganz, wie wir dies an hölzernen Brunnendeicheln und Brunnenstöcken bei vernünftiger Behandlung derselben wahrnehmen können (Fig. 15). **Halbiertes Holz, Halbholz wird an den Schnittflächen rund** (Fig. 17). **Geviertetes Holz** (Fig. 18) dagegen bleibt ziemlich gerade, hat aber, der vielfach schräg durchschnittenen Jahresringe wegen, das Bestreben, sich zu drehen.

Das Schwinden erfolgt bei neuem Holz so lange, bis es, wie schon erwähnt, den grössten Teil seiner wässerigen Bestandteile verloren hat, bis es trocken und zwar
    a) lufttrocken oder
    b) durch künstliche Wärme, Dampf etc. getrocknet ist.

Lufttrocken wird weiches Holz bei richtiger Behandlung und Aufbewahrung in $1^{1}/_{2}$ bis 3 Jahren, während harte Hölzer, wie Eiche, 4—6 Jahre Zeit hierzu nötig haben. Obgleich diese Zeit ziemlich lang ist und obschon durch den grossen Zinsenverlust die Trocknung eine verhältnismässig teure wird, so haben sich bis heute doch nur die grösseren Fabriken und Baugeschäfte mit kaufmännischem Betrieb der Trocknung durch Heizung oder Dampf zugeneigt, während die grosse Mehrzahl der Schreiner dem langsamen, die Holzfaser und Farbe weniger beeinträchtigenden Trocknen auf natürlichem Wege den Vorzug giebt. Am sichersten zum Ziel führend ist, das Holz zuerst einige Jahre in der Luft zu trocknen und es dann noch in einen Trockenofen zu bringen.

Trockenes Holz enthält noch ca. 15 bis $20^{0}/_{0}$ Wasser, während grünes, soeben gefälltes zwischen 40 bis $50^{0}/_{0}$ besitzt; das Gewicht des ersteren ist demnach wesentlich geringer und selbst dem Laien fällt der grosse Gewichtsunterschied auf, wenn er ein trockenes und ein grünes Brett zum Vergleichen in die Hand nimmt; ersteres ist ungefähr um $^{1}/_{3}$ des Grüngewichts leichter geworden. Mit dem Hammer angeschlagen, giebt gut trockenes Holz einen klirrenden Ton.

Die künstliche Trocknung geschieht in der Trockenkammer, einem nur mit Lagern zum Aufholzen des Holzes versehenen, sonst glatten, hohlen Raum, in welche die Wärme entweder durch besondere Feuerung, zweckmässiger und billiger aber durch Eisenröhren zugeführt wird, durch welche man den Abdampf der Maschine leitet. In diesen Raum wird das bereits zugeschnittene Holz eingebracht und so aufgeholzt, dass die Wärme überall und gleichmässig hingelangen kann.

Das Hochkantstellen des Holzes ist des rascheren Trocknens wegen sehr praktisch, doch schwer auszuführen, da die einzelnen Holzstösse zu schwankend werden und gern zusammenstürzen. Ausführen lässt es sich aber, wenn in der Trockenkammer quer von einer Mauer bis zur anderen Eisenschienen eingemauert sind, auf welchen die Hölzer hochkantig gestellt werden. Um sie vor dem Umkanten zu schützen, schiebt man das eine Ende in ein entsprechend gefertigtes Lattengerüste, welches an der der Eingangsthür gegenüberliegenden Wand angebracht ist.

Nach der Füllung der Kammer wird die Temperatur langsam bis auf ca. 40° R. erhöht, und das Holz je nach seiner Art 10 bis 14 Tage derselben ausgesetzt, wobei nicht versäumt werden darf, den sich bildenden Wasserdampf abzuleiten. Ist das Holz vorher schon an der Luft getrocknet, so ist das ein grosser Vorteil, da es dann nicht so leicht reisst.

Die direkte Einwirkung heissen Dampfes, welche zur Auflösung der Saftbestandteile führt, hat in Süddeutschland wenig Verbreitung gefunden. Diese Art von Trocknung wird meist von Firmen angewendet, welche einen raschen und grossen Umsatz erzielen wollen. Das in Holz angelegte Kapital wird bei der Dampftrocknung rasch nutzbar, während es bei der Lufttrocknung jahrelang totliegt. Im Dampf getrocknetes Holz bleibt zwar gut stehen, allein es hat auch durch den Verlust der Saftbestandteile — welche in Form einer bräunlich-schwärzlichen Brühe unten auslaufen — nicht nur den grössten Teil seiner relativen Festigkeit eingebüsst, sondern sich auch in der Farbe so verändert, dass es zu vielen Zwecken unbrauchbar wird. Es ist »tot« geworden. Aus diesem Grunde zieht man die oben beschriebene Trocknungsart derjenigen durch Dampf vor. Sehr gut dagegen hat sich das sog. Auslaugen, Auslohen des Holzes, das Einhängen desselben in fliessendes Wasser, bewährt, wobei dem Holz ein geringer Teil seiner Saftbestandteile entzogen wird, was die Trocknung wesentlich erleichtert. Schliesslich sei noch die letzte Trocknung in der Werkstätte selbst erwähnt, bei welcher das demnächst zur Verwendung gelangende Holz auf sog. Babelagen (an der Decke angebrachte Hängegerüste) gesetzt und nun von der stets warmen Werkstättenluft umspült wird. Diese nicht zu unterschätzende Trocknung sollte sich kein tüchtiger Meister entgehen lassen.

Sehr unangenehm kann die künstliche Trocknung werden, wenn sie sich an schon verarbeiteten Hölzern, an fertigen Arbeiten vollzieht, wenn sie eine unbeabsichtigte und unfreiwillige ist. Diese Trocknung tritt im allgemeinen überall ein, wo die betreffenden Arbeiten sich in der Nähe der Heizungseinrichtung, des Ofens befinden. Bedenkt man, dass die Luft im Neubau, selbst im Hochsommer, im August während des Tags und der Nacht einen ganz wesentlich höheren Feuchtigkeitsgehalt besitzt als die Zimmerluft im Winter in der unmittelbaren Nähe eines gut geheizten Ofens, so wird man sich nicht wundern, wenn selbst aus gut trockenem Holz und in trockener, warmer Werkstätte hergestellte Arbeiten beim Einbringen und Anschlagen im Neubau quellen, um dann im Winter darauf wieder zu schwinden. Gar zu leicht ist man in diesem Falle bereit, den Stab über den Meister zu brechen, während ihn doch eigentlich — vorausgesetzt, dass er trockenes Material verwendet — keine Schuld trifft. Denn ausser dem Töten, dem Totmachen des Holzes, giebt es kein Mittel, den Missstand des »Arbeitens« zu beseitigen. Wesentlich schlimmer als die gewöhnlichen Oefen wirken die Permanentbrenner, weil die Luft keine Gelegenheit hat, sich während der Nacht abzukühlen. Tritt hierzu noch der Umstand, dass der betreffende Raum mit Doppelfenstern versehen ist, welche eine Ausgleichung der inneren und äusseren Luft fast ganz aufheben, und wird auf dem Ofen keine Wasserverdampfschale aufgestellt, dann ist die Wirkung dieser Trocknung eine geradezu erstaunliche. Das Holz reisst unter Knall und Krach; die Täfelungen und Möbel zeigen bedenkliche Risse und Fugen. Mit dieser Bemerkung soll den ausserordentlichen Vorzügen der Amerikaner-Oefen in keiner Weise zu nahe getreten werden; sie soll aber den Schreiner veranlassen, seinen Kunden anzuraten, ständig eine Verdampfschale mit Wasser auf dem Ofen zu haben, was sich schon aus anderen Gründen empfiehlt.

b) Das Quellen.

Bei den verschiedenartigen Trocknungen des Holzes bleiben die festen Bestandteile, Zucker, Eiweiss, Gummi, Salze etc., in den Zellen zurück, woselbst sie sich, solange dasselbe in gleichmässig trockener Luft bleibt, ruhig verhalten, sofort aber vermöge ihrer hygroskopischen Eigenschaften sich regen, sobald das Holz in feuchte Luft gebracht wird. Alsbald saugen sie begierig die Feuchtigkeit derselben ein, und dies umsomehr, je mehr getrocknet oder gedörrt das Holz vorher war.

Die Folge hiervon ist eine Volumvermehrung; das Holz quillt. Das Schwinden, Quellen, Werfen, Drehen, Reissen etc. bezeichnet man allgemein mit Arbeiten: das Holz arbeitet. Wird es daran gehindert, so sprengt es die Konstruktion oder es reisst. Es ist daher Sache des Schreiners, durch zweckmässige Verbindung dieser Eigenschaft des Holzes Rechnung zu tragen. Das kann auf verschiedene Art geschehen. Nehmen wir z. B. ein Stück Holz, wie in Fig. 19, schneiden es nach a—b auseinander, kehren den einen Teil um, stürzen ihn und verleimen die beiden Teile wieder, so werden die Fasern — die ursprünglich nach einer Richtung gewirkt — nun gegeneinander wirken, wobei die Folgen sich gegenseitig aufheben. Hat man irgend welche Hölzer von aussergewöhnlicher Stärke nötig, welche gut stehen bleiben sollen, so verleimt man schwächere Hölzer auf die gleiche Weise drei-, vier- und mehrfach, indem man sie jeweils vorher stürzt. Ebenso verfahren wir nach der Breite, indem wir z. B. einen Dielen auseinanderschneiden, einen Teil desselben wenden und verleimen (Fig. 20).

Wenn man sogen. Herzdielen verarbeitet, so thut man gut, diese Dielen nicht nur zu schlitzen, sondern deren Herz auf einige cm Breite ganz herauszuschneiden und dann den Rest wieder zu verleimen; hierdurch wird verhindert, dass der Splint neben Kernholz kommt, wodurch man ein gleichmässigeres Arbeiten erzielt (Fig. 21).

Soll das Arbeiten nach allen Seiten möglichst gleichmässig erfolgen, so verleimt man die Hölzer nach zwei Richtungen und in möglichst kleinen Stücken. Nach Figur 22 fertigte man früher die Walzen für die Rollladen, um sie gegen Werfen und Verziehen zu schützen; nach Figur 23 werden aus dem gleichen Grunde die Billardstöcke verleimt. Messtisch- und Billardplatten setzte man aus möglichst vielen kleinen Stückchen zusammen. Auch durch doppeltes, d. h. beiderseitiges Furnieren (wobei das Furnier quer über die Holzfasern läuft) einer verleimten, aber gut trockenen Tafel kann das Arbeiten derselben gemildert werden.

Ausser den angeführten Vorkehrungen, bestehend in der zweckmässigen Verbindung der weicheren und festeren Holzteile, haben wir zur Verhütung des Arbeitens die Verbindungen auf Nut und Feder, das Spunden und das Stemmen, welche an anderer Stelle besprochen werden.

8. Die Zerstörung des Holzes.

Ausser dem Trocknen des Holzes ist zur Erhaltung desselben noch von grosser Wichtigkeit, es nur an trockenen, luftigen Orten zu verwenden und vor Nässe zu bewahren. Unterlassungssünden in dieser Richtung rächen sich, wenn auch manchmal langsam, doch sicher durch Zerstörung des Holzes.

Die Ursache ist fast immer in den beim Trocknen zurückgebliebenen Saftbestandteilen des Holzes zu suchen, welche hygroskopisch sind und Feuchtigkeit und Nässe aufsaugen. Geschieht dies in hohem Grade, so wird dadurch eine Gärung hervorgerufen, welche die Holzfaser in Mitleidenschaft zieht und schliesslich zerstört.

Wirkt die Nässe auf das Holz nur abwechselnd, so dass es immer wieder ganz oder teilweise zu trocknen vermag, wie dies z. B. bei Holzfussböden, in der Nähe von Balkon- oder

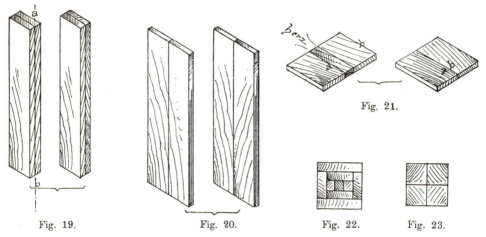

Fig. 19. Fig. 20. Fig. 22. Fig. 23.

Fig. 19 bis 23. **Das Verleimen der Hölzer.**

Verandathüren, in Badezimmern, bei Ausgussbecken an Vorplätzen, Waschtischen u. dergl. vorkommt, oder ist das Holz beständig in feuchter Luft oder direkt auf die Erde aufgelegt, so entsteht die Trockenfäule, die Vermoderung. Das Holz wird brüchig, lässt sich leicht zerbröckeln und zerreiben. Ist dagegen die Feuchtigkeit sehr bedeutend oder die Nässe anhaltend, so verläuft der Zerstörungsprozess wesentlich rascher, es tritt die Rotfäule, die nasse Fäule ein. Diese beiden Feinde des Holzes haben aber bei allem Schlimmen doch noch das Gute, dass sie sich nur soweit ausdehnen, als die Ursache reicht. Bei der Trockenfäule ist z. B. die Zerstörung in der Nähe eines Ausgussbeckens häufig nur auf einen Halbkreis von ca. 1 m Radius (dessen Mittelpunkt das Becken ist) wahrnehmbar und bei Balkonthüren schädigt die eindringende Feuchtigkeit selten mehr als 1 qm Fussboden. Aehnliches gilt auch von der nassen Fäule, wenngleich hier die Verheerungen wesentlich grösser sind. Die im Roh- und Eisenbahnbau angewendeten Mittel zur Verhütung der Fäulnis, wie das Tränken des Holzes mit Eisenvitriol, Chlorzinklösung, Quecksilberchlorid, das Imprägnieren mit Paraffin, das Karbonisieren etc., sind für die zu Schreinerarbeiten bestimmten Hölzer ungeeignet, zum Teil wegen der zu grossen Kosten, zum Teil wegen der durch die Behandlung veranlassten Verfärbung der Hölzer.

8. Die Zerstörung des Holzes.

Der schlimmste Feind des Holzes ist aber der **Hausschwamm** (Merulius lacrymans *Fr.*) Über sein Auftreten und die Gegenmittel ist schon unendlich viel behauptet und geschrieben worden, was sich als mehr oder weniger richtig erwiesen hat. Erst auf Grundlage der neueren Forschungen ist man von veralteten und unhaltbaren Ansichten abgekommen, was um so erfreulicher ist, da das Uebel gerade in neuester Zeit infolge unserer hastigen Bauweise bedenklich zugenommen hat. Bei der Wichtigkeit der Sache wollen wir das Feststehende in den wichtigsten Punkten hervorheben, wobei wir im allgemeinen den Ausführungen von Dr. Hartig in München (Der echte Hausschwamm, Berlin, Springer, 1885) folgen:

1) Der Schwamm kann sich nicht (von selbst) entwickeln, ohne dass eine Uebertragung von Schwammsporen oder Schwammteilen stattgefunden hätte.
2) An lebendem Holze kommt der Schwamm nicht vor; er wird von totem Holze auf totes Holz verschleppt (durch Bauschutt etc.).
3) Im Sommer gefälltes und richtig getrocknetes Holz erzeugt weder den Schwamm, noch ist es für sein Fortkommen mehr geeignet, als Winterholz.
4) Zum Fortkommen des Pilzes ist geschlossene, feuchte Luft erforderlich. Trockenes Holz in trockener, offener Lage schliesst die Entwickelung aus.
5) Vorhandene Alkalien, Ammoniak, wenn auch nur in geringer Menge, begünstigen die Schwammbildung; ebenso fördert Dunkelheit dieselbe, wenngleich der Pilz auch im Licht gedeiht, wenn die übrigen Bedingungen vorhanden sind.
6) Bei Temperaturen unter -5^0 und über $+50^0$ C. stirbt der Pilz ab.
7) Steinkohlenschlacke als Füllmaterial ist der Schwammbildung besonders günstig.
8) Die völlig sichere Feststellung des echten Hausschwammes ist oft nicht möglich, weil andere ähnliche Pilze das Holz ebenfalls heimsuchen. Eine solche gewährt nur die mikroskopische Untersuchung.

Das Auftreten des Hausschwammes verrät sich durch folgende Kennzeichen:

Zahlreiche, weisse oder graue, wie Spinnweben aussehende Fäden überziehen das Holz und dringen in dasselbe ein; bei genügender Feuchtigkeit wachsen sie zu lappigen Schwämmen aus, die an den Rändern tropfen und schliesslich die roten Sporen als Pulver umherstreuen. Die Farbe des Holzes verändert sich und wird gelblichbraun. Die Dielen wölben sich und ziehen die Nägel aus den bereits angegriffenen Lagern aus. Die Dielen beginnen zu schwinden; es entstehen grosse offene Fugen; das Holz wird querbrüchig und zerbröckelt in grössere und kleinere Würfel, welche sich leicht zerreiben lassen. Die Schwammbildung ist, besonders im entwickelten Zustande, mit einem eigenartigen, unangenehmen Geruche, dem charakteristischen Schwammgeruche verbunden.

In Bezug auf die Gegenmittel ist zu unterscheiden zwischen den vorbeugenden und den abhelfenden.

Als vorbeugende Mittel zur Verhütung der Schwammbildung empfehlen sich nach dem oben Ausgeführten:

1) Verwendung von möglichst trockenem, gesundem Holz.
2) Verhütung einer Ueberschleppung, sei es durch Bauschutt, Werkzeuge, Kleider etc. oder durch Zusammenlagern mit angegriffenem Holze.
3) Ausschluss von zweifelhaftem Füllmaterial (Steinkohlenlösche, Humuserde etc.) und Verwendung von reinem, gewaschenem und geröstetem Sand oder Kies. Reinhaltung in Bezug auf Urin und Exkremente.
4) Möglichste Vermeidung geschlossener, feuchter Luft durch Anbringung einer Ventilation etc.
5) Vermeidung einer Zuleitung von Feuchtigkeit, durch Teeranstrich, Isolierung etc.

6) Vernünftige, nicht überhastete Bauweise; richtiges Austrocknen des ganzen Baues und Ausfrierenlassen über Winter; genügende Trockenlegung und Entwässerung im allgemeinen; Isolierung der der Feuchtigkeit ausgesetzten Mauern u. s. w., Ersatz des Holzes durch Eisen an denjenigen Stellen, wo die Bedingungen zur Schwammbildung vorliegen und nicht beseitigt werden können.

Als Mittel zur Bekämpfung des Schwammes, wenn er bereits vorhanden ist, empfehlen sich:

1) Sofortiges Eingreifen und Aufreissen der gefährdeten Stellen.
2) Gründliche Beseitigung alles angegriffenen Holzes (auch in seinen anscheinend gesunden Teilen), des gesamten Füllmaterials und aller Schwammspuren am Boden und Mauerwerk; Vernichtung des angegriffenen Teils durch Verbrennung.
3) Sorgfältiges Abkehren der Flächen, Auskratzen der Fugen, Abwaschen mit Säure, Verputzen mit Zement etc.
4) Verwendung von völlig trockenem Holz für die Erneuerung, Anstrich desselben mit Kreosotöl, Antinonnin oder Karbolineum; Sorge für Luftzuzug.
5) Ersatz der besonders gefährdeten Teile durch Eisen.

Es ist schon ausserordentlich oft dagewesen, dass ein anscheinend beseitigter Schwammfall sich über Jahresfrist wieder eingestellt hat; deswegen kann nicht oft genug darauf hingewiesen werden, bei der Reparatur möglichst peinlich zu Werke zu gehen.

II. DIE WERKZEUGE.

1. Werkzeuge, die jedem Arbeiter gestellt werden. — 2. Allgemeine Werkzeuge, die jedem Arbeiter zur Verfügung stehen. — 3. Werkzeuge, die der Arbeiter selbst stellt.

Wenn hiermit der Versuch gemacht wird, eine Aufzählung und gedrängte Beschreibung der zur Bauschreinerei erforderlichen Werkzeuge und Hilfsvorrichtungen zu geben, so geschieht dies nicht in der Absicht, dem Schreiner zu zeigen, welche Werkzeuge er zu seinen Arbeiten braucht, oder dem Techniker begreiflich zu machen, dass der Schreiner dies oder jenes Werkzeug benützen muss, wenn die Arbeit gut werden soll. Der Schreiner muss, wenn er diesen Titel überhaupt beansprucht, es vorher schon wissen, und der Techniker wird es in dieser Kürze wohl nicht lernen. Der Zweck dieser Besprechung ist vielmehr, dem ersteren durch Vorführung neuerer bewährter Werkzeuge manches zu zeigen, was ihm vielleicht von Nutzen sein kann, dem anderen aber doch im allgemeinen einen Begriff beizubringen, womit und wie gearbeitet wird.

Die Abbildungen der Werkzeuge sind aus dem illustrierten Katalog eines der bedeutendsten und solidesten Geschäfte Süddeutschlands (der Werkzeughandlung von Ernst Straub in Konstanz, früher Karl Delisle daselbst) entnommen. Der Ruf dieser Firma ist ein derart guter und durch seine Lieferungen wohlerworbener, dass es einer besonderen Empfehlung derselben nicht bedarf. Es genügt, der Firma für die Ueberlassung des Illustrationsmaterials an dieser Stelle öffentlich zu danken.

Die Werkzeuge in einer Schreiner-Werkstätte sind dreierlei Art:

1) Werkzeuge, welche jedem Arbeiter vom Meister gestellt werden; es sind dies ausser der Hobelbank alle diejenigen, welche sich in dem zu dieser gehörigen Werkzeugkasten befinden:

1 Schichthobel, 1 Doppelhobel, 1 Putzhobel, 1 Rauhbank, 1 Schrupphobel, 1 Zahnhobel, 1 Simshobel, 1 Faustsäge, 1 Handsäge, 1 Absetzsäge, 6 Stecheisen, 1 Zirkel, 2 Streichmasse, 2 Raspeln, 2 Feilen, 1 Sägefeile, 2 Winkel, 1 Winkelmass, 1 Hammer, 1 Klöpfel, 1 Kropflade, 6 kleine, 12 mittlere und 6 grosse Schraubzwingen, 1 Oelgefäss.

2) Allgemeine Werkzeuge, welche nur in einem Exemplar oder in wenigen Exemplaren vorhanden sind und demnach sämtlichen Arbeitern der Werkstätte gemeinsam zur Verfügung stehen:

Der Nuthobel, der Grat-, der Kurven-, der Grund- und der Falzhobel, ferner die Plattbank und die verschiedenen Kehl-, Rundstab- und Gesimshobel, die Schweifsägen verschiedener Art, der Fuchsschwanz mit und ohne Rücken, der Stangenzirkel, die Loch- und die Gratsäge sowie die Feilkluppe. Ferner das Gehrmass und die Schmiege, die Winkel- und Gehrungsstosslade, die Lochbeitel, die Fischband-Einstemmeisen, die Wasser- oder Bleiwage, das Richtscheit, die Setzlatte, der Dächsel und das Handbeil. Sodann der Schraubenschlüssel, die Schraubknechte, die Schraubzwingen, die Gehrungszwingen, der Schraubstock, der Schleifstein und die verschiedenen Maschinen (Säge, Bohr- und Fräs-Maschine).

3) Werkzeuge, welche sich jeder Arbeiter selbst beschafft, welche sein Eigentum sind und verbleiben:

1 Satz Zentrumbohrer (gewöhnlich 10 Stück), 1 Satz Amerikaner Bohrer (gewöhnlich 10 Stück). Einige Nagelbohrer, 1 Ausreiber, 1 Versenker, 1 Abputzhobel mit eiserner Sohle, 1 Ziehklinge mit Stahl, 1 oder 2 Schraubenzieher, 1 Beisszange, 1 Schränkeisen, 1 Schnitzer, 1 Spitzbohrer, 1 Geisfuss, 1 Massstab, 2 Abziehsteine, 1 Stechzeug (sog. Hohleisen).

---

1. Werkzeuge, welche jedem Arbeiter vom Meister gestellt werden.

1) **Die Hobelbank.** Sie ist das bedeutendste Werkzeug des Schreiners und dient dazu, das zu bearbeitende Holz festzuhalten. Sie besteht aus dem Gestell, dem Blatt, der Vorderzange (links) und der Hinterzange (rechts). Beide Zangen werden durch Schrauben bewegt. Das Festhalten des Holzes geschieht entweder durch die Zangen direkt oder durch die eisernen Bankhaken, welche in die entsprechenden Bankhakenlöcher des Blattes und der Hinterzange eingeschoben werden. Der vertiefte Raum auf dem Blatt heisst die »Beilage«; sie dient zur Aufnahme des sonst hindernden Werkzeuges. Die Schiebladen im Gestell sind nicht absolut erforderlich, aber sehr praktisch (Fig. 24).

2) **Der Schlichthobel.** Unter einem Hobel versteht man allgemein ein Werkzeug, welches zum Glätten, zum Ebnen etc. des Holzes dient. Es besteht aus einem prismatischen länglichen Stück Holz (meist Weiss- oder Hainbuche), mit Span- und Keilloch versehen, durch welches das Hobeleisen hindurch gesteckt und durch einen Keil festgehalten wird. Am vorderen Ende hat der Hobel die sog. Nase. Hat der Hobel nur ein Eisen, so heisst er Schlichthobel; derselbe dient zum Vorarbeiten und liefert mittelfeine Späne und eine dementsprechende Arbeit. In Fig. 25 ist ein Patent-Schlichthobel mit Schraubenverschluss dargestellt, welch letzterer es ermöglicht, das Hobeleisen rascher und sicherer einzuspannen, als mittelst des Keils. Auch das Zerschlagen der Rückseite und das Aussprengen der Wände beim Herauszwängen des Hobeleisens ist hierbei vermieden.

3) **Der Doppelhobel.** Werden bei einem Hobel zwei mit den Schneiden gegeneinander liegende Hobeleisen (von denen das obere Klappe heisst) verwendet, welche durch eine Schraube mit einander verbunden sind, so heisst derselbe Doppelhobel. Derselbe liefert feine Späne.

4) **Der Putzhobel,** auch Verputz- oder Abputzhobel genannt, ist ein Doppelhobel und dient nur zu den feinsten Arbeiten, zum Glätten und namentlich zum Hobeln von Hirnholzflächen, wie bei Gehrungen von Leisten etc. Als sehr gut haben sich die amerikanischen Hobel, namentlich Baileys eiserne Patent-Hirnholzhobel bewährt. Figur 26 zeigt einen solchen mit Handgriff, Figur 27 einen andern ohne Griff mit Patentverstellung und in Figur 28 ist der Längenschnitt durch einen amerikanischen Hobel, System Bailey, wiedergegeben.

1. Werkzeuge, welche jedem Arbeiter vom Meister gestellt werden.

5) **Die Rauhbank.** Sie ist ein langer Hobel ohne Nase, aber mit einem hinteren Griff und dient hauptsächlich zum Fügen oder allgemein zur Herstellung von Ebenen. Das Eisen ist doppelt. In Figur 29 ist eine Rauhbank mit Schrauben- statt Keilverschluss dargestellt. Die

Fig. 24. Hobelbank.

zum Fügen an Fusstafeln dienende **Fügbank** ist eine Rauhbank mit beiderseitigem Anschlag und vorderem Quergriff zum Ziehen.

Fig. 25. Schlichthobel mit Schraubenverschluss.   Fig. 26. Patent-Hirnholzhobel.   Fig. 27.

6) **Der Schrupphobel.** Er ist schmäler als der Schlichthobel und dient zum rauhen Bearbeiten. Sein Eisen ist nicht gerade, sondern gewölbt geschliffen, so dass beim Hobeln flache Rinnen entstehen.

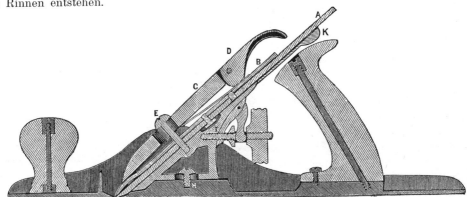

Fig. 28. Amerikanischer Hobel, Patent Bailey, Längenschnitt.

7) **Der Zahnhobel.** Er dient zum Rauhmachen der zum Verleimen bestimmten Flächen, wenn sie zu glatt sind. Sein Eisen ist auf der oberen Seite leicht gerieft, wodurch bei Anschleifen

der Schräge, des Ballens, eine sägeförmige Schneide entsteht, mittels welcher das Holz bearbeitet wird. Das Eisen steht hierbei fast senkrecht. Die Figur 30 zeigt Stanleys verstellbaren Zahn- und Furnierschabhobel.

Fig. 29. Rauhbank mit Schraubenverschluss.

Fig. 30. Stanleys verstellbarer Zahnhobel.

Fig. 31. Amerikanischer Simshobel.

8) Der Simshobel. Er ist ein schmaler Hobel, welcher dazu dient, leichte Falze herzustellen. Das Eisen nimmt die ganze Breite des Hobels ein, und der Span geht seitwärts statt oben hervor (Fig. 31).

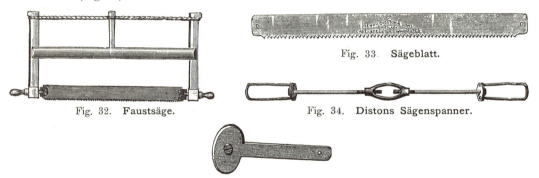

Fig. 32. Faustsäge.
Fig. 33. Sägeblatt.
Fig. 34. Distons Sägenspanner.
Fig. 35. Angel.

9) Die Faustsäge. Unter einer Säge versteht man allgemein ein Werkzeug zum Zerteilen, Zerschneiden des Holzes. Es giebt Sägen mit und ohne Gestell, mit und ohne Spannung. Zu den ersteren gehören die Faust-, Hand-, Absetz- und Schweifsägen. Die Faustsäge

Fig. 36 Stecheisen.
Fig. 37. Stanleys Massstabzirkel.

(Fig. 32) besteht aus dem Sägeblatt (Fig. 33), den Angeln (Fig. 35), den Hörnern, den Sägearmen mit Mittelsteg und Knebel samt Schnur. Die beiden letzteren werden in neuester Zeit zweckmässig durch die eisernen Sägespanner von Diston (Fig. 34) ersetzt. Ebenso haben sich die Sägeblätter, an welche man besondere Angeln beliebig an- und abschrauben kann, gut bewährt. Der Hauptvorteil besteht darin, dass man mittels dieser Vorrichtung auch ein stark ausgefeiltes Sägeblatt durch Einbohrung zweier neuer Löcher oberhalb der alten benützen kann. Auf diese Weise verzieht sich das Blatt nicht, was andernfalls unausbleiblich ist. Damit die Säge beim Schneiden sich nicht zwängt, werden die Zähne geschränkt, d. h. der eine nach

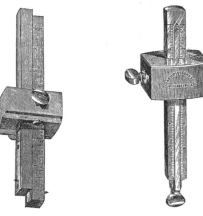

Fig. 39.   Fig. 40.
Amerikanische Präzisionsstreichmasse.

ausgebogen, wodurch der Schnitt ein breiterer wird und das Faustsäge versteht man die grösste Säge, welche der einzelne

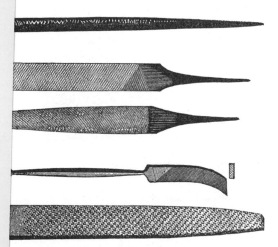

Fig. 41. Feilen und Raspeln.

ist genau wie die vorhergehende, nur kleiner als sie.
hat dieselbe Grösse und Form wie die Handsäge, dabei aber gar nicht geschränkt sind. Sie dient zum Absetzen, zum chneiden von Gehrungen etc.

II. Die Werkzeuge.

12) Das Stecheisen, auch Stechbeitel genannt, ist, wie schon sein Name sagt, ein Werkzeug zum Stechen, zum Stemmen, zum Meisseln (Fig. 36). Das Stecheisen erhält ein Heft aus Weissbuche. Die Breite des Stecheisens ist verschieden, sie wechselt von 5 bis 50 mm.

13) Der Zirkel. Er dient zum Kreisziehen, Messen und Abgreifen; er ist von Eisen und hat die gewöhnliche Form. Für grössere Arbeiten benützt man den eigentlich zur 2. Werk-

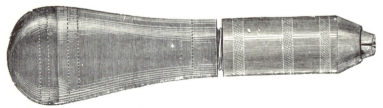

Fig. 42. Feilenheft.

zeugabteilung gehörigen Stangenzirkel »Stanley« (Fig. 38). Derselbe besteht aus 2 Teilen, welche an ein durchgehendes Lineal verstellbar befestigt werden. Sehr praktisch ist ferner Stanleys Patent-Massstabzirkel (Fig. 37).

Diese Abbildungen zeigen auch, wie man die Zirkelteile am besten befestigt. Sie können an jedem stärkeren Massstab angebracht werden und dienen als Stangen-, Teil- und Tasterzirkel. Ein praktisches Reissmass kann hergestellt werden, indem man den Massstab als Lineal benützt, wobei der eine Zirkelkopf als Spitze oder Reisskopf und der andere als Lehrenkopf dient.

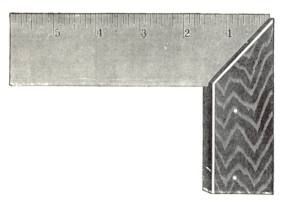 

Fig. 43. Amerikanischer Winkelhaken.   Fig. 44. Klöpfel.

14) Das Streichmass. Dasselbe, aus Holz bestehend, dient zum Vorreissen paralleler Linien. An dem einen Ende der durch den Mittelkörper durchgesteckten Stäbchen befinden sich eingeschlagene Stahlspitzen, welche die Linienzeichnung bewirken. Es giebt Streichmasse mit mehreren Stiften an ein und demselben Stäbchen. Empfehlenswert sind die amerikanischen Präzisionsstreichmasse (Fig. 39 u. 40), welche statt der glatten Stäbchen durch Schrauben verstellbare Massstäbe haben, an welchen die Vorreissstifte sitzen.

15) Raspel und Feile. Sie dienen zur Glättung und Ebnung des Holzes an solchen Arbeiten, bei welchen dies mittels des Hobels unmöglich ist. Raspel und Feile unterscheiden sich hauptsächlich dadurch, dass die angreifenden Teile bei der ersteren aus einzelnen, in regel-

mässigen Entfernungen von einander stehenden kleinen Erhöhungen oder Spitzen bestehen, während es bei der Feile quer ziehende Riefen sind, welche die glättende Wirkung hervorbringen. Die beiden werden fast immer zusammen bezw. nach einander verwendet, und zwar die Raspel zum Vorarbeiten, die Feile zur Nacharbeit. Dem Querschnitt nach giebt es vierkantige, dreikantige, halb- und ganzrunde Raspeln und Feilen (Fig. 41).

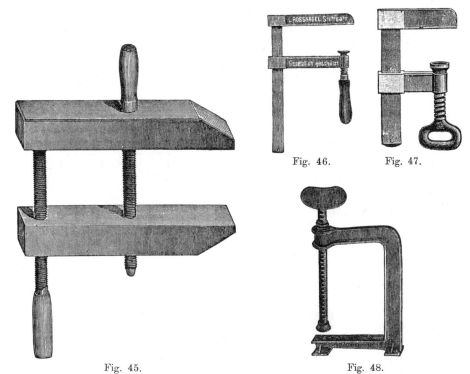

Fig. 45.
Amerikanische Schraubzwinge aus Holz.

Fig. 46. Fig. 47.

Fig. 48.
Eiserne Momentschraubzwingen.

16) Die Sägefeile. Sie ist eine Metallfeile und dient zum Feilen der Sägeblattzähne. Ihre Form ist den letzteren entsprechend dreikantig, so dass also je zwei Feilenflächen in einem Winkel von 60° zu einander stehen. In Figur 42 ist ein praktisches, leicht zu handhabendes Feilenheft dargestellt.

17) Der Winkel, auch Winkelhaken genannt, dient als Winkelmessinstrument oder als Lehre bei der Ausarbeitung rechtwinkeliger Hölzer. Er ist meist aus Holz gefertigt, doch sind die amerikanischen Winkelhaken, deren einer Schenkel aus messinggarniertem Holz besteht, während der andere aus Stahl gefertigt ist (Fig. 43), sehr zu empfehlen.

18) Das Winkelmass ist ein grosser Winkel und wird für Arbeiten benützt, für welche die kleinen Winkel nicht ausreichen.

19) Der Hammer besteht aus Stahl und hat einen Holzstiel, während

20) der Klöpfel nur aus Holz (Rotbuchen oder besser Weissbuchen) gefertigt ist. Er wird benützt zum Stemmen mit dem Lochbeitel oder in kleinerer Form zum Aushauen der Zinken mittels des Stechbeitels (Fig. 44).

21) **Die Kropflade.** Sie dient, wie der Name schon sagt, zur Herstellung von Kröpfen, d. s. nach einem bestimmten Winkel beschnittene und mit dem Hobel bestossene Gesimsstücke, welche an irgend einem Gegenstand angebracht werden sollen.

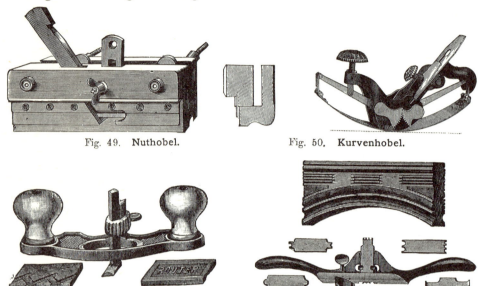

Fig. 49. Nuthobel.   Fig. 50. Kurvenhobel.

Fig. 51. Stanleys eiserner Grundhobel.   Fig. 52. Stanleys eiserner Universal-Kehler.

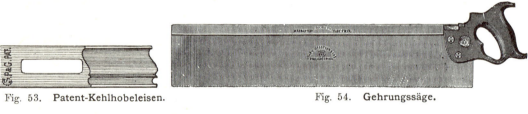

Fig. 53. Patent-Kehlhobeleisen.   Fig. 54. Gehrungssäge.

Fig. 55. Fuchsschwanzsäge.   Fig. 56. Lochsäge.   Fig. 57. Gratsäge.

22) **Die Schraubzwinge.** Sie ist meist aus Holz hergestellt, doch giebt es in neuerer Zeit auch solche von Eisen mit gewöhnlicher und Schnellspannung. Sie dient zum Zusammenschrauben zweier Holzteile und findet ihre Verwendung besonders beim Verleimen, weshalb sie auch vielfach Leimzwinge genannt wird. Je nach der Grösse der Arbeit, also nach dem Zweck verwendet man grosse, mittlere oder kleine Schraubzwingen. Fig. 45 zeigt eine amerikanische Schraubzwinge aus Holz und in den Fig. 46, 47 und 48 sind eiserne Zwingen mit Schnellspannung abgebildet.

23) **Das Oelgefäss,** ein kleines, becherartiges Blechgefäss, welches zur Aufnahme von etwas Leinöl dient.

2. Allgemeine Werkzeuge, welche sämtlichen Arbeitern gemeinsam zur Verfügung stehen.

24) **Der Nuthobel.** Er dient zur Herstellung der Nuten, ist entweder aus Holz (Fig. 49) oder in neuerer Zeit auch von Metall. Der Hobel ist durch Schraubenspindeln so verstellbar, dass schmale und breite, flache und tiefe Nuten, sowie schwache und starke Wangen damit gefertigt werden können.

25) **Der Grathobel.** Er wird benützt zur Herstellung der Grat- und Einschiebleisten; mit ihm wird der schwalbenschwanzförmige Grat an dieselben angestossen, welcher in die Gratnute eingeschoben wird.

26) **Der Kurvenhobel.** Er dient zum Aushobeln von hohlen Flächen und ist mit Ausnahme der runden Sohle konstruiert wie jeder andere Hobel. Den alten Holzkurvenhobel mit unverstellbarer Holzsohle hat längst der amerikanische, leicht verstellbare Kurvenhobel (Fig. 50) vollständig verdrängt.

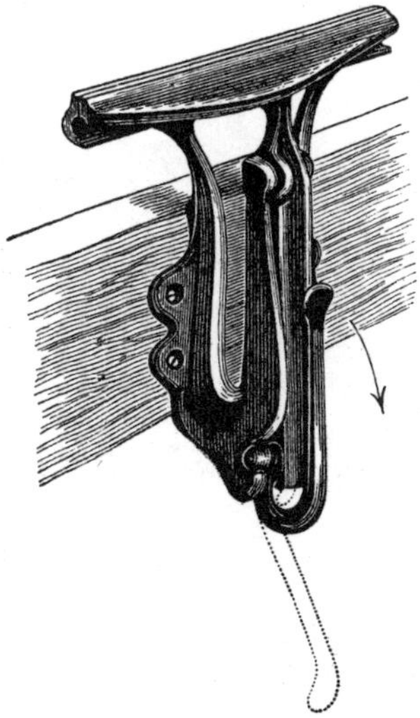

27) **Der Grundhobel.** Er wird benützt zur Glättung der durch Vorschneiden mit der Gratsäge und Ausstechen mit dem Stecheisen hergestellten Gratnute für die Einschiebleisten. Fig. 51 zeigt einen amerikanischen Grundhobel, der auch für andere Zwecke benützt werden kann.

28) **Der Falzhobel.** Man versteht darunter einen Simshobel mit Anschlag.

29) **Die Plattbank.** Sie dient dazu, die Füllungen abzuplatten, d. h. so zu verjüngen, dass sie in die Nuten passen. Sie hat einen Anschlag, welche die Breite der Abplattung bestimmt. Der Abplattung auf Querholz wegen stehen die Hobeleisen schräg und nicht winkelrecht wie bei den übrigen Hobeln.

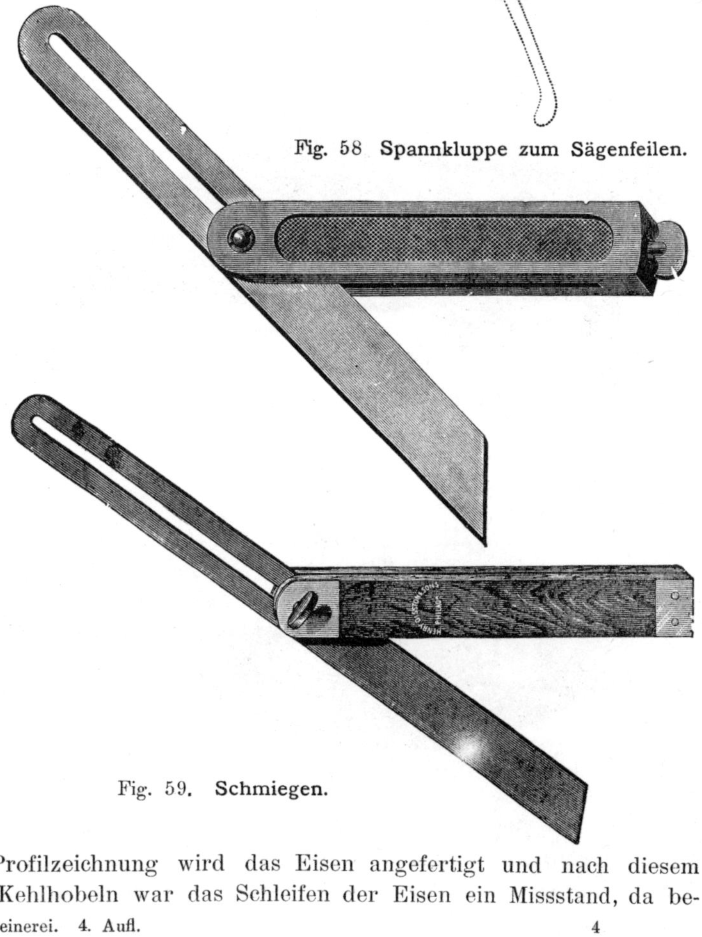

30) **Die Kehl-, Rundstab- und Gesimshobel.** Man versteht darunter Hobel mit verschiedenartig geformten Hobeleisen zur Herstellung von Hohlkehlen, Rundstäben und beliebigen Profilen. Nach der vorliegenden Profilzeichnung wird das Eisen angefertigt und nach diesem die Hobelsohle. Bei den alten Kehlhobeln war das Schleifen der Eisen ein Missstand, da be-

Fig. 58 **Spannkluppe zum Sägenfeilen.**

Fig. 59. **Schmiegen.**

sondere Geschicklichkeit dazu gehörte; bei den neuen Patent-Kehlhobeln mit glatt anschleifbarem Eisen (Fig. 53) ist derselbe beseitigt, da jedermann das Eisen nachschleifen kann, ohne das Profil im geringsten zu ändern. Bei diesem Anlass möge auch Stanleys eiserner Universal-Handkehler erwähnt sein, zu dem 7 verschiedene Eisen gehören, mit denen Rundstäbe, Kehlen, Nuten etc. hergestellt werden können (Fig. 52).

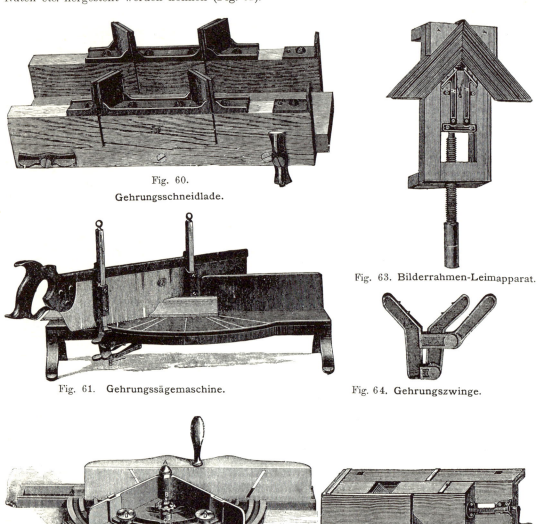

Fig. 60. Gehrungsschneidlade.

Fig. 61. Gehrungssägemaschine.

Fig. 62. Langdons Patent-Gehrungsmaschine.

Fig. 63. Bilderrahmen-Leimapparat.

Fig. 64. Gehrungszwinge.

Fig. 65. Gehrungsstosslade für Bilderrahmen.

31) **Die Schweifsäge.** Sie ist eine Säge wie die bereits besprochenen, nur ist bei ihr das Blatt schmal, da die Säge nur zum Ausschneiden von Kurven benützt wird. Nach der Grösse und der Art der Kurve richtet sich die Blattbreite.

2. Allgemeine Werkzeuge, welche sämtlichen Arbeitern gemeinsam zur Verfügung stehen.

Fig. 66. Fischband-Einstemmeisen.

Fig. 67. **Gewöhnliche Wasserwage.**

Fig. 68.

Fig. 69.

Wasserwagen in Eisenfassung, horizontal und vertikal zu gebrauchen.

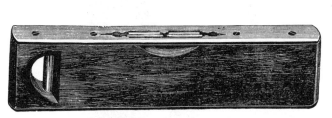

Fig. 70. Wasserwage in Holzfassung, horizontal und vertikal zu gebrauchen.

Fig. 71. Wasserwage in Eisenfassung,

Fig. 72. Dächsel.

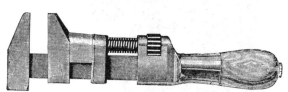

Fig. 73. Universal-Schraubenschlüssel.

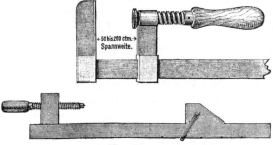

Fig. 74.

Fig. 75.

Schraubknechte aus Holz und aus Eisen.

4*

32) **Der Fuchsschwanz.** Er ist eine Säge ohne Gestell (wird also auch nicht gespannt) mit einem Handgriff. Er dient zum Sägen von Gegenständen, welchen man mit der gewöhnlichen Säge nicht beikommen kann (Fig. 55). Es giebt Fuchsschwänze mit und ohne Rückenverstärkung und solche die beiderseits verschieden gezahnt sind. Der in Figur 54 dargestellte ist für eine Gehrungssägemaschine bestimmt.

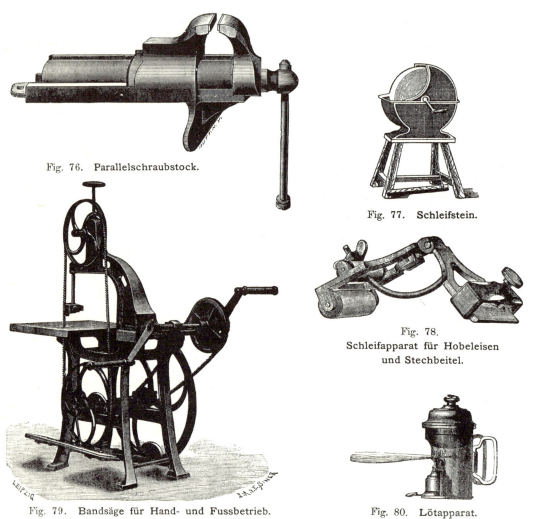

Fig. 76. Parallelschraubstock.

Fig. 77. Schleifstein.

Fig. 78. Schleifapparat für Hobeleisen und Stechbeitel.

Fig. 79. Bandsäge für Hand- und Fussbetrieb.

Fig. 80. Lötapparat.

33) **Die Lochsäge** (Fig. 56). Sie dient zum Ausschneiden von Kreisen und Figuren, welche mit einer anderen Säge des Gestells wegen nicht zu fertigen sind. Beim Beginn der Arbeit ist ein Loch zu bohren, welches der Säge als Angriff dient.

34) **Die Gratsäge** (Fig. 57). Sie ist eine kleine Säge mit Holzgriff und dient zum Einschneiden der Gräte für Einschiebleisten. Vom Fuchsschwanz unterscheidet sie sich hauptsächlich dadurch, dass sie nicht im Vorstossen, sondern im Zurückziehen schneidet.

2. Allgemeine Werkzeuge, welche sämtlichen Arbeitern gemeinsam zur Verfügung stehen.

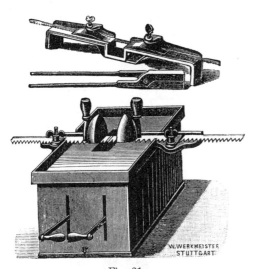

Fig. 81.
Bandsägen-Lötapparate.

Fig. 82.
Mechanischer Bandsägenschränkapparat.

Fig. 83.
Band- und Kreissäge für Hand- und Fussbetrieb.

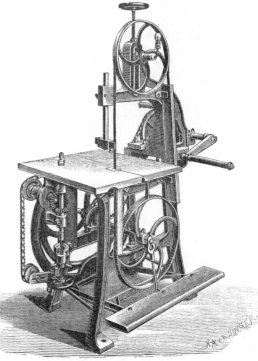

Fig. 84.
Bandsäge mit Fräsmaschine.

35) **Die Feilkluppe.** Sie ist von Holz und dient zum Einspannen der Sägeblätter beim Schärfen der Zähne. Eine eiserne Feilkluppe, die bequem anzubringen ist und für Sägeblätter aller Art passt, zeigt die Fig. 58.

36) **Das Gehrmass** gehört eigentlich zu den Winkelmassen. Während beim gewöhnlichen Winkelmass und Winkelhaken die Schenkel einen Winkel von 90° bilden, so beträgt derselbe beim Gehrmass 45° (oder auch 60°). Die beiden Schenkel sind fest mit einander verbunden. Das Gehrmass dient zum Vorreissen von Gehrungen an Gesimsen u. dergl.

37) **Die Schmiege.** Dieselbe besteht aus zwei mittels einer Schraube verstellbar miteinander verbundenen Schenkeln, welche auf einen beliebigen Winkel gestellt werden können. Sie schmiegen sich jedem Winkel an. Die Schmiege ist also eine Art Gehrmass für beliebige Winkel (Fig. 59).

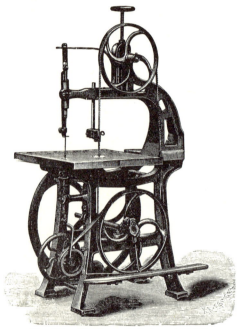

Fig. 85. Band- und Dekoupiersäge.

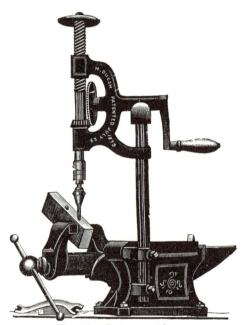

Fig. 86. Bohrmaschine.

38) **Die Winkelstosslade.** Sie dient dazu, Leisten und Profilstäbe in genauer Länge und rechtwickelig abzuschneiden und zu bestossen, d. h. an der Schnittfläche zu hobeln.

39) **Die Gehrungsstoss- und -schneidlade** ist im allgemeinen wie die vorhergehende konstruiert, nur sind die Winkel, nach welchen bei ihr die Leisten geschnitten und bestossen werden können, andere als 90°, also z. B. 45° und 60° etc. Figur 60 zeigt eine amerikanische, eisengarnierte Gehrungsschneidlade. In Figur 61 ist eine verbesserte Gehrungssägemaschine dargestellt, welche so fein schneidet, dass ein Bestossen überflüssig ist. Schliesslich zeigt Fig. 62 die amerikanische Gehrungshobelmaschine, Langdons Patent, welche sich wie die vorige auf jeden Winkel leicht stellen lässt. Beide haben sich sehr bewährt und können empfohlen werden. Wertvoll für die letztere Maschine ist, dass zu ihr alle Ersatzteile zu haben sind. Fig. 65 zeigt eine Gehrungsstosslade, speziell für Bilderrahmen. Mit erwähnt mögen hier gleich werden die Gehrungszwinge der Fig. 64 zum Festhalten der Ecke während des Verleimens

und der dem gleichen Zwecke dienende, etwas umständlicher gebaute Bilderrahmen-Leimapparat (Fig. 63).

40) **Der Lochbeitel.** Er ist ein kräftiges Stemmeisen, welches zum Stemmen der Zapfenlöcher etc. dient.

41) **Das Fischband-Einstemmeisen.** Es dient zum Einstemmen der Fischbandlappen (Fig. 66).

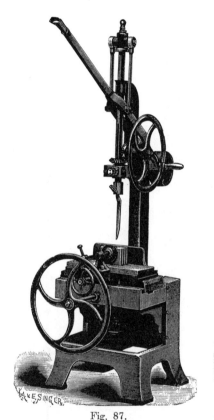

Fig. 87.
Bohr- und Stemmmaschine.

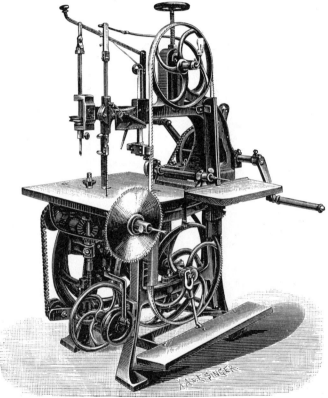

Fig. 88.
Sog. Universalmaschine.

42) **Die Wasserwage und die Bleiwage.** Sie sind Instrumente zur Bestimmung von horizontalen Flächen und Kanten. Das ältere derselben, die Bleiwage ist ein gleichschenkeliges Holzdreieck, von dessen Spitze eine unten mit einer Bleikugel beschwerte Schnur auf die Grundlinie herabhängt und durch ihre Stellung zur lotrechten Vorzeichnung angiebt, ob die Arbeit horizontal ist oder nicht. Sie ist von der Wasserwage, welche allgemein bekannt ist, vollständig verdrängt worden. In Figur 67 ist eine Wasserwage dargestellt, mit welcher man nur horizontal abwägen kann; dagegen gestatten die Wasserwagen der Fig. 68 bis 71 horizontale und vertikale Wägung. Diese Instrumente können bestens empfohlen werden.

43) **Das Richtscheit** ist ein grösseres Lineal, welches als Hilfsmittel zur Herstellung von Ebenen benützt wird.

32 II. Die Werkzeuge.

44) Die Setzlatte ist eigentlich ein grosses Richtscheit. Sie hat meist ganze Dielenlänge und halbe Dielenbreite, ist von trockenem Holz und gut gefügt. Verwendet wird sie wie das Richtscheit zur Herstellung ebener Flächen beim Boden- und Rippenlegen etc.

45) Der Dächsel ist ein Mittelding zwischen einer Hacke und einem Beil und dient zum Abdächseln der Gebälke, zur Ausgleichung, falls einzelne Balken etwas zu hoch liegen. (Fig. 72.)

46) Das Handbeil. Zweck und Konstruktion desselben ist allgemein bekannt.

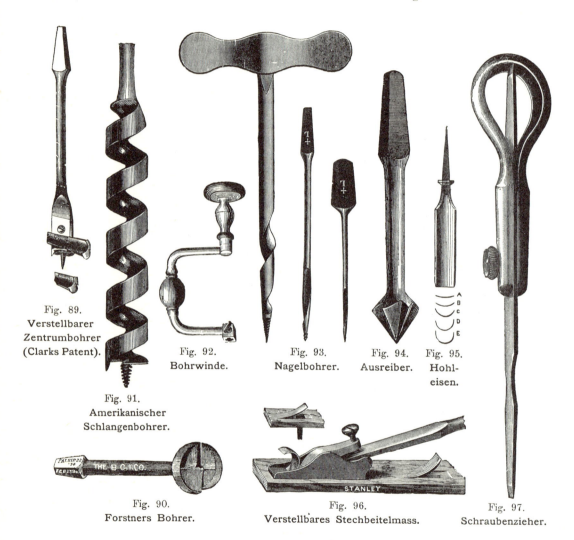

Fig. 89. Verstellbarer Zentrumbohrer (Clarks Patent).
Fig. 91. Amerikanischer Schlangenbohrer.
Fig. 90. Forstners Bohrer.
Fig. 92. Bohrwinde.
Fig. 93. Nagelbohrer.
Fig. 94. Ausreiber.
Fig. 95. Hohleisen.
Fig. 96. Verstellbares Stechbeitelmass.
Fig. 97. Schraubenzieher.

47) Der Schraubenschlüssel. Er dient zum Anziehen und Lösen der Mutterschrauben. Fig. 73 stellt eine der verschiedenen Formen dar, welche diesem Werkzeuge gegeben worden sind.

48) Die Schraubknechte. Es sind dies grosse Schraubzwingen, welche beim Verleimen grösserer Flächen Verwendung finden. Sie sind aus Holz oder Eisen, mit gewöhnlicher oder mit Schnell-Spannung etc. (Fig. 74 und 75).

2. Allgemeine Werkzeuge, welche sämtlichen Arbeitern gemeinsam zur Verfügung stehen.

49) Der Schraubstock. Obgleich eigentlich ein Werkzeug des Schlossers, ist er auch für den Schreiner, welcher viel mit Beschlägen zu thun hat, unentbehrlich, so dass er in keiner Schreinerwerkstätte fehlen sollte. Ein kleiner Parallelschraubstock nach Figur 76 ist sehr zu empfehlen.

50) Der Schleifstein. Sein Zweck ist allgemein bekannt. In dem auf dem Gestell ruhenden Kasten befindet sich Wasser, so dass der Stein beim Drehen sich selbstthätig nässt (Fig. 77). In Figur 78 ist ein sehr praktischer amerikanischer Schleifapparat gezeigt, welcher zum Egalschleifen von Hobeleisen und Stechbeiteln dient.

Schliesslich mögen noch einige sehr geeignete Maschinen für Hand- bezw. Fussbetrieb im Bilde vorgeführt werden.

51) Bandsäge mit Hand- und Fussbetrieb (Fig. 79). Mit ihr kann man Holz bis zu ca. 22 cm Stärke schneiden. Zu der Maschine werden 3 Sägeblätter und 1 Lötapparat geliefert. Preis ab Fabrik = 240 M.

Die Lötlampe, der Lötapparat (Fig. 80) dient zum Zusammenlöten des Sägeblattes, falls dieses einmal reissen sollte. Schreinermeister H. Raible in Karlsruhe schreibt über die Vornahme einer solchen Lötung in der Badischen Gewerbezeitung 1889, No. 15:

»Es kommt häufig vor, dass das Blatt einer Bandsäge springt und dass dasselbe sofort vom Schreiner selbst wieder zusammengelötet werden muss. Es geschah dies bisher mit Hilfe einer Lötzange, welche in einer Schreinerwerkstätte, in welcher man gewöhnlich über kein grosses Feuer verfügt, entsprechend glühend zu machen ziemlich umständlich ist. Schreiber dieses, welcher diesen Missstand auch vielfach zu empfinden hatte, hat nun versucht, seine Bandsägeblätter, statt mit der Zange, mit der Lötlampe zu löten und dabei gute Resultate erhalten. Er empfiehlt deshalb dieses Verfahren, bei welchem man im einzelnen wie folgt verfährt, allen Fachgenossen:

»Die zusammenzulötenden Stellen werden etwa 2 Zähne lang sauber abgefeilt, wobei man darauf zu achten hat, sie nicht zu verjüngen und sie nach dem Feilen nicht mit der Hand zu berühren. Alsdann spannt man die beiden zu vereinigenden Blattstellen in eine eiserne Kluppe (eine solche wird gewöhnlich vom Bandsägefabrikanten jeder Säge beigegeben und dürfte jedem Schreiner bekannt sein), befeuchtet die Lötstelle mit einer dünnen, wässerigen Boraxlösung, umwickelt sie mit feinem Eisendraht, legt an den Rand der Lötstelle, nicht dazwischen, feines Schlaglot und umgiebt das letztere, damit es gut liegen bleibt, mit etwas feuchtem Borax. Hierauf bringt man das zu lötende Blatt mit der Kluppe in eine kleine, mit Holzkohlen gefüllte Schüssel, umgiebt die Lötstelle gut mit Kohle und legt ein Stück von letzterer auf dieselbe. Alsdann bestreicht man die Lötstelle erst langsam und vorsichtig mit der Lampenflamme, bis der Borax angebacken ist, worauf man volle Flamme giebt. Wenn der Borax geschmolzen ist, so ist die Lötung vollendet, wobei das Lot zwischen die Lötstellen geflossen ist. Man lässt alsdann abkühlen und feilt die Lötstelle sauber, wobei man sich aber hüten muss, dieselbe zu schwächen, und schärft schliesslich die Säge nach.«

Die Fig. 81 stellt zwei Lötapparate dar, wie sie speziell für Bandsägen gebaut werden. Zum Schränken der Bandsägen hat man ebenfalls besondere Apparate hergestellt. Fig. 82 bildet einen solchen ab.

52) Bandsäge mit Kreissäge für Hand- und Fussbetrieb (Fig. 83).
53) Bandsäge mit Fräsmaschine (Fig. 84) für Hand- und Fussbetrieb.
54) Bandsäge mit Dekoupiersäge (Fig. 85).
55) Bohrmaschine (Fig. 86).
56) Bohr- und Stemmmaschine für Handbetrieb (Fig. 87).
57) Universal-Maschine (Vereinigte Bandsäge, Dekoupiersäge, vertikale Bohrmaschine, Kreissäge, Fräsmaschine und horizontale Bohrmaschine, Fig. 88).

Dieser Maschine ist folgendes beigegeben und im Preis inbegriffen:

3 scharfe Bandsägenblätter, 1 Lötapparat, 1 Feilapparat, 1 Schränkzange, 1 Fräskopf, 1 Dutzend Dekoupiersägeblätter, 1 Blasebalg für die Dekoupiersäge, 1 Dutzend rundkantige Sägefeilen, 1 Schutzvorrichtung, 2 Kreissägeblätter, 1 Parallelogramm-Linealführung für die Band- und Kreissäge, 1 Führung für die Fräsmaschine und die nötigen Mutterschlüssel. Preis 600 M.

Die Abbildungen der unter 51 bis 57 aufgeführten Maschinen sind dem Katalog der bereits erwähnten Firma Ernst Straub in Konstanz entnommen. Im übrigen verweisen wir die Interessenten für Holzbearbeitungsmaschinen noch besonders auf die illustrierten Preisverzeichnisse nachgenannter Geschäfte:

A. Goede, Spezialfabrikation von Maschinen für Holzbearbeitung, Berlin N., Chausseestr. 32.

E. Kirchner & Cie., Deutsch-amerikanische Maschinenfabrik, Leipzig-Sellerhausen.

Gebrüder Schmaltz, Maschinenfabrik in Offenbach a/M.

Teichert & Gubisch, Maschinenfabrik in Liegnitz in Schlesien; Spezialität: Säge- und Holzbearbeitungsmaschinen nach deutsch-amerikanischem System.

Karlsruher Werkzeugmaschinenfabrik, vormals Gschwind & Cie.

J. Rossnagel, Maschinen- und Werkzeugfabrik, Stuttgart.

---

3. Werkzeuge, welche sich jeder Arbeiter selbst beschafft.

58) Der Zentrumbohrer. Seine Verwendung wie seine Form ist bekannt. Noch ziemlich neu, aber sehr empfehlenswert, ist der in Figur 89 abgebildete amerikanische ausdehnbare Zentrumbohrer (Clarks Patent), mit welchem man Löcher in verschiedener Weite mit einem und demselben Bohrer bohren kann, je nachdem man den Schieber stellt. Forstners Bohrer (Fig. 90) erhält seine Führung durch den cylindrischen Rand und gestattet eine saubere und exakte Arbeit.

59) Der Amerikanerbohrer, auch Schlangenbohrer genannt, hat den Zentrumbohrer ziemlich verdrängt. Zu verwundern ist dies nicht, denn die Vorteile der Amerikanerbohrer sind den Zentrumbohrern gegenüber sehr wesentlich. In Figur 91 ist ein Amerikanerbohrer vorgeführt, bei welchem sich der Bohrspan von selbst oben ausdreht, ohne dass man also genötigt ist, den Bohrer des öfteren auszuziehen. Zu diesen beiden vorgenannten Bohrern sind sog. Bohrwinden (Fig. 92) erforderlich, wenn sie nicht in die Bohrmaschine eingespannt werden.

60) Der Nagelbohrer. Er ist ein kleiner Bohrer, welcher zum Vorbohren für Holznägel und Schrauben und unter Umständen bei Hartholz auch für eiserne Nägel dient. Hat er einen Griff wie das links dargestellte Beispiel der Figur 93, so wird er mit der Hand eingedreht; andernfalls bedient man sich ebenfalls der Bohrwinde.

61) Der Spitzbohrer, eine Art Aktenstecher, welcher zum Vorreissen dient.

62) Der Ausreiber. Er dient dazu, die mit dem Nagelbohrer vorgebohrten Löcher oben trichterförmig so zu erweitern, dass man Holzschrauben mit versenkten Köpfen so einschrauben kann, dass sie mit dem Holz bündig stehen. Er wird nur mit der Bohrwinde benützt (Fig. 94).

63) Der Geissfuss, ein Schneidwerkzeug, mit welchem man spitznutenartige Risse ziehen kann. Er hat eine V-förmige Schneide.

64) **Das Stechzeug.** Man versteht darunter eine Anzahl Hohleisen mit mehr oder weniger gebogener Schneide (Fig. 95) zum Ausstechen, zum Stechen, d. h. zum Hohlschnitzen dienend. Auch den Geissfuss kann man zum Stechzeug rechnen.

65) **Der Schnitzer** ist ein Messer mit einem langen S-förmig gebogenen Holzgriff, welcher beim Schneiden an die eine Schulter angelehnt wird.

66) **Der Versenker** ist ein kleines Werkzeug in Form eines abgekürzten Kegels, mit welchem man die Nägel unter die Holzoberfläche vertieft, ohne das Holz unnötig zu beschädigen. In Figur 96 ist ein sog. verstellbares Stechbeitelmass zum Blindnageln dargestellt.

67) **Der Schraubenzieher.** Zum Einschrauben der mit einem Schnitt versehenen Holz- oder Metallschrauben dienend. Figur 97 stellt einen praktischen amerikanischen Duplex-Schraubenzieher dar.

68) **Das Schränkeisen.** Es ist ein Instrument, mittels dessen man die Zähne der Säge schränkt, d. h. beiderseitig so ausbiegt, dass beim Sägen ein breiter Schnitt entsteht und das Blatt sich nicht klemmt.

69) **Die Ziehklinge mit Stahl.** Sie ist ein Stahlblech, dessen Kanten durch kräftiges Streichen mittels des Stahles so geschärft werden, dass es als ein vorzügliches Schabinstrument gelten kann. Die Art des Schärfens zu beschreiben ist überflüssig, da es dadurch doch niemand lernen hann.

70) **Der Abziehstein**, zum Abziehen, d. h. zum Feinschleifen der Werkzeuge dienend, nachdem dieselben bereits auf dem Schleifstein vorgeschliffen sind.

71) **Die Beisszange**, zum Ausziehen der Nägel, ist allgemein bekannt.

72) **Der Massstab** ist ein vierkantiger Stab mit Meterteilung. An seiner Stelle werden auch die bekannten zusammenlegbaren Taschenmassstäbe benützt.

III. DIE VERBINDUNGEN DER HÖLZER.

1. Die Holzverbindungen nach der Breite. — 2. Gestemmte Arbeit. — 3. Holzverbindungen nach der Länge. — 4. Eckverbindungen. — 5. Hilfsmittel zur Verbindung. Nägel, Schrauben und Verleimung.

Die Verbindungsarten der Hölzer sind der Hauptsache nach in der Schreinerei die nämlichen, wie bei den Zimmerarbeiten. Das Hauptverbindungsmittel besteht in der eigenartigen Formung der zu verbindenden Teile. Bei Schreiner- und Zimmerarbeiten treten als weitere Mittel einer gesicherten Verbindung hinzu: Holznägel, Eisennägel und Schrauben. Ein weiteres Verbindungsmittel, welches wohl die Schreinerei, nicht aber die Zimmerei benützt, ist der Leim.

Die meisten der Verbindungen sind nach der Breite des Holzes erforderlich, weil dieselbe eine verhältnismässig geringe ist im Vergleich zu den oft grössere Flächen aufweisenden Schreinerarbeiten, während anderseits die Abmessungen unserer Wohnräume die gewöhnliche Holzlänge von 4,50 m nur selten überschreiten, so dass fast sämtliche Schreinerarbeiten, auch die grössten, der Länge nach keine Zusammensetzung erfordern.

---

1. Holzverbindungen nach der Breite und Herstellung einer Fläche, einer Ebene.

Haben wir zwei oder mehrere Hölzer der Breite nach zu einer Fläche zu verbinden, so erreichen wir dies auf die einfachste Weise, indem wir die betreffenden Holzstücke stumpf neben einander legen (Fig. 98) und Leisten oder Holzstücke darüber nageln, wie es z. B. bei der Latten- und Riementhüre der Fall ist. Dabei ist aber die Verbindung nur an den Stellen erzielt, wo sich die Querleisten befinden, während an allen andern die Dielen sich beliebig werfen und drehen können. Soll dies verhindert werden, so müssen wir die Konstruktion ergänzen, d. h. runde oder viereckig geformte Holzdübel anordnen, welche in die entsprechenden Dübellöcher der Bretter eingreifen (Fig. 99 und 100).

Für viele Arbeiten genügt diese Verbindungsart, obgleich bei ihr die Fugen noch ganz offen, d. h. durchsichtig sind; es ist dies ein Nachteil, dem bei besseren Arbeiten abgeholfen werden muss. Wir erreichen dies, indem wir die Dielen an den Fugen überfälzen (Fig. 101) oder spunden (Fig. 102) oder auf Nut und Feder verbinden (Fig. 103), wobei die Dübel gespart werden.

1. Holzverbindungen nach der Breite und Herstellung einer Fläche, einer Ebene.

Zur Beseitigung von Missvertändnissen mag hier angeführt sein, dass man — sofern man überhaupt einen Unterschied macht — unter **Spunden** die Verbindung mit **angestossener**, d. h. angehobelter Feder versteht, während als Verbindung mit **Nut und Feder** diejenige mit **loser**, mit besonders gefertigter Feder bezeichnet wird. Die losen Federn müssen aus Hartholz bestehen; sie hönnen Langholz- oder Hirnholzfedern sein. Für viele Zwecke werden auch Bandeisenfedern benützt.

Bei den bis jetzt besprochenen Verbindungen ist es dem einzelnen Brett immer noch möglich, für sich zu schwinden und zu quellen; es entstehen offene Fugen, die sich beim Quellen zum Teil wieder schliessen. Ist diese Fugenbildung für eine gegebene Arbeit unzulässig, so erübrigt nur, die Hölzer zu verleimen, und wir erhalten als Verbindungsmittel:

Die Leimfuge. Dieselbe ist die dichteste Verbindung der einzelnen Holzteile. Sie muss, wenn sie gut sein soll — Trockenheit und vernünftige Behandlung vorausgesetzt — unter allen Umständen dicht bleiben. Bei guter Leimfuge bricht das Holz eher an einer andern Stelle als in der Fuge. Um diesen Zustand aber zu erreichen, ist erforderlich, dass

1) das Holz gut trocken ist und
2) die zu verleimenden Teile gut gefügt sind, damit keine Spitzfugen, d. h. an den Enden offene klaffende Fugen entstehen;
3) dass der Leim sowohl wie das Leimen selbst nichts zu wünschen übrig lassen, zwei Erfordernisse, welche an anderer Stelle eingehend besprochen werden.

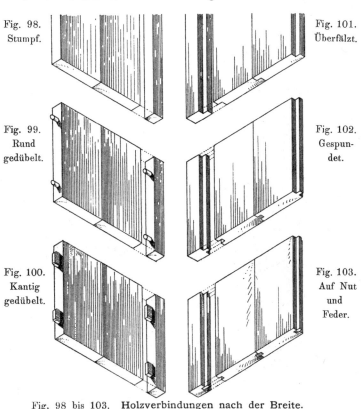

Fig. 98. Stumpf.
Fig. 99. Rund gedübelt.
Fig. 100. Kantig gedübelt.
Fig. 101. Überfälzt.
Fig. 102. Gespundet.
Fig. 103. Auf Nut und Feder.

Fig. 98 bis 103. **Holzverbindungen nach der Breite.**

Gut gefügt sind zwei Bretter oder Dielen, wenn sie mit ihren Kanten aufeinander und gegen das Licht gehalten, keinen Zwischenraum in der Fuge zeigen. Da aber die Vorrichtungen zum Zusammenpressen der beiden Teile beim Leimen derart sind, dass sie nicht an allen, sondern nur an einzelnen Punkten wirken, so fügt man die Bretter — der Länge nach — in der Mitte gern ein klein wenig hohl (Hohlfuge), womit man bezweckt, beim Anziehen der Schraubzwingen oder anderer mechanischer Mittel diese an den Enden zu sparen, ohne Spitzfugen zu erhalten. Ein Mass für die Hohlfuge zu geben, ist unmöglich; es richtet sich nach den besonderen Verhältnissen, der Grösse des zu leimenden Gegenstandes etc. und ist Gefühlssache des Arbeiters. Weiteres siehe unter Leim.

38   III. Die Verbindungen der Hölzer.

Soll eine verleimte Tafel nach der Quere Festigkeit erhalten, so versehen wir sie statt mit genagelten Querleisten:

a) mit sogen. Grat- oder Einschiebleisten;
b) mit Hirnleisten.

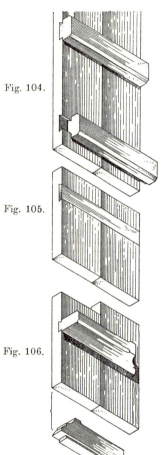

Fig. 104.

Fig. 105.

Fig. 106.

Fig. 104 bis 106.
Grat- und Einschiebleisten.

Grat- und Einschiebleisten. Die Form und Stärke derselben ist verschieden, je nach dem Zweck. Für ganz einfache Arbeiten, bei welchen ein Vorstehen nach unten nicht gewünscht wird, genügt unter Umständen eine Verbindung nach Figur 105, obgleich dieselbe nicht viel Festigkeit zu geben vermag; für gewöhnliche Arbeiten wird die Form der Figur 104 verwendet und zu besonderen Zwecken diejenige der Figur 106. Es leuchtet ohne weiteres ein, dass die letztere Form die geeignetste ist, um eine Tafel gerade zu halten.

Die Fertigung eines Grates ist folgende: Nachdem derselbe genau vorgerissen, wird mittels des Schnitzers in zwei Schnitten eine Art Spitznute hergestellt, als Führung für die nachfolgende Gratsäge. Letztere schneidet in einem Winkel von ca. 75° zwei Schnitte auf eine Tiefe von ungefähr $1/3$ der Holztärke, so dass nach Entfernung des dazwischen liegenden Holzes mittels eines Stechbeitels und des Grundhobels eine schwalbenschwanzförmige Nute, der Grat, ent-

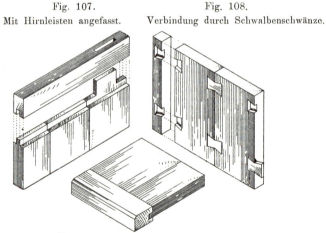

Fig. 107.
Mit Hirnleisten angefasst.

Fig. 108.
Verbindung durch Schwalbenschwänze.

Fig. 109. Mit Hirnleisten angefasst.
Fig. 107 bis 109. Holzverbindungen nach der Breite.

steht. In diesen wird die aus Hartholz gefertigte und entsprechend geformte Gratleiste fest eingeschoben, aber nicht geleimt, damit die Tafel beim Schwinden und Quellen nicht am Arbeiten gehindert ist und nicht reisst. Die beiden Enden der Gratleisten werden etwa 1 bis 2 cm abgesetzt, d. h. kürzer gemacht als die Tafel, damit sie beim Schwinden derselben nicht vorstehen, was stets schlecht aussieht. Verjüngt eingeschobene Gratleisten haben keinen besonderen Wert.

Hirnleisten. Mit Hirnleisten »fasst man eine geleimte Tafel an«, bei welcher das Hirnholz der Bretter nicht zum Vorschein kommen, die Anbringung von Einschiebleisten aber aus irgend welchem Grund unterbleiben soll.

2. Das Stemmen (Gestemmte Arbeit).

Diese Konstruktion ist eine Art Spundung, bei welcher die Feder an das Hirnholz angestossen wird, während die aus Hartholz gefertigte Hirnleiste genutet ist. Um dieselbe noch sicherer zu befestigen, lässt man einzelne Zapfen durch sie hindurchgehen und verkeilt dieselben (Fig. 107).

Hirnleisten lassen sich mit Erfolg nur bei gut trockenem Holz anwenden, da sie eingeleimt werden, mithin nichttrockenes Holz am Schwinden hindern, und das Reissen desselben befördern. Besonders vorsichtige Meister leimen dieselben nur von den Enden herein, damit das mittlere Holz der Tafel immer noch etwas arbeiten kann. Bezüglich ihrer Stärke ist zu bemerken, dass sie entweder beiderseits bündig sein (Fig. 107) oder auf einer Seite vorspringen können (Fig. 109).

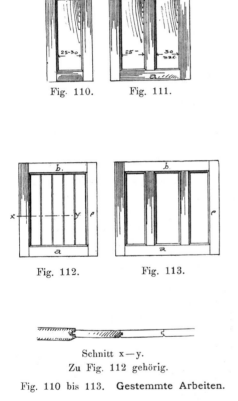

Fig. 110.   Fig. 111.

Fig. 112.   Fig. 113.

Schnitt x—y.
Zu Fig. 112 gehörig.

Fig. 110 bis 113. **Gestemmte Arbeiten.**   Fig. 114. **Einzelheiten der gestemmten Arbeit.**

In Figur 108 ist eine Verbindung dargestellt, bei welcher sogenannte Schwalbenschwänze, aus Hartholz geschnitten, in die Rückseite einer verleimten Tafel eingelassen und eingeleimt werden, um deren Zusammenhalt besser zu sichern. Für sich allein wird die Konstruktion nicht angewendet und ihr Wert ist überhaupt nicht bedeutend.

2. Das Stemmen (Gestemmte Arbeit).

Die bisher besprochenen Verbindungsweisen lassen das Holz in gewissen Grenzen frei arbeiten, d. h. die nach ihnen angefertigten Arbeiten können bei feuchter, kalter Luft quellen, bei trockener, warmer Luft schwinden, in beiden Fällen findet also eine mehr oder minder bedeutende Grössen- oder Formveränderung statt.

III. Die Verbindungen der Hölzer.

Wo es darauf ankommt, einen bei allen Witterungsverhältnissen möglichst gleichmässig dichten Verschluss zu haben, kann man die erwähnten Verbindungsweisen nicht benützen; hier bedarf es formbeständigerer Konstruktionen, als deren beste wir das Stemmen der Arbeit bezeichnen. Der Grundgedanke desselben ist: Querholz wenig — und nur in geringer Breite —,

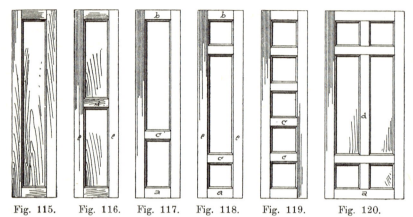

Fig. 115.   Fig. 116.   Fig. 117.   Fig. 118.   Fig. 119.   Fig. 120.

a a = Untere Friese. b b = obere Friese. c c = Querfriese. d d = Mittelfriese. e e = Höhenfriese.

Fig. 115 bis 120. **Gestemmte Arbeiten.**

Langholz dagegen in ausgedehntem Mass zur Verwendung zu bringen, da ersteres sehr, letzteres fast gar nicht schwindet. Fertigen wir daher statt einer verleimten oder auf Querleisten zusammengenagelten Tafel einen Rahmen aus vier, an den Ecken verbundenen Langhölzern (Fig. 115), so haben wir nach den Hauptausdehnungen nur Längsfasern und folglich auch fast gar keine Ausdehnung beim Quellen. Schliessen wir das noch offene Innere durch ein Füll-

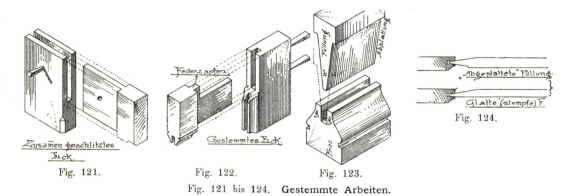

Fig. 121.   Fig. 122.   Fig. 123.

Fig. 121 bis 124. **Gestemmte Arbeiten.**

werk, welches aus auf Nut und Feder eingesteckten Brettern besteht, so haben wir — soweit überhaupt möglich — unseren Zweck erreicht, d. h. wir haben eine Arbeit erzielt, welche als Ganzes wenig wächst und schwindet. Diese Art der Zusammenstellung einer Tafel oder einer Fläche, sei es eine Brüstungslambris, eine Thür oder ein Laden, heisst Stemmen, eine derartige Arbeit gestemmte Arbeit. (Fig. 110 bis 120).

2. Das Stemmen (Gestemmte Arbeit).

Die einzelnen Teile einer solchen Konstruktion heissen:
1) Rahmenhölzer oder Friese. (Unterer und oberer etc. Fries.)
2) Füllungen.

Die Stärke der Rahmenhölzer oder Friese ist meist bedeutender, als die der gewöhnlich in Bordstärke (24 mm) ausgeführten Füllungen; die Friesbreite ist verschieden, sie kann 8 bis 12 cm und mehr betragen, je nach der Art und Grösse der zu fertigenden Arbeit.

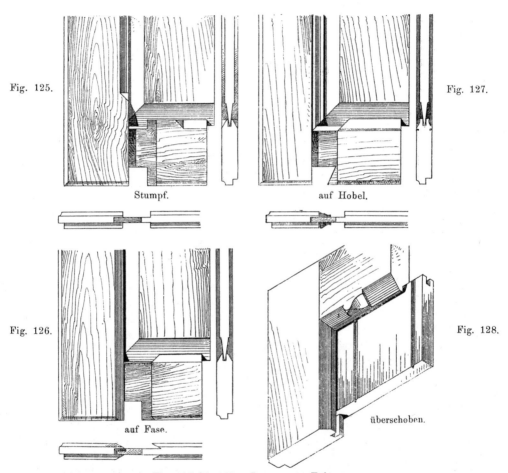

Fig. 125. Stumpf.   Fig. 127. auf Hobel.

Fig. 126. auf Fase.   Fig. 128. überschoben.

Fig. 125 bis 128. **Gestemmte Ecken.**

Die Breite der Füllungen richtet sich teils nach den Friesen und der Gesamtanordnung, hauptsächlich aber nach der gewöhnlichen Bordbreite von 25 bis 30 cm, bei welcher erfahrungsgemäss — trockenes Holz vorausgesetzt — das Schwinden der fertigen Arbeit nur ein ganz mässiges ist. Wollte man die Füllungen doppelt so breit machen, so wäre ein Schwinden in doppelter Breite zu gewärtigen, oder es könnte unter Umständen auch der umgekehrte Fall eintreten, dass die Arbeit beim Quellen auseinandergesprengt wird. Aus diesem Grunde zieht man es vor, mehrere schmälere Füllungen an Stelle von wenigen breiteren anzuordnen. Da man ferner unter gewöhnlichen Verhältnissen eine Füllung des Schwindens und des unschönen

Aussehens wegen nicht über 1,50 m hoch macht, so entstehen durch Einschaltung mittlerer Höhen- und Querfriese geometrische Muster, wie die in Figur 116 bis 120 dargestellten.

Die Friese werden an den Ecken bei einfachen Arbeiten nur zusammengeschlitzt (Fig. 121), besser aber durch verkeilte Zapfen (Fig. 122) miteinander verbunden.

Mit den Friesen ist durch Nut und Feder verbunden die Füllung (Fig. 123). Hierbei heisst f die Nute; gg sind die Wangen, und die verjüngt zugehobelten, abgeplatteten Füllungsenden werden Federn genannt.

Die Tiefe der Nute beträgt unter gewöhnlichen Verhältnissen nicht über 15 mm, ihre Breite 6 bis 8 mm.

Die Stärke der Wangen hängt von der Profilierung ab, wobei darauf zu achten ist, dass dieselben nicht zu sehr geschwächt werden; immer sollte die Wangenstärke doch mindestens gleich der Breite der Nute sein.

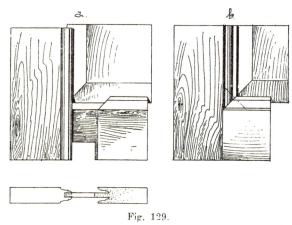

Fig. 129.

Gestemmte Ecken (auf Hobel gestemmt).

a) mit durchlaufendem Profil der Längsfriese,
b) mit eingelegtem Zinkplättchen.

Die sichtbare Abschrägung der Füllungsenden, der sog. Federn, mit dem dazugehörigen Plättchen heisst die Abplattung. Derart geformte Füllungen heissen abgeplattet zum Unterschied von glatt oder stumpf eingeschobenen (Fig. 124).

Die Abplattung muss, wenn richtig ausgeführt, so sein, dass die Feder nach dem Einsetzen in die Nute fest, aber nicht übermässig gespannt sitzt. Zu schwach gehobelte bezw. zu stark abgeplattete Füllungen klappern in der Nute, zu starke sprengen die Wangen ab. Das richtige Mass lässt sich wohl annähernd mittels einer mit demselben Nuteisen gefertigten Lehre bestimmen; immerhin bleibt es aber doch Gefühlssache des Arbeiters, das Richtige zu treffen.

Die Füllung selbst muss nach der Tiefe der Nute immer ca. 2 bis 4 mm Luft bezw. Spielraum haben, damit sie beim Quellen sich ausdehnen kann.

Wie schon ausgeführt, schwinden breite Füllungen verhältnismässig mehr als schmale; da aber viele schmale Füllungen auch viele Friese erfordern, was eine sehr wesentliche Verteuerung der Arbeit bedeutet, so behilft man sich oft, z. B. bei Brüstungslambris, so, dass man breite Füllungen verwendet, welche aus mehreren auf Nut und Feder miteinander verbundenen Riemen bestehen, die einzeln für sich schwinden (Fig. 112). Versieht man hierbei die verschiedenen Fugen der letzteren mit einem kleinen Profil, so lässt sich eine sehr hübsche Wirkung damit erzielen. (Fig. 12).

Wir unterscheiden:

a) **Stumpf gestemmt** (Fig. 125), wenn die Zapfen rechtwinkelig abgesetzt, d. h. abgeschnitten sind;

b) **auf Fase gestemmt** (Fig. 126), wenn die Zapfen schräg, der Abfasung anliegend, abgesetzt sind;

c) **auf Hobel gestemmt** (Fig. 122 und 127), wenn die Friese auf die Breite des Profilhobels in die anderen Friese eingesetzt und die Profile auf Gehrung zusammengeschnitten sind. Aus diesem Grunde ist das Stemmen auf Hobel die teuerste Konstruktion, mag die Arbeit einen Namen haben welchen sie wolle, mag sie eine Thür, eine Brüstung etc. sein.

Auf Fase — und stumpf gestemmt sind im Preise gleich, da die Mehrarbeit des schrägen Absetzens meist wieder durch die façonnierte Abfasung der Kanten beim stumpfen Stemmen ausgeglichen wird. Bei nicht ganz trockenem Holz werden beim Stemmen auf Hobel die Gehrungen leicht undicht, da die beiden Friese nur nach ihrer Breite schwinden. Diesem Uebelstand hilft man am einfachsten ab, indem man trockenes Holz statt halbtrockenes verwendet, indem man vor dem Zusammenleimen an die gefährdeten Stellen Zinkplättchen oder Furnierstücke einlegt (Fig. 129 b), oder indem man statt des angestossenen Profilhobels aufgelegte und geleimte Kehlstösse verwendet.

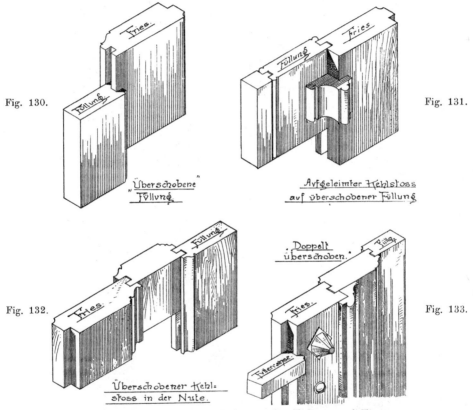

Fig. 130 bis 133. **Holzverbindungen für Thüren und Thore.**

Beim Stemmen auf Fase kommt es nicht vor, dass man durch die Thüren hindurch sehen kann, wie es bei undichten auf Hobel gestemmten Gehrungen vielfach der Fall ist. Man stellt deshalb neuerdings auch Thüren her, bei welchen der Profilhobel auf den Längsfriesen durchläuft, während die Querfriese dem Profil entsprechend am Zapfen abgesetzt werden (Fig. 129 a). Die Profile dürfen dann nicht unterschnitten sein und sind derart zu wählen, dass die falschen Gehrungen sich nicht ausfransen.

Bei allen diesen Konstruktionen kann aber die Reliefwirkung der nur wenig stärkeren Friese gegenüber den Füllungen keine sehr grosse sein, das Profil muss vielmehr sehr bescheiden

44 III. Die Verbindungen der Hölzer.

gehalten werden, damit die Wangen nicht zu schwach ausfallen. Wenn auch zugegeben werden muss, dass für die gewöhnlichen Verhältnisse die gewöhnliche Friesstärke ausreicht und tiefe

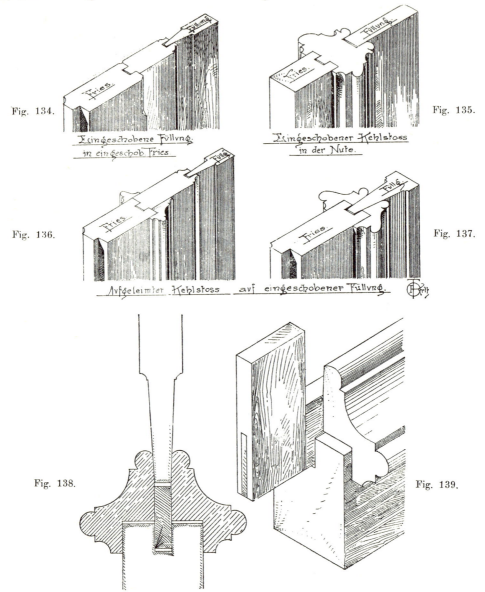

Fig. 134 bis 139. **Holzverbindungen für Thüren und Thore.**

Unterschneidungen der Profile schon im Interesse der sauberen Maler- und Tüncherarbeit besser unterbleiben, so können doch Fälle eintreten, wo die Profilierung aus irgend welchem Grunde kräftiger und reicher sein muss, und in diesem Fall hilft man sich durch Auflegen besonderer Profilleisten.

2. Das Stemmen (Gestemmte Arbeit).

Diese Profil- oder Kehlleisten dürfen aber nur an den Rahmenhölzern und nicht auch an den Füllungen befestigt werden, damit die letzteren im stande sind, sich frei zu bewegen.

Durch die Verbindung der Friese und Füllungen auf Nut und Feder ist dem Arbeiten des Holzes ziemlich vorgebeugt, leider ist aber gerade an den Verbindungsstellen auch die Festigkeit der Konstruktion sehr beeinträchtigt. Denn durch das Nuten der Friese und das Abplatten der Füllungen ist die eigentliche Holzstärke daselbst günstigsten Falls nur noch $1/4$, ein Umstand,

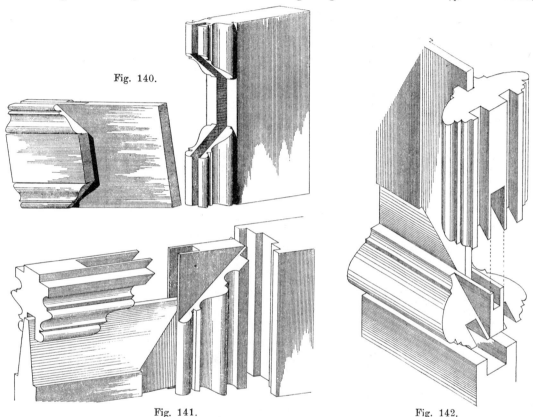

Fig. 140.

Fig. 141.   Fig. 142.
Fig. 140 bis 142. **Fries- und Kehlstossverbindungen,**
verkleinert wiedergegeben nach dem Musterbuch des Baugeschäftes von Billing & Zoller in Karlsruhe.

der diese Konstruktion überall dort als ungeeignet erscheinen lässt, wo der Schutz der Person oder des Eigentums gesichert sein soll, also bei Hauseingangsthüren und Thoren, überhaupt bei allen äusseren Thüren.

Hier wendet man vorteilhafter die sogen. überschobene Konstruktion an. Man überschiebt die Füllungen sowohl, wie allenfallsige zweite Friese und erhöht dadurch nicht nur die Festigkeit der Verbindungsstellen sehr bedeutend, sondern auch die Reliefwirkung des Ganzen, die man noch künstlich steigern kann durch sogen. eingeschobene Kehlstösse oder durch aufgelegte Profilleisten u. dergl. mehr.

Wir erhalten auf diese Weise eine Reihe von Verbindungen, deren Verwendung der grösseren Herstellungskosten wegen sich bisher auf Hausthüren, Thorwege und bessere innere Thüren beschränkte. Es sind dies:

46 III. Die Verbindungen der Hölzer.

Fig. 128 und 130. Die überschobene Füllung, und zwar nach innen oder nach aussen.

Fig. 131. Aufgeleimter Kehlstoss auf nach innen überschobener Füllung, wobei besonders und wiederholt darauf aufmerksam gemacht wird, dass die Befestigung des ersteren nur auf dem einen Teil, dem Fries oder Rahmenholz, erfolgen darf, um das Reissen zu verhindern.

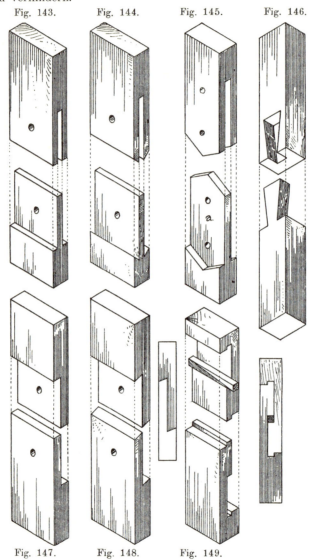

Fig. 143 bis 149. Holzverbindungen nach der Länge.

Fig. 132. Ueberschobener Kehlstoss in der Nute.

Fig. 133. Doppelt nach innen überschoben mit einem aufgelegten Kehlstoss.

Die Figuren 134 bis 139 zeigen eingeschobene Füllungen mit Konstruktionen, die von der gewöhnlichen Art ebenfalls abweichen.

Fig. 134. Eingeschobene Füllung in einem eingeschobenen, zweiten Fries.

Fig. 135. Eingeschobener Kehlstoss in der Nute, zwischen Fries und Füllung.

Fig. 136. Wie 134, aber mit aufgesetzten Kehlstössen an der Verbindungsstelle beider Friese.

Fig. 137. Eingeschobene Füllung mit beiderseits aufgesetzten Kehlstössen.

Fig. 138 und 139. Eingeschobener Federrahmen in der Nute, mit aufgesetzten Kehlstössen. Die letzteren sitzen auf dem Rahmen; Fries und Füllung sind also beim Arbeiten des Holzes von einander unabhängig. Der geschlitzte Federrahmen verhindert das Durchsichtigwerden der Gehrungen.

Die grossen Bauschreinereien haben die üblichen Verbindungen noch um einige weitere vermehrt. Der erhöhte Arbeitsaufwand, welchen diese zum Teil sehr sinnreichen und zweckmässigen Konstruktionen erfordern, kommt eben beim Dampf- und Maschinenbetrieb weniger fühlbar zur Geltung als in der Kleinschreinerei. Wir entnehmen dem Musterbuch der Firma Billing & Zoller in Karlsruhe, deren Baugeschäft weit über Baden hinaus bekannt ist, die Verbindungen der Figuren 140 bis 142. Alle drei verhindern das Durchsichtigwerden der Kehlstossgehrungen. Fig. 140 zeigt die Verbindung eines Querfrieses mit dem Höhenfries. Der Zapfen behält hierbei zum Unterschied von der gewöhnlichen Konstruktion (Fig. 127) die ganze Friesbreite. Die Solidität dieser Verbindung leuchtet ohne weiteres ein. Ebenso beachtenswert sind die Eckverbindungen der zwischen Fries und Füllung in die Nute eingesetzten, einen verzapften Rahmen bildenden Kehlstösse der Figuren 141 und 142. Die Gehrungen bleiben undurchsichtig; die Kehlstösse können für sich allein exakter gearbeitet werden; sie sind nicht an die Friesstärke gebunden und sie können aus besserem Holze hergestellt werden.

3. Holzverbindungen nach der Länge.

Dieselben werden nur selten ausgeführt, da die gewöhnliche Dielenlänge für die meisten Schreinerarbeiten ausreicht. Niemand wird es einfallen, zwei Hölzer nach der Länge künstlich miteinander zu verbinden, wenn ihm die Möglichkeit geboten ist, das Gewünschte aus einem Stück zu erhalten, indem er es von einem ganzen Dielen abschneidet. Der einzige Vorteil, der allenfalls dazu verleiten könnte, liegt in der Verwertbarkeit von Abschnitten oder Abfällen. Was dabei aber an Holz erspart wird, geht an Arbeitszeit verloren, abgesehen von der geringeren Festigkeit. Je mehr man in neuerer Zeit den Wert der Arbeit gegenüber dem des Materials schätzen gelernt hat, umsomehr ist man von diesen Verbindungen abgekommen, so dass dieselben hier nur der Vollständigkeit wegen erwähnt werden.

Fig. 143: Zusammengeschlitzt mit gerade abgesetztem Zapfen. Fig. 144: Zusammengeschlitzt mit schräg abgesetztem Zapfen. Fig. 145: Zusammengeschlitzt mit Spitzzapfen. Fig. 146: Auf Schwalbenschwanz verlängert. Fig. 147: Ueberplattung, gerade abgesetzt. Fig. 148: Ueberplattung, schräg abgesetzt. Fig. 149: Hakenblatt mit eingetriebenem Keil.

4. Eckverbindungen.

Wir unterscheiden
a) Eckverbindungen, die in einer Ebene,
b) Eckverbindungen, die in zwei verschiedenen Ebenen liegen.

Dieselben sind mannigfacher Art und finden bei allen Arbeiten Anwendung. Je nach der Bestimmung der letzteren giebt man ihnen geringere oder erhöhte Festigkeit und im Verhältnis zu dieser steht der Arbeitsaufwand, welchen dieselben verursachen.

a) Eckverbindungen in einer Ebene.

Das stumpfe Eck. Die stumpfe Ecke (auf stumpfe Gehrung Fig. 150) entsteht, wenn man die beiden zu verbindenden Hölzer auf Gehrung zusammenschneidet und leimt. Die Festigkeit dieses Verbandes ist eine verhältnismässig geringe, genügt aber immerhin für einzelne Arbeiten, wie z. B. kleine Bilderrahmen. Will man deren Festigkeit erhöhen, so bringt man auf der Gehrungsfläche Holzdübel an oder man macht mit der Säge übereck einen oder mehrere Schnitte, in welche man sog. Federn, d. h. dünne Holzplättchen, Furniere, einleimt (Fig. 158), wenn man nicht vorzieht, die Verbindung zu nageln oder zu verschrauben.

Ueberplattungen, Verplattungen.
    a) Das rechtwinkelig überplattete Eck. Für einfache Arbeiten (Fig. 151).
    b) Das auf Gehrung überplattete Eck (Fig. 157). Es findet hauptsächlich Anwendung bei Thürverkleidungen.
    c) Die kreuzweise Ueberplattung (Fig. 156).

Sämtliche drei Verbindungen werden geleimt und mit Holznägeln versehen.

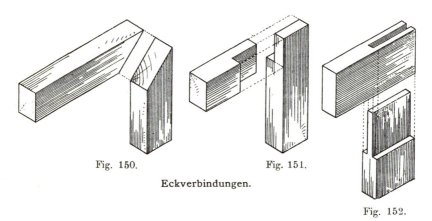

Fig. 150.   Fig. 151.

Eckverbindungen.

Fig. 152.

Fig. 150. Auf stumpfe Gehrung.   Fig. 151. Ueberplattet, verplattet.   Fig. 152. Zusammengeschlitzt.

Das zusammengeschlitzte Eck. Es bietet wesentlich grössere Festigkeit als die Ueberplattung. Wir unterscheiden:
    a) Einfach geschlitzt.

Die Zapfendicke ist gleich einem Drittel der Holzstärke, die Zapfenbreite gleich der Friesbreite (Fig. 152).
    b) Doppelt oder dreifach geschlitzt.

Es ist dies eine Verbindung, welche grosse Festigkeit giebt. Anwendung: bei Schraubzwingen etc. (Fig. 153).
    c) Auf Gehrung geschlitzt,
eine Konstruktion, die besser ist als das stumpfe, aber geringwertiger als das geschlitzte Eck. Sie wird angewendet, wo die Oberfläche ein besseres Aussehen erhalten soll (Fig. 154).

Die Zapfen und Schlitze dieser drei Verbindungen werden gut geleimt und bis zur Erhärtung des Leims mittels Schraubzwingen fest aneinander gepresst.

Weitaus die beste derartige Verbindung ist das gestemmte Eck mit Nut- oder Federzapfen (Fig. 122 und 155) und demgemäss auch die für Bauarbeiten am meisten angewendete.

4. Eckverbindungen.

Hier ist die Zapfendicke ebenfalls allgemein gleich einem Drittel der Holzstärke, richtet sich aber nach den vorhandenen Lochbeiteln, während die Breite etwa zwei Drittel der Friesbreite beträgt.

Die verminderte Zapfenbreite ist geboten:
1) um noch etwas Holz an dem Ende des mit dem Zapfenloch versehenen Friesstückes zu erhalten behufs Eintreibung der Holzkeile;
2) weil Zapfen über 8 cm Breite bei nicht ganz trockenem Holz ebenfalls wieder etwas schwinden und dadurch möglicherweise locker werden können.

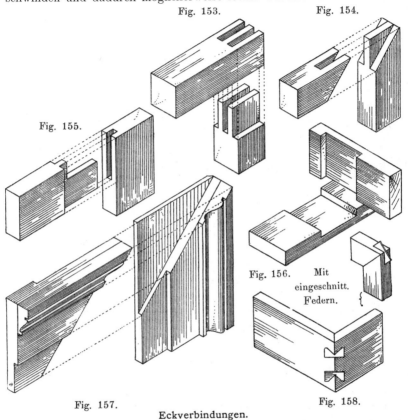

Fig. 153. Fig. 154. Fig. 155. Fig. 156. Mit eingeschnitt. Federn. Fig. 157. Fig. 158.

Eckverbindungen.
Fig. 153 und 154. Zusammengeschlitzt. Fig. 155. Gestemmt mit Nut- oder Federzapfen.
Fig. 156 und 157. Ueberplattet, verplattet. Fig. 158. Auf stumpfe Gehrung.

Um aber dem mit dem Zapfen versehenen Friesstück grössere Festigkeit gegen Abbrechen zu geben, lässt man an demselben auf den Rest der Friesbreite noch einen kleinen, etwa $1^{1}/_{2}$ bis 2 cm langen Zapfen, den sog. Nut- oder Federzapfen, stehen, welcher in eine entsprechende Nute des anderen Frieses eingreift und wie in Fig. 155, zweckmässiger wie in Fig. 159 geformt wird.

Man macht bei einer Friesbreite von 12 bis 13 cm den eigentlichen Zapfen nur etwa 8 cm breit, so dass der Nutzapfen noch 4 bis 5 cm breit wird; ist die Friesbreite wesentlich grösser, z. B. Dielenbreite, also 26 bis 29 cm (Fig. 162), so macht man zwei Zapfen zu 6 bis 8 cm und drei Nutzapfen. Figur 160 und 161 veranschaulichen die Zapfenformen bei Mittelfriesen. Das Zapfenloch wird nach aussen etwas weiter gestemmt, damit die Möglichkeit geboten ist, die Keile einzutreiben. Dieselben, je zwei auf einen Zapfen, fassen denselben von aussen und verspannen ihn (Fig. 159 und 160). Man hat auch versucht, das Gleiche zu erreichen mit einem Keil, den man in die Mitte des Zapfens bezw. des Frieses eintrieb (Fig. 161 unten); da hierbei aber nicht selten die Friese auseinandergesprengt wurden, so ist diese Keilung heute nicht mehr üblich. Dagegen hat sich eine andere Art der Verkeilung als sehr praktisch erwiesen, nämlich die, zwei Keile in den Zapfen in je einer Entfernung von etwa 1 cm von aussen einzutreiben, wodurch der Zapfen vierfach durch den Keil gefasst wird (Fig. 162). Das Mass von 1 cm genügt, um zu verhindern, dass die schmalen Teile des Zapfens, gute Arbeit natürlich vorausgesetzt, nicht abbrechen. Beim Verkeilen ist hauptsächlich deswegen Vorsicht geboten, weil das Holz des Zapfenloches leicht gesprengt wird, und praktische Leute setzen daher vorher eine Schraubzwinge mit zwei Unterlagen an die gefährdete Stelle.

Die Zapfen und Keile werden geleimt, ausnahmsweise kommen auch noch Holznägel zur Anwendung.

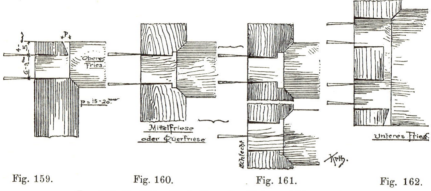

Fig. 159.   Fig. 160.   Fig. 161.   Fig. 162.
Fig. 159 bis 162. **Verkeilen der Rahmenhölzerzapfen.**

b) Eckverbindungen in zwei Ebenen.

Das Verzapfen (Fig. 163a). Eine einfache, aber häufig angewendete Verbindungsweise, wobei an das eine Holz die Zapfen angeschnitten, an das andere korrespondierend die Zapfenlöcher gestemmt werden. Die Zapfen werden geleimt und von aussen verkeilt, wobei zu beachten ist, dass die Keile so eingesetzt werden, dass sie nur nach der Längsfaserrichtung des Holzes einen Druck ausüben.

Sollen zwei Hölzer an den Enden mit einander verbunden werden, so kann man dies ebenfalls durch Verzapfen bewirken; vorteilhafter dagegen ist

das Verzinken. Diese Verbindung erfordert kaum mehr Arbeit als eine Verzapfung, erlangt aber durch die schwalbenschwanzartige Form der Zinken grössere Festigkeit. Die Schräge der Zinken ist verschieden, doch giebt die Form derselben einen Gradmesser ab für die Genauigkeit der Arbeit, ebenso wie die Zinkung überhaupt als eine Art Zeugnis für den Arbeiter betrachtet werden kann. Während nämlich Anfänger und ungenaue Arbeiter durch möglichst schräge Zinken zu ersetzen beabsichtigen, was ihrer Konstruktion an Genauigkeit abgeht, schneiden gute Arbeiter dieselben nur ganz wenig schräg, fast gerade, und erzielen damit selbst ohne Leim dennoch eine bessere Verbindung als die ersteren mit Hilfe von Leim und Holzkeilchen.

Kommt es auf das Aussehen nicht an, so verwendet man die gewöhnliche Zinkung (Fig. 163b), bei welcher auf den beiden äusseren Seiten Hirnholzflächen zu Tage treten; ist dies auf einer Seite nicht erwünscht, so wählt man die verdeckte Zinkung (Fig. 163d) und, wenn auf beiden Seiten keine Zinken bemerkt werden sollen, die

Verzinkung auf Gehrung (Fig. 163f). Die Zinkenbildung der beiden letzteren ist wie bei der gewöhnlichen Zinkung, nur bleibt bei dem einen Teil oder bei beiden Teilen ein Rest des Holzes stehen, welcher die Zinken verdeckt.

Die Figur 164 zeigt in a, b und c ebenfalls die drei Arten der Verzinkung nebst der Einteilung, nach welcher die Zinken aufgerissen werden.

Auf Grat eingeschoben ist ähnlich wie die Konstruktion der Grat- und Einschiebleiste. Die betreffenden Konstruktionen, Fig. 163c und g, werden nicht geleimt.

Auf eine anderweitige, ergänzende Befestigung angewiesen und nur für ganz bestimmte Zwecke geeignet sind die auf Nut eingeschobenen Verbindungen (Fig. 163e und h).

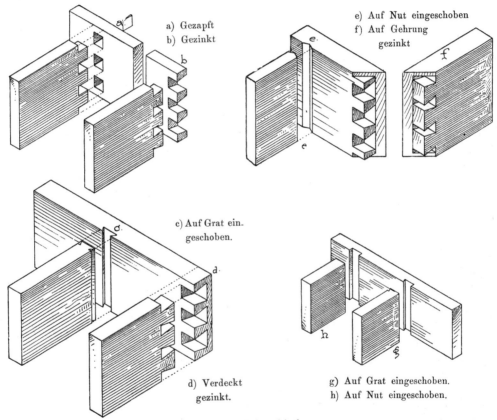

a) Gezapft
b) Gezinkt
c) Auf Grat eingeschoben.
d) Verdeckt gezinkt.
e) Auf Nut eingeschoben
f) Auf Gehrung gezinkt
g) Auf Grat eingeschoben.
h) Auf Nut eingeschoben.

Fig. 163. Eckverbindungen.

5. Hilfsmittel zur Verbindung.
Verbindung durch Holznägel.

Die Holznägel (Fig. 165), welche in der Bauschreinerei zur Verwendung kommen, werden meist aus dem Holz der Palm- oder Salweide (Salix Caprea *L.*) gefertigt, welches sehr zäh, dabei aber doch gut zu schneiden ist. Es sind dies ca. 6 bis 7 cm lange, schwach verjüngte vierseitige

Prismen von 1 cm Dicke; zwei gegenüberstehende Kanten sind gebrochen; die Spitze ist durch drei kurze Schnitte hergestellt. Nach dem Vorbohren taucht man die Nägel in heissen Leim, schlägt sie ein und hobelt das vorstehende Ende nach dem Trocknen des Leims bündig.

Verbindungen durch Metall.
(Nägel, Drahtstiften, Schrauben.)

Nägel. Geschmiedete Nägel, Bodennägel, Lattennägel (Fig. 166) kommen in der Bauschreinerei heute nur höchst selten zur Verwendung, d. h. nur bei gröberen Arbeiten, bei welchen die Köpfe sehr gross sein und die Spitzen der Nägel umgeschlagen werden sollen, wie z. B. bei Latten- und Riementhüren. Zu dieser Arbeit eignen sie sich aber ihrer Pyramidenform und ihrer feinen Spitze wegen, welche über einen vorgehaltenen Spitzbohrer umgebogen und nun quer über die Holzfaser eingeschlagen wird, vorzüglich und viel besser als die gleichdicken cylindrischen Drahtstifte.

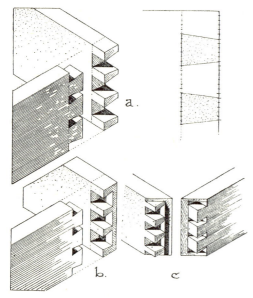

Fig. 164. **Die verschiedenen Arten der Verzinkung.**

Die Herstellung dieser Nägel geschah früher nur von Hand durch die Nagelschmiede, heute werden dieselben ebenfalls mittels Maschinen gefertigt und hauptsächlich zu Dachdecker- und Blechnerarbeiten verwendet; im übrigen sind sie durch die Drahtstifte fast vollständig verdrängt.

Drahtstifte. Unter Drahtstiften (Fig. 167) versteht man Nägel aus Eisen- oder Stahldraht, dreieckigen, quadratischen, meistens aber kreisrunden Querschnitts, an welchen unten eine Spitze und oben ein Kopf angepresst wird. Letzterer ist bei grösseren Stiften auf seiner Oberfläche geriefelt, um das Abgleiten des Hammers zu verhindern. Die Dimensionen sind sehr verschieden (je nach dem Zweck der zu fertigenden Arbeit), und jede Eisenhandlung führt Stifte von $1/2$ bis 4 mm Durchmesser und von 9 bis 150 mm Länge in unzähligen Abstufungen, welche fast ausschliesslich für Schreinerarbeiten bestimmt sind.

Die ganz feinen und kurzen Sorten bezeichnet man allgemein als Furnierstifte, während man die stärkeren einfach Stifte von bestimmter (in mm anzugebender) Länge, z. B. »Stifte, 65 mm lang« nennt. Gekauft werden die meisten Sorten in kleineren oder grösseren Packeten oder in Kistchen, im grossen aber immer nach dem Gewicht.

Nicht ganz verständlich ist die geringe Verwendung der dreikantigen kannelierten Stahldrahtstifte von J. C. Havemann in Berlin, bezw. die Abneigung vieler süddeutscher Meister gegen diese Stifte, welche doch wesentlich billiger sind als die cylindrischen. In unserer Zeit der ständigen Neuerungen und Erfindungen ist ein solches Vorurteil geradezu unerklärlich, umsomehr, als die Mehrzahl der Meister »der Spur nachredet«. Die wenigsten haben eine Probe mit denselben gemacht und sie auf ihre Brauchbarkeit geprüft; sie begnügen sich, zu sagen: »Sie biegen sich leicht um und damit fertig«. Hätten sie dieselben eines Versuchs wert gehalten, so würden sie gefunden haben, dass man bei einiger Vorsicht sie ganz wohl in das Holz einschlagen kann, während sinnlos daraufschlagende Arbeiter, wie die Erfahrung lehrt, auch cylindrische

5. Hilfsmittel zur Verbindung.

Drahtstifte gar leicht umschlagen. Vielleicht geben diese Zeilen dem einen oder anderen Meister Veranlassung, eine Probe vorzunehmen, damit er sich sein eigenes Urteil bilde.

Holzschrauben (eiserne Schrauben für Holz, Fig. 168). Während die Nägel und Stifte, welche mit dem Hammer eingeschlagen werden, die Holzfasern auseinanderpressen, die Festigkeit der Konstruktion also von der Reibung des Nagels an der Holzfaser abhängig ist, bohrt sich das Gewinde der Holzschraube gewissermassen in die Holzfaser selbst ein und fasst sie fest. Ganz abgesehen davon, dass die Verbindung eine festere wird, macht schon die Gewissheit und Sicherheit, eine Konstruktion zu haben, die nicht von Zufälligkeiten abhängig ist, und die man nötigenfalls leicht lösen kann, die Schrauben zu einem wertvollen Verbindungsmittel. Die Ausführung der Arbeit ist allerdings etwas umständlicher als die beim Nageln. Der schwach verjüngten, mehr der Cylinder- als der Kegelform sich nähernden Gestalt der Schraube wegen ist das Vorbohren eines Loches oder das Einschlagen eines solchen mit dem Spitzbohrer, einem dem Aktenstecher ähnlichen Instrument, erforderlich, damit die Schraube das Holz fassen kann. Von der Art dieses Vorbohrens hängt die Solidität der Arbeit sehr wesentlich ab. Die Vorbohrung darf nur so stark sein, als der Kern der Schraube an der Spitze.

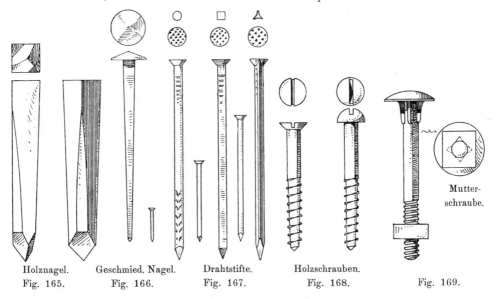

Holznagel.	Geschmied. Nagel.	Drahtstifte.	Holzschrauben.	
Fig. 165.	Fig. 166.	Fig. 167.	Fig. 168.	Fig. 169.

Mutterschraube.

Die Köpfe der Schrauben sind entweder
    flach (versenkte Schrauben) oder
    halbrund (halbrunde Schrauben).

Beide sind mit dem Einschnitt für den Schraubenzieher versehen. Vor dem Einsetzen der versenkten Holzschrauben ist das betreffende Loch auszureiben, damit der Schraubenkopf nach Vollendung der Arbeit bündig mit dem Holz ist. Es sei noch bemerkt, dass Holzschrauben, welche mit dem Hammer eingeschlagen werden, weniger Festigkeit gewähren, als Nägel, da die Holzfaser durch das Gewinde zerrissen und die durch die Spannung des Holzes verursachte Reibung geringer wird. Nicht unerwähnt darf bleiben, dass unsere heutige Industrie auch sogen. »Schrauben ohne Gewinde« auf den Markt bringt. Diese Schrauben haben nur einen Kopf mit Einschnitt, sind aber sonst glatt und müssen daher mit dem Hammer eingeschlagen werden. Sie sind allerdings billiger als Schrauben aber weiter nichts als Nägel und ihre Verwendung bezweckt nur, die Meinung zu erwecken, als seien wirkliche Schrauben verwendet. Man erkennt den Schwindel am Fehlen der Beschädigungen, welche bei wirklichen Schrauben der Angriff des Schraubenziehers an den Einschnitten hervorbringt. Der Vorwurf des Betrugs und Schwindels trifft weniger den Fabrikanten, als den Schreiner, der die gewindlose Schraube verwendet.

Mutterschrauben (Fig. 169) finden in der Bauschreinerei weniger Verwendung als Holzschrauben. Man benützt sie meistens nur bei Beschlägen, bei Befestigung der Thürbänder, und zwar derart, dass man der grösseren Solidität wegen den dem Kloben zunächst liegenden Teil des Bandes mit einem durchgehenden Schraubenbolzen versieht, dessen Mutter auf die innere Seite der Thür zu liegen kommt.

Verbindungen durch Leim.

Unter Leim versteht man im allgemeinen eine aus tierischen Abfällen der Gerbereien und Schlächtereien, aus Sehnen, Knorpel, Knochen etc., durch Kochen mit Wasser gewonnene Substanz. Die zu Schreinerzwecken verwendeten Arten sind:

1) Lederleim.
2) Knochenleim.

Von der Qualität der Rohstoffe hängt diejenige des Leims ab.

Die besten derselben für Lederleim sind:

Das sogen. Rindsleimleder (aus Vachette-Gerbereien) mit etwas Kalbsköpfen; minderwertig ist Schafs-, Ziegen- und Pferdeleimleder.

Zu Knochenleim werden Knorpel und Knochen verwendet.

Die Rohstoffe werden behufs Ausscheidung der unbrauchbaren Teile zunächst mit Kalk behandelt und kräftig gewaschen; hierauf wird das Wasser ausgepresst und das Ganze gekocht. Der flüssige Leim wird sodann in sogen. Klärbassins abgelassen; hierauf werden demselben künstlich ca. 25 bis 30 % Wasser entzogen und nun wird die schon etwas dickflüssige Masse in kleine Kästchen zwecks völliger Erkaltung abgelassen. Die hierbei gewonnenen Kuchen werden mit der Maschine in die bekannten Leimtafeln zerschnitten (woher auch der Name »Leimschnitten«) und auf Hanfnetzen zum Trocknen (auf natürlichem oder künstlichem Wege) ausgebreitet.

Lederleim aus Rindsleimleder ist der beste und, wenn auch scheinbar der teuerste, für den Schreiner doch eigentlich der billigste. Leider lässt er sich äusserlich von minderwertigen Sorten ohne Probe nicht leicht unterscheiden, ein Umstand, der dem Betrug Thür und Thor öffnet, so dass allerlei Mischungen und Zusammensetzungen als Lederleim erster Qualität in den Handel kommen. Nicht nur Leim aus Schaf- und Ziegenleimleder wird als Leim aus Rindsleimleder angeboten; auch Mischungen beider mit Knochenleim segeln unter der gleichen Flagge. Letzteren unterscheidet der erfahrene Praktiker vom guten Lederleim, indem er eine Leimschnitte mehrmals kräftig anhaucht und sie dicht vor die Nase hält. Ist der dabei entstehende Geruch ein übler, so ist gewöhnlich auf Knochen- oder doch minderwertigen Lederleim zu schliessen. Ein weiteres Merkmal ist der beim Kochen des Knochenleims hervortretende charakteristische Geruch und das geringe Wasseraufnahmevermögen.

Vor dem Kochen muss nämlich der Leim 2 bis 3 mal 24 Stunden in kaltem Wasser eingeweicht werden, wobei er jedoch nur quellen darf. Dagegen muss er, ans Feuer gesetzt, sofort sich auflösen. Bester Leim muss 5 bis 6 mal so viel Wasser einziehen, als er selbst wiegt, und dabei eine zähe, gallertartige Masse bleiben.

Guter Leim soll eine schöne hellgelbe bis gelblich-braune Farbe haben und durchscheinend, ohne Flecken, Blasen und dergl. sein. In früherer Zeit zog man den hellen Leim dem dunklen vor; heute ist dies fast umgekehrt. So ist z. B. der in Süddeutschland in sehr vielen Geschäften eingeführte Leim von F. W. Weiss & Sohn in Hilchenbach, obgleich von vorzüglicher Qualität, doch von etwas dunklerer Farbe als der früher so beliebte Kölner Leim.

Behufs Verwendung wird der Leim nach dem Abkochen in die sogen. Leimpfannen gebracht und in diesen gut warm gehalten. Dieses Warmhalten geschah früher in primitivster Weise, indem man die mit drei ziemlich hohen Füssen versehenen Gusseisenpfannen auf eine Feuerstelle brachte und mittels einiger Hobelspäne Feuer anzündete. Wenn dies in kleinen Geschäften vereinzelt auch heute noch geschieht, so ist man doch fast allgemein davon abgekommen und hat eine Einrichtung in der Weise getroffen, dass man die Leimpfannen aus Blech fertigt, diese dann (drei, vier und mehr zusammen) in einen mit Wasser gefüllten Blechkasten

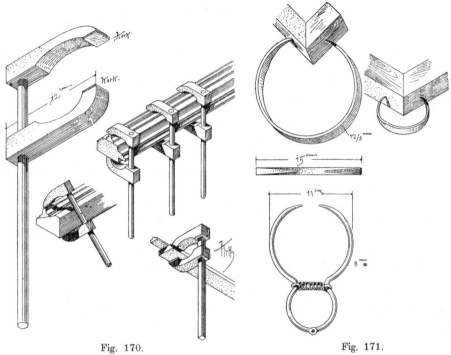

Fig. 170.
Hölzerne Knechte, sog. Sergeanten.

Fig. 171.
Spannringe aus Stahl.

einhängt, dieses Wasser und damit zugleich die Leimpfannen und den Leim erwärmt. Hierdurch wird, ganz abgesehen von dem durch das öftere Feueranmachen verursachten grossen Zeitverlust und dem Heizmaterialverbrauch, dem Verbrennen des Leims vorgebeugt. Die Heizung geschieht entweder durch Hobelspäne, welche des langsameren, stetigeren Brennens oder Glimmens wegen mit Sägespänen gemischt werden, oder einfacher, aber teurer durch Leuchtgas. Frisch abgekochter Leim ist der beste; durch das öftere Anwärmen und durch den unmöglich fernzuhaltenden Staub lässt seine Bindekraft bald wesentlich nach, worauf bei besseren Arbeiten zu achten ist. Ueberhaupt sollen in jedem Geschäft von Zeit zu Zeit sämtliche Leimpfannen ausgekocht werden, um den darin angesetzten Schmutz und verdorbenen Leim zu beseitigen, damit er den neu zugebrachten Leim nicht verderbe. Die scheinbare Verschwendung wird durch die wesentlich bessere Qualität der Leimfugen ausgeglichen.

Die Konsistenz des Leims ist je nach der zu fertigenden Arbeit verschieden. Für Fugen bei weichen Hölzern nimmt man den Leim schwach, bei mittelharten etwas stärker und bei harten

Hölzern ganz stark. Zum Verleimen grösserer Tafeln oder zum Furnieren wählt man einen mittelstarken, eher schwachen als starken Leim. Von grösster Wichtigkeit ist, dass die zu verleimenden Teile gut gefügt oder abgerichtet, schön gezahnt (d. h. auf der Oberfläche mit dem Zahnhobel rauh gemacht), und namentlich gut gewärmt sind. Grossporige Hölzer oder Hirnholzflächen tränkt man vor dem Leimen mit Leimwasser, d. h. man bestreicht sie mit einer dünnen Leimlösung (zu deren Herstellung aber nicht, wie vielfach üblich, alter verdorbener Leim benützt werden sollte), um die Poren zu füllen; nach dem Trocknen des Auftrags und nochmaligem leichten Abzahnen der Fläche wird dann geleimt.

Das Leimen selbst muss so rasch als möglich und, wo irgend thunlich, in gut gewärmtem Raum geschehen; je rascher und je wärmer dieses Geschäft erledigt wird, desto besser die Arbeit. Dem entsprechend werden in allen soliden Geschäften bei bedeutenderen Stücken alle möglichen Vorkehrungen getroffen, um jede Verzögerung zu verhindern. Die Hobelbank wird aufgeräumt und abgekehrt, wenn nötig auch der Boden davor und rings um dieselbe; die Schraubzwingen, Schraubknechte und Furnierböcke werden gerichtet und man beseitigt alle Gegenstände, welche sich allenfalls beim Leimen als Hindernis erweisen könnten. Nachdem die gut gewärmten Holzteile nicht zu stark mit entsprechendem Leim gestrichen sind, werden sie mittels der Schraubzwingen oder der Hobelbank etc. fest aneinander gepresst. Nach 4 bis 10 Stunden ist die Leimung erhärtet und fertig. Ist die Verbindung richtig, so soll nach dem Abputzen der geleimten Teile der Leim in der Fuge nicht sichtbar sein; da aber dies bei hellen oder ganz weissen Hölzern bei grösster Vorsicht nicht immer möglich ist, so benützt man in diesem Fall zweckmässig den sogen. russischen Leim, welchem durch einen chemischen Zusatz eine weisse Farbe gegeben ist.

Auf Fett haftet Leim nicht; aus diesem Grunde bestreicht man oft Schraubzwingen und Zulagen mit Seife, während man von den zu leimenden Flächen Fett sorgfältig fern hält. Ein geringer Zusatz von pulverisierter Kreide und Leinölfirniss macht den Leim wesentlich widerstandsfähiger, soweit es sich um atmosphärische Einwirkungen im Freien handelt.

Die beim Leimen zur Verwendung kommenden Schraubzwingen sind in Fig. 45 bis 48 weiter vorn abgebildet. Die hier eingereihte Figur 170 zeigt die Benützung der sogen. Sergeanten, d. s. hölzerne Schraubknechte mit Momentspannung. (Ein leichter Schlag auf den beweglichen, federnden Arm genügt zur Spannung und zur Lösung). Die Figur 171 veranschaulicht das Anlegen von Spannringen aus Stahl, üblich beim Leimen kleinerer Gegenstände, z. B. der stumpf gestossenen Ecken von Bilderrähmchen.

IV. DIE FUSSBÖDEN.

(Tafel 1 und 2.)

Allgemeines. — 1. Der Blindboden. — 2. Der rauhe Dielenboden. — 3. Der gehobelte Dielenboden. — 4. Der Tafelfussboden. — 5. Der Friesboden. — 6. Der Riemenboden. — 7. Der Fischgrat- oder Kapuzinerboden. — 8. Desgleichen in Asphalt. — 9. Tafelparketten.

Allgemeines.

Die Anforderungen, welche wir an einen guten Fussboden stellen, sind zweifacher Natur: In gesundheitlicher Beziehung soll er einen vollständig dichten Beleg bilden, welcher, ohne lange nass zu bleiben, sich feucht von Staub und Schmutz reinigen lässt, welcher keine offenen Fugen zeigt oder mit der Zeit erhält, durch welche Staub und Ausdünstung ein- und ausdringen können. In technischer Beziehung soll der Boden schön eben, horizontal, hart und widerstandsfähig sein, weder nach der einen oder anderen Seite steigen oder fallen, weder Erhöhungen noch Vertiefungen zeigen und schliesslich ein schönes Aussehen haben.

Dass die Reinhaltung des Erdbodens für die menschliche Gesundheit von grösster Bedeutung ist, weiss heute fast jedermann; weniger allgemein, aber doch immerhin den technisch und wissenschaftlich gebildeten Kreisen bekannt ist der Nachteil, welchen die Ausdünstung selbst des gewöhnlichen, nicht durch besonderen Einfluss verseuchten Erdbodens auf unsere Gesundheit auszuüben vermag, sobald seine Ausdünstungen in unsere Wohnräume eindringen, in welchen wir durchschnittlich mehr als die Hälfte des Lebens zubringen. Nicht minder schädlich kann aber auch das Füllmaterial unserer Fussböden werden, wenn es nicht frei von verweslichen Stoffen oder nicht so dicht abgeschlossen ist, dass die im Staube und in der Luft vorhandenen Mikroorganismen am Eindringen und an der Bildung von unheilvollen Brutstätten verhindert werden. Absolute Dichtigkeit ist daher Hauptbedingung für unsere Fussböden, und wenn sie sich auch nicht immer erreichen lässt, so sollte sie in allen Fällen wenigstens ernstlich angestrebt werden. Für fast ebenso wichtig zur Erhaltung der Gesundheit und für dringend nötig wird von ärztlicher Seite aber auch die Entfernung des täglich sich bildenden Staubes in unseren Wohnräumen gehalten, also die gründliche Reinigung der Fussböden. Eine solche ist aber nur möglich, wenn sie nicht mittels des Besens, sondern mit einem feuchten Tuch (welches selbstredend immer wieder gründlich in reinem Wasser auszuwaschen ist) vorgenommen wird; denn beim Kehren wird viel Staub aufgewirbelt, welcher sich in die Luft erhebt, sich auf die Wände, Decken etc. lagert, oder sich wieder auf seinen alten Platz, den Boden setzt. Nur ein gewisser Teil, und zwar der kleinere, ist also wirklich entfernt worden. Beweis hierfür ist ausser dem vorhandenen, sichtbaren Staub nach der Reinigung der Geruch der Zimmerluft. Während dieselbe bei feuchter Reinigung (unter welcher nicht das sogen. Aufwaschen zu verstehen ist) frisch und erquickend ist, riecht man nach dem Kehren auf geraume Zeit den Staub. Bei feuchter Reinigung werden täglich Bakterien aus den Zimmern entfernt, bei ansteckenden Krankheiten ein grosser Vorteil, welcher dem Kehren abgeht. Sind offene Fugen im Boden vorhanden, so gerät beim Kehren sowohl als auch sonst schon viel Staub in die Fugen und lagert sich dort. Auf diese Weise können daselbst Krankheitskeime geradezu gezüchtet werden, wenn genügende Feuchtigkeit hinzukommt, wie dies beim Aufwaschen und Scheuern der Fall ist. Das rasche Trocknen von gestrupften (nass gescheuerten) und unter Wasser gesetzten Böden ist meist auf offene Fugen zurückzuführen. Der in denselben angesammelte Staub und das feine Bodenfüllmaterial wirken in diesem Falle wie ein Saugschwamm und so sind die denkbar günstigsten Entwickelungsbedingungen für eine Bakterienkultur geschaffen.

Es muss somit unser ganzes Augenmerk darauf gerichtet sein, zu verhindern, dass Fugen entstehen, und wenn dies unmöglich, Vorkehrung zu treffen, dass sie nicht offen bleiben. Um das erstere zu erreichen, ist bei allen Bodenarten Hauptbedingung: trockenes Holz; bei ungenügend getrocknetem Holz ist die beste Konstruktion vergeblich. Im übrigen hat man bis in die Neuzeit die beste Lösung in einem Boden gesucht, welcher aus möglichst wenig Stücken bestand und also auch nicht viele Fugen hatte. Als den vollkommensten Boden in dieser Beziehung konnte man somit den in allen Fugen verleimten Patentfussboden bezeichnen, welcher seinerzeit grosses Aufsehen machte. Derselbe wurde, da er nicht genagelt werden konnte, durch an die Balken angenagelte Laufleisten, in welche er eingeschoben war, gehalten, und es wäre gegen diese Art der Befestigung nichts einzuwenden gewesen, wenn sie sich in praxi nicht als sehr schwierig und kostspielig erwiesen hätte. Man kam denn auch bald von dieser Bodenart ab und verleimte nicht mehr sämtliche Dielen eines Zimmers, sondern nur je zwei zu einer Tafel, wonach der Boden den Namen Tafelfussboden erhielt. Heute ist man auch von ihm abgekommen, denn dieser Boden muss mit einer ganz ungeheuren Sorgfalt in der Trocknung des Holzes, wie beim Legen behandelt werden, wenn keine grossen Fugen entstehen sollen; kleinere entstehen immer, und diese sind dann offen, da die Fugen stumpf sind. Heute ist man zum Extrem übergegangen; man verleimt nicht nur die Dielen nicht mehr, sondern man trennt sie sogar nochmals in der Mitte in zwei Teile, in Riemen, spundet oder verbindet sie auf Nut und Feder mit einander und erzielt so den Riemenboden. Die Riemen sind nur ca. 10 bis 15 cm breit und können vermöge ihrer geringen Breite nur wenig schwinden. Es entstehen demnach im schlimmsten Fall nur ganz schwache Fugen, welche zudem nicht offen, sondern durch die Spundung oder Federung geschlossen sind.

Um einen guten Boden zu erhalten, ist zunächst erforderlich, ein gutes, solides Auflager für denselben zu schaffen. Dieses Auflager ist in den oberen Stockwerken das Gebälke. Im Erdgeschoss dagegen fehlen die Balken überall, wo keine Balkenkeller vorhanden sind. Hier werden über den Scheitel der Gewölbe hinweg sog. Bodenrippen oder Bodenlager gelegt und befestigt. Diese Bodenrippen, auch Ripphölzer oder kurzweg Rippen genannt, werden, wo es nicht auf den Preis ankommt, am zweckmässigsten aus gut getrocknetem Eichenholz (von dem aber der Splint zu entfernen ist) gefertigt; in den weitaus meisten Fällen aber begnügt man sich mit Tannen-, Fichten- oder Forlenrippen, welche ganz wesentlich billiger sind als solche aus Eichen, und deren Dauer für diesen Zweck fast die gleiche ist, wie die der letzteren. Die Stärke der Rippen ist, wo man sehr sparen muss, die der gewöhnlichen Rahmenschenkel, 9×9 cm, besser jedoch 10×10 oder 10×12 cm. Gelegt werden sie am zweckmässigsten in reinen, gewaschenen und gedörrten Kies, auf welchen eine Schichte reinen, trockenen Sandes zum dichten Anschluss und zur Schalldämpfung aufgebracht wird. An den Enden werden sie gegen das

Mauerwerk verspannt, d. h. verkeilt. Da diese Befestigungsart aber im Widerspruch steht mit den Vorkehrungen, welche gegen Schwammbildung empfohlen werden — durch die Berührung mit der Mauer wird Feuchtigkeit in das Holz geleitet — so thut man gut, an die Mauer kleine, ca. 20 × 20 cm grosse Stücke von Asphalt-Isolierpappe anzulegen und gegen diese die Enden der Rippen zu verkeilen (Fig. 172). Eine andere Art der Befestigung ist die, in das Gewölbe verschieden lange Steinschrauben einzulassen, die Rippen zu durchbohren und mit der Unterlage zu verschrauben. Die Solidität der Konstruktion leuchtet sofort ein, doch wird sie der kostspieligen Ausführung wegen nur selten gewählt. Auf irgend welche Weise muss die Befestigung erfolgen, ein blosses Einbetten in Kies oder Sand genügt nicht, um auf ein solches Unterlager einen Boden eben legen zu können und eben zu erhalten. Die Entfernung der Bodenrippen von einander soll 60 cm von Mitte zu Mitte nicht übersteigen. Schliesslich wird hier noch besonders betont, dass im Erdgeschoss die Sorgfalt in der Ausführung die grösste sein muss, da hier die Gefahr einer Schwammbildung am nächsten liegt. Entwässerung des Hauses und Unterkellerung desselben, trockenes Mauerwerk, trockenes Füllmaterial und trockenes Holz sind unerlässliche Bedingungen, um letzteres gesund zu erhalten; will man noch ein übriges thun, so kann man die Rippen mit Kreosotöl oder Karbolineum anstreichen.

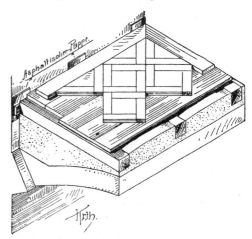

Fig. 172. **Das Legen der Bodenrippen.**

Den Holzfussboden auf die blosse Erde zu legen, ist verwerflich, da Fäulnis und Schwamm denselben in kurzer Zeit zerstören; dagegen lässt sich nichts einwenden, wenn ein ca 8 bis 10 cm starker Zementboden beschafft, der Holzboden auf diesen in Asphalt gelegt und ausserdem Vorkehrung getroffen wird, dass seitlich keine Feuchtigkeit an ihn gelangen kann. Wo es sich aber machen lässt, sollte man darauf bedacht sein, unter Räumen, welche aus irgend einem Grunde nicht unterkellert werden können, unter dem Gebälke einen zu lüftenden Hohlraum von wenigstens 50 bis 80 cm Höhe zu beschaffen, um das Holzwerk vor Zerstörung zu schützen und die schädlichen Bodendünste von den Wohnräumen abzuhalten.

Das Legen oder Verlegen der Rippen geschieht meist durch den Zimmermann, was zweckmässig ist, wenn derselbe auch den Boden zu legen hat; andernfalls ist es praktischer, dieses Geschäft dem Schreiner oder Bodenleger zuzuweisen. Denn wie in allen Dingen, wird nur derjenige, welcher auch die Folgen ungenauer Arbeit zu tragen hat, besorgt sein, seine Arbeit tadellos zu machen. Das Rippenlegen selbst wird erst unmittelbar vor dem Legen der Böden vorgenommen und bis dahin sollte auch kein Füllmaterial auf die Gewölbe gebracht werden, um sie recht austrocknen zu lassen.

Soll mit dem Legen des Bodens begonnen werden, so ist die erste Arbeit des Bauschreiners, sich genau zu vergewissern, ob das Lager, das Gebälke, genau im Blei liegt, ob aufgefüttert oder abgedächselt werden muss. Bei diesem Abbleien ist von der Treppe auszugehen; ihr Austritt ist die massgebende Stelle, nach welcher wohl oder übel der übrige Boden sich zu richten hat. Ergiebt die Untersuchung, die mit der Setzlatte und der Wasser- oder Bleiwage (woher der Ausdruck: im Blei liegen) ausgeführt wird, dass die Treppe mit der Hauptfläche des Bodens auf gleicher Höhe liegt, so wird der Rest aufgefüttert oder abgedächselt. Liegt das Gebälke unter sich richtig, der Treppenaustritt aber höher, so füttert man (wenn man den Aufwand nicht scheut) den ganzen Boden auf oder »verzieht« ihn andernfalls etwas nach der Wohnung hin, d. h. man lässt ihn in dieser Richtung leicht abfallen. Der umgekehrte Fall, in welchem das Gebälke höher liegt als die Treppe, ist weitaus schwieriger, da hier eigentlich nur Abdächseln des ganzen Gebälkes oder Steigen des Bodens nach der Wohnung Abhilfe schaffen können.

Aus diesen meist sehr mühevollen und kostspieligen Arbeiten, für deren Bezahlung ohnedies kein Mensch aufkommen will, erhellt die Wichtigkeit der richtigen Uebereinstimmung der Gebälkelage mit dem Treppenaustritt, insbesondere bei Steintreppen. Bei Holztreppen kommen Unterschiede weniger vor, indem das Mass zur Treppe im Bau selbst und zwar erst dann genommen wird, wenn das Gebälke bereits gelegt ist.

Nachdem alles im Blei ist, suchen die Bodenleger (Arbeiter, welche nur Boden legen oder wenigstens beständig in Neubauten arbeiten und die Schreinerarbeiten anschlagen), den Sand in den Balkenfächern zu ordnen, den überflüssigen zu entfernen und den fehlenden zu ergänzen. Grundsatz hierbei ist: den Sand nicht glatt über den Balkenfächern abzustreichen, sondern ihn in der Mitte etwas höher zu legen als an den Seiten, die Auffüllung also eine flachgewölbte Cylinderfläche bilden zu lassen, damit beim Nageln der Boden auch im Fach selbst in allen Punkten satt und dicht aufliegt und beim Begehen nicht hohl tönt (Fig. 173).

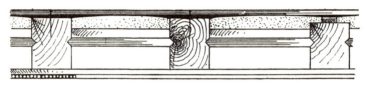

Fig. 173.
Auffüllung der Balkenfächer.

Der Wichtigkeit der Sache wegen wird hier nochmals betont, wie viel davon abhängt, dass die Stückung (Staakung) bezw. Wickelung gut trocken, das Auffüllmaterial ein reines, gut trockenes ist, also am besten gewaschener und auf Blechplatten über einem Feuer gerösteter Sand, aber keine Steinkohlenlösche und dergleichen. In die Balkenfächer soll, wie bereits gesagt, das Material erst eingebracht werden, wenn alles gut ausgetrocknet ist, also unmittelbar vor dem Bodenlegen.

Die Zeit zum Legen der Fussböden ist eine beschränkte. Wenn die Böden gut werden sollen, dürfen sie eigentlich nur in der wärmsten Jahreszeit, von Juni bis September, gelegt werden, und zwar unter allen Umständen erst dann, wenn der Verputz trocken ist und die Fenster eingesetzt und verglast sind. Zu allen andern Zeiten muss in dem bestimmten Raume mehr oder minder geheizt werden, damit das Holz nicht zu viel Feuchtigkeit aufnimmt und später schwindet. Wünschenswert ist, dass vor dem Bodenlegen die Zimmerdecken gestrichen oder gemalt sind, damit eine Beschädigung der Böden durch die Maler und ihre leider unvermeidlichen Farbentöpfe ausgeschlossen ist. Unbedingt erforderlich ist dies, wo die Böden naturfarben hell bleiben oder wo Parketten gelegt werden. In letzterem Fall legt man zuerst den Blindboden, lässt dann malen, und erst nachdem auch die Tapezierarbeiten vollendet sind, legt man das Parkett.

In Bezug auf die Konstruktion der Böden unterscheiden wir:
1) Blindboden, als Unterlage für Fischgrat- und Parkettboden;
2) Rauhen Dielenboden, stumpf oder gefälzt als Speicherboden;
3) Gehobelten Dielenboden, stumpf, gefälzt, gespundet und gefedert für Magdkammern, Speicherzimmer etc.

4) **Tafelfussboden**, gehobelt, stumpf;
5) **Friesboden**, Tafelfussboden mit Friesen;
6) **Riemenboden** (in langen Riemen), Schiffboden;
7) **Fischgrat-** oder **Kapuzinerboden** (kurze Riemen);
8) **Desgleichen in Asphalt**;
9) **Tafelparketten**.

1) **Blindboden** (Taf. 1 h und i). Er ist der einfachste Holzfussboden und dient als Unterlage für das Parkett oder die Riemen- und Friesböden. Er besteht aus 18 bis 22 cm breiten, 24 mm starken Brettern, welche weder gehobelt noch gefügt werden. Des Arbeitens wegen legt man ihn gern mit schwachen Fugen.

2) **Rauher Dielenboden** aus ordinären, 20 bis 25 cm breiten, 24 mm starken Brettern, welche entweder stumpf aneinander gestossen oder gefälzt werden, um zu verhindern, dass Staub durch die Fugen dringt und allenfalls aufzubringendes Getreide oder dergleichen durchfällt. Jeder Dielen wird auf jedem Balken mit je drei Nägeln befestigt; die einzelnen Dielen werden vorher fest aneinander gekeilt, da in den heissen Dachräumen immer ein erhöhtes Schwinden zu gewärtigen ist. Die Art der Nagelung ist in Figur 174 angedeutet.

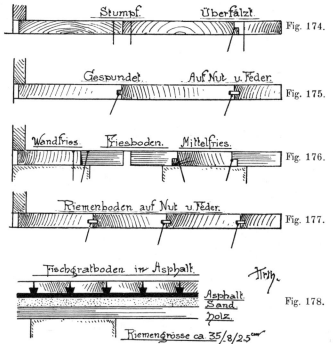

Fig. 174.
Fig. 175.
Fig. 176.
Fig. 177.
Fig. 178.

Fig. 174 bis 178. Fussboden-Konstruktionen.

3) **Gehobelter Dielenboden** (Tafel Ia 1). Er ist in Stärke und Breite wie der rauhe Dielenboden, im übrigen gehobelt und gefügt, also stumpf oder gefälzt wie Figur 174, gespundet oder gefedert wie Fig. 175. Die Dielen können ordinär oder halbrein sein, je nach dem Zweck des betreffenden Raumes. Bei der Nagelung ist darauf zu achten, dass sie des Aussehens halber in schön gerader Linie erfolgt. Die Nägel werden mit dem Versenker versenkt und die Löcher, nachdem etwaige kleine Unebenheiten an den Fugen mit dem Doppel- oder Verputzhobel ausgeglichen sind, sauber verkittet.

4) **Tafelfussboden** (Tafel 1 a 2 und b). Aus ganz- oder halbreinen, 24 mm starken und 24 bis 27 cm breiten, gehobelten, tannenen Brettern bestehend, von denen je zwei zu einer Tafel zusammengeleimt werden. Die Richtung, in welcher die Tafeln gelegt werden, ist meist durch das Gebälke bestimmt; wo dies nicht der Fall ist, wählt man sie so, dass man beim Eintritt ins Zimmer quer über die Holztafeln schreitet. Die Fuge zweier Tafeln ist eine stumpfe. Der Vorgang beim Legen ist der gleiche, wie der beim gehobelten Dielenboden: Nachdem der Sand in den Balkenfächern schön geebnet bezw. gewölbt ist, wird die erste Tafel sorgfältig aufgelegt, doch so, dass ihre Enden den Verputz der Wandflächen nicht berühren und hiernach auf ihre Breite mit 5 bis 6 Stiften genagelt. Dabei ist gleichfalls zu beachten, dass die Nägel

schön in gerader Linie und in gleichen (etwa 2½ bis 3 cm grossen) Abständen von den Fugen aus eingeschlagen werden, damit das Ganze sauber aussieht. Die Nagelköpfe werden versenkt, deren Löcher sauber verkittet. Zur besonderen Schönheit trägt es auch nicht bei, wenn die Hammerschläge auf und leider oft auch in dem Holz sichtbar sind und sich kaum durch Hobeln entfernen lassen!

Ist die erste Tafel gelegt, so wird mit dem Hammerstiel so viel Sand als möglich unter die Tafel hinuntergestopft, damit sie überall dicht aufliegt. Hierauf kommt die zweite Tafel an die Reihe, welche durch zwei Holzkeile (angelegt an eine in den Balken oder die Rippe eingeschlagene Eisenklammer) an die erst verlegte angepresst und nun wie diese genagelt und unterstopft wird (Fig. 179).

Auf diese Weise wird der ganze Boden gelegt. Neuerdings wird ein Fussboden-Legapparat empfohlen, den die Fig. 180 zur Abbildung bringt. Der Apparat wird auf den

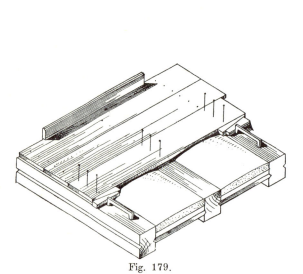

Fig. 179.
Das Festkeilen und Nageln des Fussbodens.

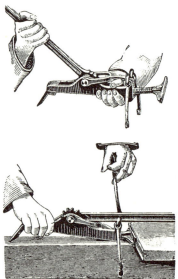

Fig. 180. Fussboden-Legapparat.

Boden aufgelegt; die seitlichen Klammern und die Klaue am hinteren Ende, auf welche mit dem Fuss getreten wird, halten ihn fest. Durch Umlegen des eisernen Hebels werden die Bretter aneinandergepresst (es werden mehrere zugleich gelegt und genagelt). Je nach der Länge der Bretter sind 2 bis 3 Apparate nötig. Die Firma E. Straub in Konstanz bietet den Apparat in zwei Ausstattungen an, zu 12,50 Mk. und 14 Mk. pro Stück.

Ist das Zimmer in seinen Dimensionen grösser als die gewöhnliche Dielenlänge von 4,50 m, also etwa 5,50 m lang, so hilft man sich, indem man an beiden Enden der 4,50 m langen Tafeln je eine solche quer legt, oder dass man die beiden Tafeln verschränkt stösst, d. h. nicht sämtliche Fugen auf einen Balken treffen lässt. Beide Legungsarten lassen in schönheitlicher Beziehung zu wünschen übrig, so dass man, wo die Mittel es irgendwie erlauben, lieber einen vollständigen Fries in Tafelbreite (Taf. 1b) herumlegt, so dass der Boden ein geschlossenes, umrahmtes Ganzes vorstellt. Es kann die letztere Art umsomehr empfohlen werden, als sich die Kosten gegenüber dem Boden mit den zwei Friesen nur unwesentlich höher belaufen. Was die Zweckmässigkeit und den Wert des Tafelfussbodens betrifft, so wurde schon eingangs des Abschnittes angedeutet, das die Glanzzeit desselben vorüber ist. Man fürchtet heute die sicher kommenden, mehr oder minder gross werdenden Fugen dieses Bodens und zahlt lieber einen etwas höheren Preis, um diese nicht zu erhalten. Betrachtet man ein solches Kunstwerk, bei dem auf je 50 bis 54 cm Holz eine Fuge von 2 bis 12 mm Breite kommt, in welche alles hineinfällt, was nicht soll, so wundert man sich nur, wie man sich so lange mit diesem Boden behelfen mochte, selbst wenn man von den eingangs angedeuteten gesundheitlichen Nachteilen absieht. Das Gleiche gilt von dem sog. ausgespänten Boden (ein Boden, bei dem Fugen künstlich durch Späne geschlossen sind, d. h, durch schmale, nach unten verjüngt gehobelte Holzleistchen, welche in die Fugen eingeleimt und genagelt werden), bei welchem die Späne beim Gehen klappern oder Anlass zum Stolpern geben.

5) **Friesboden.** Unter einem Friesboden versteht man einen Tafelfussboden, welcher von Friesen, schmäleren oder breiteren, ringsum an den Wänden oder auch kreuz und quer durch den Boden laufenden Holzstreifen eingefasst oder in regelmässige geometrische Formen geteilt ist. Die Verwendung dieser Friese ist eine sehr mannigfaltige. Man hat Friesböden mit Wand- oder Ortfriesen, also Friesen, die nur ein Gesamtfeld einfassen; sodann Böden mit Wand- und Kreuzfriesen (Taf. 1c), bei welchen der eigentliche Fond durch die beiden sich kreuzenden Friese in vier gleiche Teile zerlegt wird. Man hat ferner Böden wie Tafel 1d und 1e; ja man ging sogar zeitweise so weit, die Böden durch Friese so oft zu teilen, dass die dabei entstandenen Felder nicht mehr grösser als 70 bis 80 cm im Quadrat waren, wodurch der Uebergang zum Tafelparkett bewirkt war. Bis vor 25 Jahren zählte man den Friesboden zu den feinsten der Wohnhausböden, sein Aussehen war vielfach dem des Parketts gleich; heute hat man ihn verlassen und nur noch vereinzelt wird solch ein Boden gewünscht. Erst mit der allgemeinen Einführung neuerer, besserer Fussböden wurde das Publikum auf die Schwächen der Tafelböden wie der Friesböden aufmerksam gemacht, welche bei den letzteren allerdings anderer Natur waren als bei den ersteren. Während beim Tafelboden die grossen Fugen die Hauptursache seiner heutigen geringen Verwendung sind, ist diese beim Friesboden in dem hohen Preise, sowie in der Verschiedenartigkeit des Materials dieses Bodens zu suchen. Es leuchtet ein, dass ein Fussboden mit eichenen Friesen, dessen Felder mit reinem Tannen-Tafelholz sauber ausgelegt werden mussten, nur von den besten Arbeitern zufriedenstellend gefertigt werden konnte, da die Füllungen genau eingepasst werden mussten. Rechnet man hinzu das wertvolle Material, sowie den grossen Verschnitt (Abfälle, die nicht oder nur schwer zu verwenden sind), und schliesslich entweder ein ganzes Netz von Gebälkewechseln oder einen vollständigen Blindboden, so begreift man den hohen Preis des Bodens, welcher zu seiner Haltbarkeit in keinem richtigen Verhältnis stand. Durch die Verwendung verschieden harten Holzes (die weichen Felder traten sich sehr bald aus, und die Friese standen empor) war die Schönheit des Bodens bald dahin. Suchte man dieser wieder nachzuhelfen durch einen Ueberzug der Flächen, durch Bohnen, Wichsen oder gar durch Oelfarbenanstrich, so sah man von der schönen Arbeit so wenig wie von dem schönen Holze, und der grosse Aufwand war eigentlich umsonst. So kam es, dass auch der Friesboden heute nur noch historischen Wert hat. — Die Herstellung des Friesbodens ist im allgemeinen folgende: Nachdem das Unterlager für die sämtlichen Friese, soweit dieselben nicht auf Balken treffen, durch Einlegen von Wechseln (kleine Querbälkchen, welche von Balken zu Balken reichen), oder bei reicheren Friesböden zweckmässiger durch Legen eines Blindbodens geschaffen ist, werden die Friese genau nach Zeichnung und im Winkel gelegt, und zwar stumpf, wie Tafel 1c, d und e linke Seite, oder auf Gehrung, wie Tafel 1e rechte Seite. — Die letztere Art erfordert wesentlich mehr Holz und Arbeit, da durch die Gehrungsschnitte viel Friesholz verloren geht, ohne dass die Konstruktion solider wird. Die Nagelung erfolgt bei stumpfen Fugen von oben, bei gefälzten Friesen (Fig. 176) meist vom Falz aus. Der Zweck dabei ist, die Nagelung nicht

sehen zu lassen, doch darf man nicht vergessen dass diejenige der Füllungen auf alle Fälle von oben sichtbar bleibt, die Absicht somit nur halb erreicht wird. Ausserdem besitzt der Falz nicht genügende Festigkeit, um den Boden zu halten, und so nagelt man eben in allen Fällen, wo ein fester, sicherer Boden gewünscht wird, auch die Friese von oben. Das Zulegen der Felder geschieht in derselben Weise wie das Legen des Tafelbodens, nur ist dabei ein Augenmerk auf genaues Einpassen der Tafelstücke zu richten, deren Kanten man zweckmässig leicht unterstösst, d. h. schräg hobelt, damit nach dem Einlegen der Tafel die Fuge schön dicht wird. Der Arbeiter muss besorgt sein, die Kanten der Friese unbeschädigt zu erhalten (wobei sich das provisorische Aufnageln von Leisten bewährt), und die Nagelung der Tafeln, namentlich an den Hirnholzenden, mit der grössten Vorsicht auszuführen. Nagelt er sinnlos darauf los, so sprengt er die Tafelenden auseinander, was keineswegs zur Erhöhung der Schönheit beiträgt. Die Breite der Friese ist im allgemeinen 10 bis 15 cm; nach diesen haben sich die Lager oder Rippen zu richten, wobei bemerkt wird, dass die Tafeln doch mindestens ein Auflager von je $2^1/_2$ bis 3 cm erhalten müssen. Nachdem alle Felder zugelegt sind, werden auch die noch offenen Fensternischen mit Tafelfussboden versehen, zu dessen Nagelung, falls nicht Balken vorhanden sind, kleine Wechsel eingelegt und verspannt werden. Die Nagelköpfe werden vorsichtig versenkt, die vorhandenen Unebenheiten mit dem Verputzhobel ausgeglichen und die Nagellöcher ausgekittet. Bei den auf Tafel 1 abgebildeten Friesböden sind die Balken und Wechsel gestrichelt angedeutet.

6) **Riemenboden** (in langen Riemen), Schiffboden. Der Riemenboden ist zwar kein neu erfundener, wohl aber erst in der Neuzeit (seit etwa 30 Jahren) zur allgemeinen Verwendung gekommener Zimmerboden, welcher sich einer steigenden Beliebtheit erfreut. Die Riemen sind 10 bis 15 cm breit und 30 bis 35 mm stark; Bordstärke genügt nicht, um sie vor dem Durchbiegen oder Einschlagen zu schützen. Die Verbindung derselben ist entweder stumpf, gespundet oder auf Nut und Feder; die beiden letzteren Verbindungsarten sind vorzuziehen, weil sie den Riemen stärker und tragfähiger machen (richtiger: die Last auf verschiedene Riemen verteilen) und keinen Staub durchdringen lassen (Fig. 177).

Die Länge der Riemen ist meist 4,50 m. Ist der zu belegende oder zu dielende Raum länger als 4,50 m, z. B. 7,00 m, so werden die Riemen (sofern man nicht vorzieht, sie auf dieses Mass besonders zu bestellen) **verschränkt gestossen**, d. h. auf mindestens zwei Balken (Taf. 1f). Auf diese Weise kommen nicht nur sämtliche Fugen auf einen Balken, sondern sie werden auch fasst unsichtbar gemacht; sie verschwinden in den übrigen Fugen und Linien. Wird besonderer Wert auf schönes Aussehen gelegt, so bringt man ringsum **Wandfriese** an, die man auf Gehrung, zweckmässiger aber, aus dem beim Friesboden angegebenen Grunde, stumpf an den Ecken verbindet. Das Material ist Eichen-, Forlen- oder Tannenholz; die beiden letzteren Holzarten sind die gebräuchlichsten für Wohnhauszwecke, während die Eichenriemen sich als vorteilhaft für Wirtschaftslokale, Wartesäle etc., überhaupt für Räume bewährt haben, deren Fussböden sehr in Anspruch genommen werden. Bei Tannen- oder Forlenriemen sind diejenigen mit senkrechten Jahresringen die geeignetsten, deren Schneidart in Figur 11 dargestellt ist. Verlegt werden sie wie die Tafelfussböden. Nachdem der erste Riemen verlegt, in der Nute oder besser von oben genagelt und unterstopft ist, wird der zweite, in welchem die Feder (deren Kanten des leichteren Einstreifens wegen gebrochen sind) befestigt ist, in die entsprechende Nute des ersteren eingesteckt, mit dem Hammer und einer Zulage (ein glatt gehobeltes Brettstück, welches dazu dient, die Hammerschläge aufzunehmen) angetrieben, festgekeilt, sorgfältig genagelt und unterstopft. Nach Vollendung des ganzen Bodens und der Fensternischen (welche gewöhnlich auch mit Riemen zugelegt, in besseren Häusern auch gestemmt sind) wird der Boden mit dem Verputzhobel sauber verputzt und am besten sofort mit gekochtem Leinöl getränkt, wodurch verhindert wird, dass grober Schmutz in die Poren eindringt. Auf diese Weise bleibt der Boden schön hell und behält seine Naturfarbe, oder es kann das Oelen als Untergrund für den nachfolgenden Oelfarben- oder Lackanstrich gelten, wenn nicht vorgezogen wird, den Boden nochmals zu ölen und dann geölt zu belassen. Sämtliche drei Konservierungsarten haben für den Boden sowohl als auch für die Gesundheit der Bewohner grosse Vorteile, wenn auch der Oelfarbe und dem Lack weitaus der Vorzug über das Oel gebührt. Sämtliche gestatten eine feuchte Reinigung, ohne dass der Boden Wasser aufsaugt und längere Zeit feucht bleibt, wobei also die Holzfaser und somit der Boden selbst geschont wird. Es kann dieser Boden als einer der besten in jeder Beziehung bezeichnet und empfohlen werden.

Der Name Schiffboden stammt von den Schiffen, zu deren Deckung er sich längst bewährt hat.

7) **Fischgrat-, Kapuziner- oder Stabfussboden.** Er verlangt Blindboden und besteht aus 35 bis 60 cm langen, 6 bis 11 cm breiten und ca. 24 mm starken, auf Spundung oder Nut und Feder verbundenen Eichenholzriemen (Fig. 181), welche nicht parallel mit den Zimmer-

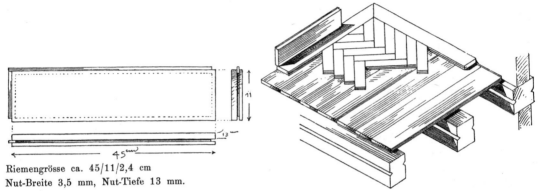

Riemengrösse ca. 45/11/2,4 cm
Nut-Breite 3,5 mm, Nut-Tiefe 13 mm.

Fig. 181. Riemen zum Fischgratboden. Fig. 182. Das Legen des Fischgratbodens.

wänden, sondern unter einem Winkel von 45° zu diesen laufen und verschränkt ineinander gebunden sind (Fig. 182 und Tafel 1h). Der Boden muss, wenn er fest werden soll, ringsum mit Wandfriesen versehen sein, welche auf Gehrung oder stumpf verbunden sind. Die Federn können an die Riemen angestossen (Spundung) oder besonders eingesetzt werden; auf alle Fälle aber müssen es Hartholzfedern sein. Hirnholzfedern (aus Querholz geschnitten) sind kräftiger aber etwas teurer als Langholzfedern. Wichtig ist, dass die Riemen von gleicher Stärke oder Dicke sind, damit keine besonderen Unterfütterungen nötig fallen und der Boden später nicht »graunzt«, d. h. beim Begehen eigentümlich knarrt. Für den Bodenleger aber gilt als Hauptregel, die Arbeit mit grösster Genauigkeit zu beginnen und unter keiner Bedingung von den mit Maschinen genau bestossenen Riemen irgend einen Hobelstoss wegzunehmen. Befolgt er diese Regel nicht und hobelt er an einem Riemen nach, so muss er, um diesen fehlenden Hobelstoss wieder auszugleichen, am zweiten Riemen mehr, am folgenden noch mehr nachhobeln und das so fort, bis der ganze Boden, wie man sagt, verhobelt, d. h. verdorben ist.

Sind die Friese verlegt, so wird mit dem Legen der Riemen und zwar in einem Eck des Zimmers begonnen (Fig. 162) und von da ab bahnweise der ganze Raum zugelegt. Die Riemen erhalten auf je einer Lang- und einer Querseite Federn, welche in die Nuten der vorher verlegten Riemen eingreifen; genagelt wird in den Nuten. Die Fensternischen werden mit einem Boden aus neben einander gelegten Friesen versehen (Taf. 1h 1) oder gestemmt (wie Taf. 1i, 10 und 12), oder der Zimmerboden greift in die Nischen direkt ein (Taf. 1h 2). Ist die letzte Bahn der

Krauth u. Meyer, Die Bauschreinerei. 4. Aufl.

Riemen von oben genagelt (oder besser geschraubt), so werden die Nagelköpfe versenkt und die dadurch entstandenen Löcher mit kleinen, sauber eingepassten und eingeleimten Holzstückchen ausgeflickt, d. h. geebnet; hierbei ist zu beachten, dass die letzteren von gleicher Farbe sind wie die Riemen, was leider sehr oft nicht berücksichtigt wird. Der Boden wird hierauf mit dem Verputzhobel verputzt und mit der Ziehklinge abgezogen. Dass hierbei jeweils mit der Holzfaser und nicht quer über sie zu fahren ist, wird wohl kaum besonders erwähnt werden müssen. Im Interesse des Bodens ist es sodann, denselben nach gründlicher Entfernung des Staubes sofort zu wachsen und zu wichsen. Zu diesem Behufe wird die ganze Fläche mit einer Mischung von Wachs und Terpentin — welche man heute auch fertig zubereitet in Droguenhandlungen kaufen kann — mittels einer Bürste oder besser eines wollenen Lappens satt eingerieben. Diese Arbeit wird nach Verlauf von 6 bis 8 Stunden nochmals wiederholt, da sehr viel Wachs in die Poren eindringt. Nachdem auch dieser zweite Anstrich 12 bis 20 Stunden getrocknet, wird mittels eines sogen. Bleistrupfers — eine grosse, möglichst rauhe, auf der Rückseite mit einem Blei- oder Eisenstück beschwerte Bürste, welche an einem Stiel befestigt ist, — oder mit einer gewöhnlichen starken Bürste so lange nach der Faserrichtung gebürstet, bis der Boden schön im Glanz erstrahlt.

Man bereitet eine einfache Bodenwichse, indem man 1 Teil weisses Wachs in einem irdenen Gefäss an geschlossenem Feuer langsam zerfliessen lässt, demselben sodann 4 Teile Terpentinöl zusetzt und die Mischung nochmals unter Beobachtung der grössten Vorsicht am Feuer schön warm werden lässt. Vorteilhaft ist, diese Wichse warm auf den Boden aufzutragen, weil sie in diesem Zustande dünnflüssig ist und leichter in die Poren eindringt.

Ausser diesen eichenen Kapuzinerböden giebt es noch schräg gelegte Riemenböden aus Forlen- oder Tannenholz, wie die auf Tafel 1g dargestellten; dieselben haben keinen Blindboden nötig, da die aus mindestens 36 mm starkem Holz (also Schleifdielen) gefertigten, ca. 15 cm breiten, gespundeten oder gefederten Riemen nur auf den Balken aufliegen, woselbst sie auf Gehrung zusammengestossen oder an die dortigen Friese angeschnitten sind. Die Arbeit solcher Böden, namentlich solcher mit Friesen auf jedem Balken (Tafel 1g 1), ist eher bedeutender wie die der Kapuzinerböden, während der erreichte Effekt kein besonderer ist. Aus diesem Grunde haben sich diese Böden auch keiner häufigen Verwendung zu erfreuen; denn für einfache Böden genügt der billigere und solidere Riemen- und Schiffboden, und in besseren Zimmern stellt der eichene Kapuzinerboden, dessen Preis nur unwesentlich höher ist, doch viel mehr vor. Verlegt werden diese Böden wie der gewöhnliche Riemenboden (doch ohne Keilung); genagelt werden sie trotz der Nuten am zweckmässigsten von oben.

8) Der Fischgrat-, Kapuziner- oder Stabfussboden in Asphalt, aus eichenen oder in neuerer Zeit buchenen, ca. 35 cm langen, 8 cm breiten und 24 mm starken Riemen bestehend, welche in heissen Asphalt verlegt werden, ist eine Erfindung der Neuzeit und der dichteste Holzfussboden, in hygienischer Beziehung der beste, den wir besitzen. Derselbe kann auf Beton wie auf Holzdielen gelegt werden, und eignet sich somit für Parterreräume sowohl als auch für obere Stockwerksböden. Seine Undurchlässigkeit macht ihn für besondere Zwecke, wie für Krankenhäuser, sehr wertvoll, doch findet er seine Hauptverwendung als Beleg für Verkaufslokale, Restaurationen, Cafés etc., überhaupt für Räume, die im Erdgeschoss liegen, und bei denen man nicht in der Lage ist, das regelrechte Austrocknen der Gewölbe abzuwarten. Die Herstellung dieses Bodens kann nur durch besonders eingeübte Arbeiter erfolgen, da es besonderer Geschicklichkeit bedarf, den rasch erkaltenden Asphalt richtig aufzutragen und die Riemen zu verlegen. Es kann somit von einer Beschreibung der Legung Umgang genommen werden. Angefügt wird noch, dass im Erdgeschoss der Asphalt direkt auf den vorher beschafften Zementbeton aufgelegt wird, während in den oberen Stockwerken auf den Holzboden zunächst eine etwa 2 cm hohe Sandschicht aufzubringen ist, welche verhindert, dass der Asphalt am Holz anklebt. Die Holz-riemen erhalten beiderseits nach unten eine schräge Ausfalzung (Fig. 178), in welche der Asphalt in Form eines Schwalbenschwanzes eingreift und so dieselben festhält, die Stärke der Asphaltschicht ist ca. 1 cm. Nach dem Legen des ganzen Bodens wird derselbe verputzt und geölt oder gewichst.

9) **Tafelparketten oder Parkettboden** (Taf. 1 i). Unter diesem Namen versteht man einen aus 24 mm starken, quadratischen Tafeln von 35 bis 40 cm Seite bestehenden Fussboden, welcher auf einen Blindboden verlegt wird. Diese Tafeln sind aus mehreren kleinen Stücken auf Nut und Feder zusammengesetzt, wodurch das Arbeiten des Holzes verhindert bezw. unschädlich gemacht, dem Boden selbst aber eine schöne Zeichnung gegeben werden soll. Dieselben werden heute nur noch aus Hartholz gefertigt und zwar nur aus einer Holzart, gewöhnlich Eichen, oder aus mehreren, namentlich verschiedenfarbigen Hölzern, wie: Eiche und Ahorn, Eiche und Nussbaum, Eiche mit Palisander-, Amarant- und Ebenholz-Einlagen oder -Adern etc.

Wo verschiedene Hölzer zu einer Tafel vereinigt werden, ist darauf zu achten, dass ihre Härte eine gleiche ist, damit sie sich gleichmässig abnützen, ein Umstand, dem bei der früheren Parkettenfabrikation oft nicht genügend Rechnung getragen wurde, indem man Tafeln konstruierte,

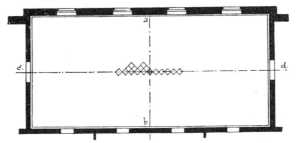

Fig. 183. Das Legen von Parkettböden in grossen Räumen.

die Hartholzfriese und Tanneneinlagen vereinigten. Eichenholz spielt seiner Härte, seines angenehmen hellgelben Tones und seines nicht übermässigen Preises wegen immer die Hauptrolle. Ahorn ist für grosse Flächen zu teuer und zu hell, Nussbaum zu teuer und zu dunkel, obgleich sonst dessen Naturton (ohne Beizung) ein schöner und angenehmer ist.

Je nach der Reinheit und Schönheit des Holzes unterscheidet man bei allen Eichenholzböden, Fischgrat- und Tafelparketten **erste, zweite und dritte Wahl**, wobei die erste die feinste ist.

Seines Preises und seiner verhältnismässig grossen Zeichnung wegen wird der Tafelparkettboden für gewöhnliche Wohnräume nur in beschränktem Masse verwendet, dagegen mit Vorliebe für bessere Wohnräume, namentlich aber Gesellschafts- und Repräsentationsräume. Räume von bescheidenen Dimensionen lässt er durch ein grosses Muster noch geringer erscheinen, während das kleine Kapuzinermuster wie kein anderes das Gegenteil bewirkt. Das Aussehen des Tafelparketts ist in grossen Räumen bei schöner Zeichnung, gutem Verlegen und tadelloser Instandhaltung entschieden ein stattliches und vornehmes, das von keinem anderen Boden erreicht wird. Unschön kann ein noch so teurer Boden wirken, wenn dessen Zeichnung unpassend oder zu gross, wenn seine Farbe zu schreiend, wenn der Boden unschön gelegt und nachlässig gehalten ist. Werden gewichste Böden nicht ununterbrochen mit grösster Sorgfalt behandelt, werden Wasser und Schmutz nicht von ihnen fern gehalten, so ist die Schönheit derselben bald dahin und sie machen statt einen vornehmen einen traurigen Eindruck. — Nicht ratsam ist die Verwendung solcher Böden für Schlafzimmer, da diese, um den Staub möglichst fern zu halten, täglich mit einem feuchten Tuch gereinigt werden müssen, was bei einem gewichsten Boden nicht angeht.

IV. Die Fussböden.

Der Tafelparkettboden erfordert wie der Kapuzinerboden, seiner Festigkeit wie auch seines Aussehens wegen, Wandfriese, in welche die ringsum genuteten Tafeln mit Hartholz- (am besten Hirnholz-) Federn eingreifen. Auch das Legen desselben geht, soweit es sich um gewöhnliche Zimmer handelt, vor sich, wie das des Kapuzinerbodens; nur gilt hier das dort Gesagte in noch viel höherem Masse. **Genauester, sorgfältigster Beginn der Arbeit und Nichtnachhobeln der Tafeln ist Hauptregel.** Vom Beginn der Arbeit hängt vielfach das Gelingen des Ganzen ab. Ist falsch oder ungenau angefangen oder wird nachgehobelt, so wird der Boden in seinen Linien krumm und verschoben, was sehr schlecht aussieht. Hat der Arbeiter den Boden in einem grösseren Raume, etwa in einem Saale, zu legen, so beginnt er nicht in einem Eck, sondern zweckmässiger, wie in Figur 183 dargestellt ist, in der Mitte des Saales, und zwar misst er sich genau die Axen ab, hängt darnach Schnüre aus und legt am Kreuzungspunkt die erste Tafel. Erst wenn diese genau und fest liegt, werden die seitlich anstossenden nach der Hauptaxe gelegt und, von dieser mittleren Tafelreihe ausgehend, wird alsdann nach allen Richtungen weiter gearbeitet. Der Vorteil, welcher aus dieser Art der Legung erwächst, ist, dass die Axen richtig und in schöner gerader Linie durch den Saal laufen, was bei jeder anderen Legungsart niemals zutrifft. Ist der zu belegende Raum nicht rechtwinkelig, so nimmt man allgemein die Fensterwand als massgebend an, mit welcher man die Tafelreihen parallel legt. Nach dem Legen, Verputzen und Abziehen des Bodens wird derselbe gewichst wie der Kapuzinerboden.

Auf Tafel 1 sind unten vier Muster der gangbarsten Parketttafeln in Ansicht und Querschnitt dargestellt.

Tafel 2 stellt den Grundriss eines Neubaues dar, wie derselbe jedem Unternehmer einer Schreinerarbeit von seiten des Architekten ausgefolgt werden sollte, wenn Irrtümer und Missverständnisse und eine Unmasse von Aerger und Unannehmlichkeiten vermieden werden sollen. Vielfach besteht diese Uebung schon, doch lange nicht im gewünschten Umfange. Nur dadurch, dass man dem Meister ganz genau, schriftlich oder durch Zeichnung angiebt, was er zu machen und wie er dies oder jenes zu behandeln hat, kann der angedeutete Zweck erreicht und unter Umständen viel Geld erspart werden. Auf dem Plan sind sämtliche Räume mit Nummern versehen und seitlich von diesem sind die für jeden Raum bestimmten Schreinerarbeiten namentlich aufgeführt und im Mass annähernd angegeben. Es bezieht sich dies auf die Zahl und Anordnung der Bodenrippen, sodann auf die Art des Fussbodens, der Friese und Fensternischen auf die Anbringung der Fussockel, Lambris und Täfelungen mit Angabe ihrer Wiederkehren, auf die Zahl der Fensterbrüstungen und der Simsbretter, die Zahl und Grösse der Thüren und die Richtung ihres Aufschlagens etc., kurzum alles, was zu wissen dem Schreinermeister wie dem Bodenleger nötig ist.

V. LAMBRIS UND TÄFELUNGEN.
(Tafel 3, 4, 5 und 6.)

1. Glatte Lambris: Sockelleisten, Fusssockel, Sockel mit Fussleiste, Sockel mit Fuss- und Deckleiste. —
2. Gestemmte Lambris: Brüstungslambris und Vertäfelungen.

---

1. Glatte Lambris.

Dieselben haben den Zweck, den Wandflächen nach unten, nach dem Boden zu, einen soliden und schönen Abschluss zu geben und die Bemalung derselben oder die aufgezogene Tapete vor Beschädigungen, besonders beim Reinigen und Scheuern, zu schützen. Die Höhe derselben braucht daher nicht bedeutend zu sein; es genügt bei Fussleisten und Sockeln meist halbe, bei Sockeln mit Fuss- und Deckleiste ganze Dielen- oder Bordbreite, also eine Breite von

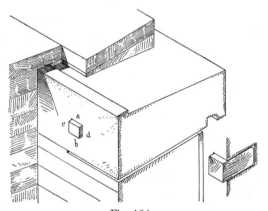

Fig. 184.
Befestigung von Holzdübeln in Fensterbänken.

ca. 9 bis 14 cm bezw. 25 bis 30 cm. Die Lambris sind oben entweder mit Fase, Hohlkehle oder Profilhobel, oder auch mit besonderer profilierter Deckleiste versehen. Am Boden sitzen sie stumpf auf oder die Fuge ist durch eine Fussleiste gedeckt. (Vergl. Fig. 12 und 13.)

Für den Fall, dass der Wandverputz, der Ersparnis halber, nicht bis auf den Boden herabgeführt ist, werden die Sockel bezw. deren Deckleisten oben nach hinten abgeschrägt, um dem Verputz einen soliden Halt zu geben.

Die Art der Befestigung ist verschieden: bei Holzfach- und Riegelwänden erfolgt sie durch direkte Annagelung an die Wandschwellen und Pfosten, bei Steinwänden dagegen durch Befestigung an in die Mauer eingesetzte Holzdübel. Die letzte Art hat den Nachteil, dass bei nicht völlig ausgetrocknetem Mauerwerk die Feuchtigkeit desselben durch die Dübel in die Holzsockel

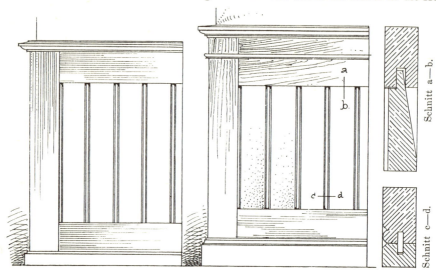

Fig. 185. Einfache, gestemmte Brüstungslambris.

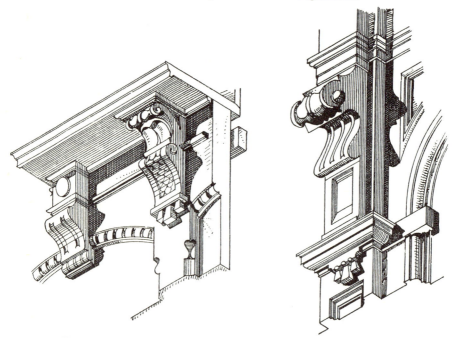

Fig. 186. Einzelheiten zu den Täfelungen b und c auf Tafel 5.

geleitet wird, wodurch einer allenfallsigen Schwammbildung bedeutend Vorschub geleistet ist. Zweckmässiger ist es daher, wenn auch etwas teurer, die Befestigung durch Annagelung an geteerte oder besser mit Kreosotöl oder Karbolineum getränkte Latten zu bewirken, welche man mit Mauerklöbchen an die Wand befestigt. Die Vorzüge dieser Konstruktion sind einleuchtend:

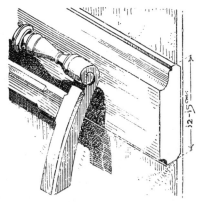

Fig. 187. Brüstungsschutzleiste.

erstens liegen nicht nur die Latten auf drei Seiten ganz frei, sondern es bleibt auch der ganze Holzsockel um die Lattendicke von der Wandfläche entfernt und ist somit von Luft umgeben.

Fig. 188. Täfelung nach dem Entwurf der Architekten Kayser und von Großheim, Berlin.

Die Vorzüge überwiegen die Nachteile dieser Konstruktion, welche in dem höheren Kostenaufwand wie in der Unmöglichkeit, die Möbel nahe an die Wand zu stellen, bestehen, ganz bedeutend, und der beste Beweis hierfür ist die täglich zunehmende Anwendung derselben. Allerdings kann man einwerfen, es sei eine der ersten Regeln der Bauschreinerei, Holzwerk nur an nachweisbar trockene Wandflächen anzuschlagen und es sei bei genauer Befolgung derselben eine solch weitgehende Vorsicht eigentlich überflüssig. Das ist richtig, aber wer kann mit Sicherheit behaupten, ob in unsern hastig hergestellten Neubauten eine Wand trocken ist oder nicht? Das Mauerwerk kann sehr schön trocken aussehen und es im Kern doch nicht sein; es kann der Verputz auf der Aussenfläche rein weiss sein — namentlich Gipsverputz — und doch noch Feuchtigkeit einschliessen. Wartet der Architekt nicht lange genug zu, so trifft ihn die Schuld, wenn die Sache verunglückt, da er den Auftrag zum Beginn der Arbeit gegeben; will er dagegen sicher gehen und zuwarten, so kann und wird man ihm den Vorwurf machen, er verschleppe die Fertigstellung der Arbeit.

Fig. 189. Brüstung der Rednertribüne des Sitzungssaales im Reichstaggebäude.

Aus dem Dargelegten dürfte hervorgehen, dass es sich empfiehlt, nur diejenige Befestigungskonstruktion zu wählen, welche den Bau nicht verzögert und sicher vor Schaden bewahrt, auch wenn sie teurer ist als die gewöhnliche.

Die besprochenen Fussleisten, Fusssockel und Lambris, wie sie auf Tafel 3, Fig. a—d gezeichnet sind, fasst man allgemein zusammen unter dem Namen »glatte Lambris«. Sie werden gewöhnlich in Tannen, 24 mm stark oder auch stärker ausgeführt. Bei a, c und d ist der Verputz bis auf den Fussboden geführt (eine grosse Erleichterung für den Schreiner beim Anschlagen); bei b reicht er nur bis Sockeloberkante. Das Anschlagen auf geteerte Latten ist ähnlich, wie es in Figur e für gestemmte Lambris angegeben ist. Werden die Lambris höher als 30 cm (gewöhnliche Dielenbreite) gemacht, so stemmt man sie und heisst sie gestemmte Lambris.

2. Gestemmte Lambris.

Ihre Höhe richtet sich nach dem Zweck, dem Stil und den verfügbaren Mitteln. Ihr Zweck ist ausser dem des Schützens noch der weitere, die Wandflächen zu zieren und warm zu halten, sowie das Zimmer wohnlich und gemütlich zu machen. Die geringe Wärmeleitungsfähigkeit des Holzes macht es hierzu ganz besonders geeignet, und auf diesen Umstand ist — im

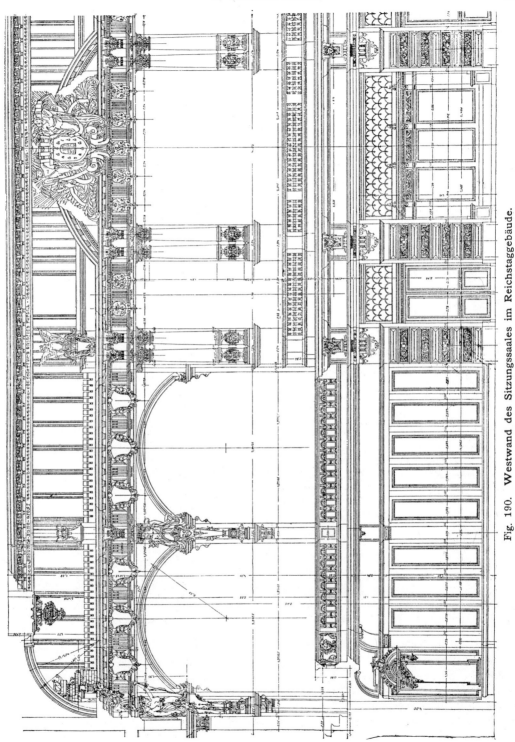

Fig. 190. Westwand des Sitzungssaales im Reichstaggebäude.

Verein mit der dekorativen Wirkung — trotz der bedeutenden Herstellungskosten hauptsächlich die grosse Beliebtheit zurückzuführen, welcher sich die Täfelungen derzeitig erfreuen. Dabei darf aber nicht unerwähnt bleiben, dass zwischen Warm- und Trockenhalten ein sehr grosser Unterschied ist, welchen leider sehr viele Leute nicht machen, da sie der irrigen Meinung sind, diese Holzbrüstungen, -Verschalungen oder -Vertäfelungen seien ein gutes Mittel gegen feuchte

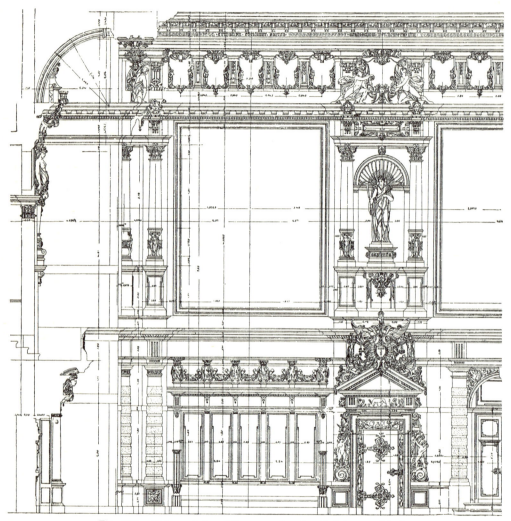

Fig. 191. Ostwand des Sitzungssaales im Reichstaggebäude.

Wände. Wenn man allerdings zufriedengestellt ist, sobald man die nasse Wand hinter der Täfelung nicht mehr sieht, so mag dies gelten; in Wirklichkeit aber ist der Zustand schlimmer geworden statt besser. Zwischen Wand und Täfelung beginnt die Schimmelbildung, welche die Zimmerluft verschlechtert und schliesslich zur Zerstörung des Holzes durch Trockenfäule führt. Auch die Isolierungen durch Asphaltpapier, durch Stanniol und selbst durch Zinktafeln helfen wenig; sie verdecken ebenfalls bloss die Feuchtigkeit, ohne sie zu beseitigen. Abhilfe kann

2. Gestemmte Lambris.

nur durch gründliche Abstellung der Ursache am Aeusseren der Wand geschaffen werden, und so lange dies nicht geschehen, ist jede Arbeit überflüssig, jede Art Täfelung eine Verschwendung, ja in gewisser Beziehung eine Gefahr für die Gesundheit. Die erste Bedingung zur Haltbarkeit jeglicher Holztäfelung ist daher vollständige Trockenheit der Wandflächen und Fernhaltung der Feuchtigkeit.

Hat die gestemmte Lambris eine Höhe von 60 cm, so heisst sie »60 cm hohe gestemmte Lambris«; erreicht sie Brüstungshöhe, also 80 bis 110 cm, so wird sie »Brüstungslambris« genannt. Wird sie noch höher gebildet, bis zu 1,80 m und mehr, so nennt man sie allgemein »Täfelung oder Vertäfelung von der und der Höhe«. Die Art des Stemmens ist verschieden: stumpf, auf Fase oder auf Hobel.

Tafel 3, Figur e stellt eine 60 cm hohe, auf Hobel gestemmte und mit aufgelegten Zierleisten versehene Lambris dar, mit abwechselnd quadratischen und rechteckigen, abgeplatteten Füllungen. Ein doppelter Sockel schliesst das Ganze unten ab und eine profilierte Deckleiste oben, auf welcher sich noch ein kleines Anschlussleistchen befindet. Die Friese sind aus 24 mm starkem Tannenholz, die Füllungen ebenso stark oder höchstens 1 bis 2 mm schwächer. Wenn man Holz sparen will, so lässt man die eigentliche Täfelung nicht bis auf den Boden reichen, sondern setzt sie hinter dem Fusssockel ab; dafür muss aber dann und wann ein Höhenfries herablaufen, auf welchem die Lambris aufruht, bis sie angeschlagen ist. Die Friese wie die Füllungen werden nur auf der sichtbaren Fläche gehobelt; die Füllungen auf der Rückseite nur abgeschrägt, nicht mit dem Platthobel abgeplattet.

Figur f und g, Tafel 3, zeigen zwei gestemmte Fensterbrüstungen,

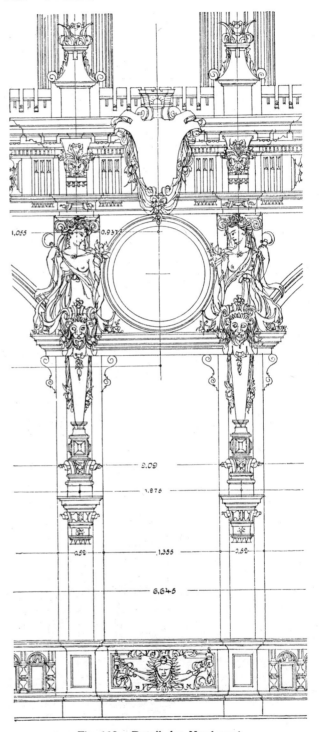

Fig. 192. Detail der Nordwand.

76   V. Lambris und Täfelungen.

Fig. 193. Figurennische in der Ostwand.

2. Gestemmte Lambris. 77

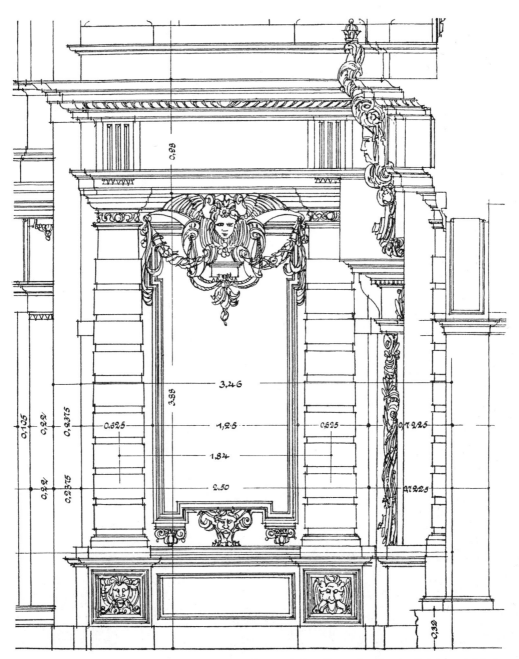

Fig. 194. Anschluss der Nordwand an die Südwand des Sitzungssaales im Reichstaggebäude.

V. Lambris und Täfelungen.

Fig. 195. Thür in der Ostwand des Sitzungssaales im Reichstaggebäude.

und zwar ist f stumpf gestemmt mit gefederter Riemenfüllung, g dagegen auf Profilhobel mit abgeplatteter Füllung. Die Friese sind 24 mm, besser 30 mm stark, die Füllungen 20 oder 24 mm;

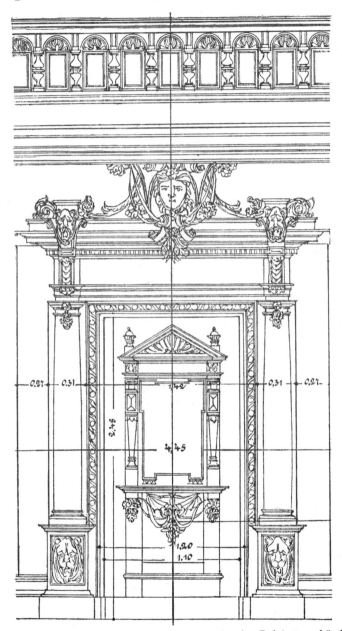

Fig. 196. Hammelsprung-Thür des Sitzungssaales im Reichstaggebäude.

die Füllungsbreite ist bei f so, dass sie noch etwas geringer ist als die Höhe, bei g gleich der gewöhnlichen Bordbreite; die Befestigung geschieht auf Holzdübel. Es sind dies 6 bis 8 cm lange

V. Lambris und Täfelungen.

Fig. 197. Reichsgerichtsgebäude in Leipzig. Empfangszimmer des Präsidenten.

2. Gestemmte Lambris.

Eichenholzklötzchen von prismatischer Form und quadratischem Querschnitt. Sie werden in die sorgfältig ausgehauenen Löcher eingepasst und eingetrieben. Dabei ist besondere Vorsicht bei Fensterbänken geboten, um Beschädigungen der letzteren zu verhüten. Es gehört nicht zu den Seltenheiten, dass grössere Stücke des Steins durch zu kräftig eingetriebene Holzdübel abgesprengt werden. Diesem unliebsamen, unter Umständen von bedenklichen Folgen begleiteten Vorkommnis kann vorgebeugt werden, indem man den Dübel nur nach der Langseite des Steines pressen lässt, während man ihm oben und unten etwas Luft lässt (vergl. Fig. 184). Der Dübel würde

Fig. 198. Reichsgerichtsgebäude in Leipzig. Südlicher Strafsenatssaal.

also an den Seiten c und d press anliegen, dagegen bei a und b einen kleinen Spielraum zeigen. Holzdübel schwalbenschwanzförmig anzuordnen, wie es viele Architekten auf ihren Zeichnungen belieben, hat keinen praktischen Wert, weil nur bei peinlichst genauer Ausführung auf diese Weise eine erhöhte Festigkeit erreicht werden kann. Die Ausfüllung einer schwalbenschwanzförmigen Vertiefung durch den Dübel wird erzielt, indem von hinten her in denselben ein Keil eingepresst wird. Ist das Loch nicht genau eingehauen und der Keil nicht genau dem Dübel angepasst, so wird beim Eintreiben der Dübel in der Mitte gesprengt, also gerade an der Stelle, wo die Nägel oder Holzschrauben ihren Halt finden sollen. Das Verkehrte dieser Einrichtung leuchtet somit ein. Anders liegt der Fall in Bezug auf schwalbenschwanzförmige Dübel, welche aus zwei Stücken bestehen, oder eiserne Dübel, welche in die Löcher eingegipst werden und in eine Mutterschraube endigen; derartige Dübel sind beispielsweise zur Befestigung der Futtertische in Pferdestallungen üblich. Wo aber Holzdübel verwendet werden, wie dies die Mehrzahl der Fälle im Neubau erfordert, benützt man am besten die gewöhnliche, prismatische Form.

Die Fuge der Brüstung am Boden deckt ein Fussleistchen; als oberer Abschluss ist das sogen. Fenstersimsbrett zu betrachten, welches aus Eichenholz gefertigt, einerseits mit Profilhobel

Fig. 199. Reichsgerichtsgebäude in Leipzig. Grosser Sitzungssaal, Stirnwand.

anderseits mit einer Feder versehen ist, mit welcher es in den Futterrahmenwetterschenkel eingreift. Oben hat es eine an beiden Enden geschlossene Hohlkehle zur Aufnahme des Regenwassers. Befestigt wird es durch Einschieben in den Futterrahmen und Aufnageln oder Aufschrauben auf die Fensterbrüstung. Die Breite des Simsbrettes wird durch die Verhältnisse bestimmt, die Stärke ist 30 mm. Die Fensterbrüstung dient als Schutz der unterhalb des Fensters befindlichen Wandfläche und läuft zu beiden Seiten stumpf in die Leibung ein. Setzt sie sich seitlich an den Leibungen und Wandflächen fort, wobei sich das Fenstersimsbrett in eine Deckleiste verwandelt, so wird dieselbe zur Brüstungslambris. Derartige Lambris, einfacher Art, in der gewöhnlichen Höhe, von ca. 80 cm, zeigt die Figur 185. Friese und Füllungen sind gehobelt

2. Gestemmte Lambris.

Fig. 200. Reichsgerichtsgebäude in Leipzig. Grosser Sitzungssaal.

V. Lambris und Täfelungen.

22 mm stark; beide sind demnach beiderseits bündig. Den Vorderkanten der Riemen ist ein Rundstäbchen angestossen.

Figur h, i, k, Tafel 3, zeigen drei Arten Brüstungslambris, wie man sie gern in Schulsälen und dergleichen Räumen verwendet, bei welchen die gewöhnliche Höhe von 80 cm nicht genügt, um die Wände vor Beschädigungen zu schützen. Das Beispiel Figur h, Tafel 3, ist stumpf gestemmt, 1,20 m hoch mit abgefasten Kanten und mit Spitznuten im mittleren Höhenfries wie in den Füllungen. Figur i unterscheidet sich von Figur h nur dadurch, dass bei ihr die horizontalen Friese Profilhobel aufweisen und die Füllungen abgeplattet sind, während sie bei Figur h stumpf eingreifen. Beide Brüstungslambris schliesst unten ein profilierter Sockel und oben eine Deckleiste ab. Das Beispiel Figur k ist 1,35 m hoch, im übrigen eine Variation der beiden vorhergehenden; ein gleiches gilt für Figur B und C, Tafel 4, wobei die Höhen 1,50 bis 1,55 m betragen. Bei diesen verschiedenen Arten von Brüstungen ist eine Abwechselung in erster Linie dadurch zu erzielen gesucht, dass die ganze Brüstung zunächst durch Höhenfriese in einzelne grosse Teile zerlegt wurde. Durch die teils oben, teils unten abgesetzten und mit einem façonnierten Kopf versehenen, etwas schmäleren Friese ist eine weitere Teilung erzielt. Diese Wirkung wird noch erhöht durch eine verschiedenartige Gestaltung der Füllungen, namentlich aber durch die Abfasungen und eine geschickte farbige Behandlung des Ganzen. Die Stärke der Friese muss mindestens 30 mm betragen, die der Füllungen 24 mm.

Tafel 4 A bringt eine Variation der Brüstungslambris Tafel 3 g und ist wie jene 80 cm hoch. Bei ihr sind je drei Füllungen zu einem Feld zusammengefast; das weitere zeigen der Schnitt und das Detail.

Tafel 4 D und E, sowie Tafel 5 a bis f zeigen sogen. hohe Täfelungen, auch Vertäfelungen, Wandvertäfelungen genannt. Ihre Bestimmung ist, wie schon erwähnt, vornehmlich, das Zimmer zu schmücken, es wohnlich zu machen und warm zu halten. Wohnlich und warm wird ein Raum schon durch die Eigenschaften des Holzes als schlechter Wärmeleiter; die Dekoration wird erzeugt durch eine schöne Gesamtzeichnung und die formale Behandlung der Einzelteile, durch Anbringung vertikaler und horizontaler Teilungen, durch Pilaster, Hermen, Karyatiden und sonstige Stützen, durch Bogen und Archivolte und die darüber befindlichen, oft sehr reichen, durch Konsolen gestützten Hauptgesimse, durch Rosetten, Profilstäbe, Abfasungen etc. etc. Hierzu kommt noch als einer der bedeutendsten Faktoren die schöne Wirkung des Holzes als solches. Während man die glatte Lambris aus Tannen fertigt, wie dies meist auch noch bei der Brüstungslambris der Fall ist, — welche man entweder naturfarben lässt, beizt und lackiert, oder aber ganz mit Oelfarbe streicht und nur in seltenen Fällen in anderem Holz herstellt — fertigt man hohe Täfelungen vornehmlich aus besseren Holzarten, aus Eichen, Nussbaum, Eschen oder Zusammenstellungen dieser Hölzer, in einfachen Fällen aber auch aus Tannenholz oder Tannen und Forlen etc. Sehr beliebt und von schöner Wirkung ist es, für die Täfelung dieselbe Holzart zu wählen wie für die Möbel des betreffenden Raumes; die Wirkung kann noch gesteigert werden, indem man für beide auch eine gleichmässige Zeichnung wählt, sie in derselben vollständig zusammenarbeitet, so dass sie als Ganzes erscheinen, dass ein Teil den anderen ergänzt und dessen Schönheit erhöht. Dass man dies kann, beweisen ausser den vielfach vorhandenen, oft geradezu mustergültig ausgestatteten Räumen des XVI. und XVII. Jahrhunderts, auch die modernen Zimmereinrichtungen. Oft sind die Schlussgesimse der Täfelungen weit vorspringend, zur Aufstellung von allerlei Prunkgefässen bestimmt (Tafel 4 D und Tafel 5 a, b und c), oft weniger ausladend, nur als Abschluss des Ganzen nach oben dienend (Tafel 4 E und Tafel 5 d, e und f).

Gelegentlich werden oben Kleider- und Huthaken angebracht (Taf. 5 a und d), sowie unten Sitzvorrichtungen (Tafel 5 d und f). Stets aber giebt sich das Bestreben kund, die vertikale Teilung überwiegen zu lassen, damit das Zimmer nicht gedrückt erscheint.

2. Gestemmte Lambris.

Tafel 4 E zeigt eine Täfelung, welche eigentlich mittelhoch und nur teilweise höher geführt ist. Die senkrechten Gliederungen haben die Form von Strebepfeilern; die Konstruktion ist die der gestemmten Lambris.

Tafel 4 D zeigt eine Täfelung, welche gleichhoch im Zimmer herumgeführt ist und daher auch in die Leibungen eingreift. Hier muss aber das Hauptgesims auf eine wenig vorspringende Profilleiste zurückgeführt werden, damit die Fensterflügel sich vollständig öffnen lassen, wie dies aus dem beigefügten Grundriss ersichtlich ist. Die Fensterbrüstungshöhe ist durch Herumführung eines besonderen Rosettenfrieses markiert.

Tafel 5a stellt zwei Muster reicherer Täfelungen mit Pilastern, Hermen, Archivolten, Konsolengesimsen und Kleiderhaken dar.

Tafel 5 b und c desgleichen. Die Formen der Konsolen der beiden Hauptgesimse sind in Figur 186 isometrisch dargestellt.

Tafel 5e zeigt eine einfache Wandtäfelung für Wirtschaftslokale, Cafés etc.

Tafel 5 d und f geben zwei etwas reichere, dem gleichen Zweck dienende Holztäfelungen wieder, von denen die letztere nur mit Sitzbank, die erstere dagegen auch noch mit Kleiderhaken versehen ist.

Sämtliche Täfelungen sind gestemmt, die Friese mindestens 30 bis 35 mm stark, die Füllungen haben gewöhnliche Bordstärke. Die Art der Befestigung — welche zweckmässig für sämtliche gestemmte Brüstungen und Täfelungen auf Latten erfolgt, da in diesem Fall das Holz $2^1/_2$ bis 3 cm von der Wand entfernt bleibt — ist aus den beigefügten Schnitten zu erkennen. Sehr zweckmässig ist es, das Holzwerk auf der Rückseite entweder tüchtig zu ölen oder besser mit Kreosotöl zu streichen.

Figur 187 veranschaulicht eine Holzleiste, welche mit Umgehung einer ganzen Holzbrüstung auf eine Höhe von 90 bis 110 cm mittels Steinschrauben an die Wand befestigt wird, um Beschädigungen des Verputzes, des Anstrichs oder der Tapete durch Stuhllehnen etc. zu verhindern. Die Breite der Leiste ist ca. 12 bis 15 cm. Derartige Vorkehrungen werden hauptsächlich in Wirtschaften und Schulzimmern getroffen.

Die auf den Tafeln 4 und 5 gebrachten Vertäfelungen entsprechen in ihrer Ausstattung dem vornehmen Wohnhaus und den neuzeitigen Wein- und Bierstuben. Die gewöhnlichen Höhen sind 1,8 oder 2 m. Für grössere Räume, für Säle und Hallen, werden gelegentlich auch grössere Verhältnisse und Abmessungen für die Täfelungen nötig, ohne dass die Konstruktion sich wesentlich ändert, so lange nicht Säulenstellungen, Nischen und andere der grossen Architektur entlehnte Formen hinzutreten. Die Tafel 6 bringt eine Täfelung, 2,5 m hoch, in welche eine Thür mit Aufsatz eingebaut ist. Wir verdanken dieses Beispiel, sowie dasjenige der Figur 188, den ausführenden Architekten Kayser und von Grofzheim in Berlin.

Um zu zeigen, wie reich und grossartig schliesslich das Täfelwerk eines Raumes gestaltet werden kann, bringen wir in den Figuren 190 bis 196 eine Anzahl der Werkzeichnungen zum Abdruck, nach welchen unter Meister Wallots Leitung die Wände des Sitzungssaales im neuen Reichstagsgebäude von den Berliner Firmen G. Olm und Gebr. Lüdtke in Eichenholz ausgeführt worden sind. Die Brüstung der Rednertribüne zeigt Fig. 189 im Detail.

Zum gleichen Zwecke sind die Figuren 197 bis 200 eingereiht. Sie beziehen sich auf die Innenausstattung des von L. Hoffmann erbauten Reichsgerichtsgebäudes in Leipzig. Die Figur 197 zeigt einen Kamin aus dem Empfangszimmer des Präsidenten, mit Holzaufsatz und anschliessender Täfelung. Figur 198 giebt das durch eine Thür unterbrochene Wandgetäfel im südlichen Strafsenatssaal. Figur 199 bringt die Ausstattung einer Stirnwand des grossen Sitzungssaales und die Figur 200 veranschaulicht die Gesamtwirkung dieses schönen Raumes mit seiner reichen Täfelung und Holzdecke nach einer photographischen Aufnahme von H. Walter.

VI. THÜREN UND THORE.

(Tafel 7 bis mit 54.)

Allgemeines. — 1. Einfache Thüren (Lattenthüre, Riementhüre, stumpf verleimte Thüre). — 2. Verdoppelte Thüren. — 3. Gestemmte Thüren für Wohnräume (Einflügelthüre, Zweiflügelthüre, Schiebthüre). — 4. Verglaste Thüren (Balkonthüre, Glasabschluss, Vorplatz- und Wartesaalthüre, Pendelthüre, Windfang). — 5. Hausthüren (Ein-, Zwei- und Dreiflügelthüren, Magazin- und Scheunenthor, Einfahrtsthor, Hofeinfriedigungsthor). — 6. Verschiedene Thüren zu bestimmten Zwecken.

Allgemeines.

Unter Thüren verstehen wir die zum Verschliessen der Thüröffnungen dienenden Vorrichtungen. Ihr Zweck ist, Unbefugten den Eintritt in einen gegebenen Raum zu verwehren und die äussere Luft, den Wind und die Kälte, sowie die atmosphärischen Niederschläge abzuhalten. Ihre Grösse hängt zumeist von dem Zweck ab, dem sie dienen, doch gehören auch ganz wesentliche Abweichungen von dieser Regel in Bezug auf architektonisch durchgebildete Innenräume und Fassaden nicht zu den Seltenheiten. So hat beispielsweise das Mittelalter die Thüren im allgemeinen nur so gross gemacht, als es das praktische Bedürfnis erforderte, während die Antike und die Renaissance die Grösse der Thüre von den übrigen Verhältnissen des Baues abhängig machten, wobei also ausser der praktischen Seite auch die ästhetisch-formale in Betracht kam. Im neuzeitigen Wohnhaus macht man die Thüren so gross, als nötig ist, d. h. dass die Bewohner desselben sie ungehindert zu passieren vermögen und alle zum Haushalt etc. gehörigen Möbel und Einrichtungsgegenstände ohne Beschädigungen hindurchtransportiert werden können. Hierfür genügt die Breite unserer gewöhnlichen einflügeligen Thüre. Erfahrungsgemäss müssen zu schmale Thüren zu weit geöffnet werden, wenn sie nicht Beschädigungen ausgesetzt sein sollen, während zu breite Thüren zu schwer sind, bei der Benützung zu viel Zugwind verursachen und die zum Stellen der Möbel benötigte Wandfläche verringern. Sollen zwei oder mehrere nebeneinander liegende Zimmer so verbunden werden, dass sie als ein Raum benützt werden können, z. B. bei Gesellschaften, oder haben wir einen Raum, in welchem eine grössere Anzahl Menschen sich aufhält, wie in Schulsälen, oder verlangt schliesslich die Art der Benützung eines Raumes eine grössere Lichtweite als die der einflügeligen Thüre, so tritt an deren Stelle die zweiflügelige oder kurzweg Flügelthüre. Diese Thüre beansprucht selbstredend viel mehr Platz als eine einflügelige und ist daher für einfache bürgerliche Wohnhausverhältnisse meist unpraktisch, zumal sie mindestens zwei- bis dreimal so teuer ist und überdies bei gewöhnlichem Gebrauch — bei dem man immer nur einen Flügel öffnet — zum Durchgehen weniger Raum bietet als eine einflügelige Thüre. Feststehende Masse giebt es zur Zeit in

Deutschland weder für einflügelige noch für Flügelthüren, obgleich solche im Interesse des Geschäftsmannes wie des Bauherrn sehr erwünscht wären. Die Verschiedenheit der Thürmasse — worunter immer die Lichtmasse derselben gemeint sind — liesse sich nur entschuldigen, wenn triftige Gründe hierfür vorlägen, was aber nicht der Fall ist. Der eine Architekt macht seine Thüren 0,90 m weit, während der andere es absolut nicht unter 0,91 m thut und der dritte sogar die ganze Wirkung der Innenräume in Frage gestellt sieht, wenn die Lichtweite der Thüren nicht 0,92 m ist! Selbst die eifrigsten Verfechter ihrer Masse werden zugeben müssen, dass sie eine Thürweiten-Differenz von 1 bis 2 cm mit dem Augenmass nicht sicher wahrzunehmen vermögen. Und wenn dem so ist, warum das unbedingte Festhalten an verschiedenen Massen, während man andererseits mit einem Einheitsmass dem Geschäftsmanne viel Unannehmlichkeiten ersparen könnte? Wie leicht liessen sich bei festen Massen im Winter oder zu anderer stiller Zeit Thüren auf Vorrat arbeiten, was bei Brüstungen, Lambris etc. nicht angeht, da deren Masse im Bau genommen werden müssen. Wer die Verhältnisse kennt, weiss, wie selten Arbeiter zu treffen sind, welche Thüren sauber zu arbeiten vermögen; er weiss auch, wie spät oft die Arbeiten bestellt werden, wie sich die Geschäfte in der Hauptzeit häufen, und wie oftmals die Termine nicht eingehalten werden können zum Nachteil des Architekten, des Schreiners und des Bauherrn. Zur Erzeugung einer heiteren Stimmung trägt ein solcher Umstand bei dem Geschäftsmanne — den noch vieles drückt, wovon andere keine Ahnung haben — nicht bei. Was nützen ihm aber zwei Dutzend vorrätige Thüren von 0,90 m Lichtweite, wenn er 6 Stück von 0,93 m machen lassen muss zu einer Zeit, wo er seine guten Schreiner zu anderer Arbeit nötig hat? Während man in England und in Amerika den grossen Vorteil fester Masse für gewöhnliche Zimmerthüren schon längst eingesehen hat und — was die Hauptsache ist — auch darnach handelt, war bei uns bisher jeder Versuch in dieser Richtung vergeblich und wird es voraussichtlich auch noch so lange bleiben, bis es zu spät ist, d. h. bis ausländische Thüren auf den Baumarkt kommen, um die inländischen überflüssig zu machen.

Allgemein macht man in Wohnräumen Thüren bis zu 1,10 m Lichtweite einflügelig; das gewöhnliche Zimmerthürmass ist in Süddeutschland 0,90 × 2,10 m. Schmälere Thüren sind höchstens für Aborte zu verwenden, weil sie dem Fremden durch ihre geringe Breite leichter auffallen. Ganz schmale, wie Schlupf- oder Tapetenthüren, müssen aber doch eine Breite von mindestens 0,60 m erhalten.

Die Breite der Flügelthüren ist 1,10 m bis ca. 1,60 m, das passendste Mass für Wohnräume ist 1,50 × 2,40 bis 2,50 m. Hierbei wird jeder Flügel 0,75 m breit, was als Minimalmass eines Thürflügels gelten muss. Schmälere Flügelthüren (weniger als 1,50 m breit) teilt man durch Anbringen zweier Schlagleisten derart, dass der aufgehende Flügel noch 0,75 m breit ist. (Tafel 15 c bis f.) Für Thüren in grossen, architektonisch durchgebildeten Räumen lassen sich bestimmte Masse nicht angeben; deren Grösse ist von dem Architekten unter Berücksichtigung der ganzen Architektur zu bestimmen. (Fig. 195 und 198.)

Schiebthüren macht man möglichst breit und hoch.

Die Abmessungen der Hausthüren und der Hausthore hängen hauptsächlich von der Fassade ab, da die Oberkante beider meist mit der Fenstersturzhöhe des Erdgeschosses abschliesst. Als Minimalmass gilt für die ersteren, 1,00 m Breite, für die letzteren 2,30 m.

Die Breite der Hofeinfriedigungsthore, sowie der Magazin- und Scheunenthore richtet sich nach den speziellen Bedürfnissen.

In Bezug auf die Konstruktion unterscheiden wir:
   1) Einfache Thüren für untergeordnete und provisorische Räume.
      a) Lattenthüre.
      b) Riementhüre mit Quer- und Bugleisten.
      c) Stumpf verleimte Thüre mit Einschiebleisten.

2) **Verdoppelte Thüren** für Kellereingänge, Waschküchen, Ställe etc. und als einfache Hausthüren.
3) **Gestemmte Thüren** für Wohnräume.
    a) **Einflügelige Zimmerthüre** (einschliesslich der Tapetenthüre).
    b) **Flügelthüre.**
    c) **Schiebthüre.**
4) **Verglaste Thüren, Glasthüren.**
    a) **Balkon- und Verandathüre.**
    b) **Glasabschluss.**
    c) **Wartesaalthüre.**
    d) **Pendelthüre.**
    e) **Windfang.**
5) **Hausthüren und Thore.**
    a) **Einflügelige, zwei- und dreiflügelige Hausthüre.**
    b) **Thor, Thorweg.**
    c) **Magazin- und Scheunenthor.**
    d) **Hofeinfriedigungsthor.**
6) **Verschiedene andere Thüren** für bestimmte Zwecke.

---

1. Einfache Thüren.

a) **Lattenthüre** (Taf. 7 a) für Holzremisen, Speicher und Kellerabteilungen. Sie ist meist rauh, d. h. nicht gehobelt und im allgemeinen die billigste der Thüren, doch wird sie nicht immer aus Sparsamkeit verwendet, sondern es ist für manche Zwecke wünschenswert, Luft und Licht in den betr. Raum zu bringen, wie z. B. in Holzremisen, Schwarzwaschkammern etc., oder beim Vorbeigehen sich durch einen Blick überzeugen zu können, ob in dem Raum noch alles in Ordnung ist. Sie besteht aus zwei senkrechten Friesen von halber Bordbreite, zwischen denen so viel Latten im ungefähren Abstand der Lattenbreite eingereiht werden, als Platz haben. Nach der Quere erhält sie ihren Zusammenhalt durch aufgenagelte Holzleisten in halber Bordbreite (Querleisten), und gegen Verschieben schützt sie die Anbringung von aufgenagelten Bugleisten (1 oder 2). Zum Nageln der Latten wie der Querleisten bedient man sich geschmiedeter Nägel (Fig. 166), welche auf der Rückseite um-, bezw. quer über die Faser in das Holz eingeschlagen werden. Die Thüre hat kein Futter, schlägt vielmehr stumpf auf den Stein- oder Holzpfosten an.

b) **Riementhüre** (Taf. 7 b).

Sie ist entweder rauh oder gehobelt und gefügt, etwas besser als die vorige und besteht aus schmalen, meist $^1/_2$ Bord breiten Riemen, welche entweder stumpf aneinander gestossen oder besser auf Nut und Feder oder durch Spundung miteinander verbunden sind. Ihre Festigkeit erhält sie ebenfalls wie die Lattenthüre durch aufgenagelte Quer- und Bugleisten; sie ist wie diese auch ohne Futter und Verkleidung.

c) **Stumpf verleimte Thüre** (Taf. 7 c).

Sie besteht aus gehobelten, schmalen oder geschlitzten Bord, welche gefügt und zu einer Tafel verleimt werden. Diese Thüre hat also keine offenen Fugen. Da sich bei dieser Konstruktion der einzelne Riemen nicht mehr frei bewegen kann, sondern nur die ganze Thüre, so ist es unzulässig, genagelte Querleisten anzubringen. An deren Stelle treten jetzt in den Grat eingeschobene Leisten, welche dem Holze gestatten, zu schwinden und zu quellen; dieselben dürfen nicht geleimt werden. Bei Verwendung als Abschluss von Dienstbotenzimmern etc. ist diese Thüre mit Futter und einfacher, glatter Verkleidung (Taf. 10 1) versehen.

Die scharfen Kanten der Quer-, Bug- und Einschiebleisten der drei besprochenen Thüren werden gebrochen oder abgefast. Das Thürbeschläge besteht aus je zwei Langbändern, Kloben in Stein, auf Platte oder in Holz, sowie einem Riegel- oder Kastenschloss mit Eisendrückern und Schliesskloben.

Soll der Abschluss dichter und zugleich fester und widerstandsfähiger werden, was besonders bei ins Freie führenden Thüren, bei Hausthüren, wünschenswert ist, so wendet man die verdoppelten Thüren an.

2. Verdoppelte Thüren.

Nicht nur für äussere, sondern auch für innere Abschlüsse, bei welchen die Thüre auf beiden Seiten verschiedener Temperatur oder mehr oder minder feuchter Luft ausgesetzt ist, wie dies z. B. bei Kellerthüren, Stall- und Waschküchenthüren der Fall ist, erfreut sich die Konstruktion allgemeiner Beliebtheit. Eine solche verdoppelte Thüre besteht im wesentlichen entweder

aus zwei einfachen Thüren, welche in allen Teilen fest aufeinander genagelt sind

oder aber aus der eigentlichen, der inneren, meist einfachen, gespundeten oder gefederten Thüre und der äusseren aufgenagelten oder geschraubten Schutzschalung, der Verdoppelung. Die letztere kann mannigfache Gestalt, vornehmlich aber eine solche erhalten, welche den Regenschlag rasch abzuleiten im stande ist.

Dabei ist zu beachten, dass die Holzfasern der äusseren Thüre quer oder schräg zu denen der inneren laufen, damit das verschiedene Arbeiten des inneren und äusseren Holzes gegenseitig ausgeglichen wird.

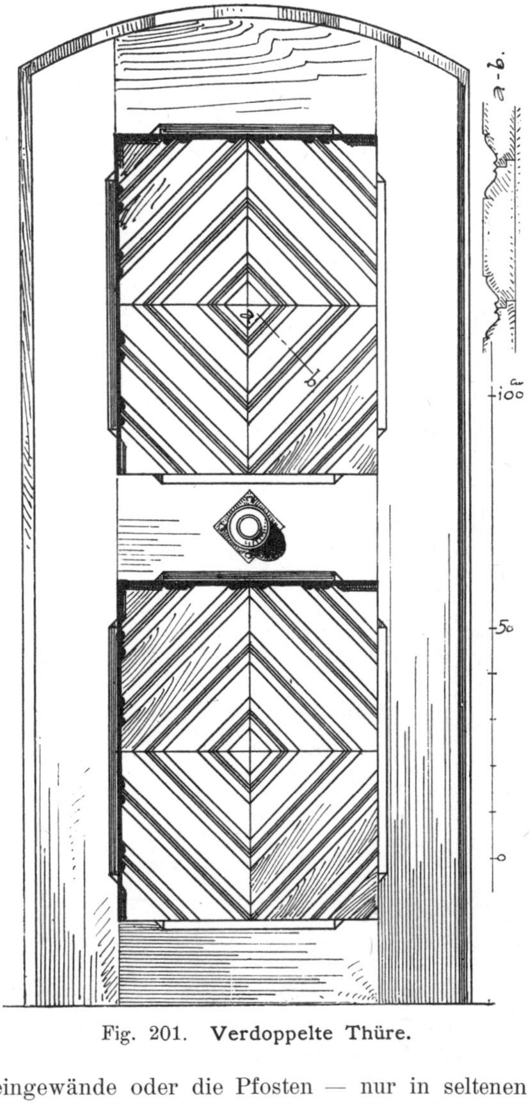

Fig. 201. **Verdoppelte Thüre.**

Behufs dichteren Anschlusses an das Steingewände oder die Pfosten — nur in seltenen Fällen ist ein Futterrahmen vorhanden — versieht man die Thüre mit einem Falz, indem man die Verdoppelung 2 bis 3 cm zu beiden Seiten und oben geringer im Mass anordnet, als die eigentliche Thüre.

Für gewöhnliche Thüren besteht die beliebteste und billigste Art der Verdoppelung in Riemen von verschiedenen Formen und Lagen, wobei entweder die Kanten leicht abgefast oder mit Profilhobel versehen sind. Durch Anordnung verschiedener Felder, welche sich leicht durch einen äusseren Rahmen mit Quer- und Mittelfriesen herstellen lassen, sowie durch Profilierung derselben sind hübsche, ja sehr reiche Muster zu erzielen (Taf. 7 f bis k und Fig. 201 u. 203).

Die Stärke der einzelnen Holzlagen ist meist diejenige gewöhnlicher Bretter, also rauh 24 mm, die der ganzen Thür somit gehobelt 42 bis 45 mm. Auf die Nagelung, welche, wie bei den besprochenen einfachen Thüren, mittels geschmiedeter Nägel geschieht, muss grosse Sorgfalt verwendet werden. Die Nägel dürfen nicht planlos in das Holz eingeschlagen werden, sondern

Fig. 202.   **Verdoppelte Hausthüren.**   Fig. 203.

es muss — damit deren sichtbare Köpfe zusammen eine schöne geometrische Figur bilden, die Anordnung mit Ueberlegung und nach einer genauen Einteilung erfolgen. Erhöht kann die Wirkung noch werden durch Verwendung verschiedener Nagelsorten, d. h. solcher mit verschiedenartig geformten Köpfen. Das Beschläge einer gewöhnlichen verdoppelten Thüre, z. B. einer Hausthüre, besteht aus zwei, oder bei schweren Thüren aus drei starken Lang- oder Winkel- und Kreuzbändern mit Kloben in Stein und einem überbauten Zweitourschloss mit starken Messing- oder Eisendrückern und Schliesskloben.

Tafel 7 d und e können als die inneren Hälften, als die Rückseiten verdoppelter Thüren gelten, während f bis k deren Aussenseiten zeigen. Die Stallthüre h ist auf halber Höhe geteilt, wodurch die Möglichkeit geboten ist, während des Sommers oben in den Stall Luft und Licht eindringen zu lassen, während gleichzeitig der untere festgeriegelte Teil noch als Abschluss des Stallraumes dient. Dieselbe Anordnung findet sich auf dem Lande auch häufig an Hausthüren.

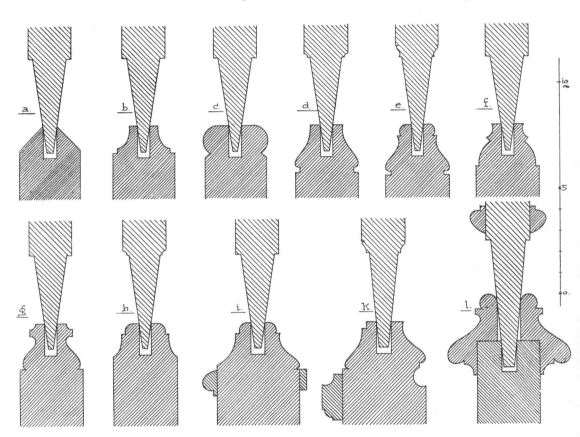

Fig. 204. **Profile von Thürfriesen.**

a auf Fase; b—h auf Hobel; i, k, l auf Hobel und mit aufgelegten Profilleisten.

3. Gestemmte Thüren.

Wie schon bei der Verbindung der Hölzer ausgeführt wurde, ändern die bis jetzt besprochenen Thüren bei Witterungswechsel ihre Form und Grösse. Sie sind daher, zumal ihnen auch meist der Futterrahmen mangelt, überall, wo ein dichter Abschluss wünschenswert erscheint, wenig geeignet. Hier wendet man, und dies gilt insbesondere für die Zimmerthüren, vorteilhafter die sogen. gestemmten Thüren mit Futter und Verkleidung an. Die Konstruktion besteht in der Bildung eines Rahmenwerks (durch Höhen-, Quer- und Mittelfriese) und Ausfüllung desselben mit Füllungen. Wir unterscheiden: stumpf gestemmt (Fig. 125), auf Fase gestemmt (Fig. 126) und auf Hobel gestemmt (Fig. 127). Nach der Zahl der Flügel

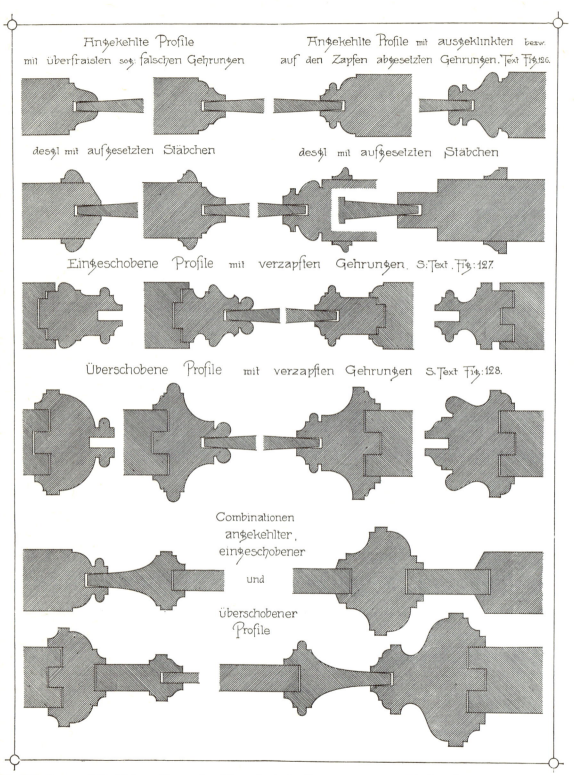

Fig. 205. Ausgewählte Thürprofile aus dem **Musterbuch** des Baugeschäftes von Billing & Zoller in Karlsruhe.

3. Gestemmte Thüren.

giebt es einflügelige und zweiflügelige Thüren; nach der Art der Thürbewegung gewöhnliche Thüren und Schiebthüren.

a) **Einflügelige Thüre.** Die gewöhnliche Grösse im Licht des Futters ist bei Wohnräumen 0,90 × 2,10 m, bei grösseren Räumen, wie Schulsälen etc., 1,00 bis 1,10 × 2,20 m. Von der Zahl der Füllungen ist die genauere Bezeichnung der Thüren abhängig. So heisst z. B. eine Thüre mit zwei Füllungen eine Zweifüllungsthüre; eine solche mit vier Füllungen eine

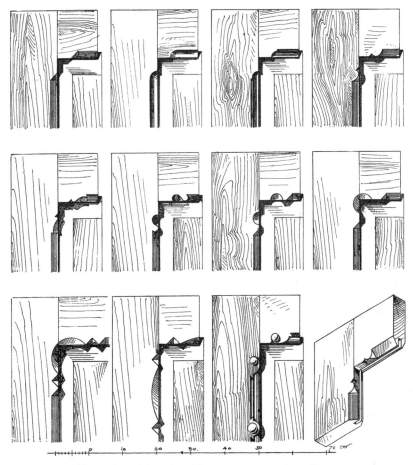

Fig. 206. **Abfasung und Kehlung von Thürfriesen.**

Vierfüllungsthüre. Es giebt Thüren mit 2, 3, 4, 5, 6, 7, 8, 9, 10 und mehr Füllungen, wie auf Tafel 8 ersichtlich. Die Breite der Thürfriese ist bei gewöhnlichen Verhältnissen gleich der halben Schleifdielenbreite, also gefügt ca. 14 cm, die Stärke derselben bei Verwendung von überbauten Schlössern gleich der Schleifdielendicke, also gehobelt ca. 33 bis 34 mm, während Einsteckschlösser mindestens 40 mm beanspruchen. Die Zapfenstärke an den Verbindungsecken ist etwa $1/3$ der Holzdicke, richtet sich aber genau nach der Stärke des benützten Lochbeitels; die Zapfenbreite beträgt 6 bis 8 cm, der Rest ist **Federzapfen**.

Die Stärke der Füllungen ist gehobelt 22 mm, die sichtbare Abplattung 30 mm breit. In der Nute muss die Füllung 2 bis 3 mm Luft haben, um quellen zu können. Die Wangen der Friese werden mit einfacherem oder reicherem Profilhobel versehen. Die Form desselben soll so gewählt werden, dass sie wirkungsvoll ist, ohne die Wangen zu sehr zu schwächen (Fig. 204 b—h). Genügt der angestossene Hobel nicht, soll die Wirkung kräftiger werden, so leimt man

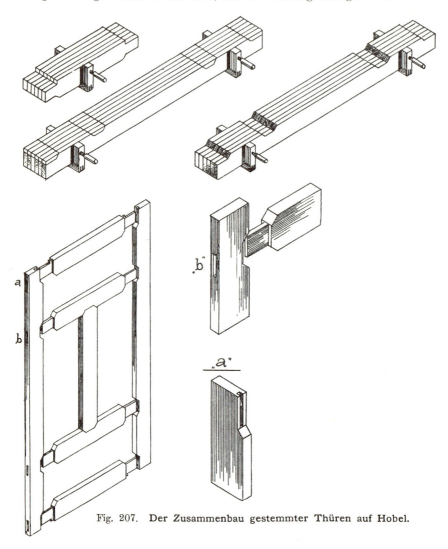

Fig. 207. **Der Zusammenbau gestemmter Thüren auf Hobel.**

sogen. Zier- oder Profilleisten auf, wobei darauf zu achten ist, dass dieselben nur am Fries befestigt sind, um die Füllung am Arbeiten nicht zu hindern (Fig. 204, i, k und l).

Wir schalten bei diesem Anlass mit der Figur 205 eine Anzahl von Thürprofilen der verschiedensten Art ein. Dieselben sind dem Musterbuch des bereits erwähnten Baugeschäftes von Billing & Zoller in Karlsruhe entnommen und von dieser Firma selbst ausgewählt (als unter vielen anderen in der Ausführung besonders wirksam und empfehlenswert).

3. Gestemmte Thüren.

Bei stumpf gestemmten Thüren fast oder kehlt man die Friese und erzielt dadurch, namentlich bei verständiger Farbenbehandlung, ganz gute Wirkungen; Figur 206 stellt einige Beispiele von Fasungen etc. dar.

Der Zusammenbau der gestemmten Thüre wird folgendermassen bewirkt, wobei vorausgesetzt wird, dass es sich empfiehlt, stets mehrere Thüren zusammen fertigen zu lassen.

Nachdem die Friese ausgehoben und gefügt sind, werden je ca. 6 Stück mit zwei Schraubzwingen zusammengeschraubt; auf der inneren Seite werden die Breiten der Querfriese mittels des Winkels und Spitzbohrers, sowie die des Profilhobels mit dem Streichmass vorgerissen und die Gehrungen mit dem Gehrmass angegeben. Nachdem hiernach die Einsätze ausgeschnitten sind, werden die Schraubzwingen abgenommen, auf jedem einzelnen Fries die Zapfenlöcher vorgerissen und mit dem Lochbeitel — nach dessen genauer Breite man sich dabei richtet — ausgestemmt. In ähnlicher Weise werden die Zapfen an den Querfriesen behandelt (Fig. 207).

Nachdem Löcher und Zapfen vollendet, werden die Rahmen zusammengesteckt und die etwa noch nicht genau passenden Gehrungen so lange mit dem Fuchsschwanz oder der Absetzsäge nachgeschnitten, bis sie dicht sind. Hierauf werden allenfallsige Unebenheiten an den Verbindungsstellen sauber verputzt, und erst dann wird die Nutung für die Füllungen sowie das Anstossen der Profile vorgenommen. Das Abplatten der Füllungen ist bei den Holzverbindungen besprochen, weshalb hier davon abgesehen werden kann. Sind diese Arbeiten sämtlich vollendet und ist alles gut passend, so werden die Thüren in den Zapfen verleimt und sorgfältig verkeilt, wobei vorsichtige Arbeiter das betreffende Eck durch Ansetzen einer Schraubzwinge gegen das Ausspringen schützen (Fig. 122, 159 bis 162, sowie Taf. 9). Nach ca. 12 bis 18 Stunden werden die Thüren verputzt und sind nun im allgemeinen fertig gestellt. Beim hierauffolgenden Bestossen der Kanten und beim Einfälzen in das Futter giebt man der Thüre 4 bis 5 mm Luft, damit sie nicht beim geringsten Witterungswechsel festquillt. Vorsichtige Schreiner verteilen diesen Spielraum auf beide Seiten so, dass sie am hinteren Thürfries oben und unten je ein 2 mm starkes Furnierstückchen aufleimen, damit der Schlosser die Thüre beim Beschlagen fest anziehen kann. Nach dem Anschlagen werden die beiden Furniere sauber entfernt.

Wie schon bemerkt, haben die Thüren behufs dichteren Abschlusses ein Futter aus Bord oder besser aus Schleifdielen, welches bis zu einer Wandstärke von 25 cm glatt bleibt, bei einer solchen über 25 cm gestemmt wird, wobei man sich mit den Querfriesen nach denjenigen der Thüre richtet. Gewöhnlich ist dasselbe von Tannenholz, der untere Teil dagegen, die Schwelle, das Schwellbrett, aus Eichen, und zwar ebenso wie das Futter bis zu 25 cm Wandstärke glatt, bei grösserer Abmessung gestemmt (Taf. 9). Verbunden sind die einzelnen Futterteile durch Verzinkung, (Taf. 9). Bei Thüren, welche vom Korridor aus in das Zimmer gehen, lässt man die Schwelle am Boden ca. 1 bis 1$^1/_2$ cm vorstehen (sogen. Anschlagsschwelle), bei solchen zwischen zwei Zimmern legt man sie bündig (Durchgangsschwelle, Taf. 9). Durch die Anschlagsschwelle soll die Luft besser abgehalten und der Thür ein guter Anschlag gegeben werden. Kommt die Schwelle nicht auf einen Balken zu liegen, auf dem sie überall dicht aufliegt, so muss ein Wechsel eingeschaltet werden (Fig. 208).

Das Futter ist befestigt an dem sogen. Thürgestell, welches bei Riegelwänden aus den Thürpfosten, dem Thürriegel und event. dem Schwellenwechsel (Fig. 208), bei 1 Stein (25 cm) starken Wänden aus einem 6 cm starken Bohlengestell (Fig. 209) besteht, dessen oberes und unteres Querholz zu sogen. Ohren verlängert ist und in die Mauer eingreift; in der Mitte der Höhe ist zum festeren Halt je noch ein weiteres Ohr angebracht, oft auch noch eine Eisen-Schlauder.

Minder gut und nicht empfehlenswert ist das Anschlagen des Thürfutters an sogen. Mauerklötzchen und obere Querdielen (Fig. 210), da die Mauerklötze, auch wenn deren Querschnitt wie a und nicht wie b ist, und man vom Quellen und Schwinden ganz absieht, beim Einschlagen der Nägel sich lösen und keinen Halt mehr bieten. Wird die Wandstärke bedeutender als 25 cm, so verwendet man sogen. Pfostengestelle aus Rahmenschenkeln, 9×9 oder bei grösseren Thüren 12×12 cm stark (Fig. 211 und 212).

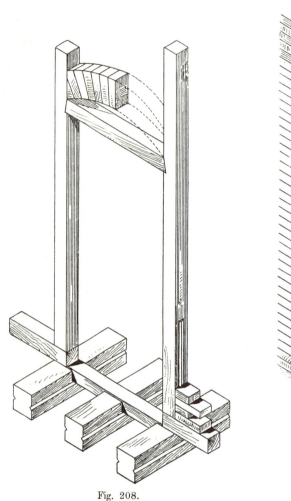

Fig. 208.
Thürgestell. Pfostengestell in Riegelwänden.

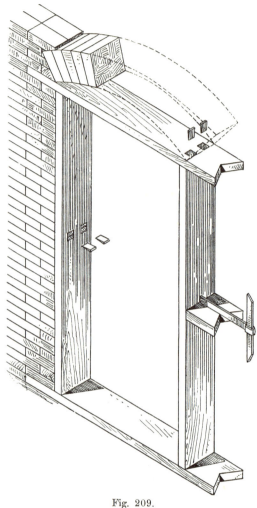

Fig. 209.
Thürgestell. Bohlengestell für 25 cm starke Wände.

Um das Futter in dem Gestell genau in Winkel und Senkel anschlagen zu können, ohne es dicht an die Wand zu bringen, ist es erforderlich, das letztere ca. 8 bis 10 cm nach beiden Richtungen grösser zu machen, als das Schreinerfutter, damit ringsum etwas Luft verbleibt (Taf. 9 und 10). In diesen Zwischenräumen werden dort, wohin die Bänder zu sitzen kommen, sogen. Hinterfütterungen (Taf. 10 aa), d. h. Brettstücke festgenagelt, in welche Schrauben eingreifen können. Die offene, beiderseits sichtbare breite Fuge wird geschlossen durch die **Verkleidung** oder **Bekleidung**, welche auf den Kanten des Futters befestigt wird (Taf. 10). Dieselbe hat

3. Gestemmte Thüren.

den weiteren Zweck, die Thüre schön zu umrahmen und dem Verputz einen sicheren Halt zu geben. Die Verkleidung kann glatt, d. h. ohne Profile (Taf. 10 1) oder profiliert sein (Taf. 10 2 bis 5).

Um einen Anschlag und einen Falz für die Thüre zu erhalten, macht man die Verkleidung auf der Thürseite etwas schmäler und schlägt sie an, wie Tafel 10 1 bis 8 es zeigen; sie heisst

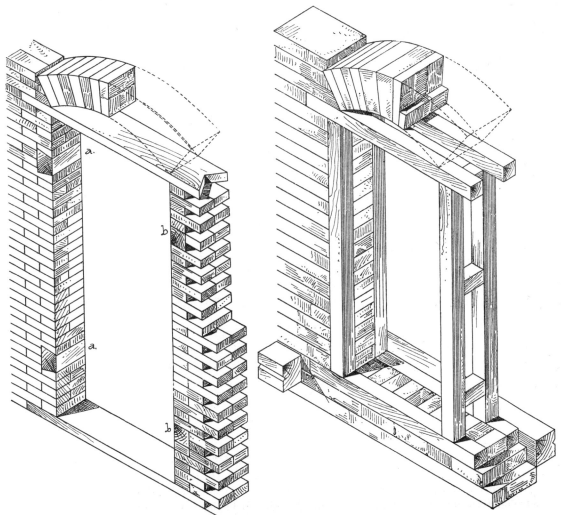

Fig. 210. Mauerklötze (a u. b) und Querdielen.   Fig. 211. Thürgestelle aus Pfosten, Schwellen etc. für mehr als 25 cm. starke Wände.

Thürgestelle.

NB.   Die Lichtmasse der Gestelle müssen um 8 bis 10 cm grösser sein, als die der Thürfutter.

dann **Falzverkleidung**, während die volle Verkleidung der anderen Wandseite **Zierverkleidung** genannt wird. Die Breite der Verkleidung ist ca. $1/7$ bis $1/8$ der Lichtweite der Thüre, die Stärke gehobelt 22 mm (Taf. 10). Bei einfachen Thüren sitzt die profilierte Verkleidung auf dem Fussboden auf, bei besseren Thüren bringt man einen mehr oder minder hohen Sockel an, dessen einfacheres Profil sich im allgemeinen dem der Verkleidung anschmiegt (Taf. 8, 9 und 10). Die Verbindung der Verkleidungsteile ist gewöhnlich stumpf auf Gehrung, oder auf Gehrung überplattet (Taf. 9), oder besser gestemmt und verkeilt, damit die Fugen sich nicht öffnen können. Einflügelige Thüren, welche mit überbauten Schlössern versehen werden, überfälzt man und lässt sie auf das Futter anschlagen (Taf. 10 2, 3, 4 und 6), solche mit Einsteckschlössern werden ganz in das Futter eingelegt (Taf. 10 5, 7 und 8). Damit die Thüren sich in geöffnetem Zustande möglichst an die Wand anlegen, muss Sorge getragen werden, dass der Drehpunkt, das Dornmittel, weit genug herausgelegt wird, wie dies auf Tafel 10 5, 7 und 8 angegeben ist. Geschieht dies nicht, so bleibt die betr. Thüre in schräger Richtung stehen (Taf. 10 2, 3, 4 und 6).

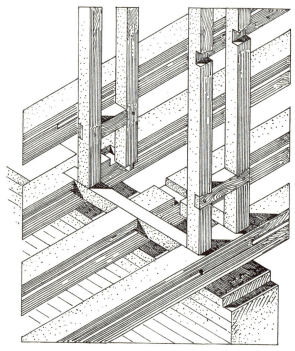

Fig. 212. Thürgestell für mehr als 25 cm starke Wände.

Zum besseren Verständnis der besprochenen und auf Tafel 10 dargestellten Konstruktionen mögen die Figuren 213 bis 216 dienen. Nach Fig. 214 und 216 (Paumelleband und Aufsatzband) tragen sich die Thüren ganz herum, d. h. sie stellen sich in geöffnetem Zustand parallel zur Wand, nach Figur 213 und 215 (Schippenband und Fischband) bleiben sie schräg stehen.

Auf Tafel 9 ist eine einflügelige Thür mit 15 und mit 35 cm tiefem Futter in Ansicht, Horizontal- und Höhenschnitt samt den zugehörigen Details abgebildet.

Tafel 10 zeigt die Einzelheiten in Bezug auf das Futter, die Verkleidung und das Anschlagen der Thüren.

Tafel 11 stellt drei einflügelige Thüren dar, deren Querfriese Profile tragen, während die Höhenfriese mit sogen. überstochenen, d. h. nicht durchlaufenden Hohlkehlen, Fasen oder Rundstäbchen versehen sind. Die Füllungen sind teils zweiseitig, teils nur einseitig abgeplattet.

Die Figuren 217 und 218 bringen zwei einflügelige Thüren aus dem Vatikan in Rom; Fig. 219 zeigt eine solche aus dem Rathaus Lindau; Fig. 220 bringt eine neuzeitige Thüre, die

3. Gestemmte Thüren.

in ihrer Ausstattung etwas über das gewöhnliche hinausgeht; Fig. 221 stellt eine eigenartige, oben verglaste Zimmerthüre gotisierenden Stils dar, welche mit Zinnen verdacht ist, und Fig. 222 giebt eine elegante Zimmerthüre nach dem Entwurf von E. Prignot wieder.

Die Tafeln 12 und 13 zeigen zwei weitere einflügelige Zimmerthüren nach den Entwürfen der Architekten Kayser und von Grofzheim in Berlin. Die Verkleidungen sind mit vorgesetzten Säulen und Pilastern und entsprechenden Verdachungen versehen. Derartige Thüren passen für architektonisch durchgeführte, grosse Räume. Die Tafel 12 zeigt gleichzeitig die anschliessende hohe Täfelung und die Thür der Tafel 13 ist in eine Wandeinrichtung eingebaut, wie sie sich in erster Linie für Bibliotheken eignen dürfte.

Hier sei auch auf die weiter oben gebrachten Figuren 195, 196 und 198 verwiesen, welche u. a. Thüren aus dem Reichstagsgebäude zu Berlin und dem Oberreichsgerichtsgebäude in Leipzig abbilden. Derartig reiche Thüren benützen für die Füllungen und Aufsätze auch die Kunst des

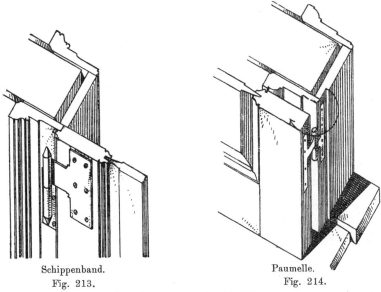

Schippenband.
Fig. 213.

Paumelle.
Fig. 214.

Das Anschlagen der Thüren.

Holzbildhauers und des Intarsienschneiders, sowie in bezug auf die Beschläge diejenige des Kunstschlossers, also Ausstattungsarten, auf welche hier nicht näher eingegangen werden kann.

Tafel 14 enthält die beiden Ansichten, die Schnitte und die Details einer Tapetenthüre, d. h. einer Thüre, die nur auf einer Seite als solche, auf der anderen aber als Wandfläche erscheinen soll. Aus diesem Grunde ist sie auf einer Seite nur stumpf und bündig, auf der anderen dagegen auf Hobel gestemmt und mit Reliefleisten versehen. Die Konstruktion ist die des Ueber- und Einschiebens. Die Thüre liegt ganz im Falz. An den äusseren Kanten ist des dichteren Anschlusses wegen eine Flacheisenschiene als Schlag- und Deckleiste eingefälzt und aussen bündig festgeschraubt; die Verkleidung liegt auf der einen Seite mit dem Verputz in einer Ebene, auf der anderen Seite ist sie wie bei jeder anderen Zimmerthüre. Um die Täuschung möglichst vollkommen zu machen, ist die in dem betr. Zimmer befindliche Lambris auf der unteren Thürhälfte ebenfalls angebracht; hierdurch wird die Verwendung stark heraustragender Bänder (sogen. Aufsatzbänder) erforderlich, damit die Thüre sich trotz der starken Profile noch weit genug zurücklegen kann. Obgleich es eigentlich nicht unmittelbar hierher gehört, so wird doch angefügt, dass direkt auf das Holz der Thüre nicht tapeziert werden darf, damit die Tapete beim Arbeiten des Holzes nicht reisst. Um dies zu verhüten, wird die Thüre zuerst mit Leinwand überspannt, und erst auf diese wird die Tapete geklebt. Die Holzstärken können den beigegebenen Einzelheiten entnommen werden.

Die einflügeligen Thüren werden angeschlagen mit je 2 Bändern (Fisch-, Schippen-, Paumelle- etc. Bändern), und mit einem überbauten oder Einsteckschloss mit Schliesskolben, oder Schliessblech und façonierten Drückern (auf beiden Seiten) versehen.

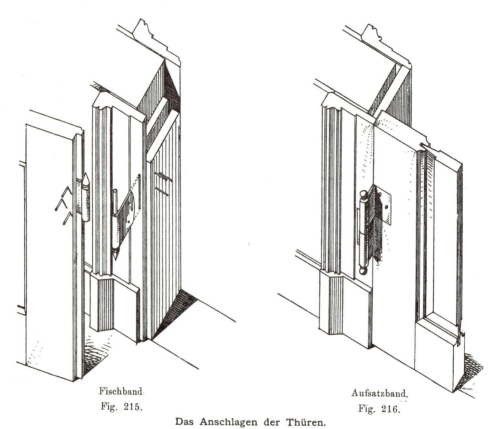

Fischband.
Fig. 215.

Aufsatzband.
Fig. 216.

Das Anschlagen der Thüren.

b) Zweiflügelige Thüre oder Flügelthüre.

Sie soll nur da angewendet werden, wo der Zweck eine breitere Thüre erfordert, nicht aber bloss deshalb, um ein mittelgrosses Zimmer grossartiger erscheinen zu lassen. Ihre Breite ist, wie schon angeführt, 110 bis 160 cm, am passendsten für Wohnhausverhältnisse 1,50 m, weil, wie schon bemerkt, bei dieser Lichtweite jeder Flügel das Minimalmass von 0,75 m erhält. Schmälere Thüren teilt man durch Anbringung zweier Schlagleisten so, dass dem aufgehenden Flügel dieses Mass noch zu teil wird. Die Höhe der Thüre soll im Verhältnis zur Breite stehen, sie soll so sein, dass das Ganze nicht gedrückt erscheint; gewöhnlich ist sie bei 1,50 m Breite gleich 2,40 bis 2,50 m. Einer etwas geringeren Höhe kann man durch Anbringung einer Verdachung mit darunter liegendem Fries nachhelfen, wie dies Tafel 15 in b und e zeigt.

3. Gestemmte Thüren.

Die Konstruktion der Thüre, des Thürgestells, des Futters und der Verkleidung ist wie bei der einflügeligen Thüre, nur sind die Friese hier 40 oder 45 mm statt 34 mm stark, während

Fig. 217 u. 218. **Zwei innere Thüren aus dem Vatikan, Rom.**

die Breite derselben bei gewöhnlichen Verhältnissen gleich der des halben Dielens bleibt. In Bezug auf die Füllungseinteilung ist allgemein zu bemerken, dass zu viele Querfriese, also zu

102   VI. Thüren und Thore.

viele Füllungen nach der Höhe, die Thüre gedrückt erscheinen lassen (Tafel 15a), während eine Thüre mit hohen, schmalen Füllungen verhältnismässig leicht und gestreckt erscheint. Die Breite der Verkleidung ist gleich $1/8$ der Lichtweite (Tafel 10).

Fig. 219. Saalthüre aus dem Rathaus zu Lindau.   Fig. 220. Zimmerthüre.

Tafel 15 stellt 6 Flügelthüren von verschiedener Breite und mit wechselnder Anordnung der Flügel und Schlagleisten dar, sowie deren Grundrisse samt Thürgestellen, Schwellen etc. etc. Während die Thüren d und f mit gewöhnlich profilierten Verkleidungen versehen sind (bei a und c sind sie Platzmangels wegen weggeblieben), haben dieselben bei b und e etwas aussergewöhnliche Formen. Bekanntlich muss man bei Anlage einer sogen. Brüstungslambris oder

Fig. 221. Kneipzimmerthüre von Prof. Direktor Theyer (Archit. Rundschau) 1885.

einer etwas höheren Täfelung die Verkleidung der Thüren ganz besonders stark vor die Wandfläche vorspringen lassen, wenn nicht die Abdeckungsgesimse vorstehen sollen, was schlecht aussieht. Allzukräftige, 60 bis 80 mm weit ausladende Verkleidungen sind aber auch unschön und machen die Thüre plump. Man sucht daher diesem Missstand abzuhelfen, indem man die Verkleidung nur an den Stellen weiter als gewöhnlich vorspringen lässt, wo es nötig ist, d. h. auf die ganze Täfelungs- oder Brüstungshöhe, oder nur da, wo die Gesimsausladungen sich befinden. Tafel 15 b und e, sowie Figur 223 zeigen dies in geometrischer Ansicht; Figur 224 dagegen veranschaulicht es in isometrischer Darstellung.

Fig. 222. Zimmerthüre nach E. Prignot.

Tafel 16 zeigt zwei Flügelthüren von 1,50 m Lichtweite, deren Flügel gleich, also je 0,75 m breit sind, während Tafel 17 zwei solche zeigt, deren aufgehender Flügel 0,84 m Breite hat. Bei sämtlichen vier Thüren ist versucht worden, durch geringe Mittel, wie durch Abfasung der Höhenfriese, Profilierung oder Fasung der Querfriese, wie auch durch verschiedenartige Behandlung der Füllungen, durch Abplatten, Einstossen von Nuten, Rundstäbchen etc. etc. die Wirkung der ganzen Thüre zu steigern. Auch die aufgesetzten Verdachungen sollen wie bei b und e der Tafel 15 hierzu beitragen, wenngleich sie noch den weiteren Zweck haben, die Thüre zu strecken, d. h. scheinbar höher zu machen. Die formale Behandlung dieser Verdachungen war bestrebt, dem Material durch thunlichste Vermeidung ausgesprochener Steinformen Rechnung

Aus der Deutschen Sattler- und Tapezierer-Zeitung.

Fig. 223. Flügelthüre.

zu tragen. Wo die Grenze zwischen Stein- und Holzformen liegt, ist schwer mit Worten zu sagen, da der stilistische Unterschied zum Teil Gefühl- und Geschmacksache ist. Die der Steinarchitektur entlehnten Gliederungen sind in der Bau- und Möbelschreinerei nun einmal so verbreitet

106 VI. Thüren und Thore.

und derart gebräuchlich geworden, dass man sich längst daran gewöhnt hat und sich über die betreffende »Stilwidrigkeit« kaum mehr aufhält. Es kommt eben alles auf das »Wie« an. Man kann eine Hängeplatte anbringen, ohne das Gefühl zu erregen, als ob notwendigerweise ein Dachkanal darauf gesetzt werden müsse. Wird die Platte, dem Material entsprechend, leicht und niedrig gehalten, mit zierlichen Untergliedern gestützt und von ebensolchen Obergliedern bekrönt, so wird bei Einhaltung der richtigen Verhältnisse aller Teile unter sich der Eindruck der Schwere und der Steinarchitektur vermieden werden können. Man vergleiche in dieser Hinsicht die Verdachungen der Thüren b und c auf Tafel 15 und ferner die Beispiele der Figur 225.

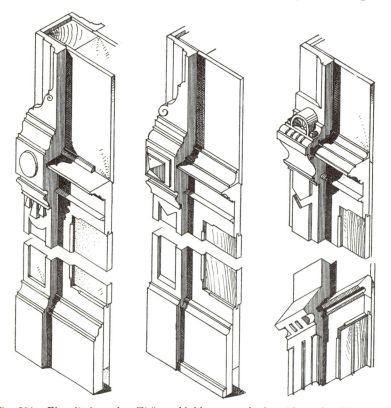

Fig. 224. Einzelheiten der Thürverkleidungen mit Anschluss der Täfelungen.

Die Konstruktion dieser Verdachungen ist, soweit sie nicht schon aus den Abbildungen ohne weiteres ersichtlich, in Figur 228 dargestellt. Es wird, um dem Arbeiten des Holzes zu begegnen, zunächst ein hohler Kasten aus Brettern oder Dielen gebildet, auf welchen die nötigen Gliederungen und Profile aufgeleimt und aufgeschraubt werden. Die Befestigung des Ganzen an der Wand erfolgt mittels Bankeisen, die in die Wand eingeschlagen oder besser mittels sauber aufgeschraubter Eisen, die in die Wand eingegipst werden.

Auf den Tafeln 16, 17 und 18 (auch Fig. 225 zählt hierher) sind einige Beispiele weit ausladender Verdachungen gegeben, zum Aufstellen von Gefässen etc. dienend. Derartige, durch entsprechende Träger gestützte Verdachungen geben der Thüre einen guten Abschluss und wahren gleichzeitig den Charakter der Holzarchitektur.

3. Gestemmte Thüren.

Die Tafeln 19 und 20 zeigen eine sogen. dreiteilige Flügelthüre von 1,35 m Lichtweite, deren Aufgehflügel 0,95 m und deren zweiter 0,40 m breit ist. Die Thüre ist mit profiliertem und dekoriertem Kämpfer versehen, über welchem sich ein zum Hereinklappen bestimmtes Oberlicht befindet. Die Höhenfriese sind gefast, die Querfriese mit Profilhobel versehen und in die Füllungen sind kleine Rundstäbchen eingestossen. Das weitere zeigen der Schnitt und die Details.

Die Tafel 21 bringt zwei Thüren nach E. Prignot in den Formen der Stilzeiten der französischen Ludwige.

Die Figuren 226 und 227 stellen moderne Flügelthüren dar, an denen ausser den geradlinig begrenzten Füllungen auch runde vorkommen und Figur 229 giebt eine Flügelthüre nach dem Entwurfe von Kayser und von Grofzheim in Berlin mit Schnitzereien in den oberen Füllungen.

Das Beschläge einer Flügelthüre besteht aus sechs starken Fisch-, Schippen- oder Paumellebändern (4 reichen nicht aus, da jeder Flügel seiner grossen

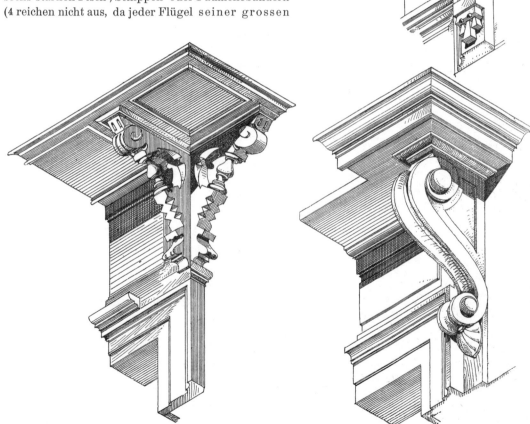

Fig. 225. **Zimmerthür-Verdachungen.**

Höhe wegen 3 Bänder erhalten muss), 2 Schieb- oder besser Kantenriegeln, einem überbauten oder Einsteckschloss mit beiderseitigen Façondrückern.

c) **Schiebthüre.**

Dieselbe hat in erhöhtem Masse den Zweck, zwei Räume so miteinander zu verbinden, dass sie zu gesellschaftlichen Zwecken möglichst wie ein einziger Raum benützt werden können. Die beiden Flügel werden hierbei nicht um eine senkrechte Axe gedreht, sondern seitlich in die hohlkonstruierte Mauer geschoben. Dass bei offener Thüre die Wandfläche nicht versperrt wird, ist der Hauptvorzug dieser Thürgattung, dem andererseits aber oftmals auch nicht unerhebliche

Fig. 226.   Fig. 227.

Flügelthüren.

Nachteile, wie schwerer Gang der Flügel, Verziehen derselben, undichter Abschluss etc. entgegenstehen. Schiebthüren macht man thunlichst gross; man nimmt gewöhnlich die halbe Zimmertiefe zur Breite und lässt sie bis zum Deckengesims hinaufreichen.

Bei der Konstruktion ist wichtig, — gut trockenes und schlichtes Holz ist dabei erste Bedingung — dass an der Thüre keine vorspringenden Profile angebracht werden, da diese einen zu breiten Thürschlitz bedingen. Im übrigen ist die Schiebthüre wie eine gewöhnliche Thüre zusammengebaut. Eine eigentliche Schlagleiste ist überflüssig, da man die Thürflügel vermittels **Wolfsrachenverschluss** (Fig. 230) ineinander greifen lässt, um einen dichten Abschluss zu erzielen.

Die Thürflügel sind nach oben und nach beiden Seiten etwas grösser als das Lichtmass der Thüre im Futter und dort durch 2 angeschraubte Festhaltungsleisten gegen zuweit gehendes Herausziehen gesichert. Das Futter, welches glatt oder gestemmt sein kann, ist in der Mitte

3. Gestemmte Thüren.

nach der Höhe in einer Weite geschlitzt, dass die Thüre sich gut bewegen lässt und überdies noch einen gewissen Spielraum hat, der sie gegen Beschädigungen schützt. Die Schwelle und die Verkleidung sind wie bei den gewöhnlichen Flügelthüren. Von besonderer Wichtigkeit ist, vor jedem Flügel je eine Seite der Wand in Holz zu konstruieren und mit Beschlägen zu versehen, damit man die Thüre bei etwaigem Steckenbleiben freilegen kann und die nötige Reparatur vorzunehmen im stande ist. Das Beschläge der Schiebthüre war seither meistens das mit Rollen,

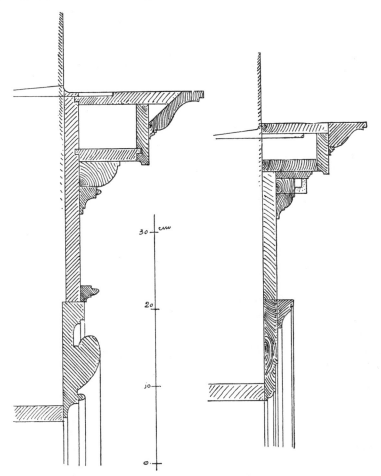

Fig. 228. **Konstruktion von Thürverdachungen.**

wie es auf Tafel 51 dargestellt ist. Die letzteren waren oben angebracht und bewegten sich auf einer zweckentsprechend befestigten Eisenschiene. Trotz mannigfacher Verbesserungen, namentlich solcher in Bezug auf leichtere Beweglichkeit, ist es aber doch nicht gelungen, das gegen die Schiebthüre gefasste Vorurteil, als bleibe eine solche, selbst im besten Falle, dann und wann einmal stecken, zu beseitigen. Mit dem auf Tafel 22 angedeuteten Weickumschen Patentbeschläge dürfte es eher gelingen, das Publikum für die Schiebthüren zu gewinnen. Hier sind die Metallrollen durch Hartgummikugeln (Taf. 22, Schnitt a—b) ersetzt, welche eine fast geräuschlose und

110 VI. Thüren und Thore.

ausserordentlich leichte Beweglichkeit der Thüre ermöglichen. Unten sind beiderseits am Futter, bezw. an der Thüre je 2, zusammen also 4, um eine senkrechte Axe drehbare Gummirollen als Leitrollen angebracht. Sollen die Kugeln nicht oben, sondern unten angebracht werden, — namentlich bei schweren Thüren und Thoren — so benützt man die Konstruktion e—f und lässt die Kugeln (Gussstahl bei schweren Thoren) auf einer flachen Eisenrinne laufen, während die Führung nach oben verlegt wird.

Zur Vervollständigung des Beschläges ist ausserdem noch erforderlich ein Einsteckschloss für Schiebthüren, wie es auf Tafel 82 dargestellt ist, wobei der Schliessriegel nicht horizontal vorwärts, sondern im Bogen bewegt wird und sich in das Schliessblech einhakt; ausserdem an jedem Flügel eine, durch einen Druck auf eine Feder selbstthätig vorschiessende Ausziehvorrichtung (Knöpfe würden zu weit vorspringen).

4. Verglaste Thüren. Glasthüren.

Die sämtlichen Glasthüren haben ausser dem Zweck, einen Raum abzuschliessen, noch den weiteren, dem dahinter liegenden Raume Licht zuzuführen. Ihre Hauptverwendung finden sie im Innern der Gebäude, da sie der Zerbrechlichkeit des Glases wegen als äussere Thüren, die Schutz geben sollen, nur unter gewissen Voraussetzungen zu gebrauchen sind. Man fertigt sie im allgemeinen

Fig. 229.
Zweiflügelthüre nach dem Entwurf der Architekten Kayser und von Grofsheim in Berlin.

Fig. 230.
Mittelfuge bei Schiebthüren. Sog. Wolfsrachenverschluss.

so, dass der untere kleinere, nur bis auf die Brüstungshöhe reichende Teil aus Holz besteht, während der obere grössere verglast, also einem Fenster ähnlich konstruiert wird.

Wir unterscheiden:
1) Glasthüren, welche am Aeusseren des Hauses angebracht sind, Balkon- und Verandathüren etc.
2) Solche im Innern der Wohnungen, Glasabschlüsse, Vorplatzthüren etc.

Die ersteren erhalten — ausgenommen bei Fachwerksbauten — einen Futterrahmen, welcher an das Steingestell befestigt wird; sie sind kräftig und solid zu konstruieren, also etwa wie Hausthüren; das Material soll wetterbeständig (Eichen, Forlen) sein. Die Verglasung gleicht meist derjenigen der Fenster, besteht also aus grossen und kleinen Scheiben, aus Tafel- oder Spiegelglas u. s. w.

Die inneren Thüren dagegen erhalten Futter und Verkleidung, ihre Konstruktion ist die der Flügelthüren gleicher Grösse, ihr Material meist Tannenholz; zur Verglasung wird gewöhnlich dekoriertes, mattiertes, geätztes oder bemaltes Glas verwendet.

Die Grösse beider richtet sich — wenn sie nicht Fassadenthüren sind, deren Form und Anlage sich nach der Architektur bestimmt — nach dem Zweck, der in den meisten Fällen es wünschenswert erscheinen lässt, sie so gross als möglich zu machen.

a) **Balkonthüre und Verandathüre.**

Man versteht darunter eine Thüre, welche in ihrem oberen Teil den Fassadenfenstern gleich konstruiert, im unteren dagegen in Holz gestemmt wird. Um das erstere zu erreichen, ohne die Thüre zu schwach zu machen, giebt man den Höhenfriesen nur unten die gewöhnliche Breite (halbe Dielenbreite) und setzt sie auf der Höhe des Brüstungsfrieses nach oben schräg ab, so dass sie nach dieser Richtung hin nur noch wie der etwas breite Höhenschenkel eines Fensters erscheinen (Fig. 231). Die Friesdicke ist 5 bis $5^{1}/_{2}$ cm, je nach der Grösse der Thüre und der Stärke der Verglasung. Am unteren Querfries befindet sich ein Wetterschenkel mit Wassernase (Fig. 232). Die Thüre hat, wie ein Fenster, einen Futterrahmen, einen Kämpfer und oberhalb desselben Fensterflügel. Von Wichtigkeit ist eine Dichtung der Thüre gegen das Eindringen von Regenwasser am Boden. Man erreicht dies am besten durch Anbringung von Fensterladen, welche den Regenschlag abhalten oder durch Anlage einer doppelten Thüre. Geht dies aus irgend welchem Grunde nicht an, so kann man das Zurücklaufen des Regenwassers durch Anbringung eines Eisenblech-Beleges (mit ringsum genieteten Winkeleisen) verhindern, welcher in den Stein eingreift und dort befestigt ist (Fig. 232). Grundregel bei allen derartigen Verwahrungen muss sein, den Anschlag an der Schwelle, welcher das Wasser am Eindringen verhindern soll, so weit nach hinten zu legen, dass noch das an der Hinterkante des Futterrahmens herablaufende, durch den Sturmwind eingepeitschte Regenwasser vor diesem Anschlag abtropft und so nach aussen abfliesst.

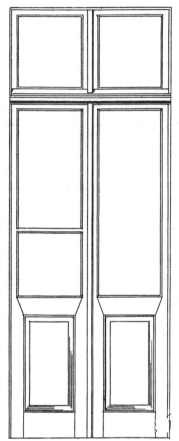

Fig. 231. Balkonthüre.

Oberbaudirektor Dr. Durm in Karlsruhe verwendet zum Zwecke der Dichtung an Stelle des Eisenblechbeleges gusseiserne Schutzschwellen (Fig. 233). Dieselben werden nach Versetzen des Thürgestelles, also nach Fertigstellung des Rohbaues von hintenher eingestreift. Einer Belastung haben sie demnach nicht zu widerstehen. Die solide und empfehlenswerte Konstruktion wird allerdings durch die Modellkosten für jeden einzelnen Fall verteuert, wenn man nicht die Balkonthürbreite dem vorhandenen Modell anpassen will oder kann.

112 VI. Thüren und Thore.

b) Glasabschluss.

Er ist eine Erfindung der Neuzeit. Sein Zweck ist, den äussersten Abschluss einer Wohnung gegen eine andere oder gegen den Vorplatz, das Treppenhaus etc. zu bilden, so dass die ganze Wohnung mit dieser einen Thür abgeschlossen ist. Zum anderen soll durch ihn dem dahinter liegenden Raum Licht zugeführt werden. Er kommt eigentlich nur im Miethaus vor, mit dessen moderner Entwickelung auch die seinige Hand in Hand ging. Während er vor etwa 30 Jahren nur ganz vereinzelt zu finden war, kann man heute kaum begreifen, wie es eine Mietwohnung geben kann ohne Glasabschluss und diese Unentbehrlichkeit ist sehr oft die Ver-

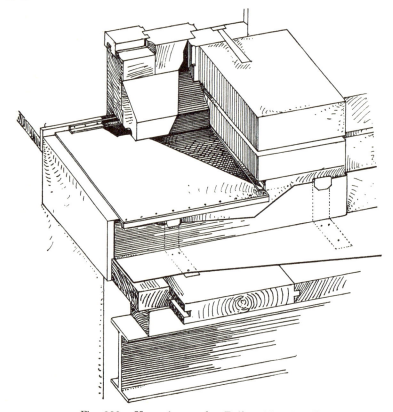

Fig. 232. **Verwahrung der Balkonthürschwelle.**

anlassung zu grossen Ausgaben, welche man zu seiner architektonischen Durchbildung macht. Im Einzelwohnhaus kommt er nicht vor, da an die Stelle des Wohnungsabschlusses die Hausthüre tritt; höchstens bringt man noch eine sog. **Pendelthüre** an, welche zwischen die Hausthüre und die Wohnung zur Abhaltung der Zugluft eingesetzt wird.

Die Grösse des Glasabschlusses richtet sich nach dem Zweck. Ueberall da, wo kein anderes Licht dem inneren Wohnungsvorplatz zugeführt werden kann, ist er so gross als möglich anzunehmen, womit jedoch nicht gesagt ist, dass der ganze Abschluss beweglich sein muss. Nur in den seltensten Fällen kommt dies vor, und zwar dort, wo Gegenstände durchtransportirt werden müssen, für deren Grösse eine gewöhnliche Thüröffnung nicht ausreicht. Da die grössten in einer Wohnung vorkommenden Gegenstände die Möbel sind und diese durch die Zimmerthüren gehen müssen, so braucht selbstredend die Thüre des Glasabschlusses auch nicht grösser zu sein. Immerhin wird das Ein- und Umziehen aber erleichtert, wenn dieser Durchlass eine reichliche Breite hat.

1) Ist ein grosser Raum durch den Glasabschluss abzusperren, so giebt man gewöhnlich der Thüre desselben die Abmessungen, welche die grösste innerhalb des Abschlusses vorhandene Thüre hat. Der verbleibende Rest wird meist beiderseits gleichartig verteilt und feststehend angeordnet.

2) Hat der Raum eine Breite von 1,2 bis 1,8 m, so bringt man wohl auch eine Flügelthüre in der ganzen Breite an und giebt dem Hauptflügel, der für gewöhnlich allein geöffnet wird, die Mindestbreite von 0,70 m. Bei einer Breite von über 1,5 m werden beide Flügel gleich.

3) Ist der Raum noch schmäler, so wird der Glasabschluss zur einflügeligen Thüre in der Breite von 0,9 bis 1,1 m.

Der Raum über dem Kämpfer der Thüre wird entweder fest verglast oder es werden bewegliche Fensterflügel zum Zwecke der Lüftung eingesetzt.

Das Material ist verschieden, gewöhnlich Tannenholz. Die Art der Verglasung hat sich in den letzten 15 Jahren ganz gewaltig geändert. Während man früher fast nur Sprosseneinteilung hatte, zu deren Verglasung man Musselinglas mit farbigen Rosetten in den Ecken verwendete, benützt man heute, dank der Hebung der Glasindustrie, meist ganze Scheiben, und zwar blanke und mattierte, oder ornamentierte, geätzte etc., oder man wählt sog. Bleiverglasung in Rauten- oder Butzenscheibenform.

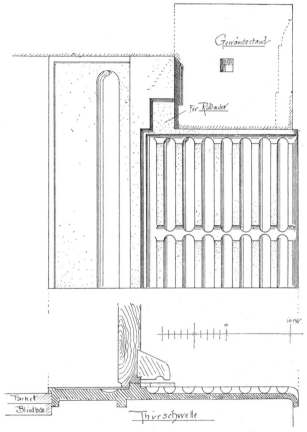

Fig. 233. Gusseiserne Balkonthürschwelle nach Oberbaudirektor Dr. Durm.

Tafel 23 zeigt einen einfachen sogen. dreiteiligen Glasabschluss von 1,44 m Lichtweite, dessen einer Flügel 1,10 m breit ist, während der zweite nur 0,34 m misst. Die eine der beiden (die rechte) über den Kämpfer sich fortsetzenden Schlagleisten ist blind.

Tafel 24 stellt ebenfalls einen Glasabschluss dar; derselbe ist vierteilig und hat 2,60 m Lichtweite. Die beiden seitlichen Teile von je 0,60 m Breite sind feststehend angeordnet, während die inmitten gelegene Thüre eine Flügelbreite von je 0,70 m bekommt. Seine Hauptfestigkeit erhält das Ganze durch starke, strebepfeilerartig geformte, senkrechte Gliederungen, welche, durch Kerbungen, Rosetten etc. geschmückt, in den starken profilierten Kämpfer eingreifen und dadurch ein kräftiges Gerippe für die Thüre bilden. Die einzelnen Konstruktionen gehen deutlich aus den beiden Schnitten, den Isometrien, sowie den auf Tafel 25 dargestellten Details hervor.

Figur 234 bringt einen Glasabschluss von 1,45 m lichter Breite. Der festgeriegelte, für gewöhnlich nicht zu öffnende Teil ist 0,55, der Thürflügel 0,90 m breit. Das Oberlicht über dem Kämpfer kann mit Vorreibern befestigt oder als Klappflügel angeschlagen werden.

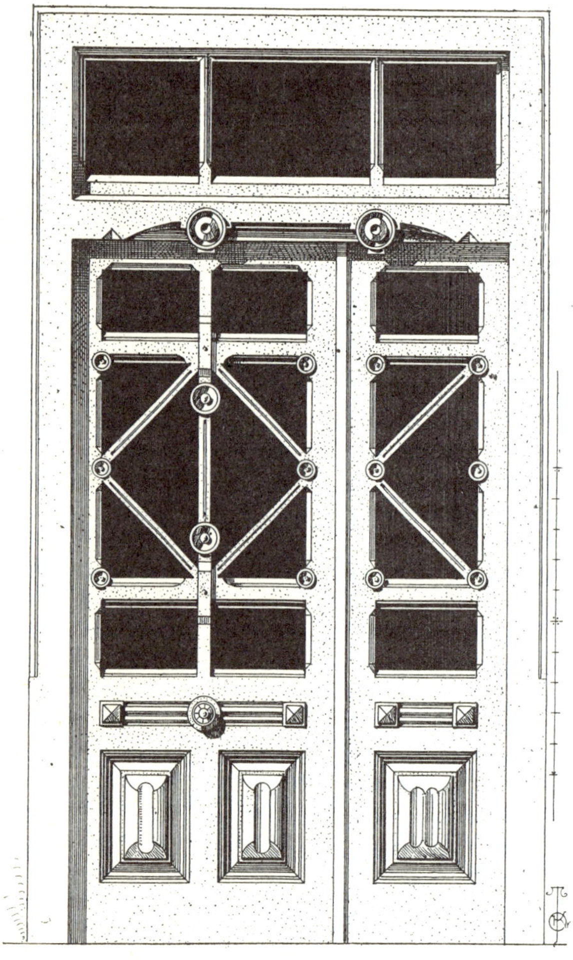

c) **Wartesaalthüre** (Vorplatzthüre).

Man versteht darunter eine Thüre, welche bestimmt ist, einen Raum zwar abzuschliessen, dabei aber doch jedermann zu gestatten, leicht in diesen Raum zu schauen. Es soll hierdurch eine Art Kontrolle geschaffen werden, welche für Wartesäle von ganz besonderem Wert ist. Die Grösse der Thüre richtet sich nach der Architektur. Die Konstruktion derselben ist wie die der gewöhnlichen Flügelthüre und des Glasabschlusses; die Verglasung kann ganz blank oder mit gemusterter Zeichnung versehen sein.

Tafel 26 zeigt eine doppelt überschobene Vorplatz- oder Wartesaalthüre, in zwei Varianten, wobei die linke Hälfte mit Sprossenwerk und mit blanker Verglasung, die rechte dagegen mit einem unteren und oberen Setzholz, sowie mit Blei-Rautenverglasung versehen ist.

Tafel 27 bringt zwei Vorplatzglasthüren, wovon die linke mit auf die Holzfüllungen aufgelegten Zierleisten versehen und mittels Stichbogen abgeschlossen ist, während die rechte unten eine äussere Verdoppelung mit Zuziehgriff zeigt und horizontal abgedeckt ist. Die oberen Flügel können zum seitlichen Aufgehen wie zum Aufklappen eingerichtet werden.

Fig. 234. **Glasabschluss.**

Das Beschläge der Glasabschlüsse wie der Vorplatzthüren besteht ausser der nötigen, von Fall zu Fall sich ändernden Anzahl starker Fischbänder, aus 2 Schieb- oder Kantenriegeln mit Schliessblechen, einem zweitourigen überbauten oder Einsteckschloss und dem Beschläge der oberen Flügel (Vorreiber oder Scharniere und Vorreiber, oder Scharniere, Scheren und Federfallen, Tafel 80 und Fig. 350 bis 357).

d) **Pendelthüre.**

Sie ist eine innere Thüre, und zwar die erste nach dem Eintritt ins Haus und hat den Zweck, die Zugluft von aussen abzuhalten und die rasche Entweichung der inneren Wärme zu verhindern. Gewöhnlich wird sie in der Ansicht wie eine zweiflügelige Glasthüre konstruiert von welcher sie sich aber hauptsächlich durch die Art ihrer Bewegung unterscheidet. Während nämlich die gewöhnliche Thüre sich nur nach einer Seite öffnen lässt, bewegen sich die Pendelthür-Flügel nach beiden Seiten und haben daher in der Mitte keinen Anschlag und Verschluss. Soll die Thür geöffnet werden, so drückt man leicht gegen einen oder beide Flügel, worauf dieselben sich öffnen, sofort nach Nachlassung des Druckes aber, durch angebrachte Federn etc. sich wieder schliessen, dabei noch einige Male nach beiden Seiten ausschlagend oder pendelnd. Der hintere Thürfries ist leicht abgerundet und greift in eine entsprechende Hohlkehle ein, teils

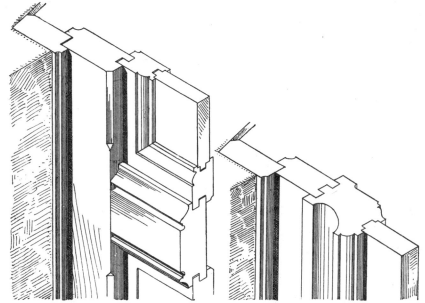

Ueberschobener Kehlstoss mit eingeschobener Füllung.   Eingeschobener Kehlstoss in der Nute.
Fig. 235. **Konstruktion von gestemmten Hausthüren.**

um einen dichten Abschluss zu bilden, teils um das Durchsehen an dieser Stelle zu verhindern. Dass diese Thüre sowohl hinten wie vorn, oben und unten Luft, d. h. Spielraum haben muss, um leicht beweglich zu sein, braucht wohl kaum besonders hervorgehoben zu werden.

Auf Tafel 28 sind zwei Pendelthüren mit hohen und niederen Glasscheiben dargestellt. Die Konstruktion der einen zeigt überschobene, die der anderen gestemmte Arbeit mit aufgeleimten Zierleisten. Man fertigt die Pendelthüren meist in besserem Material, in Mahagoni, Nussbaum etc., poliert oder reibt sie, wenn sie matt bleiben sollen, mit Politur ab; eine Behandlung derselben mit Wachs hat sich nicht bewährt, weil jeder Wassertropfen Spuren hinterlässt.

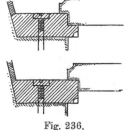

Fig. 236.
Bildung von Futterrahmen.

Das Beschläge besteht entweder:
    a) aus Zapfenbändern, von denen jeweils das untere mit der Pendelfeder verbunden ist, oder
    b) aus einem oberen Zapfenband und einem unteren Spenglerschen Rollenpendel, oder
    c) aus einem oberen Zapfenband und dem Weickumschen Patentpendelthürbeschlag.

Nach a) bewegt sich die Thüre jeweils auf gleicher Höhe, nach b) oder c) steigt dieselbe

116   VI. Thüren und Thore.

und muss daher oben genügend Luft haben. An jedem Thürflügel wird je ein Bronzeknopf zum Aufdrücken angebracht. (Vergl. Abschnitt XII. Beschläge.)

e) **Windfang.**

Man versteht darunter nicht nur eine Thüre, sondern einen auf zwei, drei oder mehreren Seiten geschlossenen Kasten, welcher mit einer Decke versehen, derart hinter der Eingangsthüre

Fig. 237. Fig. 238.
Regenschlag-Abdeckungen von nach aussen und nach innen überschobenen Füllungen.

angebracht ist, dass beim Ein- oder Ausgehen von Personen die in der Nähe Weilenden nicht durch Zugwind belästigt werden. Soll dieser Zweck erreicht werden, so muss der Windfang in seinen Dimensionen so bemessen und so gefertigt sein, dass man die innere Thüre erst dann öffnen kann, wenn die äussere bereits geschlossen ist, und umgekehrt. Hierdurch wird in dem so geschaffenen

4. Verglaste Thüren. Glasthüren.

Raum die Luft stets etwas vorgewärmt sein, so dass die Wirkung beim Oeffnen der Thüre weniger unangenehm ist. Die innere, die eigentliche Windfangthüre, ist wie eine ein- oder zweiflügelige Pendelthüre gebaut und wird wie diese angeschlagen.

Auf Tafel 29 ist ein zweiflügeliger Windfang in Hauptansicht und Grundriss dargestellt; die zugehörige Seitenansicht ist im halben Massstab beigefügt. In den Grundriss ist die Decke

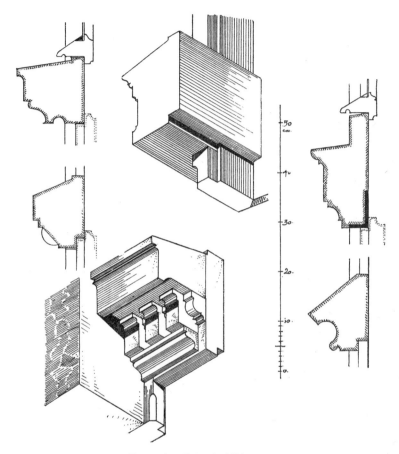

Fig. 239. **Kämpferbildungen.**

eingezeichnet. Der ganze Windfang ist im Lichten 1,92 m breit und 1,23 m tief; der Thürflügel ist 0,77 m breit. Je nach den verfügbaren Mitteln wird der Windfang in Tannen, Eichen oder Nussbaum etc. ausgeführt; das Beschläge kann das der Pendelthüre sein doch wird auch sehr häufig ein fester Verschluss gewünscht, in welchem Falle man Fischbänder oder Zapfenbänder, Kantenriegel und ein Schloss anzunehmen hat.

5. Hausthüren und Thore.
a) Gestemmte ein-, zwei- und dreiflügelige Hausthüren.

Die Hauseingangsthüre, kurzweg auch Hausthüre genannt, ist im Freien angebracht, daselbst jeder Witterung ausgesetzt und mehr als jede andere Thüre bestimmt, Unberechtigten

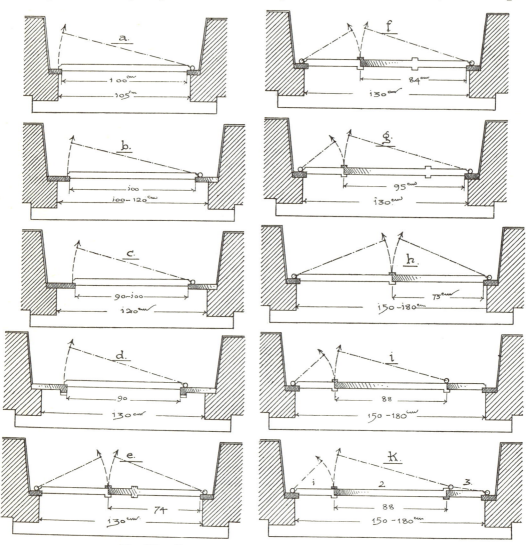

Fig. 240. Grundriss-Anlagen von Hauseingangsthüren.

den Eintritt in das Haus zu verwehren. Sie unterscheidet sich daher auch ganz wesentlich von der inneren Thüre dadurch, dass sie

1) stärkere Konstruktion, 2) wetterbeständigeres Material, 3) einen Futterrahmen, 4) einen Wetterschenkel, sowie das Regenwasser rasch ableitende Profile, 5) meist eine mit

Gitter versehene Glasfüllung, 6) Vorkehrung gegen das Aushängen und schliesslich 7) einen Kämpfer mit Oberlicht erfordert.

Was zunächst die stärkere Konstruktion anbelangt, so wurde bezüglich der Holzverbindungen betont, dass bei der gewöhnlichen Verbindung auf Nut und Feder die Festigkeit an den Verbindungsstellen am geringsten ist, weshalb man für Hausthüren die in den Figuren 130 bis 140 und in Figur 235 dargestellten Konstruktionen wählt. Es sind dies durchgängig Verbindungsarten, welche sehr stark und, wenn richtig aufgefasst und verwendet, auch geeignet sind, das Regenwasser rasch abzuleiten und für die Thüre unschädlich zu machen. Wir unterscheiden dabei:

a) die Konstruktion mit überschobenen Füllungen,
b) diejenige mit eingeschobenen Füllungen in ein- oder überschobene Kehlstösse in der Nute.

Bei beiden Konstruktionen und deren Varianten ist alle Sorgfalt darauf zu verwenden, sie so zu gestalten, dass die Festigkeit der Thüre an den Verbindungsstellen keine geringere oder doch keine wesentlich geringere ist als an anderen Stellen.

Noch widerstandsfähiger als diese Konstruktion ist diejenige der mittelalterlichen Eingangsthüren, wie wir sie an alten Baudenkmälern, namentlich an Kirchen etc., wahrnehmen. Diese Thüren waren aus starken Eichenbohlen gefertigt, gespundet, innen mit starken Querriegeln und Streben, aussen mit den die ganze Fläche überspinnenden Zierbeschlägen versehen und mittels durchgehender Schrauben fest verbunden. Eine solche Thüre macht einen sehr soliden, aber auch sehr ernsten Eindruck und ist schon aus diesem Grunde heute noch für Kirchen beliebt, obgleich bei unseren geregelten Verhältnissen eine derartige Konstruktion überflüssig stark ist.

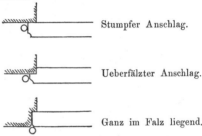

Fig. 241. Anschlag der Hausthüren.

Stumpfer Anschlag.

Ueberfälzter Anschlag.

Ganz im Falz liegend.

Schon die italienische und deutsche Renaissance haben sich von dieser, für den Profanbau weniger passenden Thüre losgesagt und die gestemmte, mit Profilleisten versehene, im übrigen sehr verschieden geformte, kräftig wirkende Hausthüre, als auch den Anforderungen genügend, angenommen. Es kann somit von einer weiteren Besprechung der gespundeten Thüre abgesehen werden.

Während das Tannenholz für Thüren im Innern sich besonders eignet, wird es am Aeussern des Hauses, wo es der Hitze wie der Kälte, dem Regen und Sonnenschein ausgesetzt ist, doch zu rasch zerstört und ist daher für Hausthüren ungeeignet. Dagegen haben sich Eichen- und nach ihm das harzreiche Kiefernholz sehr gut bewährt, während dies von Buchen, das sonst seiner Härte und Dauerhaftigkeit wegen sehr wohl geeignet wäre, nicht gesagt werden kann, da dasselbe beständig »arbeitet«.

Um die Thüre auf dem mehr oder minder rauhen Stein dicht schliessend, und die beim Schliessen der Thüre entstehenden, heftigen Erschütterungen weniger schädlich zu machen (indem man sie auf das ganze Gestell verteilt), ist die Bildung eines Futterrahmens erforderlich oder doch dringend erwünscht. Derselbe wird 4 bis 4$\frac{1}{2}$ cm stark und so breit gemacht, dass er den ganzen Gewändeanschlag deckt und noch 1$\frac{1}{2}$ bis 2 cm ins Licht der Thüre vorspringt. Befestigt wird er durch 7 bis 9 starke Steinschrauben mit versenkten Muttern (Fig. 236), wobei zwischen Stein und Rahmen eine Schichte Haarkalk (wie bei den Fenstern) eingebracht wird. Soll verhindert werden, dass das Regenwasser am Fusse der Thüre in das Haus eingepeitscht wird, so ist ausser der in Figur 232 gezeigten Schwellenkonstruktion noch die Anbringung eines Wetterschenkels erforderlich, wie dies in Fig. 237a angegeben ist. Derselbe

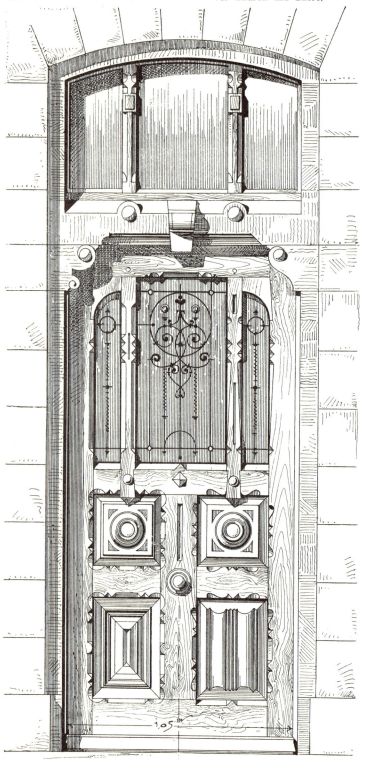

Fig. 242. Hausthüre mit Steinkämpfer und Oberlicht.

ist ein wenig in den unteren Thürfries eingesetzt, bezw. auf den Grat eingeschoben, zum besseren Halt durch Holzschrauben fest verbunden und zur Ableitung des Regenwassers mit Wasserschräge und Wassernase versehen. Aber auch bei den übrigen Profilen ist auf diesen Umstand zu achten, um einer raschen Fäulnis vorzubeugen. Die Figuren 237 und 238 zeigen an einigen Beispielen, wie man bald oben, bald unten, je nachdem die Füllung nach aussen oder innen überschoben ist, wetterschenkelartige Schutzprofile anbringen kann.

Von grossem Wert für die Sicherheit der Bewohner ist es, an der Hausthüre eine Vorrichtung zu haben, welche ermöglicht, den Einlass Begehrenden zu sehen, ohne die Thüre vorher öffnen zu müssen. Zu diesem Zwecke bringt man eine kleine und, wenn noch Licht in den Raum hinter der Hausthüre fallen soll, eine grössere Glasfüllung in der Thüre an, die man zur Sicherheit gegen Einbruch mit einem Eisengitter versieht. Die Befestigung des Gitters hat so zu geschehen, dass man es nicht leicht von aussen beseitigen, also abschrauben kann, weshalb es sich empfiehlt, dasselbe in einen Falz zu legen. Ebensowohl kann man den Rahmen des Gitters aus Winkeleisen fertigen, dessen einer Schenkel sich an die Innenseite des Frieses anlegt.

Um die Thüre gegen das unbefugte Aushängen zu

sichern, legt man sie am einfachsten in einen Falz des Futterrahmens ein, man »überfälzt« die Thüre (Fig. 236).

Gewöhnlich macht man die Hausthüre nur so hoch als nötig, giebt ihr also eine der Bedeutung der Fassade entsprechende Höhe von 2,20 bis 2,40 m. Da nun bei gleichhoher Durch-

Fig. 243. **Hausthüren.** Fig. 244.

führung des Thürsturzes und der Fensterstürze die volle Thürhöhe immer weit bedeutender wird, so benützt man das Mehrmass, um ein Oberlicht zur Beleuchtung des Flurs etc. zu erzielen; mit anderen Worten: weil unten die Brüstungshöhe hinzukommt, so vermeidet man allzu hohe Thürflügel durch Zugabe eines Oberlichts am obern Ende. Zu diesem Zwecke zapft man in die beiden Höhenfriese des Futterrahmens einen starken, entsprechend profilierten Kämpfer ein, an welchen unten die Thüre und oben der Fensterflügel des Oberlichts anschlägt. Es leuchtet ein, dass der Kämpfer zumal bei breiten, schweren Thüren sehr stark sein muss, so dass man ihn oft unter Zuhilfenahme von Eisen konstruiert. Die Figur 239 zeigt verschiedene Arten von Kämpfern, die teils aus dem vollen Holz gearbeitet, teils zusammengesetzt und verleimt sind, die entweder mit dem Futterrahmen hinten bündig gehen, oder der Verstärkung wegen über denselben vorspringen. Bei der Profilierung geht man von denselben Gesichtspunkten aus wie bei den übrigen Gliederungen; man hält auf rasche und möglichst vollständige Ableitung des

Fig. 245. **Hausthüren.** Fig. 246.

Regenwassers. Der Kämpfer erhält daher oben eine Wasserschräge mit hinterem Anschlag für die Fensterflügel und unten eine Wassernase.

In Bezug auf die Lichtbreite giebt es einflügelige, zwei- und dreiflügelige Hausthüren; bei sämtlichen geht man von der Ansicht aus, dass Flügel über 1,00 m Breite zu schwer und unpraktisch werden. Man behandelt daher Thüren von über 1,10 m Weite im allgemeinen als zweiflügelige, giebt aber dem aufgehenden Flügel stets eine Breite von mindestens 0,70 m (Fig. 240 e bis h); andernfalls muss man den Futterrahmen breiter als gewöhnlich machen und ihn in das Licht der Thüre vorspringen lassen (Fig. 240 b bis d).

5. Hausthüren und Thore.

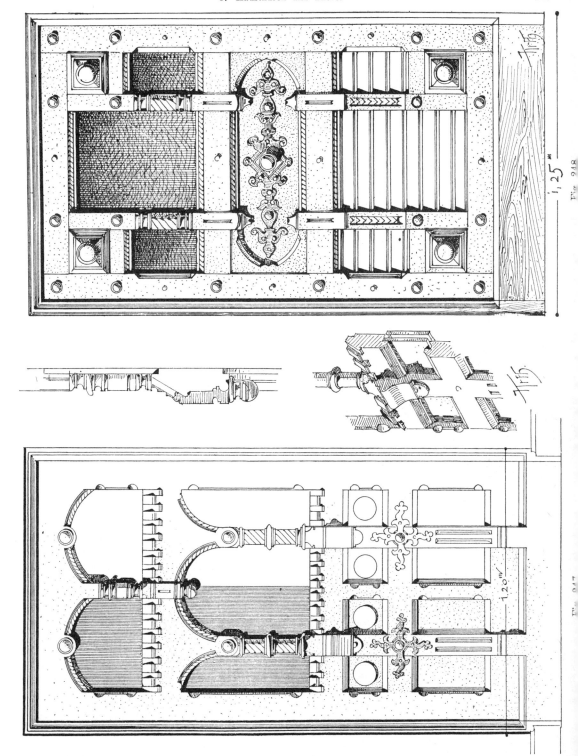

Fig. 240i zeigt eine dreiteilige Thüre, bei welcher zwei Flügel beweglich sind, während der dritte festgeriegelt oder festgeschraubt ist; nach Fig. 240k sind sämtliche drei Flügel in Bänder gehängt; für gewöhnlich steht Flügel 3 fest und an ihm ist Flügel 2 angeschlagen. Im Notfall können die drei Flügel zugleich geöffnet werden.

Fig. 249.
Hausthür mit Briefeinwurf.

Fig. 250.
Eingangsthüre, entworfen von Professor Levy in Karlsruhe.

Mit den in Fig. 240 vorgeführten Grundrissen sind die Variationen aber keineswegs erschöpft; je nach der speziellen Bestimmung der Thüre, bezw. des Hauses, der Architektur der Fassade, den verfügbaren Mitteln etc. können auch andere Anordnungen in Betracht kommen.

In Fig. 241 sind verschiedene Anschlüsse der Thüren an das Steingestell gezeigt, wenn der Futterrahmen fehlt. Der Anschlag erfolgt stumpf, mit Ueberfälzung oder ganz im Falz.

5. Haustüren und Thore.

Die Figur 242 zeigt eine einfach gehaltene Hausthüre mit Oberlicht über steinernem Kämpfer; das Beispiel 243 ist ähnlicher Art; das Oberlicht liegt hier gewissermassen in der Thüre selbst. Etwas schwerer und abweichend von der üblichen Form erscheint die Thüre der Fig. 244.

Das mit Zierbändern beschlagene Beispiel der Figur 245 hat einen Briefeinwurf in der mittleren Füllung. Aehnlich in der Konstruktion und Ausstattung sind die Thüren der Figuren

Fig. 251.

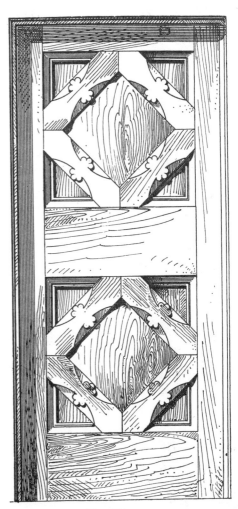
Fig. 252.

Zwei einflügelige Thüren aus München.

247 und 248 gehalten, während das weniger derbe Beispiel der Figur 246 an französische Vorbilder erinnert und in den Randfriesen mit Ziernägeln beschlagen ist.

Die Figur 249 bringt eine einfache, überschobene Thüre mit einer mittleren Ausgucköffnung nebst Verdachung und einem Briefkasten. Die Friese sind mit Nägeln geschmückt, zwei Zuziehknöpfe — von denen der eine nur der Symmetrie wegen angebracht ist — vervollständigen das Ganze.

Fig. 250 zeigt eine einfache, wirksame Thüre mit Gitterfüllung, Kämpfer und halbrundem Oberlicht, entworfen von Prof. Levy in Karlsruhe. Die Thüre ist im Licht 1,20 breit und in ihren Friesen mit Nägeln, Rosetten und einem Zuziehring verziert.

Fig. 253.
Einflügelige Hausthüre aus Basel.

Fig. 254.
Hausthüre nach E. Prignot.

In den Figuren 251 und 252 sind zwei einfachere Thüren aus München dargestellt (Synagoge), die aber erst recht zur Geltung kommen, wenn das kräftige, schwarz gehaltene Beschläge aufgelegt ist.

5. Hausthüren und Thore.

Fig. 253 bringt eine hübsche, gotische Thür aus Basel, und

Fig. 254 eine französische Thüre, die, unter einem Vordach angebracht, dem Wetter nicht ausgesetzt ist.

Auf Tafel 30 sind drei Hausthüren mit oberen Glas- und unteren Holzfüllungen dargestellt.

Die Tafel 31 zeigt links eine Thüre, deren oberer Teil nach innen überschoben und mit aufgelegten Zierstäben versehen ist; der untere Teil erhält nach aussen eine Verdoppelung mit Fuss- und Brüstungsgesims, um ihn kräftiger erscheinen und dadurch den oberen mehr zurücktreten zu lassen. Eine Verdachung schmückt das Ganze. Fig. 255 zeigt ein Stück des Brüstungsgesimses, isometrisch dargestellt.

Auf derselben Tafel ist rechts eine Thüre abgebildet, welche von einem ringsum laufenden Fries eingefasst und erst innerhalb desselben weiter gegliedert ist. Die Konstruktion zeigt der Schnitt c — d.

Tafel 32 stellt eine Thüre mit Kämpfer und Oberlicht dar. Der Schnitt und die Details zeigen die Konstruktion zur Genüge.

Die Tafeln 33 bis 35 bringen eine Anzahl von Hausthüren verschiedener Konstruktionsart. Sie können zum Teil als reicher gebildete, verdoppelte Thüren gelten; zum Teil zeigen sie den üblichen Bau mit Ueberschiebungen und Verdoppelungen. Zur Verstärkung und weiteren Ausschmückung sind teilweise schmiedeeiserne Beschläge aufgelegt und in den kleinen Lichtöffnungen, soweit solche überhaupt vorhanden sind, haben einfache Gitter Platz gefunden. Die meisten der Beispiele zeigen eine mehr strenge und ernste als zierliche Erscheinung, weshalb sie auch — besonders wenn das Motiv des Kreuzes verwertet ist — als Sakristeithüren und in ähn-

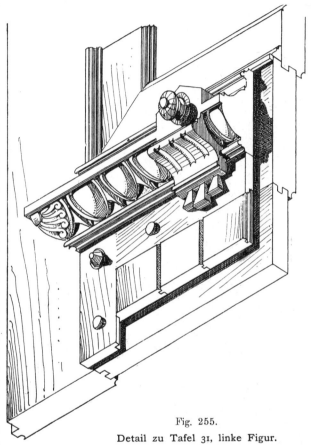

Fig. 255.
Detail zu Tafel 31, linke Figur.

lichem Sinne Verwendung finden können. Derartige Thüren schliessen nicht selten, bedingt durch den Charakter der Architektur, nach oben im Rundbogen oder Spitzbogen ab, was die Konstruktion nur insofern ändert, als eben nach oben geschweifte Friese an Stelle der geraden treten. Eine im Rundbogen endigende Einflügelthüre stellt die Tafel 46 auf der rechten Seite dar.

Tafel 36 bringt eine 1,30 m breite, einflügelige Thüre mit Oberlicht und breitem, verziertem Futterrahmen. Gestemmt mit überschobenen Füllungen, ist sie im unteren Teil aussen mit einer Verdoppelung versehen; die Querfriese haben Profil, während die Höhenfriese unten nur gefast, nach oben etwas verschmälert und gekehlt sind. Das mittlere Setzholz ist mit den

VI. Thüren und Thore.

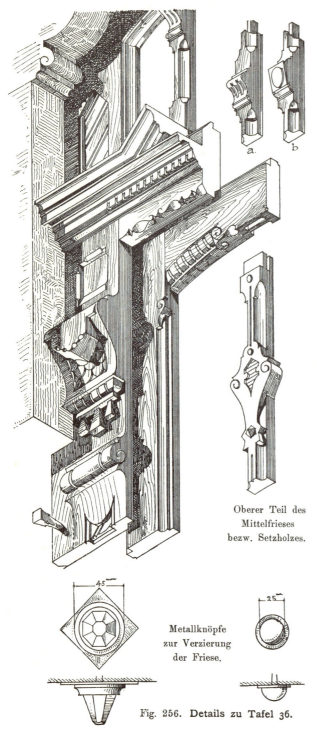

Oberer Teil des Mittelfrieses bezw. Setzholzes.

Metallknöpfe zur Verzierung der Friese.

Fig. 256. **Details zu Tafel 36.**

Querfriesen überplattet. Die Metallknöpfe (in Fig. 256 grösser gezeichnet), welche teils zum Aufputz, teils zum besseren Zusammenhalt dienen, sind aus Messing oder Bronze, der Zuziehgriff und das Ziergitter dagegen, sowie die Unterlagscheiben der Metallknöpfe aus Schmiedeeisen.

In Figur 256 ist die Ausschmückung des Futterrahmens, die Abfasung und Profilierung des Kämpfers, sowie die Form der Setzhölzer a und b angedeutet; in Fig. 257 ist die Art der Anfertigung von Füllungseinlagen für die Futterrahmen, sowie der untere Teil des Thürfrieses veranschaulicht.

Tafel 37 zeigt eine einflügelige, ähnlich wie die vorhergehende gebildete Thüre, jedoch 1,40 m breit und in zweierlei Weise durchgeführt. Während die rechte Hälfte einfach behandelt, unten mit Holz-, oben dagegen mit Glasfüllung und Gitter versehen ist, zeigt die linke, reichere Seite auf die ganze Höhe nur Holz, so dass eine allenfalls nötig fallende Beleuchtung des Flurs nur durch das niedere Oberlicht erfolgen könnte. Die zur Ausführung nötigen Profile sind rechts und links beigefügt.

Tafel 38 zeigt eine etwas reicher durchgebildete Thüre in einer Stichbogenöffnung mit breitem Futterrahmen, dekoriertem Kämpfer und dreiteiligem Oberlicht. Die auf dem Kämpfer angebrachte Verzierung ist in Schmiedeeisen gedacht.

Tafel 39 giebt eine ebenso reich behandelte, einflügelige Eingangsthüre von 1,50 m Lichtweite in einer Rundbogenöffnung mit breitem Futterrahmen, mit Kämpfer und Verdachung. Die auf den Futterrahmen, teils zur Ausschmückung, hauptsächlich aber der erhöhten Festigkeit wegen aufgelegten, strebepfeilerartigen Pfosten sind im Querschnitt angedeutet, ebenso wie die Seitenansicht eines Teils der Thüre. Sehr

5. Hausthüren und Thore.

leicht lässt sich diese einflügelige Hausthüre in eine zwei- oder dreiflügelige verwandeln, indem man die feststehenden Futterrahmenteile mit Bändern versieht und aufgehen lässt (Figur 240 i und k).

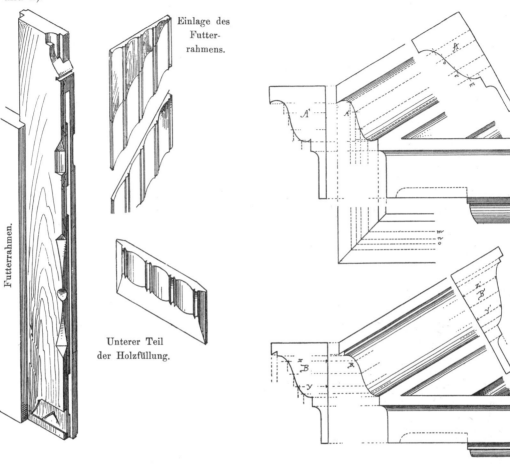

Fig. 257.
Einzelheiten zu Tafel 36.

Fig. 258.
Heraustragen der Schablonen für das Giebelgesims.

Zur Anfertigung des Giebelgesimses der Thürverdachung sind zwei Schablonen erforderlich. Würde man dem horizontalen und dem ansteigenden Teil des Gesimses das gleiche Profil geben, so würden sich beide nicht ordentlich auf Gehrung zusammenfügen lassen. Aus dem einen Profil bestimmt sich das andere. Es geschieht dies nach Figur 258. Nach dem obern Teil dieser Figur ist das schräg ansteigende Profil A angenommen oder entworfen. Wird hierauf ein Grundriss gezeichnet und werden die den Punkten m, n, o etc. des Profils entsprechenden Linien in denselben einge-

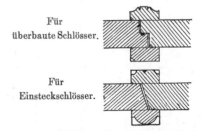

Fig. 259. Bildung der Mittelfuge bei Hausthüren.

Krauth u. Meyer, Die Bauschreinerei. 4. Aufl.

Fig. 260. Zweiflügelige Hausthüre mit Kämpfer und Oberlicht.

tragen und um das Eck herumgeführt, so ergiebt der Schnitt der zusammengehörigen Linien aus Grund- und Aufriss das neue seitliche Profil A^1.

Ein anderes, etwas abgekürztes Verfahren ist folgendes: Nachdem das seitliche Profil B entworfen, werden einzelne Punkte desselben parallel mit der Giebelschräge nach dem neu zu fertigenden Profil projiziert, wobei die entsprechenden Ausladungen y und z als y^1 und z^1 aufzutragen sind. Auf diese Weise erhält man aus dem seitlichen das schräg ansteigende Profil.

Was die zweiflügeligen Hausthüren betrifft, so verweisen wir auf die Figuren 260, 261 und 262, sowie auf die Tafeln 40 bis 44. Das gotisierende Beispiel der Fig. 260 stammt aus Kiel. Figur 261 führt den gewöhnlichen neuzeitigen Typus für derartige Thüren vor, während die Figur 262 eine Zweiflügelthüre zeigt, wie man sie häufig an den italienischen Renaissancepalästen findet, einfach aber vornehm in der Wirkung.

Tafel 40 bringt eine zweiflügelige Eingangsthüre von 1,40 m Lichtweite mit Kämpfer und Oberlicht. Die Behandlung dieser Thüre ist eine zweifache. Während der rechte Flügel ganz mit überschobenen Holzfüllungen versehen ist, zeigt der linke Glasfüllungen. Die Köpfe der beiden unteren Setzhölzer sind ähnlich gedacht, wie diejenigen des auf Tafel 56 dargestellten Fensters. Die Höhenfriese sind wie bei der Balkonthüre nach oben verjüngt, um das Aussehen leichter zu gestalten. Die mittlere, durch die innere und äussere Schlagleiste gedeckte Fuge bildet man nach der einen oder andern Art der Figur 259, je nachdem ein überbautes oder ein Einsteckschloss verwendet wird. Die erstere Bildung hat den Vorteil, dass der aufgehende Flügel beim Schliessen auf die ganze Höhe einen sicheren Anschlag findet; sie lässt sich aber des Schlossstulpes wegen bei Einsteckschlössern nicht wohl anwenden. Dass man in diesem Falle die Flügel abschrägt, geschieht, wenn man von der besseren Befestigung der Schlagleiste absieht, eigentlich mehr der lieben Gewohnheit wegen, als weil es Bedürfnis wäre. Selbst dann, wenn die Bandkloben so weit herausgesetzt werden, dass das Dornmittel mit der inneren Thürflucht bündig geht, muss doch die Thüre, damit sie bei jeder Witterung bewegbar ist, so viel Luft erhalten, dass eine Abschrägung oder richtiger ein Formen des Thürflügels nach dem Kreisumfang überflüssig ist. Immerhin aber schadet das Abschrägen nicht, solange es in mässigen Grenzen bleibt; von Unkenntnis dagegen zeigt es, wenn eine Schräge von weniger als 30, bezw. 60° gezeichnet und vom Schlosser verlangt wird, dass er danach seinen Stulp richte.

Fig. 261.
Zweiflügelige Hausthüre ohne Oberlicht.

Auf Tafel 41 sind zwei hübsche zweiflügelige Hausthüren nach den Entwürfen von Professor Max Hummel in Karlsruhe aufgezeichnet. Beide haben eine Breite von 1,75 m und sind mit Kämpfer und Oberlicht versehen. Die durch gedrehte Holzarbeit erzielten Vergitterungen dieser beiden Thüren sind in der Ausführung sehr wirksam und beide Beispiele zeigen, dass man auch ohne Bildhauerarbeit reich ausschauende Hausthüren schaffen kann.

Auf Tafel 42 sind zwei Hausthüren ohne Oberlicht dargestellt. Beide sind zweiflügelig; die eine hat eine Breite von 1,35 m, die andere von 1,50 m. An dem Beispiel der linken Seite ist die untere Partie als Verdoppelung gestaltet. Das Beispiel rechts zeigt die Anordnung zweier Schlagleisten, von denen die eine blind ist. In beiden Varianten dieses Beispiels sind die Höhenfriese in der obern Partie verschmälert und schräg abgesetzt.

Tafel 43 bringt eine reich ausgestattete, mit Schnitzerei versehene zweiflügelige Eingangsthüre mit einer lichten Breite von 1,50 m. Die Figur links oben auf der nämlichen Tafel zeigt die Gesamtanordnung der Thüre und ihre originelle Anpassung an die Architektur des Gebäudes, wobei ein grosses, halbrundes Oberlicht ermöglicht wurde. Dieses Beispiel ist nach einem Entwurf von Prof. Levy in Karlsruhe aufgezeichnet.

Auf Tafel 44 sind zwei Thüren dargestellt, die sich in ihrem Bau nicht wesentlich von einander unterscheiden. Beide sind zweiflügelig mit ungleichen Flügeln, wobei der gewöhnlich

Fig. 262. Zweiflügelige italienische Hausthüre.

aufgehende eine Breite von 90 cm hat. Diese beiden Thüren sind von Prof. Max Hummel in Karlsruhe entworfen.

Das Beschläge der Hausthüren besteht aus starken Winkel- und Kreuzbändern, Kloben auf Platten (ohne Futterrahmen in Stein), starken oberen oder unteren Schub- oder Kantenriegeln, einem Einsteck- oder überbautem Zweitourschloss mit Eisen-, Bronze- oder Messingdrückern (aussen oft auch nur Stechschlüsselvorrichtung) und der Befestigung des Futterrahmens mit 7 bis 9 starken Steinschrauben. Sind Glasflügel vorhanden, so sind sie entweder mit Fischbändern und Vorreibern oder nur mit Vorreibern anzuschlagen; das Gleiche gilt von dem Oberlicht.

b) Thor, Hausthor oder Thorweg.

Ein Thor legt man, falls nicht eine ortspolizeiliche Bestimmung ein solches verlangt, bei Wohngebäuden nur da an, wo das Bedürfnis vorliegt, in den Hofraum mit einem Wagen fahren zu können. Die polizeiliche Forderung wird aber in vielen Städten bei allen fest eingebauten Gebäuden gestellt, zu welchen grössere bewohnte Hintergebäude gehören, um die Möglichkeit zu schaffen, bei allenfallsigem Brandunglück mit der Feuerspritze in den Hofraum gelangen zu können. Zu diesem Zwecke muss das Thor, bezw. die Einfahrt mindestens 2,30 m breit sein,

ein Mass, welches im allgemeinen als Minimalmass für sämtliche Thore gelten kann. Das Wohnhausthor, das nur in seltenen Fällen als Schiebthor angelegt werden kann, ist meist als Flügelthor behandelt. In diesem Fall ist es entweder zweiflügelig oder es hat ähnlich wie das Scheunenthor noch einen dritten, kleinen Durchgehflügel, der an den einen Thorflügel angeschlagen ist. So gefällig sich eine derartige Thoranlage giebt und so praktisch sie zu sein scheint, so grosse Nachteile hat sie auch. Durch das Ausschneiden des dritten Flügels leidet nämlich die Festigkeit des Ganzen in erheblichem Grade Not und es bedarf der grössten Sorgfalt und Anstrengung seitens des Schlossers, wenn das Thor dicht schliessen und sich nicht ungleich senken soll. Es ist bei der Konstruktion darauf zu achten, dass durch Anbringung eines durchgehenden Kämpfers an dieser Stelle das Thor seinen Halt findet. Das lässt sich unschwer machen, sobald die Höhe bis zum Kämpfer als Thürhöhe genügt. Fig. 266 stellt die Skizze eines derartigen Thores dar, bei welchem der Kämpfer mit dem linken Flügel verschraubt, sich über den anderen Flügel legt und dort durch zwei starke Schraubenbolzen oder auf ähnliche Art befestigt wird. Anschlageisen an der Thorschwelle und starke, obere und untere Schubriegel vervollständigen das Ganze. Es ist vorteilhaft, statt der Riegel zwei starke Schwengelverschlüsse anzuwenden, weil man bei ihnen über eine viel grössere Kraft verfügt, um das Thor festzustellen. (Vergleiche Abschnitt XII, Beschläge.)

Auf Tafel 45 sind zwei Thore abgebildet von 2,3 und 2,4 m lichter Breite. Beide haben festen Kämpfer; das Oberlicht schliesst beim einen im Halbkreis, beim andern mit geradem Sturz ab. Die Füllungen des einen sind durch Riemen, die des andern in gewöhnlicher Art gebildet.

Auf Tafel 46 ist ein reicheres Thor nach dem Entwurf von Professor Levy in Karlsruhe dargestellt. Es hat eine Breite von 3,4 m im Licht. Die Einteilung ist derart gewählt, dass die grösseren Füllungen von schmäleren Füllungen allseitig umrahmt werden. Ausserdem zeigt dieses Thor ein reiches schmiedeeisernes Beschläge, bestehend in Bändern, Zuziehringen u. s. w. Dieselbe Tafel zeigt ferner eine zu diesem Thor passende Eingangsthüre.

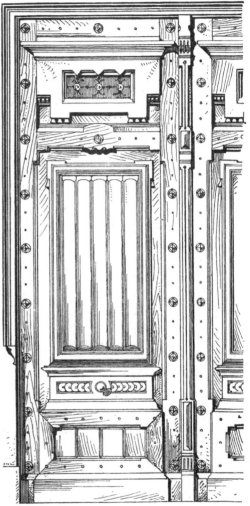

Fig. 263. **Einfahrtsthor ohne Oberlicht.**

Auf Tafel 47 ist rechts das neue Rathausthor zu Lindau dargestellt, welches unten fest in Holz geschlossen, oben mit einem durch gekreuzte Latten hergestellten Holzgitter versehen ist. Das Thor ist natureichen ausgeführt, das Beschläge und die Nagelköpfe sind verzinnt.

Die Tafel 48 bringt zwei gestemmte und ganz geschlossene Hausthüren aus Italien, welche ihrer Grösse wegen zu den Thoren gerechnet werden können. Die beigefügten Schnitte zeigen

134 VI. Thüren und Thore.

die Profilierung. Mit Ausnahme der Thürklopfer und der Nägel ist alles übrige Holz; die Schlagleisten fehlen; die Flügel sind überfälzt.

Die Tafel 49 bildet ein Zweiflügelthor, von 2,30 m Breite ab. Dasselbe hat festen Kämpfer und Oberlicht. Die Schlagleiste verstärkt sich nach unten strebepfeilerartig. Die beiden Flügel sind als Varianten mit wechselnder Ausstattung aufgezeichnet. Die rechte Seite lässt sich auch für Kirchenthüren verwerten, wobei das Oberlicht wegfällt oder anders ausgebildet wird.

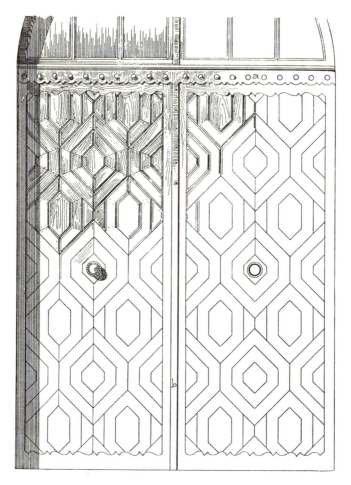

Fig. 264. Verdoppeltes Einfahrtsthor aus Augsburg.

a = Schmiedeeisen-Verzierung des Kämpfers; b = Eiserne Schlagleiste.

Die Tafel 50 (Doppeltafel) zeigt die Ansicht und den Schnitt eines zweiflügeligen Thores von 2,50 m Breite mit eisenverziertem Kämpfer und Oberlicht. Der Fuss des Thores ist zum besseren Schutz mit Eisenblech beschlagen. Die Konstruktion ist aus dem Schnitt und der isometrischen Darstellung genügend klar ersichtlich.

Fig. 263 zeigt ein modernes Thor mit kleinen, vergitterten Lichtöffnungen.

5. Hausthüren und Thore. 135

Fig. 264 giebt ein Thor aus Augsburg, wie es Ende des 18. und anfangs des 19. Jahrhunderts vielfach gefertigt wurde und heute noch vereinzelt gefertigt wird. Es lässt sich nicht leugnen, dass das Thor äusserst widerstandsfähig und fest ist, und daher in diesem Sinne bestens empfohlen werden kann, obgleich es in seiner Flachheit etwas langweilig wirkt.

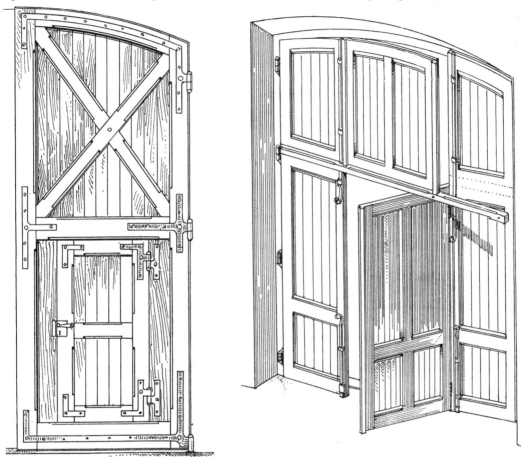

Fig. 265.
Scheunenthorflügel mit Schlupfthürchen.

Fig. 266.
Einfahrtsthor mit beweglichem Kämpfer.

c) Magazin- und Scheunenthor.

Seine Grösse hängt von dem Bedürfnis und dem Zweck, welchem es zu dienen hat, ab. So ist z. B. für ein Scheunenthor das Mass eines geladenen Heuwagens von 2,50 m Breite und 3,00 bis 3,30 m Höhe zu Grunde zu legen, welchem man den nötigen Spielraum noch zuschlägt. Bei Magazinthoren lässt sich das Mass ohne genaue Kenntnis der Bedürfnisse nicht angeben. Gewöhnlich macht man sie zweiflügelig, da die Breite zu bedeutend ist, um sie einflügelig machen zu können. Bei schwereren Thoren trifft man auch gerne Vorkehrung zum Aus- und Eingehen, ohne jeweils den ganzen Flügel in Bewegung setzen zu müssen. Dies geschieht, indem man in einem der grossen Flügel einen kleineren, leicht beweglichen einfügt (Schlupfthüre) und mit Beschlägen versieht, wie dies in Fig. 265 gezeigt ist.

VI. Thüren und Thore.

Die Bewegung der Flügel geschieht durch Drehen oder durch Schieben; in neuerer Zeit neigt man sich mehr der letzteren Art zu, da bei ihr die Flügel — falls sie einmal geöffnet — auch wirklich beseitigt sind und nicht vom Wind erfasst und zugeschlagen werden können. Wenn das Schiebthor heute noch auf ein gewisses Vorurteil stösst, so kommt dies hauptsächlich daher, dass das Beschläge zum Teil noch zu wünschen übrig lässt; übrigens kann das Schiebthor auch nicht überall angebracht werden.

Ein Thorflügel besteht aus einem gut abgesteiften, d. h. gegen das Verschieben gesicherten Pfostengerippe von 8×10, 10×10 oder 10×12 cm starkem Holz, je nach der Grösse des Flügels. Das Gerippe wird auf der Aussenseite mit senkrecht laufenden, gespundeten und gut genagelten Dielenriemen verkleidet. Der hintere Pfosten (Wendesäule) des Gerippes ist bei Flügelthoren meist etwas stärker, weil an ihm das ganze Thor seinen Halt findet; beim Schiebthor fällt dieser Unterschied weg. Die gegenseitige Entfernung der Querriegel, der Streben und Büge hängt von der Dielenstärke ab; je stärker die Dielen, desto weniger oft ist eine Nagelung erforderlich. Die übliche Entfernung ist 1,00 bis 1,30 m. Das Flügelthor und ebenso das Schiebthor hat gewöhnlich keine Schlagleiste; die Deckung der Fuge wird durch Ueberfälzung erzielt.

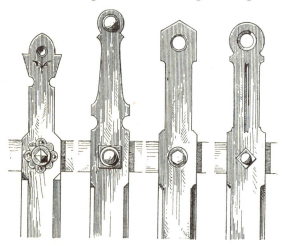

Fig. 267. **Obere Endigungen von Thorstäben.**

Tafel 51 stellt zwei Thorarten dar, ein Schiebthor und ein Flügelthor. Das Beschläge des letzteren besteht aus vier starken, eingelassenen oder aufgeschraubten Winkelbändern und zwei Kreuzbändern, Kloben in Stein, eventuell vier Eckwinkeln, einem oberen und unteren starken Schubriegel und einem überbauten Schloss, oder einem drehbaren Ueberlagseisen mit Hängeschloss. Statt der Winkel- und Kreuzbänder verwendet man auch gern starke Zapfenbänder, wie sie auf Tafel 79 abgebildet sind; ebenso statt der Riegel Schwengelverschlüsse oder Triebbasküle (Tafel 80). Das Schiebthürbeschläge besteht aus vier Winkelbändern, an deren oberen Verlängerungen sich grosse, abgedrehte Metallrollen befinden, welche auf einer durch Steinschrauben an dem Thorsturz befestigten Eisenschiene sich bewegen. Die Schiene ist an beiden Enden aufgebogen, um das Thor am Weiterlauf zu hindern; zur Feststellung beim Schliessen ist am einen Flügel unten ein Schubriegel angebracht. Damit das Thor sich bei Wind nicht bewegt, befindet sich an der unteren Kante zu beiden Enden je ein Stift, welcher in einer entsprechend geformten und in der Steinschwelle befestigten Schiene läuft und dadurch die senkrechte Haltung des Thores bewirkt. Geschlossen wird dieses Thor mittels eines Schiebthürschlosses (Tafel 82) oder durch ein Hängschloss, welches man durch zwei an den vorderen Höhenfriesen der Flügel angebrachte Oesen zieht.

Die Laufrollen unten anzubringen, hat sich nicht bewährt, da sich denselben zu leicht Hindernisse in Form von Staub und Schmutz entgegenstellen, welche den Gang zu hemmen geeignet sind. Will man am Boden die offene Fuge nicht haben, so wählt man das auf Tafel 22 in Schnitt e—f unten dargestellte Kugelbeschläge (Stahlkugeln), Patent Weickum, das sich sehr gut bewährt hat.

d) Hofeinfriedigungsthor (auch Lattenthor).

Das hölzerne Hofthor, welches, wie sein Name sagt, dazu dient, den Hofraum für Unbefugte abzuschliessen, erfreut sich heute nicht mehr so zahlreicher Verwendung wie ehedem. Seit dem Aufschwung der Eisenindustrie und dem Herabgang der Eisenpreise zieht man das wetterbeständigere Material dem Holz vor. Dies gilt sowohl für die verhältnissmässig reicheren

Fig. 268. **Polsterthüren.**

Thore als in Beziehung auf das rein zweckliche Wellenblechthor. Doch giebt es immerhin noch Ausnahmen, z. B. wenn der Ort der Herstellung in einer holzreichen Gegend liegt. Die Hof- und Garteneinfriedigungen sind dann häufig noch aus Holz und in eine derartige Einfriedigung passt ein hölzernes Thor besser, als ein solches aus Eisen.

Die Konstruktion ist derjenigen des Scheunenthores ähnlich. Man bildet ein starkes, unverschiebbares Holzgerippe, auf welchem man aussen die Dielenschalung oder die Lattung an-

VI. Thüren und Thore.

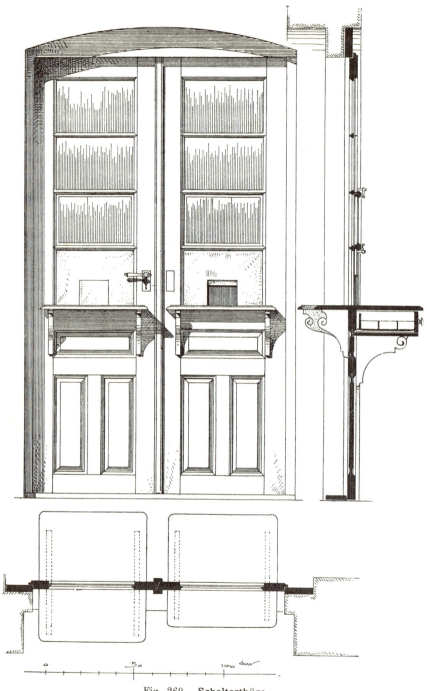

Fig. 269. Schalterthüre.

bringt, und zwar deshalb auf der Aussenseite, damit nicht Unbefugte die Querriegel als Leiter zum Darübersteigen benützen. Die Formen der Hölzer sind so zu wählen, dass das Wasser nicht stehen bleiben und eindringen kann. Man schrägt sie daher oben gern ab, fast sie und lässt das Thor nicht bis auf den Boden reichen; aus dem gleichen Grunde bringt man manchmal auf der oberen Thorkante eine breite, abgeschrägte Leiste an. Man schneidet die Pfosten, Riemen und Latten nach hinten schräg ab oder spitzt sie nach oben zu etc. Die Befestigung der Latten oder Riemen auf dem Gerippe erfolgt durch Nagelung, besser durch Mutterschrauben.

Auf Tafel 52 ist ein derartiges Holzthor dargestellt, und zwar zeigt die linke Seite das Thor von aussen, während rechts die innere, mit dem Beschläge versehene Seite gezeichnet ist. Die Schlagleiste ist weggelassen, das Thor in der Mitte überfälzt. Das Beschläge besteht, falls nicht unten je ein Zapfen- und oben ein Halsband an dem hinteren Pfosten, der Wendesäule, angebracht ist, aus starken Lang- oder Winkelbändern, unterem Anschlageisen, Schubriegel mit Schutz gegen unbefugtes Oeffnen, Spreizstange mit Hängeschloss oder Schliesse und starkem überbauten Zweitourschloss.

Figur 267 zeigt einige Muster von Lattenköpfen für Hofthore.

Auf Tafel 47 ist links ein Lattenthor dargestellt, welches im Rathaus in München als Hofabschluss dient.

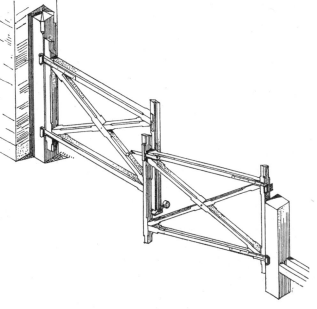

Fig. 270. Barrière.

6. Verschiedene Thüren zu bestimmten Zwecken.

1) Die Polsterthüre (Fig. 268).

Ihre Anbringung bezweckt, die Zugluft abzuhalten, das Geräusch beim Thürzuschlagen zu vermindern, sowie dem Zerbrechen der Glasscheiben vorzubeugen. Ihre Konstruktion ist die eines einfachen, abgesteiften Thürrahmens, dessen Füllungsöffnungen bis zu 12 cm mittlerer Stärke ausgepolstert sind. Ueberspannt ist das Ganze (oder nur die Füllung) mit Leder, weniger gut mit Wachstuch oder Ledertuch, wobei die sichtbaren Polsternägel genau nach einer bestimmten Zeichnung einzuschlagen sind. Um sich ausser- oder innerhalb des Abschlusses orientieren zu können, ohne ihn selbst öffnen zu müssen, bringt man zuweilen auf Gesichtshöhe eine kleine Glasscheibe an.

Fig. 268 stellt in a eine einflügelige Polsterthüre dar, welche stumpf aufschlägt, in b dagegen ist eine zweiflügelige gepolsterte Pendelthüre aufgezeichnet, deren linker Flügel fertig gepolstert erscheint, während der rechte nur das Rahmenwerk zeigt.

2) Die Schalterthüre (Fig. 269).

Sie kommt sowohl in Bahnhöfen und Postanstalten, als auch in kaufmännischen und technischen Geschäftsräumen vor und ist konstruiert wie eine ein- oder zweiflügelige Glasthüre.

18*

Auf der Höhe von ca. 1,00 m vom Boden aus ist an dem Flügel eine Zahltischplatte angebracht, die innen und aussen durch je zwei Träger unterstützt wird. Auf der inneren Seite wird zwischen der Platte und den Trägern ein mit Schieblade versehenes Kästchen angeordnet. Die erste Oeffnung oberhalb der Tischplatte ist statt mit Glas mit einem Messingblech geschlossen, in welchem sich eine durch einen Schieber verschliessbare Oeffnung zur Empfangnahme von Geld etc. befindet; über dieser Füllung ist ausserdem ein kleiner Fensterflügel eingelegt, welcher die Verständigung des Beamten mit dem Publikum erleichtert. Es kann jedoch auch (mit Weglassung der Messingplatte) ein senkrecht bewegliches Schiebfenster angeordnet werden, welches aufgezogen die unterste Abteilung im ganzen öffnet.

3) **Die Abortthüre.**

Sie ist verschieden je nach der Art und Lage des Abortes. Liegt der letztere innerhalb der Wohnung, so ist die Thüre wie eine Zimmerthüre, jedoch öfters von geringerer Breite, die aber nicht unter 0,70 m betragen soll. Hat der Abort einen Vorplatz mit einer zweiten Thüre, so erhält die letztere, um den Vorplatz etwas zu beleuchten (wenn es nicht anders geschehen kann) im obern Teil eine Verglasung; statt der obern Holzfüllung wird eine mattierte Scheibe eingesetzt. Liegt der Abort im Freien (Schüleraborte, öffentliche Aborte) oder im Erdgeschoss öffentlicher Gebäude (Bahnhöfe, Rathäuser, Gerichtsgebäude, Wirtschaften etc.), so sind die Thüren den betreffenden Verhältnissen anzupassen. Sie werden stärker und aus Forlen- oder Eichenholz gebaut in gestemmter Arbeit oder als verdoppelte Thüren etc. Wichtig sind hierbei geeignete Zuwerfungen. Zum gewöhnlichen Thürbeschlag tritt ein gutschliessender Riegel auf der Innenseite und auf dieser sollten auch einige Kleiderhaken aufgeschraubt werden. Die Abortthüren sind gewöhnlich im ganzen geschlossen. In besonderen Fällen können auch Vergitterungen erwünscht sein, wie die einfache aber hübsche Abortthüre aus dem Rathaus in München es zeigt, welche auf Tafel 53 links dargestellt ist.

4) **Die äussere Kellerthüre.**

Sie findet Anwendung, wo die innere Kellerthüre aus irgend welchem Grunde fehlt und die Benützung des Kellers von der Strasse oder dem Hofraum aus erfolgt, wie z. B. bei Weinhandlungen. Kann man in dem über dem Kellereingang gelegenen Raum im Haus einen genügend hohen »Kellerhals« anlegen, so wird die Thüre im gewöhnlichen Sinne zweiflügelig gestaltet; ist dies nicht der Fall, so muss man die Thüre vierflügelig machen, d. h. aus zwei senkrechten und zwei schräg liegenden Teilen bestehend, wie es Tafel 54 in B zeigt. Man macht die Thürflügel stark im Holz, damit sie Kälte und Hitze besser abhalten. Die liegenden Flügel werden, wie die stehenden, verdoppelt; die ersteren beschlägt man auf der Oberseite mit Zinkblech und an den Kanten zur Rechten und Linken mit Façoneisen, wie es die Schnitte e—f und g—h angeben. Auf diese Weise sind die Flügel gegen den Regen geschützt. Dass beim Oeffnen die beiden senkrechten Flügel zunächst beseitigt werden müssen, ehe die liegenden geöffnet werden können, ist wohl selbstredend. Als Unterstützung des nur an zwei Kanten aufliegenden linken Flügels bringt man gern eine am Flügel selbst befestigte eiserne Stütze an, welche sich beim Schliessen auf eine der Kellerstufen aufstellt.

Zwei Varianten gewöhnlicher zweiflügeliger Kellerthüren nach Beispielen aus Lindau zeigt die Tafel 53 rechts.

5) **Das Treppenthürchen** (Tafel 54, Fig. A).

Es dient wie das Treppengeländer als Abschluss gegen die Treppe, um das Herabstürzen der Kinder zu verhindern, und ist daher auch in Form und Abmessung wie das Geländer selbst behandelt. Wenn das letztere gedrehte Stäbe besitzt, so giebt man dem Thürchen auch solche etc. Bugleisten bringt man nicht gern an, da sie den Kindern Gelegenheit zu gefährlichen Kletterübungen geben. Besorgt man, dass das Thürchen sich einsenke, so kann man es durch Eisen-winkel versteifen. Wichtig ist, dass dasselbe sich selbstthätig schliesst, also mit einer leichten Zuwerfung versehen wird. Im übrigen erhält es Bänder und Federfalle mit möglichst verborgener Vorrichtung zum Oeffnen.

6) **Die Notausgangsthüre** (Notthüre).

Dieselbe unterscheidet sich von den anderen Thüren eigentlich nur durch die Art des Aufgehens, insofern nämlich, als sie nach aussen aufgeht, während die meisten anderen Thüren nach innen aufschlagen.

Die Notausgangsthüren sollen im Falle der Gefahr es einer grösseren Menschenmenge ermöglichen, sich rasch und rechtzeitig zu retten. Sie sind im allgemeinen nur nötig, wenn die übrigen Thüren der betreffenden Räume für eine rasche Entleerung nicht genügen. Die Erfahrung hat leider gelehrt, dass die Notthüren des öfteren verschlossen, also so gut wie nicht vorhanden waren, als das Unglück eintrat. Was der Techniker in Bezug auf Notthüren thun kann, beschränkt sich darauf, dass er sie an leicht sichtbare und leicht erreichbare Plätze verlegt, dass er sie nach aussen aufgehen lässt und dass er sie nur so stark baut und derart beschlägt, dass sie im Notfall auch vom Publikum gesprengt werden können. Alles andere ist Sache der Gebäudeverwaltung. Sie soll die Notthüren offen halten, auch wenn keine Gefahr vorhanden is' damit das Publikum die Ausgänge kennen und benutzen lernt, um sie im Falle der Gefahr u so leichter zu finden.

7) **Die Fallthüre** (Taf. 54, Fig. D).

Sie wird heutzutage nur selten angewendet, während sie früher häufig zu finden war. Verschiedenerorts ist sie baupolizeilich verboten, weil sie Anlass zu Unglücksfällen geben kann. Wir finden sie noch als Abschluss von Kehlspeichern. Sie ist gebaut als eine einfache, verleimte Thüre mit Einschiebleisten und legt sich ringsum in einen Falz oder stumpf auf. Um ihr das Gefährliche als Kehlspeicherthüre zu nehmen, versieht man sie mit einem starken Gegengewicht, so dass sie beim geringsten Druck sich öffnet, zur Schliessung aber leicht heruntergezogen werden muss. Ihr Beschläge besteht ausser dem Gegengewicht aus zwei Langbändern, und je einer Schlempe an der Thüre und an dem Futter, in welche ein Hängeschloss eingehängt wird.

8) **Die Rollenthüre** (Taf. 54, Fig. E).

Sie findet Verwendung wie die Fallthüre, doch ist sie vorteilhafter als diese. Konstruiert ist sie als verleimte Thüre mit einer ringsum ca. 6 bis 8 cm hohen, nach unten vorstehenden Zarge, an welcher 4 Holz- oder Metallrollen befestigt sind, welche in zwei auf dem Boden angebrachten Eisenlaufnuten sich bewegen. Soll die Thüre geöffnet werden, so ergreift man den unten an der Thüre angebrachten Handgriff und schiebt sie so weit als nötig zurück; soll sie geschlossen werden, so verfährt man umgekehrt und hängt dann, wie bei der Fallthüre, an zwei Schlempen ein Hängeschloss an. Gegenüber der Fallthüre hat sie den Nachteil, dass sie weniger dicht schliesst als jene.

9) **Die Dachaussteigthüre** (Taf. 54, Fig. C).

Sie ist auch eine Art Fallthüre, welche, auf dem flachen Dach angelegt, einen sicheren Abschluss in jeder Beziehung geben soll. Um das Regenwasser, namentlich aber das Schmelzwasser des Schnees abzuhalten, ist ein ca. 15 cm über Dach vorstehendes starkes Futter erforderlich, welches die gut mit Zink eingebundene Thüre oben überdeckt. Man macht die Oeffnung, bezw. die Thüre nur so gross, dass ein Erwachsener gut durchzuschlüpfen im stande ist, ca. 60$\times$60 cm

10) **Die Gartenthüre.**

In Holz und in reicher Ausstattung ist sie selten geworden, seit die Gärten eiserne Geländer und dann auch eiserne Thüren erhalten. Immerhin aber kommt sie noch vor als einfache Latten- und Riementhüre inmitten der Latten- und Plankeneinfriedigungen oder auch als verdoppelte Thüre in Gartenmauern. Da diese Thürformen in Bezug auf die Konstruktion besprochen sind, so ist hier weiter nichts beizufügen. Eine hübsche Gartenthüre, aus Eichenholz gefertigt, mit starkem Rahmengestell und gefällig ausgeschnittenen Riemen, am Bavariakeller in München ausgeführt, zeigt die Mittelfigur der Tafel 53.

11) **Die Barrière** (Fig. 270).

Sie ist ein Abschluss im Freien, für Plätze, Höfe, Gärten, Bahnhofperrons etc. und dient dazu, das Publikum abzuhalten. Sie besteht meist aus einem Gestell von quadratischen, oben abgeschrägten Hölzern, welche je nach den Abmessungen der Barrière stärkeren oder schwächeren Querschnitt haben, immer aber gut miteinander verbunden und abgesteift sind. Die Barrière kann ein-, zwei- und mehrflügelig angelegt und so angeschlagen sein, dass der Drehpunkt am Ende des Flügels sich befindet (wie bei Fig. 270), oder in der Mitte desselben, so dass die eine Seite des Flügels nach links, die andere nach rechts schlägt. Das Beschläge besteht aus Lang- und Winkel- oder Zapfen- und Halsbändern, Anschlageisen mit Festhaltung, Vorreiber oder Ueberlegeisen. (Vergl. Abschnitt XII, Beschläge.)

VII. DIE FENSTER.

(Tafel 55 bis mit 61.)

Allgemeines. — 1. Die Bildung der Futterrahmen. — 2. Die Bildung der Fensterflügel. — 3. Das einfache Fenster. — 4. Das Doppelfenster. — 5. Das Klappfenster. — 6. Das Drehfenster. — 7. Das Schiebfenster. — 8. Fenster in Fachwerkswänden. — 9. Das Schaufenster. — 10. Das Glas. — 11. Das Verglasen.

Allgemeines.

Unter Fenster verstehen wir im Sinne des Bauschreiners die zum Verschliessen der Lichtöffnungen unserer Gebäude dienenden verglasten Fensterrahmen. Das Material derselben ist hauptsächlich Eichen- und Forlen-, seltener Tannenholz, neuerdings auch Pitchpine. Die beiden ersten Holzarten haben den Vorzug, der Zerstörung durch Fäulnis grösseren Widerstand zu leisten als Tannenholz, sind aber auch im Preise wesentlich höher als letzteres. Das amerikanische Pitch-pine hält sowohl in bezug auf die Qualität wie auf den Preis ungefähr die Mitte zwischen Eiche und der heimischen Forle. Eichenholz muss nach seiner Fällung, bezw. nachdem es geschnitten ist, 3 bis 4 Wochen in fliessendes Wasser eingelegt werden, um auszulohen; wo dies versäumt wird, muss man sich darauf gefasst machen, später an den Fenstern die schwarze Lohbrühe herauslaufen zu sehen. Gelohtes Holz trocknet auch rascher als ungelohtes.

Die Grösse der Fenster hängt von der Art und Grösse des Baues ab, ist aber im allgemeinen in der Neuzeit, insbesodere seit dem Aufschwung der Glasindustrie, im Vergleich gegen früher unstreitig gewachsen. Während bis vor 3 Jahrzehnten Fassadenfenster unter 1,50 m Höhe selbst bei städtischen Wohngebäuden sehr häufig waren, sind sie heute eine Seltenheit und höchstens an untergeordneten Seiten- und Hinterbauten zu beobachten. Mit der Vergrösserung nach der Höhe hat auch die Weite zugenommen. So hat sich allmählich für die nach der Strasse gehenden Fenster unserer städtischen Wohnhäuser ein beinahe feststehendes Mass (von $1{,}00 \times 2{,}00$ m) herausgebildet, welches mit kleinen Abänderungen fast durchgehends Anwendung findet. Es bedarf wohl keiner besonderen Ausführung, dass hiermit nur die gewöhnlichen Zimmerfenster gemeint sind und dass es noch viele Fenster anderer Art giebt, kleinere und grössere, je nachdem der Zweck des Raumes und die architektonische Bedeutung des Ganzen dies verlangen.

Ein Fenster besteht aus einem oder mehreren Holzrahmen, in welchen mittels Kitt Glasscheiben eingesetzt sind. (Die Eisenrahmenfenster kommen für dieses Buch nicht in betracht.)

VII. Die Fenster.

Die Anforderungen, die wir an ein **gutes Fenster** stellen, sind: 1) möglichst luft- und wasserdichter Verschluss aller Teile; 2) möglichst schmales Rahmholz; 3) leichte und praktische Handhabung beim Oeffnen und Schliessen und schliesslich 4) zweckmässige und schöne Einteilung und Anordnung der Scheiben.

Luftdicht, oder richtiger gut schliessend gegen die Aussenluft ist ein Fenster zu machen, wenn es bloss als Lichtfenster dient, wie es z. B. in Brandgiebeln vorkommt; schwieriger dagegen wird die Aufgabe, wenn dasselbe auch als Lüftungsfenster benützt werden soll, da es dann bewegliche Flügel erhalten muss.

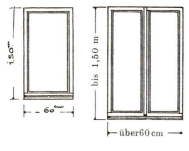

Nach Art der Bewegung der Flügel unterscheiden wir:

1) das gewöhnliche **Flügelfenster**, bei welchem die Drehung der Flügel um eine senkrechte seitliche Axe erfolgt,
2) das **Klappfenster**, bei welchem die Axe horizontal liegt, oben oder unten;
3) das **Drehfenster** mit senkrechter oder horizontaler Axe in der Mitte und schliesslich
4) das **Schiebfenster**, bei welchem die Flügel seitwärts oder auf- und abwärts geschoben werden.

Nach der Anzahl der Flügel kann man die Fenster einteilen in a) einflügelige, b) zweiflügelige, c) dreiflügelige, d) vier- und mehrflügelige Fenster.

Betrachten wir zunächst die am häufigsten vorkommenden **Flügelfenster**, so macht man allgemein, wo nicht besondere Verhältnisse eine andere Behandlung verlangen, Fenster von über 0,60 m Breite zweiflügelig. Man verhindert dadurch, dass die Scheiben zu gross werden und die Flügel beim Oeffnen über die Leibungen vorspringen, wobei sie die Vorhänge beschädigen.

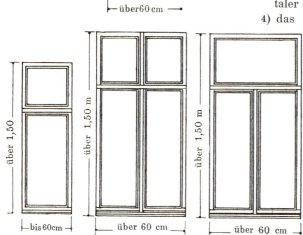

Fig. 271. **Anordnung der Fensterflügel.**

Fenster von über 1,50 m Höhe werden auch nach der Höhe geteilt, während solche unter 1,50 m ungeteilt bleiben. (Fig. 271.)

In besseren Bauten und wo die Mittel reichen, verglast man den Flügel mit einer einzigen ganzen Scheibe; wo dies aus irgend welchem Grunde nicht angeht, zerlegt man ihn wieder durch Sprossen in Unterabteilungen, in einzelne Scheiben, deren Form ein Quadrat, besser aber ein hochgestelltes Rechteck ist. Scheiben, deren Höhe geringer als ihre Breite ist, sehen unschön und gedrückt aus.

Um die Fenster luftdicht abzuschliessen, ist vor allem erforderlich, dass die Fensterflügel sich nicht direkt auf den mehr oder minder rauhen Stein auflegen. Wir nehmen daher einen besonderen, an den Ecken zusammengeschlitzten Holzrahmen und befestigen ihn mittels versenkter Steinschrauben oder Bankeisen auf die innere Kante des Fenstergestells, nachdem wir zuvor eine Zwischenlage von Haarkalk aufgetragen haben. Der Zweck des Kalks ist

1. Die Bildung der Futterrahmen.

hierbei, luftdicht abzuschliessen, derjenige der Haare, zu verhindern, dass der Kalk nach dem Trocknen und beim Zuschlagen der Fenster herausbröckelt. Ein derartiger Rahmen heisst, wie bei den Thüren, der Futterrahmen; an und in diesen schlagen die Flügel.

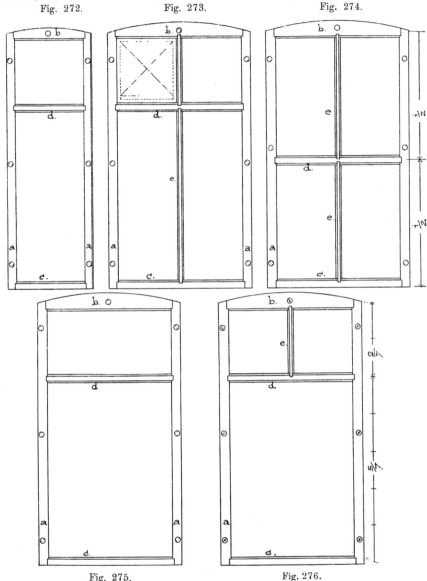

Fig. 272 bis 276. **Bildung der Futterrahmen.**
a Höhenschenkel, b Oberschenkel, c Wetterschenkel, d Kämpfer, e Setzholz.

1. Die Bildung der Futterrahmen.

Bei grösseren mehrflügeligen Fenstern genügt der Anschlag an dem Futterrahmen aber nicht, um die Fensterflügel gerade und dicht zu halten und sie gegen Winddruck und Sturm genügend widerstandsfähig zu machen. In diesem Fall bringt man zur Verstärkung noch feststehende, mit dem Futterrahmen verzapfte Mittel- und Querhölzer an. Das horizontale Querholz heisst **Kämpfer**, die senkrechten Hölzer heissen **Setzhölzer**. Die Figuren 272 bis 276 zeigen verschiedene Formen zusammengesetzter Futterrahmen. Fig. 274 stellt eine zwar sehr solide, trotzdem aber für gewöhnliche Verhältnisse ausser Gebrauch gekommene Konstruktion dar, das sogen. **Fensterkreuz**, wie es zu Anfang dieses Jahrhunderts und noch bis zur Mitte desselben sich allgemeiner Beliebtheit erfreute, so dass der Ausdruck »Kreuzstock« gleichbedeutend mit Fenster wurde. Das Fenster ist hier in 4 gleiche Teile zerlegt, wodurch der Kämpfer aber auf die menschliche Augenhöhe zu liegen kam. Diesem Missstand half man durch Hinaufschieben des Kämpfers ab (Fig. 273); immerhin aber blieb noch die mittlere Höhenteilung, das untere Setzholz, im Wege, welches beim Hinaussehen ebenfalls hinderlich war. So kam es, dass man

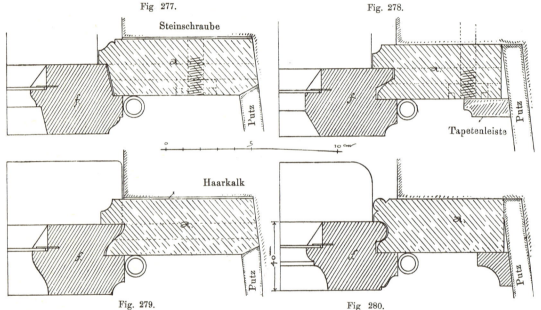

Fig. 277 bis 280. **Verbindung der Flügel mit dem Futterrahmen-Höhenschenkel.**

auch dieses für die Folge wegliess, wonach sich das Bild der Figur 276 ergab. Lässt man auch das obere Setzholz fallen, so ergiebt sich die Form der Fig. 275.

Die Anlage des Kämpfers in Bezug auf die Höhe ist

a) bei Sprossenfenstern derart, dass die beiden oberen Scheiben quadratisch, oder besser etwas überhöht werden (Fig. 273), oder aber, dass man von der lichten Höhe des Futters sämtliches Holz, Flügelwetterschenkel, Oberweitschenkel, Kämpfer und Sprossen abzieht und den Rest in gleiche Teile teilt (Taf. 55c.);

b) bei ganzen Scheiben so, dass man die Höhe der oberen Querscheibe gleich macht der Breite einer der beiden unteren, oder dass man den Kämpfer auf $2/7$ der Höhe von oben legt, wodurch für die unteren Flügel $5/7$ der Höhe verbleiben (Fig. 276).

Für Fenster von grösseren Dimensionen, wie sie bei Monumentalbauten vorkommen, lassen sich keine allgemeinen Regeln über die Anordnung der Kämpfer und Flügel geben, da bei deren Teilung ausser den zweckmässigen Holz- und Glasmassen auch noch ästhetische Rücksichten in Betracht kommen, welche nur der betr. Architekt richtig zu beurteilen im stande sein

1. Die Bildung der Futterrahmen.

wird. Die Tafel 55 zeigt neben einigen gewöhnlichen Fensterteilungen auch Anordnungen der letztgenannten Art.

Die seitlichen Teile des Futterrahmens a a (Fig. 272 bis 280) heissen Futterrahmenhöhenschenkel, das obere Querholz b heisst Futterweitschenkel oder Futteroberschenkel.

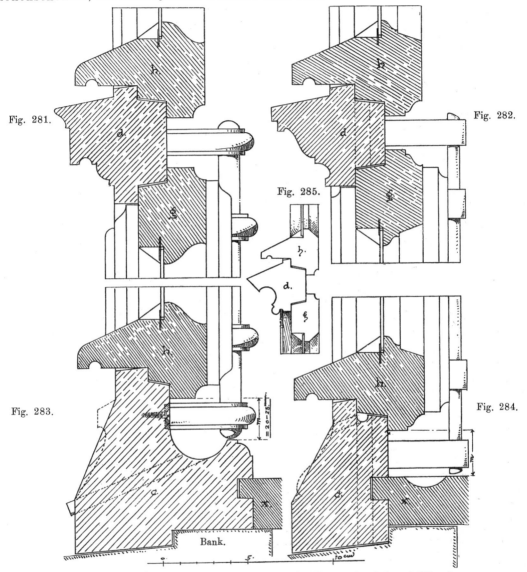

Fig. 281 bis 285. **Verbindung der Flügel mit Wetterschenkel und Kämpfer.**

Die Stärke derselben ist 3,0 bis 3,5 cm gehobelt, bei besseren Fenstern die gleiche wie die des Flügelholzes. Die erstgenannte Stärke reicht meist aus, weil die Breite 7,5 bis 8 cm beträgt und überdies der Rahmen auf die ganze Fensterhöhe mindestens 3 mal mittels Eisen an den Stein befestigt wird.

Das untere Querholz c heisst **Futterrahmenwetterschenkel**, dessen Dicke ca. 5 bis 9 cm beträgt. Seine Höhe ist so zu bemessen, dass der Zwischenraum m (Fig. 283 und 284) noch 2,0 bis 3,0 cm beträgt, um einen starken Schliesskloben anbringen zu können, beträgt also im ganzen mindestens 8 cm.

Das mittlere Querholz d, an welchem die unteren Flügel, sowohl wie die oberen ihren Anschlag finden, wird **Kämpfer, Losholz**, auch **Stab** genannt. Seine Höhe ist 6,5 bis 8 cm, seine Dicke hängt von der beabsichtigten Profilierung ab, ist aber im allgemeinen gleich der Höhe, also 6 bis 8 cm.

Sind senkrechte Mittelteilungen e vorhanden, welche in den Oberschenkel, Wetterschenkel und Kämpfer eingezapft werden, so heissen sie **Setzhölzer**. Ihre Dicke ist gleich der des Futterrahmens, ihre Breite 4 bis 4,5 cm (Fig. 286). Die Futterrahmenhöhenschenkel und der Oberschenkel legen sich gegen das Mauerwerk der Leibung stumpf an, werden aber besser abgefast, damit der Verputz einen soliden Anschluss erhält (Fig. 277 und 279); oft wird die Fuge auch mit einer Leiste, der sogen. **Tapetenleiste** (Fig. 278 und 280), gedeckt.

Nach der Lichtöffnung zu sind die Höhenschenkel, wie auch der Kämpfer-, Wetter- und Weitschenkel mit **schrägem Falz** (Fig. 277) versehen; wesentlich dichter wird aber der Verschluss, wenn an die ersteren, statt des Falzes, die sogen. **Hinternute** (Fig. 278) oder ein **S-Falz** angestossen wird (Fig. 279).

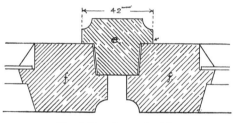

Fig. 286. Setzholz (e).

Der Kämpfer wird nach aussen meist profiliert, wobei dem raschen Abfluss des Regenwassers Rechnung zu tragen ist; er erhält eine **Wasserschräge** und eine **Wassernase** (Fig. 281 und 285). Die Form der Profilierung richtet sich nach dem Stil des ganzen Bauwerks. Werden Rollladen angebracht, so ist zur Vermeidung starker Auffütterung das Kämpferprofil nur schwach auszuladen.

Beim Wetterschenkel sind Profile überflüssig, da sie nicht gesehen werden; sollen dennoch welche angebracht werden, so gilt bezüglich deren Form das über die Kämpferprofile Gesagte. An der Innenseite des Wetterschenkels wird eine ca. 10 bis 12 mm starke Nute eingestossen, in welche das Fenstersimsbrett x (Fig. 283 und 284) eingreift.

Die einzelnen Teile des Rahmens werden zusammengeschlitzt und gestemmt, wobei die Zapfenstärke sich nach dem Falz und dem betr. Lochbeitel von ca. 9 bis 12 mm Breite richtet. Man nimmt im allgemeinen an, dass die Vorderkante der Zapfen bündig geht mit der Aussenflucht der Hinternute oder des S-Falzes (Fig. 278 u. 279 und Fig. 295 u. 297). Die Zapfen werden geleimt, verbohrt und mit Holznägeln versehen; verkeilt werden sie nicht.

Um die Lichtweite der Flügelöffnungen genau einzuhalten und dem Rahmen festeren Halt zu geben, setzt man die drei horizontalen Hölzer 8 bis 10 mm, bezw. bis zur Tiefe der Hinternute oder des S-Falzes in die Höhenschenkel ein (Fig. 272 bis 276 und Fig. 295 u. 297).

2. Die Bildung der Fensterflügel.

Die in den Futterrahmen einschlagenden Glasrahmen heissen **Flügel, Fensterflügel** und zwar untere und obere. Sie bestehen aus den senkrechten Hölzern, **Flügelhöhenschenkel** genannt (Fig. 277 bis 280 f.), dem oberen Querholz, **Flügelweitschenkel** oder **Flügeloberschenkel** geheissen (Fig. 281 u. 282 g), und dem unteren Querholz oder dem **Flügelwetter-**

2. Die Bildung der Fensterflügel.

schenkel (Fig. 281 bis 284 h). Die Stärke oder Dicke des Flügelholzes ist gehobelt, gewöhnlich 3,5 bis 4,5 cm, je nach der Grösse der Flügel und der Stärke des Glases; die Breite desselben ist 5,5 bis 6 cm. Der hintere Höhenschenkel ist, entsprechend dem Futterrahmen, bei einfachen Fenstern hinten mit schrägem Falz, bei besseren mit Hinternute oder S-Falz versehen.

Die Flügelwetter- und Weitschenkel schlagen mit Falz in den Futterrahmen ein und ebenso legen sich die beiden mittleren Höhenschenkel übereinander, wobei eine Schlagleiste die Fuge deckt; oder sie greifen mittels eines Wulstes ineinander, der sogen. Wolfsrachenkonstruktion (Fig. 287 bis 290). Die letztere Verbindung ist zwar dicht schliessend, dennoch aber nicht sehr

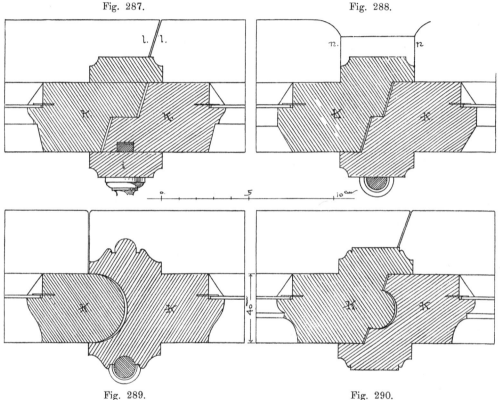

Fig. 287 bis 290. **Bildung der Schlagleisten.**

beliebt, da man bei ihrer Anwendung jeweils die beiden Flügel zusammen öffnen muss.

Bei ganz gewöhnlichen Fenstern und bei solchen, an denen die Triebstange des Verschlusses unsichtbar sein soll, wird die Schlagleiste i besonders gefertigt und auf den vorderen Höhenschenkel aufgeleimt und aufgeschraubt (Fig. 287); solider dagegen ist es, sie an den Höhenschenkel anzuarbeiten, d. h. mit ihm aus einem Stück Holz zu hobeln (Fig. 288 bis 290). In diesem Fall nennt man das Ganze, also Höhenschenkel und Schlagleiste zusammen, Schlagleiste (k). Nach der Glasseite zu sind die Höhen-, Wetter- und Weitschenkel, wie auch die etwa vorhandenen Sprossen aussen mit dem Kittfalz, auch Glasfalz genannt, versehen, welcher mindestens 12 bis 15 mm tief und 7 bis 9 mm breit sein muss; nach innen dagegen sind sie entweder gefast und es heissen die betreffenden Fenster Fasefenster (Fig. 291) und zwar:

a) Fenster mit **Halbfase**,
b) Fenster mit **Spitzfase**,
c) Fenster mit **eingreifender Fase**, event. mit Karnies und Stäbchen,

oder sie sind profiliert und heissen **profilierte Fenster, Fenster auf Hobel, auf Profilhobel** (Fig. 292 e u. f).

Sind die Flügel mit **Holzsprossen** versehen, welche in der Ansicht meist 2,5 cm stark gemacht werden, so können dieselben entweder auf die ganze Holzdicke durchgreifen (Fig. 291) oder nur zum Teil, wie bei Fig. 292 a. Scheinen für einen bestimmten Zweck die Holzsprossen zu stark, oder befürchtet man ein zu rasches Faulen derselben, so kann man sie aus Profileisen herstellen (Fig. 292 b bis d).

Der Oberweitschenkel g ist, wenn das Fenster hinten im Falz liegt, genau in Stärke und Profil wie die Höhenschenkel; er verändert seine Form aber auch dann nicht und behält den Falz bei, wenn letztere mit Hinternute versehen sind, da sonst das Fenster nicht zu öffnen wäre.

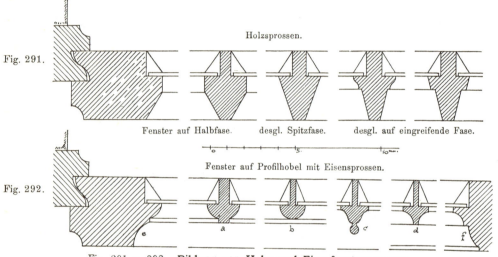

Fig. 291 u. 292. **Bildung von Holz- und Eisenfenstersprossen.**

Der **Flügelwetterschenkel** h ist, von innen betrachtet, genau wie ein umgekehrter Oberweitschenkel, nach aussen dagegen hat er einen Ansatz, welcher das Eindringen des Regenschlags verhindert. Dieser Ansatz mit ca. 4 cm Ausladung ist oben als Wasserschräge behandelt, während unten meist durch ein halbrundes Kehlchen eine Wassernase gebildet ist. Eine anderweitige Profilierung des Wetterschenkels ist überflüssig, ja schädlich, da sie den Wasserabfluss hindert. Von Wichtigkeit ist, dass die Flügelwetterschenkel in der Mitte dicht aneinander schliessen, wie es die Figuren 287, 289 und 290 im Grundriss darstellen. (Reichen dieselben nur bis zur Aussenkante der Schlagleiste (Fig. 288), so dringt das Regenwasser an diesen Stellen sehr leicht ein.) Ebenso wird der Abschluss gegen Regen ein dichterer, wenn man den Flügelwetterschenkel seitlich bis ans Gewände gehen lässt, ihn also zum Teil in den Futterrahmen einsetzt (Fig. 279 und 294), statt ihn abzusetzen, wie Fig. 277, 278 und 280. Die oberen Flügel sind, wenn sie **beweglich**, also zum Oeffnen eingerichtet werden, in Konstruktion, Holzstärke und Profil ganz wie die unteren. Sie erhalten entweder das nämliche Beschläge wie diese, oder sie werden unten mit Bändern, seitlich mit Scheren und oben mit Federfalle versehen (Klappfenster). Anders verhält es sich, wenn die obern Flügel **feststehend** angeordnet werden. Feststehende Flügel sind solche

2. Die Bildung der Fensterflügel.

Flügel, welche zwar nicht festgeschraubt sind, aber doch nicht so leicht geöffnet werden können, wie bewegliche. Die Beweglichkeit hat in vielen Fällen gar keinen Sinn, inbesondere wenn innen angebrachte Rouleaux oder Stores ein Oeffnen unmöglich machen. Im Interesse der Reinigung ist es aber geboten, die Flügel wenigstens herausnehmen und putzen zu können, und man er-

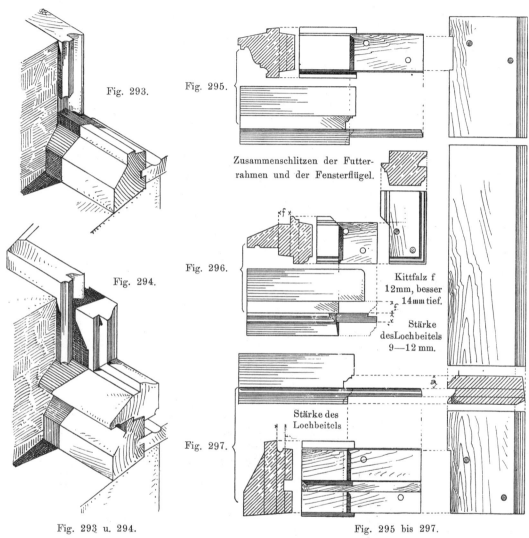

Fig. 293 u. 294.
Einsetzen des Flügelwetterschenkels in den Futterrahmen.

Fig. 295 bis 297.
Konstruktion der Futterrahmen und Flügel.

reicht diesen Zweck, indem man entweder die Flügel in eine obere, besonders vertiefte Nute hinaufschiebt und dann auf die am Kämpfer angestossene Feder herabzieht, wobei man das Beschläge spart, Taf. 56 links, oder indem man sie ringsum in den Falz legt und je nach ihrer Grösse mit 4 bis 6 Vorreibern befestigt.

VII. Die Fenster.

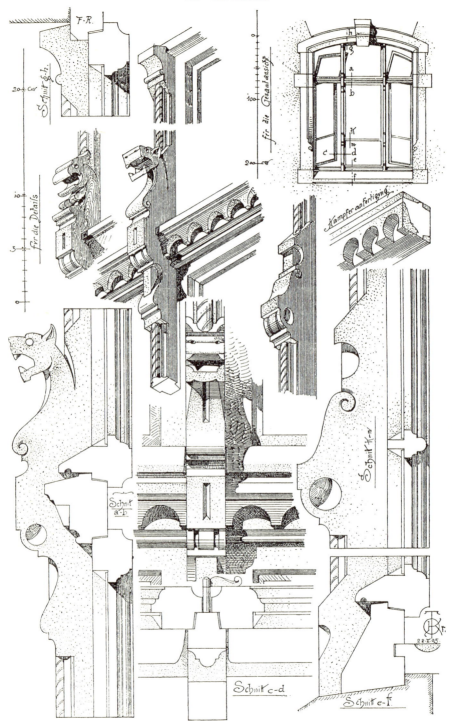

Fig. 298. Dreiteiliges Fenster.

3. Das einfache Fenster.

Die Art der Anfertigung eines Fensterflügels ist im allgemeinen derjenigen der Thüre ähnlich: Nachdem die sämtlichen Hölzer ausgehobelt, d. h. in der richtigen Breite und Dicke gearbeitet sind, werden dieselben vorgerissen, geschlitzt, gestemmt; die Gehrungen werden nach der Profilhobelbreite angeschnitten, und zwar entweder ganz auf Gehrung (bei Profilfenstern) oder nur zusammengestochen (bei Fasefenstern). Alsdann werden, ohne den Rahmen vorher noch provisorisch zusammenzustecken, der Kittfalz und das Profil angestossen und das Fenster ist im allgemeinen fertig. Beim späteren Leimen werden die Zapfen gut vorgewärmt und das Ganze wird mit Holznägeln verbohrt (Fig. 296).

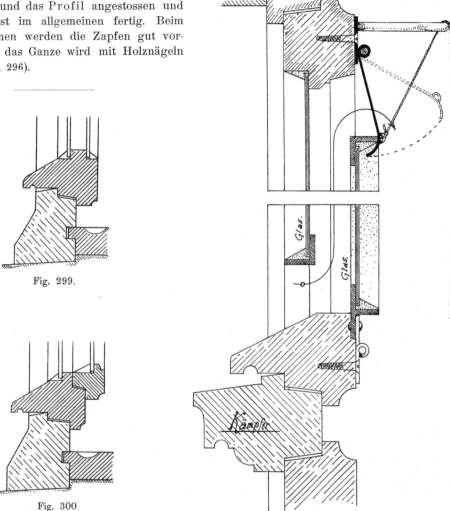

Fig. 299.

Fig. 300.
Doppelte Verglasung.

Fig. 301. Lüftungs-Flügel.

3. Das einfache Fenster.

Das einfache Fenster (im Gegensatz zum Doppelfenster) ist in seiner gewöhnlichen Art mit dem Vorausgegangenen vollständig beschrieben und es erübrigt nur noch, über reichere und aussergewöhnliche Anlagen einige Worte hinzuzufügen. Die Konstruktionen bleiben auch in diesem Fall dem Prinzipe nach dieselben; aber mit den vergrösserten Abmessungen der Lichtöffnungen verstärken sich die Hölzer entsprechend und wo neben der praktischen Seite auch die ästhetische in Betracht kommt, da tritt an Stelle der glatten Arbeit die Verzierung durch Kannelierung, Schnitzerei etc. Die Kämpfer können reicher gegliedert werden nach Art von Gesimsen; die Setzhölzer nehmen die Formen von Lesinen und Pilastern an. Wo die Setzhölzer sich mit dem Kämpfer oder mit anderen Querhölzern im Rahmen kreuzen, werden sie gerne verstärkt und nach vorn vorgesetzt; in ähnlichem Sinne können sie nach unten sich strebepfeilerartig verstärken. Die Kämpfermitten werden bei reichen Beispielen auch mit Krönungen und Giebeln versehen. In diesem Sinne verweisen wir auf die Tafeln 55 und 56, sowie auf die Fig. 298.

Die Tafel 55 zeigt neben der gewöhnlichen Fensterform im Stich- und im Rundbogen abschliessende Fenster, deren Rahmen- und Flügel-Oberschenkel sich dementsprechend ändern. Neben Vierflügelfenstern sind solche mit 6, mit 8, mit 9, 12 und 16 Flügeln aufgezeichnet, die selbstverständlich nicht alle beweglich zu sein brauchen. Ein grosses, reiches Fenster zeigt das mittlere Beispiel der unteren Reihe. Der Kämpfer ist in seiner Mitte vorgebaut, mit Konsolen und Giebelkrönung versehen.

Die Tafel 56 bringt ein Sechsflügelfenster mit Kämpfer- und Setzholzverzierung. Wie die Schnitte zeigen, sind die unteren Flügel in gewöhnlicher Art angeschlagen. Sie sind alle beweglich, was sich jedoch auch auf den Mittelflügel oder auf die Seitenflügel beschränken könnte. Die oberen Flügel werden durch Höhenverschiebung in der bereits beschriebenen Art ohne Beschläge festgestellt. Die Tafel zeigt gleichzeitig die Anbringung des Rollladens, wovon später zu reden sein wird.

Ein weiteres dreiteiliges (Sechsflügel-) Fenster ist in Figur 298 gegeben. Der Gesamtansicht sind so viele Schnitte und Einzelheiten beigezeichnet, dass die Konstruktion weiter keiner Schilderung bedarf.

Das einfache Fenster ist, trockenes Holz und gute Arbeit vorausgesetzt, soweit wetterdicht, als man es eben von Holz und Glas überhaupt verlangen kann. Auch das beste Holz quillt und schwindet und die dabei entstehenden Fugen werden immer etwas Luft durchlassen. Im Sommer stört dies auch nicht; die unbeabsichtigte, selbstthätige Ventilation ist sogar im Interesse der Gesundheit zu begrüssen. Anders ist die Sache aber im Winter. Wer zu dieser Zeit gezwungen ist, in der Nähe des Fensters zu sitzen, der wird die eindringende Luft als leichten Zug wahrnehmen und das Glas, welches ja an sich keine Luft durchlässt, ist zu dünn, um das Eindringen der Kälte zu verhindern. Es kühlt sich an der Aussenluft derart ab, dass auf der Innenseite die Feuchtigkeit der Zimmerluft sich als Kondensationswasser niederschlägt. Die Scheiben »schwitzen« und bei grosser Kälte bilden sich sogar Eisblumen. Das sind Missstände, welche oft geradezu gesundheitsschädlich werden. Eine gründliche Abhilfe schaffen nur die Doppelfenster, von denen im nächsten Absatz zu reden sein wird. Da dieselben eine teure und teilweise auch unbequeme Einrichtung sind, so hat man längst versucht, das einfache Fenster derart umzugestalten, dass es den Anforderungen der Luft- und Wetterdichtigkeit an sich genüge. Man hat komplizierte Falzungen angebracht, Filz- und Selbendstreifen, Gummiröhrchen etc. eingelegt und dickes Glas verwendet. Dadurch lässt sich schon etwas erzielen aber nicht viel; mit der zunehmenden Kompliziertheit wird das Fenster empfindlicher und ebenfalls teurer, ohne dass das Anlaufen der Gläser vermieden wäre. Bezüglich des letzteren ist man auf die Anwendung der sogen. stehenden Luftschicht verfallen. Man hat die einfachen Fenster doppelt verglast, also innen und aussen mit Kittfalz versehen (Fig. 299). Da hat sich gezeigt, dass im Laufe der Zeit der Staub in das Innere dringt und die Möglichkeit des Reinigens nicht vorhanden ist. Um diesem Uebel abzuhelfen, hat man in den Flügel einen zweiten leichteren eingesetzt, zum Oeffnen oder Fortnehmen eingerichtet (Fig. 300). Ein derartiges Fenster hat gewisse Vorteile. Die Scheiben beschlagen sich weniger; die Kälte dringt weniger ein. Man kann den inneren Flügel bunt verglasen und hat dann je nach Wunsch (bei Oeffnung oder Schluss des Innenflügels) ein durchsichtiges oder undurchsichtiges Fenster. Etwaige Glasgemälde des inneren Flügels sind durch die Aussenscheiben geschützt. Trotz der doppelten Verglasung ist das Fenster aber nur ein verbessertes einfaches und die Dichtung hat wenig gewonnen; die Luft zieht nach wie vor durch die Fugen des Hauptflügels.

Bei diesem Anlass möge noch ein Lüftungsflügel erwähnt sein, den wir nach dem Prinzip konstruiert haben, welches der Oberstabsarzt Castaing für die Ventilation der Kasernenbauten in La Rochelle aufgestellt hat (Fig. 301). Der Fensteroberflügel ist doppelt verglast. Die Scheiben nehmen jedoch das volle Feld nicht ein; die äussere Scheibe lässt unten, die innere oben einen etwa 5 cm breiten Schlitz. Die äussere Scheibe liegt auf beiden Seiten und oben im Falz des Holzes, während sie unten durch Sprossenwinkeleisen gefasst ist. Die innere Scheibe liegt allseitig in einem schwachen Eisenrahmen, welcher der Reinigung wegen beweglich in den Holzrahmen eingesetzt ist und mittels Vorreiber oder auf andere Weise festgehalten wird.

Die Luft zieht zwischen den beiden Scheiben langsam durch, wobei sich die Temperaturen etwas ausgleichen, so dass der Zug weniger schroff zur Geltung kommt. Zur Abschwächung oder Einstellung der Lüftung ist der innere Schlitz mit einer Blechkappe versehen, welche, leicht beweglich befestigt, den Schlitz durch ihr Eigengewicht verschliesst, so lange nicht ein Emporheben mittels der durch eine Ringschraube geführten Schnur stattfindet.

---

4. Das Doppelfenster.

a) Das Vorfenster oder Winterfenster.

Bis vor kurzer Zeit waren diese Vorfenster oder Winterfenster in Süddeutschland die einzige Art der Doppelfenster. Sie sind ausserhalb der eigentlichen feststehenden Fenster angebracht, werden im Herbst bei Eintritt der rauhen Witterung eingesetzt und bleiben bis zum Frühjahr, also nur die Hälfte des Jahres, im Gebrauch. Man fertigt sie daher meist auch nur aus Forlen- oder Tannenholz und streicht sie beiderseits gut mit Oelfarbe. Die Dimensionen der einzelnen Teile sind geringer als die der anderen Fenster (Holzstärke 30 mm), teils um sie recht billig herstellen zu können, teils um viel Licht einzulassen. Sie liegen behufs dichteren Anschlusses entweder in einem an das Steingestell angearbeiteten Falz oder Spunden, oder sie sind an der scharfen Kante desselben überfälzt; ihre Befestigung an das innere, bleibende Fenster geschieht mittels Einhänghaken und Ringschrauben. Die Flügel dieser Vorfenster sind entweder feststehend, d. h. mit Vorreibern versehen, wobei zur Lüftung nur ein kleiner Flügel mit Bändern angeschlagen wird, oder aber es werden die beiden unteren Flügel mit vollständigem Beschläge versehen, welches ihr Oeffnen und Schliessen leicht ermöglicht. In beiden Fällen gehen die Flügel entweder nach aussen — in welchem Fall sie mit einer sogen. Festhaltung versehen sein müssen — oder sie schlagen nach innen.

Die Flügelanordnung und Scheibenteilung richtet sich nach derjenigen der inneren Fenster. Sind die letztern durch Sprossen geteilt, so teilt man die Vorfenster ebenso; haben sie dagegen ganze Scheiben, so versieht man die Vorfenster entweder auch mit solchen oder man bringt auf der halben Höhe des Flügels einen Zwischensprossen an. Besondere Vorsicht ist beim Einstellen und Herausnehmen der Vorfenster zur Verhütung eines Unglückes nötig.

Durch diese Vorfenster, deren praktischen Nutzen wohl niemand bezweifelt, wird das Aeussere des Hauses jedoch nicht verschönert. Am meisten verlieren dabei diejenigen Häuser, die an sich schon infolge sparsamer Bauweise wenig Relief zeigen; ihre Fassaden erscheinen nach dem Einsetzen der Winterfenster glatt und schlagschattenlos. Die besser gearbeiteten und verglasten Hauptfenster werden durch die weniger schönen Vorfenster auf die Hälfte des Jahres verdeckt und unwirksam gemacht.

Dieser Missstand zeitigte den Wunsch nach einem zweckmässigen Ersatz der Winterfenster und so hat sich denn auch in Süddeutschland das im Norden längst verwendete feststehende Doppelfenster oder Kastenfenster eingebürgert.

b) Das Kastenfenster (Tafel 57).

Dasselbe unterscheidet sich von dem gewöhnlichen Vorfenster dadurch, dass es

1) im Sommer wie im Winter eingesetzt bleibt,
2) dass das bessere Fenster sich nicht innen, sondern aussen befindet,
3) dass das geschützte innere Fenster doch immerhin wesentlich besser konstruiert ist, als ein gewöhnliches Winterfenster, und
4) dass das Relief der Fassade nicht Not leidet, indem das äussere Fenster an der gewöhnlichen Stelle des einfachen angebracht wird, also unmittelbar hinter dem Steingestell, während das eigentliche Doppelfenster nach innen springt.

Was die Konstruktion des einzelnen Fensters betrifft, so ist sie im allgemeinen die gleiche wie die der einfachen Fenster, nur fehlen dem inneren Fenster die Wetterschenkel, und der Kämpfer bleibt unprofiliert. Bei der Bestimmung der Holzstärken und der Abmessungen der Flügel für den inneren Futterrahmen ist darauf zu achten, dass die Flügel des äusseren Fensters sich leicht und ganz nach innen öffnen lassen, ohne durch den Kämpfer etc. daran gehindert zu werden. Unter sich verbunden sind die beiden Fenster durch einen zusammengezinkten Holzrahmen, dessen unterer Teil — an welchem der innere Flügel seinen Anschlag findet — das Fenstersimsbrett ist. Die Tiefe des Rahmens richtet sich nach dem Abstand, welchen man beiden Fenstern von einander geben will; als Minimalmass gilt 10 cm, so dass die Entfernung der Gläser von einander ca. 13 bis 15 cm beträgt. Die Befestigung des Futterrahmens an das Steingewände kann so erfolgen, dass entweder beide Fenster mit einer durchgehenden Steinschraube, oder aber, dass nur das äussere mit einer Steinschraube, das innere dagegen mittels Bankeisen an die Leibung angemacht wird. Will man der Ersparnis halber nicht beide Fenster aus Eichenholz fertigen, so wählt man für das innere Forlen- oder Tannenholz. Die Verglasung der beiden Fenster muss des guten Aussehens halber miteinander übereinstimmen. Hat das Aussenfenster ganze Scheiben, so erhält auch das innere solche und umgekehrt.

Das Beschläge ist für beide Fenster gleich, wenn nicht etwa die Olive des äusseren Verschlusses etwas einfacher gehalten wird, als die dem Auge näher liegende des inneren Fensters.

c) Das Blumenfenster oder Fenster-Glashäuschen. (Tafel 58).

Es ist dies ebenfalls ein Doppelfenster, welches sich in der Konstruktion von dem gewöhnlichen wenig, im Zweck dagegen sehr wesentlich unterscheidet. Dasselbe dient zur Ueberwinterung von Blumen im Zimmer und muss so eingerichtet sein, dass die Pflanzen, ohne ihnen das zu ihrem Gedeihen nötige Licht entziehen zu müssen, durch die Kälte nicht Not leiden. Will man nicht ein feststehendes Doppelfenster, also ein Kastenfenster in der benötigten Tiefe (von ca. 40 cm) herstellen, was das Einfachste ist, so kann man sich ein Blumenfenster schaffen, indem man

1) ausserhalb, d. h. vor dem feststehenden Fenster ein provisorisches, nur für den Winter dienendes Vorfenster anbringt (Tafel 58 B), wie es gewöhnlich geschieht, oder besser indem man

2) innerhalb ein gewöhnliches zweites Fenster anfügt (Tafel 58 A).

Das äussere Fenster, das sogen. Glashäuschen (Tafel 58B), konstruiert man so, dass dessen unterer Teil auch im Sommer als Blumenbänkchen belassen bleiben kann, während im Spätjahr nur der eigentliche Kasten darüber gestülpt wird. Derselbe besteht aus zwei geschlossenen paralleltrapezförmigen Seiten und einem ringsum vorspringenden, mit Wassernase versehenen Oberteil. Diese 3 Teile werden gut mit starkem Zinkblech eingebunden und so gegen das Wetter geschützt. Die Flügel sind nach aussen gehende Klappfenster — die Bänder befinden sich oben, — und zum Aufstellen dient ein handlicher, mit Löchern versehener Griff, welcher eingehakt wird, oder ein Kniehebel. Obgleich der letztere nur ein mässiges Oeffnen gestattet, so ist dieses doch genügend, um frische Luft einzulassen. Dass das Ganze durch Anhänghaken und Ringschrauben festgestellt werden muss, braucht wohl nicht besonders hervorgehoben zu werden.

Der untere Fries des eigentlichen Futterrahmens ist ziemlich hoch angenommen, um den Blumen Schutz gegen Kälte zu gewähren, da andernfalls die Wurzeln gerne Not leiden. Zur besseren Warmhaltung ist schliesslich auf das Blumenbänkchen ein Zinkeinsatz gefertigt, welcher, ca. 5 bis 6 cm hoch mit Torfmull ausgefüllt, den Töpfen einen warmen Stand giebt. Des Ver-

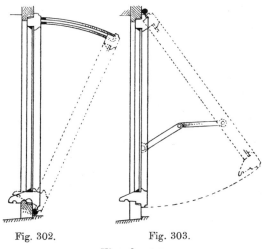

Fig. 302.   Fig. 303.
Klappfenster.

ziehens wegen sind die beiden oberen Zwischenteilungen nicht aus je einem vollen Brett gefertigt, sondern als eine Art Lattenrost behandelt. Dieser Rost kann mit Vorteil durch einen Rahmen ersetzt werden, der mit einem starken Drahtgeflecht überspannt ist. Es kann dann mehr Licht einfallen.

Das Blumenfenster (Tafel 58 A) besteht aus einem gewöhnlichen Fenster, welches sich in ein zwischen die Fenterleibung eingepasstes und eingestemmtes Futter einlegt und nach dem Zimmer mit einer ringsum laufenden Verkleidung abschliesst. Futter und Verkleidung samt dem breiten Fenstersimsbrett und dem Kämpfer bilden ein Ganzes, in welches sich die Flügel einlegen. Ebenfalls damit verbunden, oder auch getrennt, kann unterhalb des Simsbrettes ein mit Thüren versehenes Schränkchen angebracht sein, welches zur Aufbewahrung verschiedener Gerätschaften dient. Das Ganze kann provisorisch befestigt werden, so dass es im Frühjahr ganz zu beseitigen ist, aus welchem Grund zwei Simsbretter angenommen sind.

5. Das Klappfenster.

Klappfenster, d. h. Fenster, welche sich um eine horizontale Axe drehen und zum Hereinklappen konstruiert sind, haben wir als solche allein und in Holzausführung sehr selten, dagegen sehr oft kombiniert mit gewöhnlichen Fenstern. In diesem Fall sind meist die Flügel oberhalb des Kämpfers zum Klappen eingerichtet, und zwar so, dass der Klappflügel unten mit zwei

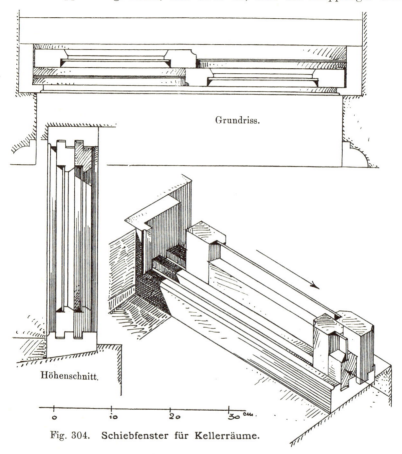

Fig. 304. Schiebfenster für Kellerräume.

Scharnier- oder Fischbändern, oben mit Schere, Federfalle und Zugkettchen angeschlagen wird (Fig. 302). In geöffnetem Zustand strömt nun die Luft schräg nach oben, nach der Decke aufsteigend, ein, vermischt sich dort mit der warmen Zimmerluft und sinkt allmählich herab, die günstigste und für die Bewohner unschädlichste Art der Lufterneuerung. Weniger vorteilhaft ist das umgekehrte Anschlagen des Klappflügels, wobei die Bänder obenhin zu sitzen kommen und das Fenster, nachdem es aufgehoben ist, durch Kniegelenkhebel in geöffnetem Zustand erhalten wird (Fig. 303).

Die Konstruktion der Klappflügel ist entweder die der gewöhnlichen im Falz liegenden Flügel oder es kann der Klappflügel nur an drei Seiten im Falz liegen, an der unteren Seite dagegen mit Bändern versehen und so konstruiert sein wie der Wetterschenkel des oberen

Flügels des einfachen Fensters auf Tafel 56. Klappflügel werden angewendet in Wohnzimmern, Schulsälen (woselbst sie seitlich mit Blechbacken versehen werden), bei Glasabschlüssen, in den meisten Hauswirtschaftsräumlichkeiten und Ställen.

Klappfenster, die sich nach aussen öffnen, kommen an Veranden, in Treppenhäusern etc. vor. Sie werden angeschlagen und aufgestellt wie an dem Fensterglashäuschen Tafel 58 B.

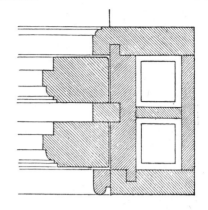

6. Drehfenster.

Die Flügel drehen sich um die horizontale oder um die vertikale Mittelaxe und sind an den Drehpunkten mit Zapfen versehen. Derartige Fenster werden kaum mehr in Holz ausgeführt, wohl aber in Eisen für Fabriken, Schlachthäuser, Ställe u. s. w. Sie können demnach hier übergangen werden.

7. Schiebfenster (Tafel 59).

Die Schiebfenster in Holz sind heute weniger häufig in Anwendung als ehedem; sie werden meistens nur da angebracht, wo drehbare Flügel aus irgend einem Grunde unmöglich sind oder wo man versichert sein will, dass die Fenster nicht vom Winde erfasst und zertrümmert werden können. Aus dem letzteren Grunde werden die Schiebfenster (wie die Klappfenster) gerne für Veranden, Erker und Treppenhäuser benützt; auch als Kellerfenster sind sie vielfach im Gebrauch. Im ersteren Falle werden sie gewöhnlich zum Schieben nach der Höhe, im letzteren Falle zum Schieben nach der Seite eingerichtet. Die Konstruktion ist in beiden Fällen ziemlich dieselbe und der Bau ist demjenigen anderer Fenster ähnlich. Die Schiebfenster haben Futterrahmen und die Höhenschenkel oder die Querschenkel haben Federn angearbeitet und laufen in Nuten oder umgekehrt. Das zu verwendende Holz muss gut trocken, schlicht und geradfaserig sein, damit die Flügel sich nicht verziehen und die Bewegung erschweren. Den

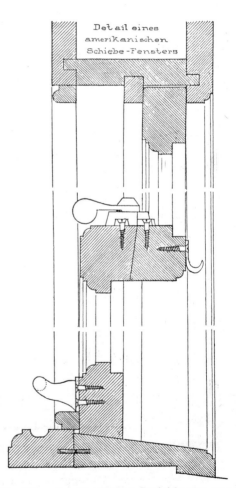

Fig. 305. Amerikanisches Schiebfenster von Billing & Zoller in Karlsruhe.

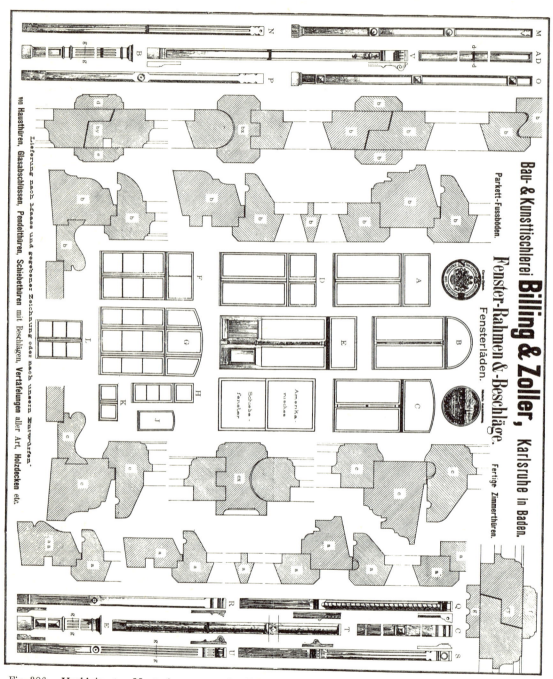

Fig. 306. Verkleinerter Musterbogen aus der Fenster-Preisliste der Bau-, Kunst- und Parkettschreinerei von Billing & Zoller in Karlsruhe i. B.

7. Schiebfenster.

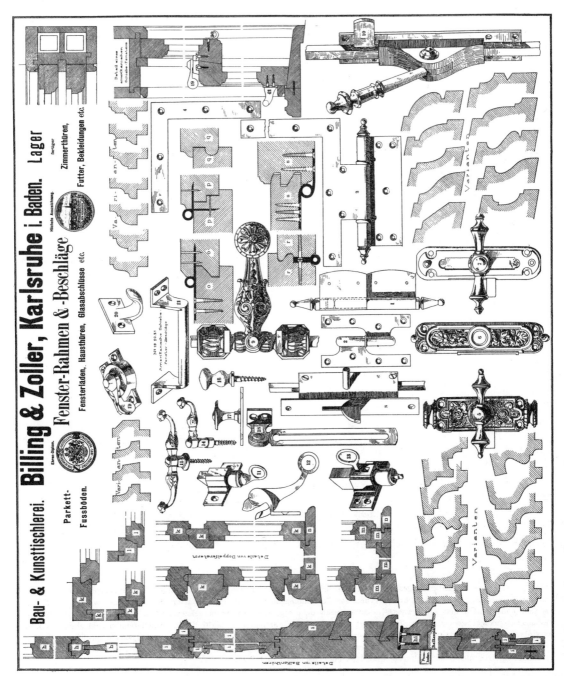

Fig. 307. Verkleinerter Musterbogen der Fenster-Preisliste der Bau-, Kunst- und Parkettschreinerei von Billing & Zoller in Karlsruhe i. B.

162 VII. Die Fenster.

Federn ist in den Nuten genügend Spielraum zu geben, damit die Fenster sich nicht klemmen und stecken bleiben, wenn das Holz quillt. Das richtige Mass zu treffen, ist nicht leicht. Wird sehr genau gearbeitet, so ist die Handhabung der Fenster erschwert; wird reichlich Spielraum gegeben, dann klappern die Flügel im Winde. Diese Unzuverlässigkeit hat die Schiebfester beim Publikum in Misskredit gebracht und es müssen erst wesentliche Verbesserungen angebracht werden, wenn deren Verwendung allgemeiner werden soll. Ein verbessertes Schiebfenster scheint sich von Amerika aus einbürgern zu sollen. Das Prinzip ist folgendes: Die Laufnuten des Futterrahmens sind nicht gleichbreit, sondern verjüngen sich nach den Enden hin, während sie an der Stelle, wo die Flügel geschlossen übereinander greifen, einige mm weiter sind. Die Flügel laufen dabei leicht; in der Nutenenge sitzen sie ohne weiteres fest, am anderen Ende besorgt ein einfacher Hebelverschluss die Feststellung, so dass das Klappern vermieden ist (Fig. 305).

Grosse und schwere Schiebfester, welche nach der Höhe laufen, versieht man beiderseits mit Gegengewichten, welche in geschlossenen, in die Leibung eingebauten Kästen hängen und an Schnüren über Rollen geführt, sich auf- und abbewegen können. Ist dabei alles in Ordnung, so laufen die Flügel sehr leicht und bleiben ausserdem in jeder Höhe stehen, sobald die Hand zu schieben oder zu ziehen aufhört, und die andernfalls zur Festhaltung in den Nuten anzubringenden Schleppfedern kommen in Wegfall. Mit Blei ausgegossene Gasrohrstücke eignen sich zu Gegengewichten besonders, da sie wenig Platz einnehmen und sich in den Kästchen gut führen. Selbstredend sollen letztere etwaiger Reparaturen wegen zum Oeffnen eingerichtet sein.

Auf Tafel 59 ist ein Fenster dargestellt, dessen mittlerer unterer Flügel zum Aufwärtsschieben eingerichtet ist, Der mittlere obere Flügel ist nach der Zeichnung als Klappfenster behandelt, während die vier kleinen seitlichen Flügel wie gewöhnliche Fenster angeschlagen sind. Selbstredend können die fünf letztgenannten Stücke auch mit Vorreibern befestigt werden.

Die Figur 304 zeigt ein Schiebfenster für Kellerräume mit seitlich verschiebbaren Flügeln. Die letzteren beiden laufen in Nuten des Futterrahmens, welcher erst zusammengesetzt wird, nachdem die Flügel eingebracht sind.

Bei den Schiebfenstern liegen die Glasflächen der einzelnen Flügel nicht in einer Ebene, was nicht zu umgehen ist. Wenn die Symmetrie hierdurch gestört wird, wie bei dem Kellerfenster, so ist das nicht gerade schön; weniger auffällig ist die Sache, wenn die Symmetrie gewahrt werden kann, wie bei dem Fenster der Tafel 59 und der Figur 305.

Die letztere Figur bezieht sich auf das amerikanische Schiebfenster, dessen Konstruktionsprinzip bereits erwähnt wurde. Die Abbildung ist einem Musterbogen entnommen, mit welchem die rühmlichst bekannte Karlsruher Bau-, Kunst- und Parkettschreinerei von Billing & Zoller ihre Fenster-Preisliste illustriert.*) Da dieser Musterbogen wohl geeignet erscheint, das über die Fenster Vorgebrachte bildlich zu ergänzen, so geben die Figuren 306 und 307 denselben, auf $\frac{1}{3}$ verkleinert, wieder.

---

*) Diese sorgfältig ausgearbeitete Preisliste wird von der Firma den Interessenten auf Ersuchen gerne zugestellt. Bemerkenswert ist der neueingeführte, verbesserte Berechnungsmodus. Die (unverglasten) Fensterrahmen werden nicht, wie seither üblich, nach dem Quadratmass, sondern nach dem sog. Eckmeter berechnet. Höhe und Breite, an der Futterrahmenaussenkante gemessen, werden addiert und mit den Einheitspreisen multipliziert. Die letzteren sind abgestuft nach 3 Holzarten (Eiche, Pitchpine, Forle), 4 Holzstärken (31, 36, 41 und 46 mm) und 10 Ausstattungsarten (profilierte Vorderfrontfenster ohne Sprossen und gefaste Hinterfrontfenster mit Sprossen, je 1, 2, 3, 4 oder 6 flügelig). Für aussergewöhnliche Ausstattung, Grundieren mit Oelfarbe, Beschläge etc. Zuschlags- und Sonderpreise.

8. Fenster in Fachwerkswänden.

Bis jetzt war nur vom Anschlagen der Fenster die Rede, soweit dieselben in Wänden aus Stein eingesetzt werden. Nun giebt es aber auch Fachwerksbauten und selbst beim massiven Haus sind nicht selten Dach- und Gaupenfenster in Holzgestelle einzusetzen.

Im allgemeinen sind, auch wenn das Pfosten- und Riegelholz gehobelt ist, Futterrahmen zu beschaffen und die Befestigung derselben geschieht vermittelst beiderseitiger Deckleisten (Fig. 309), welche die Dichtung besorgen. Der Futterrahmen selbst soll nämlich im Gestell Luft haben, d. h. nicht dicht anliegen, damit das Fenster nicht Not leidet, wenn beim Setzen des Baues, beim Arbeiten des Holzes, bei Erschütterungen etc. das Gestell seine Form verändert.

Die Figuren 308 bis 311 zeigen die Konstruktion in der Abbildung. Nach den Figuren 308 und 310 sind die Brustriegel der Fenster, die man gerne aus Eichenholz macht, nach aussen abgeschrägt und mit einem Falz versehen, an welchen sich der Futterrahmen des Fensters anlegt, wobei ein Zinkblech den Brustriegel abdeckt und schützt. Man kann aber auch zwischen Brustriegel und Futterrahmen ein eichenes Simsbrett einfügen, welches ebenfalls mit Zink abgedeckt wird (Fig. 311). Die Figur 309 zeigt einen Horizontalschnitt durch den Flügel- und Futterrahmenhöhenschenkel samt Pfosten und den bereits erwähnten Dichtungsleisten.

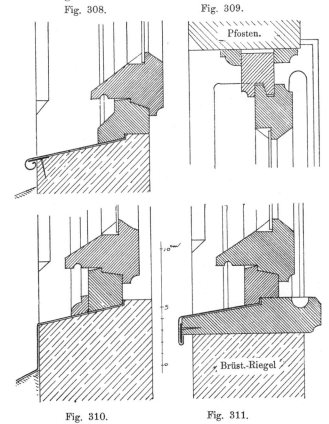

Fig. 308 bis 311. Befestigung der Futterrahmen in Fachwerkswänden und Dachgaupen.

9. Schaufenster, Auslage- oder Ladenfenster (Tafel 60 und 61).

Schliesslich ist noch zu erwähnen das sogen. Schaufenster der Kaufhäuser. Dasselbe besteht aus einem starken Futterrahmen, in welchem das Glas, 6 bis 8 mm starkes, unbelegtes Spiegelglas mittels Holzleisten befestigt ist (Fig. 312). Dieses Fenster, nur aus einer Scheibe bestehend, ist dicht und undurchlässig für Regen und Luft, hat aber, wie alle einfachen Fenster, den Nachteil, anzulaufen und zu schwitzen, sobald die Temperatur im Laden eine höhere wird, als im Freien. Die Klagen über diesen Missstand, welcher der eigentlichen Bestimmung des Schau- oder Auslagefensters zuwiderläuft, sind so alt wie das Fenster selbst, und man war seit seiner Einführung bestrebt, Mittel zur Abhilfe zu entdecken. Was hilft das schönste Anordnen der Waren, wenn man sie von aussen nicht sehen kann? Alle Mühe aber war bisher ziemlich vergeblich, sofern man Abhilfe bei einem einfachen Fenster erhoffte. Wohl sucht man das Schwitzen zu verhindern, indem man am Fuss der Scheibe und oben mittels durchbrochener Metallfüllungen Luft ein-, bezw. auslässt, welches Mittel allerdings hilft, wenn die Temperatur des Ladens dadurch auf diejenige im Freien reduziert wird (Fig. 312), oder man bringt am Fuss der Spiegelscheibe Gasflämmchen an, welche durch Erzeugung einer warmen, aufsteigenden Luftschichte die Scheibe erwärmen und dadurch die Bildung feuchter Niederschläge verhindern. Dieses Verfahren ist, abgesehen vom Gasverbrauch, nicht unzweckmässig, aber es ist seiner Gefährlichkeit halber nicht überall anwendbar, so dass sogar ein Verbot desselben gerechtfertigt wäre. Gründliche Abhilfe verschafft nur die Erstellung eines doppelten Abschlusses nach Art der Doppelfenster durch Bildung einer stehenden Luftschichte, d. h. der zur Auslage dienende Raum muss durch ein zweites Fenster oder einen Glasabschluss von dem Ladenraum getrennt werden. Hierdurch wird die Möglichkeit geschaffen, den Laden zu heizen, ohne dass die Fenster anlaufen; ausserdem werden die Waren am Schaufenster staubfrei gehalten. Wie dieser innere Abschluss zu fertigen ist, hängt von den jeweiligen Umständen und Verhältnissen ab. Im allgemeinen wird derselbe der Form nach den Glasabschlüssen ähnlich behandelt, doch nicht befestigt, sondern vielmehr beweglich auf Rollen verschiebbar gebaut, um die Anordnung der Waren leicht bewirken zu können. Der Unterbau nimmt die Treppen und Gestelle für die auszulegenden Waren auf. Im oberen Teil des Apparates erfolgt die Verglasung bei grösserer Tiefe auf drei Seiten, bei geringerer Tiefe nur nach hinten. Bei feststehender Anordnung muss der Auslageraum durch Anbringung entsprechender Flügel zum Oeffnen zugänglich gemacht werden.

Auf den Tafeln 60 und 61 bringen wir ein Schaufenster zur Abbildung. Tafel 60 zeigt die äussere Ausstattung, Tafel 61 die innere Einrichtung. Während die das Schaufenster aufnehmende Architektur durchweg im Material des Steins gehalten sein kann, so ist bei diesem Beispiel bloss die Sockelpartie aus Stein, dagegen sind die Pfeiler oder Ständer, sowie der Architrav aus Eisen konstruiert und mit Holz verkleidet, wie es neuerdings in steinarmen Gegenden oft gemacht wird, wenn man nicht vorzieht, die Ständer samt ihren Verzierungen in Eisen zu giessen oder dieselben, wie auch den Architrav, aus Mannstaedtschem Walzeisen zu bilden.

Die Art und Weise, wie die Verkleidung der Eisenteile und der Leibungen erfolgt, ist aus Tafel 60 und aus den Schnitten der Tafel 61 genügend ersichtlich. Als Material dient am besten Eichenholz oder, wenn die Holzfarbe umgangen werden soll, Forlenholz, welches der übrigen Architektur entsprechend oder zu ihr passend, einen Oelfarbanstrich erhält. Zwischen die beiden Schaufenster ist die Ladenthür mit Kämpfer und Oberlicht eingebaut. Der Grösse und Schwere der Thüre entsprechend muss das Thürfutter genügend stark und solid gehalten werden.

Der Auslagekasten im Innern des Fensters (Taf. 61) besteht zunächst aus zwei Teilen, von denen der eine mitsamt der Rollladeneinrichtung in die Fensterleibung eingebaut ist und

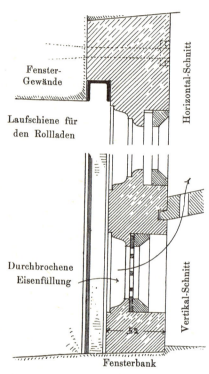

Fig. 312. **Konstruktion des Schaufensters.**

in Bezug auf die Konstruktion bereits an anderer Stelle beschrieben ist, während der zweite fahrbare Teil stumpf oder auf Nut und Feder an den ersten anstösst. Dieser bewegliche Teil setzt sich zusammen aus einem unteren und oberen »Kranz«, den gestemmten oder mit Glasfüllungen versehenen Seiten und der verglasten Rückwand. Die Verbindung der Seiten und der Rückwand ist durch angeleimte Lesinen bewirkt. Ein Kämpfer, von der einen Lesine zur andern laufend, trennt den fest verglasten oberen Teil der Rückwand von den Fensterflügeln des unteren Teils, an deren Stelle auch Schiebfenster treten können. Der obere Kranz ist ebenfalls unter Anwendung von Sprosseneisen fest verglast. Der auf dem unteren Kranz liegende Boden des Auslagekastens muss wegnehmbar sein, um bei etwaigen Störungen bequem zu den Laufrollen gelangen zu können. Die letzteren sind gegossen, fein abgedreht, mit einer Rinne versehen und mit ihren Lagern seitlich im unteren Kranz befestigt, wie es die Zeichnung zeigt. Die Rollen laufen auf $\perp$-Eisen-Schienen, welche versenkt im Fussboden befestigt sind.

Alles weitere dürfte zur Genüge aus den Abbildungen erhellen.

Wir sind wiederholt darum angegangen worden, gerade dieses Kapitel mit weiteren Abbildungen zu belegen. Wir haben jedoch davon abgesehen, weil derartige Anlagen, wie bereits erwähnt, zu sehr von den gegebenen Verhältnissen abhängen und von Fall zu Fall durchdacht und entworfen werden müssen.

10. Das Glas.

Glas entsteht durch Zusammenschmelzen von Quarz mit Kali oder Natron und Kalk. Von der Wahl und Reinheit des Rohmaterials hängt die Qualität ab. Metalloxyde färben die an sich farblose Masse; zur Erzeugung farbigen Glases werden sie in bestimmter Weise zugesetzt; zufällig im Rohmaterial vorhanden, inbesondere in der Form von Eisenverbindungen, machen sie das Glas unrein und geben ihm farbige Stiche ins Grüne, Gelbe etc.

Das Glas wird technisch ausserordentlich vielseitig verwendet. Hier kommt jedoch nur das tafelförmige Glas in Betracht, wie es zum Verglasen der Fenster und Thüren dient, im übrigen aber auch für Spiegel, Dacheindeckungen, Oberlichter u. a. m. verwendet wird.

In Bezug auf die Herstellung kann man drei Arten tafelförmigen Glases unterscheiden:
    a) **Geblasenes Glas, Walzenglas, kurzweg Tafelglas genannt.**
        Die mit der Pfeife erblasenen Cylinder werden aufgeschnitten, tafelförmig auseinandergelegt, gebügelt u. s. w.
    b) **Gegossenes Glas oder Rohglas.**
        Die glühend aufgegossene Masse wird mit Walzen zu Tafeln ausgebreitet, in gemusterte Formen gepresst u. s. w.
    c) **Geschliffenes Glas oder Spiegelglas.**
        Das gegossene Rohglas wird in vier Folgen geschliffen (Rauhschliff, Klarschliff, Feinschliff und Polierung).

Das Tafelglas ist durchsichtig wie das geschliffene Glas, aber nicht so vollständig eben, was beim Hindurchsehen weniger auffällt, als wenn beide Arten von Gläsern als Spiegel verglichen werden. Das Rohglas ist je nach der Rauheit seiner Oberfläche mehr oder weniger durchsichtig oder nur durchscheinend.

Für die gewöhnlichen Fälle der Bauschreinerei wird das billigere Tafelglas benützt. Das bedeutend mehrwertige Spiegelglas kommt hauptsächlich für Schaufenster in Anwendung, für ganz feine Fenster überhaupt, für Thürschutzplatten etc.

In Bezug auf die Stärke des tafelförmigen Glases ist zu bemerken, dass das Rohglas 4 bis 30 mm stark, das Spiegelglas 4 bis 8 mm stark (für Aquarien etc. auch stärker) hergestellt wird, wobei die Stärken in mm angegeben werden, während für das Tafelglas andere Bezeichnungen üblich sind. Man unterscheidet in dieser Hinsicht:

    a) einfaches oder $^4/_4$-Glas, ca. 2 bis $2^1/_2$ mm stark;
    b) anderthalbfaches oder $^6/_4$-Glas, ca. $2^1/_2$ bis $3^1/_2$ mm stark;
    c) Doppelglas oder $^8/_4$-Glas, ca. $3^1/_2$ bis $4^1/_2$ mm stark;
    d) dreifaches oder $^{12}/_4$-Glas, ca. $4^1/_2$ bis 6 mm stark.

Für gewöhnliche Sprossenfenster genügt $^4/_4$-Glas; zur Verglasung ganzer Flügel mit einer Scheibe dient $^6/_4$ oder besser $^8/_4$-Glas; $^{12}/_4$-Glas kommt nur ausnahmsweise in Anwendung.

In Bezug auf die Reinheit unterscheidet man ordinäres oder grünes, halbweisses, $^3/_4$ weisses und ganz weisses Glas. Der Grad des farbigen Stiches kommt beim Hindurchsehen wenig, deutlicher aber auf dem Schnitt zur Geltung, der sich stets mehr oder weniger grünlich zeigt. Die Reinheit ist ausserdem abhängig von Blasen, Flecken, Unebenheiten etc. Die Glashütten machen in ihren Verzeichnissen meist drei Abstufungen: erste, zweite und dritte Sorte oder Wahl.

Ausser dem tafelförmigen, durchsichtigen und farblosen Glas finden auch andere Formen, wie die sogen. Butzenscheiben, bloss durchscheinende oder künstlich mattierte, in der Masse gefärbte oder auf der Oberfläche bemalte oder anderweitig verzierte Gläser Verwendung zur Verglasung von Fenstern, Thüren etc. für bestimmte Zwecke.

Tafelglas.

Das gewöhnliche Tafelglas wird direkt von den Hütten, bezw. von den Generalagenten derselben bezogen, nach deren von Zeit zu Zeit zur Ausgabe gelangenden Preis-Verzeichnissen und Bedingungen die Besteller sich zu richten haben.

Es möge hier ein Auszug aus dem Verzeichnis der vereinigten Tafelglashütten Saarbrücken folgen:

Das Glas ist rechtwinklig nach cm geschnitten. Die ungeraden cm werden wie die nächst höheren geraden berechnet, z. B. $26^1/_2 \times 35$ als 28×36. Breite und Länge einer Tafel addiert, ergiebt die vereinigten cm oder die Stufe der Preisliste. Tafeln von über 160 cm Länge zählen jedoch zur nächst höheren Stufe (z. B. 60×162 gleich 222 vereinigte cm zählen zur 9. statt zur 8. Stufe der folgenden Tabelle):

Feste Masse				Preis für 1 qm					
				1. Sorte		2. Sorte		3. Sorte	
				ℳ	₰	ℳ	₰	ℳ	₰
1. Stufe		bis 60	ver. cm	2	—	1	80	1	60
2. »	von 62	» 90	» »	2	30	1	90	1	70
3. »	» 92	» 122	» »	2	60	2	—	1	80
4. »	» 122	» 150	» »	3	—	2	30	2	—
5. »	» 152	» 180	» »	3	50	2	50	2	10
6. »	» 182	» 200	» »	4	—	2	60	2	20
7. »	» 202	» 220	» »	4	50	3	—	2	30
8. »	» 222	» 240	» »	5	—	3	40	2	60
9. »	» 242	» 260	» »	6	—	4	30	3	50
10. »	» 262	» 280	» »	7	50	5	50	4	40
11. »	» 282	» 300	» »	8	50	7	30	6	—

Diese Preise gelten für $4/_4$-Stärke; $6/_4$-Glas kostet das $1\frac{1}{2}$ fache, $8/_4$ das doppelte und $12/_4$ das vierfache.

Unter Normalkisten werden solche verstanden, welche Glas von einer Grösse, Sorte und Stärke enthalten und zwar:

20 qm $4/_4$, Mindestgewicht 120 kg.
20 » $6/_4$, » 140 »
15 » $8/_4$, » 180 »

Halbe Kisten, zu 10 qm Inhalt, enthalten die Hälfte der Tafelzahl der Normalkisten oder — wenn die letztere ungerade ist — die Hälfte weniger 1. Für Normalkisten und Nichtnormalkisten, welche mehr Inhalt gleicher Grösse, Sorte und Stärke als jene haben, sowie für halbe Kisten der 1. Stufe wird die Verpackung nicht berechnet.

Unter freien Massen werden solche verstanden, bei welchen die Wahl der Breite und Länge den Hütten überlassen ist und der Besteller nur die Stufe bestimmt. Unter 62 cm Breite werden freie Masse nicht geliefert. Der ungefähre Inhalt der Freimasskisten beträgt 30, 20 und 15 qm für $4/_4$, $6/_4$ und $8/_4$-Glas.

Streifen werden nur in $4/_4$-Stärke 2. und 3. Sorte und nicht über 160 cm lang geliefert, in Kisten von 25 qm Inhalt bis 50 cm breit, in Kisten von 30 qm Inhalt 52 bis 60 cm breit. Die Verpackung ist frei, wenn nicht mehr als drei Breiten vertreten sind.

106 bis 112 cm breite Tafeln erleiden einen Aufschlag von 10%, über 112 cm breite Tafeln von 20%, sowohl bei festen als freien Massen.

Das Abschleifen der Kanten wird mit 1 ℳ oder 1 ℳ 50 ₰ pro lfd. m berechnet, je nachdem die Seitenlänge oder Breite weniger oder mehr als 100 cm misst.

Die Preisschwankungen werden durch Rabattsätze auf die Grundpreise geregelt. Dieselben sind verschieden je nach Sorte, Grösse und Stärke und werden von Zeit zu Zeit bekannt gegeben.

Für aussergewöhnliche Anforderungen, für das Biegen der Scheiben etc., sind besondere Vereinbarungen zu treffen.

Für eine genaue Stärke der Tafeln wird keine Garantie geleistet, ebensowenig für überall gleichmässige Stärke derselben. Ganz abgesehen von der Unmöglichkeit, eine solche bei der derzeitigen Fabrikationsweise des Walzenglases einzuhalten, besitzen die Glashütten durch das gemeinsame, geeinigte Vorgehen ihrer Direktorien — das manchen Körperschaften als Muster dienen könnte — eine Art Monopol, welches sie in den Stand setzt, Vorschriften zu machen. Auch bezüglich der Farbe, ob etwas mehr gelblich oder grünlich oder gar bläulich, übernehmen die Hütten keine Garantie. Es kommt nämlich vor, dass in einer und derselben Kiste im Bruch ganz verschieden schillernde Glassorten sich befinden, ein Umstand, der, wenn das Glas sonst rein, d. h. ohne Blasen, Flecken oder andere Unreinigkeiten, ohne Unebenheiten und dergl. ist, welche das Durchsehen erschweren oder verhindern, von geringer Bedeutung ist, da man, sobald das Glas eingesetzt ist, die Farbe nicht mehr sieht.

Spiegelglas: unbelegtes und belegtes.

Für den Bezug von geschliffenem Spiegelglas (gegossenes Tafelglas, welches geschliffen und poliert wird), auch unbelegtes Spiegelglas von 4 bis 8 mm Stärke genannt, haben die Fabriken ähnliche Bedingungen aufgestellt und veröffentlicht. Das dem Verfasser vorliegende Preis-Verzeichnis ist dasjenige des Vereins Deutscher Spiegelglasfabriken, Köln a. Rh., Hermannstr. 1, welches bestimmt, dass die Scheiben von 3 zu 3 cm wachsen, sowohl nach der Höhe wie nach der Breite. Dazwischen fallende Abmessungen werden wie das nächst höhere Mass berechnet, z. B. 156×64 für 156×66 cm, $157\frac{1}{2} \times 64\frac{1}{2}$ für 159×66 cm u. s. w.

Die Grundpreise sind nach einem Tarif vom 1. Januar 1884 geregelt; die jeweiligen Preise werden durch Rabattgewährung festgesetzt. Der Rabatt beträgt zur Zeit (1899):

für Gläser 1. Wahl zum Belegen . . . . 25 %
» » 2. » » » . . . . 35 %
» » 3. » » » . . . . 45 %
» » zu Verglasungen im Betrage bis 80 ℳ . . . . 50 %
» » » » » » von 80 bis 160 ℳ 50+10 %
» » » » » » » 160 » 240 ℳ 50+5+15 %
» » » » » » » über 240 ℳ 50+5+20 %

Für gleichmässige Dicke der Gläser einer Sendung wird nicht garantiert; sie schwankt zwischen 4 und 8 mm. Genau vorgeschriebene Stärken werden zum Rabattsatze der nächst höheren Wahl berechnet.

Für das Justieren und Polieren der Kanten, für das Facettieren, für das Abrunden der Ecken, für das Bohren von Löchern, sowie für das Mattschleifen von Gläsern enthält das Verzeichnis besondere Zuschlagstabellen.

Nicht rechteckige Formen werden wie das kleinste umschriebene Rechteck berechnet, wenn nicht ein besonders schwieriger Umriss die Kosten erhöht.

Die Verpackung wird mit 4 ℳ für den Quadratmeter der Kistendeckeloberfläche berechnet. Der Kistendeckel gilt nach beiden Richtungen um je 12 cm grösser als das Glas. Aussergewöhnliche Verpackungen nach Vereinbarung. Die Kisten werden nicht zurückgenommen. Die Anfuhr der Kisten zur Bahn wird mit 20 ₰ für 100 kg berechnet. Die Sendungen reisen auf Kosten und Gefahr des Bestellers. Frachtgutsendungen können beim Verein gegen Bruch auf dem Transport versichert werden, Eilgutsendungen nicht.

Streifen bis zu 18 cm Breite gehen nach dem Gewicht, zu 40 ℳ für 100 kg; Streifen von 18 bis 30 Breite cm kosten pro Quadratmeter 10 ℳ.

Für Glasthürplatten mit einfachem Schliff regelt ein besonderes Verzeichnis die Preise für je 100 Stück.

Silberbelegte Gläser (Spiegel) werden ähnlich wie die unbelegten berechnet. Einfache Gläser zahlen einen Aufschlag von 12 %; facettierte von 16 %. Kleine, fein weiss silberbelegte Spiegelgläser ohne Wahl, nicht über 84 cm und nicht über 40 qdm messend, haben einen besonderen Tarif und werden in Kisten zu 15 qm Inhalt geliefert. Unbelegt 20 % billiger.

Rohglas.

Die dem erwähnten Verein angehörigen Glasmanufakturen liefern auch Rohglas, d. h. gegossenes, nicht geschliffenes Glas zu Bedingungen, die den vorgebrachten ähnlich sind und zu Preisen, die sich in ähnlicher Weise berechnen. Von den Grundpreisen kommen wieder veränderliche Rabatte in Abzug. Für Kantenjustierung, Eckenabrundung, Löcherbohren, Mattschleifen, Biegen etc. sind Aufschläge zu bezahlen.

Es ist zu unterscheiden zwischen folgenden Arten:
1) Gewöhnliches Rohglas, 10 bis 13 mm stark, zur Verglasung und Verdachung von Bahnhöfen, Fabriken, Magazinen, Passagen etc.
2) Glatte Fussbodenplatten, 20 bis 25 mm stark.
3) Dünnes weisses Rohglas, 4 bis 6 mm stark, glatt, gerippt, kleingerautet, grossgerautet (Diamantglas), für Dächer, Treibhäuser, Kirchenfenster etc.
4) Gemustertes Patent-Rohglas, 3 bis 4 mm stark, mit verschiedenerlei Zeichnung und Relief, für Korridorthüren, Treppen- und Gangfenster, Veranden etc.

Kathedralglas.

Dieses Glas ist schwach in der Masse gefärbt und zwar meistens gelblich, grünlich oder bläulich. Es hat eine eigentümlich rauhe Oberfläche, durch deren Runzeln das durchfallende

Licht gedämpft und zerstreut wird. Für Kirchenfenster, Treppenhaus- und Gangfenster ist das Glas vielfach in Anwendung; Rauten- und andere einfache Muster mit Verbleiung sind eine bekannte Erscheinung und wirken mit Beizug geeigneter farbiger Borden sehr gut. Mit Schwarzlot aufgemalte oder schablonierte Muster (Grisaille) geben auf diesem Glase ebenfalls eine gute Wirkung.

Farbiges Hüttenglas.

Als solches bezeichnet man das kräftig in der Masse gefärbte Tafelglas. Es kann durchsichtig sein oder opak, wie das bekannte weisse Milchglas.

Ueberfangglas.

Das ist gewöhnliches, einerseits gefärbtes oder überfangenes Glas. Nach seiner Herstellung könnte man dieses Glas plattiert nennen. Der gewöhnliche Glasklumpen wird in eine farbige Glasmasse getaucht und Grundmasse und Ueberzug werden beim Ausblasen in eins vereinigt. Durch teilweises Ausschleifen oder Wegätzen des Ueberfanges können zweifarbige Muster mit Uebergängen erzielt werden.

Antikglas

ist eine besondere Art von Ueberfangglas für die Zwecke der Glasmalerei. Mit Hilfe der vier letztgenannten Glasarten lassen sich reich wirkende Kunstverglasungen erzielen, wenn die einzelnen Stücke mosaikartig nach hübschem Muster und in guter Farbenverteilung miteinander verbleit werden. Man bezeichnet derartige Verglasungen, insbesondere wenn noch eine Verzierung mit Schwarzlot dazutritt, als Teppiche. Die monumentale Glasmalerei benützt dieselben Mittel auch in Bezug auf figürliche Darstellungen zum Unterschied von der Kabinettglasmalerei, welche auf ein und derselben Scheibe verschiedene aufgemalte Farben einbrennt. Beide Methoden können sich selbstredend vereinigen und verbinden.

Mattglas.

Das gewöhnliche Glas wird mattiert. Das kann auf verschiedene Weise geschehen: durch Rauhschleifen der Oberfläche, durch Anätzen derselben mit Flusssäure (Fluorwasserstoff), durch Sandgebläse, durch Einbrennen eines mattwirkenden Ueberzuges u. s. w. Je nach der Herstellung sind Preis und Wirkung der mattierten Gläser verschieden. Sie dienen da zu Verglasungen, wo man Licht einlassen will, ohne durchsehen zu können.

Geätztes Glas.

Statt die ganze Fläche einer Tafel zu ätzen, kann sich die Aetzung auf bestimmte Stellen verteilen. Man schützt durch geeignete Ueberzüge die blank bleibenden Stellen (durch Aufmalung, Bedrucken oder Schablonieren), ätzt das Freibleibende an und entfernt den Schutzlack. Auf diese Weise lassen sich schöne Wirkungen erzielen vom einfachen Ornament bis zur reichen künstlerischen Darstellung. Benützt man im ersteren Falle schwach farbig überfangenes Glas, so entstehen zweifarbige Muster, während gewöhnliches Glas nur durch den Unterschied von blank und matt wirkt. Wird andernfalls das Aetzverfahren in verschiedener Weise wiederholt, so lassen sich Abstufungen erzielen und die Darstellungen können sich wie zarte Gemälde von grau in grau geben.

Für gröbere Arbeiten dieser Art eignet sich auch das Sandblasverfahren; man schützt dann die Blankstellen durch aufgelegte Kupfer-, Gummi- oder Kautschukschablonen.

Musselinglas.

Man versteht darunter mattgeätzte Tafelgläser mit Blankstellen und kleinen, einfachen Mustern, sowie blanke oder matte Tafelgläser mit eingebrannten Musterungen. Gepulverte und geschlemmte Glasflüsse oder Emailfarben werden aufschabloniert, aufgedruckt oder aufgesiebt und eingeschmolzen. Ein umgekehrtes Verfahren besteht darin, die ganze Tafel zu bestreichen, über einer aufgelegten Schablone die Blankstellen auszubürsten und den verbleibenden, nachkorrigierten Rest einzubrennen, welch letzteres nur wenige Minuten beansprucht. Für Glasabschlüsse und ähnliches waren die Musselingläser häufig benützt, solange es nichts besseres gab. Das Hereinsehen war beschränkt oder aufgehoben, während das Hinaussehen durch die Blankstellen einen Ueberblick zuliess. Immerhin werden hübsche Damastmuster sowohl, wie abgepasste Rand-, Eck- und Mittelstücke noch verwendet, wo man auf eine billige Zierverglasung Wert legt.

Butzenscheiben.

Das sind durch Blasen mit der Pfeife hergestellte kreisrunde Scheiben mit verstärktem Rand und verdickter Mitte (Butzen). Sie haben verschiedene Farbe und verschiedene Grösse. Die grösseren Butzenscheiben aus gewöhnlichem, grünem Glas waren früher als Fensterverglasung zwischen Bleifassungen eine häufige Erscheinung. Später ganz ausser Mode gekommen, sind sie in anderem Sinne neuerdings wieder beliebt. Kleine, gelbe oder andersfarbige Butzenscheiben werden in die Kunstverglasungen an passender Stelle eingereiht und bringen gewissermassen mit ihrer eigentümlichen Wirkung Relief in die flachen Teppiche.

11. Das Verglasen.

Vor dem Verglasen, also vor dem Einsetzen der Glasscheiben, sind die Holzrahmen gut mit Oelfarbe zu grundieren oder doch gut mit Leinöl zu ölen. Es soll dadurch verhindert werden, dass dem später anzubringenden Kitt durch das trockene, ungeölte Holz zu viel Oel entzogen wird, was ihn minderwertig macht. Nach dem Grundieren wird bei guter Arbeit zuerst leicht und gleichmässig Fensterkitt oder Glaserkitt — ein aus Leinöl, bezw. Leinölfirnis und geschlemmter Kreide bestehendes inniges Gemenge — aufgetragen und erst hierauf die Scheibe aufgelegt und mit kleinen Stiftchen, besser aber mit kleinen dreieckig geschnittenen Blechstückchen welche mit der einen Spitze in das Flügelholz eingeschlagen werden, befestigt. Bei Verglasung von Eisenfenstern ist es nötig, dass in die Rahmen, etwa in Entfernungen von 30 bis 40 cm kleine Löcher gebohrt werden, durch welche man kleine Nieten oder Drahtstückchen als Halt für die Scheibe steckt.

Es folgt sodann die im Querschnitt als Dreieck erscheinende eigentliche Verkittung, welche den Zweck hat, die Scheibe gleichmässig zu pressen, hauptsächlich aber zu dichten. Den Kitt selbst fertigen heute nur noch sehr wenige Meister, die grosse Mehrzahl bezieht ihn billiger fertig aus den Kittfabriken.

Die verbleiten Kunstverglasungen werden ähnlich behandelt, wie die glatten Scheiben. In dem Falze der Holzrahmen werden sie ebenfalls mit Stiften befestigt, worauf Verkittung wie gewöhnlich erfolgt. Ausserdem werden in passenden Abständen sogen. Windstangen aus dünnem Rundeisen von Höhenschenkel zu Höhenschenkel geführt. An diesen Rundeisen ist die Verbleiung an geeigneten Stellen befestigt; die Enden der Stangen sind platt geschlagen, durchlocht und dem Rahmenholz aufgeschraubt. Bei einigermassen grossen Flügeln wird diese Verstärkung nötig, weil aus einzelnen Stücken zusammengesetzte Tafeln selbstredend nicht den Halt und die Festigkeit haben, wie ganze Scheiben. Grosse Kunstverglasungen erhalten eiserne Rahmen, oder sie werden mit Deckleisten in starken Holzrahmen befestigt, denen wiederum Windstangen aufzuschrauben sind.

VIII. FENSTERLÄDEN.

(Tafel 62, 63 und 64.)

A. Aeussere Läden: 1. Klappläden (Glatte Läden, gestemmte Läden). — 2. Rollläden. — 3. Schiebläden. — 4. Zugjalousien. — B. Innere Läden.

D er Zweck derselben ist
1) bei den Licht- und Luftöffnungen einfacher, meist landwirtschaftlicher Gebäude, welche nicht mit Fenstern versehen sind, den einzigen Abschluss gegen die atmosphärischen Niederschläge zu bilden;
2) von den Fenstern unserer Wohnräume die Kälte, Hitze und den Regen abzuhalten, sowie das Hineinsehen zu verhindern und schliesslich
3) Schutz gegen Einbruch zu gewähren.

Angebracht werden sie sowohl ausserhalb wie innerhalb der Fenster und heissen danach äussere oder innere Läden.

In Bezug auf die Art ihrer Bewegbarkeit unterscheiden wir:
1) Klappläden oder Flügelläden, wenn die Drehung der Flügel um eine vertikale oder horizontale Axe vor sich geht;
2) Rollläden, wenn der Laden aufwärts (oder abwärts) bewegt wird und sich dabei auf eine Walze aufrollt; hierher gehören auch die Rolljalousien;
3) Schiebläden, wenn dieselben seitlich in den Mauerkern eingeschoben werden und
4) Zugjalousien, nach Grösse, Form und Stellung veränderliche, zusammenschiebbare Läden.

A. ÄUSSERE LÄDEN.
1. Klappläden.
(Tafel 62).

Dieselben legen sich entweder stumpf an das Fenstergestell an (Fig. 313 a), oder sie liegen halb oder ganz im Falz (Spunden) (Fig. 313 b—d). Die beiden letzten Arten bieten grössere Sicherheit gegen unbefugtes Oeffnen, da sie sich in geschlossenem Zustande nicht aus den Bandkloben heben lassen, was bei dem stumpfen Anschlag ohne grosse Mühe möglich ist.

VIII. Fensterläden.

Die Grösse der Ladenflügel entspricht im allgemeinen derjenigen der Fenster. Fensterläden über 0,60 m Breite macht man zweiflügelig; unter diesem Mass einflügelig; nach der Höhe der Fenster wird der Laden selten geteilt.

Nach der Konstruktion unterscheiden wir:

1) **Glatte Läden.**

a) Aus einzelnen Dielen bestehend, welchen aufgenagelte Quer- und Bugleisten Zusammenhalt geben. Sind die Dielen rauh, so sind die Fugen stumpf; gehobelte Läden werden gespundet oder gefedert. Die Art der Anfertigung ist die gleiche, wie die der einfachen Thüren.

b) Gehobelt und verleimt, mit Einschiebleisten versehen oder mit Hirnleisten angefasst. Der Schluss der Mittelfuge kann durch Ueberfälzung oder durch Anbringung einer aufgeschraubten Schlagleiste erfolgen (Fig. 316 bis 319). Die Stärke des Holzes ist hierbei gehobelt 22 bis 28 mm. Die verleimten Läden verändern bei Witterungswechsel mehr oder minder ihre Grösse und ihr Aussehen ist ein sehr gewöhnliches. Bei städtischen Wohngebäuden finden dieselben daher weniger Verwendung, desto mehr dagegen an Hintergebäuden, Werkstätten etc. Wesentlich besser im Aussehen sind

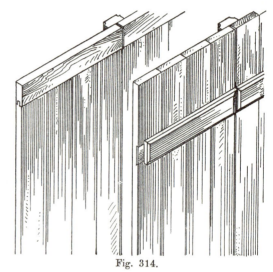

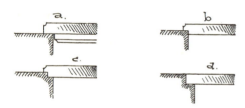

Fig. 313.
Anschlag der Fensterläden.

Fig. 314.
Fensterläden mit Hirn- und Einschiebleisten.

2) **Gestemmte Läden.**

Die Konstruktion derselben ist die der gestemmten Thüren und besteht in der Herstellung eines Langholzrahmens aus stärkeren Dielen (30 bis 35 mm), welcher durch Füllungen geschlossen wird. Die Verbindung des Rahmens und der Füllungen ist in Fig. 122 dargestellt. Die Zapfen werden in gestemmte Zapfenlöcher eingesteckt, geleimt, verkeilt und meist auch noch mit Holznägeln verbohrt. Die Art des Zusammenbaues kann sein:

stumpf mit gebrochenen Kanten oder Fasen (Fig. 125);

auf Fase (Fig. 126);

auf Hobel (Fig. 127), genau wie bei den Thüren.

Die Breite der Friese ist 9 bis 10 cm, deren Stärke wie oben angegeben. Die Füllungen, meist aus 24 mm starken Dielen bestehend, sind, wenn die gewöhnliche Dielenbreite ausreicht, aus einem Stück gefertigt. Ist die Breite bedeutender, so thut man gut, sie aus einzelnen Riemen zu fertigen und diese zu spunden, da geleimte Füllungen, dem Sonnenbrande und der Winterkälte ausgesetzt, erfahrungsgemäss auf die Dauer nicht halten.

Die Verbindung der Füllungen mit den Friesen ist verschieden: Die Füllungen sind entweder an den Enden zu einer Feder abgeplattet und in die Nuten der Friese eingesteckt oder

1. Klappläden.

sie sind, wenn die Konstruktion stärker sein soll, überschoben (Fig. 326), oder schliesslich, wie in Tafel 62 links dargestellt, mit Nuten versehen, während die entsprechenden Federn an die Friese angestossen sind. Bei dieser auf den ersten Blick etwas ungewöhnlichen, aber trotzdem gediegenen Art der Verbindung kann das Regenwasser weniger wie bei der gewöhnlichen Zusammensetzung in die Nuten der Friese eindringen.

Fig. 315 bis 324. **Glatte (verleimte) und gestemmte Fensterläden.**

Läden, welche nur Schutz gewähren sollen, schliesst man am einfachsten mit ganzen Füllungen, wobei der abgeschlossene Raum auch bei Tag vollständig dunkel wird. Um sich in demselben aber doch noch orientieren zu können, bringt man am oberen Teil der Füllung kleine Ausschnitte an (Fig. 322) oder man versieht das obere Feld mit feststehenden sogen. Jalousiebrettchen (Fig. 319). In neuerer Zeit bevorzugt man Läden, welche auch an Stelle der unteren geschlossenen Holzfüllung mit Jalousiebrettchen versehen sind (Fig. 320 und 323), da man

174 VIII. Fensterläden.

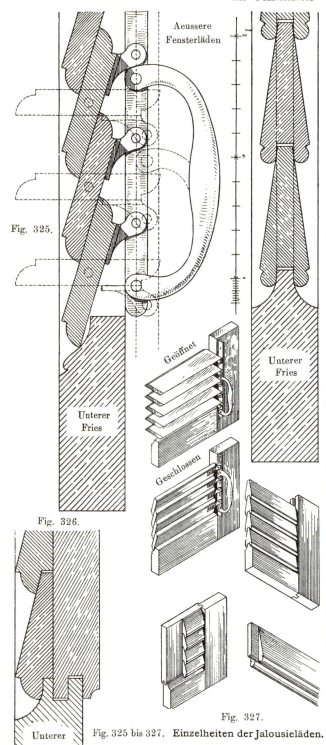

Fig. 325 bis 327. **Einzelheiten der Jalousieläden.**

vom Zimmer aus das ausserhalb des Hauses Vorfallende beobachten kann, ohne den Schutz gegen Regen und Sonnenschein aufgeben zu müssen. Noch vorteilhafter sind solche Läden, bei welchen die unteren Jalousien beweglich sind, so dass man sie verschiedenartig stellen kann, um Licht und Luft einzulassen. Zu diesem Zwecke werden die Jalousiebrettchen entweder fest in bewegliche Rahmen gefasst, die, um eine obere Axe drehbar, unten nach aussen gestellt werden können (Fig. 321); oder die Brettchen greifen mit Eisenzäpfchen an beiden Enden in entsprechende Oesen der Ladenfriese ein und sind auf diese Weise einzeln beweglich. Gewöhnlich verbindet man dieselben dann durch eine Eisenstange zu einem beweglichen System, das sich mittels Griff im gesamten öffnen und schliessen lässt (Fig. 325).

Die Brettchen selbst fertigt man am zweckmässigsten aus Hartholz, um sie gegen Zerstörung widerstandsfähig zu machen; ihre Stärke beträgt 11 bis 15 mm, ihre Breite richtet sich nach der Konstruktion. Bezüglich der feststehenden und in bewegliche Flügel gefassten Jalousiebrettchen gilt als allgemeine Regel, dieselben so zu gestalten und zu einander zu stellen, dass die hintere Oberkante des unteren Brettchens (Tafel 62 rechts) mindestens 13 bis 16 mm über der vorderen Unterkante des oberen Brettchens liegt, wodurch das Hereinsehen von aussen verhindert wird. Die Brettchen können auf beiden Seiten mit den Friesen bündig sein oder nur auf einer Seite und an der anderen um ihr Profil vorstehen, oder sie sind auf beiden Seiten profiliert und stehen beiderseits vor. Hiernach richtet sich die Art ihrer Befestigung an den Friesen, in deren schräge Nuten sie eingeschoben sind. Haben die Brettchen an beiden Langseiten oder nur hinten Profil, so genügt zu ihrem Halt die einfache, bereits erwähnte Nute, indem das hintere Profil sie vor dem Herausfallen schützt (Fig. 329). Sind sie dagegen hinten bündig, so muss je ein Zäpfchen angesetzt werden, welches in ein entsprechend gestemmtes Loch der Friese eingreift und das Brettchen dadurch festhält (Fig. 328).

Fig. 327 stellt eine Konstruktion dar, welche das Aussehen eines geschlossenen Jalousieladens ergiebt und Vorsorge für die rasche und gründliche Ableitung des Regenwassers trifft. Die Füllung besteht hier aus einzelnen, quer stehenden, durch Nut und Feder miteinander verbundenen und mit Zapfen in die Friese eingesetzten profilierten Brettchen. Ein nach Figur 326 gebildeter Laden ist dem vorerwähnten ähnlich, was das Aussehen betrifft, aber stärker und solider bezüglich der Konstruktion. Die Füllungen sind den Friesen überschoben und zum Schutze derselben ist eine Verdoppelung aus einzelnen profilierten, schuppenartig übereinander greifenden Brettchen angebracht.

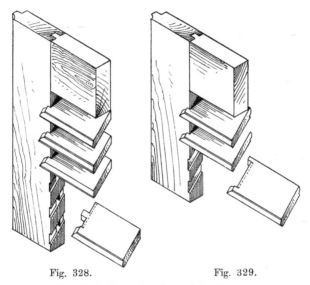

Fig. 328.   Fig. 329.

Das Befestigen der Jalousiebrettchen.

Das Beschläge der Fensterläden besteht im allgemeinen aus starken Winkel- oder Schippenbändern mit Kloben in Stein oder Holz (Tafel 78), einer Vorrichtung zum Feststellen der geöffneten Läden, bestehend aus Federfallen, Riegeln oder Schlempen mit den dazu gehörigen Kloben, sowie der Schliessvorrichtung, gebildet aus Einhängehaken, Riegeln, Basküllen, Anschlageisen etc. (Vergl. Abschnitt: Beschläge.)

2. Rollläden.
(Tafel 63.)

Sie besitzen die Vorzüge der Jalousieläden in noch höherem Masse: ausserdem aber gestattet deren Konstruktion das Oeffnen und Schliessen der Läden bei geschlossenem Fenster, ein Vorzug, welchen hauptsächlich schwächliche Personen zu schätzen wissen, für welche das Schliessen der äusseren Läden in höheren Stockwerken unter Umständen nicht ungefährlich ist. Auch das

Aussehen der Fassade wird bei geöffneten Läden nicht verändert. Auf diese Vorteile, zu denen noch der durch die Massenherstellung in Fabriken ermöglichte, niedere Preis der Rollläden kommt, ist im allgemeinen wohl auch die heutige, massenhafte Verwendung derselben zurückzuführen. Während vor 30 Jahren die Rollläden ganz vereinzelt und nur an Schaufenstern zu treffen waren, alle übrigen Gebäude aber Klappläden zeigten, sind die ersteren heute ganz allgemein verwendet. Im Interesse der Bauherren ist dies zu begrüssen, doch kann nicht verhehlt werden, dass mit der allgemeinen Einführung der Rollläden dem Gewerbetreibenden, dem Schreiner, ein grosses Arbeitsgebiet für immer abgenommen und den Fabriken zugeteilt wurde. Aeussere Klappläden werden nur noch selten und dann zu den niedersten Preisen begehrt und bezüglich der Rollläden kann der ohne Maschinen arbeitende Schreinermeister mit den Fabriken nicht in Konkurrenz treten.

Die Rollläden bestehen im wesentlichen aus einzelnen schmalen, eigenartig profilierten Holzstäbchen, welche zu einem Ganzen verbunden das Auf- und Abrollen über eine Walze ermöglichen. Als **Verbindungsmittel** dieser Stäbchen können dienen:

 a) **Leinwand**, auf welche sie aufgeleimt werden;
 b) **Leinwandgurten**, welche durch die gelochten Stäbchen gezogen und mit diesen verschraubt werden;
 c) **Stahlbänder**, welche an Stelle der Gurten treten;
 d) **Stahlplättchen**, welche, unter sich und mit den Stäben verbunden, eine Art Kette bilden, und schliesslich
 e) **Stahldrahtschnüre**.

Sämtliche fünf Verbindungsarten haben Vorteile, je nach dem Zweck der Rollläden. Als dauernd gut haben sich im Laufe der Jahre die Verbindungen mit Leinwand und Gurtendurchzug bewährt, während die Stahlverbindungen verhältnismässig neueren Datums sind und ihre Dauerhaftigkeit noch beweisen müssen. Auf Tafel 63 sind die Profile verschiedener Arten von Rollläden abgebildet.

Die **Holzwalze**, auf welche sich der Laden aufrollt, ist innerhalb des Fenstersturzes angebracht und mit zwei an den Enden befestigten Zapfen in Lagern beweglich, welche der Mauer eingegipst sind. Das Aufziehen erfolgt durch eine starke Gurte, welche sich beim Herablassen des Ladens auf die an einem Ende der Walze befindliche **Gurten- oder Riemenscheibe** aufwickelt.

Neben dieser gewöhnlichen Anordnung findet sich auch folgende im Gebrauch: Man bringt unterhalb der Ballenwalze eine für sich gelagerte Riemenscheibe an, welche mit einem Zahnrad versehen ist, das in ein zweites an der Walze befestigtes eingreift, wobei gleichzeitig eine Uebersetzung stattfinden und das Aufziehen und Herablassen durch eine Kurbel erfolgen kann.

Um den Laden in der richtigen Bahn zu halten und zu leiten, ist seitlich je eine eiserne **Laufschiene** angebracht, die, wenn der Laden nicht ausstellbar sein soll, aus einem Stück besteht und hinter das Gewände springen kann, im anderen Falle jedoch vor das Gewände zu stehen kommt. Im letzteren Falle, d. h. wenn der Laden zum Hinausstellen eingerichtet werden soll, ist jede der Schienen aus zwei, durch Scharniere miteinander verbundenen Stücken zu fertigen, an deren unterem die Ausstellvorrichtung (Taf. 56) angebracht wird. Die Laufschienen selbst müssen so weit vor das Fenster zu liegen kommen, dass der Laden in seinen Bewegungen durch vorspringende Fensterprofile nicht gehemmt ist. Will man daher auf die Ausladung der Profile nicht verzichten, so ist eine mehr oder minder starke Auffütterung nötig, wie dies auf Tafel 56 ersichtlich ist. Es ist wichtig, diesen Punkt hier zu erwähnen, da es öfter vorzukommen pflegt, dass die Auffütterungen im Ueberschlag vergessen werden und dann, wenn

2. Rollläden.

sie niemand umsonst fertigen will, den Anlass zu Streitigkeiten geben. Also entweder die Kämpferprofile nur schwach ausladen oder die Aufbesserung für Auffütterung nicht vergessen!

Zur Verhinderung des Abschürfens des Ladens bei nahezu vollendeter Abwickelung ist unterhalb der Ballenwalze eine **zweite dünnere Holzwalze**, die sogen. **Leitwalze**, angebracht, über welche der Laden hinwegrollt, um in die Laufschiene einzuleiten (Taf. 56). Eine ähnliche Vorrichtung hat den Zweck, die **Zuggurte** richtig zu leiten und zu schonen. An beiden Enden des untersten, etwas stärker konstruierten Ladenbrettchens befindet sich je eine aufgeschraubte **Nase**, welche beim Oeffnen des Ladens oben an dem Gewändesturz anschlägt und dadurch verhindert, dass derselbe höher hinaufgezogen wird, als nötig ist. Befindet der Laden sich auf der richtigen Höhe, so wird die Gurte mittels des **Schnur- oder Gurthalters** (Taf. 64) festgeklemmt, und der Laden ist gehalten. Soll derselbe herabgelassen werden, so wird der Verschluss gelöst und das Gewicht des Ladenunterteils lässt ihn von selbst sich abwickeln, vorausgesetzt, dass alles in Ordnung ist. Zum Schutz gegen unbefugtes Oeffnen ist an der unteren, inneren Seite des Ladens ein **Riegel** angebracht, welcher in ein entsprechendes Schliessblech eingreift und dadurch einen sicheren Verschluss bewirkt.

Bei Anlage der Rollläden ist vor allem Sorge zu tragen, dass der nötige Raum für den **Ladenballen** geschaffen wird. Nur in ganz seltenen Fällen wird in dieser Beziehung nicht gesündigt, in weitaus den meisten Fällen ist der Raum ungenügend und alle möglichen Nacharbeiten sind das Resultat. Das nötige Mass lässt sich nicht unter eine gewisse Grenze herabdrücken, wenn der Laden richtig funktionieren soll; obgleich die Fabriken ihre warnende Stimme erheben und es an der nötigen Belehrung nicht fehlen lassen, kommt es immer und immer wieder vor, dass die Monteure ratlos vor einem Fenster stehen und nicht wissen, wie sie den umfangreichen Ballen in dem ungenügenden Raume anbringen sollen. Es kann nicht genug empfohlen werden, lieber ein oder zwei Centimeter mehr Raum zu lassen, als $1/2$ cm zu wenig.

Die folgende Tabelle hat den Zweck, die richtigen Abmessungen festzustellen und damit vor Schaden und Nacharbeiten zu bewahren:

Ballendurchmesser verschiedener Rollläden.

Ladenhöhe	160	180	200	220	240	260	280	300 cm
Rollläden, auf Leinwand geleimt	22	23	24	25	26	27	28	29 cm
» mit Gurtendurchzug	21	23	24	25	26	27	28	29
Rolljalousien	16	17	18	19	19	20	20	21

Der Rollkasten ist ca. 4 cm. weiter zu machen als die angegebenen Durchmesser betragen.

Diese Angaben sind, wie auch zum grossen Teile die Konstruktionen dem Musterbuche von C. Leins & Co. in Stuttgart, einem der ersten Geschäfte dieses Artikels in Süddeutschland, dessen Fabrikate bestens empfohlen werden können, entnommen. Das Musterbuch wird von der Firma jedem Interessenten auf Ansuchen gratis verabfolgt, so dass eine Besprechung weiterer Arten von Läden, deren Bewegungsarten und Vorrichtungen u. s. w. hier überflüssig sein dürfte.

Reicht die verfügbare Höhe des Ballenraumes nicht aus, so dass der Ballen unter die Sturzkante in das Licht des Fensters einspringt, so kann man sich durch Anbringen sogen. **Lambrequins** (ausgeschweifte und gepresste Schutzbleche) helfen, welche das Mehrmass decken. Der ganze Ballenraum wird zum Schluss an seinen freien Seiten durch einen **Holzkasten** abgeschlossen, dessen Einzelteile glatt gestemmt und so mit Scharnieren und Vorreibern angeschlagen sind, dass sie ermöglichen, bei etwaigen Reparaturen den ganzen Raum für den Monteur freizulegen.

Die Rolljalousie ist eine leichte Art von Rollladen, ein Mittelding zwischen Roll- und Jalousieladen, welches weniger das Wetter, als vielmehr die Sonnenstrahlen und die Hitze abzuhalten bestimmt ist. Ihre Konstruktion ist fast die gleiche wie die des Rollladens. Die Rolljalousien unterscheiden sich von den Rollläden eigentlich nur dadurch, dass bei den letzteren dem Zimmer etwas Licht durch die aus den einzelnen Teilen ausgefrästen oder mittels Einlegen

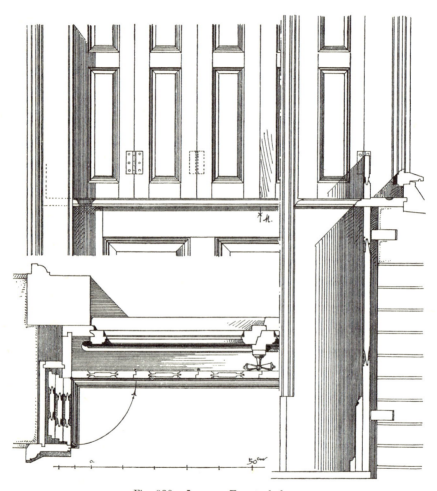

Fig. 330. Innerer Fensterladen.

von Rundstäbchen gebildeten sogen. Lichtschlitze zugeführt wird, während bei den Rolljalousien die einzelnen Leisten in gewissen Entfernungen voneinander geheftet sind, wobei sich durchgehende Lichtschlitze bilden. Im übrigen laufen sie in eisernen Laufschienen, können hinausgestellt werden und rollen sich auf wie die Rollläden; der für sie benötigte Raum ist in der obigen Tabelle vermerkt und das Profil ist auf Tafel 63 als letztes rechts den übrigen beigefügt.

3. Die Schiebläden

unterscheiden sich von den Klappläden nur durch ihr Beschläge, welches in kleinerem Massstabe genau das der Schiebthüren ist und somit füglich hier übergangen werden kann. Die Vor- und Nachteile der Schiebthüren sind auch diejenigen der Schiebläden.

---

4. Die Zugjalousie.
(Tafel 64.)

Sie unterscheidet sich von den Rollläden und Rolljalousien hauptsächlich dadurch, dass sie, statt aufgerollt zu werden, mittels Zugschnüren nach oben zusammengezogen wird. Ihr eigentlicher Zweck ist, die Sonnenstrahlen und die Hitze von den Zimmern abzuhalten und nicht, wie auch irrtümlicher Weise angenommen wird, gegen Regen und Kälte Schutz zu gewähren. Für die letztgenannte Bestimmung ist die Konstruktion viel zu leicht und die Zugjalousie wird in kurzer Zeit zerstört, wenn sie dem Sturm und Regen ausgesetzt ist. Die in der Neuzeit so beliebte Verwendung als Ersatz für Roll- oder andere Läden ist zunächst dem billigen Preis, sodann aber auch dem gefälligen Aussehen (im neuen Zustand) zuzuschreiben. Die gemachte Ersparnis ist nur eine scheinbare; denn sobald die Jalousien abwechselnd dem Regen und Sonnenschein und ständig dem Winde ausgesetzt sind, werden Reparaturen nötig, deren Kosten kapitalisiert bald höher sind, als die Mehrausgabe bei Beschaffung von Rollläden. Auch in dieser Hinsicht haben die betr. Fabriken es nicht an Belehrung fehlen lassen, leider umsonst, und so sahen sie sich schliesslich gezwungen, ihrerseits alles zu thun, um die nachteilige Wirkung der verkehrten Verwendung möglichst abzuschwächen oder unschädlich zu machen. Zum Teil ist dies auch gelungen. Die im Freien bald Not leidenden Leinwandgurten sind durch verzinkte Eisenkettchen und die Hanfschnüre durch verzinkte Stahldrahtschnüre ersetzt; an den inneren Jalousien sind sogen. Sturmführungen angebracht worden (Tafel 64), welche verhindern, dass der Wind den schwankenden Laden beschädigt. Auch die Rolljalousie, wie schon erwähnt, ein Mittelding zwischen Laden und Jalousie, ist ein Ergebnis derartiger Bemühungen.

Bezüglich der Konstruktion der Zugjalousien ist zu bemerken:

Sie bestehen aus einzelnen ca. 3 mm starken und 60 mm breiten Brettchen aus schön geradfaserigem Tannenholz, welche durch angebrachte Gurten oder Kettchen in bestimmter Entfernung voneinander gehalten werden, durch Zugschnüre zusammengezogen und in verschiedenartige Stellungen zu einander gebracht werden können, so dass der Laden bald mehr, bald weniger geschlossen ist. Das Ganze ist an ein 30 mm starkes und 60 mm breites Dielenstück befestigt, welches im Fensterlicht unmittelbar unter dem Sturz so angebracht ist, dass es jederzeit leicht abgeschraubt werden kann.

Auf diesem lattenartigen Dielen befindet sich auf zwei Holzlagern eine Holzwalze, auf welcher die verschiedenen Zugschnüre sich aufwickeln. Es sind dies die eigentliche Hanfzugschnur m, mittels welcher die Walze in Umdrehung gesetzt wird, und die beiden verzinkten Stahldrahtschnüre $n\,n$, welche, am untersten Brett angemacht, dieses beim Drehen der Walze und Aufwickeln der Schnüre langsam heraufziehen und, die einzelnen Brettchen dabei mitnehmend, den Laden öffnen. Festgestellt wird die geöffnete Jalousie durch Andrücken des Schnurhalters. Beim Lösen derselben sinkt sie durch das Gewicht des schweren untersten Brettchens herab, worauf die beiden an demselben seitlich angebrachten kleinen Riegel in entsprechende, in das Steingestell eingegipste Oesen eingreifen. Zieht man hierauf die Aufziehschnur kräftig an und klemmt sie fest, so ist der ganze Laden gespannt. Vermittels des links angebrachten Kettchens o, welches oben in zwei über Holzrollen laufende Hanfschnüre sich verzweigt, reguliert man die Stellung der Brettchen. Einfach und sinnreich ist ferner die Vorkehrung für das schöne Aufwickeln der Zugschnur. Dieselbe besteht ausser dem Eisen p, welches die Schnur im allgemeinen leitet, darin, dass das eine Walzenlager ein Schraubengewinde hat, mittels dessen die Walze sich beim Drehen seitlich bewegt, so dass die Zugschnur sich schön glatt aufrollen kann; beim Oeffnen des Ladens geht dann die Walze im Gewinde wieder auf die andere Seite zurück. Das Hinausstellen der Jalousie wird durch das Einstecken von zwei seitlich am Gewände angebrachten, beweglichen Eisenstäbchen bewirkt, welche in entsprechende Oesen am unteren Brettchen eingreifen.

Die Tafel 64 zeigt links die Sturmführungen, rechts die Schnurhalter.

B. INNERE LÄDEN.
(Fig. 330.)

Sie werden seit der allgemeinen Einführung der Rollläden und Jalousien nur noch selten ausgeführt. Sie dienen dazu, Sonne und Hitze abzuhalten; Schutz gegen Einbruch gewähren sie wenig.

Sie bestehen aus schmalen, gestemmten und in den Fugen überfälzten Ladenflügeln, welche miteinander durch Scharniere verbunden, so in die Leibung der Fenster eingelegt sind, dass das Ganze wie ein gewöhnliches gestemmtes Futter aussieht. Die Art der Zusammenlegung des Ladens hängt von der Leibungstiefe ab; je schwächer dieselbe, desto öfter muss der Laden gebrochen werden, desto mehr Teile erhält er. Nach Fig. 330 ist er z. B. in sechs Teile zerlegt, von denen je drei zusammenhängend sich nach rechts und links legen. Der Laden selbst erhält ein glattes Futter, welches mit dem Fensterfutterrahmen verbunden ist. Nach dem Zimmer zu schliesst eine profilierte Verkleidung die Oeffnung ab. Unten am Simsbrett und oben am Futter erhält der Laden einen Anschlag, damit er feststeht. Der Verschluss kann durch in der Mitte angebrachte Schlempen oder besser durch ein Baskülenschloss bewirkt werden.

IX. HOLZDECKEN.

(Tafel 65 bis 69.)

Allgemeines. — 1. Die Balkendecke. — 2. Die Kassettendecke. — 3. Die Felderdecke.

Allgemeines.

Die Holzdecken, im Sinne der Bauschreinerei, können verschiedener Art sein. Sie werden gebildet, indem man entweder die Konstruktion, die tragenden und ausfüllenden Teile sichtbar lässt, dieselben in richtiger und verständiger Weise schmückt und dadurch erst recht zum Ausdruck zu bringen sucht. Oder man verkleidet diese Teile mit einem feineren, edleren Material, wobei aber noch die Konstruktion zu erkennen ist, oder man fertigt schliesslich eine blinde Decke, d. h. eine Art Täfelung, und schraubt sie einfach unten an die Deckenbalken, bezw. die allenfalls nötig werdende Auffütterung fest. Im ersten Fall ist das zu verwendende Material meist das gewöhnliche Bauholz, Tanne, Fichte, Forle. In den beiden anderen Fällen wählt man gerne ein besseres Holz, Eichen oder Nussbaum, vielleicht mit Eschen- oder Ahornfüllungen.

Von besonderer Wichtigkeit ist die richtige Trocknung des Holzes, da ungenügend trockenes Holz an der warmen Luft der Decke schwindet und reisst. Aber selbst bei trockenem Holz sind Risse und Sprünge nicht ganz ausgeschlossen, wenn die Hitze in dem Raum zu bedeutend wird. Die Ofenwärme, namentlich die der Füllöfen, und die von den Gaskronen ausgestrahlte Hitze sind schlimme Feinde der Decken und man thut gut daran, alles vorzusehen, was geeignet ist, bei allenfallsigem Reissen dasselbe erträglich zu machen. Zu diesem Zweck richtet man die Dekoration der Hölzer so ein, dass durch die Risse die Gesamtwirkung nicht beeinträchtigt oder gestört wird. Bei sichtbaren Balkendecken fast und kehlt man daher die Balken und Unterzüge, versieht sie mit Riefen, welche mit den Holzfasern parallel laufen und vermeidet möglichst alle quergehenden Verzierungen, welche bei etwaigem Reissen unschön aussehen (Tafel 65, a, b und c). Können aus irgend welchem Grund die Balken an ihren sichtbaren Flächen nicht gehobelt werden, so verkleidet man sie mit gehobelten und profilierten Brettern und Kehlleisten (Tafel 65, e und f). Konstruiert man eine blinde Balkendecke, so bildet man die Balken aus hohlen Kästchen, wie Fig. d auf Tafel 65 zeigt, und befestigt sie an der eigentlichen, den Raum nach oben abschliessenden Decke.

Die Füllungen macht man schmal oder fertigt sie, wenn dies nicht angehen sollte, aus schmalen gespundeten Riemen, bei welchen man die Fugen absichtlich durch Anstossen eines kleinen Kehlhobels hervorhebt. Schwinden dann die Füllungen, so werden höchstens die Fugen etwas grösser, was nicht auffällt; im übrigen sind aber Risse vermieden. Auf Tafel 66 sind, Decke III und IV, derartig behandelte Füllungen verwendet. Sind die Felder sehr breit, so muss man sie stemmen und die Füllungen in die Nuten von Friesen einschieben, damit sie arbeiten können.

Bei fast allen Balkendecken schliesst man die Fugen zwischen Füllungen und Balken mit mehr oder minder starken, profilierten Leisten und bildet dadurch sanfte Uebergänge vom Senkrechten in das Wagrechte (Tafel 66, IV). Sind die nötigen Mittel vorhanden, so lassen sich auch ganze Gesimse mit Zahnschnitten, Konsolen etc. einlegen (Taf, 65, m bis p). Seltener kommt es vor, dass die den Uebergang bildenden Leisten wegbleiben, und es ist hauptsächlich dann der Fall, wenn Balkendecken älterer Zeit nachgeahmt werden (Taf. 66, I, II und III).

Die Höhe, auf welche die Füllungen einer blinden Decke zu liegen kommen, ist gewöhnlich die untere Deckenfläche, und zwar liegen sie dann entweder in einer Ebene, wie dies z. B. bei den beiden Decken der Tafel 69 der Fall ist, oder man legt bei reicheren Decken ein Feld oder einige bedeutendere Felder tiefer an, als die übrigen und erzielt dadurch eine viel lebhaftere Wirkung des Ganzen. Die Tafel 67 zeigt in b, d, f und g, die Tafel 68 in f Decken mit verschieden hohen Feldern. Um sie solid herstellen zu können, ist eine Auffütterung erforderlich. Man versteht darunter ein in Holz ausgeführtes Gerippe, welches fest mit der Tragdecke verbunden, den Grund bildet, auf welchem die blinde Decke befestigt wird. Wie die Auffütterung konstruiert werden muss, lässt sich allgemein nicht sagen, da sie von der Zeichnung der Decke, wie von dem Profil derselben abhängt. Darf ein wohlgemeinter Rat aber hier angebracht werden, so sei es der, an Eisenwerk, namentlich an durchgehenden Mutterschrauben nicht zu sparen, da nur sie allein die erwünschte Sicherheit zu geben im stande sind — und ferner besorgt zu sein, dass für alle schwereren, an der Decke zu befestigenden Teile oder Gegenstände, wie schwere Rosetten, Gaskronen etc., starke Wechsel in das Hauptgebälke eingelassen werden, woran jene dann zu befestigen sind.

Beim Entwurf einer Decke ist zu beachten:
    a) die Grösse des Raumes. Von den Abmessungen desselben, namentlich von der Höhe, kann zuweilen die ganze Wirkung abhängen. Das Deckenmotiv muss mit denselben im Einklang stehen: für grosse Räume grosse Muster, für kleinere umgekehrt; für bedeutende Höhen kräftige Gesimse, für niedrige Räume feinere und zierliche Gliederungen;
    b) die Form der Decke. Dieselbe ist bedingt durch die Grundform des betr. Raumes. Die gewöhnliche, das Rechteck, ergiebt ein schönes Gesamtbild, wenn sein Verhältnis $1:1\frac{1}{2}$ bis $1:2$ ist. Anderen Verhältnissen, wie $1:3$ oder $1:4$, welche zu lang, zu riemenartig erscheinen, nimmt man das Unschöne, indem man sie durch wirkliche oder blinde Unterzüge in zwei oder mehrere kleine Decken teilt und jede für sich behandelt. Die Quadratform ist, wenn die Abmessungen nicht zu bedeutend sind, für Holzdecken immer eine schöne Form. Das Gleiche gilt vom Sechs- und Achteck;
    c) die Einteilung der Decke. Man kann dem Gebälk entsprechend, die Decke in schmale Rechtecke zerlegen. Man kann viele gleichartige Felder, die sog. Kassetten anordnen. Man kann schliesslich eine beliebige Einteilung machen mit ausgesprochener Mitte, mit grossen und kleinen Feldern verschiedener Form und Lage. Darnach kann man unterscheiden:
        1) Die Balkendecke, auch Fachdecke genannt; 2) die Kassettendecke 3) die Felderdecke.

1. Die Balkendecke.

Sie ist gewöhnlich Konstruktionsdecke. Bei ihr sind die Balken echt oder doch nur verkleidet, die Füllungen sind zugleich die untere Deckenschalung (Taf. 65, a, b u. c) oder besonders angebracht (Taf. 65 h). Die Balken liegen an den Wänden meist auf einem sie zusammenfassenden architrav- oder balkenartigen Gesims oder auf Konsolen und Tragsteinen auf, deren Zweck ist, teils den Uebergang von der Wand nach der Decke weniger hart erscheinen zu lassen, teils die Tragfähigkeit derselben wirklich oder nur scheinbar für das Auge zu erhöhen. Bei Fig. I und II, Tafel 66 (Rathaus zu Lindau) ist dieses Wandgesims um jedes Balkenende herumgekröpft, während es bei den Fig. III und IV sich jeweils an den Balkenkonsolen »totläuft«, d. h. ohne Verkröpfung endigt.

Die Balkendecke beginnt entweder mit einem ganzen Balken oder aber mit einem Teil eines solchen, welcher ermöglicht, ihn dekorativ so zu behandeln, wie die ganzen Zwischenbalken. Wie dies zu geschehen hat, sowie auch die Füllungsgestaltung zeigen die Details auf Tafel 66. Ausserdem sind auf Tafel 65 in Fig. i bis l einige Unterzugsbildungen mit Konsolen etc. dargestellt für grosse Decken, welche in Einzeldecken zerlegt werden.

Die Figur 331 bringt eine alte Decke aus Schloss Reifenstein in Tirol, aufgenommen von F. Paukert, zur Abbildung. Die an sich höchst einfache Balkendecke hat ihren Schmuck durch reiche Bemalung gefunden.

---

2. Die Kassettendecke.

Sie ist zwar keine Konstruktionsdecke, sucht aber den Schein einer solchen zu wahren. Die Kassette kann quadratische, vieleckige oder kreisrunde Form haben; ebenso kann sie nur in einerlei oder in mehrerlei Form an einer und derselben Decke vorkommen. In der Mitte der Kassette befindet sich gewöhnlich eine Rosette; absolut erforderlich ist sie nicht. Die Verbindungspunkte des Gerippes werden durch Rosetten, Diamantquader oder Knöpfe hervorgehoben und ausgezeichnet (Tafel 65, d und m). Die Profilierung der Rippen ist verschieden; sie richtet sich sowohl nach den vorhandenen Mitteln, wie nach der Entfernung, aus welcher die Decke gesehen wird. Man kann die Deckenprofile eines kleinen Raumes nicht ohne weiteres für einen doppelt so grossen verwenden. Was dort passend und wirkungsvoll war, würde hier zu zierlich und fein erscheinen.

Im allgemeinen werden die Rippen so gebildet, dass man sie unten mit einem kleinen seitlichen Profil versieht, welches sie leichter erscheinen lässt; dies zeigen die durch die Figuren n, o, p, q und r auf Tafel 65 verzeichneten Kassettenrippen-Profile, während Tafel 68 in Figur g, h und i einige ganze Kassettendecken darstellt. Von den Raumverhältnissen wird es auch abhängen, ob man — die gewöhnliche Kassettendecke mit ihren quadratischen Feldern ausgenommen — mit halben oder ganzen Kassetten beginnen wird. Für die erstere Art spricht der Umstand, dass die mittleren Kassetten freier erscheinen und besser zur Geltung kommen. Auf die Ausschmückung der Füllungen wird grosses Gewicht gelegt. Für sie eignet sich sowohl Malerei, als auch Intarsia und reiche Stemmarbeit, drei Arten, mit welchen prächtige Wirkungen erzielt werden können.

Die Figur 332 bringt eine alte Holzdecke aus dem Stifte Sekkau in Steiermark, aufgenommen von R. Bakalowits. Diese Decke beginnt einerseits mit halben, andererseits mit ganzen Kassetten.

184 IX. Holzdecken.

3. Die Felderdecke.

Während die Balkendecke und die Kassettendecke sich beliebig fortsetzen und erweitern lassen, ist dies bei der Felderdecke nur unter bestimmten Voraussetzungen der Fall; sie ist eine für sich abgeschlossene, abgepasste blinde Decke. Mit wenigen Ausnahmen ist bei ihr ein

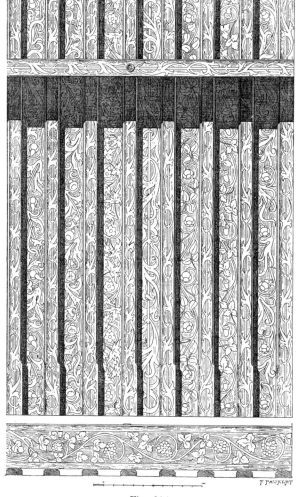

Fig. 331.

Haupt- oder Mittelfeld geschaffen, an welches sich die kleineren Felder anschliessen. Dieses Mittelfeld kann dominierend geformt sein, wie bei den Beispielen Tafel 67 b, d und e, so dass es sofort kräftig in die Augen fällt, oder es hält sich mehr in der Art und Weise der übrigen Felder, wie bei dem Beispiel Tafel 68 a, wodurch die Decke ruhiger wirkt.

Seiner Bedeutung entsprechend, ist dieses Mittelfeld entweder reicher dekoriert, mit Konsolengesims, Zahnschnitten, Eierstäben oder kräftigen Profilstäben versehen, oder es ist

3. Die Felderdecke.

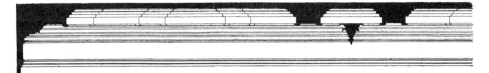

HOLZ PLAFOND IM STIFTE SEKKAU.

Fig. 332. Holzdecke aus dem Stifte Sekkau in Steiermark.
Aufgenommen von R. Bakalowits.

186　　　　　　　　　　　IX. Holzdecken.

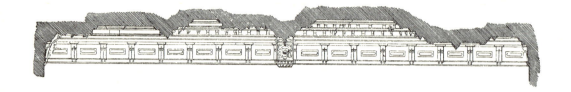

Fig. 333.　Holzdecke aus dem Rathause zu Augsburg.
Aufgenommen von L. Leybold.

3. Die Felderdecke.

tiefer gehalten, d. h. seine Decke liegt nicht bündig mit denen der kleinen Felder, oder es hat nur allein eine Rosette oder die bedeutendste der sämtlichen; kurzum: man ist nach jeder Weise bedacht, es auszuzeichnen. Wie die Profile in jedem einzelnen Fall sein müssen, lässt sich zum voraus ebensowenig, wie bei den Kassettendecken sagen. Die Rippen können hier die gleichen Profile erhalten, wie bei den Kassetten (Tafel 65, m bis r); sie können unten glatt, profiliert oder mit Rosetten und Knöpfen dekoriert sein.

Die Figur 333 bringt eine alte Decke aus dem Rathaus in Augsburg, aufgenommen von L. Leybold. Diese reiche Holzdecke mit ihren Schnitzereien hat gar kein eigentliches Mittelfeld, sondern an dessen Stelle eine Rosette auf pyramidenartigem Vorsprung. Die um diese herum liegenden 8 Felder bilden zusammen die ausgesprochene Mitte.

Zwei einfachere Felderdecken sind auf Tafel 69 dargestellt. Sie werden sich hauptsächlich für das Herrenzimmer oder das Speisezimmer des besseren Wohnhauses, sowie für kleine Kneipzimmer in Wirtschaften eignen. Die Herstellung aus Friesen, Füllungen und aufgesetzten oder eingeschobenen Leisten gleicht genau der Arbeit, wie sie für Wandtäfelungen üblich ist. Ein passender Wechsel im Holz wird die Wirkung wesentlich erhöhen.

X. DIE HOLZTREPPEN.

(Tafel 70 bis 76).

Allgemeines. — 1. Die Führung der Treppe: a) die gewöhnliche, b) die gemischte, c) die gewundene Treppe. — 2. Treppenarme und Treppenbenennung. — 3. Die Konstruktion der Treppen: a) die eingeschobene, b) die gestemmte, c) die aufgesattelte Treppe. — 4. Das Treppengeländer.

Allgemeines.

Unter einer Treppe versteht man die gangbare Verbindung eines tiefer gelegenen Fussbodens mit einem höher gelegenen. Die Haupterfordernisse einer solchen sind Sicherheit und Bequemlichkeit.

Jede Treppe besteht aus einer Anzahl aufeinanderfolgender Stufen von gleicher Höhe. Ebenso ist die Breite der Stufen entweder überall, oder aber doch in der Mittellinie des Grundrisses die gleiche. Die einzelne Stufe heisst Trittstufe oder Tritt; die horizontale Breite derselben, von der Vorderkante der einen Stufe bis zur Vorderkante der nächstfolgenden gemessen, heisst Auftritt, die senkrechte Höhe der Stufe heist Steigung. Tafel 70 (Fig. XXIII und XXIV). Die erste Stufe einer Treppe, auch die eines jeden Stockwerks, bezeichnet man als Antrittsstufe oder Antritt, die letzte Stufe als Austrittsstufe oder Austritt. Eine Anzahl in gerader Richtung aufeinanderfolgender Stufen wird Treppenarm oder Treppenlauf genannt, daher ein-, zwei- und mehrarmige Treppen. Die seitlichen Begrenzungen der Trittstufen — falls solche überhaupt vorhanden sind — heissen Zargen oder Wangen. Unter Laufbreite oder Breite des Treppenarms versteht man die Länge der Trittstufen nebst den Zargenstärken. Die Laufbreite beträgt bei Haupttreppen mindestens 1,00 m, bei Neben- oder Diensttreppen mindestens 0,75 m. Die Mittellinie des Grundrisses der Treppe, auf welcher die Auftritte eingeteilt werden, nennt man Teilungs- oder Lauflinie und die Summe der Auftritte eines Stockwerkes, auf dieser Lauflinie gemessen, heisst Grund. Die Vorkehrung zur Verhinderung des seitlichen Herabstürzens von der Treppe nennt man Geländer, bestehend aus a) dem Geländerpfosten, b) den Staketen, Docken, Stäben etc. und c) dem Handgriff. Oft ist auch ein Handgriff an der Wandseite angebracht.

Das Begehen einer Treppe, selbst der bestangelegten, ermüdet mehr als das einer horizontalen Ebene. Nehmen wir den mittleren menschlichen Geheschritt zu 60 bis 65 cm an, so muss derselbe beim Treppensteigen — da hier auch die Steigung zu überwinden ist — wesentlich verringert werden, soll er nicht den Begehenden ausserordentlich ermüden. Diese Reduktion ist erfahrungsgemäss genügend, wenn man die Steigung, also die Höhe, auf welche man den Fuss senkrecht heben muss, doppelt in Anrechnung bringt, so dass sich z. B. bei einem Auftritt von 30 cm und einer Schrittgrösse von 62 cm eine Steigung von 16 cm ergibt. Hierauf beruht die Formel: **zwei Steigungen und ein Auftritt sind gleich 60 bis 65 cm**. Bei Haupttreppen wählt man die Steigung nicht über 17,5 cm, die Auftrittsbreite nicht unter 22 cm; bei Keller- und Speichertreppen soll der Neigungswinkel nicht über 45° betragen. Erfahrungsgemäss steigen sich gut: 15,5 : 29, 16 : 29, 16,5 : 28, 17 : 26, 17 : 29 und 17 : 31 cm.

Um das Begehen der Treppe zu erleichtern, legt man nach einer gewissen Anzahl Stufen einen sogen. Podest an, auf welchem man zwei oder mehrere Schritte in horizontaler Richtung machen und sich etwas erholen kann. Bei zwei- oder mehrarmigen Treppen erhält der Podest meist die Rechtecksform und zwar nimmt man die Podestbreite allgemein gleich der Laufbreite an, obgleich ein breiterer Podest viel weniger Anlass zu Beschädigungen beim Transport von Möbeln etc. ergiebt. Die Zahl der Trittstufen eines Laufes soll nicht mehr als 12 bis 15 und nicht weniger als 3 betragen. Eine Treppe mit Podest heisst Podesttreppe.

---

1. Die Führung der Treppen.

In Bezug auf die Führung der Treppe, die sich nach der Form der aufeinanderfolgenden Trittstufen richtet, unterscheiden wir:

a) Die gewöhnliche Treppe, welche nur aus gleich breiten Trittstufen besteht, die mit oder ohne Unterbrechung durch einen Podest aufeinanderfolgen. Bei richtigem Steigungsverhältnis ist sie die bequemste aller Treppen.

b) Die gemischte Treppe mit verschiedenen Stufen. Reicht bei einer Treppenanlage der vorhandene Grund nicht aus, um eine schöne Podesttreppe anzulegen, so kann man zur Not den Podest etwas verringern und einige Wendelstufen, d. h. ungleich breite Trittstufen anbringen, wie Tafel 70, Fig. XI es zeigt. Zu gering darf aber der Podest auch nicht werden und eine Grenze in dieser Richtung ist geboten. Kann derselbe nicht so gross wie nach der oben genannten Figur gemacht werden, so entspricht er seiner eigentlichen Bestimmung nicht mehr. Man verzichtet in diesem Fall dann besser ganz auf den Podest und verwendet den dadurch gewonnenen Grund zu einem besseren Steigungsverhältnis der ganzen Treppe, wonach sich die Form XII ergiebt. Wenn die Wendelstufen dieser Treppe nach einem gemeinsamen Mittelpunkt laufen, so werden sie am inneren Ende sehr schmal und spitz, so dass ein Begehen derselben an diesen Stellen unmöglich oder doch nicht gefahrlos ist. Da aber die Erfahrung lehrt, wie nur alte Leute sich ihre Lauflinie den Wandzargen entlang wählen, während Kinder an den mittleren, freien Zargen, also an den gefährlichsten Stellen, auf und ab gehen, so erwächst hieraus für den Techniker die Verpflichtung, dafür zu sorgen, dass diese Stufen verbreitert werden, um Unglück zu verhüten. Zu diesem Zweck »verzieht« man die Treppe, d. h. die spitz zulaufenden Tritte werden auf Kosten der gleichbreiten verbreitert. Diese Verbreiterung hat nach bestimmter Proportion zu erfolgen, damit die Form der Zargen, welche sich nach den Stufen richtet, eine schön geschwungene wird. Je weiter eine Treppe verzogen ist, um so besser ist sie zu begehen, um so schwieriger wird aber auch ihre Anfertigung, da die Zargen geschweift, d. h. gekrümmt und nicht mehr gerade sind.

Auf Tafel 72 ist die Konstruktion des Verziehens dargestellt: Nachdem der Grundriss der Treppe im allgemeinen angelegt ist, die Laufbreite und Zargenstärke bestimmt und die Trittstufen auf der Lauflinie eingeteilt sind — wozu bemerkt wird, dass wir aus Schönheits- wie aus technischen Gründen die Treppen, wo es irgend angeht, symmetrisch anlegen — konstruiert man nach Fig. F mit Hilfe der seitlich aufgetragenen Steigungen zunächst ein schematisches Profil der gleich breit bleibenden Stufen, welches bis a reicht. Hierauf verbindet man die

Vorderkanten dieser erhaltenen Stufen, errichtet auf dem Endpunkte dieser Linie eine Senkrechte und ermittelt den letzten Punkt f im Aufriss. Derselbe liegt auf der Höhe zwischen 9 und 10, im übrigen aber horizontal soweit von a entfernt, als die Strecke a f im Grundriss, verstreckt aufgetragen, beträgt. Verbindet man a mit f und errichtet in der Mitte derselben eine Senkrechte, so erhält man den Mittelpunkt g, von welchem aus der Kreisbogen a f gezogen wird. Derselbe schneidet die einzelnen Steigungen in Punkten, deren Horizontalprojektion, an den betreffenden Stellen der Zarge im Grundriss abgetragen, die Breite der Stufen daselbst ergiebt. Man hat nun für jede Wendelstufe zwei Punkte und kann sie somit konstruieren; für den oberen Teil des Treppengrundrisses, welcher symmetrisch mit dem unteren ist, braucht die Konstruktion selbstredend nicht wiederholt zu werden.

c) Die gewundene Treppe, auch gewendelte Treppe oder Wendeltreppe genannt, mit unter sich gleichen, aber ungleich breiten Stufen, entsteht, wenn die Richtungsänderung stetig und gleichmässig von Stufe zu Stufe statt von Podest zu Podest erfolgt. (Tafel 70, XIX, XXI und XXII.) Dieselbe lässt sich sowohl in kreisrunden, wie in quadratischen und vieleckigen Räumen anbringen und gestattet den Austritt an beliebigen Stellen des Umfangs. Sie nimmt den geringsten Raum ein und passt auf jede Schrittgrösse, hat aber den Nachteil, dass man bei geringem Durchmesser des Treppenhauses eine ziemlich hohe Steigung zu nehmen gezwungen ist, um nicht den Kopf an den Hinterkanten der darüber liegenden Stufen anzuschlagen und dass man der fortwährend sich ändernden Richtung wegen leicht schwindelig wird. Bei grösseren Anlagen fällt dieser Missstand weg. Erfahrungsgemäss kann man zwei volle Umdrehungen durchlaufen, ohne von Schwindel befallen zu werden.

Man unterscheidet zwei Arten von Wendeltreppen:

α) Die Spindeltreppe. Sie ist die eigentliche Wendeltreppe. Sie besteht aus Trittstufen und Futterbrettern, welche, wie bei den übrigen Treppen, einerseits in eine Zarge (Wandzarge), andererseits in die sogen. Spindel, eine Holzsäule, auch Mönch genannt, eingestemmt werden. Der Zusammenhalt des Ganzen erfolgt durch durchgehende Mutterschrauben. Zu den Spindeltreppen gehören die Beispiele XXI und XXII.

β) Die Hohltreppe. Sie ist eine Wendeltreppe mit zwei gewundenen Zargen, einer inneren und einer äusseren, im übrigen aber konstruiert wie eine geradarmige Zargen- oder Wangentreppe. (XIX.)

Ist der zur Verfügung stehende Raum nicht dazu geeignet, um eine volle oder ganz gewundene Wendeltreppe anzuordnen, so wird man in vielen Fällen eine halbgewundene Treppe anlegen können (Taf. 70, XVII). Sie hat lauter gleiche Stufen, die nach einem gemeinsamen Mittelpunkt gerichtet sind. Auch die Ellipse und ihre Näherungskonstruktion, der Korbbogen, lassen sich als Grundrissform verwerten (Taf. 70, XX). Die Stufen sind dann nicht alle gleich und richten sich nach verschiedenen Punkten. Es können ferner Zusammensetzungen von geradläufigen und gewundenen Anlagen gemacht werden (Taf. 70, XVI und XVIII). Man wird diese gemischten Formen als gemischte Treppen kurzweg oder als gemischte Wendeltreppen bezeichnen können, je nachdem die gleichbreiten oder die ungleichbreiten in der Ueberzahl sind.

2. Treppenarme und Treppenbenennung.

In Bezug auf die Anzahl der Arme unterscheidet man einarmige, zwei- und mehrarmige Treppen; ein Mittelding zwischen den ein- und zweiarmigen Treppen sind die rechtoder schiefwinklig gebrochenen (VI u. VII). Den auf Tafel 70 zusammengestellten Grundformen entsprechen nach dem Vorausgeschickten folgende Bezeichnungen:

I. geradläufig, einarmig
II. » mit oberer Viertelswendung;
III. » mit oberer und unterer Viertelswendung nach einer Seite;
IV. » » » » » » » zwei Seiten;
V. » mit leichter, zum Begehen einladender Wendelung am Beginn:
VI. rechtwinkelig gebrochen mit Wendelstufen im Eck;
VII. » » » Podest im Eck;
VIII. und IX. zweiarmig, mit parallelen, gleichlangen Armen, ohne und mit Krümmling;
X. zweiarmig mit parallelen, ungleichlangen Armen;
XI. zweiarmig, gemischt, mit Podest;
XII. » » ohne Podest;
XIII. » mit schräg zu einander liegenden Armen und mit Podest;
XIV. und XV. dreiarmig, ohne Podeste und mit Podesten, gemischt und geradläufig;
XVI. gemischt, ähnlich wie III;
XVII. halb gewunden, halbrunde Wendeltreppe;
XVIII. gemischt, an den Enden halb gewendelt, in der Mitte geradläufig;
XIX. Wendeltreppe; Hohltreppe;
XX. gemischt-gewendelt; Korbbogentreppe mit drei Mittelpunkten;
XXI. Spindelwendeltreppe auf kreisrundem Grundriss;
XXII. » » vieleckigem Grundriss;

3. Die Konstruktion der Treppen.

In Bezug auf die Konstruktion der Treppen unterscheidet man:
    a) die eingeschobene Treppe;
    b) die gestemmte Treppe;
    c) die aufgesattelte Treppe;

a) **Die eingeschobene Treppe.** Sie ist die einfachste Treppe und wird nur zu untergeordneten Zwecken, als Speichertreppe etc., verwendet. Sie besteht aus zwei, ca. 6 cm. starken, 22 bis 27 cm breiten Zargen oder Wangen, in welche von vorne die etwa 5 cm starken Trittstufen in den Grat eingeschoben werden. Die Treppe hat keine Futterbretter und bringt daher beim Begehen ein Gefühl von Unsicherheit hervor, da man durch die Stufen hindurch nach unten sieht. Figur 334 B zeigt eine derartige Treppe. In Figur 334 A ist eine ältere Konstruktion dargestellt, bei welcher die Stufen nur in Nuten eingesetzt waren. Um aber der Treppe den nötigen Zusammenhalt zu geben, wurden einzelne Stufen mit Zapfen versehen, welche man durch die Zargen durchgehen liess und von aussen verkeilte.

b) **Die gestemmte Treppe,** auch Zargen- oder Wangentreppe genannt (Tafel 71, I), ist die gebräuchlichste Wohnhaustreppe. Sie hat Zargen, Auftritte und Futterbretter. Die ersteren können gleichstark sein; zweckmässiger ist es jedoch, die innere, sich freitragende Zarge etwas stärker zu machen als die Wandzarge, welche man so oft als nötig, mittels Eisenwerks an der Wand befestigen und unterstützen kann. Die gewöhnliche Stärke der Wandzarge ist 6 cm, die der mittleren Zarge 7,5 cm. Die Höhe oder Breite derselben wird allgemein so bestimmt, dass man das Treppenprofil, bestehend aus Auftritt und Futterbrett, konstruiert und von der Vorderkante eines Trittes senkrecht auf die Treppenschräge 4 cm aufwärts und von der betreffenden Hinterkante 4 cm nach abwärts aufträgt und durch die beiden so gewonnenen Punkte Parallelen mit der Treppenschräge zieht. (Tafel 70, Figur XXIV.) Eine solche Treppe bietet bei guter Ausführung und gesundem Holz hinreichend Festigkeit, um auf ihr alle zu einem Haushalt gehörigen Gegenstände transportieren zu können. Nur bei aussergewöhnlichen Verhältnissen, bei aussergewöhnlicher Lauflänge oder wo der Transport besonders schwerer Lasten in Betracht kommt, ist eine besondere Verstärkung der Zargen erforderlich.

Auf die gleiche Weise bestimmt man auch die Form der geschweiften Zargen oder Wangen der auf Tafel 72 dargestellten gemischten Treppe. Auch hier wird aus dem Grundriss unter Zuhilfenahme der seitlich aufgetragenen Steigungen ebenfalls das Profil zusammengestellt, von jeder Vorder- bezw. Hinterkante der Stufen werden 4 cm auf- bezw. abwärts getragen und die hierdurch erhaltenen Punkte durch schöne Kurven verbunden. Um den letzten Punkt im

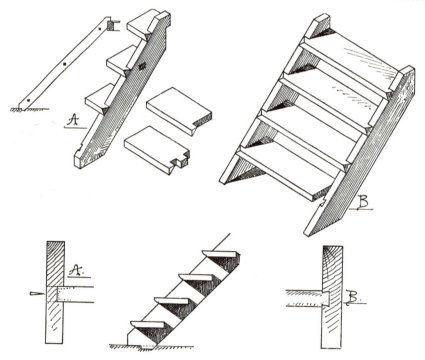

Fig. 334. Eingeschobene Treppe.

Eck genau zu erhalten, muss man den einen Teil des Ecktrittes herumklappen, da die Steigung sich auf dessen ganze Breite gleichmässig verteilt. Die Zargen-Ober- und Unterkanten sind stets rechtwinkelig gearbeitet. Allgemein bemerkt sei noch, dass man der soliden Konstruktion wegen gewöhnlich eine Trittstufe ins Eck legt und dass man, wenn irgend möglich, bestrebt ist, die mittleren Zargen durch Pfosten zu unterstützen, damit die Treppe sich nicht einsenkt oder einsackt.

Die Auftritte werden 5 bis 6 cm stark gemacht, ihre Breite richtet sich nach dem Profil und dem Auftrittsmass. Die Futterbretter sind gehobelt $1\frac{1}{2}$ bis 2 cm stark und in die Nuten der Auftritte eingeschoben oder von hinten angenagelt. (Taf. 70, Fig. XXVIII.) In die Zargen stemmt man die Futterbretter nur ca. 2 cm tief ein, während man die Auftritte 3 cm eingreifen lässt. Zusammengehalten wird der Treppenlauf durch eiserne Mutterschrauben, welche alle 5 bis 6 Stufen eingelegt werden, so dass auf den Lauf einer gewöhnlichen Stockwerks-

Podesttreppe 3 Schrauben kommen. Dieselben können durchgehend und mit versenkten Muttern versehen sein, oder sie reichen nur auf eine Länge von ca. 25 bis 30 cm von aussen herein, so dass der Tritt selbst als Zugschlauder dient. (Taf. 73.)

Der Anfall der Zargen am Antritt, am Podest und am Austritt ist auf Taf. 72 ersichtlich und ausserdem in der Textfigur 335 dargestellt. Soll das Geländer, z. B. bei der zweiarmigen, gestemmten Treppe auf Tafel 71, fortlaufend herumgeführt werden, so ist es nötig, einen sogen. **Krümmling** anzubringen, d. h. ein Holz, welches so geformt ist, dass es die obere Fläche der unteren Zarge in schöner Form auf die obere Zarge überleitet. Diese Form ist, da der Grundriss des Krümmlings meist die Halbkreisform zeigt, die einer gewöhnlichen windschiefen Schraubenfläche. In diesen Krümmling sind die beiden Zargen eingezapft, er selbst wird dem Podest-

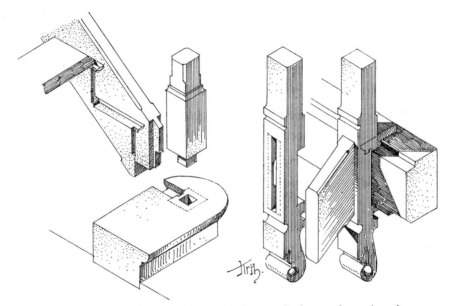

Fig. 335. **Anfall der Zargen am Antritt, am Podest und am Austritt.**

balken eingelassen und aufgesetzt; alle drei zusammen sind durch starke Mutterschrauben miteinander verbunden. Die Konstruktion des Krümmlings und der zu seiner Herstellung erforderlichen Schablonen ist die gleiche, wie die in Fig. 337 dargestellte und weiter unten beschriebene.

Der **Podest** (Fig. 336) besteht aus mindestens zwei **Podestbalken**, welche, ca. 20 cm in die Mauer eingreifend, durch eine entsprechende Anzahl **Podestwechsel** oder **Stiche** versteift werden, so dass ein Nachgeben des Ganzen unmöglich ist. Eine Stärke für diese Balken hier als Regel anzugeben, ist unthunlich, da dieselbe von den Verhältnissen abhängt. Dagegen kann nicht genug auf die Wichtigkeit und Sorgfalt dieser Konstruktion hingewiesen werden. Bedenkt man, dass der eine dieser Balken an zwei Stellen sehr stark belastet wird und zwar gerade da, wo er durch Zapfenlöcher geschwächt ist, so wird man diese Vorsicht begreifen. Sehr zu empfehlen ist die Anlage eines sehr starken, über das gewöhnliche Mass hinausgehenden Podestbalkens, oder eine Verstärkung desselben durch einen **Unterzug**. Die gleiche Vorsicht ist beim Austritt auf Stockhöhe geboten, sofern sich ein neuer Lauf des Stockwerkes dort aufstützt. Die

194 X. Die Holztreppen.

Balken selbst sowie die Wechsel sind entweder gehobelt und gefast oder profiliert oder mit Brettern und Profilleisten verkleidet.

Der Podestboden muss stark und undurchlässig sein, weshalb man ihn aus starken Dielen fertigt und ihn spundet oder federt. Auch legt man oft bei besseren Ausführungen einen Blindboden und auf diesen ein Parkett (Fig. 336), während man die Untersicht für sich als blinde Decke fertigt und anschraubt. Auf Tafel 71 sind zwei solche Decken dargestellt und einige weitere Beispiele finden sich auf Tafel 74.

Die Austrittsstufe auf dem Podest oder auf der Stockhöhe wird der Ersparnis halber meist nur in einer Breite von 10 bis 12 cm hergestellt.

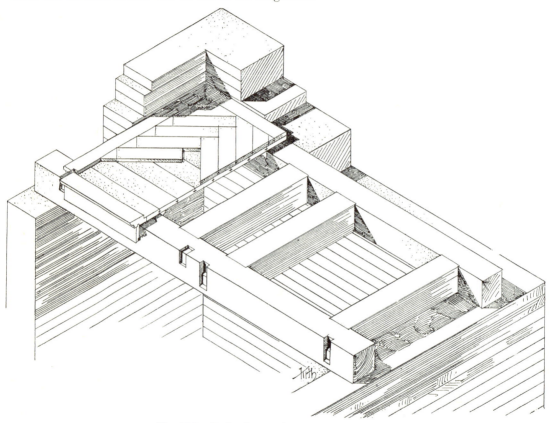

Fig. 336. **Podestkonstruktion mit Bodenbeleg.**

c) **Die aufgesattelte Treppe.** (Tafel 71 II.) Sie ist unstreitig schöner und eleganter, aber auch kostspieliger als die gestemmte oder Wangentreppe. Von dieser unterscheidet sie sich nur durch die Form und Befestigung der mittleren, freitragenden Zarge, während die Auftritte, Futterbretter und Wandzargen die gleichen sind wie bei der gestemmten Treppe. Die mittlere Zarge, gewöhnlich 8 bis 10 cm stark, befindet sich nur unterhalb der Trittstufen, weshalb sie treppenförmig ausgeschnitten ist. Die Stufen sind auf dieselbe aufgeschraubt, während sie in die Wandzarge eingestemmt sind. Um das Hirnholz des Futterbretts aussen nicht sichtbar zu lassen, setzt man es, wie die Zarge selbst, auf Gehrung ab. Weniger ratsam ist das Ansetzen von Langholzprofilen an Stelle des Hirnholzes der Auftritte, da bei nicht völlig übereinstimmender Färbung der Hölzer die Zusammensetzung von oben sichtbar wird.

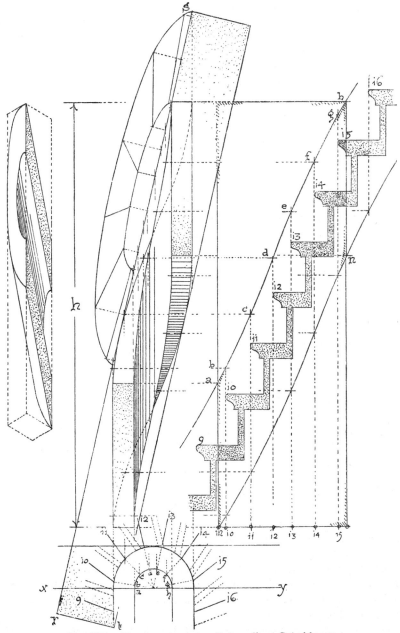

Fig. 337. Heraustragen der Krümmlings-Schablonen.

Wesentlich anders als bei der Wangentreppe ist die Art der Befestigung der oberen Zargenenden an den Podestbalken. Während die Wandzargen sich mit einer Art Klaue über die Podestbalken legen, lehnen die Mittelzargen der aufgesattelten Treppe, weil nur unterhalb der Stufen befindlich, sich günstigsten Falles nur an die Podestbalken an. Ein Blick auf das Detail, Tafel 71, Fig. II, genügt, um zu zeigen, dass hier ein aussergewöhnlich hoher Podestbalken oder zwei übereinanderliegende und miteinander verschraubte Balken nötig sind, um die Zarge aufzunehmen. Um die Konstruktion solid herzustellen, fertigt man daher das die Zargen aufnehmende Podestholz aus Eichen und lehnt es gegen den eigentlichen Podest an. Die Art der Zapfen ist aus dem Detail ersichtlich.

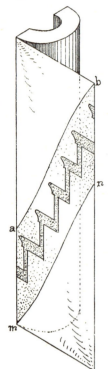

Fig. 338.
Heraustragen der Krümmlings-Schablonen.

Schliesslich erübrigt noch, die Konstruktion des Krümmlings Fig. 337 (zu Tafel 70, Fig. XII gehörig), welche sich mit derjenigen aller gewundenen Zargen deckt, zu besprechen. Es gilt als Grundsatz, die Holzfaser mit der Längsrichtung des Krümmlings laufen zu lassen. Zu allen diesen Konstruktionen ist die Abwickelung des äusseren oder inneren Cylindermantels (jeweils desjenigen, auf welchem die Stufen zum Einstemmen vorgerissen werden) erforderlich.

Nachdem der Grundriss mit den Vorder- und Futterbrettkanten gezeichnet ist, wird (im Fall der Fig. 337) der äussere Halbkreis nebst der Horizontalprojektion der Stufen seitlich verstreckt aufgetragen, worauf mit Hilfe der Steigungen das Profil der Treppe gezeichnet wird. Hierauf wird wie bei der Zargenbestimmung der gemischten Treppe die obere und untere Begrenzungslinie des Krümmlings gesucht. Aus dem Grundriss und der erhaltenen Abwickelungsfläche lässt sich nunmehr leicht der Aufriss des Krümmlings konstruieren und zwar, indem man durch die Vorderkanten der Trittstufen im Grundriss radiale, in der Abwickelung dagegen senkrechte Schnitte legt, wodurch man eine Anzahl Rechtecke erhält, deren Endpunkte, miteinander verbunden, den Aufriss des Krümmlings angeben.

Von da ab giebt es zwei Konstruktionen. Die erstere und einfachere ist die, aus einem Stück Holz einen halben Hohlcylinder auszuschneiden, dessen Querschnitt gleich dem Grundriss des Krümmlings und dessen Länge gleich der Höhe desselben, gleich h ist. Legt man an diesen Hohlcylinder (Fig. 338) die zuerst ermittelte und auf Papier aufgezeichnete Abwickelung so an, dass der Punkt m an das untere Eck, die Kante m a bündig mit der Krümmlingskante zu liegen kommt, schneidet man ferner nach der Kurve der Abwickelung das Holz winkelrecht ab und stemmt die Stufen ein, so ist der Krümmling im allgemeinen fertig. Bei dieser ersten Konstruktion liegen die Holzfasern lotrecht.

Anders dagegen verhält sich dies bei der zweiten Konstruktion (Fig. 337). Nachdem der Grundriss, die Abwickelung und mit diesen beiden der Aufriss gefunden sind, wird die Holzstärke bestimmt. Die Dicke ist einfach im Grundriss abzugreifen; sie ist bestimmt durch die beiden an die äusseren Punkte angelegten Parallellinien. Die Breite findet man, indem man beiderseits an die Kurven des Aufrisses parallele Tangenten so zieht, dass der ganze Krümmling innerhalb derselben zu liegen kommt. Die Länge wird schliesslich durch die Verlängerung der äusseren senkrechten Aufrisskanten bis zu deren Schnitt mit der Holzkante bestimmt (Punkt s und t).

Aus dem Dargelegten geht hervor, dass die Holzfaser nicht senkrecht, sondern mit der Längsrichtung des Krümmlings läuft und der Grundriss nicht kurzweg zum Ausschneiden des Hohlcylinders benützt werden kann, sondern zuvor der Umänderung bedarf. Diese Aenderung ist eine Dehnung, ein Strecken nach der Länge. Zu diesem Zwecke bringen wir sämtliche äusseren und inneren Punkte des Grundrisses hinauf auf die Holzkante r s und tragen die betreffenden Entfernungen derselben von der Grundrisskante x y senkrecht daselbst auf und erhalten eine Anzahl neuer Punkte, welche, miteinander verbunden, die Schablone ergeben. Legen wir dieselbe nun auf das Holz so auf, dass sie auf der einen Seite oben, auf der anderen dagegen unten bündig liegt, und verbinden wir die Punkte miteinander, so ist der ganze Krümmling vorgerissen und wir können ihn als halben Hohlcylinder ausschneiden. Nachdem dies geschehen, wird die ursprüngliche Abwickelung auch auf diese Mantelfläche aufgelegt, vorgerissen und das Ganze rechtwinkelig nach der erhaltenen Kurve ausgeschnitten.

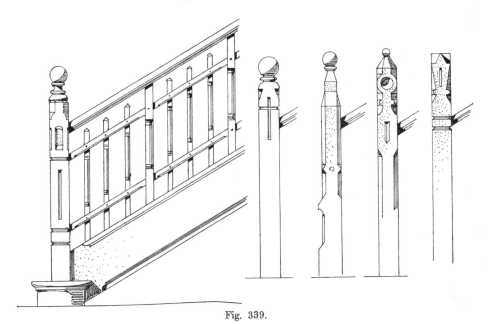

Fig. 339.
Treppengeländer und Treppengeländerpfosten.

In ähnlicher Weise wird die Schablone für das Zargenstück C der Hohltreppe auf Tafel 73 konstruiert. Nachdem ein Stück des Grundrisses — dessen Länge man nach der verfügbaren Holzstärke so bemisst, dass der Stoss auf die Mitte eines Trittes zu liegen kommt — gezeichnet hat und nachdem die Abwickelung sowie der Aufriss gefunden ist, wird die Schablone genau ebenso ermittelt wie die des vorbesprochenen Krümmlings. Hierbei ist zu beachten, dass die abzuwickelnde Mantelfläche nicht die äussere, sondern die innere ist, während bei der inneren Zarge sich dies wieder umgekehrt verhält.

Die Verbindung der Zargen durch Ueberplattung, Federung und durchgehende Mutterschrauben ist in den Figuren F—J auf Tafel 73 dargestellt.

Auf Taf. 74 sind verschiedene Podestdecken, zu geradläufigen und gemischten Treppen gehörig, in der Untersicht dargestellt.

X. Die Holztreppen.

4. Das Treppengeländer.

Wie oben bemerkt, bringt man zum Schutz gegen seitliches Herabfallen an der Aussenseite der Treppe das Geländer an, welches genügend hoch und stark sein muss. Es besteht aus:

a) dem Geländer-Pfosten;
b) den Geländer-Stäben oder Staketen und
c) dem Geländergriff, dem Handgriff.

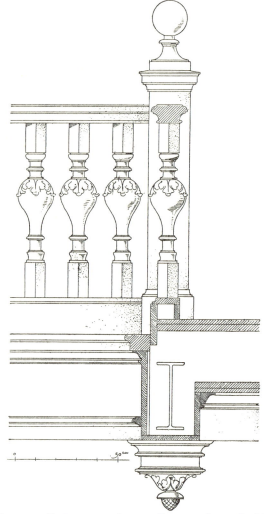

Fig. 340. **Brüstungsgeländer**, entworfen von Kayser und von Grofzheim in Berlin.

Die Höhe des Geländers nimmt man allgemein zu 85 cm von Vorderkante Trittstufe bis Oberkante Handgriff an.

Der Geländerpfosten ist meist höher und entweder mit einer schönen Endigung nach oben versehen oder zu einem Kandelaber etc. gearbeitet. Seine Form kann verschieden sein, ebenso wie sein Querschnitt; massgebend ist nur, dass er stark genug ist, dem Geländer Festigkeit zu geben. Seine Befestigung am steinernen Antritt ist durch einen unteren Zapfen, der in den Stein eingreift (Fig. 335), sowie durch Verzapfen und Verschrauben mit der Zarge bewirkt. Beim Austritt muss man — falls nicht die beste, in Fig. 335 und auf Taf. 72 dargestellte Konstruktion gewählt wird, bei welcher der ganze Treppenarm sich gegen den Pfosten stemmt und

Fig. 341.
Aufgang zur Redner- und Präsidententribüne im Sitzungssaal des deutschen Reichstags. Unter Wallots Leitung ausgeführt von Gebr. Lüdtke, Berlin.

ihn festhält — zu Eisenwinkeln greifen, da der gewöhnliche in den Boden eingreifende Zapfen nicht ausreicht, um genügende Festigkeit zu geben.

Die Geländerstäbe, entweder einfach vierkantig gehobelte und übereck gestellte oder gedrehte Stäbe, sind entweder unten und oben in die Zarge bezw. den Griff eingestemmt oder eingebohrt oder sie sind nur unten auf diese Art festgemacht, oben dagegen mittels einer nach der Oberfläche der Zargen abgebogenen Flacheisenschiene zu einem Ganzen verbunden. Die Staketen werden im letzteren Fall nach der Schmiege schräg abgeschnitten und durch eine Holzschraube mit der Schiene verschraubt. Die Entfernung der Stäbe voneinander ist allgemein so, dass zwei Stäbe auf einen Tritt kommen, also ca. 13 bis 15 cm von Mitte zu Mitte, so dass kleine Kinder nicht durchfallen können. Bei der aufgesattelten Treppe können die Staketen von oben eingebohrt oder seitlich angebracht sein, wie die auf Taf. 75 in A und B gezeichneten Beispiele zeigen.

Der Handgriff, meist aus Hartholz bestehend, erhält gewöhnlich eine Stärke von 5 auf 6 cm und ein leicht mit der Hand zu umfassendes Profil. Befestigt wird er an den Geländerpfosten und den Staketen oder aber, wenn eine Eisenschiene vorhanden ist, mit dieser. Die Schiene wird auf der Unterseite des Handgriffes eingelassen und mit ihm verschraubt. Die Oberfläche des Griffes wird poliert oder doch gut geglättet. Treten an Stelle der schlanken Stäbe für die Geländerbildung runde oder kantige Docken, dann kommt auf jeden Tritt eine solche aufzustehen und die Abdeckleiste wird gesimsartig verstärkt und verbreitert.

Auf Tafel 75 sind einige gedrehte Geländerpfosten, Stäbe und Griffe dargestellt, welche das Gesagte illustrieren: dieselben sind dem Schulwerk: Kircher, Vorlagen für den gewerblichen Fachunterricht (Karlsruhe, Bielefeld) entnommen. Die Tafel 76 giebt eine Anzahl von Geländerstäben und Docken zur Auswahl im Bedarfsfall. Einige weitere Einzelheiten einfacher Art bringt die Fig. 339, während Fig. 340 ein Brüstungsgeländer nach dem Entwurf der Architekten Kayser und von Grofzheim in Berlin dargestellt, welches ebensowohl als Treppengeländer Verwendung finden kann. Schliesslich zeigt die Fig. 341, wie die Geländerbildung bei sehr reich gehaltenen Treppen sich etwa gestaltet. Derartige Ausstattungen waren im 17. und 18. Jahrhundert keine Seltenheit; heute sind sie die seltene Ausnahme von der Regel.

XI. ABORTSITZE.
(Tafel 77.)

Die grosse Mehrzahl der Techniker, wie das wirklich gebildete Publikum überhaupt, weiss eine gediegene Aborteinrichtung aus Schicklichkeits- wie Gesundheitsgründen zu schätzen. Es ist bekannt, dass selbst der rohe Mensch bei Benützung eines hellen, sauber gehaltenen Aborts bemüht ist, denselben rein zu halten; ebenso steht fest, dass selbst sonst wohlerzogene Menschen — aus Furcht, sich und ihre Kleider zu beschmutzen — einen schlecht beleuchteten und unsauberen Abort unter Umständen noch mehr verunreinigen. Zu welchen Bildern die letztere Benützungsart führt, zeigen die unter keiner besonderen Aufsicht stehenden öffentlichen Aborte. Dass diese Bilder nicht veredelnd wirken können, ist klar, wie für den umgekehrten Fall die Erfahrung den Beweis liefert.

Bedenkt man ferner, wie nachteilig mangelhaft konstruierte Aborte in Wohnungen — von den Abortgruben ganz abgesehen — für die Gesundheit werden können, so werden hier einige Worte über die Einrichtung, soweit sie den Schreiner betrifft, gerechtfertigt sein.

Der Abortsitz darf, wenn er seinen Zweck erfüllen und eine bequeme Benützung gestatten soll, nicht zu hoch angebracht sein (für Erwachsene 0,40 bis höchstens 0,45 m vom Boden aus); er muss ferner wenigstens 0,60 m tief sein, damit man die Kleider nicht an der Wand verdirbt. Die Breite des Sitzes ist meist gleich der Abortbreite. Der Sitz wird am zweckmässigsten aus hellem, hartem Holz, das sich gut glätten, bezw. polieren lässt, hergestellt, oder doch hell in Oelfarbe angestrichen (aber nicht mit bleiweisshaltiger Farbe, weil der Schwefelwasserstoff dieselbe schwärzt), damit jede, auch die geringste Verunreinigung sichtbar wird und somit eine stete Warnung zur Vorsicht für die Benützenden ist.

In Anbetracht des Umstandes, dass die Luft im Abort oder wenigstens in dem Abtrittstrichter immer etwas feucht ist, soll der Sitz so konstruiert sein, dass er sich nicht wirft, nicht reisst oder seine Form sonst verändert, weshalb man einfache Sitze fest verleimt, mit Einschiebleisten versieht und die Oeffnung ausschneidet. Besser ist es jedoch, die Sitze zu stemmen. Das Sitzbrett darf nicht auf dem Trichter oder der Porzellanschüssel aufliegen, vielmehr muss ein Zwischenraum von ca. 1 cm zwischen beiden bleiben. Unterhalb des Sitzes befindet sich gewöhnlich als Abschluss nach dem Abtrittsraum das sogen. Vorbrett, welches den Apparat verdeckt und dem Sitz als Auflager dient. Bei aussergewöhnlichen Aborten, wo der Sitz im Eck oder frei im Raume steht, kann das Vorbrett auf zwei, bezw. drei Seiten nötig werden. Dasselbe ist glatt verleimt, oder besser auch gestemmt.

Als Auflager des Sitzes dient ausser dem Vorbrett entweder ein Rahmenschenkelgestell, wie es an Sitz A, Tafel 77, angedeutet ist, und in diesem Falle muss die oben zwischen Sitz und Wand verbleibende Fuge durch ein mit Messingschrauben zu befestigendes Leistchen gedeckt werden, worauf der Verputz angeschlossen wird — oder es werden zu beiden Seiten des Sitzes und hinten an der Wand starke Leisten, nach den Figuren B, C und D, sowie g und h, mittels Steinschrauben angebracht, in welche der Sitz entweder von oben eingelegt und durch ein Leistchen gedeckt (g) oder von vorn eingeschoben (h) wird. Die Leisten sind oben nach hinten abgeschrägt, um dem Verputz einen sicheren Halt zu geben. Auf diese Weise ist es möglich, den Sitz bei allenfalls nötig fallenden Reparaturen des Apparates leicht zu beseitigen, ohne den Verputz und dessen Anstrich beschädigen zu müssen, was bei A fast nicht zu verhüten ist. Bezüglich der Wahl zwischen der Leiste g oder h sei bemerkt, dass es zweckmässiger ist, für Sitz C (mit Einschiebleisten) und D (mit Zuggriff) die Leiste g zu wählen, womit jedoch nicht gesagt sein soll, dass h für diese Fälle unbrauchbar sei.

Noch vorteilhafter gestaltet sich das Ganze, wenn man weitere Leisten derart anbringt, dass auch das Vorbrett von oben eingeschoben werden kann, wodurch man im Notfalle den ganzen Sitz frei zu legen im stande ist.

Die Tafel 77 zeigt in A und C zwei gefederte und verleimte, aus Schleifdielen bestehende Sitze, wovon der erstere eine kreisrunde, der letztere eine eiförmige Sitzöffnung zeigt; bei beiden ist darauf Rücksicht genommen, dass die Exkremente direkt in die Oeffnung fallen, ohne die Schüssel zu beschmutzen. Bei Bestimmung der Oeffnung ist zu beachten, dass dieselbe mindestens 28 cm Durchmesser hat und der Abstand von Vorderkante-Sitz bis Vorderkante-Oeffnung nicht mehr als 8 cm beträgt. Sitz A zeigt einen befestigten Drehdeckel, C hat einen kreisrunden, gedrehten Deckel zum Wegnehmen, auf dessen unterer Seite eine in die Oeffnung des Sitzes passende Verdoppelung in Eiform aufgeschraubt ist. Sitz B besteht aus zwei Teilen, aus dem eigentlichen Sitzbrett und der übrigen Fläche des Sitzes, in welchen das Sitzbrett in den Falz eingelegt ist; hinten ist dasselbe mit zwei Messingscharnierbändern angeschlagen, vorn liegt es auf dem Vorbrett auf. Die Sitzöffnung ist kreisrund und mit einem gedrehten Deckel mit Knopf geschlossen. Der Schnitt zeigt die vorerwähnte Verdoppelung, die in diesem Falle kreisrund ist. Sitz D ist ebenfalls gestemmt und für einen Zugapparat bestimmt. Der Deckel liegt hier in einer Ebene mit der Sitzoberfläche und ist hinten mit Bändern angeschlagen. Das eigentliche Sitzbrett wird erst sichtbar, sobald der Deckel nach hinten aufgeklappt ist. Bei sämtlichen Sitzen muss die Oeffnung für das Abtritts- oder Dunstrohr so ausgeschnitten sein, dass die Sitze bequem eingeschoben werden können. Die dadurch entstehende grössere Oeffnung an den Rohren wird durch schön angepasste und festgeschraubte Holzleisten gedeckt.

XII. DIE BESCHLÄGE.

(Tafel 78 bis mit 82.)

1. Steinschrauben, Bankeisen, Eckwinkel. — 2. Bänder. — 3. Riegel, Vorreiber, Ruder-, Baskülen-, Schwengel- und Espagnolettverschlüsse: Aufstellvorrichtungen, Festhaltungen und Zuwerfungen; Schlösser und andere Thürsicherungen.

Die in Betracht kommenden Beschläge der Bauschreinereiarbeiten haben dreierlei Zwecke: Sie dienen

1) zur Befestigung und Verbindung einzelner Teile der Schreinerarbeiten;
2) zur Bewegung derselben, und
3) zum Festhalten in bestimmten Lagen, zum Verschliessen.

Ausserdem dienen die Beschläge in vielen Fällen auch zur Ausschmückung und Verzierung. Das Material derselben ist Schmiedeeisen, Stahl oder schmiedbares Gusseisen. Das letztere wird besonders zu kleineren und schwierig in Schmiedeeisen herzustellenden Arbeiten (Schlossteile, Schlüssel, Drücker etc.) verwendet. Die gegossenen Teile werden durch Tempern oder Glühfrischen kohlenstoffärmer und dadurch dem Schmiedeeisen ähnlich gemacht, so dass sie sich wie dieses bearbeiten lassen. Auch Bronze, Messing und sonstige Legierungen werden zu sichtbaren Beschlägeteilen, wie Thürdrückern, Oliven, Knöpfen etc., benützt.

Die Beschläge werden meistens mit Holzschrauben auf das Holzwerk befestigt. Wichtig ist dabei, dass die Beschläge, wo es irgend angeht, sauber in das Holz eingelassen und befestigt werden, da ein eingelassenes Band ohne Schrauben unter Umständen den Thür- oder Fensterflügel besser zu tragen im stande ist, als ein nur aufgelegtes und aufgeschraubtes Band.

Im allgemeinen sind die Beschläge für Thüren, Fenster und Läden die gleichen; sie unterscheiden sich von einander nur in ihren Abmessungen.

---

1. Beschläge zur Befestigung und Verbindung einzelner Teile.

a) **Die Steinschraube** (Fig. 342 und 343).

Sie dient hauptsächlich zur Befestigung der Futterrahmen und hat entweder die Form Fig 342b, wenn der Gewändeanschlag breit genug ist, um sie genügend befestigen zu können, oder die Form Fig. 342a, wenn dies nicht der Fall ist. Die Mutter der Steinschraube ist versenkt, so dass sie über das zu befestigende Holz nicht vorsteht.

XII. Die Beschläge.

Eine besondere Art von Steinschraube zeigt die Figur 343. Sie wird von Bauinspektor Kredell in Baden-Baden vorgeschlagen zur sichern Befestigung der Holzrosetten, an welche die Vorhanghalter aufgehängt werden. Wer die Missstände der üblichen Paterenbefestigung kennt, wird diese Errungenschaft mit Freuden begrüssen.

Dem nämlichen Zwecke dient die von Werkmeister Ad. Bodenmüller erfundene Steinschraube der Figur 344. Dieselbe endigt in eine Metallscheibe, in welche ein nach oben offener Grat mit Mittelsteg eingeschnitten ist. Die Holzrosette des Vorhanghalters wird mit einer flachköpfigen Schraube zentrisch angebohrt, deren Kopf dann in den erwähnten Grat eingeführt wird.

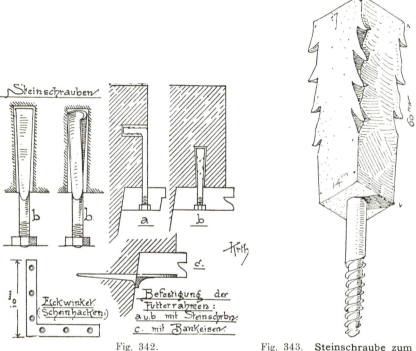

Fig. 342.
Befestigung der Futterrahmen.

Fig. 343. Steinschraube zum Befestigen von Vorhanghaltern.

Dreht man hierauf die Holzrosette nach rechts herum, bis die Schraube fest angezogen ist, so sitzt auch der Vorhanghalter genügend fest, kann aber ebenso leicht durch die umgekehrten Bewegungen wieder ausgelöst werden.

Bei dieser Gelegenheit mag auch eine andere zweckmässige Vorrichtung erwähnt sein, die von demselben Erfinder herrührt, eine Steinschraube zum Befestigen der Vorhanggalerien in Mietwohnungen dienend. Dieser paarweise zu verwendende Galeriekloben ist durch die Figur 345 abgebildet. Die Steinschraube endigt in einen doppelt durchbohrten »Kreuzkopf« mit Stellschraube. Mittels der letzteren wird im Kreuzkopf ein Rundeisendoppelwinkel festgestellt, nach oben, nach unten, nach rechts oder links gerichtet, je nach dem gegebenen Fall. An dem längeren Arm des Doppelwinkels ist ein zweiter Kreuzkopf verschiebbar und in diesem der Vierkantwinkel d. Auf diese Weise passt sich die vielseitige Vorrichtung jeder Vorhanggalerie und den an ihr schon befestigten Oesen an und erleichtert das Aufmachen der Galerie ganz wesentlich. (Bestellungen sind zu richten an den Erfinder, Karlsruhe, Werderstrasse 11. Ein Rosettenhalter kostet 20, ein Galeriehalter 40 Pfg.)

b) **Das Bankeisen** (Fig. 342c).

Es findet Verwendung wie die gewöhnliche Steinschraube, ist aber weniger sicher als diese. Befestigt wird es, indem man es in eine Mauerfuge einschlägt; das vorstehende Ende wird entweder eingelassen oder auf den Futterrahmen aufgeschraubt.

c) **Der Eckwinkel** (Fig. 342),

auch Scheinhaken genannt, wird aus starkem Schwarzblech gefertigt und dient zur Verstärkung von Eckverbindungen, besonders solcher von Fensterflügeln. Soll er wirklichen Wert haben, so muss er eingelassen und verschraubt werden.

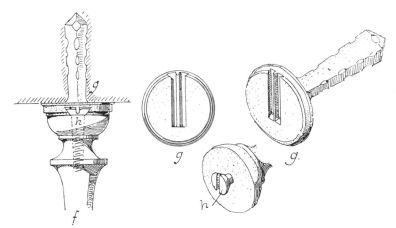

Fig. 344. Steinschraube zum Befestigen von Vorhanghaltern.

2. Beschläge zur Bewegung einzelner Konstruktionsteile.

Die Bänder (Thür-, Fenster- und Ladenbänder).

Sie geben dem beweglichen Flügel eine feste Drehaxe und bestehen aus zwei Teilen: aus dem Kloben, d. h. dem an dem Thürfutter oder Gewände befestigten und feststehenden Teil, und dem eigentlichen Band, dem beweglichen Teil, welcher an dem Flügel angebracht wird; oder sie bestehen aus zwei gleichen Bandlappen. Beide Lappen verbindet miteinander der Dorn, auch Kegel oder Stift genannt, welcher in den meisten Fällen mit dem Kloben vernietet ist. Der Kloben hat, wenn er in Stein eingesetzt werden soll, zur Befestigung an der Wand entweder einen kräftigen Ansatz, oder er wird bei Verwendung in Holz nach hinten zugespitzt oder auf eine Eisenplatte aufgenietet. Je nachdem heisst er dann 1) **Kloben in Stein**, 2) **Spitzkloben** oder 3) **Kloben auf Platte**. Der letztere ist dem Spitzkloben entschieden vorzuziehen; er wird sauber in das Holz eingelassen und aufgeschraubt. Ein Mittelding zwischen 2 und 3 ist der **Stützkloben** (Tafel 78). Das eigentliche Band, der Bandlappen, wird aus starkem Schwarzblech oder Schmiedeeisen angefertigt, welches mit dem einen Ende um einen passenden provisorischen Dorn warm herumgebogen (herumgewunden), mit dem anderen an dem Flügel befestigt wird; die herumgebogene Hülse nennt man das **Gewinde**. Wird dieser Bandlappen über den Dorn hereingestülpt, so legt er sich auf den Kloben auf, die Dornspitze steht oben vor und man sagt: das Band läuft auf dem Gewinde. Schraubt oder nietet man einen Stift an das obere Ende des Bandgewindes, so dass der Kloben und das Bandlappengewinde nicht aufeinander aufsitzen, so läuft das Band auf dem Dorn.

Wir unterscheiden:

a) **Das Langband und das Kurzband** (Tafel 78).

Ersteres, über 30 cm lang, wird meist für Latten- und Riementhüren, für Fensterläden, überhaupt für solche Flügel verwendet, welche aus mehreren, nebeneinander liegenden Teilen

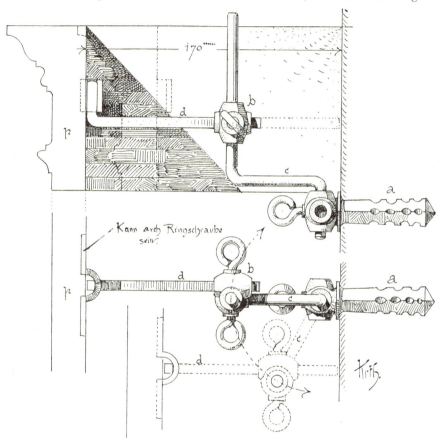

Fig. 345. Steinschraube mit Zubehör, zum Befestigen von Vorhanggalerien.

bestehen und daher durch das Band fest zusammengehalten werden sollen. Es ist meist aufgelegt und entweder in seiner Form nur glatt und gleich breit oder nach vorn verjüngt oder formal besser durchgebildet. Befestigt wird es durch Holzschrauben; die erste Schraube beim Gewinde ist eine durchgehende Mutterschraube. Das Kurzband unterscheidet sich von dem Langband nur durch seine Abmessungen, es ist unter 30 cm lang und daher nur für leichtere Flügel zu verwenden.

b) **Das Schippenband** (Tafel 78 und Fig. 346).

Es ist ein häufig benütztes Band für einfache Thüren und Läden. Der Kloben ist Stein-, Spitz-, Platten- oder Stützkloben; der sichtbare, meist aufgelegte Bandlappen ist entweder glatt mit gebrochenen Kanten, oder er ist verschiedenartig gebildet, wie Fig. 346 zeigt; bei schweren Thüren bringt man ausser den Holzschrauben noch eine durchgehende Mutterschraube an. Das Band wird auf Gewinde oder Dorn laufend gefertigt. In Fig. 213, sowie auf Tafel 10 1 u. 2 ist das angeschlagene Schippenband dargestellt. Dem Schippenband ähnlich ist

c) **Das Winkelband** (Tafel 78).

Für Thüren, Fenster und Läden gleichwohl geeignet, hat es gegenüber dem Schippenband den Vorteil, dass sich an dem Bandlappen, rechtwinkelig abgebogen, noch ein weiterer, über einen grossen Teil des Flügels hinweggreifender Lappen befindet, welcher, gut aufgeschraubt, der betreffenden Eckverbindung grosse Festigkeit verleiht und die Eckwinkel an dieser Stelle überflüssig macht. Das Band wird eingelassen oder aufgelegt; bei schweren Flügeln ist die erste, dem Kloben zunächst sitzende Schraube wieder eine durchgehende Mutterschraube.

d) **Das Kreuzband** (Tafel 78).

Dasselbe ist von dem Winkelband, bezw. dem Schippenband, welch letzteres man in einzelnen Gegenden irrtümlich mit diesem Namen belegt, abgeleitet. Bei schweren Thorflügeln z. B., bei welchen das Aushauen der Schippenbänder wie der entsprechend starken Winkelbänder aus den Blechtafeln grosse Mühe und Kosten verursachen würde, zieht man es vor, die sonst aus einem Stück bestehenden Bandlappen aus zwei Teilen zu fertigen und sie fest miteinander zu verbinden. Es geschieht dies entweder dadurch, dass man auf dem in das Holz einzulassenden Bandlappen durch Annieten oder Aufschweissen zweier Leistchen ein sicheres Lager für den zweiten Lappen schafft, oder dass man den ersten Lappen über den zweiten hinwegkröpft. Beide Teile werden vernietet und mit einer durchgreifenden Mutterschraube versehen. In das Gewinde des Klobens ist der Kegel von unten eingesteckt und durch zwei starke Nieten befestigt; umgekehrt ist in den Bandlappen von oben ein kleiner Kegel eingepasst und eingeschraubt, so dass die beiden aufeinander laufen und das Gewinde frei bleibt. Das Kreuzband, wie

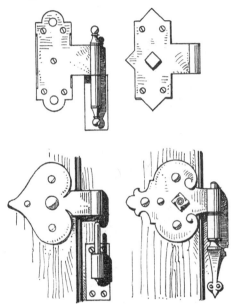

Fig. 346. Schippenbänder.

das kombinierte Kreuz- und Winkelband kommt nur bei schweren Flügeln vor.

e) **Das Fischband** (Tafel 78).

Es ist das häufigst verwendete Band für Bauschreinerarbeiten und verdankt diesen Vorzug dem Umstand, dass von dem ganzen Band nur der fischförmige Körper, aus Kegel und Gewinde bestehend, sichtbar ist. Das Band besteht aus zwei fast gleichen Teilen, deren einer — wie bei dem Kreuzband — mit dem von unten eingesteckten und geschweissten oder genieteten Kegel versehen ist, während der obere Teil einen ebenfalls festgenieteten kürzeren Kegel besitzt, der auf dem unteren aufsitzt. Weniger gut ist, wenn das Band auf dem Gewinde läuft, da dieses sich rasch abnützt, wobei die Thüre sich senkt. Man kann zwar diesem letzten Missstand durch Unterlegen von Eisenringen abhelfen; doch ist immerhin, wo es angeht, das Laufen auf dem Kegel anzustreben. Verwendung findet das Band bei Thüren wie bei Fenstern, und zwar wird es beiderseits eingestemmt, oder nur mit einem Lappen, während der andere eingelassen und aufgeschraubt wird, wie dies die Grundrissskizzen a, b u. c (Taf. 78), sowie die

Figuren 3 und 4 auf Tafel 10 zeigen. Was die Grösse der Bänder anbelangt, so sind allgemein gebräuchlich für Fenster (wobei 3 Stück des Verziehens wegen auf den unteren Flügel kommen) solche von ca. 12 mm Durchmesser; für Zimmerthüren bis 2,10 m Höhe 2 (über dieses Mass 3 Bänder pro Flügel) solche von 18 mm Durchmesser. Bei den letzteren beträgt die Lappen- oder Gewindstärke ca. 3½ bis 4 mm. Fig. 215 zeigt die Art des Einstemmens und das Befestigen des Bandes.

Für den einzelnen ist es bekanntlich keine leichte Arbeit, Thür- und Fensterflügel einzuhängen, die mit 2 oder 3 Bändern beschlagen sind. Während man dem einen Dorn seine Aufmerksamkeit zuwendet, hat sich die Sache am anderen Ende wieder verschoben u. s. w. Schliesslich gelingt es durch Zufall. Vor vielen Jahren bereits wurde in der Leipziger Illustrierten Zeitung vorgeschlagen, diesem Uebelstand dadurch abzuhelfen, dass die Dornlängen der 2 oder 3 Bänder verschieden gehalten werden, so dass das Einpassen erst an dem einen und nachher an dem andern Ende erfolgen könne. Es ist bedauerlich, dass dieses einfache Mittel sich nicht eingeführt hat, und dass die betreffenden Bänderpaare nicht schon mit ungleichen Kegeln geliefert werden.

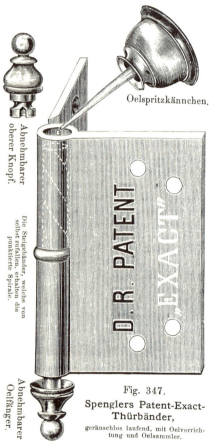

Fig. 347.
Spenglers Patent-Exact-Thürbänder,
geräuschlos laufend, mit Oelvorrichtung und Oelsammler.

f) **Das Aufsatzband** (Taf. 79).

Unter Aufsatzband versteht man in Süddeutschland im allgemeinen jedes Band, dessen Drehpunkt weiter als gewöhnlich vor der Thüre liegt und welches demgemäss geeigenschaftet ist, dieselbe schön herumzutragen, damit sie sich parallel an die Wand legen kann. Dasselbe wird nicht eingestemmt, sondern höchstens um seine Dicke eingelassen und festgeschraubt. Man verwendet es gerne für Tapeten- und andere Thüren, welche aus irgend welchem Grunde des Heraustragens benötigt sind. Zu den Aufsatzbändern gehört auch

g) **Das Paumelleband** (Taf. 79).

Dasselbe ist ein starkes, vollgeschmiedetes Aufsatzband, das auf eingelegten Bronzeringen und nicht auf dem Kegel läuft, obgleich die Hülse des oberen Bandlappens oben geschlossen ist. Es wird auf den Kanten der Thüre und des Futters angeschlagen, eingelassen und festgeschraubt. Fig. 214 zeigt dasselbe angeschlagen, ebenso die Tafel 10 in den Figuren 5 und 7.

h) **Das Scharnierband** (Taf. 79).

Es wird im allgemeinen für Bauarbeiten weniger verwendet, da die damit beschlagenen Flügel sich nicht ohne weiteres aushängen lassen, dagegen ist es aus demselben Grunde gerade für solche Arbeiten beliebt, welche man entweder gar nicht aushängen kann, oder welche nicht ausgehängt werden sollen, z. B. eine inmitten des Futters liegende Thüre oder zum Hereinklappen eingerichtete Glasabschluss- und Fensterflügel. Es wird gefertigt mit festem und losem Dorn, und zwar bei einfachen Verhältnissen in der Weise, dass man das entsprechend ausgekerbte Blech um den Dorn windet, so dass es am Lappen doppelt aufeinander liegt. Besser dagegen ist, die Lappen aus dem Vollen zu arbeiten, wie es das zu empfehlende Spenglersche

Exakt-Scharnierband zeigt. Der Dorn desselben ist beweglich und das Ganze schön zum Oelen eingerichtet. Die Anzahl der Kerbungen ist verschieden, sie kann zwei-, drei- und vierfach sein; in der Möbelschreinerei giebt es Scharnierbänder, die zwanzig- und mehrfach gekerbt sind und nach dem laufenden Meter verkauft werden (Klavierscharnierne).

Das Fig. 347 und auch auf Taf. 79 dargestellte Exaktband von Spengler ist ein Mittelding zwischen den Scharnier- und Aufsatzbändern.

i) **Das Zapfenband,**
auch Stift- oder Dornband genannt. Unter dieser Bezeichnung versteht man solche Bänder, welche nicht an der hinteren, sondern an der unteren und oberen Thürkante befestigt sind. Dabei ist wichtig, dass das Auflager des unteren Bandes ein solides ist, da das ganze Gewicht der Thüre auf ihm allein ruht. Das Band besteht aus zwei nach hinten verstärkten Lappen, deren einer den Dorn trägt, während der andere mit dem Dornloch versehen ist. Bei schweren Thüren oder Thoren biegt man den Bandlappen rechtwinkelig auf und setzt in diesen Winkel das Thor fest ein, wie dies auf Tafel 79 dargestellt ist. Verwendet wird es für Pendel- und Windfangthüren (da es eine Bewegung derselben nach zwei Seiten gestattet) und für schwere Thore.

---

3. Beschläge zum Festhalten einzelner Konstruktionsteile in bestimmten Lagen und zum Verschliessen.

a) Riegel. b) Vorreiber. c) Ruderverschluss. d) Baskülenverschluss. e) Schwengelverschluss. f) Espagnolettstangenverschluss. g) Aufstellvorrichtungen für Fensterflügel etc. h) Festhaltungen. i) Zuwerfungen. k) Pendelthürbeschläge. l) Schlösser. m) Thürsicherungen.

a) **Der Riegel** (Taf. 79).

1) Der Schieb- oder Schubriegel, auch Lang- und Kurzriegel genannt, wird vielfach bei gewöhnlichen Bauarbeiten verwendet. Er besteht aus einem Stück Blech, auf welchem sich in zwei angenieteten Ueberkloben der an einem Ende zu einem Griff umgebogene Riegel bewegt. Begrenzt wird diese Bewegung durch je eine ober- und unterhalb der Ueberkloben befindliche angeschmiedete Nase. Der Riegel greift in das in der Schwelle befindliche Schliessblech ein; ist für dieses und seine Befestigung nicht genügend Platz vorhanden, so kröpft man den Riegel, d. h. man biegt ihn zweimal rechtwinkelig ab. Je nach der Länge des Riegelschaftes unterscheidet man zwischen Lang- oder Kurzriegel. Massgebend für die Länge ist, dass man den Riegel bequem mit der Hand erfassen und bewegen kann. Werden grössere Riegel, Lang- oder Kurzriegel in senkrechter Lage verwendet, so ist es nötig, ihnen eine sogen. Blatt- oder Schleppfeder zu geben, damit sie sich in der gegebenen Höhe zu halten im stande sind und nicht von selbst herabfallen. Bei schwereren unteren Thorriegeln bringt man statt der Federn vorteilhafter eine Aufhängevorrichtung an. Der Riegel wird in diesem Fall mit seinem ringförmigen Griff in einem am Thor angebrachten Haken eingehängt und kann nun nicht mehr herabfallen. Für Zimmerthüren und Fenster verwendet man sogen. façonnierte Schubriegel, welche sich von den einfacheren durch ihre schönere Ausstattung unterscheiden. Wird der etwas dünne Schaft dieser Riegelart bei einer hohen Thüre allenfalls zu schwach, so

dass zu befürchten steht, er könnte sich bei der Bewegung seitlich ausbiegen, so kann man ihn ganz wesentlich versteifen durch Anbringung eines weiteren mittleren Ueberklobens. Sämtliche Schiebriegel werden sichtbar auf das Holzwerk aufgesetzt.

2) Der Kantenriegel (Taf. 79).

Derselbe ist ein verdeckter, d. h. bei geschlossenen Flügeln nicht sichtbarer Riegel. Sein Vorteil besteht darin, dass er in diesem Falle nicht nur nicht sichtbar, sondern auch nicht zu öffnen ist. Er wird, wie sein Name schon besagt, auf der Thürkante eingelassen und aufgeschraubt und greift in Schliessbleche ein, die am Boden und oben am Futter befestigt sind. Besser als der gewöhnliche ist der ebenfalls auf Tafel 79 abgebildete Spenglersche Kantenriegel, welcher so konstuiert ist, dass, des herausgelegten Griffes wegen, an ein Schliessen des zweiten Thürflügels nicht gedacht werden kann, bevor der Kantenriegel wirklich geschlossen ist. Wer da weiss, wie schwer das Dienstpersonal daran zu gewöhnen ist, die Kantenriegel jeweils einzuriegeln, wird diese sinnreiche Vorrichtung begrüssen.

b) **Vorreiber** (Taf. 80).

1) Einfacher Vorreiber. Derselbe ist der einfachste Fensterverschluss, erfüllt aber seinen Zweck bei kleineren, nicht zu hohen und nicht zu hoch angelegten Flügeln ganz wohl. Er wird in neuerer Zeit meist von Gusseisen gefertigt. Man befestigt ihn mittels einer starken Holzschraube auf den Futterrahmen, Setzhölzern oder dergleichen. Vom Drehpunkt aus legt er sich über den Flügel und presst ihn fest. Um ein Beschädigen des Flügelholzes zu verhüten, bringt man sogen. Streicheisen (Taf. 80 J), kleine, an den Enden zugespitzte Eisendrähte an, welche mit den Spitzen in das Holz eingeschlagen werden und auf deren Oberfläche nun der Riegel läuft. Besser und solider als diese Drähte sind kleine hälftig eingelassene und aufgeschraubte Eisenplättchen. Auf Tafel 80 ist bei der Skizze A die Verwendung des Vorreibers, zur Befestigung eines oberen, feststehenden Fensterflügels angegeben.

2) Doppelter Vorreiber. Derselbe wird natürlich nur da verwendet, wo zwei Flügel so nebeneinander liegen, dass sie miteinander geschlossen werden können. In diesem Falle ist er zwei einfachen Vorreibern entschieden vorzuziehen, da man zu seiner Bewegung nur einer Handbewegung bedarf. Die Skizze B auf Tafel 80 zeigt zwei obere Fensterflügel, welche mit Bändern angeschlagen und mit einem doppelten Vorreiber geschlossen sind.

c) **Der Ruderverschluss** (Taf. 80).

Er wird verwendet wie der doppelte Vorreiber. Von diesem unterscheidet er sich hauptsächlich dadurch, dass sich der Drehpunkt am einen Ende befindet, während er beim doppelten Vorreiber in der Mitte liegt. (Einarmiger und zweiarmiger Hebel.) Zum Ruderverschluss ist gleichfalls ein Streicheisen und weiter ein auf das Setzholz zu befestigender Schliesskloben erforderlich. Im allgemeinen verwendet man Vorreiber und Ruder für Fenster mit Setzhölzern, während für solche ohne Setzhölzer die nachstehend beschriebenen Basküle-, Schwengel- und Espagnolettverschlüsse benützt werden.

d) **Der Basküleverschluss.**

Man versteht darunter einen doppelten Riegelverschluss, welcher durch Umdrehung eines die beiden Riegel verbindenden Griffs, Taf. 80 m, nach der früher üblichen Form Olive genannt, bewirkt wird. Diese Doppelbewegung erfolgt z. B. bei Basküle E auf Tafel 80 durch ein mit der Olive verbundenes Zahnrädchen, welches in die entsprechend angebrachten Zähne der beiden Riegel (hier Triebstangen genannt) eingreift und den einen derselben aufwärts, den anderen dagegen abwärts in die an dem Kämpfer, bezw. Futterrahmen angebrachten Ueberkloben oder Schliesskloben einschiebt und den Verschluss bewirkt. Bringt man an der einen Triebstange eine sogen. Nase an, welche in einen entsprechenden dritten Schliesskloben eingreift, so ist der Verschluss ein dreifacher. Man fertigt die Triebstangen entweder aus Rund-, Halbrund-

3. Beschläge zum Festhalten einzelner Konstruktionsteile in bestimmten Lagen und zum Verschliessen.

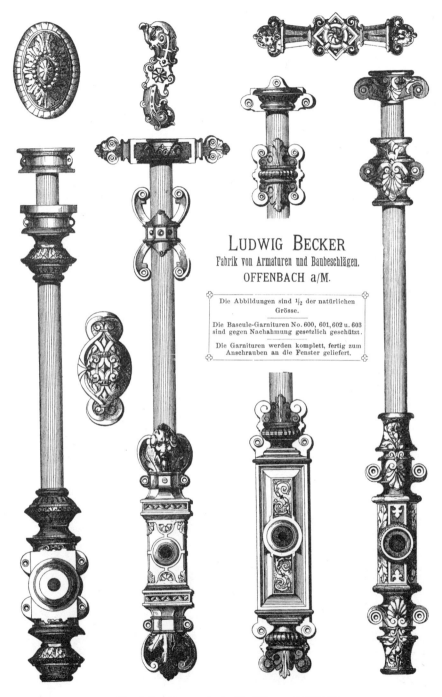

Fig. 348. Façonnierte Baskülenverschlüsse.

212   XII. Die Beschläge.

oder Flacheisen an. Das Zahnrädchen mit den Zahnstangen sitzt in dem Baskülekasten, dessen Seitenwandungen den Triebstangen gleichzeitig als Führung dienen. Oben und unten am Flügel

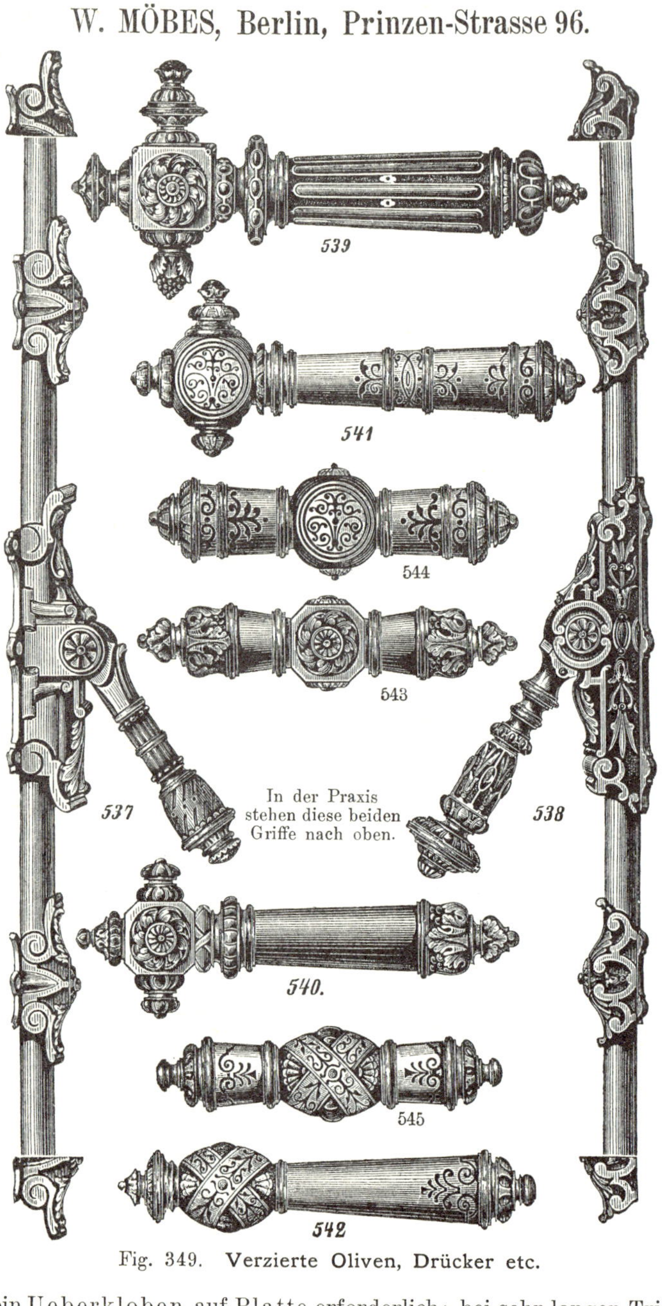

W. MÖBES, Berlin, Prinzen-Strasse 96.

Fig. 349. Verzierte Oliven, Drücker etc.

In der Praxis stehen diese beiden Griffe nach oben.

ist dann noch je ein Ueberkloben auf Platte erforderlich; bei sehr langen Triebstangen werden zwischen hinein weitere Ueberkloben zur Führung und gegen das Ausbiegen eingeschoben.

3. Beschläge zum Festhalten einzelner Konstruktionsteile in bestimmten Lagen und zum Verschliessen. 213

Bei genauer Ausführung ist die Konstruktion gut, dagegen lässt sie, wo diese nicht vorhanden ist, vielfach zu wünschen übrig. So begegnet man öfters dem Missstand, dass der Apparat klappert und rasselt, oder ein mangelhaftes Triebwerk hat, so dass man fast eine halbe Umdrehung auszuführen hat, bevor die Triebstangen in Bewegung gesetzt werden. Wesentlich besser und daher auch in neuerer Zeit sehr beliebt ist der Basküleverschluss F auf Tafel 80,

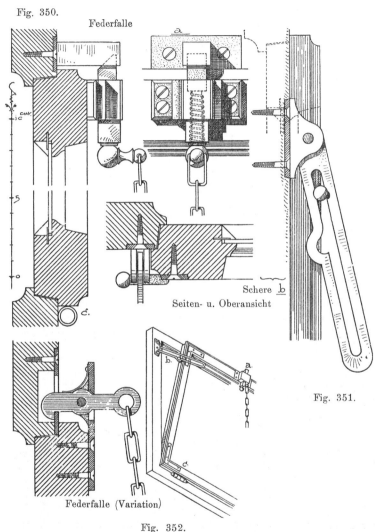

Fig. 350 bis 352. **Federfallen und Scheren.**

das sogen. Scheibenbasküle. Die Konstruktion ist aus der Zeichnung genau ersichtlich. An einer durch die Olive zu drehenden Scheibe sind um zwei Stifte beweglich die beiden Triebstangen befestigt, und zwar bei f direkt, bei g unter Zuhilfenahme eines kleinen eingeschobenen Gelenkes. Durch das letztere wird die Möglichkeit geschaffen, bei der Drehung die Triebstange in senkrechter Lage zu erhalten, was bei f nicht nötig ist, da die Schrägstellung so gering ist, dass sie kaum stört (weil die obere Stange länger ist, als die untere). An der Scheibe ist eine Nase oder Zunge angebracht, welche den dritten Verschluss bewirkt; im übrigen sind der Baskülekasten, die Ueberkloben und Schliesskloben dieselben wie beim Zahnradbasküle. Die Skizze C auf Tafel 80 stellt ein vollständig beschlagenes Fenster mit Basküleverschluss dar.

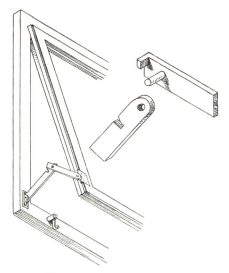

Fig. 353. Klappfenster.

e) **Der Schwengelverschluss** (Tafel 80).

Verbindet man die beiden Triebstangen des vorerwähnten Verschlusses zu einer einzigen, bringt ebenfalls in der Mitte derselben eine Nase, sowie auf der Vorderkante eine Anzahl Zähne

Fig. 354. Klappfenster-Verschluss.

an, in welche ein mit einem Hebel verbundenes halbes Zahnrad eingreift, so wird bei einer Bewegung des Hebels nach aufwärts die Stange abwärts bewegt und dadurch der Verschluss unten und in der Mitte bewirkt. Gestaltet man ferner den oberen Schliesskloben zu einer Art

3. Beschläge zum Festhalten einzelner Konstruktionsteile in bestimmten Lagen und zum Verschliessen. 215

Gabel, in welche das mit einem kleinen Querstück versehene Ende der Triebstange eingreift und sich beim Herabbewegen derselben festhakt, so ist auch an dieser Stelle der Verschluss hergestellt. Der Schwengelverschluss ist überall, wo grosse, schwere Flügel fest verschlossen werden sollen, zu empfehlen. Er kommt nicht nur für Fenster, sondern auch für Magazin- und andere Thore in Betracht. Im letzteren Falle macht man den Schwengel bis zu 50 cm lang.

Die Figuren 348 und 349 zeigen Basküle- und Schwengelverschlüsse nebst Ueberkloben, Drückern etc. in besserer Ausstattung, entnommen den Musterbüchern von L. Becker in Offenbach und von W. Möbes in Berlin.

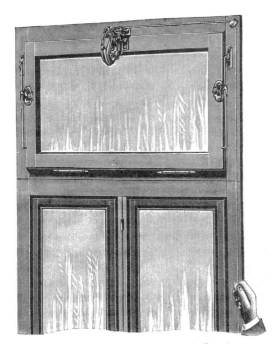

Fig. 355. Rechtecksflügel.

Fig. 356. Rundfenster.

Klappfensterverschlüsse von Seilnacht in Baden-Baden.

f) **Der Espagnolettstangen-Verschluss** (Taf. 80).

Derselbe ist sowohl ein Fenster- wie Thürverschluss; er hat ebenfalls nur eine Triebstange aus starkem Rundeisen, welche auf die Höhe des Flügels mehrmals durch Ueberkloben so gefasst ist, dass sie sich nicht ausbiegen, wohl aber drehen kann. Die beiden Enden derselben sind zu kleinen abgerundeten Widerhaken ausgeschmiedet und in der Mitte ist ein ca. 10 bis 12 cm langer Hebel befestigt. Bewegt man den letzteren horizontal, so macht die Stange eine Drehung um ihre senkrechte Axe, wodurch die beiden Haken in entsprechend geformte und am Kämpfer oder dem Futterrahmen befestigte Schliesskloben eingreifen und den Flügel schliessen. Um noch einen dritten Verschluss in der Mitte oder — was bei unseren dermaligen Wohnhausfenstern praktischer ist — auf ca. 1,40 m Höhe vom Boden aus zu erhalten, macht man den Hebel nach oben drehbar und legt ihn dann wie ein Ruder von oben in einen besonders angebrachten dritten Schliesskloben. Man hat hier also nicht wie bei dem Basküle- und Schwengelverschluss nur eine, sondern zwei verschiedene Bewegungen auszuführen, bis der Verschluss erfolgt oder gelöst ist. Soll z. B. das letztere ausgeführt werden, so hebt man zunächst den Hebel aufwärts und damit aus dem mittleren Schliesskloben heraus, und erst dann wird die Drehung und damit die vollständige Oeffnung des Verschlusses bewirkt. Beim Schliessen verfährt man umgekehrt. Skizze G auf Tafel 80 zeigt den mittleren Teil eines mit Espagnolettverschluss versehenen Fensters.

g) **Aufstellvorrichtungen** für obere Fensterflügel.

Dieselben sind meist derart konstruiert, dass die damit versehenen Flügel als Klappflügel sich bewegen. Die Flügel haben oben oder unten Fisch- oder Scharnierbänder und klappen entweder von oben herab, wie bei Fig. 352, oder von unten herauf, wie bei Fig. 353. Im ersten Fall werden sie in bestimmter, geöffneter Lage gehalten durch seitlich angebrachte sogen.

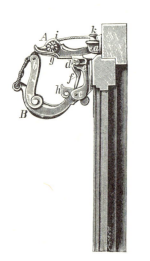
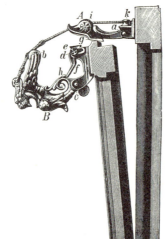
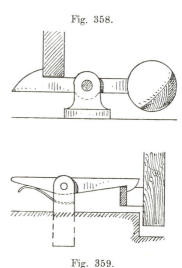

Fig. 357.

Klappfensterverschluss von Seilnacht in Baden-Baden.

Fig. 359.

Festhaltungen für Thore.

Scheren (Fig. 351), beim letzteren durch Kniehebel (Patent Leins). Verschlossen wird der Klappflügel der Fig. 352 durch eine oben angebrachte Federfalle mit Schliesskloben und Kettchen, nach Fig. 350. Bei diesem Verschluss zieht man an dem Kettchen und löst die schliessende Falle oder den Widerhaken aus, worauf das Fenster mit Hilfe eines kleinen Gegengewichts entweder von selbst sich öffnet oder doch mittels des Kettchens sich leicht aufziehen lässt. Da aber die Fenster sich gelegentlich festklemmen und sich trotz alles Zerrens nicht gutwillig öffnen wollen, so war man auf einen Verschluss bedacht, welcher diesem Uebelstand abhelfen soll. Das schwierige Oeffnen rührt hauptsächlich daher, dass der Kräfteangriff unter einem zu kleinen Winkel erfolgt, was bei dem Maraskyschen Klappfensterverschluss (Fig. 354) vermieden ist. Auf eine sinnreiche Weise ist nämlich bei ihm an dem Verschlusshebel eine Nase angebracht, welche bei der Bewegung des ersteren sich gegen den Futterrahmen stemmt und so das Fenster aus dem Falz herauszwängt. Es wird dabei aber hervorgehoben, dass die Bewegung nicht durch ein Kettchen ausgeführt werden kann, sondern durch eine Stange mit einem oben angebrachten Haken.

3. Beschläge zum Festhalten einzelner Konstruktionsteile in bestimmten Lagen und zum Verschliessen.

Gut bewährt hat sich ferner der Universalverschluss für Klappfenster, Patent Seilnacht, den die Figuren 355 bis 357 vorführen. Die sinnreiche, scheinbar komplizierte, aber sichere und bequeme Konstruktion wird durch eine Schnur bewegt, sowohl beim Oeffnen als beim Schliessen des Flügels. Während die Figuren 355 und 356 die Gesamtanordnung zeigen, giebt Figur 357 das Hauptbeschläge in zwei verschiedenen Ausstattungen wieder. Zieht man bei geschlossenem Flügel an der Schnur, so hebt die Rolle b des Hebels B die Falle A etwas in die Höhe, wobei gleichzeitig der Flügel aus dem Falz gezogen wird. Sobald die Spaltöffnung ca. 8 cm breit ist,

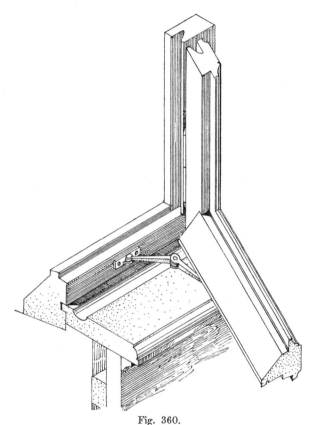

Fig. 360.
Fenstersteller von W. Kinzinger in Heidelberg.

lässt man die Schnur nach und der Flügel senkt sich durch sein Eigengewicht, soweit es die angebrachten Scheren gestatten. Wird hierauf die Schnur freigegeben, so senkt sich der Hebel B, bis das Pendel f an die Nase h anstösst. Soll der Flügel geschlossen werden, so zieht man an der Schnur; während der Flügel unter der Falle A hingleitet, hebt er dieselbe und sobald der Flügel sich in den Falz legt, fällt die Falle durch ihr Gewicht und die Spannung der Schnur herab und bewirkt den Verschluss, indem der untere Vorsprung der Falle sich an den Vorsprung e des Flügelbeschläges anlegt. Der letztere ist verschiebbar und wird mittels Stellschraube ein für allemal der Rahmenstärke entsprechend für die Falle angepasst. Der Preis der Garnitur samt Schrauben und Schnur beträgt 2,50 Mark.

218　　　　　　　　　　　XII. Die Beschläge.

h) **Festhaltungen** (für geöffnete Flügel).

1) für Thüren und Thore: Man benützt allgemein Vorkehrungen nach Fig. 358 und 359, welche so deutlich in der Zeichnung sind, dass sie einer Erklärung nicht bedürfen;

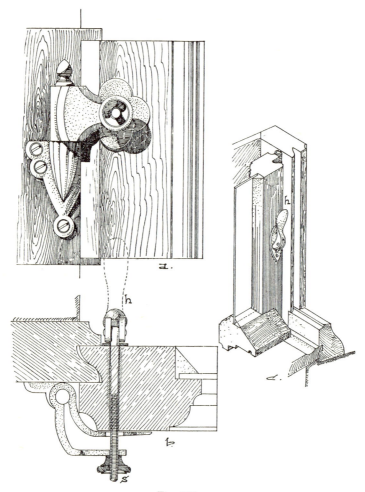

Fig. 361.

Fensterhalter der Gesellschaft für Fabrikation von Fensterverschlüssen in Baden-Baden.

2) für Fenster: Einhänghaken mit Ringschraube (Fig. 364) werden trotz der neuerdings so vielfach angepriesenen, komplizierten und patentierten Fensteraufstellvorrichtungen immer noch angewendet und zählen auch fernerhin wegen ihrer Einfachheit und Billigkeit zu den besten Vorrichtungen dieser Art; immerhin mögen einige der sogen. Fenstersteller hier Erwähnung und Abbildung finden:

3. Beschläge zum Festhalten einzelner Konstruktionsteile in bestimmten Lagen und zum Verschliessen.

Die Baubeschlägfabrik von W. Kinzinger in Heidelberg liefert einen einfachen Fenstersteller nach Figur 360, der sich deshalb empfiehlt, weil man das Fenster in gewöhnlicher Weise öffnet und schliesst, ohne einen weiteren Griff nötig zu haben. Will man der vollständigen Sicherheit halber das Knie ganz gerade stellen, anstatt es im stumpfen Winkel stehen zu lassen, so genügt ein leichter Druck auf das Gelenk beim Oeffnen und beim Schliessen. Je nach der Ausstattung kostet das Paar dieser Fenstersteller 1,20 bis 1,80 Mk.

Die Gesellschaft zur Fabrikation von Fensterverschlüssen in Baden-Baden, die insbesondere den bereits erwähnten Seilnachtschen Verschluss vertreibt, fertigt auch den in Fig. 361 dargestellten Fenster- und Thürensteller. Das einem Schippenband ähnliche Beschläge wird auf beliebiger Höhe befestigt, wie es die Figur zeigt. Der den Dorn umfassende Doppellappen wird vermittels der Stellschraube s soweit angezogen, dass der Flügel sich leicht bewegen lässt, so lange der Hebel h herabhängt, dass die Beweglichkeit der vermehrten Reibung wegen aber aufhört oder wenigstens sehr erschwert ist, sobald der Hebel h nach oben umgelegt wird, wobei sich der Raum zwischen den beiden Lappen verengert, so dass der Dorn geklemmt wird. Das Beschläg kostet für Fenster 0,90 bis 1,10 Mk., für Thüren etwas mehr.

Die Firma Nagel & Weber in Karlsruhe fertigt einen automatischen Fenstersteller (D.R.-P. Nr. 62563), den die Figur 362 abbildet. In dem am Fensterrahmen befestigten Gehäuse liegen zwei durch Federkraft aufeinander gedrükte, radial gewellte Stahlwellblechscheiben. Die eine Scheibe ist zu einem Hebelarm verlängert, welcher sich in einem auf dem Flügel befestigten Lager verschiebt, so dass beim Oeffnen die beiden Scheiben übereinander weggleiten müssen. Diese ruckweise verstärkte Reibung, regulierbar durch eine Stellschraube, verhindert das Zuwerfen des Flügels durch den Wind und ermöglicht seine Einstellung in beliebiger Oeffnungsweite. Der Apparat arbeitet mit etwas Geräusch, ist aber gut. Die Preise sind je nach Ausstattung und Grösse 1 bis 2,50 Mk. für das Stück.

3) für **Fensterläden** dient als Festhaltung: Die Schlempe (Fig. 365). Sie war seither ausser dem Riegel die gewöhnlichste Art der Feststellung von Läden; besser dagegen ist die selbstthätig wirkende **Federfalle** (Fig. 363), da man bei ihr den Laden nur aufzuschlagen und an die Wand zu drücken hat, was beim Riegel und bei der Schlempe bekanntlich nicht genügt.

i) **Zuwerfungen** (für Thüren und Thore).

Dieselben bezwecken, wie ihr Name sagt, das selbstthätige Schliessen der Thüren; sie können auf verschiedene Weise konstruiert werden. Die älteste und einfachste derselben dürfte die sein, das **untere Thürband etwas weiter herauszusetzen**, so dass also der Dorn desselben weiter vom Thürfutter entfernt ist als der obere. Hierdurch wird die Drehaxe der Thüre eine geneigte und durch die Schwerpunktslage wird dieselbe geschlossen — vorausgesetzt, dass die Thüre nicht bis 90° geöffnet ist. Das Mothessche **Gabelband** ist nach diesem Grundgedanken konstruiert. Eine weitere alte, aber nichtsdestoweniger immer noch gediegene Zuwerfung ist diejenige mit **Gegengewicht**. Dieselbe hat den Vorzug, dass sie fast nie versagt, wenn sie gut ausgeführt, mit abgedrehten Messingrollen und Darmsaiten versehen ist. Des besseren Aussehens wegen ist das Gegengewicht in einem in der Leibung anzubringenden Kasten versteckt.

Ausser den angeführten giebt es noch:

1) Zuwerfungen durch Heben in den Bändern.

α) **Band mit Schraubengewinde.** Man versieht den Bandkloben oder Banddorn mit einem Schraubengewinde, auf welchem mit Hilfe des oberen entsprechend geformten Bandlappens beim Oeffnen die Thüre sich hebt und durch ihr eigenes Gewicht herabsinkend sich wieder

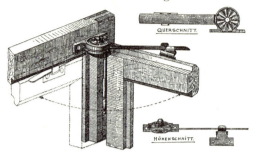

Fig. 362. Fenstersteller von Nagel & Weber in Karlsruhe (Baden).

schliesst. Die an sich gute Einrichtung hat den Nachteil, dass die Bänder sich bald auslaufen und locker werden, wenn sie nicht sehr gut befestigt sind. Ein hierher zu rechnendes Steigband zeigt die Figur 347.

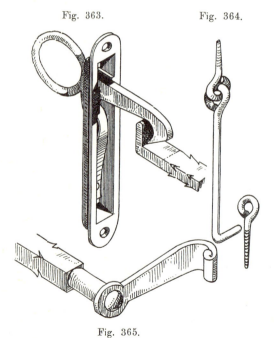

Fig. 363. Fig. 364.

Fig. 365.

Fig. 363 bis 365. Festhaltungen für Fensterladen.

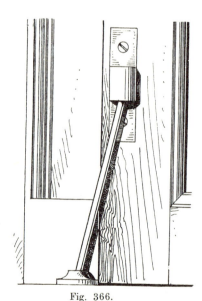

Fig. 366. Strebespindel.

β) **Die Strebespindel** (Fig. 366). Ein Eisenstab wird unten schräg gegen den hinteren Thürfries angelehnt und unten und oben in eine Pfanne eingesetzt; derselbe hebt nun beim Oeffnen die Thüre, um sie beim Loslassen sofort zu schliessen. Die Konstruktion ist gut und für einfache Thüren sehr praktisch.

3. Beschläge zum Festhalten einzelner Konstruktionsteile in bestimmten Lagen und zum Verschliessen. 221

Auf demselben Prinzip beruht die Fig 367 dargestellte Zuwerfung. Was die Strebespindel als Stütze leistet, das besorgt diese Vorrichtung in Hinsicht auf Zug. Werden an entsprechenden Stellen der Thüre und des Futters Ringe eingeschraubt und durch ein dem Distonschen Sägespanner (Fig. 33) ähnliches Zwischenglied verbunden, so hebt sich beim Oeffnen die Thüre um ein Weniges und schliesst sich infolge der veränderten Schwerpunktslage von selbst.

Fig. 367. **Ketten-Zuwerfung.**

γ) Der Weickumsche Thürselbstschliesser (Fig. 368). Unten an der Thüre ist eine Nase und entsprechend derselben am Boden eine kreisbogenförmige, schiefe Ebene angebracht. Eine zwischen beide gelegte Gussstahl- oder Hartgummikugel lässt beim Oeffnen die Thüre auf der schiefen Ebene hinauf-, beim Loslassen sofort wieder herabgleiten. Der Grundgedanke ist nicht neu, aber gut verwertet.

Angefügt wird noch, dass die Thüren nicht in Falz gelegt werden dürfen, sobald die Zuwerfungen der einen oder anderen Art eine Hebung der Thüre bedingen.

2) Zuwerfungen durch Federtrieb.

Hierher gehören:

α) die amerikanische Thürzuwerffeder (Fig. 369), sowie

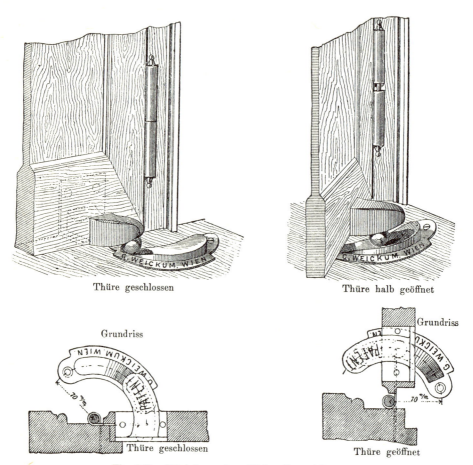

Fig. 368. **Weickumscher Thürselbstschliesser.**

β) die gewöhnliche deutsche Zuwerfungsfeder (Fig. 370), die in verschiedenen Stärken für verschieden schwere Thüren gefertigt wird. Bei ihr werden eine Anzahl Stahlstreifen, welche, in einer Hülse befindlich, am einen Ende befestigt sind, am anderen Ende durch das Oeffnen der Thüre gedreht, so dass sie nach Loslassen derselben das Bestreben haben, in ihre richtige Form zurückzukehren und dadurch die Thüre zu schliessen. Allgemein gilt, dass bei Feder- und Torsionszuwerfungen die Kraft am stärksten ist, wenn die Spannung am grössten ist, also am Anfang der Bewegung, während bei allen Zuwerfungen, die auf Hebung der Thüre basieren, nach dem Fallgesetz eine beschleunigte Bewegung eintritt.

γ) der Hartungsche selbstthätige, geräuschlose Thürschliesser (D. R.-P. Nr. 35 601), welcher ausser der starken, aufgerollten Feder noch eine Vorrichtung zum sanften Schliessen der Thüre besitzt. Das letztere geschieht dadurch, dass in einen Cylinder ein Kolben eingeführt wird, welcher die darin befindliche Luft zu einer kleinen Oeffnung hinauspresst und dadurch die Bewegung verlangsamt (Fig. 372).

Das Publikum konnte im grossen Ganzen sich jedoch nicht daran gewöhnen, einen derartigen Thürschliesser ruhig ausarbeiten zu lassen. Es drückte die Thüren voreilig mit Gewalt

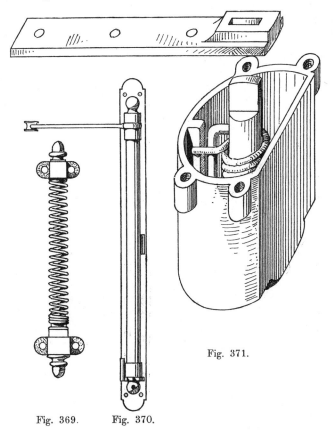

Fig. 369. Fig. 370. Fig. 371.

Fig. 369 bis 371. **Zuwerfungsfedern.**

zu, was die Apparate ruinierte. Darauf hat die Verbesserung von Schubert & Werth, Berlin C, Prenzlauerstr. 41, Rücksicht genommen, indem sie ein nachgiebiges Glied (Rohr mit Bolzen und Feder) einschob (Fig. 373), so dass die Thüre auch ohne Schaden gewaltsam geschlossen werden kann. Diese zu empfehlenden, verbesserten Thürschliesser (D. R.-P. 49 615) kosten je nach der Grösse 16 bis 27 Mk. ohne Anschlag. »Spielende Schliessfallen«, das sind solche, welche das Zufallen der Thüre ins Schloss ohne weiteres ermöglichen, kosten ohne Einsetzen 50 Pfg.

Die verschiedenen Federzuwerfungen haben Vorteile, wenn sie richtig verwendet werden, d. h. wenn die Thüre sofort wieder geschlossen werden kann. Anders dagegen liegen die Verhältnisse, wenn die Thüre längere Zeit geöffnet bleiben soll, wie dies ja zuweilen vorkommt.

224 XII. Die Beschläge.

In diesem Falle leiden die Federn sehr Not und sind bald ruiniert, und man wählt dann zweckmässiger Zuwerfungen ohne Federn.

Die hier aufgeführten Zuwerfungen sind nur eine Auswahl der zur Zeit üblichen; ausser ihnen giebt es noch eine ganze Reihe anderer, deren Erwähnung hier zu weit führen würde.

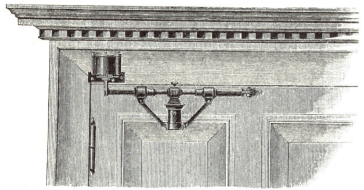

Fig. 372.
Hartungscher Thürschliesser.

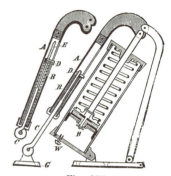

Fig. 373.
Konstruktion der Thürschliesser von Schubert & Werth in Berlin.

k) Windfang- und Pendelthürbeschläge.

Dieselben sind eigentlich nach zwei Seiten wirkende Zuwerfungen, ein Umstand, welcher fast jedes Pendelthürbeschläge, sofern man es halbieren kann, als Zuwerfung zu benützen

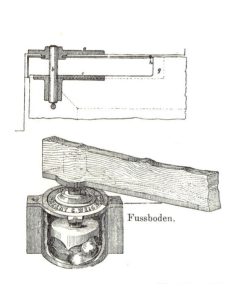

Fig. 374. Weickumscher Pendelthür-Apparat.

gestattet. Auch hier haben wir Federn als treibende Kraft und Thürhebungen auf schiefen Ebenen. Doch gewinnen die letzteren nach und nach die Oberhand, da ihre Dauer eben eine weit grössere, als die der besten Federn ist. Auf dem erstgenannten Prinzip beruht u. a. die englische Windfangfeder (Fig. 371). Das Federgehäuse wird in den Boden versenkt und das Zapfenband an die Thüre befestigt. Sehr haltbar und empfehlenswert ist der Weickumsche Pendelthür-Apparat (Fig. 374). Die Thüre wird hier durch eine auf einer schiefen Ebene laufende Kugel nach beiden Richtungen um 18 bis 22 mm gehoben, durch ihre eigene Schwere

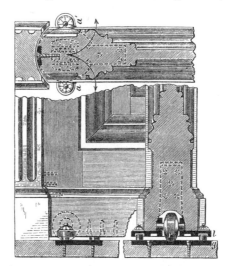

Fig. 375. Spenglers Exact-Rollen-Pendel.

geschlossen und durch die Kugel in der Ruhelage fixiert. Die Kugel nebst der schiefen Ebene befindet sich in einem in den Boden zu versenkenden Gehäuse. Für schwere Thore ebenfalls zu empfehlen ist das sogen. Spenglersche Rollenpendel (Fig. 375), welches auf Grund des Mothesschen Gabelbandes konstruiert ist. Die Thüre läuft hier auf zwei Rollen, die sich auf einer schiefen Ebene bewegen, wobei sie einen festen Anschlag bei a oder a[1] hat; durch ihr Eigengewicht kehrt sie, nachdem sie losgelassen, wieder in ihre ursprüngliche Lage zurück.

Fig. 376. Zu- und Aufziehknopf.

$\frac{1}{4}$ natürlicher Grösse

Fig. 377. Thürzuziehgriff.

Bei den sämtlichen angeführten Pendelthürbeschlägen sind die oberen Bänder Zapfenbänder. Ausser den Bändern sind zu einem vollständigen Beschläge aber noch erforderlich sogen. Aufziehknöpfe (Fig. 376, aus dem Musterbuch von Spengler), oder Aufziehgriffe (Fig. 377, von W. Möbes in Berlin).

Krauth u. Meyer, Die Bauschreinerei. 4. Aufl.

1) Die Schlösser.

Sie dienen dazu, die Thür- oder sonstigen Flügel an dem Gewände oder Futterrahmen festzuhalten und Schutz gegen unbefugtes Oeffnen zu gewähren.

Wir unterscheiden:

1) das Fallenschloss,
2) » » mit Nachtriegel,
3) » Riegelschloss,
4) » » mit Nachtriegel,
5) » eintourige Kastenschloss mit hebender Falle,
6) » zweitourige » » » » mit oder ohne Nachtriegel,
7) » zweitourige Kastenschloss mit Ueberbau, hebender Falle und Nachtriegel,
8) » gewöhnliche Einsteckschloss a) mit hebender Falle,
  » » » b) » schiessender Falle,
9) » Einsteckschloss für Schiebthüren,
10) » Chubb-Schloss,
11) » Stangenschloss,
12) » Baskülen- oder Doppelstangenschloss.

1) Das Fallenschloss. Es ist das einfachste Schloss und hat, wie schon sein Name sagt, nur eine Falle, die meist eine hebende, seltener eine schiessende ist. Verwendung findet es nur bei solchen Thüren, die man geschlossen wünscht, damit sie der Sturm nicht erfasst oder damit Geflügel und andere Haustiere abgehalten werden; Sicherheit gegen unbefugtes Oeffnen bietet es nicht.

2) Das Fallenschloss mit Nachtriegel. Es ist ein Fallenschloss, dem noch ein Nachtriegel beigegeben ist. Verwendung findet es zuweilen bei Abortthüren.

3) Das Riegelschloss. Es ist meist ein eintouriges Schloss, verwendet für einfache Thüren, wie Kellerabteilungs-, Speicherraumthüren etc., weniger für Wohnräume geeignet. Auf Tafel 81 ist dasselbe ohne und mit Schlossdecke dargestellt. Es besteht aus dem Schlosskasten, welcher wieder aus dem Boden oder Schlossblech und den seitlichen Umfassungen, dem Umschweif und Stulp zusammengesetzt ist. Sämtliche drei sind aus starkem Schwarzblech gefertigt und mittels der Umschweifstifte miteinander vernietet. Im Schlosskasten, geführt durch die Stulpöffnung und den Riegelstift, befindet sich der sogen. Schliessriegel a. Derselbe ist vornen zum Riegelkopf verstärkt und hat ausser dem Schlitz für den Riegelstift auf seiner oberen Kante die beiden Zuhaltungseinschnitte, unten dagegen die Toureinfeilung, die Angriffe. Festgehalten wird der Schliessriegel in geöffnetem wie geschlossenem Zustand durch die Zuhaltung, ein Eisen, welches an einem Ende um den Zuhaltungsstift drehbar befestigt, mit einem am anderen Ende befindlichen Zäpfchen in die Zuhaltungseinschnitte des Schliessriegels eingreift und ihn feststellt. Als Fortsetzung hat die Zuhaltung nach der einen Seite die Zuhaltungsfeder, nach der anderen den Zuhaltungbogen. Unterhalb des Riegels (in der Zeichnung auf dem Schlossboden) befindet sich das Schlüsselloch und demselben gegenüber an der Schlossdecke das Schlüsselrohr, durch welches der Schlüssel eingeführt wird. Der letztere besteht aus dem Griff (der Räute oder Raute) dem Rohr und dem Bart. Das Rohr wird nur so lang gemacht, als es der Zweck erfordert; den Bart kann man verschieden behandeln. Im allgemeinen richtet man sich nach der alten Schlosserregel, nach welcher alle Teile des Schlüssels in einem bestimmten Verhältnis zu einander stehen sollen, d. h. man nimmt den Rohrdurchmesser als Einheit an, macht den Bart an seinem Ende, dem

3. Beschläge zum Festhalten einzelner Konstruktionsteile in bestimmten Lagen und zum Verschliessen.

Eingriff, ebenso stark und giebt ihm zur Höhe und Länge die doppelte Rohrdicke. Von der Höhe des Bartes und der Entfernung des Rohres von dem Schliessriegel hängt die Schliesslänge, d. h. die Entfernung ab, um welche der Riegel bei einmaligem Umdrehen vorwärts oder rückwärts bewegt wird. Als Regel gilt, die Entfernung des Schlüsselrohrs von der Unterkante des Riegels gleich der Dicke des ersteren zu machen, so dass der Bart um eine Rohrdicke in den Riegel eingreift und die Schliesslänge gleich der Bartlänge wird, wie es die Konstruktion auf Tafel 81 zeigt. Es bedarf keines besonderen Beweises, dass bei veränderter Schlüsselstellung auch die Schliesslänge eine andere wird. Wie schon angedeutet, fasst der Schlüsselbart den Riegel an der Tour oder dem Angriff — nachdem er zuvor den Zuhaltungsbogen so weit in die Höhe gehoben, bis die Zuhaltung oben ausgelöst ist und der Riegel frei wird — und schiebt ihn so lange vorwärts, bis der Bart aus dem Riegel wieder heraustritt. In demselben Augenblick hört die Bewegung auf, gleichzeitig aber ist von oben die Zuhaltung wieder herabgekommen und hat sich in den anderen Zuhaltungseinschnitt eingehakt und den Riegel festgestellt. Beim Zurückschliessen des Riegels wird umgekehrt verfahren. Der Bart hebt die Zuhaltung auf, fasst den Riegel am Angriff und schiebt ihn zurück, worauf die Zuhaltung sich wieder in den anderen Einschnitt einhakt. Nimmt man ein starkes Blechstück und giebt ihm an einem Ende die Länge und Höhe des Bartes, so dass es im stande ist, die Zuhaltung auf die richtige Höhe zu heben und den Riegel zu bewegen, so ist der Verschluss auch geöffnet. Die Sicherheit, welche ein solches Schloss bietet, ist daher gering, da ein Nachschlüssel leicht gefertigt werden kann. Will man dieselbe erhöhen, so bringt man auf dem Schlossblech ein sogen. Reifchen an und verhindert dadurch, dass ein gewöhnlich geformter Schlüsselbart eingeführt wird. Erst wenn derselbe einen diesem Reifchen entsprechenden Einschnitt hat, kann er vollständig eingesteckt werden. Ein solches Reifchen heisst eine Besatzung oder Reifchenbesatzung. Hält man ein Reifchen für ungenügend, so lassen sich auch zwei und mehrere anbringen, ebenso schrägstehende, und es leuchtet ein, dass mit jedem weiteren die Nachbildung des Schlüssels schwerer wird. Ausserdem kann man dem Schlüsselbart (und dem Schlüsselloch) eine geschweifte, gebogene Form geben und den Schlüssel bei einseitig zu schliessenden Schlössern als Hohlschlüssel konstruieren. Jedoch macht man heute kunstvoll geschweifte Schlüsselbärte, wie man sie früher fertigte, nicht mehr, da der Aufwand nicht im richtigen Verhältnis zum erzielten Gewinn steht. Das einzige, was man in dieser Richtung thut, ist, sie bei Zimmerthüren, um Abwechselung zu erhalten, leicht zu schweifen; im übrigen wird die Sicherheit auf andere Weise zu erhöhen gesucht.

An der Schlossdecke, welche auf den sogen. Schenkelfüsschen aufliegt, ist eine Schleppfeder angebracht, welche dem Riegel einen steten, ruhigen Gang verschaffen soll. Angeschlagen wird das Schloss so, dass die Decke gegen das Holz zu liegen kommt. Das Schlüsselrohr und der Stulp werden in das Holz eingelassen. In der auf Tafel 81 dargestellten Weise ist das Schloss nur von einer Seite schliessbar, da von der anderen der Schlüssel nicht eingeführt werden kann.

4) **Das Riegelschloss mit Nachtriegel** wird bei Abortthüren verwendet und ist dem Fallenschloss mit Nachtriegel vorzuziehen, weil man den Abort abschliessen kann.

5) **Das eintourige Kastenschloss mit hebender Falle.** Es ist dies ein Riegelschloss mit hebender Falle, wie sie in Bezug auf das Zweitourschloss beschrieben werden wird.

6) **Das zweitourige Kastenschloss ohne Ueberbau mit hebender Falle**, auch Mauskastenschloss genannt. Unter gewöhnlichen Verhältnissen ist die Rohrdicke ca. 7 mm, die Schlusslänge somit gleich 14 mm, ein Mass, welches bei breiten, gestemmten Thüren (des Schwindens wegen) ungenügend ist, den Verschluss der Thüre zu sichern. Will man in dieser Beziehung sicher gehen, so empfiehlt es sich, eine grössere Schliesslänge anzunehmen und

das zweitourige Schloss zu wählen, welches in der unter 7. zu besprechenden Form Tafel 81 abgebildet ist. Dasselbe ist, wenn wir von der Falle d, dem Nachtriegel h und dem Eingerichte c absehen, dem Riegelschloss sehr ähnlich. Nur hat hier der Schliessriegel, der zwei Touren wegen, unten zwei Toureinfeilungen und oben drei Zuhaltungseinschnitte. Die Riegelführung ist bei dem aufgezeichneten Beispiel nicht durch einen Stift, sondern durch eine Art Leistchen und den Schlosstulp bewirkt, die Zuhaltung b mit ihrer Feder nicht aus einem Stück gefertigt; dagegen ist das Schloss von zwei Seiten schliessbar. Das einfache Reifchen ist hier durch ein doppeltes ersetzt und überdies ist noch auf zwei Schenkelfüsschen eine sogen.

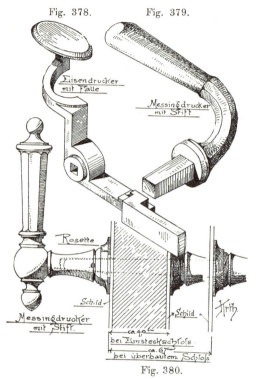

Fig. 378 bis 380. **Thürdrücker.**

Mittelbruchplatte oder kurzweg ein Mittelbruch angebracht. Hierdurch wird die Sicherheit abermals erhöht, wenn auch nur in geringem Masse, denn alle die so kompliziert erscheinenden Einschnitte des Schlüsselbartes, welche als Hindernisse beim unbefugten Oeffnen erscheinen, lassen sich sehr leicht durch Herausfeilen der Mittelpartie des Bartes umgehen, wie dies am Schlüssel rechts unten durch die Punktierung angedeutet ist. Man wählt aber dennoch diese Art Eingerichte für gewöhnliche Zimmerthüren, da man durch Verstellung der Reifchen zu einander doch vielerlei Schlüssel, bezw. Schlösser ausführen kann, von denen sich keines mit dem Schlüssel des anderen öffnen lässt, während sämtliche mit einem sogen. Hauptschlüssel von der angedeuteten Form zu öffnen sind.

Oberhalb des Riegels befindet sich die Falle d, welche aus dem Fallenkopf, dem Fallenschaft und der Nuss besteht. Ersterer ist geführt durch den Stulp, während die Nuss im Schlossblech und der Schlossdecke ihre Befestigung findet und drehbar gemacht ist.

3. Beschläge zum Festhalten einzelner Konstruktionsteile in bestimmten Lagen und zum Verschliessen.

Auf die Falle — hier eine hebende — drückt die Fallenfeder; bewegt wird die Falle durch den Drücker, welcher entweder mit ihr aus einem Stück besteht (Fig. 378) oder besonders gefertigt ist und mit dem Drückerstift in die Nussöffnung eingreift (Fig. 379 und 380). Die Drückergriffe in Wohnräumen, meist aus Messing oder Bronze (Fig. 381), Horn oder Bein bestehend, sind auf Eisenstifte montiert und werden paarweise zusammengepasst. Beim Anbringen an die Thüre wird der eine Drücker mit dem Stift durch die Nuss hindurchgesteckt, worauf dann an dessen Ende der zweite Drücker darüber gestülpt und durch ein Stiftchen befestigt wird, welches in eine Durchbohrung von Stift und Drücker eingreift. Dem des sorgfältigen Einpassens wegen etwas mühsamen Befestigen der Drücker suchte man in neuerer Zeit vielfach durch besonders konstruierte, für jede Thürdicke passende Drücker abzuhelfen. In Figur 382 ist ein solcher dargestellt, wozu der Erfinder folgende Erläuterung giebt:

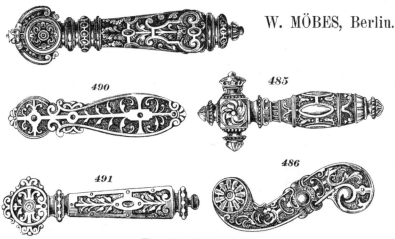

Fig. 381. Thürdrücker.

Die bisherige Befestigungsweise der Thürdrücker war insofern eine ungenügende, als der Zusammenhalt durch einen vierkantigen Stift sich alsbald lockerte, die Drückerführung sich in den Rosetten nach kurzer Zeit auslieerte, und demzufolge die Drücker in den Schildern wackelten. Ferner geschah die Verstiftung der Drücker vor der Rosette, der Stift wurde mit Leichtigkeit entfernt und die Drücker entwendet, auch mussten die Drücker stets nach der Thürstärke eingepasst werden — All diese Uebelsände fallen bei der neuen Befestigung fort, indem dieselbe durch ein Schraubengewinde geschieht, welches den Drücker mit der Nuss verbindet, sich über dieselbe in verjüngtem Massstabe fortsetzt und jenseits der Nuss das entgegengesetzte Gewinde führt, auf welches der Hohldrücker aufgeschraubt wird. Hierdurch bildet das Ganze eine Welle, und werden die Drücker bei jeder Benutzung nur noch fester angezogen.

Die Rosetten, an den Schildern befestigt, sind auf den Drückerhälsen verstellbar und demzufolge ist das Zusammenschrauben der Drücker unabhängig von der Thürstärke, wodurch Zeit und Geld gespart wird.

Um das Entwenden der Drücker zu verhindern, findet die Verstiftung hinter der Rosette statt, was besonders bei Haushürdrückern von grossem Vorteil ist.

Die Anwendung dieser Befestigung wird überall da empfohlen, wo man eine präzise Funktion der Drücker und Wegfall von Reparaturen liebt.

Ueberpreis für Stubenthür-Garnituren M. 0.70.

Jeder bessere Drücker hat ferner eine meist aus dem gleichen Material gefertigte Rosette, in welcher er sich dreht. Dieselbe sitzt einerseits auf dem Schlosskasten, andererseits auf einem Messing- oder Eisenblechschild, welches eingelassen oder aufgelegt sein kann und auf welchem sich das Schlüsselloch befindet. Die Falle greift in den Schliesskloben (Fig. 383) ein, welcher entweder einfach in die Thürverkleidung eingeschlagen oder besser auf eine Platte aufgesetzt und aufgeschraubt wird. Der untere Teil des Schliesskloben ist für den Schliessriegel bestimmt und kann beim einfachen Fallenschloss fortbleiben.

7) **Das überbaute zweitourige Kastenschloss mit hebender Falle und Nachtriegel**. Bei dem auf Tafel 81 abgebildeten Zweitourschloss ist bei geschlossener Thüre der Schliesskloben durch den vor dem Schlossstulp befindlichen Teil des Kastens, den sogen. Ueber-

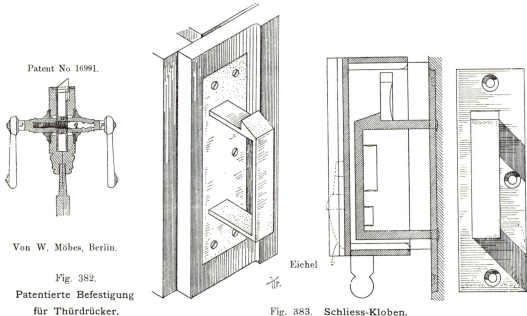

Patent No 16991.

Von W. Möbes, Berlin.

Fig. 382.
Patentierte Befestigung für Thürdrücker.

Eichel

Fig. 383. Schliess-Kloben.

bau, verdeckt, weshalb es überbautes Zweitourschloss heisst. Dasselbe hat gegenüber dem Kastenschloss ohne Ueberbau, dem sogen. Mauskastenschloss den Vorteil, dass man bei ihm die Falle ohne Drücker nicht heben und den Schliessriegel nicht zurückschieben kann. In Wohnzimmern versieht man die Thürschlösser zur Erhöhung der persönlichen Sicherheit noch mit einem nur auf der Zimmerseite zu bewegenden Nachtriegel h.

Die auf der Aussenseite des Schlossblechs angebrachte Façonleiste hat nur dekorativen Zweck; dagegen soll das über dem Schlüsselloch befestigte, aber doch drehbare Vorhängerle oder die Eichel das Eindringen von Staub in das Innere des Schlosses verhindern (Fig. 383).

8) **Das gewöhnliche Einsteckschloss**, a) mit hebender Falle, b) mit schiessender Falle. Die bisher besprochenen Schlösser werden auf der Zimmerseite auf die Holzfläche der Thüren aufgesetzt, stehen somit vor und bilden dadurch gerade keine besondere Zierde für dieselben. Findet man sich aber mit dieser unsymmetrischen Anbringung bei einflügeligen Thüren zur Not noch ab, — da man sie eben von jeher so zu sehen gewohnt ist — so kann doch nicht geleugnet werden, dass sie bei Flügelthüren, in besseren Zimmern verwendet, sehr störend wirkt.

3. Beschläge zum Festhalten einzelner Konstruktionsteile in bestimmten Lagen und zum Verschliessen. 231

Diesem Missstand lässt sich allenfalls abhelfen durch Anbringung eines zweiten blinden Schlosskastens mit Drücker auf dem zweiten Thürflügel, wodurch die Symmetrie hergestellt ist. Viel einfacher liegt aber die Sache, wenn man das Schloss so baut und anbringt, dass es unsichtbar ist, wenn man es in das Friesholz einstemmt und einsteckt, wenn man das Einsteckschloss verwendet. Ein weiterer Vorzug, der das letztere empfiehlt, ist der kurze Schlüssel, ein Vorzug, welcher es namentlich für stark konstruierte, also dicke Thüren, wie Hausthüren, als ganz besonders geeigenschaftet erscheinen lässt. Vergleicht man die Schlüssel eines überbauten und

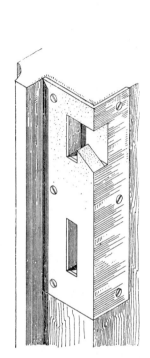

Fig. 384. Schliessblech für ein Einsteckschloss mit hebender Falle.

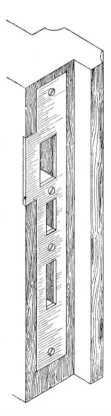

Fig. 385. Schliessblech für ein Einsteckschloss mit schiessender Falle.

eines eingesteckten Schlosses einer Hausthüre miteinander, so wird, wenn der eine oder der andere in der Tasche mitgetragen werden muss, die Wahl niemand besonders schwer fallen.

In Bezug auf die Falle unterscheiden wir:

    das Einsteckschloss mit hebender Falle, Tafel 82 links,

    das Einsteckschloss mit schiessender Falle, Tafel 82 rechts.

Die letztere Art ist die meist angewendete, während die erstere sich durch grosse Solidität auszeichnet.

Die Konstruktion ist bei beiden, mit Ausnahme der Falle, die gleiche und derjenigen des Kastenschlosses sehr ähnlich.

a) **Das Einsteckschloss mit hebender Falle** zeigt, wenn von dem fehlenden Umschweif des Kastens abgesehen wird (das Schloss hat nur einen Messingstulp), den Bau eines zweitourigen Kastenschlosses. Der Schliessriegel mit Zuhaltung und Federn, sowie die Einführung des Schlüssels und die Toureinfeilung sind dieselben; der Unterschied ist nur in der Falle zu suchen, welche hier mit der Nuss nicht aus einem, sondern aus zwei Stücken besteht. Das Schliessblech wird auf die Thürfrieskante des zweiten Flügels aufgeschraubt und durch die Schlagleiste gedeckt (Fig. 384).

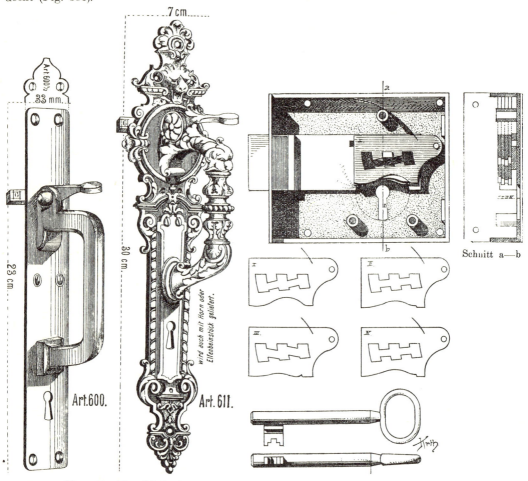

Fig. 386. Thürklinkgriffe von Gebr. Graeff in Elberfeld.

Fig. 387. Chubb-Schloss mit 4 Zuhaltungen.

b) **Das Einsteckschloss mit schiessender Falle** unterscheidet sich von dem soeben besprochenen zunächst dadurch, dass die Falle nicht gehoben, sondern nach vorn bewegt wird. Durch die Feder g stets nach vornen gedrückt und im Stulp geführt, gleitet sie beim Schliessen der Thüre mit ihrem schrägen Kopf über die Kante des Schliessblechs, um dann sofort in dasselbe einzuschiessen. Die Wirkung der Feder g wird unterstützt durch die Feder h, welche gegen die Nase der beweglichen Nuss drückt; die Führung der Falle ist hinten durch einen

3. Beschläge zum Festhalten einzelner Konstruktionsteile in bestimmten Lagen und zum Verschliessen. 233

oberhalb derselben angenieteten Stift und die unterhalb befindliche Nuss bewirkt. Was den Schliessriegel anbelangt, so ist derselbe genau so gefertigt, wie der des vorher besprochenen Schlosses. Unterhalb desselben ist auf der Abbildung noch ein Nachtriegel befestigt, dessen Führung durch den Stulp und einen Riegelstift gesichert ist, während die Bewegung durch die Nuss i erfolgt, in welche von der einen Seite aus ein Knöpfchen oder eine Olive zum Drehen eingesteckt ist. Die Schleppfedern für die Schliessriegel befinden sich an der Schlossdecke und sind auf der Zeichnung nicht ersichtlich.

Fig. 388. Standard-Schiebthürschloss.   Fig. 389. Chubbschlüssel aus Stahlblech.

Das Einsteckschloss wird auf der Kante des einen Thürfrieses eingestemmt und so weit versenkt, dass der Schlossstulp bündig sitzt. Die Höhe, auf welche derselbe an der Thüre zu sitzen kommt, hängt von der Art der Benützung ab. Wenn viele Kinder die Thüre benützen, so wird man den Drücker und somit das Schloss tiefer setzen, als wenn dies nicht der Fall ist. Im allgemeinen nimmt man 1,10 m Drückerhöhe an, wobei man aber Rücksicht auf die Querfriese nimmt, damit kein Zapfen derselben abgestemmt wird. Die Höhenangabe gilt selbstredend auch für die anderen Schlösser. Dem Stulp des Schlosses gegenüber befindet sich, auf der Kante des zweiten Thürflügels bündig eingelassen, das Schliessblech, welches hier die Form annimmt, wie sie Fig. 385 zeigt.

Bei diesem Anlasse machen wir auf die Einsteckschlösser der Gebr. Graeff in Elberfeld aufmerksam. Ihre Konstruktion unterscheidet sich von der gewöhnlichen Art dadurch, dass an Stelle der Thürdrücker sog. Klinkgriffe treten, die verschiedene Vorteile gegenüber jenen haben. Die Figur 386 bildet zwei Klinkgriffe ab, einen einfachen und einen reich ausgestatteten. Die Preise der Garnituren, bestehend aus zwei Griffen mit Schildern und Einsteckschloss, wechseln nach der Art der Ausstattung; Art. 600 kostet beispielsweise in Messing Mk. 6.90, während Art. 611 in vergoldeter Bronze mit Mk. 19.50 berechnet ist.

9) **Das Einsteckschloss für Schiebthüren** (Taf. 82 rechts unten). Es unterscheidet sich von den gewöhnlichen Einsteckschlössern dadurch, dass es keine Falle hat und dass der Schliessriegel nicht horizontal vorgeschoben wird, sondern sich in einem Bogen durch den Stulp des Schlosses herausbewegt und in denjenigen des anderen Thürflügels einhakt, so dass der

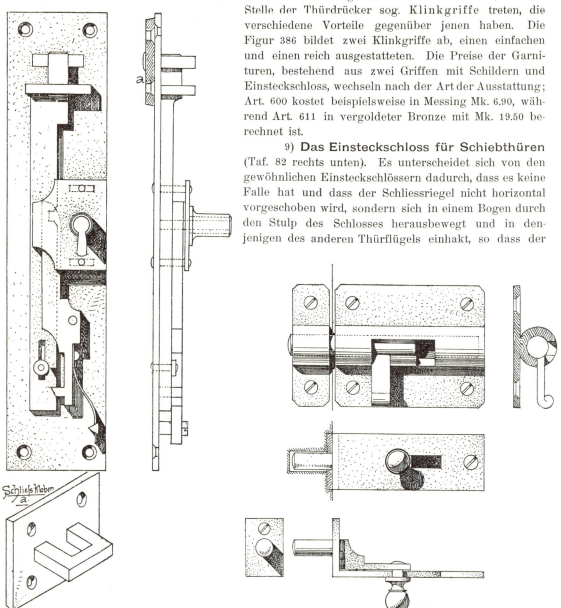

Fig. 390. Stangenriegelschloss.   Fig. 391. Thürriegel. Selbständige Nachtriegel.

Verschluss gesichert ist. Die Konstruktion ist aus der Zeichnung genau ersichtlich. Das Schloss hat Mittelbruch mit Reifchenbesatzung, und der Schlüssel desselben ist zum Umklappen eingerichtet, damit er nicht weit vorsteht, also beim vollständigen Einschieben der Thüre in den

3. Beschläge zum Festhalten einzelner Konstruktionsteile in bestimmten Lagen und zum Verschliessen. 235

Schlitz nicht beschädigt werden kann. In neuerer Zeit zieht man Beschläge, welche die gänzliche Beseitigung der Thüre ermöglichen, den älteren vor, bei denen die Thüre immer noch einige cm, d. h. so weit vor das Futter vorstand, dass man die Handgriffe der Thüre noch erfassen konnte. In dem vorliegenden Falle ist eine Vorkehrung zum vollständigen Einschieben getroffen. Drückt man nämlich leicht mit dem Finger von unten auf den kleinen Riegel a am Stulp des Schlosses, so dreht sich der Hebel m hinten etwas nach abwärts und löst den als schiessende Falle konstruierten Griff n aus, welcher nun durch den von einer Feder gedrückten Hebelarm k nach vorn geschoben wird. Beim Zurückdrücken des Griffes hakt sich der von einer weiteren Feder stets nach oben gedrückte Hebel m wieder ein. Der Stulp besteht des

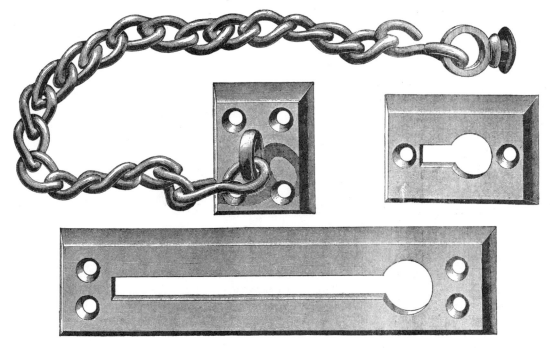

Fig. 392. Thürsperrkette.

besseren Aussehens wegen aus zwei Stücken, aus einem hinteren, starken Eisenstulp und einem feinen Zierstulp aus Messing. Das auf Tafel 82 links abgebildete Gegenschloss des zweiten Thürflügels hat nur die soeben beschriebene Herauszieh-Vorrichtung und dient im übrigen als Schliessblech. (Wird von Gebr. Wenner in Schwelm, Westfalen, geliefert.)

10) **Das Chubb-Schloss.** Ein Schloss wird seinen Zweck um so besser erfüllen, je weniger leicht es unbefugter Weise durch Nachschlüssel etc. geöffnet werden kann. Die Reifbesatzung, die Eingerichte, die Hohlschlüssel, die geschweiften Bärte u. a. m. erreichen diesen Zweck jedoch nur unvollkommen; erfahrungsgemäss öffnen geschickte Leute derartig ausgestattete Schlösser mit einfachen Haken verhältnismässig rasch und leicht. Sobald es gelingt, die Zuhaltung zu heben und den Riegel zu schieben, ist die Sache gemacht. Das Prinzip des Chubb-Schlosses besteht nun darin, an Stelle der einen Zuhaltung deren mehrere zu setzen und je mehr Zuhaltungen angebracht werden, je schwieriger wird das Oeffnen des Schlosses ohne passenden Schlüssel sein. Figur 387 zeigt ein Chubbschloss mit 4 verschiedenen Zuhaltungen,

die als sogen. Fenster aus Messingblech ausgeschnitten sind. Der Bart des Schlüssels ist abgetreppt und jeder Stufe entspricht eine anders geformte Schweifung der zugehörigen Zuhaltung. Die Ausschnitte müssen so beschaffen sein, dass sie gleichzeitig den Riegelstift fassen oder freigeben, wenn die Schlüsseldrehung die einzelnen Zuhaltungsplatten senkt oder hebt. Da das letztere nicht gleichmässig erfolgt, so hat jede Zuhaltung eine getrennte Feder aus Stahlblech.

Das im Prinzipe gekennzeichnete Schloss ist dann mannigfach verändert und den verschiedensten Zwecken angepasst worden. Alle Schlösser mit einfacher Zuhaltung sind auch als Chubbschlösser zu bauen. Um die Zuhaltungsplatten genügend stark machen zu können (was beim gewöhnlichen Schlüssel einer Bartvergrösserung gleichkommen würde), hat man den Schlüssel auch derart umgemodelt, dass er, aus Stahlblech hergestellt, das Aussehen der Figur 389 annimmt. Dabei wird das Schlüsselloch zum schmalen Schlitz, der an sich schon das Einbringen der Nachschlüssel erschwert. Eine solche Schlüsselöffnung zeigt das Standard-Schiebthürschloss der Fig. 388, welches als Chubb-Schloss mit 3 Zuhaltungen gebaut ist.

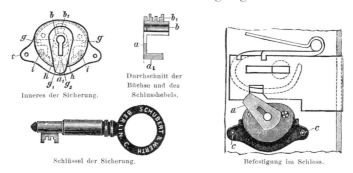

Fig. 393. Thürschlosssicherung von Schubert & Werth in Berlin.

11) **Das Stangenschloss.** Es ist ein Riegelschloss mit stangenartig verlängertem Schliessriegel und ist überall da am Platze, wo das Schloss, bezw. das Schlüsselloch nicht in der Nähe des Schliesshakens angebracht werden kann. Wird der Riegel dabei sehr lang, so muss er durch Ueberkloben geführt werden. Ein derartiges Schloss mit nach oben schliessendem Riegel stellt Fig. 390 dar.

12) **Das Baskülen- oder Doppelstangenschloss** (Tafel 82). Es schliesst nach oben und unten und wird sowohl für leichte als schwere Thüren gefertigt. Es besteht aus einem Baskülenverschluss, bei dem an Stelle der Scheibe oder des Zahnrädchens eine Schliessriegelkonstruktion mit Toureneinfeilung und Zuhaltung tritt. Die Enden der Riegel greifen in Schliesskloben oder Schliessbleche ein. Lange Stangen müssen wiederum durch Ueberkloben am Ausweichen verhindert werden.

Damit ist die Aufzählung neuzeitiger Schlösser nicht erschöpft. Es existieren noch verschiedene Sicherheits- und Kombinationsschlösser von sehr sinnreicher Erfindung, wie das Yaleschloss und das Bramahschloss. Ihre kleinen Schlüssel bewegen die Riegel nicht unmittelbar, sondern geben nur durch Auslösen der Zuhaltungen die im Innern des Schlosses liegenden Teile frei, welche die Schliessriegel bewegen. Diese Schlösser werden hauptsächlich für Kassenschränke, seltener für gewöhnliche Thüren gebaut, weshalb die kurze Erwähnung genügen mag.

m) Thürsicherungen anderer Art.

Die gewöhnlichste Thürsicherung, welche angebracht wird, um die Thür auch dann gesperrt zu halten, wenn das Schloss von aussen geöffnet werden sollte, ist der auf der Innenseite angebrachte Riegel. In seiner Form als Kurz- und Langriegel ist er bereits beschrieben und der mit besseren Schlössern in Verbindung stehenden Nachtriegel ist ebenfalls schon gedacht. Es erübrigt demnach nur noch, den selbständigen Nachtriegel zu erwähnen, der übrigens auch bei Tage und bei offenem Schloss als Schutz gegen unbefugtes Eindringen benützt wird. Die Figur 391 stellt zwei Arten derartiger Riegel dar, von denen besonders die obenan gezeichnete sich empfiehlt, weil sie beliebig stark zur Ausführung kommen kann. Der Riegel ist ein Rundeisen. Die Riegelhülse und der Schliesskloben sind rohrartig. Der Riegel ist nicht nur der Länge nach verschiebbar, sondern ausserdem um seine Axe drehbar. Er hat einen seitlichen Flansch oder Knopf, in passende Ausfeilungen des Rohrschlitzes einzuhaken, am einen Ende bei geöffnetem, am andern bei geschlossenem Zustande.

Eine weitere Thürsicherung ist die Sperrkette. Sie ist erfunden worden auf Grund des Wunsches, die Thüre behufs Hinaussehens etwas öffnen zu können, ohne einen etwa vorkommenden gewaltthätigen Eintritt zu gestatten. Die bekannte, einfache aber gute Einrichtung ist durch Fig. 392 veranschaulicht. Am Futterrahmen ist die Kette mit Kloben auf Platte befestigt, während auf der Innenseite der Thüre die Einhakplatte für den Kettenknopf aufgeschraubt ist. Sind beide Teile in richtiger Art und Entfernung befestigt, so kann man die Thüre auf etwa 5 cm Weite öffnen, ohne dass es von aussen gelingt, die Kette auszuhängen.

Die Thürsicherung von Schubert & Werth in Berlin ist ebenfalls zu empfehlen (Fig. 393). Sie kann mit 2 Schrauben in jedem einfachen Thürschloss befestigt werden und gestaltet es damit zu einem Chubbschloss mit vielen Zuhaltungen. Der Schlusshebel des Apparates vertritt die Stelle des ursprünglichen Schlüsselbartes und wird durch einen zugehörigen kleinen Chubbschlüssel bewegt. Bei Wohnungswechsel kann die Einrichtung wieder abgeschraubt werden und das Schloss verbleibt in seiner ehemaligen Art. Der in den Eisenhandlungen zu habende Apparat kostet samt Schlüssel 5 ℳ. Nachstehend folgt die dem Prospekt entnommene Beschreibung:

Der Schlusshebel a, welcher mit der Büchse b fest verbunden ist, hat Führung in den beiden Platten, welche durch Schrauben und Wände fest verbunden sind. Die Hebel g g, welche auf die Stifte h h lose gesteckt sind, und deren Zapfen bei g^1 und g^2 ineinander greifen, werden von den Federn i i gegen die Büchse b gedrückt. Letztere hat bei b^1 verschiedene Erhöhungen, deren Stärke mit der der Hebel g g gleich ist. Um das Schliessen zu bewirken, wird der Schlüssel in die Oeffnung der Büchse b gesteckt; durch Drehung des Schlüssels werden die Hebel g g so weit seitwärts gedrückt, dass die Büchse b mit ihren Erhöhungen bei b^1 an den Hebeln g g vorbei kann. Werden die Hebel g g durch einen nicht passenden Schlüssel nicht genug seitwärts gedrückt, so lassen dieselben die Büchse b bei b^1 nicht vorbei; ebenfalls lassen, wenn die Hebel g g durch einen nicht passenden Schlüssel zu weit seitwärts gedrückt werden, dieselben den Schlusshebel a bei a^1 nicht vorbei. Bei Oeffnungsversuchen mit einem falschen Schlüssel wird mindestens einer der vielen Hebel nicht genau gehoben und somit das Oeffnen unmöglich.

DAS SCHREINERBUCH

VON

THEODOR KRAUTH UND FRANZ SALES MEYER

I.

DIE BAUSCHREINEREI

II. BAND: TAFELN.

DIE GESAMTE
BAUSCHREINEREI

EINSCHLIESSLICH

DER HOLZTREPPEN, DER GLASERARBEITEN

UND DER BESCHLÄGE

HERAUSGEGEBEN

VON

THEODOR KRAUTH

ARCHITEKT, GROSSH. PROFESSOR UND REGIERUNGSRAT IN KARLSRUHE.

VIERTE, DURCHGESEHENE UND VERMEHRTE AUFLAGE

MIT 82 VOLLTAFELN UND 393 WEITEREN FIGUREN IM TEXT

ZWEITER BAND: TAFELN

LEIPZIG

VERLAG VON E. A. SEEMANN

1899.

VERZEICHNIS DER TAFELN.

Tafel 1. Fussböden.
„ 2. Wohnhausgrundriss mit eingezeichneten Schreinerarbeiten.
„ 3 bis 6. Lambris und Wandtäfelungen.
„ 7. Einfache und verdoppelte Thüren.
„ 8 bis 12. Gestemmte Zimmerthüren, einflügelig.
„ 13. Einflügelthüre, eingebaut in eine Bibliothek.
„ 14. Tapetenthüre.
„ 15 bis 17. Zweiflügelige Zimmerthüren.
„ 18. Thürverdachungen.
„ 19 und 20. Dreiteilige Flügelthüre.
„ 21. Flügelthüren nach E. Prignot.
„ 22. Schiebthüre.
„ 23 bis 25. Glasabschlüsse.
„ 26 und 27. Abschlussthüren für Vorplätze, Wartesäle, Gänge etc.
„ 28 und 29. Windfangthüren.
„ 30 bis 44. Ein- und zweiflügelige Hausthüren.
„ 45 bis 50. Haus- und Hofthore.
„ 51. Magazin- und Scheunenthore.
„ 52. Lattenthor.

Tafel 53. Abort-, Keller- und Gartenthüren.
„ 54. Treppen-, Kellerhals-, Dachaussteig-, Fall- und Rollenthüren.
„ 55. Fenster.
„ 56. Fenster mit Rollladen.
„ 57. Kastenfenster.
„ 58. Blumenfenster, Fenster-Glashäuschen.
„ 59. Schiebfenster.
„ 60 und 61. Schau-, Auslage- oder Ladenfenster.
„ 62. Läden.
„ 63. Rollladen.
„ 64. Zugjalousie.
„ 65 bis 69. Holzdecken.
„ 70 bis 73. Treppenkonstruktionen.
„ 74. Treppenpodestdecken.
„ 75 und 76. Treppengeländer.
„ 77. Abortsitze.
„ 78 und 79. Bänder, Kloben und Riegel.
„ 80. Baskülen-, Schwengel-, Ruder-, Espagnolett- und Vorreiberverschlüsse.
„ 81 und 82. Schlösser.

Tafel 1.

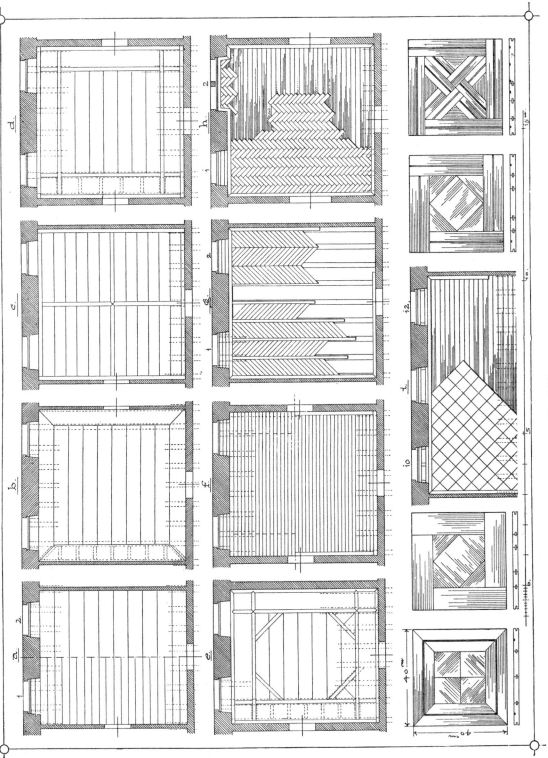

Fussböden.

Tafel 2.

Neubau des Herrn N. N. in O.
Grundriss für die Schreinerarbeiten des Erdgeschosses.

Die Maße sind am Bau zu nehmen!

R. 1.
Bodenrippen ca. 31 lfd. m.
Blindboden ca. 17 qm.
Eich. Fischgratboden 17 qm.
Brüstungslambris (Fig. „g", Taf. 3) ca. 13 lfd. m.
Fenstersimsbrett 1 Stück.
Vierfüllungsthüre 1 Stück.

R. 2.
Bodenrippen ca. 50 lfd. m.
Blindboden ca. 27 qm.
Eich. Tafelparketten 27 qm.
Täfelung (Fig. „c", Taf. 5) 14 lfd. m.
Flügelthür 1 Stück.
Vierfüllungsthüren 2 Stück.
Fenstersimsbretter 2 Stück.

R. 3.
Bodenrippen ca. 43 lfd. m.
Tann. Schiffboden ca. 22 qm.
Lambris mit Fuss- u. Deckleiste (Fig. „d", Taf. 3) 17 lfd. m.
Fenstersimsbrett 1 Stück.
Vierfüllungsthür 1 Stück.

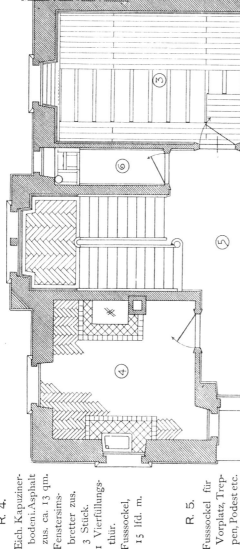

R. 4.
Eich. Kapuzinerbodeni. Asphalt zus. ca. 13 qm.
Fenstersimsbretter zus. 3 Stück.
1 Vierfüllungsthür.
Fussockel, 15 lfd. m.

R. 5.
Fussockel für Vorplatz, Treppen, Podest etc. zus. ca. 20 lfd. m.
Fenstersimsbrett f. Doppelf. 1 St.
1 Hauseingangsthür.

R. 6.
Vierfüllungsthüre 1 Stück
Fussockel ca. 3 lfd. m.
Fenstersimsbrett 1 Stück.
Abtrittsitz 1 Stück.

Grundriss mit eingezeichneten Schreinerarbeiten.

Tafel 3.

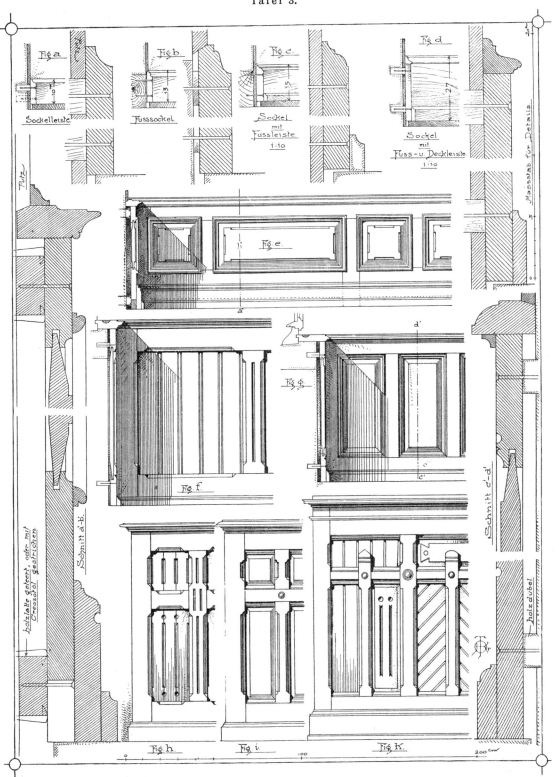

Lambris.

Tafel 4.

Lambris und Täfelungen.

Tafel 5.

Täfelungen.

Tafel 6.

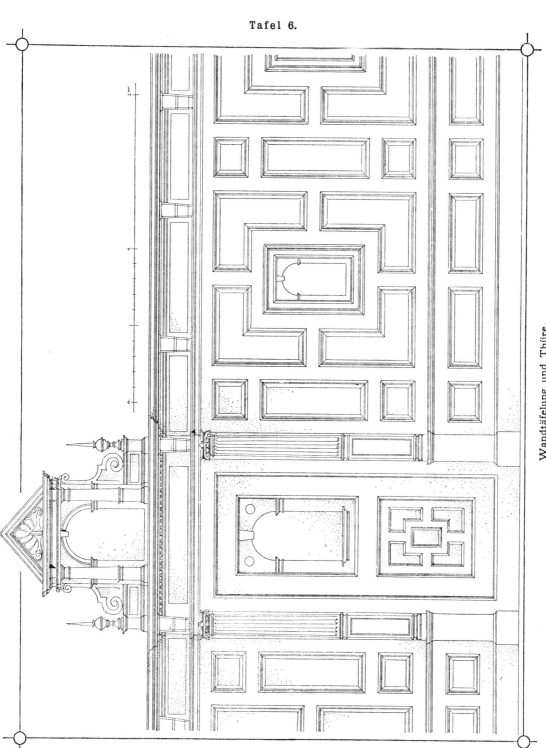

Wandtäfelung und Thüre.
Nach dem Entwurf von Kayser und von Grofzheim in Berlin.

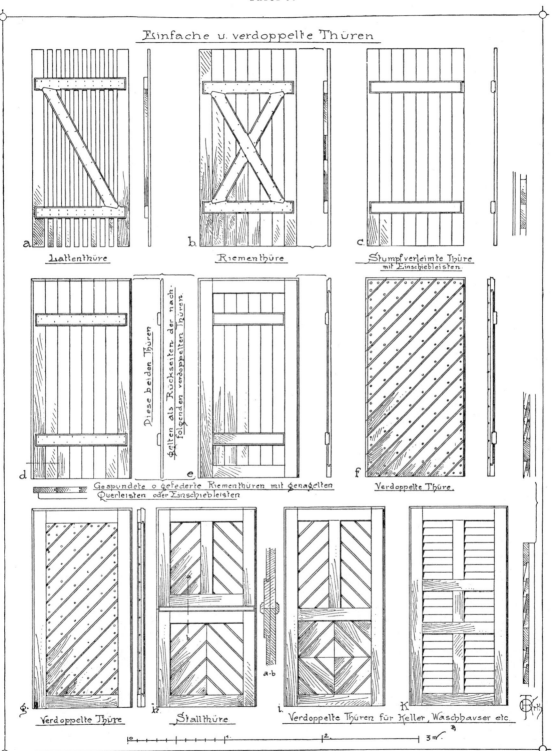

Einfache und verdoppelte Thüren.

Tafel 8.

Gestemmte Zimmerthüren.

Tafel 9.

Gestemmte einflügelige Zimmerthüre.

Tafel 10.

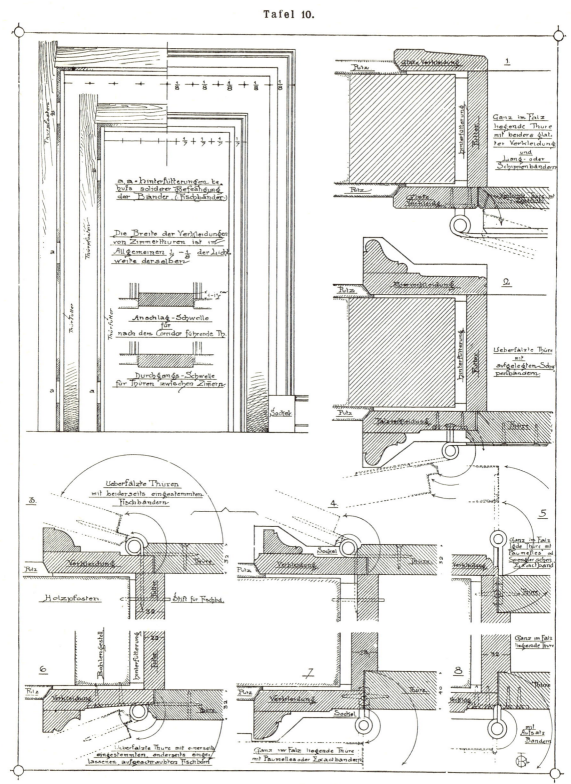

Thürverkleidungen etc.

Tafel 11.

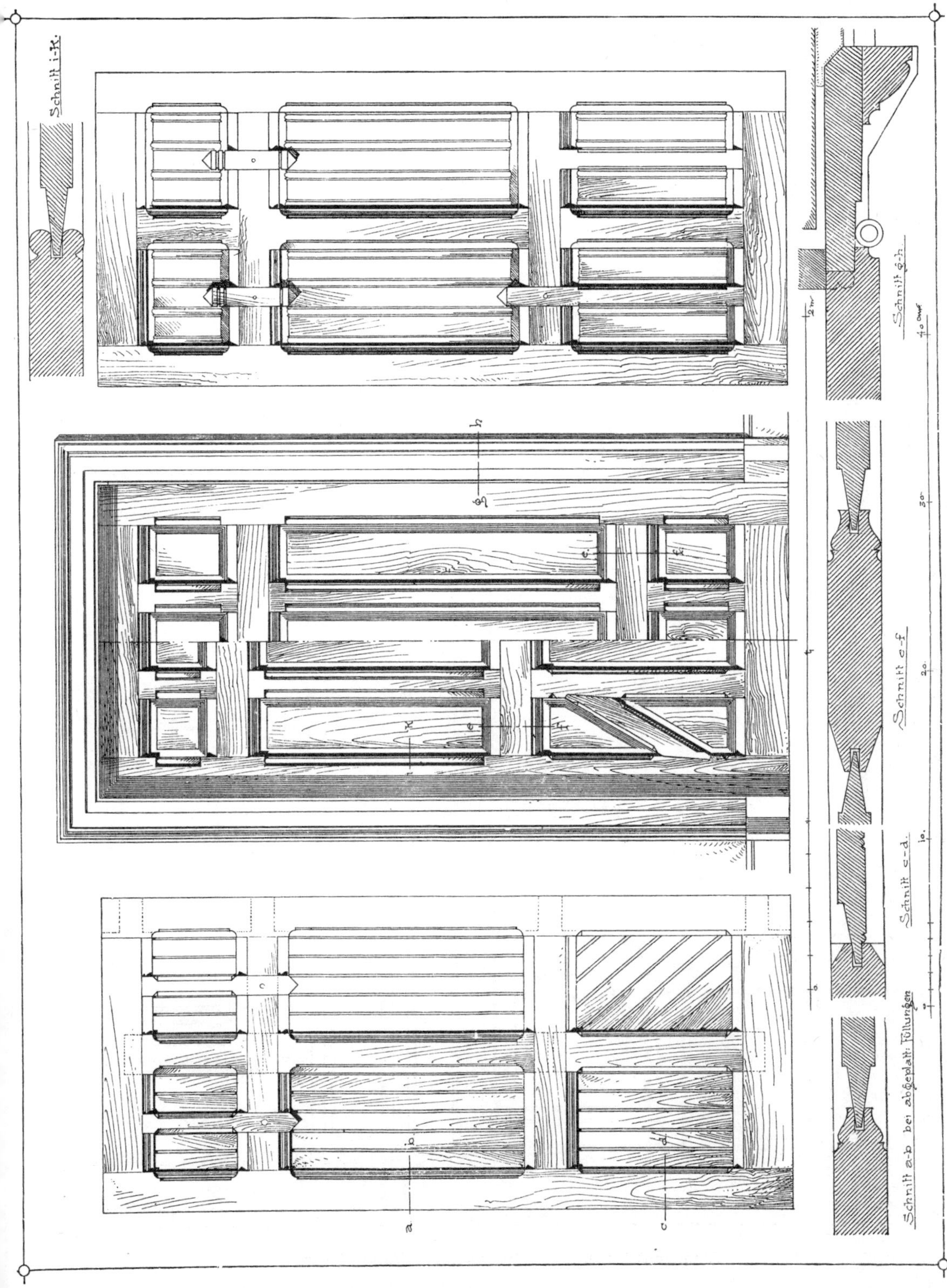

Einflügelige Zimmerthüren mit Einzelheiten.

Einflügelthüre mit anschliessender Wandtäfelung.
Nach den Entwürfen von Kayser und von Grofzheim in Berlin.

Tafel 13.

Einflügelthüre, eingebaut in eine Bibliothekeinrichtung.
Nach den Entwürfen von Kayser und von Grofsheim in Berlin.

Tafel 14.

Schnitt a-b.

Tapetenthüre.

Tapetenthüre.

Tafel 15.

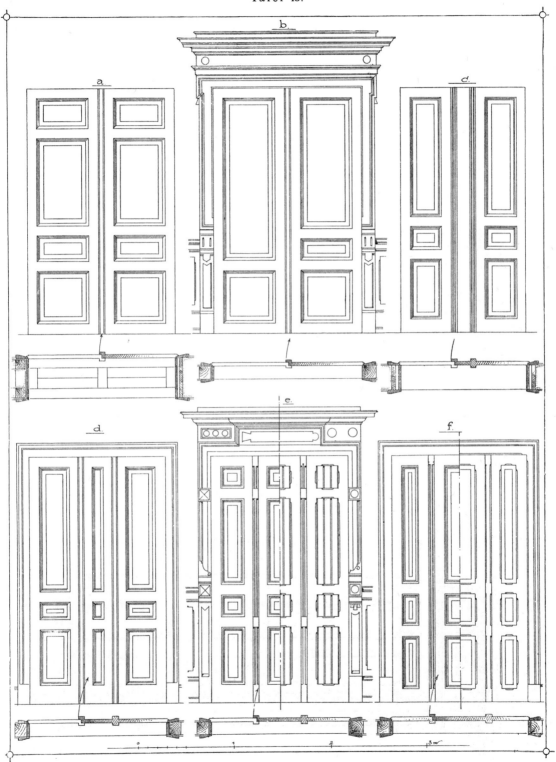

Zweiflügelige Zimmerthüren.

Tafel 16.

Zweiflügelige Zimmerthüren.

Tafel 17.

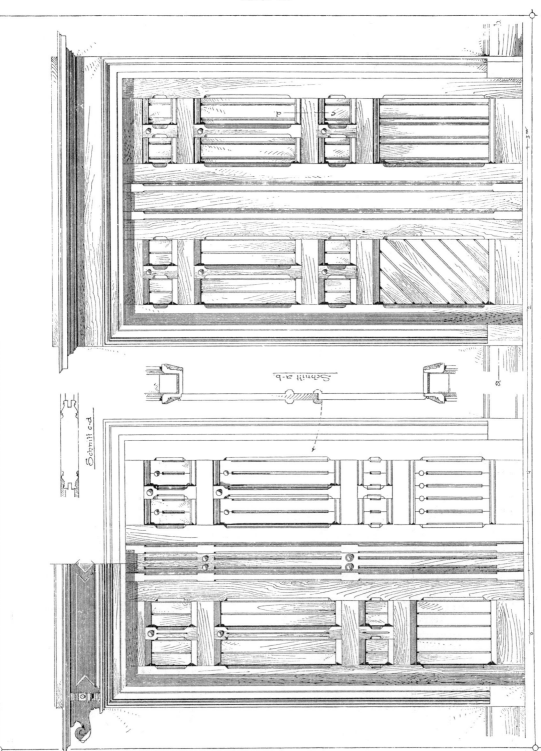

Zweiflügelige Zimmerthüren.

Tafel 18.

Thürverdachungen.

Tafel 19.

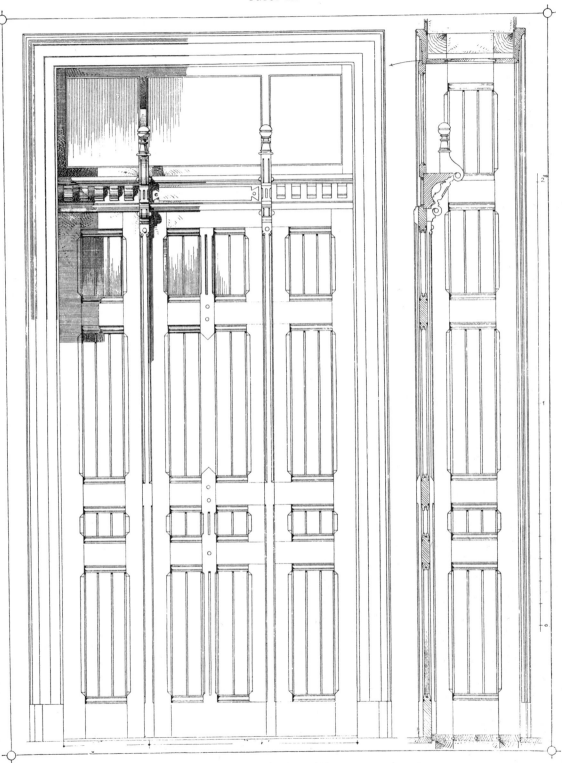

Dreiteilige Flügelthüre.

Tafel 20.

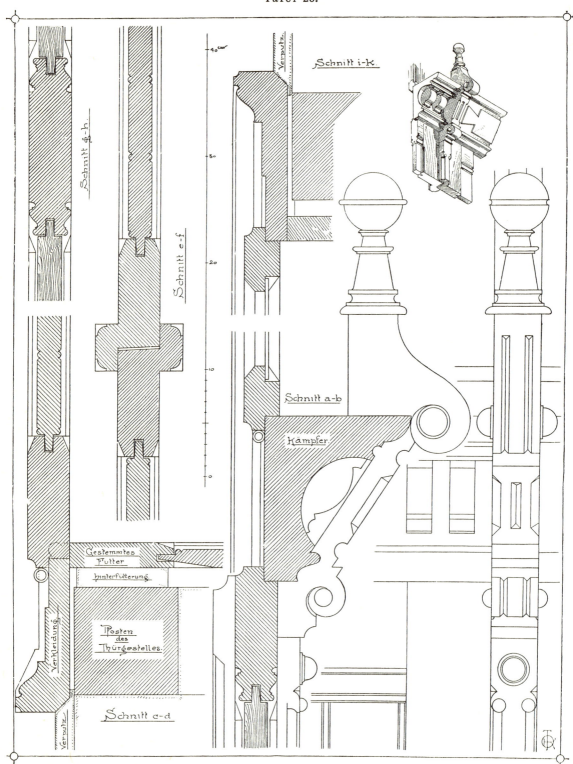

Einzelheiten zu Tafel 19.

Tafel 21.

Flügelthüren nach E. Prignot.

Tafel 22.

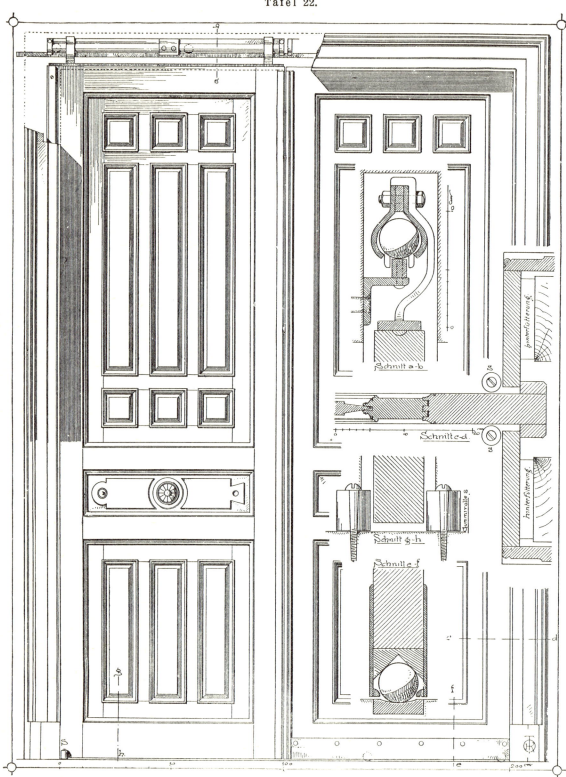

Schiebthüre.

Tafel 23.

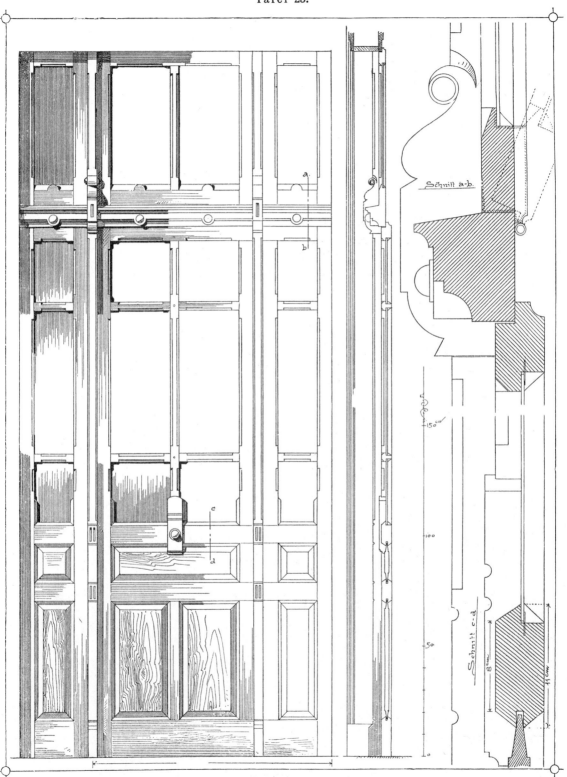

Glasabschluss.

Tafel 24.

Glasabschluss.

Tafel 25.

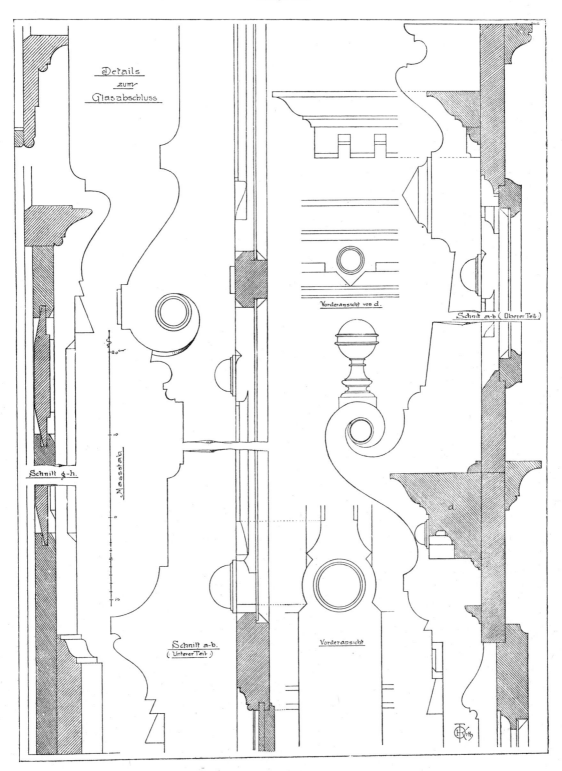

Einzelheiten zu Tafel 24.

Abschlussthüren für Vorplätze, Wartesäle etc.

Tafel 27.

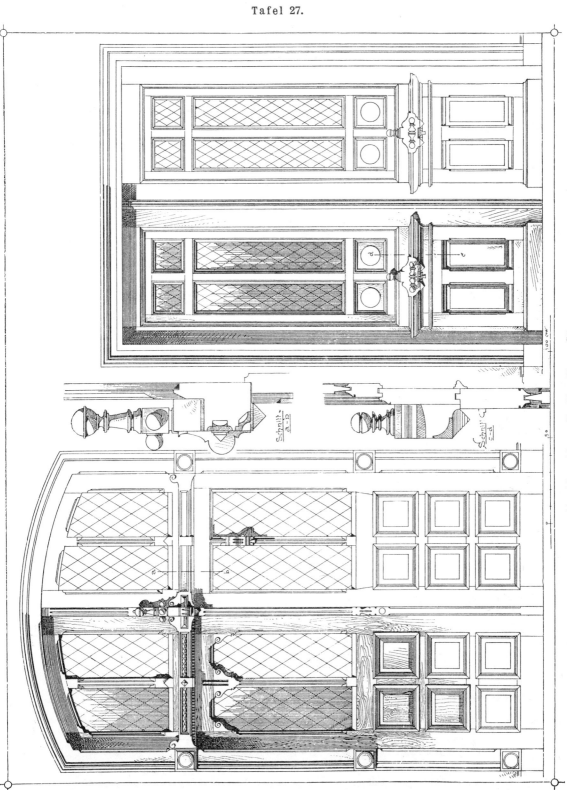

Abschlussthüren für Vorplätze, Gänge etc.

Tafel 28.

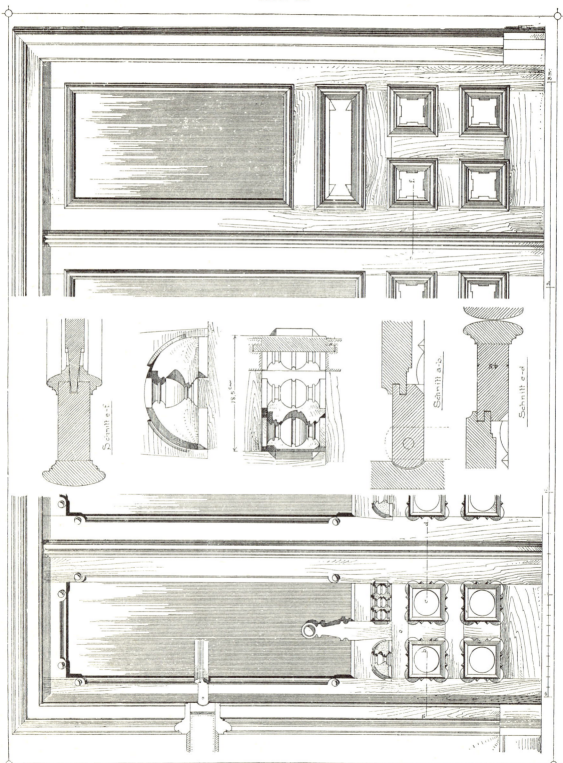

Windfang- oder Pendelthüren.

Zweiflügel-Windfang.

Tafel 30.

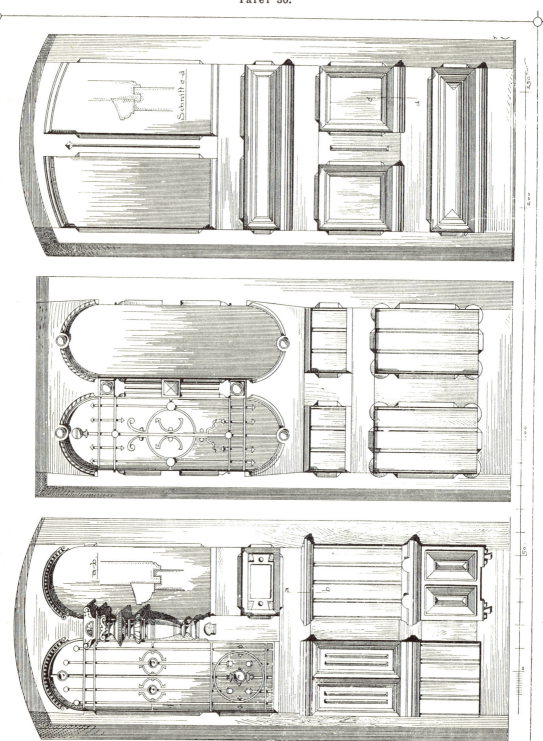

Einflügelige Hausthüren.

Tafel 31.

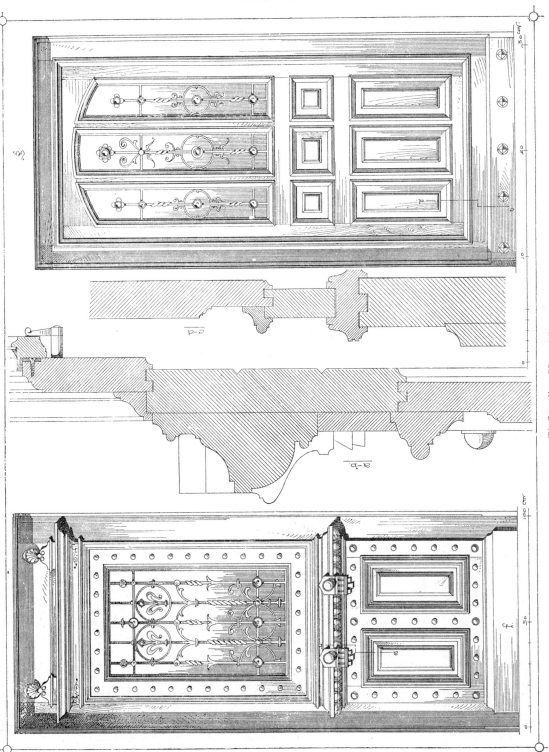

Einflügelige Hausthüren.

Tafel 32.

Einflügelige Hausthüre mit Oberlicht.

Tafel 33.

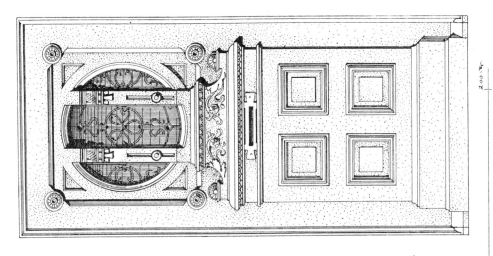

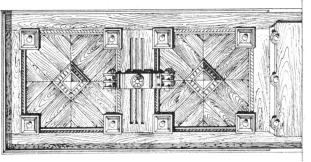

Einflügelige Hausthüren.

Tafel 34.

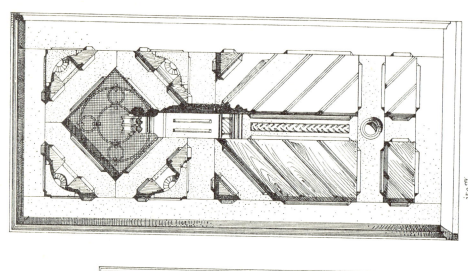
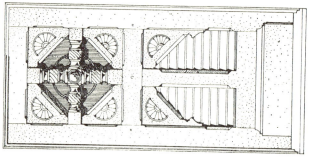
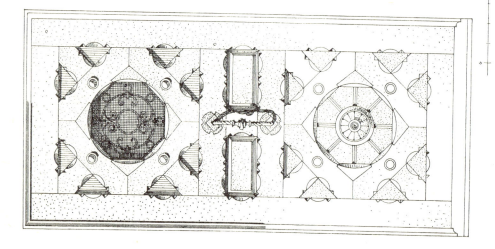

Einflügelige Hausthüren.

Tafel 35.

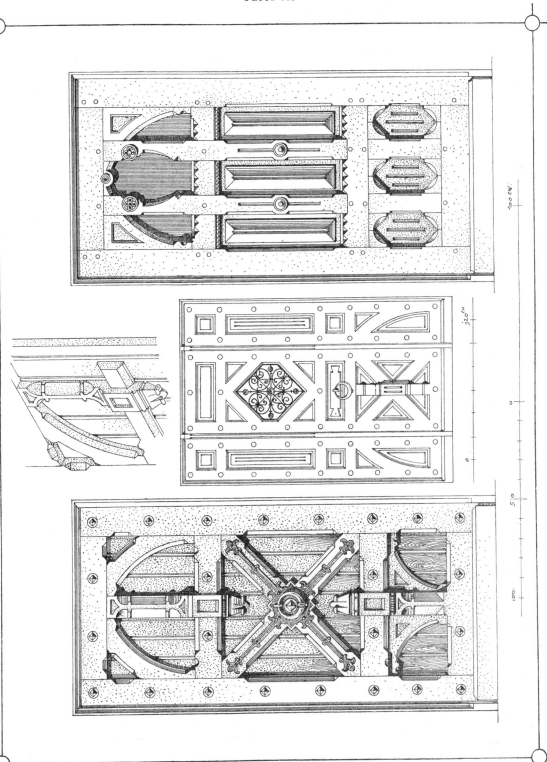

Hausthüren.

Tafel 36.

Einflügelige Hausthüre mit Kämpfer und Oberlicht.

Tafel 37.

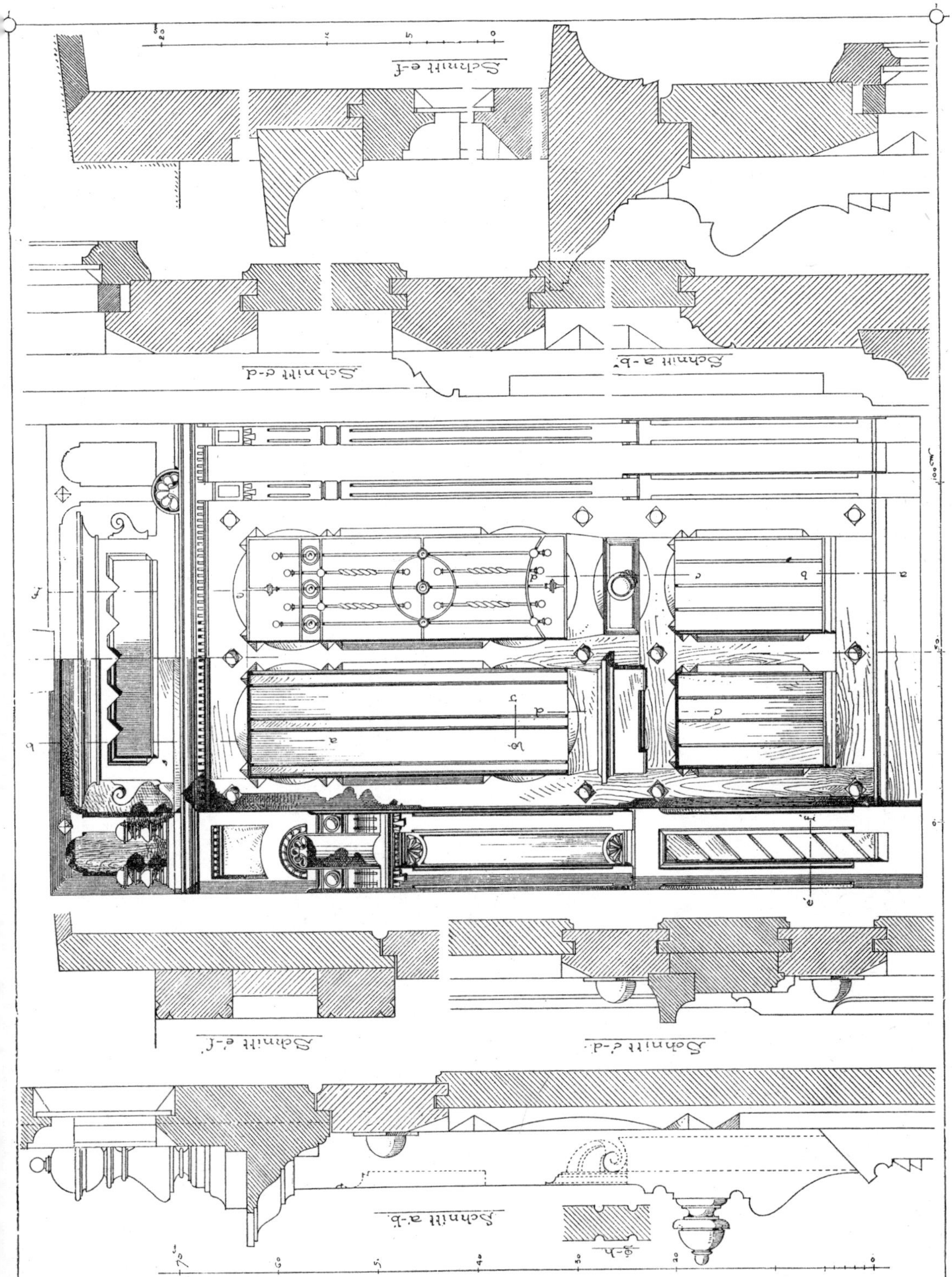

Einflügelige Hausthüre mit breitem Futterrahmen und Oberlicht.

Tafel 38.

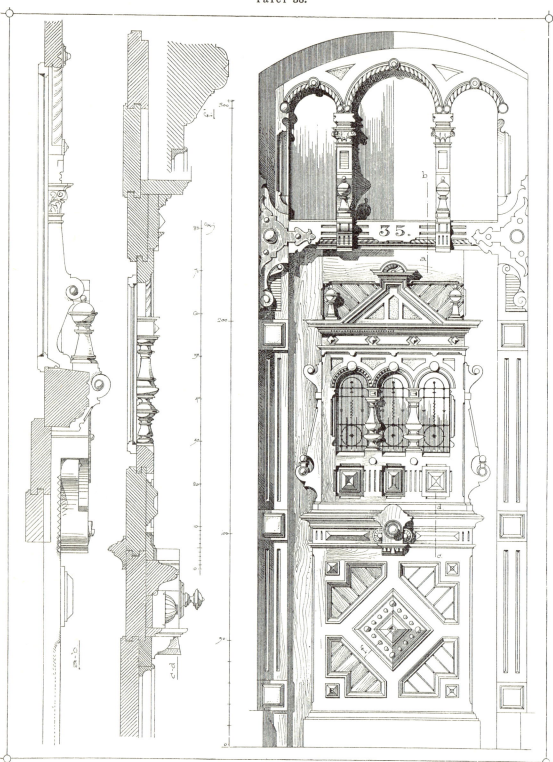

Einflügelige Hausthüre mit Kämpfer und Oberlicht.

Tafel 39.

Einflügelige Hausthüre mit breitem Futterrahmen, Kämpfer und Oberlicht.

Tafel 40.

Zweiflügelige Hausthüre mit Kämpfer und Oberlicht.

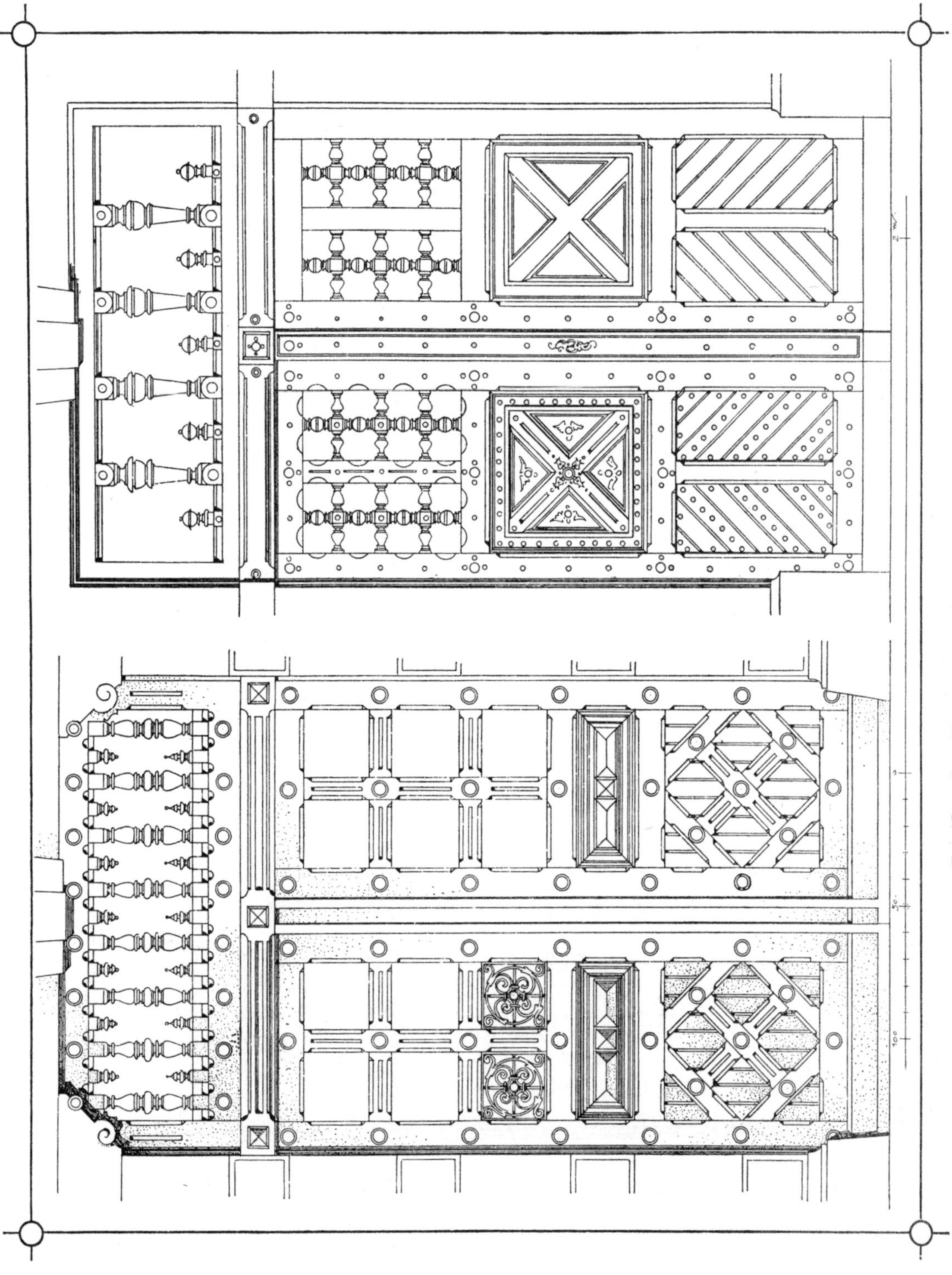

Tafel 41.

Zweiflügelige Haustüren mit Kämpfer und Oberlicht.
Entworfen von Professor Max Hummel in Karlsruhe.

Tafel 42.

Zweiflügelige Hausthüren ohne Oberlicht.

Tafel 43.

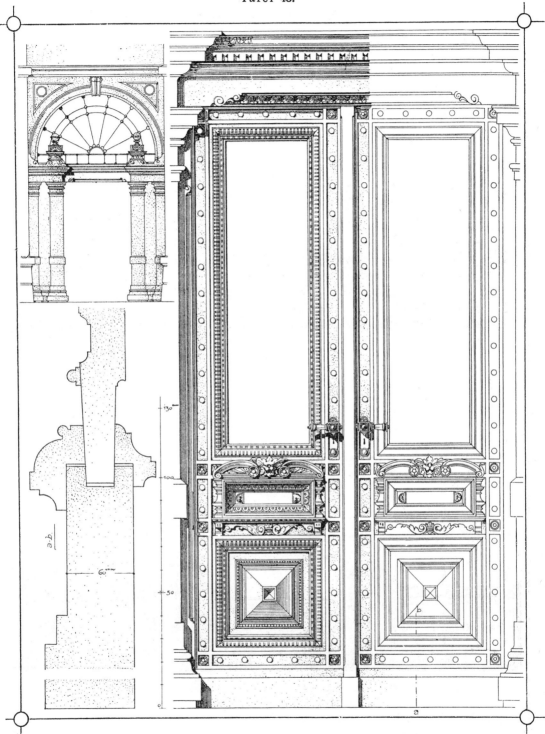

Reiche, zweiflügelige Hausthüre mit grossem Oberlicht.
Entworfen von Professor Levy in Karlsruhe.

Tafel 44.

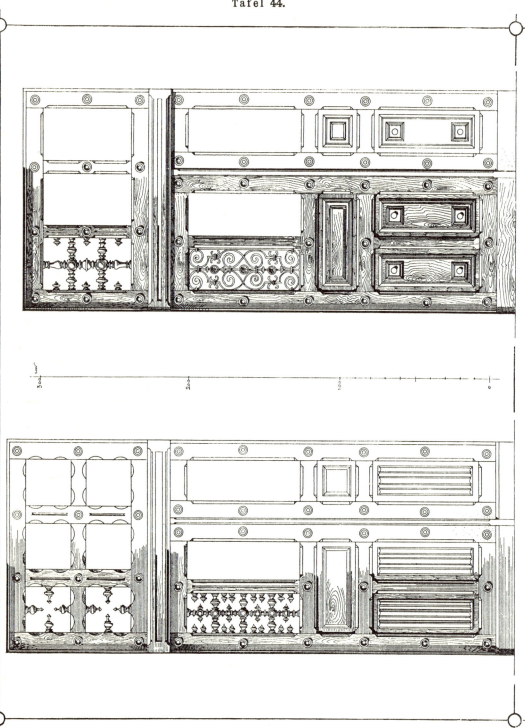

Zweiflügelige Hausthüren mit Oberlicht und ungleichen Flügeln.

Tafel 45.

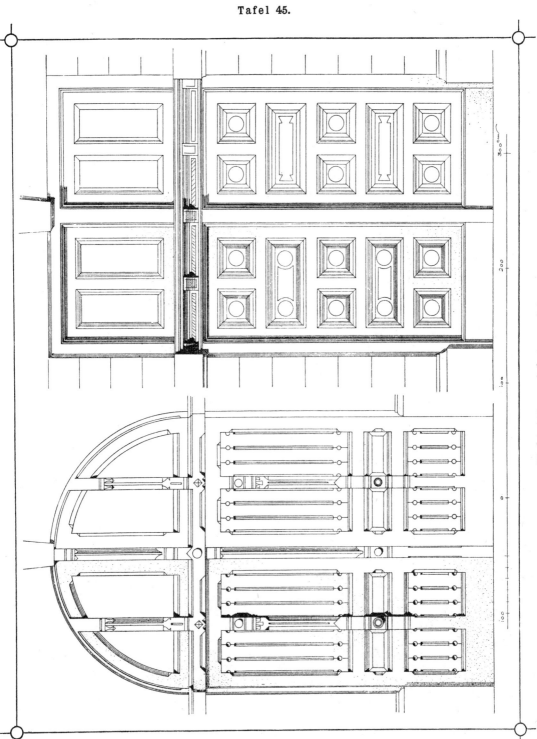

Einfahrtsthore einfacherer Art.

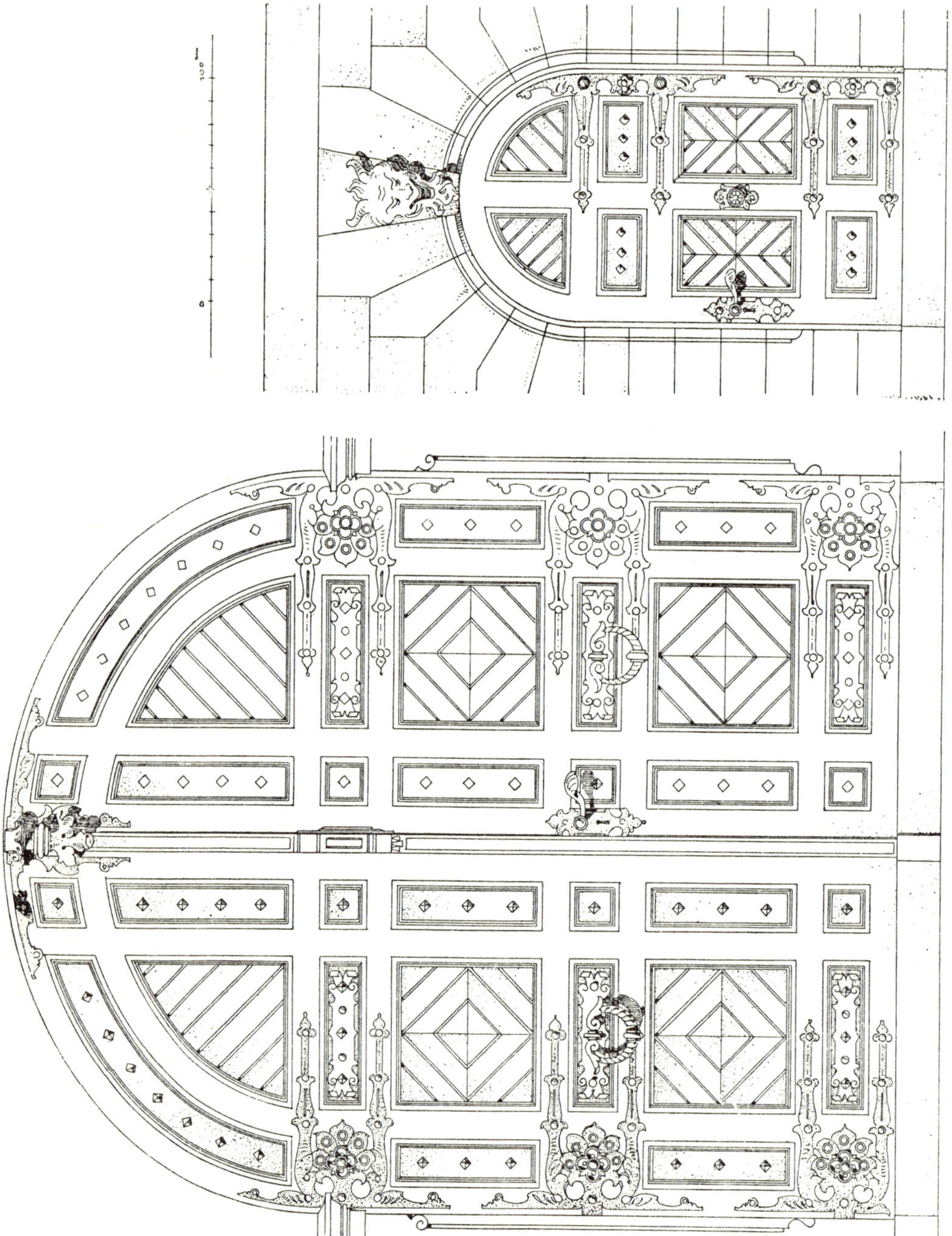

Tafel 46.

Einfahrtsthor und Hausthüre.
Entworfen von Professor Levy in Karlsruhe.

Tafel 47.

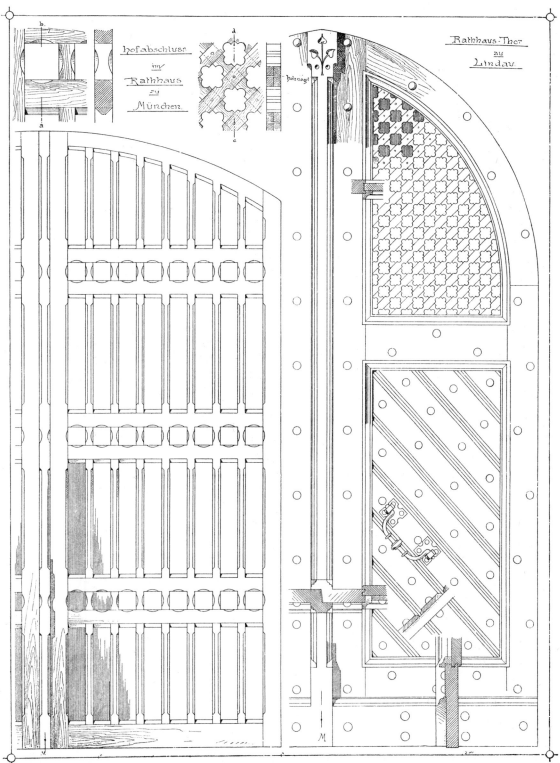

Hofthore.

Tafel 48.

Hausthüren oder Hausthore aus Pisa.

Tafel 49.

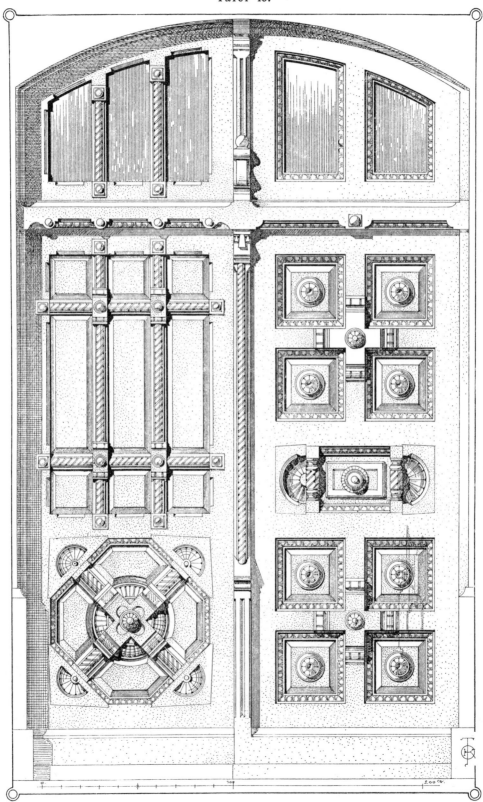

Hausthor.

Tafel 50.

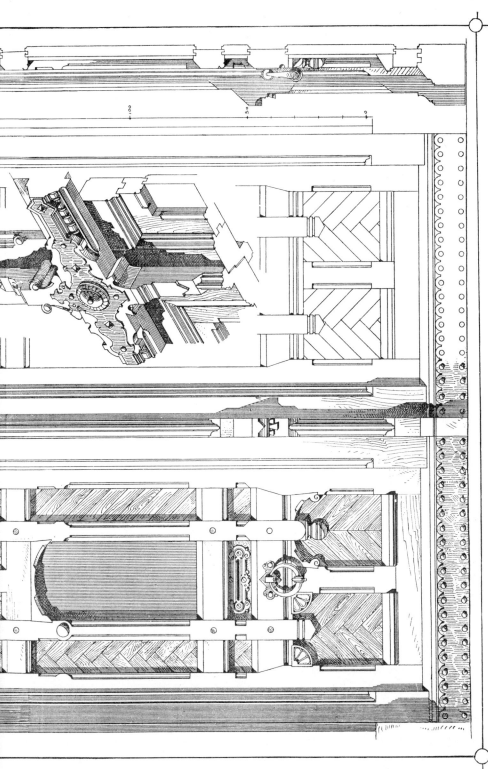

Einfahrtsthor.

Tafel 51.

Magazin- und Scheunenthore.

Tafel 52.

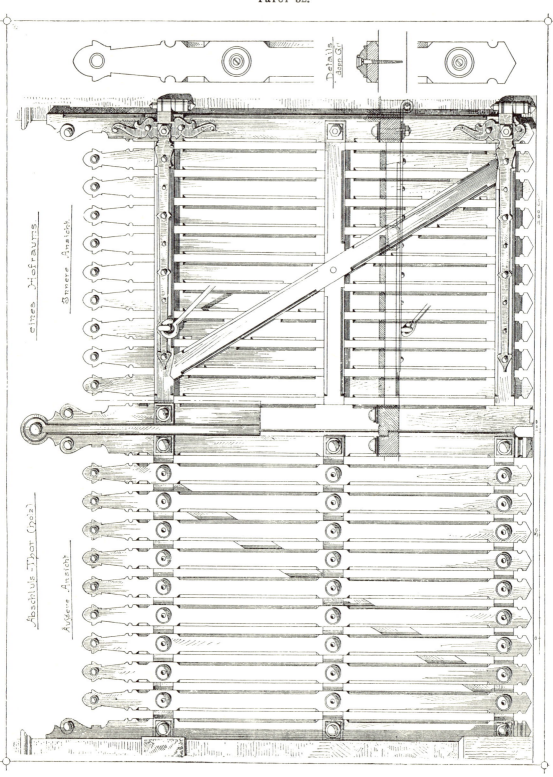

Tafel 53.

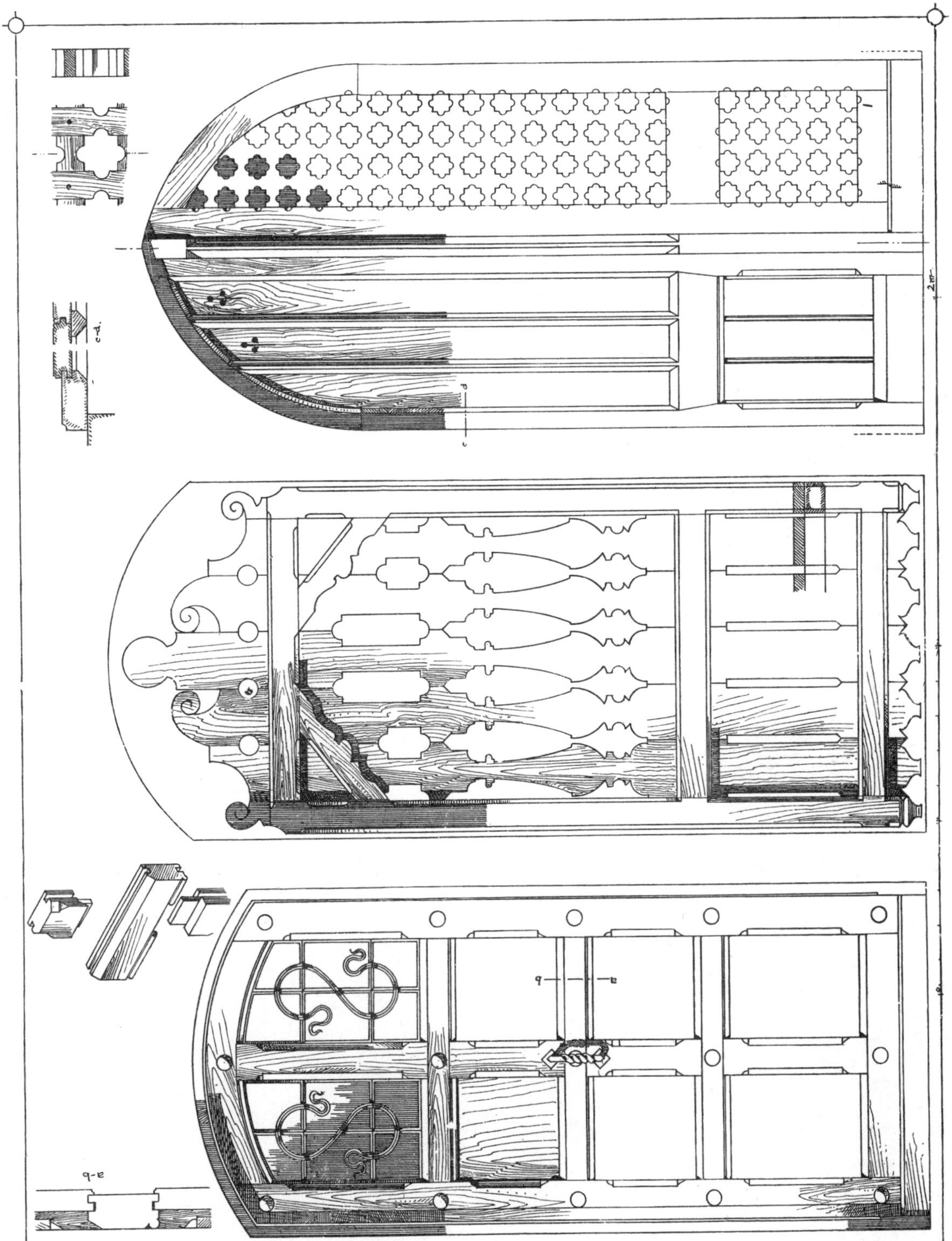

Aborttüre. — Gartenthüre. — Kellerthüre.

Tafel 54.

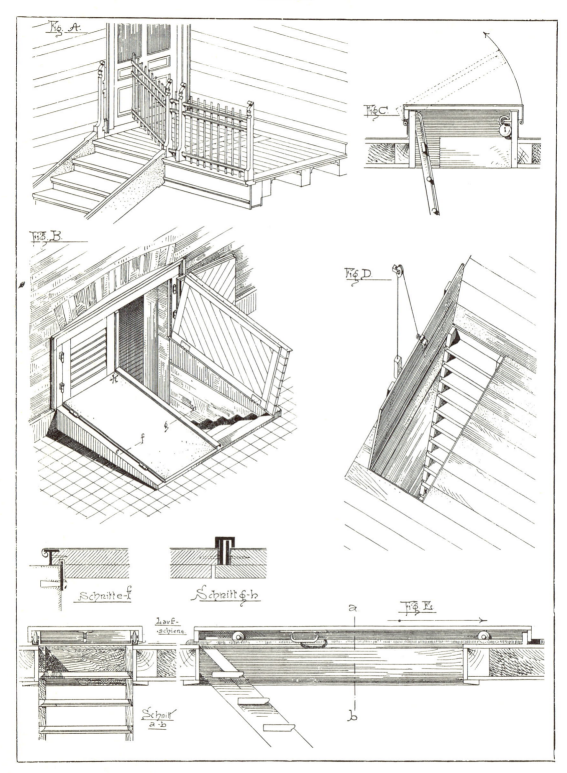

Treppenthüre, Kellerhalsthüre, Dachaussteigthüre, Fallthüre und Rollenthüre.

Tafel 55.

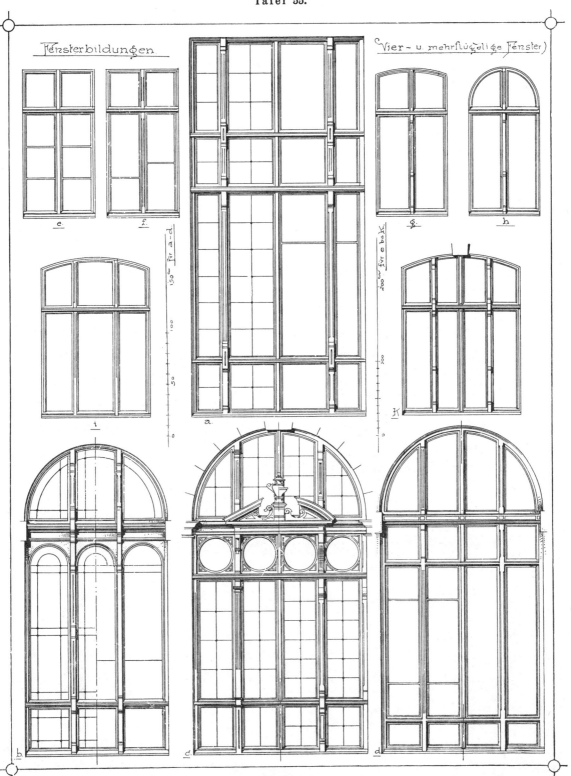

Fensterbildungen.

Tafel 56.

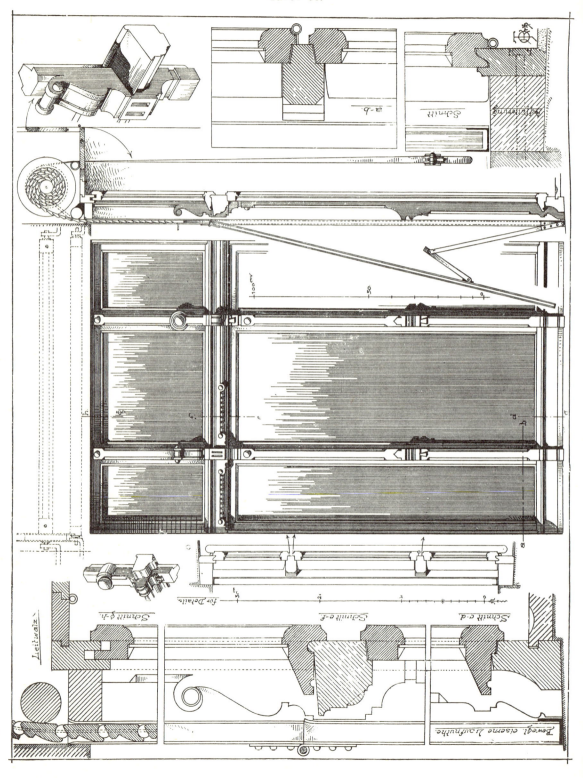

Tafel 57.

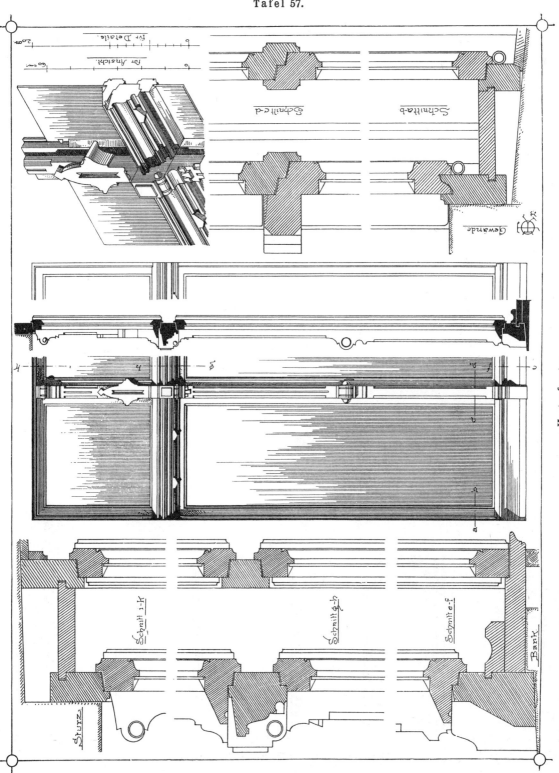

Tafel 58.

Blumenfenster oder Fenster-Glashäuschen.

Tafel 59.

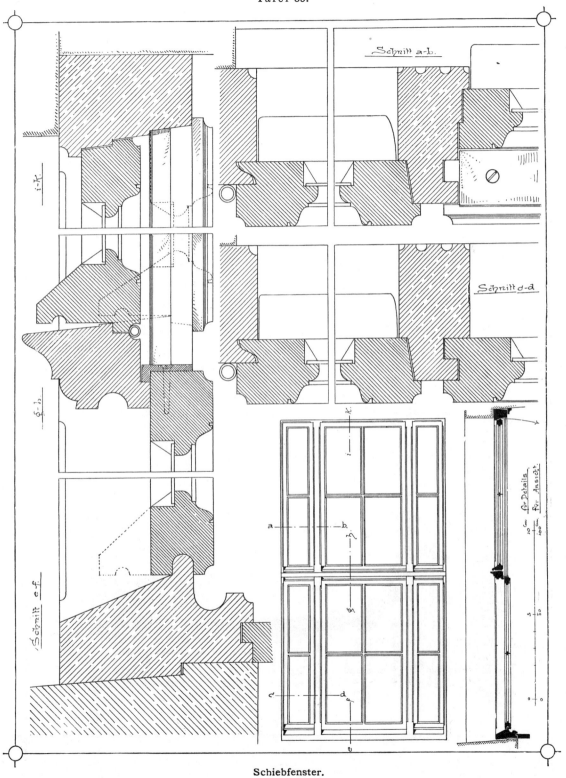

Schiebfenster.

Tafel 60.

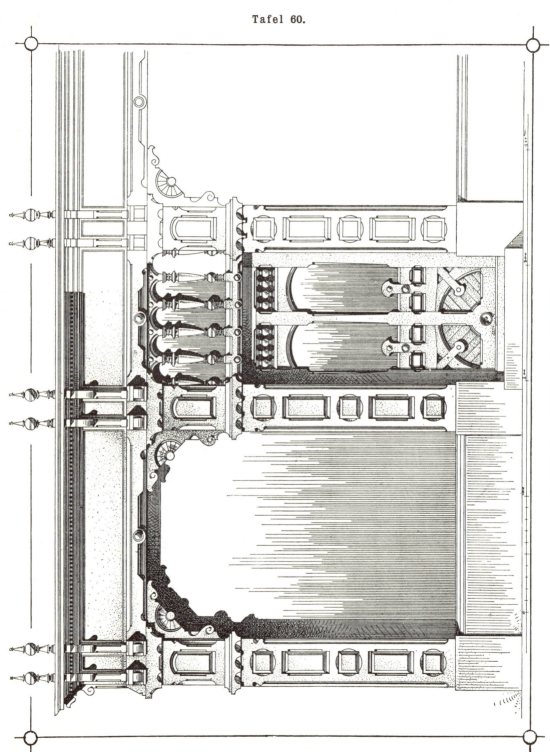

Schaufenster, Auslage- oder Ladenfenster.

Tafel 61.

Auslagekasten, zum Schaufenster auf Tafel 60 gehörig.

Tafel 62.

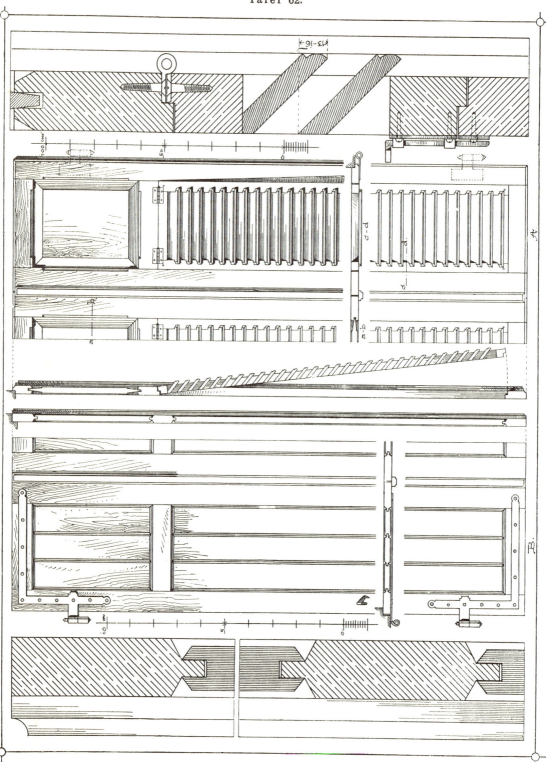

Laden.

Tafel 63.

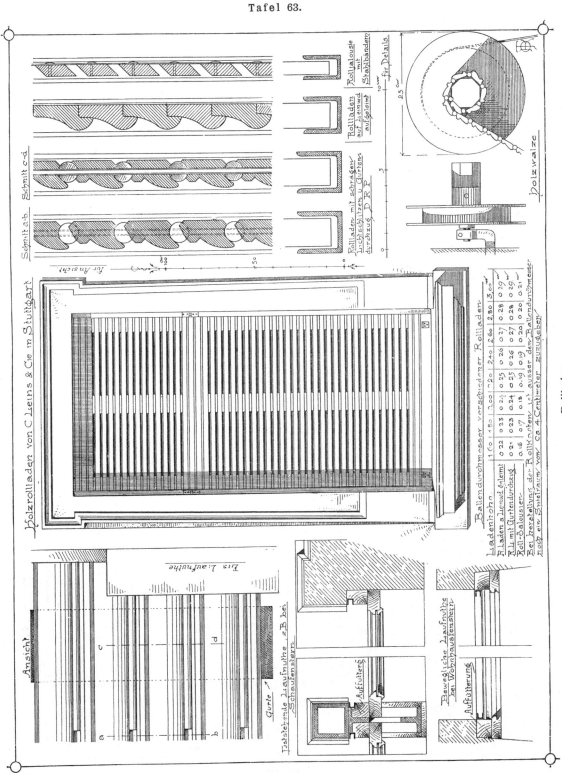

Rollladen.

Tafel 64.

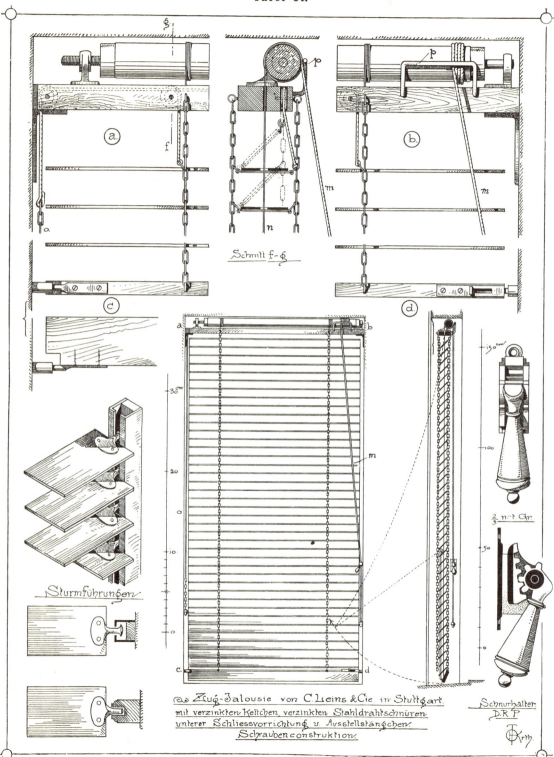

Zugjalousie.

Tafel 65.

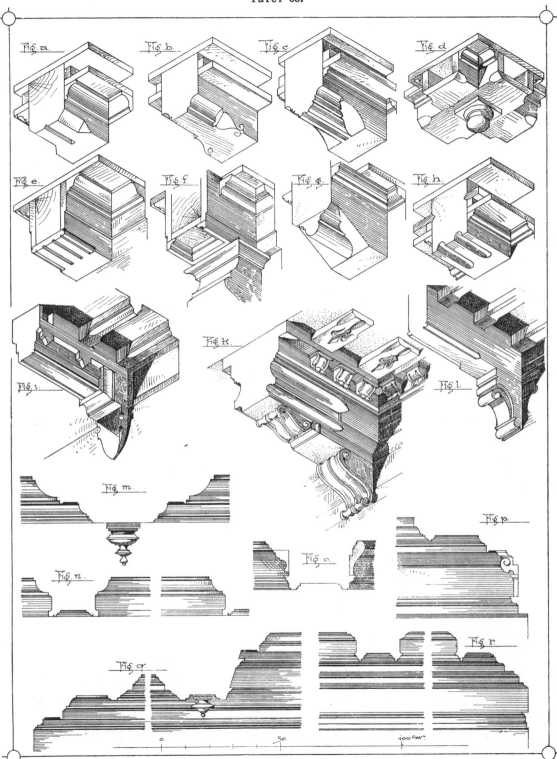

Einzelheiten der Holzdecken.

Tafel 66.

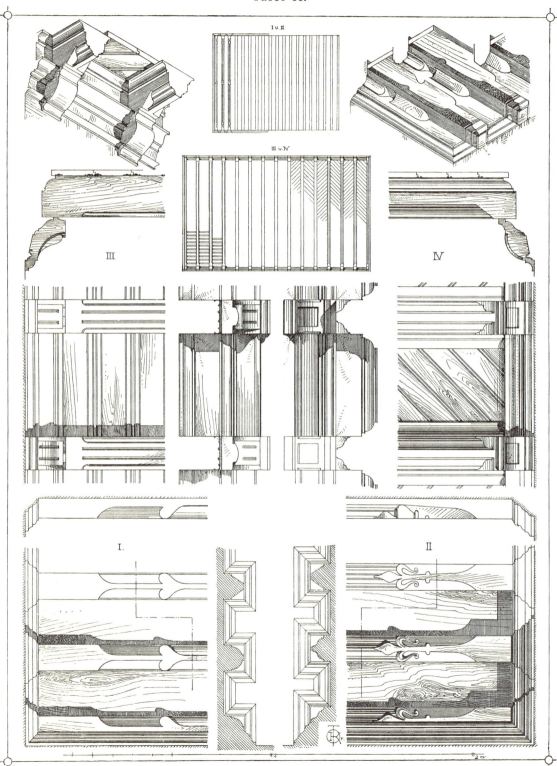

Einzelheiten der Holzdecken.

Tafel 67.

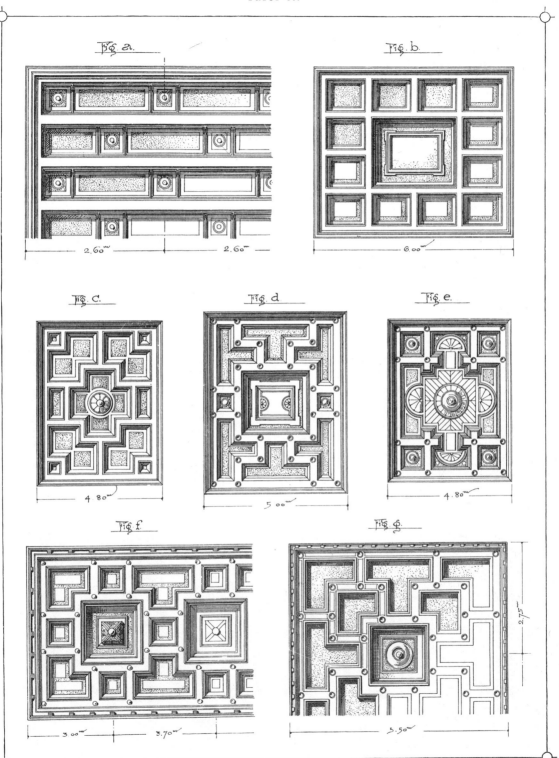

Holzdecken.

Tafel 68.

Holzdecken.

Tafel 69.

Holzdecken.

Tafel 70.

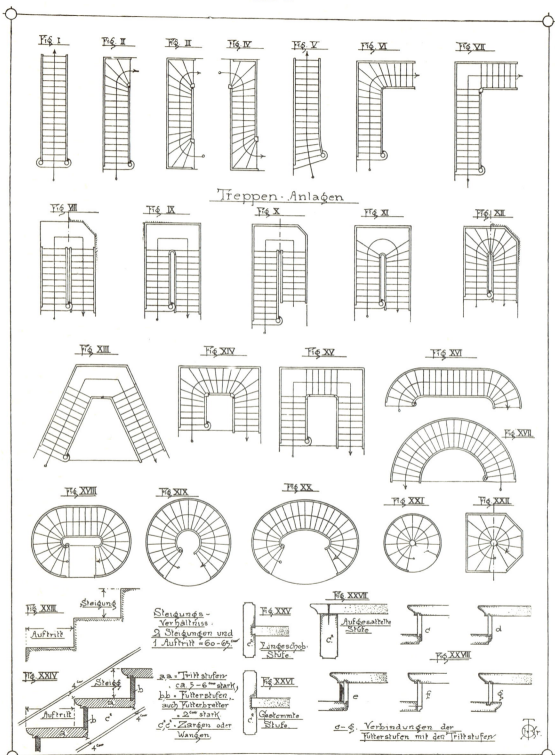

Treppenanlagen und Einzelheiten.

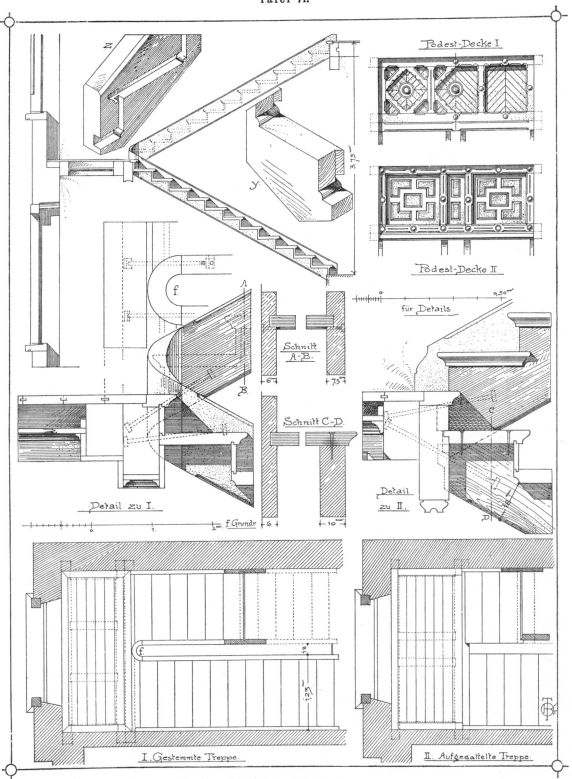

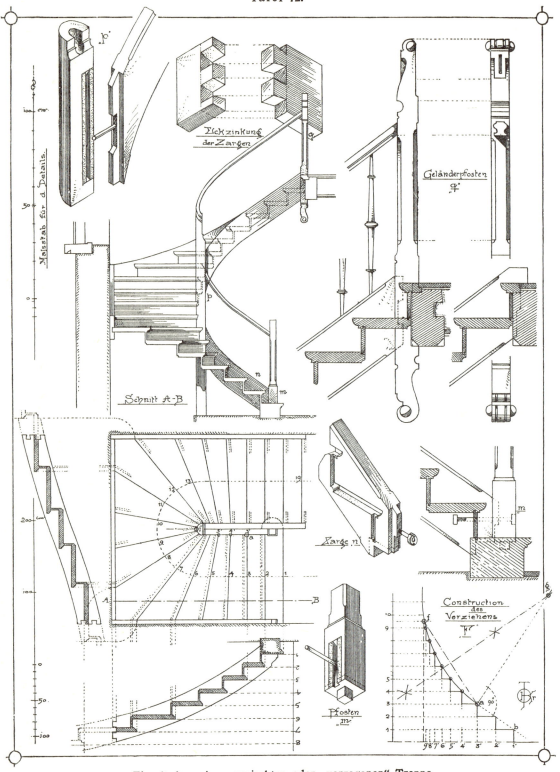

Einzelheiten einer gemischten oder „verzogenen" Treppe.

Tafel 73.

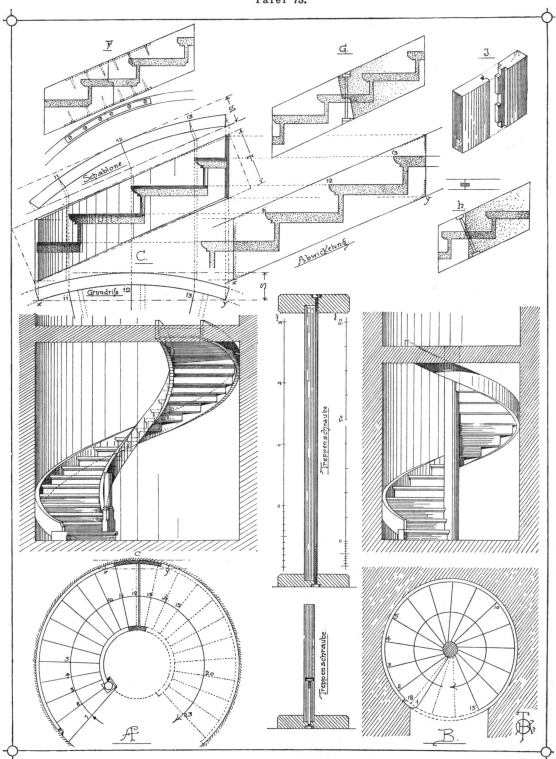

Einzelheiten der Wendeltreppe.

Tafel 74.

Treppenpodestdecken.

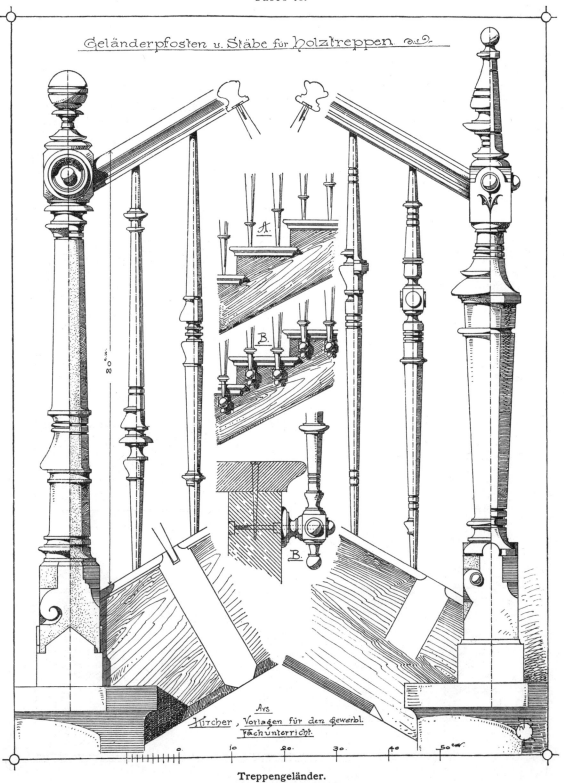
Treppengeländer.

Tafel 76.

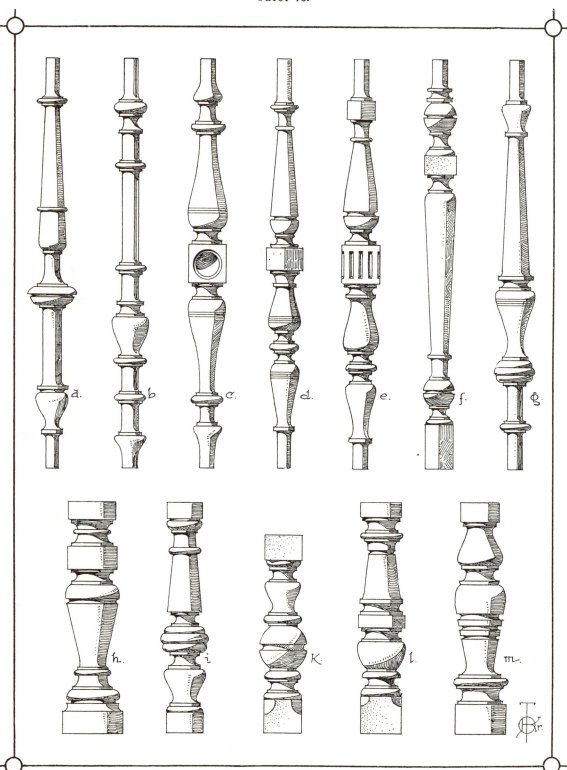

Geländerstäbe und Docken.

Tafel 77.

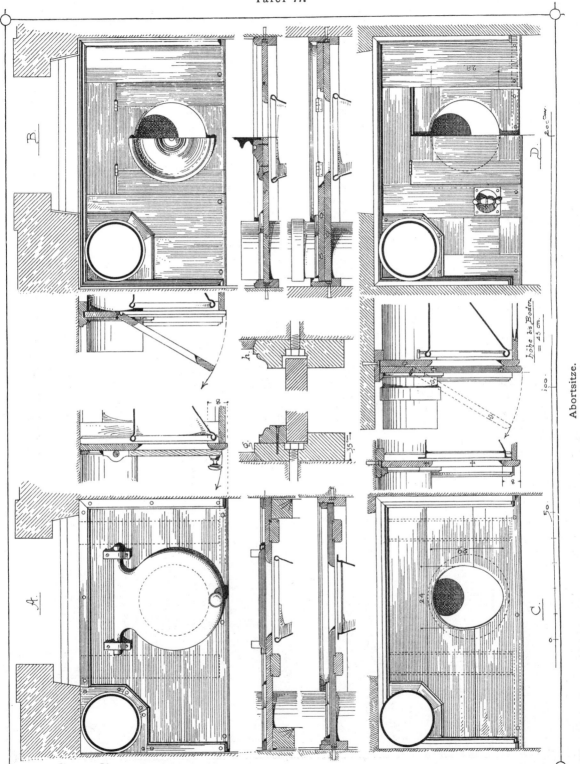

Abortsitze.

Tafel 78.

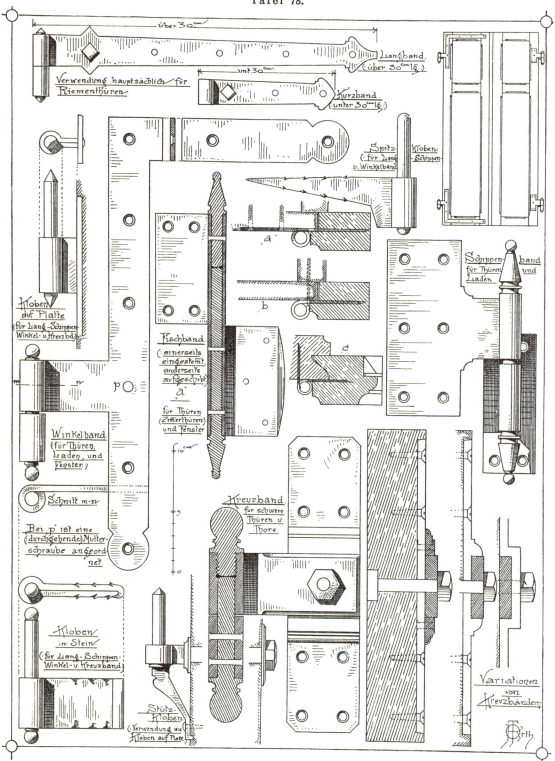

Bänder und Kloben.

Tafel 79.

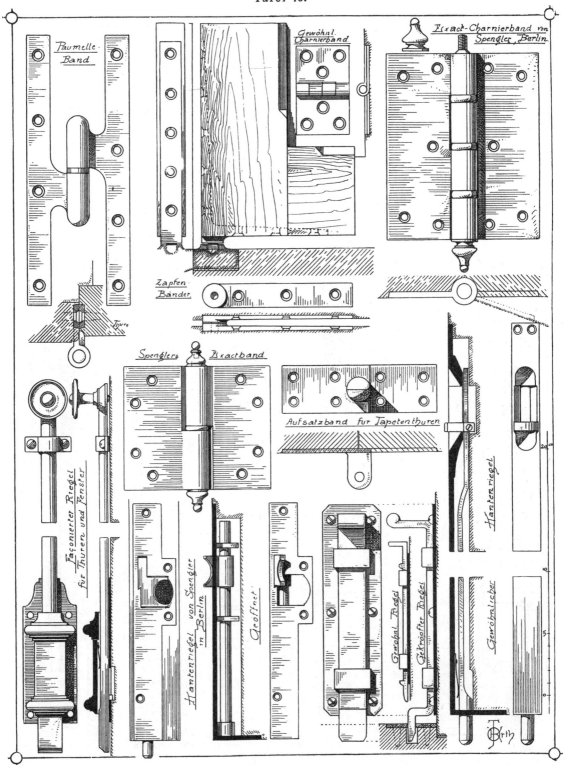

Bänder und Riegel.

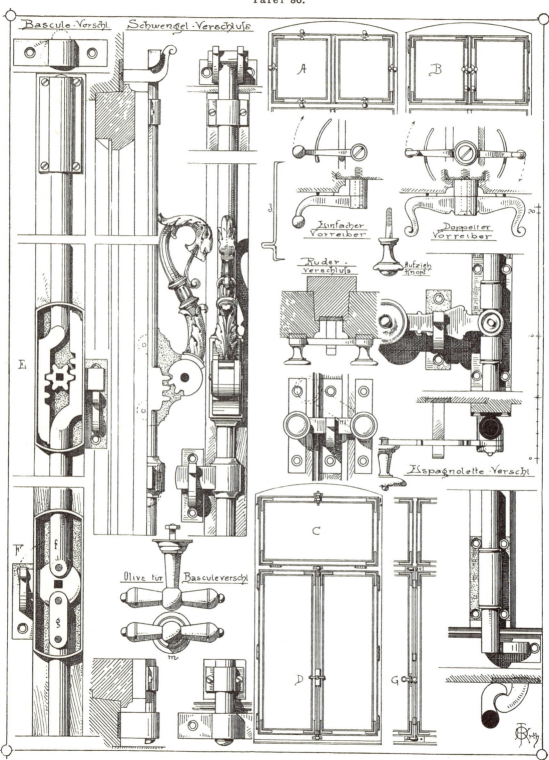

Baskülen-, Schwengel-, Ruder-, Espagnolett- und Vorreiberverschlüsse.

Tafel 81.

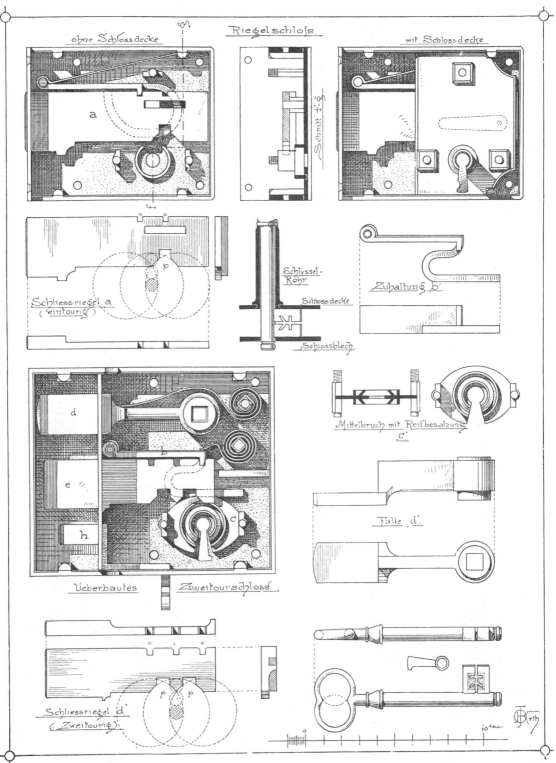

Schlösser.

Tafel 82.

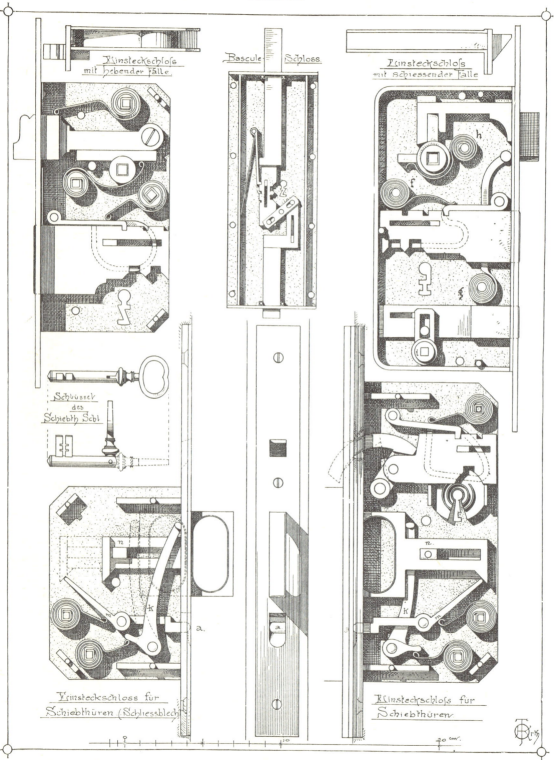

Schlösser.

Über **120** Fach- und Sachbücher aus vergangenen Jahrhunderten

EDITION libri rari